SOURCE BOOK *of* ENZYMES

John Stephen White
Dorothy Chong White

WHITE Technical Research GROUP

CRC Press
Boca Raton   London   New York   Washington, D.C.

Library of Congress Cataloging-in-Publication Data

White, John Stephen
    Source book of enzymes/ John Stephen White and Dorothy Chong White
        p. cm.
    Includes index.
    ISBN 0-8493-9470-8
    1. Enzymes—sources. 2. Biochemistry. 3. Molecular biology—biotechnology. I. White, John Stephen. II. Title.
QP749.G78 1997
616′.0108849—dc21
                                                                                                                                                                 97-10241
                                                                                                                                                                CIP

    This book contains information obtained from authentic and highly regarded sources. Reprinted material is quoted with permission, and sources are indicated. A wide variety of references are listed. Reasonable efforts have been made to publish reliable data and information, but the author and the publisher cannot assume responsibility for the validity of all materials or for the consequences of their use.

    Neither this book nor any part may be reproduced or transmitted in any form or by any means, electronic or mechanical, including photocopying, microfilming, and recording, or by any information storage or retrieval system, without prior permission in writing from the publisher.

    The consent of CRC Press LLC does not extend to copying for general distribution, for promotion, for creating new works, or for resale. Specific permission must be obtained in writing from CRC Press LLC for such copying.

    Direct all inquiries to CRC Press LLC, 2000 Corporate Blvd., N.W., Boca Raton, Florida 33431.

© 1997 by CRC Press LLC

No claim to original U.S. Government works
International Standard Book Number 0-8493-9470-8
Library of Congress Card Number 97-10241
Printed in the United States of America   2 3 4 5 6 7 8 9 0
Printed on acid-free paper

Table of Contents

PREFACE v

PART 1. INTRODUCTION 1

CHAPTER 1. USER'S GUIDE 2
CHAPTER 2. SYSTEM OF ENZYME COMMISSION NUMBERS 5
CHAPTER 3. GLOSSARY 7

PART 2. NUMERICAL LISTING OF ENZYMES ASSIGNED UNIQUE EC CLASSIFICATIONS 13

CHAPTER 4. SINGLE ENZYME PREPARATIONS 13
    Oxidoreductases (EC 1.X.X.X) 14
    Transferases (EC 2.X.X.X) 181
    Hydrolases (EC 3.X.X.X) 299
    Lyases (EC 4.X.X.X) 663
    Isomerases (EC 5.X.X.X) 702
    Ligases (EC 6.X.X.X) 720

CHAPTER 5. MULTIPLE ENZYME PREPARATIONS 737
    Carbohydrase Mixtures 738
    Protease Mixtures 748
    Carbohydrase and Protease Mixtures 750
    Carbohydrase, Lipase and Protease Mixtures 753
    Carbohydrase, Lipase, Nucleic Acid-Active Enzyme and Protease Mixtures 754
    Nucleic Acid-Active Enzyme Mixtures 756
    Miscellaneous Enzyme Mixtures 757

PART 3. ALPHABETICAL LISTING OF ENZYMES SHARING COMMON EC CLASSIFICATIONS — 761

CHAPTER 6. RESTRICTION ENDONUCLEASES — 761
CHAPTER 7. DNA METHYLTRANSFERASES — 1221

PART 4. UNCLASSIFIED ENZYMES — 1239

CHAPTER 8. ENZYMES NOT ASSIGNED AN EC NUMBER — 1239
    Carbohydrases — 1240
    Nucleic Acid-Active Enzymes — 1242
    Proteases — 1243
    Macerating and Lytic Enzymes — 1245
    Bakery Enzymes — 1248
    Miscellaneous Enzymes — 1249

PART 5. INDEXES — 1251

INDEX A. ENZYME SUPPLIER INDEX — 1251
INDEX B. GENERAL INDEX OF EC NUMBERS AND ENZYME NAMES, SYNONYMS AND ABBREVIATIONS — 1273

Preface

Source Book of Enzymes was inspired by our difficulty in locating a suitable replacement for a depleted enzyme in the midst of an urgent research project. To our dismay, we found the enzyme supplier out of business and the product line no longer available. The reagent catalogues on our library bookshelf offered a limited selection and incomplete functional information. We were ultimately able to locate a satisfactory alternative only by making countless inquiries and leafing through numerous product catalogues and technical data sheets. We needed—but could not find—a resource that catalogued available enzymes, provided critical technical information to distinguish one from another and told us where we could buy them.

We conceived *Source Book of Enzymes* as a unique reference tool to satisfy this need. It has been carefully designed to inform users of the broad variety of enzymes available from commercial suppliers worldwide. *Source Book of Enzymes* facilitates the selection process by providing systematic and comparative functional information about each enzyme. The task of maintaining and searching scores of current supplier catalogues and product data bulletins is eliminated.

Source Book of Enzymes will be a practical tool for researchers in academia, industry and government. Students and educators will regularly refer to its concise information and helpful indexes. Enzyme suppliers and marketers will find *Source Book of Enzymes* an invaluable resource in conducting competitive assessments and identifying new product trends and opportunities. And to speed their identification and ordering of specific enzymes, purchasing agents will rely on *Source Book of Enzymes*' distillation of critical product information into thorough reference tables and a comprehensive index of worldwide suppliers.

Source Book of Enzymes is organized in a user-friendly format which combines the following unique features in one expansive volume:
- an encyclopedic comparison of over 7000 commercially-available enzymes, restriction endonucleases, DNA methyltransferases and enzyme mixtures, blends and complexes
- a worldwide listing of nearly 100 enzyme suppliers, including local brokers and distributors
- critical functional information and applications for each enzyme
- biological sources, pH and temperature optima, additional (i.e., contaminating) enzymes, activity definitions and specific activity comparisons to facilitate selection of compatible enzymes of appropriate purity
- complete indexes of Enzyme Commission numbers, enzyme names, synonyms and abbreviations

It is our hope that *Source Book of Enzymes* finds a place on your bookshelf as a valued research tool, and that it eases the process of locating the enzyme and supplier best suited to your needs and geographic location, as freeing your time for more productive uses is our goal.

We made a concerted effort to make *Source Book of Enzymes* an exhaustive and global compendium of all commercially available enzymes. That we fell short of this goal is due in part to the reluctance of a few suppliers to participate and in part to our ignorance of additional suppliers. If we overlooked you, please accept our apology and invitation to contribute to the next edition.

Because of the continuing discovery, preparation and commercialization of new enzymes and sources, the information in this book will quite naturally become dated. To address this shortcoming, the authors and publisher are committed to updating *Source Book of Enzymes* on a regular basis. We welcome your suggestions for corrections and additions to the book. Particularly valued are your suggestions for improving the content, readability, accessibility and organization of *Source Book of Enzymes*.

About the Authors

Dorothy Chong White has worked for fifteen years in the corn wet-milling industry, holding varied positions in research and development for the A.E. Staley Manufacturing Company in Decatur, Illinois. She has developed considerable expertise in manufacturing process trouble-shooting and product characterization. She has the distinction of being a member of the research team which won technical approval for 100% use of high fructose corn syrup-55 in the carbonated beverage industry. Ms. White holds two US patents and is actively engaged in new product development from biotechnology. Her scientific training was taken at the Universities of Illinois at Champaign-Urbana and Springfield.

John Stephen White developed an enduring fascination for the use of enzymes through practical experience in the isolation and purification of enzymes for academic research, development of a multi-enzyme *in vitro* enzyme system to predict the *in vivo* digestibility of carbohydrate polymers, and large-scale applications of enzymes for food and industrial ingredient manufacture. His training in biochemistry was taken at the University of California at San Diego and the University of Utah. Dr. White held joint appointments at the University of Illinois, Champaign-Urbana, as postdoctoral fellow and visiting assistant professor of biochemistry. He spent thirteen years in research and management in the corn wet milling industry, one of the largest consumers of commercial enzymes, and is the author of over 20 journal articles and book chapters.

Dr. White has worked extensively with scientists from the food and beverage industries, academia and government to investigate nutritional issues surrounding traditional and novel foods and ingredients. He has enjoyed professional affiliations with the Calorie Control Council, the Corn Refiners Association, the Institute of Food Technologists and the International Life Sciences Institute, which he has served as trustee and scientific advisor. Dr. White is founder and principal in the consulting firm of WHITE *Technical Research* GROUP, established in 1994 to serve scientific research and management.

WHITE *Technical Research* GROUP
Rural Route *One*, Box *Twenty-nine*
Argenta, Illinois 62501 USA
Phone: 217-795-4437
E-mail: wtrg@webmart.net

Acknowledgments

We gratefully acknowledge permission granted by Academic Press (Orlando, Florida) to draw upon the recommendations of the Nomenclature Committee of the International Union of Biochemistry and Molecular Biology published in *Enzyme Nomenclature 1992*. We are likewise indebted to the pioneering work of Dr. Richard J. Roberts in classifying restriction endonucleases and methyl transferases. His REBASE database is updated annually in *Nucleic Acids Research* (24: 223-235, 1996).

Finally, we express our gratitude to the many marketing, technical service, customer service, research and public affairs personnel from enzyme suppliers who gave so generously of their time to provide us with technical information about their enzyme products.

Dedication

To Helen Chong and R. Stephen & Freda White for the possibility,
to Drs. Paul Saltman, Clark Gubler, Hans Rilling and Lowell Hager for kindling the enthusiasm,
to Chasib N. Khalaf for the encouragement,
to Paul Petralia and Harvey Kane at CRC Press, Inc. for embracing the concept,
and to Dorothy White for the belief and the partnership...

Part 1. Introduction

Chapter 1. User's Guide

Source Book of Enzymes is designed chiefly for two purposes: to increase accessibility to globally-available enzymes and to provide information to assist the enzyme selection process. In this chapter we will guide you through the enzyme selection process, using NADPH Dehydrogenase as an example and suggesting selection criteria for your consideration. The next chapter will review the system of Enzyme Commission (EC) numbers.

If you know the EC number...

locating an enzyme is easy since *Source Book of Enzymes* is organized by EC classification. Simply turn to the page containing the desired EC number in the page heading. The EC number for NADPH Dehydrogenase is 1.6.99.1 and it can be found on pages 125-127, between NAD(P)H Dehydrogenase (FMN) (EC 1.6.8.1) and NAD(P)H Dehydrogenase (Quinone) (EC 1.6.99.2). EC numbers are also listed in numerical sequence in the General Index (Index B), beginning on page 1274.

If you know the enzyme name...

consult the General Index. Enzymes are listed alphabetically beginning on page 1278 by systematic name, synonym, trade name and abbreviation; so you are likely to find the enzyme you want. NADPH dehydrogenase is listed among other index entries as follows:

> NADH-FMN oxidoreductase, *128*
> NADP-cytochrome reductase, *122*
> NADPH-dependent dehydrogenase, *133*
> →NADPH dehydrogenase, *125*
> NADPH diaphorase, *125*
> NADPH:(acceptor) oxidoreductase, *125*
> NADPH:ferricytochrome oxidoreductase, *122*

Note in this example that synonyms NADPH diaphorase and NADPH:(acceptor) oxidoreductase are also indexed for completeness of identification.

If you don't know the enzyme name or EC number...

don't be discouraged. You can browse the General Index alphabetically (begin on page 1278) for names or synonyms that describe your application. Alternatively, novel applications may occur to you while browsing broad EC classifications such as oxidoreductases, transferases, hydrolases, etc. Enzymes with like specificities are grouped together for simple comparison of functionality. Their location in *Source Book of Enzymes* is provided in the detail under Chapter 4 in the Table of Contents (1.X.X.X through 6.X.X.X).

Confirming the enzyme selection...

is the next step once the enzyme is located. *Source Book of Enzymes* makes this straightforward by providing broad functional information in columnar form.

Columnar information is useful for confirming the enzyme selection by:
- verifying the reaction catalyzed or unique recognition sequence
- viewing alternative names
- reviewing sample applications
- obtaining practical information about specific activators, inhibitors, cofactors and stabilizers

> **Please turn to page 125 now and review the columnar information for NADPH Dehydrogenase.**

User's Guide

Selecting the right enzyme for your application...

is the final step after confirming the enzyme selection. *Source Book of Enzymes* simplifies this task by providing product-specific technical data in a tabular format which encourages product comparisons.

Use the **tabular data** as selection criteria for comparing and contrasting enzyme products. Select that product which most nearly matches your requirements for pH and temperature optima, specific activity, preparation form and purity.

- biological source—responsible for the wondrous variety in pH and temperature optima, stability, molecular weight and subunit structure
- specific activity, concentration or volume activity—gives the relative potency/purity/value, in units of enzyme activity per unit weight (or volume) of enzyme, total protein or diluent (solid or liquid)
- unit definition—levels the playing field between suppliers by disclosing how the specific activity is measured
- preparation form or reaction buffer—describes the physical form of the product (crystallized, lyophilized, etc.) and any added buffers, salts, stabilizers or preservatives
- additional activities—lists other enzyme activities present (***Note***: an empty column may indicate no contaminating activities or the failure of the supplier to test for additional activities; contact the supplier for more information)

> *Please return to pages 125-127 now and review the tabular data and selection criteria for NADPH Dehydrogenase enzyme products. Note the variety in biological sources, functional properties and forms from which to choose.*

Confused about...

a term or abbreviation used in the data tables? Chapter 3 (see page 7) contains an extensive Glossary of terms used in *Source Book of Enzymes* that will clarify those with which you are unfamiliar.

> *Please turn to the Glossary now and find an explanation for the abbreviation, NADPH.*

To order the selected enzyme...

simply record the enzyme supplier's catalog number from the data table and locate their office, broker or distributor closest to you from the Enzyme Supplier Index (Index A), beginning on page 1251.

> *Please note a supplier or two now from the tabular data for NADPH Dehydrogenase, turn to the* **Enzyme Supplier Index** *and find the supplier's representative most convenient to your location.*

Source Book of Enzymes specifications...

- 500 conventional enzymes (6500 products), catalogued alphabetically by EC number in Chapter 4
- over 350 unique restriction enzymes and DNA methyltransferases (1500 products), presented alphabetically in Chapters 6 and 7
- more than 100 enzyme mixtures (170 products) and enzymes with no EC numbers assigned (80 products), detailed in Chapters 5 and 8, respectively
- nearly 100 worldwide suppliers, brokers and distributors of enzymes are listed in Index A

Conventions

Verbal descriptions of product attributes have been condensed and standardized in order to simplify reader comparisons and conserve space in data tables. The authors have endeavored to do this without compromising supplier product claims.

Enzyme specific activities and unit definitions vary considerably from supplier to supplier. Supplier-defined activities and unit definitions were incorporated into data tables with only minor changes to improve readability.

The authoritative source for Enzyme Commission numbers is *Enzyme Nomenclature 1992*: Recommendations of the Nomenclature Committee of the International Union of Biochemistry and Molecular Biology, published by Academic Press, Inc.

Restriction endonucleases and methyl transferases are organized according to guidelines promoted by Dr. Richard J. Roberts and published annually in *Nucleic Acids Research* (Oxford University Press).

The abbreviations used in this work are those in common use. Since abbreviations differ somewhat around the world, we adopted for use those which seemed most descriptive and straightforward. New abbreviations were coined only where necessary to avoid confusion with another enzyme.

The authors have made every attempt to ensure that the information contained in this work is accurate. Since the field of enzymes is changing so rapidly, readers are encouraged to verify enzyme information with suppliers before ordering.

Information was gathered from supplier catalogues and electronic data bases. Obvious discrepancies between suppliers were resolved where possible. The sheer volume of data in this work, however, made it impossible to verify the veracity of each supplier's claim.

Because of their large number, restriction endonucleases and methyl transferases have been removed from the EC numerical sequence of Part 2. They follow alphabetically in Chapters 6 and 7 of Part 3. We believe that organizing the material in this way makes it easier for a user to find a specific restriction endonuclease or methyl transferase, while preventing this large number of enzymes from disrupting the flow of conventional enzymes presented by EC number.

Chapter 2. System of Enzyme Commission Numbers

The organization of enzymes in *Source Book of Enzymes* follows recommendations of the Enzyme Commission (EC) of the International Union of Biochemistry (now called the International Union of Biochemistry and Molecular Biology). EC involvement began at a time when systematic naming conventions were non-existent and new enzymes were being discovered at an accelerating rate. Names coined for new discoveries were inadvertently repeated for enzymes with unrelated activities or the new name very often failed to adequately describe the reaction catalyzed. Equally confusing, the same enzyme was known by different names in different laboratories.

EC began developing a code of systematic rules for the consistent naming of new enzymes in 1956. Their recommendations were first published in 1961 and have been updated five times since. The most recent edition appeared in 1992 under the title *Enzyme Nomenclature*[1]. It is significant to note that over these three decades, the number of entries in *Enzyme Nomenclature* increased from 712 to 3196. With the widespread application of powerful molecular cloning, culture and isolation technology, it is likely that the rapid pace of enzyme discovery, development and commercialization will continue.

---

[1] Webb, Edwin C., for the Nomenclature Committee of the International Union of Biochemistry and Molecular Biology. *Enzyme Nomenclature 1992*: Recommendations of the Nomenclature Committee of the International Union of Biochemistry and Molecular Biology Academic Press, Inc., San Diego.

The following sections present in condensed form the general nomenclature principles and classification/numbering scheme adopted by EC.

General Nomenclature Principles

- Names of enzymes ending in –ase are used only for enzymes with a single catalytic activity. Enzymes with more than one catalytic activity should contain the word *system* in the name. Examples of the latter include the succinate oxidase system and the pyruvate dehydrogenase system.

- Enzymes are classified and named according to the reaction they catalyze.

- Enzymes are divided into groups on the basis of the type of reaction catalyzed. Individual enzymes within each group are classified, assigned code numbers and named using preferred substrate(s).

Classification and Numbering Scheme

The Enzyme Commission (EC) classifies enzymes in six main divisions:

Class 1	Oxidoreductases
Class 2	Transferases
Class 3	Hydrolases
Class 4	Lyases
Class 5	Isomerases
Class 6	Ligases

Each enzyme is assigned a unique EC number consisting of four parts, each separated by a period:

X.__.__.___	main division (above) in which an enzyme is classified
_.XX.__.___	enzyme subclass
_.__.XX.___	enzyme sub-subclass
_.__.__.XXX	serial number of the enzyme within its sub-subclass

Systematic and Recommended Names

Two types of names are given and indexed in *Source Book of Enzymes*:

Systematic Name	highly descriptive and lengthy, this name is given in strict accordance with EC naming conventions
Recommended or Trivial Name	not particularly descriptive but succinct, this name is often in general use

In addition to the naming and classification rules summarized above, EC developed 32 specific rules for classification and nomenclature. Readers wishing to learn more about these enzyme naming and classification rules are referred to *Enzyme Nomenclature*. It should be noted that IUBMB continues to convene expert scientific panels in an on-going effort to clarify the often murky waters of enzyme nomenclature.

Restriction Endonucleases and Methyl Transferases

While the EC system for naming and classifying enzymes suffices for the majority of conventional enzymes, it does not for restriction endonucleases and methyl transferases. Although nearly five hundred of these important enzymes have already been commercialized, they are classified under just five EC numbers (below).

Schemes have been developed to further organize enzymes within these classifications. Highly regarded among these is one promoted by Dr. Richard J. Roberts which is updated annually in *Nucleic Acids Research*, published by Oxford University Press.

3.1.21.3	Type I Site-specific deoxyribonuclease
3.1.21.4	Type II Site-specific deoxyribonuclease
3.1.21.5	Type III Site-specific deoxyribonuclease
2.1.1.72	Site-specific DNA-methyltransferase (adenine-specific)
2.1.1.73	Site-specific DNA-methyltransferase (cytosine-specific)

Chapter 3. Glossary

A

A	adenine
AAO	ascorbic acid oxidase, ascorbate oxidase
Ab	antibody
ABDP	N-(4-aminobenzoyl)-N'-(pyridyldithioisopropionyl-hydrazine)
ABTS	2,2-azino-di-[3-ethylbenzthiazoline sulfonate (6)]
Ac	N-acetyl-
Acetyl-CoA	acetyl-coenzyme A
AChE	acetyl cholinesterase
ACOD	acyl-CoA oxidase
ACP	acyl carrier protein
ACS	long-chain-fatty-acid-CoA ligase
ADA	adenosine deaminase
ADH	alcohol dehydrogenase
AdK	adenylate kinase
ADP	adenosine diphosphate
AEBSF	4-(2 aminoethyl)-benzenesulfonyl-fluoride (Pefabloc® SC)
Ag	antigen
AGH	α-glucosidase
AICAR	5-amino-imidazole-4-carboxamide-1-riboside
AIDS	acquired immunodeficiency syndrome
AK	acetate kinase
Ala	alanine
AlaDH	alanine dehydrogenase
AlDH	aldehyde dehydrogenase
AlcDH	alcohol dehydrogenase
ALOD	alcohol oxidase
ALP	alkaline phosphatase
ALT	alanine amino transferase
AMG	amyloglucosidase, glucoamylase
AMP	adenosine monophosphate
cAMP	adenosine 3′,5′-cyclic monophosphate
AMV	avian myeloblastosis virus
AO	ascorbate oxidase
AOD	amino acid oxidase
AP	alkaline phosphatase
APMSF	(4-amidinophenyl)-methane-sulfonyl fluoride
Arg	arginine
Arg-C	arginine specific endoproteinase
ARS	arylsulfatase
Asn	asparagine
ASOD	ascorbate oxidase
ASOM	ascorbate oxidase
Asp	aspartate
AST	aspartate amino transferase
Asx	asparagine or aspartate
ATEE	acetyl-L-tyrosine ethylester
ATP	adenosine triphosphate
ATPase	adenosine triphosphatase

B

B	C or G or T
BAEE	N-benzoyl-L-arginine ethyl ester
BC	butyryl cholinesterase
BCECF	2′,7′-bis-(carboxyethyl)-5(6)-carboxyfluorescein
BME	β-mercaptoethanol
BMME	bis-(maleimido)-methyl ether
Bp	base pair(s)
BSA	bovine serum albumin
BTEE	N-benzoyl-L-tyrosine ethyl ester

C

C	cytosine
C4S	chondro-4-sulfatase
C6S	chondro-6-sulfatase
CA	carbonic anhydrase
CaM	calmodulin
CaOAc	calcium acetate
CAT	chloramphenicol acetyltransferase
CCK	cholecystokinin
cDNA	complementary DNA
CDP	cytidine diphosphate
CE	cholesterol esterase
CHAPS	3-[(3-cholamidopropyl)-dimethyl-ammonio]-1-propane sulfonate
ChO	cholesterol oxidase
CK	creatine kinase
CL	citrate lyase
CMC	carboxymethyl cellulose
CMCase	carboxymethyl cellulase
CMP	cytidine monophosphate
CNPase	2′,3′-cyclic nucleotide 3′-phosphodiesterase
CO	cytochrome oxidase
CoA	coenzyme A
COD	choline oxidase
CoQ	coenzyme Q

COMT	catechol o-methyl transferase	dTTP	2′-deoxythymidine-5′-triphosphate	**E**	
CON	cholesterol oxidase	dU	deoxyuridine	E	extinction (absorbance)
ConA	concanavalin A	dUTP	2′-deoxyuridine-5′-triphosphate	EC	Enzyme Commission (IUB System)
CPase Y	carboxypeptidase Y	D	A or G or T	EDTA	ethylenediamine tetracetate
CPK	creatine phosphokinase	Da	dalton (1.66×10^{-24} g)	EGTA	ethyleneglycol-bis-(β-aminoethyl)-N,N,N',N'-tetraacetate
CS	citrate synthase	*dam*	DNA adenine methylase; mutation blocks methylation of adenine residues in the recognition sequence 5′-GmATC-3′	EIA	enzyme immunoassay
CTP	cytidine triphosphate			ELISA	enzyme-linked immunoabsorbent assay
Cys	cysteine			EtOH	ethanol
Cyt-c	cytochrome c	DAAO	D-amino acid oxidase		
		dcm	DNA cytosine methylase; mutation blocks methylation of internal cytosine residues in the recognition sequences 5′-CmCAGG-3′ or 5′-CmCTGG-3′	**F**	
D				F6P	fructose-6-phosphate
d	2′-deoxyribo			F6P-K	fructose-6-phosphate kinase
dATP	2′-deoxyadenosine-5′-triphosphate			FA	fatty acid
dCDP	2′-deoxycytidine-5′-diphosphate			Fab	antigen-binding fragment
dCMP	2′-deoxycytidine-5′-monophosphate	DEAE	diethylaminoethyl	F(ab′)$_2$	divalent antigen-binding fragment
dCTP	2′-deoxycytidine-5′-triphosphate	DFP	diisopropyl fluorophosphate	FAD	flavin adenine dinucleotide (oxidized form)
dGDP	2′-deoxyguanosine-5′-diphosphate	DHAP	dihydroxyacetone phosphate	FADH$_2$	flavin adenine dinucleotide (reduced form)
dGMP	2′-deoxyguanosine-5′-monophosphate	DIPF	diisopropylphosphofluoridate	Fc	constant sequence fragment of immunoglobulin
dGTP	2′-deoxyguanosine-5′-triphosphate	DMF	dimethyl formamide	FDNB	fluorodinitrobenzene
DI	diaphorase, NADPH dehydrogenase	DMSO	dimethyl sulfoxide	FMN	flavin mononucleotide (oxidized form)
		DNA	deoxyribonucleic acid		
		DNase	dexoxyribonuclease	FMNH$_2$	flavin mononucleotide (reduced form)
dITP	2′-deoxyinosine-5′-triphosphate	DNP	dinitrophenol		
		DPCC	diphenyl carbamyl chloride		
dNTP	deoxynucleoside triphosphate	DPN$^+$	see NAD$^+$	Fru	fructose
		DPNH	see NADH		
		ds	double-stranded (DNA)	**G**	
dTDP	2′-deoxythymidine-5′-diphosphate	DTE	dithioerythritol		
		DTNB	5,5′-dithio-bis-(2-nitrobenzoic acid)	G	guanine
dTMP	2′-deoxythymidine-5′-monophosphate	DTT	dithiothreitol	G3P	glycerol-3-phosphate

G3PDH	glycerol-3-phosphate dehydrogenase		GOT	glutamic oxaloacetic transaminase		

Glossary

G3PDH	glycerol-3-phosphate dehydrogenase
GABA	γ-aminobutyric acid
GAD	glutamic acid decarboxylase
GAH	β-galactosidase
Gal	galactose
α-, β-Gal	α-, β-galactosidase
Gal1P	galactose-1-phosphate
GalO	galactose oxidase
GAM	glucoamylase, amyloglucosidase
GAP	glyceraldehyde 3-phosphate
GAPDH	glyceraldehyde 3-phosphate dehydrogenase
GAT	glutamic aspartate transaminase
GDH	glycerol dehydrogenase
GDP	guanosine diphosphate
GK	glycerokinase, glycerol kinase
Glc	glucose
GlcDH	glucose dehydrogenase
Glc1P	glucose-1-phosphate
Glc6P	glucose-6-phosphate
Glc6PDH	glucose-6-phosphate dehydrogenase
GlcK	glucokinase
GlcNAc	N-acetylglucosamine
GlDH	glutamate dehydrogenase
Gln	glutamine
Glu	glutamate
Glu-C	glutamic acid specific endoproteinase
αGluD	α-glucosidase
Glx	glutamine or glutamate
Gly	glycine
GMP	guanosine monophosphate
GO	glucose oxidase
GOD	glucose oxidase
GOT	glutamic oxaloacetic transaminase
GPCP	glycerophosphocholine phosphodiesterase
GPDH	glycerol phosphate dehydrogenase
GPO	glycerol 3-phosphate oxidase
GPT	glutamic pyruvic transaminase
GR	glutathione reductase
GRS	glucuronidase
GS	glutamine synthetase
GSH	glutathione (reduced form)
GSSG	glutathione (oxidized form)
γGT	γ-glutamyl transferase
GTP	guanosine triphosphate
GTPase	guanosine triphosphatase

H

H	A or C or T
H_2O_2	hydrogen peroxide
HAT	hypoxanthine/aminopterin/thymidine
Hb	hemoglobin
3HBDH	3-hydroxybutyrate dehydrogenase
HBc	hepatitis B core protein
HBsAg	hepatitis B surface antigen
HCl	hydrochloric acid, hydrochloride
HCV	hepatitis C virus
HEPES	4-(2-hydroxyethyl)-1-piperazineethanesulfonic acid
His	histidine
HIV	human immunodeficiency virus
HK	hexokinase
HNMT	histamine N-methyl transferase
HPETE	hydroperoxy eicosatetraenoic acid
HPLC	high pressure (performance) liquid chromatography
HRP	horse radish peroxidase
HSA	human serum albumin
3αHSD	3α-hydroxysteroid dehydrogenase (B-specific)
7αHSD	7α-hydroxysteroid dehydrogenase
12αHSD	12α-hydroxysteroid dehydrogenase
HSDH	hydroxysteroid dehydrogenase
hsdM	E. coli (or EcoK) DNA methylase; mutation blocks sequence specific adenine methylation in the sequence A^{N6}mACNNNNNNNGTGC or GC^{N6}mACNNNNNNNGTT; DNA isolated from a HsdM⁻ strain will be restricted by a HsdR⁺ host
hsdR	E. coli (or EcoK) restriction endonuclease; absence of this activity permits the introduction of DNA propagated from non -E. coli sources
hsdS	Specificity determinant for hsdM and hsdR; mutation eliminates HsdM and HsdR activity

I

ICDH	isocitrate dehydrogenase
IDP	inosine diphosphate

Ig	immunoglobulin	
IgG	immunoglobulin G	
Ile	isoleucine	
IMP	inosine monophosphate	
IPTG	isopropyl-β-D-thiogalactopyranoside	
ITP	inosine triphosphate	
IU	international units	
IUB	International Union of Biochemists	

K

α-Kg	α-ketoglutarate
K	G or T
K^+	potassium ion
kBp	kilobase pair(s)
KCl	potassium chloride
KOAc	potassium acetate
K_M	Michaelis constant, the substrate concentration at which the reaction rate is half its maximal value
KPO_4	potassium phosphate

L

LAOD	L-amino-acid oxidase
LAP	leucine aminopeptidase
LD	lipoamide dehydrogenase
LDH	lactate dehydrogenase
Leu	leucine
LeuDH	leucine dehydrogenase
LO	lactate oxidase
LOX	lactate 2-monooxygenase
LPO	lactoperoxidase
LYPL	lysophospholipase
Lys	lysine
Lys-C	lysine specific endoproteinase

M

M	A or C
M	molar
MAO	monoamine oxidase
Mb	myoglobin
MCA	4-methyl-coumaryl-7-amide
mcrA	$E.\ coli$ restriction system; mutation prevents McrA restriction of methylated DNA of sequence 5′-CmCGG; formerly known as rglA
mcrCB	$E.\ coli$ restriction system; mutation prevents McrCB restriction of methylated DNA of sequence 5′-G^{5}mC, 5′-G^{5h}mC or 5′-G^{N4}mC; formerly known as rglB
MDH	malate dehydrogenase
MeOH	methanol
MES	2-(N-morpholino)ethane-sulfonic acid
Met	methionine
mg	milligram
Mg^{2+}	magnesium ion
$MgSO_4$	magnesium sulfate
min	minutes
MK	myokinase
mL	milliliter
mM	millimolar
MP	malate phosphorylase
mRNA	messenger RNA
mrr	$E.\ coli$ restriction system; mutation prevents Mrr restriction of methylated DNA of sequence 5′-GmAC or CmAG; mutation also prevents McrF restriction of methylated cystosine sequences
β-MSH	β-mercaptoethanol, 2-mercaptoethanol
MW	molecular weight

N

N	A or C or G or T
N-Ac	N-acetyl-
NAD^+	nicotinamide adenine dinucleotide (oxidized form)
NADH	nicotinamide adenine dinucleotide (reduced form)
$NADP^+$	nicotinamide adenine dinucleotide phosphate (oxidized form)
NADPH	nicotinamide adenine dinucleotide phosphate (reduced form)
NAG	N-acetyl-D-galactosaminidase
NAGlc	N-acetyl-glucosaminidase
NAH	N-acetyl-hexosaminidase
$NaHCO_3$	sodium bicarbonate
NaN_3	sodium azide
NANA	N-acetyl-neuraminic acid aldolase
NaOAc	sodium acetate
NaOH	sodium hydroxide
$NaPO_4$	sodium phosphate
Na_2S	sodium sulfide
NH_3	ammonia
NH_4HCO_3	ammonium bicarbonate

Glossary

(NH₄)₂SO₄	ammonium sulfate
nM	nanomolar
nmol	nanomole
2NNap	β-naphthylamide
NO	nitric oxide
NOS	nitric oxide synthase
NT	nitrate reductase
NTP	nucleoside triphosphate

O

OAA	oxaloacetic acid
ODH	oligosaccharide DH
OMP	oligo-*N*-methylmorpholinium propylene oxide

P

PAGE	polyacrylamide gel electrophoresis
PAP	prostatic acid phosphatase
PCA	pyrrolidine carboxylate
PCNB	pentachloronitrobenzene
PCR	polymerase chain reaction
PDE	phosphodiesterase
PDH	pyruvate dehydrogenase
PEG	polyethylene glycol
PEP	phospho(enol)pyruvate
PEPC	phospho(enol)pyruvate carboxylase
PFK	phosphofructokinase
3PGDH	3-phosphoglycerate dehydrogenase, glycerate 3-phosphate dehydrogenase
6PGDH	6-phosphoglycerate dehydrogenase
6PGluDH	6-phosphogluconate dehydrogenase
PGI	phosphoglycerate isomerase
PGlcI	phosphoglucose isomerase
PGlcM	phosphoglucomutase
PGlyM	phosphoglycerate mutase
PGK	3-phosphoglycerate kinase
PGM	phosphoglucomutase
3PGPK	3-phosphoglycerate phosphokinase
pH	acid-base scale; reciprocal log of hydrogen ion concentration
Phe	phenylalanine
PHI	phosphohexose isomerase
pI	isoelectric point, the pH at which there is no net electric charge on a protein
P_i	inorganic phosphate (orthophosphate)
pK	reciprocal log of dissociation constant
PK	pyruvate kinase
PKA	protein kinase A
PKC	protein kinase C
PKG	protein kinase G
PL	pancreatic lipase
PLA₂	phospholipase A₂
PLC	phospholipase C
PLD	phospholipase D
PME	phosphomonoesterase
PMI	phosphomannose isomerase
PMS	phenazine methosulfate
PMSF	phenyl-methane-sulfonyl fluoride
PNGase F	*N*-glycosidase F
PNMT	phenylethanolamine *N*-methyl-transferase
PNPase	polynucleotide phosphorylase
PNPG	α-*p*-nitrophenylglycerine
PO	pyruvate oxidase
PO₄	phosphate
POD	peroxidase
PP	protein phosphatase
PP$_i$	inorganic pyrophosphate
PPase	inorganic pyrophosphatase
ppb	parts per billion
ppm	parts per million
Pro	proline
Prot	protein
PRPP	phosphoribosylpyrophosphate
PTA	phosphotransacetylase
PTK	protein tyrosine kinase
Pu	purine
Py	pyrimidine

R

R	A or G
RBC	red blood cell
RNA	ribonucleic acid
rRNA	ribosomal RNA
tRNA	transfer RNA
RNase	ribonuclease
RT	reverse transcriptase

S

S	C or G
SAM	*S*-adenosyl-L-(methyl)methionine
SCS	succinyl-CoA synthetase
SDH	sorbitol dehydrogenase
SDS	sodium dodecyl sulfate
SDS-PAGE	sodium dodecyl sulfate polyacrylamide gel electrophoresis
Ser	serine
SH	sulfhydryl
SOD	superoxide dismutase

T

SOX	sarcosine oxidase
SP	sucrose phosphorylase
ss	single-stranded (DNA)
T	thymine
TBS	Tris-buffered saline
TCA	trichloroacetic acid
TD	tyrosine decarboxylase
TEA	triethanolamine
TEMED	N,N,N',N'-tetramethylethylenediamine
TFA	trifluoroacetic acid
TH	tyrosine hydroxylase
Thr	threonine
TIM	triose-phosphate isomerase
TK	tyrosine kinase
TLCK	tosyl-L-lysine chloromethyl ketone
TOD	amine oxidase (flavine-containing), tyramine oxidase
TPI	triose-phosphate isomerase
TPCK	N-tosyl-L-phenylalanine-chloromethyl ketone
TPN$^+$	see NADP$^+$
TPNH	see NADPH
TPP	thymidine pyrophosphate
Tris	tris (hydroxymethyl)-aminomethane
Trp	tryptophan
TTP	thymidine triphosphate
Tyr	tyrosine

U

U	unit of enzyme activity
U	uracil
UDP	uridine diphosphate
UDPGal	uridine diphosphogalactose
UDPGlc	uridine diphosphoglucose
UDPGlcDH	uridine 5'-diphosphoglucose dehydrogenase
UMP	uridine monophosphate
UNG	uracil-DNA-glycosylase
UOD	urate oxidase
USP	*United States Pharmacopoeia*
UTP	uridine triphosphate
UV	ultraviolet
μg	microgram
μmol	micromole
μM	micromolar

V

V	A or C or G
Val	valine

W

W	A or T

X

XO	xanthine oxidase
XOD	xanthine oxidase

Y

Y	C or T

Z

ZnOAc	zinc acetate
ZPCK	carbobenzoxy-L-phenylalanine chloromethyl ketone

Part 2. Numerical Listing of Enzymes Assigned Unique EC Classifications

Chapter 4. Single Enzyme Preparations

Alcohol Dehydrogenase

1.1.1.1

REACTION CATALYZED
Alcohol + NAD$^+$ ↔ Aldehyde or Ketone + NADH

SYSTEMATIC NAME
Alcohol:NAD$^+$ oxidoreductase

SYNONYMS
Aldehyde reductase, ADH, AlcDH

REACTANTS
Primary alcohol, secondary alcohol, cyclic secondary alcohol (animal enzyme), aldehyde, hemiacetal, ketone, NAD$^+$, NADH

APPLICATIONS
- Enzymatic determination of primary alcohols, aldehydes, β-NAD and β-NADH
- Synthesis of chiral compounds
- Spectrophotometric assay of plasmalogenase
- Enzymatic catalysis in organic solvents
- Studies of NAD$^+$, NADH, NADP$^+$ and NADPH

NOTES
- Classified as an oxidoreductase acting on the CH-OH group of donors with NAD$^+$ or NADP$^+$ as acceptor
- A component of NADH recycling systems
- A metalloprotein; yeast enzyme contains four tightly bound zinc atoms per molecule
- Inhibited by heavy metals and -SH binding reagents
- Stabilized in dilute solution by serum albumin and glutathione or cysteine

SPECIFIC ACTIVITY	UNITS DEFINITION	PREPARATION FORM	ADDITIONAL ACTIVITIES	SUPPLIER CATALOG NO.
Equine liver (MW = 80,000 Da [two subunits])				
2.7 U/mg	1 unit converts 1 μmol EtOH to product/min at 25°C	Suspension in 20 mM KPO$_4$ buffer and 10% EtOH, pH 7	<0.1% LDH	Boehringer 102741
3 U/mg protein	1 unit oxidizes 1 μmol EtOH/min at pH 8.8, 25°C	Crystallized; suspension in 20 mM KPO$_4$ buffer and 10% EtOH	<0.1% LDH	Fluka 05645
3 U/mg protein	1 unit oxidizes 1 μmol EtOH/min at pH 8.8, 25°C	Crystallized; suspension in 20 mM KPO$_4$ buffer		Fluka 05646
2 U/mg	1 unit oxidizes 1 μmol EtOH/min at pH 8.8, 25°C	Lyophilized		Fluka 05648
1-2 U/mg protein	1 unit converts 1.0 μmol EtOH to acetaldehyde/min at pH 8.8, 25°C	1X Crystallized; lyophilized containing 98% protein, <2% PO$_4$ buffer salts, <0.1% EtOH		Sigma A6128
Lactobacillus kefir				
0.4 U/mg	1 unit reduces 1 μmol acetophenone to phenyl-EtOH/min at pH 7.0, 25°C	Powder		Fluka 05643

1.1.1.1 Alcohol Dehydrogenase *continued*

SPECIFIC ACTIVITY	UNITS DEFINITION	PREPARATION FORM	ADDITIONAL ACTIVITIES	SUPPLIER CATALOG NO.
Yeast, bakers (optimum pH = 8.6-9.0, pI = 5.4, T = 25°C, MW = 141,000 Da)				
≥400 U/mg	1 unit converts 1 μmol EtOH to acetaldehyde/min at pH 8.8, 25°C	Lyophilized		AMRESCO 9432
≥300 U/mg protein; 200-300 U/mg solid	1 unit reduces 1 μmol NAD$^+$/min at pH 8.8, 25°C	Highly purified; lyophilized containing 35% sucrose		Biozyme ADH1
400 U/mg protein	1 unit converts 1 μmol EtOH to product/min at 25°C	Lyophilized containing 30 mg protein, 15 mg sucrose, 5 mg KPO$_4$	<0.01% LDH, MDH	Boehringer 102717
300 U/mg	1 unit converts 1 μmol EtOH to product/min at 25°C	Suspension in 3.2 M (NH$_4$)$_2$SO$_4$, pH 6	<0.01% LDH, MDH	Boehringer 126900 127540 127558
300 U/mg protein	1 unit reduces 1 μmol NAD$^+$/hr at pH 8.8, 25°C	Lyophilized containing 90-95% protein		Calzyme 054A0300
250 U/mg	1 unit oxidizes 1 μmol EtOH/min at pH 8.8, 25°C	Powder		Fluka 05635
300 U/mg protein	1 unit oxidizes 1 μmol alcohol/min at pH 8.8, 25°C	Lyophilized containing 35% sucrose as stabilizer		Fluka 05640
>200 U/mg protein; >100 U/mg solid	1 unit reduces 1.0 μmol NAD$^+$/min at pH 8.8, 25°C	Lyophilized	<0.01% PK, LDH <0.02% aldolase <0.03% MK <0.05% G3PDH, PGK, carboxylase	Genzyme 1091
300 U/mg	1 unit converts 1.0 μmol NAD/min at 25°C	2X Crystallized; lyophilized		ICN 100161
>300 U/mg	1 unit reduces 1 μmol NAD/min at 25°C	Recrystallized by ion-exchange chromatography; lyophilized		ICN 151430
≥300 IU/mg protein	1 IU transforms 1 μmol substrate/min under standard IUB conditions at 25°C	Lyophilized; solution containing 50% glycerol	<0.01% LDH, PK <0.02% aldolase <0.03% MK <0.05% carboxylase, PGK, G3PDH	OYC
75-150 U/g polyacrylamide	1 unit converts 1.0 μmol EtOH to acetaldehyde/min at pH 8.8, 30°C	Insoluble enzyme attached to polyacrylamide containing 30% borate buffer salts		Sigma A0761
900-1200 U/g agarose 1mL gel yields 25-40 U	1 unit converts 1.0 μmol EtOH to acetaldehyde/min at pH 8.8, 30°C	Insoluble enzyme attached to beaded agarose; lyophilized containing lactose and citrate stabilizers		Sigma A2529

Alcohol Dehydrogenase continued — 1.1.1.1

SPECIFIC ACTIVITY	UNITS DEFINITION	PREPARATION FORM	ADDITIONAL ACTIVITIES	SUPPLIER CATALOG No.
Yeast, bakers *continued*				
200-400 U/mg protein	1 unit converts 1.0 µmol EtOH to acetaldehyde/min at pH 8.8, 25°C	Crystallized; lyophilized containing 90% protein, <2% citrate buffer salts, <0.005 mol β-NAD and β-NADH/mol ADH		Sigma A3263
300-500 U/mg protein	1 unit converts 1.0 µmol EtOH to acetaldehyde/min at pH 8.8, 25°C	Crystallized; lyophilized containing 90% protein and <2% citrate buffer salts		Sigma A7011
Yeast (optimum pH = 8.6-9 [EtOH oxidation], 7.0 [acetaldehyde reduction], pI = 5.4, MW = 141,000 Da; unstable at pH <6.0 and >8.5)				
≥300 U/mg protein	1 unit reduces 1 µmol NAD/min at pH 8.8, 25°C	2X Crystallized; lyophilized		Worthington LS01069 LS01070 LS01071
≥300 U/mg protein	1 unit reduces 1 µmol NAD/min at pH 8.8, 25°C	2X Crystallized; suspension in 2.4 M $(NH_4)_2SO_4$, 3% pyrophosphate, 1% Gly		Worthington LS01087 LS01088 LS01089
Zymomonas mobilis (optimum pH = 9.5-10.0, MW = 148,000 Da [37,000/subunit]; K_M = 110 mM [EtOH], 350 mM [MeOH], 0.12 mM [NAD^+]; stable pH 7.0-9.0, T < 40°C)				
>400 U/mg protein	1 unit forms 1 µmol NADH/min at 30°C	Lyophilized		Unitika
>400 U/mg protein	1 unit forms 1 µmol NADH/min at 30°C	Suspension in $(NH_4)_2SO_4$		Unitika

1.1.1.2 Alcohol Dehydrogenase (NADP$^+$)

REACTION CATALYZED
  An alcohol + NADP$^+$ ↔ An aldehyde + NADPH

SYSTEMATIC NAME
  Alcohol:NADP$^+$ oxidoreductase

SYNONYMS
  Aldehyde reductase (NADPH); (R)-aromatic alcohol dehydrogenase, NADP$^+$-dependent, (S)-aromatic alcohol dehydrogenase (NADP$^+$-dependent), ADH

REACTANTS
  Primary alcohols (some enzymes), secondary alcohols (some enzymes), aldehydes, acetaldehyde, ketones, 8-methyldec-2-yl propanoate, NADP$^+$, NADPH

APPLICATIONS
- Synthesis of bifunctional chirons and chiral furan derivatives. *Lactobacillus* enzyme is highly enantiospecific for many (R)-aromatic alcohols. *Thermoanaerobium* enzyme is highly enantiospecific for many (S)-aromatic alcohols
- Conversion of secondary alcohols to corresponding lactones

NOTES
- Classified as an oxidoreductase acting on the CH-OH group of donors with NAD$^+$ or NADP$^+$ as acceptor
- A zinc protein
- A-specific with respect to NADPH
- *Thermoanaerobium* enzyme is extremely thermostable
- May be identical with glucuronate reductase (EC 1.1.1.19), mevaldate reductase (NADPH) (EC 1.1.1.33) and lactaldehyde reductase (NADPH) (EC 1.1.1.55)

SPECIFIC ACTIVITY	UNITS DEFINITION	PREPARATION FORM	ADDITIONAL ACTIVITIES	SUPPLIER CATALOG NO.
Lactobacillus kefiranofaciens				
1-5 U/mg solid	1 unit oxidizes 1.0 μmol 2-hexanol to 2-hexanone/min at pH 7.8, 50°C	Lyophilized containing 5% protein; balance primarily HEPES, MgCl$_2$, DTT		Sigma A9560
Thermoanaerobium brockii				
300-1200 U/g agarose	1 unit oxidizes 1.0 μmol 2-propanol to acetone/min at pH 7.8, 40°C in the presence of NADP$^+$	Insoluble enzyme attached to 4% cross-linked beaded agarose, PNP chloroformate activated; lyophilized containing lactose stabilizer		Sigma A2809
30-90 U/g acrylic beads	1 unit oxidizes 1.0 μmol 2-propanol to acetone/min at pH 7.8, 40°C in the presence of NADP$^+$	Insoluble enzyme attached to macroporous acrylic beads; lyophilized containing 50% glucose stabilizer		Sigma A6184

Alcohol Dehydrogenase (NADP$^+$) continued   1.1.1.2

SPECIFIC ACTIVITY	UNITS DEFINITION	PREPARATION FORM	ADDITIONAL ACTIVITIES	SUPPLIER CATALOG NO.
Thermoanaerobium brockii continued				
30-90 U/mg protein	1 unit oxidizes 1.0 μmol 2-propanol to acetone/min at pH 7.8, 40°C in the presence of NADP$^+$	Purified; lyophilized containing 50% protein; balance primarily Na citrate and DTT		Sigma A9287
35 U/mg	1 unit oxidizes 1 μmol 2-butyl alcohol/min at pH 7.8, 40°C	Powder	NADH-linked alcohol-aldehyde/ketone oxidoreductase	Fluka 05655
5-15 U/mg protein	1 unit oxidizes 1.0 μmol 2-propanol to acetone/min at pH 7.8, 40°C in the presence of NADP$^+$	Lyophilized containing 60% protein; balance primarily NaPO$_4$ buffer salts and DTT		Sigma A8435
Thermoanaerobium species				
15 U/mg		Powder		Biocatalysts S300
1-5 U/mg solid	1 unit oxidizes 1.0 μmol 2-propanol to acetone/min at pH 9.0, 50°C	Lyophilized containing 5% protein; balance primarily raffinose, DTT, KPO$_4$ buffer salts		Sigma A9685

Acetoin Dehydrogenase   1.1.1.5

REACTION CATALYZED
   Acetoin + NAD$^+$ ↔ Diacetyl + NADH

SYSTEMATIC NAME
   Acetoin:NAD$^+$ oxidoreductase

SYNONYMS
   Diacetyl reductase, (S)-diacetyl reductase

REACTANTS
   Acetoin, diacetyl, NAD$^+$, NADH, NADP$^+$

NOTES
- Classified as an oxidoreductase acting on the CH-OH group of donors with NAD$^+$ or NADP$^+$ as acceptor

SPECIFIC ACTIVITY	UNITS DEFINITION	PREPARATION FORM	ADDITIONAL ACTIVITIES	SUPPLIER CATALOG NO.
Chicken liver				
12-30 U/mg protein	1 unit oxidizes 1.0 μmol NADH/min at pH 6.1, 25°C using diacetyl as substrate	Lyophilized		Sigma D5044
Lactobacillus kefir				
0.1 U/mg	1 unit oxidizes 1 μmol 2,3-butanedione/min at pH 7.5, 25°C	Powder		Fluka 00545

1.1.1.5 Acetoin Dehydrogenase continued

SPECIFIC ACTIVITY	UNITS DEFINITION	PREPARATION FORM	ADDITIONAL ACTIVITIES	SUPPLIER CATALOG NO.
Yeast, bakers				
0.25-0.5 U/mg solid	1 unit converts 1 μmol diacetyl and NADPH to (S)-acetoin and NADP/min at pH 6.9, 25°C	Lyophilized containing Tris and DTT		Sigma D1918

1.1.1.6 Glycerol Dehydrogenase

REACTION CATALYZED
Glycerol + NAD$^+$ ↔ Glycerone + NADH

SYSTEMATIC NAME
Glycerol:NAD$^+$ 2-oxidoreductase

SYNONYMS
GDH

REACTANTS
Glycerol, propane-1,2-diol, glycerone, triacylglycerols, NAD$^+$, NADH, dihydroxyacetone, 1,2-propandiol

APPLICATIONS
- Determination of glycerol and triacylglycerols in biological fluids

NOTES
- Classified as an oxidoreductase acting on the CH-OH group of donors with NAD$^+$ or NADP$^+$ as acceptor
- Activated by NH_3, K^+, Rb^+ and Mn^{2+}
- Inhibited by Ba^{2+}, Ca^{2+}, Co^{2+}, Li^+, Na^+, Ni^{2+}, Cu^{2+}, Cd^{2+}, Mg^{2+}, Zn^{2+}, PCMB, o-phenanthroline, monoiodoacetate, and high ionic strength solutions
- Stabilized by BSA

SPECIFIC ACTIVITY	UNITS DEFINITION	PREPARATION FORM	ADDITIONAL ACTIVITIES	SUPPLIER CATALOG NO.
Bacillus megaterium (optimum pH = 8.0-9.0, pI = 4.0, T = 40°C, MW = 230,000 Da [Sephadex G-200]; K_M = 2.9 mM [glycerol], 0.23 mM [NAD$^+$]; stable pH 7.0-9.0 [37°C, 60 min], T < 50°C [pH 8.0, 10 min])				
20-60 U/mg solid	1 unit oxidizes 1 μmol glycerol to dihydroxyacetone/min at pH 9, 25°C	Lyophilized containing 80% protein	<0.01% NADH oxidase <0.1% LDH 1.0-2.0% G3PDH	Asahi GDH T-03
>25 U/mg protein; >20 U/mg solid	1 unit oxidizes 1.0 μmol glycerol/min at pH 9.0, 25°C	Lyophilized	<0.01% NADH oxidase <0.1% LDH 1-2% G3PDH	Genzyme 1281
15 U/mg solid	1 unit oxidizes 1.0 μmol glycerol to dihydroxyacetone/min at pH 8.8, 25°C	Lyophilized containing 70% protein; balance Tris-HCl and stabilizer		Sigma G6267

SPECIFIC ACTIVITY	UNITS DEFINITION	PREPARATION FORM	ADDITIONAL ACTIVITIES	SUPPLIER CATALOG NO.
Bacteria				
≥50 IU/mg protein	1 IU transforms 1 μmol substrate/min under standard IUB conditions at 25°C	Lyophilized		OYC
***Cellulomonas* species** (optimum pH = 10.0-10.5, pI = 4.4, T = 50°C, MW = 390,000 Da [10 subunits per mol enzyme at 42,000 Da each]; K_M = 11 mM [glycerol], 89 μM [NAD^+]; stable pH 7.5-10.5 [25°C, 20 hr], T < 55°C [pH 7.5, 15 min])				
50-125 U/mg protein	1 unit oxidizes 1.0 μmol glycerol to dihydroxyacetone/min at pH 10.0, 25°C	Lyophilized containing 70% protein; balance primarily gluconate and KPO_4 buffer salts		Sigma G3512
≥50 U/mg solid	1 unit forms 1 μmol NADH/min at pH 10.5, 25°C	Lyophilized containing 50% BSA as stabilizer	<0.001% NADH oxidase	Toyobo GYD-301
Enterobacter aerogenes (optimum pH = 9.0; nearly inactive at pH 11)				
25 U/mg	1 unit converts 1 μmol glycerol to product/min at 25°C	Suspension in 3.2 M $(NH_4)_2SO_4$, pH 7.5	<0.01% NADH oxidase	Boehringer 258555
	1 unit oxidizes 1.0 μmol glycerol to dihydroxyacetone/min at 25°C	Lyophilized containing 75% protein; balance primarily KPO_4 buffer salts		Sigma G3755
≥5 U/mg solid	1 unit reduces 1 μmol NAD/min at pH 10.0, 25°C	Partially purified; lyophilized		Worthington LS04532 LS04533 LS04535
Rabbit muscle				
40-100 U/mg protein	1 unit oxidizes 1 μmol glycerol to dihydroxyacetone/min at pH 10.0, 25°C	Lyophilized containing 70% protein; balance primarily gluconate and KPO_4 buffer salts		ICN 151195

1.1.1.8 Glycerol-3-Phosphate Dehydrogenase (NAD$^+$)

REACTION CATALYZED

    sn-Glycerol 3-phosphate + NAD$^+$ ↔ Glycerone phosphate + NADH

SYSTEMATIC NAME

    sn-Glycerol 3-phosphate:NAD$^+$ 2-oxidoreductase

SYNONYMS

    Glycerol dehydrogenase, α-glycerophosphate dehydrogenase, GPDH, G3PDH

REACTANTS

    sn-Glycerol 3-phosphate, glycerone, propane-1,2-diol phosphate (much lower affinity), glycerone sulfate (much lower affinity), NAD$^+$, NADH, dihydroxyacetone phosphate, L-glycerol-3-P

APPLICATIONS

- Detection of serum triglycerides and glycerol
- Reduced enzyme levels are found in chronic multiple sclerosis plaques

NOTES

- Classified as an oxidoreductase acting on the CH-OH group of donors with NAD$^+$ or NADP$^+$ as acceptor
- Highly dependent on substrate and sulfate ion concentration
- Specific for L-glycerol-3-P and dihydroxyacetone phosphate
- Reaction velocity for NADH is 10X that for NADPH
- Inhibited by sulfhydryl reagents, dihydroxyacetone phosphate and fructose 1,6-diphosphate
- Stabilized by NAD$^+$ and lactose; 0.5% BSA in dilute form

SPECIFIC ACTIVITY	UNITS DEFINITION	PREPARATION FORM	ADDITIONAL ACTIVITIES	SUPPLIER CATALOG NO.
Rabbit muscle (optimum pH = 7.0-9.0, T = 45-50°C, MW = 78,000 Da; K_M = 0.77 mM [G3P], 0.16 mM [NAD$^+$], 0.45 mM [DHAP], 13 μM [NADH]; stable pH 5.5-8.5 [25°C, 21 hr] and T < 40°C [pH 8.5, 15 min])				
>300 U/mg protein	1 unit converts 1 μmol DHAP to G3P/min at pH 7.6, 25°C	Highly purified; suspension in 70% (NH$_4$)$_2$SO$_4$ containing 2 mM EDTA	<0.001% ALD, GK <0.01% PK <0.02% LDH <0.03% TIM	Biozyme G3PD3
250 U/mg protein; ≥55 U/mg solid	1 unit converts 1 μmol DHAP to G3P/min at pH 7.6, 25°C	Highly purified; salt-free; lyophilized	<0.001% ALD, GK <0.01% PK <0.02% LDH <0.03% TIM	Biozyme G3PD3F
170 U/mg	1 unit converts 1 μmol DHAP to product/min at 25°C	Suspension in 3.2 M (NH$_4$)$_2$SO$_4$, pH 6	<0.001% aldolase, GAPDH <0.01% LDH, TIM	Boehringer 127124 127752 127779
200-250 U/mg protein	1 unit converts 1 μmol DHAP to α-glycerophosphate/min at pH 7.9, 25°C	Crystallized; suspension in 2.5 M (NH$_4$)$_2$SO$_4$ containing NaPO$_4$ buffer and 10 mg/mL protein, pH 6.5		Calzyme 059B0200

Glycerol-3-Phosphate Dehydrogenase (NAD$^+$) continued 1.1.1.8

SPECIFIC ACTIVITY	UNITS DEFINITION	PREPARATION FORM	ADDITIONAL ACTIVITIES	SUPPLIER CATALOG NO.
Rabbit muscle *continued*				
270 U/mg protein	1 unit reduces 1 µmol DHAP/min at 25°C with NADH as second substrate	Crystalline suspension in 3.2 M (NH$_4$)$_2$SO$_4$, pH 6	<0.001% aldolase, GAPDH <0.01% LDH, TPI	Fluka 50013
120 U/mg	1 unit reduces 1 µmol DHAP/min at 25°C with NADH as second substrate	Lyophilized	<0.001% aldolase, GK <0.01% LDH <0.02% TPI	Fluka 50015
>180 U/mg protein; >45 U/mg solid	1 unit oxidizes 1.0 µmol NADH/min at pH 7.9, 25°C	Lyophilized	<0.005% aldolase <0.01% GK <0.05% LDH <0.1% TPI	Genzyme 1301
>180 U/mg protein	1 unit oxidizes 1.0 µmol NADH/min at pH 7.9, 25°C	Suspension in (NH$_4$)$_2$SO$_4$	<0.005% aldolase <0.01% GK <0.05% LDH <0.1% TPI	Genzyme 1302
300 U/mg protein	1 unit converts 1.0 µmol DHAP to G3P/min at pH 7.6, 25°C	Highly purified; crystallized; suspension in 2.4 M (NH$_4$)$_2$SO$_4$, pH 6.3	<0.001% aldolase, GK <0.03% TPI	ICN 104938
80 U/mg solid	1 unit converts 1.0 µmol DHAP to G3P/min at pH 7.6, 25°C	(NH$_4$)$_2$SO$_4$-free; lyophilized		ICN 151197
≥120 IU/mg protein	1 IU transforms 1 µmol substrate/min under standard IUB conditions at 25°C	Suspension in (NH$_4$)$_2$SO$_4$	<0.001% GK <0.005% aldolase <0.05% LDH <0.1% TPI	OYC
100-300 U/mg protein	1 unit converts 1.0 µmol DHAP to α-glycerophosphate/min at pH 7.4, 25°C	Crystallized; suspension in 2.0 M (NH$_4$)$_2$SO$_4$ and 100 µg/mL EDTA, pH 6.0	<0.01% LDH, PK, aldolase, GAPDH <0.02% TPI	Sigma G6751 (Type I) G4381 (Type III)
100-300 U/mg protein	1 unit converts 1.0 µmol DHAP to α-glycerophosphate/min at pH 7.4, 25°C	Sulfate-free; lyophilized containing 80% protein; balance primarily citrate buffer salts and EDTA	<0.01% LDH, PK, aldolase, GAPDH <0.02% TPI	Sigma G6880
≥15 U/mg solid	1 unit forms 1 µmol NADH/min at pH 9.0, 37°C	Lyophilized containing NAD$^+$ and lactose as stabilizers	<0.01% GK <0.05% aldolase <0.3% LDH, PK <10% TPI	Toyobo G3D-301

1.1.1.10 L-Xylulose Reductase

REACTION CATALYZED
Xylitol + NADP$^+$ ↔ L-Xylulose + NADPH

SYSTEMATIC NAME
Xylitol:NADP$^+$ 4-oxidoreductase (L-xylulose-forming)

REACTANTS
Xylitol, L-xylulose, NADP$^+$, NADPH

NOTES
- Classified as an oxidoreductase acting on the CH-OH group of donors with NAD$^+$ or NADP$^+$ as acceptor

SPECIFIC ACTIVITY	UNITS DEFINITION	PREPARATION FORM	ADDITIONAL ACTIVITIES	SUPPLIER CATALOG NO.
Pigeon liver				
5-15 U/mg protein	1 unit oxidizes 1.0 μmol xylitol to L-xylulose/min at pH 10.0, 25°C	Suspension in 3 M (NH$_4$)$_2$SO$_4$, 0.01 M KPO$_4$, 5.0 mM DTT, 3.0 mM EDTA, pH 8.5		Sigma X5250

1.1.1.14 L-Iditol 2-Dehydrogenase

REACTION CATALYZED
L-Iditol + NAD$^+$ ↔ L-Sorbose + NADH

SYSTEMATIC NAME
L-Iditol: NAD$^+$ 2-oxidoreductase

SYNONYMS
Polyol dehydrogenase, sorbitol dehydrogenase, SDH

SUBSTRATES
L-Iditol, D-glucitol, related sugar alcohols, L-sorbose, D-fructose, D-sorbitol, NAD$^+$, NADH

NOTES
Classified as an oxidoreductase acting on the CH-OH group of donors with NAD$^+$ or NADP$^+$ as acceptor

SPECIFIC ACTIVITY	UNITS DEFINITION	PREPARATION FORM	ADDITIONAL ACTIVITIES	SUPPLIER CATALOG NO.
Candida utilis				
10-20 U/mg protein	1 unit converts 1.0 μmol xylitol to D-xylulose/min at pH 8.6, 25°C	Lyophilized containing Na citrate		Sigma P3759
Ovine liver (MW = 115,000 Da)				
40 U/mg protein	1 unit converts 1 μmol D-fructose to product/min at 25°C	Lyophilized containing 2 mg protein, 10 mg maltose; 60 mg containing 10 mg protein, 50 mg maltose	<0.01% ADH <0.02% GlDH, GlcDH <0.05% LDH, MDH	Boehringer 109312 109339

L-Iditol 2-Dehydrogenase continued — 1.1.1.14

SPECIFIC ACTIVITY	UNITS DEFINITION	PREPARATION FORM	ADDITIONAL ACTIVITIES	SUPPLIER CATALOG NO.
Ovine liver continued				
6 U/mg	1 unit converts 1 µmol D-fructose to D-sorbitol/min at pH 7.6, 25°C	Lyophilized containing 17% protein	<0.01% ADH <0.02% GlDH, GlcDH <0.05% LDH, MDH 83% D-maltose	Fluka 84535
20-40 U/mg protein	1 unit converts 1.0 µmol D-fructose to D-sorbitol/min at pH 7.6, 25°C	Lyophilized containing 15% protein; balance primarily maltose		ICN 156642
20-40 U/mg protein	1 unit converts 1.0 µmol D-fructose to D-sorbitol/min at pH 7.6, 25°C	Lyophilized containing 15% protein; balance primarily maltose		Sigma S1128
40-80 U/mg protein	1 unit converts 1.0 µmol D-fructose to D-sorbitol/min at pH 7.6, 25°C	Lyophilized containing 15% protein; balance primarily maltose	No detectable GlDH, LDH, MDH	Sigma S3764

myo-Inositol 2-Dehydrogenase — 1.1.1.18

REACTION CATALYZED
myo-Inositol + NAD$^+$ ↔ 2,4,6/3,5-Pentahydroxycyclohexanone + NADH

SYSTEMATIC NAME
myo-Inositol:NAD$^+$ 2-oxidoreductase

SUBSTRATES
myo-Inositol, 2,4,6/3,5-pentahydroxycyclohexanone, NAD$^+$, NADH

NOTES
- Classified as an oxidoreductase acting on the CH-OH group of donors with NAD$^+$ or NADP$^+$ as acceptor

SPECIFIC ACTIVITY	UNITS DEFINITION	PREPARATION FORM	ADDITIONAL ACTIVITIES	SUPPLIER CATALOG NO.
Enterobacter aerogenes				
10-20 U/mg protein	1 unit converts 1.0 µmol myo-inositol and β-NAD to scyllo-inosose and β-NADH/min at pH 9.0, 25°C	Lyophilized containing 5% protein, buffer salts, stabilizers		Sigma I0255
25-50 U/mg protein	1 unit converts 1.0 µmol myo-inositol and β-NAD to scyllo-inosose and β-NADH/min at pH 9.0, 25°C	Suspension in 3.1 M (NH$_4$)$_2$SO$_4$, 1% BSA, 2% trehalose, pH 7.3		Sigma I3517
25-50 U/mg protein	1 unit converts 1.0 µmol myo-inositol and β-NAD to scyllo-inosose and β-NADH/min at pH 9.0, 25°C	Highly purified; lyophilized containing 35% protein, buffer salts, stabilizers		Sigma I5010

1.1.1.21 Aldehyde Reductase

REACTION CATALYZED
Alditol + NAD(P)$^+$ ↔ Aldose + NAD(P)H

SYSTEMATIC NAME
Alditol:NAD(P)$^+$ 1-oxidoreductase

SYNONYMS
Aldose reductase, polyol dehydrogenase (NADP$^+$)

REACTANTS
Alditol, NAD(P)$^+$, aldose, NAD(P)H

APPLICATIONS
- Study of sorbitol effects on the body

NOTES
- Classified as an oxidoreductase acting on the CH-OH group of donors with NAD$^+$ or NADP$^+$ as acceptor
- Primary catalyst in aldo-ketose reduction of glyceraldehydes
- Wide specificity

SPECIFIC ACTIVITY	UNITS DEFINITION	PREPARATION FORM	ADDITIONAL ACTIVITIES	SUPPLIER CATALOG NO.
Human muscle cell, recombinant (MW = 36,000 Da)				
1.4-1.6 U/mg; 1 U/mL	1 unit depletes 1 µmol NADPH/min at pH 6.2, 25°C	≥95% purity by SDS-PAGE and silver stain; solution containing 5 mM DTT, 50% glycerin, 50 mM NaPO$_4$ buffer, pH 7.0		Wako 012-13991

1.1.1.22 UDP Glucose 6-Dehydrogenase

REACTION CATALYZED
UDP-Glucose + 2 NAD$^+$ + H$_2$O ↔ UDP-Glucuronate + 2 NADH

SYSTEMATIC NAME
UDP-Glucose:NAD$^+$ 6 oxidoreductase

SYNONYMS
UDP-Glucose dehydrogenase, uridine-5´-diphosphoglucose dehydrogenase, UDPGlcDH

SUBSTRATES
UDP-Glucose, UDP-2-deoxyglucose, UDP-glucuronate, NAD$^+$, NADH, H$_2$O

APPLICATIONS
- Determination of inorganic pyrophosphate, UDP-glucose and UTP

NOTES
- Classified as an oxidoreductase acting on the CH-OH group of donors with NAD$^+$ or NADP$^+$ as acceptor

UDP Glucose 6-Dehydrogenase continued

1.1.1.22

SPECIFIC ACTIVITY	UNITS DEFINITION	PREPARATION FORM	ADDITIONAL ACTIVITIES	SUPPLIER CATALOG NO.
Bovine liver (optimum T = 50-55°C)				
0.6 U/mg	1 unit converts 1 μmol UDP-glucose to product/min at 25°C	Suspension in 3.2 M $(NH_4)_2SO_4$ and BSA as stabilizer, pH 6	<1% LDH	Boehringer 110078
0.02-0.08 U/mg protein	1 unit oxidizes 1.0 μmol UDP-glucose to UDP-glucuronic acid/min at pH 8.7, 25°C	Lyophilized containing 40% protein, 40% $(NH_4)_2SO_4$, 20% Na citrate, pH 6.0	Substantially pyrophosphatase-free β-Glucuronidase, catalase, LDH, GlcDH, MK, nucleotide 5´-diP kinase	Sigma U5500
0.03-0.06 U/mg protein	1 unit oxidizes 1.0 μmol UDP-glucose to UDP-glucuronic acid/min at pH 8.7, 25°C	Lyophilized containing 50% protein, 30% KPO_4, 20% Na citrate	<1% β-glucuronidase, LDH <5% catalase, GlcDH 5% MK, pyrophosphatase 10% nucleotide 5´-diphosphate kinase	Sigma U7251

Glyoxylate Reductase

1.1.1.26

REACTION CATALYZED

Glycolate + NAD^+ ↔ Glyoxylate + NADH

SYSTEMATIC NAME

Glycolate:NAD^+ oxidoreductase

SUBSTRATES
- Glycolate, D-glycerate, glyoxylate, hydroxypyruvate, NAD^+, NADH

NOTES
- Classified as an oxidoreductase acting on the CH-OH group of donors with NAD^+ or $NADP^+$ as acceptor

SPECIFIC ACTIVITY	UNITS DEFINITION	PREPARATION FORM	ADDITIONAL ACTIVITIES	SUPPLIER CATALOG NO.
Spinach leaves				
50 U/mg protein	1 unit converts 1.0 μmol glyoxylate to glycolate/min at pH 6.4, 25°C	Suspension in 3.0 M $(NH_4)_2SO_4$ and 0.01 M KPO_4, pH 7.0	100-300 U/mg protein glycerate dehydrogenase	Sigma G5259

L-Lactate Dehydrogenase

1.1.1.27

REACTION CATALYZED
(S)-Lactate + NAD$^+$ ↔ Pyruvate + NADH

SYSTEMATIC NAME
(S)-Lactate:NAD$^+$ oxidoreductase

SYNONYMS
L-Lactic acid dehydrogenase, L-lactic dehydrogenase, lactate dehydrogenase, lactate dehydrogenase NAD-dependent, lactic acid dehydrogenase, lactic dehydrogenase, NAD-lactate dehydrogenase, LDH

REACTANTS
(S)-Lactate, (S)-2-hydroxymonocarboxylic acids, pyruvate, NAD$^+$, NADH, NADP$^+$ (animal enzyme, slowly)

APPLICATIONS
- Determination of ATP, ADP, glucose, creatinine, pyruvate, lactate and glycerol
- Determination of GPT, PK and CPK by coupling with related enzymes
- Used clinically as a marker for incidence of myocardial infarction

NOTES
- Classified as an oxidoreductase acting on the CH-OH group of donors with NAD$^+$ or NADP$^+$ as acceptor
- Mammalian enzyme exists as 5 tetrameric isozymes composed of combinations of 2 different subunits; molecular weight ranges 134,600-138,800 Da
- Mammalian isozymes differ in catalytic, physical and immunological properties:
  - H promotes aerobic oxidation of pyruvate
  - M promotes anaerobic metabolism and pyruvate reduction
  - H (heart) and M (muscle) combine to form H$_4$, M$_4$, H$_3$M, H$_2$M$_2$ and HM$_3$
  - H$_4$ is most negatively charged
- Activated by dimethyl sulfoxide, ethanol and methanol
- Stabilized by BSA, 2-mercaptoethanol, sucrose, diethylstilbestrol and NADH
- Inhibited by Ag$^+$, Cu^{2+}, Hg^{2+}, hydroxylamine, I$^-$, p-chloromercuribenzoate, oxalate and oxamic acid

SPECIFIC ACTIVITY	UNITS DEFINITION	PREPARATION FORM	ADDITIONAL ACTIVITIES	SUPPLIER CATALOG NO.
Bacillus species (optimum pH = 7.5, pI = 4.87, T = 65°C, MW = 105,000 Da [gel filtration]; K_M = 60 μM [NADH], 12.2 mM [pyruvate]; stable pH 5.5-9.0 [65°C, 10 min], T < 65°C [pH 5.5-9.0, 10 min])				
	1 unit produces 1 μmol lactate/min at 37°C	Lyophilized containing BSA and sucrose as stabilizers		Asahi LDHS T-49

L-Lactate Dehydrogenase continued — 1.1.1.27

SPECIFIC ACTIVITY	UNITS DEFINITION	PREPARATION FORM	ADDITIONAL ACTIVITIES	SUPPLIER CATALOG NO.
Bacillus stearothermophilus (optimum pH = 6.0, MW = 140,000 Da [35,000/subunit]; K_M = 2.6 mM [pyruvate], 48 μM [NADH]; stable pH 8-9.5 [4°C, 24 hr], T < 55°C [pH 7.5, 15 min])				
200 U/mg protein	1 unit reduces 1.0 μmol pyruvate to L-lactate/min at pH 7.5, 37°C	Lyophilized containing 60% protein; balance KPO_4 buffer salts		Sigma Type XXXVII L3514
>300 U/mg protein	1 unit forms 1 μmol of NAD^+/min at 30°C	Lyophilized	<0.01% GOT, GPT, NADH oxidase	Unitika
>300 U/mg protein	1 unit forms 1 μmol of NAD^+/min at 30°C	50% glycerol solution	<0.01% GOT, GPT, NADH oxidase	Unitika
200-600 U/mg protein	1 unit reduces 1.0 μmol pyruvate to L-lactate/min at pH 7.5, 37°C	Suspension in 2.7 M $(NH_4)_2SO_4$		Sigma L5275
Bovine adrenal glands				
300-500 U/mg protein; 90% activity as H4 isozyme	1 unit reduces 1.0 μmol pyruvate to L-lactate/min at pH 7.5, 37°C	Suspension in 3.2 M $(NH_4)_2SO_4$, pH 6.0	No detectable PK	Sigma Type XXIV L5008
Bovine heart				
300 U/mg protein	1 unit oxidizes 1 μmol NADH/min at pH 7.4, 25°C	2X Crystallized; suspension in 60% $(NH_4)_2SO_4$, pH 7.2	<0.01% PK, GPT <0.05% MDH	Biozyme LDHB2
300 U/mg protein	1 unit oxidizes 1 μmol NADH/min at pH 7.4, 25°C	2X Crystallized; lyophilized; $(NH_4)_2SO_4$-free	<0.01% PK, GPT <0.05% MDH	Biozyme LDHB2F
300 U/mg protein	1 unit oxidizes 1 μmol NADH/min at pH 7.4, 25°C	2X Crystallized; 50% glycerol; NH_3-free	<0.01% PK, GPT <0.05% MDH	Biozyme LDHB2G
250 U/mg at 25°C 620 U/mg at 37°C	1 unit converts 1 μmol pyruvate to product/min	Suspension in 3.2 M $(NH_4)_2SO_4$, pH 6.5	<0.01% MK, PK <0.05% MDH	Boehringer 106984
300 U/mg protein	1 unit reduces 1 μmol of pyruvate to L-lactate/min at 25°C in 0.1 M PO_4 buffer, pH 7.0	Crystallized; suspension in 70% $(NH_4)_2SO_4$, PO_4 buffer, 10 mg/mL protein, pH 7		Calzyme 195B0300
250 U/mg protein	1 unit reduces 1 μmol of pyruvate/min at pH 7, 25°C	2x Crystallized; suspension in 3.5 M $(NH_4)_2SO_4$, pH 7.2	<0.01% PK <0.5% MDH	Fluka 61310
300-400 U/mg protein	1 unit reduces 1 μmol of pyruvate to L-lactate/min at pH 7.4, 25°C	Highly purified; 2X crystallized; suspension in 60% saturated $(NH_4)_2SO_4$, pH 7.2	<0.01% GPT, PK <0.05% MDH	ICN 151531
1000 U/mL	1 unit reduces 1.0 μmol pyruvate to L-lactate/min at pH 7.5, 37°C	Crystallized; suspension in $(NH_4)_2SO_4$		Sigma Type III 826-6
500-700 U/mg protein	1 unit reduces 1.0 μmol pyruvate to L-lactate/min at pH 7.5, 37°C	Crystallized; suspension in 2.1 M $(NH_4)_2SO_4$, pH 6.0	< 0.03% PK	Sigma Type III L2625

SPECIFIC ACTIVITY	UNITS DEFINITION	PREPARATION FORM	ADDITIONAL ACTIVITIES	SUPPLIER CATALOG NO.
Bovine heart *continued*				
500-700 U/mg protein	1 unit reduces 1.0 μmol pyruvate to L-lactate/min at pH 7.5, 37°C	Crystallized; suspension in 2.1 M $(NH_4)_2SO_4$, pH 6.0	<0.01% PK, MK, α-GPDH, GOT <0.02% GPT	Sigma Type XV L7755
400-600 U/mg protein	1 unit reduces 1.0 μmol pyruvate to L-lactate/min at pH 7.5, 37°C	Solution containing 50% glycerol, 0.025 M KPO_4 buffer, pH 7.5 <10 mg/mg protein NH_3	<0.01% PK, MK, α-GPDH <0.02% GPT, GOT	Sigma Type XVII L1006
≥200 U/mg protein	1 unit oxidizes 1 μmol NADH/min at pH 7.3, 25°C	Chromatographically purified; crystallized; suspension in 2.2 M $(NH_4)_2SO_4$	<0.01% PK	Worthington LS01222 LS01223 LS01224 LS01228
Bovine heart, LDH-1 (H4) Isozyme				
200 U/mg		Crystallized; suspension in 0.55 M saturated $(NH_4)_2SO_4$	Substantially free of PK	ICN 100833
500 U/mg protein	1 unit reduces 1.0 μmol pyruvate to L-lactate/min at pH 7.5, 37°C	Crystallized; suspension in 2.1 M $(NH_4)_2SO_4$, pH 6.0	<0.01% PK, MK, α-GPDH, GOT <0.02% GPT	Sigma Type IX L0377
Bovine heart, LDH-2 (H3M) Isozyme				
400-700 U/mg protein	1 unit reduces 1.0 μmol pyruvate to L-lactate/min at pH 7.5, 37°C	Suspension in 2.8 M $(NH_4)_2SO_4$, pH 6.0	<0.01% PK	Sigma Type XXV L0133
Bovine muscle				
450 U/mg	1 unit converts 1 μmol pyruvate to product/min at 25°C	Suspension in 3.2 M $(NH_4)_2SO_4$, pH 6.5	<0.01% GPT, MK <0.1% MDH, PK	Boehringer 106992
600-900 U/mg protein	1 unit reduces 1.0 μmol pyruvate to L-lactate/min at pH 7.5, 37°C	Crystalline suspension in 2.4 M $(NH_4)_2SO_4$, pH 6.0	< 0.01% PK	Sigma Type X L1378
Bovine muscle, LDH-5 (M4) isozyme				
600-900 U/mg protein	1 unit reduces 1.0 μmol pyruvate to L-lactate/min at pH 7.5, 37°C	Suspension in 2.6 M $(NH_4)_2SO_4$, pH 6.0	< 0.01% PK	Sigma Type XXVI L0508
Bovine semen, X isozyme				
60-120 U/mg protein	1 unit reduces 1.0 μmol pyruvate to L-lactate/min at pH 7.5, 37°C	Lyophilized containing 0.5% protein and 2% DTT; balance Ficoll and Tris buffer salts		Sigma L7892
Chicken heart (optimum pH = 8.9, pI = 4.8, MW = 140,000 Da)				
300 U/mg solid	1 unit reduces 1 μmol of pyruvate to L-lactate/min at 25°C in 0.1 M PO_4 buffer, pH 7.0	Lyophilized powder		Calzyme 127A0300

SPECIFIC ACTIVITY	UNITS DEFINITION	PREPARATION FORM	ADDITIONAL ACTIVITIES	SUPPLIER CATALOG NO.
Chicken heart *continued*				
300 U/mg solid		Lyophilized		ICN 153883
500-800 U/mg protein	1 unit reduces 1.0 μmol pyruvate to L-lactate/min at pH 7.5, 37°C	Crystallized; suspension in 1.3 M $(NH_4)_2SO_4$, pH 6.0	No detectable PK	Sigma Type VIII L9126
≥300 U/mg		Lyophilized		Wako 126-03411
Chicken liver				
400-800 U/mg protein	1 unit reduces 1.0 μmol pyruvate to L-lactate/min at pH 7.5, 37°C	Suspension in 3.2 M $(NH_4)_2SO_4$ and 0.005 M KPO_4, pH 6.5	No detectable PK	Sigma Type XXXIII L6504
Chicken muscle				
800-1000 U/mg protein	1 unit reduces 1.0 μmol pyruvate to L-lactate/min at pH 7.5, 37°C	Suspension in 2.1 M $(NH_4)_2SO_4$, pH 6.0	< 0.01% PK	Sigma Type XXXIV L9887
Dog muscle				
400-900 U/mg protein	1 unit reduces 1.0 μmol pyruvate to L-lactate/min at pH 7.5, 37°C	Crystallized; suspension in 2.1 M $(NH_4)_2SO_4$, pH 6.0	< 0.01% PK	Sigma Type XIV L0755
Escherichia coli*, recombinant *Bacillus (optimum pH = 6.0-6.4; relative activity: 25°C = 1.0, 30°C = 1.25, 37°C = 1.68)				
>400 U/mg protein; >150 U/mg solid	1 unit oxidizes 1.0 μmol NADH/min at pH 6.0, 30°C	Lyophilized	<0.003% GlDH <0.01% CK, GAT, PGM, enolase, AK, PK, GOT	Genzyme 1561
Human				
				Vital 17234
				Vital 17234
Human erythrocytes				
>50 IU/mg protein	Sigma LDH kit (lactate to pyruvate), 37°C	30% purity by SDS-PAGE; 80-100% purity with respect to other LDH isozymes by agarose electrophoresis; solution containing >0.4 mg/mL protein and Tris buffer, pH 6.8-7.2	No detectable HBsAg, HCV, HIV	Aalto 1026
>50 IU/mg protein	Sigma LDH kit (lactate to pyruvate), 37°C	30% purity by SDS-PAGE; 80-100% purity with respect to other LDH isozymes by agarose electrophoresis; solution containing >0.4 mg/mL protein and Tris buffer, pH 6.8-7.2	No detectable HBsAg, HCV, HIV	Aalto 1028

SPECIFIC ACTIVITY	UNITS DEFINITION	PREPARATION FORM	ADDITIONAL ACTIVITIES	SUPPLIER CATALOG No.
Human erythrocytes *continued*				
300-500 U/mg protein	1 unit reduces 1.0 μmol pyruvate to L-lactate/min at pH 7.5, 37°C	Suspension in 2.1 M $(NH_4)_2SO_4$, pH 6.0	No detectable PK LDH-1 (H4) & LDH-2 (H3M) Isozymes	Sigma Type XII L3379
Human erythrocytes, LDH-1 (H4) isozyme				
200-400 U/mg	LDH-1 activity determined using lactate as a substrate, measuring the appearance of NADH at A_{340}, 37°C	>95% purity; solution containing 0.5-1.0 mg/mL protein	<0.1% PK	Cortex CP9062 LDH-1
250-450 U/mg protein	1 unit reduces 1.0 μmol pyruvate to L-lactate/min at pH 7.5, 37°C	≥98% homogeneous; suspension in 2.1 M $(NH_4)_2SO_4$, pH 6.0	< 0.1% PK	Sigma Type XIX L3632
Human erythrocytes, LDH-2 (H3M) isozyme				
100-200 U/mg	LDH-2 activity determined enzymatically	>95% purity; solution containing 0.5-1.0 mg/mL protein	<0.1% PK	Cortex CP9062 LDH-2
150 U/mg protein	1 unit reduces 1.0 μmol pyruvate to L-lactate/min at pH 7.5, 37°C	98% homogeneous; suspension in 2.1 M $(NH_4)_2SO_4$, pH 6.0	No detectable PK	Sigma Type XX L3757
Human erythrocytes, LDH-3 (H2M2) isozyme				
250 U/mg protein	1 unit reduces 1.0 μmol pyruvate to L-lactate/min at pH 7.5, 37°C	≥95% homogeneous; suspension in 2.1 M $(NH_4)_2SO_4$, pH 6.0	No detectable PK	Sigma Type XXI L3882
Human heart				
50-100 U/mg protein		Crystallized; suspension in 70% $(NH_4)_2SO_4$, PO_4 buffer, 10 mg/mL protein, pH 7		ICN 153882
Human heart, LDH-1				
50-100 U/mg protein	1 unit reduces 1 μmol of pyruvate to L-lactate/min at 25°C in 0.1 M PO_4 buffer, pH 7.0	Crystallized; suspension in 70% $(NH_4)_2SO_4$, PO_4 buffer, 10 mg/mL protein, pH 7		Calzyme 060B0050
>150 U/mg	LDH-1 activity is determined using lactate as substrate, measuring the appearance of NADH at A_{340}, 37°C	35-60% purity; suspension in 20 mM Tris-HCl, 1 mM EDTA, 80% $(NH_4)_2SO_4$, 0.5-1.0 mg/mL protein, , pH 7.5		Scripps L1223
>300 U/mg	LDH-1 activity is determined using lactate as substrate, measuring the appearance of NADH at A_{340}, 37°C	>99% purity; suspension in 20 mM Tris-HCl, 1 mM EDTA, 80% $(NH_4)_2SO_4$, 0.5-1.0 mg/mL protein, pH 7.5	No detectable contaminants	Scripps L1224
Human heart, LDH-1 & -2				
100-150 U/mg	Activity determined using lactate as a substrate, measuring the appearance of NADH at A_{340}, 37°C	80% purity; solution containing 0.5-1.0 mg/mL protein		Cortex CP9062 LDH-1 & -2

L-Lactate Dehydrogenase continued 1.1.1.27

SPECIFIC ACTIVITY	UNITS DEFINITION	PREPARATION FORM	ADDITIONAL ACTIVITIES	SUPPLIER CATALOG No.
Human heart, LDH-2				
350 U/mg protein	1 unit reduces 1 μmol of pyruvate to L-lactate/min at pH 7.0, 25°C in 0.1 M PO_4 buffer	Crystallized; suspension in 70% $(NH_4)_2SO_4$, PO_4 buffer, 10 mg/mL protein, pH 7		Calzyme 170B0350
Human heart, LDH-3				
200 U/mg	1 unit reduces 1 μmol of pyruvate to L-lactate/min at pH 7.0, 25°C in 0.1 M PO_4 buffer	Crystallized; suspension in 70% $(NH_4)_2SO_4$ and PO_4 buffer, pH 7		Calzyme 175B0200
Human liver, LDH-4 isozyme				
400 U/g	1 unit reduces 1.0 μmol of pyruvate to L-lactate/min at pH 7.5, 37°C	$(NH_4)_2SO_4$ suspension		Sigma L2526
200 U/mg	1 unit reduces 1 μmol of pyruvate to L-lactate/min at pH 7.0, 25°C in 0.1 M PO_4 buffer	Crystallized; suspension in 70% $(NH_4)_2SO_4$ and PO_4 buffer, pH 7		Calzyme 193B0200
Human muscle				
50-100 U/mg protein		Crystallized; suspension in 70% $(NH_4)_2SO_4$, PO_4 buffer, 10 mg/mL protein, pH 7		ICN 153881
Human muscle, LDH-5				
200 U/mg	1 unit reduces 1 μmol of pyruvate to L-lactate/min at pH 7.0, 25°C in 0.1 M PO_4 buffer	Crystallized; suspension in 70% $(NH_4)_2SO_4$ and PO_4 buffer, pH 7		Calzyme 194B0200
Human placenta, LDH-5 (M4) isozyme				
300-600 U/mg protein	1 unit reduces 1.0 μmol pyruvate to L-lactate/min at pH 7.5, 37°C	Suspension in 3.2 M $(NH_4)_2SO_4$, pH 6.0	No detectable PK	Sigma Type XXVIII L6508
Human tissue				
>50 IU/mg protein	Sigma LDH kit (lactate to pyruvate), 37°C	30% purity by SDS-PAGE; 80-100% purity with respect to other LDH isozymes by agarose electrophoresis; solution containing >0.4 mg/mL protein and Tris buffer, pH 6.8-7.2	No detectable HBsAg, HCV, HIV	Aalto 1030
>50 IU/mg protein	Sigma LDH kit (lactate to pyruvate), 37°C	30% purity by SDS-PAGE; 80-100% purity with respect to other LDH isozymes by agarose electrophoresis; solution containing >0.4 mg/mL protein and Tris buffer, pH 6.8-7.2	No detectable HBsAg, HCV, HIV	Aalto 1032
***Lactobacillus* species (optimum pH = 6.0, pI = 4.96, T = 60°C, MW = 136,000 Da; K_M = 50 mM [NADH], 0.89 mM [pyruvate]; stable pH 6.0-8.0 [50°C, 30 min], T < 65°C [pH 6.2-7.5, 30 min])**				
	1 unit produces 1 μmol lactate to pyruvate/min at pH 8, 37°C	Lyophilized		Asahi LDH T-54

1.1.1.27 L-Lactate Dehydrogenase continued

SPECIFIC ACTIVITY	UNITS DEFINITION	PREPARATION FORM	ADDITIONAL ACTIVITIES	SUPPLIER CATALOG NO.
Leuconostoc mesenteroides				
800-1200 U/mg	1 unit oxidizes 1 μmol of NADH/min at pH 7.4, 25°C	Lyophilized		ICN 151534
Lobster tail				
400-900 U/mg protein	1 unit reduces 1.0 μmol pyruvate to L-lactate/min at pH 7.5, 37°C	Crystallized; suspension in 3.2 M $(NH_4)_2SO_4$ and 0.005 M KPO_4, pH 6.5	< 0.1% PK	Sigma Type XXIII L0883
Microbial (optimum pH = 6.0-7.0, T = 35-40°C)				
100 U/mg protein	1 unit oxidizes 1 μmol NADH/min	Lyophilized		AMRESCO 0951
Pigeon breast muscle				
400-800 U/mg protein	1 unit reduces 1.0 μmol pyruvate to L-lactate/min at pH 7.5, 37°C	Suspension in 3.2 M $(NH_4)_2SO_4$, pH 6.0	< 0.02% PK	Sigma Type XXII L9757
Porcine heart (optimum pH = 6.0-7.4, T = 60°C, MW = 115,000 ± 6,500 Da [4 subunits/mol enzyme]; K_M = 25 mM [lactate], 0.1 mM [pyruvate]; stable pH 6.0-10.0 [23°C, 22 hr], T < 50°C [pH 7.4, 10 min])				
5000 U/mL	1 unit oxidizes 1 μmole of NADH/min under assay conditions at 25°C	$(NH_4)_2SO_4$ suspension		AMRESCO 0253
350 U/mg protein	1 unit oxidizes 1 μmol NADH/min	Lyophilized		AMRESCO 0367
300 U/mg protein	1 unit oxidizes 1 μmol NADH/min at pH 7.4, 25°C	2X Crystallized; suspension in 70% $(NH_4)_2SO_4$, pH 7.2	<0.01% PK, GPT <0.05% MDH	Biozyme LDHP2
300 U/mg protein	1 unit oxidizes 1 μmol NADH/min at pH 7.4, 25°C	2X Crystallized; lyophilized; $(NH_4)_2SO_4$-free	<0.01% PK, GPT <0.05% MDH	Biozyme LDHP2F
300 U/mg protein	1 unit oxidizes 1 μmol NADH/min at pH 7.4, 25°C	2X Crystallized; 50% glycerol; NH_3-free	<0.01% PK, GPT <0.05% MDH	Biozyme LDHP2G
300 U/mg at 25°C; 750 U/mg at 37°C	1 unit converts 1 μmol pyruvate to product/min	Suspension in 3.2 M $(NH_4)_2SO_4$, pH 6.0	<0.01% MDH, MK, PK <0.02% GOT	Boehringer 107034 107069
300 U/mg	1 unit reduces 1 μmol L-lactate to pyruvate/min at pH 7.4, 25°C	3X Crystallized; suspension in 70% $(NH_4)_2SO_4$, pH 7.2	<0.01% GPT, PK <0.05% MDH	Calbiochem 42721
350 U/mg protein	1 unit reduces 1 μmol of pyruvate to L-lactate/min at pH 7.0, 25°C in 0.1 M PO_4 buffer	Lyophilized containing PO_4 buffer and 80-90% protein, pH 7		Calzyme 092A0350
>280 U/mg protein; >150 U/mg solid	1 unit oxidizes 1 μmol of NADH/min at 25°C	Lyophilized powder	<0.01% GOT, GPT, PK <0.05% MDH, MK	Genzyme 1391
>280 U/mg protein; >150 U/mg solid	1 unit oxidizes 1 μmol of NADH/min at 25°C	$(NH_4)_2SO_4$ suspension	<0.01% GOT, GPT, PK <0.05% MDH, MK	Genzyme 1392

L-Lactate Dehydrogenase continued — 1.1.1.27

SPECIFIC ACTIVITY	UNITS DEFINITION	PREPARATION FORM	ADDITIONAL ACTIVITIES	SUPPLIER CATALOG NO.
Porcine heart *continued*				
350-400 U/mg protein	1 unit reduces 1 μmol of pyruvate to L-lactate/min at pH 7.4, 25°C	Highly purified; 2X crystallized; suspension in 70% saturated $(NH_4)_2SO_4$, pH 7.2	<0.01% GPT, PK <0.05% MDH	ICN 151532
>300 IU/mg protein	1 IU transforms 1 μmol of substrate/min under standard conditions defined by the IUB	Lyophilized	<0.003% PK <0.01% MK <0.03% MDH, GPT, GOT	OYC
>300 IU/mg protein	1 IU transforms 1 μmol of substrate/min under standard conditions defined by the IUB	$(NH_4)_2SO_4$ suspension	<0.003% PK <0.01% MK <0.03% MDH, GPT, GOT	OYC
>300 IU/mg protein	1 IU transforms 1 μmol of substrate/min under standard conditions defined by the IUB	50% glycerol solution	<0.003% PK <0.01% MK <0.03% MDH, GPT, GOT	OYC
400-600 U/mg protein	1 unit reduces 1.0 μmol pyruvate to L-lactate/min at pH 7.5, 37°C	Crystallize; suspension in 3.2 M $(NH_4)_2SO_4$, pH 6.0	<0.01% PK, MK, MDH, GOT <0.03% GPT	Sigma Type XVIII L2881
500 U/mg protein	1 unit reduces 1.0 μmol pyruvate to L-lactate/min at pH 7.5, 37°C	Solution containing 50% glycerol and KPO_4 buffer, pH 7.5	<0.01% PK, MK, MDH, GOT <0.03% GPT	Sigma Type XXXV L9889
≥350 U/mg protein	1 unit oxidizes 1 μmol of NADH/min at pH 7.4, 25°C	Crystallized; suspension in 1.6 M $(NH_4)_2SO_4$	<0.001% PK <0.01% MK <0.02% MDH <0.03% GPT	Toyobo LCD-209
Porcine heart, LDH-1 (H4) isozyme				
500 U/mg protein	1 unit reduces 1.0 μmol pyruvate to L-lactate/min at pH 7.5, 37°C	Crystallize; suspension in 3.2 M $(NH_4)_2SO_4$, pH 6.0	<0.01% PK, MK, MDH, GOT <0.03% GPT	Sigma Type VII-S L9382
Porcine muscle				
400 U/mg protein	1 unit oxidizes 1 μmol NADH/min at pH 7.4, 25°C	3X Crystallized; suspension in 70% $(NH_4)_2SO_4$, pH 7.2	<0.01% MDH, GPT <0.05% PK	Biozyme LDHPM2
400 U/mg protein	1 unit oxidizes 1 μmol NADH/min at pH 7.4, 25°C	3X Crystallized; lyophilized; $(NH_4)_2SO_4$-free	<0.01% MDH, GPT <0.05% PK	Biozyme LDHPM2F
400 U/mg protein	1 unit oxidizes 1 μmol NADH/min at pH 7.4, 25°C	3X Crystallized; 50% glycerol; NH_3-free	<0.01% MDH, GPT <0.05% PK	Biozyme LDHPM2G
550 U/mg at 25°C 1000 U/mg at 37°C	1 unit converts 1 μmol pyruvate to product/min	Suspension in 3.2 M $(NH_4)_2SO_4$, pH 6.5	<0.001% PK, aldolase <0.01% GlDH, GOT, GPT, MDH, MK	Boehringer 107077
550 U/mg at 25°C 850 U/mg at 37°C	1 unit converts 1 μmol pyruvate to product/min	50% glycerol solution, pH 6.5	<0.001% PK, aldolase <0.005% GlDH <0.01% GOT, GPT, MDH, MK	Boehringer 127221 127868

1.1.1.27 L-Lactate Dehydrogenase continued

SPECIFIC ACTIVITY	UNITS DEFINITION	PREPARATION FORM	ADDITIONAL ACTIVITIES	SUPPLIER CATALOG NO.
Porcine muscle *continued*				
550 U/mg protein	1 unit reduces 1 μmol of pyruvate to L-lactate/min at pH 7.0, 25°C in 0.1 M PO$_4$ buffer	50% glycerol solution containing 10 mg/mL protein, pH 7		Calzyme 051C0550
400-450 U/mg protein	1 unit reduces 1 μmol of pyruvate to L-lactate/min at pH 7.4, 25°C	Purified; 3X crystallized; suspension in 65% saturated (NH$_4$)$_2$SO$_4$, pH 7.2	<0.01% GPT <0.015% MDH <0.05% PK	ICN 151533
1000 U/mg protein	1 unit reduces 1.0 μmol pyruvate to L-lactate/min at pH 7.5, 37°C	Solution containing 50% glycerol and KPO$_4$ buffer, pH 7.0	<0.01% PK, MK, MDH, GOT, GPT	Sigma Type XXIX-S L0762
800-1000 U/mg protein	1 unit reduces 1.0 μmol pyruvate to L-lactate/min at pH 7.5, 37°C	Suspension in 2.1 M (NH$_4$)$_2$SO$_4$, pH 6.0	<0.01% PK, MK, MDH, GOT, GPT	Sigma Type XXX-S L4387
Porcine muscle, LDH-5 (M4) isozyme				
1000 U/mg protein	1 unit reduces 1.0 μmol pyruvate to L-lactate/min at pH 7.5, 37°C	Suspension in 2.1 M (NH$_4$)$_2$SO$_4$, pH 6.0	<0.01% PK, MK, MDH, GOT, GPT	Sigma Type XXXII-S L5762
Rabbit heart, LDH-1 (H4) isozyme				
400-600 U/mg protein	1 unit reduces 1.0 μmol pyruvate to L-lactate/min at pH 7.5, 37°C	Suspension in 3.2 M (NH$_4$)$_2$SO$_4$, pH 6.0	PK, PGluM activity < 0.01% of l-LDH activity	Sigma Type XXXVI L2889
Rabbit muscle (optimum pH = 6.0-7.0; relative activity: 25°C = 1.0, 30°C = 1.27, 37°C = 1.74)				
550 U/mg protein	1 unit oxidizes 1 μmol NADH/min at pH 7.4, 25°C	2X Crystallized; suspension in 65% (NH$_4$)$_2$SO$_4$, pH 7.2	<0.01% MDH, GPT, PK	Biozyme LDHR2
550 U/mg protein	1 unit oxidizes 1 μmol NADH/min at pH 7.4, 25°C	2X Crystallized; lyophilized; (NH$_4$)$_2$SO$_4$-free	<0.01% MDH, GPT, PK	Biozyme LDHR2F
550 U/mg protein	1 unit oxidizes 1 μmol NADH/min at pH 7.4, 25°C	2X Crystallized; 50% glycerol; NH$_3$-free	<0.01% MDH, GPT, PK	Biozyme LDHR2G
550 U/mg at 25°C 1100 U/mg at 37°C	1 unit converts 1 μmol pyruvate to product/min	Suspension in 3.2 M (NH$_4$)$_2$SO$_4$, pH 7	<0.001% PK, aldolase <0.01% GOT, GPT, MDH, MK	Boehringer 127230 127876 127884
200 U/mg protein	1 unit converts 1 μmol L-lactate to pyruvate in 0.1 M PO$_4$ buffer/min at pH 7.0, 30°C	Lyophilized	<0.001% GOT <0.005% GPT	Calbiochem 427217
350 U/mg protein	1 unit reduces 1 μmol of pyruvate to L-lactate/min at pH 7.0, 25°C in 0.1 M PO$_4$ buffer	Lyophilized containing PO$_4$ buffer, mannitol, 90% protein, pH 7		Calzyme 127A0350
140 U/mg	1 unit reduces 1 μmol of pyruvate/min at pH 7, 25°C	Lyophilized	<0.01% MK, enolase, PGM, CPK, GAT <0.02% GOT <0.05% PK	Fluka 61309

L-Lactate Dehydrogenase continued   1.1.1.27

SPECIFIC ACTIVITY	UNITS DEFINITION	PREPARATION FORM	ADDITIONAL ACTIVITIES	SUPPLIER CATALOG NO.
Rabbit muscle *continued*				
500 U/mg protein	1 unit reduces 1 µmol of pyruvate/min at pH 7, 25°C	Aqueous suspension in 2.1 M $(NH_4)_2SO_4$, pH 6	<0.01% CPK, GOT, GPT, GDPDH, MDH, GAPDH, PGluM, N5DP, MK <0.02% PK	Fluka 61311
>400 U/mg protein; >300 U/mg solid	1 unit oxidizes 1 µmol of NADH/min at 25°C	Lyophilized powder	<0.01% MDH, GOT, GAT, MK <0.05% PK	Genzyme 1371
>400 U/mg protein; >300 U/mg solid	1 unit oxidizes 1 µmol of NADH/min at 25°C	$(NH_4)_2SO_4$ suspension	<0.01% MDH, GOT, GAT, MK <0.05% PK	Genzyme 1372
>400 U/mg protein; >300 U/mg solid	1 unit oxidizes 1 µmol of NADH/min at 25°C	50% glycerol solution	<0.01% GOT, GPT, PK <0.05% MDH, MK	Genzyme 1373
600 - 1000 U/mg protein	1 unit reduces 1 µmol of pyruvate to L-lactate/min at pH 7.4, 25°C	2X Crystallized; suspension in 65% saturated $(NH_4)_2SO_4$, pH 7.2	<0.01% MDH, GPT <0.05% PK	ICN 151530
>400 IU/mg protein	1 IU transforms 1 µmol of substrate/min under standard conditions defined by the IUB	$(NH_4)_2SO_4$ suspension	<0.005% GPT, GOT <0.01% MK, PK <0.03% MDH	OYC
300-500 U/g of beaded agarose; 1 mL yields 9-16 U	1 unit reduces 1.0 µmol pyruvate to L-lactate/min at pH 7.5, 30°C	Insoluble enzyme attached to beaded agarose; suspension in 2.0 M $(NH_4)_2SO_4$, pH 7.0		Sigma L8011
25-40 U/g solid	1 unit reduces 1.0 µmol pyruvate to L-lactate/min at pH 7.5, 30°C	Insoluble enzyme attached to polyacrylamide containing 30% borate buffer salts		Sigma L8253
40-100 U/mg protein	1 unit converts 1.0 µmol PEP to pyruvate/min at pH 7.6, 37°C	Crude; suspension in 3.2 M $(NH_4)_2SO_4$, pH 6.0	40-100 U/mg protein PK	Sigma Type I L2375
800-1200 U/mg protein	1 unit reduces 1.0 µmol pyruvate to L-lactate/min at pH 7.5, 37°C	Crystallized; suspension in 3.2 M $(NH_4)_2SO_4$, pH 6.0	<0.01% PK, MK, MDH, GPT, GOT, α-GPDH	Sigma Type II L2500
700-1200 U/mg protein	1 unit reduces 1.0 µmol pyruvate to L-lactate/min at pH 7.5, 37°C	Salt-free; lyophilized	<0.01% PK, MK, MDH, GPT, GOT, α-GPDH	Sigma Type XI L1254
700-1000 U/mg protein	1 unit reduces 1.0 µmol pyruvate to L-lactate/min at pH 7.5, 37°C	Solution containing 50% glycerol, 10 mM KPO_4 buffer, pH 7.5 <1 mg/mg protein free ammonium ions	<0.01% PK, MK, MDH, GPT, GOT, α-GPDH	Sigma Type XXXIX L2518
≥250 U/mg protein	1 unit oxidizes 1 µmol NADH/min at pH 7.3, 25°C	Chromatographically purified; 2.2 M $(NH_4)_2SO_4$		Worthington LS02750 LS02751 LS02752

SPECIFIC ACTIVITY	UNITS DEFINITION	PREPARATION FORM	ADDITIONAL ACTIVITIES	SUPPLIER CATALOG NO.
Rabbit muscle *continued*				
≥250 U/mg protein	1 unit oxidizes 1 μmol NADH/min at pH 7.3, 25°C	Chromatographically purified; lyophilized		Worthington LS02755 LS02756 LS02757
Rabbit muscle, LDH-5 (M4) isozyme				
800 U/mg protein	1 unit reduces 1.0 μmol pyruvate to L-lactate/min at pH 7.5, 37°C	Crystallized; suspension in 2.1 M $(NH_4)_2SO_4$, pH 6.0	<0.01% PK, MK, MDH, GPT, GOT, α-GPDH	Sigma Type V-S L5132
Staphylococcus epidermidis (optimum pH = 6.0-7.0, T = 35-40°C)				
10-20 U/mg	1 unit reduces 1 μmol pyruvate to L-lactate/min at pH 7.5, 37°C	Lyophilized		ICN 190108
Trout muscle				
600-900 U/mg protein	1 unit reduces 1.0 μmol pyruvate to L-lactate/min at pH 7.5, 37°C	Crystallized; suspension in 3.2 M $(NH_4)_2SO_4$, pH 6.5	< 0.4% PK	Sigma Type XXVII L6383
Yeast				
0.1-0.6 U/mg	1 unit reduces 1 μmol of ferricyanide/min under assay conditions	Lyophilized		ICN 154604

D-Lactate Dehydrogenase 1.1.1.28

REACTION CATALYZED
(R)-Lactate + NAD$^+$ ↔ Pyruvate + NADH

SYSTEMATIC NAME
(R)-Lactate: NAD$^+$ oxidoreductase

SYNONYMS
Lactic acid dehydrogenase, LDH

REACTANTS
(R)-Lactate, pyruvate, NAD$^+$, NADH

APPLICATIONS
- Determination of glutamate pyruvate transaminase

NOTES
- Classified as an oxidoreductase acting on the CH-OH group of donors with NAD$^+$ or NADP$^+$ as acceptor
- Inhibited by Ag$^+$, Hg^{2+} and sulfhydryl reagents

SPECIFIC ACTIVITY	UNITS DEFINITION	PREPARATION FORM	ADDITIONAL ACTIVITIES	SUPPLIER CATALOG NO.
Lactobacillus leichmanii				
300 U/mg	1 unit converts 1 μmol pyruvate to product/min at 25°C	Suspension in 3.2 M (NH$_4$)$_2$SO$_4$ and BSA as stabilizer, pH 6	<0.01% ADH, GlDH, SDH <0.1% MDH	Boehringer 106941 736970 1585436
300 U/mg	1 unit converts 1 μmol pyruvate to product/min at 25°C	Lyophilized	<0.01% ADH, GlDH, SDH <0.1% MDH	Boehringer 732737
300 U/mg protein	1 unit reduces 1 μmol pyruvate to D-lactate/min at pH 7.0, 25°C	Suspension in 3.2 M (NH$_4$)$_2$SO$_4$, pH 6.0	<0.005% L-(+)-LDH, <0.01% ADH, β-NADH oxidase, sorbitol dehydrogenase, GlDH <0.5% MDH	Fluka 61306
250-500 U/mg protein	1 unit reduces 1 μmol pyruvate to D-lactate/min at pH 7.0, 25°C	Suspension in 3.2 M (NH$_4$)$_2$SO$_4$, pH 6.0	MDH <0.5% D-LDH activity	Sigma L2011
150-300 U/mg protein	1 unit reduces 1 μmol pyruvate to D-lactate/min at pH 7.0, 25°C	Lyophilized containing 50% protein; balance PO$_4$ buffer salts	MDH <0.5% D-LDH activity	Sigma L3888
Leuconostoc mesenteroides				
≥1000 IU/mg protein	1 IU transforms 1 μmol substrate/min under standard IUB conditions at 25°C	Lyophilized; suspension in (NH$_4$)$_2$SO$_4$	<0.001% MDH, GPT, GOT <0.003% PK <0.02% MK	OYC
1000-3000 U/mg protein	1 unit reduces 1 μmol pyruvate to D-lactate/min at pH 7.0, 25°C	Suspension in 3.2 M (NH$_4$)$_2$SO$_4$ and 0.1 M KPO$_4$, pH 7.0		Sigma L2395
Microbial (optimum pH = 6.0-7.0 and 7.8 [Tris-HCl], pI = 4.0, T = 35-40°C, MW = 140,000 Da [gel filtration]; K_M = 0.16 mM [pyruvate, pH 7.0]; stable pH 5.0-9.0 [25°C, 48 hr], T < 45°C [pH 7.0, 15 min])				
>100 U/mg solid	1 unit oxidizes 1.0 μmol NADH/min at pH 7.8, 25°C	Lyophilized	<0.01% GPT, enolase <0.05% PK <0.3% CK, AK	Genzyme 1411

1.1.1.28 D-Lactate Dehydrogenase continued

SPECIFIC ACTIVITY	UNITS DEFINITION	PREPARATION FORM	ADDITIONAL ACTIVITIES	SUPPLIER CATALOG NO.
Microbial continued				
≥400 U/mg solid	1 unit oxidizes 1 μmol NADH/min at pH 7.4, 25°C	Lyophilized	<0.001% NADH oxidase, PK <0.005% GOT, GPT <0.01% MDH, MK	Toyobo LCD-211
Staphylococcus epidermidis				
10-20 U/mg solid	1 unit reduces 1.0 μmol pyruvate to D-lactate/min at pH 7.0, 25°C	Lyophilized containing 2% protein; balance primarily dextran		Sigma L9636
Staphylococcus species (optimum pH = 5.0, pI = 4.3, T = 40°C, MW = 64,000 Da; stable pH 5.5-8.5 [30°C, 1hr], T < 35°C [pH 7, 1hr])				
≥100 U/mg	1 unit reduces 1 μmol pyruvate to D-lactate/min at pH 7.0, 25°C	Powder		Amano Lactate dehydrogenase

1.1.1.30 3-Hydroxybutyrate Dehydrogenase

REACTION CATALYZED

(R)-3-Hydroxybutanoate + NAD$^+$ ↔ Acetoacetate + NADH

SYSTEMATIC NAME

(R)-3-Hydroxybutanoate:NAD$^+$ oxidoreductase

SYNONYMS

β-Hydroxybutyrate dehydrogenase, D-3-hydroxybutyrate dehydrogenase, 3HBDH

REACTANTS

(R)-3-Hydroxybutanoate, 3-hydroxymonocarboxylic acids, acetoacetate, NAD$^+$, NADH

APPLICATIONS
- Determination of acetoacetate and D-(-)-3-hydroxybutyrate

NOTES
- Classified as an oxidoreductase acting on the CH-OH group of donors with NAD$^+$ or NADP$^+$ as acceptor
- Selectively oxidizes (R)-3-hydroxymonocarboxylic acids or reverse
- Stabilized by sucrose, mannitol and BSA
- Inhibited by Ag$^+$, Hg^{2+}, PCMB, SDS, MIA, IAA and DAC

SPECIFIC ACTIVITY	UNITS DEFINITION	PREPARATION FORM	ADDITIONAL ACTIVITIES	SUPPLIER CATALOG NO.
Microbial (optimum pH = 8.5, T = 37°C)				
>70 U/mg protein	1 unit forms 1 μmol NADH/min at pH 8.5, 37°C	Solution		GDS Tech HR-100

3-Hydroxybutyrate Dehydrogenase continued — 1.1.1.30

SPECIFIC ACTIVITY	UNITS DEFINITION	PREPARATION FORM	ADDITIONAL ACTIVITIES	SUPPLIER CATALOG NO.
Pseudomonas lemoignei				
50-200 U/mg protein	1 unit oxidizes 1.0 μmol D-β-hydroxybutyrate to acetoacetate/min at pH 7.8, 37°C	Chromatographically purified; lyophilized containing 10% protein; balance sucrose and Tris buffer salts	<0.05% LDH <0.1% MDH	Sigma H5132
50-100 U/mL	1 unit oxidizes 1.0 μmol D-β-hydroxybutyrate to acetoacetate/min at pH 7.8, 37°C	Suspension in 3.1 M $(NH_4)_2SO_4$, 1% BSA, 2% trehalose, pH 7.3		Sigma H6523
Pseudomonas species (optimum pH = 8.3, pI = 5.6, T = 55°C, MW = 130,000 Da [gel filtration]; K_M [D-3-hydroxybutyrate] = 0.42 mM [25°C, pH 8.3], 0.7 mM [37°C, pH 8.3]; K_M [NAD$^+$] = 49 μM [25°C, pH 8.3], 72 μM [37°C, pH 8.3]; K_M [acetoacetate] = 81 μM [25°C, pH 7.1], 0.24 mM [37°C, pH 7.1]; K_M [NADH] = 8.4 μM [25°C, pH 7.1], 15 μM [37°C, pH 7.1]; stable pH 5.0-8.5 [25°C, 20 hr], T < 40°C [pH 6.5, 15 min])				
≥100 U/mg solid	1 unit forms 1 μmol NADH/min at pH 8.5, 37°C	Lyophilized		ICN 153484
≥100 U/mg solid	1 unit forms 1 μmol NADH/min at pH 8.5, 37°C	Lyophilized containing sucrose, BSA, mannitol as stabilizers	<0.002% MDH, LDH, NADH oxidase	Toyobo HBD-301
100 U/mg				Wako 086-05441
Rhodobacter sphaeroides (Rhodopseudomonas spheroides)				
12 U/mg	1 unit converts 1 μmol 3-hydroxybutyrate to product/min at 25°C	Suspension 3.2 M $(NH_4)_2SO_4$ and BSA as stabilizer, pH 6	<0.05% LDH <1% MDH	Boehringer 106577
3 U/mg	1 unit converts 1 μmol 3-hydroxybutyrate to product/min at 25°C	Suspension 3.2 M $(NH_4)_2SO_4$ and BSA as stabilizer, pH 6	<0.1% LDH <5% MDH	Boehringer 127191 127833 127841 737054
3 U/mg protein	1 unit oxidizes 1.0 μmol D-β-hydroxybutyrate to acetoacetate/min at pH 7.8, 37°C	Aqueous solution containing 3.2 M $(NH_4)_2SO_4$, pH 6.0	<0.1% LDH, MDH <0.5% Glc6PDH, HK	Fluka 54975
20-30 U/mg protein	1 unit oxidizes 1.0 μmol D-β-hydroxybutyrate to acetoacetate/min at pH 7.8, 37°C	Suspension in 3.2 M $(NH_4)_2SO_4$ and BSA, pH 6.0		Sigma H4005
5-10 U/mg protein	1 unit oxidizes 1.0 μmol D-β-hydroxybutyrate to acetoacetate/min at pH 7.8, 37°C	Suspension in 3.2 M $(NH_4)_2SO_4$, pH 6.0	<1% LDH, MDH	Sigma H6126
250-500 U/mg protein	1 unit oxidizes 1.0 μmol D-β-hydroxybutyrate to acetoacetate/min at pH 7.8, 37°C	Chromatographically purified; lyophilized containing 10% protein; balance primarily Tris buffer salts	<0.05% LDH <0.1% MDH	Sigma H8509

1.1.1.35 3-Hydroxyacyl-CoA Dehydrogenase

REACTION CATALYZED

(S)-3-Hydroxyacyl-CoA + NAD$^+$ ↔ 3-Oxoacyl-CoA + NADH

SYSTEMATIC NAME

(S)-3-Hydroxyacyl-CoA: NAD$^+$ oxidoreductase

SYNONYMS

β-Hydroxyacyl dehydrogenase, β-hydroxyacyl-CoA dehydrogenase, β-keto-reductase

REACTANTS

(S)-3-Hydroxyacyl-CoA, (S)-3-hydroxyacyl-N-acylthioethanolamine, (S)-3-hydroxyacyl-hydrolipoate, 3-oxoacyl-CoA, NAD$^+$, NADH; NADP$^+$ (slowly, some enzymes)

NOTES

- Classified as an oxidoreductase acting on the CH-OH group of donors with NAD$^+$ or NADP$^+$ as acceptor
- Broad specificity for acyl chain-length

SPECIFIC ACTIVITY	UNITS DEFINITION	PREPARATION FORM	ADDITIONAL ACTIVITIES	SUPPLIER CATALOG NO.
Bovine liver				
150 U/mg protein	1 unit converts 1.0 μmol acetoacetyl-CoA to β-hydroxybutyryl-CoA/min at pH 7.3, 37°C in the presence of β-NADH	Lyophilized containing 10% protein; balance primarily KPO$_4$, sucrose, GSH as stabilizers		Sigma H7384
Porcine heart				
150-200 U/mg protein	1 unit converts 1.0 μmol acetoacetyl-CoA to β-hydroxybutyryl-CoA/min at pH 7.3, 37°C in the presence of β-NADH	Suspension in 3.2 M (NH$_4$)$_2$SO$_4$, 2 mM β-MSH, 1 mM EDTA, pH 6.0		Sigma H3516

Malate Dehydrogenase

1.1.1.37

REACTION CATALYZED
(S)-Malate + NAD$^+$ ↔ Oxaloacetate + NADH

SYSTEMATIC NAME
(S)-Malate: NAD$^+$ oxidoreductase

SYNONYMS
Malic dehydrogenase, MDH

REACTANTS
(S)-Malate, 2-hydroxydicarboxylic acids, oxaloacetate, NAD$^+$, NADH

APPLICATIONS
- Determination of CO_2/bicarbonate levels in biological fluids, with PEP-carboxylase
- Determination of oxaloacetate and malate
- Determination of aspartame in carbonated beverages
- Indicator enzyme for determination of GOT [2.6.1.1], citrate synthase [4.1.3.7] and ATP-citrate lyase [4.1.3.8]

NOTES
- MDH levels in serum and cerebrospinal fluid are clinically significant
- Classified as an oxidoreductase acting on the CH-OH group of donors with NAD$^+$ or NADP$^+$ as acceptor
- All eukaryotic cells contain mitochondrial (M) and cytoplasmic (S) isozymes; both are dimers with one coenzyme binding site per subunit
- Prokaryotic cells contain only a single form
- Specific for the L-configuration
- Activated by phosphate, arsenate and zinc ions
- Inhibited by thyroxine, iodine, cyanide, molecular iodine, 2-thenoyltrifluoroacetone (TTFA) and chlorothricin

SPECIFIC ACTIVITY	UNITS DEFINITION	PREPARATION FORM	ADDITIONAL ACTIVITIES	SUPPLIER CATALOG NO.
Bovine heart				
1250 U/mg protein	1 unit oxidizes 1.0 μmol NADH/min at pH 7.4, 25°C	Highly purified; suspension of mitochondrial MDH in 70% (NH$_4$)$_2$SO$_4$, pH 7.4	<0.005% GOT, GPT, LDH	ICN 151578
1250 U/mg protein		(NH$_4$)$_2$SO$_4$-free; lyophilized	<0.005% GOT, GPT, LDH	ICN 151579
2000-4000 U/mg protein	1 unit converts 1.0 μmol oxalacetate and β-NADH to L-malate and β-NAD/min at pH 7.5, 25°C	Suspension in 3 M (NH$_4$)$_2$SO$_4$ and 0.01 M KPO$_4$, pH 7.3	Low LDH and transaminase	Sigma M9004
Chicken heart				
600 U/mg protein	1 unit converts 1.0 μmol oxalacetate and β-NADH to L-malate and β-NAD/min at pH 7.5, 25°C	Suspension in 3.0 M (NH$_4$)$_2$SO$_4$, pH 7.0		Sigma M2262

1.1.1.37 Malate Dehydrogenase continued

SPECIFIC ACTIVITY	UNITS DEFINITION	PREPARATION FORM	ADDITIONAL ACTIVITIES	SUPPLIER CATALOG No.
Escherichia coli, recombinant (optimum pH = 7.8-8.0)				
>1000 U/mg protein; >300 U/mg solid	1 unit oxidizes 1.0 μmol NADH/min at pH 8.0, 30°C	Lyophilized	<0.005% GOT	Genzyme 1571
1200-1700 U/mg protein	1 unit converts 1.0 μmol oxalacetate and β-NADH to L-malate and β-NAD/min at pH 7.5, 25°C	Solution containing 50% glycerol and 5 mM KPO$_4$ buffer, pH 7.2		Sigma M1037
Human erythrocytes				
100-200 U/mg protein	1 unit converts 1.0 μmol oxalacetate and β-NADH to L-malate and β-NAD/min at pH 7.5, 25°C	Suspension in 3.6 M (NH$_4$)$_2$SO$_4$, pH 7.0	<0.2% GPT, GPT <0.5% LDH	Sigma M6137
Human heart, cytoplasmic (MW = 70,000 Da)				
50 U/mg protein	1 unit converts 1 μmol oxaloacetate and NADH/min at pH 7.5, 25°C in 0.1 M NaPO$_4$ buffer	Suspension in 70% (NH$_4$)$_2$SO$_4$ containing 10 mg/mL protein and PO$_4$ buffer, pH 7	<0.1% GPT, GOT	Calzyme 067B0050
Human heart, mitochondrial (MW = 70,000 Da)				
200 U/mg protein	1 unit converts 1 μmol oxaloacetate and NADH/min at pH 7.5, 25°C in 0.1 M NaPO$_4$ buffer	Suspension in 70% (NH$_4$)$_2$SO$_4$ containing 10 mg/mL protein and PO$_4$ buffer, pH 7	<0.1% GPT, GOT	Calzyme 068B0200
Human placenta				
200-300 U/mg protein	1 unit converts 1.0 μmol oxalacetate and β-NADH to L-malate and β-NAD/min at pH 7.5, 25°C	Suspension in 3.2 M (NH$_4$)$_2$SO$_4$ and 0.1 M KPO$_4$, pH 6.5	<0.1% transaminase	Sigma M5888
Microbial (optimum pH = 8.0, pI = 4.8, T = 70°C, MW = 140,000 Da [4 subunits/mol enzyme]; K_M = 54 μM [L-malate], 5 μM [oxaloacetate], 8.1 μM [NADH]; stable pH 3.0-9.0 [25°C, 20 hr], T < 70°C [pH 7.5, 15 min])				
>100 U/mg solid	1 unit oxidizes 1.0 μmol NADH/min at pH 8.0, 25°C	Lyophilized	<0.003% GOT	Genzyme 1521
≥400 U/mg solid	1 unit oxidizes 1 μmol NADH/min at pH 7.5, 30°C	Lyophilized	<0.001% GOT, LDH, fumarase <0.01% NADH oxidase	Toyobo MAD-211
Pigeon breast muscle				
4000 U/mg protein	1 unit converts 1.0 μmol oxalacetate and β-NADH to L-malate and β-NAD/min at pH 7.5, 25°C	Crystallized; suspension in 2.8 M (NH$_4$)$_2$SO$_4$ and 0.005 M KPO$_4$, pH 7.5	Low LDH and transaminase	Sigma M7508
Porcine heart (optimum pH = 7.8, pI = 6.1-6.4, T = 45-50°C, MW = 70,000 Da [2 subunits/mol enzyme]; K_M = 0.4 mM [L-malate], 33 μM [oxaloacetate]; stable pH 7.3-8.5 [25°C, 20 hr], T < 30°C [pH 7.5, 10 min])				
12,280 U/mL	1 unit oxidizes 1 μmol NADH/min at pH 7.4, 25°C	Solution containing 3.2 M (NH$_4$)$_2$SO$_4$, pH 6.0		AMRESCO 0373 0901(enzyme III)

Malate Dehydrogenase continued

1.1.1.37

SPECIFIC ACTIVITY	UNITS DEFINITION	PREPARATION FORM	ADDITIONAL ACTIVITIES	SUPPLIER CATALOG NO.
Porcine heart continued				
≥1250 U/mg protein	1 unit oxidizes 1 µmol NADH/min at pH 7.4, 25°C	Highly purified; suspension of mitochondrial MDH in 80% $(NH_4)_2SO_4$	<0.001% fumarase, GLDH <0.005% CK, GPT, LDH, GOT	Biozyme MDHP2
≥1250 U/mg protein	1 unit oxidizes 1 µmol NADH/min at pH 7.4, 25°C	Salt-free; lyophilized	<0.001% fumarase, GLDH <0.005% CK, GPT, LDH, GOT	Biozyme MDHP2F
≥1250 U/mg protein	1 unit oxidizes 1 µmol NADH/min at pH 7.4, 25°C	50% glycerol; $(NH_4)_2SO_4$-free	<0.001% fumarase, GLDH <0.005% CK, GPT, LDH, GOT	Biozyme MDHP2G
≥1500 U/mg protein	1 unit oxidizes 1 µmol NADH/min at pH 7.4, 25°C	Highly purified; suspension of mitochondrial MDH in 80% $(NH_4)_2SO_4$	<0.001% CK, fumarase, GLDH, GPT, LDH, GOT	Biozyme MDHP3
1500 U/mg protein	1 unit oxidizes 1 µmol NADH/min at pH 7.4, 25°C	Salt-free; lyophilized	<0.001% CK, fumarase, GLDH, GPT, LDH, GOT	Biozyme MDHP3F
≥1500 U/mg protein	1 unit oxidizes 1 µmol NADH/min at pH 7.4, 25°C	50% glycerol; $(NH_4)_2SO_4$-free	<0.001% CK, fumarase, GLDH, GPT, LDH, GOT	Biozyme MDHP3G
1000 U/mg protein	1 unit oxidizes 1 µmol NADH/min at pH 7.5, 25°C in $NaPO_4$ buffer	Salt-free; lyophilized	<0.0015% GPT <0.005% GOT <0.01% LDH	Calbiochem 442610
1200 U/mg protein	1 unit reduces 1 µmol oxaloacetate/min at pH 7.5, 25°C	Crystallized; suspension in 3.2 M $(NH_4)_2SO_4$, pH 6	<0.01% fumarase, LDH	Fluka 63170
>120 U/mg protein; >100 U/mg solid	1 unit oxidizes 1.0 µmol NADH/min at pH 7.4, 25°C	Lyophilized	<0.002% GOT	Genzyme 1501
>120 U/mg protein	1 unit oxidizes 1.0 µmol NADH/min at pH 7.4, 25°C	Suspension in $(NH_4)_2SO_4$	<0.002% GOT	Genzyme 1502
>120 U/mg protein	1 unit oxidizes 1.0 µmol NADH/min at pH 7.4, 25°C	Solution containing 50% glycerol	<0.002% GOT	Genzyme 1503
1500 U/mg protein	1 unit oxidizes 1.0 µmol NADH/min at pH 7.4, 25°C	Highly purified; suspension of mitochondrial MDH in 70% $(NH_4)_2SO_4$, pH 7.4	<0.001% GOT, GPT, LDH	ICN 151580
1200-1300 U/mg protein	1 unit oxidizes 1.0 µmol NADH/min at pH 7.4, 25°C	Highly purified; suspension of mitochondrial MDH in 70% $(NH_4)_2SO_4$, pH 7.4	<0.005% GOT, GPT, LDH	ICN 151581
1500 U/mg protein	1 unit oxidizes 1.0 µmol NADH/min at pH 7.4, 25°C	$(NH_4)_2SO_4$-free; lyophilized	<0.001% GOT, GPT, LDH	ICN 151582
1200-1300 U/mg protein	1 unit oxidizes 1.0 µmol NADH/min at pH 7.4, 25°C	$(NH_4)_2SO_4$-free; lyophilized	<0.005% GOT, GPT, LDH	ICN 151583
≥1100 IU/mg protein	1 IU transforms 1 µmol substrate/min under standard IUB conditions at 25°C	Lyophilized; suspension in $(NH_4)_2SO_4$; 50% glycerol solution	<0.001% NADH oxidase <0.002% GlDH (NAD) <0.01% fumarase (L-malate), L-LDH (NADH), GOT	OYC

1.1.1.37 Malate Dehydrogenase continued

SPECIFIC ACTIVITY	UNITS DEFINITION	PREPARATION FORM	ADDITIONAL ACTIVITIES	SUPPLIER CATALOG NO.
Porcine heart continued				
400 U/mg protein	1 unit converts 1.0 μmol oxalacetate and β-NADH to L-malate and β-NAD/min at pH 7.5, 25°C	Suspension in 2.8 M $(NH_4)_2SO_4$, pH 6.0	<0.1% transaminase	Sigma 410-12
≥400 U/mg solid	1 unit oxidizes 1 μmol NADH/min at pH 7.5, 25°C	Lyophilized	<0.005% GOT <0.01% LDH, fumarase, NADH oxidase	Toyobo MAD-201
Porcine heart, cytoplasmic (MW = 70,000 Da)				
250 U/mg protein	1 unit converts 1 μmol oxalacetate and NADH/min at pH 7.5, 25°C in 0.1 M NaPO₄ buffer	Suspension in 70% $(NH_4)_2SO_4$ containing 10 mg/mL protein, pH 7	<0.01% GPT, GOT	Calzyme 069B0250
40 U/mg protein	1 unit converts 1.0 μmol oxalacetate and β-NADH to L-malate and β-NAD/min at pH 7.5, 25°C	Suspension in 3.2 M $(NH_4)_2SO_4$ and 0.1 M KPO₄, pH 7.0	<0.01% transaminase	Sigma M7383
Porcine heart, mitochondrial (optimum pH = 7.4, MW = 70,000 Da)				
1200 U/mg	1 unit converts 1 μmol oxaloacetate to product/min at 25°C	50% glycerol solution, pH 7	<0.002% GOT <0.01% fumarase, LDH	Boehringer 127248 127892 127906
1200 U/mg	1 unit converts 1 μmol oxaloacetate to product/min at 37°C	Suspension in 3.2 M $(NH_4)_2SO_4$, pH 6	<0.002% GOT <0.01% fumarase, LDH	Boehringer 127256 127914 127922
1000-1800 U/mg protein	1 unit converts 1 μmol oxaloacetate and NADH/min at pH 7.5, 25°C in 0.1 M NaPO₄ buffer	50% glycerol solution containing 10 mg/mL protein, pH 7	<0.001% fumarase, GPT, GOT, LDH	Calzyme 054C1800
1000-1800 U/mg protein	1 unit converts 1 μmol oxaloacetate and NADH/min at pH 7.5, 25°C in 0.1 M NaPO₄ buffer	Suspension in 70% $(NH_4)_2SO_4$ containing 10 mg/mL protein	<0.001% fumarase, GPT, GOT, LDH	Calzyme 071B1700
50-100 U/mg solid	1 unit converts 1 μmol oxaloacetate and NADH/min at pH 7.5, 25°C in 0.1 M NaPO₄ buffer	Lyophilized containing 10% protein/mg solid, PO₄ buffer, mannitol, pH 7	<0.01% fumarase, GPT, GOT, LDH	Calzyme 098A0050
1200 U/mg protein	1 unit oxidizes 1.0 μmol NADH/min at pH 7.4, 25°C	Lyophilized	<0.002% GOT <0.01% fumarase, LDH	ICN 160049
1000 U/mg protein	1 unit converts 1.0 μmol oxalacetate and β-NADH to L-malate and β-NAD/min at pH 7.5, 25°C	Suspension in 2.8 M $(NH_4)_2SO_4$, pH 6.0	<0.01% transaminase	Sigma 410-13
1000 U/mg protein	1 unit converts 1.0 μmol oxalacetate and β-NADH to L-malate and β-NAD/min at pH 7.5, 25°C	Solution containing 50% glycerol and 0.05 M KPO₄ buffer, pH 7.5 <10 mg/mg protein free ammonium ions	<0.01% transaminase	Sigma M2634

Malate Dehydrogenase continued
1.1.1.37

SPECIFIC ACTIVITY	UNITS DEFINITION	PREPARATION FORM	ADDITIONAL ACTIVITIES	SUPPLIER CATALOG NO.
Porcine heart, mitochondrial *continued*				
≥1000 U/mg protein	1 unit oxidizes 1 μmol NADH/min at pH 7.4, 25°C	Chromatographically purified; suspension in 3.2 M $(NH_4)_2SO_4$	<0.01% transaminase	Worthington LS05569 LS05570 LS05572
≥400 U/mg protein	1 unit oxidizes 1 μmol NADH/min at pH 7.4, 25°C	Chromatographically purified; lyophilized		Worthington LS05575 LS05577 LS05573
Thermus flavua				
5-16 U/mg solid	1 unit converts 1.0 μmol oxalacetate and β-NADH to L-malate and β-NAD/min at pH 7.5, 25°C	Lyophilized containing 10% protein; balance primarily dextrin		ICN 190113
	1 unit converts 1.0 μmol oxalacetate and β-NADH to L-malate and β-NAD/min at pH 7.5, 25°C	Lyophilized containing dextrin		Sigma M7032
Thermus species (optimum pH = 8.0, pI = 5.05, T = 90°C, MW = 54,000 Da; stable pH 6.5-8.0 [80°C, 30 min], T < 70°C [pH 7.8, 30 min])				
≥100 U/mg	1 unit reduces 1 μmol NAD^+/min at pH 7.8, 25°C	Powder		Amano
Thermus species, recombinant expressed in *E. coli*				
100-200 U/mg protein	1 unit converts 1.0 μmol oxalacetate and β-NADH to L-malate and β-NAD/min at pH 7.5, 25°C	Suspension in 3.5 M $(NH_4)_2SO_4$		Sigma M1035
Yeast				
≥1000 IU/mg protein	1 IU transforms 1 μmol substrate/min under standard IUB conditions at 25°C	Lyophilized; suspension in $(NH_4)_2SO_4$; 50% glycerol solution	<0.001% NADH oxidase, GlDH (NAD) <0.01% fumarase (L-malate), L-LDH (NADH), GOT	OYC
Yeast, bakers (MW = 70,000 Da)				
1000 U/mg protein	1 unit converts 1 μmol oxaloacetate and NADH/min at pH 7.5, 25°C in 0.1 M $NaPO_4$ buffer	Lyophilized containing 95% protein		Calzyme 200A1000

1.1.1.40 Malate Dehydrogenase (Oxaloacetate-Decarboxylating) (NADP$^+$)

REACTION CATALYZED

(S)-Malate + NADP$^+$ ↔ Pyruvate + CO$_2$ + NADPH

SYSTEMATIC NAME

(S)-Malate:NADP$^+$ oxidoreductase (oxaloacetate-decarboxylating)

SYNONYMS

Malic enzyme, pyruvic-malic carboxylase

REACTANTS

(S)-Malate, oxaloacetate, NADP$^+$, pyruvate, CO$_2$, NADPH

NOTES

- Classified as an oxidoreductase acting on the CH-OH group of donors with NAD$^+$ or NADP$^+$ as acceptor

SPECIFIC ACTIVITY	UNITS DEFINITION	PREPARATION FORM	ADDITIONAL ACTIVITIES	SUPPLIER CATALOG NO.
Chicken liver				
1200 U/mg protein	1 unit converts 1 μmol L-malate and NADP to pyruvate, CO$_2$ and NADPH/min at pH 7.4, 25°C	Suspension in 2.9 M (NH$_4$)$_2$SO$_4$ solution, 10 mM KPO$_4$, 0.5 mM β-MSH, 10 mM MnCl$_2$, 3 mM Na$_4$ EDTA, pH 6		Fluka 63172

1.1.1.41 Isocitrate Dehydrogenase (NAD$^+$)

REACTION CATALYZED

Isocitrate + NAD$^+$ ↔ 2-Oxoglutarate + CO$_2$ + NADH

SYSTEMATIC NAME

Isocitrate: NAD$^+$ oxidoreductase (decarboxylating)

SYNONYMS

Isocitric dehydrogenase, β-ketoglutaric-isocitric carboxylase, ICDH

REACTANTS

Isocitrate, 2-oxoglutarate, CO$_2$, NAD$^+$, NADH

NOTES

- Classified as an oxidoreductase acting on the CH-OH group of donors with NAD$^+$ or NADP$^+$ as acceptor
- The isomer of isocitrate involved is $(1R,2S)$-1-hydroxypropane-1,2,3-tricarboxylate, formerly termed threo-D$_S$-isocitrate
- Does not decarboxylate oxalosuccinate

SPECIFIC ACTIVITY	UNITS DEFINITION	PREPARATION FORM	ADDITIONAL ACTIVITIES	SUPPLIER CATALOG NO.
Yeast (MW = 380,000 Da)				
30 U/mg	1 unit converts 1 μmol isocitrate to α-ketoglutarate/min at pH 8.0, 25°C	50% glycerol solution		Calbiochem 420125
≥30 IU/mg protein	1 IU transforms 1 μmol substrate/min under standard IUB conditions at 25°C	50% glycerol solution	<0.2% ICDH (NADP$^+$)	OYC

Isocitrate Dehydrogenase (NADP$^+$)

1.1.1.42

REACTION CATALYZED
Isocitrate + NADP$^+$ ↔ 2-Oxoglutarate + CO$_2$ + NADPH

SYSTEMATIC NAME
Isocitrate: NADP$^+$ oxidoreductase (decarboxylating)

SYNONYMS
Isocitric dehydrogenase (NADP), oxalosuccinate decarboxylase, ICDH

REACTANTS
Isocitrate, oxalosuccinate, 2-oxoglutarate, CO$_2$, NADP$^+$, NADPH

NOTES
- Classified as an oxidoreductase acting on the CH-OH group of donors with NAD$^+$ or NADP$^+$ as acceptor
- The isomer of isocitrate involved is *(1R,2S)*-1-hydroxypropane-1,2,3-tricarboxylate, formerly termed *threo-Ds*-isocitrate

SPECIFIC ACTIVITY	UNITS DEFINITION	PREPARATION FORM	ADDITIONAL ACTIVITIES	SUPPLIER CATALOG NO.
Porcine heart				
3 U/mg solid	1 unit converts 1 μmol isocitrate to product/min at 25°C	Lyophilized	<3% MDH	Boehringer 236322
10 U/mg	1 unit converts 1 μmol isocitrate to α-ketoglutarate at pH 7.4, 37°C	Powder		Fluka 58774
6 U/mg protein	1 unit converts 1 μmol isocitrate to α-ketoglutarate at pH 7.4, 37°C	50% glycerol solution containing EDTA and buffer, pH 6		Fluka 58775
0.1-1.0 U/mg solid	1 unit converts 1.0 μmol isocitrate to α-ketoglutarate/min at pH 7.4, 37°C	Crude	Numerous enzyme activities associated with porcine heart	Sigma I1877
3-8 U/mg protein	1 unit converts 1.0 μmol isocitrate to α-ketoglutarate/min at pH 7.4, 37°C	Solution containing 50% glycerol, EDTA buffer salts, pH 6.0		Sigma I2002
20-60 U/mg protein	1 unit converts 1.0 μmol isocitrate to α-ketoglutarate/min at pH 7.4, 37°C	Highly purified; lyophilized containing 20% protein; balance primarily Na citrate and MnSO$_4$		Sigma I2516
Yeast				
≥8 IU/mg protein	1 IU transforms 1 μmol substrate/min under standard IUB conditions at 25°C	Lyophilized; 50% glycerol solution	<0.5% ICDH (NAD$^+$)	OYC

1.1.1.44 Phosphogluconate Dehydrogenase (Decarboxylating)

REACTION CATALYZED

6-Phospho-D-gluconate + NADP$^+$ ↔ D-Ribulose 5-phosphate + CO_2 + NADPH

SYSTEMATIC NAME

6-Phospho-D-gluconate: NADP$^+$ 2-oxidoreductase (decarboxylating)

SYNONYMS

Phosphogluconic acid dehydrogenase, 6-phosphogluconic dehydrogenase, 6-phosphogluconic carboxylase, 6PGluDH

REACTANTS

6-Phospho-D-gluconate, D-ribulose 5-phosphate, CO_2

NOTES

- Classified as an oxidoreductase acting on the CH-OH group of donors with NAD$^+$ or NADP$^+$ as acceptor
- Certain preparations reduce both NADP$^+$ and NAD$^+$

SPECIFIC ACTIVITY	UNITS DEFINITION	PREPARATION FORM	ADDITIONAL ACTIVITIES	SUPPLIER CATALOG NO.
Human erythrocytes				
15-50 U/mg protein	1 unit oxidizes 1.0 μmol 6-phospho-D-gluconate to D-ribulose 5-phosphate and CO_2/min at pH 7.4, 37°C in the presence of NADP$^+$	Lyophilized containing 10% protein, 80% lactose, 1% 2',5'-ADP, DTT; balance primarily NaPO$_4$ buffer salts	<0.5% MDH <5% Glc6PDH, <10% LDH, GR	Sigma P7533
Leuconostoc mesenteroides				
≥50 IU/mg protein	1 IU transforms 1 μmol substrate/min under standard IUB conditions at 25°C	Suspension in (NH$_4$)$_2$SO$_4$	<0.005% NADH oxidase, GR <0.01% PGI <0.02% HK <0.05% G6PDH	OYC
30-70 U/mg protein	1 unit oxidizes 1.0 μmol 6-phospho-D-gluconate to D-ribulose 5-phosphate and CO_2/min at pH 7.5, 25°C in the presence of NAD$^+$	Suspension in 3.0 M (NH$_4$)$_2$SO$_4$ and 0.1 M KPO$_4$ buffer, pH 6.5		Sigma P7281
Ovine liver				
2-10 U/mg protein	1 unit oxidizes 1.0 μmol 6-phospho-D-gluconate to D-ribulose 5-phosphate and CO_2/min at pH 7.4, 37°C in the presence of NADP$^+$	Lyophilized containing 80% protein; balance primarily Tris buffer salts	<0.3% Glc6PDH, HK	Sigma P8406
Yeast				
12 U/mg	1 unit converts 1 μmol gluconate-6-P to product/min at 25°C	Suspension in 2.5 M (NH$_4$)$_2$SO$_4$, 0.5 M NaCl, 0.05 M KPO$_4$, BSA as stabilizer, pH 7.5	<0.01% G6PDH, GR, HK <0.03% PGI	Boehringer 108391 108405
≥1.4 IU/mg protein	1 IU transforms 1 μmol substrate/min under standard IUB conditions at 25°C	Suspension in (NH$_4$)$_2$SO$_4$	<0.01% NADH oxidase, <0.05% G6PDH, GR	OYC

Phosphogluconate Dehydrogenase (Decarboxylating) continued — 1.1.1.44

SPECIFIC ACTIVITY	UNITS DEFINITION	PREPARATION FORM	ADDITIONAL ACTIVITIES	SUPPLIER CATALOG NO.
Yeast continued				
3-6 U/mg protein	1 unit oxidizes 1.0 µmol 6-phospho-D-gluconate to D-ribulose 5-phosphate and CO_2/min at pH 7.4, 37°C in the presence of $NADP^+$	Lyophilized containing 15% protein		Sigma P4553
Yeast, bakers (MW = 101,000–111,000 Da)				
10 U/mg protein	1 unit oxidizes 1.0 µmol 6-phospho-D-gluconate to D-ribulose 5-phosphate and CO_2/min at pH 7.4, 37°C in the presence of $NADP^+$	Lyophilized containing 95% protein		Calzyme 203A0003
Yeast, Torula				
20 U/mg protein	1 unit oxidizes 1 µmol 6-phospho-D-gluconate to D-ribulose 5-phosphate and CO_2/min at pH 7.4, 37°C	Crystallized; suspension in 3.2 M $(NH_4)_2SO_4$, pH 7.0	<0.93% glycolytic activity	ICN 100982
20 U/mg protein	1 unit oxidizes 1.0 µmol 6-phospho-D-gluconate to D-ribulose 5-phosphate and CO_2/min at pH 7.4, 37°C in the presence of $NADP^+$	Crystallized; suspension in 3.1 M $(NH_4)_2SO_4$ and 0.2 M Gly-Gly, pH 7.6	<0.1% Glc6PDH, 2% HK	Sigma P0507
50 U/mg protein	1 unit oxidizes 1.0 µmol 6-phospho-D-gluconate to D-ribulose 5-phosphate and CO_2/min at pH 7.4, 37°C in the presence of $NADP^+$	Crystallized; suspension in 3.1 M $(NH_4)_2SO_4$ and 0.2 M Gly-Gly, pH 7.6	<0.1% Glc6PDH, HK	Sigma P0632

Glucose 1-Dehydrogenase — 1.1.1.47

REACTION CATALYZED

β-D-Glucose + $NAD(P)^+$ ↔ D-Glucono-1,5-lactone + NAD(P)H

SYSTEMATIC NAME

β-D-Glucose:$NAD(P)^+$ 1-oxidoreductase

SYNONYMS

Glucose dehydrogenase

SUBSTRATES

β-D-Glucose, D-xylose, D-glucono-1,5-lactone, $NAD(P)^+$, NAD(P)H

APPLICATIONS
- Determination of glucose in urine
- Manufacture of valine from ketoisovalerate

NOTES
- Classified as an oxidoreductase acting on the CH-OH group of donors with NAD^+ or $NADP^+$ as acceptor
- Inhibited by Ag^+, Hg^{2+} and monoiodoacetate

1.1.1.47 Glucose 1-Dehydrogenase continued

SPECIFIC ACTIVITY	UNITS DEFINITION	PREPARATION FORM	ADDITIONAL ACTIVITIES	SUPPLIER CATALOG NO.
Bacillus megaterium				
50-150 U/mg solid	1 unit oxidizes 1.0 μmol β-D-glucose to D-glucono-δ-lactone/min at pH 7.6, 25°C	Chromatographically purified; lyophilized		Sigma G7653
33 U/mg	1 unit oxidizes 1 μmol β-D-glucose to D-glucono-δ-lactone/min at pH 7.6, 25°C			Fluka 49165
Bacillus species (optimum pH = 8.0-8.5, T = 37°C, MW = 116,000-120,000 Da))				
>30 U/mg solid	1 unit reduces 1.0 μmol NAD^+/min at pH 8.0, 25°C	Lyophilized	<0.002% LDH (pyruvate), LDH (lactate), NADH oxidase	Genzyme 1191
30 U/mg		Lyophilized		Wako 071-02411
100-300 U/mg protein	1 unit oxidizes 1.0 μmol β-D-glucose to D-glucono-δ-lactone/min at pH 7.6, 25°C	Chromatographically purified; lyophilized		ICN 157199
Calf liver				
15 U/g solid	1 unit oxidizes 1.0 μmol β-D-glucose to D-glucono-δ-lactone/min at pH 7.6, 25°C	Crude; powder		Sigma G5625
Microbial (optimum pH = 9.0, pI = 4.5, T = 55°C, MW = 101,000 Da [gel filtration]; K_M [NAD^+-linked] = 13.8 mM [D-glucose], 0.309 mM [NAD^+]; K_M [$NADP^+$-linked] = 12.5 mM [D-glucose], 40.7 μM [$NADP^+$]; stable pH 6.0-7.5 [20°C, 16 hr], T < 45°C [pH 7.0, 15 min])				
>50 U/mg protein	1 unit converts 1 μmol NAD to NADH/min at pH 8.25, 37°C	Lyophilized		GDS Tech G-020
≥250 U/mg solid	1 unit forms 1 μmol NADH/min at pH 8.0, 37°C	Lyophilized	<0.001% NADH oxidase, α-glucosidase, G6PDH	Toyobo GLD-311

1.1.1.48 Galactose 1-Dehydrogenase

REACTION CATALYZED

D-Galactose + NAD^+ ↔ D-Galactono-1,4-lactone + NADH

SYSTEMATIC NAME

D-Galactose:NAD^+ 1-oxidoreductase

SYNONYMS

β-Galactose dehydrogenase, galactose dehydrogenase

REACTANTS

D-Galactose, D-galactono-1,4-lactone

APPLICATIONS

- Determination of D-galactose

NOTES

- Classified as an oxidoreductase acting on the CH-OH group of donors with NAD^+ or $NADP^+$ as acceptor

Galactose 1-Dehydrogenase continued — 1.1.1.48

SPECIFIC ACTIVITY	UNITS DEFINITION	PREPARATION FORM	ADDITIONAL ACTIVITIES	SUPPLIER CATALOG No.
Microbial (optimum pH = 8.5, T = 37°C)				
>3 U/mg protein	1 unit converts 1 μmol NAD to NADH/min at pH 8.5, 37°C	Lyophilized		GDS Tech G-030
Pseudomonas fluorescens				
5 U/mg protein	1 unit converts 1 μmol D-galactose to product/min at 25°C	Suspension in 3.2 M $(NH_4)_2SO_4$, 1 mM EDTA, BSA as stabilizer, pH 6	<0.01% ADH, β-galactosidase <0.1% NADH oxidase <0.5% LDH	Boehringer 104973 104981
3-6 U/mg protein	1 unit converts 1.0 μmol D-galactose to D-galactonate/min at pH 8.6, 25°C	Suspension in 3.2 M $(NH_4)_2SO_4$, 1 mM EDTA, BSA, pH 6.0		Sigma G2004
5 U/mg protein	1 unit converts 1 μmol D-galactose to D-galactonate/min at pH 8.6, 25°C	Suspension in 1 M $(NH_4)_2SO_4$	<0.05% ADH, LDH, MDH, β-NADH oxidase, β-galactosidase	Fluka 48265
Pseudomonas fluorescens gene, recombinant from E. coli				
80 U/mg protein	1 unit converts 1 μmol galactose to product/min at 25°C	Suspension in 3.2 M $(NH_4)_2SO_4$, pH 6	<0.01% ADH, β-galactosidase <0.05% NADH oxidase <0.1% LDH	Boehringer 662046
50-200 U/mg protein	1 unit converts 1.0 μmol D-galactose to D-galactonate/min at pH 8.6, 25°C	Suspension in 3.2 M $(NH_4)_2SO_4$, pH 6		Sigma G6637

Glucose-6-Phosphate 1-Dehydrogenase — 1.1.1.49

REACTION CATALYZED

D-Glucose 6-phosphate + $NADP^+$ ↔ D-Glucono-1,5-lactone 6-phosphate + NADPH

SYSTEMATIC NAME

D-Glucose 6-phosphate:$NADP^+$ 1-oxidoreductase

SYNONYMS

Glucose-6-phosphate dehydrogenase, Glc6PDH

REACTANTS

D-Glucose 6-phosphate, β-D-glucose, sugars, D-glucono-1,5-lactone 6-phosphate, $NADP^+$, NADPH; NAD^+ (some enzymes)

APPLICATIONS

- Useful in systems in which NADPH or NADH production is measured; e.g., pregnanediol 3α-glucuronide by fluoroenzyme immunoassay
- Determination of glucose and ATP (when coupled with HK), fructose, glucose-6-phosphate, $NAD(P)^+$ and CK
- Deficient in 10-13% of American Black males; susceptible to severe hemolytic anemia without proper diagnosis

1.1.1.49 Glucose-6-Phosphate 1-Dehydrogenase continued

NOTES

- Classified as an oxidoreductase acting on the CH-OH group of donors with NAD^+ or $NADP^+$ as acceptor
- Obtained from yeast, *E. coli* and various mammalian tissues
- Composed of 2 subunits of molecular weight 55,000 Da
- A component of the cofactor recycling system for NADH and NADPH
- Activity with NAD is 1.6X greater than with NADP
- Unusual in that it contains no cysteine or cystine residues
- Stabilized by BSA
- Inhibited by acetyl-CoA, ATP, Al^{2+}, Cu^{2+}, Mn^{2+}, metal ions and sulfate
- *Leuconostoc* enzyme does not absolutely require Mg^{2+}; is modestly stimulated by HCO_3^- ($\leq 0.3\ M$); and is inhibited by high concentrations of palmitoyl-CoA, various acyl-CoA, ATP, 1-fluoro-2,4-dinitrobenzene (irreversibly) and pyridoxal 5´-phosphate with respect to glucose-6-phosphate (competitively) and NAD or NADP (noncompetitively)
- Phosphate buffer yields 70% of the activity with Tris buffer

SPECIFIC ACTIVITY	UNITS DEFINITION	PREPARATION FORM	ADDITIONAL ACTIVITIES	SUPPLIER CATALOG NO.
Bacillus species (optimum pH = 8.4, pI = 6.13, T = 75°C, MW = 342,000 Da; K_M = 8.3 μM [$NADP^+$], 0.12 mM [Glc6P]; stable pH 6.0-8.0 [75°C, 30 min], T < 65°C [pH 7.5, 10 min])				
100-200 U/mg solid	1 unit oxidizes 1 μmol D-glucose-6-phosphate to D-glucono-δ-lactone-6-phosphate/min at pH 7.5, 37°C	Lyophilized containing 80% protein		Asahi G6PDHII T-51
Bacillus stearothermophilus (optimum pH = 8.7, pI = 6.5-6.8, T = 70°C, MW = 195,000 Da [53,000/subunit]; K_M = 0.16 mM [G6P], 16 μM [$NADP^+$], 1.64 mM [NAD^+]; stable pH > 7.5, T < 60°C)				
> 150 U/mg protein	1 unit forms 1 μmol NADPH/min at 30°C	Lyophilized	<0.01% GR, 6PHI, PGluM <0.02% GlcK, 6PGDH	Unitika
> 150 U/mg protein	1 unit forms 1 μmol NADPH/min at 30°C	50% glycerol solution	<0.01% GR, 6PHI, PGluM <0.02% GlcK, 6PGDH	Unitika
100-200 U/mg protein	1 unit oxidizes 1.0 μmol Glc6P to 6-phosphogluconate/min at pH 7.8, 30°C in the presence of NADP	Lyophilized containing 75% protein; balance primarily Tris buffer, pH 8.5		ICN 190482
Leuconostoc mesenteroides (optimum pH = 7.4-8.0, pI = 4.6, T = 50°C, MW = 104,000 Da [55,000/subunit]; K_M [NAD^+-linked] = 0.106 mM [NAD^+], 52.7 μM [G6P]; K_M [$NADP^+$-linked] = 5.69 μM [$NADP^+$], 81 μM [G6P]; stable pH 5.5-7.5 [30°C, 17 hr], T < 37°C [pH 8.0, 30 min])				
2250-2750 IU/mL	1 unit reduces 1 μmol NAD^+ to $NADH_2$/min at 30°C	30% glycerol solution, free of particulate matter	<0.01% 6PGDH, CPK, GR, PGluM, MK <0.05% HK, PGI	Beckman 682082

Glucose-6-Phosphate 1-Dehydrogenase continued

1.1.1.49

SPECIFIC ACTIVITY	UNITS DEFINITION	PREPARATION FORM	ADDITIONAL ACTIVITIES	SUPPLIER CATALOG NO.
Leuconostoc mesenteroides continued				
400 IU/mg protein	1 unit reduces 1 μmol NAD/min at 30°C	Lyophilized	<0.01% 6PGDH, CPK, GR, PGluM, MK <0.05% HK, PGI <0.085% 6PGDH	Beckman 682370
≥500 U/mg protein	1 unit reduces 1 μmol NAD^+/min at pH 7.8, 25°C	Crystallized; suspension in 70% $(NH_4)_2SO_4$	<0.01% CPK, 6PGDH, PGI, PGM <0.05% HK	Biozyme GPDHLM2
≥500 U/mg protein; 400 U/mg solid	1 unit reduces 1 μmol NAD^+/min at pH 7.8, 25°C	Salt-free; lyophilized	<0.01% CPK, 6PGDH, PGI, PGM <0.05% HK	Biozyme GPDHLM2F
550 U/mg at 25°C; 650 U/mg at 30°C	1 unit converts 1 μmol Glc6P and NAD to product/min	Suspension in 3.2 M $(NH_4)_2SO_4$, pH 6	<0.001% CK, 6PGDH <0.01% GR, PGI <0.02% NADH oxidase <0.05% HK	Boehringer 165875 165883 737186
150 U/mg	1 unit oxidizes 1 μmol Glc6P/min at pH 7.6, 25°C	Lyophilized	<0.01% CPK, MK, ATPase, PGI <0.02% PGluM <0.05% HK	Fluka 49275
260 U/mg protein	1 unit oxidizes 1 μmol Glc6P/min at pH 7.6, 25°C	Crystallized; suspension in 3.2 M $(NH_4)_2SO_4$, pH 6.0	<0.01% HK, NADH-oxidase, NADPH-oxidase, 6PGDH <0.02% PGI	Fluka 49276
>300 U/mg protein; >170 U/mg solid	1 unit reduces 1.0 μmol NAD^+/min at pH 7.8, 25°C	Lyophilized	<0.002% CK including AK, AK <0.005% ATPase, 6PGDH <0.01% PGI, GR <0.02% PGluM <0.05% HK	Genzyme 1201
>300 U/mg protein	1 unit reduces 1.0 μmol NAD^+/min at pH 7.8, 25°C	Suspension in $(NH_4)_2SO_4$	<0.002% CK including AK <0.005% ATPase, 6PGDH <0.01% PGI, GR <0.02% PGluM <0.05% HK	Genzyme 1202
>300 U/mg protein	1 unit reduces 1.0 μmol NAD^+/min at pH 7.8, 25°C	50% glycerol solution	<0.002% CK including AK <0.005% ATPase, 6PGDH <0.01% PGI, GR <0.02% PGluM <0.05% HK	Genzyme 1203
400 U/mg protein	1 unit reduces 1.0 μmol NAD^+/min at pH 7.8, 25°C	Crystallized; suspension in 70% $(NH_4)_2SO_4$	<0.01% CPK, 6PGDH, PGI, PGM <0.05% HK	ICN 105623
250 U/mg protein	1 unit reduces 1 μmol $NADP^+$/min at pH 7.8, 30°C	Suspension in 3 M $(NH_4)_2SO_4$		ICN 151186
400 U/mg protein	1 unit reduces 1.0 μmol NAD^+/min at pH 7.8, 25°C	Salt-free; lyophilized	<0.01% CPK, 6PGDH, PGI, PGM <0.05% HK	ICN 151187

1.1.1.49 Glucose-6-Phosphate 1-Dehydrogenase continued

SPECIFIC ACTIVITY	UNITS DEFINITION	PREPARATION FORM	ADDITIONAL ACTIVITIES	SUPPLIER CATALOG NO.
Leuconostoc mesenteroides continued				
≥400 IU/mg protein	1 IU transforms 1 μmol substrate/min under standard IUB conditions at 25°C	Lyophilized; suspension in $(NH_4)_2SO_4$	<0.001% CK, MK, GR, PGluM, 6PGDH <0.005% PGI <0.01% HK, LDH	OYC
≥400 U/mg solid	1 unit forms 1 μmol NADH/min at pH 7.8, 30°C	Lyophilized	<0.001% CPK, PGluM, GR <0.005% 6PGDH <0.01% PGI, HK, MK, NADH oxidase, NADPH oxidase	Toyobo G6D-311
250-380 U/mg		Lyophilized; optimum pH = 7.8, pI = 6.2 and 6.9, T = 30°C, MW = 35,000 Da		Wako 079-02451
≥200 NADP U/mg protein	1 unit reduces 1 μmol pyridine nucleotide/min at pH 7.8, 30°C	Chromatographically purified; lyophilized	<0.002% AdK, CPK <0.003% 6PGDH <0.02% PHI	Worthington LS03981 LS03980 LS03982
≥200 NADP U/mg protein	1 unit reduces 1 μmol pyridine nucleotide/min at pH 7.8, 30°C	Chromatographically purified; suspension in 2.4 M $(NH_4)_2SO_4$	<0.002% AK, CPK <0.003% 6PGDH <0.02% PHI	Worthington LS03983 LS03985 LS03987
≥360 NAD U/mg protein	1 unit reduces 1 μmol pyridine nucleotide/min at pH 7.8, 30°C	Chromatographically purified; suspension in 2.4 M $(NH_4)_2SO_4$	<0.002% AK, CPK <0.003% 6PGDH <0.02% PHI	Worthington LS03992 LS03993 LS03994
≥360 NAD U/mg protein	1 unit reduces 1 μmol pyridine nucleotide/min at pH 7.8, 30°C	Chromatographically purified; lyophilized	<0.002% AK, CPK <0.003% 6PGDH <0.02% PHI	Worthington LS03997 LS03998 LS03999
Saccharomyces cereviseae (optimum pH = 9.2)				
400 U/mg protein	1 unit oxidizes 1 μmol Glc6P/min at pH 8.0, 30°C in the presence of NADP	Suspension in 3.3 M $(NH_4)_2SO_4$	No detectable GR, PGM <0.001% ATPase, 6PDG, PGI <0.05% HK	Calbiochem 34678
Yeast (optimum pH = 7.5-9.0, MW = 128,000 Da; relative activity: 25°C = 1.0, 30°C = 1.4, 37°C = 2.2)				
≥250 U/mg protein	1 unit reduces 1 μmol $NADP^+$/min at pH 8.0, 25°C	Crystallized; suspension in 80% $(NH_4)_2SO_4$	<0.001% ATPase, PGM, PGI, 6PGDH <0.005% CPK <0.05% HK	Biozyme GPDH2
≥250 U/mg protein; 150 U/mg solid	1 unit reduces 1 μmol $NADP^+$/min at pH 8.0, 25°C	Salt-free; lyophilized	<0.001% ATPase, PGM, PGI, 6PGDH <0.005% CPK <0.05% HK	Biozyme GPDH2F

Glucose-6-Phosphate 1-Dehydrogenase continued

SPECIFIC ACTIVITY	UNITS DEFINITION	PREPARATION FORM	ADDITIONAL ACTIVITIES	SUPPLIER CATALOG NO.
Yeast *continued*				
350 U/mg	1 unit converts 1 μmol Glc6P to product/min at 25°C	Suspension in 3.2 M $(NH_4)_2SO_4$, pH 6	<0.001% CK <0.002% PGI <0.01% GR, HK, 6PGDH, PGluM	Boehringer 127035 127655 737224
140 U/mg	1 unit converts 1 μmol Glc6P to product/min at 25°C	Suspension in 3.2 M $(NH_4)_2SO_4$ and BSA as stabilizer, pH 6	<0.001% CK <0.002% PGI <0.01% 6PGDH, PGluM <0.02% HK <0.05% GR	Boehringer 127043 127663 127671 737232
140 U/mg protein; 15 U/mg solid	1 unit converts 1 μmol Glc6P to product/min at 25°C	Lyophilized	<0.001% CK <0.002% PGI <0.01% 6PGDH, PGluM <0.02% HK <0.05% GR	Boehringer 197726 737178
140 U/mg protein	1 unit oxidizes 1 μmol Glc6P /min at pH 7.6, 25°C	Crystallized; suspension in 3.2 M $(NH_4)_2SO_4$, BSA as stabilizer, pH 6	<0.02% HK <0.05% GR	Fluka 49270
160 U/mg	1 unit oxidizes 1 μmol Glc6P /min at pH 7.6, 25°C	Lyophilized	<0.005% ATPase, PGI, PGluM <0.01% CPK, MK <0.2% GR <0.3% HK	Fluka 49273
>200 U/mg protein; >120 U/mg solid	1 unit reduces 1.0 μmol $NADP^+$/min at pH 8.0, 25°C	Lyophilized	<0.001% PGluM <0.005% ATPase, 6PGDH, AK, PGI <0.2% GR <0.3% HK	Genzyme 1211
>200 U/mg protein	1 unit reduces 1.0 μmol $NADP^+$/min at pH 8.0, 25°C	Suspension in $(NH_4)_2SO_4$	<0.001% PGluM <0.005% ATPase, 6PGDH, AK, PGI <0.2% GR <0.3% HK	Genzyme 1212
250 U/mg protein	1 unit reduces 1.0 μmol $NADP^+$/min at pH 8.0, 25°C	Crystallized; suspension in 3.2 M $(NH_4)_2SO_4$, pH 6.8	<0.001% ATPase <0.005% CPK <0.3% HK	ICN 100677
200 U/mg protein	1 unit reduces 1.0 μmol $NADP^+$/min at pH 8.0, 25°C	Salt-free; lyophilized	<0.001% ATPase <0.005% CPK <0.3% HK	ICN 151188
≥250 IU/mg protein	1 IU transforms 1 μmol substrate/min under standard IUB conditions at 25°C	Lyophilized; suspension in $(NH_4)_2SO_4$	<0.001% ATPase, CK <0.01% MK, PGluM, 6PGDH, PGI <0.02% HK <0.2% GR	OYC

1.1.1.49 Glucose-6-Phosphate 1-Dehydrogenase continued

SPECIFIC ACTIVITY	UNITS DEFINITION	PREPARATION FORM	ADDITIONAL ACTIVITIES	SUPPLIER CATALOG NO.
Yeast continued				
160-320 U/mg		Lyophilized		Wako 076-02461
Yeast, bakers				
20 U/mg solid	1 unit reduces 1 µmol NADP/min at pH 7.6, 25°C	Lyophilized containing 50% protein	0.002% HK 0.001% CK 0.01% 6PGDH, PGluM	Calzyme 078A0020
240 U/mg protein	1 unit oxidizes 1 µmol Glc6P /min at pH 7.6, 25°C	Crystallized; suspension in 3.2 M $(NH_4)_2SO_4$, pH 7.0	<0.01% GR, GlcDH, NAd-act. G6PDH, NADPH-oxidase, PGluM, 6PGDH, PGI <0.05% HK, MK	Fluka 49271
210 U/mg	1 unit oxidizes 1 µmol Glc6P /min at pH 7.6, 25°C	Lyophilized <0.01% sulfate 20% Na citrate	<0.01% NAd-act., G6PDH, NADH-oxidase, PGluM, 6PGDH <0.05% MK, HK, GlcDH <0.1% GR, PGI	Fluka 49272
150 U/mg protein	1 unit oxidizes 1.0 µmol D-Glc6P to 6-phospho-D-gluconate/min at pH 7.4, 25°C in the presence of NADP	Dry $(NH_4)_2SO_4$ cake	Very low to absent levels of HK, 6PGDH	Sigma G7750
Yeast, Torula				
420 U/mg	1 unit oxidizes 1 µmol Glc6P/min at pH 7.6, 25°C	Powder containing 15% Na citrate and <0.01% sulfate	<0.05% GR, HK, MK, NAd-act. G6PDH, NADPH-oxidase, PGluM, 6PGDH, PGI	Fluka 49278
180 U/mg protein	1 unit oxidizes 1 µmol Glc6P/min at pH 7.6, 25°C	Crystallized; suspension in 3.2 M $(NH_4)_2SO_4$, pH 7.5; <0.01% sulfate	<0.05% GR, HK, MK, NAD-act. G6PDH, NADPH-oxidase, PGluM, 6PGDH, PGI	Fluka 49279
Zymomonas mobilis (optimum pH = 8.0, MW = 208,000 Da [52,000/subunit]; K_M = 0.14 mM [G6P, NAD$^+$], 20 μM [NADP$^+$]; stable pH 5.0-10.0, T < 50°C)				
>300 U/mg protein	1 unit forms 1 µmol NADH/min at 30°C	Lyophilized		Unitika
>300 U/mg protein	1 unit forms 1 µmol NADH/min at 30°C	Suspension in $(NH_4)_2SO_4$		Unitika

3α-Hydroxysteroid Dehydrogenase (B-specific)

1.1.1.50

Reaction Catalyzed
Androsterone + NAD(P)$^+$ ↔ 5α-androstane-3,17-dione + NAD(P)H

Systematic Name
3α-Hydroxysteroid:NAD(P)$^+$ oxidoreductase (B-specific)

Synonyms
Hydroxyprostaglandin dehydrogenase, 3α-hydroxysteroid dehydrogenase, HSDH, 3αHSD

Reactants
Androsterone, 3α-hydroxysteroids, 9-hydroxyprostaglandin, 11-hydroxyprostaglandin, 15-hydroxyprostaglandin, 5α-androstane-3,17-dione, NAD(P)$^+$, NAD(P)H

Applications
- Determination of steroids in drug and pharmaceutical processing, and lipid and steroid research
- Diagnostic determination of bile acids in blood

Notes
- Classified as an oxidoreductase acting on the CH-OH group of donors with NAD$^+$ or NADP$^+$ as acceptor
- B-specific with respect to NAD$^+$ or NADP$^+$
- Oxidizes only 3 α hydroxysteroids of the C_{19}, C_{21} and C_{24} series
- Inhibited by heavy metals and sulfhydryl-binding agents

SPECIFIC ACTIVITY	UNITS DEFINITION	PREPARATION FORM	ADDITIONAL ACTIVITIES	SUPPLIER CATALOG NO.
Pseudomonas species (optimum pH = 8.0-10.0, pI = 4.6, MW = 46,000 Da; K_M = 31 μM [cholic acid]; stable pH 6-10 [37°C, 1 hr], T < 45°C [pH 8.0, 10 min])				
5-20 U/mg solid	1 unit oxidizes 1 μmol androsterone to 3-oxoandrosterone/min at pH 8.0, 37°C	Lyophilized containing 30% protein	<0.05% MDH <0.6% LDH	Asahi 3αHSD T-27
15 U/mg protein	1 unit converts 1 μmol androsterone to product/min at 37°C	Lyophilized	<0.05% ADH <0.1% NADH oxidase <0.2% 3β-HSDH	Boehringer 980552
Pseudomonas testosteroni (optimum pH = 10.2-10.5, MW = 47,000 Da)				
≥95 U/mg protein	1 unit oxidizes 1 μmol androsterone/min at pH 8.9, 25°C in the presence of β-NAD	Lyophilized containing 27% protein	No detectable LDH, NADH oxidase, ADH, L-MDH 0.5% β-HSDH	Nacalai 189-49 GR
20-50 U/mg protein	1 unit oxidizes 1.0 μmol androsterone/min at pH 8.9, 25°C in the presence of β-NAD$^+$	Chromatographically purified; lyophilized containing 50% protein; balance primarily KPO$_4$ buffer salt and EDTA	<0.05% ADH <0.5% β-HSDH	Sigma H1506
50-100 U/mL	1 unit oxidizes 1.0 μmol androsterone/min at pH 8.9, 25°C in the presence of β-NAD$^+$	Suspension in 3.1 M (NH$_4$)$_2$SO$_4$, 1% BSA, 2% trehalose, pH 7.3		Sigma H6398

1.1.1.50 3α-Hydroxysteroid Dehydrogenase (B-specific) continued

SPECIFIC ACTIVITY	UNITS DEFINITION	PREPARATION FORM	ADDITIONAL ACTIVITIES	SUPPLIER CATALOG NO.
Pseudomonas testosteroni continued				
9 U/mg	1 unit oxidizes 1 μmol androsterone/min at pH 8.9, 25°C in the presence of NAD	Lyophilized	<0.05% ADH <0.1% NADH oxidase <0.5% 3β-HSDH	Fluka 56460
95 U/mg protein	1 unit oxidizes 1 μmol androsterone as substrate/min at pH 8.9, 25°C in the presence of β-NAD	Lyophilized containing 27% protein	0.5% β-HSDH	ICN 153849

1.1.1.51 3 (or 17) β-Hydroxysteroid Dehydrogenase

REACTION CATALYZED

Testosterone + NAD(P)$^+$ ↔ Androst-4-ene-3,17-dione + NAD(P)H

SYSTEMATIC NAME

3 (or 17) β-Hydroxysteroid:NAD(P)$^+$ oxidoreductase

SYNONYMS

β-Hydroxysteroid dehydrogenase, hydroxysteroid dehydrogenase, HSDH

REACTANTS

Testosterone, 3β- or 17β-hydroxysteroids, androst-4-ene-3,17-dione, NAD(P)$^+$, NAD(P)H, 4-androsterone-3,17-dione

APPLICATIONS

- Determination of steroids in drug and pharmaceutical processing, and lipid and steroid research
- Diagnostic determination of bile acids in blood

NOTES

- Classified as an oxidoreductase acting on the CH-OH group of donors with NAD$^+$ or NADP$^+$ as acceptor
- Oxidizes 3 β-steroids of the C_{19} and C_{21} series, 17 β-hydroxysteroids of the C_{18}, C_{19} and C_{21} series, and certain 16 β-hydroxysteroids
- Inhibited by heavy metals and reducing agents
- Testosterone oxidation is inhibited by 3,17-α-estradiol and other 1,3,5-estradiene derivatives

SPECIFIC ACTIVITY	UNITS DEFINITION	PREPARATION FORM	ADDITIONAL ACTIVITIES	SUPPLIER CATALOG NO.
Pseudomonas testosteroni (optimum pH = 10.1-10.3, MW = 100,000 Da)				
≥25 U/mg protein	1 unit oxidizes 1 μmol testosterone/min at pH 8.9, 25°C in the presence of β-NAD	Lyophilized containing 40% protein	No detectable LDH, NADH oxidase, ADH, L-MDH 0.1% 3α-HSDH	Nacalai 189-50 GR

3 (or 17) β-Hydroxysteroid Dehydrogenase continued

SPECIFIC ACTIVITY	UNITS DEFINITION	PREPARATION FORM	ADDITIONAL ACTIVITIES	SUPPLIER CATALOG NO.
Pseudomonas testosteroni continued				
20-50 U/mg protein	1 unit oxidizes 1.0 μmol testosterone/min at pH 8.9, 25°C in the presence of β-NAD$^+$	Chromatographically purified; lyophilized containing 15% protein; balance primarily KPO$_4$ buffer salt and EDTA	<0.01% ADH <0.5% 3α-HSDH	Sigma H5133
≥0.5 U/mg solid	1 unit reduces 1 μmol NAD/min at pH 9.0, 25°C using androsterone as substrate	Purified; powder	No detectable activity on testosterone Activity on androsterone	Worthington LS04908 LS04910 LS04911
≥0.03 U/mg solid	1 unit reduces 1 μmol NAD/min at pH 9.0, 25°C using androsterone as substrate	Dried cell preparation	α- and β- activities	Worthington LS04915 LS04916 LS04918
≥0.5 U/mg solid	1 unit reduces 1 μmol NAD/min at pH 9.0, 25°C using androsterone as substrate	Purified; powder	α- and β- activities	Worthington LS04919 LS04920 LS04921 LS04922
24 U/mg protein	1 unit oxidizes 1 μmol testosterone as a substrate/min at pH 8.9, 25°C in the presence of β-NAD	Lyophilized containing 40% protein	0.1% 3α-HSDH	ICN 153850

1.1.1.53 3α (or 20β)-Hydroxysteroid Dehydrogenase

REACTION CATALYZED

Androstan-3α,17β-diol + NAD$^+$ ↔ 17β-Hydroxyandrostan-3-one + NADH

SYSTEMATIC NAME

3α (or 20β)-Hydroxysteroid:NAD$^+$ oxidoreductase

SYNONYMS

Cortisone reductase, (R)-20-hydroxysteroid dehydrogenase, 3α-20β-hydroxysteroid dehydrogenase, HSDH

REACTANTS

Androstan-3α,17β-diol, 17β-hydroxyandrostan-3-one, 3α-hydroxypregnane, 3α-hydroxyandrostane, 20β-hydroxypregnane, 20β-hydroxyandrostane

NOTES

- Classified as an oxidoreductase acting on the CH-OH group of donors with NAD$^+$ or NADP$^+$ as acceptor
- Oxidizes only 3-α hydroxysteroids of the C_{19}, C_{21} and C_{24} series
- The 3α- or 20β-hydroxy groups of pregnane and androstane steroids can act as donors
- Inhibited by heavy metals and sulfhydryl-binding agents

SPECIFIC ACTIVITY	UNITS DEFINITION	PREPARATION FORM	ADDITIONAL ACTIVITIES	SUPPLIER CATALOG NO.
Bacteria				
≥50 IU/mg protein	1 IU transforms 1 μmol substrate/min under standard IUB conditions at 25°C	Lyophilized; suspension in $(NH_4)_2SO_4$	<0.01% ADH, NADH oxidase <0.5% β-HSDH	OYC
Streptomyces hydrogenans				
10 U/mg protein	1 unit reduces 1.0 μmol cortisone to 20-dihydrocortisone/min at pH 7.6, 25°C in the presence of β-NADH	Crystallized; suspension in 3.2 M $(NH_4)_2SO_4$ and BSA, pH 6.0	No detectable NADH oxidase	Sigma H7252
5-8 U/mg protein	1 unit reduces 1.0 μmol cortisone to 20-dihydrocortisone/min at pH 7.6, 25°C in the presence of β-NADH	Partially purified; lyophilized containing 30% protein; balance primarily Tris buffer salts and EDTA	No detectable NADH oxidase	Sigma H2267

Mannitol 2-Dehydrogenase 1.1.1.67

REACTION CATALYZED
    D-Mannitol + NAD$^+$ ↔ D-Fructose + NADH
SYSTEMATIC NAME
    D-Mannitol:NAD$^+$ 2-oxidoreductase
SYNONYMS
    Mannitol dehydrogenase
REACTANTS
    D-Mannitol, D-fructose, NAD$^+$, NADH

APPLICATIONS
- Determination of mannitol in urine

NOTES
- Classified as an oxidoreductase acting on the CH-OH group of donors with NAD$^+$ or NADP$^+$ as acceptor

SPECIFIC ACTIVITY	UNITS DEFINITION	PREPARATION FORM	ADDITIONAL ACTIVITIES	SUPPLIER CATALOG NO.
Actinobacillus species				
10 U/mg solid	1 unit converts 1.0 μmol D-mannitol to D-fructose/min at pH 7.6, 37°C	Lyophilized containing 50-70% protein; balance primarily KPO$_4$ and DTT buffer salts		Sigma M3154
Leuconostoc mesenteroides				
10-50 U/mg protein	1 unit reduces 1.0 μmol D-fructose/min at pH 5.3, 30°C in the presence of NADH	Lyophilized		Sigma M9532
Pseudomonas species (optimum pH = 9.5-11.0, T = 40°C)				
>5 U/mg		Powder		Biocatalysts M093P

1.1.1.72 Glycerol Dehydrogenase (NADP$^+$)

REACTION CATALYZED
    Glycerol + NADP$^+$ ↔ D-Glyceraldehyde + NADPH

SYSTEMATIC NAME
    Glycerol:NADP$^+$ oxidoreductase

REACTANTS
    Glycerol, NADP$^+$, D-glyceraldehyde, NADPH

NOTES
- Classified as an oxidoreductase acting on the CH-OH group of donors with NAD$^+$ or NADP$^+$ as acceptor

SPECIFIC ACTIVITY	UNITS DEFINITION	PREPARATION FORM	ADDITIONAL ACTIVITIES	SUPPLIER CATALOG NO.
Aspergillus niger				
10-20 U/mg protein	1 unit reduces 1.0 μmol NADP/min at pH 9.0, 25°C in the presence of glycerol	Lyophilized containing 70% protein; balance primarily citrate buffer salt		sigma G9509

1.1.1.83 D-Malate Dehydrogenase (Decarboxylating)

REACTION CATALYZED
    (R)-Malate + NAD$^+$ ↔ Pyruvate + CO$_2$ + NADH

SYSTEMATIC NAME
    (R)-Malate:NAD$^+$ oxidoreductase (decarboxylating)

REACTANTS
    (R)-Malate, NAD$^+$, pyruvate, CO$_2$, NADH

APPLICATIONS
- Determination of D-malic acid, e.g., in wine, juice, fruit and other foods

NOTES
- Classified as an oxidoreductase acting on the CH-OH group of donors with NAD$^+$ or NADP$^+$ as acceptor

SPECIFIC ACTIVITY	UNITS DEFINITION	PREPARATION FORM	ADDITIONAL ACTIVITIES	SUPPLIER CATALOG NO.
Escherichia coli				
≥18 U/mg protein; ≥1.6 U/mg solid	1 unit converts 1 μmol D-malic acid to product/min at 25°C	Lyophilized	<0.01% GalDH, GlDH, LDH, MDH, NADH oxidase	Boehringer 1544853

Phosphoglycerate Dehydrogenase

REACTION CATALYZED
3-Phosphoglycerate + NAD^+ ↔ 3-Phosphohydroxypyruvate + NADH

SYSTEMATIC NAME
3-Phosphoglycerate:NAD^+ 2-oxidoreductase

SYNONYMS
3-Phosphoglycerate dehydrogenase, 3PGDH

REACTANTS
3-Phosphoglycerate, 3-phosphohydroxypyruvate

NOTES
- Classified as an oxidoreductase acting on the CH-OH group of donors with NAD^+ or $NADP^+$ as acceptor

SPECIFIC ACTIVITY	UNITS DEFINITION	PREPARATION FORM	ADDITIONAL ACTIVITIES	SUPPLIER CATALOG NO.
Chicken liver				
1.5-7.0 U/mg protein	1 unit converts 1.0 μmol 3-phosphoglycerate to phosphohydroxypyruvate/min at pH 9.0, 25°C	Lyophilized containing 60% protein		sigma P4579

Glucose 1-Dehydrogenase (NADP$^+$)

REACTION CATALYZED

D-Glucose + NADP$^+$ ↔ D-Glucono-1,4-lactone + NADPH

SYSTEMATIC NAME

D-Glucose:NADP$^+$ 1-oxidoreductase

SYNONYMS

Glucose dehydrogenase, GlcDH

REACTANTS

D-Glucose, D-mannose, 2-deoxy-D-glucose, 2-amino-2-deoxy-D-mannose, D-glucono-1,4-lactone

APPLICATIONS

- Determination of glucose in serum or urine

NOTES

- Classified as an oxidoreductase acting on the CH-OH group of donors with NAD$^+$ or NADP$^+$ as acceptor
- Activated by non-ionic detergents
- Stabilized by BSA, EDTA and NADP$^+$

SPECIFIC ACTIVITY	UNITS DEFINITION	PREPARATION FORM	ADDITIONAL ACTIVITIES	SUPPLIER CATALOG NO.
Cryptococcus uniguttulatus (optimum pH = 6.0-7.0, pI = 4.9, MW = 110,000 Da; K_M = 2.6 mM [glucose], 4.2 µM [NADP$^+$]; stable pH 6-7, T < 50°C [pH 6.0, 10 min])				
	1 unit oxidizes 1 µmol β-D-glucose to glucono-δ-lactone/min at pH 8.0, 37°C	Lyophilized containing 40% protein		Asahi GLCDH T-43
> 15 U/mg solid	1 unit oxidizes 1.0 µmol β-D-glucose/min at pH 8.0, 37°C	Lyophilized		Genzyme 6351
10 U/mg solid	1 unit oxidizes 1.0 µmol β-D-glucose to D-glucono-δ-lactone/min at pH 8.0, 37°C in the presence of NADP	Lyophilized		Sigma G0649

D-threo-Aldose 1-Dehydrogenase

1.1.1.122

REACTION CATALYZED

D-threo-aldose + NAD$^+$ ↔ D-threo-aldono-1,5-lactone + NADH

SYSTEMATIC NAME

D-threo-Aldose:NAD$^+$ 1-oxidoreductase

SYNONYMS

L-Fucose dehydrogenase, (2S,3R)-aldose dehydrogenase

REACTANTS

D-threo-Aldose, L-fucose, L-xylose, D-arabinose, NAD$^+$, D-threo-aldono-1,5-lactone, NADH; L-glucose (*Pseudomonas caryophylli* enzyme), L-arabinose (animal enzyme)

NOTES

- Classified as an oxidoreductase acting on the CH-OH group of donors with NAD$^+$ or NADP$^+$ as acceptor

SPECIFIC ACTIVITY	UNITS DEFINITION	PREPARATION FORM	ADDITIONAL ACTIVITIES	SUPPLIER CATALOG NO.
Pseudomonas species				
≥20 U/mg solid	1 unit oxidizes 1.0 μmol L-fucose to L-fucono-1,5-lactone/min at pH 9.5, 37°C in the presence of NADP	Lyophilized		Sigma F0400

7α-Hydroxysteroid Dehydrogenase

REACTION CATALYZED

3α,7α,12α-Trihydroxy-5β-cholanate + NAD$^+$ ↔ 3α,12α-Dihydroxy-7-oxo-5β-cholanate + NADH

SYSTEMATIC NAME

7α-Hydroxysteroid:NAD$^+$ 7-oxidoreductase

SYNONYMS

HSDH, 7αHSD

REACTANTS

3α,7α,12α-Trihydroxy-5β-cholanate, NAD$^+$, 3α,12α-dihydroxy-7-oxo-5β-cholanate, chenodeoxycholic acid, NADH; NADP$^+$ (*Bacteroides fragilis* and *Clostridium* enzymes)

APPLICATIONS

- Oxidizes the 7α-hydroxyl group of bile acids and alcohols in both free and conjugated forms

NOTES

- Classified as an oxidoreductase acting on the CH-OH group of donors with NAD$^+$ or NADP$^+$ as acceptor
- No activity with deoxycholic acid, androsterone or testosterone

SPECIFIC ACTIVITY	UNITS DEFINITION	PREPARATION FORM	ADDITIONAL ACTIVITIES	SUPPLIER CATALOG No.
Escherichia coli				
5-20 U/mg protein	1 unit oxidizes 1.0 μmol chenodeoxycholic acid/min at pH 8.9, 25°C in the presence of β-NAD$^+$	Partially purified; lyophilized containing buffer salts	No detectable ADH or NADH oxidase	Sigma H9506
Pseudomonas species (optimum pH = 8.0-9.0, pI = 4.6, MW = 94,000 Da; K_M = 0.25 *mM* [cholic acid], 0.27 *mM* [chenodeoxycholic acid], 1.2 *mM* [glycocholic acid]; stable pH 7-11 [37°C, 1 hr], T < 50°C [pH 8, 10 min])				
5-20 U/mg solid	1 unit oxidizes 1 μmol cholic acid to 7-oxocholic acid/min at pH 8.0, 37°C	Lyophilized		Asahi 7αHSD T-28
3-5 U/mg protein	1 unit oxidizes 1.0 μmol chenodeoxycholic acid/min at pH 8.9, 25°C in the presence of β-NAD$^+$	Lyophilized		Sigma H2140

12α-Hydroxysteroid Dehydrogenase

1.1.1.176

REACTION CATALYZED
  3α,7α,12α-Trihydroxy-5β-cholanate + NADP$^+$ ↔ 3α,7α-Dihydroxy-12-oxo-5β-cholanate + NADPH

SYSTEMATIC NAME
  12α-Hydroxysteroid:NADP$^+$ 12-oxidoreductase

SYNONYMS
  HSDH, 12αHSD

REACTANTS
  3α,7α,12α-Trihydroxy-5β-cholanate, NADP$^+$, 3α,7α-dihydroxy-12-oxo-5β-cholanate, NADPH

APPLICATIONS
- Oxidizes the 12α-hydroxyl group of bile acids in both free and conjugated forms

NOTES
- Classified as an oxidoreductase acting on the CH-OH group of donors with NAD$^+$ or NADP$^+$ as acceptor

SPECIFIC ACTIVITY	UNITS DEFINITION	PREPARATION FORM	ADDITIONAL ACTIVITIES	SUPPLIER CATALOG NO.
Bacillus sphaericus (optimum pH = 8.8-9.5, pI = 4.77, MW = 250,000 Da; K_M = 0.1 mM [cholic acid]; stable pH 7.0-8.5 [37°C, 1 hr])				
	1 unit oxidizes 1 μmol deoxycholic acid to 12-oxodeoxycholic acid/min at pH 8.0, 37°C	Lyophilized containing 60% protein		Asahi 12αHSD T-29
150-350 U/mg protein	1 unit oxidizes 1.0 μmol deoxycholic acid to 12-ketodeoxycholic acid/min at pH 8.0, 37°C	Lyophilized		Sigma H2265

L-Lactate Dehydrogenase (Cytochrome)

1.1.2.3

REACTION CATALYZED
  (S)-Lactate + 2 ferricytochrome c ↔ Pyruvate + 2 ferrocytochrome c

SYSTEMATIC NAME
  (S)-Lactate:ferricytochrome c 2-oxidoreductase

SYNONYMS
  Lactic acid dehydrogenase, L-lactate dehydrogenase (cytochrome b$_2$)

REACTANTS
  (S)-Lactate, ferricytochrome c, pyruvate, ferrocytochrome c, ferricyanide, phenazine methosulfate, redox dyes, quinone

APPLICATIONS
- Specific determination of L-lactate

1.1.2.3 L-Lactate Dehydrogenase (Cytochrome) continued

NOTES
- Classified as an oxidoreductase acting on the CH-OH group of donors with a cytochrome as acceptor
- Identical with cytochrome b_2
- A flavohemoprotein (FMN)
- Acts on L-lactate, but not the D-isomer or α-hydroxybutyrate
- EDTA prevents metal inhibition

SPECIFIC ACTIVITY	UNITS DEFINITION	PREPARATION FORM	ADDITIONAL ACTIVITIES	SUPPLIER CATALOG NO.
Yeast, bakers (optimum pH = 5.5-9.0 [cytochrome c], 7-8.5 [lactate, ferricyanide]; MW = 228,000 Da)				
0.1-0.6 U/mg protein	1 unit reduces 1.0 μmol ferricyanide and oxidizes 0.5 μmol L-lactate to pyruvate/min at pH 8.4, 37°C	Suspension in 3.2 M $(NH_4)_2SO_4$, pH 6.0		Sigma L4506
≥6 U/vial	1 unit reduces 1 μmol ferricyanide/min at pH 8.4, 25°C	Partially purified; suspension in 2.6 M $(NH_4)_2SO_4$		Worthington LS03970 LS03968 LS03971

1.1.2.4 D-Lactate Dehydrogenase (Cytochrome)

REACTION CATALYZED
    (R)-Lactate + 2 ferricytochrome c ↔ Pyruvate + 2 ferrocytochrome c

SYSTEMATIC NAME
    (R)-Lactate:ferricytochrome-c 2-oxidoreductase

SYNONYMS
    Lactic acid dehydrogenase

REACTANTS
    (R)-Lactate, ferricytochrome c, pyruvate, ferrocytochrome c

NOTES
- Classified as an oxidoreductase acting on the CH-OH group of donors with a cytochrome as acceptor

See Chapter 5. Multiple Enzyme Preparations

Glucose Oxidase 1.1.3.4

REACTION CATALYZED

β-D-Glucose + O_2 ↔ D-Glucono-1,5-lactone + H_2O_2

SYSTEMATIC NAME

β-D-Glucose:oxygen 1-oxidoreductase

SYNONYMS

Glucose oxyhydrase, GO, GOD

REACTANTS

β-D-Glucose, O_2, D-glucono-1,5-lactone, H_2O_2

APPLICATIONS

- Determination of glucose in serum and urine, with peroxidase
- Determination of amylase activity, with α-glucosidase
- (Bio)chemical synthesis in bioelectrochemical cells
- A labeling enzyme for chemiluminescence immunoassays
- Removes glucose from mixtures of saccharides
- Biosensors
- Improves baking quality

NOTES

- Classified as an oxidoreductase acting on the CH-OH group of donors with oxygen as acceptor
- Composed of 2 identical subunits, covalently linked by disulfide bonds; each subunit contains 1 mol each of iron and FAD
- Oxidizes free and terminal-bound glucose
- Saturating the reaction mixture with oxygen increases activity to 100%
- Inhibited by Ag^+, Hg^{2+}, Cu^{2+}, 4-chloromercuribenzoate and D-arabinose
- Discovered initially as an antibiotic; activity subsequently shown to be due to peroxide formation

SPECIFIC ACTIVITY	UNITS DEFINITION	PREPARATION FORM	ADDITIONAL ACTIVITIES	SUPPLIER CATALOG No.
Aspergillus niger (optimum pH = 5.5 (4-7), pI = 4.2, T = 30-40°C, MW = 160,000 Da; stable pH 4.0-7.0, T = 15-60°C)				
≥750 IU/mL ≥2000 IU/mL ≥5000 IU/mL	1 unit liberates 1 μmol H_2O_2/min at 25°C	Chromatographically purified; solution containing buffer and preservative, pH 5.0-7.0	catalase	ABM-RP Glucox-P75 Glucox-P200 Glucox-P500
75,000 IU/mL	1 unit liberates 1 μmol H_2O_2/min at 25°C	Chromatographically purified; lyophilized containing $NaPO_4$ buffer	100-400 BU/mL catalase	ABM-RP Glucox-PS
≥100 U/mg	1 unit oxidizes 1 μmol β-D-glucose to D-gluconic acid and H_2O_2/min at pH 5.1, 35°C	Solution		AMRESCO 0243
>2625 IU/mL	1 unit oxidizes 1 μmol β-D-glucose/min at pH 6.0, 37°C	Solution containing 0.2 M KPO_4 and parabens, pH 6.0	1:1 GO to catalase ratio	Beckman 682726

Glucose Oxidase continued

SPECIFIC ACTIVITY	UNITS DEFINITION	PREPARATION FORM	ADDITIONAL ACTIVITIES	SUPPLIER CATALOG NO.
Aspergillus niger continued				
300 U/mg protein; 250 U/mg solid	1 unit oxidizes 1 μmol glucose/min at pH 7.0, 25°C	Salt-free; lyophilized	<0.2% amylase, saccharase, maltase <10 U/mg catalase	Biozyme GO2A
300 U/mg protein; 170 U/mg solid	1 unit oxidizes 1 μmol glucose/min at pH 7.0, 25°C	Salt-free; lyophilized	<0.2% amylase, saccharase, maltase <3 U/mg catalase	Biozyme GO2AS
360 U/mg protein; 270 U/mg solid	1 unit oxidizes 1 μmol glucose/min at pH 7.0, 25°C	Highly purified; salt-free; lyophilized	<0.05% amylase, saccharase, maltase	Biozyme GO3A
360 U/mg protein	1 unit oxidizes 1 μmol glucose/min at pH 7.0, 25°C	Highly purified; solution containing 20 mg/mL protein	<0.05% amylase, saccharase, maltase	Biozyme GO3AS
300 U/mg protein	1 unit oxidizes 1 μmol glucose/min at pH 7.0, 25°C	Purified; solution containing 10-50 mg/mL protein	<0.05% amylase, saccharase, maltase <2.5 U/mg catalase	Biozyme GO4
300 U/mg protein; 250 U/mg solid	1 unit oxidizes 1 μmol glucose/min at pH 7.0, 25°C	Salt-free; lyophilized	<0.05% amylase, saccharase, maltase <2.5 U/mg catalase	Biozyme GO4F
140 U/mg protein; 20 U/mg solid	1 unit oxidizes 1 μmol glucose/min at pH 7.0, 25°C	Lyophilized	<0.2% amylase, saccharase, maltase <10 U/mg catalase	Biozyme GO5
250 U/mg protein; 1500 U/mL	1 unit oxidizes 1 μmol glucose/min at pH 7.0, 25°C	Purified; solution	<0.05% amylase, saccharase, maltase <5 U/mg catalase	Biozyme GO6
250 U/mg solid	1 unit converts 1 μmol glucose to product/min at 25°C; Q_{O_2} = 180,000 (Kusai)	Lyophilized	<0.01% amylase, saccharase <10 U/mg catalase	Boehringer 105139 105147 737194
200 U/mg solid	1 unit converts 1 μmol glucose to product/min at 25°C; Q_{O_2} = 100,000 (Kusai)	Lyophilized	<0.03% amylase, saccharase <200 U/mg catalase	Boehringer 646431
200-250 U/mg	1 unit oxidizes 1 μmol glucose/min at pH 7.0, 25°C	Lyophilized containing 90-95% protein		Calzyme 077A0250
200 U/mg protein	1 unit oxidizes 1 μmol glucose/min at pH 7.0, 25°C	Solution containing 0.1 M NaOAc buffer, 0.002% thimerosal as preservative, pH 4.0	<0.01% α-amylase, β-amylase <0.05% invertase, glycogenase <2% maltase, GalO <8% catalase	Fluka 49177
20 U/mg	1 unit oxidizes 1 μmol glucose/min at pH 7.0, 25°C	Lyophilized	<50% catalase	Fluka 49178
200 U/mg	1 unit oxidizes 1 μmol glucose/min at pH 7.0, 25°C	Lyophilized	<4% catalase	Fluka 49180

SPECIFIC ACTIVITY	UNITS DEFINITION	PREPARATION FORM	ADDITIONAL ACTIVITIES	SUPPLIER CATALOG No.
Aspergillus niger continued				
130 U/mg	1 unit oxidizes 1 μmol glucose/min at pH 7.0, 25°C	Powder	<0.001% protease <0.05% α-amylase, β-amylase <0.1% invertase, glycogenase, GalO <3% maltase <4% catalase	Fluka 49181
150 U/mg	1 unit oxidizes 1 μmol glucose/min at pH 7.0, 25°C	Powder	<0.001% protease <0.05% invertase, α-amylase, β-amylase, <0.1% glycogenase <1% GalO, catalase <2% maltase	Fluka 49182
4000-6000 Titrimetric U/mL	1 unit oxidizes 3.0 mg glucose to gluconic acid/15 min at 35°C	Food grade; Kosher certified; solution containing 20% protein		Genencor OxyGO^R HP L5000
120-175 Titrimetric U/mg	1 unit oxidizes 3.0 mg glucose to gluconic acid/15 min at 35°C	Food grade; Kosher certified; lyophilized containing 90% protein		Genencor OxyGO^R HP S200
5000-5500 Titrimetric U/mL	1 unit oxidizes 3.0 mg glucose to gluconic acid/15 min at 35°C	Food grade; Kosher certified; solution		Genencor OxyGO^R L5
5-5.5 Titrimetric U/mg	1 unit oxidizes 3.0 mg glucose to gluconic acid/15 min at 35°C	Food grade; Kosher certified; lyophilized containing soya flour and PEG as stabilizer		Genencor OxyGO^R S5
30-75 Titrimetric U/mg	1 unit oxidizes 3.0 mg glucose to gluconic acid/15 min at 35°C	Food grade; Kosher certified; lyophilized containing 70% protein		Genencor OxyGO^R S75
20-30 Titrimetric U/mg	1 unit oxidizes 3.0 mg glucose to gluconic acid/15 min at 35°C	Food grade; Kosher certified; lyophilized containing microcrystalline cellulose and PEG as stabilizer		Genencor OxyGO^R WS25
>250 U/mg protein; >200 U/mg solid	1 unit oxidizes 1.0 μmol glucose/min at pH 7.0, 25°C	Lyophilized	>10:1 GO:catalase ratio	Genzyme 1171
>5000 U/mL	1 unit oxidizes 1.0 μmol glucose/min at pH 7.0, 25°C; 1 kinetic U (Genzyme) = 0.67 titration U (Genencor); 70 Bergmeyer U (Genzyme) = 1 Baker U (Genencor U)	Solution	<2000 U/mL catalase	Genzyme RDX 6424
>190 U/mg protein; >180-260 U/mg solid	1 unit oxidizes 1.0 μmol glucose/min at pH 7.0, 25°C; 1 kinetic U (Genzyme) = 0.67 titration U (Genencor); 70 Bergmeyer U (Genzyme) = 1 Baker U (Genencor U)	Lyophilized	>2 GO:catalase ratio (100 by Genencor method)	Genzyme HP S100 6451

1.1.3.4 Glucose Oxidase continued

SPECIFIC ACTIVITY	UNITS DEFINITION	PREPARATION FORM	ADDITIONAL ACTIVITIES	SUPPLIER CATALOG NO.
Aspergillus niger continued				
>220 U/mg protein; >180-260 U/mg solid	1 unit oxidizes 1.0 μmol glucose/min at pH 7.0, 25°C; 1 kinetic U (Genzyme) = 0.67 titration U (Genencor); 70 Bergmeyer U (Genzyme) = 1 Baker U (Genencor U)	Lyophilized	>5 GO:catalase ratio (250 by Genencor method)	Genzyme HP S120 6461
15-20 U/mg solid	1 unit liberates 1.0 μmol H_2O_2/min at pH 7.0, 25°C			ICN 100289
30-40 U/mg solid		Lyophilized		ICN 100330
0.62 U/mg	pH 5.9, 25°C	Immobilized on DEAE-cellulose		ICN 100331
>200 U/mg protein; >200 U/mg solid	1 unit oxidizes 1 μmol glucose/min at pH 7.0, 25°C	Lyophilized	>10:1 GO:catalase	Randox GO 833L
2000-10,000 U/g solid (without O_2)	1 unit oxidizes 1.0 μmol β-D-glucose to D-gluconic acid and H_2O_2/min at pH 5.1, 35°C (22.4 μL/min O_2 uptake)	Crude	May contain catalase, amylase, maltase, glycogenase, invertase, GalO	Sigma G1262
100,000-200,000 U/g solid (without O_2)	1 unit oxidizes 1.0 μmol β-D-glucose to D-gluconic acid and H_2O_2/min at pH 5.1, 35°C (22.4 μL/min O_2 uptake)	Lyophilized containing 80% protein; balance PO_4 buffer salts and NaCl	10 Sigma U/mg protein catalase Trace amylase, maltase, glycogenase, invertase, GalO	Sigma G2133
15,000-25,000 U/g solid (without O_2)	1 unit oxidizes 1.0 μmol β-D-glucose to D-gluconic acid and H_2O_2/min at pH 5.1, 35°C (22.4 μL/min O_2 uptake)	Lyophilized containing K gluconate	2 Sigma U/mg solid catalase 2% amylase, maltase, glycogenase, invertase, GalO	Sigma G6125 G7773
15,000-25,000 U/g solid (without O_2)	1 unit oxidizes 1.0 μmol β-D-glucose to D-gluconic acid and H_2O_2/min at pH 5.1, 35°C (22.4 μL/min O_2 uptake)	Lyophilized containing 20% protein; balance primarily K gluconate	10 Sigma U/mg protein catalase 2% maltase, glycogenase May contain amylase, invertase, GalO	Sigma Type II-S G6641
2000-10,000 U/g solid (without O_2)	1 unit oxidizes 1.0 μmol β-D-glucose to D-gluconic acid and H_2O_2/min at pH 5.1, 35°C (22.4 μL/min O_2 uptake)	Crude	May contain catalase, amylase, maltase, glycogenase, invertase, GalO	Sigma G6766
1000 U/mL (without O_2)	1 unit oxidizes 1.0 μmol β-D-glucose to D-gluconic acid and H_2O_2/min at pH 5.1, 35°C (22.4 μL/min O_2 uptake)	Solution containing 0.5 M NaOAc buffer, 0.002% thimerosal as preservative, pH 4	No detectable amylase, maltase, invertase, glycogenase 30 Sigma U/mL protein catalase 2% GalO	Sigma G6891
100,000-200,000 U/g solid (without O_2)	1 unit oxidizes 1.0 μmol β-D-glucose to D-gluconic acid and H_2O_2/min at pH 5.1, 35°C (22.4 μL/min O_2 uptake)	Lyophilized containing 80% protein; balance PO_4 buffer and NaCl	10 Sigma U/mg protein catalase Trace amylase, maltase, glycogenase, invertase, GalO	Sigma G7016
100,000-200,000 U/g solid (without O_2)	1 unit oxidizes 1.0 μmol β-D-glucose to D-gluconic acid and H_2O_2/min at pH 5.1, 35°C (22.4 μL/min O_2 uptake)	Lyophilized containing 75% protein	5 Sigma U/mg protein catalase Trace amylase, maltase, glycogenase, invertase, GalO	Sigma G7141

SPECIFIC ACTIVITY	UNITS DEFINITION	PREPARATION FORM	ADDITIONAL ACTIVITIES	SUPPLIER CATALOG NO.
Aspergillus niger continued				
200 U/mg protein	1 unit oxidizes 1.0 μmol β-D-glucose to D-gluconic acid and H_2O_2/min at pH 5.1, 35°C (22.4 μL/min O_2 uptake)	Solution containing 0.1 M NaOAc buffer, 0.002% thimerosal as preservative, pH 4	0.1 Sigma U/mg protein catalase	Sigma G9010
150-250 U/mg		Lyophilized		Wako 074-02401
≥110 U/mg solid	1 unit oxidizes 1 μmol o-dianisidine/min at pH 6.0, 25°C	Chromatographically purified; lyophilized	<0.1% cellobiase, lactase, amylase, sucrase <0.2% maltase	Worthington LS04571 LS04572 LS04573
≥1.25 U/mg solid	1 unit oxidizes 1 μmol o-dianisidine/min at pH 6.0, 25°C	Powder containing gluconate as carrier		Worthington LS08495 LS08497
Aspergillus species (optimum pH = 4.5-7.0, pI < 4.5, T = 40-50°C, MW = 153,000-160,000 Da; K_M = 33 mM [β-D-glucose], 61 mM [2-deoxyglucose]; stable pH 4.5-6.0 [30°C, 20 hr] or 4-8 [50°C, 15 min], T < 50°C [pH 5.7, 1 hr] or < 40°C [pH 7, 30 min])				
≥100 U/mg	1 unit oxidizes 1 μmol glucose/min at pH 7.0, 25°C	Powder		Amano Glucose oxidase
≥2000 U/mL	1 unit oxidizes 1 μmol glucose/min at pH 7.0, 25°C	Solution		Amano Glucose oxidase L
100 GU/g				Danisco GRINDAMYL™ G 55
≥100 U/mg solid	1 unit forms 1 μmol H_2O_2 (0.5 μmol quinoneimine dye)/min at pH 5.7, 37°C	Lyophilized containing 50% K gluconate and Na Glu as stabilizers	<3% catalase	Toyobo GLO-201
Microbial (optimum pH = 5.5, T = 25°C)				
>80 U/mg solid	1 unit oxidizes 1 μmol β-D-glucose to D-gluconic acid and H_2O_2/min at pH 5.5, 30°C	Lyophilized		GDS Tech G-010
≥300 IU/mg protein	1 IU transforms 1 μmol substrate/min under standard IUB conditions at 25°C	Lyophilized	<0.01% amylase, saccharase <0.05% catalase	OYC

1.1.3.6 Cholesterol Oxidase

REACTION CATALYZED
Cholesterol + O_2 ↔ Cholest-4-en-3-one + H_2O_2

SYSTEMATIC NAME
Cholesterol:oxygen oxidoreductase

SYNONYMS
ChO

REACTANTS
Cholesterol, O_2, cholest-4-en-3-one, H_2O_2

APPLICATIONS
- Determination of free and bound cholesterol in serum; total cholesterol, with cholesterol esterase
- Determination of cholesterol, cholesterol esters and cholesterol-associated lipoproteins in lipid extracts
- Chemiluminescence analysis
- Enzyme electrode analysis

NOTES
- Classified as an oxidoreductase acting on the CH-OH group of donors with oxygen as acceptor
- Hygroscopic; may lose activity as it absorbs moisture
- Activated by Triton X-100, Adekatol, DOC and hydroxypolyethoxydodecane
- Inhibited by $ZnCl_2$, Hg^{2+}, ionic detergents, Brij 35, Tween 60, Tween 40, SDS and LBS
- Stabilized by BSA and sugars

SPECIFIC ACTIVITY	UNITS DEFINITION	PREPARATION FORM	ADDITIONAL ACTIVITIES	SUPPLIER CATALOG NO.
Brevibacterium species				
≥4 U/mg	1 unit liberates 1 μmol H_2O_2/min at 37°C with cholesterol as substrate	Lyophilized	<0.005% L-AOD <0.2% GO, uricase <0.5% catalase	Beckman 682840
20 U/mg solid	1 unit decomposes 1 μmol cholesterol ester/min at pH 7.0, 37°C	Lyophilized	<0.1% GO, uricase <0.4% catalase	Scripps C1614
4 U/mg solid	1 unit converts 1.0 μmol cholesterol to 4-cholesten-3-one/min at pH 7.0, 37°C	Lyophilized		Sigma C8153
Cellulomonas species				
20-60 U/mg protein	1 unit converts 1.0 μmol cholesterol to 4-cholesten-3-one/min at pH 7.5, 25°C	Solution containing 10 m*M* Tris-HCl, pH 8.0		Sigma C5421
Microbial (optimum pH = 7.0, pI = 4.8, T = 37°C, MW = 58,000 Da; K_M = 25 μ*M* [cholesterol]; stable pH 4.0-8.5 [37°C, 1 hr], T < 60°C [pH 7.0, 10 min])				
1-5 U/mg solid	1 unit generates 1 μmol Δ⁴-cholesten-3-one/min at pH 7.0, 37°C	Lyophilized containing 20% protein	<0.0002% CE <2.0% catalase	Asahi CON T-19
25 U/mg at 25°C; 45 U/mg at 37°C	1 unit converts 1 μmol cholesterol to product/min	Solution containing 1 *M* NaCl and BSA as stabilizer, pH 7	<0.005% CE <0.01% uricase, GO, NADH oxidase	Boehringer 126934 737658

Cholesterol Oxidase continued — 1.1.3.6

SPECIFIC ACTIVITY	UNITS DEFINITION	PREPARATION FORM	ADDITIONAL ACTIVITIES	SUPPLIER CATALOG No.
Microbial continued				
>5 U/mg protein	1 unit oxidizes 1 μmol cholesterol to cholestenone/min at pH 7.0, 37°C	Lyophilized		GDS Tech C-040 C-061
Nocardia erythropolis				
25 U/mg at 25°C; 45 U/mg at 37°C	1 unit converts 1 μmol cholesterol to product/min	Solution containing 3 M NaCl and BSA as stabilizer, pH 6	<0.005% CE <0.01% uricase, GO, NADH oxidase	Boehringer 393924 396818
25 U/mg protein	1 unit converts 1 μmol cholesterol to 4-cholesten-3-one/min at pH 7.5, 25°C	Solution containing 1 M $(NH_4)_2SO_4$, pH 6		Fluka 26746
30-50 U/mg protein	1 unit oxidizes 1 μmol cholesterol/min at pH 7.0, 30°C	Lyophilized		ICN 150672
25-45 U/mg		Solution containing 1 M $(NH_4)_2SO_4$ and BSA, pH 6.0		ICN 150673
20 U/mg protein	1 unit converts 1.0 μmol cholesterol to 4-cholesten-3-one/min at pH 7.5, 25°C	Solution containing 1 M $(NH_4)_2SO_4$ and BSA, pH 6		Sigma C1512
25-50 U/mg protein	1 unit converts 1.0 μmol cholesterol to 4-cholesten-3-one/min at pH 7.5, 25°C	Chromatographically purified; lyophilized containing 5% protein; balance primarily Tris buffer salts and deoxycholate as stabilizer		Sigma C1638
Nocardia species (optimum pH = 5.0-8.0, pI = 4.85, MW = 59,000 Da; relative activity: 25°C = 1.0, 30°C = 1.2, 37°C = 1.35; stable pH 5.5-6.5, T < 50°C)				
>16 U/mg protein	1 unit oxidizes 1 μmol cholesterol/min at pH 7.5, 25°C	Lyophilized	<0.1% catalase, GO	Calbiochem 228230
>15 U/mg protein; >10 U/mg solid	1 unit oxidizes 1.0 μmol cholesterol/min at pH 7.0, 37°C	Lyophilized	<0.05% GO, uricase <3.00% catalase	Genzyme 1101
>15 U/mg protein	1 unit oxidizes 1.0 μmol cholesterol/min at pH 7.0, 37°C	Solution containing Tris buffer	<0.05% GO, uricase <3.00% catalase	Genzyme 1104
>10 U/mg solid	1 unit converts 1 μmol cholesterol to cholestenone/min at pH 7.0, 37°C	Lyophilized	<0.05% GO, uricase <3% catalase	Randox CO 958L
Pseudomonas fluorescens				
10-50 U/mg protein	1 unit converts 1.0 μmol cholesterol to 4-cholesten-3-one/min at pH 7.5, 25°C	Lyophilized containing 15% protein; balance primarily KPO_4		Sigma C7149
Pseudomonas species (optimum pH = 5.0-6.0, pI = 6.2 and 6.9, T = 50°C, MW = 35,000 Da; stable pH 6-8.5 [37°C, 1 hr], T < 60°C [pH 7, 1 hr])				
≥5 U/mg	1 unit oxidizes 1 μmol cholesterol/min at pH 7.0, 37°C	Powder		Amano Cholesterol oxidase

SPECIFIC ACTIVITY	UNITS DEFINITION	PREPARATION FORM	ADDITIONAL ACTIVITIES	SUPPLIER CATALOG NO.
Pseudomonas* species *continued				
5 U/mg	1 unit converts 1 μmol cholesterol to 4-cholesten-3-one/min at pH 7.5, 25°C	Powder		Fluka 26747
>5 U/mg protein; >4 U/mg solid	1 unit oxidizes 1.0 μmol cholesterol/min at pH 7.0, 37°C	Lyophilized		Genzyme 1111
3-10 U/mg solid	1 unit oxidizes 1.0 μmol cholesterol/min at pH 7.0, 37°C	Lyophilized		ICN 190006
3-5 U/mg		Lyophilized		Wako 033-11201
Schizophyllum commune				
10 U/mg protein	1 unit converts 1.0 μmol cholesterol to 4-cholesten-3-one/min at pH 5.0, 37°C	Lyophilized containing 75% protein; balance primarily borate buffer salts		Sigma C7274
Streptomyces cinnamomeus (optimum pH = 6.0-8.0, pI = 6.1 and 7.3, MW = 38,000 Da; K_M = 36 μM [cholesterol]; stable pH 6.0-8.5 [37°C, 10 min], T < 40°C [pH 7.0, 10 min])				
15-40 U/mg solid	1 unit produces 1 μmol Δ⁴-cholesten-3-one/min at pH 7, 37°C	Lyophilized containing 80% protein	<0.01% GO <1.0% catalase	Asahi CO T-04
>20 U/mg protein; >15 U/mg solid	1 unit oxidizes 1.0 μmol cholesterol/min at pH 7.0, 37°C	Lyophilized	<0.01% GO <1.0% catalase	Genzyme 6231
***Streptomyces* species** (optimum pH = 6.5-7.0, pI = 5.1 and 5.4, T = 45-50°C, MW = 34,000 Da [Sephadex G-200]; K_M = 43 μM [cholesterol]; stable pH 5.0-10.0 [25°C, 20 hr], T < 45°C [pH 7.0, 15 min])				
20 U/mg	1 unit converts 1 μmol cholesterol to 4-cholesten-3-one/min at pH 7.5, 25°C	Powder		Fluka 26748
15 U/mg solid	1 unit forms 1 μmol H_2O_2/min at pH 7.0, 37°C	Lyophilized		ICN 150674
	1 unit converts 1.0 μmol cholesterol to 4-cholesten-3-one/min at pH 7.5, 25°C	Lyophilized containing 60% protein; balance Na cholate, borate, BSA		Sigma C8649
≥15 U/mg solid	1 unit forms 1 μmol H_2O_2 (0.5 μmol quinoneimine dye)/min at pH 7.0, 37°C	Lyophilized containing 40% BSA and sugars as stabilizers	<0.01% CE <1% catalase	Toyobo COO-311
Streptomyces*, recombinant *Nocardia (optimum pH = 6.5-7.0; relative activity: 25°C = 1.0, 30°C = 1.2, 37°C = 1.35)				
>25 U/mg protein; >25 U/mg solid	1 unit oxidizes 1.0 μmol cholesterol/min at pH 7.0, 37°C	Lyophilized	<0.01% GO, uricase <1% catalase	Genzyme 1221

Galactose Oxidase

REACTION CATALYZED
$$D\text{-Galactose} + O_2 \leftrightarrow D\text{-Galacto-hexodialdose} + H_2O_2$$

SYSTEMATIC NAME
  D-Galactose:oxygen 6-oxidoreductase

SYNONYMS
  GalO

REACTANTS
  D-Galactose, O_2, D-galacto-hexodialdose, H_2O_2

APPLICATIONS
- Determination of lactose and galactose
- Structural investigations of poly- and oligo-saccharides and glycoproteins
- Detect and distinguish galactose and glycoproteins histochemically
- Cell surface research

NOTES
- Classified as an oxidoreductase acting on the CH-OH group of donors with oxygen as acceptor
- A copper protein (1 per molecule)
- Oxidizes galactose and some galactose derivatives at the C_6 position in both free and polymeric forms
- D-Glucose, L-galactose, L-arabinose and D-glucuronate and D-galactose substituted at the C_4-OH *are not oxidized*
- 2-Deoxy-D-galactose, lactose, melibiose, raffinose and stachyose react with galactose oxidase in the peroxidase/O-tolidine system
- Inhibited by cyanide, diethyldithiocarbamate, azide and hydroxylamine
- Stabilized by sucrose and Cu^{2+}

SPECIFIC ACTIVITY	UNITS DEFINITION	PREPARATION FORM	ADDITIONAL ACTIVITIES	SUPPLIER CATALOG NO.
Dactylium dendroides (optimum pH = 7.0, MW = 65,000-71,000 Da)				
250 U/mg solid	1 unit produces a ΔA_{425} of 1.0/min at pH 6.0, 25°C in the peroxide/2-toluidine assay	DFP-treated; lyophilized		Boehringer 1213784
55 U/mg	1 unit oxidizes 1 μmol D-galactose/min at pH 7.0, 25°C	Lyophilized		Fluka 48267
500-1500 U/mg protein	1 unit produces a ΔA_{425} of 1.0/min at pH 6.0, 25°C in a peroxidase and o-tolidine system in a 3.4 mL reaction vol.; light path=1cm	Partially purified; lyophilized containing 25% protein; balance is primarily buffer salts and stabilizer	No detectable lactase, invertase, GO <50 Sigma U/mg protein catalase	Sigma G3385
	1 unit produces a ΔA_{425} of 1.0/min at pH 6.0, 25°C in a peroxidase and o-tolidine system in a 3.4 mL reaction vol.; light path=1cm	Crude; lyophilized		Sigma G7400

1.1.3.9 Galactose Oxidase continued

SPECIFIC ACTIVITY	UNITS DEFINITION	PREPARATION FORM	ADDITIONAL ACTIVITIES	SUPPLIER CATALOG NO.
Dactylium dendroides continued				
≥30 U/mg solid	1 unit has a ΔA_{425} of 1.0/min at pH 6.0, 25°C in a coupled peroxidase-*o*-tolidine system, using galactose as substrate; 1 absorbance U = 0.54 μmol galactose oxidized/min	Partially purified; lyophilized containing 50-60% sucrose		Worthington LS04520 LS04522 LS04524 LS04523
Dactylium species (optimum pH = 6.8-7.0, T = 40°C, MW = 65,000-71,000 Da [1 copper atom per molecule]; K_M = 14 mM [D-galactose]; stable pH 4.0-10.0 [25°C, 20 hr], T < 60°C [pH 5.0, 10 min])				
≥15 U/mg solid	1 unit forms 1 μmol H_2O_2 (0.5 μmol quinoneimine dye)/min at pH 7.0, 25°C	Lyophilized containing 70% sucrose and Cu^{2+} as stabilizers	<0.1% GO <1% catalase	Toyobo GAO-201

1.1.3.13 Alcohol Oxidase

REACTION CATALYZED
Primary alcohol + O_2 ↔ Aldehyde + H_2O_2

SYSTEMATIC NAME
Alcohol:oxygen oxidoreductase

SYNONYMS
Methanol oxidase

REACTANTS
Primary alcohols, unsaturated alcohols, O_2, aldehydes, H_2O_2

APPLICATIONS
- Determination of alcohol levels in biological fluids

NOTES
- Classified as an oxidoreductase acting on the CH-OH group of donors with oxygen as acceptor
- A flavoprotein (FAD)
- Does not act on branched-chain and secondary alcohols
- Inhibited by PCMB, *N*-ethyl maleinimide, Cu^{2+}, Ag^+, Hg^{2+}, hydroxylamine, NaF and DAC

SPECIFIC ACTIVITY	UNITS DEFINITION	PREPARATION FORM	ADDITIONAL ACTIVITIES	SUPPLIER CATALOG NO.
Candida boidinii				
5-15 U/mg protein	1 unit oxidizes 1.0 μmol MeOH to formaldehyde/min at pH 7.5, 25°C	Lyophilized containing 10% protein; balance KPO_4 buffer salts, DTE and stabilizer		Sigma A6941

SPECIFIC ACTIVITY	UNITS DEFINITION	PREPARATION FORM	ADDITIONAL ACTIVITIES	SUPPLIER CATALOG NO.
Candida species (optimum pH = 7.5-9.0, pI = 4.1, MW = 520,000 Da [gel filtration], 75,000 Da [SDS-PAGE]; K_M = 2.9 mM [MeOH], 8.2 mM [EtOH]; stable pH 6.0-9.5 [37°C, 1 hr], T < 40°C [pH 7.5, 10 min])				
7-20 U/mg solid	1 unit generates 1 μmol H_2O_2/min at pH 8.0, 37°C	Lyophilized		Asahi ALOD T-38
>14 U/mg protein; >7 U/mg solid	1 unit oxidizes 1.0 μmol MeOH/min at pH 8.0, 37°C	Lyophilized		Genzyme 6031
Hansenula polymorpha strain Q3N				
4.4 U/mg protein	1 unit oxidizes 1 μmol MeOH/min at pH 7.0, 37°C	Suspension in 50 mM KPO_4 and 0.38 M $(NH_4)_2SO_4$, pH 7.5		Fluka 64253
Hansenula species				
20-40 U/mg protein	1 unit oxidizes 1.0 μmol MeOH to formaldehyde/min at pH 7.5, 25°C	Lyophilized		Sigma A0438
Pichea pastoris				
10-20 U/mg protein	1 unit oxidizes 1.0 μmol MeOH to formaldehyde/min at pH 7.5, 25°C	Solution containing 35% sucrose		ICN 190155
10-40 U/mg protein	1 unit oxidizes 1.0 μmol MeOH to formaldehyde/min at pH 7.5, 25°C	Solution containing PO_4-buffered 60% sucrose		Sigma A2404
Yeast (optimum pH = 7.0-8.5, pI = 4.8, T = 45°C, MW = 500,000 Da [7-8 subunits at 75,000 Da each]; K_M = 4.0 mM [MeOH], 72 mM [EtOH]; stable pH 5.5-9.0 [25°C, 20 hr], T < 40°C [pH 7.5, 15 min])				
≥10 U/mg protein; ≥15 mg/mL protein	1 unit forms 1 μmol H_2O_2 (0.5 μmol quinoneimine dye)/min at pH 7.5, 30°C	Solution containing 35% sucrose	<1% catalase	Toyobo AOO-309

1.1.3.15 (S)-2-Hydroxy-acid Oxidase

REACTION CATALYZED
(S)-2-Hydroxy acid + O_2 ↔ 2-Oxo acid + H_2O_2

SYSTEMATIC NAME
(S)-2-Hydroxy acid:oxygen 2-oxidoreductase

SYNONYMS
Glycolate oxidase, hydroxy-acid oxidase A, hydroxy-acid oxidase B

REACTANTS
(S)-2-Hydroxy acid, O_2, 2-oxo acid, H_2O_2

NOTES
- Classified as an oxidoreductase acting on the CH-OH group of donors with oxygen as acceptor
- A flavoprotein (FMN)
- The A isozyme preferentially oxidizes short-chain aliphatic hydroxy acids
- The B isozyme preferentially oxidizes long-chain and aromatic hydroxy acids
- Rat isozyme B also acts as L-amino-acid oxidase (EC 1.4.3.2)

SPECIFIC ACTIVITY	UNITS DEFINITION	PREPARATION FORM	ADDITIONAL ACTIVITIES	SUPPLIER CATALOG NO.
Spinach oleracea (Spinach)				
6 U/mg	1 unit produces 1 µmol glyoxylate from glycolate/min at 37°C	Powder	<0.1% glycerate dehydrogenase, glyoxylate reductase, peroxidase, catalase	Fluka 50581
5-20 U/mg protein	1 unit produces 1.0 µmol glyoxylate from glycolate/min at pH 7.8, 25°C in the presence of phenylhydrazine	Lyophilized containing 50% protein; balance primarily Tris buffer salts, $(NH_4)_2SO_4$, FMN	May contain glyoxylate reductase, glycerate dehydrogenase, peroxidase, catalase	Sigma G1886
3-12 U/mg protein	1 unit produces 1.0 µmol glyoxylate from glycolate/min at pH 7.8, 25°C in the presence of phenylhydrazine	Suspension in 3.2 M $(NH_4)_2SO_4$ and 2 mM FMN, pH 7.4	May contain glyoxylate reductase, glycerate dehydrogenase, peroxidase, catalase	Sigma G8260
Beta vulgaris vulgaris (Sugar beet)				
10-30 U/mg protein	1 unit produces 1.0 µmol glyoxylate from glycolate/min at pH 8.3, 25°C in the presence of phenylhydrazine	Suspension in 2.4 M $(NH_4)_2SO_4$, 10 mM Tris, 5 mM FMN, pH 8.3	May contain glyoxylate reductase, glycerate dehydrogenase, peroxidase, catalase	Sigma G4136

Choline Oxidase 1.1.3.17

REACTION CATALYZED
Choline + O_2 ↔ Betaine aldehyde + H_2O_2

SYSTEMATIC NAME
Choline:oxygen 1-oxidoreductase

REACTANTS
Choline, betaine aldehyde, betaine aldehyde, betaine

APPLICATIONS
- Determination of lecithins in biological fluids
- Determination of phospholipids, with phospholipase D

NOTES
- Classified as an oxidoreductase acting on the CH-OH group of donors with oxygen as acceptor
- A flavoprotein which covalently binds 1 mol FAD per mol enzyme
- Inhibited by Cu^{2+}, Co^{2+}, Hg^{2+}, Ag^+ and PCMB
- Stabilized by EDTA, BSA and amino acids

SPECIFIC ACTIVITY	UNITS DEFINITION	PREPARATION FORM	ADDITIONAL ACTIVITIES	SUPPLIER CATALOG No.
Alcaligenes species (optimum pH = 8.0-8.5, pI = 4.1, T = 40-45°C, MW = 95,000 Da; K_M = 2.84 mM [choline], 5.33 mM [betaine aldehyde]; stable pH 7.0-9.0 [30°C, 2 hr], T < 37°C [pH 7.5, 10 min])				
10 U/mg solid	1 unit forms 1 µmol H_2O_2/min	Lyophilized		AMRESCO E220
11 U/mg	1 unit produces 1 µmol H_2O_2/min at pH 8.0, 37°C with choline as substrate	Powder		Fluka 26986
≥10 U/mg solid	1 unit forms 1 µmol H_2O_2/min at pH 8.0, 37°C	Lyophilized containing 20% stabilizers		ICN 153492
10 U/mg solid	1 unit forms 1.0 µmol H_2O_2 from choline and H_2O/min at pH 8.0, 37°C	Lyophilized		Sigma C5896
≥10 U/mg solid	1 unit forms 1 µmol H_2O_2 (0.5 µmol quinoneimine dye)/min at pH 8.0, 37°C	Lyophilized containing 20% BSA, EDTA, amino acids as stabilizers	<100% catalase	Toyobo CHO-301
10 U/mg		Lyophilized		Wako 037-14401
Arthrobacter globiformis (optimum pH = 7.5-8.0, pI = 4.5, MW = 83,000 Da; K_M = 1.2 mM [choline], 8.7 mM [betaine aldehyde]; stable pH 7.5-9 [37°C, 30 min], T < 40°C [pH 7.5, 10 min])				
8-20 U/mg solid	1 unit generates 1 µmol H_2O_2/min at pH 8, 37°C	Lyophilized containing 90% protein	<0.01% GO <10% catalase/CO	Asahi COD T-05
10 U/mg protein	1 unit forms 1 µmol H_2O_2/min at 37°C	Solution containing 4 M NaCl and 10 mM EDTA	<0.02% GO	Boehringer 430323

1.1.3.17 Choline Oxidase continued

SPECIFIC ACTIVITY	UNITS DEFINITION	PREPARATION FORM	ADDITIONAL ACTIVITIES	SUPPLIER CATALOG NO.
Arthrobacter globiformis continued				
>9 U/mg protein; >8 U/mg solid	1 unit forms 1.0 μmol H_2O_2/min at pH 8.0, 37°C	Lyophilized	<0.01% GO <10 catalase:GO	Genzyme 1121
8-20 U/mg solid	1 unit forms 1.0 μmol H_2O_2 from choline and H_2O/min at pH 8.0, 37°C	Lyophilized		Sigma C4405

1.1.3.21 Glycerol-3-Phosphate Oxidase

REACTION CATALYZED

sn-Glycerol 3-phosphate + O_2 ↔ Glycerone phosphate + H_2O_2

SYSTEMATIC NAME

sn-Glycerol 3-phosphate:oxygen 2-oxidoreductase

SYNONYMS

GPO

REACTANTS

sn-Glycerol 3-phosphate, O_2, glycerone phosphate, H_2O_2

APPLICATIONS
- Determination of triacylglycerols
- Determination of triglycerides, with lipoprotein lipase and glycerol kinase

NOTES
- Classified as an oxidoreductase acting on the CH-OH group of donors with oxygen as acceptor
- A flavoprotein (FAD)
- Stabilized by FAD, ammonium sulfate, amino acids and sucrose
- Inhibited by ionic detergents (SDS, LBS), Ag^+, Hg^{2+} and sulfhydryl reagents

SPECIFIC ACTIVITY	UNITS DEFINITION	PREPARATION FORM	ADDITIONAL ACTIVITIES	SUPPLIER CATALOG NO.
Aerococcus viridans (optimum pH = 7.5-8.5, pI = 4.2, T = 30°C, MW = 74,000 Da [Sephadex G-150], 75,000 Da [SDS-PAGE]; K_M = 3.2 mM [L-α-glycero-3-phosphate]; stable pH 6-8 [45°C, 10 min], T < 40°C [0.2 M dimethylglutarate buffer, pH 7, 10 min])				
75-150 U/mg solid	1 unit generates 1 μmol H_2O_2/min at pH 8.0, 37°C	Lyophilized containing 80% protein	<0.01% LO <0.5% catalase	Asahi GPO T-15
40 U/mg	1 unit oxidizes 1 μmol sn-G3P/min at 37°C	Powder	<0.01% LO	Fluka 50030
>100 U/mg protein; >70 U/mg solid	1 unit oxidizes 1.0 μmol L-α-glycerol phosphate/min at pH 8.0, 37°C	Lyophilized	<0.002% LO	Genzyme 1321

SPECIFIC ACTIVITY	UNITS DEFINITION	PREPARATION FORM	ADDITIONAL ACTIVITIES	SUPPLIER CATALOG NO.
Aerococcus viridans continued				
120 U/mg solid	1 unit oxidizes 1.0 μmol L-G3P to DHAP with the formation of H_2O_2/min at pH 8.1, 37°C	Lyophilized containing 70% protein; balance primarily sucrose		ICN 157239
70 U/mg solid	1 unit oxidizes 1.0 μmol L-G3P to DHAP with the formation of H_2O_2/min at pH 8.1, 37°C	Lyophilized containing 70% protein; balance primarily sucrose		Sigma G9888
Microbial (optimum pH = 6.5-8.0, pI = 4.6, T = 40°C, MW = 75,000 Da, 93,000 Da [gel filtration]; K_M = 2.3 mM [L-α-glycerophosphate]; stable pH 5.0-7.5 [25°C, 60 min], T < 45°C [pH 6.5, 10 min])				
40 U/mg solid or 50 U/mg protein (L-G3P); 65 U/mg solid or 82 U/mg enzyme protein (D-, L-G3P)	1 unit converts 1 μmol L-G3P (25°C) or D-, L-G3P (37°C) to product/min	Lyophilized	<0.002% LO	Boehringer 775797
>50 U/mg protein; >40 U/mg solid	1 unit oxidizes 1.0 μmol L-α-glycerol phosphate/min at pH 6.5, 37°C	Lyophilized	<0.001% LO <0.1% AK	Genzyme 6061
>40 U/mg protein; >5 U/mg solid	1 unit oxidizes 1 μmol G3P/min at pH 7.5, 25°C	Lyophilized	<0.002% ChO, LO, uricase	Randox GC 947L
≥15 U/mg solid	1 unit forms 1 μmol H_2O_2 (0.5 μmol quinoneimine dye)/min at pH 6.5, 30°C	Lyophilized containing 60% amino acids as stabilizer	<0.0002% LO <0.001% phosphatase	Toyobo G30-311
Pediococcus species (optimum pH = 8.0-8.5, pI = 4.1, T = 35-40°C, MW = 76,000 Da [gel filtration]; K_M 3.2 mM [L-α-glycerophosphate], 6.8 mM [D-, L-glycerophosphate]; stable pH 6.5-8.5 [25°C, 20 hr], T < 40°C [pH 7.0, 15 min])				
>50 U/mg protein; >40 U/mg solid	1 unit oxidizes 1.0 μmol L-α-glycerol phosphate/min at pH 8.1, 37°C	Lyophilized	<0.001% LO	Genzyme 6051
40-80 U/mg solid	1 unit oxidizes 1.0 μmol L-G3P to DHAP with the formation of H_2O_2/min at pH 8.1, 37°C	Lyophilized containing 70% protein; balance stabilizers		ICN 151198
40-80 U/mg solid	1 unit oxidizes 1.0 μmol L-G3P to DHAP with the formation of H_2O_2/min at pH 8.1, 37°C	Lyophilized containing 70% protein; balance stabilizers		Sigma G9637
≥40 U/mg solid	1 unit forms 1 μmol H_2O_2 (0.5 μmol quinoneimine dye)/min at pH 8.1, 37°C	Lyophilized containing 40% sucrose and FAD as stabilizers	<0.001% LO	Toyobo G30-301

1.1.3.21 Glycerol-3-Phosphate Oxidase continued

SPECIFIC ACTIVITY	UNITS DEFINITION	PREPARATION FORM	ADDITIONAL ACTIVITIES	SUPPLIER CATALOG NO.
Streptococcus species (optimum pH = 6.5-7.0 [PIPES buffer] and 7.0-7.5 [KPO_4 buffer], pI = 4.4 and 4.03, T = 37°C, MW = 170,000 Da, 190,000 Da [Sephacryl S-200]; K_M = 5.4 mM [L-α-glycerophosphate]; stable pH 6-7 [37°C, 30 min], T < 37°C [pH 7, 10 min])				
≥15 U/mg	1 unit oxidizes 1 μmol L-α-G3P/min at pH 7.0, 37°C	Powder		Amano
40-80 U/mg solid	1 unit generates 1 μmol H_2O_2/min at pH 6.5, 37°C	Lyophilized containing 80% protein	<0.001% LO <0.1% AK	Asahi GPOS T-40
>25 U/mg protein; >20 U/mg solid	1 unit oxidizes 1.0 μmol L-α-glycerol phosphate/min at pH 7.0, 37°C	Lyophilized	<0.01% LO	Genzyme 1341
>30 U/mg protein; >15 U/mg solid	1 unit oxidizes 1.0 μmol L-α-glycerol phosphate/min at pH 6.5, 30°C	Lyophilized	<0.0002% LO <0.001% phosphatase	Genzyme 6101
Streptococcus thermophilus				
10-20 U/mg solid	1 unit oxidizes 1.0 μmol L-G3P to DHAP with the formation of H_2O_2/min at pH 7.0, 37°C	Lyophilized		ICN 157240
10 U/mg solid	1 unit oxidizes 1.0 μmol L-G3P to DHAP with the formation of H_2O_2/min at pH 7.0, 37°C	Lyophilized		Sigma G4388

1.1.3.22 Xanthine Oxidase

REACTION CATALYZED

Xanthine + H_2O + O_2 ↔ Urate + H_2O_2

SYSTEMATIC NAME

Xanthine:oxygen oxidoreductase

SYNONYMS

Hypoxanthine oxidase, XO, XOD

REACTANTS

Xanthine, hypoxanthine, some purines and pterins, aldehydes, hydrated derivatives, H_2O, O_2, urate, H_2O_2

APPLICATIONS

- Determination of superoxide dismutase activity
- Determination of purine nucleoside phosphorylase activity
- Determination of xanthine and hypoxanthine in biological fluids
- Determination of inorganic phosphate
- Catalyzes binding of DNA to L-nitropyrene
- Immobilized enzyme is useful in organic synthesis, e.g., the regiospecific introduction of an oxygen atom into an azaheterocycle

Xanthine Oxidase continued — 1.1.3.22

NOTES

- Healthy individuals have appreciable amounts only in the liver and jejunum. In liver disorders, enzyme is released to the circulation
- Classified as an oxidoreductase acting on the CH-OH group of donors with oxygen as acceptor
- A complex enzyme containing flavin (FAD), molybdenum, iron and sulfide cofactors
- Activity is 50% with hypoxanthine as substrate
- Under some conditions, the product is mainly superoxide rather than peroxide:
  $$R\text{-}H + H_2O \leftrightarrow ROH + 2\,O_2^-$$
- Enzyme from animal sources can be converted to xanthine dehydrogenase (EC 1.1.1.204)
- Dehydrogenase enzyme can be converted into the oxidase by enzyme-thiol transhydrogenase (oxidized-glutathione) (EC 1.8.4.7) in the presence of oxidized glutathione
- Liver enzyme exists mainly as the dehydrogenase, but is converted to the oxidase by:
  - storage at -20°C
  - treatment with proteolytic enzymes
  - treatment with organic solvents
  - treatment with thiol reagents like Cu^{2+}, N-ethylmaleimide or 4-mercuribenzoate
- Thiol effects can be reversed by reagents like 1,4-dithioerythritol
- *Micrococcus* enzyme can use ferredoxin as acceptor
- Reported to have antitumor effects in mice
- Participates in the release of iron from hepatic ferritin stores in the plasma
- Stabilized by salicylate, cysteine, histamine, glutamate, BSA and versenate
- Inhibited by reducing agents, urea, purine 6-aldehyde, 2-amino-4-hydroxypteridine 6-aldehyde, Hg^{2+}, Ag^+ and MIA

SPECIFIC ACTIVITY	UNITS DEFINITION	PREPARATION FORM	ADDITIONAL ACTIVITIES	SUPPLIER CATALOG NO.
Bovine milk (optimum pH = 4.6, MW = 275,000 Da)				
0.1 U/mg solid	1 unit converts 1 μmol xanthine to product/min at 25°C	Lyophilized	<0.005% guanase, NP, uricase <0.05% ADA, AP(4-nitrophenyl phosphate as substrate) <10 U/mg proteases at 37°C with azocoll as substrate	Boehringer 1048180
1 U/mg	1 unit converts 1 μmol xanthine to product/min at 25°C	Chromatographically purified; suspension in 3.2 M $(NH_4)_2SO_4$ and 10 mM EDTA, pH 8	<0.005% guanase, NP, uricase <0.05% ADA, AP(4-nitrophenyl phosphate as substrate)	Boehringer 110434 110442
0.5 U/mg protein	1 unit forms 1 μmol uric acid/min at pH 7.5, 25°C	Suspension in 0.60 saturated $(NH_4)_2SO_4$ and 0.02% Na salicylate		ICN 101229
	1 unit converts 1.0 μmol xanthine to uric acid/min at pH 7.5, 25°C	Suspension in 2.3 M $(NH_4)_2SO_4$ containing 1 mM Na salicylate	Protease	Sigma X4875

1.1.3.22 Xanthine Oxidase continued

SPECIFIC ACTIVITY	UNITS DEFINITION	PREPARATION FORM	ADDITIONAL ACTIVITIES	SUPPLIER CATALOG NO.
Bovine milk *continued*				
≥0.04 U/mg protein	1 unit forms 1 μmol urate from hypoxanthine/min at pH 7.5, 25°C	Purified; suspension in 0.6 saturated $(NH_4)_2SO_4$		Worthington LS01150 - 2 LS01154
Buttermilk (optimum pH = 7.5-8.3, pI = 5.3, MW = 275,000-362,000 Da)				
≥0.5 U/mg protein	1 unit oxidizes 1 μmol xanthine/min at pH 7.5, 25°C	Purified; suspension in 60% $(NH_4)_2SO_4$ containing Na salicylate and EDTA		Biozyme XO1
≥0.5 U/mg protein; 0.3 U/mg solid	1 unit oxidizes 1 μmol xanthine/min at pH 7.5, 25°C	Salt-free; lyophilized containing Na salicylate and EDTA		Biozyme XO1F
1.0-1.5 U/mg protein	1 unit oxidizes 1 μmol xanthine/min at pH 7.5, 25°C	Purified; suspension in 60% $(NH_4)_2SO_4$ containing Na salicylate and EDTA		Biozyme XO2
1.0-1.5 U/mg protein	1 unit oxidizes 1 μmol xanthine/min at pH 7.5, 25°C	Chromatographically prepared; highly purified; suspension in 60% $(NH_4)_2SO_4$ containing Na salicylate and EDTA	No detectable LPO	Biozyme XO2B
1.0 U/mg protein; 0.4 U/mg solid	1 unit oxidizes 1 μmol xanthine/min at pH 7.5, 25°C	Salt-free; lyophilized containing Na salicylate and EDTA		Biozyme XO2F
1.0-4.0 U/mg protein	1 unit oxidizes 1 μmol xanthine to uric acid/min at pH 7.5, 25°C	Lyophilized containing 95% protein		Calzyme 076A0001
1.0-4.0 U/mg protein	1 unit oxidizes 1 μmol xanthine to uric acid/min at pH 7.5, 25°C	Suspension in 60% $(NH_4)_2SO_4$ containing 0.02% Na salicylate and 10 mg/mL protein, pH 7.5		Calzyme 076B0001
1 U/mg protein	1 unit oxidizes 1 μmol xanthine/min at pH 7.8, 30°C	Suspension in 2.3 *M* $(NH_4)_2SO_4$ and 1 *mM* Na salicylate	<0.1% uricase 15% protease	Fluka 95492
0.4 U/mg	1 unit oxidizes 1 μmol xanthine/min at pH 7.8, 30°C	Lyophilized		Fluka 95493
1.25 U/mg protein	1 unit oxidizes 1 μmol xanthine/min at pH 7.8, 30°C	Suspension in 60% saturated $(NH_4)_2SO_4$, Na salicylate, EDTA	<0.01% uricase 15% protease	Fluka 95495
>0.5 U/mg protein; >0.3 U/mg solid	1 unit oxidizes 1.0 μmol xanthine/min at pH 7.5, 25°C	Lyophilized		Genzyme 1751
>1.0 U/mg protein; >0.4 U/mg solid	1 unit oxidizes 1.0 μmol xanthine/min at pH 7.5, 25°C	Lyophilized		Genzyme 1761
0.5 U/mg protein	1 unit converts 1.0 μmol xanthine to uric acid/min at pH 7.5, 25°C	Suspension in 2.3 *M* $(NH_4)_2SO_4$ and 1 *mM* Na salicylate	No detectable uricase Protease	Sigma X1875
0.4-1.0 U/mg protein	1 unit converts 1.0 μmol xanthine to uric acid/min at pH 7.5, 25°C	Lyophilized containing 15% protein, 0.5% Na salicylate, <50 ppm PO_4	Protease	Sigma X4376
1-2 U/mg protein	1 unit converts 1.0 μmol xanthine to uric acid/min at pH 7.5, 25°C	Chromatographically purified; suspension in 2.3 *M* $(NH_4)_2SO_4$, 10 *mM* $NaPO_4$ buffer, 1 *mM* EDTA, 1 *mM* Na salicylate, pH 7.8	No detectable uricase Protease	Sigma X4500

Xanthine Oxidase continued   1.1.3.22

SPECIFIC ACTIVITY	UNITS DEFINITION	PREPARATION FORM	ADDITIONAL ACTIVITIES	SUPPLIER CATALOG No.
Buttermilk *continued*				
0.5 U/mg		Suspension		Wako 247-00571
Microbial (optimum pH = 7.5-8.0, pI = 4.0, T = 65°C, MW = 160,000 Da; K_M = 45 μM [xanthine], 76 μM [hypoxanthine]; stable pH 6.5-9.0 [25°C, 15 hr], T < 55°C [pH 8.0, 30 min])				
10 U/mg solid	1 unit converts 1.0 μmol xanthine to uric acid/min at pH 7.5, 25°C	Lyophilized containing BSA and Na Glu as stabilizers	Protease	sigma X2252
≥15 U/mg solid	1 unit oxidizes 1 μmol uric acid/min at pH 7.5, 37°C	Lyophilized containing EDTA, Na Glu as stabilizers	<0.001% phosphatase, uricase, ADA <0.005% purine-nucleoside phosphorylase <20% catalase	Toyobo XTO-211

Gluconate 2-Dehydrogenase   1.1.99.3

REACTION CATALYZED

    D-Gluconate + Acceptor ↔ 2-Dehydro-D-gluconate + Reduced acceptor

SYSTEMATIC NAME

    D-Gluconate:(acceptor) 2-oxidoreductase

SYNONYMS

    Gluconate dehydrogenase (NAD(P) independent)

REACTANTS

    D-Gluconate, 2-dehydro-D-gluconate

NOTES

- Classified as an oxidoreductase acting on the CH-OH group of donors with other acceptors
- A flavoprotein (FAD)

SPECIFIC ACTIVITY	UNITS DEFINITION	PREPARATION FORM	ADDITIONAL ACTIVITIES	SUPPLIER CATALOG No.
***Pseudomonas* species**				
50-150 U/mg protein	1 unit reduces 1.0 μmol 2,6-dichlorophenol-indophenol/min at pH 6.0, 25°C in the presence of Na gluconate	Solution containing 50% glycerol containing $MgCl_2$, Na gluconate, 1% Triton X-100		sigma G7275

1.1.99.10 Glucose Dehydrogenase (Acceptor)

REACTION CATALYZED
D-Glucose + acceptor ↔ D-Glucono-1,5-lactone + reduced acceptor

SYSTEMATIC NAME
D-Glucose:(acceptor) 1-oxidoreductase

SYNONYMS
Glucose dehydrogenase (*Aspergillus*)

REACTANTS
D-Glucose, 2,6-dichloroindophenol, D-glucono-1,5-lactone

NOTES
- Classified as an oxidoreductase acting on the CH-OH group of donors with other acceptors
- A glycoprotein containing one mole of FAD per mole of enzyme

SPECIFIC ACTIVITY	UNITS DEFINITION	PREPARATION FORM	ADDITIONAL ACTIVITIES	SUPPLIER CATALOG NO.
Bacillus species (optimum pH = 8.0)				
≥30 U/mg	1 unit produces 1 μmol NADH/min at pH 8.0, 25°C	Powder		Amano Glucose DH

1.1.99.11 Fructose 5-Dehydrogenase

REACTION CATALYZED
D-Fructose + Acceptor ↔ 5-Dehydro-D-fructose + reduced acceptor

SYSTEMATIC NAME
D-Fructose:(acceptor) 5-oxidoreductase

SYNONYMS
D-Fructose dehydrogenase

REACTANTS
D-Fructose, 5-dehydro-D-fructose, 2,6-dichloroindophenol

APPLICATIONS
- Determination of D-fructose

NOTES
- Classified as an oxidoreductase acting on the CH-OH group of donors with other acceptors
- Inhibited by Ag^+, Hg^{2+} and SDS
- Stabilized by sugars, amino acids and BSA

Fructose 5-Dehydrogenase continued

1.1.99.11

SPECIFIC ACTIVITY	UNITS DEFINITION	PREPARATION FORM	ADDITIONAL ACTIVITIES	SUPPLIER CATALOG NO.
Gluconobacter species (optimum pH = 4.0, pI = 5.0, T = 37°C, MW = 140,000 Da [gel filtration]; K_M = 5 mM [D-Fru]; stable pH 4.0-6.0 [25°C, 16 hr], T < 40°C [pH 4.5, 15 min])				
20 U/mg solid	1 unit oxidizes 1 μmol D-fructose (formation of 2 μmol Prussian blue)/min at pH 4.5, 25°C	Lyophilized containing 70% stabilizers		ICN 191505
20 U/mg solid	1 unit converts 1.0 μmol D-fructose to 5-keto-D-fructose/min at pH 4.5, 37°C	Lyophilized		Sigma F4892
400-1200 U/mg protein	1 unit converts 1.0 μmol D-fructose to 5-keto-D-fructose/min at pH 4.5, 37°C	Lyophilized containing 5% protein; balance citrate-PO_4 buffer salts, Triton X-100 and stabilizer		Sigma F5152
≥20 U/mg solid	1 unit oxidizes 1 μmol D-fructose (2 μmol Prussian blue)/min at pH 4.5, 37°C	Lyophilized containing 70% BSA, sugars, amino acids as stabilizers		Toyobo FCD-301

Formaldehyde Dehydrogenase (Glutathione)

1.2.1.1

REACTION CATALYZED
Formaldehyde + glutathione + NAD^+ ↔ S-Formylglutathione + NADH

SYSTEMATIC NAME
Formaldehyde:NAD^+ oxidoreductase (glutathione-formylating)

SYNONYMS
Formic dehydrogenase

REACTANTS
Formaldehyde, 2-oxoaldehydes, NAD^+, S-formylglutathione, NADH

NOTES
- Classified as an oxidoreductase acting on the aldehyde or oxo group of donors with NAD^+ or $NADP^+$ as acceptor
- NADPH can replace NADH in the reverse reaction

SPECIFIC ACTIVITY	UNITS DEFINITION	PREPARATION FORM	ADDITIONAL ACTIVITIES	SUPPLIER CATALOG NO.
Candida boidinii				
7-20 U/mg protein	1 unit oxidizes 1.0 μmol formaldehyde to formic acid/min at pH 7.5, 37°C in the presence of GSH	Lyophilized containing 10% protein; balance PO_4 buffer salts and sucrose as stabilizer		Sigma F4136

1.2.1.2 Formate Dehydrogenase

REACTION CATALYZED
Formate + NAD$^+$ ↔ CO$_2$ + NADH

SYSTEMATIC NAME
Formate:NAD$^+$ oxidoreductase

REACTANTS
Formate, NAD$^+$, CO$_2$, NADH

APPLICATIONS
- Determination of formate
- Widely used in cofactor recycling systems for NADH

NOTES
- Classified as an oxidoreductase acting on the aldehyde or oxo group of donors with NAD$^+$ or NADP$^+$ as acceptor
- Together with hydrogen dehydrogenase (EC 1.12.1.2), forms a system previously known as formate hydrogenlyase

SPECIFIC ACTIVITY	UNITS DEFINITION	PREPARATION FORM	ADDITIONAL ACTIVITIES	SUPPLIER CATALOG No.
Candida boidinii (Xylaria digitata)				
3 U/mg protein; 0.4 U/mg solid	1 unit converts 1 μmol formate to product/min at 25°C	Lyophilized	<0.05% LDH, ADH <0.1% MDH	Boehringer 244678 837016
1.5-4.5 U/mg protein	1 unit oxidizes 1.0 μmol formate to CO$_2$/min at pH 7.5, 37°C in the presence of β-NAD	Lyophilized containing 20% protein; balance PO$_4$ buffer salts and sucrose		Sigma F5632
Pseudomonas oxalaticus				
0.4-1.2 U/mg protein	1 unit oxidizes 1.0 μmol formate to CO$_2$/min at pH 7.0, 37°C in the presence of β-NAD	Lyophilized containing 60% protein; balance primarily buffer salts		Sigma F3753
Yeast				
0.7 U/mg	1 unit oxidizes 1 μmol Na formate/min at pH 7.6, 25°C	Lyophilized	<0.05% LDH, ADH	Fluka 47711
5-15 U/mg protein	1 unit oxidizes 1.0 μmol formate to CO$_2$/min at pH 7.6, 37°C in the presence of β-NAD	Lyophilized containing 15% protein		Sigma F8649

Aldehyde Dehydrogenase (NADP⁺)

1.2.1.4

REACTION CATALYZED
An aldehyde + $NADP^+$ + H_2O ↔ An acid + NADH

SYSTEMATIC NAME
Aldehyde:$NADP^+$ oxidoreductase

SYNONYMS
AlDH

REACTANTS
Aldehyde, $NADP^+$, H_2O, acid, NADH

NOTES
- Classified as an oxidoreductase acting on the aldehyde or oxo group of donors with NAD^+ or $NADP^+$ as acceptor

SPECIFIC ACTIVITY	UNITS DEFINITION	PREPARATION FORM	ADDITIONAL ACTIVITIES	SUPPLIER CATALOG NO.
Yeast				
≥5 IU/mg protein	1 IU transforms 1 μmol substrate/min under standard IUB conditions at 25°C	Lyophilized	<0.01% NADPH oxidase, ADH, LDH	OYC

Aldehyde Dehydrogenase [NAD(P)⁺]

1.2.1.5

REACTION CATALYZED
An aldehyde + $NAD(P)^+$ + H_2O ↔ An acid + $NAD(P)H$

SYSTEMATIC NAME
Aldehyde:$NAD(P)^+$ oxidoreductase

REACTANTS
Aldehyde, $NAD(P)^+$, H_2O, acid, $NAD(P)H$

APPLICATIONS
- Determination of acetaldehyde
- Component of NADH and NADPH recycling systems

NOTES
- Classified as an oxidoreductase acting on the aldehyde or oxo group of donors with NAD^+ or $NADP^+$ as acceptor
- Activated by potassium, glutathione, 2-mercaptoethanol and cysteine
- Inhibited by trace heavy metals, especially Cu^{2+}

SPECIFIC ACTIVITY	UNITS DEFINITION	PREPARATION FORM	ADDITIONAL ACTIVITIES	SUPPLIER CATALOG NO.
Yeast				
5 U/mg protein; ≥1 U/mg solid	1 unit oxidizes 1 μmol acetaldehyde to acetic acid/min at pH 8.0, 25°C in the presence of K^- and thiols	Lyophilized	<0.1% ADH	Biozyme ALDH1

1.2.1.5 Aldehyde Dehydrogenase [NAD(P)$^+$] continued

SPECIFIC ACTIVITY	UNITS DEFINITION	PREPARATION FORM	ADDITIONAL ACTIVITIES	SUPPLIER CATALOG No.
Yeast continued				
5 U/mg protein; ≥1 U/mg solid	1 unit oxidizes 1 μmol acetaldehyde to acetic acid/min at pH 8.0, 25°C in the presence of K$^-$ and thiols	Lyophilized	<0.01% ADH	Biozyme ALDH2
12 U/mg protein; 1.2 U/mg solid	1 unit oxidizes 1 μmol acetaldehyde to acetic acid/min at pH 8.0, 25°C in the presence of K$^-$ and thiols	Chromatographically prepared; lyophilized	<0.002% ADH	Biozyme ALDH3
20 U/mg protein	1 unit converts 1 μmol acetaldehyde to product/min at 25°C	Lyophilized containing KPO$_4$ buffer as stabilizer, pH 6	<0.01% NADH oxidase, ADH, LDH	Boehringer 171832
5 U/mg protein; 1 U/mg solid	1 unit oxidizes 1 μmol acetaldehyde to acetic acid/min at pH 8.0, 25°C in the presence of K$^-$ and thiols	Lyophilized	<0.1% ADH	ICN 190679
Yeast, bakers (MW = 200,000)				
10 U/mg protein	1 unit oxidizes 1 μmol acetaldehyde to acetic acid/min at pH 8.0, 25°C in the presence of K$^-$, thiols and β-NAD	Lyophilized containing 50% protein and KPO$_4$, pH 7		Calzyme 056A0010
2-10 U/mg protein	1 unit oxidizes 1.0 μmol acetaldehyde to acetic acid/min at pH 8.0, 25°C in the presence of β-NAD$^+$, K$^-$ and thiols	Lyophilized containing 50% protein; balance primarily mannitol, KPO$_4$ and citrate buffer salts, DTT, traces of β-NAD and propionic acid		Sigma A0826
20-40 U/mg protein	1 unit oxidizes 1.0 μmol acetaldehyde to acetic acid/min at pH 8.0, 25°C in the presence of β-NAD$^+$, K$^-$ and thiols	Chromatographically purified; lyophilized containing 50% protein; balance primarily mannitol, KPO$_4$ and citrate buffer salts, DTT, traces of β-NAD and propionic acid	<0.1% ADH 1% β-NADP$^+$-linked AlDH	Sigma A5550

1.2.1.12 Glyceraldehyde-3-Phosphate Dehydrogenase (Phosphorylating)

REACTION CATALYZED

D-Glyceraldehyde 3-phosphate + orthophosphate + NAD$^+$ ↔ 3-Phospho-D-glyceroyl phosphate + NADH

SYSTEMATIC NAME

D-Glyceraldehyde 3-phosphate:NAD$^+$ oxidoreductase (phosphorylating)

SYNONYMS

Triosephosphate dehydrogenase, GAPDH

REACTANTS

D-Glyceraldehyde 3-phosphate, orthophosphate, NAD$^+$, 3-phospho-D-glyceroyl phosphate, NADH; D-glyceraldehyde and other aldehydes (very slowly)

APPLICATIONS

- Determination of ATP and 3-phospho-D-glycerate, with phosphoglycerate kinase

Glyceraldehyde-3-Phosphate Dehydrogenase (Phosphorylating) continued   1.2.1.12

NOTES

- Synthesis of radioactive 8-azidoadenosine 5'-triphosphate
- Classified as an oxidoreductase acting on the aldehyde or oxo group of donors with NAD^+ or $NADP^+$ as acceptor
- A key enzyme in intermediary metabolism, catalyzing the oxidation and subsequent phosphorylation of aldehydes to acyl phosphate
- Rabbit muscle enzyme is comprised of four identical subunits, each of molecular weight 36,000 Da and containing 1 molecule of NAD^+
- Thiols and arsenate can replace phosphate

SPECIFIC ACTIVITY	UNITS DEFINITION	PREPARATION FORM	ADDITIONAL ACTIVITIES	SUPPLIER CATALOG NO.
Bacillus stearothermophilus (optimum pH = 8.5, pI = 4.2, MW = 140,000 Da [36,000/subunit]; K_M = 0.22 mM [GAP], 1.0 mM [disodium arsenate]; stable pH 6.5-8.5, T < 40°C)				
50-200 U/mg protein	1 unit reduces 1.0 μmol 3-phosphoglycerate to D-GAP/min at pH 7.6, 30°C in a coupled system	Lyophilized containing 85% protein; balance primarily Tris-HCl buffer salts		Sigma G5892
>50 U/mg protein	1 unit forms 1 μmol NADH/min at 30°C	Lyophilized	<0.01% PGK, G3PDH, PGluM, TPI	Unitika
>50 U/mg protein	1 unit forms 1 μmol NADH/min at 30°C	50% glycerol solution	<0.01% PGK, G3PDH, PGluM, TPI	Unitika
Chicken muscle				
40-100 U/mg protein	1 unit reduces 1.0 μmol 3-phosphoglycerate to D-GAP/min at pH 7.6, 25°C in a coupled system	Lyophilized containing 95% protein; balance primarily citrate buffer salts	<0.01% 3-phosphoglyceric phosphokinase <0.1% TPI	Sigma G9263
Human erythrocytes				
50-150 U/mg protein	1 unit reduces 1.0 μmol 3-phosphoglycerate to D-GAP/min at pH 7.6, 25°C in a coupled system	Lyophilized containing 60% protein; balance primarily Na citrate buffer salts	<2% 3-phosphoglyceric phosphokinase	Sigma G6019
Porcine muscle				
75-110 U/mg protein	1 unit reduces 1.0 μmol 3-phosphoglycerate to D-GAP/min at pH 7.6, 25°C in a coupled system	Lyophilized containing 85% protein; balance primarily citrate buffer salts	<0.05% 3-phosphoglyceric phosphokinase <0.01% TPI, LDH, MK	Sigma G6517
40-100 U/mg protein	1 unit reduces 1.0 μmol 3-phosphoglycerate to D-GAP/min at pH 7.6, 25°C in a coupled system	Crystallized; suspension in 2.6 M $(NH_4)_2SO_4$, 1 mM EDTA, 0.02 M β-MSH, pH 7.5	<0.05% 3-phosphoglyceric phosphokinase <0.01% TPI, LDH, MK	Sigma G9013
Rabbit muscle (MW = 144,000 Da)				
60 U/mg protein	1 unit oxidizes 1 μmol NADH/min at pH 7.6, 25°C	Crystallized; suspension in 75% $(NH_4)_2SO_4$ and 1 mM EDTA	<0.005% LDH, G3PDH <0.01% PGM, 3PGK	Biozyme GAPD1

1.2.1.12 Glyceraldehyde-3-Phosphate Dehydrogenase (Phosphorylating) *continued*

SPECIFIC ACTIVITY	UNITS DEFINITION	PREPARATION FORM	ADDITIONAL ACTIVITIES	SUPPLIER CATALOG NO.
Rabbit muscle *continued*				
80 U/mg	1 unit converts 1 μmol glycerate-1,3-phosphate to product/min at 25°C	Suspension in 3.2 M $(NH_4)_2SO_4$ and 0.1 mM EDTA, pH 7.5	<0.01% GDH, LDH, PGK, PGM <0.03% TIM	Boehringer 105686 105694
80 U/mg protein	1 unit converts 1 μmol glycerate 1,3-diphosphate to D-GAP/min at pH 7.6, 25°C	Crystalline suspension in 3.2 M $(NH_4)_2SO_4$ and 0.1 mM EDTA, pH 7.5		Fluka 49830
60 U/mg protein	1 unit oxidizes 1.0 μmol NADH/min at pH 7.6, 20°C	Crystallized; suspension in 75% $(NH_4)_2SO_4$, pH 7.5		ICN 100693
90-120 U/mg protein	1 unit reduces 1.0 μmol 3-phosphoglycerate to D-GAP/min at pH 7.6, 25°C in a coupled system	Crystallized; suspension in 2.6 M $(NH_4)_2SO_4$, 1 mM EDTA, 0.02 M β-MSH, pH 7.5	<0.05% 3-phosphoglyceric phosphokinase <0.01% TPI, LDH, MK, PK	Sigma G0763
80 U/mg protein	1 unit reduces 1.0 μmol 3-phosphoglycerate to D-GAP/min at pH 7.6, 25°C in a coupled system	Lyophilized containing 85% protein; balance primarily citrate buffer salts	<0.05% 3-phosphoglyceric phosphokinase <0.01% TPI, LDH, MK, PK	Sigma G2267
≥25 U/mg protein	1 unit reduces 1 μmol NAD/min at pH 8.5, 25°C	Crystallized; suspension in 0.72 saturated $(NH_4)_2SO_4$, pH 8.3		Worthington LS02330 LS02332 LS02334
Yeast, bakers				
	1 unit reduces 1.0 μmol 3-phosphoglycerate to D-GAP/min at pH 7.6, 25°C in a coupled system	Insoluble enzyme attached to beaded agarose; lyophilized containing lactose, citrate and Cys stabilizers		Sigma G0389
100 U/mg protein	1 unit reduces 1.0 μmol 3-phosphoglycerate to D-GAP/min at pH 7.6, 25°C in a coupled system	Solution containing 50% glycerol and 1.5 M $(NH_4)_2SO_4$, pH 8.0	<0.01% 3-phosphoglyceric phosphokinase	Sigma G2647
70-140 U/mg protein	1 unit reduces 1.0 μmol 3-phosphoglycerate to D-GAP/min at pH 7.6, 25°C in a coupled system	Sulfate-free; lyophilized containing 75% protein; balance primarily citrate buffer salts	<0.01% 3-phosphoglyceric phosphokinase	Sigma G8380

Succinate-Semialdehyde Dehydrogenase [NAD(P)$^+$]

1.2.1.16

REACTION CATALYZED
Succinate semialdehyde + NAD(P)$^+$ + H$_2$O ↔ Succinate + NAD(P)H

SYSTEMATIC NAME
Succinate-semialdehyde:NAD(P)$^+$ oxidoreductase

REACTANTS
Succinate semialdehyde, NAD(P)$^+$, H$_2$O, succinate, NAD(P)H

NOTES
- Classified as an oxidoreductase acting on the aldehyde or oxo group of donors with NAD$^+$ or NADP$^+$ as acceptor

SPECIFIC ACTIVITY	UNITS DEFINITION	PREPARATION FORM	ADDITIONAL ACTIVITIES	SUPPLIER CATALOG NO.
Pseudomonas fluorescens				
10-30 U/mg protein	1 unit converts 1.0 μmol succinic semialdehyde to succinate with a stoichiometric reduction of 1.0 μmol NADP$^+$/min at pH 8.6, 25°C	Lyophilized containing 10% protein; balance KPO$_4$ buffer salts, NaCl, stabilizers	<1 U/mg protein 4-aminobutyrate aminotransferase; No detectable 4-aminobutanal dehydrogenase	Sigma S3907

Formaldehyde Dehydrogenase

1.2.1.46

REACTION CATALYZED
Formaldehyde + NAD$^+$ + H$_2$O ↔ Formate + NADH

SYSTEMATIC NAME
Formaldehyde:NAD$^+$ oxidoreductase

REACTANTS
Formaldehyde, NAD$^+$, H$_2$O, formate, NADH

APPLICATIONS
- Determination of formaldehyde and H$_2$O$_2$, with catalase

NOTES
- Classified as an oxidoreductase acting on the aldehyde or oxo group of donors with NAD$^+$ or NADP$^+$ as acceptor
- This *Pseudomonas* enzyme does not require reduced glutathione, unlike EC 1.2.1.1
- Comprised of two subunits
- Inhibited by chelating agents, Ni^{2+}, Pb^{2+}, Hg^{2+}, PCMB and ionic detergents
- Stabilized by Mg^{2+}, Ca^{2+}, BSA, glycine and lysine

1.2.1.46 Formaldehyde Dehydrogenase continued

SPECIFIC ACTIVITY	UNITS DEFINITION	PREPARATION FORM	ADDITIONAL ACTIVITIES	SUPPLIER CATALOG NO.
Pseudomonas putida				
1.0 U/mg solid	1 unit oxidizes 1.0 μmol formaldehyde to formic acid/min at pH 7.5, 37°C	Lyophilized		ICN 190668
1-6 U/mg solid	1 unit oxidizes 1.0 μmol formaldehyde to formic acid/min at pH 7.5, 37°C	Lyophilized		Sigma F1879
Pseudomonas species (optimum pH = 9.0, pI = 5.25, T = 40°C, MW = 150,000 Da [gel filtration], 2 subunits at 75,000 Da per mol enzyme; K_M = 80 μ*M* [HCHO], 0.12 m*M* [NAD$^+$]; stable pH 8.0-10.0 [30°C, 16 hr], T < 40°C [pH 7.5, 30 min])				
≥1 U/mg solid	1 unit forms 0.5 μmol diformazan/min at pH 7.5, 37°C	Lyophilized containing 70% BSA, Mg^{2+}, Ca^{2+}, Gly, Lys as stabilizers	<0.1% NADH oxidase	Toyobo FRD-201

1.2.1.51 Pyruvate Dehydrogenase (NADP$^+$)

REACTION CATALYZED

Pyruvate + CoA + NADP$^+$ ↔ Acetyl-CoA + CO$_2$ + NADPH

SYSTEMATIC NAME

Pyruvate:NADP$^+$ 2-oxidoreductase (CoA-acetylating)

SYNONYMS

PDH

REACTANTS

Pyruvate, CoA, methyl viologen, FAD, NADP$^+$, acetyl-CoA, CO$_2$, NADPH

APPLICATIONS

- Determination of ADP, pyruvate, Pi, sialic acid, GOT and GPT, with related enzymes

NOTES

- Classified as an oxidoreductase acting on the aldehyde or oxo group of donors with NAD$^+$ or NADP$^+$ as acceptor
- *Euglena* enzyme also uses FAD and methyl viologen as acceptors, but more slowly
- Stabilized by mannitol and FAD
- Inhibited by oxygen and Hg^{2+}

SPECIFIC ACTIVITY	UNITS DEFINITION	PREPARATION FORM	ADDITIONAL ACTIVITIES	SUPPLIER CATALOG NO.
Microbial (optimum pH = 6.2-6.3, pI = 4.4, T = 55°C, MW = 160,000 Da; K_M = 0.44 m*M* [pyruvate], 14 m*M* [K$^+$]; stable pH 5.0-8.0 [25°C, 20 hr], T < 50°C [pH 6.3, 15 min])				
≥2 U/mg solid	1 unit forms 0.5 μmol diformazan/min at pH 6.3, 37°C	Lyophilized containing 50% mannitol and FAD as stabilizers		Toyobo PYD-301

Pyruvate Dehydrogenase (NADP⁺) continued — 1.2.1.51

SPECIFIC ACTIVITY	UNITS DEFINITION	PREPARATION FORM	ADDITIONAL ACTIVITIES	SUPPLIER CATALOG NO.
Porcine heart				
1-5 U/mg protein	1 unit converts 1.0 μmol β-NAD to β-NADH/min at pH 7.4, 30°C in the presence of saturating levels of CoA	50% glycerol, 30% sucrose, 2.5 mM EDTA, 2.5 mM EGTA, 2.5 mM β-MSH, 0.5% Triton X-100, 0.005% NaN₃, 25 mM KPO₄	<15% α-ketoglutarate dehydrogenase	Sigma P5194
0.1-1.0 U/mg protein	1 unit converts 1.0 μmol β-NAD to β-NADH/min at pH 7.4, 30°C in the presence of saturating levels of CoA	50% glycerol, 10 mg/mL BSA, 30% sucrose, 2.5 mM EDTA, 2.5 mM EGTA, 2.5 mM β-MSH, 0.5% Triton X-100, 0.005% NaN₃, 25 mM KPO₄, traces of PEG, pH 6.8	<15% α-ketoglutarate dehydrogenase	Sigma P7032

Pyruvate Oxidase — 1.2.3.3

REACTION CATALYZED

Pyruvate + orthophosphate + O_2 + H_2O ↔ Acetyl phosphate + CO_2 + H_2O_2

SYSTEMATIC NAME

Pyruvate:oxygen 2-oxidoreductase (phosphorylating)

SYNONYMS

Pyruvic oxidase G, PO

REACTANTS

Pyruvate, orthophosphate, O_2, H_2O, acetyl phosphate, CO_2, H_2O_2

APPLICATIONS
- Determination of ALT and AST in biological fluids
- Determination of pyruvate, GOT and GPT in clinical analyses

NOTES
- Classified as an oxidoreductase acting on the aldehyde or oxo group of donors with oxygen as acceptor
- A flavoprotein (FAD)
- Requires thiamin diphosphate
- Activated by Mn^{2+}, Mg^{2+}, Ca^{2+} and Co^{2+}
- Stabilized by FAD and sugars
- Inhibited by Hg^{2+}, Ag^+, Cu^{2+}, EDTA and inorganic phosphate

SPECIFIC ACTIVITY	UNITS DEFINITION	PREPARATION FORM	ADDITIONAL ACTIVITIES	SUPPLIER CATALOG NO.
Aerococcus viridans (optimum pH = 6.5-7.0, pI = 4.0, T = 40°C, MW = 70,000 Da [SDS-PAGE], 155,000 [gel filtration]; K_M [pyruvate] = 5.9 mM [Mn^{2+}], 41 mM [Mg^{2+}]; K_M [K^+] = 2.0 mM [Mn^{2+}], 4.5 mM [Mg^{2+}]; stable pH 6-7 [37°C, 1hr, 10 μM FAD], T < 45°C [KPO_4 buffer, pH 6.5, 10 min, 10 μM FAD])				
25-60 U/mg solid	1 unit generates 1 μmol H_2O_2/min at 37°C	Lyophilized containing 70% protein	No detectable catalase <0.002% GOT, LO <0.006% GPT	Asahi POPG T-45
>35 U/mg protein; >25 U/mg solid	1 unit forms 1.0 μmol H_2O_2/min at 37°C	Lyophilized	No detectable catalase <0.002% LO, total AST (GOT) <0.006% total ALT (GPT)	Genzyme 1691
Bacterial				
1.5 U/mg solid	1 unit produces 1.0 μmol H_2O_2 during conversion of pyruvate and PO_4 to acetylphosphate and CO_2/min at pH 5.7, 37°C	Lyophilized containing glucose as stabilizer		Sigma P3673
Microbial (optimum pH = 5.7, pI = 4.9, T = 50°C, MW = 230,000 Da; K_M = 50 mM [pyruvate], 2 mM [K^+]; stable pH 4.5-6.5 [25°C, 20 hr], T < 40°C [pH 6.0, 15 min])				
≥1.5 U/mg solid	1 unit forms 1 μmol H_2O_2 (0.5 μmol quinoneimine dye)/min at pH 5.7, 37°C	Lyophilized		ICN 153486
≥1.5 U/mg solid	1 unit forms 1 μmol H_2O_2 (0.5 μmol quinoneimine dye)/min at pH 5.9, 37°C	Lyophilized containing sugars as stabilizers	<0.01% LO <0.05% GPT, GOT, ATPase	Toyobo PYO-301
Pediococcus pseudomonas (MW = 150,000 Da)				
20 U/mg solid	1 unit oxidizes 1 μmol pyruvate/min at pH 6.7, 37°C	Lyophilized	<0.01% GOT, GPT, LO	Calbiochem 550700
Pediococcus species (optimum pH = 6.5-7.5, pI = 4.0, MW = 150,000 Da [Sephacryl S-200]; K_M = 1.7 mM [pyruvate], 0.5 mM [K^+]; stable pH 5.5-7.0 [40°C, 10 min], T < 40°C [pH 7.0, 10 min])				
25-60 U/mg solid	1 unit generates 1 μmol H_2O_2/min at 37°C	Lyophilized containing 80% protein	No detectable catalase <0.02% LO, GOT <0.06% GPT	Asahi POP T-12
>30 U/mg protein; >25 U/mg solid	1 unit forms 1.0 μmol H_2O_2/min at 37°C	Lyophilized	No detectable catalase <0.02% LO, total AST (GOT) <0.06% total ALT (GPT)	Genzyme 1611
	1 unit produces 1.0 μmol H_2O_2 during conversion of pyruvate and PO_4 to acetylphosphate and CO_2/min	Lyophilized containing 75% protein; balance primarily buffer and sugar		Sigma P6779

Oxalate Oxidase 1.2.3.4

REACTION CATALYZED

Oxalate + O_2 ↔ 2 CO_2 + H_2O_2

SYSTEMATIC NAME

Oxalate:oxygen oxidoreductase

REACTANTS

Oxalate, O_2, CO_2, H_2O_2

NOTES
- Classified as an oxidoreductase acting on the aldehyde or oxo group of donors with oxygen as acceptor
- A flavoprotein

SPECIFIC ACTIVITY	UNITS DEFINITION	PREPARATION FORM	ADDITIONAL ACTIVITIES	SUPPLIER CATALOG NO.
Barley seedlings				
5 U/mg protein; 0.25 U/mg solid	1 unit converts 1 μmol oxalate to product/min at 37°C	Lyophilized		Boehringer 567698
0.5-1.5 U/mg solid	1 unit forms 1.0 μmol H_2O_2 from oxalate/min at pH 3.8, 37°C	Partially purified; lyophilized	No detectable catalase	Sigma O4127

Pyruvate Dehydrogenase (lipoamide) 1.2.4.1

REACTION CATALYZED

Pyruvate + lipoamide ↔ S-Acetyldihydrolipoamide + CO_2

SYSTEMATIC NAME

Pyruvate:lipoamide 2-oxidoreductase (decarboxylating and acceptor-acetylating)

SYNONYMS

Pyruvate decarboxylase, pyruvate dehydrogenase, pyruvic dehydrogenase

REACTANTS

Pyruvate, lipoamide, S-acetyldihydrolipoamide, CO_2

NOTES
- Classified as an oxidoreductase acting on the aldehyde or oxo group of donors with a disulfide as acceptor
- Requires thiamin diphosphate
- Component of the multienzyme complex pyruvate dehydrogenase

SPECIFIC ACTIVITY	UNITS DEFINITION	PREPARATION FORM	ADDITIONAL ACTIVITIES	SUPPLIER CATALOG NO.
Lactobacillus delbrueckii				
3 U/mg solid	1 unit produces 1.0 μmol acetyl PO_4 and CO_2 from pyruvate and P_i/min at pH 6.3, 37°C	Lyophilized containing 50% mannitol and FAD as stabilizers		Sigma P3798

1.2.4.2 Oxoglutarate Dehydrogenase (Lipoamide)

REACTION CATALYZED
2-Oxoglutarate + lipoamide ↔ S-Succinyldihydrolipoamide + CO_2

SYSTEMATIC NAME
2-Oxoglutarate:lipoamide 2-oxidoreductase (decarboxylating and acceptor-succinylating)

SYNONYMS
Oxoglutarate decarboxylase, α-ketoglutaric dehydrogenase, α-ketoglutarate dehydrogenase

REACTANTS
2-Oxoglutarate, lipoamide, S-succinyldihydrolipoamide, CO_2

NOTES
- Classified as an oxidoreductase acting on the aldehyde or oxo group of donors with a disulfide as acceptor
- Requires thiamin diphosphate
- Component of the multienzyme 2-oxoglutarate dehydrogenase complex

SPECIFIC ACTIVITY	UNITS DEFINITION	PREPARATION FORM	ADDITIONAL ACTIVITIES	SUPPLIER CATALOG NO.
Porcine heart				
0.1-1 U/mg protein	1 unit converts 1.0 μmol β-NAD to β-NADH/min at pH 7.4, 30°C in the presence of saturating levels of CoA	50% glycerol, 10 mg/mL BSA, 30% sucrose, 2.5 mM EDTA, 2.5 mM EGTA, 2.5 mM β-MSH, 0.5% Triton X-100, 0.005% NaN_3, 25 mM KPO_4, trace PEG, pH 6.8	<10% PDH	Sigma K1502

1.3.1.14 Orotate Reductase (NADH)

REACTION CATALYZED
(S)-Dihydroorotate + NAD^+ ↔ Orotate + NADH

SYSTEMATIC NAME
(S)-Dihydroorotate:NAD^+ oxidoreductase

SYNONYMS
Dihydroorotate dehydrogenase

REACTANTS
(S)-Dihydroorotate, NAD^+, orotate, NADH

NOTES
- Classified as an oxidoreductase acting on the CH-CH group of donors with NAD^+ or $NADP^+$ as acceptor
- A flavoprotein (FAD, FMN)

Orotate Reductase (NADH) continued   1.3.1.14

SPECIFIC ACTIVITY	UNITS DEFINITION	PREPARATION FORM	ADDITIONAL ACTIVITIES	SUPPLIER CATALOG NO.
Zymobacterium oroticum				
7.0 U/mg protein	1 unit converts 1.0 μmol orotic acid to dihydroorotate/min at pH 6.5, 25°C	Lyophilized containing 25% protein; balance primarily PO_4 buffer salts	2 U/mg protein NADH oxidase (intrinsic)	Sigma D6384

Bilirubin Oxidase   1.3.3.5

REACTION CATALYZED
  Bilirubin + O_2 ↔ Biliverdin + H_2O

SYSTEMATIC NAME
  Bilirubin:oxygen oxidoreductase

REACTANTS
  Bilirubin, O_2, biliverdin, H_2O

APPLICATIONS
- Determination of bilirubin in biological fluids

NOTES
- Increases the accuracy of oxidase/peroxidase colorimetric assays by eliminating bilirubin interference
- Classified as an oxidoreductase acting on the CH-CH group of donors with oxygen as acceptor

SPECIFIC ACTIVITY	UNITS DEFINITION	PREPARATION FORM	ADDITIONAL ACTIVITIES	SUPPLIER CATALOG NO.
Myrothecium verrucaria				
15-65 U/mg protein	1 unit oxidizes 1.0 μmol bilirubin/min at pH 8.4, 37°C	Lyophilized containing 20% protein		Sigma B0390
1-5 U/mg protein	1 unit oxidizes 1.0 μmol bilirubin/min at pH 8.4, 37°C	Crude		Sigma B1515
Myrothecium species (optimum pH = 6.0 or 8.0 [Tris], pI = 4.1, MW = 52,000 Da; stable pH 7-11 [37°C, 1 hr], T < 45°C [pH 7, 30 min])				
≥0.8 U/mg	1 unit oxidizes 1 μmol albumin bound bilirubin/min at pH 7.0, 37°C	Powder		Amano Bilirubin oxidase
>30 U/mg protein; >10 U/mg solid	1 unit oxidizes 1.0 μmol bilirubin/min at pH 8.0, 37°C	Lyophilized		Genzyme 1021
8-15 U/mg solid	1 unit oxidizes 1 μmol bilirubin/min at pH 8.4, 37°C	Lyophilized		ICN 190000

Acyl-CoA Oxidase

1.3.3.6

REACTION CATALYZED
Acyl-CoA + O_2 ↔ *Trans*-2,3-dehydroacyl-CoA + H_2O_2

SYSTEMATIC NAME
Acyl-CoA:oxygen 2-oxidoreductase

SYNONYMS
Acyl coenzymeA oxidase, ACOD

REACTANTS
Acyl-CoA, O_2, *trans*-2,3-dehydroacyl-CoA, **palmitoyl-CoA**

APPLICATIONS
- Determination of free fatty acids in serum or plasma, with acyl-CoA synthetase

NOTES
- Classified as an oxidoreductase acting on the CH-CH group of donors with oxygen as acceptor
- A flavoprotein (FAD)
- Acts on CoA derivatives of fatty acids with 8-18 chain length
- Activated by Triton X-100
- Stabilized by FAD and sugars
- Inhibited by Hg^{2+} and Ag^+

SPECIFIC ACTIVITY	UNITS DEFINITION	PREPARATION FORM	ADDITIONAL ACTIVITIES	SUPPLIER CATALOG NO.
Arthrobacter species (optimum pH = 7.5 [serum acyl-CoA], 8.5 [palmitoyl-CoA], pI = 4.7, MW = 210,000 Da [Sephadex G-150]; K_M = 20 μM [palmitoyl-CoA]; stable pH 6-7.5 [37°C, 1 hr], T < 40°C [pH 7.0, 10 min])				
20-40 U/mg solid	1 unit generates 1 μmol H_2O_2/min at pH 8.0, 37°C	Lyophilized containing 60% protein		Asahi ACOD T-17
120 U/tablet	1 unit converts 1 μmol palmitoyl-CoA to product/min at 37°C	Tablet		Boehringer 1521195
20-40 U/mg protein	1 unit forms 1.0 μmol H_2O_2 and hexadecenoyl-CoA from palmitoyl-CoA/min at pH 8.0, 30°C in a peroxidase coupled system	Lyophilized containing 50% protein; balance primarily FAD as stabilizer		Sigma A2167
Candida species (optimum pH = 7.5-8.0, pI = 5.5, T = 35-40°C, MW = 600,000 [8 subunits/molecule]; K_M = 33 μM [palmitoyl-CoA]; stable pH 6.5-7.5 [25°C, 16 hr], T < 45°C [pH 7.8, 15 min])				
≥5 U/mg solid	1 unit forms 1 μmol H_2O_2 (0.5 μmol quinoneimine dye)/min at pH 7.5, 30°C	Lyophilized		ICN 190677
≥5 U/mg solid	1 unit forms 1 μmol H_2O_2 (0.5 μmol quinoneimine dye)/min at pH 8.0, 30°C	Lyophilized containing 50% FAD and sugars as stabilizers	<0.1% lipase <100% catalase	Toyobo ACO-201
2-8 U/mg				Wako 019-10841

Alanine Dehydrogenase

1.4.1.1

REACTION CATALYZED
  L-Alanine + H_2O + NAD^+ ↔ Pyruvate + NH_3 + NADH

SYSTEMATIC NAME
  L-Alanine:NAD^+ oxidoreductase (deaminating)

SYNONYMS
  L-Alanine dehydrogenase, AlaDH

SUBSTRATES
  L-Alanine, H_2O, NAD^+, pyruvate, NH_3, NADH

NOTES
- Classified as an oxidoreductase acting on the CH-NH_2 group of donors with NAD^+ or $NADP^+$ as acceptor

SPECIFIC ACTIVITY	UNITS DEFINITION	PREPARATION FORM	ADDITIONAL ACTIVITIES	SUPPLIER CATALOG NO.
Bacillus subtilis				
20 U/mg	1 unit converts 1 μmol L-Ala to product/min at 25°C	Suspension in 2.4 M $(NH_4)_2SO_4$, pH 7	<0.01% LDH (L-lactate as substrate), MDH (L-malate as substrate)	Boehringer 102636
30-50 U/mg protein	1 unit converts 1 μmol L-Ala to pyruvate and NH_3/min at pH 10.0, 25°C with agar as substrate	Suspension in 2.4 M $(NH_4)_2SO_4$, pH 7.0		Fluka 05192
	1 unit converts 1.0 μmol L-Ala to pyruvate and NH_3/min at pH 10.0, 25°C	Suspension in 2.4 M $(NH_4)_2SO_4$, pH 7.0		Sigma A7189
30 U/mg protein	1 unit converts 1.0 μmol L-Ala to pyruvate and NH_3/min at pH 10.0, 25°C	Solution containing 50% glycerol and 10 mM KPO_4 buffer, pH 7.7	1% LDH	Sigma A7653
Sporolactobacillus species (optimum pH = 8.5-9.5, pI = 4.57, T = 65°C, MW = 245,000 Da [gel filtration]; K_M = 11.8 mM [L-Ala], 36 μM [NAD^+], 0.84 mM [pyruvate], 20 mM [NH_4^+], 0.11 μM [NADH]; stable pH 6-9 [80°C, 15 min], T < 75°C [pH 8, 15 min])				
200-450 U/mg solid	1 unit reduces 1 μmol pyruvate to L-Ala/min at pH 8.5, 37°C	Lyophilized		Asahi ALADH T-52

Glutamate Dehydrogenase [NAD(P)$^+$]

1.4.1.3

REACTION CATALYZED
  L-Glutamate + H_2O + NAD(P)$^+$ ↔ 2-Oxoglutarate + NH_3 + NAD(P)H

SYSTEMATIC NAME
  L-Glutamate:NAD(P)$^+$ oxidoreductase (deaminating)

SYNONYMS
  Glutamic dehydrogenase, L-glutamic dehydrogenase, L-glutamate dehydrogenase, GlDH

REACTANTS
  L-Glutamate, H_2O, NAD(P)$^+$, 2-oxoglutarate, NH_3, NAD(P)H

1.4.1.3 Glutamate Dehydrogenase [NAD(P)$^+$] continued

APPLICATIONS

- Determination of serum transaminases, 5'-nuclease, ammonia, urea, L-glutamate, 2-ketoglutarate and 2-oxoglutarate in biological fluids
- Significant role in the determination of amino acids, especially in organisms that excrete ammonia

NOTES

- Classified as an oxidoreductase acting on the CH-NH$_2$ group of donors with NAD$^+$ or NADP$^+$ as acceptor
- Activated by ADP (3-fold stimulation with 0.4 mM), EDTA and β-mercaptoethylamine
- Inhibited by Cu^{2+}, Ag$^+$, Hg^{2+}, Zn^{2+}, ferrous ions, sodium azide, orthophosphate and 4-chloromercuribenzoate
- Bovine liver enzyme undergoes reversible association-dissociation reactions; rat liver reportedly does not
- Comprised of eight subunits

SPECIFIC ACTIVITY	UNITS DEFINITION	PREPARATION FORM	ADDITIONAL ACTIVITIES	SUPPLIER CATALOG NO.
Bovine liver (optimum pH = 7.0-8.0, T = 37°C, MW = 2,200,000 Da [8 X 280,000]; relative activity: 25°C = 1.0, 30°C = 1.27, 37°C = 1.65; stable pH 5-8, T < 60°C)				
40 U/mg protein; ≥10 U/mg solid	1 unit transforms 1 μmol 2-oxoglutarate/min at pH 7.3, 25°C under conditions NOT containing ADP	Salt-free; lyophilized containing <0.001 μmol/U ammonium ions	<0.01% LDH, MDH	Biozyme GDHB3
40 U/mg protein	1 unit transforms 1 μmol 2-oxoglutarate/min at pH 7.3, 25°C under conditions NOT containing ADP	Suspension in 50% (NH$_4$)$_2$SO$_4$, pH 7.6	<0.01% LDH, MDH	Biozyme GDHB5
40 U/mg protein	1 unit transforms 1 μmol 2-oxoglutarate/min at pH 7.3, 25°C under conditions NOT containing ADP	50% glycerol solution containing <0.001 μmol/U ammonium ions, pH 7.3	<0.01% LDH, MDH	Biozyme GDHB5G
120 U/mg	1 unit converts 1 μmol 2-oxoglutarate to product/min at 25°C with ADP as activator	3X Crystallized; solution containing 50% glycerol and <10 mg/mL NH$_3$, pH 7	<0.005% ADH, LDH, MDH	Boehringer 127086 127710 737119
120 U/mg	1 unit converts 1 μmol 2-oxoglutarate to product/min at 25°C with ADP as activator	Suspension in 2.0 M (NH$_4$)$_2$SO$_4$, pH 7	<0.005% ADH, LDH, MDH	Boehringer 127701
120 U/mg enzyme protein; 10 U/mg solid	1 unit converts 1 μmol 2-oxoglutarate to product/min at 25°C with ADP as activator	Lyophilized	<0.005% ADH, LDH, MDH	Boehringer 197734
>140 U/mg protein	1 unit oxidizes 1 μmol α-ketoglutarate in the presence of ADP at pH 7.8, 25°C	50% glycerol solution	<0.02% LDH, MDH	Calbiochem 35134

Glutamate Dehydrogenase [NAD(P)$^+$] continued 1.4.1.3

SPECIFIC ACTIVITY	UNITS DEFINITION	PREPARATION FORM	ADDITIONAL ACTIVITIES	SUPPLIER CATALOG NO.
Bovine liver *continued*				
10-50 U/mg solid	1 unit converts 1 μmol α-ketoglutarate to L-Glu/min at pH 7.5, 25°C without ADP	Lyophilized containing 90-95% protein		Calzyme 080B0010 080A0010
>50 U/mg protein; >10 U/mg solid	1 unit oxidizes 1 μmol NADH/min at pH 7.5, 37°C in the absence of ADP	Lyophilized containing stabilizers, buffer salts and <0.0001 μmol/U NH$_3$	<0.006% MDH	Diagnostic G-042
>50 U/mg protein; >10 U/mg solid	1 unit oxidizes 1 μmol NADH/min at pH 7.5, 37°C in the absence of ADP	Lyophilized containing stabilizers, buffer salts and <0.0001 μmol/U NH$_3$	<0.006% MDH	Diagnostic G-042B
>30 U/mg protein; >6 U/mg solid	1 unit oxidizes 1 μmol NADH/min at 25°C	Lyophilized containing stabilizers and buffer salts	<0.002% ADA <0.01% MDH	Diagnostic G-042D
40 U/mg	1 unit reduces 1 μmol 2-oxoglutarate/min at pH 7.9, 25°C	Solution containing 50% NaPO$_4$ buffer, 50% glycerol, pH 7.3		Fluka 49390
10 U/mg	1 unit reduces 1 μmol 2-oxoglutarate/min at pH 7.9, 25°C	Lyophilized containing BSA as stabilizer; Fraction V containing buffer salts		Fluka 49392
>50 U/mg protein	1 unit oxidizes 1 μmol NADH to NAD/min at 37°C	Lyophilized		GDS Tech G-042
>40 U/mg protein; >10 U/mg solid	1 unit oxidizes 1.0 μmol NADH/min at pH 7.5, 25°C	Lyophilized containing <0.0005 μmol/U NH$_3$	<0.01% Asp dehydrogenase, MDH, LDH	Genzyme 1251
40 U/mg protein	1 unit transforms 1.0 μmol α-ketoglutarate/min at pH 7.3, 25°C	Suspension in 50% (NH$_4$)$_2$SO$_4$, pH 7.6	<0.005% GOT, GPT <0.01% LDH, MDH	ICN 151189
>40 U/mg protein; >10 U/mg solid	1 unit transforms 1 μmol oxoglutarate/min at pH 7.3, 25°C in the absence of ADP	Lyophilized containing <0.1 μg/mg NH$_3$	<0.005% MDH, GOT, GPT <0.01% LDH	Randox GD 825L
40 U/mg protein	1 unit reduces 1.0 μmol α-ketoglutarate to L-Glu/min at pH 7.3, 25°C in the presence of ammonium ions	Lyophilized containing 40% protein; balance primarily sucrose and PO$_4$ buffer salts; <0.5 μg/mg protein NH$_3$		Sigma G2009
40 U/mg protein	1 unit reduces 1.0 μmol α-ketoglutarate to L-Glu/min at pH 7.3, 25°C in the presence of ammonium ions	Suspension in 2.0 M (NH$_4$)$_2$SO$_4$, pH 7.0		Sigma G2501
40 U/mg protein	1 unit reduces 1.0 μmol α-ketoglutarate to L-Glu/min at pH 7.3, 25°C in the presence of ammonium ions	3X Crystallized; solution containing 50% glycerol and NaPO$_4$ buffer, pH 7.3		Sigma G2626
40 U/mg protein	1 unit reduces 1.0 μmol α-ketoglutarate to L-Glu/min at pH 7.3, 25°C in the presence of ammonium ions	Suspension in 2.0 M (NH$_4$)$_2$SO$_4$ and 0.05 M NaPO$_4$, pH 7.5		Sigma G4008

1.4.1.3 Glutamate Dehydrogenase [NAD(P)$^+$] continued

SPECIFIC ACTIVITY	UNITS DEFINITION	PREPARATION FORM	ADDITIONAL ACTIVITIES	SUPPLIER CATALOG NO.
Bovine liver continued				
40 U/mg protein	1 unit reduces 1.0 μmol α-ketoglutarate to L-Glu/min at pH 7.3, 25°C in the presence of ammonium ions	Lyophilized containing 70% protein; balance primarily citrate buffer salts; <0.5 μg/mg protein NH$_3$		Sigma G7882
Ovine liver				
40 U/mg protein	1 unit reduces 1.0 μmol α-ketoglutarate to L-Glu/min at pH 7.3, 25°C in the presence of ammonium ions	Crystallized; suspension in 3.2 M (NH$_4$)$_2$SO$_4$ and 0.01 M KPO$_4$, pH 7.5		Sigma G4393
Rat liver				
30 U/mg protein	1 unit reduces 1.0 μmol α-ketoglutarate to L-Glu/min at pH 7.3, 25°C in the presence of ammonium ions	Crystallized; solution containing 50% glycerol, 12.5 mM KPO$_4$ buffer, 0.5 mM EDTA, pH 7.6; 1 μg/mg protein NH$_3$		Sigma G5636
Yeast, bakers				
10 U/mg protein	1 unit converts 1 μmol α-ketoglutarate to L-Glu/min at pH 7.5, 25°C without ADP	Lyophilized		Calzyme 190A0010

1.4.1.4 Glutamate Dehydrogenase (NADP$^+$)

REACTION CATALYZED

L-Glutamate + H$_2$O + NADP$^+$ ↔ 2-Oxoglutarate + NH$_3$ + NADPH

SYSTEMATIC NAME

L-Glutamate:NADP$^+$ oxidoreductase (deaminating)

SYNONYMS

Glutamic dehydrogenase, L-glutamic dehydrogenase

REACTANTS

L-Glutamate, H$_2$O, NADP$^+$, 2-oxoglutarate, NH$_3$, NADPH

NOTES

- Classified as an oxidoreductase acting on the CH-NH$_2$ group of donors with NAD$^+$ or NADP$^+$ as acceptor
- Inhibited by Hg^{2+}, Cd^{2+}, PCMB, pyridine, 4,4´-dithiopyridine, 2,2´-dithiopyridine
- Stabilized by EDTA

Glutamate Dehydrogenase (NADP$^+$) continued — 1.4.1.4

SPECIFIC ACTIVITY	UNITS DEFINITION	PREPARATION FORM	ADDITIONAL ACTIVITIES	SUPPLIER CATALOG NO.
Candida utilis				
50-200 U/mg protein	1 unit reduces 1.0 μmol α-ketoglutarate to L-Glu/min at pH 8.3, 30°C in the presence of ammonium ions and NADPH	Lyophilized containing KPO$_4$ buffer salt		Sigma G5027
Proteus species (optimum pH = 9.5 [forward], 8.5 [reverse], T = 45-55°C [forward], 45°C [reverse], MW = 300,000 Da [6 subunits at 50,000 Da]; K_M = 1.1 mM [NH$_3$], 0.34 mM [α-KG], 1.2 mM [L-Glu], 14 μM [NADH], 15 μM [NADP$^+$]; stable pH 6.0-8.5 [25°C, 20 hr], T < 50°C [pH 7.4, 10 min])				
400 U/mg protein	1 unit reduces 1.0 μmol α-ketoglutarate to L-Glu/min at pH 8.3, 30°C in the presence of ammonium ions and NADPH	Solution containing 50 mM Tris HCl, 5 mM Na$_2$ EDTA, 0.05% NaN$_3$, pH 7.8		Sigma G4387
≥300 U/mg protein; ≥9000 U/mL	1 unit oxidizes 1 μmol NADPH/min at pH 8.3, 30°C	Solution containing 50 mM Tris-HCl, 0.05% NaN$_3$, 5 mM EDTA, pH 7.8	<0.01% NADPH oxidase, GR	Toyobo GTD-209
Yeast				
≥10 IU/mg protein	1 IU transforms 1 μmol L-Glu/min under standard IUB conditions at 25°C	Lyophilized	<0.01% NADPH oxidase <0.1% G6PDH, GlDH, GR <0.5% 6PGDH	OYC

Leucine Dehydrogenase — 1.4.1.9

REACTION CATALYZED

  L-Leucine + H$_2$O + NAD$^+$ ↔ 4-Methyl-2-oxopentanoate + NH$_3$ + NADH

SYSTEMATIC NAME

  L-Leucine:NAD$^+$ oxidoreductase (deaminating)

SYNONYMS

  LeuDH

REACTANTS

  L-Leucine, isoleucine, valine, norvaline, norleucine, H$_2$O, NAD$^+$, 4-methyl-2-oxopentanoate, NH$_3$, NADH

NOTES

- Classified as an oxidoreductase acting on the CH-NH$_2$ group of donors with NAD$^+$ or NADP$^+$ as acceptor
- Stabilized by 2-mercaptoethanol, L-cysteine, dithiothreitol, EDTA
- Inhibited by Na$_2$S, Hg^{2+}, Cu^{2+}, Co^{2+}, Mg^{2+} and PCMB

1.4.1.9 Leucine Dehydrogenase continued

SPECIFIC ACTIVITY	UNITS DEFINITION	PREPARATION FORM	ADDITIONAL ACTIVITIES	SUPPLIER CATALOG NO.
Bacillus species (optimum pH = 10.5-10.8 [forward], 7.4 [reverse], T > 70°C, MW = 245,000 Da [6 subunits/mol enzyme]; K_M = 1.0 mM [L-Leu], 0.39 mM [NAD^+], 35 μM [NADH], 0.31 mM [α-ketoisocaproate], 0.2 M [NH_3]; stable pH 5.5-10.5 [25°C, 20 hr], T < 60°C [pH 6.9, 10 min])				
60-120 U/mg protein	1 unit converts 1.0 μmol L-Leu to α-ketoisocaproate/min at pH 10.5, 37°C	Lyophilized containing 40% protein; balance primarily Lys		Sigma L5135
≥20 U/mg solid	1 unit forms 1 μmol NADH/min at pH 10.5, 37°C	Lyophilized containing 70% DTT, EDTA, β-MSH, L-Cys as stabilizers	<0.01% NADH oxidase, Leu-Val, Leu-Gly-Gly	Toyobo LED-201

1.4.1.20 Phenylalanine Dehydrogenase

REACTION CATALYZED
  L-Phenylalanine + H_2O + NAD^+ ↔ Phenylpyruvate + NH_3 + NADH

SYSTEMATIC NAME
  L-Phenylalanine:NAD^+ oxidoreductase (deaminating)

SYNONYMS
  L-Phenylalanine dehydrogenase

REACTANTS
  L-Phenylalanine, L-tyrosine, H_2O, NAD^+, phenylpyruvate, NH_3, NADH

NOTES
- Classified as an oxidoreductase acting on the $CH-NH_2$ group of donors with NAD^+ or $NADP^+$ as acceptor
- *Bacillus badius* and *Sporosarcina ureae* enzymes are highly specific for L-phenylalanine
- *Bacillus sphaericus* enzyme also acts on L-tyrosine

SPECIFIC ACTIVITY	UNITS DEFINITION	PREPARATION FORM	ADDITIONAL ACTIVITIES	SUPPLIER CATALOG NO.
Microbial (optimum pH = 10.5, T = 37°C)				
>20 U/mg protein	1 unit converts 1 μmol NAD to NADH/min at pH 10.5, 37°C	Lyophilized		GDS Tech P-010
***Sporosarcina* species**				
20 U/mg protein	1 unit oxidizes 1.0 μmol L-Phe/min at pH 10.5, 30°C in the presence of β-NAD	Lyophilized		Sigma P4798
***Sporosarcinae* (optimum pH = 10.0-11.0, T < 45°C)**				
20-30 U/mg protein		Powder		Biocatalysts P098P

L-Amino-Acid Oxidase — 1.4.3.2

REACTION CATALYZED

An L-amino acid + H_2O + O_2 ↔ A 2-Oxo acid + NH_3 + H_2O_2

SYSTEMATIC NAME

L-Amino-acid:oxygen oxidoreductase (deaminating)

SYNONYMS

Ophio-amino-acid oxidase, AOD, LAOD

REACTANTS

L-Amino acid, H_2O, O_2, 2-oxo acid, NH_3, H_2O_2

APPLICATIONS

- Oxidation reactions
- Diagnostic component
- Purification and determination of certain amino acids
- Preparation of α-keto acids

NOTES

- Classified as an oxidoreductase acting on the $CH-NH_2$ group of donors with oxygen as acceptor
- A glycoprotein composed of 2 subunits combined in unequal amounts; 3 electrophoretically distinct isozymes occur as different combinations of the 2 subunits
- Contains approximately 2 mol FAD/mol holoenzyme
- Has absolute specificity for the L-isomer of a number of amino acids
- Inhibited by cyanide; competitively inhibited by aromatic carboxylates

SPECIFIC ACTIVITY	UNITS DEFINITION	PREPARATION FORM	ADDITIONAL ACTIVITIES	SUPPLIER CATALOG NO.
Bothrops atrox venom				
0.2-0.5 U/mg solid	1 unit oxidatively aminates 1.0 μmol L-Phe/min at pH 6.5, 37°C	Crude		Sigma A4257
Colletotrichum species (optimum pH = 7.0-8.5, pI = 3.9, MW = 200,000 Da [gel filtration]; K_M = 0.17 mM [L-Leu]; stable pH 8-10.5, T < 70°C)				
5-15 U/mg solid	1 unit oxidizes 1 μmol L-Leu to 2-oxo-4-methyl-valeric acid/min at pH 7.5, 37°C	Lyophilized		Asahi LAOD T-26
Crotalus adamanteus venom (MW = 70,000 Da; reversibly inactivated at 38°C [pH 7.5] and upon freezing)				
0.5 U/mg	1 unit deaminates 1 μmol L-Phe/min at pH 6.5, 37°C	Powder	<6% phosphodiesterase	Fluka 09417
4-8 U/mg protein	1 unit oxidizes 1 μmol L-Leu/min at pH 7.6, 25°C	Aqueous solution containing toluene as preservative		ICN 100272
1-3 U/mg protein	1 unit oxidatively aminates 1.0 μmol L-Phe/min at pH 6.5, 37°C	Lyophilized containing 60% protein; balance Gly buffer salts		Sigma A3016
0.3 U/mg solid	1 unit oxidatively aminates 1.0 μmol L-Phe/min at pH 6.5, 37°C	Crude		Sigma A9253
4-10 U/mg protein	1 unit oxidatively aminates 1.0 μmol L-Phe/min at pH 6.5, 37°C	Aqueous suspension containing toluene as preservative		Sigma A9378

1.4.3.2 L-Amino-Acid Oxidase continued

SPECIFIC ACTIVITY	UNITS DEFINITION	PREPARATION FORM	ADDITIONAL ACTIVITIES	SUPPLIER CATALOG NO.
Crotalus adamanteus venom *continued*				
≥4 U/mg protein	1 unit oxidizes 1 µmol L-Leu/min at pH 7.6, 25°C	Solution containing toluene		Worthington LS02763 LS02764 LS02766
Crotalus atrox venom				
0.3 U/mg	1 unit deaminates 1 µmol L-Phe/min at pH 6.5, 37°C	Powder		Fluka 09419
0.2-0.3 U/mg solid	1 unit oxidatively aminates 1.0 µmol L-Phe/min at pH 6.5, 37°C	Crude		Sigma A5147
1-3 U/mg protein	1 unit oxidatively aminates 1.0 µmol L-Phe/min at pH 6.5, 37°C	Lyophilized containing 45% protein; balance Gly buffer salts		Sigma A8390
Crotalus durissus venom				
7 U/mg at 25°C; 12 U/mg at 37°C	1 unit converts 1 µmol L-Leu to product/min	Suspension in 3.2 M $(NH_4)_2SO_4$, pH 6	<0.01 U/mg proteases (Kunitz)	Boehringer 102792
3-8 U/mg protein	1 unit oxidatively aminates 1.0 µmol L-Phe/min at pH 6.5, 37°C	Suspension in 3.2 M $(NH_4)_2SO_4$, pH 6		Sigma A2805

1.4.3.3 D-Amino-Acid Oxidase

REACTION CATALYZED

A D-amino acid + H_2O + O_2 ↔ A 2-Oxo acid + NH_3 + H_2O_2

SYSTEMATIC NAME

D-Amino-acid:oxygen oxidoreductase (deaminating)

SYNONYMS

DAAO

REACTANTS

D-Amino acid, glycine, H_2O, O_2, 2-oxo acid, NH_3, H_2O_2

APPLICATIONS

- Determination of FAD and D-alanine
- Separation of L-amino acids from racemic mixtures
- Preparation of keto acids
- Oxidation-reduction studies
- Usefulness is increased through immobilization

NOTES

- Classified as an oxidoreductase acting on the $CH-NH_2$ group of donors with oxygen as acceptor
- Oxidatively deaminates a wide variety of D-amino acids to the corresponding α-keto acid
- A monomer which can undergo concentration-dependent dimerization; monomer is more active than dimer
- A flavoprotein containing 1 mol FAD/monomer
- Cysteine and tyrosine are involved at the active site

D-Amino-Acid Oxidase continued — 1.4.3.3

SPECIFIC ACTIVITY	UNITS DEFINITION	PREPARATION FORM	ADDITIONAL ACTIVITIES	SUPPLIER CATALOG No.
Porcine kidney (optimum pH = 9 [D-Ala], MW = 38,000-39,000 Da)				
	1 unit transforms 1 μmol D-Ala to pyruvate/min at pH 8.3, 25°C in the presence of catalase	Salt-free; lyophilized		Biozyme APO DOX
≥40 U/g solid	1 unit transforms 1 μmol D-Ala to pyruvate/min at pH 8.3, 25°C in the presence of catalase	Salt-free; lyophilized		Biozyme DOX1
≥7000 U/g protein; ≥6000 U/g solid	1 unit transforms 1 μmol D-Ala to pyruvate/min at pH 8.3, 25°C in the presence of catalase	Salt-free; lyophilized		Biozyme DOX2
15,000 U/g protein	1 unit transforms 1 μmol D-Ala to pyruvate/min at pH 8.3, 25°C in the presence of catalase	Highly purified suspension in 3.2 M $(NH_4)_2SO_4$		Biozyme DOX3S
15 U/mg	1 unit converts 1 μmol D-Ala to product/min at 25°C	Suspension in 3.2 M $(NH_4)_2SO_4$, pH 6.5	<0.005 U/mg proteases (Kunitz)	Boehringer 102784
15-20 U/mg protein	1 unit oxidatively deaminates 1 μmol D-Ala to pyruvate/min at pH 8.3, 37°C in the presence of catalase	Lyophilized containing 90% protein		Calzyme 052A0015
15-20 U/mg protein	1 unit oxidatively deaminates 1 μmol D-Ala to pyruvate/min at pH 8.3, 37°C in the presence of catalase	Suspension in $(NH_4)_2SO_4$ containing 10 mg/mL protein		Calzyme 052B0015
25-30 U/mg protein	1 unit oxidatively deaminates 1 μmol D-Ala to pyruvate/min at pH 8.3, 37°C in the presence of catalase	Lyophilized containing 90% protein	No detectable FAD, AP	Calzyme 176A0025
10-15 U/mg protein	1 unit deaminates 1 μmol D-Ala/min at pH 8.3, 25°C in the presence of catalase	Suspension in 3.6 M $(NH_4)_2SO_4$, pH 6.5		Fluka 09412
0.5-1.5 U/mg	1 unit deaminates 1 μmol D-Ala/min at pH 8.3, 25°C in the presence of catalase	Powder		Fluka 09413
15-20 U/mg protein	1 unit deaminates 1.0 μmol D-Ala to pyruvate/min at pH 8.3, 20°C in the presence of catalase	Crystallized; suspension in 3.2 M $(NH_4)_2SO_4$, pH 6.5		ICN 190681
15 U/mg protein	1 unit oxidatively aminates 1.0 μmol D-Ala to pyruvate/min at pH 8.3, 25°C in the presence of catalase	Crystallized; suspension in 3.6 M $(NH_4)_2SO_4$ solution, pH 6.5		Sigma A1789
0.5-1.5 U/mg solid	1 unit oxidatively aminates 1.0 μmol D-Ala to pyruvate/min at pH 8.3, 25°C in the presence of catalase	Lyophilized containing 85% protein; balance Tris buffer salts		Sigma A1914
10 U/mg protein	1 unit oxidatively aminates 1.0 μmol D-Ala to pyruvate/min at pH 8.3, 25°C in the presence of catalase	Lyophilized containing 85% protein; balance Tris buffer salts		Sigma A5418

1.4.3.3 D-Amino-Acid Oxidase continued

SPECIFIC ACTIVITY	UNITS DEFINITION	PREPARATION FORM	ADDITIONAL ACTIVITIES	SUPPLIER CATALOG NO.
Porcine kidney *continued*				
	1 unit oxidatively aminates 1.0 µmol D-Ala to pyruvate/min at pH 8.3, 25°C in the presence of catalase	Crude		Sigma A9128
≥2 U/mg solid	1 unit oxidizes 1 µmol D-Ala/min at pH 8.3, 37°C	Chromatographically purified		Worthington LS06310 LS06308 LS06311

1.4.3.4 Amine Oxidase (Flavin-Containing)

REACTION CATALYZED

$$RCH_2NH_2 + H_2O + O_2 \leftrightarrow RCHO + NH_3 + H_2O_2$$

SYSTEMATIC NAME

Amine:oxygen oxidoreductase (deaminating) (flavin-containing)

SYNONYMS

Monoamine oxidase, tyramine oxidase, tyraminase, amine oxidase, adrenalin oxidase, MAO, TOD

REACTANTS

RCH_2NH_2, H_2O, O_2, $RCHO$, NH_3, H_2O_2

NOTES

- Classified as an oxidoreductase acting on the CH-NH_2 group of donors with oxygen as acceptor
- A flavoprotein (FAD)
- Acts on primary amines, and some secondary and tertiary amines

SPECIFIC ACTIVITY	UNITS DEFINITION	PREPARATION FORM	ADDITIONAL ACTIVITIES	SUPPLIER CATALOG NO.
***Arthrobacter* species** (optimum pH = 7.0, pI = 4.23, MW = 48,000 [gel filtration]; K_M = 27 µM [tyramine]; stable pH 6-8 [37°C, 1 hr], T < 50°C [pH 7.5, 5 min])				
3-7 U/mg solid	1 unit oxidizes 1 µmol tyramine to p-hydroxybenzaldehyde/min at pH 7.5, 37°C	Lyophilized containing 70% protein	<0.1% peptidase	Asahi TOD T-25
3 U/mg solid	1 unit oxidizes 1.0 µmol tyramine to p-hydroxyphenylacetaldehyde/min at pH 7.5, 37°C			Sigma T0905
Bovine plasma				
40-100 U/g protein	1 unit oxidizes 1.0 µmol benzylamine to benzaldehyde/min at pH 7.4, 25°C	Suspension in 2.0 M $(NH_4)_2SO_4$, pH 6.0		Sigma M4636
2-6 U/g protein	1 unit oxidizes 1.0 µmol benzylamine to benzaldehyde/min at pH 7.4, 25°C	Lyophilized containing 80% protein; balance PO_4 buffer salts		Sigma M9643

Amine Oxidase (Copper-Containing) 1.4.3.6

REACTION CATALYZED

$RCH_2NH_2 + H_2O \leftrightarrow RCHO + NH_3 + H_2O_2$

SYSTEMATIC NAME

Amine:oxygen oxidoreductase (deaminating) (copper-containing)

SYNONYMS

Diamine oxidase, diamino oxhydrase, histaminase, amine oxidase, plasma amine oxidase, PAO

REACTANTS

RCH_2NH_2, H_2O, RCHO, NH_3, H_2O_2, catecholamines, tryptamine derivatives, benzyl amine

NOTES

- Classified as an oxidoreductase acting on the $CH-NH_2$ group of donors with oxygen as acceptor
- A group of copper quinoproteins oxidizing primary monoamines, diamines and histamine
- Plasma enzyme is composed of 2 identical chains, 2 pyridoxal phosphates and 2 Cu^{2+} atoms
- Inhibited by copper chelating agents, many carboxyl reagents (e.g., cuprizone), hydroxylamine and cyanide; benzoic acid and benzyl alcohol are non-competitive inhibitors
- Orotate reductase (NADPH) (EC 1.3.1.15) from rat kidney also catalyzes this reaction

SPECIFIC ACTIVITY	UNITS DEFINITION	PREPARATION FORM	ADDITIONAL ACTIVITIES	SUPPLIER CATALOG NO.
Bovine plasma (optimum pH = 6.2 [spermine], 7.2 [spermidine], MW = 170,000 Da)				
≥17 Tabor U/mg solid	1 IU oxidizes 1 μmol benzylamine/min at pH 7.2, 25°C; 1 IU = 4330 Tabor U	Chromatographically prepared; lyophilized		Worthington LS03113 LS03114 LS03110
Porcine kidney				
0.05-0.15 U/mg solid	1 unit oxidizes 1.0 μmol putrescine/hr at pH 7.2, 37°C	Salt-free; lyophilized		ICN 100589
0.05-0.15 U/mg solid	1 unit oxidizes 1.0 μmol putrescine/hr at pH 7.2, 37°C		No detectable MAO	Sigma D7876

1.4.3.10 Putrescine Oxidase

REACTION CATALYZED
Putrescine + O_2 + H_2O ↔ 4-Aminobutanal + NH_3 + H_2O_2

SYSTEMATIC NAME
Putrescine:oxygen oxidoreductase (deaminating)

SUBSTRATES
Putrescine, O_2, H_2O, 4-aminobutanal, NH_3, H_2O_2, 1-pyrroline, spermidine, cadaverine

APPLICATIONS
- Determination of polyamine in clinical analyses

NOTES
- Classified as an oxidoreductase acting on the $CH-NH_2$ group of donors with oxygen as acceptor
- A flavoprotein (FAD)
- 4-Aminobutanal condenses non-enzymatically to 1-pyrroline
- Stabilized by sugars
- Inhibited by MIA, Cd^{2+}, Cu^{2+}, Pb^{2+} and Hg^{2+}
- Comprised of 2 subunits of 50,000 Da each and 1 mol FAD/mol enzyme

SPECIFIC ACTIVITY	UNITS DEFINITION	PREPARATION FORM	ADDITIONAL ACTIVITIES	SUPPLIER CATALOG NO.
Microbial (optimum pH = 8.0, pI = 4.0, T = 55°C, MW = 130,000 Da [gel filtration, 50,000/subunit, 1 mol FAD/mol enzyme]; K_M = 0.13 mM [putrescine], 0.14 mM [spermidine], 0.24 mM [cadaverine]; stable pH 5.0-9.0 [25°C, 20 hr], T < 45°C [pH 8.0, 15 min])				
≥15 U/mg solid	1 unit forms 1 μmol H_2O_2 (0.5 μmol quinoneimine dye)/min at pH 8.0, 30°C	Lyophilized containing sugars as stabilizers	<0.0005% AlaDH, NADH oxidase <0.5% catalase	Toyobo PUO-301

1.4.3.11 L-Glutamate Oxidase

REACTION CATALYZED
L-Glutamate + O_2 + H_2O ↔ 2-Oxoglutarate + NH_3 + H_2O_2

SYSTEMATIC NAME
L-Glutamate:oxygen oxidoreductase (deaminating)

REACTANTS
L-Glutamate, O_2, H_2O, 2-oxoglutarate, NH_3, H_2O_2

NOTES
- Classified as an oxidoreductase acting on the $CH-NH_2$ group of donors with oxygen as acceptor
- A flavoprotein (FAD)

L-Glutamate Oxidase continued   1.4.3.11

SPECIFIC ACTIVITY	UNITS DEFINITION	PREPARATION FORM	ADDITIONAL ACTIVITIES	SUPPLIER CATALOG NO.
Streptomyces species (optimum pH = 7.0-9.0, T = 30°C)				
Relative reaction rates with 10 mM substrate: 100% for L-Glu, 0.6% for L-Asp, 0% for other amino acids	1 unit consumes 1 μmol O_2/min at pH 7.4, 30°C in the absence of catalase	Lyophilized		Seikagaku 100645-1
5 U/mg solid	1 unit forms 1.0 μmol α-ketoglutaric acid from L-Glu/min at pH 7.4, 30°C	Lyophilized containing 20% protein; balance KPO_4 buffer salts, lactose		Sigma G0400

L-Lysine Oxidase   1.4.3.14

REACTION CATALYZED

L-Lysine + O_2 + H_2O ↔ 6-Amino-2-oxohexanoate + NH_3 + H_2O_2

SYSTEMATIC NAME

L-Lysine:oxygen 2-oxidoreductase (deaminating)

SYNONYMS

Lysine oxidase, L-lysine α-oxidase

REACTANTS

L-Lysine, and more slowly on L-ornithine, L-phenylalanine, L-arginine, L-histidine; O_2, H_2O, 6-amino-2-oxohexanoate, NH_3, H_2O_2

NOTES

- Classified as an oxidoreductase acting on the $CH-NH_2$ group of donors with oxygen as acceptor

SPECIFIC ACTIVITY	UNITS DEFINITION	PREPARATION FORM	ADDITIONAL ACTIVITIES	SUPPLIER CATALOG NO.
Trichoderma viride (optimum pH = 7.5-8.5, T = 30°C)				
Relative reaction rates with 10 mM substrate; 100% for L-lysine, 18.2% for L-ornithine, 8.3% for L-Phe, 6.1% for L-Arg, 5.5% for L-Tyr, 3.8% for L-His	1 unit consumes 1 μmol O_2/min at pH 7.4, 30°C in the absence of catalase	Lyophilized		Seikagaku 100932-1 100932-2
20-40 U/mg protein	1 unit forms 1 μmol 6-amino-2-oxohexanoic acid from L-Lys/min at pH 8.0, 37°C	Lyophilized containing 10% protein; balance PO_4 buffer salts and stabilizer		Sigma L6150

1.5.1.3 Dihydrofolate Reductase

REACTION CATALYZED
5,6,7,8-Tetrahydrofolate + NADP$^+$ ↔ 7,8-Dihydrofolate + NADPH

SYSTEMATIC NAME
5,6,7,8-Tetrahydrofolate:NADP$^+$ oxidoreductase

SYNONYMS
Tetrahydrofolate dehydrogenase

REACTANTS
5,6,7,8-Tetrahydrofolate, NADP$^+$, 7,8-dihydrofolate, NADPH

APPLICATIONS
- Determination of dihydrofolic acid

NOTES
- Classified as an oxidoreductase acting on the CH-NH group of donors with NAD$^+$ or NADP$^+$ as acceptor
- Animal and (some) microorganism enzyme also slowly reduce folate to 5,6,7,8-tetrahydrofolate

SPECIFIC ACTIVITY	UNITS DEFINITION	PREPARATION FORM	ADDITIONAL ACTIVITIES	SUPPLIER CATALOG NO.
Bovine liver				
8 U/mg protein	1 unit converts 1 μmol 7,8-dihydrofolate and NADPH to 5,6,7,8-tetrahydrofolate and NADP/min at pH 6.5, 25°C with starch as substrate	Suspension in 3.6 M (NH$_4$)$_2$SO$_4$, pH 7.0		Fluka 37294
8 U/mg protein	1 unit converts 1.0 μmol 7,8-dihydrofolate and NADPH to 5,6,7,9-tetrahydrofolate and NADP/min at pH 6.5, 25°C	Suspension in 3.6 M (NH$_4$)$_2$SO$_4$, pH 8		Sigma D6385
Chicken liver				
2-5 U/mg protein	1 unit converts 1.0 μmol 7,8-dihydrofolate and NADPH to 5,6,7,9-tetrahydrofolate and NADP/min at pH 6.5, 25°C	Solution containing 50% glycerol, 1.0 M (NH$_4$)$_2$SO$_4$, 0.1 M KPO$_4$, pH 6.4		Sigma D2768
Rat liver				
2-5 U/mg protein	1 unit converts 1.0 μmol 7,8-dihydrofolate and NADPH to 5,6,7,9-tetrahydrofolate and NADP/min at pH 6.5, 25°C	Solution containing 50% glycerol, 0.7 M (NH$_4$)$_2$SO$_4$, 0.05 M KPO$_4$, pH 6.5		Sigma D5519

Methylenetetrahydrofolate Dehydrogenase (NADP$^+$) 1.5.1.5

REACTION CATALYZED
  5,10-Methylenetetrahydrofolate + NADP$^+$ ↔ 5,10-Methenyltetrahydrofolate + NADPH

SYSTEMATIC NAME
  5,10-Methylenetetrahydrofolate:NADP$^+$ oxidoreductase

SYNONYMS
  5,10-Methylenetetrahydrofolate dehydrogenase

REACTANTS
  5,10-Methylenetetrahydrofolate, NADP$^+$, 5,10-methenyltetrahydrofolate, NADPH

APPLICATIONS
- Determination of tetrahydrofolic acid

NOTES
- Classified as an oxidoreductase acting on the CH-NH group of donors with NAD$^+$ or NADP$^+$ as acceptor
- A trifunctional enzyme in eukaryotes also having methenyltetrahydrofolate cyclohydrolase (EC 3.5.4.9) and formate-tetrahydrofolate ligase (EC 6.3.4.3) activities
- A bifunctional enzyme in some prokaryotes having methenyltetrahydrofolate cyclohydrolase activity (EC 3.5.4.9)

SPECIFIC ACTIVITY	UNITS DEFINITION	PREPARATION FORM	ADDITIONAL ACTIVITIES	SUPPLIER CATALOG NO.
Yeast				
0.5-1.0 U/mg protein	1 unit converts 1.0 μmol 5,10-methylenetetrahydrofolate and NADP to 5,10-methenyltetrahydrofolate and NADPH/min at pH 7.5, 25°C	Suspension in 2.8 M (NH$_4$)$_2$SO$_4$, 0.003 M KPO$_4$, 0.03 M KCl, 0.003 M EDTA Na$_4$, 0.01 M MgCl$_2$, 0.01 M DTT, pH 7.5		Sigma M4882

Saccharopine Dehydrogenase (NAD$^+$, L-Lysine-Forming) 1.5.1.7

REACTION CATALYZED
  N^6-(L-1,3-Dicarboxypropyl)-L-lysine + NAD$^+$ + H$_2$O ↔ L-Lysine + 2-oxoglutarate + NADH

SYSTEMATIC NAME
  N^6-(L-1,3-Dicarboxypropyl)-L-lysine:NAD$^+$ oxidoreductase (L-lysine-forming)

SYNONYMS
  Lysine-2-oxoglutarate reductase

REACTANTS
  N^6-(L-1,3-Dicarboxypropyl)-L-lysine, NAD$^+$, H$_2$O, L-lysine, 2-oxoglutarate, NADH

APPLICATIONS
- Determination of L-lysine

NOTES
- Classified as an oxidoreductase acting on the CH-NH group of donors with NAD$^+$ or NADP$^+$ as acceptor

1.5.1.7 Saccharopine Dehydrogenase (NAD$^+$, L-Lysine-Forming) continued

SPECIFIC ACTIVITY	UNITS DEFINITION	PREPARATION FORM	ADDITIONAL ACTIVITIES	SUPPLIER CATALOG NO.
Yeast, bakers				
40-80 U/mg protein	1 unit converts 1.0 μmol L-Lys and α-ketoglutaric acid to saccharopine/min at pH 6.8, 25°C	Affinity purified; lyophilized containing 5-15% protein; balance primarily KPO$_4$ and EDTA	<0.1% NADH oxidase, GlDH	Sigma S9383

1.5.1.11 D-Octopine Dehydrogenase

REACTION CATALYZED

N^2-(D-1-Carboxyethyl)-L-arginine + NAD$^+$ + H$_2$O ↔ L-Arginine + pyruvate + NADH

SYSTEMATIC NAME

N^2-(D-1-Carboxyethyl)-L-arginine:NAD$^+$ oxidoreductase (L-arginine forming)

SYNONYMS

D-Octopine synthase

REACTANTS

N^2-(D-1-Carboxyethyl)-L-arginine, NAD$^+$, H$_2$O, L-ornithine, L-lysine, L-histidine, L-arginine, pyruvate, NADH

NOTES

- Classified as an oxidoreductase acting on the CH-NH group of donors with NAD$^+$ or NADP$^+$ as acceptor

SPECIFIC ACTIVITY	UNITS DEFINITION	PREPARATION FORM	ADDITIONAL ACTIVITIES	SUPPLIER CATALOG NO.
Scallops				
5-15 U/mg protein	1 unit forms 1.0 μmol octopine from 1.0 μmol of pyruvate and 1.0 μmol L-Arg/min at pH 6.6, 25°C	Lyophilized containing 90% protein; balance primarily Tris buffer salts		Sigma O1252

Sarcosine Oxidase 1.5.3.1

REACTION CATALYZED
Sarcosine + H_2O + O_2 ↔ Glycine + formaldehyde + H_2O_2

SYSTEMATIC NAME
Sarcosine:oxygen oxidoreductase (demethylating)

SYNONYMS
Sarcosine oxidase G, SOX

REACTANTS
Sarcosine, H_2O, O_2, glycine, formaldehyde, H_2O_2

APPLICATIONS
- Determination of creatine, creatinine and sarcosine in biological fluids

NOTES
- Classified as an oxidoreductase acting on the CH-NH group of donors with oxygen as acceptor
- A flavoprotein (FAD)
- The flavin is both covalently and non-covalently bound in a molar ratio of 1:1
- Stabilized by potassium gluconate
- Inhibited by PCMB, SDS, heavy metal ions, dimethyl-benzylalkyl-ammonium chloride, MIA and N-ethylmaleimide

SPECIFIC ACTIVITY	UNITS DEFINITION	PREPARATION FORM	ADDITIONAL ACTIVITIES	SUPPLIER CATALOG No.
Arthrobacter nicotianae				
5 U/mg protein; 1.5 U/mg solid	1 unit converts 1 μmol sarcosine to product/min at 25°C	Lyophilized	<0.01% uricase, creatininase <2% catalase	Boehringer 789798
Arthrobacter species (optimum pH = 8.0, pI = 4.0, T = 35°C, MW = 160,000 Da [gel filtration]; K_M = 4.1 mM [sarcosine]; stable at pH 6.5-8.5 [25°C, 20 hr] and < 37°C [pH 7.0, 10 min])				
5-10 U/mg solid	1 unit forms 1 μmol formaldehyde from sarcosine/min at pH 8.3, 37°C	Lyophilized containing 40% stabilizer		ICN 152047
5-15 U/mg protein	1 unit forms 1.0 μmol formaldehyde from sarcosine/min at pH 8.3, 37°C	Lyophilized containing 50% protein; balance primarily K gluconate and EDTA		Sigma S8764
≥5 U/mg solid	1 unit forms 1 μmol H_2O_2 (0.5 μmol quinoneimine dye)/min at pH 8.0, 37°C	Lyophilized containing 40% K gluconate as stabilizer	<0.001% NADH oxidase <10% catalase	Toyobo SAO-311
Bacillus species (optimum pH = 9.0, pI = 4.7, MW = 40,000 Da [gel filtration]; K_M = 50 mM [sarcosine]; stable pH 6-11 [37°C, 1 hr], T < 40°C [pH 7.5, 10 min])				
30-50 U/mg solid	1 unit oxidizes 1 μmol sarcosine to Gly/min at pH 8.0, 37°C	Lyophilized containing 95% protein		Asahi SOXG T-46
>10 U/mg protein; >10 U/mg solid	1 unit oxidizes 1.0 μmol sarcosine/min at pH 8.0, 37°C	Lyophilized		Genzyme 1671
25-50 U/mg solid	1 unit forms 1.0 μmol formaldehyde from sarcosine/min at pH 8.3, 37°C	Lyophilized		Sigma S7897
Corynebacterium species				
5-10 U/mg protein	1 unit forms 1.0 μmol formaldehyde from sarcosine/min at pH 8.3, 37°C	Lyophilized containing 50% protein; balance primarily PO_4 buffer and lactose		Sigma S7759

1.5.3.1 Sarcosine Oxidase *continued*

SPECIFIC ACTIVITY	UNITS DEFINITION	PREPARATION FORM	ADDITIONAL ACTIVITIES	SUPPLIER CATALOG NO.
Microbial (optimum pH = 7.5-8.5, pI = 4.9, T = 40-50°C, MW = 65,000 Da [gel filtration]; K_M = 2.8 *mM* [sarcosine]; stable at pH 6.5-9.0 [25°C, 24 hr], T < 55°C [pH 7.5, 10 min])				
≥8 U/mg solid	1 unit forms 1 μmol H_2O_2 (0.5 μmol quinoneimine dye)/min at pH 8.0, 37°C	Lyophilized containing K gluconate as stabilizer	<1% catalase	Toyobo SAO-321
***Pseudomonas* species**				
2-4 U/mg	1 unit produces 1 μmol formaldehyde from sarcosine/min at pH 8.3, 37°C	Lyophilized		Fluka 84580
5 U/mg protein	1 unit forms 1.0 μmol formaldehyde from sarcosine/min at pH 8.3, 37°C	Lyophilized containing 30% protein		Sigma S5896

1.5.99.1 Sarcosine Dehydrogenase

REACTION CATALYZED

Sarcosine + acceptor + H_2O ↔ Glycine + formaldehyde + reduced acceptor

SYSTEMATIC NAME

Sarcosine:(acceptor) oxidoreductase (demethylating)

REACTANTS

Sarcosine, acceptor, H_2O, glycine, formaldehyde, reduced acceptor

APPLICATIONS

- Determination of sarcosine, creatinine and creatine, with creatinine amidohydrolase and creatine amidinohydrolase

NOTES

- Classified as an oxidoreductase acting on the CH-NH group of donors with other acceptors
- A flavoprotein (FMN)
- Inhibited by PCMB, Cu^{2+}, Hg^{2+} and Ag^+

SPECIFIC ACTIVITY	UNITS DEFINITION	PREPARATION FORM	ADDITIONAL ACTIVITIES	SUPPLIER CATALOG NO.
***Pseudomonas* species** (optimum pH = 8.8, pI = 5.9, T = 45°C, MW = 170,000 Da [4 subunits/molecule and 1 mol FMN/subunit]; K_M = 10 *mM* [sarcosine], 0.38 *mM* [PMS]; stable pH 7.0-8.0 [30°C, 3.5 hr], T < 40°C [pH 7.6, 30 min])				
0.5-1.5 U/mg protein	1 unit converts 1 μmol sarcosine to Gly and formaldehyde/min at pH 7.5, 37°C	Lyophilized containing 30% protein; balance 60% sucrose, 10% KPO_4 buffer salts and trace EDTA		ICN 152046
0.5-1.5 U/mg protein	1 unit converts 1.0 μmol sarcosine to Gly and formaldehyde/min at pH 7.5, 37°C	Lyophilized containing 30% protein; balance 60% sucrose, 10% KPO_4 buffer salts, trace EDTA		Sigma S4384
≥0.15 U/mg solid	1 unit forms 0.5 μmol diformazan/min at pH 7, 37°C	Lyophilized containing 80% sucrose, lactose, EDTA, nonionic detergents as stabilizers	<0.1% NADH oxidase, formaldehyde dehydrogenase	Toyobo SAD-301

NADPH-Ferrihemoprotein Reductase

1.6.2.4

REACTION CATALYZED
  NADPH + 2 ferricytochrome ↔ NADP$^+$ + 2 ferrocytochrome

SYSTEMATIC NAME
  NADPH:ferricytochrome oxidoreductase

SYNONYMS
  NADP-cytochrome reductase, TPNH$_2$-cytochrome c reductase, ferrihemoprotein P-450 reductase, NADPH P-450 oxidoreductase

REACTANTS
  NADPH, ferricytochrome, NADP$^+$, ferrocytochrome, cytochrome c, cytochrome b$_5$

APPLICATIONS
- Reconstituted P450-dependent biotransformation reactions
- *In vitro* drug metabolism and toxicology research

NOTES
- Classified as an oxidoreductase acting on NADH or NADPH with NAD$^+$ or NADP$^+$ as acceptor
- A flavoprotein (1 mol each of FMN and FAD)
- Catalyzes reductions of heme-thiolate-dependent, unspecific monooxygenases (EC 1.14.14.1)
- Part of the microsomal hydroxylating system

SPECIFIC ACTIVITY	UNITS DEFINITION	PREPARATION FORM	ADDITIONAL ACTIVITIES	SUPPLIER CATALOG NO.
Rat, recombinant expressed in *E. coli*				
45-70 μmol/min/mg protein; 1-5 mg/mL	45-70 μmol cytochrome c reduced/min/mg protein	>90% purity by Coomassie-stained SDS gel		PanVera P2196 P2197

Glutathione Reductase (NADPH)

1.6.4.2

REACTION CATALYZED
  NADPH + oxidized glutathione ↔ NADP$^+$ + 2 glutathione

SYSTEMATIC NAME
  NADPH:oxidized-glutathione oxidoreductase

SYNONYMS
  GR

REACTANTS
  NADPH, oxidized glutathione, NADP$^+$, glutathione

NOTES
- Classified as an oxidoreductase acting on NADH or NADPH with a disulfide as acceptor
- A dimeric flavoprotein (FAD)
- Activity is dependent on a redox-active disulfide in each active center
- Activated by phosphate
- Inhibited by KCl

1.6.4.2 Glutathione Reductase (NADPH) continued

SPECIFIC ACTIVITY	UNITS DEFINITION	PREPARATION FORM	ADDITIONAL ACTIVITIES	SUPPLIER CATALOG NO.
Bovine intestinal mucosa				
50-100 U/mg protein	1 unit reduces 1.0 μmol GSSG/min at pH 7.6, 25°C	Lyophilized containing 85% protein; balance citrate buffer salts, pH 7.0	2-10 U/mg protein CoA-GR	Sigma G1762
Spinach				
30-100 U/mg protein	1 unit reduces 1.0 μmol GSSG/min at pH 7.6, 25°C	Suspension in 3.6 M $(NH_4)_2SO_4$ and 10 mM KPO_4, pH 7.0	<0.5 U/mg protein CoA-GR	Sigma G3011
Wheat germ				
0.08 U/mg protein	1 unit reduces 1.0 μmol GSSG/min at pH 7.6, 25°C	Crude		Sigma G6004
Yeast				
120 U/mg	1 unit converts 1 μmol GSSG to product/min at 25°C	Suspension in 3.2 M $(NH_4)_2SO_4$ and BSA as stabilizer, pH 6	<0.01% NADPH oxidase, 6PGDH <0.1% G6PDH	Boehringer 105678 697788
≥120 IU/mg protein	1 IU transforms 1 μmol substrate/min under standard IUB conditions at 25°C	50% glycerol solution	<0.01% NADPH oxidase, 6PGDH <0.1% G6PDH	OYC
Yeast, bakers (MW = 118,000-124,000 Da)				
150 U/mg protein	1 unit reduces 10 mM GSSG/min at pH 7.6, 25°C	Lyophilized containing 90% protein		Calzyme 191A0001
200 U/mg protein	1 unit reduces 1 μmol GSSG/min at pH 7.6, 25°C	Aqueous suspension in 3.6 M $(NH_4)_2SO_4$, pH 7.0	<0.01% 6PGDH, Glc6PDH, ICDH, LD, NADPH oxidase 15 U/mg protein CoA-GR-activity (CoASSG, pH 5.5, 25°C)	Fluka 49755
150 U/mg protein	1 unit reduces 1 μmol GSSG/min at pH 7.6, 25°C	Aqueous suspension in 3.6 M $(NH_4)_2SO_4$, pH 7.0	<0.05% 6PGDH, Glc6PDH, ICDH, LD, NADPH oxidase 15 U/mg protein CoA-GR-activity (CoASSG, pH 5.5, 25°C)	Fluka 49756
120 U/mg	1 unit oxidizes 1 μmol GSH/min at pH 5.5, 25°C	Highly purified; crystallized; suspension in 3.2 M $(NH_4)_2SO_4$, pH 6.0	No detectable NADPH-oxidase, 6PGDH	ICN 100689
100-200 U/mg protein	1 unit reduces 1.0 μmol GSSG/min at pH 7.6, 25°C	Highly purified; suspension in 3.6 M $(NH_4)_2SO_4$, pH 7.0	<0.01% G6PDH, 6PGDH, NADPH oxidase <0.1% LD 10-30 U/mg protein CoA-GR	Sigma G4751
100-300 U/mg protein	1 unit reduces 1.0 μmol GSSG/min at pH 7.6, 25°C	Affinity chromatographically purified, suspension in 3.6 M $(NH_4)_2SO_4$, pH 7.0	<0.01% G6PDH, 6PGDH, NADPH oxidase <0.1% LD 10-30 U/mg protein CoA-GR	Sigma G4759

Nitrate Reductase (NADH)

1.6.6.1

REACTION CATALYZED
  NADH + nitrate ↔ NAD$^+$ + nitrite + H$_2$O

SYSTEMATIC NAME
  NADH:nitrate oxidoreductase

SYNONYMS
  Assimilatory nitrate reductase, NT

REACTANTS
  NADH, nitrate, NAD$^+$, nitrite

NOTES
- Classified as an oxidoreductase acting on NADH or NADPH with a nitrogenous group as acceptor
- A flavoprotein (FAD or FMN)
- Contains molybdenum

SPECIFIC ACTIVITY	UNITS DEFINITION	PREPARATION FORM	ADDITIONAL ACTIVITIES	SUPPLIER CATALOG NO.
Corn seedling				
5 U/mL	1 unit reduces 1.0 μmol nitrate to nitrite/min at pH 7.3, 30°C in NADH system	Solution containing 0.01 M Gly, 50 mM MOPS, 20% glycerol, 1 mg/mL BSA, pH 7.2		sigma N2397

Nitrate Reductase [NAD(P)H]

1.6.6.2

REACTION CATALYZED
  NAD(P)H + nitrate ↔ NAD(P)$^+$ + nitrite + H$_2$O

SYSTEMATIC NAME
  NAD(P)H:nitrate oxidoreductase

SYNONYMS
  Assimilatory nitrate reductase

REACTANTS
  NAD(P)H, nitrate, NAD(P)$^+$, nitrite, H$_2$O

NOTES
- Classified as an oxidoreductase acting on NADH or NADPH with a nitrogenous group as acceptor
- A flavoprotein (FAD or FMN)

SPECIFIC ACTIVITY	UNITS DEFINITION	PREPARATION FORM	ADDITIONAL ACTIVITIES	SUPPLIER CATALOG NO.
***Aspergillus* species**				
10 U/mg protein; 0.4 U/mg solid	1 unit converts 1 μmol nitrate to product/min at 25°C	Lyophilized	<0.15% nitrite reductase <0.5% NADPH oxidase <0.8% NAD(P)H dependent ADH	Boehringer 981249
300 U/g solid	1 unit reduces 1.0 μmol nitrate/min at pH 7.5, 25°C in presence of β-NADPH	Lyophilized		sigma N7265

1.6.8.1 NAD(P)H Dehydrogenase (FMN)

REACTION CATALYZED
  NAD(P)H + FMN ↔ NAD(P)$^+$ + FMNH$_2$

SYSTEMATIC NAME
  NAD(P)H:FMN oxidoreductase

SYNONYMS
  FMN reductase

REACTANTS
  NAD(P)H, FMN, riboflavin, FAD, NAD(P)$^+$, FMNH$_2$

NOTES
- Classified as an oxidoreductase acting on NADH or NADPH with a flavin as acceptor
- Luminescent bacterial enzyme also slowly reduces riboflavin and FAD

SPECIFIC ACTIVITY	UNITS DEFINITION	PREPARATION FORM	ADDITIONAL ACTIVITIES	SUPPLIER CATALOG No.
Photobacterium fischeri				
100 U/mg protein; 7 U/mg solid	1 unit converts 1 μmol FMN to product/min at pH 7.0, 25°C with NADH as coenzyme	Lyophilized containing BSA as stabilizer		Boehringer 476480

1.6.99.1 NADPH Dehydrogenase

REACTION CATALYZED
  NADPH + acceptor ↔ NADP$^+$ + reduced acceptor

SYSTEMATIC NAME
  NADPH:(acceptor) oxidoreductase

SYNONYMS
  NADPH diaphorase, diaphorase, diaphorase I, NADPH-FMN oxidoreductase, DI

REACTANTS
  NADPH, acceptor, NADP$^+$, reduced acceptor, dichlorophenol-indophenol (DCPIP)

APPLICATIONS
- Colorimetric determination of dehydrogenases and ethanol, when coupled with dyes which accept hydrogen from NAD(P)H

NOTES
- Classified as an oxidoreductase acting on NADH or NADPH with other acceptors
- Ubiquitous flavoprotein (FAD in plants, FMN in yeast)
- Catalyzes the reduction of various dyes which accept hydrogen from the reduced form of di- and triphosphopyridine nucleotides
- Activated by K$^+$, Na$^+$, NH$_3$, FMN, FAD and non-ionic detergents
- Stabilized by BSA and sugars
- Inactivated by N-ethylmaleimide (<5 mM)
- Basic solution and 0.5% BSA are required for full activity
- Oxygen and cytochrome c are not substrates

SPECIFIC ACTIVITY	UNITS DEFINITION	PREPARATION FORM	ADDITIONAL ACTIVITIES	SUPPLIER CATALOG NO.
Bacillus megaterium (optimum pH = 7.0-9.0; K_M = 16 μM [NADH], 0.29 mM [NADPH])				
30-60 U/mg solid	1 unit oxidizes 1 μmol NADH to NAD^+/min at pH 8, 37°C	Lyophilized containing 90% protein optimum pH = 7.5-8.5, pI = 4.2, MW = 58,000 Da; stable pH 6-9 [50°C, 10 min], T < 50°C [pH 8, 10 min]	No detectable MK <6% in relative activity NADH oxidase	Asahi DI T-06
10-20 U/mg solid	1 unit oxidizes 1 μmol NADPH to $NADP^+$/min at pH 8, 30°C	Lyophilized containing 80% protein optimum pH = 7-9, pI = 3.0, MW = 48,000 Da; stable pH 6.5-9 [37°C, 1 hr], T < 60°C [pH 7.5, 10 min]		Asahi DIP T-10
Bacillus species (optimum pH = 8.0-9.0, pI = 4.38, T = 65°C, MW = 82,000 Da [gel filtration]; K_M = 35 μM [NADH], 0.53 mM [NADPH]; stable pH 6-9.5, T < 70°C [pH 7.5, 1 hr])				
180-340 U/mg solid	1 unit oxidizes 1 μmol NADH to NAD^+/min at pH 8.0, 37°C	Lyophilized		Asahi DIS T-44
Bacillus stearothermophilus (optimum pH = 8.0, pI = 4.7, T = 70°C, MW = 30,000 Da; stable pH > 7.5, T < 50°C)				
>1000 U/mg protein	1 unit reduces 1 μmol DCPIP /min at 30°C	Lyophilized	<0.01% AK, NADH oxidase	Unitika
>1000 U/mg protein	1 unit reduces 1 μmol DCPIP /min at 30°C	50% glycerol solution	<0.01% AK, NADH oxidase	Unitika
Clostridium kluyveri (optimum pH = 7.5-8.5, T = 25°C, MW = 24,000 Da; relative activity: 25°C = 1.0, 30°C = 1.7, 37°C = 2.8)				
≥96 U/mg	1 unit reduces 1 μmol thiazolyl blue tetrazolium bromide (MTT) to MTT formazen/min at 30°C	Lyophilized		AMRESCO 0273 9693
>5.0 U/mg solid (INT assay); >70.0 U/mg solid (DCPIP assay)	1 unit oxidizes 1.0 μmol NADH/min at pH 8.0, 25°C	Lyophilized	<0.7% (of INT activity) CK, AK <1.3% (of INT activity) ATPase <2.5% (of INT activity) GR	Genzyme 1151
600 U/vial		Salt-free; lyophilized	No detectable Py	ICN 100592
25 U/mg; 1800 U/vial	1 unit reduces 1.0 μmol DCPIP/min at pH 8.5, 20°C	Salt-free; lyophilized	No detectable Py	ICN 150843
≥30 U/mg solid	1 unit decreases A_{600} by 1.0/min at pH 7.5, 25°C	Salt-free; lyophilized	No detectable Py	Worthington LS04327 LS04326
≥25 U/mg solid	1 unit reduces 1 μmol DCPIP/min at pH 8.5, 25°C	Salt-free; lyophilized	No detectable Py	Worthington LS04330 LS04333

1.6.99.1 NADPH Dehydrogenase continued

SPECIFIC ACTIVITY	UNITS DEFINITION	PREPARATION FORM	ADDITIONAL ACTIVITIES	SUPPLIER CATALOG NO.
Clostridium species (optimum pH = 8.5, T = 50°C, MW = 24,000 Da; K_M = 20 μM [NADH], 6 μM [NADPH]; stable pH 7.5 [30°C, 3 hr], T < 30°C [pH 7.5, 30 min])				
≥30 U/mg solid	1 unit decreases DCPIP absorbance by 1.0/min at pH 7.5, 25°C	Lyophilized containing 15% FMN and NAD(P)H as stabilizers	<0.5% MK, NAD(P)H oxidase	Toyobo DAD-301
Porcine heart				
80 U/mg protein (lipoamide as substrate, 20°C); 10 U/mg (6,8-dithio-*n*-octanoate as substrate)		Lyophilized		ICN 150844
Vibrio harveyi (optimum pH = 5.5, T = 55°C, MW = 60,000 Da; K_M = 22.2 μM [NADPH, FMN]; stable pH 6.5-9.5 [25°C, 6 hr], T < 40°C [pH 7.0, 10 min])				
0.01-0.02 U/mg solid	1 unit reduces 1.0 μmol FMN to FMNH$_2$/min at pH 7.0, 30°C	Lyophilized containing BSA and raffinose		Sigma N4392
≥10 mU/mg solid	1 unit forms 1 μmol FMNH$_2$/min at pH 7.0, 30°C	Lyophilized containing 80% BSA and sugars as stabilizers		Toyobo NPR-201

1.6.99.2 NAD(P)H Dehydrogenase (Quinone)

REACTION CATALYZED
   NAD(P)H + acceptor ↔ NAD(P)$^+$ + reduced acceptor

SYSTEMATIC NAME
   NAD(P)H:(quinone-acceptor) oxidoreductase

SYNONYMS
   Menadione reductase, phylloquinone reductase, quinone reductase

REACTANTS
   NAD(P)H, acceptor, NAD(P)$^+$, reduced acceptor

NOTES
- Classified as an oxidoreductase acting on NADH or NADPH with other acceptors
- A flavoprotein
- Inhibited by dicoumarol

See Chapter 5. Multiple Enzyme Preparations

NADH Dehydrogenase

1.6.99.3

REACTION CATALYZED
NADH + acceptor ↔ NAD^+ + reduced acceptor

SYSTEMATIC NAME
NADH:(acceptor) oxidoreductase

SYNONYMS
Cytochrome c reductase, type 1 dehydrogenase, NADH-FMN oxidoreductase

SUBSTRATES
NADH, acceptor, NAD^+, reduced acceptor

APPLICATIONS
- Determination of NADH, with luciferase

NOTES
- Classified as an oxidoreductase acting on NADH or NADPH with other acceptors
- A flavoprotein containing iron-sulfur centers
- Cytochrome c may act as acceptor in certain preparations
- Present in the mitochondrial complex of NADH dehydrogenase (ubiquinone) (EC 1.6.5.3)
- Stabilized by BSA and sugars

SPECIFIC ACTIVITY	UNITS DEFINITION	PREPARATION FORM	ADDITIONAL ACTIVITIES	SUPPLIER CATALOG NO.
Porcine heart				
1-3 U/mg protein	1 unit reduces 1.0 μmol oxidized cytochrome c/min at pH 8.5, 25°C	Crude; lyophilized containing 50% protein; balance KPO_4, pH 7		Sigma C3381
Vibrio harveyi (optimum pH = 5.5-7.0, T = 55°C, MW = 20,000 Da; K_M = 23.4 μM [NADH], 2.24 μM [FMN]; stable pH 6.5-8.0 [25°C, 6 hr], T < 45°C [pH 7.0, 10 min])				
≥10 mU/mg solid	1 unit forms 1 μmol $FMNH_2$/min at pH 7.0, 30°C	Lyophilized containing 80% stabilizers		ICN 153485
0.01-0.05 U/mg solid	1 unit reduces 1.0 μmol FMN/min at pH 7.0, 30°C in the presence of β-NADH	Lyophilized containing BSA and sugar as stabilizers		Sigma N3517
≥10 mU/mg solid	1 unit forms 1 μmol $FMNH_2$/min at pH 7.0, 30°C	Lyophilized containing 80% BSA and sugars as stabilizers		Toyobo NHR-201

1.6.99.7 Dihydropteridine Reductase

REACTION CATALYZED
  NAD(P)H + 6,7-dihydropteridine ↔ NAD(P)$^+$ + 5,6,7,8-tetrahydropteridine

SYSTEMATIC NAME
  NAD(P)H:6,7-dihydropteridine oxidoreductase

REACTANTS
  NAD(P)H, 6,7-dihydropteridine, NAD(P)$^+$, 5,6,7,8-tetrahydropteridine

NOTES
- Classified as an oxidoreductase acting on NADH or NADPH with other acceptors
- The substrate is the quinoid form of dihydropteridine
- Different from dihydrofolate reductase (EC 1.5.1.3)

SPECIFIC ACTIVITY	UNITS DEFINITION	PREPARATION FORM	ADDITIONAL ACTIVITIES	SUPPLIER CATALOG NO.
Ovine liver				
30-75 U/mg protein	1 unit oxidizes 1.0 μmol NADH to NAD with 6,7-dimethyldihydropterine (quinonoid isomer)/min at pH 7.2, 25°C	Solution containing 60% glycerol, 10 mM KPO$_4$, pH 7.0; may contain up to 0.5 M (NH$_4$)$_2$SO$_4$		Sigma D6888

1.7.3.3 Urate Oxidase

REACTION CATALYZED
  Urate + O$_2$ + H$_2$O ↔ Allantoin + H$_2$O$_2$ + CO$_2$ ↔ Unknown products

SYSTEMATIC NAME
  Urate:oxygen oxidoreductase

SYNONYMS
  Uricase, UOD

REACTANTS
  Urate, O$_2$, H$_2$O, allantoin, H$_2$O$_2$, CO$_2$

APPLICATIONS
- Determination of uric acid in biological fluids, with HRP or catalase and ALDH
- Spectrophotometric determination of purines and purine nucleosides

NOTES
- Classified as an oxidoreductase acting on other nitrogenous compounds as donors with oxygen as acceptor
- The termination of purine catabolism in all mammals except for man, higher apes and the Dalmatian dog
- A copper protein highly specific for urate
- Initial products decompose to form allantoin
- Activated by high concentrations of acetate and phosphate
- Stabilized by borate, EDTA and nonionic detergents
- Inhibited by excess urate, cyanide, Cu^{2+}, Hg^{2+}, Ag$^+$ and urate analogs

Urate Oxidase continued

SPECIFIC ACTIVITY	UNITS DEFINITION	PREPARATION FORM	ADDITIONAL ACTIVITIES	SUPPLIER CATALOG NO.
Arthrobacter globiformis (optimum pH = 8.5-9.5 [8.9-9.2, borate buffer], pI = 4.64, MW = 56,000 Da [Sephadex G-100]; K_M = 0.1 mM [uric acid]; stable pH 6-9.5 [37°C, 1 hr], T < 55°C [pH 8.0, 10 min])				
10-30 U/mg solid	1 unit oxidizes 1 μmol urate to allantoin/min at pH 9.0, 25°C	Lyophilized containing 90% protein		Asahi UODN T-34
>10 U/mg protein; >10 U/mg solid	1 unit oxidizes 1.0 μmol uric acid/min at pH 9.1, 25°C	Lyophilized		Genzyme 1721
15-30 U/mg protein	1 unit converts 1.0 μmol uric acid to allantoin/min at pH 9.0, 25°C	Lyophilized		Sigma U7128
Arthrobacter protophormiae				
3-10 U/mg protein	1 unit converts 1.0 μmol uric acid to allantoin/min at pH 8.5, 25°C	Lyophilized		Sigma U4004
Bacillus fastidiosus (optimum pH = 8.9-9.2, T = 40°C, MW = 93,000 Da; relative activity: 25°C = 1.0, 30°C = 1.5, 37°C = 2.0; stable pH 6-11, T < 40°C)				
>10 U/mg protein; >5 U/mg solid	1 unit oxidizes 1.0 μmol uric acid/min at pH 9.1, 37°C	Lyophilized	<0.005% ChO, GO <1% catalase	Genzyme 1701
9 U/mg	1 unit oxidizes 1 μmol uric acid/min at pH 9.1, 37°C	Lyophilized		Fluka 94310
>10 U/mg protein; >5 U/mg solid	1 unit oxidizes 1 μmol uric acid/min at pH 9.1, 37°C	Lyophilized	<0.005% ChO, GO <1% catalase	Randox UA 864L
Bacillus species (optimum pH = 8.5, pI = 4.7, T = 45°C, MW = 150,000 Da [4 subunits/molecule]; K_M = 13.6 μM [uric acid]; stable pH 6.0-9.5 [25°C, 20 hr], T < 60°C [pH 8.0, 10 min])				
≥1.5 U/mg solid	1 unit oxidizes 1 μmol uric acid/min at pH 8.0, 37°C	Lyophilized containing EDTA, borate, nonionic detergents as stabilizers	<1% catalase	Toyobo UAO-211
4.0 U/mg solid	1 unit oxidizes 1 μmol uric acid/min at pH 8.5, 25°C	Lyophilized		ICN 191498
Bovine kidney				
15 U/g protein	1 unit converts 1.0 μmol uric acid to allantoin/min at pH 8.5, 25°C	Powder		Sigma U3500
4.5 U/g	1 unit oxidizes 1 μmol uric acid/min at pH 8.5, 25°C	Powder		ICN 101202
Candida species (optimum pH = 8.5, pI = 5.4, T = 40°C, MW = 120,000 Da [2 subunits/molecule]; K_M = 25 μM [uric acid]; stable pH 7.0-11.0 [25°C, 20 hr], T < 50°C [pH 8.5, 10 min])				
≥4 U/mg solid	1 unit oxidizes 1 μmol uric acid/min at pH 8.5, 25°C	Lyophilized containing 20% EDTA, borate, nonionic detergents as stabilizers	<1% catalase	Toyobo UAO-201

SPECIFIC ACTIVITY	UNITS DEFINITION	PREPARATION FORM	ADDITIONAL ACTIVITIES	SUPPLIER CATALOG NO.
Candida species continued				
3.0-6.5 U/mg		Lyophilized		Wako 218-00721
Candida utilis (optimum pH = 8.5 [8-9, borate], pI = 5.6, T = 40°C, MW = 70,000-76,000 Da; stable at pH 8.5-9.0 and T < 35°C)				
0.9-2.0 IU/mg protein; 0.2-0.4 IU/mg powder	1 unit oxidizes 1 μmol uric acid/min at 25°C	Lyophilized		Beckman 682087
≥1000 IU/g protein	1 unit oxidizes 1 μmol uric acid/min at 25°C	Lyophilized	ADH <0.15% uricase	Beckman 684169
>5 U/mg protein; >4 U/mg solid	1 unit oxidizes 1.0 μmol uric acid/min at pH 9.1, 25°C	Lyophilized	<1% catalase	Genzyme 1711
≥3.0 U/mg protein	1 unit oxidizes 1 μmol uric acid/min at pH 8.5, 25°C			ICN 101203
≥6 IU/mg protein	1 IU transforms 1 μmol substrate/min under standard IUB conditions at 25°C	Lyophilized		OYC
≥4 U/mg	1 unit oxidizes 1 μmol uric acid/min at pH 8.5, 25°C	Lyophilized		Seikagaku 120451-1 120451-2
3-10 U/mg protein	1 unit converts 1.0 μmol uric acid to allantoin/min at pH 8.5, 25°C	50% glycerol solution		Sigma U1377
20 U/mg protein	1 unit converts 1.0 μmol uric acid to allantoin/min at pH 8.5, 25°C	Affinity purified; lyophilized containing 70% protein; balance primarily Tris buffer salts		Sigma U1878
3-10 U/mg protein	1 unit converts 1.0 μmol uric acid to allantoin/min at pH 8.5, 25°C	Partially purified; lyophilized containing 75% protein		Sigma U8500
≥2 U/mg solid	1 unit converts 1 μmol uric acid to allantoin/min at pH 8.5, 25°C	Lyophilized		Worthington LS03857 LS03855
Corynebacterium species				
50-150 U/g solid	1 unit converts 1.0 μmol uric acid to allantoin/min at pH 8.5, 25°C	Lyophilized		Sigma U8002
Microbial (optimum pH = 7.0-8.0, 9.0-9.5, T = 25°C)				
>3.5 U/mg protein	1 unit hydrolyzes 1 μmol uric acid to allantoin/min at pH 9.1, 25°C	Lyophilized		GDS Tech U-010
>5 U/mg protein	1 unit hydrolyzes 1 μmol uric acid to allantoin/min at pH 8.5, 37°C	Lyophilized		GDS Tech U-020

SPECIFIC ACTIVITY	UNITS DEFINITION	PREPARATION FORM	ADDITIONAL ACTIVITIES	SUPPLIER CATALOG NO.
Porcine liver (optimum pH = 9.0, MW = 125,000 Da)				
≥1000 U/g protein; 500 U/g solid	1 unit oxidizes 1 μmol uric acid to allantoin/min at pH 8.5, 25°C	Lyophilized containing Na_2CO_3		Biozyme U3
≥1000 U/g protein	1 unit oxidizes 1 μmol uric acid to allantoin/min at pH 8.5, 25°C	50% glycerol solution, pH 9.3		Biozyme U3G
4000-7000 U/g protein; 1000 U/g solid	1 unit oxidizes 1 μmol uric acid to allantoin/min at pH 8.5, 25°C	Highly purified; lyophilized containing Na_2CO_3		Biozyme U4
4000-7000 U/g protein	1 unit oxidizes 1 μmol uric acid to allantoin/min at pH 8.5, 25°C	Highly purified; 50% glycerol solution, pH 9.3		Biozyme U4G
9 U/mg at 25°C; 22 U/mg at 37°C	1 unit converts 1 μmol urate to product/min at O_2 saturation	Solution containing 50% glycerol, 50 mM Gly, 0.13 M Na_2CO_3, pH 10.2		Boehringer 127469
1000 U/mg protein	1 unit converts 1 μmol uric acid to allantoin/min at pH 8.5, 25°C	Lyophilized containing 90% protein and Na_2CO_3		Calzyme 117A1000
0.2-0.4 U/mL	1 unit converts 1.0 μmol uric acid to allantoin/min at pH 8.5, 25°C	Purified; suspension in $(NH_4)_2SO_4$		Sigma 292-8
2 U/mL; 4000-8000 U/g protein	1 unit converts 1.0 μmol uric acid to allantoin/min at pH 8.5, 25°C	Purified; suspension in 2.0 M $(NH_4)_2SO_4$		Sigma U3250
5-10 U/mg protein	1 unit converts 1.0 μmol uric acid to allantoin/min at pH 8.5, 25°C	Solution containing 50% glycerol, 0.13 M Na_2CO_3, 5 mM Gly buffer, pH 10.2		Sigma U3377
20 U/g protein	1 unit converts 1.0 μmol uric acid to allantoin/min at pH 8.5, 25°C	Powder		Sigma U9375
≥600 U/mg protein	1 unit converts 1 μmol uric acid to allantoin/min at pH 8.5, 25°C	Highly purified; suspension in 0.4 M $(NH_4)_2SO_4$		Worthington LS03826 LS03827 LS03838
Yeast				
6 U/mg protein	1 unit converts 1 μmol uric acid to allantoin/min at pH 8.5, 25°C	Lyophilized containing 90% protein		Calzyme 117A0006

1.8.1.4 Dihydrolipoamide Dehydrogenase

REACTION CATALYZED

Dihydrolipoamide + NAD^+ ↔ Lipoamide + NADH

SYSTEMATIC NAME

Dihydrolipoamide:NAD^+ oxidoreductase

SYNONYMS

Diaphorase, Straub diaphorase, diaphorase II, lipoamide reductase (NADH), lipoyl dehydrogenase, lipoamide dehydrogenase, LD

REACTANTS

Dihydrolipoamide, NAD^+, lipoamide, NADH

APPLICATIONS

- Determination of NADH- and NADPH-dependent dehydrogenases
- Determination of L- and D-lactic acid
- Determination of total bile acids
- Enzymatic amplification systems
- Conversion of dehydrogenase to colorimetric reactions

NOTES

- Classified as an oxidoreductase acting on a sulfur group of donors with NAD^+ or $NADP^+$ as acceptor
- A flavoprotein (FAD)
- A component of the multienzyme pyruvate and 2-oxoglutarate dehydrogenase complexes
- The name diaphorase has been loosely applied to several enzymes which oxidize β-NAD(P)H in the presence of an electron acceptor, such as methylene blue or 2,6-dichlorophenol-indophenol
- Activity with oxygen is 5% that with 6,8-dithio-n-octanate
- Pig heart enzyme has native diaphorase as well as lipoic and lipoamide dehydrogenase activity. Diaphorase has been reported to be a denatured lipoamide dehydrogenase. Preincubation of enzyme with Cu^{2+} reduces lipoamide dehydrogenase activity and proportionately increases β-NADH diaphorase activity

SPECIFIC ACTIVITY	UNITS DEFINITION	PREPARATION FORM	ADDITIONAL ACTIVITIES	SUPPLIER CATALOG NO.
Bacillus stearothermophilus (optimum pH = 6.5, pI = 4.8, MW = 110,000 Da [50,000/subunit]; K_M = 2.0 mM [lipoate], 10 μM [NADH]; stable T < 60°C)				
>180 U/mg protein	1 unit forms 1 μmol NAD^+/min at 30°C	Lyophilized	<0.01% AK <0.1% NADH oxidase	Unitika
>180 U/mg protein	1 unit forms 1 μmol NAD^+/min at 30°C	50% glycerol solution	<0.01% AK <0.1% NADH oxidase	Unitika
Bovine heart (optimum pH = 7.0)				
>15 U/mg protein; >150 U/mg solid	1 unit oxidizes 1.0 μmol NADH/min at 25°C	Suspension in $(NH_4)_2SO_4$	3 U/mg protein NADH	Genzyme 1492

Dihydrolipoamide Dehydrogenase continued   1.8.1.4

SPECIFIC ACTIVITY	UNITS DEFINITION	PREPARATION FORM	ADDITIONAL ACTIVITIES	SUPPLIER CATALOG NO.
Bovine intestinal mucosa				
100-200 U/mg protein; 3-15 U/mg protein lipoic DH; 2-10 U/mg protein NADH	1 unit reduces 1.0 μmol DL-lipoamide to DL-dihydrolipoamide/min at pH 6.5, 25°C	Suspension in 3.2 M $(NH_4)_2SO_4$, pH 6	<1 U/mg protein NADH oxidase, NADPH diaphorase, NADPH oxidase, NADPH LD	Sigma L6777
Clostridium kluyveri				
6 U/mg	1 unit reduces 1 μmol INT/min at pH 8.0, 25°C with NADH as second substrate	Crystallized		Fluka 33460
4-15 U/mg protein using DCPIP	1 unit oxidizes 1.0 μmol β-NADH/min at pH 7.5, 25°C	Solution containing 50% glycerol, Tris buffer, stabilizers, FMN, pH 7.6		Sigma D1404
5-20 U/mg protein using DCPIP	1 unit oxidizes 1.0 μmol β-NADH/min at pH 7.5, 25°C	Lyophilized containing 80% protein		Sigma D2381
5-20 μM U/mg protein using DCPIP	1 unit oxidizes 1.0 μmol β-NADH/min at pH 7.5, 25°C	Lyophilized		Sigma D5540
Microbial				
15 U/mg solid	1 unit converts 1 μmol NADH and INT to product/min at pH 8.8, 25°C	Lyophilized		Boehringer 411558
Porcine heart (MW = 110,000-114,000 Da)				
25 U/mg protein	1 unit oxidizes 1 μmol NADH/min at pH 5.65, 25°C	Highly purified; suspension in 70% $(NH_4)_2SO_4$	3 U/mg diaphorase	Biozyme LIP3
≥20 U/mg protein	1 unit oxidizes 1 μmol NADH/min at pH 5.65, 25°C	Highly purified; salt-free; lyophilized	4 U/mg diaphorase	Biozyme LIP3F
80 U/mg protein (lipoamide as substrate); 10 U/mg protein (6,8-dithio-n-octanoate as substrate)	1 unit converts 1 μmol substrate to product/min at 25°C	Lyophilized containing 10 mg protein, 20 mg sucrose, BSA as stabilizer		Boehringer 104353
25 U/mg protein; 50 U/mg protein diaphorase	1 unit oxidizes 1 μmol NADH/min at pH 5.65, 25°C	Suspension in 3.2 M $(NH_4)_2SO_4$ containing 90-95% protein		Calzyme 066B0025
25 U/mg protein	1 unit oxidizes 1 μmol NADH/min at pH 5.65, 25°C	Lyophilized containing 90% protein		Calzyme 153A0025

1.8.1.4 Dihydrolipoamide Dehydrogenase continued

SPECIFIC ACTIVITY	UNITS DEFINITION	PREPARATION FORM	ADDITIONAL ACTIVITIES	SUPPLIER CATALOG NO.
Porcine heart continued				
>5 U/mg protein; >3 U/mg solid	1 unit reduces DCPIP/min at 25°C	Lyophilized		Diagnostic S-060
β-NADH = 2-5 β-NADPH = 0.1 U/mg protein using 2,6-dichlorophenol-indophenol	1 unit oxidizes 1.0 μmol β-NADH/min at pH 7.5, 25°C	Suspension in 3.2 M $(NH_4)_2SO_4$, pH 6	<0.1 U/mg protein β-NADPH oxidase <1 U/mg β-NADH oxidase 100-200 U/gm protein LD using DL-lipoamide	Sigma D3752
100-200 U/mg protein; 5-15 U/mg protein lipoic dehydrogenase; 2-5 U/mg protein NADH	1 unit reduces 1.0 μmol DL-lipoamide to DL-dihydrolipoamide/min at pH 6.5, 25°C	Suspension in 3.2 M $(NH_4)_2SO_4$, pH 6	<1 U/mg protein NADH oxidase, NADPH diaphorase, NADPH oxidase, NADPH LD	Sigma L2002
Yeast, Torula				
25-50 U/mg protein	1 unit reduces 1.0 μmol DL-lipoamide to DL-dihydrolipoamide/min at pH 6.5, 25°C	Lyophilized containing 75% protein; balance EDTA, PO_4, citrate buffer salts	Lipoic dehydrogenase, NADH, NADPH diaphorase, NADH oxidase, NADPH LD, GR	Sigma L1634

1.8.3.1 Sulfite Oxidase

REACTION CATALYZED
$$\text{Sulfite} + O_2 + H_2O \leftrightarrow \text{Sulfate} + H_2O_2$$

SYSTEMATIC NAME
  Sulfite:oxygen oxidoreductase

REACTANTS
  Sulfite, O_2, H_2O, sulfate, H_2O_2

NOTES
- Classified as an oxidoreductase acting on a sulfur group of donors with oxygen as acceptor
- A molybdohemoprotein

Sulfite Oxidase continued   1.8.3.1

SPECIFIC ACTIVITY	UNITS DEFINITION	PREPARATION FORM	ADDITIONAL ACTIVITIES	SUPPLIER CATALOG NO.
Chicken liver				
30-70 U/mg protein	1 unit oxidizes 1.0 μmol sulfite to sulfate during the reduction of cytochrome c/min at pH 8.5, 25°C	Crystallized; suspension in 3.2 M $(NH_4)_2SO_4$, pH 7.5		ICN 156725
20-70 U/mg protein	1 unit oxidizes 1.0 μmol sulfite to sulfate during the reduction of cytochrome c/min at pH 8.5, 25°C	Suspension in 3.2 M $(NH_4)_2SO_4$ containing 1.6 mM molybdic acid, pH 7.5		Sigma S9526

Glutathione Oxidase   1.8.3.3

REACTION CATALYZED

2 Glutathione + O_2 ↔ Oxidized glutathione + H_2O_2

SYSTEMATIC NAME

Glutathione:oxygen oxidoreductase

SYNONYMS

Glutathione sulfhydryl oxidase

REACTANTS

Glutathione, L-cysteine, O_2, oxidized glutathione, H_2O_2

NOTES
- Classified as an oxidoreductase acting on a sulfur group of donors with oxygen as acceptor
- A flavoprotein (FAD)
- Acts slowly on L-cysteine and several other thiols

SPECIFIC ACTIVITY	UNITS DEFINITION	PREPARATION FORM	ADDITIONAL ACTIVITIES	SUPPLIER CATALOG NO.
***Penicillium* species K-6-5 (optimum pH = 7-8, T = 30°C)**				
Relative reaction rates with 10 mM substrate: 100% for L-GSH, 59.1% for L-Cys, 22.8% for DTT, 3.5% for cysteamine	1 unit consumes 1 μmol O_2 /min at pH 7.4, 30°C in the absence of catalase	Lyophilized		Seikagaku 100662-1 100662-2

1.9.3.1 Cytochrome-c Oxidase

REACTION CATALYZED
    4 Ferrocytochrome c + O_2 ↔ 4 Ferricytochrome c + 2 H_2O

SYSTEMATIC NAME
    Ferrocytochrome-c:oxygen oxidoreductase

SYNONYMS
    Cytochrome oxidase, cytochrome a_3, cytochrome aa_3, CO

REACTANTS
    Ferrocytochrome c, O_2, ferricytochrome c, H_2O

NOTES
- Classified as an oxidoreductase acting on a heme group of donors with oxygen as acceptor
- A cytochrome of the *a* type containing copper

SPECIFIC ACTIVITY	UNITS DEFINITION	PREPARATION FORM	ADDITIONAL ACTIVITIES	SUPPLIER CATALOG NO.
Bovine heart				
3.5 U/mg protein; 0.3 U/mg solid	1 unit takes up 1 μmol O_2/min at pH 7.4, 25°C	Lyophilized		Biozyme CO2
25 U/vial	1 unit takes up 1 μmol O_2/min at pH 7.4, 25°C	Lyophilized		Biozyme CO2V
25 U/vial	1 unit causes the uptake of 1 μmol O_2/min at pH 7.4, 25°C	Lyophilized	No detectable cytochrome c	ICN 190317
5 U/mg solid	1 unit reduces 1.0 μmol reduced cytochrome c to oxidized cytochrome c/min at pH 7.0, 37°C using 0.5 μmol O_2	Lyophilized containing BSA and buffer salts		Sigma C5771

1.9.6.1 Nitrate Reductase (Cytochrome)

REACTION CATALYZED
    Ferrocytochrome + nitrate ↔ Ferricytochrome + nitrite

SYSTEMATIC NAME
    Ferrocytochrome:nitrate oxidoreductase

REACTANTS
    Ferrocytochrome, nitrate, ferricytochrome, nitrite

NOTES
- Classified as an oxidoreductase acting on a heme group of donors with a nitrogenous group as acceptor
- Made up of a 3 polypeptide complex (155,000, 64,000 and 19,000 Da); may exist as a dimer (498,000 Da) and a tetramer (1,000,000 Da)
- Contains iron and molybdenum; 4 atoms of Mo and 4 active sites per enzyme
- Inhibited by cyanide (competitive), azide, chelating agents, *o*-phenanthroline and 8-hydroxyquinone
- Not inhibited by chloromercuribenzoate or iodoacetate

Nitrate Reductase (Cytochrome) continued — 1.9.6.1

SPECIFIC ACTIVITY	UNITS DEFINITION	PREPARATION FORM	ADDITIONAL ACTIVITIES	SUPPLIER CATALOG NO.
***Aspergillus* species**				
200-600 U/g solid	1 unit reduces 1.0 μmol nitrate to nitrite/min at pH 7.0, 30°C in a methylviologen system	Lyophilized		ICN 155840
***Escherichia coli* (MW = 498,000 Da)**				
25-50 U/g solid	1 unit reduces 1.0 μmol nitrate to nitrite/min at pH 7.0, 30°C in a methylviologen system	Lyophilized containing 5% protein; balance primarily PO_4 buffer salts		ICN 155841
5 U/g solid	1 unit reduces 1.0 μmol nitrate to nitrite/min at pH 7.0, 30°C in methylviologen system	Lyophilized		Sigma N0519
≥5 U/mg solid	1 unit reduces 1 μmol nitrate to nitrite/min at pH 7.0, 30°C	Lyophilized containing stabilizing salts and buffers		Worthington LS04806 LS04807

Catechol Oxidase — 1.10.3.1

REACTION CATALYZED

2 Catechol + O_2 ↔ 2 1,2-Benzoquinone + 2 H_2O

SYSTEMATIC NAME

1,2-Benzenediol:oxygen oxidoreductase

SYNONYMS

Diphenol oxidase; *o*-diphenolase, phenolase, polyphenol oxidase, tyrosinase

REACTANTS

Catechol, O_2, 1,2-benzoquinone, H_2O

NOTES

- Classified as an oxidoreductase acting on diphenols and related substances as donors with oxygen as acceptor
- A group of copper proteins acting also on a variety of substituted catechols
- Tyrosinase and others catalyze the monophenol monooxygenase (EC 1.14.18.1) reaction

SPECIFIC ACTIVITY	UNITS DEFINITION	PREPARATION FORM	ADDITIONAL ACTIVITIES	SUPPLIER CATALOG NO.
Potato				
300-1000 U/mg protein	1 unit produces a ΔA_{265} of 0.01/min at pH 7.5, 30°C with chlorogenic acid as substrate in 3 mL volume	Lyophilized containing 60% protein; balance KPO_4 buffer salts, DTT, $MgCl_2$, EDTA		Sigma P4788

1.10.3.2 Laccase

REACTION CATALYZED
  4 Benzenediol + O_2 ↔ 4 Benzosemiquinone + 2 H_2O
SYSTEMATIC NAME
  Benzenediol:oxygen oxidoreductase
SYNONYMS
  Urishiol oxidase
REACTANTS
  Benzenediol, o-quinols and p-quinols, aminophenols, phenylenediamine, O_2, benzosemiquinone, H_2O

NOTES
- Classified as an oxidoreductase acting on diphenols and related substances as donors with oxygen as acceptor
- A group of multi-copper enzymes of low specificity
- The semiquinone may react further enzymatically or chemically

SPECIFIC ACTIVITY	UNITS DEFINITION	PREPARATION FORM	ADDITIONAL ACTIVITIES	SUPPLIER CATALOG NO.
Pyricularia oryzae				
50-150 U/mg solid	1 unit causes a ΔA_{530} of 0.001/min at pH 6.5, 30°C in a 3 mL reaction volume using syringaldazine as substrate	Crude		ICN 155158
100-300 U/mg solid	1 unit produces a ΔA_{530} of 0.001/min at pH 6.5, 30°C using syringaldazine as substrate	Crude		Sigma L5510

1.10.3.3 *L-Ascorbate Oxidase*

REACTION CATALYZED
  2 L-Ascorbate + O_2 ↔ 2 Dehydroascorbate + 2 H_2O
SYSTEMATIC NAME
  L-Ascorbate:oxygen oxidoreductase
SYNONYMS
  Ascorbate oxidase, AAO, AO, ASOD, ASOM

REACTANTS
  L-Ascorbate, O_2, 2 dehydroascorbate, H_2O
APPLICATIONS
- Determination of ascorbic acid in foods and biological materials
- Elimination of ascorbic acid interference in clinical analyses

L-Ascorbate Oxidase continued

NOTES

- Classified as an oxidoreductase acting on diphenols and related substances as donors with oxygen as acceptor
- A multi-copper protein
- Stabilized by BSA, borax, basic amino acids and mannitol
- Inhibited by azide, cyanide, Na_2S, NaF, SDS, diethyldithiocarbamate, sodium laurylbenzene, sulfonate, PCMB

SPECIFIC ACTIVITY	UNITS DEFINITION	PREPARATION FORM	ADDITIONAL ACTIVITIES	SUPPLIER CATALOG NO.
Acremonium species (optimum pH = 4.0-4.5, pI = 4.0, MW = 80,000 Da [gel filtration]; K_M [ascorbic acid] = 0.1 mM [pH 7], 0.38 mM [pH 4]; stable pH 6-10 [30°C, 24 hr], T < 50°C [pH 7.0, 10 min])				
200-450 U/mg solid	1 unit oxidizes 1 μmol ascorbate to dehydroascorbate/min at pH 5.6, 30°C	Lyophilized containing 40% protein		Asahi ASOM T-53
Cucumis species (cucumber) (optimum pH = 5.6-5.8, pI = 6.0-8.2, T = 30-40°C, MW = 132,000-140,000 Da; K_M = 0.25 mM [ascorbic acid]; stable pH 7.0-10.0 [25°C, 17 hr], T < 40°C [pH 8.0, 30 min])				
≥230 U/mg	1 unit oxidizes 1 μmol L-ascorbic acid/min at 30°C	Powder		Amano Ascorbate oxidase
200 U/mg solid	1 unit oxidizes 1.0 μmol L-ascorbate to dehydroascorbate/min at pH 5.6, 20°C	Lyophilized		ICN 190075
≥100 IU/mg protein	1 IU transforms 1 μmol substrate/min under standard IUB conditions at 25°C	Lyophilized		OYC
≥200 U/mg solid	1 unit decreases 1 μmol ascorbic acid/min at pH 5.6, 30°C	Lyophilized containing 70% BSA, borax, basic amino acids as stabilizers	<0.02% phosphatase <0.1% catalase	Toyobo ASO-301
Cucurbita pepo medullosa (zucchini squash)				
500 U/mg protein	1 unit oxidizes 1 μmol ascorbic acid/min at pH 5.6, 25°C	Lyophilized containing 20% protein and 80% stabilizers	<0.02% phosphatase <0.1% catalase	Calzyme 061A0500
Cucurbita species (optimum pH = 5.5-6.5, pI = 5-6, T = 45°C, MW = 140,000 Da; K_M = 0.27 mM [ascorbic acid]; stable pH 6.0-10.0 [25°C, 20 hr], T < 45°C [pH 7.0, 30 min])				
2000 U/mg protein; ≥300 U/mg solid	1 unit oxidizes 1 μmol L-ascorbic acid/min at pH 5.6, 30°C	Salt-free; lyophilized		Biozyme AO1
2500 U/mg protein; ≥450 U/mg solid	1 unit oxidizes 1 μmol L-ascorbic acid/min at pH 5.6, 30°C	Highly purified; lyophilized	<0.0005% GOT, GPT <0.005% phosphatase <0.05% CAT	Biozyme AO2
1700 U/mg protein; 170 U/mg solid	1 unit converts 1 μmol L-ascorbate to product/min at 25°C	Lyophilized		Boehringer 236314

1.10.3.3 L-Ascorbate Oxidase continued

SPECIFIC ACTIVITY	UNITS DEFINITION	PREPARATION FORM	ADDITIONAL ACTIVITIES	SUPPLIER CATALOG No.
Cucurbita species continued				
17 U/spatula	1 unit converts 1 μmol L-ascorbate to product/min at 25°C	Spatula		Boehringer 736619
270 U/mg	1 unit oxidizes 1 μmol L-ascorbic acid to dehydroascorbate/min at pH 5.6, 30°C	Powder		Fluka 11136
>300 U/mg protein; >200 U/mg solid	1 unit oxidizes 1.0 μmol ascorbic acid/min at pH 5.6, 37°C	Lyophilized	<0.15 U/mg solid catalase <0.5 U/mg solid AK	Genzyme AAO 6141
>1200 U/mg protein; >120 U/mg solid	1 unit oxidizes 1 μmol L-ascorbate/min at pH 5.6, 25°C	Lyophilized	<0.001% GOT, GPT <0.2% catalase	Randox AO 811L
1000-3000 U/mg protein	1 unit oxidizes 1.0 μmol L-ascorbate to dehydroascorbate/min at pH 5.6, 25°C	Lyophilized containing 10% protein; balance primarily buffers and sucrose as stabilizer		Sigma A0157
≥200 U/mg solid	1 unit decreases 1 μmol ascorbic acid/min at pH 5.6, 30°C	Lyophilized containing BSA and borate as stabilizers	<0.02% phosphatase <0.1% catalase	Toyobo ASO-311
150-300 U/mg		Lyophilized		Wako 016-10851

1.11.1.1 NADH Oxidase

REACTION CATALYZED
$$NADH + H_2O_2 \leftrightarrow NAD^+ + 2\ H_2O$$

SYSTEMATIC NAME
NADH:hydrogen-peroxide oxidoreductase

SYNONYMS
NADH peroxidase

REACTANTS
$NADH$, H_2O_2, NAD^+, H_2O

APPLICATIONS
- Determination of NADH dehydrogenases

NOTES
- Classified as an oxidoreductase acting on a peroxide as acceptor
- A flavoprotein (FAD)
- Ferricyanide, quinones, etc. can replace H_2O_2
- Addition of 10 mM sodium azide to reaction mixtures is recommended for complete elimination of catalase
- Inhibited by KCN and PCMB

NADH Oxidase continued

SPECIFIC ACTIVITY	UNITS DEFINITION	PREPARATION FORM	ADDITIONAL ACTIVITIES	SUPPLIER CATALOG No.
Bacillus licheniformis (optimum pH = 6.5-7.5, T = 37°C, MW = 240,000 Da)				
>50 U/mg protein	1 unit oxidizes 1 µmol NADH/min at pH 7.0, 30°C	Lyophilized	Possible trace catalase	Calbiochem 481925
50 U/mg protein		Lyophilized		ICN 153847
>50 U/mg protein	1 unit oxidizes 1 µmol NADH/min at pH 7.0, 30°C	Chromatographically purified; lyophilized	Catalase	Nacalai 236-26
Streptococcus faecalis				
45 U/mg protein	1 unit decomposes 1.0 µmol H_2O_2/min at pH 6.0, 25°C	Suspension in 3.0 M $(NH_4)_2SO_4$, pH 7		ICN 155766
20-50 U/mg protein	1 unit decomposes 1.0 µmol H_2O_2/min at pH 6.0, 25°C	Suspension in 3.0 M $(NH_4)_2SO_4$, pH 7		sigma N6755
Streptococcus faecalis ATCC 11700, recombinant expressed in *E. coli*				
50-200 U/mg protein	1 unit oxidizes 1.0 µmol β-NADH/min at pH 5.4, 25°C	>95% (SDS-PAGE); solution containing 0.05 M KPO_4 and 0.6 mM EDTA, pH 7		sigma N0895

Catalase

REACTION CATALYZED

$$2 H_2O_2 \leftrightarrow O_2 + 2 H_2O$$

SYSTEMATIC NAME

Hydrogen-peroxide:hydrogen-peroxide oxidoreductase

REACTANTS

H_2O_2, O_2, H_2O

APPLICATIONS

- Of commercial interest wherever H_2O_2 is used as a germicide
- Milk preservative, with peroxidase
- Disposes of H_2O_2 used in milk pasteurization prior to cheese making
- Increases synthesis and stability of diacetyl in cultured milk
- Egg desugaring
- Free radical research
- Oxygen radical scavenger
- Decomposes residual H_2O_2 after bleaching woven and knitted cotton fabrics before drying
- Fabrication of porous materials
- Deodorization
- Determination of uric acid, with aldehyde dehydrogenase
- Determination of cysteamine
- H_2O_2 detection from activated phagocytes by a peroxidase-independent method
- Biocatalytic production of glycolic acid
- Production of gluconic acid in a disk reactor, with immobilized glucose oxidase

1.11.1.6 Catalase continued

- Contact lens solution
- Removal of oxidation products
- Color removal in herring processing
- Cleaning silicon semiconductor plates

NOTES

- Classified as an oxidoreductase acting on a peroxide as acceptor
- A hemoprotein with 4 subunits of equal size
- Found in nearly all animal cells and organs, and aerobic microorganisms
- H_2O_2 acts as both substrate and hydrogen acceptor; ethanol and other organics can act as hydrogen donor
- Pseudocatalase is the name given to the enzyme containing Mn(III) in the resting state
- Catalase, superoxide dismutase and glutathione peroxidase control levels of oxygen-derived free radicals in mammalian cells and function as somatic oxidant defense
- Heat stability is reduced by H_2O_2
- *Aspergillus* enzyme is more heat stable and more active in NaCl solution than liver catalase
- Inhibited by ascorbate alone or with Cu^{2+}
- Inactivated by freezing, lyophilization, peroxidase and sunlight under aerobic conditions

SPECIFIC ACTIVITY	UNITS DEFINITION	PREPARATION FORM	ADDITIONAL ACTIVITIES	SUPPLIER CATALOG NO.
Aspergillus niger (optimum pH = 7 [4.5-8.5], T = 55-65°C, MW = 250,000 Da; stable to T < 65°C)				
430 Baker U/mL 1000 Baker U/mL	1 Baker unit decomposes 264 mg H_2O_2/min	Food grade; Solution containing 50% glycerol and PO_4 buffer, pH 5.0-6.0		ABM-RP Catalase L43 Catalase L100
≥300,000 U/g	1 unit reduces 1 μmol H_2O_2/min at pH 7.0, 30°C	Powder		Amano Catalase N
≥50,000 U/mL	1 unit reduces 1 μmol H_2O_2/min at pH 7.0, 30°C	Food grade; solution		Amano Catalase NL
> 3000 U/mg protein; ≥2000 U/mg solid	1 unit decomposes 1 μmol H_2O_2/min at pH 7.0, 25°C	Salt-free; lyophilized		Biozyme CATNIF
10,000 U/mg protein	1 unit decomposes 1 μmol H_2O_2/min at pH 7.0, 30°C	Lyophilized	<0.001% GO	Calbiochem 219261
2000 U/mg protein	1 unit decomposes 1 μmol H_2O_2/min at pH 7.00, 25°C	Lyophilized containing 95% protein		Calzyme 052A2000
7000 U/mg protein	1 unit decomposes 1 μmol H_2O_2/min at pH 7.0, 25°C	Suspension in 3.2 M $(NH_4)_2SO_4$, pH 6	<0.01% GO	Fluka 60628
≥5000 Baker U/mL	1 unit decomposes 264 mg H_2O_2/hr at 25°C	Food grade; Kosher certified; solution containing 2% protein		Genencor CAT HP L5000
240-290 Baker U/mg; ≥220 Baker U/mL	1 unit decomposes 264 mg H_2O_2/hr at 25°C	Food grade; Kosher certified; lyophilized containing 90% protein		Genencor CAT HP S200

Catalase continued

SPECIFIC ACTIVITY	UNITS DEFINITION	PREPARATION FORM	ADDITIONAL ACTIVITIES	SUPPLIER CATALOG NO.
Aspergillus niger continued				
≥1000 Baker U/mL	1 unit decomposes 264 mg H_2O_2/hr at pH 7.0, 25°C	Food grade; Kosher certified; solution, pH 5.25		Genencor Fermcolase^R 1000
>1700 U/mg protein; >1200 U/mg solid	1 unit decomposes 1.0 μmol H_2O_2/min at pH 7.0, 25°C	Lyophilized		Genzyme 1031
100 U/mL	1 unit decomposes 1.0 μmol H_2O_2/min at pH 7.0, 25°C	Fungal suspension		ICN 100402
2000 U/mg solid	1 unit decomposes 1.0 μmol H_2O_2/min at pH 7.0, 25°C	Lyophilized		ICN 190311
50,000 CIU/mL		GRAS; solution		Novo Nordisk Novozym^R 355
4000-8000 U/mg protein	1 unit decomposes 1.0 μmol of H_2O_2/min at pH 7.0, 25°C while H_2O_2 falls from 10.3 to 9.2 mM	Suspension in 3.2 M $(NH_4)_2SO_4$, pH 6.0		Sigma C3515
Bison liver				
2000-5000 U/mg protein	1 unit decomposes 1.0 μmol of H_2O_2/min at pH 7.0, 25°C while H_2O_2 falls from 10.3 to 9.2 mM			Sigma C9447
Bovine liver (optimum pH = 7.0, pI = 5.4, MW = 250,000 Da [4 identical subunits])				
40,000 U/mg protein	1 unit decomposes 1 μmol H_2O_2/min at pH 7.0, 25°C	Suspension in 80% $(NH_4)_2SO_4$		Biozyme CAT1A
40,000 U/mg protein; 7000 U/mg solid	1 unit decomposes 1 μmol H_2O_2/min at pH 7.0, 25°C	Salt-free; lyophilized		Biozyme CAT1F
40,000 U/mg protein; 13,000 U/mg solid	1 unit decomposes 1 μmol H_2O_2/min at pH 7.0, 25°C	Salt-free; lyophilized		Biozyme CAT2A
65,000 U/mg	1 unit converts 1 μmol H_2O_2 to product/min at 25°C	Suspension in water and 0.01% alkylbenzyldimethyl ammonium chloride as stabilizer		Boehringer 106810 106828
260,000 U/mL	1 unit converts 1 μmol H_2O_2 to product/min at 25°C	Solution containing 30% glycerol and 10% EtOH		Boehringer 106836
50,000 U/mg protein	1 unit decomposes 1 μmol H_2O_2/min at pH 7.00, 25°C	Crystallized; suspension in distilled water containing 0.1% thymol and 10 mg/mL protein		Calzyme 052D50000
20,000 U/mg	1 unit decomposes 1 μmol H_2O_2/min at pH 7.0, 25°C	Powder containing 0.1% thymol stabilizer		Fluka 60629

1.11.1.6 Catalase continued

SPECIFIC ACTIVITY	UNITS DEFINITION	PREPARATION FORM	ADDITIONAL ACTIVITIES	SUPPLIER CATALOG NO.
Bovine liver *continued*				
65,000 U/mg protein	1 unit decomposes 1 µmol H_2O_2/min at pH 7.0, 25°C	Crystallized; aqueous suspension containing 0.01% alkylbenzyldimethyl ammonium chloride stabilizer, pH 6		Fluka 60630
1300 U/mg	1 unit decomposes 1 µmol H_2O_2/min at pH 7.0, 25°C	Powder		Fluka 60632
2500 U/mg	1 unit decomposes 1 µmol H_2O_2/min at pH 7.0, 25°C	Lyophilized		Fluka 60635
260,000 U/mL	1 unit decomposes 1 µmol H_2O_2/min at pH 7.0, 25°C	30% glycerol solution containing 10% (v/v) EtOH stabilizer		Fluka 60640
>3000 U/mg	1 unit decomposes 1.0 µmol H_2O_2/min at pH 7.0, 25°C			ICN 100428
>40,000 U/mg solid; 30,000 U/mL	1 unit decomposes 1.0 µmol H_2O_2/min at pH 7.0, 25°C	Sterile; aqueous solution		ICN 100429
≥260 Keil U/mL		Crude; solution containing NaCl, propylene glycol, pH 7.0-7.5		Intergen 8450-00
>50,000 IU/mg; ≥260,000 IU/mL		Purified; solution containing $NaPO_4$ buffer, glycerol, EtOH, NaCl, EDTA, pH 6.5-7.5		Intergen 8452-00
40,000-60,000 U/mg protein	1 unit decomposes 1.0 µmol of H_2O_2/min at pH 7.0, 25°C while H_2O_2 falls from 10.3 to 9.2 mM	2X Crystallized; suspension in water and 0.1% thymol		Sigma C-100
10,000-30,000 U/mg protein	1 unit decomposes 1.0 µmol of H_2O_2/min at pH 7.0, 25°C while H_2O_2 falls from 10.3 to 9.2 mM	Crystallized; suspension in water and 0.1% thymol		Sigma C-30
10,000-25,000 U/mg protein	1 unit decomposes 1.0 µmol of H_2O_2/min at pH 7.0, 25°C while H_2O_2 falls from 10.3 to 9.2 mM	Thymol-free		Sigma C-40
50,000 U/mg protein	1 unit decomposes 1.0 µmol of H_2O_2/min at pH 7.0, 25°C while H_2O_2 falls from 10.3 to 9.2 mM	2X Crystallized; aseptically filled; solution containing <0.01 mg/mL thymol		Sigma C3155
2000-5000 U/mg protein	1 unit decomposes 1.0 µmol H_2O_2/min at pH 7.0, 25°C while H_2O_2 conc. falls from 10.3 to 9.2 mM	Cell culture tested		Sigma C6665
60,000-120,000 U/g agarose	1 unit decomposes 1.0 µmol of H_2O_2/min at pH 7.0, 25°C while H_2O_2 falls from 10.3 to 9.2 mM	Attached to 4% beaded agarose; suspension in 3.2 M $(NH_4)_2SO_4$ and 0.02% thimerosal, pH 7.0		Sigma C9284
2000-5000 U/mg protein	1 unit decomposes 1.0 µmol of H_2O_2/min at pH 7.0, 25°C while H_2O_2 falls from 10.3 to 9.2 mM			Sigma C9322

SPECIFIC ACTIVITY	UNITS DEFINITION	PREPARATION FORM	ADDITIONAL ACTIVITIES	SUPPLIER CATALOG No.
Bovine liver *continued*				
40,000 U/mg		Lyophilized		Wako 039-12901
≥3000 U/mg protein	1 unit decomposes 1 µmol H_2O_2/min at pH 7.0, 25°C	Partially purified; lyophilized containing 1/3 sucrose by weight		Worthington LS01847 LS01849 LS01851
≥20,000 U/mg protein; 6 mg/mL	1 unit decomposes 1 µmol H_2O_2/min at pH 7.0, 25°C	2X Crystallized; suspension saturated with thymol as preservative	<0.04 U/U catalase SOD	Worthington LS01872 LS01873 LS01874
≥40,000 U/mg protein; ≥30,000 U/mL; 10 mL/vial	1 unit decomposes 1 µmol H_2O_2/min at pH 7.0, 25°C	2X Crystallized; sterile-filtered; solution		Worthington LS01896 LS01898
Dog liver				
900 U/mg protein	1 unit decomposes 1.0 µmol of H_2O_2/min at pH 7.0, 25°C while H_2O_2 falls from 10.3 to 9.2 mM	Suspension in 2.1 M $(NH_4)_2SO_4$, pH 6.5		Sigma C4138
Human erythrocyte (MW = 256,000 Da [4 identical subunits])				
		≥95% purity by electrophoresis; frozen in 50 mM Tris-HCl, pH 8.0	No detectable HBsAg, anti-HCV, anti-HBc, anti-HIV	ART 16-05-030000
>50,000 U/mg protein	1 unit decomposes 1 µmol H_2O_2/min at pH 7.0, 25°C	>95% purity; frozen in 50 mM Tris, pH 8.0	No detectable HBsAg, HIV Ab	Calbiochem 219008
		>95% purity by electrophoresis; frozen in 50 mM Tris-HCl, pH 8.0	No detectable HIV I, HIV II, HCV Ab, HBsAg	Cortex CP3008
160,000 U/mg protein	1 unit decomposes 1.0 µmol H_2O_2/min at pH 7.0, 25°C	Solution containing 1.5 mM $NaPO_4$ and 20 mM NaCl, pH 7.0	No detectable HBsAg, HIV Ab	ICN 191341
30,000 U/mg protein	1 unit decomposes 1.0 µmol of H_2O_2/min at pH 7.0, 25°C while H_2O_2 falls from 10.3 to 9.2 mM	95% purity by SDS-PAGE; solution containing 50 mM Tris, pH 8.0		Sigma C3556
Micrococcus lysodeicticus (optimum pH = 5.0-8.0, T = 15-30°C [pH 7.0, 5 min], 15-40°C [pH 7.0, 1 min]; stable pH 5-11 [30°C, 1 hr], T < 50°C [pH 7.0, 30 min])				
				EnzymeDev Enzeco[R] Catalase L
>50,000 CtUN/mL	1 CtUN decomposes 1 µmol H_2O_2/min at pH 7.0, 30°C	Solution		Nagase Reyonet
>50,000 CtUN/mL	1 CtUN decomposes 1 µmol H_2O_2/min at pH 7.0, 30°C	Solution, pH 6.0-8.0		Nagase Reyonet S

1.11.1.6 Catalase continued

SPECIFIC ACTIVITY	UNITS DEFINITION	PREPARATION FORM	ADDITIONAL ACTIVITIES	SUPPLIER CATALOG NO.
Murine liver				
3000-6000 U/mg protein	1 unit decomposes 1.0 μmol of H_2O_2/min at pH 7.0, 25°C while H_2O_2 falls from 10.3 to 9.2 mM	Suspension in 2.7 M $(NH_4)_2SO_4$, pH 6.5		Sigma C8531

1.11.1.7 Peroxidase

REACTION CATALYZED
Donor + H_2O_2 ↔ Oxidized donor + 2 H_2O

SYSTEMATIC NAME
Donor:hydrogen-peroxide oxidoreductase

SYNONYMS
Lactoperoxidase (milk, mammalian glands), bromoperoxidase (*Corallina officinalis*), microperoxidase (equine heart), myeloperoxidase (human leukocytes, neutrophils, tissue), LAP, HRP, LPO, POD

REACTANTS
Donor, H_2O_2, oxidized donor, H_2O, I^-, I_2, phenols, aromatic amines, pyrogallol, ascorbate, guaiacol, ferrocyanide, 4-aminoantipyrine, cytochrome c

APPLICATIONS
- Chemiluminescent determination of H_2O_2 with luminol
- Enhanced chemiluminescent determination of membrane-bound nucleic acid sequences
- Oxidation of N-substituted aromatic amines
- A gentle, specific alternative to chloramine-T for radioiodination of proteins
- A tracer and marker enzyme when coupled to other proteins, e.g., histochemistry and cytochemistry
- Cell permeability studies, immunocytochemistry, neurophysiology and electron microscopy
- Staining of biological specimens
- Indicator for chemical reactions where peroxide is produced
- Routinely conjugated to monoclonal or polyclonal antibodies for use in EIA, based on ability to yield chromogenic products and its relatively good stability
- Determination of glucose and galactose in fluids
- Determination of plasma uric acid, with uricase
- Epoxidation of double bonds
- Oxidation of phenols and aromatics in the presence of H_2O_2
- Enzyme-chloride-mediated killing of microbes and tumor cells
- Inactivation of chemotactic factors

Peroxidase continued — 1.11.1.7

NOTES

- Classified as an oxidoreductase acting on a peroxide as acceptor
- A hemoprotein, which may occur as a mixture of 2 isozymes
- Specific for hydrogen acceptors: H_2O_2, methyl and ethyl peroxides
- Not specific for the hydrogen donor: a large number of compounds react
- *Arthromyces* enzyme has no isozymes
- Human neutrophil activity decreases over extended periods
- Seven isozymes of horseradish peroxidase have been described. Reversibly inhibited by cyanide and sulfide at 10^{-5} M. Inhibited by fluoride and azide

SPECIFIC ACTIVITY	UNITS DEFINITION	PREPARATION FORM	ADDITIONAL ACTIVITIES	SUPPLIER CATALOG NO.
Arthromyces ramosus (optimum pH = 6-8, T = 35-45°C, MW = 41,000 Da)				
3000 U/mg	1 unit oxidizes 1 µmol ABTS/min at pH 5.0, 25°C			Fluka 77325
250 U/mg protein	1 unit reduces 1 µmol H_2O_2/min at pH 7.0, 37°C	Chromatographically purified; lyophilized; RZ: 2.5+		Nacalai 266-47
Bovine milk (MW = 77,500 Da)				
200 U/mg protein	1 unit decomposes 1 µmol peroxide/min at pH 5.5, 25°C	Highly purified; suspension in 75% $(NH_4)_2SO_4$, pH 7.0	<0.0001% XO	Biozyme LP2
200 U/mg protein; 200 U/mg solid	1 unit decomposes 1 µmol peroxide/min at pH 5.5, 25°C	Highly purified; lyophilized		Biozyme LP2F
6-10 mg enzyme/mL gel; 300-400 U/mL		Chromatographically purified; attached to beaded agarose in 0.02 M NaOAc and 0.01% NaN_3, pH 5		Elastin A689
6-10 mg enzyme/mL gel; 300-400 U/mL		Chromatographically purified; attached to beaded agarose in 0.02 M NaOAc and 0.01% NaN_3, pH 5		Elastin AD689
6-10 mg enzyme/mL gel; 300-400 U/mL		Chromatographically purified; attached to beaded agarose in 0.02 M NaOAc and 0.01% NaN_3, pH 5		Elastin AE689
250 U/mg	1 unit oxidizes 1 µmol 2,2´-azino-bis(3-ethylbezothiazoline-6-sulfonic acid) diammonium salt/min at pH 6.0, 25°C	Lyophilized		Fluka 61328
200 U/mg protein	1 unit oxidizes 1 µmol 2,2´-azino-bis(3-ethylbezothiazoline-6-sulfonic acid) diammonium salt/min at pH 6.0, 25°C	Crystallized; suspension in 3.2 M $(NH_4)_2SO_4$ and 0.01 M KPO_4, pH 7		Fluka 61330

1.11.1.7 Peroxidase continued

SPECIFIC ACTIVITY	UNITS DEFINITION	PREPARATION FORM	ADDITIONAL ACTIVITIES	SUPPLIER CATALOG No.
Bovine milk continued				
250 U/mg	1 unit oxidizes 1 µmol 2,2´-azino-bis(3-ethylbezothiazoline-6-sulfonic acid) diammonium salt/min at pH 6.0, 25°C	Lyophilized		Fluka 61331
80-100 U/mg protein	1 unit forms 1.0 mg purpurogallin from pyrogallol/20 sec at pH 6.0, 20°C	Lyophilized; A_{412}/A_{280} = 0.88 to 0.92		ICN 195271
60-80 U/mg protein	1 unit forms 1.0 mg purpurogallin from pyrogallol/20 sec at pH 6.0, 20°C	Lyophilized; A_{412}/A_{280} = 0.70 to 0.85		ICN 195272
>200 U/mg protein	1 unit forms 1.0 mg purpurogallin from pyrogallol/20 sec at pH 6.0, 20°C	Lyophilized containing 90-95% protein; A_{412}/A_{280} = >0.9		ICN 195273
75-150 U/mg protein	1 unit forms 1.0 mg purpurogallin from pyrogallol/20 sec at pH 6.0, 20°C	Suspension in 3.2 M $(NH_4)_2SO_4$ and 20 mM NaOAc, pH 7		Sigma L0515
80-120 U/mg protein	1 unit forms 1.0 mg purpurogallin from pyrogallol/20 sec at pH 6.0, 20°C	Lyophilized		Sigma L2005
50-150 U/mg protein	1 unit forms 1.0 mg purpurogallin from pyrogallol/20 sec at pH 6.0, 20°C	Suspension in 3.2 M $(NH_4)_2SO_4$, pH 7		Sigma L2130
80-150 U/mg protein	1 unit forms 1.0 mg purpurogallin from pyrogallol/20 sec at pH 6.0, 20°C	Lyophilized		Sigma L8257
≥300 U/mg solid	1 unit reduces 1 µmol H_2O_2/min at pH 6.0, 25°C	Chromatographically purified; bound to Sepharose^R; suspension in 0.1 M acetate buffer, pH 4.5		Worthington LS00140 LS00142
≥35 U/mg solid	1 unit reduces 1 µmol H_2O_2/min at pH 6.0, 25°C	Chromatographically purified; lyophilized		Worthington LS00150 LS00151 LS00152
Bovine raw skim milk (MW = 77,500 Da)				
≥200 U/mg protein	1 unit forms 1 µmol triiodide/min at pH 7.0, 25°C in 0.033 M NaPO$_4$ buffer	Lyophilized containing PO$_4$ buffer and 90% protein		Calzyme 095A0200
Corallina officinalis				
50-150 U/mg protein	1 unit converts 1.0 µmol monochloro-dimedon to monobromomono-chlorodimedon/min at pH 6.4, 25°C	Partially purified; lyophilized containing 10% protein; balance primarily Tris buffer salts	<1% chloroperoxidase	Sigma B2404
Equine heart				
		SDS-PAGE single band; chromatographically purified; lyophilized		Biozyme MPO1
		>95% purity; lyophilized	<2% 8-microperoxidase	Cortex CP9001
1 mg/vial		80% purity	<2% 11-microperoxidase	Cortex CP9001

Peroxidase continued — 1.11.1.7

SPECIFIC ACTIVITY	UNITS DEFINITION	PREPARATION FORM	ADDITIONAL ACTIVITIES	SUPPLIER CATALOG NO.
Horseradish (optimum pH = 6.0-7.0, pI = 7.2, T = 45°C, MW = 40,000 Da; stable pH 5.0-10.0 [25°C, 20 hr], T < 50°C [pH 6.0, 10 min])				
≥180 U/mg	1 unit decomposes 1 µmol H_2O_2/min at pH 7.0, 37°C	Powder		Amano Peroxidase
≥100 U/mg powder	1 unit decomposes 1 µmol peroxidase/min at 25°C with 4-AAP-phenol as chromogen	>1.0 purity number; powder or flake		Beckman 667965
	1 unit produces 1 mg purpurogallin from pyrogallol/20 sec at pH 6.0, 20°C	Salt-free; lyophilized; RZ > 0.05		Biozyme APO HRP
≥60 U/mg solid	1 unit produces 1 mg purpurogallin from pyrogallol/20 sec at pH 6.0, 20°C	Salt-free; lyophilized; RZ > 0.6		Biozyme HRP2
≥150 U/mg solid	1 unit produces 1 mg purpurogallin from pyrogallol/20 sec at pH 6.0, 20°C	Salt-free; lyophilized; RZ = 2.0		Biozyme HRP3C
≥250 U/mg solid	1 unit produces 1 mg purpurogallin from pyrogallol/20 sec at pH 6.0, 20°C	Chromatographically prepared; salt-free; lyophilized; RZ > 3.0		Biozyme HRP4
≥250 U/mg solid	1 unit produces 1 mg purpurogallin from pyrogallol/20 sec at pH 6.0, 20°C	90% isozyme C; chromatographically prepared; salt-free; lyophilized; RZ > 3.0		Biozyme HRP4B
≥250 U/mg solid	1 unit produces 1 mg purpurogallin from pyrogallol/20 sec at pH 6.0, 20°C	Chromatographically prepared; salt-free; lyophilized; RZ > 3.0		Biozyme HRP4C
80 U/mg solid	1 unit produces 1 mg purpurogallin from pyrogallol/20 sec at pH 6.0, 20°C	Mainly acidic isozymes; chromatographically prepared; salt-free; lyophilized; RZ = 3.5		Biozyme HRP5
250 U/mg	1 unit converts 1 µmol guaiacol and H_2O_2 to product/min at 25°C	Suspension in 3.2 M $(NH_4)_2SO_4$, pH 6		Boehringer 108073 108081
200 U/mg solid	1 unit converts 1 µmol guaiacol and H_2O_2 to product/min at 25°C	Purity 2.0 at A_{403}/A_{275}; lyophilized		Boehringer 127361 128066
800 U/mg protein	1 unit converts 1 µmol ABTS and H_2O_2 to product/min at pH 5.0, 25°C	Purity 3.0-3.5 at A_{403}/A_{275}; lyophilized	>90% Isozyme C (HPLC)	Boehringer 1428861
250 U/mg solid	1 unit converts 1 µmol guaiacol and H_2O_2 to product/min at 25°C	Purity 3.0 at A_{403}/A_{275}; salt-free; lyophilized	<0.001% ATPase, acid phosphatase <0.7% catalase	Boehringer 413470 108090
250 U/mg solid (guaiacol as substrate); 1000 U/mg solid (ABTS as substrate, pH 5.0); 3500 U/mg solid (tetra-methylbenzidine as substrate)	1 unit converts 1 µmol substrate and H_2O_2 to product/min at 25°C	EIA grade; purity 3.2-3.3 at A_{403}/A_{275}; salt-free; lyophilized containing 12.0-14.5% carbohydrates and 2-3 moles/mole enzyme amino groups	>90% isoenzyme C (HPLC)	Boehringer 814393 814407

1.11.1.7 Peroxidase continued 151

SPECIFIC ACTIVITY	UNITS DEFINITION	PREPARATION FORM	ADDITIONAL ACTIVITIES	SUPPLIER CATALOG NO.
Horseradish *continued*				
250 U/mg solid	by Guiacol	Salt-free; lyophilized; RZ: 3.0		Crystal Chem
180 U/mg solid	by Guiacol	Salt-free; lyophilized; RZ: 2.0		Crystal Chem
100 U/mg solid	by Guiacol	Salt-free; lyophilized; RZ: 1.0		Crystal Chem
100 U/mg	1 unit oxidizes 1 μmol ABTS/min at pH 6.0, 25°C	Salt-free; lyophilized		Fluka 77330
50 U/mg	1 unit oxidizes 1 μmol ABTS/min at pH 6.0, 25°C	Salt-free; lyophilized		Fluka 77332
700 U/mg	1 unit oxidizes 1 μmol ABTS/min at pH 6.0, 25°C	Lyophilized		Fluka 77333
850 U/mg	1 unit oxidizes 1 μmol ABTS/min at pH 6.0, 25°C	Lyophilized		Fluka 77334
900 U/mg	1 unit oxidizes 1 μmol ABTS/min at pH 6.0, 25°C	Lyophilized		Fluka 77336
0.05 U/mg peroxidase activity	1 unit oxidizes 1 μmol ABTS/min at pH 6.0, 25°C	Salt-free; powder		Fluka 77338
>150 U/mg solid	1 unit converts 1 μmol pyrogallol to purpurogallin/20 sec at pH 6.0, 20°C	Lyophilized		GDS Tech P-011
>180 U/mg solid	1 unit produces 1.0 mg purpurogallin/20 sec at pH 6.0, 20°C	Lyophilized; RZ: >2.0		Genzyme 1551
>8 U/mg solid	1 unit forms 1.0 mg purpurogallin/20 sec at pH 6.0, 20°C	Lyophilized; RZ: >1.0		Genzyme 1591
>300 U/mg solid	1 unit forms 1.0 mg purpurogallin/20 sec at pH 6.0, 20°C	Lyophilized; RZ: >3.0		Genzyme 1981
45 purpurogallin U/mg solid	1 unit produces 1 mg purpurogallin/20 sec at 20°C	Salt-free; lyophilized; RZ: 1.0-1.5		ICN 150035
60 purpurogallin U/mg solid	1 unit produces 1 mg purpurogallin/20 sec at 20°C	Salt-free; lyophilized; RZ: >0.60		ICN 150036
100-150 purpurogallin U/mg solid	1 unit produces 1 mg purpurogallin/20 sec at 20°C	Salt-free; lyophilized; RZ: 1.0-2.0		ICN 150037
300 purpurogallin U/mg solid	1 unit produces 1 mg purpurogallin/20 sec at 20°C	Salt-free; lyophilized; RZ: >3.0		ICN 150038
300 U/mL	1 mg activated enzyme yields 0.5 mL conjugate, working dilution 1:2000	Conjugation grade; activated; lyophilized		ICN 191398
400 U/mg solid	1 unit produces 1 mg purpurogallin from pyrogallol/20 sec at pH 6.0, 20°C	Crude; salt-free; powder; RZ: 0.3		ICN 195370

SPECIFIC ACTIVITY	UNITS DEFINITION	PREPARATION FORM	ADDITIONAL ACTIVITIES	SUPPLIER CATALOG NO.
Horseradish *continued*				
80 U/mg solid		Salt-free; powder; RZ: 0.6		ICN 195371
150-200 U/mg solid	1 unit produces 1 mg purpurogallin from pyrogallol/20 sec at pH 6.0, 20°C	Salt-free; powder; RZ: 1.0-1.5		ICN 195372
250-330 U/mg solid	1 unit produces 1 mg purpurogallin from pyrogallol/20 sec at pH 6.0, 20°C	Salt-free; powder; RZ: 3.0		ICN 195373
250 U/mg protein		Crystallized; suspension in 3.2 M $(NH_4)_2SO_4$ with KPO_4 buffer, pH 6.0; RZ:3.0		ICN 195374
>250 U/mg protein; >150 U/mg solid	1 unit produces 1 mg purpurogallin/20 sec at 20°C	Lyophilized	<0.01% phosphatase <0.1% cellulase	Randox PO 883L
225-275 guaiacol U/mg	1 unit converts 1 μmol peroxide/min at 25°C; 1.0 guaiacol U at 25°C = 1.0 purpurogallin U at 20°C = 13.0 IUB U at 25°C = 24.0 4-aminoantipyrine U at 25°C	Lyophilized; RZ>3.0		Scripps H0214
10,000-20,000 U/g agarose; 1 mL gel yields 60-100 U	1 unit forms 1.0 μmol purpurogallin from pyrogallol/min at pH 6.0, 30°C; 1/10th of the purpurogallin[20 sec] unit	Insoluble enzyme attached to beaded agarose; suspension in 2.0 M $(NH_4)_2SO_4$ and 0.02% thimerosal, pH 7.0		Sigma P3912
1000 U/mg solid using ABTS; 250-330 U/mg solid using pyrogallol	1 unit oxidizes 1.0 μmol 2,2´-Azino-bis(3-ethylbenzthiazoline-6-sulfonic acid)/min at pH 5.0, 25°C using ABTS	Salt-free; powder		Sigma P6782
1000 U/mg solid (ABTS); 250-330 U/mg solid (pyrogallol)	1 ABTS unit oxidizes 1 μmol 2,2´-azino-bis(3-ethylbenzthiazoline-6-sulfonic acid)/min at pH 5.0, 25°C	Salt-free; powder		Sigma P6782
≥250 purpurogallin U/mg solid	1 purpurogallin unit forms 1 mg purpurogallin/20 sec at 20°C	Electrophoretically homogeneous; chromatographically purified; salt-free; lyophilized	<0.001% phosphatase	Toyobo PEO-131
≥250 purpurogallin U/mg solid	1 purpurogallin unit forms 1 mg purpurogallin/20 sec at 20°C	Mixture of basic isozymes; partially purified; salt-free; lyophilized	<0.001% phosphatase	Toyobo PEO-201
≥110 purpurogallin U/mg solid	1 purpurogallin unit forms 1 mg purpurogallin/20 sec at 20°C	Basic and acidic isozymes; partially purified; lyophilized containing 30% stabilizers	<0.001% phosphatase	Toyobo PEO-301
250-350 U/mg		Lyophilized		Wako 166-12881 162-12883

1.11.1.7 Peroxidase continued

SPECIFIC ACTIVITY	UNITS DEFINITION	PREPARATION FORM	ADDITIONAL ACTIVITIES	SUPPLIER CATALOG NO.
Horseradish *continued*				
≥100 U/mg				Wako 169-10791 165-10793
Horseradish root (optimum pH = 6.0-7.0, T = 45°C, MW = 40,000 Da)				
≥185 U/mg; 20 g/vial ≥1000 U/mg; 20 g/vial	1 unit incorporates 1 μmol H_2O_2/min at 25°C using 4-amino antipyrin as substrate	2 basic (B & C) and no acidic isozymes; ≥98% purity by SDS-PAGE; lyophilized; RZ: 0.5 and 3.0		Adv Immuno HrPx1
250 U/mg	1 unit converts 1 μmol peroxide/min at 25°C	Lyophilized; RZ: 3		AMRESCO 0343 (RZ-3)
≥100 U/mg	1 unit converts 1 μmol peroxide/min at 25°C	Lyophilized; RZ: 1		AMRESCO 0417 (RZ-1)
600 U/mg	1 unit decomposes 1 μmol H_2O_2/min at pH 7, 25°C	Several isozymes; lyophilized containing 95% protein; RZ value > 3.00		Calzyme 100A0600
400 U/mg	1 unit decomposes 1 μmol H_2O_2/min at pH 7, 25°C	Lyophilized containing 75% protein; RZ value > 2.00		Calzyme 101A0400
200 U/mg	1 unit decomposes 1 μmol H_2O_2/min at pH 7, 25°C	Lyophilized containing 60% protein; RZ value > 1.00		Calzyme 102A0200
250 U/mg	1 unit converts 1 μmol peroxide/min at pH 7.0, 25°C in guaiacol assay; 1 guaiacol unit at 25°C = 0.95 purpurogallin units at 20°C = 13 o-dianisidine units at 25°C	Labeling grade; Lyophilized; RZ: 3.0		ICN 364511 364512 364514
		Lyophilized; RZ: >3		ProZyme HP10 000 HP10 010
>100 U/mg	1 unit forms 1.0 mg purpurogallin from pyrogallol/20 sec at pH 6.0, 20°C	Powder		STC LABS STHRP-1
>200 PPU/mg	1 unit forms 1.0 mg purpurogallin from pyrogallol/20 sec at pH 6.0, 20°C	Powder		STC LABS STHRP-2
>300 PPU/mg	1 unit forms 1.0 mg purpurogallin from pyrogallol/20 sec at pH 6.0, 20°C	Powder		STC LABS STHRP-3
>300 PPU/mg	1 unit forms 1.0 mg purpurogallin from pyrogallol/20 sec at pH 6.0, 20°C	Powder		STC LABS STHRP-C
≥85 U/mg solid	1 unit decomposes 1 μmol H_2O_2/min at pH 7.0, 25°C using aminoantipyrine and phenol	Diagnostic grade; lyophilized; RZ: ≥1		Worthington LS02559 LS02560 LS02561

Peroxidase continued   1.11.1.7

SPECIFIC ACTIVITY	UNITS DEFINITION	PREPARATION FORM	ADDITIONAL ACTIVITIES	SUPPLIER CATALOG NO.
Horseradish root *continued*				
≥500 U/mg protein	1 unit decomposes 1 μmol H_2O_2/min at pH 7.0, 25°C using aminoantipyrine and phenol	Basic isozyme; chromatographically purified; lyophilized; RZ: 2.9-3.0		Worthington LS06474 LS06476 LS06472
Human leukocytes (MW = 118,000 Da)				
150-200 U/mg protein	1 unit decomposes 1 μmol H_2O_2/min at pH 6.0, 25°C	≥95% purity; lyophilized from buffer containing 50 mM NaOAc and 100 mM NaCl, pH 6.0	No detectable HBsAg, HIV Ab	Calbiochem 475911
60 U/mg protein	1 unit increases A_{470} by 1.0/min at pH 7.0, 25°C, calculated from the initial rate of reaction using guaiacol as substrate	Lyophilized		Fluka 70021
50 U/mg protein	1 unit increases ΔA_{470} by 1.0/min at pH 7.0, 25°C calculated from initial reaction rate with guaiacol as substrate	Lyophilized containing 0.02 M NaOAc buffer, pH 6.0		Sigma M6908
Human neutrophils (MW = 150,000 Da)				
150-200 U/mg protein	1 unit decomposes 1 μmol H_2O_2/min at pH 6.0, 25°C	≥95% purity by electrophoresis; lyophilized containing 50 mM NaOAc and 100 mM NaCl, pH 6.0	No detectable HBsAg, anti-HCV, anti-HBc, anti-HIV	ART 16-14-130000
200 U/mg protein; 20 U/vial	1 unit decomposes 1.0 μmol H_2O_2/min at pH 6.0, 25°C	>95% purity by electrophoresis; lyophilized containing 50 mM NaOAc and 100 mM NaCl, pH 6.0	No detectable HIV Ab, HBsAg	Cortex CP3028
200-1000 U/mg protein; 10-15 U/vial	1 unit decomposes 1.0 μmol H_2O_2/min at pH 6.0, 25°C	>98% purity by electrophoresis; lyophilized containing 50 mM NaOAc, 100 mM NaCl, pH 6.0	No detectable HBsAg, HIV Ab	ICN 191336
Human sputum, purulent (leucocyte) (MW = 120,000 Da)				
28-33 U/mg protein	1 unit decomposes 1 μmol H_2O_2/min at pH 7.0, 25°C using 4-amino-antipyrine as hydrogen donor	98% purity by PAGE and gel diffusion; salt-free; lyophilized		Elastin MY862
Human tissue culture cells (MW = 140,000-150,000 Da)				
	1 unit generates an absorbance of 1.2 when using the standard protocol for enzyme immunoassay	Affinity purified; chromatographically heterogeneous; solution containing 10 mM Tris buffer and 0.02% NaN_3 as preservative		ImmunoVision MPO-3000 MPO-3010 MPO-3020

1.11.1.8 Iodide Peroxidase

REACTION CATALYZED
  Iodide + H_2O_2 ↔ Iodine + 2 H_2O
SYSTEMATIC NAME
  Iodide:hydrogen-peroxide oxidoreductase
SYNONYMS
  Iodotyrosine deiodase, iodinase

REACTANTS
  Iodide, H_2O_2, iodine, H_2O
NOTES
- Classified as an oxidoreductase acting on a peroxide as acceptor
- A hemoprotein

> See Chapter 5. Multiple Enzyme Preparations

1.11.1.9 Glutathione Peroxidase

REACTION CATALYZED
  2 Glutathione + H_2O_2 ↔ Oxidized glutathione + 2 H_2O
SYSTEMATIC NAME
  Glutathione:hydrogen-peroxide oxidoreductase
REACTANTS
  Glutathione, H_2O, oxidized glutathione, H_2O
APPLICATIONS
- Determination of lipid hydroperoxidase

NOTES
- Classified as an oxidoreductase acting on a peroxide as acceptor
- A protein containing a selenocysteine residue
- Steroid and lipid hydroperoxides can act as acceptors, but more slowly than H_2O_2
- Activity at pH 8.8 is 10X that at pH 7.0
- A mechanism for protecting cell membranes against peroxidative damage
- Inhibited by sodium sulfite, mercuric chloride and nucleotides
- Stabilized by DTT, glutathione, glycerol, sucrose, potassium phosphate

SPECIFIC ACTIVITY	UNITS DEFINITION	PREPARATION FORM	ADDITIONAL ACTIVITIES	SUPPLIER CATALOG NO.
Bovine erythrocytes (optimum pH = 8.8, T = 42°C, MW = 84,000 Da [4 subunits per mol enzyme; 1 mol selenium per subunit]; K_M = 3.0 mM [GSH], 1-10 μM [H_2O_2, lipid hydroperoxide]; stable at pH 6.0-6.5 [25°C, 20 hr] and < 45°C [pH 7.0, 10 min])				
50 U/mg solid	1 unit oxidizes 1 μmol GSH/min at pH 7.0, 25°C	Lyophilized	<0.02% catalase	Calbiochem 353916

SPECIFIC ACTIVITY	UNITS DEFINITION	PREPARATION FORM	ADDITIONAL ACTIVITIES	SUPPLIER CATALOG NO.
Bovine erythrocytes *continued*				
60 U/mg solid	1 unit oxidizes 1 μmol GSH (0.5 μmol NADPH)/min at pH 7, 25°C	Hemoglobin (unreacted)-free; lyophilized containing 85% stabilizer	0.002% catalase	Calzyme 167A0060
100 U/mg	1 unit catalyzes the oxidation by H_2O_2 of 1 μmol GSH to GSSG/min at pH 7.0, 25°C	Lyophilized		Fluka 49753
30-100 U/mg protein	1 unit oxidizes 1.0 μmol GSH to GSSG by H_2O_2/min at pH 7.0, 25°	Lyophilized containing 20% protein, 25% sucrose, 3% DTT; balance primarily $NaPO_4$ buffer salts		Sigma G4013
300-700 U/mg protein	1 unit oxidizes 1.0 μmol GSH to GSSG by H_2O_2/min at pH 7.0, 25°C	Lyophilized containing 20% protein, 25% sucrose, 2.5% DTT; balance primarily $NaPO_4$ buffer salts		Sigma G6137
≥50 U/mg solid	1 unit oxidizes 1 μmol GSH (0.5 μmol NADPH)/min at pH 7.0, 25°C	Lyophilized containing 85% DTT, GSH, glycerol, sucrose, KPO_4 as stabilizers	<0.02% catalase	Toyobo GSP-301

Chloride Peroxidase

REACTION CATALYZED
$$2\ RH + 2\ Cl^- + H_2O_2 \leftrightarrow 2\ RCl + 2\ H_2O$$

SYSTEMATIC NAME
  Chloride:hydrogen-peroxide oxidoreductase

SYNONYMS
  Chloroperoxidase

REACTANTS
  Chloride, bromide, iodide, H_2O_2, H_2O

APPLICATIONS
- A reagent in asymmetric synthesis
- A useful alternative to lactoperoxidase for [^{31}I]labeling studies, bromination of proteins and [^{36}Cl]-labeling of macromolecules in long-term isolation procedures

NOTES
- Classified as an oxidoreductase acting on a peroxide as acceptor
- Probably a heme-thiolate protein (*P*-450)
- Chlorinates a range of organic molecules, forming stable C-Cl bonds

SPECIFIC ACTIVITY	UNITS DEFINITION	PREPARATION FORM	ADDITIONAL ACTIVITIES	SUPPLIER CATALOG NO.
Caldariomyces fumago (MW = 42,000 Da)				
1000-2000 U/mg protein	1 unit converts 1 μmol monochlorodimedon to dichlorodimedon/min at pH 2.75, 25°C in the presence of H_2O_2 and KCl	Aqueous solution containing 1-3 m*M* NaOAc buffer, pH 4.0		Calbiochem 220490
1300 U/mg protein	1 unit converts 1 μmol monochlorodimedon to dichlorodimedon/min at pH 2.75, 25°C in the presence of KCl and H_2O_2	Suspension in 0.1 *M* NaPO$_4$, pH 4.0		Fluka 25810
≥3000 U/mL	1 unit converts 1.0 μmol monochlorodimedon to dichlorodimedon/min at pH 2.75, 25°C in the presence of KCl and H_2O_2	Crude; suspension in 0.1 *M* NaPO$_4$, pH 4.0		Sigma C0278
1000-2000 U/mg protein	1 unit converts 1.0 μmol monochlorodimedon to dichlorodimedon/min at pH 2.75, 25°C in the presence of KCl and H_2O_2	Purified; suspension in 0.1 *M* NaPO$_4$, pH 4.0		Sigma C0887
1000-2000 U/mg protein	1 unit converts 1.0 μmol monochlorodimedon to dichlorodimedon/min at pH 2.75, 25°C in the presence of KCl and H_2O_2	Partially purified; lyophilized containing KPO$_4$ buffer salts and stabilizer		Sigma C8902

Protocatechuate 3,4-Dioxygenase 1.13.11.3

REACTION CATALYZED
3,4-Dihydroxybenzoate + O_2 ↔ 3-Carboxy-cis,cis-muconate

SYSTEMATIC NAME
Protocatechuate:oxygen 3,4-oxidoreductase (decyclizing)

SYNONYMS
Protocatechuate oxygenase

REACTANTS
3,4-Dihydroxybenzoate, O_2, 3-carboxy-cis,cis-muconate, protocatechuate

APPLICATIONS
- Determination of choline esterase, with p-hydroxybenzoate hydroxylase

NOTES
- Classified as an oxidoreductase acting on single donors with incorporation of molecular oxygen (oxygenases), and incorporation of two atoms of oxygen
- Requires Fe^{3+}
- Stabilized by sugars
- Inhibited by Ag^+, Hg^{2+} and PCMB

SPECIFIC ACTIVITY	UNITS DEFINITION	PREPARATION FORM	ADDITIONAL ACTIVITIES	SUPPLIER CATALOG No.
Pseudomonas species (optimum pH = 9.0, T = 60-65°C, MW = 700,000 Da; K_M = 18.5 μM [protocatechuate]; stable at pH 7-9 [25°C, 72 hr], < 50°C [pH 6.0, 1hr])				
3-5 U/mg solid	1 unit oxidizes 1 μmol protocatechuate to 3-carboxy-cis,cis-muconate/min at pH 7.5, 37°C	Lyophilized containing 40% stabilizers		ICN 151975
3-5 U/mg solid	1 unit oxidizes 1.0 μmol protocatechuate to 3-carboxy-cis,cis-muconate/min at pH 7.5, 37°C	Lyophilized containing 65% protein; balance primarily sugar stabilizer		Sigma P8279
≥3 U/mg solid	1 unit forms 1 μmol protocatechuate/min at pH 7.5, 37°C	Lyophilized containing 40% sugars as stabilizers	<0.1% NADPH oxidase	Toyobo PCO-301

Lipoxygenase

1.13.11.12

Reaction Catalyzed
Linoleate + O_2 ↔ (9Z,11E)-(13S)-13-hydroperoxyoctadeca-9,11-dienoate

Systematic Name
Linoleate:oxygen 13-oxidoreductase

Synonyms
Lipoxidase, carotene oxidase

Reactants
Linoleate, methylene-interrupted polyunsaturated fatty acids, O_2, (9Z,11E)-(13S)-13-hydroperoxyoctadeca-9,11-dienoate

Applications
- Enzyme-catalyzed hydroperoxide formation in microemulsions containing nonionic surfactants
- Degradation of hydroperoxide by lipoxygenase prevents the breakdown of amino acids and proteins associated with odorous carbyl compounds (i.e., the beany flavor in legumes)

Notes
- Classified as an oxidoreductase acting on single donors with incorporation of molecular oxygen (oxygenases), and incorporation of two atoms of oxygen
- An iron protein
- Isozymes have been reported in cow peas

SPECIFIC ACTIVITY	UNITS DEFINITION	PREPARATION FORM	ADDITIONAL ACTIVITIES	SUPPLIER CATALOG NO.
Soybeans				
≥100,000 U/mg solid	1 unit increases E_{234} by 0.001/min at pH 9.0, 25°C	Salt-free; lyophilized		Biozyme LPX1
≥250,000 U/mg protein	1 unit increases E_{234} by 0.001/min at pH 9.0, 25°C	Chromatographically prepared; suspension in 60% $(NH_4)_2SO_4$		Biozyme LPX2
≥200,000 U/mg solid	1 unit increases E_{234} by 0.001/min at pH 9.0, 25°C	Salt-free; lyophilized		Biozyme LPX2F
10 U/mg protein	1 unit increase $A_{232.5}$ by 0.001/min at pH 9.0, 25°C using linoleic acid as substrate in 3.0 mL volume (1 cm light path); 1 $A_{232.5}$ unit = oxidation of 0.12 μmol linoleic acid	60% purity; lyophilized		Calzyme 198A0010
9 U/mg	1 unit forms 1 μmol linoleic acid hydroperoxide/min at pH 9.0, 25°C with linoleate as substrate	Lyophilized		Fluka 62340
250,000 U/mg	1 unit increases A_{234} by 0.001/min at pH 9.0, 25°C due to linoleate oxidation	Purified; salt-free; lyophilized		ICN 100825
≥10 IU/mg protein	1 IU transforms 1 μmol substrate/min under standard IUB conditions at 25°C	Lyophilized		OYC

Lipoxygenase continued 1.13.11.12

SPECIFIC ACTIVITY	UNITS DEFINITION	PREPARATION FORM	ADDITIONAL ACTIVITIES	SUPPLIER CATALOG No.
Soybeans *continued*				
300,000-600,000 U/mg protein	1 unit increases A_{234} 0.001/min at pH 9.0, 25°C using linoleic acid as substrate in 3.0 volume (1 cm light path); 1 A_{234} U equals oxidation of 0.12 µmol linoleic acid	2X Crystallized; suspension in 2.3 M $(NH_4)_2SO_4$, pH 6		sigma L3004
500,000-1,000,000 U/mg protein	1 unit increases A_{234} 0.001/min at pH 9.0, 25°C using linoleic acid as substrate in 3.0 volume (1 cm light path); 1 A_{234} U equals oxidation of 0.12 µmol linoleic acid	Affinity purified; suspension in 2.3 M $(NH_4)_2SO_4$, pH 6		sigma L6632
150,000-350,000 U/mg protein	1 unit increases A_{234} 0.001/min at pH 9.0, 25°C using linoleic acid as substrate in 3.0 volume (1 cm light path); 1 A_{234} U equals oxidation of 0.12 µmol linoleic acid	Lyophilized containing 60% protein; balance stabilizer and NaCl		sigma L7395
60,000-120,000 U/mg protein	1 unit increases A_{234} 0.001/min at pH 9.0, 25°C using linoleic acid as substrate in 3.0 volume (1 cm light path); 1 A_{234} U equals oxidation of 0.12 µmol linoleic acid	Lyophilized containing 60% protein		sigma L8383

Arachidonate 12-Lipoxygenase 1.13.11.31

REACTION CATALYZED
  Arachidonate + O_2 ↔ (5Z,8Z,10E,14Z)-(12S)-12-Hydroperoxyicosa-5,8,10,14-tetraenoate

SYSTEMATIC NAME
  Arachidonate:oxygen 12-oxidoreductase

SYNONYMS
  12-Lipoxygenase

REACTANTS
  Arachidonate, O_2, (5Z,8Z,10E,14Z)-(12S)-12-Hydroperoxyicosa-5,8,10,14-tetraenoate

NOTES
- Classified as an oxidoreductase acting on single donors with incorporation of molecular oxygen (oxygenases), and incorporation of two atoms of oxygen
- Product is rapidly reduced to the corresponding 12S-hydroxy compound

1.13.11.31 Arachidonate 12-Lipoxygenase continued

SPECIFIC ACTIVITY	UNITS DEFINITION	PREPARATION FORM	ADDITIONAL ACTIVITIES	SUPPLIER CATALOG No.
Porcine leukocytes				
50 U/mg protein	1 unit converts 1 nmol arachidonic acid to 12-HPETE/min at 20°C	Partially purified; 20 mM Tris-HCl, 0.5 mM EDTA, PMSF, DTT, pH 7.2		BIOMOL PL-012
300 U/mL	1 unit consumes 1 nmol O_2/min in 0.1 M Tris-HCl, 5 mM EDTA, 0.03% Tween 20, 100 mM arachidonate at pH 7.5, 30°C	20 mM phosphate buffer, pH 7.5		Cayman 60300
50 U/mg protein	1 unit converts 1 nmol arachidonic acid to 12-HPETE/min at 20°C	20 mM Tris-HCl, 0.5 mM EDTA, PMSF, DTT, pH 7.2		ICN 159179

1.13.11.33 Arachidonate 15-Lipoxygenase

REACTION CATALYZED
Arachidonate + O_2 ↔ (5Z,8Z,11Z,13E)-(15S)-15-Hydroperoxyicosa-5,8,11,13-tetraenoate

SYSTEMATIC NAME
Arachidonate:oxygen 15-oxidoreductase

SYNONYMS
15-Lipoxygenase, lipoxygenase P4

SUBSTRATES
Arachidonate, O_2, (5Z,8Z,11Z,13E)-(15S)-15-Hydroperoxyicosa-5,8,11,13-tetraenoate

NOTES
- Classified as an oxidoreductase acting on single donors with incorporation of molecular oxygen (oxygenases), and incorporation of two atoms of oxygen
- Product is rapidly reduced to the corresponding 15S-hydroxy compound

SPECIFIC ACTIVITY	UNITS DEFINITION	PREPARATION FORM	ADDITIONAL ACTIVITIES	SUPPLIER CATALOG No.
Rabbit reticulocytes				
8000 U/mg protein	1 unit converts 1 nmol linoleic acid/min at 20°C	Partially purified; 10 mM KPO_4 buffer, pH 6.3		BIOMOL PL-015
8 KU/mg protein	1 unit converts 1 nmol linoleic acid/min at 20°C	10 mM KPO_4, pH 6.3		ICN 159180
Soybean (Provar) (optimum pH = 7.0, MW = 94,000 Da)				
1,021,277 U/mg; 1,440,000 U/mL	1 unit increases A_{234} by 0.001 at pH 9.0, 25°C in 0.02% linoleate and 0.1 M borate buffer in a 3.0 mL total volume	≥98% purity; 0.1 M Tris-HCl, 10% glycerol, 1.41 mg/mL protein, pH 7.0		Cayman 60700

Arachidonate 15-Lipoxygenase continued

1.13.11.33

SPECIFIC ACTIVITY	UNITS DEFINITION	PREPARATION FORM	ADDITIONAL ACTIVITIES	SUPPLIER CATALOG NO.
Soybean (Provar) *continued*				
531,417 U/mg	1 unit increases A_{234} by 0.001 at pH 7.0, 25°C in 0.28% linoleate, 0.14 M PO_4 buffer, 0.1% Tween 20 in a 1.0 mL total volume	≥98% purity; 0.1 M Tris-HCl, 10% glycerol, 132 µg/mL protein, pH 7.0		Cayman 60710

Arachidonate 5-Lipoxygenase

1.13.11.34

REACTION CATALYZED
Arachidonate + O_2 ↔ (6E,8Z,11Z,14Z)-(5S)-5-Hydroperoxyicosa-6,8,11,14-tetraenoate

SYSTEMATIC NAME
Arachidonate:oxygen 5-oxidoreductase

SYNONYMS
Leukotriene-A_4 synthase, 5-lipoxygenase

REACTANTS
Arachidonate, O_2, (6E,8Z,11Z,14Z)-(5S)-5-Hydroperoxyicosa-6,8,11,14-tetraenoate, (7E,9E,11Z,14Z)-(5S,6S)-5,6-epoxyicosa-7,9,11,14-tetraenoate (leukotriene A_4)

NOTES
- Classified as an oxidoreductase acting on single donors with incorporation of molecular oxygen (oxygenases), and incorporation of two atoms of oxygen

SPECIFIC ACTIVITY	UNITS DEFINITION	PREPARATION FORM	ADDITIONAL ACTIVITIES	SUPPLIER CATALOG NO.
Potato				
24 U/mg protein (linoleic acid); 8 U/mg protein (arachidonic acid)	1 unit consumes 1 µmol oxygen/min at 30°C	98% purity; 2 M $(NH_4)_2SO_4$, 0.15 M KPO_4, pH 6.3		BIOMOL PL-005
60,400 U/mg; 11,000 U/mL	1 unit consumes 1 nmol oxygen/min at pH 6.3, 25°C in 0.1 M PO_4 buffer and 200 µM linoleate (arachidonate as substrate gives 3 times lower activity)	≥98% purity; suspension in 0.1 M PO_4 and 2 M $(NH_4)_2SO_4$, 0.182 mg/mL protein, pH 6.3		Cayman 60400

1.13.11.34 Arachidonate 5-Lipoxygenase continued

SPECIFIC ACTIVITY	UNITS DEFINITION	PREPARATION FORM	ADDITIONAL ACTIVITIES	SUPPLIER CATALOG NO.
Potato continued				
8 U/mg protein with arachidonic acid; 24 U/mg protein with linoleic acid	1 unit consumes 1 µmol O_2/min at 30°C	2 M $(NH_4)_2SO_4$ and 0.15 M KPO_4, pH 6.3		ICN 159178

1.13.12.4 Lactate 2-Monooxygenase

REACTION CATALYZED
    (S) Lactate + O_2 ↔ Acetate + CO_2 + H_2O

SYSTEMATIC NAME
    (S)-Lactate:oxygen 2-oxidoreductase (decarboxylating)

SYNONYMS
    Lactate oxidative decarboxylase, lactate oxidase, lactate oxidase II, LO, LOX

REACTANTS
    (S) Lactate, O_2, acetate, CO_2, H_2O

APPLICATIONS
- Determination of lactate and transaminases in biological fluids

NOTES
- Classified as an oxidoreductase acting on single donors with incorporation of molecular oxygen (oxygenases), and incorporation of one atom of oxygen (internal monooxygenases or internal mixed function oxidases)
- A flavoprotein (FMN)

SPECIFIC ACTIVITY	UNITS DEFINITION	PREPARATION FORM	ADDITIONAL ACTIVITIES	SUPPLIER CATALOG NO.
Aerococcus viridans (optimum pH = 6.0-7.0, pI = 4.6, T = 35°C, MW = 80,000 Da [gel filtration]; K_M = 0.7 mM [L-lactate]; stable pH 6-9 [50°C, 10 min], T < 50°C [pH 7, 10 min])				
20-60 U/mg solid	1 unit generates 1 µmol H_2O_2/min at pH 6.5, 37°C	Lyophilized containing 20% protein		Asahi LOXII T-47
Aerococcus viridans, recombinant (optimum pH = 6.0-7.0)				
>50 U/mg protein; >20 U/mg solid	1 unit oxidizes 1.0 µmol lactate/min at pH 6.5, 37°C	Lyophilized		Genzyme 1381

SPECIFIC ACTIVITY	UNITS DEFINITION	PREPARATION FORM	ADDITIONAL ACTIVITIES	SUPPLIER CATALOG NO.
Pediococcus species (optimum pH = 6.0-7.0, pI = 4.6, MW = 80,000 Da [gel filtration]; K_M = 0.7 mM [L-lactate]; stable pH 7-9 [37°C, 1 hr], T < 40°C [pH 7, 10 min])				
20-60 U/mg solid	1 unit generates 1 μmol H_2O_2/min at 37°C	Lyophilized containing 40% protein	<0.00002% uricase <0.0004% pyruvate oxidase <0.0006% GO <0.001% ChO	Asahi LOX T-13
55 U/mg protein; 20 U/mg solid	1 unit converts 1 μmol L-lactate to product/min at 25°C	Lyophilized		Boehringer 981222
30 U/mg	1 unit oxidizes 1 μmol L-lactate/min at pH 6.5, 37°C	Lyophilized		Fluka 61315
>50 U/mg protein; >20 U/mg solid	1 unit oxidizes 1.0 μmol lactate/min at pH 6.5, 37°C	Lyophilized	<0.00002% uricase <0.0004% pyruvate oxidase <0.0006% GO <0.001% ChO	Genzyme 1361
20-40 U/mg solid	1 unit oxidizes 1.0 μmol L-lactate to pyruvate and H_2O_2/min at pH 6.5, 37°C	Lyophilized		Sigma L0638

Photinus-Luciferin 4-Monooxygenase (ATP-Hydrolyzing)   1.13.12.7

REACTION CATALYZED

Photinus luciferin + O_2 + ATP ↔ Oxidized *Photinus* luciferin + CO_2 + H_2O + AMP + pyrophosphate + hν

SYSTEMATIC NAME

Photinus luciferin:oxygen 4-oxidoreductase (decarboxylating, ATP-hydrolyzing)

SYNONYMS

Firefly luciferase, ciferase

REACTANTS

Photinus luciferin, O_2, ATP, oxidized *Photinus* luciferin, CO_2, H_2O, AMP, pyrophosphate, hν

APPLICATIONS

- Determination of ATP in ATP-generating or consuming reactions
- Determination of ATPase and creatine kinase
- Determination of GTP, ITP and CTP in luciferase-luciferin light emission systems, with nucleoside diphosphokinase and ADP
- Detection of small molecular weight antigens like methotrexate, trinitrotoluene and dinitrophenol

1.13.12.7 *Photinus*-Luciferin 4-Monooxygenase (ATP-Hydrolyzing) *continued*

NOTES

- Classified as an oxidoreductase acting on single donors with incorporation of molecular oxygen (oxygenases), and incorporation of one atom of oxygen (internal monooxygenases or internal mixed function oxidases)
- *Photinus* luciferin is (S)-4,5-dihydro-2-(6-hydroxy-2-benzothiazoloyl)-4-thiazolecarboxylic acid
- An acid anhydride is formed between the carboxylic group and AMP in the first reaction step, with the release of pyrophosphate
- The enzyme may be measured by light emission

SPECIFIC ACTIVITY	UNITS DEFINITION	PREPARATION FORM	ADDITIONAL ACTIVITIES	SUPPLIER CATALOG No.
Photinus pyralis (American firefly)				
8 mU/mg protein	1 unit converts 0.1 mM luciferin and 0.7 mM ATP to product/min at pH 7.75, 25°C with 0.1 M Tris-acetate	2X Crystallized; lyophilized	No detectable luciferin, arsenate <1 mU/mg solid ATPase <10 mU/mg solid nucleoside-diphosphatase kinase	Boehringer 411523 634409
30-50 U/mg protein	1 unit causes light emission of 1 μmol photon/min at pH 7.0, 25°C	Lyophilized containing 90% protein		Calzyme 199A0030
		Solid containing 0.005% Na arsenate and NaPO$_4$, 0.009% NaCl, 0.1% Tris, 0.3% D-luciferin, 4% EDTA, 10% MgSO$_4$, 19% HAS, 66% Gly	0.2% luciferase	Fluka 62646
	Sensitivity: <1 femtomol ATP can be determined using 20 μg luciferase	Lyophilized containing 90% protein	<0.005 U/mg ATPase <0.02 U/mg nucleoside-diphosphokinase	Fluka 62647
	Sensitivity: <1 femtomol ATP can be determined using 20 μg luciferase	Powder containing 15% protein and 85% Tris succinate	<0.005 U/mg ATPase <0.02 U/mg MK, nucleoside-diphosphokinase	Fluka 62648
10-25 X 10^6 light U/mg protein	1 light unit produces a biometer peak height equivalent to 0.02 μCi of ^{14}C in PPO/POPOP cocktail in 50 mL assay mixture at pH 7.4, 25°C	Crystallized; lyophilized containing 15% protein; balance primarily Tris and Asp		Sigma L1759
15-30 X 10^6 light U/mg protein	1 light unit produces a biometer peak height equivalent to 0.02 μCi of ^{14}C in PPO/POPOP cocktail in 50 mL assay mixture at pH 7.4, 25°C	Chromatographically purified; crystallized; lyophilized containing 90% protein; balance primarily NaPO$_4$, potassium arsenate, NaCl, EDTA	<5 nM U/mg protein ATPase 20 nM U/mg protein nucleoside-diphosphokinase	Sigma L5256

Photinus-Luciferin 4-Monooxygenase (ATP-Hydrolyzing) continued

1.13.12.7

SPECIFIC ACTIVITY	UNITS DEFINITION	PREPARATION FORM	ADDITIONAL ACTIVITIES	SUPPLIER CATALOG NO.
Photinus pyralis continued				
10-25 × 10^6 light U/mg protein	1 light unit produces a biometer peak height equivalent to 0.02 μCi of ^{14}C in PPO/POPOP cocktail in 50 mL assay mixture at pH 7.4, 25°C	Crystallized; lyophilized containing 15% protein; balance primarily Tris and succinic acid		Sigma L9009
Photinus pyralis, recombinant				
30 U/mg protein	1 unit releases 1 nmol pyrophosphate from ATP/min at 25°C in the presence of 540 μM ATP and 100 μM D-luciferin	Crystallized > 95% purity (SDS-PAGE)	No detectable ATP, nucleoside-5´-diphosphate kinase	Calbiochem 440325

Salicylate 1-Monooxygenase

1.14.13.1

REACTION CATALYZED

Salicylate + NADH + O$_2$ ↔ Catechol + NAD$^+$ + H$_2$O + CO$_2$

SYSTEMATIC NAME

Salicylate,NADH:oxygen oxidoreductase (1-hydroxylating, decarboxylating)

SYNONYMS

Salicylate hydroxylase

REACTANTS

Salicylate, NADH, O$_2$, catechol, NAD$^+$, H$_2$O

NOTES

- Classified as an oxidoreductase acting on paired donors with incorporation of molecular oxygen, with NADH or NADPH as one donor, and incorporation of one atom of oxygen
- A flavoprotein (FAD)

SPECIFIC ACTIVITY	UNITS DEFINITION	PREPARATION FORM	ADDITIONAL ACTIVITIES	SUPPLIER CATALOG NO.
Microbial (optimum pH = 7.6, T = 45°C)				
> 7 U/mg protein	1 unit converts 1 μmol NADH/min at pH 7.6, 37°C	Lyophilized		GDS Tech SR-100
Pseudomonas species				
0.5-3 U/mg protein	1 unit converts 1.0 μmol salicylate and β-NADH to catechol and β-NAD/min at pH 7.6, 30°C	Lyophilized containing 45% protein; balance primarily Tris PO$_4$ and FAD		Sigma S2907

4-Hydroxybenzoate 3-Monooxygenase

1.14.13.2

REACTION CATALYZED

4-Hydroxybenzoate + NADPH + O_2 ↔
Protocatechuate + $NADP^+$ + H_2O

SYSTEMATIC NAME

4-Hydroxybenzoate,NADPH:oxygen oxidoreductase (3-hydroxylating)

SYNONYMS

p-Hydroxybenzoate hydrolyase

REACTANTS

4-Hydroxybenzoate, NADPH, O_2, protocatechuate, $NADP^+$, H_2O

NOTES

- Classified as an oxidoreductase acting on paired donors with incorporation of molecular oxygen, with NADH or NADPH as one donor, and incorporation of one atom of oxygen
- A flavoprotein; 1 mol FAD per mol enzyme
- The *Pseudomonas* enzyme is highly specific for NADPH
- Inhibited by Ag^+, Hg^{2+}, PCMB and SDS
- Stabilized by sugars and FAD

SPECIFIC ACTIVITY	UNITS DEFINITION	PREPARATION FORM	ADDITIONAL ACTIVITIES	SUPPLIER CATALOG No.
Pseudomonas species (optimum pH = 7.7-7.9, T = 35°C, MW = 55,000-60,000 Da; K_M = 20 μM [p-hydroxybenzoate], 40 μM [NADPH]; stable pH 5.0-7.5 [25°C, 72 hr], T < 40°C [pH 6.0, 15 min])				
20 U/mg solid	1 unit hydroxylates 1.0 μmol p-hydroxybenzoate to protocatechuate/min at pH 8.2, 37°C in the presence of NADPH	Lyophilized containing 40% protein; balance mannitol and buffer salts		ICN 151286
20 U/mg solid	1 unit hydroxylates 1.0 μmol p-hydroxybenzoate to protocatechuate/min at pH 8.2, 37°C in the presence of NADPH	Lyophilized containing 40% protein; balance mannitol and buffer salts		Sigma H9886
≥20 U/mg solid	1 unit oxidizes 1 μmol NADPH/min at pH 8.2, 37°C	Lyophilized containing 40% sucrose and FAD as stabilizers	<0.1% NADPH oxidase	Toyobo HBH-301

Cyclopentanone Monooxygenase 1.14.13.16

REACTION CATALYZED

Cyclopentanone + NADPH + O_2 ↔ 5-Valerolactone + $NADP^+$ + H_2O

SYSTEMATIC NAME

Cyclopentanone,NADPH:oxygen oxidoreductase (5-hydroxylating, lactonizing)

REACTANTS

Cyclopentanone, NADPH, O_2, 5-valerolactone, $NADP^+$, H_2O

NOTES

- Classified as an oxidoreductase acting on paired donors with incorporation of molecular oxygen, with NADH or NADPH as one donor, and incorporation of one atom of oxygen

SPECIFIC ACTIVITY	UNITS DEFINITION	PREPARATION FORM	ADDITIONAL ACTIVITIES	SUPPLIER CATALOG NO.
Pseudomonas species				
0.1 U/mg	1 unit catalyzes the cyclopentanone-stimulated oxidation of 1 μmol NADPH/min at pH 7.7, 30°C	Powder		Fluka 29800

1.14.13.22 Cyclohexanone Monooxygenase

REACTION CATALYZED
Cyclohexanone + NADPH + O_2 ↔ 6-Hexanolide + $NADP^+$ + H_2O

SYSTEMATIC NAME
Cyclohexanone,NADPH:oxygen oxidoreductase (6-hydroxylating, 1,2-lactonizing)

REACTANTS
Cyclohexanone, NADPH, O_2, 6-hexanolide, $NADP^+$, H_2O

APPLICATIONS
- Introduction of enantioselectivity in the classical Baeyer-Villiger oxidation, e.g., enantioselective synthesis of lactones from achiral cyclohexanones and bicyclo(3.2.0)heptanones

NOTES
- Classified as an oxidoreductase acting on paired donors with incorporation of molecular oxygen, with NADH or NADPH as one donor, and incorporation of one atom of oxygen
- A flavoprotein (FAD)
- Acts on other cyclic ketones

SPECIFIC ACTIVITY	UNITS DEFINITION	PREPARATION FORM	ADDITIONAL ACTIVITIES	SUPPLIER CATALOG NO.
Acinetobacter species				
0.5 U/mg	1 unit catalyzes the cyclohexanone-stimulated oxidation of 1 μmol NADPH/min at pH 9.0, 30°C	Powder		Fluka 29170

1.14.13.39 Nitric-Oxide Synthase

REACTION CATALYZED
L-Arginine + ηNADH + $m$$O_2$ ↔ Citrulline + nitric oxide + $\eta$$NADP^+$

SYSTEMATIC NAME
L-Arginine,NADPH:oxygen oxidoreductase (nitric-oxide-forming)

SYNONYMS
NOS

REACTANTS
L-Arginine, NADH, O_2, Citrulline, nitric oxide, $NADP^+$

APPLICATIONS
- Study of enzyme regulation and kinetics

NOTES
- Classified as an oxidoreductase acting on paired donors with incorporation of molecular oxygen, with NADH or NADPH as one donor, and incorporation of one atom of oxygen
- Brain enzyme requires Ca^{2+}
- The stoichiometry may involve two-electron and one-electron oxidation steps

Nitric-Oxide Synthase continued

1.14.13.39

SPECIFIC ACTIVITY	UNITS DEFINITION	PREPARATION FORM	ADDITIONAL ACTIVITIES	SUPPLIER CATALOG NO.
Bovine brain				
10-50 U/mg protein	1 unit forms 1.0 nmol citrulline from Arg/min at pH 7.3, 37°C	Solution containing 1 mM DTT, 50 mM Tris, 50% glycerol, 1 mM CaCl$_2$, 0.5 mM EDTA, 0.1 mM EGTA, 0.5% Triton X-100, pH 7.5		Sigma N4523
Murine macrophage, isolated from baculovirus (MW = 130,000 Da)				
92 U/mg	1 unit produces 1 nmol nitric oxide/hr at pH 7.4, 37°C in 50 mM HEPES	Purity: 100,000 X g supernatant; 50 mM HEPES, 10% glycerol, 24 mg/mL protein, pH 7.4		Cayman 60862
Rat brain, recombinant				
0.6 μmol L-citrulline/min/mg	pH 7.0, 37°C	cDNA isotype; >95% purity by SDS-PAGE		BIOMOL SE-126
Rat neuronal, overexpressed in SF9 cells, isolated from baculovirus (MW = 150,000 Da)				
2100 U/mg	1 unit produces 1 nmol nitric oxide/hr at pH 7.4, 37°C in 50 mM HEPES	Purity: 100,000 X g supernatant; 50 mM HEPES, 10% glycerol, 100 μM DTT, 1.7 mg/mL protein, pH 7.4		Cayman 60870

Tyrosine *N*-Monooxygenase

1.14.13.41

REACTION CATALYZED

L-Tyrosine + NADPH + O$_2$ ↔ *N*-Hydroxy-L-tyrosine + NADP$^+$ + H$_2$O

SYSTEMATIC NAME

L-Tyrosine,NADPH:oxygen oxidoreductase (*N*-hydroxylating)

SYNONYMS

Tyrosine *N*-hydroxylase, tyrosine hydroxylase, TH

REACTANTS

L-Tyrosine, NADPH, O$_2$, *N*-hydroxy-L-tyrosine, H$_2$O, NADP$^+$

NOTES

- Classified as an oxidoreductase acting on paired donors with incorporation of molecular oxygen, with NADH or NADPH as one donor, and incorporation of one atom of oxygen
- A heme-thiolate protein (*P*-450)

SPECIFIC ACTIVITY	UNITS DEFINITION	PREPARATION FORM	ADDITIONAL ACTIVITIES	SUPPLIER CATALOG NO.
Bovine adrenal glands				
10-100 U/mg protein	1 unit forms 1.0 nmol L-DOPA from Tyr/min at pH 7.0, 37°C	Solution containing 50% glycerol, 20 mM Tris, 1 mM DTT, 2% sucrose, 2 mg/mL BSA, pH 7.0		Sigma T8799

Alkanal Monooxygenase (FMN-linked)

REACTION CATALYZED
 $RCHO + FMNH_2 + O_2 \leftrightarrow RCOOH + FMN + H_2O + h\nu$

SYSTEMATIC NAME
 Alkanal,reduced-FMN:oxygen oxidoreductase (1-hydroxylating, luminescing)

SYNONYMS
 Bacterial luciferasealdehyde monooxygenase, LU

REACTANTS
 $RCHO$, $FMNH_2$, O_2, $RCOOH$, FMN, H_2O

APPLICATIONS
- Determination of FAD, FMN and trace metabolic intermediates in pyridine nucleotide-dependent dehydrogenase reactions

NOTES
- Classified as an oxidoreductase acting on paired donors with incorporation of molecular oxygen, with reduced flavin or flavoprotein as one donor, and incorporation of one atom of oxygen
- Bacterial luciferase (LU) is composed of distinct a and b subunits, each of molecular weight 40,000 Da; there is one reduced FMN binding site per a-b dimer
- Highly specific for reduced FMN and long-chain aliphatic aldehydes with eight carbons or more
- Reaction sequence:
  - incorporation of oxygen into reduced FMN
  - reaction with the aldehyde to form an activated $FMN.H_2O$ complex
  - breakdown of the activated complex with emission of light
- Aldehyde activates luminescence, but is not essential
- NADH is preferred in analytical applications since $FMNH_2$ is so rapidly oxidized by air
- Inhibited by PCMB, riboflavin and low concentrations of oxidase inhibitors

SPECIFIC ACTIVITY	UNITS DEFINITION	PREPARATION FORM	ADDITIONAL ACTIVITIES	SUPPLIER CATALOG NO.
Bacterial				
3 U/mg protein	1 unit causes light emission of 1 μmol photon/min at pH 7.0, 25°C	Lyophilized containing 95% protein		Calzyme 199A0003
***Photobacterium fischeri (Vibrio fischeri)* (optimum pH = 6.8, MW = 2 x 40,000 Da)**				
15 mU/mg protein	1 unit converts 1 μmol FMN and myristine aldehyde to product/min at 25°C, 50 mU NAD(P)H:FMN oxidoreductase/mL assay solution (based on light emission in ATP system under formation of pyrophosphate with 1 mU luciferase from *Photinus pyralis*)	Lyophilized	<1 mU/mg protein NADH oxidase <10 mU /mg protein MK, NADH:FMN oxidoreductase	Boehringer 476498
15 U/g protein		Lyophilized containing 8% protein		Fluka 62641
	1 mg/mL produces 80% transmittance	Partially purified; powder		ICN 100826

SPECIFIC ACTIVITY	UNITS DEFINITION	PREPARATION FORM	ADDITIONAL ACTIVITIES	SUPPLIER CATALOG NO.
Photobacterium fischeri (Vibrio fischeri) continued				
	Produces light in a system containing FMN, NADH or NADPH and n-decyl aldehyde	Partially purified; lyophilized containing 50% protein; balance primarily buffer salts and stabilizer	FMN-dependent luciferase and NADH- and NADPH-dependent FMN reductases	Sigma L8507
	Gives light response with reduced flavin mononucleotide and long chain aldehyde	Partially purified; powder		Worthington LS04686 LS04689
Vibrio harveyi (optimum pH = 6.8, pI = 4.3, T = 30°C, MW = 79,000 Da; K_M = 0.3 μM [FMNH$_2$]; stable pH 6.5-7.5 [30°C, 6 hr], T < 35°C [pH 7.0, 10 min])				
	Produces light in a system containing FMN, NADH or NADPH and n-decyl aldehyde	Partially purified; soluble; lyophilized containing 50% protein; balance KPO$_4$ buffer salts and stabilizer	FMN-dependent luciferase and NADH- and NADPH-dependent FMN reductases	Sigma L1637
≥0.03 U/mg solid	1 unit causes the light emission of 1 μmol photon/min at pH 7.0, 25°C	Lyophilized containing sugar and amino acid as stabilizers		Toyobo LUC-201

Dopamine β-Monooxygenase

REACTION CATALYZED

3,4-Dihydroxyphenethylamine + ascorbate + O_2 ↔ Noradrenaline + dehydroascorbate + H_2O

SYSTEMATIC NAME

3,4-Dihydroxyphenethylamine,Ascorbate:oxygen oxidoreductase (β-hydroxylating)

SYNONYMS

Dopamine β-hydroxylase

REACTANTS

3,4-Dihydroxyphenethylamine, ascorbate, O_2, noradrenaline, dehydroascorbate, H_2O

NOTES

- Classified as an oxidoreductase acting on paired donors with incorporation of molecular oxygen, with ascorbate as one donor, and incorporation of one atom of oxygen
- A copper protein
- Stimulated by fumarate

SPECIFIC ACTIVITY	UNITS DEFINITION	PREPARATION FORM	ADDITIONAL ACTIVITIES	SUPPLIER CATALOG NO.
Bovine adrenal glands				
2 U/mg	1 unit converts 1 μmol tyramine to octopamine/min at pH 5.0, 37°C with starch as substrate	Powder		Fluka 44248
2-5 U/mg protein	1 unit converts 1.0 μmol tyramine to octopamine/min at pH 5.0, 37°C	Lyophilized containing 95% protein; balance $NaPO_4$ buffer salt		Sigma D1147
1-3 U/mg protein	1 unit converts 1.0 μmol tyramine to octopamine/min at pH 5.0, 37°C	Suspension in 3.6 M $(NH_4)_2SO_4$, pH 7		Sigma D1893

Monophenol Monooxygenase 1.14.18.1

REACTION CATALYZED

L-Tyrosine + L-dopa + O_2 ↔ L-Dopa + dopaquinone + H_2O

SYSTEMATIC NAME

Monophenol,L-dopa:oxygen oxidoreductase

SYNONYMS

Tyrosinase, phenolase, monophenol oxidase, cresolase, polyphenol oxidase

REACTANTS

L-Tyrosine, L-dopa, 1,2-benzenediols, O_2, L-dopa, dopaquinone, H_2O

APPLICATIONS

- Causes cut surfaces of many fruits and vegetables to darken
- Oxidizes tyrosine to melanin pigment in mammals
- Suicide inactivation of tyrosinase
- Browning reactions

NOTES

- Classified as an oxidoreductase acting on paired donors with incorporation of molecular oxygen, with another compound as one donor, and incorporation of one atom of oxygen
- A bifunctional copper oxidase having catecholase and cresolase activity
- A tetramer containing 4 gram atoms of copper and 2 binding sites for substrate per molecule, and a distinct binding site for oxygen
- Substrates are aromatic and phenolic compounds
- Also catalyzes catechol oxidase reaction (EC 1.10.3.1) if 1,2-benzenediols are only available substrates
- Four isozymes are reported in tissues of mice with malignant melanoma
- Inhibited by compounds that complex with copper, and competitively inhibited by benzoate (catechol) and cyanide (oxygen)

SPECIFIC ACTIVITY	UNITS DEFINITION	PREPARATION FORM	ADDITIONAL ACTIVITIES	SUPPLIER CATALOG NO.
Mushrooms (optimum pH = 6-7, pI = 5, MW = 128,000 Da)				
2000 U/mg solid	1 unit increases E_{280} by 0.001/min at pH 6.5, 25°C	Lyophilized containing 90% protein		Calzyme 115A2000
500 U/mg solid	1 unit increases E_{280} by 0.001/min at pH 6.5, 25°C	Lyophilized containing 90% protein		Calzyme 159A0500
125 U/mg	1 unit oxidizes 1 μmol 4-methyl-catechol to 4-methyl-1,2-benzoquinone/min at pH 6.5, 25°C	Lyophilized		Fluka 93898
2000 U/mg solid	1 unit yields ΔA_{280} of 0.001/min at pH 6.5, 25°C in 3 mL reaction mix containing L-Tyr		Contains polyphenol oxidase and catechol oxidase activity where 1 U = ΔA_{265} of 0.001/min at pH 6.5, 25°C in 3 mL reaction mix containing L-dihydroxy-Phe or catechol and L-ascorbic acid	Sigma T7755

1.14.18.1 Monophenol Monooxygenase continued

SPECIFIC ACTIVITY	UNITS DEFINITION	PREPARATION FORM	ADDITIONAL ACTIVITIES	SUPPLIER CATALOG No.
Mushrooms continued				
2000 U/mg		Lyophilized		Wako 205-11541
500 U/mg		Lyophilized		Wako 208-11531
≥500 U/mg solid	1 unit causes ΔA_{280} of 0.001/min at pH 6.5, 25°C with L-Tyr as substrate	Dialyzed; lyophilized		Worthington LS03789 LS03792 LS03793 LS03791

1.14.99.1 Prostaglandin-Endoperoxide Synthase

REACTION CATALYZED
Arachidonate + AH_2 + 2 O_2 ↔ Prostaglandin H_2 + A + H_2O

SYSTEMATIC NAME
(5Z,8Z,11Z,14Z)-Icosa-5,8,11,14-tetraenoate,hydrogen-donor:oxygen oxidoreductase

SYNONYMS
Prostaglandin synthase, prostaglandin G/H synthase, cyclooxygenase

REACTANTS
Arachidonate, AH_2, O_2, prostaglandin H_2, A, H_2O

NOTES
- Classified as an oxidoreductase acting on paired donors with incorporation of molecular oxygen, with *miscellaneous*
- Acts as both a dioxygenase and a peroxidase
- Catalyzes the sequential bis-oxygenation of arachidonate to prostaglandin G_2, which is reduced to prostaglandin H_2 through enzyme hydroperoxidase activity
- Contains Fe^{3+} protoporphyrin IX (hematin) as cofactor, which dissociates during enzyme purification giving apo- and holo-enzymes. Added hematin gives maximal activity

SPECIFIC ACTIVITY	UNITS DEFINITION	PREPARATION FORM	ADDITIONAL ACTIVITIES	SUPPLIER CATALOG No.
Ovine placenta				
6202 U/mg; 2852.8 U/mL	1 unit consumes 1 nmol O_2/min at pH 8.0, 37°C in 0.1 M Tris-HCl, 100 μM arachidonate and 2 mM phenol	70% purity; 80 mM Tris-HCl, 0.1% Tween 20, 300 μM diethyldithiocarbamate, 0.46 mg/mL protein, pH 8.0		Cayman 60120

SPECIFIC ACTIVITY	UNITS DEFINITION	PREPARATION FORM	ADDITIONAL ACTIVITIES	SUPPLIER CATALOG NO.
Ram seminal vesicles (MW = 70,000 Da)				
24 U/mg protein (arachidonic acid)	1 unit consumes 1 µmol O_2/min at 30°C	>95% purity; 0.1 M Tris-HCl, 1% Tween 80, pH 8.0		BIOMOL EH-001
>20 U/mg protein	1 unit consumes 1 µmol O_2/min at pH 7.0, 30°C with 0.1 mM arachidonic acid, 1 µM hematin, 0.24 µM O_2, 5 µM Trp	>95% purity; 100 mM Tris-HCl and 5 mM Trp, pH 7.4		Calbiochem 434738
30,600 U/mg; 10,800 U/mL	1 unit consumes 1 nmol O_2/min at pH 8.0, 37°C in 0.1 M Tris-HCl, 100 µM arachidonate and 2 mM phenol	≥95% purity; 80 mM Tris-HCl, 0.1% Tween 20, 300 µM diethyldithio-carbamate, 354 µg/mL protein, pH 8.0		Cayman 60100
24 U/mg protein (arachidonic acid)	1 unit consumes 1 µmol O_2/min at 30°C	>95% purity; containing 0.1 M Tris-HCl and 1% Tween 80, pH 8.0		ICN 154277

Superoxide Dismutase 1.15.1.1

REACTION CATALYZED

$$2O_2^- + 2H^+ \leftrightarrow O_2 + H_2O_2$$

SYSTEMATIC NAME

Superoxide:superoxide oxidoreductase

SYNONYMS

Erythrocuprein, hemocuprein, cytocuprein, orgotein (copper-zinc enzyme with strong anti-inflammatory properties and no toxic side effects), SOD

REACTANTS

O_2^-, H^+, O_2, H_2O_2, nitro blue tetrazolium

APPLICATIONS
- Toxic radical scavenger in anaerobe culture
- Biochemical electron transfer studies
- Copper and zinc enzyme is used in cell immunology, gene technology and cancer research

NOTES
- Classified as an oxidoreductase acting on superoxide radicals as acceptor
- Catalyzes the removal of the O_2^- free radical, protecting oxygen-metabolizing cells against the harmful effects of superoxide free radicals
- A metalloprotein: bovine and human erythrocytes contain copper and zinc; chicken and rat liver contain manganese; *E. coli* contains iron
- Catalyzes lipid peroxidation and peroxidative hemolysis of erythrocytes
- High levels are found in brain, liver, heart, erythrocytes and kidney
- Inactivated by hydrogen peroxide
- Inhibited by cyanide and diethyldithiocarbamate

1.15.1.1 Superoxide Dismutase continued

SPECIFIC ACTIVITY	UNITS DEFINITION	PREPARATION FORM	ADDITIONAL ACTIVITIES	SUPPLIER CATALOG No.
Bacillus species (optimum pH = 6.0-9.0, pI = 4.5, MW = 50,000 Da)				
10,000 U/mg		Lyophilized		Wako 195-10291
Bacillus stearothermophilus (optimum pH = 9.5, pI = 4.5, MW = 50,000 Da [25,000/subunit]; stable pH 6-9, T < 60°C)				
2000-6000 U/mg protein	1 unit inhibits the rate of reduction of cytochrome c by 50% in a coupled system with xanthine and xanthine oxidase at pH 7.8, 25°C (3.0 mL reaction volume)	Lyophilized containing 70% protein; balance primarily KPO_4 buffer salts		Sigma S4525
>10,000 U/mg protein	1 unit inhibits the rate of reduction of cytochrome c by 50% at 30°C	Lyophilized	<0.01% catalase	Unitika
>10,000 U/mg protein	1 unit inhibits the rate of reduction of cytochrome c by 50% at 30°C	50% glycerol solution	<0.01% catalase	Unitika
Bacillus stearothermophilus; recombinant expressed in E. coli				
2000-6000 U/mg protein	1 unit inhibits the rate of reduction of cytochrome c by 50% in a coupled system with xanthine and XO at pH 7.8, 25°C	Lyophilized containing Tris-HCl and lactose		Sigma S8151
Bovine erythrocyte (optimum pH = 9.0, pI = 4.95, T = 30°C, MW = 32,500 Da [2 identical subunits joined by a disulfide bond]; stable pH 7.0-8.5 [25°C, 20 hr], T < 70°C [pH 7.0, 30 min])				
≥100 U/mg solid	1 unit inhibits the rate of reduction cytochrome c by 50%	Salt-free; lyophilized		Biozyme SOD1
3500 U/mg protein; ≥3000 U/mg solid	1 unit inhibits the rate of reduction cytochrome c by 50%	Salt-free; lyophilized	<0.01% carbonic dehydratase	Biozyme SOD2
5000 U/mg solid	1 unit inhibits the auto-oxidation of pyrogallol at 50%	Lyophilized		Boehringer 567680 837113
3000 U/mg solid	1 unit causes a 50% inhibition in the rate of reduction of cytochrome c at pH 7.8, 25°C	Salt-free; lyophilized	<30 U/mg carbonic anhydrase	Calbiochem 574594
3000 U/mg protein	1 unit inhibits the rate of reduction of ferricytochrome c by 50%	Lyophilized containing 95% protein		Calzyme 205A3000
3000 U/mg	1 unit inhibits the autooxidation of pyrogallol by 50% at pH 8.2, 25°C	Powder		Fluka 86200
3500 U/mg	1 unit causes a 50% inhibition in the rate of cytochrome c	Salt-free; lyophilized		ICN 190117

SPECIFIC ACTIVITY	UNITS DEFINITION	PREPARATION FORM	ADDITIONAL ACTIVITIES	SUPPLIER CATALOG NO.
Bovine erythrocyte *continued*				
3000 U/mg	1 unit causes a 50% inhibition in the rate of cytochrome c	Salt-free; lyophilized containing traces of albumin		ICN 835001 835002
2500-7000 U/mg protein	1 unit inhibits the rate of reduction of cytochrome c by 50% in a coupled system with xanthine and XO at pH 7.8, 25°C	Lyophilized containing 98% protein; balance primarily KPO_4 buffer salts		Sigma S2515
2500-7000 U/mg protein	1 unit inhibits the rate of reduction of cytochrome c by 50% in a coupled system with xanthine and XO at pH 7.8, 25°C	Cell culture tested; lyophilized containing 98% protein; balance primarily KPO_4 buffer salts		Sigma S5395
≥3000 U/mg solid	1 unit causes half maximum inhibition of cytochrome c reduction at pH 7.8, 25°C	Lyophilized	<0.01% catalase	Toyobo SOD-301
3000-4000 U/mg		Lyophilized optimum pH = 7.8, T = 25°C		Wako 190-08771
≥1400 U/mg solid	1 unit inhibits 50% maximum reduction of nitro blue tetrazolium	Highly purified; lyophilized		Worthington LS03540 LS03541 LS03542
Bovine kidney				
3000 U/mg protein	1 unit inhibits the rate of reduction of ferricytochrome c by 50%	Lyophilized containing 95% protein		Calzyme 205A3000
4000-10,000 U/mg protein	1 unit inhibits the rate of reduction of cytochrome c by 50% in a coupled system with xanthine and XO at pH 7.8, 25°C	Suspension in 3.8 M $(NH_4)_2SO_4$, pH 7.0		Sigma S2139
Bovine liver				
3000 U/mg protein	1 unit inhibits the rate of reduction of ferricytochrome c by 50%	Lyophilized containing 95% protein		Calzyme 205A3000
2000-6000 U/mg protein	1 unit inhibits the rate of reduction of cytochrome c by 50% in a coupled system with xanthine and XO at pH 7.8, 25°C	Lyophilized containing 80% protein; balance primarily KPO_4 buffer salts		Sigma S4761
3000 U/mg protein	1 unit inhibits the rate of reduction of cytochrome c by 50% in a coupled system with xanthine and XO at pH 7.8, 25°C	Suspension in 3.8 M $(NH_4)_2SO_4$, pH 7.0		Sigma S7008

1.15.1.1 Superoxide Dismutase *continued*

SPECIFIC ACTIVITY	UNITS DEFINITION	PREPARATION FORM	ADDITIONAL ACTIVITIES	SUPPLIER CATALOG No.
Dog erythrocytes				
2000-6000 U/mg protein	1 unit inhibits the rate of reduction of cytochrome c by 50% in a coupled system with xanthine and XO at pH 7.8, 25°C (3.0 mL reaction volume)	Lyophilized containing 90% protein and KPO_4 buffer salts		Sigma S9511
Escherichia coli				
3000-6000 U/mg protein	1 unit inhibits the rate of reduction of cytochrome c by 50% in a coupled system with xanthine and XO at pH 7.8, 25°C	Lyophilized containing 70% protein; balance primarily Tris buffer salts		Sigma S5389
2500-5000 U/mg protein	1 unit inhibits the rate of reduction of cytochrome c by 50% in a coupled system with xanthine and XO at pH 7.8, 25°C	Lyophilized containing 90% protein; balance primarily Tris buffer salts		Sigma S5639
Horseradish				
1000-3000 U/mg protein	1 unit inhibits the rate of reduction of cytochrome c by 50% in a coupled system with xanthine and XO at pH 7.8, 25°C	Lyophilized containing 70% protein; balance primarily KPO_4 buffer salts		Sigma S4636
Human erythrocytes				
5000 U/mg	1 unit inhibits the autooxidation of pyrogallol by 50% at pH 8.2, 25°C	Powder	No detectable hepatitis B virus, HIV	Fluka 86198
3000-6000 U/mg protein	1 unit inhibits the rate of reduction of cytochrome c by 50% in a coupled system with xanthine and XO at pH 7.8, 25°C	Lyophilized containing 98% protein; balance primarily KPO_4 buffer salts		Sigma S9636
Yeast, bakers				
3000 U/mg protein	1 unit inhibits the rate of reduction of ferricytochrome C by 50%	Lyophilized containing 95% protein		Calzyme 205A3000

Ferroxidase

1.16.3.1

REACTION CATALYZED
$$4\ Fe(II) + 4\ H^+ + O_2 \leftrightarrow 4\ Fe(III) + 2\ H_2O$$

SYSTEMATIC NAME
Fe(II):oxygen oxidoreductase

SYNONYMS
Ceruloplasmin (animals), rusticyanin (*Thiobacillus ferroxidans*)

REACTANTS
Fe(II), H^+, O_2, Fe(III), H_2O

NOTES
- Classified as an oxidoreductase oxidizing metal ions with oxygen as acceptor
- A multi-copper protein

SPECIFIC ACTIVITY	UNITS DEFINITION	PREPARATION FORM	ADDITIONAL ACTIVITIES	SUPPLIER CATALOG NO.
Bovine				
20-60 U/mg protein; 2000-6000 U/mL	1 unit causes a ΔA_{550} of 0.01/min using N,N-dimethyl-*p*-phenylenediamine as substrate at pH 6.4, 37°C in a 7 mL reaction volume	Buffered in 0.25 M NaCl and 0.05 M NaOAc, pH 7; 80-120 µg/mL Cu		sigma C2026
Human				
30-50 U/mg protein	1 unit causes a ΔA_{550} of 0.01/min using N,N-dimethyl-*p*-phenylenediamine as substrate at pH 5.5, 37°C in a 7 mL reaction volume	Lyophilized containing 30% protein and 1.0-2.5 mg /mg protein copper; balance primarily NaCl and NaOAc		sigma C4519
25-75 U/mg protein; 2000-6000 U/mL	1 unit causes a ΔA_{550} of 0.01/min using N,N-dimethyl-*p*-phenylenediamine as substrate at pH 5.5, 37°C in a 7 mL reaction volume	Buffered in 0.25 M NaCl, 100-150 mg/mL Cu, 0.05 M NaOAc, pH 7		sigma C4770
Porcine				
50-150 U/mg protein; 6000-18,000 U/mL	1 unit causes a ΔA_{550} of 0.01/min using N,N-dimethyl-*p*-phenylenediamine as substrate at pH 5.5, 37°C in a 7 mL reaction volume	Buffered in 0.25 M NaCl, 100-150 mg/mL Cu, 0.05 M NaOAc, pH 7		sigma C3938

1.18.1.2 Ferredoxin-NADP$^+$ Reductase

REACTION CATALYZED
　Reduced ferredoxin + NADP$^+$ ↔ Oxidized ferredoxin + NADPH

SYSTEMATIC NAME
　Ferredoxin:NADP$^+$ oxidoreductase

SYNONYMS
　Adrenodoxin reductase

REACTANTS
　Reduced ferredoxin, NADP$^+$, oxidized ferredoxin, NADPH

NOTES
- Classified as an oxidoreductase acting on reduced ferredoxin as donor with NAD$^+$ or NADP$^+$ as acceptor
- A flavoprotein

SPECIFIC ACTIVITY	UNITS DEFINITION	PREPARATION FORM	ADDITIONAL ACTIVITIES	SUPPLIER CATALOG NO.
Spinach leaves				
2-6 U/mg solid	1 unit reduces 1.0 μmol NADP/min at pH 7.6, 25°C	Lyophilized	≤2.5 U/mg solid NADPH-diaphorase 15 U/mg solid NADH-diaphorase	Sigma F0628

2.1.1.6 Catechol O-Methyltransferase

REACTION CATALYZED
　S-Adenosyl-L-methionine + a catechol ↔ S-Adenosyl-L-homocysteine + a guaiacol

SYSTEMATIC NAME
　S-Adenosyl-L-methionine:catechol O-methyltransferase

SYNONYMS
　COMT

REACTANTS
　S-Adenosyl-L-methionine, catechol, catecholamine, adrenaline, noradrenaline, S-Adenosyl-L-homocysteine, guaiacol, epinephrine, norepinephrine, metanephrine, normetanephrine, dopamine

NOTES
- Classified as a transferase, transferring one-carbon groups: methyltransferase (or transmethylase)
- Mammalian enzyme acts more rapidly on catecholamines like adrenaline or noradrenaline than on catechols
- Activated by Mg^{2+}
- Enzyme inhibition increases catecholamine levels, leading to hyperexcitation

Catechol O-Methyltransferase continued 2.1.1.6

SPECIFIC ACTIVITY	UNITS DEFINITION	PREPARATION FORM	ADDITIONAL ACTIVITIES	SUPPLIER CATALOG NO.
Porcine liver				
2000 U/mg solid	1 unit methylates 1 nmol dihydroxybenzoic acid by S-adenosyl-L-(methyl-^{14}C)-Met/hr at pH 8.0, 37°C	Lyophilized		Calbiochem 219352
4000 U/mg protein	1 unit methylates 1 nmol dihydroxybenzoic acid/hr at pH 8.0, 37°C using S-adenosyl-L-(methyl-^{14}C)-Met as the methyl donor	Lyophilized in the presence of Tris buffer containing DTT as stabilizer and 90% protein	<0.01% PNMT	Calzyme 064A4000
4000 U/mg protein		Lyophilized containing Tris buffer, DTT as stabilizer and 90% protein	<0.01% PNMT	ICN 153880
1000-2000 U/mg protein	1 unit methylates 1.0 nmol protocatechuic acid/hr at pH 7.9, 37°C using S-adenosyl-L-methyl-Met	Chromatographically purified; lyophilized containing 70% protein; balance Tris buffer and DTT		Sigma C4512

Histamine N-Methyltransferase 2.1.1.8

REACTION CATALYZED

S-Adenosyl-L-methionine + histamine ↔ S-Adenosyl-L-homocysteine + N^t-methylhistamine

SYSTEMATIC NAME

S-Adenosyl-L-methionine:histamine N-tele-methyltransferase

SYNONYMS

HNMT

REACTANTS

S-Adenosyl-L-methionine, histamine, S-Adenosyl-L-homocysteine, N^t-methylhistamine

APPLICATIONS

- Determination of histamine by radio-enzymology methods

NOTES

- Classified as a transferase, transferring one-carbon groups: methyltransferase (or transmethylase)
- Responsible for inactivation of histamine (via methylation) in mammals

SPECIFIC ACTIVITY	UNITS DEFINITION	PREPARATION FORM	ADDITIONAL ACTIVITIES	SUPPLIER CATALOG NO.
Bovine kidney				
50-100 U/mg protein	1 unit converts 1 nmol histamine to methyl histamine/hr at pH 8.5, 37°C	Lyophilized containing 90% protein		Calzyme 157A0050
Rat kidney (optimum pH = 8.00-8.25, MW = 33,400 Da)				
50-100 U/mg protein	1 unit converts 1 nmol histamine to methyl histamine/hr at pH 8.56, 37°C	Lyophilized containing 90% protein		Calzyme 160A0050

2.1.1.28 Phenylethanolamine N-Methyltransferase

REACTION CATALYZED
S-Adenosyl-L-methionine + phenylethanolamine ↔ S-Adenosyl-L-homocysteine + N-methylphenylethanolamine

SYSTEMATIC NAME
S-Adenosyl-L-methionine:phenylethanolamine N-methyltransferase

SYNONYMS
Noradrenaline N-methyltransferase, PNMT

REACTANTS
S-Adenosyl-L-methionine, noradrenaline, phenylethanolamine, S-Adenosyl-L-homocysteine, adrenaline, N-methylphenylethanolamine, norepinephrine

NOTES
- Classified as a transferase, transferring one-carbon groups: methyltransferase (or transmethylase)
- Acts on various phenylethanolamines.
- Localized in adrenal medulla; also found in mammalian brain and heart tissue

SPECIFIC ACTIVITY	UNITS DEFINITION	PREPARATION FORM	ADDITIONAL ACTIVITIES	SUPPLIER CATALOG NO.
Bovine adrenal medulla (MW = 37,000-38,000 Da)				
50-100 U/mg protein	1 unit converts 1 nmol normetanephrine to metanephrine/hr at pH 8.5, 37°C	Lyophilized		Calzyme 103A0050
50 U/mg protein	1 unit converts 1.0 nmol normetanephrine to metanephrine/hr at pH 8.5, 37°C in the presence of β-NAD	Chromatographically purified; lyophilized containing 70% protein; balance primarily PO_4 buffer salts, pH 7.5		Sigma P8924

Protein-L-Isoaspartate (D-Aspartate) O-Methyltransferase   2.1.1.77

REACTION CATALYZED

S-Adenosyl-L-methionine + protein L-isoaspartate ↔ S-Adenosyl-L-homocysteine + protein L-isoaspartate a-methyl ester

SYSTEMATIC NAME

S-Adenosyl-L-methionine:protein-L-isoaspartate O-methyltransferase

SYNONYMS

Protein-L-isoaspartate O-methyltransferase, protein-β-aspartate O-methyltransferase

REACTANTS

S-Adenosyl-L-methionine, protein L-isoaspartate, protein D-aspartate, S-adenosyl-L-homocysteine, protein L-isoaspartate a-methyl ester

NOTES

- Classified as a transferase, transferring one-carbon groups: methyltransferase (or transmethylase)
- L-Aspartate residues in proteins cannot act as acceptors

See Chapter 5. Multiple Enzyme Preparations

Protein-Glutamate O-Methyltransferase   2.1.1.80

REACTION CATALYZED

S-Adenosyl-L-methionine + protein L-glutamate ↔ S-Adenosyl-L-homocysteine + protein L-glutamate methyl ester

SYSTEMATIC NAME

S-Adenosyl-L-methionine:protein L-glutamate O-methyltransferase

SYNONYMS

Methyl-accepting chemotaxis protein O-methyltransferase

SUBSTRATES

S-Adenosyl-L-methionine, protein L-glutamate

PRODUCTS

S-Adenosyl-L-homocysteine, protein L-glutamate methyl ester

NOTES

- Classified as a transferase, transferring one-carbon groups: methyltransferase (or transmethylase)
- Forms ester groups with L-glutamate residues in a number of membrane proteins

See Chapter 5. Multiple Enzyme Preparations

2.1.2.5 Glutamate Formiminotransferase

REACTION CATALYZED
    5-Formiminotetrahydrofolate + L-glutamate ↔
    Tetrahydrofolate + N-formimino-L-glutamate

SYSTEMATIC NAME
    5-Formiminotetrahydrofolate:L-glutamate N-formiminotransferase

SYNONYMS
    Glutamate formyltransferase, formimino-L-glutamic acid transferase

REACTANTS
    5-Formiminotetrahydrofolate, 5-formyltetrahydrofolate, L-glutamate, tetrahydrofolate, N-formimino-L-glutamate

NOTES
- Classified as a transferase, transferring one-carbon groups: hydroxymethyl-, formyl- and related transferase
- A pyridoxal-phosphate protein
- A bifunctional enzyme in eukaryotes also having formiminotetrahydrofolate cyclodeaminase activity (EC 4.3.1.4)

SPECIFIC ACTIVITY	UNITS DEFINITION	PREPARATION FORM	ADDITIONAL ACTIVITIES	SUPPLIER CATALOG NO.
Porcine liver				
1 U/mg protein	1 unit converts 1.0 μmol FIGLU and THF to L-Glu and 5-formimino-THF/min at pH 7.2, 25°C (measured as 5,10-methenyl-THF after perchloric acid treatment)	Lyophilized containing 30% protein; balance primarily KPO$_4$ buffer salts		Sigma F0777

2.1.3.2 Aspartate Carbamoyltransferase

REACTION CATALYZED
    Carbamoyl phosphate + L-aspartate ↔
    Orthophosphate + N-carbamoyl-L-aspartate

SYSTEMATIC NAME
    Carbamoyl-phosphate:L-aspartate carbamoyltransferase

SYNONYMS
    Carbamylaspartotranskinase, aspartate transcarbamylase

REACTANTS
    Carbamoyl phosphate, L-aspartate, orthophosphate, N-carbamoyl-L-aspartate

NOTES
- Classified as a transferase, transferring one-carbon groups: carboxyl- and carbamoyltransferase

SPECIFIC ACTIVITY	UNITS DEFINITION	PREPARATION FORM	ADDITIONAL ACTIVITIES	SUPPLIER CATALOG NO.
Streptococcus faecalis				
10-50 U/mg protein	1 unit forms 1.0 μmol N-carbamyl-L-Asp/min at pH 8.5, 37°C	Partially purified; lyophilized containing 40% protein; balance primarily KPO_4 buffer salts		Sigma A8886

Ornithine Carbamoyltransferase 2.1.3.3

REACTION CATALYZED
  Carbamoyl phosphate + L-ornithine ↔ Orthophosphate + L-citrulline

SYSTEMATIC NAME
  Carbamoyl phosphate:L-ornithine carbamoyltransferase

SYNONYMS
  Citrulline phosphorylase, ornithine transcarbamylase

REACTANTS
  Carbamoyl phosphate, L-ornithine, orthophosphate, L-citrulline

NOTES
- Classified as a transferase, transferring one-carbon groups: carboxyl- and carbamoyltransferase
- The plant enzyme also acts as putrescine synthase, converting agmatine into putrescine and ornithine into citrulline (see EC 2.1.3.6, 2.7.2.2, 3.5.3.12)

SPECIFIC ACTIVITY	UNITS DEFINITION	PREPARATION FORM	ADDITIONAL ACTIVITIES	SUPPLIER CATALOG NO.
Streptococcus faecalis				
600 U/mg protein	1 unit forms 1.0 μmol citrulline from ornithine and carbamyl PO_4/min at pH 8.5, 37°C	Lyophilized containing 80% protein; balance Tris buffer salts		Sigma O2501

2.2.1.1 Transketolase

REACTION CATALYZED
: Sedoheptulose 7-phosphate + D-glyceraldehyde 3-phosphate ↔ D-Ribose 5-phosphate + D-xylulose 5-phosphate

SYSTEMATIC NAME
: Sedoheptulose 7-phosphate:D-glyceraldehyde 3-phosphate glycolaldehydetransferase

SYNONYMS
: Glycolaldehydetransferase

REACTANTS
: Sedoheptulose 7-phosphate, D-glyceraldehyde 3-phosphate, D-ribose 5-phosphate, D-xylulose 5-phosphate, hydroxypyruvate, R-CHO, CO_2, R-CHOH-CO-CH_2OH

NOTES
- Classified as a transferase, transferring aldehyde or ketone residues: transketolase and transaldolase
- A thiamin-diphosphate protein
- *Alkaligenes faecalis* enzyme shows high activity with D-erythrose as acceptor

SPECIFIC ACTIVITY	UNITS DEFINITION	PREPARATION FORM	ADDITIONAL ACTIVITIES	SUPPLIER CATALOG NO.
Yeast, bakers				
15 U/mg	1 unit produces 1 μmol GAP from xylulose-5-phosphate/min at pH 7.7, 25°C in the presence of ribose-5-phosphate, thiamine pyrophosphate, Mg^{2+}	Powder containing 70% protein	<0.1% ADH, GAPDH	Fluka 90197
15-25 U/mg protein	1 unit produces 1.0 μmol GAP from xylulose 5-phosphate/min at pH 7.7, 25°C in a coupled system with α-GDH/TPI	Sulfate-free; lyophilized containing 30% buffer salts		Sigma T6133

Transaldolase 2.2.1.2

REACTION CATALYZED
    Sedoheptulose 7-phosphate + D-glyceraldehyde 3-phosphate ↔ D-Erythrose 4-phosphate + D-fructose 6-phosphate

SYSTEMATIC NAME
    Sedoheptulose 7-phosphate:D-glyceraldehyde 3-phosphate glyceronetransferase

SYNONYMS
    Dihydroxyacetonetransferase

REACTANTS
    Sedoheptulose 7-phosphate, D-glyceraldehyde 3-phosphate, D-erythrose 4-phosphate, D-fructose 6-phosphate

NOTES
- Classified as a transferase, transferring aldehyde or ketone residues: transketolase and transaldolase

SPECIFIC ACTIVITY	UNITS DEFINITION	PREPARATION FORM	ADDITIONAL ACTIVITIES	SUPPLIER CATALOG NO.
Yeast, bakers				
10-30 U/mg protein	1 unit produces 1.0 μmol D-GAP from D-F6P/min at pH 7.7, 25°C in the presence of D-erythrose 4-phosphate in a coupled system with GDH/TPI and β-NADH	Sulfate-free; lyophilized containing 5% citrate buffer salts		Sigma T6008

Choline O-Acetyltransferase 2.3.1.6

REACTION CATALYZED
    Acetyl-CoA + choline ↔ CoA + O-acetylcholine

SYSTEMATIC NAME
    Acetyl-CoA:choline O-acetyltransferase

SYNONYMS
    Choline acetyltransferase, choline acetylase

REACTANTS
    Acetyl-CoA, propanoyl-CoA (more slowly), choline, CoA, O-acetylcholine

NOTES
- Classified as an acyltransferase, transferring acyl groups and forming either esters or amides

SPECIFIC ACTIVITY	UNITS DEFINITION	PREPARATION FORM	ADDITIONAL ACTIVITIES	SUPPLIER CATALOG NO.
Bovine brain				
10 U/mg protein	1 unit transfers 1.0 nmol acetate from ^{14}C-acetyl CoA to choline/min at pH 7.5, 37°C	Lyophilized containing 5-20% protein; balance primarily $NaPO_4$, NaCl, EDTA		Sigma C3388

2.3.1.6 Choline O-Acetyltransferase continued

SPECIFIC ACTIVITY	UNITS DEFINITION	PREPARATION FORM	ADDITIONAL ACTIVITIES	SUPPLIER CATALOG No.
Human placenta				
20 U/mg protein	1 unit transfers 1.0 nmol acetate from ^{14}C-acetyl CoA to choline/min at pH 7.5, 37°C	Suspension in 2.8 M $(NH_4)_2SO_4$, 5 mM KPO_4, 0.5 mM EDTA, 0.5 mM DTT, pH 7.4		Sigma C2898
Rat, recombinant				
		≥95% purity		Chemicon AG280

2.3.1.7 Carnitine O-Acetyltransferase

REACTION CATALYZED
    Acetyl-CoA + carnitine ↔ CoA + O-acetylcarnitine

SYSTEMATIC NAME
    Acetyl-CoA:carnitine O-acetyltransferase

REACTANTS
    Acetyl-CoA, propanoyl-CoA, butanoyl-CoA, carnitine, CoA, O-acetylcarnitine

NOTES
- Classified as an acyltransferase, transferring acyl groups and forming either esters or amides

SPECIFIC ACTIVITY	UNITS DEFINITION	PREPARATION FORM	ADDITIONAL ACTIVITIES	SUPPLIER CATALOG No.
Pigeon breast muscle				
80 U/mg	1 unit converts 1 μmol acetyl-D,L-carnitine and CoA to product/min at 25°C	Suspension 2.9 M $(NH_4)_2SO_4$, pH 7.5	<0.01% acetyl-CoA-deacylase	Boehringer 103241 1693654
100 U/mg protein	1 unit converts 1 μmol acetyl-L-carnitine and CoA to L-carnitine and acetyl-CoA/min at pH 8.0, 25°C	Crystallized; suspension in 2.9 M $(NH_4)_2SO_4$, pH 7.5		Fluka 22017
80 U/mg protein	1 unit converts 1 μmol acetyl-L-carnitine and CoA to L-carnitine and acetyl-CoA/min at pH 8.0, 25°C	Crystallized; suspension in 3.2 M $(NH_4)_2SO_4$, 50 mM KPO_4, 1 mM DTT, pH 7.0		Fluka 22019
40-100 U/mg protein	1 unit converts 1.0 μmol acetyl-L-carnitine and CoA to L-carnitine and acetyl-CoA/min at pH 8.0, 25°C	Lyophilized containing 50% protein, 30% lactose; balance KPO_4 salts, EDTA, DTT		Sigma C4430

SPECIFIC ACTIVITY	UNITS DEFINITION	PREPARATION FORM	ADDITIONAL ACTIVITIES	SUPPLIER CATALOG NO.
Pigeon breast muscle *continued*				
80 U/mg protein	1 unit converts 1.0 μmol acetyl-L-carnitine and CoA to L-carnitine and acetyl-CoA/min at pH 8.0, 25°C	Crystallized; suspension in 3.2 M $(NH_4)_2SO_4$, 50 mM KPO_4, 1 mM DTT, pH 7.0		Sigma C4899
80 U/mg protein	1 unit converts 1.0 μmol acetyl-L-carnitine and CoA to L-carnitine and acetyl-CoA/min at pH 8.0, 25°C	Crystallized; suspension in 2.9 M $(NH_4)_2SO_4$, pH 7.5		Sigma C8757

Phosphate Acetyltransferase

2.3.1.8

REACTION CATALYZED
Acetyl-CoA + orthophosphate ↔ CoA + acetyl phosphate

SYSTEMATIC NAME
Acetyl-CoA:orthophosphateacetyltransferase

SYNONYMS
Phosphotransacetylase, phosphoacylase, PTA

REACTANTS
Acetyl-CoA, short-chain acyl-CoA, orthophosphate, CoA, acetyl phosphate

NOTES
- Classified as an acyltransferase, transferring acyl groups and forming either esters or amides

SPECIFIC ACTIVITY	UNITS DEFINITION	PREPARATION FORM	ADDITIONAL ACTIVITIES	SUPPLIER CATALOG NO.
Bacillus stearothermophilus (optimum pH = 7.5, pI = 4.5, MW = 70,000 Da [35,000/subunit]; K_M = 0.4 mM [CoA], 1.1 mM [acetyl phosphate]; stable pH > 7, T < 50°C)				
11,600 U/mg protein; 3000 U/mg solid	1 unit converts 1 μmol acetyl-CoA to product/min at 30°C	Lyophilized	<0.01% acetyl-CoA-deacylase, AK, LDH	Boehringer 525995
4000–10,000 U/mg protein	1 unit converts 1.0 μmol CoA to acetyl CoA/min at pH 7.4, 25°C using acetyl phosphate as substrate	Lyophilized containing 30% protein; balance Tris buffer salts		Sigma P2783
>5000 U/mg protein	1 unit forms 1 μmol acetyl-CoA/min at 30°C	Lyophilized	<0.01% LDH, AK, AdK	Unitika
>5000 U/mg protein	1 unit forms 1 μmol acetyl-CoA/min at 30°C	50% glycerol solution	<0.01% LDH, AK, AdK	Unitika

2.3.1.8 Phosphate Acetyltransferase continued

SPECIFIC ACTIVITY	UNITS DEFINITION	PREPARATION FORM	ADDITIONAL ACTIVITIES	SUPPLIER CATALOG NO.
Clostridium kluyveri				
20-80 U/mg protein	1 unit converts 1.0 μmol CoA to acetyl CoA/min at pH 7.4, 25°C using acetyl phosphate as substrate	Lyophilized containing 50% protein; balance Tris buffer salts, EDTA, stabilizer		Sigma P2669
Leuconostoc mesenteroides				
≥5000 IU/mg protein	1 IU transforms 1 μmol substrate/min under standard IUB conditions at 25°C	Suspension in $(NH_4)_2SO_4$	<0.005% MK, LDH, AK	OYC
5000-10,000 U/mg protein	1 unit converts 1.0 μmol CoA to acetyl CoA/min at pH 7.4, 25°C using acetyl phosphate as substrate	Suspension in 3.0 M $(NH_4)_2SO_4$, pH 7		Sigma P7156

2.3.1.16 Acetyl-CoA C-Acyltransferase

REACTION CATALYZED
  Acyl-CoA + acetyl-CoA ↔ CoA + 3-oxoacyl-CoA

SYSTEMATIC NAME
  Acyl-CoA:acetyl-CoA C-acyltransferase

SYNONYMS
  β-Ketothiolase, 3-ketoacyl-CoA thiolase

REACTANTS
  Acyl-CoA, acetyl-CoA, CoA, 3-oxoacyl-CoA

NOTES
- Classified as an acyltransferase, transferring acyl groups and forming either esters or amides

---

See Chapter 5. Multiple Enzyme Preparations

Chloramphenicol O-Acetyltransferase 2.3.1.28

REACTION CATALYZED
  Acetyl-CoA + chloramphenicol ↔ CoA + chloramphenicol 3-acetate

SYSTEMATIC NAME
  Acetyl-CoA:chloramphenicol 3-O-acetyltransferase

SYNONYMS
  Chloramphenicol acetyltransferase, CAT

REACTANTS
  Acetyl-CoA, chloramphenicol, CoA, chloramphenicol 3-acetate

NOTES
  - Classified as an acyltransferase, transferring acyl groups and forming either esters or amides

SPECIFIC ACTIVITY	UNITS DEFINITION	PREPARATION FORM	ADDITIONAL ACTIVITIES	SUPPLIER CATALOG NO.
Escherichia coli				
100,000 U/mg; 1500 U/mL	1 unit transfers 1 nmol and acetyl residues from acetyl-CoA to chloramphenicol/min at 25°C	>90% purity by SDS-PAGE; 50% glycerol solution, pH 7.8		Boehringer 874434
≥1000 U/mL	1 unit acetylates 1 nmol chloramphenicol/min at pH 7.8, 37°C	Solution containing 100 mM Tris-HCl, 2 mg BSA/mL, 50% glycerol, pH 7.8		Pharmacia 27-0847-01 27-0847-02
50,000-150,000 U/mg protein	1 unit converts 1.0 nmol chloramphenicol and acetyl-CoA to chloramphenicol 3-acetate and CoA/min at pH 7.8, 25°C	Partially purified; lyophilized containing Tris buffer salts		Sigma C2900
50,000-150,000 U/mg protein	1 unit converts 1.0 nmol chloramphenicol and acetyl-CoA to chloramphenicol 3-acetate and CoA/min at pH 7.8, 25°C			Sigma C8413
Escherichia coli, recombinant				
≥10 U/μL	1 unit transfers 1 nmol acetate to chloramphenicol/min at pH 8.0, 37°C	>85% purity by SDS gel		Promega E1051 E1052
Tn9 infected E. coli				
	pH 7.4, 37°C	Affinity purified; 50 mM Tris-HCl, 50 μM DTT, 50 μM chloramphenicol, 50% glycerol, 200 μg/mL protein, pH 7.8		5'→3' 5306-567890
>30,000 U/mg; 200 μg/mL	1 unit acetylates 1 nmol chloramphenicol/min at pH 7.8, 37°C	Solution containing 50% glycerol, 50 mM Tris-HCl, 50 μM β-MSH, 50 μM chloramphenicol, pH 7.8		ICN 153900

2.3.1.43 Phosphatidylcholine-Sterol O-Acyltransferase

REACTION CATALYZED
Phosphatidylcholine + sterol ↔ 1-Acylglycerolphosphocholine + sterol ester

SYSTEMATIC NAME
Phosphatidylcholine:sterol O-acyltransferase

SYNONYMS
Lecithin-cholesterol acyltransferase, phospholipid-cholesterol acyltransferase

REACTANTS
Phosphatidylcholine, cholesterol, sterol, 1-acylglycerolphosphocholine, sterol ester

NOTES
- Classified as an acyltransferase, transferring acyl groups and forming either esters or amides
- Transfers palmitoyl, oleoyl and linoleoyl residues
- A variety of sterols act as acceptors
- Activated by apo-A1
- Liver enzyme esterifies free cholesterol acquired by HDL particles in the plasma, resulting in a progressive increase in HDL size and decrease in HDL density
- Bacterial enzyme also catalyzes phospholipase A_2 (EC 3.1.1.4) and lysophospholipase (EC 3.1.1.5) reactions

SPECIFIC ACTIVITY	UNITS DEFINITION	PREPARATION FORM	ADDITIONAL ACTIVITIES	SUPPLIER CATALOG No.
Human plasma				
		>99% purity; 10 mM Tris buffer solution containing 10 mM NaCl and 1 mM EDTA, pH 7.6	No detectable HBsAg, HIV Ab	Calbiochem 428875

2.3.1.55 Kanamycin 6'-N-Acetyltransferase

REACTION CATALYZED
Acetyl-CoA + kanamycin ↔ CoA + $N^{6'}$-acetylkanamycin

SYSTEMATIC NAME
Acetyl-CoA:kanamycin $N^{6'}$-acetyltransferase

REACTANTS
Acetyl-CoA, kanamycin, CoA, $N^{6'}$-acetylkanamycin; antibiotics kanamycin A, kanamycin B, neomycin, gentamicin C_{1a}, gentamicin C_2, sisomicin

NOTES
- Classified as an acyltransferase, transferring acyl groups and forming either esters or amides

Kanamycin 6'-*N*-Acetyltransferase continued — 2.3.1.55

SPECIFIC ACTIVITY	UNITS DEFINITION	PREPARATION FORM	ADDITIONAL ACTIVITIES	SUPPLIER CATALOG NO.
Escherichia coli				
100-200 U/mg protein	1 unit converts 1.0 nmol kanamycin and acetyl-CoA to kanamycin 6-acetate and CoA/min at pH 5.7, 37°C	Lyophilized containing 5% protein; balance PO$_4$ buffer salts and stabilizer		Sigma K0377

γ-Glutamyltransferase — 2.3.2.2

REACTION CATALYZED
  (5-L-Glutamyl)-peptide + an amino acid ↔ Peptide + 5-L-glutamyl amino acid

SYSTEMATIC NAME
  (5-L-Glutamyl)-peptide:amino-acid 5-glutamyltransferase

SYNONYMS
  Glutamyl transpeptidase, γ-glutamyl transpeptidase, γGT

REACTANTS
  (5-L-Glutamyl)-peptide, amino acid, peptide, 5-L-glutamyl amino acid

NOTES
- Classified as an aminoacyltransferase
- Found in small intestine microvilli and kidney brush border plasma membrane
- Naturally-occurring donors and acceptors are unknown
- High serum levels are associated with liver disorders
- Activated by Mg^{2+} and EDTA
- Inhibited by excess glutathione, bromosulphthalein, *N*-ethylmaleimide, L-glutamine, iodoacetate, Hg^{2+} and Mn^{2+}

SPECIFIC ACTIVITY	UNITS DEFINITION	PREPARATION FORM	ADDITIONAL ACTIVITIES	SUPPLIER CATALOG NO.
Bovine kidney				
5 U/mg solid	1 unit releases 1 μmol *p*-nitroaniline/min at pH 8.2, 25°C	Salt-free; lyophilized	<0.5% GOT, GPT, LDH, AP	Biozyme GGTB1
30 U/mg protein; ≥15 U/mg solid	1 unit releases 1 μmol *p*-nitroaniline/min at pH 8.2, 25°C	Highly purified; lyophilized	<0.01% GOT, GPT, AP <0.1% LDH	Biozyme GGTB2
20-50 U/mg solid	1 unit releases 1 μmol *p*-nitroaniline/min at pH 8.2, 25°C	Lyophilized containing 90% protein		Calzyme 076A0020
3-9 U/mg solid	1 unit liberates 1.0 μmol *p*-nitroaniline from L-γ-glutamyl-*p*-nitroanilide/min at pH 8.5, 25°C	Crude		Sigma G4135

2.3.2.2 γ-Glutamyltransferase continued

SPECIFIC ACTIVITY	UNITS DEFINITION	PREPARATION FORM	ADDITIONAL ACTIVITIES	SUPPLIER CATALOG NO.
Bovine kidney continued				
15-30 U/mg solid	1 unit liberates 1.0 μmol p-nitroaniline from L-γ-glutamyl-p-nitroanilide/min at pH 8.5, 25°C	Lyophilized containing 85% protein	<0.5% CPK, LDH, GOT, GPT, AP	Sigma G4756
10 U/mg		Lyophilized		Wako 079-02831
Equine kidney				
5-12 U/mg solid	1 unit liberates 1.0 μmol p-nitroaniline from L-γ-glutamyl-p-nitroanilide/min at pH 8.5, 25°C	Crude		Sigma G9270
Porcine kidney (optimum pH = 8.0-9.0, MW = 95,000 Da; stable pH 5.0-11.0, T < 50°C)				
≥50 U/mg protein; ≥45 U/mg solid	1 unit releases 1 μmol p-nitroaniline/min at pH 8.2, 25°C	Salt-free; lyophilized	<0.005% LDH, GPT <0.05% GOT <0.2% LAP	Biozyme GGTB3
20-50 U/mg solid	1 unit releases 1 μmol p-nitroaniline/min at pH 8.2, 30°C	Lyophilized containing 90% protein		Calzyme 130A0020
3 U/mg	1 unit liberates 1 μmol 4-nitroanilide/min at pH 8.5, 25°C	Powder		Fluka 49640
>10 U/mg solid	1 unit releases 1 μmol 5-amino-2-nitrobenzoate/min at pH 8.25, 25°C	Lyophilized	<0.1% LDH <0.05% GPT <0.5% ALP, GOT	Randox GT 822L
2-6 U/mg solid	1 unit liberates 1.0 μmol p-nitroaniline from L-γ-glutamyl-p-nitroanilide/min at pH 8.5, 25°C	Crude		Sigma G2262

Protein-Glutamine γ-Glutamyltransferase 2.3.2.13

REACTION CATALYZED

Protein glutamine + alkylamine ↔ Protein N^5-alkylglutamine + NH_3

SYSTEMATIC NAME

Protein-glutamine:amine γ-glutamyltransferase

SYNONYMS

Transglutaminase, Factor XIIIa, fibrinoligase

REACTANTS

Protein glutamine, alkylamine, protein N^5-alkylglutamine

NOTES

- Classified as an aminoacyltransferase
- Requires Ca^{2+}
- Formed by proteolytic cleavage of plasma Factor XIII
- Inter-molecular N^6-(5-glutamyl)-lysine crosslinks are formed from donor γ-carboxamide groups of peptide-bound glutamine residues and acceptor protein- and peptide-bound lysine residues

SPECIFIC ACTIVITY	UNITS DEFINITION	PREPARATION FORM	ADDITIONAL ACTIVITIES	SUPPLIER CATALOG NO.
Guinea pig liver				
2 U/mg	1 unit forms 1 μmol hydroxamate/min from N-α-CBZ-Glu-Gly and hydroxylamine at pH 6.0, 37°C compared with L-Glu γ-hydroxamate standard	Powder containing 10% Tris and DTE		Fluka 90195
1.5-3 U/mg protein	1 unit forms 1.0 μmol hydroxamate from N-α-CBZ-Glu-Gly and hydroxylamine/min at pH 6.0, 37°C	Lyophilized containing 90% protein; balance primarily Tris and DTT		Sigma T5398

Phosphorylase 2.4.1.1

REACTION CATALYZED

$(1,4-\alpha-D-Glucosyl)_n$ + orthophosphate ↔ $(1,4-\alpha-D-Glucosyl)_{n-1}$ + α-D-glucose 1-phosphate

SYSTEMATIC NAME

$(1,4-\alpha-D-Glucosyl)_n$:orthophosphate α-D-glucosyltransferase

SYNONYMS

Muscle phosphorylase *a* and *b*, amylophosphorylase, polyphosphorylase

REACTANTS

$(1,4-\alpha-D-Glucosyl)_n$, maltodextrin, starch, glycogen, orthophosphate, $(1,4-\alpha-D-glucosyl)_{n-1}$, α-D-glucose 1-phosphate

NOTES

- Classified as a glycosyltransferase: hexosyltransferase
- Acts on a variety of α-1,4-linked glucose polymers
- Skeletal muscle enzyme exists in two interconvertible forms: active (phosphorylase *a*) and less active (phosphorylase *b*)

2.4.1.1 Phosphorylase continued

SPECIFIC ACTIVITY	UNITS DEFINITION	PREPARATION FORM	ADDITIONAL ACTIVITIES	SUPPLIER CATALOG No.
Rabbit muscle (MW = 185,000 Da)				
60 U/mg protein	1 unit liberates 1 μmol Glc1P from glycogen and orthophosphate/min at pH 6.8, 37°C in the presence of 5'-AMP	Lyophilized containing 90% protein		Calzyme 104A0060
10 U/mg	1 unit forms 1 μmol α-D-Glc1P from glycogen and PO_4/min at pH 6.8, 25°C	Lyophilized containing <0.5% MgOAc, 35% protein, 64% lactose	<0.002% AMP phosphatase, ATP phosphatase, phosphorylase phosphatase <0.02% debranching enzyme <0.1% phosphorylase kinase, 5'-AMP, PGluM <5% phosphorylase *a*	Fluka 79700
		Single-band purity by SDS-PAGE; lyophilized		ProZyme MW10 000 MW10 010 MW10 100
25 U/mg protein	1 unit forms 1.0 μmol α-D-Glc1P from glycogen and orthophosphate/min at pH 6.8, 30°C	2X Crystallized; lyophilized containing 40% protein; balance primarily β-glycerophosphate buffer salts and EDTA	<0.1% AMP-phosphatase, ATP-phosphatase, debrancher enzyme, PGluM, phosphorylase phosphatase <20% phosphorylase *b*	Sigma P1261
30 U/mg protein	1 unit forms 1.0 μmol α-D-Glc1P from glycogen and orthophosphate in the presence of 5'-AMP/min at pH 6.8, 30°C	2X Crystallized; lyophilized containing 35% protein; balance primarily buffer salts as lactose, 10 μmol/100 mg protein MgOAc, 1 μmol/100 mg protein 5'-AMP	<0.1% phosphorylase kinase, phosphorylase phosphatase, debrancher enzyme, PGluM, AMPase, ATPase <5% phosphorylase *a*	Sigma P6635

2.4.1.5 Dextransucrase

REACTION CATALYZED

Sucrose + (1,6-α-D-glucosyl)$_n$ ↔ D-Fructose + (1,6-α-D-glucosyl)$_{n+1}$

SYSTEMATIC NAME

Sucrose:1,6-α-D-glucan 6-α-D-glucosyltransferase

SYNONYMS

Sucrose 6-glucosyltransferase

REACTANTS

Sucrose, (1,6-α-D-glucosyl)$_n$, D-fructose, (1,6-α-D-glucosyl)$_{n+1}$

NOTES

- Classified as a glycosyltransferase: hexosyltransferase

Dextransucrase continued

SPECIFIC ACTIVITY	UNITS DEFINITION	PREPARATION FORM	ADDITIONAL ACTIVITIES	SUPPLIER CATALOG NO.
Leuconostoc mesenteroides				
150-450 U/mg protein	1 unit liberates 1.0 μmol fructose from sucrose/min at pH 5.2, 30°C	Lyophilized containing 15% protein; balance primarily dextran, MES buffer salts, $CaCl_2$		sigma D9909

Sucrose Phosphorylase

2.4.1.7

REACTION CATALYZED
    Sucrose + orthophosphate ↔ D-Fructose + α-D-glucose 1-phosphate

SYSTEMATIC NAME
    Sucrose:orthophosphate α-D-glucosyltransferase

SYNONYMS
    Sucrose glucosyltransferase, SP

REACTANTS
    Sucrose, orthophosphate, arsenate, D-fructose, ketoses (various), L-arabinose, α-D-glucose 1-phosphate

NOTES
- Classified as a glycosyltransferase: hexosyltransferase
- Arsenate may replace phosphate in the forward reaction
- Various ketoses and L-arabinose may replace D-fructose in the reverse reaction

SPECIFIC ACTIVITY	UNITS DEFINITION	PREPARATION FORM	ADDITIONAL ACTIVITIES	SUPPLIER CATALOG NO.
Leuconostoc mesenteroides				
≥20 IU/mg protein	1 IU transforms 1 μmol substrate/min under standard IUB conditions at 25°C	Lyophilized		OYC
3-7 U/mg protein	1 unit converts 1.0 μmol sucrose and PO_4 to Glc1P and fructose/min at pH 7.0, 30°C	Partially purified; lyophilized containing 40% protein, 10% Tris-HCl buffer salts, stabilizer		sigma S7760

2.4.1.8 Maltose Phosphorylase

REACTION CATALYZED
Maltose + orthophosphate ↔ D-Glucose + β-D-glucose 1-phosphate

SYSTEMATIC NAME
Maltose:orthophosphate 1-β-D-glucosyltransferase

REACTANTS
Maltose, orthophosphate, D-glucose, β-D-glucose 1-phosphate

NOTES
- Classified as a glycosyltransferase: hexosyltransferase

SPECIFIC ACTIVITY	UNITS DEFINITION	PREPARATION FORM	ADDITIONAL ACTIVITIES	SUPPLIER CATALOG NO.
Bacterial				
≥5 IU/mg protein	1 IU transforms 1 μmol substrate/min under standard IUB conditions at 25°C	Suspension in $(NH_4)_2SO_4$	<0.05% α-glucosidase, α-amylase	OYC

2.4.1.11 Glycogen (Starch) Synthase

REACTION CATALYZED
UDPglucose + (1,4-α-D-glucosyl)$_n$ ↔ UDP + (1,4-α-D-glucosyl)$_{n+1}$

SYSTEMATIC NAME
UDPglucose:glycogen 4-α-D-glucosyltransferase

SYNONYMS
UDPglucose-glycogen glucosyltransferase

REACTANTS
UDPglucose, (1,4-α-D-glucosyl)$_n$, UDP, (1,4-α-D-glucosyl)$_{n+1}$

NOTES
- Classified as a glycosyltransferase: hexosyltransferase
- Animal tissue enzyme is a complex of a catalytic subunit and the protein glycogenin. Glycosylated glycogenin is required as a primer and is produced by glycogenin glucosyltransferase (EC 2.4.1.186)

SPECIFIC ACTIVITY	UNITS DEFINITION	PREPARATION FORM	ADDITIONAL ACTIVITIES	SUPPLIER CATALOG NO.
Rabbit muscle				
2-7 U/mg protein without Glc6P	1 unit incorporates 1.0 μmol glucose from UDP-glucose into glycogen/min at pH 8.2, 30°C yielding 1.0 μmol UDP, measured in a PK/LDH/NADH system	Suspension in 60% glycerol, 5 mM β-MSH, 5 mM EDTA, 1.0 M NaCl, 2.8 mM maltose, 0.05 M Tris, pH 7.5	Both D- and L- forms of glycogen synthase <5% phosphorylase	Sigma G2259

Sucrose Synthase 2.4.1.13

REACTION CATALYZED
UDPglucose + D-fructose ↔ UDP + sucrose

SYSTEMATIC NAME
UDPglucose:D-fructose 2-α-D-glucosyltransferase

SYNONYMS
UDPglucose-fructose glucosyltransferase, sucrose-UDP glucosyltransferase, sucrose synthetase

REACTANTS
UDPglucose, D-fructose, UDP, sucrose

NOTES
- Classified as a glycosyltransferase: hexosyltransferase

SPECIFIC ACTIVITY	UNITS DEFINITION	PREPARATION FORM	ADDITIONAL ACTIVITIES	SUPPLIER CATALOG NO.
Wheat germ				
5-20 U/mg protein	1 unit converts 1.0 µmol UDP-glucose and D-fructose to UDP and sucrose/30 min at pH 7.5, 37°C in a coupled system with PK and LDH	Solution containing 60% glycerol, 1.0 M $(NH_4)_2SO_4$, 5 mM KPO_4, 15 mM β-MSH, 1 mM $MgCl_2$, 5 mM GSH, 0.1 mM EDTA, pH 7.0	-	Sigma S6379

Sucrose-Phosphate Synthase 2.4.1.14

REACTION CATALYZED
UDPglucose + D-fructose 6-phosphate ↔ UDP + sucrose 6-phosphate

SYSTEMATIC NAME
UDPglucose:D-fructose-6-phosphate 2-α-D-glucosyltransferase

SYNONYMS
UDPglucose-fructose-phosphate glucosyltransferase, sucrose phosphate-UDP glucosyltransferase, sucrose phosphate synthetase

REACTANTS
UDPglucose, D-fructose 6-phosphate, UDP, sucrose 6-phosphate

NOTES
- Classified as a glycosyltransferase: hexosyltransferase

2.4.1.14 Sucrose-Phosphate Synthase continued

SPECIFIC ACTIVITY	UNITS DEFINITION	PREPARATION FORM	ADDITIONAL ACTIVITIES	SUPPLIER CATALOG No.
Wheat germ				
3-10 U/mg protein	1 unit converts 1.0 μmol UDP-glucose and D-F6P to UDP and sucrose 6-phosphate/30 min at pH 7.5, 37°C in a coupled system with PK and LDH	Solution containing 60% glycerol, 1.0 M (NH$_4$)$_2$SO$_4$, 5 mM KPO$_4$, 15 mM β-MSH, 1 mM MgCl$_2$, 5 mM GSH, 0.1 mM EDTA, pH 7.0		Sigma S6254

2.4.1.17 Glucuronosyltransferase

REACTION CATALYZED
UDPglucuronate + acceptor ↔ UDP + acceptor β-D-glucuronoside

SYSTEMATIC NAME
UDPglucuronate:(acceptor) β-D-glucuronosyltransferase

SYNONYMS
UDPglucuronosyltransferase, UDPglucuronyltransferase

REACTANTS
UDPglucuronate, phenols, alcohols, amines, fatty acids, acceptor, UDP, acceptor β-D-glucuronoside

NOTES
- Classified as a glycosyltransferase: hexosyltransferase
- A family of enzymes accepting a wide range of substrates including phenols, alcohols, amines and fatty acids

SPECIFIC ACTIVITY	UNITS DEFINITION	PREPARATION FORM	ADDITIONAL ACTIVITIES	SUPPLIER CATALOG No.
Bovine liver				
1-4 U/g solid	1 unit transfers 1.0 μmol glucuronic acid from uridine 5'-diphosphoglucuronic acid to phenolphthalein/min at pH 8.0, 37°C	Crude; microsomal; lyophilized containing 40% protein; balance KCl, GSH and sucrose	β-glucuronidase (20,000 Sigma U/g at pH 4.5; <1000 Sigma U/g at pH 8.0)	Sigma U4502
Human DNA, recombinant overexpressed in insect cells using *Baculovirus*				
		Cell extract		PanVera
Rabbit liver				
1-3 U/g solid	1 unit transfers 1.0 μmol glucuronate from uridine 5'-diphosphoglucuronate to phenolphthalein/min at pH 8.0, 37°C	Crude; microsomal; lyophilized containing 50% protein; balance KCl and GSH	β-glucuronidase (20,000 Sigma U/g at pH 4.5; <1000 Sigma U/g at pH 8.0)	Sigma U3626

Cyclomaltodextrin Glucanotransferase 2.4.1.19

REACTION CATALYZED
Cyclizes part of a 1,4-α-D-glucan chain by formation of a 1,4-α-D-glucosidic bond

SYSTEMATIC NAME
1,4-α-D-Glucan 4-α-D-(1,4-α-D-glucano)-transferase (cyclizing)

SYNONYMS
Bacillus macerans amylase, cyclodextrin glucanotransferase, CGTase

REACTANTS
Starch, maltodextrin, cyclomaltodextrins (Schardinger dextrins)

APPLICATIONS
- Synthesis of cyclodextrins and transglycosylation products

NOTES
- Classified as a glycosyltransferase: hexosyltransferase
- Reversibly forms cyclomaltodextrins (Schardinger dextrins) of glucose chain length 6, 7, 8, etc.
- Also disproportionates linear maltodextrins without cyclizing

SPECIFIC ACTIVITY	UNITS DEFINITION	PREPARATION FORM	ADDITIONAL ACTIVITIES	SUPPLIER CATALOG No.
Bacillus macerans (optimum pH = 6, T = 65°C)				
≥600 U/g		Solution		Amano CGTase
Bacillus stearothermophilus (optimum pH = 5.0-6.0)				
1400 U/g	1 unit decreases iodine stain in 15 mg soluble starch by 10%/min at 40°C	Crude; solution		Hayashibara EN301

Lactose Synthase

REACTION CATALYZED

UDPgalactose + D-glucose → UDP + lactose

SYSTEMATIC NAME

UDPgalactose:D-glucose 4-β-D-galactotransferase

SYNONYMS

UDPgalactose-glucose galactosyltransferase, N-acetyllactosamine synthase, galactosyl transferase

REACTANTS

UDPgalactose, D-glucose, UDP, lactose, N-acetylglucosamine

APPLICATIONS

- Determination of α-lactalbumin, UDP-gal, N-acetylglucosamine, N-acetyl-mannose amine and glucosamine

NOTES

- Classified as a glycosyltransferase: hexosyltransferase
- An essentially irreversible reaction which proceeds extremely slowly; the addition of α-lactalbumin increases the reaction rate
- A complex of proteins A and B (α-lactalbumin)
- In the absence of protein B, the enzyme acts as N-acetyllactosamine synthase (EC 2.4.1.90) and transfers galactose from UDPgalactose to N-acetylglucosamine

SPECIFIC ACTIVITY	UNITS DEFINITION	PREPARATION FORM	ADDITIONAL ACTIVITIES	SUPPLIER CATALOG NO.
Bovine milk (MW = 70,000 Da)				
3 U/mg	1 unit transfers 1 µmol galactose from UDP-galactose to D-glucose/min at pH 8.4, 30°C in the presence of α-lactalbumin	Lyophilized containing 50% protein (including 5% α-lactalbumin); 50% Tris, EDTA, $(NH_4)_2SO_4$		Fluka 48279
5-15 U/mg protein with added α-lactalbumin; <0.3 U/mg protein	1 unit transfers 1.0 µmol galactose from UDP-galactose to D-glucose/min at pH 8.4, 30°C in presence of α-lactalbumin	Lyophilized containing 50% protein (<10% α-lactalbumin); balance primarily Tris, EDTA, $(NH_4)_2SO_4$ buffer salts		Sigma G5507
≥2 U/mg protein	1 unit transfers 1.0 µmol galactose from UDP-galactose to D-glucose/min at pH 8.4, 37°C in the presence of α-lactalbumin	Chromatographically purified; lyophilized containing <5% α-lactalbumin		Worthington LM02020 LM02022 LM02024
Human milk				
	1 unit transfers 1.0 µmol galactose from UDP-galactose to D-glucose/min at pH 8.4, 30°C in presence of α-lactalbumin	Lyophilized containing Tris-HCl, EDTA, KCl, NaN_3, BSA		Sigma G1024

1,4-α-Glucan 6-α-Glucosyltransferase 2.4.1.24

REACTION CATALYZED
: Transfers an α-D-glucosyl residue in a 1,4-α-D-glucan to the primary hydroxyl group of glucose, free or combined in a 1,4-α-D-glucan

SYSTEMATIC NAME
: 1,4-α-D-Glucan:1,4-α-D-glucan (D-glucose) 6-α-D-glucosyltransferase

SYNONYMS
: Oligoglucan-branching glycosyltransferase

REACTANTS
: 1,4-α-D-Glucan, glucose

NOTES
- Classified as a glycosyltransferase: hexosyltransferase

See Chapter 5. Multiple Enzyme Preparations

4-α-Glucanotransferase 2.4.1.25

REACTION CATALYZED
: Transfers a segment of a 1,4-α-D-glucan to a new position in an acceptor, which may be glucose or a 1,4-α-D-glucan

SYSTEMATIC NAME
: 1,4-α-D-Glucan:1,4-α-D-glucan 4-α-D-glycosyltransferase

SYNONYMS
: Disproportionating enzyme, dextrin glycosyltransferase, D-enzyme (plants), maltogenic amylase

REACTANTS
: Glucose, 1,4-α-D-glucan, glycogen

APPLICATIONS
- Used in baking to prolong shelf-life and retard staling without producing gumminess or affecting dough handling. Modifies flour starch to reduce retrogradation and create low molecular weight sugars and dextrins. The latter improve the water retention capacity of baked goods

NOTES
- Classified as a glycosyltransferase: hexosyltransferase
- Forms part of mammalian and yeast glycogen debranching system

2.4.1.25 4-α-Glucanotransferase continued

SPECIFIC ACTIVITY	UNITS DEFINITION	PREPARATION FORM	ADDITIONAL ACTIVITIES	SUPPLIER CATALOG NO.
Bacillus subtilis (optimum pH = 4.8-6.0, T = 45-75°C)				
10,000 MANU/g	1 MANU hydrolyzes 1 μmol maltotriose/min	Food grade; purified; powder at 150 micron particle size		Novo Nordisk Novamyl™ 10.000 BG
1500 MANU/g	1 MANU hydrolyzes 1 μmol maltotriose/min	Food grade; purified; granulate at 350 micron particle size; wheat grits as carrier		Novo Nordisk Novamyl™ 1500 MG

2.4.1.38 β-*N*-Acetylglucosaminylglycopeptide β-1,4-Galactosyltransferase

REACTION CATALYZED
UDPGalactose + *N*-acetyl-β-*D*-glucosaminyl-glycopeptide ↔ UDP + β-*D*-galactosyl-1,4-*N*-acetyl-β-*D*-glucosaminylglycopeptide

SYSTEMATIC NAME
UDPgalactose:*N*-acetyl-β-*D*-glucosaminyl-glycopeptide β-1,4-galactosyltransferase

SYNONYMS
UDPgalactose-glycoprotein galactosyltransferase, glycoprotein 4-β-galactosyl-transferase, β-*N*-acetyl-*D*-glucosaminide β-1,4-galactosyltransferase

REACTANTS
UDPGalactose, *N*-acetyl-β-*D*-glucosaminyl-glycopeptide, *N*-acetyl-β-*D*-glucosaminyl-polysaccharide, *N*-acetyl-β-*D*-glucosaminyl-glycoprotein, UDP, β-*D*-galactosyl-1,4-*N*-acetyl-β-*D*-glucosaminylglycopeptide

NOTES
- Classified as a glycosyltransferase: hexosyltransferase
- High activity toward *N*-acetyl-β-*D*-glucosaminyl-residues linked β-1,6- to galactose; lower activity toward β-1,3- links

See Chapter 5. Multiple Enzyme Preparations

Purine-Nucleoside Phosphorylase — 2.4.2.1

REACTION CATALYZED
Purine nucleoside + orthophosphate ↔ Purine + α-D-ribose 1-phosphate

SYSTEMATIC NAME
Purine nucleoside:orthophosphate ribosyltransferase

SYNONYMS
Inosine phosphorylase, nucleoside phosphorylase

REACTANTS
Purine nucleoside orthophosphate, purine, α-D-ribose 1-phosphate

APPLICATIONS
- Component of a double enzyme reactor system for determination of fish freshness
- Determination of inorganic phosphate, 5′-nucleosidase and adenosine deaminase, with related enzymes

NOTES
- Classified as a glycosyltransferase: pentosyltransferase
- Incompletely defined specificity
- Also catalyzes nucleoside ribosyltransferase (EC 2.4.2.5) -type reactions
- Inhibited by PCMB, SDS, Hg^{2+} and Ag^+

SPECIFIC ACTIVITY	UNITS DEFINITION	PREPARATION FORM	ADDITIONAL ACTIVITIES	SUPPLIER CATALOG NO.
Bacterial				
15-30 U/mg protein	1 unit phosphorylates 1.0 µmol inosine to hypoxanthine and ribose 1-phosphate/min at pH 7.4, 25°C	Lyophilized containing 60% protein; balance K gluconate, mannitol, EDTA		Sigma N8264
Calf spleen				
20 U/mg	1 unit converts 1 µmol inosine and P_i to product/min at 25°C	Suspension in 3.2 M $(NH_4)_2SO_4$, pH 6	<0.01% AP, guanase <0.5% ADA	Boehringer 107956 107964
20 U/mg protein	1 unit converts 1 µmol inosine to hypoxanthine and ribose-1-phosphate/min at pH 7.4, 25°C	Crystallized; suspension in 3.2 M $(NH_4)_2SO_4$, pH 6	<0.01% AP <0.5% ADA	Fluka 74678
20 U/mg protein	1 unit phosphorylates 1.0 µmol inosine to hypoxanthine and ribose-1-phosphate/min at pH 7.4, 25°C	Crystallized; suspension in 3.2 M $(NH_4)_2SO_4$, pH 6.0	<0.01% AP, guanase, XO <0.5% ADA	ICN 100889
20 U/mg protein	1 unit phosphorylates 1.0 µmol inosine to hypoxanthine and ribose 1-phosphate/min at pH 7.4, 25°C	Crystallized; suspension in 3.2 M $(NH_4)_2SO_4$, pH 6.0	<0.01% AP, guanase, XO May contain up to 0.5% ADA	Sigma N3003

2.4.2.1 Purine-Nucleoside Phosphorylase continued

SPECIFIC ACTIVITY	UNITS DEFINITION	PREPARATION FORM	ADDITIONAL ACTIVITIES	SUPPLIER CATALOG NO.
Calf spleen continued				
20 U/mg protein	1 unit phosphorylates 1.0 μmol inosine to hypoxanthine and ribose 1-phosphate/min at pH 7.4, 25°C	Crystallized; suspension in 3.2 M $(NH_4)_2SO_4$ and 50 mM KPO_4, pH 6.5	<0.01% AP, guanase, XO May contain up to 0.5% ADA	Sigma N4262
1.5-3.5 U/mg solid; 40-60 U/vial	1 unit phosphorylates 1.0 μmol inosine to hypoxanthine and ribose 1-phosphate/min at pH 7.4, 25°C	Lyophilized; <0.1% PO_4		Sigma N5643
Human blood				
20 U/mg protein	1 unit phosphorylates 1.0 μmol inosine to hypoxanthine and ribose 1-phosphate/min at pH 7.4, 25°C	Lyophilized containing 95% protein; balance primarily KPO_4 buffer salts and $MgCl_2$	<0.01% guanase, XO < 0.5% ADA, AP	Sigma N3514
Microbial (optimum pH = 7.5-8.0, pI = 4.1, T = 65°C, MW = 120,000 Da; K_M = 64 μM [inosine], 0.32 mM [P_i]; stable pH 6.0-9.0 [30°C, 16 hr], T < 60°C [pH 7.7, 30 min])				
≥15 U/mg solid	1 unit forms 1 μmol uric acid/min at pH 7.7, 37°C	Lyophilized containing K gluconate, mannitol, EDTA as stabilizers	<0.001% 5'-nucleotidase, ADA <0.01% ATPase <20% catalase	Toyobo PNP-301

2.4.2.4 Thymidine Phosphorylase

REACTION CATALYZED
    Thymidine + orthophosphate ↔ Thymine + 2-deoxy-D-ribose 1-phosphate

SYSTEMATIC NAME
    Thymidine:orthophosphate deoxy-D-ribosyl-transferase

SYNONYMS
    Pyrimidine phosphorylase

REACTANTS
    Thymidine, orthophosphate, thymine, 2-deoxy-D-ribose 1-phosphate

APPLICATIONS
- Preparation of media for testing sulfonamide and trimethoprim susceptibility

NOTES
- Classified as a glycosyltransferase: pentosyltransferase
- *E. coli* enzyme is inhibited by thymidine, a product of the forward reaction; the reverse reaction is specific for 2-deoxy-D-ribose 1-phosphate
- Enzyme from some tissue sources also catalyzes nucleoside deoxyribosyltransferase (EC 2.4.2.6)-type reactions

Thymidine Phosphorylase continued

2.4.2.4

SPECIFIC ACTIVITY	UNITS DEFINITION	PREPARATION FORM	ADDITIONAL ACTIVITIES	SUPPLIER CATALOG NO.
Escherichia coli				
1000 U/mL	1 unit phosphorylates 1 mole thymidine to thymine and 2-deoxyribose 1-phosphate/min at pH 7.4, 25°C	Solution containing 0.5 M KPO$_4$, 2 mM uracil, 0.02% NaN$_3$, BSA		ICN 156915
500 U/mL	1 unit converts 1.0 μmol thymidine and phosphate to thymine and 2-deoxyribose 1-phosphate/min at pH 7.4, 25°C	Solution containing 0.5 M KPO$_4$, 2 mM uracil, 0.02% NaN$_3$, BSA		Sigma T7006
≥0.1 U/mg protein		Chromatographically purified; lyophilized		Worthington LM01520 LM01522 LM01524

Hypoxanthine Phosphoribosyltransferase

2.4.2.8

REACTION CATALYZED

IMP + pyrophosphate ↔ Hypoxanthine + 5-phospho-α-D-ribose 1-diphosphate

SYSTEMATIC NAME

IMP:pyrophosphate phospho-D-ribosyltransferase

SYNONYMS

IMP pyrophosphorylase, transphosphoribosidase, hypoxanthine-guanine phosphoribosyltransferase, guanine phosphoribosyltransferase

REACTANTS

IMP, pyrophosphate, hypoxanthine, guanine, 6-mercaptopurine, 5-phospho-α-D-ribose 1-diphosphate

APPLICATIONS

- Determination of phosphoribosyl pyrophosphate

NOTES

- Classified as a glycosyltransferase: pentosyltransferase

SPECIFIC ACTIVITY	UNITS DEFINITION	PREPARATION FORM	ADDITIONAL ACTIVITIES	SUPPLIER CATALOG NO.
Yeast, bakers				
300-600 U/mg protein	1 unit forms 1.0 nmol GMP from guanine and PRPP/min at pH 7.5, 37°C	Lyophilized containing 80% protein and 5% MgCl$_2$; balance primarily Tris buffer salts		Sigma H3389

2.4.2.10 Orotate Phosphoribosyltransferase

REACTION CATALYZED
Orotidine 5'-phosphate + pyrophosphate ↔ Orotate + 5-phospho-α-D-ribose 1-diphosphate

SYSTEMATIC NAME
Orotidine 5'-phosphate:pyrophosphate phospho-α-D-ribosyltransferase

SYNONYMS
Orotidylic acid phosphorylase, orotidine-5'-phosphate pyrophosphorylase

REACTANTS
Orotidine 5'-phosphate, pyrophosphate, orotate, 5-phospho-α-D-ribose 1-diphosphate

APPLICATIONS
- Determination of orotidine 5'-monophosphate
- Production of OMP analogs from the corresponding orotic acid

NOTES
- Classified as a glycosyltransferase: pentosyltransferase
- The eukaryotic enzyme also acts as orotidine-5'-phosphate decarboxylase (EC 4.1.1.23)

SPECIFIC ACTIVITY	UNITS DEFINITION	PREPARATION FORM	ADDITIONAL ACTIVITIES	SUPPLIER CATALOG NO.
Yeast, brewer's bottom				
25 U/mg protein	1 unit converts 1.0 μmol orotidine 5'-monophosphate to orotic acid/hr at pH 8.0, 30°C	Lyophilized containing 50% buffer salts	<1% OMP decarboxylase; 5% inorganic pyrophosphatase	Sigma O1501

2.4.2.30 NAD$^+$ ADP-Ribosyltransferase

REACTION CATALYZED
NAD$^+$ + (ADP-D-ribosyl)$_n$-acceptor ↔ Nicotinamide + (ADP-D-ribosyl)$_{n+1}$-acceptor

SYSTEMATIC NAME
NAD$^+$:poly(adenine-diphosphate-D-ribosyl)-acceptor ADP-D-ribosyl-transferase

SYNONYMS
Poly(ADP-ribose) synthase, ADP-ribosyltransferase (polymerizing), *botulinum* neurotoxin C3, ADP-ribosyltransferase C3, exoenzyme C3

REACTANTS
NAD$^+$, (ADP-D-ribosyl)$_n$-acceptor, nicotinamide, (ADP-D-ribosyl)$_{n+1}$-acceptor

APPLICATIONS
- Selective ADP-ribosylation of low molecular weight G proteins of the *Rho* subfamily at Asn^{41}
- Studying the role of *Rho* and related proteins in lymphocyte-mediated cytotoxicity, cell mobility and thrombin-induced platelet aggregation

NAD⁺ ADP-Ribosyltransferase continued 2.4.2.30

NOTES
- Classified as a glycosyltransferase: pentosyltransferase
- The ADP-D-ribosyl group of NAD⁺ is transferred to an acceptor carboxyl group on the enzyme or a histone; further ADP-ribosyl groups are transferred to the 2'-position of the terminal adenosine moiety, building up a polymer with an average chain length of 20-30 units
- More potent than *botulinum* neurotoxin C1. ADP-ribosylates 21-24 kDa proteins in platelets and brain membranes similar to *botulinum* neurotoxins C1 and D

SPECIFIC ACTIVITY	UNITS DEFINITION	PREPARATION FORM	ADDITIONAL ACTIVITIES	SUPPLIER CATALOG NO.
Clostridium botulinum (MW = 25,000 Da)				
		98% purity; lyophilized		BIOMOL G-130
		Single band purity; lyophilized		Calbiochem 341208
1 pmol/mg P2/μg C3				Wako 011-14441

β-Galactoside α-2,6-Sialyltransferase 2.4.99.1

REACTION CATALYZED

CMP-*N*-acetylneuraminate + β-D-galactosyl-1,4-*N*-acetyl-β-D-glucosamine ↔ CMP + α-*N*-acetylneuraminyl-2,6-β-D-galactosyl-1,4-*N*-acetyl-β-D-glucosamine

SYSTEMATIC NAME

CMP-*N*-acetylneuraminate:β-D-galactosyl-1,4-*N*-acetyl-β-D-glucosamine α-2,6-*N*-acetylneuraminyltransferase

SYNONYMS

Sialyltransferase, 2,6-sialyl transferase

REACTANTS

CMP-*N*-acetylneuraminate, β-D-galactosyl-1,4-*N*-acetyl-β-D-glucosamine, CMP, α-*N*-acetylneuraminyl-2,6-β-D-galactosyl-1,4-*N*-acetyl-β-D-glucosamine

APPLICATIONS
- *In vitro* sialylation of Galb(1,4)GlcNAc structures in *N*-glycans via α-2,6 linkages

NOTES
- Classified as a glycosyltransferase, transferring other glycosyl groups
- The terminal β-D-galactosyl residue of lactose and the oligosaccharide of glycoproteins can act as acceptors

2.4.99.1 β-Galactoside α-2,6-Sialyltransferase continued

SPECIFIC ACTIVITY	UNITS DEFINITION	PREPARATION FORM	ADDITIONAL ACTIVITIES	SUPPLIER CATALOG NO.
Rat liver (optimum pH = 6.0, MW = 40,500 Da)				
8 U/mg protein (substrate A); 2.5 U/mg protein (substrate B)	1 unit converts 1 μmol substrate A (CMP-N-Ac-neuraminic acid) or substrate B (CMP-9-(3-fluoresceinyl-thioureido)-9-deoxy-N-Ac-neuraminic acid) to product/min at 37°C	Solution containing 35 mM Mes-NaOH, 0.45 M NaCl, 0.08% Triton CF-54, 50% glycerol, pH 6.0	No detectable proteases <0.4% Galβ(1-3(4))GlcNAcα(2-3)-sialyltransferase	Boehringer 981583
2 U/mL	1 unit transfers 1.0 μmol N-Ac-neuraminic acid from CMP-N-Ac-neuraminic acid to asialomucin/min at pH 6.5, 37°C	Solution containing 50% glycerol, 0.5 M NaCl, 0.10% Triton, 35 mM Na cacodylate, pH 6		ICN 156622
2 U/mL	1 unit transfers 1.0 μmol N-Ac-neuraminic acid from CMP-Ac-neuraminic acid to asialomucin/min at pH 6.5, 37°C	Solution containing 50% glycerol, 0.45 M NaCl, 0.08% Triton CF-54, 35 mM Na cacodylate, pH 6		Sigma S2769
2.5 U/mg protein		Solution		Wako 196-10581

2.4.99.4 β-Galactoside α-2,3-Sialyltransferase

REACTION CATALYZED
CMP-N-acetylneuraminate + β-D-galactosyl-1,3-N-acetyl-α-D-galactosaminyl-R ↔ CMP + α-N-acetylneuraminyl-2,3-β-D-galactosyl-1,3-N-acetyl-α-D-galactosaminyl-R

SYSTEMATIC NAME
CMP-N-acetylneuraminate:β-D-galactoside α-2,3-N-acetylneuraminyltransferase

SYNONYMS
α-2,3-Sialyltransferase

SUBSTRATES
CMP-N-acetylneuraminate, β-D-galactosyl-1,3-N-acetyl-α-D-galactosaminyl-R, lactose, CMP, α-N-acetylneuraminyl-2,3-β-D-galactosyl-1,3-N-acetyl-α-D-galactosaminyl-R

NOTES
- Classified as a glycosyltransferase, transferring other glycosyl groups
- Acceptor is Galβ(1,3)GalN-acetyl-R, where R is H, a threonine or serine residue in a glycoprotein or glycolipid
- Lactose can act as acceptor
- May be identical with monosialoganglioside sialyltransferase (EC 2.4.99.2)

SPECIFIC ACTIVITY	UNITS DEFINITION	PREPARATION FORM	ADDITIONAL ACTIVITIES	SUPPLIER CATALOG NO.
Porcine liver				
	1 unit sialylates 1.0 µmol N-glycosidase F-treated asialofetuin/min at pH 6.5, 37°C	Solution containing 50% glycerol, 0.8 M NaCl, 1% Triton X-100, 10 mM Na cacodylate buffer, pH 6.5		Sigma S0403

Glutathione Transferase 2.5.1.18

REACTION CATALYZED
RX + glutathione ↔ HX + R-S-G

SYSTEMATIC NAME
RX:glutathione R-transferase

SYNONYMS
Glutathione S-alkyltransferase, glutathione S-transferase, glutathione S-aryltransferase, S-(hydroxyalkyl)glutathione lyase, glutathione S-aralkyltransferase

REACTANTS
RX, glutathione, polyol nitrate, aliphatic epoxides, arene oxides, HX, R-S-G, polyol, nitrile

APPLICATIONS
- In vitro drug metabolism and toxicology research
- Pro-carcinogen activation studies

NOTES
- Classified as a transferase, transferring alkyl or aryl groups, other than methyl groups
- A group of enzymes of broad specificity
- R may be an aliphatic, aromatic or heterocyclic group
- X may be a sulfate, nitrile or halide group
- Also adds aliphatic epoxides and arene oxides to glutathione, reduces polyol nitrate and glutathione to polyol and nitrile, and catalyzes isomerization reactions and disulfide interchanges
- Type A1-1 forms thioether conjugates between glutathione and reactive xenobiotic compounds, acting as defense against electrophilic chemical species formed through cellular oxidations catalyzed by cytochrome P-450. Comprised of four structural classes: alpha, mu, pi and theta
- Type P1-1 is abundant in most tumor cells and is widely distributed throughout the body (except the liver)
- Type M1-1 is the b allelic variant corresponding to ν purified from human liver

SPECIFIC ACTIVITY	UNITS DEFINITION	PREPARATION FORM	ADDITIONAL ACTIVITIES	SUPPLIER CATALOG No.
Bovine liver				
10-20 U/mg protein; 0.2-1.0 U/mg protein (*p*-nitrobenzyl chloride); 0.02-0.5 U/mg protein (1,2-epoxy-3-(*p*-nitrophenoxy)propane)	1 unit conjugates 1.0 µmol 1-chloro-2,4-dinitrobenzene with GSH/min at pH 6.5, 25°C	Lyophilized containing 60% protein; balance primarily Tris and PO_4 buffer salts, GSH, EDTA		Sigma G4385
Equine liver				
50-100 U/mg protein; 0.2-1.0 U/mg protein (*p*-nitrobenzyl chloride); 0.02-0.1 U/mg protein (1,2-epoxy-3-(*p*-nitrophenoxy)propane)	1 unit conjugates 1.0 µmol 1-chloro-2,4-dinitrobenzene with GSH/min at pH 6.5, 25°C	Lyophilized containing 75% protein; balance primarily Tris and PO_4 buffer salts, GSH, EDTA		Sigma G6511
Human placenta				
25-125 U/mg protein; 0.2-1.0 U/mg protein (*p*-nitrobenzyl chloride); 0.02-0.5 U/mg protein (1,2-epoxy-3-(*p*-nitrophenoxy)propane)	1 unit conjugates 1.0 µmol 1-chloro-2,4-dinitrobenzene with GSH/min at pH 6.5, 25°C	Lyophilized containing 40% protein; balance primarily Tris buffer salts, GSH, EDTA		Sigma G8642
Porcine liver				
50-150 U/mg protein; 0.2-1.5 U/mg protein (*p*-nitrobenzyl chloride); 0.02-0.5 U/mg protein (1,2-epoxy-3-(*p*-nitrophenoxy)propane)	1 unit conjugates 1.0 µmol 1-chloro-2,4-dinitrobenzene with GSH/min at pH 6.5, 25°C	Lyophilized containing 75% protein; balance primarily Tris and PO_4 buffer salts, GSH, EDTA		Sigma G6636
Rabbit liver				
50-150 U/mg protein; 0.2-1.5 U/mg protein (*p*-nitrobenzyl chloride); 0.02-0.5 U/mg protein (1,2-epoxy-3-(*p*-nitrophenoxy)propane)	1 unit conjugates 1.0 µmol 1-chloro-2,4-dinitrobenzene with GSH/min at pH 6.5, 25°C	Lyophilized containing 75% protein; balance primarily Tris and PO_4 buffer salts, GSH, EDTA		Sigma G8261
Rat liver				
25-50 U/mg protein	1 unit conjugates 1.0 µmol 1-chloro-2,4-dinitrobenzene with GSH/min at pH 6.5, 25°C	Lyophilized containing 75% protein; balance primarily Tris and PO_4 buffer salts, GSH, EDTA		Sigma G8386

Glutathione Transferase continued

2.5.1.18

SPECIFIC ACTIVITY	UNITS DEFINITION	PREPARATION FORM	ADDITIONAL ACTIVITIES	SUPPLIER CATALOG NO.
Human DNA, recombinant expressed in *E. coli*				
1-5 mg/mL	≥50 µmol/min/mg	>95% purity by Coomassie-stained SDS gel; 50 m*M* Tris-HCl, 50 m*M* NaCl, 1 m*M* DTT, 1 m*M* EDTA, 50% glycerol, pH 7.5		PanVera P2175 P2176
1-5 mg/mL	≥75 µmol/min/mg	>95% purity by Coomassie-stained SDS gel; 50 m*M* Tris-HCl, 50 m*M* NaCl, 1 m*M* DTT, 1 m*M* EDTA, 50% glycerol, pH 7.5		PanVera P2177 P2178
1-5 mg/mL	≥100 µmol/min/mg	>95% purity by Coomassie-stained SDS gel; 50 m*M* Tris-HCl, 50 m*M* NaCl, 1 m*M* DTT, 1 m*M* EDTA, 50% glycerol, pH 7.5		PanVera P2192 P2193

Aspartate Transaminase

2.6.1.1

REACTION CATALYZED
    L-Aspartate + 2-oxoglutarate ↔ Oxaloacetate + L-glutamate

SYSTEMATIC NAME
    L-Aspartate:2-oxoglutarate aminotransferase

SYNONYMS
    Glutamic-oxaloacetic transaminase, glutamic-aspartic transaminase, transaminase A, aspartate aminotransferase, AST, GOT

REACTANTS
    L-Aspartate, L-tyrosine, L-phenylalanine, L-tryptophan, 2-oxoglutarate, oxaloacetate, L-glutamate, NADH

APPLICATIONS
- Synthesis of unnatural L-amino acids from α-keto acids
- A component of QC sera for diagnosis of liver and heart disorders

NOTES
- Classified as a transferase, transferring nitrogenous groups: transaminase
- A pyridoxal-phosphate protein
- Controlled proteolyis of aromatic-amino-acid transaminase (EC 2.6.1.57) forms this activity
- Widely distributed in plants and animals, but is concentrated in mammalian heart and liver
- Serum levels are elevated in acute and chronic hepatitis, obstructive jaundice, liver carcinoma, myocardial infarction and muscular dystrophy
- Exists in two isozyme forms: M (mitochondrial) and S (cytoplasmic)
- Activated by pyridoxal or pyridoxamine phosphate
- Stabilized by α-ketoglutarate and EDTA in maleate or succinate buffers
- Inhibited by Hg^{2+}, cyanide, PMCB, maleate, succinate, glutarate, adipate, hydroxyaspartate and fluoro oxaloacetate

2.6.1.1 Aspartate Transaminase continued

SPECIFIC ACTIVITY	UNITS DEFINITION	PREPARATION FORM	ADDITIONAL ACTIVITIES	SUPPLIER CATALOG NO.
Human heart				
50 U/mg protein	1 unit transaminates 1 μmol L-Asp/min at pH 7.5, 25°C	Suspension in 70% $(NH_4)_2SO_4$ containing 10 mg/mL protein		Calzyme 055B0050
Porcine heart (optimum pH = 8.0-8.5, pI = 7.5, MW = 91,000-110,000 Da; stable pH 4.5-8.0, T < 65°C [with 10 mM 2-oxoglutarate])				
180 U/mg protein	1 unit transaminates 1 μmol L-Asp/min at pH 7.4, 25°C	Suspension in 75% $(NH_4)_2SO_4$ containing 0.0025 M 2-oxoglutarate	<0.01% GLDH, LDH, MDH <0.05% GPT	Biozyme GOT1
180 U/mg protein; ≥25 U/mg solid	1 unit transaminates 1 μmol L-Asp/min at pH 7.4, 25°C	Salt-free; lyophilized	<0.01% GLDH, LDH, MDH <0.05% GPT	Biozyme GOT1FS
200 U/mg at 25°C; 380 U/mg at 37°C	1 unit converts 1 μmol L-Asp and 2-oxoglutarate to product/min	Suspension in 3.2 M $(NH_4)_2SO_4$, 2.5 mM 2-oxoglutarate, 50 mM maleate, 1 mM pyridoxal phosphate, pH 6	<0.01% GIDH, LDH, MDH, GPT, oxaloacetate decarboxylase	Boehringer 105546 105554 737046
200 U/mg protein	1 unit transaminates 1 μmol L-Asp/min at pH 7.5, 25°C	Suspension in 70% $(NH_4)_2SO_4$ containing 10 mg/mL protein	<0.01% GPT, LDH, MDH, GlDH <0.05% OAA decarboxylase	Calzyme 056B0200
100 U/mg	1 unit transaminates 1 μmol L-Asp/min at pH 7.5, 25°C	Lyophilized		Calzyme 126A0100
50 U/mg protein	1 unit transaminates 1 μmol L-Asp/min at pH 7.5, 25°C	Suspension in 70% $(NH_4)_2SO_4$ containing 90-95% protein		Calzyme 197B0050
280 U/mg protein	1 unit converts 1 μmol 2-oxoglutarate to L-Glu/min at pH 7.5, 37°C in the presence of L-Asp	Suspension in 3 M $(NH_4)_2SO_4$, 0.05 M malate, 2.5 mM 2-oxoglutarate solution, pH 7.0	<0.01% GAPDH, GlDH, GPT, LDH, MDH, NADH oxidase	Fluka 49396
180 U/mg protein	1 unit transaminates 1.0 μmol L-Asp/min at pH 7.4, 25°C	Salt-free; lyophilized	<0.01% GLDH, LDH, MDH	ICN 104830
180 U/mg protein	1 unit transaminates 1.0 μmol L-Asp/min at pH 7.4, 25°C	Suspension in 3.0 M $(NH_4)_2SO_4$	<0.01% GLDH, LDH, MDH	ICN 151190
>150 U/mg protein; >30 U/mg solid	1 unit transaminates 1 μmol L-Asp/min at pH 7.4, 25°C	Lyophilized	<0.01% MDH, GLDH <0.05% LDH, GPT	Randox GO 900L
200-500 U/mg protein	1 unit converts 1.0 μmol α-ketoglutarate to L-Glu/min at pH 7.5, 37°C in the presence of L-Asp; 1 U = 2000 O.D. U at 25°C	Suspension in 3.0 M $(NH_4)_2SO_4$, 0.05 M maleate, 2.5 mM α-ketoglutarate, pH 6.0	<0.01% LDH, MDH <0.03% GPT	Sigma G2751
100-400 U/mg protein	1 unit converts 1.0 μmol α-ketoglutarate to L-Glu/min at pH 7.5, 37°C in the presence of L-Asp; 1 U = 2000 O.D. U at 25°C	Lyophilized containing 80% protein and 20% Na citrate buffer salt	<0.01% LDH, MDH <0.03% GPT	Sigma G7005

Aspartate Transaminase continued 2.6.1.1

SPECIFIC ACTIVITY	UNITS DEFINITION	PREPARATION FORM	ADDITIONAL ACTIVITIES	SUPPLIER CATALOG No.
Porcine heart continued				
≥50 U/mg solid	1 unit decreases A_{340} by 0.001/min at 25°C using a coupled reaction with MDH in the presence of NADH	Chromatographically purified; dialyzed; lyophilized	<0.01% MDH, LDH, GPT	Worthington LS05266 LS05267 LS05268

Alanine Transaminase 2.6.1.2

REACTION CATALYZED
    L-Alanine + 2-oxoglutarate ↔ Pyruvate + L-glutamate

SYSTEMATIC NAME
    L-Alanine:2-oxoglutarate aminotransferase

SYNONYMS
    Glutamic-pyruvic transaminase, glutamic-alanine transaminase, alanine aminotransferase, ALT, GPT

SUBSTRATES
    L-Alanine, 2-oxoglutarate, 2-aminobutanoate (acts slowly), pyruvate, L-glutamate

APPLICATIONS
- Synthesis of unnatural L-amino acids from α-keto acids
- Human enzyme is used in the diagnosis of liver and heart disease

NOTES
- Classified as a transferase, transferring nitrogenous groups: transaminase
- A pyridoxal-phosphate protein
- 2-Aminobutanoate acts slowly in place of alanine
- Found in plants and animals, but most concentrated in mammalian heart and liver
- Exists as two isozymes: M (mitochondrial) and S (cytoplasmic)

SPECIFIC ACTIVITY	UNITS DEFINITION	PREPARATION FORM	ADDITIONAL ACTIVITIES	SUPPLIER CATALOG No.
Human heart				
10 U/mg protein	1 unit converts 1 μmol α-ketoglutarate to L-Glu/min at pH 7.5, 25°C in PO_4 buffer	Suspension in 70% $(NH_4)_2SO_4$ containing 10 mg/mL protein		Calzyme 057B0010
Porcine heart (optimum pH = 7.0-9.0, T = 37°C, MW = 100,000-116,000 Da)				
≥80 U/mg protein	1 unit transaminates 1 μmol L-Ala/min at pH 7.5, 25°C	Purified; suspension in 75% $(NH_4)_2SO_4$	<0.01% GLDH, GOT, LDH, MDH	Biozyme GPT1

2.6.1.2 Alanine Transaminase continued

SPECIFIC ACTIVITY	UNITS DEFINITION	PREPARATION FORM	ADDITIONAL ACTIVITIES	SUPPLIER CATALOG No.
Porcine heart continued				
≥80 U/mg protein; ≥35 U/mg solid	1 unit transaminates 1 μmol L-Ala/min at pH 7.5, 25°C	Salt-free; lyophilized	<0.01% GLDH, GOT, LDH, MDH	Biozyme GPT1F
80 U/mg at 25°C; 140 U/mg at 37°C	1 unit converts 1 μmol L-Ala and 2-oxoglutarate to product/min	Suspension in 3.2 M $(NH_4)_2SO_4$, pH 6	<0.01% GlDH, LDH, MDH <0.03% GOT	Boehringer 105562 105589 737127
100 U/mg protein	1 unit converts 1 μmol α-ketoglutarate to L-Glu/min at pH 7.5, 25°C in PO_4 buffer	Suspension in 70% $(NH_4)_2SO_4$ containing 10 mg/mL protein	<0.001% GOT, GAPDH <0.002% MDH <0.005% LDH	Calzyme 078B0025
100 U/mg protein	1 unit converts 1 μmol α-ketoglutarate to L-Glu/min at pH 7.5, 25°C in PO_4 buffer	Lyophilized containing 40% protein	<0.001% GOT, GAPDH <0.002% MDH <0.005% LDH	Calzyme 081A0100
60 U/mg	1 unit converts 1 μmol 2-oxoglutarate to L-Glu/min at pH 7.5, 37°C in the presence of L-Asp	Powder	<0.01% GlDH, LDH, MDH <0.03% GOT	Fluka 49399
100 U/mg protein	1 unit converts 1 μmol α-ketoglutarate to L-Glu/min at pH 7.6, 37°C in the presence of L-Ala	Aqueous suspension in 1.8 M $(NH_4)_2SO_4$, pH 6.0	<0.01% GlDH, MDH, LDH <0.03% GOT	Fluka 49400
80 U/mg protein	1 unit transaminates 1.0 μmol L-Ala/min at pH 7, 25°C	Purified; suspension in 1.8 M $(NH_4)_2SO_4$, pH 6.0	<0.01% GLDH, LDH, MDH, GOT	ICN 104879
80 U/mg protein	1 unit transaminates 1.0 μmol L-Ala/min at pH 7, 25°C	Salt-free; lyophilized	<0.01% GLDH, LDH, MDH, GOT	ICN 151191
80 U/mg protein	1 unit converts 1.0 μmol α-ketoglutarate to L-Glu/min at pH 7.6, 37°C in the presence of L-Ala	Lyophilized containing 85% protein; balance primarily citrate and acetate buffer salts	<0.01% L-GlDH, LDH, MDH <0.03% GOT	Sigma G8255
80 U/mg protein	1 unit converts 1.0 μmol α-ketoglutarate to L-Glu/min at pH 7.6, 37°C in the presence of L-Ala	Suspension in 1.8 M $(NH_4)_2SO_4$, pH 6.0	<0.01% L-GlDH, LDH, MDH <0.03% GOT	Sigma G9880
30-50 U/mg		Lyophilized		Wako 015-11801

Hexokinase 2.7.1.1

REACTION CATALYZED
ATP + D-hexose ↔ ADP + D-hexose 6-phosphate

SYSTEMATIC NAME
ATP:D-hexose 6-phosphotransferase

SYNONYMS
Glucokinase, hexokinase type IV, hexokinase D, hexokinase II, HK

REACTANTS
ATP, ITP, dATP, D-hexose, D-glucose, D-mannose, D-fructose, sorbitol, D-glucosamine, ADP, D-hexose 6-phosphate, 5-keto-D-fructose, 2-deoxy-D-glucose

APPLICATIONS
- Determination of fructose, mannose and other saccharides
- Preparation of glucose-6-phosphate
- Determination of ATP and glucose in biological fluids, with glucose-6-phosphate dehydrogenase

NOTES
- Classified as a transferase, transferring phosphorus-containing groups: phosphotransferases with an alcohol group as acceptor
- Four distinct mammalian enzymes have been identified in all tissues. Types I and II predominate in adipose tissue, Type I in skeletal muscle and Type IV in the liver
- Two enzymes (P-I and -II) have been isolated from yeast
- Requires Mg^{2+} for catalytic activity; Ca^{2+} do not affect activity
- Activated by Mg^{2+}, catecholamine and related compounds
- Stabilized by ATP, albumin, KCl and NaCl
- Inhibited by -SH-reactive compounds, PCMB, Hg^{2+}, Ag^+, sorbose-1-phosphate, polyphosphates, 6-deoxy-6-fluoro-glucose, 2-C-hydromethyl-glucose, xylose and lyxose

SPECIFIC ACTIVITY	UNITS DEFINITION	PREPARATION FORM	ADDITIONAL ACTIVITIES	SUPPLIER CATALOG NO.
Bacillus species (optimum pH = 7.5-8.0, pI = 5.64, T = 50°C, MW = 68,000 Da [gel filtration]; K_M = 0.82 mM [glucose], 87 μM [ATP], 1.6 mM [$MgCl_2$]; stable pH 7-8.5 [phosphate buffer, 55°C, 10 min], T < 60°C [pH 8.0, 10 min])				
250-400 U/mg solid	1 unit generates 1 μmol NADPH/min at pH 8.0, 37°C	Lyophilized containing 70% protein		Asahi [HKII] T-50
Bovine heart				
10-30 U/mg protein	1 unit phosphorylates 1.0 μmol glucose/min at pH 7.6, 25°C	Lyophilized containing 5% protein; balance primarily Tris, EDTA, DTT, Ficoll		Sigma H8755

SPECIFIC ACTIVITY	UNITS DEFINITION	PREPARATION FORM	ADDITIONAL ACTIVITIES	SUPPLIER CATALOG No.
Saccharomyces species (optimum pH = 7.5, T = 40°C, MW = 100,000 Da [2 subunits/mol enzyme]; K_M = 0.1 mM [Glc, ATP], 0.7 mM [Fru], 0.6 mM [5-keto-D-Fru], 50 μM [mannose], 0.3 mM [2-deoxyGlc], 0.15 mM [glucosamine, pH 7.5], 1 mM [N-Ac-glucosamine], 20 μM [glucosone], > 50 mM [Gal], 5mM [6-deoxy-6-fluoroGlc], 3.7 mM [ITP], 2.6 [Mg^{2+}]; stable pH 5.5-7.5 [20°C, 17 hr], T < 30°C [pH 8.0, 30 min])				
≥150 U/mg solid	1 unit forms 1 μmol NADH/min at pH 8.0, 30°C	Lyophilized	<0.01% 6PGDH, G6PDH, MK <0.1% PGI <0.5% GR	Toyobo HXK-301
Yeast (optimum pH = 7.5-9.0, MW = 100,000 Da; relative activity: 25°C = 1.0, 30°C = 1.32, 37°C = 1.86)				
80 U/mg protein	1 unit forms 1 μmol NADH/min at pH 8.2, 25°C	Salt-free; lyophilized		Biozyme HK2B
130-250 U/mg protein	1 unit forms 1 μmol NADH/min at pH 8.2, 25°C	Crystallized; suspension in 75% $(NH_4)_2SO_4$	<0.005% ATPase, 6PGDH <0.05% PGI	Biozyme HK3
130-250 U/mg protein; 120 U/mg solid	1 unit forms 1 μmol NADH/min at pH 8.2, 25°C	Salt-free; lyophilized	<0.005% ATPase, 6PGDH <0.05% PGI	Biozyme HK3F
120 U/mg protein	1 unit phosphorylates 1 μmol D-hexose/min at pH 7.5, 30°C in the presence of ATP	Salt-free; lyophilized	<0.01% 6-PGDH	Calbiochem 376811
140 U/mg protein	1 unit converts 1 μmol glucose and ATP/min at pH 7.6, 25°C	Crystalline suspension in 3.2 M $(NH_4)_2SO_4$, pH 6	<0.05% PGI	Fluka 53110
25 U/mg	1 unit converts 1 μmol glucose and ATP/min at pH 7.6, 25°C	Lyophilized		Fluka 53115
>150 U/mg protein; >100 U/mg solid	1 unit forms 1.0 μmol Glc6P/min at pH 8.2, 25°C	Lyophilized containing <10 pmol/U glucose	<0.002% CK (including AK), AK <0.005% ATPase, GR, 6PGDH, PGI	Genzyme 1351
>150 U/mg protein	1 unit forms 1.0 μmol Glc6P /min at pH 8.2, 25°C	Suspension in $(NH_4)_2SO_4$ and <10 pmol/U glucose	<0.002% CK, AK <0.005% ATPase, GR, 6PGDH, PGI	Genzyme 1352
>150 U/mg protein	1 unit forms 1.0 μmol Glc6P /min at pH 8.2, 25°C	Solution containing 50% glycerol and <10 pmol/U glucose	<0.002% CK, AK <0.005% ATPase, GR, 6PGDH, PGI	Genzyme 1353
20-25 U/mg solid	1 unit forms 1.0 μmol NADH/min at pH 8.2, 25°C	Purified; salt-free; lyophilized		ICN 100716
140 U/mg protein	1 unit forms 1.0 μmol NADH/min at pH 8.2, 25°C	Crystallized; suspension in 3.2 M $(NH_4)_2SO_4$ and PO_4 buffer, pH 7.0		ICN 151253
140 U/mg protein	1 unit forms 1.0 μmol NADH/min at pH 8.2, 25°C	Purified; salt-free; lyophilized		ICN 151254

Hexokinase continued 2.7.1.1

SPECIFIC ACTIVITY	UNITS DEFINITION	PREPARATION FORM	ADDITIONAL ACTIVITIES	SUPPLIER CATALOG NO.
Yeast *continued*				
≥140 IU/mg protein	1 IU transforms 1 μmol substrate/min under standard IUB conditions at 25°C	Lyophilized; suspension in $(NH_4)_2SO_4$	<0.001% ATPase <0.005% GR, CK, 6PGDH <0.01% G6PDH, PGluM, MK <0.1% PGI	OYC
≥180 IU/mg protein	1 IU transforms 1 μmol substrate/min under standard IUB conditions at 25°C	Suspension in $(NH_4)_2SO_4$	<0.001% ATPase, PGluM, 6PGDH, MK <0.003% PGI <0.005% GR, G6PDH, CK	OYC
≥150 U/mg protein	1 unit reduces 1 μmol NAD/min at pH 8.0, 30°C in a coupled assay system with G6PDH	Chromatographically purified; suspension in 3.0 M $(NH_4)_2SO_4$		Worthington LS02498 LS02499 LS02500
≥150 U/mg protein	1 unit reduces 1 μmol NAD/min at pH 8.0, 30°C in a coupled assay system with G6PDH	Chromatographically purified; dialyzed; lyophilized	<0.005% AdK, CPK <0.003% 6-PGDH <0.05% PHI	Worthington LS02511 LS02512 LS02514
Yeast, bakers (optimum pH = 7.5, MW = 100,000 Da)				
150 U/mg solid	1 unit phosphorylates 1 μmol D-glucose/min at pH 7.6, 25°C	Lyophilized containing 95% protein		Calzyme 088A0150
150 U/mg solid	1 unit phosphorylates 1 μmol D-glucose/min at pH 7.6, 25°C	Suspension in 3.2 M $(NH_4)_2SO_4$ containing 95% protein		Calzyme 116B0150
	1 unit phosphorylates 1 μmol D-glucose/min at pH 7.6, 25°C	Meets reference specifications for use in FDA glucose measurement method		Sigma H1131
1200-2000 U/g beaded agarose; 1 mL yields 35-65 U	1 unit phosphorylates 1 μmol D-glucose/min at pH 7.6, 30°C	Insoluble enzyme attached to beaded agarose; suspension in 2.0 M $(NH_4)_2SO_4$, pH 7.0		Sigma H2005
	1 unit phosphorylates 1.0 μmol glucose/min at pH 8.5, 25°C	Biotin-labeled; lyophilized containing 80% protein; balance primarily Na citrate		Sigma H2884
200 U/mg protein	1 unit phosphorylates 1 μmol D-glucose/min at pH 7.6, 25°C	Crystallized; lyophilized containing 15% Na citrate	No detectable ATPase, MK, Glc6PDH, 6PGDH, PGI	Sigma H4502
15-24 U/mg protein	1 unit phosphorylates 1.0 μmol D-glucose/min at pH 7.6, 25°C	Lyophilized	Glc6PDH, PGI	Sigma H5000
25-40 U/mg protein	1 unit phosphorylates 1.0 μmol D-glucose/min at pH 7.6, 25°C	Lyophilized	Glc6PDH, PGI	Sigma H5125
40-60 U/mg protein	1 unit phosphorylates 1.0 μmol D-glucose/min at pH 7.6, 25°C	Lyophilized	Glc6PDH, PGI	Sigma H5250

SPECIFIC ACTIVITY	UNITS DEFINITION	PREPARATION FORM	ADDITIONAL ACTIVITIES	SUPPLIER CATALOG No.
Yeast, bakers *continued*				
60-130 U/mg protein	1 unit phosphorylates 1.0 µmol D-glucose/min at pH 7.6, 25°C	Lyophilized	Glc6PDH, PGI	Sigma H5375
130-250 U/mg protein	1 unit phosphorylates 1.0 µmol D-glucose/min at pH 7.6, 25°C	Crystallized; suspension in 3.2 M $(NH_4)_2SO_4$, 2.5 mM PO_4 buffer, 0.25 mM EDTA, pH 7; a mixture of isozymes	No detectable ATPase, MK, Glc6PDH, 6PGDH, PGI	Sigma H5500
250 U/mg protein	1 unit phosphorylates 1.0 µmol D-glucose/min at pH 7.6, 25°C	Crystallized; suspension in 3.2 M $(NH_4)_2SO_4$, 2.5 mM PO_4 buffer, 0.25 mM EDTA, pH 7; a mixture of isozymes	No detectable ATPase, MK, Glc6PDH, 6PGDH, PGI	Sigma H5625
250 U/mg protein	1 unit phosphorylates 1.0 µmol D-glucose/min at pH 7.6, 25°C	Crystalline suspension in 3.2 M $(NH_4)_2SO_4$, 2.5 mM PO_4 buffer, 0.25 mM EDTA, pH 7; Kaji Fraction I	No detectable ATPase, MK, Glc6PDH, 6PGDH, PGI	Sigma H5750
250 U/mg protein	1 unit phosphorylates 1.0 µmol D-glucose/min at pH 7.6, 25°C	Crystallized; suspension in 3.2 M $(NH_4)_2SO_4$, 2.5 mM PO_4 buffer, 0.25 mM EDTA, pH 7; Kaji Fraction II	No detectable ATPase, MK, Glc6PDH, 6PGDH, PGI	Sigma H5875
10-15 U/g solid	1 unit phosphorylates 1.0 µmol D-glucose/min at pH 7.6, 25°C	Insoluble enzyme attached to polyacrylamide containing 30% borate buffer salts		Sigma H8254
Yeast, overproducing strain				
450 U/mg protein; 70 U/mg solid	1 unit converts 1 µmol glucose and ATP to product/min at 25°C	Lyophilized	<0.0005% G6PDH, GR <0.001% CK, MK, 6PGDH ADH <0.002% PGI <0.05% GlDH	Boehringer 1317466
450 U/mg protein	1 unit converts 1 µmol glucose and ATP to product/min at 25°C	Suspension in 3.2 M $(NH_4)_2SO_4$ and <30 µg/mL glucose, pH 6.5	<0.0005% G6PDH, GR <0.001% CK, MK, 6PGDH ADH <0.002% PGI <0.05% GlDH, invertase (β-fructosidase)	Boehringer 1426362 1426389
350 U/mg protein	1 unit phosphorylates 1.0 µmol D-glucose/min at pH 7.6, 25°C	Lyophilized containing 15% protein and stabilizer		Sigma H6380

Glucokinase 2.7.1.2

REACTION CATALYZED
  ATP + D-glucose ↔ ADP + D-glucose 6-phosphate
SYSTEMATIC NAME
  ATP:D-glucose 6-phosphotransferase
SYNONYMS
  GlcK
REACTANTS
  ATP, D-glucose, ADP, D-glucose 6-phosphate

NOTES
- Classified as a transferase, transferring phosphorus-containing groups: phosphotransferases with an alcohol group as acceptor
- A group of enzymes found in invertebrates and microorganisms
- Highly specific for glucose.
- Activated by P_i

SPECIFIC ACTIVITY	UNITS DEFINITION	PREPARATION FORM	ADDITIONAL ACTIVITIES	SUPPLIER CATALOG No.
Bacillus stearothermophilus (optimum pH = 8.5, pI = 5.0, T = 65°C, MW = 68,000 Da [32,000/subunit]; K_M = 0.1 mM [Glc], 50 μM [ATP]; stable pH > 8.0, T < 60°C)				
300 U/mg solid	1 unit phosphorylates 1.0 μmol D-glucose to D-Glc6P/min at pH 9.0, 30°C	Lyophilized containing 75% protein; balance primarily Tris buffer salts		ICN 190481
100-500 U/mg protein	1 unit phosphorylates 1.0 μmol D-glucose to D-Glc6P/min at pH 9.0, 30°C	Lyophilized containing 75% protein; balance primarily Tris buffer salts		Sigma G8887
>350 U/mg protein	1 unit forms 1 μmol Glc6P/min at 30°C	Lyophilized	<0.01% G6DH, PGluM, 6PGDH, PHI, GR	Unitika
>350 U/mg protein	1 unit forms 1 μmol Glc6P/min at 30°C	50% glycerol solution	<0.01% G6PDH, PGluM, 6PGDH, PHI, GR	Unitika
Zymomonas mobilis (optimum pH = 7.0-8.0, MW = 66,000 Da [33,000/subunit]; K_M = 0.1 mM [Glc], 0.65 mM [ATP]; stable pH 6.0-8.0, T < 40°C)				
>200 U/mg protein	1 unit forms 1 μmol Glc6P/min at 30°C	Lyophilized		Unitika
>200 U/mg protein	1 unit forms 1 μmol Glc6P/min at 30°C	Suspension in $(NH_4)_2SO_4$		Unitika

2.7.1.6 Galactokinase

REACTION CATALYZED
ATP + D-galactose ↔ ADP + α-D-galactose 1-phosphate

SYSTEMATIC NAME
ATP:D-galactose 1-phosphotransferase

REACTANTS
ATP, D-galactose, D-galactosamine, ADP, α-D-galactose 1-phosphate

NOTES
- Classified as a transferase, transferring phosphorus-containing groups: phosphotransferases with an alcohol group as acceptor

SPECIFIC ACTIVITY	UNITS DEFINITION	PREPARATION FORM	ADDITIONAL ACTIVITIES	SUPPLIER CATALOG NO.
Yeast, galactose adapted				
25-50 U/mg protein	1 unit converts 1.0 μmol D-galactose to Gal1P /min at pH 7.0, 25°C	Lyophilized in vials containing 50% protein; balance primarily citrate buffer salts	<0.1% UDP-GlcDH, UDP-glucose pyrophosphorylase, UDP-galactose 4-epimerase 1% HK 2% Gal1P uridyl transferase	Sigma G0130

2.7.1.11 6-Phosphofructokinase

REACTION CATALYZED
ATP + D-fructose 6-phosphate ↔ ADP + D-fructose 1,6-bisphosphate

SYSTEMATIC NAME
ATP:D-fructose 6-phosphate 1-phosphotransferase

SYNONYMS
Phosphohexokinase, phosphofructokinase, fructose-6-phosphate kinase, PFK

REACTANTS
ATP, UTP, CTP, ITP, D-fructose 6-phosphate, D-tagatose 6-phosphate, sedoheptulose 7-phosphate, ADP, D-fructose 1,6-bisphosphate

NOTES
- Classified as a transferase, transferring phosphorus-containing groups: phosphotransferases with an alcohol group as acceptor
- Skeletal muscle enzyme is reportedly 50X more active than liver enzyme. The two differ significantly in regulatory properties
- Inhibited by PEP, citrate
- Activated by K^+, $(NH_4)_2SO_4$

SPECIFIC ACTIVITY	UNITS DEFINITION	PREPARATION FORM	ADDITIONAL ACTIVITIES	SUPPLIER CATALOG NO.
Bacillus stearothermophilus (optimum pH = 9.0, pI = 6.0-6.2, MW = 74,000 Da [34,000/subunit]; K_M = 1.6 mM [F6P], 35 μM [ATP]; stable pH 5.5-10.5, T < 50°C)				
100-250 U/mg protein at pH 9.0, 30°C	1 unit converts 1.0 μmol F6P and ATP to fructose 1,6-diphosphate and ADP/min at pH 8.0, 37°C	Lyophilized containing 70% protein; balance KPO$_4$		ICN 158196
100-250 U/mg protein	1 unit converts 1.0 μmol F6P and ATP to fructose 1,6-diphosphate and ADP/min at pH 9.0, 30°C	Lyophilized containing PO$_4$ buffer salt		Sigma F0137
>100 U/mg protein	1 unit forms 1 μmol fructose 1,6-bisphosphate/min at 30°C	Lyophilized	<0.01% AdK, ATPase, 6PGDH, GR, PGluM, PGI	Unitika
>100 U/mg protein	1 unit forms 1 μmol fructose 1,6-bisphosphate/min at 30°C	50% glycerol solution	<0.01% AdK, ATPase, 6PGDH, GR, PGluM, PGI	Unitika
Rabbit liver				
5-20 U/mg protein	1 unit converts 1.0 μmol fructose 6-diphosphate ATP to fructose 1,6-diphosphate and ADP/min at pH 8.0, 30°C in a coupled system	Lyophilized containing 2% protein, 1% ATP, 5% DTT, 10% (NH$_4$)$_2$SO$_4$; balance primarily stabilizers		Sigma F8134
Rabbit muscle				
100-200 U/mg protein	1 unit converts 1.0 μmol F6P and ATP to fructose 1,6-diphosphate and ADP/min at pH 8.0, 37°C	Suspension in 3.2 M (NH$_4$)$_2$SO$_4$, 0.01 M PO$_4$, 0.001 M adenosine PO$_4$, pH 7.5		ICN 158195
300-500 U/g agarose; 1 mL yields 10-15 U	1 unit converts 1.0 μmol F6P and ATP to fructose 1,6-diphosphate and ADP/min at pH 8.0, 37°C	Insoluble enzyme attached to beaded agarose; suspension in 2.0 M (NH$_4$)$_2$SO$_4$, 0.004 M ATP, 0.05 M β-glycerophosphate, 0.001 M DTT, 0.002 M EDTA, pH 7.0		Sigma F2129
200 U/mg protein	1 unit converts 1.0 μmol F6P and ATP to fructose 1,6-diphosphate and ADP/min at pH 8.0, 37°C	Crystallized; suspension in 1.4 M (NH$_4$)$_2$SO$_4$, 0.05 M β-glycerophosphate, 0.004 M ATP, 0.002 M EDTA, 0.001 M DTT, pH 7.2	<0.01% PGI, GlcK, aldolase	Sigma F6877

2.7.1.12 Gluconokinase

REACTION CATALYZED
  ATP + D-gluconate ↔ ADP + 6-phospho-D-gluconate
SYSTEMATIC NAME
  ATP:D-gluconate 6-phosphotransferase
SYNONYMS
  Gluconate kinase

REACTANTS
  ATP, D-gluconate, ADP, 6-phospho-D-gluconate
NOTES
- Classified as a transferase, transferring phosphorus-containing groups: phosphotransferases with an alcohol group as acceptor

SPECIFIC ACTIVITY	UNITS DEFINITION	PREPARATION FORM	ADDITIONAL ACTIVITIES	SUPPLIER CATALOG NO.
Escherichia coli				
40 U/mg	1 unit converts 1 μmol gluconate and ATP to product/min at 25°C	Suspension in 3.2 M $(NH_4)_2SO_4$ and BSA as stabilizer, pH 6	<0.05% G6PDH, HK, NADPH oxidase	Boehringer 105058
40 U/mg protein	1 unit converts 1.0 μmol D-gluconate to 6PGlu/min at pH 8.0, 30°C	Suspension in 3.2 M $(NH_4)_2SO_4$, pH 6		ICN 157192
40 U/mg protein	1 unit converts 1.0 μmol D-gluconate to 6PGlu/min at pH 8.0, 30°C	Suspension in 3.2 M $(NH_4)_2SO_4$, pH 6		Sigma G6380

2.7.1.15 Ribokinase

REACTION CATALYZED
  ATP + D-ribose ↔ ADP + D-ribose 5-phosphate
SYSTEMATIC NAME
  ATP:D-ribose 5-phosphotransferase
REACTANTS
  ATP, D-ribose, 2-deoxy-D-ribose, ADP, D-ribose 5-phosphate

NOTES
- Classified as a transferase, transferring phosphorus-containing groups: phosphotransferases with an alcohol group as acceptor

SPECIFIC ACTIVITY	UNITS DEFINITION	PREPARATION FORM	ADDITIONAL ACTIVITIES	SUPPLIER CATALOG NO.
Bovine liver				
1-10 U/mg protein	1 unit converts 1.0 μmol D-ribose to D-ribose-5-phosphate/min at pH 7.7, 37°C in the presence of ATP and Mg^{2+}	Solution containing 50% glycerol, 10 mM KPO_4, 1 mM DTT, pH 8.0		Sigma R9013

Phosphoribulokinase 2.7.1.19

REACTION CATALYZED
 ATP + D-ribulose 5-phosphate ↔ ADP + D-ribulose 1,5-bisphosphate

SYSTEMATIC NAME
 ATP:D-ribulose 5-phosphate 1-phosphotransferase

SYNONYMS
 Phosphopentokinase, ribulose-5-phosphate kinase

REACTANTS
 ATP, D-ribulose 5-phosphate, ADP, D-ribulose 1,5-bisphosphate

NOTES
- Classified as a transferase, transferring phosphorus-containing groups: phosphotransferases with an alcohol group as acceptor

SPECIFIC ACTIVITY	UNITS DEFINITION	PREPARATION FORM	ADDITIONAL ACTIVITIES	SUPPLIER CATALOG NO.
Spinach				
2-8 U/mg protein	1 unit transfers 1.0 μmol PO_4 from ATP to D-ribulose 5-phosphate/min at pH 7.9, 37°C	Partially purified; powder containing 50% protein; balance primarily $(NH_4)_2SO_4$		sigma P9877

NAD^+ Kinase 2.7.1.23

REACTION CATALYZED
 ATP + NAD^+ ↔ ADP + $NADP^+$

SYSTEMATIC NAME
 ATP:NAD^+ 2'-phosphotransferase

SYNONYMS
 DPN Kinase

REACTANTS
 ATP, NAD^+, ADP, $NADP^+$

NOTES
- Classified as a transferase, transferring phosphorus-containing groups: phosphotransferases with an alcohol group as acceptor

SPECIFIC ACTIVITY	UNITS DEFINITION	PREPARATION FORM	ADDITIONAL ACTIVITIES	SUPPLIER CATALOG NO.
Chicken liver				
10-20 U/mg solid	1 unit phosphorylates 1.0 nmol β-NAD to β-NADP/min at pH 7.5, 37°C, with ATP	Lyophilized containing 85% protein; balance primarily citrate buffer		sigma N8882

Glycerol Kinase

REACTION CATALYZED
ATP glycerol → ADP + sn-glycerol 3-phosphate

SYSTEMATIC NAME
ATP:glycerol 3-phosphotransferase

SYNONYMS
Glycerokinase, GK

REACTANTS
ATP, UTP, ITP, GTP, glycerol, glycerone, L-glyceraldehyde, dihydroxyacetone, ADP, sn-glycerol 3-phosphate

APPLICATIONS
- Determination of glyceraldehyde, dihydroxyacetone, triglycerides and glycerol in plasma, serum and tissues
- Determination of serum creatine kinase, with peroxidase

NOTES
- Classified as a transferase, transferring phosphorus-containing groups: phosphotransferases with an alcohol group as acceptor
- Makes possible microbial utilization of glycerol as a carbon source
- The interfacing link between carbohydrate and lipid metabolism in mammals
- ITP and GTP are donors for the yeast enzyme
- Activated by Mg^{2+}; substitution of Mn^{2+} reduces activity to one-third
- Stabilized by Na^+, K^+, glycerol and ATP
- Inhibited by p-chloromercuribenzoate and heavy metal ions
- Slowed by fructose diphosphate

SPECIFIC ACTIVITY	UNITS DEFINITION	PREPARATION FORM	ADDITIONAL ACTIVITIES	SUPPLIER CATALOG NO.
Arthrobacter species (optimum pH = 9.7, pI = 4.3, T = 50°C, MW = 82,000 Da; stable pH 5.5-9.5 [37°C, 1 hr], T < 45°C [pH 7, 15 min])				
≥10 U/mg	1 unit produces 1 μmol NADH/min at pH 9.8, 25°C	Powder		Amano Glucose kinase
25 U/mg	1 unit phosphorylates 1 μmol glycerol/min at pH 9.8, 25°C		<0.001% NADH oxidase <0.01% HK, ATPase <0.5% TPI	Fluka 49950
>35 U/mg protein; >20 U/mg solid	1 unit phosphorylates 1.0 μmol glycerol/min at pH 9.8, 25°C	Lyophilized	<0.01% NADH oxidase <0.02% ATPase, HK	Genzyme 1291
Bacillus stearothermophilus (optimum pH = 10.0-10.5, T = 50-55°C, MW = 232,000 Da [4 subunits of 58,000 Da each]; stable pH 5.0-11.0, T < 65°C)				
85 U/mg at 25°C; 410 U/mg at 50-55°C	1 unit converts 1 μmol glycerol and ATP to product/min	Solution containing Tris buffer, pH 7.3	<0.01% NADH oxidase, HK	Boehringer 691836
>60 U/mg protein; >10 U/mg solid	1 unit phosphorylates 1 μmol glycerol/min at pH 9.8, 25°C	Lyophilized	<0.01% HK, NADH oxidase	Randox GK 945L

SPECIFIC ACTIVITY	UNITS DEFINITION	PREPARATION FORM	ADDITIONAL ACTIVITIES	SUPPLIER CATALOG NO.
Bacillus stearothermophilus continued				
75 U/mg protein	1 unit converts 1.0 µmol glycerol and ATP to L-α-glycerophosphate and ADP/min at pH 9.8, 25°C in a coupled system with PK/LDH	Solution containing Tris buffer, pH 7.3		Sigma G0774
15 U/mg solid	1 unit converts 1.0 µmol glycerol and ATP to L-α-glycerophosphate and ADP/min at pH 9.8, 25°C in a coupled system with PK/LDH	Lyophilized containing Tris buffer salts		Sigma G6402
Candida mycoderma				
80 U/mg at 25°C; 160 U/mg at 37°C	1 unit converts 1 µmol glycerol and ATP to product/min	Suspension in 3.2 M $(NH_4)_2SO_4$, 1% ethylene glycol, BSA as stabilizer, pH 6	<0.01% NADH oxidase, HK <0.02% MK	Boehringer 127159 127795 737267
85 U/mg protein	1 unit phosphorylates 1 µmol glycerol/min at pH 9.8, 25°C	Aqueous suspension in 3.2 M $(NH_4)_2SO_4$ and 1% ethylene glycol, pH 6.0		Fluka 49953
85 U/mg protein	1 unit converts 1.0 µmol glycerol to L-α-glycerophosphate/min at pH 9.8, 25°C	Crystallized; suspension in 2.4 M $(NH_4)_2SO_4$ and 1% ethylene glycol, pH 6.0		ICN 151196
80-100 U/mg protein	1 unit converts 1.0 µmol glycerol to L-α-glycerophosphate/min at pH 9.8, 25°C	Crystallized; suspension in 3.2 M $(NH_4)_2SO_4$ and 1% ethylene glycol, pH 6		ICN 157238
80-100 U/mg protein	1 unit converts 1.0 µmol glycerol and ATP to L-α-glycerophosphate and ADP/min at pH 9.8, 25°C in a coupled system with PK/LDH	Crystallized; suspension in 3.2 M $(NH_4)_2SO_4$ and 1% ethylene glycol, pH 6		Sigma G5751
Candida utilis				
30-150 U/mg protein	1 unit converts 1.0 µmol glycerol and ATP to L-α-glycerophosphate and ADP/min at pH 9.8, 25°C in a coupled system with PK/LDH	Lyophilized containing 50% protein; balance PO_4-citrate salts, EDTA and DTT		Sigma G8771
Cellulomonas species (optimum pH = 7.8 [GPO] or 9.8 [G3PDH], pI = 4.2, T = 50°C, MW = 128,000 Da [gel filtration]; K_M = 44 µM [glycerol], 0.43 mM [ATP]; stable pH 5.5-10.0 [25°C, 20 hr], T < 40°C [pH 7.5, 15 min])				
20 U/mg	1 unit forms 1 µmol NADH/min	Lyophilized		AMRESCO 0651
>35 U/mg protein; >20 U/mg solid	1 unit phosphorylates 1.0 µmol glycerol/min at pH 9.8, 25°C	Lyophilized	<0.002% phosphatase (pH 6.0) <0.01% NADH oxidase <0.1% catalase	Genzyme 6491
25-75 U/mg protein	1 unit converts 1.0 µmol glycerol and ATP to L-α-glycerophosphate and ADP/min at pH 9.8, 25°C in a coupled system with PK/LDH	Lyophilized containing 70% protein; balance PO_4 buffer salts and Na gluconate		Sigma G6142

SPECIFIC ACTIVITY	UNITS DEFINITION	PREPARATION FORM	ADDITIONAL ACTIVITIES	SUPPLIER CATALOG NO.
Cellulomonas species _continued_				
≥20 U/mg solid	1 unit forms 1 μmol NADH/min at pH 9.8, 25°C	Lyophilized containing 50% stabilizer	<0.002% phosphatase <0.01% NADH oxidase <0.1% catalase	Toyobo GYK-301
Escherichia coli (optimum pH = 9.0-9.8, MW = 217,000 Da; maximum stability at pH 7.0)				
≥30 U/mg protein; 12-36 IU/mg powder	1 unit catalyzes 1 μmol glycerol to G1P/min at 30°C	Lyophilized	<0.01% GPDH, MK, NADH oxidase <0.02% ATPase, HK <0.025% TPI	Beckman 682820
50 U/mg	1 unit phosphorylates 1 μmol glycerol/min at pH 9.8, 25°C	Lyophilized	<0.05% GDH, HK, MK, NADH oxidase, TPI	Fluka 49955
50-100 U/mg protein	1 unit converts 1.0 μmol glycerol and ATP to L-α-glycerophosphate and ADP/min at pH 9.8, 25°C in a coupled system with PK/LDH	Partially purified; lyophilized containing 90% protein; balance primarily salts and EDTA	<0.05% TPI, HK, Mk, GDH, NADH oxidase	Sigma G4509
300-600 U/mL	1 unit converts 1.0 μmol glycerol and ATP to L-α-glycerophosphate and ADP/min at pH 9.8, 25°C in a coupled system with PK/LDH	Suspension in 3.1 M $(NH_4)_2SO_4$, 1% BSA, 2% trehalose, pH 7.3		Sigma G6278
≥30 U/mg solid	1 unit oxidizes 1 μmol NADH/min at pH 8.9, 25°C in a coupled assay system with PK and LDH	Highly purified by chromatography on DEAE-Sephadex^R; lyophilized		Worthington LS05865 LS05866 LS05875
≥250 U/mL	1 unit oxidizes 1 μmol NADH/min at pH 8.9, 25°C in a coupled assay system with PK and LDH	Chromatographically purified; suspension in 0.65 M $(NH_4)_2SO_4$		Worthington LS05877 LS05878
Streptomyces canus (optimum pH = 9.0-10.0, pI = 4.5, MW = 72,000 Da [Sephadex G-100]; K_M = 37 μM [ATP], 48 μM [glycerol]; stable pH 5.5-10 [37°C, 1 hr], T < 40°C [pH 7.8, 10 min])				
40-80 U/mg solid	1 unit converts 1 μmol glycerol to G3P/min at pH 9.8, 37°C	Lyophilized containing 100% protein	<0.01% NADH oxidase <0.02% MK, HK	Asahi GK T-09
>40 U/mg protein; >40 U/mg solid	1 unit phosphorylates 1.0 μmol glycerol/min at pH 9.8, 37°C	Lyophilized	<0.01% NADH oxidase <0.02% AK, HK	Genzyme 6481
10-30 U/mg protein	1 unit converts 1.0 μmol glycerol and ATP to L-α-glycerophosphate and ADP/min at pH 9.8, 25°C in a coupled system with PK/LDH	Lyophilized		Sigma G4147

Glycerol Kinase continued

2.7.1.30

SPECIFIC ACTIVITY	UNITS DEFINITION	PREPARATION FORM	ADDITIONAL ACTIVITIES	SUPPLIER CATALOG No.
Yeast				
100-300 U/mg solid	1 unit produces 1 μmol G3P from glycerol/min at pH 9.8, 37°C in the presence of ATP	Lyophilized containing 50% protein		Calzyme 192A0100

Choline Kinase

2.7.1.32

REACTION CATALYZED
    ATP + choline ↔ ADP + O-phosphocholine

SYSTEMATIC NAME
    ATP:choline phosphotransferase

REACTANTS
    ATP, choline, ethanolamine, ADP, O-phosphocholine

NOTES
- Classified as a transferase, transferring phosphorus-containing groups: phosphotransferases with an alcohol group as acceptor
- Ethanolamine and its methyl- and ethyl-derivatives also act as acceptors

SPECIFIC ACTIVITY	UNITS DEFINITION	PREPARATION FORM	ADDITIONAL ACTIVITIES	SUPPLIER CATALOG No.
Saccharomyces cerevisiae				
0.5-1.5 U/mg protein	1 unit phosphorylates 1.0 μmol choline to choline PO_4 by ATP/min at pH 8.5, 25°C	Lyophilized containing 10% protein; balance Tris buffer salts and stabilizer		sigma C7635
Yeast				
0.04 U/mg solid; 0.5 U/mg enzyme protein	1 unit converts 1 μmol choline to product/min at 25°C	Lyophilized		Boehringer 348651

Protein Kinase

REACTION CATALYZED
ATP + a protein ↔ ADP + a phosphoprotein

SYSTEMATIC NAME
ATP:protein phosphotransferase

SYNONYMS
Phosphorylase b kinase kinase, glycogen synthase α kinase, hydroxyalkyl-protein kinase, serine (threonine) protein kinase, protein kinase, protein kinase A, protein kinase C, protein kinase G, cAMP-dependent protein kinase, cGMP-dependent protein kinase, protein kinase II (Ca^{2+}/CaM-dependent), casein kinase I, casein kinase II, $p34^{cdc2}$/cyclinB, PKA, PKC, PKG

REACTANTS
ATP, protein, ADP, phosphoprotein

APPLICATIONS
- In vitro phosphorylation of target enzymes, proteins and cell lysates, especially those involved in gene expression, protein synthesis and signaling pathways. Used in intact cells if the enzyme is microinjected or cells are permeabilized
- Phosphorylation of target substrates in enzyme regulation and kinetics studies
- Microassay of cAMP

NOTES
- Classified as a transferase, transferring phosphorus-containing groups: phosphotransferases with an alcohol group as acceptor
- Many cAMP-dependent protein kinases have been reported. They contain at least two subunits, identified as regulatory and catalytic. Activity is inhibited when both are linked. When cAMP binds the regulatory subunit, the catalytic subunit is released and transfers phosphate from ATP to various proteins. Activity is expressed both as phosphorylating and cAMP binding
- Mediates the actions of cAMP
- Regulates cellular functions, including gene expression, carbohydrate metabolism, calcium homeostasis and ion channel function
- Involved in the regulation of transfer factor CREB at Ser^{133} and functional modulation of Glu^{R6}

SPECIFIC ACTIVITY	UNITS DEFINITION	PREPARATION FORM	ADDITIONAL ACTIVITIES	SUPPLIER CATALOG NO.
Bovine heart				
1500 U/mg	1 unit transfers 1 µmol ^{32}P to hydrolyzed, partially dephosphorylated casein/min at pH 6.5, 30°C; γ-^{32}P-ATP as substrate	Powder containing 80% protein, 10% EDTA, KPO_4 (pH 7)	<0.01% ATPase, 3'5'-cyclic nucleotide PDE	Fluka 82523

SPECIFIC ACTIVITY	UNITS DEFINITION	PREPARATION FORM	ADDITIONAL ACTIVITIES	SUPPLIER CATALOG NO.
Bovine heart continued				
Phosphorylating: 1-2 U/μg protein; Binding: 40 nM cyclic AMP bound at 0.1 pmol/mg protein and 160 nM cyclic AMP bound at 0.2 pmol/mg protein	Phosphorylating: 1 unit transfers 1.0 pmol PO_4 from γ-^{32}P-ATP to hydrolyzed, partially dephosphorylated casein/min at pH 6.5, 30°C; binding: assays carried out in 0.05 M acetate buffer, pH 4.0, 0°C	Crude; lyophilized containing 80% protein, 10% EDTA, 10% KPO_4, pH 7.0	<0.025 μM U/mg protein ATPase <0.1 μM U/mg protein 3′:5′-cyclic nucleotide PDE	Sigma P5511
Bovine heart, catalytic subunit (MW = 40,000 Da)				
>6 U/mg	1 unit transfers 1 μmol PO_4 from ATP to the synthetic substrate kemptide/min at 30°C; 400 mU (1 μmol P_i/min, kemptide as substrate) = 2500 U (1 pmol P_i/min, casein as substrate)	>95% purity by SDS-PAGE; solution filled under N_2	No detectable protease	Boehringer 1529307
750 U/μg protein	1 unit transfers 1 pmol PO_4 to histone H1/min at 30°C	>95% purity; solution containing 100 mM NaCl, 20 mM MES, 50% ethylene glycol, 20 mM β-MSH, 100 μM EDTA, pH 6.5		Calbiochem 539486
		≥95% purity		Chemicon AG630
40,000 U/mg protein	1 unit transfers 1 μmole of ^{32}P to hydrolyzed, partially dephosphorylated casein/min at pH 6.5, 30°C	Lyophilized containing 99.9% sucrose, PO_4 buffer salts and 0.1% protein		Fluka 82501
Phosphorylating: 30-65 U/μg protein	1 unit transfers 1.0 pmol PO_4 from γ-^{32}P-ATP to hydrolyzed, partially dephosphorylated casein/min at pH 6.5, 30°C	Lyophilized containing <1% protein; balance primarily sucrose and PO_4 buffer salts as stabilizer	Binding: 320 nM cyclic AMP at 0.01 pmol/μg protein	Sigma P2645
Bovine heart, regulatory subunit				
80,000 U/mg protein	1 unit transfers 1 μmol ^{32}P to hydrolyzed, partially dephosphorylated casein/min at pH 6.5, 30°C, in the absence of cyclic AMP	99.5% sucrose, PO_4 buffer salts and 0.5% protein	Phosphorylase activity: 80 U/mg protein	Fluka 82499
Inhibitory: 50-100 U/μg protein	1 unit inhibits 1 unit of catalytic subunit phosphorylating activity in the absence of cyclic AMP	Lyophilized containing <1% protein; balance primarily sucrose and PO_4 buffer salts as stabilizer	Binding: 64 nM cyclic-AMP at 5-10 pmol/μg protein, pH 7.0, 0 °C Phosphorylating: 0.08 U/μg protein at pH 6.5, 30 °C	Sigma P6022

SPECIFIC ACTIVITY	UNITS DEFINITION	PREPARATION FORM	ADDITIONAL ACTIVITIES	SUPPLIER CATALOG NO.
Bovine lung α isozyme (MW = 78,000 Da)				
2500 U/μg protein	1 unit transfers 1 pmol PO_4 to RKRSRAE peptide substrate/min at 30°C	>90% purity by SDS-PAGE; 0.1 mg/mL in 10 mM KPO_4, 0.15 M NaCl, 1 mM EDTA, 25 mM β-MSH, 12% sucrose, pH 6.8	No detectable kinase	BIOMOL SE-118
Bovine Iα isozyme, recombinant expressed in *E. coli*				
10,000 U/mg protein	1 unit transfers 1 pmol PO_4 to GRTGRRNSI/min at 30°C	>95% purity; 40% glycerol		Calbiochem 370350
Human glioblastoma α and β isozymes, recombinant expressed in *E. coli* (tetrameric, MW = 44,000 [catalytic], 26,000 Da [regulatory])				
700,000 U/mg protein	1 unit transfers 1 pmol PO_4 from ATP to peptide substrate (RRREEETEEE)/min at pH 7.5, 30°C	0.7 mg/mL in 20 mM Tris, 0.35 M NaCl, 1 mM EDTA, 10 mM β-MSH, 0.1% Triton X-100, pH 7.5	No detectable protease, phosphatase, RNase, DNase	BIOMOL SE-124
500,000 U/mL	1 unit transfers 1 pmol PO_4 to peptide substrate RRREEETEEE/min at 30°C in CKII buffer in a 25 μL reaction volume	>90% homogeneous; 20 mM Tris-HCl, 350 mM NaCl, 1 mM Na_2EDTA, 10 mM β-MSH, 0.1% Triton X-100, pH 7.5	No detectable protease, phosphatase, nuclease, kinase	NE Biolabs 6010S 6010L
Human $p34^{cdc2}$ and human cyclin, recombinant expressed in *Spodoptera frugiperda* using *Baculovirus* (MW = 34,000 [catalytic], 55,000 Da [regulatory])				
2000 U/mL	1 unit transfers 1 pmol PO_4 to histone H1/min at 30°C in cdc2 buffer in a 30 μL reaction volume	>90% homogeneous; 20 mM Tris-HCl, 10 mM NaCl, 1 mM Na_2EDTA, 1 mM DTT, 0.01% Brij, 10% glycerol, pH 7.5	No detectable protease, phosphatase, nuclease, kinase	NE Biolabs 6020S 6020L
Human, recombinant expressed in *E. coli*				
>1 U/mg	1 unit transfers 1 μmol PO_4 from ATP to the synthetic casein kinase II substrate/min at 37°C	>90% by SDS-PAGE; solution filled under N_2	No detectable protease	Boehringer 1500767
1 U/mg protein	1 unit transfers 1 μmol PO_4 from ATP to a synthetic casein kinase II substrate/min at 37°C	>95% purity; solution containing 25 mM Tris-HCl, 500 mM NaCl, 7 mM β-MSH, 0.5 mM PMSF, pH 8.5	No detectable proteases	Calbiochem 218698
Human, recombinant expressed in *Spodoptera frugiperda*				
2000 U/mL	1 unit phosphorylates 1 pmol histone H1/min at pH 7.5, 30°C	>90% purity; solution containing 100 mM NaCl, 20 mM Tris-HCl, 1 mM DTT, 1 mM EDTA, 0.01% Brij, 10% glycerol, pH 7.5	No detectable DNase, other kinase, protease, phosphatase, RNase	Calbiochem 506116
Human, recombinant overexpressed in insect cells using *Baculovirus*				
>1000 nmol PO_4 transferred to substrate/min/mg		>95% purity by SDS-PAGE		PanVera PKC α, βI, βII, γ, δ, ε, η, ζ, σ, ι

SPECIFIC ACTIVITY	UNITS DEFINITION	PREPARATION FORM	ADDITIONAL ACTIVITIES	SUPPLIER CATALOG NO.
Murine PKA catalytic subunit, recombinant expressed in *E. coli*				
1000 U/mL	1 unit transfers 1 nmol PO_4 to kemptide (LRRASLG) substrate/min at 30°C in PKA buffer in a 25 µL reaction volume	>90% homogeneous; 50 mM MES, 50 mM NaCl, 1 mM Na_2EDTA, 1 mM β-MSH, 50% glycerol, pH 6.5	No detectable protease, phosphatase, nuclease, kinase	NE Biolabs 6000S 6000L
Murine, recombinant expressed in *E. coli* (MW = 38,000 Da)				
1250 U/mg protein	1 unit adds 1 nmol PO_4 to Kemptide substrate/min at pH 7.5, 30°C	0.8 mg/mL in 50 mM MES, 50 mM NaCl, 1 mM EDTA, 1 mM β-MSH, 50% glycerol, pH 6.5	No detectable protease, phosphatase, RNase, DNase	BIOMOL SE-122
Porcine heart (MW = 38,000 Da)				
30-65 U/µg protein	1 unit adds 1 pmol PO_4 to dephosphorylated casein/min at pH 6.5, 30°C	Lyophilized containing 1% protein, PO_4, sucrose as stabilizer		BIOMOL SE-110
Phosphorylating: 1.0 U/µg protein; Binding: 40 nM cyclic AMP bound at 0.05 pmol/mg protein and 160 nM cyclic AMP bound at 0.15 pmol/mg protein	Phosphorylating: 1 unit transfers 1.0 pmol PO_4 from γ-^{32}P-ATP to hydrolyzed, partially dephosphorylated casein/min at pH 6.5, 30°C; binding: assays carried out in 0.05 M acetate buffer, pH 4.0, 0°C	Crude; lyophilized containing 80% protein, 10% EDTA, 10% KPO_4, pH 7.0	<0.01 µMU/mg protein ATPase, 3′:5′-cyclic nucleotide PDE	Sigma P8164
Porcine heart, catalytic subunit				
Phosphorylating: 30-65 U/µg protein	1 unit transfers 1.0 pmol PO_4 from γ-^{32}P-ATP to hydrolyzed, partially dephosphorylated casein/min at pH 6.5, 30°C	Lyophilized containing <1% protein; balance primarily sucrose and PO_4 buffer salts as stabilizer	Binding: 320 nM cyclic AMP at 0.01 pmol/µg protein	Sigma P8289
Rabbit muscle				
Phosphorylating: 0.3 U/µg protein; Binding: 50 nM cyclic AMP bound at 0.04 pmol/mg protein	Phosphorylating: 1 unit transfers 1.0 pmol PO_4 from γ-^{32}P-ATP to hydrolyzed, partially dephosphorylated casein/min at pH 6.5, 30°C; binding: assays carried out in 0.05 M acetate buffer, pH 4.0, 0°C	Crude; lyophilized containing 85% protein, 15% KPO_4, pH 7.0		Sigma P3891
Phosphorylating: 0.2 U/µg protein; Binding: 50 nM cyclic AMP bound at 0.02 pmol/mg protein	Phosphorylating: 1 unit transfers 1.0 pmol PO_4 from γ-^{32}P-ATP to hydrolyzed, partially dephosphorylated casein/min at pH 6.5, 30°C; binding: assays carried out in 0.05 M acetate buffer, pH 4.0, 0°C	Crude; lyophilized containing 98% protein, 2% KPO_4, pH 7.0		Sigma P4890

SPECIFIC ACTIVITY	UNITS DEFINITION	PREPARATION FORM	ADDITIONAL ACTIVITIES	SUPPLIER CATALOG NO.
Rabbit skeletal muscle 3β isozyme, recombinant expressed in *E. coli*				
40,000 U/mg protein	1 unit transfers 1 pmol PO_4 to Protein Phosphatase Inhibitor 2/min at pH 7.5, 30°C	>90% purity; solution containing 20 mM Tris-HCl, 10 mM β-MSH, 1 mM EDTA, 1 mM EGTA, 50% glycerol, pH 7.5	No detectable DNase, other kinase, phosphatase, protease, DNase	Calbiochem 361525
5000 U/mL	1 unit transfers 1 pmol PO_4 to protein substrate I-2 (phosphatase inhibitor 2)/min at 30°C in GSK-3 buffer in a 25 μL reaction volume	>90% homogeneous; 20 mM Tris-HCl, 1 mM Na_2EDTA, 1 mM EGTA, 10 mM β-MSH, 50% glycerol, pH 7.5	No detectable protease, phosphatase, nuclease, kinase	NE Biolabs 6040S 6040L
Rat brain (MW = 80,000–82,000 Da)				
1530 U/mg	1 unit transfers 1 nmol PO_4 from ATP to H1 histone/min at pH 7.4, 22°C	>95% purity by SDS-PAGE; 35 μg/mL in 20 mM Tris, 0.5 mM EDTA, 0.5 mM EGTA, 1 mM DTT, 0.05% Triton X-100, 10% glycerol, pH 7.5		BIOMOL SE-111
800 U/mg	1 unit transfers 1 nmol PO_4 from ATP to H1 histone/min at pH 7.4, 30°C	>95% purity by SDS-PAGE; 5 μg/mL in 20 mM Tris, 100 mM NaCl, 1 mM EDTA, 1 mM EGTA, 15 mM DTT, 10% glycerol, pH 7.5		BIOMOL SE-133
>1 U/mg	1 unit transfers 1 μmol PO_4 from ATP to histone H1/min at 30°C	>90% by SDS-PAGE; solution filled under N_2	No detectable protease	Boehringer 1459651
1000 U/mg protein	1 unit phosphorylates H1-histone at 1 nmol/min at pH 7.4, 22°C	>95% purity; solution containing 20 mM Tris-HCl, 100 mM NaCl, 10% glycerol, 500 μM EDTA, 500 μM EGTA, 1 mM DTT, pH 7.5		Calbiochem 539494
		≥95% purity		Chemicon AG605
		≥95% purity		Chemicon AG610
		≥95% purity		Chemicon AG625
	1 unit transfers 1 nmol PO_4/min from ATP to histone H1 at pH 7.4, 30°C	Solution containing 50% glycerol, 20 mM Tris, 0.5 mM EDTA, 0.5 mM EGTA, 5 mM DTT, 0.02% Tween 20, 100 mM NaCl, 1 μg/mL leupeptin		Sigma P0329
	1 unit transfers 1 nmol PO_4/min from ATP to histone H1 at pH 7.4, 30°C	Lyophilized containing Tris buffer salts, EDTA, EGTA, DTT, Tween 20, NaCl, leupeptin and sucrose as stabilizer		Sigma P7956
Rat brain, catalytic subunit (MW = 50,000–55,000 Da)				
750 U/mg protein	1 unit transfers 1 nmol PO_4 to histone H1/min at 30°C	>95% purity; solution containing 100 mM NaCl, 20 mM Tris-HCl, 15 mM DTT, 1 mM EGTA, 10% glycerol, pH 7.5		Calbiochem 539513

Protein Kinase continued — 2.7.1.37

SPECIFIC ACTIVITY	UNITS DEFINITION	PREPARATION FORM	ADDITIONAL ACTIVITIES	SUPPLIER CATALOG NO.
Rat liver (MW = 35,000-37,000 Da)				
17.5 kU/mg	1 unit transfers 1 pmol PO_4 from ATP to partially dephosphorylated casein/min at pH 7.4, 37°C	>90% purity by SDS-PAGE; 0.6 mg/mL in 25 mM Tris, 1 mM EDTA, 1 mM DTT, 50% glycerol, pH 7.4	No detectable kinase	BIOMOL SE-121
Rat testis, recombinant				
4 MU/mg	1 unit transfers 1 pmol PO_4 from ATP to peptide substrate (KRRRALpSVASLPGL)/min at 30°C	1 MU/mL in 20 mM Tris, 0.25 M NaCl, 1 mM EDTA, 1 mM EGTA, 10 mM β-MSH, 0.1% Triton X-100, 50% glycerol, pH 7.0	No detectable protease, phosphatase	BIOMOL SE-125
Rat testis CKI, recombinant expressed in *E. coli* (MW = 36,000 Da)				
1,000,000 U/mL	1 unit transfers 1 pmol PO_4 to peptide substrate KRRRALS(P)VASLPGL/ min at 30°C in CKI buffer in a 25 μL reaction volume	>90% homogeneous; 20 mM Tris-HCl, 250 mM NaCl, 1 mM Na_2EDTA, 1 mM EGTA, 10 mM β-MSH, 0.1% Triton X-100, 50% glycerol, pH 7.0	No detectable protease, phosphatase, nuclease, kinase	NE Biolabs 6030S 6030L
Starfish oocytes				
4000 U/mL	1 unit phosphorylates 1 pmol histone H1/min at pH 7.2, 30°C	Solution containing 200 mM Na_2CO_3, 200 mM NaCl, 10% glycerol, pH 8.2		Calbiochem 506118

Phosphorylase Kinase — 2.7.1.38

REACTION CATALYZED

4 ATP + 2 phosphorylase *b* ↔ 4 ADP + phosphorylase *a*

SYSTEMATIC NAME

ATP:phosphorylase-*b* phosphotransferase

SYNONYMS

Dephosphophosphorylase kinase

REACTANTS

ATP, phosphorylase *b*, ADP, phosphorylase *a*

NOTES
- Classified as a transferase, transferring phosphorus-containing groups: phosphotransferases with an alcohol group as acceptor
- Cyclic GMP, but not cyclic AMP, activates some enzymes

2.7.1.38 Phosphorylase Kinase continued

SPECIFIC ACTIVITY	UNITS DEFINITION	PREPARATION FORM	ADDITIONAL ACTIVITIES	SUPPLIER CATALOG NO.
Rabbit muscle				
200-500 U/mg protein	1 unit forms 1.0 μM unit phosphorylase a from phosphorylase b/min at pH 7.7, 30°C in the presence of ATP	Lyophilized containing 30% protein; balance primarily $(NH_4)_2SO_4$, β-glycerophosphate, DTT buffer salts, sucrose	<0.2% ATPase, phosphorylase a <1% phosphorylase b	Sigma P2014

2.7.1.40 Pyruvate Kinase

REACTION CATALYZED
   ATP + pyruvate ↔ ADP + phosphoenolpyruvate

SYSTEMATIC NAME
   ATP:pyruvate 2-O-phosphotransferase

SYNONYMS
   Phosphoenolpyruvate kinase, phosphoenol transphosphorylase, PK

REACTANTS
   ATP, UTP, GTP, CTP, ITP, dATP, pyruvate, hydroxylamine, fluoride, ADP, phosphoenolpyruvate, CO_2

APPLICATIONS
- Determination of phosphoenol pyruvate, glycerol, 2-phosphoglycerate, ADP, and ATP; with lactate dehydrogenase, e.g., in UV triglyceride methods

NOTES
- Classified as a transferase, transferring phosphorus-containing groups: phosphotransferases with an alcohol group as acceptor
- A key enzyme in glycogen metabolism
- Phosphorylates hydroxylamine and fluoride in the presence of CO_2
- Rabbit muscle enzyme is tetrameric with:
   - 4 metal binding sites;
   - no covalently bound prosthetic group;
   - an absolute requirement for a divalent and monovalent metal ion;
   - 4 integral sulfhydryl groups at the active site
- Activated by Mg^{2+} and K^+; activity is considerably lower without fructose 1,6-diphosphate
- Competitively inhibited by fluorophosphate (vs. PEP), creatine phosphate (vs. PEP) and ATP (vs. ADP and PEP if $[Mg^{2+}] > [ATP]$)
- Inhibited by PCMB, SDS, DAC, Ca^{2+}, Zn^{2+}, Cd^{2+}, Cu^{2+}, Ag^+, Hg^{2+}, N-ethylmaleimide, ATP (removes Mg^{2+} from substrate $MgADP^-$) and ADP (>6 mM)
- Inactivated by pyridoxal 5'-phosphate

SPECIFIC ACTIVITY	UNITS DEFINITION	PREPARATION FORM	ADDITIONAL ACTIVITIES	SUPPLIER CATALOG NO.
Bacillus stearothermophilus (optimum pH = 7.0, pI = 5.2, MW = 260,000 Da [68,000/subunit]; K_M = 0.6 mM [PEP], 0.9 mM [ATP]; stable pH 8-10, T < 55°C)				
100-200 U/mg protein		Lyophilized containing 90% protein; balance primarily Tris buffer salts, pH 8.5		ICN 152001
100-300 U/mg protein	1 unit converts 1.0 μmol PEP to pyruvate/min at pH 7.2, 30°C	Lyophilized containing Tris buffer salts, pH 8.5		Sigma P1903
>400 U/mg protein	1 unit forms 1 μmol pyruvate/min at 30°C	Lyophilized	<0.01% AdK, LDH	Unitika
>400 U/mg protein	1 unit forms 1 μmol pyruvate/min at 30°C	50% glycerol solution	<0.01% AdK, LDH	Unitika
Bovine pancreas (optimum pH = 7.5, T = 25°C)				
35 U/mg solid	1 unit oxidizes 1 μmol NADH/min at 25°C	Lyophilized		AMRESCO 0514 94036 94037
Chicken muscle				
300-800 U/mg protein	1 unit converts 1.0 μmol PEP to pyruvate/min at pH 7.2, 37°C	Crystallized; suspension in 3.2 M $(NH_4)_2SO_4$, pH 6.0	No detectable CPK, PGluM, MK, LDH	Sigma P6406
Dog muscle				
350-500 U/mg protein	1 unit converts 1.0 μmol PEP to pyruvate/min at pH 7.2, 37°C	Crystallized; suspension in 2.2 M $(NH_4)_2SO_4$, pH 6.0	No detectable CPK, PGluM, MK 0.03% LDH	Sigma P7889
Human muscle				
50-100 U/mg protein	1 unit produces 1 μmol pyruvate from PEP/min at pH 7.5, 25°C in the presence of ADP	Suspension in 65% $(NH_4)_2SO_4$ containing 10 mg/mL protein		Calzyme 106B0050
Porcine heart				
≥200 U/mg protein	1 unit oxidizes 1 μmol NADH/min at pH 7.4, 25°C	Highly purified; suspension in 80% $(NH_4)_2SO_4$	<0.001% LDH	Biozyme HPK1
250 U/mg protein	1 unit produces 1 μmol pyruvate from PEP/min at pH 7.5, 25°C in the presence of ADP	Suspension in 65% $(NH_4)_2SO_4$ containing 10 mg/mL protein	<0.001% GPDH, CPK, ATPase, NADH oxidase, PGM <0.005% MK	Calzyme 131B0250
>200 U/mg protein; >3000 U/mL	1 unit dephosphorylates 1.0 μmol pyruvate/min at pH 7.5, 25°C	Suspension in 80% saturated $(NH_4)_2SO_4$, pH 6.5	<0.003% HK <0.01% PGM, enolase, AK, GAPDH, CK, GPDH, ATPase <0.02% LDH	Genzyme 1772
200-300 U/mg solid	1 unit oxidizes 1 μmol NADH/min at pH 7.4, 25°C	Lyophilized		ICN 152002

SPECIFIC ACTIVITY	UNITS DEFINITION	PREPARATION FORM	ADDITIONAL ACTIVITIES	SUPPLIER CATALOG NO.
Porcine heart *continued*				
≥200 IU/mg protein	1 IU transforms 1 μmol substrate/min under standard IUB conditions at 25°C	Lyophilized; suspension in $(NH_4)_2SO_4$	<0.003% HK <0.01% PGM, G3PDH, enolase, MK, GPDH, CK, ATPase <0.02% LDH	OYC
300-500 U/mg protein	1 unit converts 1.0 μmol PEP to pyruvate/min at pH 7.2, 37°C	Suspension in 3.2 M $(NH_4)_2SO_4$, pH 7	No detectable CPK, PGluM, MK, LDH, MDH	Sigma P2040
Rabbit liver				
10-20 U/mg protein	1 unit converts 1.0 μmol PEP to pyruvate/min at pH 7.6, 37°C in the presence of 1.0 mM fructose 1,6-diphosphate activator	Suspension in 3.5 M $(NH_4)_2SO_4$, pH 6.0		Sigma P3267
20-40 U/mg protein	1 unit converts 1.0 μmol PEP to pyruvate/min at pH 7.6, 37°C in the presence of 1.0 mM fructose 1,6-diphosphate activator	Salt-free; lyophilized	<0.01% LDH, CPK, MK	Sigma P7286
Rabbit muscle (optimum pH = 7.0, pI = 5.98, T = 40°C, MW = 237,000 Da [4 subunits of 57,000 Da/mol enzyme]; K_M = 0.86 mM [ATP], 10 mM [pyruvate], 0.3 mM [ADP], 70 μM [PEP]; stable pH 5.3-8.0 [25°C, 20 hr], T < 45°C [pH 7.6, 10 min])				
350 U/mg protein	1 unit oxidizes 1 μmol NADH/min at pH 7.4, 25°C	3X Crystallized; suspension in 70% $(NH_4)_2SO_4$	<0.001% GK <0.005% CK, LDH	Biozyme PK2
350 U/mg protein; 150 U/mg solid	1 unit oxidizes 1 μmol NADH/min at pH 7.4, 25°C	Salt-free; lyophilized	<0.001% GK <0.005% CK, LDH	Biozyme PK3
200 U/mg	1 unit converts 1 μmol PEP to product/min at 25°C	50% glycerol solution, pH 6	<0.001% GK <0.002% HK, NADH oxidase, ATPase <0.01% enolase, LDH, MK	Boehringer 109037 109045 109053
200 U/mg at 25°C; 500 U/mg at 37°C	1 unit converts 1 μmol PEP to product/min	Suspension in 3.2 M $(NH_4)_2SO_4$, pH 6	<0.001% GK <0.002% HK, NADH oxidase, ATPase <0.01% enolase, LDH, MK	Boehringer 127418 128155 128163
300 U/mg protein	1 unit hydrolyzes 1.0 μmol PEP/min at pH 7.4, 25°C	3X Crystallized; suspension in 3 M $(NH_4)_2SO_4$	<0.001% GK <0.005% LDH, CK	Calbiochem 5506
250 U/mg protein	1 unit produces 1 μmol pyruvate from PEP/min at pH 7.5, 25°C in the presence of ADP	Suspension in 65% $(NH_4)_2SO_4$ containing 10 mg/mL protein	<0.001% GPDH, CPK, ATPase, NADH oxidase, PGM <0.005% MK	Calzyme 073B0250
250 U/mg protein	1 unit produces 1 μmol pyruvate from PEP/min at pH 7.5, 25°C in the presence of ADP	Lyophilized containing 90-95% protein	<0.001% GPDH, CPK, ATPase, NADH oxidase, PGM <0.005% MK	Calzyme 107A0250

SPECIFIC ACTIVITY	UNITS DEFINITION	PREPARATION FORM	ADDITIONAL ACTIVITIES	SUPPLIER CATALOG NO.
Rabbit muscle *continued*				
200 U/mg	1 unit catalyzes 1 μmol PEP/min at pH 7.4, 25°C	Lyophilized	<0.01% CPK, GOT, GPT, LDH, MDH, MK, nucleoside diphosphate kinase <0.02% PGluM	Fluka 83328
320 U/mg protein	1 unit catalyzes 1 μmol PEP/min at pH 7.4, 25°C	Crystallized; suspension in 3.8 M $(NH_4)_2SO_4$, pH 6	<0.01% LDH	Fluka 83330
>250 U/mg protein; >200 U/mg solid	1 unit dephosphorylates 1.0 μmol pyruvate/min at pH 7.1, 25°C	Lyophilized	<0.01% enolase, AK, ATPase, GPDH <0.02% CK <0.05% PGM, LDH	Genzyme 1601
>250 U/mg protein	1 unit dephosphorylates 1.0 μmol pyruvate/min at pH 7.1, 25°C	Suspension in $(NH_4)_2SO_4$	<0.01% enolase, AK, ATPase, GPDH <0.02% CK <0.05% PGM, LDH	Genzyme 1602
350 U/mg protein	1 unit oxidizes 1 μmol NADH/min at pH 7.4, 25°C	Salt-free; lyophilized		ICN 151999
350 U/mg protein	1 unit oxidizes 1 μmol NADH/min at pH 7.4, 25°C	Highly purified; suspension in 70% $(NH_4)_2SO_4$, pH 6.0		ICN 152000
≥200 IU/mg protein	1 IU transforms 1 μmol substrate/min under standard IUB conditions at 25°C	Suspension in $(NH_4)_2SO_4$	<0.01% PGM, G3PDH, enolase, MK, GPDH <0.02% LDH	OYC
40-100 U/mg protein	1 unit converts 1.0 μmol PEP to pyruvate/min at pH 7.6, 37°C	Crude; suspension in 3.2 M $(NH_4)_2SO_4$, pH 6.0	40-100 U/mg protein LDH	Sigma P1381
400-600 U/mg protein	1 unit converts 1.0 μmol PEP to pyruvate/min at pH 7.6, 37°C	Crystallized; suspension in 3.2 M $(NH_4)_2SO_4$, pH 6.0	<0.01% LDH, CPK, MK, PGluM	Sigma P1506
20-30 U/g solid	1 unit converts 1.0 μmol PEP to pyruvate/min at pH 7.6, 30°C	Insoluble enzyme attached to polyacrylamide; containing 30% borate buffer salts		Sigma P4010
400-600 U/mg protein	1 unit converts 1.0 μmol PEP to pyruvate/min at pH 7.6, 37°C	Solution containing 50% glycerol and 0.01 M PO_4, pH 7.0	<0.01% LDH, CPK, PGluM, MK	Sigma P7768
400-600 U/mg protein	1 unit converts 1.0 μmol PEP to pyruvate/min at pH 7.6, 37°C	Salt-free; lyophilized	<0.01% LDH, CPK, MK, PGluM	Sigma P9136
≥100 U/mg solid	1 unit oxidizes 1 μmol NADH/min at pH 7.6, 25°C	Lyophilized	<0.01% LDH, enolase, MK, NADH oxidase	Toyobo PYK-301
200 U/mg protein		Suspension		Wako 162-15161

2.7.1.40 Pyruvate Kinase continued

SPECIFIC ACTIVITY	UNITS DEFINITION	PREPARATION FORM	ADDITIONAL ACTIVITIES	SUPPLIER CATALOG NO.
Rabbit muscle *continued*				
≥160 U/mg protein	1 unit produces 1 μmol pyruvate from PEP/min at pH 7.6, 25°C in the presence of ADP	Purified; lyophilized containing 33% sucrose as stabilizer		Worthington LS03278 LS03279 LS03282 LS03281
Zymomonas mobilis (optimum pH = 7.0-8.0, MW = 114,000 Da; K_M = 60 *mM* [PEP], 0.13 *mM* [ADP]; stable pH 6-7 [4°C, 24 hr], T < 40°C [pH 6.0])				
>200 U/mg protein	1 unit forms 1 μmol pyruvate/min at 30°C	Lyophilized		Unitika
>200 U/mg protein	1 unit forms 1 μmol pyruvate/min at 30°C	Suspension in $(NH_4)_2SO_4$		Unitika

2.7.1.78 Polynucleotide 5′-Hydroxyl-Kinase

REACTION CATALYZED
ATP + 5′-dephospho-DNA ↔ ADP + 5′-phospho-DNA

SYSTEMATIC NAME
ATP:5′-dephosphopolynucleotide 5′-phosphotransferase

SYNONYMS
T4 Polynucleotide kinase

REACTANTS
ATP, 5′-dephospho-DNA, 5′-dephospho-RNA 3′-mononucleotides, ADP, 5′-phospho-DNA

APPLICATIONS
- Adds phosphate groups to synthetic DNA linkers and fragments that lack the 5′-phosphate termini needed for ligation

NOTES
- Classified as a transferase, transferring phosphorus-containing groups: phosphotransferases with an alcohol group as acceptor
- Transfers the terminal phosphate of ATP (γ-orthophosphate) to the 5′-hydroxyl termini of ribo- and deoxyribonucleotides
- In the presence of ADP, the 5′ terminal phosphate is transferred from phosphorylated DNA to ADP. Dephosphorylated DNA is then rephosphorylated by transfer of the γ-phosphate from ATP
- This enzyme is encoded by the bacteriophage T4 and is one of the earliest induced activities

Polynucleotide 5′-Hydroxyl-Kinase continued — 2.7.1.78

SPECIFIC ACTIVITY	UNITS DEFINITION	PREPARATION FORM	ADDITIONAL ACTIVITIES	SUPPLIER CATALOG NO.
T4 am N81pseT1 phage-infected *E. coli* B.B.				
>40,000 U/mg; 10,000 U/mL	1 unit incorporates 10 nmol [^{32}P] into acid-precipitable products/30 min at 37°C	50 mM Tris-HCl, 1 mM DTT, 0.1 mM EDTA, 1 μM ATP, 50% glycerol, pH 7.5	No detectable 3′-phosphatase	Boehringer 709557 838292
T4 gene in an λ vector, recombinant in *E. coli*				
10 U/μL	1 unit transfers 1 nmol PO$_4$ of ATP to the 5′-OH termini of micrococcal nuclease-treated DNA/30 min at pH 7.6, 37°C	350 mM Tris-HCl, 50 mM MgCl$_2$, 500 mM KCl, 5 mM β-MSH, 350 μM ADP, pH 6.4		Life Technol 18004-010 18004-028
T4 gene in an overexpressing plasmid, recombinant in *E. coli*				
10 U/μL	1 unit produces 1 nmol acid insoluble [^{32}P]/30 min at 37°C	50 mM KCl, 10 mM Tris-HCl, 0.1 mM EDTA, 1 mM DTT, 0.1 μM ATP, 200 μg/mL BSA, 50% glycerol, pH 7.4	No detectable non-specific endonuclease, exonuclease, RNase	Ambion 2310 2312
T4 phage				
	1 unit transfers 1 nmol γ-PO$_4$ from ATP to the 5′-OH termini of salmon sperm DNA fragments/30 min at 37°C	50 mM Tris-HCl, 1.0 mM DTT, 0.1 μM ATP, 0.1 mM EDTA, 25 mM KCl, 50% glycerol, pH 7.6	No detectable endonuclease, RNase, exonuclease	CHIMERx 1261-01 1261-02
T4 phage, cloned				
35,000 U/mg	1 unit transfers 1 nmol PO$_4$ from γ-[^{32}P]-labeled ATP to the 5′-OH ends of micrococcal nuclease-treated calf thymus DNA/30 min at pH 7.5, 37°C	Solution containing 50% glycerol, 50 mM Tris-HCl, 0.1 M NaCl, 0.1 mM EDTA, 1 mM DTT, 0.1% Triton X-100, pH 7.5	No detectable exonuclease, endonuclease, RNase (by agarose gel electrophoresis)	Epicentre P0505H P0501K P0503K P0510K
30,000-60,000 U/mg protein; 5000-10,000 U/mL	1 unit transfers 1 nmol PO$_4$ from ATP to polynucleotide/30 min at pH 7.6, 37°C using duplex DNA partially digested with micrococcal nuclease as substrate	Molecular biology grade; homogeneous purity; solution containing 20 mM KPO$_4$, 25 mM KCl, 10 mM β-MSH, ATP, 50% glycerol, pH 7.0	No detectable RNase, DNase, nickase	Pharmacia 27-0736-01 27-0736-02
T4 phage-infected *E. coli*				
	1 unit incorporates 1 nmol acid-insoluble [^{32}P]/30 min at pH 7.6, 37°C	50 mM KCl, 10 mM Tris-HCl, 0.1 mM EDTA, 1 mM DTT, 0.1 μM ATP, 50% glycerol, pH 7.4	Exonuclease: releases <0.05% of radioactivity from [^{3}H] DNA; Endonuclease: 5% φX174 RFI DNA converted to RFII	Adv Biotech AB-0269 AB-0269b
20-50 U/μL	1 Richardson unit incorporates 1 nmol [^{32}P] from [γ^{32}P] ATP into micrococcal nuclease treated DNA/30 min at 37°C	>99% purity by SDS-PAGE; 50 mM Tris-HCl, 0.1 mM EDTA, 1 mM DTT, 50% glycerol, pH 7.5	No detectable exonuclease, endonuclease, ribonuclease	Amersham E 70031Y E 70031Z E 70013X
5-20 U/μL	1 unit incorporates 1 nmol [^{32}P] in acid precipitable material/30 min at pH 7.5, 37°C	10 mM Tris-HCl, 0.1 mM EDTA, 1 mM DTT, 50 mM KCl, 50% glycerol, pH 7.5	No detectable exonuclease, endonuclease, RNase	ACS Heidelb A00810S A00810M

SPECIFIC ACTIVITY	UNITS DEFINITION	PREPARATION FORM	ADDITIONAL ACTIVITIES	SUPPLIER CATALOG No.
T4 phage-infected *E. coli* continued				
>40,000 U/mg; 10,000 U/mL	1 unit incorporates 1 nmol [^{32}P]/30 min at 37°C	50 mM Tris-HCl, 1 mM DTT, 0.1 mM EDTA, 1 μM ATP, 50% glycerol, pH 7.5		Boehringer 174645 633542
500 and 2500 U	1 unit incorporates 1 nmol [^{32}P] from [γ-^{32}P]ATP into micrococcal nuclease-treated calf thymus DNA/30 min at pH 7.6, 37°C	500 mM Tris-HCl, 100 mM MgCl$_2$, 100 mM β-MSH, pH 8.0 for protruding ends (pH 9.5 for blunt or recessed ends)		CLONTECH 8408-1 8408-2
	1 unit transfers 1 nmol γ-PO$_4$ from ATP to the 5'-OH terminus of salmon sperm DNA fragments/30 min at 37°C	Solution containing 25 mM KCl, 50 mM Tris-HCl, 0.1 μM ATP, 1 mM DTT, 0.1 mM EDTA, 50% glycerol, pH 7.6		ICN 151935
	1 unit transfers 1 nmol γ-PO$_4$ from ATP to the 5'-OH termini of salmon sperm DNA fragments/30 min at 37°C	Solution containing 25 mM KCl, 50 mM Tris-HCl, 0.1 μM ATP, 1 mM DTT, 0.1 mM EDTA, 50% glycerol, pH 7.6	3'-phosphatase	ICN 800708
≥30,000 Richardson U/mg protein	1 unit transfers 1 nmol PO$_4$ from ATP onto the 5'-OH end of a polynucleotide/30 min at 37°C	50 mM Tris-HCl, 1 μM ATP, 1 mM DTT, 0.1 mM EDTA, 50% glycerol, pH 7.5		Oncor 120104 120105
5-10 U/μL	1 unit transfers 1 nmol PO$_4$ to the 5'-OH end of an oligonucleotide from [γ-^{32}P]ATP/30 min at pH 7.5	Cloning Qualified; ≥90% purity by SDS gel; 10X reaction buffer: 700 mM Tris-HCl, 100 mM MgCl$_2$, 50 mM DTT, pH 7.6	<1% DNase <3% RNase ≥90% supercoiled plasmid	Promega M4101 M4102
10,000 U/mL	1 unit transfers 1 nmol [^{32}P] to the 5'-end of micrococcal nuclease-treated DNA/30 min at 37°C			Sigma P4390
1000-12,000 U/mL	1 unit transfers 1 nmol γ-PO$_4$ from ATP to the 5'-OH terminus of random length oligonucleotides/30 min at 37°C	10 mM Tris-HCl, 50 mM KCl, 1 mM TdT, 0.1 mM EDTA, 0.01 mM ATP, 50% glycerol, pH 7.4	No detectable endonuclease, exonuclease	Stratagene 600103 600104
≥10,000 U/mg protein	1 unit incorporates 1 nmol γ-[^{32}P] from [γ-^{32}P]ATP into micrococcal nuclease treated calf thymus DNA/30 min at 37°C	Solution containing 50 mM Tris-HCl, 1 mM DTT, 15% glycerol, pH 7.5		Worthington LM01050 LM01052 LM01054
T4 phage-infected *E. coli* W3110 (IRC4), overexpressed				
10,000 U/mL	1 Richardson unit incorporates 1 nmol acid insoluble [^{32}P]/30 min at 37°C	50 mM KCl, 10 mM Tris-HCl, 0.1 mM EDTA, 1 mM DTT, 50% glycerol, pH 7.4	No detectable exonuclease, endonuclease, RNase	NE Biolabs 201S 201L
T4 phage-infected *E. coli*, overproduced				
5-15 U/μL	1 unit transfers 1 nmol PO$_4$ from [γ-^{32}P]-ATP to 5'-OH DNA/30 min at 37°C	20 mM Tris-HCl, 0.1 mM EDTA, 2 mM DTT, 25 mM KCl, 0.1 mM ATP, 50% glycerol, pH 7.5	No detectable endo-, exodeoxyribonucleases, RNase	Fermentas EK0031 EK0032

SPECIFIC ACTIVITY	UNITS DEFINITION	PREPARATION FORM	ADDITIONAL ACTIVITIES	SUPPLIER CATALOG NO.
T4 pseT phage-infected *E. coli* (contains the T4 PNK gene) (4 identical subunits of 33,000 Da each)				
5-10 U/μL	1 unit transfers 1 nmol PO_4 from [γ-^{32}P]-ATP to the 5'-OH end of the polynucleotide/30 min at pH 7.5, 37°C	25 mM Tris-HCl, 10 mM β-MSH, 50% glycerol, pH 7.5		NBL Gene 020903 020905
10 U/μL; 1000 U and 5000 U	1 unit incorporates 1 nmol [^{32}P] from [γ-^{32}P]ATP into acid insoluble form/30 min at pH 7.6, 37°C with calf thymus DNA activated by micrococcal nuclease as substrate	Solution containing 50 mM Tris-HCl, 50 mM KCl, 1 mM DTT, 0.1 μM ATP, 50% glycerol, pH 7.5	No detectable nuclease Contains 3'-phosphatase activity	TaKaRa 2021
T7 phage				
				Oncogene XME10

Pyrophosphate-Fructose-6-Phosphate 1-Phosphotransferase

2.7.1.90

REACTION CATALYZED
  Pyrophosphate + D-fructose 6-phosphate ↔ Orthophosphate + D-fructose 1,6-bisphosphate

SYSTEMATIC NAME
  Pyrophosphate:D-fructose 6-phosphate 1-phosphotransferase

SYNONYMS
  6-Phosphofructokinase (pyrophosphate), fructose-6-phosphate kinase

REACTANTS
  Pyrophosphate, D-fructose 6-phosphate, orthophosphate, D-fructose 1,6-bisphosphate

APPLICATIONS
- Microassay of fructose 2,6-diphosphate
- Determination of enzymatically-generated pyrophosphate

NOTES
- Classified as a transferase, transferring phosphorus-containing groups: phosphotransferases with an alcohol group as acceptor

2.7.1.90 Pyrophosphate-Fructose-6-Phosphate 1-Phosphotransferase continued

SPECIFIC ACTIVITY	UNITS DEFINITION	PREPARATION FORM	ADDITIONAL ACTIVITIES	SUPPLIER CATALOG NO.
Mung bean				
5-20 U/mg protein	1 unit converts 1.0 μmol pyrophosphate and F6P to fructose 1,6-diphosphate and P$_i$/min at pH 7.6, 30°C in coupled system	Lyophilized containing 40% protein; balance imidazole salts and stabilizers		sigma F8757
Potato tuber				
2-6 U/mg protein	1 unit converts 1.0 μmol pyrophosphate and F6P to fructose 1,6-diphosphate and P$_i$/min at pH 8.0, 30°C in coupled system	Lyophilized containing 30% protein; balance primarily NaPO$_4$ buffer and stabilizers with trace PP$_i$ and DTT		sigma F2258
Propionibacterium freudenreichii (shermanii)				
150-250 U/mg protein	1 unit converts 1.0 μmol pyrophosphate and F6P to fructose 1,6-diphosphate and P$_i$/min at pH 7.4, 30°C in coupled system	Lyophilized containing 5% protein; balance primarily imidazole salts and stabilizer		sigma F4384
4-8 U/mg protein	1 unit converts 1.0 μmol pyrophosphate and F6P to fructose 1,6-diphosphate and P$_i$/min at pH 7.4, 30°C in a coupled system	Lyophilized containing 15% protein; balance primarily imidazole salts and stabilizer		sigma F8381

2.7.1.95 Kanamycin Kinase

REACTION CATALYZED
ATP + kanamycin ↔ ADP + kanamycin 3′-phosphate

SYSTEMATIC NAME
ATP:kanamycin 3′-O-phosphotransferase

SYNONYMS
Neomycin-kanamycin phosphotransferase, neomycin phosphotransferase II

REACTANTS
ATP, kanamycin, neomycin, paromomycin, neamine, paromamine, vistamycin, gentamicin A, bitirosin, ADP, kanamycin 3′-phosphate

APPLICATIONS
- Standard in sandwich-type ELISAs for determination of NPT II in crude cell lysates

NOTES
- Classified as a transferase, transferring phosphorus-containing groups: phosphotransferases with an alcohol group as acceptor
- *Pseudomonas aeruginosa* enzyme also acts on bitirosin

Kanamycin Kinase continued

2.7.1.95

SPECIFIC ACTIVITY	UNITS DEFINITION	PREPARATION FORM	ADDITIONAL ACTIVITIES	SUPPLIER CATALOG NO.
Escherichia coli (MW = 29,000 Da)				
	1 unit converts 1.0 nmol ATP and kanamycin to ADP and phosphorylated kanamycin/min at pH 7.1, 37°C	>98% purity by SDS-PAGE; 50 mM Tris-HCl, 200 mM NaCl, 3 mM KCl, 1 mM β-MSH, 10 μg/mL neomycin, 50% glycerol, 0.5 mg/mL protein, pH 7.5		5′→3′ 5306-639134
Tn5-infected *E. coli*				
10,000 U/mg; 1 μg/mL	1 unit converts 1 nmol ATP and kanamycin to ADP and phosphorylated kanamycin/min at 37°C	Solution containing 50% glycerol, 150 mM NaCl, 8 mM Na$_2$HPO$_4$, 2 mM KH$_2$PO$_4$, 3 mM KCl, 5 mM DTT, 0.5 mM PMSF, 10 μg neomycin, pH 7.4		ICN 153899

Diacylglycerol Kinase

2.7.1.107

REACTION CATALYZED
    ATP + 1,2-diacylglycerol ↔ ADP + 1,2-diacyl-*sn*-glycerol 3-phosphate

SYSTEMATIC NAME
    ATP:1,2-diacylglycerol 3-phosphotransferase

SYNONYMS
    Diglyceride kinase, *sn*-1,2-diacylglycerol kinase

REACTANTS
    ATP, 1,2-diacylglycerol, ADP, 1,2-diacyl-*sn*-glycerol 3-phosphate, ceramide, monoglycerol

APPLICATIONS
- Determination of diacylglycerol in crude lipid extracts

NOTES
- Classified as a transferase, transferring phosphorus-containing groups: phosphotransferases with an alcohol group as acceptor
- Specific for the *sn*-1,2 isomer of diacylglycerol, but will phosphorylate ceramide and monoglycerol at lower rates

SPECIFIC ACTIVITY	UNITS DEFINITION	PREPARATION FORM	ADDITIONAL ACTIVITIES	SUPPLIER CATALOG NO.
Escherichia coli (optimum pH = 6.3-8.3, MW = 13,700 Da)				
6.0 U/mg protein	1 unit phosphorylates 1.0 μmol diacylglycerol/min	Turbid membrane suspension in 10 mM KPO$_4$, 20% glycerol, 1 mM DTT, 1 mg/mL protein, pH 7.0		BIOMOL SE-100

2.7.1.107 Diacylglycerol Kinase continued

SPECIFIC ACTIVITY	UNITS DEFINITION	PREPARATION FORM	ADDITIONAL ACTIVITIES	SUPPLIER CATALOG NO.
Escherichia coli, recombinant (optimum pH = 6.3-8.3, MW = 13,700 Da)				
2 U/mg protein	1 unit phosphorylates 1 μmol DAG/min at pH 6.6, 25°C	Solution containing 20 mM NaCl, 10 mM KPO$_4$, 1 mM DTT, 20% glycerol, pH 7.0		Calbiochem 266724
12 U/mg protein	1 unit phosphorylates 1 μmol DAG/min at pH 6.8, 25°C	>90% purity; solution containing 300 mM NaCl, 250 mM imidazole, 0.5% n-decyl-β-D-maltopyranoside, 50 mM NaPO$_4$, pH 7.5		Calbiochem 266725

2.7.1.112 Protein-Tyrosine Kinase

REACTION CATALYZED
    ATP + protein tyrosine ↔ ADP + protein tyrosine phosphate

SYSTEMATIC NAME
    ATP:protein tyrosine *O*-phosphotransferase

SYNONYMS
    Tyrosylprotein kinase, protein kinase (tyrosine), hydroxyaryl-protein kinase, p60^{c-src} enzyme, PTK, TK

REACTANTS
    ATP, protein tyrosine, ADP, protein tyrosine phosphate

NOTES
- Classified as a transferase, transferring phosphorus-containing groups: phosphotransferases with an alcohol group as acceptor
- A glyco- and receptor protein
- Some of the many eukaryotic genes for this enzyme are closely related
- Other nucleotides may act as donors
- Some are activated by cyclic GMP but not by cyclic AMP; some are activated by neither

SPECIFIC ACTIVITY	UNITS DEFINITION	PREPARATION FORM	ADDITIONAL ACTIVITIES	SUPPLIER CATALOG NO.
Human A431 cells (MW = 170,000 Da)				
100 U/mL	1 unit adds 1 nmol PO$_4$ to angiotensin II/min at pH 7.4, 30°C	>90% purity by SDS-PAGE; 3.7 μg/mL in 20 mM HEPES, 130 mM NaCl, 1 mM EDTA, 1 mM DTT, 10% glycerol, 0.05% Triton X-100 buffer containing 5 mM EGF	No detectable serine kinase	BIOMOL SE-116
Human platelets				
	1 unit incorporates 1 nmol PO$_4$ from ATP into tyrosyl residues/min at 30°C	Chromatographically purified		Oncogene PK03

Ca^{2+}/Calmodulin-Dependent Protein Kinase 2.7.1.123

REACTION CATALYZED

ATP + protein ↔ ADP + O-phosphoprotein

SYSTEMATIC NAME

ATP:protein O-phosphotransferase (calmodulin-dependent)

SYNONYMS

Microtubule-associated protein 2 kinase, CaM kinase II, Ca^{2+}/calmodulin kinase II, protein kinase II (Ca^{2+}/CaM dependent)

REACTANTS

ATP, protein, ADP, O-phosphoprotein, calcineurin, CRE-binding protein, IP3 receptor, intermediate filaments, ribosomal protein S6, synapsin I

APPLICATIONS
- Phosphorylates target sites

NOTES
- Classified as a transferase, transferring phosphorus-containing groups: phosphotransferases with an alcohol group as acceptor
- Regulates cellular function, including neurotransmitter synthesis and release, gene expression, ion channel function, carbohydrate metabolism, cytoskeletal function and calcium homeostasis
- Requires calmodulin and Ca^{2+}
- Vimentin, synapsin, glycogen synthase, myosin light chains and microtubule-associated *tau* protein are among the variety of proteins which act as acceptor
- Composed of 60,000 and 50,000 Da subunit in the ratios of 1:3 and 1:4
- Not identical with myosin-light-chain kinase (EC 2.7.1.117), caldesmon kinase (EC 2.7.1.120) or *tau*-protein kinase (EC 2.7.1.135)

SPECIFIC ACTIVITY	UNITS DEFINITION	PREPARATION FORM	ADDITIONAL ACTIVITIES	SUPPLIER CATALOG NO.
Rat brain (MW = 650,000 Da)				
900 U/mg	1 unit transfers 1 nmol PO$_4$ from ATP to synapsin I/min at pH 7.4, 30°C	>95% purity by SDS-PAGE; 20 µg/mL in 25 m*M* Tris, 10 m*M* β-MSH, 10 µ*M* AEBSF, 0.5 m*M* EGTA, 0.1 mg/mL leupeptin, 50% glycerol, pH 7.5		BIOMOL SE-134
1 U/mg protein	1 unit transfers 1 µmol PO$_4$ to synapsin I/min at 35°C	>95% purity; solution containing 25 m*M* Tris-HCl, 0.5 m*M* EDTA, 10 m*M* β-MSH, 100 µ*M* AEBSF, 1 mg/mL leupeptin, 50% glycerol, pH 7.5		Calbiochem 208707
		≥95% purity		Chemicon AG625

Acetate Kinase

REACTION CATALYZED
ATP + acetate ↔ ADP + acetyl phosphate

SYSTEMATIC NAME
ATP:acetate phosphotransferase

SYNONYMS
Acetokinase, AK

REACTANTS
ATP, acetate, propanoate (slowly), ADP, acetyl phosphate

NOTES
- Classified as a transferase, transferring phosphorus-containing groups: phosphotransferases with a carboxyl group as acceptor
- Activated by fructose 1,6-bisphosphate

SPECIFIC ACTIVITY	UNITS DEFINITION	PREPARATION FORM	ADDITIONAL ACTIVITIES	SUPPLIER CATALOG NO.
Bacillus stearothermophilus (optimum pH = 7.2, pI = 4.8, MW = 160,000 Da [40,000/subunit]; K_M = 120 mM [acetate], 2.3 mM [acetylphosphate], 1.2 mM [ATP], 0.8 mM [ADP]; stable pH 7.0-8.0, T < 65°C)				
>1350 U/mg protein	1 unit forms 1.0 μmol ADP/min at pH 2.6	Lyophilized containing 40% protein; balance primarily KPO₄ buffer		ICN 151421
400-1200 U/mg solid at pH 7.2, 30°C	1 unit phosphorylates 1.0 μmol acetate to acetyl PO₄/min at pH 7.6, 25°C	Lyophilized containing KPO₄ buffer		Sigma A6781
>1400 U/mg protein	1 unit forms 1 μmol ADP/min at 30°C	Lyophilized	<0.01% LDH, AdK, NADH oxidase, GOT, GPT	Unitika
>1400 U/mg protein	1 unit forms 1 μmol ADP/min at 30°C	50% glycerol solution	<0.01% LDH, AdK, NADH oxidase, GOT, GPT	Unitika
Escherichia coli				
250 U/mg	1 unit converts 1 μmol acetate and ATP to product/min at 25°C	Suspension in 3.2 M (NH₄)₂SO₄, pH 6	0.01% GPT, GOT, LDH, MK, NADH oxidase	Boehringer 101834
150-300 U/mg protein	1 unit phosphorylates 1.0 μmol acetate to acetyl PO₄/min at pH 7.6, 25°C	Suspension in 3.2 M (NH₄)₂SO₄, pH 6	<0.01% GOT, LDH, MK, NADH oxidase	ICN 154678
150-300 U/mg protein	1 unit phosphorylates 1.0 μmol acetate to acetyl PO₄/min at pH 7.6, 25°C	Suspension in 3.2 M (NH₄)₂SO₄ and BSA, pH 6	<0.01% GOT, GPT, LDH, MK, NADH oxidase	Sigma A2384
300-800 U/mg protein	1 unit phosphorylates 1.0 μmol acetate to acetyl PO₄/min at pH 7.6, 25°C	Lyophilized containing 10% protein, trehalose, KPO₄, MgCl₂, DTT	<0.01% ATPase, NADH oxidase	Sigma A7437
Methanosarcina thermophila, recombinant as *E. coli*				
	1 unit phosphorylates 1.0 μmol acetate to acetyl PO₄/min at pH 7.6, 25°C	Lyophilized containing Tris buffer salt, KCl, DTT, Ficoll		Sigma A9701

Carbamate Kinase 2.7.2.2

REACTION CATALYZED
  ATP + NH_3 + CO_2 ↔ ADP + carbamoyl phosphate
SYSTEMATIC NAME
  ATP:carbamate phosphotransferase
REACTANTS
  ATP, NH_3, CO_2, ADP, carbamoyl phosphate

NOTES
- Classified as a transferase, transferring phosphorus-containing groups: phosphotransferases with a carboxyl group as acceptor

SPECIFIC ACTIVITY	UNITS DEFINITION	PREPARATION FORM	ADDITIONAL ACTIVITIES	SUPPLIER CATALOG No.
Streptococcus faecalis				
2000-4000 U/mg protein	1 unit forms 1.0 μmol ATP from ADP and carbamyl PO_4/min at pH 8.3, 37°C	Chromatographically purified; lyophilized containing 40% protein; balance succinate buffer salts and stabilizer		sigma C5408
400-900 U/mg protein	1 unit forms 1.0 μmol ATP from ADP and carbamyl PO_4/min at pH 8.3, 37°C	Lyophilized containing 40% protein; balance MES buffer salt		sigma C8398

Phosphoglycerate Kinase 2.7.2.3

REACTION CATALYZED
  ATP + 3-phospho-D-glycerate ↔ ADP + 3-phospho-D-glyceroyl phosphate
SYSTEMATIC NAME
  ATP:3-phospho-D-glycerate 1-phosphotransferase
SYNONYMS
  3-Phosphoglyceric phosphokinase, 3-phosphoglycerate kinase, PGK, 3PGPK
REACTANTS
  ATP, 3-phospho-D-glycerate, ADP, 3-phospho-D-glyceroyl phosphate; comparative rates: lysophosphatidylcholine (100%), phosphatidylcholine (87%), sphingosylphosphorylcholine (22%)

APPLICATIONS
- Determination of choline phospholipids, with choline oxidase

NOTES
- Classified as a transferase, transferring phosphorus-containing groups: phosphotransferases with a carboxyl group as acceptor
- Does not react with AMP, phosphoenolpyruvate or glycerate-2,3-diphosphate

SPECIFIC ACTIVITY	UNITS DEFINITION	PREPARATION FORM	ADDITIONAL ACTIVITIES	SUPPLIER CATALOG No.
Bacillus stearothermophilus (optimum pH = 7, pI = 4.4, MW = 40,000 Da; K_M = 0.7 mM [glycerate 3-P, ATP]; stable pH > 7.5, T < 50°C)				
500 U/mg protein	1 unit converts 1.0 μmol 1,3-diphosphoglycerate to 3-phosphoglycerate/min at pH 7.2, 30°C	Lyophilized containing 70% protein; balance primarily Tris buffer salts, pH 8.5		Sigma P5663
>400 U/mg protein	1 unit forms 1 μmol 1,3-diphosphoglycerate/min at 30°C	Lyophilized	<0.01% AdK, 3PGDH, GAPDH, TPI, NADH oxidase	Unitika
>400 U/mg protein	1 unit forms 1 μmol 1,3-diphosphoglycerate/min at 30°C	50% glycerol solution	<0.01% AdK, 3PGDH, GAPDH, TPI, NADH oxidase	Unitika
Rabbit muscle				
500 U/mg protein	1 unit converts 1.0 μmol 1,3-diphosphoglycerate to 3-phosphoglycerate/min at pH 6.9, 25°C	Sulfate-free; lyophilized containing 50% protein; balance primarily citrate buffer salts		Sigma P2399
Yeast (MW = 47,000 Da)				
450 U/mg	1 unit converts 1 μmol glycerate-3-phosphate and ATP to product/min at 25°C	Suspension in 3.2 M $(NH_4)_2SO_4$ and 1 mM EDTA, pH 7	<0.001% NADH oxidase <0.01% GAPDH, GDH, MK <0.1% TIM	Boehringer 108430 108448
400 U/mg protein	1 unit converts 1.0 μmol 1,3-diphosphoglycerate to 3-phosphoglycerate/min at pH 6.9, 25°C	Lyophilized containing 95% protein		Calzyme 204A0400
450 U/mg		Crystallized; suspension in 0.80 $(NH_4)_2SO_4$, 40 mM pyrophosphate, pH 7.0	<0.01% MK, glycolytic enzymes	ICN 100984
Yeast, bakers				
1000 U/mg protein	1 unit converts 1.0 μmol 1,3-diphosphoglycerate to 3-phosphoglycerate/min at pH 6.9, 25°C	Sulfate-free; lyophilized containing 50% protein; balance primarily citrate buffer salts		Sigma P1136
1500-2500 U/mg protein	1 unit converts 1.0 μmol 1,3-diphosphoglycerate to 3-phosphoglycerate/min at pH 6.9, 25°C	Crystallized; suspension in 3.0 M $(NH_4)_2SO_4$ and 0.04 M tetra sodium pyrophosphate, pH 8.0	<0.05% GAPDH	Sigma P7634

Creatine Kinase — 2.7.3.2

REACTION CATALYZED
ATP + creatine ↔ ADP + phosphocreatine

SYSTEMATIC NAME
ATP:creatine N-phosphotransferase

SYNONYMS
Creatine phosphokinase, CK, CPK

REACTANTS
ATP, creatine, N-ethylglycocyamine, ADP, phosphocreatine, glycocyamine

APPLICATIONS
- Marker for cancer of breast, ovary, prostate, and colon and other gastrointestinal carcinomas
- Sensitive diagnostic marker for myocardial infarction
- MB2 is a non-invasive marker for coronary artery perfusion, eliminating surgery
- Determination of creatine
- Immunogen for mono- and polyclonal antibodies
- Used as an ATPase and ATP kinase

NOTES
- Classified as a transferase, transferring phosphorus-containing groups: phosphotransferases with a nitrogenous group as acceptor
- Dimeric enzyme found in three isozyme forms: MM (muscle), MB (brain/muscle hybrid) and BB (brain)
- Widely distributed and represents 10-20% of muscle cytoplasmic protein
- Regenerates ATP in contractile and transport systems
- Normally absent in serum of normal adults, but may increase with severe damage to tissues containing the BB isozyme (i.e., brain)
- Rabbit muscle enzyme is activated by divalent cations like Mg^{2+}, Ca^{2+} and Mn^{2+}
- Rabbit muscle enzyme is inhibited by various sulfhydryl-binding reagents, chelating agents, some adenosine phosphate compounds, orthophosphate, pyro- and tripolyphosphate, adenosine, Cl^-, SO_4^{2-}, acetate, dibenamine, phenothiazone and 3,5-dinitro-O-cresol; ADP competitively inhibits the forward reaction with respect to ATP and noncompetitively with respect to creatine
- Addition of 0.1% albumin to reaction buffer prevents dilution-related inactivation

SPECIFIC ACTIVITY	UNITS DEFINITION	PREPARATION FORM	ADDITIONAL ACTIVITIES	SUPPLIER CATALOG NO.
Bovine heart (optimum pH = 8.0-9.0, MW = 81,000 Da)				
25 U/mg solid	1 unit phosphorylates 1 μmol creatine/min at pH 9.0, 25°C	Salt-free; lyophilized		Biozyme CKB1
25-50 U/mg solid	1 unit phosphorylates 1.0 μmol creatine/min at pH 9.0, 25°C	Salt-free; lyophilized		ICN 105057
150-300 U/mg protein	1 unit transfers 1.0 μmol PO_4 from phosphocreatine to ADP/min at pH 7.4, 30°C	Salt-free; lyophilized containing MM isoenzyme	<0.01% ATPase, LDH <0.02% PK <0.03% HK <0.07% MK	Sigma C7886

SPECIFIC ACTIVITY	UNITS DEFINITION	PREPARATION FORM	ADDITIONAL ACTIVITIES	SUPPLIER CATALOG NO.
Bovine heart *continued*				
50 U/mg		Lyophilized		Wako 036-13393
Human brain				
600 U/mg protein	1 unit transfers 1 μmol PO_4 from phosphocreatine to ADP/min at pH 6.9, 25°C	Single band (PAGE); lyophilized containing 90-95% protein		Calzyme 068A0600
600 U/mg protein	1 unit transfers 1 μmol PO_4 from phosphocreatine to ADP/min at pH 6.9, 25°C	Single band (PAGE); 50% glycerol solution containing 90-95% protein		Calzyme 068C0600
Human brain BB isozyme (MW = 84,000 Da)				
>700 U/mg protein	1 unit converts 1 μmol creatine to creatine PO_4/min at pH 6.6, 30°C	>98% purity; solution containing 50% glycerol, 10 mM NaCl, 1 mM EDTA, 5 mM succinate, 5 mM β-MSH, pH 7.0	No detectable HBsAg, HIV Ab	Calbiochem 238397
500-600 U/mg protein; 1 mg/mL	pH 6.8, 37°C	Frozen solution containing 10 mM Tris-HCl, 1 mM EDTA, 10 mM β-MSH, 150 mM NaCl, 50% glycerol, pH 7.4	No detectable HIV I and II Ab, HBsAg, anti-HCV	Cortex CP3013
≥400 IU/mg	1 unit converts 1 μmol creatine to creatine PO_4/min at pH 6.6, 30°C	>99% purity by SDS-PAGE; solution containing 50% glycerol, 5 mM succinate, 1 mM EDTA, 5 mM β-MSH, 10 mM NaCl, 0.5-2 mg/mL protein, pH 7.0	No detectable CK isozymes, HGsAg, HCV, HIV-1 Ab	Scripps C1124
Human heart				
600 U/mg protein	1 unit transfers 1 μmol PO_4 from phosphocreatine to ADP/min at pH 6.9, 25°C	Single band (PAGE); 50% glycerol solution containing 90-95% protein		Calzyme 128C0600
Human heart MB isozyme (MW = 80,000 Da)				
600 U/mg protein	1 unit converts 1 μmol creatine from phosphocreatine/min at pH 6.6, 30°C	Single band observed on agarose electrophoresis when stained for CPK-MB activity; buffered solution containing 50% glycerol and 10 mM Tris-HCl, pH 7.5	No detectable HBsAg, HIV Ab	Calbiochem 238399
1007 IU/mg; 2114 IU/mL		>95% purity by SDS-PAGE; solution containing 10 mM Tris, 5 mM β-MSH, 50% glycerol, pH 7.2	No detectable HIV I, HIV II, HCV Ab, HBsAg	Cortex CP3014
300-1000 IU/mg		80-98% purity; 10 mM β-MSH, 2 mM EDTA, 50% glycerol, 50 mM Tris buffer, pH 7.2		Genzyme 5503
≥100 U/mg	1 unit converts 1 μmol creatine to creatine PO_4/min at pH 6.6, 30°C	≥90% purity by SDS-PAGE; Spiking Grade A; solution containing 50% glycerol, 5 mM succinate, 1 mM EDTA, 5 mM β-MSH, 10 mM NaCl, pH 7.0	No detectable isozymes, HGsAg, HCV, HIV-1 Ab <10% CK-BB isoenzyme	Scripps C1222

SPECIFIC ACTIVITY	UNITS DEFINITION	PREPARATION FORM	ADDITIONAL ACTIVITIES	SUPPLIER CATALOG NO.
Human heart MB isozyme continued				
≥300 U/mg	1 unit converts 1 μmol creatine to creatine PO_4/min at pH 6.6, 30°C	≥99% purity by SDS-PAGE; Spiking Grade B; solution containing 50% glycerol, 5 mM succinate, 1 mM EDTA, 5 mM β-MSH, 10 mM NaCl, pH 7.0	No detectable CK isozymes, HGsAg, HCV, HIV-1 Ab	Scripps C1223
≥400 IU/mg; ≥700 IU/mg (37 C)	1 unit converts 1 μmol creatine to creatine PO_4/min at pH 6.6, 30°C	>99% purity by SDS-PAGE; solution containing 50% glycerol, 5 mM succinate, 1 mM EDTA, 5 mM β-MSH, 10 mM NaCl, 0.5-2 mg/mL protein, pH 7.0	No detectable CK isozymes, HGsAg, HCV, HIV-1 Ab	Scripps C1224
≥400 IU/mg; ≥700 IU/mg (37°C)	1 unit converts 1 μmol creatine to creatine PO_4/min at pH 6.6, 30°C	Purity by SDS-PAGE; solution containing 50% glycerol, 5 mM succinate, 1 mM EDTA, 5 mM β-MSH, 10 mM NaCl, 0.5-2 mg/mL protein, pH 7.0	No detectable CK isozymes, HGsAg, HCV, HIV-1 Ab; 50% CK-MB_2; 50% CK-MB_1; ≥99% CK-MB	Scripps C1723
≥400 IU/mg; ≥700 IU/mg (37°C)	1 unit converts 1 μmol creatine to creatine PO_4/min at pH 6.6, 30°C	Purity by SDS-PAGE; solution containing 50% glycerol, 5 mM succinate, 1 mM EDTA, 5 mM β-MSH, 10 mM NaCl, 0.5-2 mg/mL protein, pH 7.0	No detectable HGsAg, HCV, HIV-1 Ab; <10% CK-MB_2; ≥90% CK-MB_1, CK-MB	Scripps C1724
≥300 U/mg	1 unit converts 1 μmol creatine to creatine PO_4/min at pH 6.6, 30°C	≥99% purity by SDS-PAGE; solution containing 50% glycerol, 5 mM succinate, 1 mM EDTA, 5 mM β-MSH, 10 mM NaCl, pH 7.0	No detectable CK isozymes, HGsAg, HCV, HIV-1 Ab; <3% LD-1, LD-2	Scripps C4023
Human muscle				
600 U/mg protein	1 unit transfers 1 μmol PO_4 from phosphocreatine to ADP/min at pH 6.9, 25°C	Single band (PAGE); lyophilized containing 90-95% protein		Calzyme 069A0600
600 U/mg protein	1 unit transfers 1 μmol PO_4 from phosphocreatine to ADP/min at pH 6.9, 25°C	Single band (PAGE); 50% glycerol solution containing 90-95% protein		Calzyme 069C0600
Human skeletal muscle MM isozyme (MW = 80,000 Da)				
400 IU/mg protein	1 unit converts 1 μmol creatine from phosphocreatine/min at pH 7.4, 37°C	Buffered solution containing 50% glycerol, 5 mM β-MSH, 50 μM EDTA, pH 7.0	No detectable HBsAg, HIV Ab	Calbiochem 238407
910 IU/mg; 4548 IU/mL		>95% purity; solution containing 25% ethylene glycol, 10 mM Tris, 1 mM EDTA, 5 mM β-MSH, pH 7.5	No detectable HIV I, HIV II, HCV Ab, HBsAg	Cortex CP4012
≥400 IU/mg	1 unit converts 1 μmol creatine to creatine PO_4/min at pH 6.6, 30°C	>99% purity by SDS-PAGE; solution containing 50% glycerol, 5 mM succinate, 1 mM EDTA, 5 mM β-MSH, 10 mM NaCl, 0.5-2 mg/mL protein, pH 7.0	No detectable CK isozymes, HGsAg, HCV, HIV-1 Ab	Scripps C1324

2.7.3.2 Creatine Kinase continued

SPECIFIC ACTIVITY	UNITS DEFINITION	PREPARATION FORM	ADDITIONAL ACTIVITIES	SUPPLIER CATALOG NO.
Human tissue BB isozyme				
>200 IU/mg protein	CK-NAC activated assay, 37°C	50% purity by SDS-PAGE; 1 CK isozyme band by agarose electrophoresis; solution containing >0.4 mg/mL protein and Tris buffer, pH 6.8-7.2	No detectable HBsAg, HCV, HIV	Aalto 1020
>700 IU/mg protein	CK-NAC activated assay, 37°C	Immunopure; >95% purity by SDS-PAGE; 1 CK isozyme band by agarose electrophoresis; solution containing >4.0 mg/mL protein and Tris buffer, pH 6.8-7.2	No detectable HBsAg, HCV, HIV	Aalto 1021
Human tissue MB isozyme				
>300 IU/mg protein	CK-NAC activated assay, 37°C	50% purity by SDS-PAGE; 1 CK isozyme band by agarose electrophoresis; solution containing >0.5 mg/mL protein and Tris buffer, pH 6.8-7.2	No detectable HBsAg, HCV, HIV	Aalto 1016 1052
>700 IU/mg protein	CK-NAC activated assay, 37°C	Immunopure; >95% purity by SDS-PAGE; 1 CK isozyme band by agarose electrophoresis; solution containing >4.0 mg/mL protein and Tris buffer, pH 6.8-7.2	No detectable HBsAg, HCV, HIV	Aalto 1056
Human tissue MM isozyme				
>600 IU/mg protein	CK-NAC activated assay, 37°C	Immunopure; >95% purity by SDS-PAGE; 1 CK isozyme band by agarose electrophoresis; solution containing >1.0 mg/mL protein and Tris buffer, pH 6.8-7.2	No detectable HBsAg, HCV, HIV	Aalto 1011
>600 IU/mg protein	CK-NAC activated assay, 37°C	50% purity by SDS-PAGE; 1 CK isozyme band by agarose electrophoresis; solution containing >1.0 mg/mL protein and Tris buffer, pH 6.8-7.2	No detectable HBsAg, HCV, HIV	Aalto 1011
Porcine heart				
30 U/mg solid	1 unit phosphorylates 1 µmol creatine/min at pH 9.0, 25°C	Salt-free; lyophilized	<0.01% PK, LDH <0.02% ATPase	Biozyme CKP2
≥300 U/mg protein	1 unit transfers 1 µmol PO_4 from phosphocreatine to ADP/min at pH 6.9, 25°C	Lyophilized containing 95% protein		Calzyme 053A0300
15 U/mg solid	1 unit phosphorylates 1.0 µmol creatine/min at pH 9.0, 25°C	Salt-free; lyophilized		ICN 150714
Rabbit brain				
40-80 U/mg protein	1 unit transfers 1.0 µmol PO_4 from phosphocreatine to ADP/min at pH 7.4, 30°C	Lyophilized containing 55% protein; balance primarily Gly buffer salts, pH 9.0	<0.05% LDH, MK <0.1% ATPase, HK, PK	Sigma C6638

SPECIFIC ACTIVITY	UNITS DEFINITION	PREPARATION FORM	ADDITIONAL ACTIVITIES	SUPPLIER CATALOG NO.
Rabbit heart				
150-300 U/mg protein	1 unit transfers 1.0 µmol PO_4 from phosphocreatine to ADP/min at pH 7.4, 30°C	Lyophilized containing 90% protein; balance primarily Gly buffer salts	Contains MM isoenzyme	Sigma C3792
Rabbit muscle (optimum pH = 6.0-7.0 [reverse] or 9.0 [forward], MW = 81,000 Da)				
25 U/mg solid	1 unit phosphorylates 1 µmol creatine/min at pH 9.0, 25°C	Salt-free; lyophilized	<0.05% ATPase, LDH, MDH	Biozyme CK1
40 U/mg solid	1 unit phosphorylates 1 µmol creatine/min at pH 9.0, 25°C	Salt-free; lyophilized	<0.001% ATPase, LDH, MDH, PK, MK, HK	Biozyme CK2
350 U/mg solid at 25°C; 800 U/mg solid at 37°C	1 unit converts 1 µmol creatine PO_4 and ADP to product/min with N-Ac-L-Cys as activator	Lyophilized	<0.001% ATPase, HK, MK	Boehringer 126969 127566 736988
40 U/mg solid	1 unit transfers 1 µmol PO_4 from phosphocreatine to ADP/min at pH 6.9, 25°C	Lyophilized containing 90% protein		Calzyme 149A0040
14 U/mg	1 unit phosphorylates 1 µmol creatine/min at pH 9.0, 25°C	Lyophilized		Fluka 27957
20 U/mg	1 unit phosphorylates 1 µmol creatine/min at pH 9.0, 25°C	Salt-free; lyophilized		Fluka 27960
25 U/mg solid	1 unit phosphorylates 1.0 µmol creatine/min at pH 9.0, 20°C	Salt-free; lyophilized		ICN 100509
40 U/mg solid	1 unit phosphorylates 1.0 µmol creatine/min at pH 9.0, 20°C	Salt-free; lyophilized from 2X crystallized material		ICN 150715
≥250 IU/mg protein	1 IU transforms 1 µmol substrate/min under standard IUB conditions at 25°C	Suspension in $(NH_4)_2SO_4$	<0.001% ATPase <0.01% HK, MK, LDH, MDH, PK	OYC
150-250 U/mg protein	1 unit transfers 1.0 µmol PO_4 from phosphocreatine to ADP/min at pH 7.4, 30°C	Salt-free; lyophilized	<0.01% ATPase, LDH, HK, MK, PK	Sigma C3755
≥30 U/mg protein	1 unit converts 1 µmol creatine to creatine PO_4/min at pH 8.9, 25°C	Lyophilized containing 0.01 M Gly buffer, pH 9.0		Worthington LS01822 LS01824 LS01828
Rabbit skeletal muscle (optimum pH = 9.0 [forward] and 6.0-7.0 [reverse], MW = 81,000 Da)				
140 U/mg, creatine PO_4 as substrate; 25 U/mg, creatine as substrate	1 unit converts 1 µmol substrate/min at pH 6.8 or 8.9, 30°C	Salt-free; lyophilized	<0.01% ATPase, HK, MK	Calbiochem 2384

2.7.3.3 Arginine Kinase

REACTION CATALYZED
ATP + L-arginine ↔ ADP + N-phospho-L-arginine
SYSTEMATIC NAME
ATP:L-arginine N-phosphotransferase
REACTANTS
ATP + L-arginine, ADP + N-phospho-L-arginine

NOTES
- Classified as a transferase, transferring phosphorus-containing groups: phosphotransferases with a nitrogenous group as acceptor

SPECIFIC ACTIVITY	UNITS DEFINITION	PREPARATION FORM	ADDITIONAL ACTIVITIES	SUPPLIER CATALOG No.
Lobster tail muscle				
75-125 U/mg protein	1 unit converts 1.0 μmol L-Arg and ATP to N-ω-phospho-L-Arg and ADP/min at pH 8.6, 30°C	Suspension in 3.2 M $(NH_4)_2SO_4$, 0.01 M Gly buffer, 0.01 M β-MSH, 1 mM EDTA, pH 8.0		Sigma A3389

2.7.4.3 Adenylate Kinase

REACTION CATALYZED
ATP + AMP ↔ ADP + ADP
SYSTEMATIC NAME
ATP:AMP phosphotransferase
SYNONYMS
Myokinase, AdK, MK
SUBSTRATES
ATP, dATP, ADP, dADP, AMP, dAMP
APPLICATIONS
- Facile, selective synthesis of diadenosine polyphosphate, with leucyl t-RNA synthetase and ATP regeneration

NOTES
- Classified as a transferase, transferring phosphorus-containing groups: phosphotransferases with a phosphate group as acceptor
- Specific for substrates listed; does not react with any other mono-, di- or triphosphates

Adenylate Kinase continued — 2.7.4.3

SPECIFIC ACTIVITY	UNITS DEFINITION	PREPARATION FORM	ADDITIONAL ACTIVITIES	SUPPLIER CATALOG NO.
Bacillus stearothermophilus (optimum pH = 6.5, pI = 5.0, MW = 20,000 Da; K_M = 20 μM [AMP], 40 μM [ATP], 50 μM [ADP]; stable pH > 8.0, T < 65°C)				
200 U/mg protein	1 unit converts 2.0 μmol ADP to ATP and AMP/min at pH 7.6, 37°C	Lyophilized containing 60% protein; balance primarily Tris buffer salts, pH 8.5		Sigma M3266
>200 U/mg protein	1 unit forms 2 μmol ADP/min at 30°C	Lyophilized	<0.01% ATPase, PGK	Unitika
>200 U/mg protein	1 unit forms 2 μmol ADP/min at 30°C	50% glycerol solution	<0.01% ATPase, PGK	Unitika
Chicken muscle				
1500-3000 U/mg protein	1 unit converts 2.0 μmol ADP to ATP and AMP/min at pH 7.6, 37°C	Salt-free; lyophilized	<0.01% LDH, PK <0.1% 3-phosphoglyceric phosphokinase	Sigma M5520
Porcine muscle				
360 U/mg	1 unit converts 1 μmol AMP and ATP to product/min at 25°C	Suspension in 3.2 M (NH₄)₂SO₄ and 30 mM KPO₄, pH 7	<0.01% ATPase	Boehringer 107506
1500-3000 U/mg protein	1 unit converts 2.0 μmol ADP to ATP and AMP/min at pH 7.6, 37°C	Suspension in 3.2 M (NH₄)₂SO₄ and 0.001 M EDTA, pH 6.0	<0.01% LDH, PK <0.1% 3-phosphoglyceric phosphokinase	Sigma M3382
Rabbit muscle (MW = 21,000 Da)				
360 U/mg	1 unit converts 1 μmol AMP and ATP to product/min at 25°C	Suspension in 3.2 M (NH₄)₂SO₄ and 30 mM KPO₄, pH 7	<0.01% ATPase	Boehringer 127272 127949
360 U/mg protein	1 unit converts 2.0 mM ADP to ATP and AMP/min at pH 7.6, 25°C	Suspension in 3.2 M (NH₄)₂SO₄ containing 10 mg/mL protein, pH 6		Calzyme 202B0360
360 U/mg protein	1 unit converts 1 μmol AMP and ATP to ADP/min at pH 7.6, 25°C	Crystallized; suspension in 3.2 M (NH₄)₂SO₄, pH 6	<0.01% PGK	Fluka 70040
1000-1500 U/mg protein	1 unit converts 2.0 μmol ADP to ATP and AMP/min at pH 7.6, 37°C	Crystallized; suspension in 3.2 M (NH₄)₂SO₄	<0.03% PGK <0.1% LDH, GLDH	ICN 100866
1500-3000 U/mg protein	1 unit converts 2.0 μmol ADP to ATP and AMP/min at pH 7.6, 37°C	Suspension in 3.2 M (NH₄)₂SO₄ and 0.001 M EDTA, pH 6.1		Sigma M3003
Rabbit muscle subfragment 1				
ATPase: 1-4 U/mg protein	1 unit liberates 1.0 μmol Pi from ATP/min at pH 9.0, 24°C in the presence of Ca^{2+}	Produced from a chymotryptic digest of myosin in the presence of EDTA; 50% glycerol, 0.5 M KCl and 0.025 M KPO₄, pH 6.2		Sigma M5772
Yeast				
≥200 IU/mg protein	1 IU transforms 1 μmol substrate/min under standard IUB conditions at 25°C	50% glycerol solution	<0.01% ATPase <0.1% PGK	OYC

2.7.4.3 Adenylate Kinase continued

SPECIFIC ACTIVITY	UNITS DEFINITION	PREPARATION FORM	ADDITIONAL ACTIVITIES	SUPPLIER CATALOG NO.
Yeast, bakers				
120 U/mg protein	1 unit converts 2.0 mM ADP to ATP and AMP/min at pH 7.6, 25°C	Lyophilized containing 90% protein		Calzyme 201A0200

2.7.4.4 Nucleoside-Phosphate Kinase

REACTION CATALYZED
    ATP + nucleoside phosphate ↔ ADP + nucleoside diphosphate

SYSTEMATIC NAME
    ATP:nucleoside-phosphate phosphotransferase

SYNONYMS
    Nucleoside monophosphate kinase

REACTANTS
    ATP, nucleoside phosphate, ADP, nucleoside diphosphate

NOTES
- Classified as a transferase, transferring phosphorus-containing groups: phosphotransferases with a phosphate group as acceptor
- Many nucleotides act as acceptors
- Other nucleoside triphosphates may act in place of ATP

SPECIFIC ACTIVITY	UNITS DEFINITION	PREPARATION FORM	ADDITIONAL ACTIVITIES	SUPPLIER CATALOG NO.
Bovine liver				
1 U/mg protein	1 unit converts 1 μmol UMP and ATP to product/min at 25°C	Lyophilized containing 10 mg protein and 50 mg saccharose	<0.2% NADH oxidase <1% ATPase	Boehringer 107948
0.5 U/mg protein	1 unit converts 1.0 μmol UMP and ATP to UDP and ADP/min at pH 7.6, 25°C in a coupled system with PK/LDH	Lyophilized containing 80% saccharose		ICN 155962
0.5 U/mg protein	1 unit converts 1.0 μmol UMP and ATP to UDP and ADP/min at pH 7.6, 25°C in a coupled system with PK/LDH	Lyophilized containing 80% saccharose		Sigma N4379

Nucleoside-Diphosphate Kinase — 2.7.4.6

REACTION CATALYZED
   ATP + nucleoside diphosphate ↔ ADP + nucleoside triphosphate

SYSTEMATIC NAME
   ATP:nucleoside-diphosphate phosphotransferase

SYNONYMS
   Nucleoside 5'-diphosphate kinase

REACTANTS
   ATP, nucleoside diphosphate, ADP, nucleoside triphosphate

NOTES
- Classified as a transferase, transferring phosphorus-containing groups: phosphotransferases with a phosphate group as acceptor
- Many nucleoside diphosphates act as acceptors
- Many ribo- and deoxyribonucleoside triphosphates act as donors

SPECIFIC ACTIVITY	UNITS DEFINITION	PREPARATION FORM	ADDITIONAL ACTIVITIES	SUPPLIER CATALOG NO.
Bovine liver				
80-120 U/mg protein	1 unit converts 1.0 μmol each of TDP and ATP to TTP and ADP/min at pH 7.6, 25°C in a coupled system with PK/LDH	50% glycerol solution, pH 6		ICN 155961
1000 U/mg protein	1 unit converts 1.0 μmol TDP and ATP to TTP and ADP/min at pH 7.6, 25°C in a coupled system with PK/LDH	50% glycerol solution, pH 8	<0.1% LDH, nucleoside monophosphokinase, MK, ATPase, β-NADH oxidase	Sigma N2635
80-120 U/mg protein	1 unit converts 1.0 μmol TDP and ATP to TTP and ADP/min at pH 7.6, 25°C in a coupled system with PK/LDH	50% glycerol solution, pH 6; may contain BSA		Sigma N7130
Human erythrocytes				
400-800 U/mg protein	1 unit converts 1.0 μmol TDP and ATP to TTP and ADP/min at pH 7.6, 25°C in a coupled system with PK/LDH	50% glycerol solution, pH 8	<0.1% LDH, nucleoside monophosphokinase	Sigma N8646
Yeast, bakers				
1000-2000 U/mg protein	1 unit converts 1.0 μmol TDP and ATP to TTP and ADP/min at pH 7.6, 25°C in a coupled system with PK/LDH	Sulfate-free; lyophilized containing 20% protein; balance primarily Na citrate, traces of Mg^{2+} and EDTA salts		Sigma N0379

Guanylate Kinase

REACTION CATALYZED
    ATP + GMP ↔ ADP + GDP

SYSTEMATIC NAME
    ATP:(d)GMP phosphotransferase

SYNONYMS
    Deoxyguanylate kinase

REACTANTS
    ATP, dATP, GMP, dGMP, ADP, GDP

NOTES
- Classified as a transferase, transferring phosphorus-containing groups: phosphotransferases with a phosphate group as acceptor

SPECIFIC ACTIVITY	UNITS DEFINITION	PREPARATION FORM	ADDITIONAL ACTIVITIES	SUPPLIER CATALOG NO.
Bovine brain				
10-30 U/mg protein	1 unit converts 1.0 µmol each of GMP and ATP to GDP and ADP/min at pH 7.5, 30°C	Lyophilized containing 50% protein; balance primarily KPO_4 buffer salts	5% MK	Sigma G7510
20-60 U/mg protein	1 unit converts 1.0 µmol each of GMP and ATP to GDP and ADP/min at pH 7.5, 30°C	Solution containing 50% glycerol, 1 mM KPO_4, 0.1 mM EDTA, pH 7.0	5% MK	Sigma G9385

Ribose-Phosphate Pyrophosphokinase 2.7.6.1

REACTION CATALYZED
   ATP + D-ribose 5-phosphate ↔ AMP + 5'-phospho-α-D-ribose 1-diphosphate

SYSTEMATIC NAME
   ATP:D-ribose-5-phosphate pyrophosphotransferase

SYNONYMS
   Phosphoribosylpyrophosphate synthetase

REACTANTS
   ATP, dATP, D-ribose 5-phosphate, AMP, 5'-phospho-α-D-ribose 1-diphosphate

NOTES
- Classified as a transferase, transferring phosphorus-containing groups: diphosphotransferases

SPECIFIC ACTIVITY	UNITS DEFINITION	PREPARATION FORM	ADDITIONAL ACTIVITIES	SUPPLIER CATALOG NO.
Escherichia coli				
1-5 U/mg protein	1 unit forms 1.0 μmol AMP from ATP and ribose 5-phosphate/min at pH 7.6, 37°C	Solution containing 55% glycerol and 0.06 M NaPO$_4$, pH 7.6		Sigma P0287

Nicotinamide-Nucleotide Adenylyltransferase 2.7.7.1

REACTION CATALYZED
   ATP + nicotinamide ribonucleotide ↔ Pyrophosphate + NAD$^+$

SYSTEMATIC NAME
   ATP:nicotinamide-nucleotide adenylyltransferase

SYNONYMS
   NAD$^+$ pyrophosphorylase

REACTANTS
   ATP, nicotinamide ribonucleotide, nicotinate nucleotide, pyrophosphate, NAD$^+$

NOTES
- Classified as a transferase, transferring phosphorus-containing groups: nucleotidyltransferases

SPECIFIC ACTIVITY	UNITS DEFINITION	PREPARATION FORM	ADDITIONAL ACTIVITIES	SUPPLIER CATALOG NO.
Porcine liver				
1.5-3 U/vial	1 unit forms 1.0 μmol NAD$^+$ from nicotinamide mononucleotide and ATP/min at pH 7.4, 37°C	Lyophilized containing 30% protein; balance primarily sucrose		Sigma N5268

DNA-Directed RNA Polymerase

REACTION CATALYZED
Nucleoside triphosphate + RNA_n ↔ Pyrophosphate + RNA_{n+1}

SYSTEMATIC NAME
Nucleoside-triphosphate:RNA nucleotidyltransferase (DNA-directed)

SYNONYMS
RNA polymerase, RNA nucleotidyltransferase (DNA-directed), RNA polymerase I, RNA polymerase II, RNA polymerase III, poly(A) polymerase, T7 RNA polymerase

REACTANTS
Nucleoside triphosphate, RNA_n, pyrophosphate, RNA_{n+1}

APPLICATIONS
- T7 enzyme produces large amounts of specific RNA by transcribing DNA ligated to a promoter for T7 RNA polymerase. This RNA is useful as a hybridization probe, messenger for *in vitro* translation and as substrate for process reaction analysis

NOTES
- Classified as a transferase, transferring phosphorus-containing groups: nucleotidyltransferases
- Catalyzes DNA-template-directed extension of the 3'-end of an RNA strand one nucleotide at a time
- Can initiate a chain *de novo*
- Three enzyme forms have been distinguished in eukaryotes on the basis of sensitivity to α-amanitin and the type of RNA synthesized
- T7 enzyme is produced early in the T7 infection and has a stringent specificity for its own promoters. Unlike bacterial and eukaryotic enzyme, T7 enzyme consists of a single chain with molecular weight 100,000 Da. RNA synthesis proceeds in the 5'→3' direction, transcribing double-stranded DNA carrying the bacteriophage-specific promoters

SPECIFIC ACTIVITY	UNITS DEFINITION	PREPARATION FORM	ADDITIONAL ACTIVITIES	SUPPLIER CATALOG NO.
Escherichia coli				
200-1000 U/mg protein	1 unit incorporates 1 nmol AMP into acid-insoluble material/min at 37°C with poly(dA)·poly(dT) as a template	10 m*M* Tris-HCl, 100 m*M* NaCl, 0.1 m*M* EDTA, 0.1 m*M* DTT, 50% glycerol, pH 7.9	No detectable non-specific exonuclease, RNase	Amersham E 78022Y
	1 unit incorporates 1 nmol AMP into acid-insoluble form/10 min at 37°C using tRNA as substrate	25 m*M* Tris-HCl, 0.1 m*M* DTT, 0.01% Triton X-100, 1.0 m*M* EDTA, 0.5 *M* NaCl, 50% glycerol, pH 7.9	No detectable RNase	CHIMERX 1240-01 1240-02
1000-5000 U/mg protein; 500-1000 U/mL	1 unit incorporates 1 nmol AMP into acid-insoluble material/10 min at pH 7.9, 37°C	Solution containing 25 m*M* Tris-HCl, 500 m*M* NaCl, 1 m*M* EDTA, 0.1 m*M* DTT, 100 μg/mL BSA, 50% glycerol, pH 7.9	No detectable RNase, DNase	Pharmacia 27-0206-01 27-0206-02
200-1000 U/mg; 4000-10,000 U/mL	1 unit incorporates 1 nmol AMP into acid-insoluble product/10 min at pH 7.9, 37°C using poly(dA)·poly(dT) as template	Solution containing 10 m*M* Tris-HCl, 100 m*M* NaCl, 0.1 m*M* EDTA, 0.1 m*M* DTT, 50% glycerol, pH 7.9	No detectable RNase, DNase	Pharmacia 27-0916-01 27-0916-02

SPECIFIC ACTIVITY	UNITS DEFINITION	PREPARATION FORM	ADDITIONAL ACTIVITIES	SUPPLIER CATALOG NO.
Escherichia coli continued				
100-300 U/mg protein (ATP substrate)	1 unit incorporates 1.0 nmol labeled nucleoside triphosphate into an acid-insoluble product/10 min at pH 7.9, 37°C using calf thymus DNA as template	Solution containing 60% glycerol, 0.05-0.45 M $(NH_4)_2SO_4$, 50 mM Tris buffer, 2.0 mM DTT, 0.1 mM EDTA, 10 mM $MgCl_2$, pH 8.0	No detectable DNase, RNase	Sigma R5376
Escherichia coli strain B (optimum pH = 8.0, MW = 58,000 Da)				
1 U/µL	1 unit incorporates 1 nmol AMP into tRNA/10 min at pH 7.9, 37°C with ATP as substrate		No detectable RNase	Amersham E 2180Y
20 U and 100 U	1 unit incorporates 1 nmol AMP into tRNA/10 min at pH 7.9, 37°C with ATP as substrate	Solution containing 25 mM Tris-HCl, 0.1 mM DTT, 500 mM NaCl, 1 mM EDTA, 50% glycerol, pH 7.9	No detectable RNase	TaKaRa 2180
Escherichia coli strain BL21 holoenzyme				
1200 U/mg	1 unit incorporates 1 nmol radiolabeled ribonucleoside triphosphates into RNA/10 min at pH 7.9, 37°C	Solution containing 50% glycerol, 50 mM Tris-HCl, 250 mM NaCl, 0.1 mM EDTA, 1 mM DTT, pH 7.5	No detectable DNase, RNase	Epicentre S90050 S90100 S90250
Escherichia coli strain K-12				
750-3000 U/mg protein	1 unit releases 1.0 nmol PPi/10 min at pH 7.9, 37°C using calf thymus DNA as template	Affinity purified; solution containing 50% glycerol, 50 mM Tris-HCl, 50 mM KCl, 1.0 mM EDTA, 1.0 mM DTT, pH 7.9		Sigma R1510
Escherichia coli strain MRE 600				
1-5 U/µL	1 unit incorporates 1 nmol AMP into tRNA/10 min at pH 7.9, 37°C			Life Technol 18032-011 18032-029
20,000-40,000 U/mL	1 unit incorporates 1 nmol AMP in acid-precipitable RNA fraction/10 min at 37°C	10 mM Tris-HCl, 100 mM KCl, 10 mM β-MSH, 1 mM EDTA, 50% glycerol, pH 7.5		Boehringer 109304
Escherichia coli strain Q13				
≥200 U/mg protein	1 unit incorporates 1 nmol AMP into acid-insoluble product/10 min at 37°C in presence of calf thymus DNA as template	Highly purified containing sigma factor; solution containing 10 mM Tris-HCl, 1 mM DTT, 100 mM NaCl, 15% glycerol, pH 7.9	No detectable RNase	Worthington LM03003 LM03006 LM03000
Escherichia coli strain Ymal (MW = 450,000 Da [5 subunits])				
1-3 U/µL	1 unit incorporates 1 nmol NTP into TCA-insoluble products/10 min at pH 7.9, 37°C	>95% purity by SDS-PAGE; >50% σ factor content; 10 mM Tris-HCl, 100 mM NaCl, 0.1 mM EDTA, 0.1 mM DTT, 50% glycerol, pH 7.9	No detectable RNase, DNase	NBL Gene 020402 020404

2.7.7.6 DNA-Directed RNA Polymerase continued

SPECIFIC ACTIVITY	UNITS DEFINITION	PREPARATION FORM	ADDITIONAL ACTIVITIES	SUPPLIER CATALOG NO.
SP6 phage				
	1 unit incorporates 1 nmol labeled UTP into acid insoluble material/hr at 37°C	>95% purity by SDS-PAGE; 10 mM KPO$_4$, 0.2 M KCl, 0.1 mM EDTA, 1.0 mM DTT, 50% glycerol, pH 7.9	No detectable endonuclease, RNase, exonuclease	CHIMERx 1280-01 1280-02
				Oncogene XME09
SP6 phage, recombinant expressed in *E. coli* carrying plasmids with the gene for phage SP6 infecting *Salmonella typhimurium* LT2				
3000 U and 15,000 U	1 unit incorporates 1 nmol [^{3}H]GMP into acid-insoluble fraction/hr at pH 7.5, 37°C	>95% homogeneous by SDS-PAGE; solution containing 10 mM KPO$_4$, 1 mM DTT, 0.1 mM EDTA, 150 mM NaCl, 50% glycerol, pH 7.9	No detectable nuclease	TaKaRa 2520
SP6 phage-infected *E. coli*				
	1 unit incorporates 1 nmol labeled nucleoside triphosphates into acid-insoluble form/hr at pH 7.9, 37°C	10 mM KPO$_4$, 0.1 mM EDTA, 10 mM DTT, 50% glycerol, 250 µg/mL BSA, pH 7.9	No detectable 5'-exonuclease, non-specific endonuclease, RNase	Adv Biotech AB-0327 AB-0327b
100,000 U/mg; 2 U/µL	1 unit incorporates 1 nmol nucleoside triphosphate into acid-insoluble material/hr at pH 7.9	Highly purified; 100 mM NaCl, 50 mM Tris-HCl, 0.1 mM EDTA, 0.1% Triton X-100, 1 mM DTT, 50% glycerol, pH 7.9	No detectable non-specific endonuclease, exonuclease, RNase	Ambion 2071 2073
100,000 U/mg; 200 U/µL	1 unit incorporates 1 nmol nucleoside triphosphate into acid-insoluble material/hr at pH 7.9	Highly purified; 100 mM NaCl, 50 mM Tris-HCl, 0.1 mM EDTA, 0.1% Triton X-100, 1 mM DTT, 50% glycerol, pH 7.9	No detectable non-specific endonuclease, exonuclease, RNase	Ambion 2075
30 U/µL	1 unit incorporates 1 nmol [^{3}H]GMP into acid-insoluble RNA products/hr at 37°C	>95% homogeneous by SDS-PAGE; 10 mM KPO$_4$, 150 mM NaCl, 0.1 mM EDTA, 1 mM DTT, 50% glycerol, pH 7.9	No detectable non-specific endonuclease, exonuclease, RNase	Amersham E 2520Y
600,000 U/mg	1 unit converts 1 nmol labeled ribonucleoside triphosphate into acid-insoluble material/hr at pH 7.5, 37°C	Solution containing 50% glycerol, 50 mM Tris-HCl, 0.1 M NaCl, 0.1 mM EDTA, 1 mM DTT, 0.1% Triton X-100, pH 7.5	No detectable DNA exo- and endonuclease, *E. coli* RNA polymerase, phosphatase	Epicentre SP6 RNA Polymerase
10-20 U/µL; 40-80 U/µL; >100 U/µL	1 unit incorporates 1 nmol ribonucleotides into DE-81 absorbable form/hr at 37°C	Overexpressed; 50 mM Tris-HCl, 0.01 mM EDTA, 2 mM DTT, 150 mM NaCl, 0.5 mM ELUGENT detergent, 40% ethylene glycol, pH 8.0	No detectable endo-, exodeoxyribonucleases, RNases	Fermentas EP0131 EP0132 EP0133 EP0134
15 U/µL	1 unit incorporates 1 nmol ribonucleotide into acid-precipitable material/hr at pH 7.9, 37°C using an SP6 transcription vector as template	200 mM Tris-HCl, 30 mM MgCl$_2$, 10 mM spermidine, pH 7.9 and 10 mM DTT		Life Technol 18018-010 18018-069
40,000-60,000 U/mL	1 unit incorporates 1 nmol AMP into acid-insoluble product/hr at pH 7.9, 37°C using a plasmid containing SP6 promoter as template	Molecular biology grade; homogeneous purity; solution containing 10 mM KPO$_4$, 200 mM KCl, 0.1 mM EDTA, 5 mM DTT, 100 µg/mL BSA, 50% glycerol, pH 7.9	No detectable RNase, DNase, nickase	Pharmacia 27-0808-01 27-0808-02

DNA-Directed RNA Polymerase continued — 2.7.7.6

SPECIFIC ACTIVITY	UNITS DEFINITION	PREPARATION FORM	ADDITIONAL ACTIVITIES	SUPPLIER CATALOG NO.
SP6 phage-infected *E. coli* continued				
10-20 U/µL and 80 U/µL	1 unit incorporates 1 nmol nucleotide triphosphate into acid-insoluble product/hr at pH 7.9, 37°C	≥90% purity by SDS gel; 5X reaction buffer: 200 mM Tris-HCl, 50 mM NaCl, 30 mM $MgCl_2$, 100 mM DTT, 10 mM spermidine, pH 7.5	<1% DNase <3% RNase	Promega P1085 P1086 P4084
50,000 U/mL	1 unit incorporates 1 nmol nucleoside triphosphate into acid-insoluble form/hr at 37°C	Overproduced; 90% purity by SDS-PAGE; 20 mM KPO_4, 1 mM EDTA, 100 mM NaCl, 10 mM DTT, 50% glycerol, pH 7.7	No detectable DNase, RNase	Stratagene 600151 600152
SP6 phage-infected *Salmonella typhimurium*				
	1 unit incorporates 1.0 nmol ATP into acid-insoluble form/hr at 37°C	Solution containing 10 mM NaCl, 50 mM Tris-HCl, 1 mM EDTA, 20 mM β-MSH, 50% glycerol, pH 7.9		ICN 800709 800710
10,000-50,000 U/mL	1 unit incorporates 1.0 nmol rNTP into acid-precipitable material/hr at 37°C	50% glycerol, 100 mM NaCl, 50 mM Tris-HCl, 0.1 mM EDTA, 0.1% Triton X-100, 1 mM DTT, pH 7.9	No detectable DNase, RNase, protease	Sigma R0759
SP6 phage-infected *Salmonella typhimurium* LT2				
10-20 U/µL	1 unit incorporates 1 nmol GTP into TCA-insoluble RNA products/hr at pH 7.2, 37°C	>90% purity by SDS-PAGE; 10 mM KPO_4, 200 mM KCl, 0.1 mM EDTA, 10 mM β-MSH, 50% glycerol, 100 µg/mL BSA, pH 7.9	No detectable RNase, DNase	NBL Gene 020806 020803
SP6 phage-infected *E. coli* BL21pSP3/BL21 *Salmonella typhimurium* LT2				
10,000-20,000 U/mL	1 unit incorporates 1 nmol CMP in acid-precipitable RNA products/hr at 37°C	10 mM KPO_4, 200 mM KCl, 0.1 mM EDTA, 10 mM β-MSH, 100 µg/mL BSA, 50% glycerol, pH 7.9		Boehringer 810274 1487671
SP6 phage-infected *Salmonella typhimurium* LT2				
20,000 U/mL	1 unit incorporates 1 nmol GMP into acid-precipitable RNA/hr at pH 7.9, 37°C	50 mM Tris-HCl, 10 mM β-MSH, 100 mM KCl, 0.1 mM EDTA, 50% glycerol, pH 7.9	No detectable RNase	AGS Heidelb A00780S A00780M
50,000 U/mL	1 unit incorporates 1 nmol ATP into acid-insoluble form/hr at 37°C	100 mM NaCl, 50 mM Tris-HCl, 1 mM EDTA, 20 mM β-MSH, 0.1% Triton X-100, 50% glycerol, pH 7.9	No detectable RNA polymerase, DNase, RNase	NE Biolabs 207S 207L
T3 phage-infected *E. coli*				
	1 unit incorporates 1 nmol labeled nucleoside triphosphates into acid-insoluble form/hr at pH 7.9, 37°C	50 mM Tris-HCl, 0.1 mM NaCl, 0.1 mM EDTA, 1 mM DTT, 0.1% Triton X-100, 50% glycerol, pH 7.9	No detectable 5'-exonuclease, non-specific endonuclease, RNase	Adv Biotech AB-0328 AB-0328b
100,000 U/mg; 2 U/µL	1 unit incorporates 1 nmol nucleoside triphosphate into acid-insoluble material/hr at pH 7.9	100 mM NaCl, 50 mM Tris-HCl, 0.1 mM EDTA, 0.1% Triton X-100, 1 mM DTT, 50% glycerol, pH 7.9	No detectable non-specific endonuclease, exonuclease, RNase	Ambion 2060 2062
100,000 U/mg; 200 U/µL	1 unit incorporates 1 nmol nucleoside triphosphate into acid-insoluble material/hr at pH 7.9	Highly purified; 100 mM NaCl, 50 mM Tris-HCl, 0.1 mM EDTA, 0.1% Triton X-100, 1 mM DTT, 50% glycerol, pH 7.9	No detectable non-specific endonuclease, exonuclease, RNase	Ambion 2063

T3 phage-infected E. coli continued

SPECIFIC ACTIVITY	UNITS DEFINITION	PREPARATION FORM	ADDITIONAL ACTIVITIES	SUPPLIER CATALOG NO.
	1 unit incorporates 1 nmol nucleoside triphosphate into acid-insoluble material/hr at 37°C using a T3 transcription vector as template	Overproduced; 20 mM KPO$_4$, 100 mM NaCl, 1.0 mM EDTA, 1.0 mM DTT, 100 µg/mL BSA, 50% glycerol, pH 7.5	No detectable non-specific nuclease, endonuclease, RNase	Amersham E 70051Y E 70051Z
20-100 U/µL	1 unit incorporates 1 nmol ribonucleoside monophosphate into DE-81 absorbable form/hr at 37°C	Overexpressed; 10 mM Tris-HCl, 2.5 mM MgCl$_2$, 0.1 mM EDTA, 2 mM DTT, 50% glycerol, pH 8.0	No detectable unspecific endonuclease, nickase, RNase, host polymerase	AGS Heidelb F00787S F00787M
1,000,000 U/mg	1 unit converts 1 nmol labeled ribonucleoside triphosphate into acid-insoluble material/hr at pH 7.5, 37°C	Solution containing 50% glycerol, 50 mM Tris-HCl, 0.1 M NaCl, 0.1 mM EDTA, 1 mM DTT, 0.1% Triton X-100, pH 7.5	No detectable DNA exo- and endonuclease, E. coli RNA polymerase, phosphatase	Epicentre T3 RNA Polymerase
10-20 U/µL; 40-80 U/µL; >100 U/µL	1 unit incorporates 1 nmol ribonucleotides into DE-81 absorbable form/hr at 37°C	Overexpressed; 50 mM Tris-HCl, 0.01 mM EDTA, 2 mM DTT, 150 mM NaCl, 0.5 mM ELUGENT detergent, 40% ethylene glycol, pH 8.0	No detectable endo-, exodeoxyribonucleases, RNases	Fermentas EP0101 EP0102 EP0103 EP0104
50 U/µL	1 unit incorporates 1 nmol ribonucleotide into acid-precipitable material/hr at pH 8.0, 37°C using a T3 transcription vector as template primer	200 mM Tris-HCl, 40 mM MgCl$_2$, 10 mM spermidine, pH 8.0 and 10 mM DTT		Life Technol 18036-012 18036-046
50 U/µL	1 unit incorporates 1 nmol ribonucleoside into acid-precipitable RNA/hr at pH 7.5, 37°C		No detectable E. coli RNA polymerase, RNase, phosphatase, endonuclease, exonuclease, nickase	NBL Gene 021807 021808
40,000 U/mL	1 unit incorporates 1 nmol ribonucleotide into DE-81 adsorbable form/hr at 37°C	Overproduced; 50 mM Tris-HCl, 2.5 mM MgCl$_2$, 2 mM DTT, 0.1 mM EDTA, 50% glycerol, pH 8.0	No detectable RNase, phosphatase, endo- and exonuclease	Oncor 120191 120192
40,000-60,000 U/mL	1 unit incorporates 1 nmol AMP into acid-insoluble product/hr at pH 8.0, 37°C using a plasmid containing T3 promoter as template	Molecular biology grade; homogeneous purity; solution containing 20 mM KPO$_4$, 100 mM NaCl, 1.0 mM EDTA, 10 mM DTT, 0.1% Triton X-100, 50% glycerol, pH 7.7	No detectable RNase, DNase, nickase	Pharmacia 27-0802-02 27-0802-03
10-20 U/µL and 80 U/µL	1 unit incorporates 1 nmol nucleotide triphosphate into acid-insoluble product/hr at pH 7.9, 37°C	≥90% purity by SDS gel; 5X reaction buffer: 200 mM Tris-HCl, 50 mM NaCl, 30 mM MgCl$_2$, 100 mM DTT, 10 mM spermidine, pH 7.5	<1% DNase <3% RNase	Promega P2083 P2084 P4024
10,000-50,000 U/mL	1 unit incorporates 1.0 nmol rNTP into acid-precipitable material/hr at 37°C	50% glycerol, 100 mM NaCl, 50 mM Tris-HCl, 0.1 mM EDTA, 0.1% Triton X-100, 1 mM DTT, pH 7.9	No detectable DNase, RNase, protease	Sigma R1009
50,000 U/mL	1 unit incorporates 1.0 nmol nucleoside triphosphate into acid-insoluble form/hr at 37°C	Overproduced; 99% purity by SDS-PAGE; 20 mM KPO$_4$, 1 mM EDTA, 100 mM NaCl, 10 mM DTT, 50% glycerol, pH 7.7	No detectable DNase, RNase	Stratagene 600111 600112

DNA-Directed RNA Polymerase continued — 2.7.7.6

SPECIFIC ACTIVITY	UNITS DEFINITION	PREPARATION FORM	ADDITIONAL ACTIVITIES	SUPPLIER CATALOG NO.
T3 phage-infected *E. coli* HB 101/pCM 56				
20,000–40,000 U/mL	1 unit incorporates 1 nmol CMP in acid-precipitable RNA products/hr at 37°C	10 mM KPO$_4$, 200 mM KCl, 0.1 mM EDTA, 10 mM β-MSH, 100 μg/mL BSA, 50% glycerol, pH 7.9		Boehringer 1031163 1031171
T4 phage-infected *E. coli*				
20,000 U/mg protein	1 unit incorporates 10 nmol nucleotide into DE-81 adsorbable form/30 min at 37°C with calf thymus DNA as substrate	200 mM KPO$_4$, 2 mM DTT, 50% glycerol, pH 6.5		Oncor 120041 120042
T7 phage				
	1 unit incorporates 1 nmol labeled UTP into acid-insoluble material/hr at 37°C	>95% purity by SDS-PAGE; 20 mM KPO$_4$, 0.1 M NaCl, 1.0 mM EDTA, 1.0 mM DTT, 50% glycerol, 100 μg/mL BSA, pH 7.5	No detectable endonuclease, RNase, exonuclease	CHIMERX 1290-01 1290-02
	1 unit incorporates 1 nmol labeled UTP into acid-insoluble material/hr at 37°C	Solution containing 0.1 M NaCl, 50 mM Tris-HCl, 0.1 mM EDTA, 1.0 mM DTT, 50% glycerol, pH 7.9		ICN 152031
				Oncogene XME11
T7 phage-infected *E. coli* BL21/pAR1219				
50,000 U/mL	1 unit incorporates 1 nmol ATP into acid-insoluble form/hr at 37°C	100 mM NaCl, 50 mM Tris-HCl, 1 mM EDTA, 20 mM β-MSH, 0.1% Triton X-100, 50% glycerol, pH 7.9	No detectable RNA polymerase, DNase, RNase	NE Biolabs 251S 251L
T7 phage-infected *E. coli*, overproduced				
	1 unit incorporates 1 nmol labeled nucleoside triphosphates into acid-insoluble form/hr at pH 7.9, 37°C		No detectable 5'-exonuclease, non-specific endonuclease, RNase	Adv Biotech AB-0329 AB-0329b
100,000 U/mg; 2 U/μL	1 unit incorporates 1 nmol nucleoside triphosphate into acid-insoluble material/hr at pH 7.9	Highly purified; 100 mM NaCl, 50 mM Tris-HCl, 0.1 mM EDTA, 0.1% Triton X-100, 1 mM DTT, 50% glycerol, pH 7.9	No detectable non-specific endonuclease, exonuclease, RNase	Ambion 2080 2084
50,000 U/mL	1 unit incorporates 1 nmol GMP into acid-precipitable RNA/hr at pH 7.9, 37°C	50 mM Tris-HCl, 0.1 mM EDTA, 10 mM β-MSH, 100 mM KCl, 50% glycerol, pH 7.9	No contaminating nuclease	AGS Heidelb A00840S A00840M
100,000 U/mg; 200 U/μL	1 unit incorporates 1 nmol nucleoside triphosphate into acid-insoluble material/hr at pH 7.9	Highly purified; 100 mM NaCl, 50 mM Tris-HCl, 0.1 mM EDTA, 0.1% Triton X-100, 1 mM DTT, 50% glycerol, pH 7.9	No detectable non-specific endonuclease, exonuclease, RNase	Ambion 2085
>80 U/μL	1 unit incorporates 1 nmol labeled nucleoside triphosphate into acid-insoluble material/hr at 37°C	>95% purity by SDS-PAGE; 20 mM KPO$_4$, 0.1 mM EDTA, 0.1 mM DTT, 50% glycerol, pH 7.5	No detectable non-specific endonuclease, exonuclease, RNase	Amersham E 70001Y E 70001Z
20 U/μL	1 unit incorporates 1 nmol labeled nucleoside triphosphate into acid-insoluble material/hr at 37°C	>95% purity by SDS-PAGE; 20 mM KPO$_4$, 0.1 mM EDTA, 0.1 mM DTT, 50% glycerol, pH 7.5	No detectable non-specific endonuclease, exonuclease, RNase	Amersham E 70047Y E 70047Z

2.7.7.6 DNA-Directed RNA Polymerase continued

SPECIFIC ACTIVITY	UNITS DEFINITION	PREPARATION FORM	ADDITIONAL ACTIVITIES	SUPPLIER CATALOG NO.
T7 phage-infected *E. coli*, overproduced *continued*				
500,000 U/mg	1 unit converts 1 nmol labeled ribonucleoside triphosphate into acid-insoluble material/hr at pH 7.5, 37°C	Solution containing 50% glycerol, 50 mM Tris-HCl, 0.1 M NaCl, 0.1 mM EDTA, 1 mM DTT, 0.1% Triton X-100, pH 7.5	No detectable DNA exo- and endonuclease, *E. coli* RNA polymerase, phosphatase	Epicentre T7 RNA Polymerase
10-20 U/µL; 40-80 U/µL; >100 U/µL	1 unit incorporates 1 nmol ribonucleotides into DE-81 absorbable form/hr at 37°C	50 mM Tris-HCl, 0.01 mM EDTA, 2 mM DTT, 150 mM NaCl, 0.5 mM ELUGENT detergent, 40% ethylene glycol, pH 8.0	No detectable endo-, exodeoxyribonucleases, RNases	Fermentas EP0111 EP0112 EP0113 EP0114
	1 unit incorporates 1 nmol labeled ribonucleotide into acid-precipitable material/hr at 37°C	Solution containing 20 mM KPO$_4$, 0.1 mM DTT, 0.1 mM EDTA, 50% glycerol, pH 7.5		ICN 800711
50 U/µL	1 unit hydrolyzes 1 nmol ribonucleotide into acid-precipitable material/hr at pH 8.0, 37°C using T7 transcription vector as template	200 mM Tris-HCl, 40 mM MgCl$_2$, 10 mM spermidine, 125 mM NaCl, pH 8.0 and 10 mM DTT		Life Technol 18033-019 18033-100
50 U/µL	1 unit incorporates 1 nmol ribonucleoside triphosphate into acid-precipitable RNA/hr at pH 7.5, 37°C	50 mM Tris-HCl, 100 mM NaCl, 0.1 mM EDTA, 0.1% Triton X-100, 1 mM DTT, 50% glycerol, pH 7.9	No detectable *E. coli* RNA polymerase, RNase, phosphatase, endonuclease, exonuclease, nickase	NBL Gene 021907 021908
	1 unit incorporates 1 nmol ribonucleotide in an acid-insoluble form/hr at 37°C	50 mM Tris-HCl, 100 mM KCl, 10 mM β-MSH, 0.1 mM EDTA, pH 8		Oncor 120161 120162
40,000-70,000 U/mL and 10,000-15,000 U/mL	1 unit incorporates 1 nmol AMP into acid-insoluble product/hr at pH 8.0, 37°C using a 3-kb plasmid containing T7 promoter as template	Molecular biology grade; homogeneous purity; solution containing 20 mM KPO$_4$, 100 mM NaCl, 1 mM EDTA, 1 mM DTT, 100 µg/mL BSA, 50% glycerol, pH 7.5	No detectable RNase, DNase, nickase	Pharmacia 27-0801-01 27-0801-02 27-0805-01
10-20 U/µL and 80 U/µL	1 unit incorporates 1 nmol nucleotide triphosphate into acid-insoluble product/hr at pH 7.9, 37°C	≥90% purity by SDS gel; 5X reaction buffer: 200 mM Tris-HCl, 50 mM NaCl, 30 mM MgCl$_2$, 100 mM DTT, 10 mM spermidine, pH 7.5	<1% DNase <3% RNase	Promega P2075 P2076 P4074
10,000-50,000 U/mL	1 unit incorporates 1.0 nmol rNTP into acid-precipitable material/hr at 37°C	50% glycerol containing 100 mM NaCl, 50 mM Tris-HCl, 0.1 mM EDTA, 0.1% Triton X-100, 1 mM DTT, pH 7.9	No detectable DNase, RNase, protease	Sigma R0884
50,000 U/mL	1 unit incorporates 1.0 nmol nucleoside triphosphate into acid-insoluble form/hr at 37°C	99% purity by SDS-PAGE; 20 mM KPO$_4$, 1 mM EDTA, 100 mM NaCl, 10 mM DTT, 50% glycerol, pH 7.7	No detectable DNase, RNase	Stratagene 600123 600124
5000 U and 25,000 U	1 unit incorporates 1 nmol [^{3}H]GMP into acid-insoluble fraction/hr at pH 8.0, 37°C	>95% homogeneous by SDS-PAGE; solution containing 20 mM KPO$_4$, 1 mM DTT, 0.1 mM EDTA, 100 mM NaCl, 50% glycerol, pH 7.9	No detectable nuclease	TaKaRa 2540

DNA-Directed RNA Polymerase continued

2.7.7.6

SPECIFIC ACTIVITY	UNITS DEFINITION	PREPARATION FORM	ADDITIONAL ACTIVITIES	SUPPLIER CATALOG No.
T7 phage-infected *E. coli*, overproduced *continued*				
≥8000 U/mg protein	1 unit incorporates 1 nmol labeled nucleoside monophosphate into acid-insoluble product/hr at 37°C	Highly purified; solution containing 50 mM KPO$_4$, 1 mM DTT, 20% glycerol, pH 7.4		Worthington LM03010 LM03012 LM03014
T7 phage-infected *E. coli* HMS 174/pAR 1219, overproduced				
20,000–40,000 U/mL	1 unit incorporates 1 nmol CMP in acid-precipitable RNA products/hr at 37°C	10 mM KPO$_4$, 200 mM KCl, 0.1 mM EDTA, 10 mM β-MSH, 100 µg/mL BSA, 50% glycerol, pH 7.9		Boehringer 881767 881775
Thermophilic bacterium				
	1 unit incorporates 1 nmol radiolabeled ribonucleotide triphosphate into RNA/hr at pH 8.2, 65°C	Solution containing 50% glycerol, 50 mM Tris-HCl, 0.1 M NaCl, 0.1 mM EDTA, 1 mM DTT, 0.1% Triton X-100, pH 7.5	No detectable DNA exo- and endonuclease, RNase, protease	Epicentre T90050 T90100 T90250
Wheat germ				
5–20 U/mg protein	1 unit releases 1.0 nmol PPi/10 min at pH 7.9, 37°C using heat-denatured calf thymus DNA as template	Solution containing 50% glycerol, 50 mM Tris-HCl, 50 mM KCl, 1.0 mM EDTA, 1.0 mM DTT, pH 7.9		Sigma R1635
Yeast				
600 U/µL	1 unit incorporates 1 pmol AMP into acid-insoluble material/min at 30°C	20 mM Tris-HCl, 50 mM KCl, 0.5 mM DTT, 50% glycerol, pH 8.0	No detectable RNase	Amersham E 74225Y E 74225Z

DNA-Directed DNA Polymerase

2.7.7.7

REACTION CATALYZED

Deoxynucleoside triphosphate + DNA$_n$ ↔ Pyrophosphate + DNA$_{n+1}$

SYSTEMATIC NAME

Deoxynucleoside-triphosphate:DNA deoxynucleotidyltransferase (DNA-directed)

SYNONYMS

DNA polymerase I, II, III, DNA polymerase α, β, γ, DNA nucleotidyltransferase (DNA-directed)

SUBSTRATES

Deoxynucleoside triphosphate, DNA$_n$

PRODUCTS

Pyrophosphate, DNA$_{n+1}$

APPLICATIONS

- High percentage incorporation of radioactivity for nick translation assays

2.7.7.7 DNA-Directed DNA Polymerase continued

NOTES

- Classified as a transferase, transferring phosphorus-containing groups: nucleotidyltransferases
- In the presence of template and primer DNA, transfers dNMPs from dNTPs to the 3'-OH terminus of primer complementary to the template. Contains integral activities: 3'→5' exonuclease specific to single-stranded DNA and 5'→3' exonuclease activity specific to double-stranded DNA. Cannot initiate a chain *de novo*. Requires a DNA or RNA primer. Molecular weight is 109,000 Da
- The Klenow fragment of DNA polymerase transfers dNMPs from dNTPs to the 3'-OH termini of DNA primer that is complementary to the template. It is prepared by subtilisin proteolysis of native DNA polymerase I. It retains the 3'→5' polymerase and 3'→5' exonuclease activities of native enzyme, but lacks 5'→3' exonuclease activity. Molecular weight is 62,000 Da
- T4 DNA polymerase is a single chain of molecular weight 114,000 Da. It polymerizes dNTPs in the 5'→3' direction and requires DNA template and primer. T4 enzyme lacks the 5'→3' exonuclease activity, but possesses a potent 3'→5' exonuclease. It has a strong preference for single-stranded DNA
- T7 DNA polymerase is comprised of two subunits: the catalytic T7 gene 5 protein (molecular weight 85,000 Da) and thioredoxin, required to keep the enzyme attached to the template (molecular weight 12,000 Da). In the presence of DNA template and primer, it polymerizes dNTPs in the 5'→3' direction. *In vivo*, T7 DNA polymerase replicates T7 phage DNA during infection. T7 enzyme lacks the 5'→3' exonuclease, but possesses potent 5'→3' polymerase and 3'→5' exonuclease activities
- Taq DNA polymerase (from *Thermus aquaticus*) consists of a single chain of molecular weight 95,000 Da. In the presence of DNA template and primer, it polymerizes dNTPs in the 5'→3' direction. The enzyme is thermostable, with optimal activity at 75°C
- Tbr DNA polymerase (from *Thermus brockianus*) has an error frequency 5X lower than Taq DNA polymerase

SPECIFIC ACTIVITY	UNITS DEFINITION	PREPARATION FORM	ADDITIONAL ACTIVITIES	SUPPLIER CATALOG NO.
Bacillus stearothermophilus (Bst DNA polymerase) (inactivated at 75°C, 15 min)				
	1 unit incorporates 10 nmol total nucleotide into acid-insoluble form/30 min at 60°C	20 mM KPO$_4$, 1.0 mM DTT, 50% glycerol, pH 6.8	No detectable endonuclease, 3'-exonuclease, 5'-exonuclease/5'-phosphatase, nonspecific RNase, ss- and ds-DNase	CHIMERX 1078-01 1078-02

SPECIFIC ACTIVITY	UNITS DEFINITION	PREPARATION FORM	ADDITIONAL ACTIVITIES	SUPPLIER CATALOG NO.
Bacillus stearothermophilus (Bst DNA polymerase) continued				
600,000 U/mg	1 unit converts 10 nmol deoxyribonucleoside triphosphates into acid-insoluble form/30 min at pH 9.3, 65°C	Solution containing 50% glycerol, 50 mM Tris-HCl, 0.1 M NaCl, 0.1 mM EDTA, 1 mM DTT, 0.1% Triton X-100, pH 7.5	No detectable endonuclease, RNase, protease	Epicentre IsoTherm™ BL9100 BL9500 BL901K
Calf thymus (DNA polymerase-α)				
	1 unit incorporates 1 nmol total nucleotide into acid-insoluble form/hr at 37°C	20 mM Tris-HCl, 1.0 mM DTT, 0.25 mM EDTA, 50 mM KCl, 50% glycerol, pH 8.5	No detectable endo- and exonuclease	CHIMERx 1074-01 1074-02
Escherichia coli (optimum pH = 7.5, MW = 75,000 Da (Klenow fragment); inactivated at 75°C, 15 min)				
7813 U/mg protein; 1.0 U/μL	1 unit incorporates 10 nmol deoxynucleoside triphosphate into acid-insoluble product/30 min at 37°C with poly (dA-dT) as template	Klenow Fragment; 50 mM KPO$_4$, 1 mM DTT, 50% glycerol, pH 7.0		5'→3' 5306-332565
	1 unit incorporates 10 nmol deoxynucleotides into acid-insoluble form/30 min at pH 7.5, 37°C	Recombinant; 25 mM Tris-HCl, 0.1 mM EDTA, 1 mM DTT, 5% glycerol, pH 7.5	No detectable non-specific endonuclease	Adv Biotech AB-0331 AB-0331b
	1 unit incorporates 10 nmol total deoxyribonucleotide into acid-insoluble form/30 min at 37°C with DNase I-activated DNA as template primer	>95% purity by SDS-PAGE; 50 mM KPO$_4$, 0.25 mM DTT, 50% glycerol, pH 7.0	No detectable endonuclease	CHIMERx 1080-01 1080-02
	1 unit incorporates 10 nmol total nucleotide into acid-insoluble form/30 min at 37°C	Klenow Fragment; >95% purity by SDS-PAGE; 50 mM KPO$_4$, 0.25 mM DTT, 50% glycerol, pH 7.0	No detectable endonuclease, 5'-exonuclease	CHIMERx 1091-01 1091-02
	1 unit incorporates 10 nmol total nucleotide into acid-insoluble form/30 min at 37°C	Klenow Fragment; 50 mM KPO$_4$, 0.25 mM DTT, 50% glycerol, pH 7.0	No detectable endonuclease, 3'- and 5'-exonuclease, nonspecific ss- and ds-DNase	CHIMERx 1092-01 1092-02
150,000 U/mg	1 unit converts 10 nmol deoxyribonucleoside triphosphates into acid-insoluble form/30 min at pH 9.3, 65°C	rBst DNA Polymerase; recombinant; solution containing 50% glycerol, 50 mM Tris-HCl, 0.1 M NaCl, 0.1 mM EDTA, 1 mM DTT, 0.1% Triton X-100, pH 7.5	No detectable endonuclease, RNase, protease	Epicentre BH1100 BH1500 BH101K
5-15 U/μL	1 unit incorporates 10 nmol deoxyribonucleotides into DE-81 absorbable form/30 min at 37°C	Overproducer; 50 mM KPO$_4$, 1 mM DTT, 50% glycerol, pH 7.4	No detectable endodeoxyribonucleases	Fermentas EP0041 EP0042
1-2 U/μL; 5-15 U/μL	1 unit incorporates 10 nmol deoxyribonucleotides into DE-81 absorbable form/30 min at 37°C	Klenow Fragment; overproducer; 50 mM KPO$_4$, 1 mM DTT, 50% glycerol, pH 7.4	No detectable endo- and 5'→3' exodeoxyribonucleases	Fermentas EP0054 EP0051 EP0052

SPECIFIC ACTIVITY	UNITS DEFINITION	PREPARATION FORM	ADDITIONAL ACTIVITIES	SUPPLIER CATALOG NO.
Escherichia coli continued				
3-10 U/μL	1 unit incorporates 10 nmol deoxyribonucleotides into DE-81 absorbable form/30 min at 37°C	T4 DNA Polymerase; overproducer; 20 mM KPO$_4$, 200 mM KCl, 2 mM DTT, 50% glycerol, pH 7.5	No detectable endodeoxyribonucleases	Fermentas EP0061 EP0062
5-20 U/μL	1 unit incorporates 10 nmol deoxyribonucleotides into DE-81 absorbable form/30 min at 37°C	T7 DNA Polymerase; overproducer; 20 mM KPO$_4$, 0.1 mM EDTA, 1 mM DTT, 50% glycerol, pH 7.4	No detectable endodeoxyribonucleases	Fermentas EP0081 EP0082
4-5 U/μL; 1 U/μL	1 unit incorporates 10 nmol deoxyribonucleotides into DE-81 absorbable form/30 min at 70°C	Taq DNA Polymerase; overproducer; 20 mM Tris-HCl, 100 mM KCl, 0.5% NP40, 1 mM DTT, 0.1 mM EDTA, 0.5% Tween 20, 50% glycerol, pH 8.0	No detectable endo-, exodeoxyribonucleases, RNases	Fermentas EP0401 EP0402 EP0403 EP0404
	1 unit incorporates 10 nmol deoxyribonucleotide into acid-insoluble material/30 min at 37°C with DNase I-activated DNA as template primer	Solution containing 50 mM KPO$_4$, 0.025 mM DTT, 50% glycerol, pH 7.0		ICN 151008
	1 unit incorporates 10 nmol total nucleotide into acid-insoluble form/30 min at 37°C	Klenow Fragment; solution containing 50 mM KPO$_4$ buffer, 0.025 mM DTT, 50% glycerol, pH 7.5		ICN 151009
	1 unit converts 10 nmol deoxyribonucleotides to acid-insoluble form/30 min at 37°C	Klenow Fragment; recombinant; >99% purity by SDS-PAGE; 0.1 M KPO$_4$, 1 mM β-MSH, 50% glycerol, pH 6.5	No detectable 5'→3' exonuclease, endonuclease	MINOTECH 205
5-10 U/μL	1 unit incorporates 10 nmol deoxyribonucleotides into TCA-insoluble form/30 min at pH 7.5, 37°C	Recombinant; >90% purity by SDS gel; 10X reaction buffer: 500 mM Tris-HCl, 100 mM MgSO$_4$, 1 mM DTT, pH 7.2	≥90% supercoiled plasmid	Promega M2051 M2052
5-10 U/μL	1 unit incorporates 10 nmol total deoxyribonucleotides into TCA-insoluble form/30 min at pH 7.5, 37°C	Klenow Fragment; recombinant; ≥90% purity by SDS gel; 10X Reaction buffer: 500 mM Tris-HCl, 100 mM MgSO$_4$, 1 mM DTT, pH 7.2	≥90% supercoiled plasmid	Promega M2201 M2202 M2205
5-10 U/μL	1 unit incorporates 10 nmol total nucleotide into acid-insoluble form/30 min at pH 8.8, 37°C	T4 DNA Polymerase; recombinant; ≥90% purity by SDS gel	≥90% supercoiled plasmid	Promega M4211 M4212
1000-10,000 U/mL	1 unit converts 10 nmol total deoxyribonucleotides into acid-insoluble form/30 min at 37°C	Klenow Fragment; overproducer, recombinant; >99% purity by SDS-PAGE; 50 mM KPO$_4$, 1 mM DTT, 50% glycerol, pH 7.0	No detectable endonuclease	Stratagene 600071
Escherichia coli BNN 50-594, a lysogen carrying pol A transducing phage (MW = 109,000 Da)				
5-15 U/μL	1 unit converts 10 nmol total deoxyribonucleotides to acid-insoluble form/30 min at 37°C using poly [d(A-T)] as template primer	>95% purity by SDS-PAGE; 50 mM KPO$_4$, 1.0 mM DTT, 50% glycerol, pH 7.0	No detectable ss- and ds-DNA endonuclease	Amersham E 70010Y

DNA-Directed DNA Polymerase continued   2.7.7.7

SPECIFIC ACTIVITY	UNITS DEFINITION	PREPARATION FORM	ADDITIONAL ACTIVITIES	SUPPLIER CATALOG No.
Escherichia coli BNN 50-594 *continued*				
5-10 U/μL	1 unit incorporates 10 nmol dNTP into a TCA-insoluble form/30 min at pH 7.5, 37°C	>95% purity by SDS-PAGE; 100 mM KPO$_4$, 1 mM DTT, 50% glycerol, pH 6.5	No detectable endonuclease, nickase	NBL Gene 020203 020205
Escherichia coli CJ 375				
≥2000 U/mg protein		Klenow Fragment; purified; solution containing 50 mM KPO$_4$, 1 mM DTT, 20% glycerol, pH 7.4	No detectable RNase, DNase, endonuclease, exonuclease	Worthington LM01070 LM01072 LM01074
Escherichia coli CM 5197				
4000-10,000 U/mg protein; 3000-10,000 U/mL	1 unit incorporates 10 nmol total deoxynucleotide into acid-insoluble product/30 min at pH 7.4, 37°C with poly(dA-dT) as template	Solution containing 50 mM KPO$_4$, 0.025 mM DTT, 50% glycerol, pH 7.0	No detectable nickase activity	Pharmacia 27-0926-01 27-0926-02
5000-10,000 U/mg protein; 5000-10,000 U/mL and 1000 U/mL	1 unit incorporates 10 nmol total deoxynucleotide into acid-insoluble product/30 min at pH 7.4, 37°C with poly(dA-dT) as template	Klenow Fragment; molecular biology grade; homogeneous purity; solution containing 50 mM KPO$_4$, 1.0 mM DTT, 50% glycerol, pH 7.0	No detectable nickase activity	Pharmacia 27-0928-01 27-0928-02 27-0929-01
Escherichia coli CM 5199 (Inactivated at 75°C, 20 min)				
10,000 U/mL	1 unit converts 10 nmol dNTPs to acid-insoluble form/30 min at 37°C	0.1 M KPO$_4$, 1 mM DTT, 50% glycerol, pH 6.5	No detectable endonuclease	NE Biolabs 209S 209L
Escherichia coli pol A gene, genetic fusion (Inactivated at 75°C, 20 min)				
20,000 U/mg; 5000 U/mL	1 unit converts 10 nmol dNTPs to acid-insoluble form/30 min at 37°C	Klenow Fragment; 0.1 M KPO$_4$, 1 mM DTT, 50% glycerol, pH 6.5	No detectable exonuclease, endonuclease <1% uncleaved DNA polymerase I (SDS-PAGE)	NE Biolabs 210S 210L
20,000 U/mg; 5000 U/mL	1 unit converts 10 nmol dNTPs to acid-insoluble form/30 min at 37°C	Klenow Fragment; 0.1 M KPO$_4$, 1 mM DTT, 50% glycerol, pH 7.5	No detectable exonuclease, endonuclease	NE Biolabs 212S 212L
Escherichia coli, carrying a modified 9°N DNA polymerase gene from *Thermococcus* species (strain 9°N-7) ($T_{1/2}$ = 6.7 hr at 95°C)				
2000 U/mL	1 unit incorporates 10 nmol dNTP into acid-insoluble material/30 min at 75°C	Purified; 100 mM KCl, 10 mM Tris-HCl, 1 mM DTT, 0.1 mM EDTA, 0.1% Triton X-100, 50% glycerol, pH 7.4	No detectable endonuclease, exonuclease	NE Biolabs 9° Nm™ 260S 260L
Escherichia coli, carrying a plasmid containing phage T4 DNA polymerase gene (T4 DNA Polymerase) (optimum pH = 8-9)				
	1 unit incorporates 10 nmol deoxynucleotides into acid-insoluble form/30 min at pH 8.8, 37°C	20 mM KPO$_4$, 5 mM DTT, 50% glycerol, pH 6.5	No detectable 5'-exonuclease, non-specific endonuclease	Adv Biotech AB-0330 AB-0330b

SPECIFIC ACTIVITY	UNITS DEFINITION	PREPARATION FORM	ADDITIONAL ACTIVITIES	SUPPLIER CATALOG NO.
Escherichia coli, carrying a plasmid containing phage T4 DNA polymerase gene (T4 DNA Polymerase) *continued*				
100 U and 500 U	1 unit incorporates 10 nmol total nucleotides into acid-insoluble products/30 min at pH 8.8, 37°C with heat-denatured calf thymus DNA as template-primer	Solution containing 200 mM KPO$_4$, 10 mM β-MSH, 50% glycerol, pH 6.5	No detectable endonuclease	TaKaRa 2040
Escherichia coli, carrying a plasmid encoding *E. coli* DNA polymerase I (MW = 76,000 Da [Klenow Fragment])				
1 U/μL and 5 U/μL	1 unit incorporates 10 nmol dNTP into a TCA-insoluble form/30 min at pH 7.5, 37°C	Klenow Fragment; 5 mM KPO$_4$, 1 mM DTT, 50% glycerol, pH 6.5	No detectable endonuclease, exonuclease, nickase	NBL Gene 020302 020304 020352 020354
>8000 U/mg; 500 U and 2500 U	1 unit incorporates 10 nmol total nucleotides into acid-insoluble products/30 min at pH 7.4, 37°C with poly d(A-T) as template primer	>90% homogeneous by SDS-PAGE; solution containing 50 mM KPO$_4$, 10 mM β-MSH, 50% glycerol, pH 6.5	No detectable endonuclease	TaKaRa 2130
Escherichia coli, carrying a plasmid encoding the C-terminal two-thirds of *Bacillus caldotenax* DNA polymerase gene (*Bac* DNA Polymerase) (optimum pH = 7.5)				
	1 unit incorporates 10 nmol total nucleotides into acid-insoluble products/30 min at pH 7.4, 60°C with poly d(A-T) as template primer	>95% homogeneous by SDS-PAGE; 50 mM Tris-HCl, 2 mM DTT, 0.1% gelatin, 50% ethylene glycol, pH 7.0	No detectable endonuclease	PanVera TAK 2710B
Escherichia coli, carrying a plasmid encoding the C-terminal two-thirds of *E. coli* DNA polymerase I gene (optimum pH = 7.4)				
1000-4000 U/mg; 5 U/μL	1 unit incorporates 10 nmol total nucleotides into acid-insoluble material/30 min at 37°C in 50 μL total volume	Klenow Fragment; >95% homogeneous by SDS-PAGE; 50 mM KPO$_4$, 10 mM β-MSH, 50% glycerol, pH 6.5	No detectable non-specific nuclease	Amersham E 2141Y E 2141Z
2-10 U/μL	1 unit incorporates 10 nmol total deoxyribonucleotides into acid-precipitable material/30 min at 37°C using poly [d(A-T)] as template primer	Klenow Fragment; >98% purity by SDS-PAGE; 50 mM KPO$_4$, 1.0 mM DTT, 0.1 mM EDTA, 50% glycerol, pH 7.0	No detectable ss- and ds-DNA endonuclease, exonuclease	Amersham E 70057Y E 70057Z
	1 unit incorporates 10 nmol total nucleotides into acid-insoluble products/30 min at pH 7.4, 37°C with poly d(A-T) as template primer	Klenow Fragment; >95% homogeneous by SDS-PAGE; solution containing 50 mM KPO$_4$, 10 mM β-MSH, 50% ethylene glycol, pH 6.5	No detectable endonuclease	PanVera TAK 2140MK
200 U and 1000 U	1 unit incorporates 10 nmol total nucleotides into acid-insoluble products/30 min at pH 7.4, 37°C with poly d(A-T) as template primer	Klenow Fragment; >95% homogeneous by SDS-PAGE; solution containing 50 mM KPO$_4$, 10 mM β-MSH, 50% glycerol, pH 6.5	No detectable endonuclease, nickase	TaKaRa 2140

SPECIFIC ACTIVITY	UNITS DEFINITION	PREPARATION FORM	ADDITIONAL ACTIVITIES	SUPPLIER CATALOG NO.
Escherichia coli, carrying a plasmid encoding the *Thermus aquaticus* DNA polymerase gene (rTaq DNA polymerase)				
50 µL; 5 U/µL	1 unit incorporates 10 nmol dNTP into acid-insoluble products/30 min at 74°C with activated salmon sperm DNA as template primer	Purified	No detectable nickase, endonuclease, exonuclease	PanVera TAK R001
5 U/µL	1 unit incorporates 10 nmol dNTP into acid-insoluble products/30 min at pH 9.3, 74°C with activated salmon sperm DNA as template primer	20 mM Tris-HCl, 100 mM KCl, 0.1 mM EDTA, 1 mM DTT, 0.5% Tween 20, 0.5% NP-40, 50% glycerol, pH 8.0	No detectable exonuclease, endonuclease, nickase	TaKaRa R001
Escherichia coli, carrying an overexpressing plasmid for T4 DNA polymerase (T4 DNA polymerase) (inactivated at 75°C, 10 min)				
10,000 U/mg; 2 U/µL	1 unit incorporates 10 nmol dNTP into acid-precipitable material/30 min at 37°C	100 mM KPO$_4$, 10 mM β-MSH, 50% glycerol, pH 6.5	No detectable non-specific endonuclease, exonuclease	Ambion 2026 2028
3000 U/mL	1 unit incorporates 10 nmol dNTP into acid-precipitable material/30 min at 37°C	Purified; 100 mM KPO$_4$, 10 mM β-MSH, 50% glycerol, pH 6.5	No detectable endonuclease	NE Biolabs 203S 203L
Escherichia coli, carrying clones that overproduce T7 gene 5 protein and thioredoxin (T7 DNA polymerase)				
5-20 U/µL	1 unit incorporates 10 nmol total nucleotides into acid-insoluble material/30 min at 37°C	>95% purity by SDS-PAGE; 20 mM KPO$_4$ buffer, 1 mM DTT, 0.1 mM EDTA, 50% glycerol, pH 7.4	No detectable endonuclease	Amersham E 70017Y E 70017Z
8-25 U/µL	1 unit incorporates 1 nmol total nucleotide into acid insoluble form/min at 37°C using 5 µg primed M13mp18 DNA as template	Solution containing 20 mM KPO$_4$ buffer, 1 mM DTT, 0.1 mM EDTA, 50% glycerol, pH 7.4	No detectable exonuclease	Amersham Sequenase™ E 70775Y E 70775Z
8-30 U/µL	1 unit incorporates 10 nmol total nucleotide into acid-insoluble form/30 min at 37°C	Solution containing 20 mM KPO$_4$ buffer, 1.0 mM DTT, 0.1 mM EDTA, 50% glycerol, pH 7.4	<1% exonuclease activity of native T7 DNA polymerase	Amersham Sequenase™ US70722-200 units US70722-1000 units
5000 U/mL	1 unit incorporates 10 nmol dNTP into acid-precipitable material/30 min at pH 7.5, 37°C with calf thymus DNA as template	20 mM KPO$_4$, 0.1 mM EDTA, 1 mM DTT, 50% glycerol, pH 7.4	No detectable endonuclease	AGS Heidelb F00855S F00855M
≥2000 U/mg protein		Chromatographically purified; solution containing 50 mM KPO$_4$, 1 mM DTT, 20% glycerol, pH 7.4	No detectable nuclease, endonuclease	Worthington LM01080 LM01082 LM01084

2.7.7.7 DNA-Directed DNA Polymerase continued

SPECIFIC ACTIVITY	UNITS DEFINITION	PREPARATION FORM	ADDITIONAL ACTIVITIES	SUPPLIER CATALOG NO.
Escherichia coli, carrying plasmid pCJ122				
1600 U/mg; 5 U/μL	1 unit incorporates 10 nmol deoxynucleotides into acid-insoluble material/30 min at pH 7.4, 37°C	Klenow Fragment; single band of 51 U visible by SDS-PAGE; 50 mM Tris-HCl, 0.5 mM DTT, 50% glycerol, pH 7.5	No detectable non-specific endonuclease, exonuclease	Ambion 2012 2014
Escherichia coli, carrying plasmid pCJ374				
30,000 U/mg; 5 U/μL	1 unit incorporates 10 nmol deoxynucleotides into acid-insoluble material/30 min at pH 7.4, 37°C	Klenow Fragment; >98% purity by SDS-PAGE; 50 mM Tris-HCl, 0.5 mM DTT, 50% glycerol, pH 7.5	No detectable ss- and ds-endonuclease, exonuclease	Ambion 2006 2008
Escherichia coli, carrying the Deep Vent DNA polymerase gene from *Pyrococcus* species GB-D ($T_{1/2}$ = 23 hr at 95°C, 8 hr at 100°C)				
2000 U/mL	1 unit incorporates 10 nmol dNTP into acid-insoluble material/30 min at 75°C	Purified; 100 mM KCl, 10 mM Tris-HCl, 1 mM DTT, 0.1 mM EDTA, 0.1% Triton X-100, 50% glycerol, pH 7.4	No detectable endonuclease, exonuclease	NE Biolabs Deep Vent$_R^R$ 258S 258L
2000 U/mL	1 unit incorporates 10 nmol dNTP into acid-insoluble material/30 min at 75°C	Purified; 100 mM KCl, 10 mM Tris-HCl, 1 mM DTT, 0.1 mM EDTA, 0.1% Triton X-100, 50% glycerol, pH 7.4	No detectable endonuclease, exonuclease	NE Biolabs Deep Vent$_R^R$ 259S 259L
Escherichia coli, carrying the DTth-1 gene from *Thermus thermophilus* HB8 (ΔTth-1 DNA Polymerase)				
100 and 500 U	1 unit incorporates 10 nmol total nucleotide into acid-precipitable form/30 min at pH 8.8, 75°C	100 mM Tris-HCl, 800 mM KCl, 15 mM MgCl$_2$, 5 mg/mL BSA, 1% Na cholate, pH 8.9	No detectable RT	CLONTECH 8414-1 8414-2
Escherichia coli, carrying the gene from *Thermus aquaticus* (Taq DNA Polymerase)				
5 U/μL	1 unit incorporates 10 nmol total nucleotides into acid-insoluble material/30 min at 70°C in 50 μL total volume	50 mM Tris-HCl, 0.1 mM EDTA, 5 mM DTT, 50% glycerol, stabilizers, pH 7.5	No detectable non-specific nuclease, Taq I restriction endonuclease	Amersham T 0303Y T 0303Z T 0303X RPN 0303Y RPN 0303Z RPN 0303X
Escherichia coli, carrying the Klenow fragment on a plasmid				
	1 unit incorporates 10 nmol total nucleotide into TCA-precipitable form/30 min at pH 7.5, 37°C	Klenow Fragment; 50 mM KPO$_4$, 1 mM DTT, 50% glycerol, pH 7.0	No detectable 5'-exonuclease, non-specific endonuclease	Adv Biotech AB-0268 AB-0268b
3-9 U/μL	1 unit incorporates 10 nmol total deoxyribonucleotide into acid-precipitable material/30 min at 37°C using poly(dA-dT) as template primer	Klenow Fragment; purity by SDS-PAGE; 500 mM Tris-HCl, 100 mM MgCl$_2$, 500 mM NaCl, pH 8.0 and dilution buffer		Life Technol 18012-021 18012-039 18012-096

DNA-Directed DNA Polymerase continued   2.7.7.7

SPECIFIC ACTIVITY	UNITS DEFINITION	PREPARATION FORM	ADDITIONAL ACTIVITIES	SUPPLIER CATALOG NO.
Escherichia coli, carrying the T4 DNA polymerase gene on a plasmid (T4 DNA Polymerase)				
1-8 U/μL	1 unit incorporates 10 nmol total deoxyribonucleotide into acid-precipitable material/30 min at pH 8.8, 37°C using DNase I-nicked DNA as template primer	165 mM Tris acetate, 330 mM NaOAc, 50 mM MgOAc, 500 μg/mL BSA, 2.5 mM DTT, pH 7.9		Life Technol 18005-017 18005-025
Escherichia coli, carrying the Vent DNA polymerase gene from *Thermococcus litoralis* ($T_{1/2}$ = 6.7 hr at 95°C, 1.8 hr at 100°C)				
2000 U/mL	1 unit incorporates 10 nmol dNTP into acid-insoluble material/30 min at 75°C	Purified; 100 mM KCl, 10 mM Tris-HCl, 1 mM DTT, 0.1 mM EDTA, 0.1% Triton X-100, 50% glycerol, pH 7.4	No detectable endonuclease, exonuclease	NE Biolabs Vent$_R^R$ 254S 254L
2000 U/mL	1 unit incorporates 10 nmol dNTP into acid-insoluble material/30 min at 75°C	Purified; 100 mM KCl, 10 mM Tris-HCl, 1 mM DTT, 0.1 mM EDTA, 0.1% Triton X-100, 50% glycerol, pH 7.4	No detectable endonuclease, exonuclease	NE Biolabs Vent$_R^R$ 257S 257L
Escherichia coli, lysogenic				
5000 U/mg; 5000 U/mL	1 unit incorporates 10 nmol total nucleotide into acid-precipitable fraction/30 min	Nick translation grade; 50 mM KPO$_4$, 0.25 mM DTT, 50% glycerol, pH 7.0		Boehringer 104485 104493
6000 U/mg; 5000 U/mL	1 unit incorporates 10 nmol total nucleotide into acid-precipitable fraction/30 min	50 mM KPO$_4$, 0.25 mM DTT, 50% glycerol, pH 7.0	No detectable endonucleases	Boehringer 642711 642720
Escherichia coli, lysogenic for CM 5199 (MW = 62,000 Da [Klenow Fragment], 109,000 Da [native])				
≥2000 U/mg protein	1 unit converts 10 nmol total deoxyribonucleotides to acid-insoluble product/30 min at 37°C using poly[d(A-T)] as template primer	Solution containing 50 mM KPO$_4$, 1 mM DTT, 20% glycerol, pH 7.4		Worthington LM01003 LM01006 LM01000
≥2000 U/mg protein	1 unit incorporates 10 nmol total deoxynucleotides to acid-insoluble product/30 min at 37°C using poly[d(A-T)] as template primer	Klenow Fragment; chromatographically purified; solution containing 50 mM KPO$_4$, 1 mM DTT, 20% glycerol, pH 7.4	No detectable endonuclease, DNase, RNase	Worthington LM01013 LM01016 LM01010
Escherichia coli, lysogenic for λ-phage NM964 (Inactivated at 75°C, 10 min)				
20,000 U/mg protein	1 unit incorporates 10 nmol total deoxynucleotide into acid-insoluble form/30 min at pH 7.4, 37°C	>99% purity by electrophoresis; 100 mM KPO$_4$, 1 mM DTT, 50% glycerol, 1 mg/mL BSA, pH 7.4	No detectable 5'-exonuclease, non-specific endonuclease, nicking activity	Adv Biotech AB-0333 AB-0333b
13,500 U/mL	1 unit converts 10 nmol dNTPs to acid-precipitable form/30 min at pH 7.4, 37°C using poly(dA-dT) template primer	100 mM KPO$_4$, 1 mM DTT, 50% glycerol, pH 7.4	No detectable endonuclease	AGS Heidelb A00700S A00700M

SPECIFIC ACTIVITY	UNITS DEFINITION	PREPARATION FORM	ADDITIONAL ACTIVITIES	SUPPLIER CATALOG NO.
Escherichia coli, lysogenic for λ-phage NM964 *continued*				
5-50 U/μL	1 unit incorporates 10 nmol total nucleotides into acid-precipitable form/30 min at pH 7.5, 37°C using poly(dA-dT) template-primer	Klenow Fragment; subtilisin-treated DNA polymerase I; 50 mM KPO$_4$, 1 mM DTT, 50% glycerol, pH 7.4	No detectable 5'-exonuclease, endonuclease	AGS Heidelb A00740S A00740M
> 3000 U/mg; 2000 U/mL	1 unit incorporates 10 nmol total nucleotides into acid-precipitable products/30 min at 37°C with poly [d(A-T)] as primer	Klenow Fragment; labeling grade; solution containing 50 mM potassium phosphate KPO$_4$, 0.25 mM DTT, 50% glycerol, pH 7.0		Boehringer 1008404 1008412
100 and 500 U	1 unit incorporates 10 nmol total nucleotide into acid-precipitable form/30 min at pH 7.4, 37°C	Klenow Fragment; 500 mM Tris-HCl, 1 mM DTT, 100 mM MgCl$_2$, 500 μg/mL BSA, pH 7.4		CLONTECH 8411-1 8411-2
5-10 U/μL	1 unit incorporates 10 nmol total deoxyribonucleotide into acid-precipitable material/30 min at pH 7.5, 37°C using poly(dA-dT) as template primer			Life Technol 18010-017 18010-025
Escherichia coli, T4 gene 43 (T4 DNA Polymerase) (MW = 114,000 Da)				
≥2000 U/mg protein	1 unit incorporates 10 μmol total deoxynucleotides to acid-precipitable product with denatured nuclease digested salmon sperm DNA as template primer	Chromatographically purified to >95% by SDS-PAGE; solution containing 50 mM KPO$_4$, 1 mM DTT, 20% glycerol, pH 7.4	No detectable endonuclease, RNase	Worthington LM01030 LM01032 LM01034
Escherichia coli, T4 gene 43, modified (T4 DNA Polymerase)				
≥2000 U/mg protein	1 unit incorporates 10 μmol total deoxynucleotides to acid-precipitable product with denatured nuclease digested salmon sperm DNA as template primer	Chromatographically purified to >95% by SDS-PAGE; solution containing 50 mM KPO$_4$, 1 mM DTT, 20% glycerol, pH 7.4	No detectable exonuclease	Worthington LM01040 LM01042 LM01044
Human				
	1 unit incorporates 1 nmol total nucleotide into acid-insoluble form/hr at 37°C	DNA polymerase-α; 20 mM Tris-HCl, 1.0 mM DTT, 0.25 mM EDTA, 50 mM KCl, 50% glycerol, pH 8.5	No detectable endo- and exonuclease	CHIMERX 1075-01 1075-02
	1 unit incorporates 1 nmol total nucleotide into acid-insoluble form/hr at 37°C	DNA polymerase-β; 20 mM Tris-HCl, 1.0 mM DTT, 0.1 mM EDTA, 0.2 M NaCl, 50% glycerol, pH 8.0	No detectable endonuclease, 3'-exonuclease, 5'-exonuclease/5'-phosphatase, nonspecific RNase, ss- and ds-DNase	CHIMERX 1077-01 1077-02

DNA-Directed DNA Polymerase continued — 2.7.7.7

SPECIFIC ACTIVITY	UNITS DEFINITION	PREPARATION FORM	ADDITIONAL ACTIVITIES	SUPPLIER CATALOG No.
Micrococcus luteus (M. lysodeikticus)				
5-20 U/mg protein	1 unit incorporates 10 nmol total nucleotide into acid-insoluble product/30 min at pH 7.0, 37°C using DNA from calf thymus as template	Chromatographically prepared on DEAE cellulose and gel filtration media; lyophilized containing DTT		ICN 157557
	1 unit polymerizes 20 nmol deoxynucleotide/hr at pH 7.0, 37°C using heat-denatured DNA as template primer	Highly purified; frozen solution containing 20 mM Tris-HCl, 100 mM NaCl, 1 mM MgCl2, 1 mM β-MSH, pH 8.0		Midland D-0700
Pyrococcus furiosus, native (Pfu DNA polymerase) (95% active at 95°C, 1 hr)				
2500 U/mL	1 unit incorporates 10 nmol [^{3}H]dTTP/30 min at 72°C	50 mM Tris-HCl, 1 mM DTT, 0.1 mM EDTA, 0.1% NP40, 0.1% Tween 20, 50% glycerol, pH 8.2		Stratagene 600135 600136
Pyrococcus furiosus, recombinant (Pfu DNA polymerase) (95% active at 95°C, 1 hr)				
2500 U/mL	1 unit incorporates 10 nmol [^{3}H]dTTP/30 min at 72°C	50 mM Tris-HCl, 1 mM DTT, 0.1 mM EDTA, 0.1% NP40, 0.1% Tween 20, 50% glycerol, pH 8.2		Stratagene 600153 600154 600159
2500 U/mL	1 unit incorporates 10 nmol [^{3}H]dTTP/30 min at 72°C	98% purity by silver-stained SDS-PAGE; 50 mM Tris-HCl, 1 mM DTT, 0.1 mM EDTA, 0.1% NP40, 0.1% Tween 20, 50% glycerol, pH 8.2	No detectable nonspecific nuclease	Stratagene 600163
Pyrococcus woesei, recombinant from *E. coli* (Pwo DNA polymerase) (MW = 90,000 Da)				
5 U/μL	1 unit incorporates 10 nmol total dNTPs into acid-precipitable DNA/30 min at 70°C	Highly purified	No detectable unspecific endo- or exonuclease	Boehringer 1644947 1644955
T4 am-infected *E. coli* B (T4 DNA polymerase) (optimum pH = 8-9)				
5 U/μL	1 unit incorporates 10 nmol total nucleotides into acid-insoluble material/30 min at 37°C in 50 μL total volume	200 mM KPO$_4$, 10 mM β-MSH, 50% glycerol, pH 6.5	No detectable endonuclease	Amersham E 2040Y E 2040Z
T4 amN82-infected *E. coli* (T4 DNA polymerase)				
4000 U/mL	1 unit incorporates 10 nmol dNTP into acid-precipitable DNA/30 min at pH 8.8, 30°C	20 mM KPO$_4$, 200 mM KCl, 1 mM DTT, 50% glycerol, pH 7.5	No detectable endonuclease	AGS Heidelb F00830S F00830M
T4 Phage (T4 DNA polymerase)				
	1 unit incorporates 10 nmol total nucleotide into acid-insoluble form/30 min at 37°C	20 mM KPO$_4$, 5.0 mM DTT, 50% glycerol, pH 6.5	No detectable endonuclease	CHIMERX 1100-01 1100-02

SPECIFIC ACTIVITY	UNITS DEFINITION	PREPARATION FORM	ADDITIONAL ACTIVITIES	SUPPLIER CATALOG NO.
T4 Phage (T4 DNA polymerase) *continued*				
6000 U/mg	1 unit converts 10 nmol dNTP into acid-insoluble material/30 min at pH 7.8, 37°C	Solution containing 50% glycerol, 50 mM Tris-HCl, 0.1 M NaCl, 0.1 mM EDTA, 1 mM DTT, 0.1% Triton X-100, pH 7.5	No detectable endonuclease (by agarose gel electrophoresis)	Epicentre D0602H D0605H D0610H
5 U/µL	1 unit incorporates 10 nmol dNTP into a TCA-insoluble product/10 min at pH 8.3, 37°C	50 mM Tris-HCl, 100 mM NaCl, 0.1 mM EDTA, 0.1% Triton X-100, 1 mM DTT, 50% glycerol, pH 7.6	No detectable endonuclease	NBL Gene 021706 021703
T4, cloned (T4 DNA polymerase)				
30,000-65,000 U/mg protein; 5000-10,000 U/mL	1 unit incorporates 10 nmol total nucleotide into acid-insoluble product/30 min at pH 8.8, 37°C	Molecular biology grade; homogeneous purity; solution containing 200 mM KPO_4, 2 mM DTT, 50% glycerol, pH 6.5	No detectable nickase activity	Pharmacia 27-0718-01 27-0718-02
T4-infected *E. coli* (T4 DNA polymerase)				
100 and 500 U	1 unit incorporates 10 nmol total nucleotide into acid-precipitable form/30 min at pH 8.5, 37°C	500 mM Tris-HCl, 1 mM EDTA, 70 mM $MgCl_2$, 150 mM $(NH_4)_2SO_4$, 100 mM β-MSH, pH 8.5		CLONTECH 8412-1 8412-2
	1 unit incorporates 10 nmol total nucleotide into acid-insoluble product/30 min at 37°C	Solution containing 0.2 M KPO_4, 2 mM DTT, 50% glycerol, pH 6.5		ICN 151007
1000-5000 U/mL	1 unit converts 10 nmol deoxyribonucleoside triphosphate into acid-insoluble material/30 min at 37°C		No detectable endonuclease <2% degradation/U exonuclease or phosphatase	Sigma D0410
1000-10,000 U/mL	1 unit incorporates 10 nmol deoxyribonucleotides into acid-insoluble form/30 min at 37°C	200 mM KPO_4, 2 mM DTT, 50% glycerol, pH 6.5	No detectable endonuclease	Stratagene 600091 600092
T4-infected *E. coli* 71-18/pTL43W (T4 DNA polymerase)				
≥5000 U/mg; ≥1000 U/mL	1 unit incorporates 10 nmol dNTP into acid-precipitable DNA products/30 min at 37°C	100 mM KPO_4, 10 mM β-MSH, 50% glycerol, pH 8.5		Boehringer 1004786 1004794
T7, carrying clones that overproduce T7 gene 5 protein and *E. coli* thioredoxin (T7 DNA polymerase)				
4500 U/mg	1 unit converts 10 nmol dNTP into acid-insoluble material/30 min at pH 7.8, 37°C	Solution containing 50% glycerol, 50 mM Tris-HCl, 0.1 M NaCl, 0.1 mM EDTA, 1 mM DTT, 0.1% Triton X-100, pH 7.5	No detectable endonuclease (by agarose gel electrophoresis), RNase	Epicentre D07250 D07500 D0701K
5000-15,000 U/mL	1 unit converts 10 nmol deoxynucleotides to acid-insoluble form/30 min at 37°C	50 mM KPO_4, 1 mM DTT, 0.1 mM EDTA, 50% glycerol, pH 7.0	No detectable endonuclease	NE Biolabs 256S 256L

DNA-Directed DNA Polymerase continued — 2.7.7.7

SPECIFIC ACTIVITY	UNITS DEFINITION	PREPARATION FORM	ADDITIONAL ACTIVITIES	SUPPLIER CATALOG No.
T7, cloned (T7 DNA polymerase)				
8000-12,000 U/mL	1 unit incorporates 50 nmol total nucleotide into acid-insoluble product/30 min at pH 7.5, 37°C using M13mp19(+) DNA as template	Molecular biology grade; homogeneous purity; solution containing 25 mM Tris-HCl, 0.25 M NaCl, 5 mM DTT, 50% glycerol, pH 7.5	No detectable nickase, DNase, exonuclease	Pharmacia 27-0985-03 27-0985-04
***Thermococcus litoralis* (*Tli* DNA polymerase) (MW = 85,000 Da)**				
1 U/μL	1 unit incorporates 10 nmol dNTP into acid-insoluble form/30 min at pH 9.0, 74°C	>90% purity by SDS gel	≥90% supercoiled plasmid	Promega M7101 M7102
Thermophilic bacteria				
	1 unit converts 10 nmol dNTP into acid-insoluble material/30 min at pH 9.3, 74°C	Solution containing 50% glycerol, 50 mM Tris-HCl, 0.1 M NaCl, 0.1 mM EDTA, 1 mM DTT, 0.1% Triton X-100, pH 7.5	No detectable DNase, RNase, protease	Epicentre Replitherm™ R04100 R04250 R04500 R0401K R0405K
***Thermus aquaticus* (Taq DNA polymerase) (optimum T = 75°C, MW = 85,000 [SDS-PAGE] to 95,000 Da)**				
4000 U/mL	1 unit incorporates 10 nmol dNTP into acid-insoluble material/30 min at pH 8.55, 70°C	20 mM Tris-HCl, 7 mM β-MSH, 100 mM KCl, 0.2 mM EDTA, 55% glycerol, stabilizers, pH 8	No detectable nuclease	AGS Heidelb
5 U/μL	1 unit incorporates 10 nmol dNTP into acid-insoluble form/30 min at 74°C		No detectable non-specific nuclease, Taq I endonuclease	ICN 152489
5 U/μL	1 unit incorporates 10 nmol dNTP into acid-insoluble form/30 min at 74°C	Sequencing grade	No detectable non-specific nuclease, Taq I endonuclease	ICN 152490
5 U/μL	1 unit incorporates 10 nmol nucleoside into DE-81 adsorbable product/30 min	20 mM Tris-HCl, 1 mM DTT, 1 mM EDTA, pH 8.0	No detectable nickase, endonuclease, 3' exonuclease, RNase	Oncor 120181 120182 120183
50 μL		Purified		PanVera TAK RR001 TAK RR009
5000 U/mL	1 unit incorporates 10 nmol total nucleotide into acid-insoluble product/30 min at pH 9.3, 70°C using M13mp18(+) DNA as template	Solution containing 50 mM Tris-HCl, 0.1 mM EDTA, 5 mM DTT, 50% glycerol, stabilizers, pH 7.5	No detectable nickase, DNase, endonuclease	Pharmacia 27-0799-01 27-0799-02 27-0799-03
5000 U/mL	1 unit incorporates 10 nmol [^{3}H]dTTP/30 min at 80°C	50 mM Tris-HCl, 1 mM DTT, 0.1 mM EDTA, 87.5 mM KCl, 50% glycerol, stabilizers, pH 8.2		Stratagene 600203 600204

SPECIFIC ACTIVITY	UNITS DEFINITION	PREPARATION FORM	ADDITIONAL ACTIVITIES	SUPPLIER CATALOG NO.
Thermus aquaticus (*Taq* DNA polymerase) *continued*				
5 U/μL	1 unit incorporates 10 nmol dNTP into acid-insoluble products/30 min at pH 9.3, 74°C with activated salmon sperm DNA as template primer	20 mM Tris-HCl, 100 mM KCl, 0.1 mM EDTA, 1 mM DTT, 0.5% Tween 20, 0.5% NP40, 50% glycerol, pH 8.0	No detectable exonuclease, endonuclease, nickase	TaKaRa RR001A
5 U/μL	1 unit incorporates 10 nmol dNTP into acid-insoluble products/30 min at pH 9.3, 74°C with activated salmon sperm DNA as template primer	20 mM Tris-HCl, 100 mM KCl, 0.1 mM EDTA, 1 mM DTT, 0.5% Tween 20, 0.5% NP40, 50% glycerol, pH 8.0	No detectable exonuclease, endonuclease, nickase	TaKaRa RR002A
≥2000 U/mg protein	1 unit incorporates 10 nmol total deoxynucleotides to acid-precipitable product/30 min at 37°C with activated calf thymus DNA as template primer	Chromatographically purified to >95% by SDS-PAGE; solution containing 50 mM KPO$_4$, 1 mM DTT, 20% glycerol, pH 7.4		Worthington LM01060 LM01062 LM01064
Thermus aquaticus BM (*Taq* DNA polymerase)				
1000-5000 U/mL	1 unit incorporates 10 mM total deoxyribonucleoside-3P into acid-precipitable DNA/30 min at 75°C	20 mM Tris-HCl, 1 mM DTT, 0.1 mM EDTA, 0.1 M KCl, 0.5% NP40, 0.5% Tween 20, 50% glycerol, pH 8.0		Boehringer 1146165 1146173 1418432 1435094 1596594
Thermus aquaticus BM, recombinant in *E. coli* (*Taq* DNA polymerase) (optimum pH = 9, T = 75°C, MW = 95,000 Da)				
1 U/μL	1 unit incorporates 10 nmol total dNTPs into acid-precipitable DNA/30 min at 75°C	Highly purified	No detectable unspecific endo- or exonuclease	Boehringer 1647679 1647687
Thermus aquaticus, recombinant in *E. coli* (*Taq* DNA polymerase) (MW = 94,000 Da)				
5 U/μL	1 unit incorporates 10 nmol deoxyribonucleotide into acid-precipitable material/30 min at pH 9.3, 74°C	PCR qualified; 200 mM Tris-HCl, 500 mM NaCl, pH 8.4 and 50 mM MgCl$_2$		Life Technol 10342-053 10342-020 10342-046
Thermus aquaticus, strain YT1 (*Taq* DNA polymerase) (optimum pH = 8.3, MW = 80,000-94,000 Da; stable to 95°C)				
2-5 U/μL	1 unit incorporates 10 nmol deoxyribonucleotides into DE-81 absorbable form/30 min at 70°C	20 mM Tris-HCl, 100 mM KCl, 0.5% NP40, 1 mM DTT, 0.2 mg/mL BSA, 50% glycerol, pH 7.5	No detectable endo- and exodeoxyribonucleases, RNases	Fermentas EP0071 EP0072
4-5 U/μL; 1 U/μL	1 unit incorporates 10 nmol deoxyribonucleotides into DE-81 absorbable form/30 min at 70°C	20 mM Tris-HCl, 100 mM KCl, 0.5% NP40, 1 mM DTT, 0.1 mM EDTA, 0.5% Tween 20, 50% glycerol, pH 8.0	No detectable endo- and exodeoxyribonucleases, RNases	Fermentas EP0281 EP0282 EP0283 EP0284

SPECIFIC ACTIVITY	UNITS DEFINITION	PREPARATION FORM	ADDITIONAL ACTIVITIES	SUPPLIER CATALOG NO.
Thermus aquaticus, strain YT1 (Taq DNA polymerase) continued				
5 U/μL	1 unit incorporates 10 nmol deoxyribonucleotide into acid-precipitable material/30 min at pH 9.3, 74°C	PCR qualified; 200 mM Tris-HCl, 500 mM NaCl, pH 8.4 and 50 mM MgCl$_2$		Life Technol 18038-018 18038-042 18038-067
	1 unit incorporates 10 nmol ^{3}H-dTTP/30 min at 70°C	20 mM Tris-HCl, 0.1 mM EDTA, 50 mM KCl, 1 mM DTT, 200 μg/mL BSA, 50% glycerol, 0.5% NP40, 0.5% Tween 20, pH 8.0	No detectable non-specific nuclease, Taq I restriction endonuclease	MINOTECH 204
2-5 U/μL	1 unit incorporates 10 nmol dNTP into acid-insoluble form/30 min at pH 9.0, 74°C	>90% purity by SDS gel; 20 mM Tris-HCl, 100 mM KCl, 0.1 mM EDTA, 1 mM DTT, 50% glycerol, 0.5% NP40, 0.5% Tween 20, pH 8.0	≥90% supercoiled plasmid	Promega M1661-M1669
2-5 U/μL	1 unit incorporates 10 nmol dNTP into acid-insoluble form/30 min at pH 9.0, 74°C	>90% purity by SDS gel; 50 mM Tris-HCl, 100 mM NaCl, 0.1 mM EDTA, 5 mM DTT, 50% glycerol, 1.0% Triton X-100, pH 8.0	≥90% supercoiled plasmid	Promega M1861-M1869
2-5 U/μL	1 unit incorporates 10 nmol dNTP into acid-insoluble form/30 min at pH 9.0, 74°C	Sequencing grade; ≥90% purity by SDS gel	≥90% supercoiled plasmid	Promega M2031-M2039
2500-5000 U/mL	1 unit incorporates 10 nmol [^{3}H]dTTP/30 min at 80°C	100 mM KCl, 10 mM DTT, 0.1 mM EDTA, 0.5% NP40, 0.5% Tween 20, 20 mM Tris-HCl, 50% glycerol, pH 8.0		Stratagene 600131 600132 600139
Thermus brockianus ($T_{1/2}$ = 2.5 hr, 96°C)				
2-5 U/μL	1 unit incorporates 10 nmol dNTP into acid-precipitable DNA/30 min at pH 8.8, 72°C	20 mM Tris-HCl, 100 mM KCl, 0.1 mM EDTA, 0.1% Triton X-100, 1 mM DTT, 50% glycerol, 160 μg/mL BSA, pH 7.4	No detectable endonuclease, exonuclease, nickase	NBL Gene 021111
Thermus flavus (Tfl DNA polymerase) (MW = 82,000 Da)				
	1 unit incorporates 10 nmol total nucleotide into acid-insoluble form/30 min at 70°C	50 mM Tris-HCl, 5.0 mM DTT, 0.1 mM EDTA, 50% glycerol, stabilizers, pH 7.5	No detectable endonuclease, 3'-exonuclease, nonspecific ss- and ds-DNase	CHIMERx 1112-01 1112-02
	1 unit incorporates 10 nmol total nucleotides into acid-insoluble material/30 min at 70°C	50 mM Tris-HCl, 5.0 mM DTT, 0.1 mM EDTA, 50% glycerol, stabilizers, pH 7.5	No detectable endonuclease, 3'-exonuclease, nonspecific ss- and ds-DNase	CHIMERx 8010-01
	1 unit incorporates 10 nmol total nucleotide into acid-insoluble material/30 min at 70°C	50 mM Tris-HCl, 5.0 mM DTT, 0.1 mM EDTA, 50% glycerol, stabilizers, pH 7.5	No detectable endonuclease, 3'-exonuclease, nonspecific ss- and ds-DNase	CHIMERx 8020-01

SPECIFIC ACTIVITY	UNITS DEFINITION	PREPARATION FORM	ADDITIONAL ACTIVITIES	SUPPLIER CATALOG NO.
Thermus flavus (*Tfl* DNA polymerase) *continued*				
70,000 U/mg	1 unit converts 10 nmol deoxyribonucleoside triphosphates into acid-insoluble form/30 min at pH 9.3, 74°C	Solution containing 50% glycerol, 50 mM Tris-HCl, 0.1 M NaCl, 0.1 mM EDTA, 1 mM DTT, 0.5% Tween 20, 0.5% NP40, pH 7.5	No detectable DNA exo- and endonuclease, RNase, protease	Epicentre F10100 F10250 F10500 F1001K F1005K
2-6 U/μL	1 unit incorporates 10 nmol total nucleotide into TCA-precipitable material/30 min at pH 9.0, 74°C	>90% purity by SDS gel	≥90% supercoiled plasmid	Promega M1941 M1942
Thermus thermophilus (*Tth* DNA Polymerase) (optimum T = 75°C)				
	1 unit incorporates 10 nmol total nucleotide into acid-insoluble form/30 min at 70°C	50 mM Tris-HCl, 5.0 mM DTT, 0.1 mM EDTA, 50% glycerol, stabilizers, pH 7.5	No detectable endonuclease, 3'-exonuclease, nonspecific ss- and ds-DNase	CHIMERX 1115-01 1115-02
	1 unit converts 10 nmol deoxyribonucleotide triphosphates into acid-insoluble material/30 min at pH 9.3, 74°C	Solution containing 50% glycerol, 50 mM Tris-HCl, 0.1 M NaCl, 0.1 mM EDTA, 1 mM DTT, 0.5% Tween 20, 0.5% NP40, pH 7.5	No detectable DNA exo- and endonuclease, protease, RNase	Epicentre TTH19100 TTH19250 TTH19500 TTH1901K TTH1905K
5000 U/mL	1 unit incorporates 10 nmol total nucleotide into acid-insoluble product/30 min at pH 9.3, 70°C using M13mp18(+) DNA as template	Solution containing 10 mM Tris-HCl, 50 mM KCl, 1 mM DTT, 0.1 mM EDTA, 50% glycerol, 100 μg/mL gelatin, stabilizers, pH 7.5	No detectable nickase, DNase, endonuclease	Pharmacia 27-0800-01
Thermus thermophilus HB-8 (*Tth* DNA Polymerase) (optimum T = 75°C, MW = 94,000 Da)				
1500 U/mL	1 unit incorporates 10 nmol deoxyribonucleoside-3P into acid-precipitable DNA/30 min at 70°C	10 mM Tris-HCl, 1 mM DTT, 0.1 mM EDTA, 300 mM KCl, 0.1% Triton X-100, 50% glycerol, pH 7.5		Boehringer 1480014 1480022
100 and 500 U	1 unit incorporates 10 nmol total nucleotide into acid-precipitable form/30 min at pH 8.8, 75°C	100 mM Tris-HCl, 800 mM KCl, 15 mM MgCl$_2$, 5 mg/mL BSA, 1% Na cholate, pH 8.9		CLONTECH 8413-1 8413-2
4-6 U/μL	1 unit incorporates 10 nmol total nucleotide into TCA-precipitable material/30 min at pH 9.0, 70°C using calf thymus as substrate	10X Reaction buffer: 500 mM KCl, 100 mM Tris-HCl, 1.0% Triton X-100, pH 9.0	≥90% supercoiled plasmid	Promega M2101 M2102

DNA-Directed DNA Polymerase continued   2.7.7.7

SPECIFIC ACTIVITY	UNITS DEFINITION	PREPARATION FORM	ADDITIONAL ACTIVITIES	SUPPLIER CATALOG NO.
Thermus ubiquitous				
5 U/μL	1 unit incorporates 10 nmol total nucleotides into acid-insoluble material/30 min at 70°C in 50 μL total volume	>95% purity by SDS-PAGE; 50 mM Tris-HCl, 0.1 mM EDTA, 5 mM DTT, 50% glycerol, stabilizers, pH 7.5	No detectable non-specific nuclease	Amersham Hot Tub™ T 0333Y T 0333Z RPN 0333Y RPN 0333Z
Unspecified				
	1 unit converts 10 nmol deoxyribonucleoside triphosphates into acid-insoluble material/30 min at pH 9.3, 74°C	Solution containing 50% glycerol, 50 mM Tris-HCl, 0.1 M NaCl, 0.1 mM EDTA, 1 mM DTT, 0.5% Tween 20, 0.5% NP40, pH 7.5	No detectable DNA exo- and endonuclease, RNase, phosphatase	Epicentre SequiTherm™ S3305H S3301K
	1 unit incorporates 10 nmol deoxynucleotides into DE-81 adsorbable form/30 min at 37°C	Klenow Fragment; 100 mM KPO$_4$, 1 mM DTT, 0.1 mg/mL BSA, 50% glycerol, pH 7.0	No detectable 5'→3' exonuclease, endonuclease	Oncor 120061 120062

Polyribonucleotide Nucleotidyltransferase   2.7.7.8

REACTION CATALYZED
   RNA$_{n+1}$ + orthophosphate ↔ RNA$_n$ + a nucleoside diphosphate

SYSTEMATIC NAME
   Polyribonucleotide:orthophosphate nucleotidyltransferase

SYNONYMS
   Polynucleotide phosphorylase, PNPase

REACTANTS
   RNA$_{n+1}$, orthophosphate, ADP, IDP, GDP, UDP, CDP, RNA$_n$, a nucleoside diphosphate

NOTES
- Classified as a transferase, transferring phosphorus-containing groups: nucleotidyltransferases

SPECIFIC ACTIVITY	UNITS DEFINITION	PREPARATION FORM	ADDITIONAL ACTIVITIES	SUPPLIER CATALOG NO.
Bacillus stearothermophilus (optimum pH = 9-9.5, pI = 4.0, MW = 300,000-340,000 Da [85,000/subunit]; K_M = 0.27 mM [poly A], 3.0 mM [KH$_2$PO$_4$]; stable pH 9-11, T < 55°C)				
>2000 U/mg protein	1 unit forms 1 μmol ADP/hr at 60°C by depolymerizing of Poly A; 1 unit polymerizes 1 μmol ADP/hr at 60°C	Lyophilized	<0.0001% RNase	Unitika

SPECIFIC ACTIVITY	UNITS DEFINITION	PREPARATION FORM	ADDITIONAL ACTIVITIES	SUPPLIER CATALOG NO.
Bacillus stearothermophilus continued				
>2000 U/mg protein	1 unit forms 1 μmol ADP/hr at 60°C by depolymerizing of Poly A; 1 unit polymerizes 1 μmol ADP/hr at 60°C	50% glycerol solution	<0.0001% RNase	Unitika
Escherichia coli				
10-30 U/mg protein	1 unit polymerizes 1.0 μmol ADP releasing 1.0 μmol Pi/15 min at pH 9.0, 37°C	Solution containing 50% glycerol, 5 mM Tris and 0.5 mM DTT, pH 8.0		Sigma P6264
15-60 U/mg gel	1 unit polymerizes 1.0 μmol ADP releasing 1.0 μmol Pi/15 min at pH 9.0, 37°C	Attached to 4% cross-linked beaded agarose; suspension in 50% glycerol, 20 mM Tris, 0.5 mM DTT, 1 mM EDTA, pH 8.2		Sigma P9686
Micrococcus lysodeikticus				
40-90 U/mg protein	1 unit incorporates 1.0 μmol labeled UDP into poly-UMP/hr min at pH 8.8, 37°C using GpU as primer	Lyophilized containing 50% protein and 50% Tris buffer salts		Sigma P1384
	1 unit liberates 1 μmol Pi from ADP/15 min at pH 9.0, 37°C with no template primer added	Highly purified; frozen solution containing 20 mM Tris-HCl, 100 mM NaCl, 1 mM MgCl2, 1 mM β-MSH, pH 8.0		Midland P-1200

2.7.7.9 UTP-Glucose-1-Phosphate Uridylyltransferase

REACTION CATALYZED
UTP + α-D-glucose 1-phosphate ↔ Pyrophosphate + UDPglucose

SYSTEMATIC NAME
UTP:α-D-glucose-1-phosphate uridylyltransferase

SYNONYMS
UDPglucose pyrophosphorylase, glucose-1-phosphate uridylyltransferase, uridine-5′-diphosphoglucose pyrophosphorylase

REACTANTS
UTP, α-D-glucose 1-phosphate, pyrophosphate, UDPglucose

NOTES
- Classified as a transferase, transferring phosphorus-containing groups: nucleotidyltransferases
- Inhibited by sulfate ions

UTP-Glucose-1-Phosphate Uridylyltransferase continued   2.7.7.9

SPECIFIC ACTIVITY	UNITS DEFINITION	PREPARATION FORM	ADDITIONAL ACTIVITIES	SUPPLIER CATALOG No.
Bovine liver				
100 U/mg protein; 4 U/mg solid	1 unit converts 1 μmol UDPG and PPi to product/min at 25°C	Lyophilized	<0.1% NDP kinase, LDH	Boehringer 411507
50–150 U/mg protein	1 unit forms 1.0 μmol Glc1P from uridine 5′-diphosphoglucose and PPi/min at pH 8.5, 25°C	Lyophilized containing 10% protein, 80% sucrose, 5% DTT; balance primarily Tris buffer salts	<0.1% LDH, nucleoside diphosphate kinase	Sigma U5877
Yeast, bakers				
50–250 U/mg protein	1 unit forms 1.0 μmol Glc1P from uridine 5′-diphosphoglucose and PPi/min at pH 8.5, 25°C	Sulfate-free; lyophilized containing 50% protein; balance primarily citrate buffer salts	<0.1% inorganic pyrophosphatase, UDP-GlcDH, UDP-galactose-4-epimerase, Gal1P uridyltransferase	Sigma U8501

UTP-Hexose-1-Phosphate Uridylyltransferase   2.7.7.10

REACTION CATALYZED
    UTP + α-D-galactose 1-phosphate ↔ Pyrophosphate + UDPgalactose

SYSTEMATIC NAME
    UTP:α-D-hexose-1-phosphate uridylyltransferase

SYNONYMS
    Galactose-1-phosphate uridylyltransferase

REACTANTS
    UTP, α-D-galactose 1-phosphate, α-D-glucose 1-phosphate, pyrophosphate, UDPgalactose

NOTES
- Classified as a transferase, transferring phosphorus-containing groups: nucleotidyltransferases
- α-D-Glucose 1-phosphate acts slowly as substrate

SPECIFIC ACTIVITY	UNITS DEFINITION	PREPARATION FORM	ADDITIONAL ACTIVITIES	SUPPLIER CATALOG No.
Yeast, galactose-adapted				
20–60 U/mg protein	1 unit forms 1.0 μmol Glc1P from UDP-glucose, Gal1P and NADP$^+$/min at pH 8.7, 25°C as detected by a coupled system using PGluM	Lyophilized containing 25% protein; balance is buffer salts as citrate and GSH	<0.2% UDP glucose pyrophosphorylase, galactokinase, UDP galactose-4-epimerase, 6PGDH	Sigma G4256

2.7.7.12 UDP-Glucose-Hexose-1-Phosphate Uridylyltransferase

REACTION CATALYZED

UDPglucose + α-D-galactose 1-phosphate ↔ α-D-glucose 1-phosphate + UDPgalactose

SYSTEMATIC NAME

UDPglucose:α-D-galactose-1-phosphate uridylyltransferase

SYNONYMS

Uridyl transferase, hexose-1-phosphate uridylyltransferase

REACTANTS

UDPglucose, α-D-galactose 1-phosphate, α-D-glucose 1-phosphate, UDPgalactose

NOTES

- Classified as a transferase, transferring phosphorus-containing groups: nucleotidyltransferases

SPECIFIC ACTIVITY	UNITS DEFINITION	PREPARATION FORM	ADDITIONAL ACTIVITIES	SUPPLIER CATALOG No.
Calf liver				
1.5 U/mg protein	1 unit converts 1 μmol Gal1P and UDP-glucose to product/min at 25°C	Lyophilized containing 5 mg protein and 45 mg sucrose	<0.01% 6PGDH <0.1% NADPH oxidase <1% GlcDH <5% UDPG-pyrophosphorylase	Boehringer 110230
1-3 U/mg protein	1 unit oxidizes 1.0 μmol UDP-galactose and Glc1P from UDP-glucose and Gal1P/min at pH 8.0, 25°C in a coupled system	Lyophilized containing 10% protein; balance sucrose		Sigma U4880

2.7.7.31 DNA Nucleotidylexotransferase

REACTION CATALYZED

Deoxynucleoside triphosphate + DNA_n ↔ Pyrophosphate + DNA_{n+1}

SYSTEMATIC NAME

Nucleoside-triphosphate:DNA deoxynucleotidylexotransferase

SYNONYMS

Terminal deoxyribonucleotidyltransferase, terminal addition enzyme, terminal transferase

REACTANTS

Deoxynucleoside triphosphate, DNA_n, pyrophosphate, DNA_{n+1}

APPLICATIONS

- Addition of homopolymer tails to vector DNA fragments
- Insertion of DNA for *in vitro* construction of recombinant DNA
- 3'-End labeling of DNA

DNA Nucleotidylexotransferase continued — 2.7.7.31

NOTES

- Classified as a transferase, transferring phosphorus-containing groups: nucleotidyltransferases
- Catalyzes template-independent extension of the 3'-end of single- or double-stranded DNA one nucleotide at a time
- Cannot initiate a chain *de novo*
- Uses ribo- or deoxyribonucleoside
- Does not require a template
- Accepts radiolabeled nucleotides and nucleotides labeled with haptens like digoxigenin or biotin.
- Requires an oligonucleotide of 3+ bases as substrate
- Activated by Co^{2+} for addition of dATP or dTTP; Mn^{2+} for addition of dCTP or dGTP
- Inhibited by Na^+ and EDTA
- May lose activity at low protein concentration; ≥ 10 mg/mL recommended

SPECIFIC ACTIVITY	UNITS DEFINITION	PREPARATION FORM	ADDITIONAL ACTIVITIES	SUPPLIER CATALOG NO.
Bovine thymus				
20,000 U/mg protein	1 unit adds 1 nmol deoxyadenylate onto a p(dT)$_6$ oligodeoxyribonucleotide/hr at 37°C	100 mM KPO$_4$, 1 mM DTT, 50% glycerol, pH 7.2		Oncor 120151 120152
Calf thymus (MW = 32,000 Da)				
	1 unit incorporates 1 nmol dATP into acid-insoluble form/hr at pH 7.0, 37C	50 mM KPO$_4$, 1 mM β–MSH, 50% glycerol, pH 7.2		Adv Biotech AB-0334 AB-0334b
10,000 U/mg	1 unit incorporates 1 nmol [³H]dATP into DNA/hr at pH 7.2, 37°C with calf thymus DNA treated with DNase and heat-denatured (activated) as initiator	60 mM KPO$_4$, 150 mM KCl, 1 mM β–MSH, 50% glycerol, pH 7.2	No detectable nuclease	Amersham E 2230Y E 2230Z
15,000 U/mL	1 unit incorporates 1 nmol dTMP into acid-precipitable material/30 min at pH 7.5, 37°C	50 mM K cacocylate, 2 mM MgCl$_2$, 0.02 mM EDTA, 2 mM DTT, 50% glycerol, pH 7.5	No detectable nuclease	AGS Heidelb F00850S F00850M
10,000-50,000 U/mL	1 unit incorporates 1 nmol dAMP into acid-insoluble products/hr at 37°C using d(pT)$_6$ as primer	200 mM Na cacodylate, 200 mM NaCl, 1 mM EDTA, 4 mM β–MSH, 50% glycerol, pH 6.5		Boehringer 220582
200 and 1000 U	1 unit incorporates 1 nmol dATP into acid-insoluble form/hr at pH 7.2, 37°C	2 M Na cacodylate, 10 mM MgCl$_2$, 10 mM β-MSH, pH 7.2		CLONTECH 8415-1 8415-2
20-40 U/μL	1 unit incorporates 1 nmol deoxythymidylate into DE-81 absorbable form/hr at 37°C	200 mM Tris-KPO$_4$, 0.01 mM EDTA, 2 mM DTT, 0.01% Triton X-100, 40% ethylene glycol, pH 6.8	No detectable endo-, exodeoxyribonucleases, RNases	Fermentas EP0161 EP0162
3000-5000 U/mg	1 unit incorporates 1 nmol deoxyadenylic acid into polymer/hr at pH 7.0, 37°C	Lyophilized		ICN 152098

SPECIFIC ACTIVITY	UNITS DEFINITION	PREPARATION FORM	ADDITIONAL ACTIVITIES	SUPPLIER CATALOG NO.
Calf thymus *continued*				
10-20 U/μL	1 unit incorporates 1 nmol dATP into acid-precipitable material/hr at pH 7.2, 37°C using d(pA)$_{50}$ as primer	500 mM K cacodylate, 10 mM CoCl$_2$, 1 mM DTT, pH 7.2		Life Technol 18008-011 18008-060
	1 unit polymerizes 1 nmol deoxyadenylate/hr at pH 7.0, 37°C using p(dT)$_4$ as primer	Purified; solution containing 20 mM Tris-HCl, 200 mM KCl, 50% glycerol, pH 7.6	No detectable endonuclease, exonuclease	Midland T-1201
10,000 U/mg	1 unit incorporates 1 nmol [^{3}H]dATP into activated calf thymus DNA as initiator/hr at pH 7.2, 37°C	Solution containing 60 mM KPO$_4$, 150 mM KCl, 1 mM β-MSH, 50% glycerol, pH 7.2	No detectable nuclease	PanVera TAK 2230
20,000-50,000 U/mg protein; 15,000-25,000 U/mL	1 unit adds 1 nmol deoxyadenylate to p(dT)$_6$ primer/hr at pH 6.8, 37°C	Molecular biology grade; homogeneous purity; solution containing 100 mM KPO$_4$, 10 mM β-MSH, 50% glycerol, pH 6.9	No detectable RNase, DNase, nickase	Pharmacia 27-0730-01 27-0730-02
15-30 U/μL	1 unit transfers 1 nmol dATP to p(dT)$_{12}$/hr at 37°C	5X Reaction buffer: 500 mM cacodylate buffer, 5 mM CoCl$_2$, 0.5 mM DTT, pH 6.8	No detectable RNase H <1% DNase <3% RNase ≥90% supercoiled plasmid	Promega M1871 M1872
5000 U/mL	1 unit incorporates 1 nmol dATP into acid-precipitable material/hr at 37°C with d(pT)$_6$ as primer		No detectable endonuclease <2% degradation/U exonuclease or phosphatase	Sigma T6511
≥5000 U/mL	1 unit incorporates 1.0 nmol dATP into acid-precipitable material/1 hr at 37°C using d(pT)$_6$ as primer		No detectable endonuclease (nickase) <2% degradation/U exonuclease or phosphatase	Sigma T6511
10,000-50,000 U/mL	1 unit transfers 1.0 nmol dAMP from dATP to the 3'-OH terminus of the oligonucleotide initiator p(dT)$_6$/hr at 37°C	100 mM KPO$_4$, 1 mM DTT, 50% glycerol, pH 7.2	No detectable endonuclease, exonuclease	Stratagene 600137 600138
10,000 U/mg; 12 U/μL; 300 U and 1500 U	1 unit incorporates 1 nmol [^{3}H]dATP into primer DNA/hr at pH 7.2, 37°C with calf thymus DNA that is treated with DNA and heat-denatured as initiator	Solution containing 60 mM KPO$_4$, 150 mM KCl, 1 mM β-MSH, 50% glycerol, pH 7.2	No detectable nuclease	TaKaRa 2230B
Unspecified				
	1 unit transfers 1 nmol dAMP from dATP to the 3'-OH terminus of the oligodeoxynucleotide initiator p(dA)$_{50}$/hr at 37°C	100 mM KPO$_4$, 1.0 mM DTT, 50% glycerol, pH 7.2	No detectable endonuclease, exonuclease, RNase	CHIMERx 1390-01 1390-02
				Oncogene XME12

RNA-Directed DNA Polymerase   2.7.7.49

REACTION CATALYZED
Deoxynucleoside triphosphate + DNA_n ↔ Pyrophosphate + DNA_{n+1}

SYSTEMATIC NAME
Deoxynucleoside-triphosphate:DNA deoxynucleotidyltransferase (RNA-directed)

SYNONYMS
DNA nucleotidyltransferase (RNA-directed), reverse transcriptase, revertase, RT

REACTANTS
Deoxynucleoside triphosphate, DNA_n, pyrophosphate, DNA_{n+1}

APPLICATIONS
- AIDS research
- Preparation of cDNA from mRNA

NOTES
- Classified as a transferase, transferring phosphorus-containing groups: nucleotidyltransferases
- Catalyzes RNA-template-directed extension of the 3'-end of a DNA strand one deoxynucleotide at a time
- Cannot initiate a chain *de novo*
- Requires RNA or DNA primer
- DNA also serves as template
- Also has RNase H activity
- Reverse transcriptases most often used are isolated from avian myeloblastosis virus (AMV)
- Cloned HIV enzyme is primarily used for AIDS research. It is less efficient than AMV enzyme in applications like preparation of cDNA from mRNA

SPECIFIC ACTIVITY	UNITS DEFINITION	PREPARATION FORM	ADDITIONAL ACTIVITIES	SUPPLIER CATALOG NO.
Avian myeloblastosis virus (optimum pH = 7.2, T = 50-60°C, MW = 68,000 Da [α-subunit], 92,000 Da [β-subunit])				
	1 unit incorporates 1 nmol TTP into acid-insoluble form/10 min at 35°C using poly(A)·oligo dT_{12-18} as substrate	0.2 M KPO_4, 2 mM DTT, 0.2% Triton X-100, 50% glycerol, pH 7.2	No detectable RNase, exonuclease	Adv Biotech AB-0321 AB-0321b
26,700 U/mL	1 unit incorporates 1 nmol dTMP into acid-precipitable form/10 min at pH 8.3, 37°C	200 mM KPO_4, 2 mM DTT, 0.2% Triton X-100, 50% glycerol, pH 7.2	No detectable endonuclease, RNase	AGS Heidelb F00750S F00750M
10-20 U/μL	1 unit incorporates 1.0 nmol [³H]TTP into acid-insoluble products/10 min at 37°C	0.2 M KPO_4, 2.0 mM DTT, 0.2% Triton X-100, 50% glycerol, pH 7.2	No detectable endonuclease, exonuclease, RNase	Amersham E 70041Y E 70041Z
> 50,000 U/mg; > 20,000 U/mL	1 unit incorporates 1 nmol [³H]-dTMP into acid-precipitable products/10 min at 37°C using poly(A)·d(pT)$_{15}$ as template primer	200 mM KPO_4, 2 mM DTT, 0.2% Triton X-100, 50% glycerol, pH 7.2	No detectable nonspecific RNases, nonspecific DNases (gel electrophoresis)	Boehringer 109118 1495062
	1 unit incorporates 1 nmol dTTP into acid-insoluble form/10 min at 37°C	200 mM KPO_4, 2.0 mM DTT, 0.2% Triton X-100, 50% glycerol, pH 7.2	No detectable endonuclease, exonuclease	CHIMERX 1372-01 1372-02

2.7.7.49 RNA-Directed DNA Polymerase continued

SPECIFIC ACTIVITY	UNITS DEFINITION	PREPARATION FORM	ADDITIONAL ACTIVITIES	SUPPLIER CATALOG NO.
Avian myeloblastosis virus *continued*				
30,000 U/mL	1 unit incorporates 1 nmol (^{3}H)-TMP into nucleic acid product/10 min at 37°C	Solution containing 0.2 M KPO$_4$, 2.0 mM DTT, 0.2% Triton X-100, 50% glycerol, pH 7.2	No detectable nonspecific nuclease	ICN 855928 855929
1-5 U/μL	1 unit incorporates 1 nmol deoxynucleotide into acid-precipitable material/10 min at pH 8.3, 37°C using poly(A)·oligo(dT)$_{12-18}$ as template primer	500 mM Tris-HCl, 50 mM MgCl$_2$, 50 mM DTT, pH 8.3 and 250 mM KCl		Life Technol 18020-016 18020-024
13 U/μL	1 unit incorporates 1 nmol dTNP into a TCA-insoluble product/10 min at pH 8.3, 37°C	0.2 M KPO$_4$, 2 mM DTT, 0.2% Triton X-100, 50% glycerol, pH 7.2	No detectable exogenous RNase, nicking or degradation of RNA	NBL Gene 020604
30 U/μL	1 unit incorporates 1 nmol dTNP into a TCA-insoluble product/10 min at pH 8.3, 37°C	0.2 M KPO$_4$, 2 mM DTT, 0.2% Triton X-100, 50% glycerol, pH 7.2; for high efficiency synthesis of full length cDNA in the 6-10 kilobase range	No detectable RNase, exonuclease, endonuclease, nicking	NBL Gene 020703
> 20,000 U/mg; 10,000-20,000 U/mL	1 unit incorporates 1 nmol dNTP into DE-81 adsorbable form/10 min at 37°C	200 mM KPO$_4$, 2 mM DTT, 0.2% Triton X-100, 50% glycerol, pH 7.2	No detectable RNase, DNase	Oncor 120111 120112
25,000-50,000 U/mg protein; 10,000-20,000 U/mL	1 unit incorporates 1 nmol dTMP into acid-insoluble product/10 min at pH 8.3, 37°C using poly(rA)·p(dT)$_{12-18}$ as template primer	Molecular biology grade; homogeneous purity; solution containing 0.2 M KPO$_4$, 2.0 mM DTT, 0.2% Triton X-100, 50% glycerol, pH 7.2	No detectable RNase, DNase, nickase	Pharmacia 27-0922-01 27-0922-02
5-10 U/μL and 20-25 U/μL	1 unit incorporates 1 nmol dTTP into acid-insoluble form/10 min at pH 8.3, 37°C	> 95% purity by SDS gel; 10X Reaction buffer: 250 mM Tris-HCl, 250 mM KCl, 50 mM MgCl$_2$, 2.5 mM spermidine, 50 mM DTT, pH 8.3	<1% DNase <3% RNase ≥90% supercoiled plasmid	Promega M5101 M5102 M9004
80 U/μg protein	1 unit incorporates 1.0 nmol dTMP into acid-insoluble products/10 min at 37°C	Solution containing 50% glycerol, 0.2 M KPO$_4$, 2 mM DTT, 0.2% Triton X-100 optimum T = 37°C		Seikagaku 120248-0 120248-1 120248-2
10,000-70,000 U/mL	1 unit incorporates 1.0 nmol [^{3}H]dTTP into acid-insoluble product/10 min at 37°C	Purified; 20 mM KPO$_4$, 2 mM DTT, 0.2% Triton X-100, 50% glycerol, pH 7.2	No detectable endonuclease, nonspecific RNase	Stratagene 600081 600082
***Escherichia coli* strain NPR, recombinant**				
3000-5000 U/mg protein	1 unit incorporates 1 nmol dTTP into acid-insoluble form/hr at 37°C with poly(rA) dT 12-18 as template	≥95% purity; 50 mM Tris-HCl, 100 mM NaCl, 0.10 mM EDTA, 1.0 mM DTT, 0.1% NP40, 50% glycerol, pH 7.0	No detectable endo- and exonuclease activities except intrinsic RNase H	Adv Immuno HIVRT1

SPECIFIC ACTIVITY	UNITS DEFINITION	PREPARATION FORM	ADDITIONAL ACTIVITIES	SUPPLIER CATALOG NO.
Escherichia coli HB101/pB6B 15.23, recombinant (MW = 71,000 Da)				
>40,000 U/mg; >20,000 U/mL	1 unit incorporates 1 nmol TMP in acid-insoluble product/10 min at 37°C with poly(A)·(dT)$_{15}$ as substrate	Coomassie Blue shows a single band purity; 50 mM Tris-HCl, 10 mM DTT, 100 mM NaCl, 0.05% polydocanol, 1 mM EDTA, 50% glycerol, pH 8.4	No detectable nonspecific RNases, nonspecific DNases (gel electrophoresis)	Boehringer 1062603
Escherichia coli, carrying a clone of M-MLV reverse transcriptase (optimum T = 50°C, MW = 71,000 Da)				
50-250 U/μL	1 unit incorporates 1 nmol deoxynucleotides into acid-precipitable material/10 min at 37°C using poly(rA)·oligo(dT)$_{12-18}$ as template primer	>90% purity by SDS-PAGE; 20 mM Tris-HCl, 0.1 M NaCl, 0.1 mM EDTA, 1 mM DTT, 0.01% NP40, 50% glycerol, pH 7.5	No detectable non-specific ss- and ds-endonuclease, exonuclease, RNase	Amersham E 70456Y E 70456Z
Escherichia coli, carrying modified M-MLV reverse transcriptase on a plasmid				
50,000 U/mL	1 unit incorporates 1 nmol TTP into acid-insoluble form/10 min at pH 8.0, 37°C using poly(rA)-oligo(dT) as template-primer	50 mM Tris-HCl, 0.1 mM DTT, 100 mM NaCl, 1 mM EDTA, 0.1% NP40, 50% glycerol, pH 8.3	No detectable endonuclease, RNase	AGS Heidelb F00755S F00755M
Escherichia coli, carrying the M-MLV pol gene				
200 U/μL	1 unit incorporates 1 nmol deoxyribonucleotide into acid-precipitable material/10 min at pH 8.3, 37°C using poly(A)·oligo(dT)$_{12-18}$ as template primer	Purity by SDS-PAGE; 250 mM Tris-HCl, 15 mM MgCl$_2$, 375 mM KCl, pH 8.3 and 100 mM DTT	No detectable RNase H	Life Technol 18053-017
200 U/μL	1 unit incorporates 1 nmol deoxyribonucleotide into acid-precipitable material/10 min at pH 8.3, 37°C using poly(A)·oligo(dT)$_{12-18}$ as template primer	Purity by SDS-PAGE; 250 mM Tris-HCl, 15 mM MgCl$_2$, 375 mM KCl, pH 8.3 and 100 mM DTT	No detectable RNase H	Life Technol 18064-014 18064-071
200 U/μL	1 unit incorporates 1 nmol deoxyribonucleotide into acid-precipitable material/10 min at pH 8.3, 37°C using poly(A)·oligo(dT)$_{12-18}$ as template primer			Life Technol 28025-013 28025-021
Escherichia coli, carrying the plasmid pB6B15.23				
25,000 U/mL	1 unit incorporates 10 nmol TTP into acid-insoluble material/10 min at 37°C using poly(rA)-oligo(dT) as template primer	0.1 mM NaCl, 50 mM Tris-HCl, 5 mM DTT, 1 mM EDTA, 0.1% NP40, 50% glycerol, pH 7.6	No detectable endonuclease, RNase	NE Biolabs 253S 253L

SPECIFIC ACTIVITY	UNITS DEFINITION	PREPARATION FORM	ADDITIONAL ACTIVITIES	SUPPLIER CATALOG NO.
Escherichia coli, plasmid pRC-RT (MW = 66,000 and 51,000 Da [dimeric])				
≥10,000 U/mg protein	1 unit incorporates 1 nmol [^{3}H]dTMP into acid-insoluble product/10 min at 37°C using polyA/dp(T$_P$-T$_{12}$) as template primer	Extensively purified by chromatography to single band by SDS-PAGE; solution containing 10 mM KPO$_4$, 1 mM DTT, 20% glycerol, pH 7.4		Worthington LM05003 LM05006 LM05000
Escherichia coli, recombinant				
≥5000 U/mg protein	1 unit incorporates 1.0 nmol [^{3}H]-TMP into acid-insoluble products/10 min at 37°C using poly(A)·d(pT)$_{15}$ as substrate	Recombinant; 99% by HPLC, SDS-PAGE; lyophilized containing 0.2% BSA as stabilizer	No detectable nuclease	Boehringer 1465333
20-40 U/μL	1 unit incorporates 1 nmol deoxyribonucleotide into DE-81 absorbable form/10 min at 37°C	Overproducer; 50 mM Tris-HCl, 0.1 M NaCl, 0.1% Triton X-100, 1 mM EDTA, 5 mM DTT, 50% glycerol, pH 8.3	No detectable endo- and exodeoxyribonucleases, RNases	Fermentas EP0351 EP0352
100-200 U/μL	1 unit incorporates 1 nmol dTTP into acid-insoluble form/10 min at pH 8.3, 37°C	Recombinant; ≥90% purity by SDS gel; 5X reaction buffer: 250 mM Tris-HCl, 375 mM KCl, 15 mM MgCl$_2$, 50 mM DTT, pH 8.3	No detectable RNase H <1% DNase <3% RNase ≥90% supercoiled plasmid	Promega M5301 M5302
HIV				
10 U/μL	1 unit incorporates 1 nmol labeled dTTP into acid-insoluble material/10 min at pH 8.3, 37°C	100 mM KPO$_4$, 1 mM DTT, 50% glycerol, pH 7.1		Calbiochem 382129
	1 unit incorporates 1 nmol dTTP into acid-insoluble form/10 min at 37°C	20 mM KPO$_4$, 1.0 mM DTT, 0.02% Triton X-100, 50% glycerol, pH 7.1	No detectable endonuclease, exonuclease	CHIMERX 1373-01 1373-02
Moloney murine leukemia virus (MW = 71,000 Da; stable at 37°C, low activity at 42°C)				
	1 unit incorporates 1 nmol TTP into acid-insoluble form/10 min at 37°C using poly(A)-oligo dT$_{12-18}$ as substrate	50 mM Tris-HCl, 0.1 M NaCl, 1 mM EDTA, 5 mM DTT, 0.1% Triton X-100, 50% glycerol, pH 8.3	No detectable RNase, exonuclease	Adv Biotech AB-0322 AB-0322b
	1 unit incorporates 1 nmol dTTP into acid-insoluble form/10 min at 37°C	50 mM Tris-HCl, 5.0 mM DTT, 1.0 mM EDTA, 0.1 M NaCl, 0.1% NP40, 50% glycerol, pH 8.0	No detectable endonuclease, RNase	CHIMERX 1375-01 1375-02
35,000 U/mg	1 unit incorporates 10 nmol dTTP into acid-insoluble material/10 min at pH 8.6, 37°C using oligo(dT)$_{12-18}$-primed poly(A)$_n$ as template	Solution containing 50% glycerol, 50 mM Tris-HCl, 0.1 M NaCl, 0.1 mM EDTA, 1 mM DTT, 0.1% Triton X-100, pH 7.5	No detectable RNase, endonuclease, exonucleolytic DNase, protease	Epicentre M4425H M4410H
	1 unit incorporates 1 nmol labeled dATP into acid-insoluble material/10 min at 37°C	Solution containing 0.1 mM NaCl, 50 mM Tris-HCl, 1 mM EDTA, 5 mM DTT, 0.1% NP40, 50% glycerol, pH 8.0		ICN 152020
	1 unit incorporates 1 nmol TMP into DE-81 adsorbable form/10 min at 37°C using polyA-oligodT$_{12-18}$ as substrate	50 mM Tris-HCl, 0.1 M NaCl, 5 mM DTT, 1 mM EDTA, 0.1% Triton X-100, 50% glycerol, pH 8.3	No detectable RNase, DNase	Oncor 120301 120302

RNA-Directed DNA Polymerase continued — 2.7.7.49

SPECIFIC ACTIVITY	UNITS DEFINITION	PREPARATION FORM	ADDITIONAL ACTIVITIES	SUPPLIER CATALOG NO.
Moloney murine leukemia virus *continued*				
50,000–95,000 U/mg protein; 10,000–20,000 U/mL	1 unit incorporates 1 nmol dTMP into acid-insoluble product/10 min at pH 8.3, 37°C using poly(rA)·p(dT)$_{12-18}$ as template primer	Molecular biology grade; homogeneous purity; solution containing 50 mM Tris-HCl, 0.1 M NaCl, 1 mM EDTA, 5 mM DTT, 0.1% Triton X-100, 50% glycerol, pH 8.3	No detectable RNase, DNase, nickase	Pharmacia 27-0925-01 27-0925-02
50,000 U/mL	1 unit incorporates 1.0 nmol [^{3}H]TTP into acid-insoluble product/10 min at 37°C	50 mM Tris-HCl, 5 mM DTT, 1 mM EDTA, 100 mM NaCl, 0.1% NP40, 50% glycerol, pH 8.0	No detectable RNase H, DNase, nonspecific RNase	Stratagene 600085
Rous associated virus 2				
10–30 U/μL	1 unit incorporates 1 nmol [^{3}H]dTMP/10 min at 37°C with poly(rA)-oligo(dT) as template primer	200 mM KPO$_4$, 2 mM DTT, 0.2% NP40, 50% glycerol, pH 7.2	No detectable non-specific nuclease	Amersham E 2610Y E 2610Z
400 U and 1600 U	1 unit incorporates 1 nmol [^{3}H]dTMP/10 min at pH 8.3, 37°C with poly(rA)·oligo(dT) as template primer	Solution containing 200 mM KPO$_4$, 2 mM DTT, 0.2% NP40, 50% glycerol, pH 7.2	No detectable nuclease	TaKaRa 2610
Thermus thermophilus 111				
	1 unit incorporates 1 nmol dTTP into acid-insoluble form/10 min at 50°C	50 mM Tris-HCl, 5.0 mM DTT, 0.1 mM EDTA, 50% glycerol, stabilizers, pH 7.5	No detectable endonuclease, 3′-exonuclease, 5′-exonuclease/5′-phosphatase, nonspecific RNase, ss- and ds-DNase	CHIMERX 1374-01 1374-02

mRNA Guanylyltransferase — 2.7.7.50

REACTION CATALYZED

GTP + (5′)pppPur-mRNA ↔ Pyrophosphate + G(5′)pppPur-mRNA (mRNA containing a guanosine residue linked 5′ through three phosphates to the 5′ position of the terminal residue)

SYSTEMATIC NAME

GTP:mRNA guanylyltransferase

SYNONYMS

mRNA capping enzyme, guanylyltransferase

REACTANTS

GTP, (5′)pppPur-mRNA, poly(A), poly(G), pyrophosphate, G(5′)pppPur-mRNA, m^{7}G(5′)pppAn, m^{7}G(5′)pppGn

APPLICATIONS

- Capping *in vitro* transcripts for labeling or *in vitro* translation

NOTES

- Classified as a transferase, transferring phosphorus-containing groups: nucleotidyltransferases

2.7.7.50 mRNA Guanylyltransferase continued

SPECIFIC ACTIVITY	UNITS DEFINITION	PREPARATION FORM	ADDITIONAL ACTIVITIES	SUPPLIER CATALOG NO.
Saccharomyces cerevisiae pep4				
25 U	1 unit incorporates 1 pmol GMP into ppG-terminated RNA/30 min at pH 7.0, 30°C			CLONTECH 8416-1
Vaccinia virus				
1-5 U/μL	1 unit incorporates 1 pmol GMP into yeast 5S RNA/30 min at pH 7.9, 37°C			Life Technol 18024-018

2.8.1.1 Thiosulfate Sulfurtransferase

REACTION CATALYZED
    Thiosulfate + cyanide ↔ Sulfite + thiocyanate
SYSTEMATIC NAME
    Thiosulfate:cyanide sulfurtransferase
SYNONYMS
    Thiosulfate cyanide transsulfurase, thiosulfate thiotransferase, rhodanese

REACTANTS
    Thiosulfate, cyanide, sulfite, thiocyanate
NOTES
- Classified as a transferase, transferring sulfur-containing groups: sulfurtransferases
- A few other sulfur compounds act as donors

SPECIFIC ACTIVITY	UNITS DEFINITION	PREPARATION FORM	ADDITIONAL ACTIVITIES	SUPPLIER CATALOG NO.
Bovine liver				
100-300 U/mg solid	1 unit converts 1.0 μmol cyanide to thiocyanate/min at pH 8.6, 25°C	Highly purified; salt-free; lyophilized		Sigma R1756
10-20 U/mg solid	1 unit converts 1.0 μmol cyanide to thiocyanate/min at pH 8.6, 25°C	Purified; powder		Sigma R4125

3-Oxoacid CoA-Transferase

2.8.3.5

REACTION CATALYZED
Succinyl-CoA + a 3-oxo acid ↔ Succinate + a 3-oxoacyl-CoA

SYSTEMATIC NAME
Succinyl-CoA:3-oxo-acid CoA-transferase

SYNONYMS
Succinyl coenzyme A transferase

REACTANTS
Succinyl-CoA, malonyl-CoA, 3-oxo acid, acetoacetate, 3-oxopropanoate, 3-oxopentanoate, 3-oxo-4-methylpentanoate, 3-oxohexanoate, succinate, 3-oxoacyl-CoA

NOTES
- Classified as a transferase, transferring sulfur-containing groups: CoA-transferases
- Acting more slowly are 3-oxopropanoate, 3-oxopentanoate, 3-oxo-4-methylpentanoate, 3-oxohexanoate

SPECIFIC ACTIVITY	UNITS DEFINITION	PREPARATION FORM	ADDITIONAL ACTIVITIES	SUPPLIER CATALOG No.
Porcine heart				
2-4 U/mg protein	1 unit forms 1.0 μmol succinyl CoA from acetoacetyl CoA and succinic acid/min at pH 8.1, 25°C	Suspension in 3.2 M $(NH_4)_2SO_4$, 0.01 M KPO_4 buffer, 1 mM EDTA, pH 7		Sigma S1259

3.1.1.1 Carboxylesterase

REACTION CATALYZED
A carboxylic ester + H_2O ↔ An alcohol + a carboxylate

SYSTEMATIC NAME
Carboxylic-ester hydrolase

SYNONYMS
Ali-esterase, B-esterase, monobutyrase, cocaine esterase, procaine esterase, methylbutyrase, esterase, lipase

REACTANTS
Carboxylic ester, vitamin A esters, H_2O, alcohol, carboxylate

APPLICATIONS
- Lactose reduction and flavor modification in dairy applications. Develops strong piccanted flavor in Romano and other aged Italian cheeses

NOTES
- Enantioselective hydrolysis of monoesters
- Selective hydrolysis of racemic or mesodiesters
- Classified as a hydrolase, acting on ester bonds: carboxylic ester hydrolases
- Wide specificity
- Microsomal enzymes also catalyzes the reactions of arylesterase (EC 3.1.1.2), lysophospholipase (EC 3.1.1.5), acetylesterase (EC 3.1.1.6), acylglycerol lipase (EC 3.1.1.23), acylcarnitine hydrolase (EC 3.1.1.28), palmitoyl-CoA hydrolase (EC 3.1.2.2), amidase (EC 3.5.1.4), aryl-acylamidase (EC 3.5.1.13)

SPECIFIC ACTIVITY	UNITS DEFINITION	PREPARATION FORM	ADDITIONAL ACTIVITIES	SUPPLIER CATALOG No.
Bacillus species				
0.1 U/mg	1 unit hydrolyzes 1 µmol ethyl valerate/min at pH 8.0, 25°C			Fluka 46062
Equine liver				
0.7 U/mg	1 unit hydrolyzes 1 µmol ethyl butyrate/min at pH 8.0, 25°C	Lyophilized		Fluka 46069
Mucor miehei				
1 U/mg	1 unit hydrolyzes 1 µmol ethyl valerate/min at pH 8.0, 25°C	Powder		Fluka 46059
Natural pregastric				
48 U/g		Powder containing salt and nonfat dry milk		ChrHansens Romase 500
Porcine liver				
130 U/mg	1 unit converts 1 µmol butyric acid ethylester to product/min at 25°C	Suspension 3.2 M $(NH_4)_2SO_4$, pH 6	<0.01% AP	Boehringer 104698
220 U/mg	1 unit hydrolyzes 1 µmol ethyl valerate/min at pH 8.0, 25°C	Lyophilized		Fluka 46058

SPECIFIC ACTIVITY	UNITS DEFINITION	PREPARATION FORM	ADDITIONAL ACTIVITIES	SUPPLIER CATALOG NO.
Porcine liver *continued*				
130 U/mg protein	1 unit hydrolyzes 1 μmol ethyl butyrate/min at pH 8.0, 25°C	Suspension in 3.2 M $(NH_4)_2SO_4$, pH 6.0	<0.01% AP	Fluka 46063
2300 U/g solid; 800 U/g moistened	1 unit hydrolyzes 1 μmol ethyl valerate/min at pH 8.0, 25°C	Moist pearls immobilized on copolymer of methacrylamide, allyl-glycidylether, methylene-bis-acrylamide	13 mg protein/g dried material 4 mg protein/g wet material	Fluka 46064
150 U/mg protein	1 unit hydrolyzes 1.0 μmol ethyl butyrate to butyric acid and EtOH/min at pH 8.0, 25°C	Suspension in 3.2 M $(NH_4)_2SO_4$, pH 8		Sigma E2884
15 U/mg solid	1 unit hydrolyzes 1.0 μmol ethyl butyrate to butyric acid and EtOH/min at pH 8.0, 25°C	Lyophilized containing <5% buffer salts		Sigma E3019
200 U/mg protein	1 unit hydrolyzes 1.0 μmol ethyl butyrate to butyric acid and EtOH/min at pH 8.0, 25°C	Suspension in 3.2 M $(NH_4)_2SO_4$, pH 8		Sigma E3128
Rabbit liver				
100 U/mg protein	1 unit hydrolyzes 1.0 μmol o-nitrophenyl butyrate to butyric acid and o-nitrophenol/min at pH 7.5, 25°C	Lyophilized containing 85% protein; balance primarily Tris buffer salts		Sigma E0887
80-120 U/mg protein	1 unit hydrolyzes 1.0 μmol o-nitrophenyl butyrate to butyric acid and o-nitrophenol/min at pH 7.5, 25°C	Crystallized; suspension in 3.6 M $(NH_4)_2SO_4$ and 0.001 M Tris, pH 8.5		Sigma E9636
Rhizomucor miehei (optimum pH = 6.0-7.0)				
		Food grade		EnzymeDev Enzeco[R] Esterase Lipase
		Food grade		EnzymeDev Enzeco[R] Esterase Lipase
Streptomyces rochei (optimum pH = 5-6, T = 30-35°C)				
10,000 U/g wet		Immobilized		Wako 032-14571
500 U/mg		Purified		Wako 035-14561
50 U/mg		Crude		Wako 038-14551
Thermoanaerobium brockii				
0.1 U/mg	1 unit hydrolyzes 1 μmole of ethyl valerate/min at pH 8.0, 25°C	Powder		Fluka 46061

3.1.1.3 Triacylglycerol Lipase

REACTION CATALYZED
   Triacylglycerol + H_2O ↔ Diacylglycerol + a carboxylate

SYSTEMATIC NAME
   Triacylglycerol acylhydrolase

SYNONYMS
   Alkaline lipase, colipase, lipase, lipase A, lipase B, pancrelipase, phospholipase C, PL, tributyrase, triglyceride lipase

REACTANTS
   Triacylglycerol, H_2O, diacylglycerol, carboxylate; comparative reactivity: triolein (100%), tripalmitin (22%), trilaurin (103%), trimyristin (53%), tricuprulin (312%), tricaprin (166%), tripropionin (22%), triacetin (38%)

APPLICATIONS
- Preparation of fatty acids and mono- and diglycerides from fats and oils (e.g., olive, butter, lard, cow tallow, soybean, cotton seed, sesame and rape seed)
- Low levels increase palatability and greater levels improve digestibility of fatty animal feeds, especially reduced calorie 'finishing' rations
- Addition of mild piccanted flavor to parmesan, provolone, feta, mozzarella, blue cheese and Italian-style cheeses
- Acylation and deacylation of furanose and pyranose derivatives
- Enantioselective hydrolysis of 2-methyl 3-acetoxyesters
- Preparative resolution of racemic acids and alcohols in chemical syntheses
- Interesterification of 1- and 3-ester bonds in triglycerides
- Synthesis of fatty hydroxamic acids
- Selective acylation of primary alcohols in organic solvents
- Chemoenzymatic synthesis of (-)-carbocyclic 7-deazoxetanocin G
- Tandem evaluation of serum lipase and pancreatic amylase is a valuable diagnostic tool for acute pancreatitis
- Determination of triacylglycerols
- Triglyceride hydrolysis in fabric-based soils and stains, releasing fatty acids into the wash solution. Removes fat and oil stains like cosmetics, frying fats, salad oils, butter, fat-based sauces, soups and human sebum
- Enhanced cleaning efficiency of surfactants and other chemicals
- Removal of interfering triglycerides in the electroimmunoassay of apolipoprotein B
- Hydrolysis of N-benzyloxycarbonyl-cis-2,6-(acetoxymethyl)piperidine

NOTES
- Classified as a hydrolase, acting on ester bonds: carboxylic ester hydrolases
- Yeast enzyme completely hydrolyzes triglycerides to fatty acids and glycerol, with little apparent specificity. Glycerol α- and β-positions are similarly hydrolyzed, with no isomerase requirement. It is activated by taurocholate
- Two exocrine lipase isozymes are found in pancreatic juice: lipase A is more acidic than lipase B and both contain a carbohydrate moiety. Pancreatic lipase (PL) aids in digestion by hydrolyzing emulsified esters of glycerol and fatty acids at the ester-water interface, preferentially hydrolyzing outer ester links. It is activated by Ca^{2+} (required for activity), Sr^{2+},

Mg^{2+} and detergents like sodium taurocholate. PL is inhibited by lipstatin, monoglycerides, tetrahydrolipstatin, apoproteins, benzene boronic acid, PCMB (10 μM), detergents (20 μM), fluoride, DPF (20 μM), Hg^{2+} (10 mM), Zn^{2+}, Cu^{2+}, iodine, PCMB and conjugated bile salts. Elevated serum levels are found in pancreatic and renal disease and carcinoma

SPECIFIC ACTIVITY	UNITS DEFINITION	PREPARATION FORM	ADDITIONAL ACTIVITIES	SUPPLIER CATALOG NO.
Achromobacter species (optimum pH = 10.0, T = 45°C)				
≥300 U/mg protein	1 unit liberates 1 µmol fatty acid/min at pH 10.0, 37°C	Lyophilized		Seikagaku 100148-1
Animal pregastric tissue				
17.0-20.0 U/g	1 unit liberates 5 µmol butyric acid from tributyrin substrate/min at pH 6.2, 40°C	Powder containing salt, nonfat dry milk, silicon dioxide		SBI CAPALASE™ K 01011
17.0-20.0 U/g	1 unit liberates 5 µmol butyric acid from tributyrin substrate/min at pH 6.2, 40°C	Powder containing salt, nonfat dry milk, silicon dioxide		SBI CAPALASE™ KL 01012
14.2-16.5 U/g	1 unit liberates 5 µmol butyric acid from tributyrin substrate/min at pH 6.2, 40°C	Powder containing salt, nonfat dry milk, silicon dioxide		SBI CAPALASE™ L 01013
14.2-17.3 U/g	1 unit liberates 5 µmol butyric acid from tributyrin substrate/min at pH 6.2, 40°C	Powder containing salt, nonfat dry milk, silicon dioxide		SBI ITALASE™ C 01002
14.2-17.3 U/g	1 unit liberates 5 µmol butyric acid from tributyrin substrate/min at pH 6.2, 40°C	Kosher certified; powder containing salt, soy protein concentrate, silicon dioxide		SBI KSHRASE™ C 01008
Animal pregastric tissue & Microbial lipases, produced by fermentation				
9.5-10.5 U/g	1 unit liberates 5 µmol butyric acid from tributyrin substrate/min at pH 6.2, 40°C	Kosher certified; powder containing salt, soy protein concentrate, silicon dioxide		SBI KSHRASE™ KLM 01005
9.5-10.5 U/g	1 unit liberates 5 µmol butyric acid from tributyrin substrate/min at pH 6.2, 40°C	Kosher certified; powder containing water, whey, salt, corn oil		SBI KSHRASE™ KM 01007
Aspergillus niger (optimum pH = 5.0-7.0, T = 30-40°C; stable pH 2-10 [60 min])				
60,000 U/g; 120,000 U/g	pH 6.0	Powder		Amano Lipase AP6 Lipase AP12

SPECIFIC ACTIVITY	UNITS DEFINITION	PREPARATION FORM	ADDITIONAL ACTIVITIES	SUPPLIER CATALOG NO.
Aspergillus niger continued				
		Food grade		EnzymeDev Enzeco[R] Microbial Lipase
1 U/mg	1 unit liberates 1 μmol oleic acid/min at pH 8.0, 40°C with triolein as substrate	Powder		Fluka 62294
4 U/mg	1 unit liberates 1 μmol oleic acid/min at pH 8.0, 40°C with triolein as substrate	Powder		Fluka 62301
Aspergillus oryzae (optimum pH 10.5, T = 35°C; stable pH 6-8 [40-60°C, 2 hr])				
		Granulate or solution		Novo Nordisk Lipolase™
Bacillus cereus				
≥50 U/mg solid	1 unit hydrolyzes 1 μmol lecithin/min at pH 8.0, 37°C	Lyophilized containing 10 mM Tris-HCl, pH 8.0	No detectable GO <0.3 U/mg catalase	Calbiochem 525186
Calf				
Standardized to specific enzyme level		Powder containing salt, non-fat dry milk, edible animal tissue		ChrHansens Calf Rennet Extract 961020210
Calf tongue root & salivary gland (optimum pH = 4-5, T = 37°C; stable pH = 5.5-7.5 [37°C, 30 min])				
≥40 U/g		Powder		Amano Lipase PGE
Candida antarctica				
3 U/mg	1 unit liberates 1 μmol oleic acid/min at pH 8.0, 40°C with triolein as substrate	Powder		Fluka 62299
Candida lipolytica (optimum pH = 4.0 [A] and 8.0 [B], T = 20°C [A] and 30°C [B]; stable pH 4-7, T < 40°C)				
		Powder		Amano Lipase L
1 U/mg	1 unit liberates 1 μmol oleic acid/min at pH 8.0, 40°C with triolein as substrate	Powder		Fluka 62303
Candida rugosa (formerly C. cylindraccae) (pH = 5.0-8.0, T = 30-55°C, MW = 100,000-120,000 Da; active pH 3-9)				
30,000 U/g	pH 7.0	Powder		Amano Lipase AY 30
115,000 U/g lipase		Powder		Biocatalysts L034P
105,000 U/g lipase		Food grade; powder	60,000 U/g esterase	Biocatalysts Lipomod 34P L034P

SPECIFIC ACTIVITY	UNITS DEFINITION	PREPARATION FORM	ADDITIONAL ACTIVITIES	SUPPLIER CATALOG NO.
Candida rugosa continued				
2000-5000	1 unit liberates 1 µmol glycerol/min at pH 7.0, 37°C	Lyophilized containing 90% protein		Calzyme 141A2000
		Industrial grade		EnzymeDev Enzeco[R] Lipase Concentrate
		Industrial grade		EnzymeDev Enzeco[R] Lipase
30 U/mg	1 unit liberates 1 µmol oleic acid/min at pH 8.0, 40°C with triolein as substrate	Lyophilized		Fluka 62302
20 U/mg	1 unit liberates 1 µmol oleic acid/min at pH 8.0, 37°C with olive oil as substrate	Powder containing 40% lactose	<0.001% protease <0.01% α-amylase	Fluka 62316
>600 U/mg protein; >400 U/mg solid	1 unit forms 1.0 µmol fatty acid/min at pH 7.0, 37°C	Lyophilized		Genzyme 1451
≥10,000 U/mg protein	1 unit liberates 1 µmol fatty acid/min at pH 7.0, 37°C	Lyophilized		Seikagaku 100878-1
300-1200 U/g solid	1 unit hydrolyzes 1.0 µeq fatty acid from triglyceride/hr at pH 7.2, 37°C, olive oil as substrate; equivalent to 10 µL CO_2/30 min	Insoluble enzyme attached to macroporous acrylic beads containing 50% glucose as stabilizer		Sigma L1150
700-1500 U/mg solid	1 unit hydrolyzes 1.0 µeq fatty acid from triglyceride/30 min at pH 7.2, 37°C with olive oil as substrate; equivalent to 10 µL CO_2/30 min	Contains lactose as extender	No detectable α-amylase and protease	Sigma L1754
100,000-400,000 U/mg protein	1 unit hydrolyzes 1.0 µeq fatty acid from triglyceride/30 min at pH 7.2, 37°C with olive oil as substrate; equivalent to 10 µL CO_2/30 min	Lyophilized containing 20% protein and traces of Ca^{2+}	No detectable α-amylase and protease	Sigma L8525
≥300 U/mg solid	1 unit releases 1 µmol fatty acid from emulsified olive oil/min at pH 8.0, 25°C	Purified; dialyzed; lyophilized		Worthington LS09615 LS09616 LS09617
Candida utilis				
0.1 U/mg	1 unit liberates 1 µmol oleic acid/min at pH 8.0, 40°C with triolein as substrate	Powder		Fluka 62307
Chromobacterium viscosum (optimum pH = 3.0-10.0, T = 50-60°C, MW = 120,000 [pI 3.7] and 30,000 [pI 7.3] Da; stable pH 4-10 [50°C, 1 hr], T < 70°C [pH 7.0, 10 hr])				
≥2000 U/mg	1 unit produces 1 µmol fatty acid from olive oil emulsion/min at 37°C	Lyophilized		AMRESCO 0753

SPECIFIC ACTIVITY	UNITS DEFINITION	PREPARATION FORM	ADDITIONAL ACTIVITIES	SUPPLIER CATALOG NO.
Chromobacterium viscosum continued				
2500-4500 U/mg solid	1 unit liberates 1 μmol fatty acid/min at 37°C	Lyophilized containing 80% protein	<0.01% NADH oxidase, ChO, catalase	Asahi [LP] T-01
25,000 U/g lipase		Powder		Biocatalysts L050P
2500 U/mg solid	1 unit liberates 1 μmol fatty acid from olive oil/min at pH 8.0, 37°C	Lyophilized		Calbiochem 437707
2000-8000 U/mg protein	1 unit hydrolyzes 1.0 μeq fatty acid from triglyceride/30 min at pH 7.7, 37°C with olive oil as substrate; equivalent to 10 μL CO_2/30 min	Lyophilized containing 65% protein		Sigma L0763
Fungal (optimum pH = 5.0-7.0, T = 40-50°C)				
11,000 U/g esterase	1 unit hydrolyzes glycerol tributyrate releasing 1 μmol butyric acid/min at pH 6.9, 30°C	Food grade; powder		Biocatalysts Lipomod 187P L187P
Fungal & animal (optimum pH = 7.0-9.0, T = 35-45°C)				
200 U/g lipase		Food grade; powder		Biocatalysts Lipomod 41P L041P
Human				
				Vital 17236
Human pancreas (optimum pH = 5-6, MW = 48,000 Da)				
1100-1200 U/mg protein	1 unit produces a ΔA_{340} of 1/min at pH 9.0, 25°C using triolein as substrate under colipase saturation	≥95% purity; frozen in 20 mM Tris-HCl, pH 7.8	No detectable HBsAg, anti-HCV, anti-HBc, anti-HIV	ART 16-19-120916
1100-1200 U/mg protein	1 unit changes A_{340} by 1.0/min at 25°C with triolein as substrate under colipase saturation, pH 9.2	>95% purity; solution containing 20 mM Tris-HCl, pH 7.8	No detectable HBsAg, HIV Ab	Calbiochem 437709
30-60 U/mg protein	1 unit releases 1 μmol fatty acid from olive oil emulsion/min at pH 8.0, 25°C	Chromatographically purified; lyophilized containing 98% protein, 2% salts, 0.01% NaN_3 as preservative, pH 7.5		Elastin HL938
100-150 U/mg		Solution containing 50 mM Tris-HCl, 100 mM NaCl, 1 mM benzamidine, pH 8.0	No detectable HGsAg, HCV, HIV-1 Ab <0.2% amylase	Scripps L0513
25,000 U/mg protein	1 unit hydrolyzes 1.0 μeq fatty acid from triglyceride/hr at pH 7.4, 37°C with triacetin as substrate; equivalent to 10 μL CO_2/30 min	Solution containing 0.1 M Tris, 0.1 M NaCl, serine protease inhibitor		Sigma L9780

SPECIFIC ACTIVITY	UNITS DEFINITION	PREPARATION FORM	ADDITIONAL ACTIVITIES	SUPPLIER CATALOG No.
Kid goat				
Standardized to specific enzyme level		Powder containing salt, non-fat dry milk, edible animal tissue		ChrHansens Calf Rennet Extract 960820210
Microbial (optimum pH = 5.0-8.0)				
≥1000 U/mg	1 unit produces 1 μmol fatty acid from olive oil emulsion/min at 37°C	Lyophilized		AMRESCO 0424
***Mucor javanicus* (optimum pH = 5.5-8.5, T = 30-45°C)**				
5 U/mg	1 unit liberates 1 μmol oleic acid/min at pH 8.0, 40°C with triolein as substrate	Powder		Fluka 62304
***Mucor miehei* (optimum pH = 7.0-8.0, T = 45°C)**				
4500 U/g lipase		Food grade; powder		Biocatalysts L054P
***Mucor miehei* (optimum pH = 5.0-8.0, T = 35-50°C)**				
24 U/mg	1 unit liberates 1 μmol oleic acid/min at pH 8.0, 40°C with triolein as substrate	Powder		Fluka 62298
10,000 LU/g	1 LU liberates 1 μmol butyric acid from emulsified tributyrin substrate in a pH-stat/min at pH 7.0, 30°C	Food grade; solution		Novo Nordisk Lipozyme^R 10.000 L
5-6 BAUN/g		Immobilized on phenolic anion exchange resin; granulate at 0.2-0.6 mm particle size with no crosslinking agents	No detectable side activities	Novo Nordisk Lipozyme^R IM
***Mucor* species (optimum pH = 7.0, T = 30-45°C; stable pH 4-8 [37°C, 1 hr])**				
10,000 U/g	pH 7.0	Powder		Amano Lipase MAP
Natural pregastric				
17 U/g		Kosher certified (including Passover); powder containing salt		ChrHansens Kosher Lipase-KLC
Pancreas (optimum pH = 7.0-7.5)				
21,000 U/g esterase	1 unit hydrolyzes glycerol tributyrate releasing 1 μmol butyric acid/min at pH 6.9, 30°C	Food grade; powder		Biocatalysts Lipomod 224P L224P
2000 U/g esterase	1 unit hydrolyzes glycerol tributyrate releasing 1 μmol butyric acid/min at pH 6.9, 30°C	Food grade; powder		Biocatalysts Lipomod 299P L299P
≥24 USP U/mg		Powder	≥100 USP U/mg amylase, protease	SPL 0610

3.1.1.3 Triacylglycerol Lipase continued

SPECIFIC ACTIVITY	UNITS DEFINITION	PREPARATION FORM	ADDITIONAL ACTIVITIES	SUPPLIER CATALOG No.
Pancreas continued				
≥30 USP U/mg	1 g digests 531.4 g dietary fat (triolein)/hr at pH 9.0, 37.5°C	Powder		SPL 0710
≥48 USP U/mg	1 g digests 850.2 g dietary fat (triolein)/hr at pH 9.0, 37.5°C	Powder prepared without activation of other naturally occurring zymogens		SPL 0770
≥75 USP U/mg	Pancreatin USP converts ≥250 times its weight of USP Potato Starch Reference Standard into soluble carbohydrates and ≥275 times its weight of casein into proteases; 1 g digests 1327 g tristearin/hr at pH 9.0, 37.5°C	Powder	≥250 USP U/mg amylase ≥275 USP U/mg protease	SPL 0820
Pancreas & fungal (optimum pH = 5.0-8.0, T = 40-50°C)				
15,000 U/g esterase	1 unit hydrolyzes glycerol tributyrate releasing 1 μmol butyric acid/min at pH 6.9, 30°C	Food grade; powder	7500 U/g lipase	Biocatalysts Lipomod 29P L029P
2000 U/g esterase	1 unit hydrolyzes glycerol tributyrate releasing 1 μmol butyric acid/min at pH 6.9, 30°C	Food grade; powder		Biocatalysts Lipomod 309P L309P
Penicillium cyclopium (optimum pH = 4.5-7.5, T = 30-50°C)				
50,000 U/g	pH 5.6	Powder		Amano Lipase G 50
Penicillium roqueforti (optimum pH = 6.0-8.0, T = 30°C; stable < 30°C)				
900-1100 U/g	pH 7.0	Powder		Amano Lipase R 10
4500 U/g esterase	1 unit hydrolyzes glycerol tributyrate releasing 1 μmol butyric acid/min at pH 6.9, 30°C	Food grade; powder		Biocatalysts Lipomod 338P L338P
2 U/mg	1 unit liberates 1 μmol oleic acid/min at pH 8.0, 40°C with triolein as substrate	Powder		Fluka 62308
Phycomyces nitens NRRL 2444 (optimum pH = 6.0 and 7.0 [PVA], pI = 5.9, T = 40°C, MW = 26,000-27,000 Da)				
100-200 U/mg		Lyophilized		Wako 122-02651
Porcine pancreas (optimum pH = 6.0-9.0, pI = 5.0, T = 37-55°C, MW = 11,000 Da [1 chain, 5 disulfides]; inactivated at 75°C when 4 > pH > 11.5)				
8, 16, 24 and 30 USP U/mg		Powder		Am Labs
≥24 USP U/mg lipase		USP powder	100 USP U/mg protease 100 USP U/mg amylase	Am Labs

SPECIFIC ACTIVITY	UNITS DEFINITION	PREPARATION FORM	ADDITIONAL ACTIVITIES	SUPPLIER CATALOG No.
Porcine pancreas continued				
42,000 U/g esterase	1 unit hydrolyzes glycerol tributyrate releasing 1 µmol butyric acid/min at pH 6.9, 30°C	Food grade; powder	4000 U/g lipase	Biocatalysts L115P
300 U/mg solid	1 unit converts 1 µmol triolein to product/min at 25°C under colipase saturation	Purified; lyophilized	No detectable protease	Boehringer 644072
70,000 U/mg solid		Lyophilized		Boehringer 644099
20,000-50,000	1 unit liberates 1 µmol glycerol/min at pH 7.0, 37°C	Lyophilized containing 90% protein		Calzyme 140A20000
4000-12,000 U/mg solid	1 unit releases 1 µmol fatty acid from tributyrin/min at pH 6.5, 25°C	Lyophilized containing 95% protein	No detectable lipase	Calzyme 142A4000
		Partially purified (50%); lyophilized containing 2% NaCl	No detectable lipase	Elastin CL644
120,000-145,000 U/g protein; 16,000 U/g solid	1 unit releases 100 µmol fatty acid from olive oil emulsion/hr at pH 7.8, 37°C	Purified; lyophilized containing lactose	0.02-0.05 U/mg α-amylase	Elastin IC398
50-80 U/mg protein	1 unit releases 1 µmol fatty acid from olive oil emulsion/min at pH 8.0, 25°C	Purified; lyophilized		Elastin L397
30,000-45,000 U/mg protein	1 unit releases 1 µmol fatty acid from olive oil emulsion/hr at pH 7.8, 37°C	Chromatographically purified; salt-free; lyophilized	No detectable elastase, protease, α-amylase, carboxypeptidase	Elastin LC298
3.5 U/mg	1 unit liberates 1 µmol oleic acid/min at pH 8.0, 37°C with olive oil as substrate	Powder		Fluka 62300
100 U/mg	1 unit liberates 1 µmol oleic acid/min at pH 8.0, 37°C with olive oil as substrate	Lyophilized		Fluka 62313
>12,000 U/mg protein; >500 U/mg solid	1 unit catalyzes a 1.0 µmol increase in the release of fatty acid from tributyrin/min at pH 6.5, 20°C	Lyophilized		Genzyme 6251
>12,000 U/mg protein	1 unit releases 1.0 µmol fatty acid from tributyrin/min at pH 6.5, 20°C	Solution		Genzyme 6254
16,000 U/g	1 unit liberates 100 µmol fatty acid/hr at pH 7.8, 37°C	Purified		ICN 100817
12,000 U/g	1 unit liberates 100 µmol fatty acid/hr at pH 7.8, 37°C		No detectable diastase	ICN 100823
Lipase value is 3.5 times the USP specification				ICN 102189

3.1.1.3 Triacylglycerol Lipase continued

SPECIFIC ACTIVITY	UNITS DEFINITION	PREPARATION FORM	ADDITIONAL ACTIVITIES	SUPPLIER CATALOG NO.
Porcine pancreas continued				
20,000-100,000 U/mg protein	1 unit hydrolyzes 1.0 µeq fatty acid from triglyceride/hr at pH 7.7, 37°C with olive oil as substrate; equivalent to 10 µL CO_2/30 min	Lyophilized containing 70% protein		Sigma L0382
35-70 U/mg protein (triacetin); 110-220 U/mg protein (olive oil)	1 unit hydrolyzes 1.0 µeq fatty acid from triglyceride/hr at pH 7.4, 37°C with triacetin as substrate; 30 min incubation time at pH 7.7 using olive oil as substrate; equivalent to 10 µL CO_2/30 min	Crude; containing 25% protein	Amylase and protease	Sigma L3126
≥100 U/mg solid	1 unit releases 1 µmol fatty acid from emulsified olive oil/min at pH 8.0, 25°C	Lyophilized		Worthington LS03295 LS03296 LS03297
Pseudomonas cepacia				
30 U/mg	1 unit liberates 1 µmol oleic acid/min at pH 8.0, 40°C with triolein as substrate	Powder		Fluka 62309
Pseudomonas fluorescens (optimum pH = 5.0-9.0, T = 37-65°C)				
10,000 U/g lipase		Powder		Biocatalysts L056P
42 U/mg	1 unit liberates 1 µmol oleic acid/min at pH 8.0, 40°C with triolein as substrate	Powder		Fluka 62312
3500 U/mg	1 unit liberates 1 µmol oleic acid/min at pH 8.0, 40°C with triolein as substrate	Lyophilized		Fluka 62321
Pseudomonas mendocina expressed in Bacillus (optimum pH = 7-10, T = 50-60°C [pH 8.0, p-nitrophenyl butyrate substrate])				
≥20,000 GLU/kg	Activity is determined spectrophotometrically against an internal standard using p-nitrophenyl-butyrate as substrate	Coated granulate at 0.3-0.85 mm particle size		Genencor Lumafast™ 2000G with Enzoguard^R
Pseudomonas species (optimum pH = 7.0-9.0, pI = 5.95, T = 45-50°C, MW = 134,000 Da; stable at pH 7.0-9.0 [25°C, 20 hr] and T < 55°C [pH 7.0, 10 min])				
20,000 U/g	pH 7.0	Powder; optimum T = 60°C; stable pH 4-10, T < 70°C		Amano Lipase AK
30,000 U/g; 800,000 U/g	pH 7.0	Powder; optimum T = 40-60°C; active pH 3-11		Amano Lipase PS 30 Lipase PS 800
25 U/mg solid	1 unit hydrolyzes 1.0 µeq fatty acid from triglyceride/hr at pH 7.7, 37°C; equivalent to 10 µL CO_2/30 min	Lyophilized containing 50% BSA and enzyme protein; balance primarily Na cholate, sucrose, MgCl, KPO_4 buffer		Sigma L9518

SPECIFIC ACTIVITY	UNITS DEFINITION	PREPARATION FORM	ADDITIONAL ACTIVITIES	SUPPLIER CATALOG NO.
Pseudomonas species continued				
≥0.5 U/mg solid	1 unit forms 1 μmol ethyloctanoate/min at 25.5°C	Immobilized on Hyflo Super-Cel; powder containing sugars as stabilizers		Toyobo LIP-301
≥0.5 U/mg		Immobilized		Wako 124-04171
***Rhizomucor miehei* (optimum pH = 5-7, T < 50°C)**				
1000 LU/g	1 LU liberates 1 μmol butyric acid from emulsified tributyrin substrate in a pH-stat/min at pH 7.0, 30°C	Food grade; solution		Novo Nordisk Palatase[R] M 1000 L
200 LU/g	1 LU liberates 1 μmol butyric acid from emulsified tributyrin substrate in a pH-stat/min at pH 7.0, 30°C	Food grade; solution		Novo Nordisk Palatase[R] M 200 L
Rhizopus arrhizus				
14,000 U/mg	1 unit converts 1 μmol olive oil to product/min at 37°C with Rhodoviol as detergent	Suspension 3.2 M $(NH_4)_2SO_4$, 0.01 M KPO_4, BSA as stabilizer, pH 6	<0.00001% GK, AP, acid phosphatase, HK <0.001% ATPase	Boehringer 414590
2 U/mg	1 unit liberates 1 μmol oleic acid/min at pH 8.0, 40°C with triolein as substrate	Powder		Fluka 62305
400,000 U/mg protein	1 unit hydrolyzes 1.0 μeq fatty acid from triglyceride/30 min at pH 7.7, 37°C with olive oil as substrate; equivalent to 10 μL CO_2/30 min	Suspension in 3.2 M $(NH_4)_2SO_4$ and 10 mM KPO_4, pH 6.0		Sigma L4384
***Rhizopus delemar* (optimum pH = 5.6, T = 40°C)**				
60 U/mg	1 unit liberates 1 μmol oleic acid/min at pH 8.0, 37°C with olive oil as substrate	Powder		Fluka 62311
≥600 U/mg	1 unit releases fatty acids corresponding to 0.2 mL of 0.05 N NaOH solution/30 min at pH 5.6, 30°C using standard olive oil as substrate	Powder		Seikagaku 100890-1
***Rhizopus japonicus* (optimum pH = 7.0, T = 40°C; stable pH 3-8, T < 40°C [pH 7.0, 30 min])**				
150,000 U/g	pH 7.0	Powder		Amano Lipase FAP 15
500,000 U/g solid	1 unit liberates 1 μmol fatty acid/hr at pH 7.0, 37°C	Powder; enzyme preparation		Nagase
1,000,000 U/g solid	1 unit liberates 1 μmol fatty acid/hr at pH 7.0, 37°C	Powder; enzyme preparation		Nagase Lilipase A-10
100,000 U/g solid	1 unit liberates 1 μmol fatty acid/hr at pH 7.0, 37°C	Powder; whole cell preparation		Nagase Lilipase B-2
200,000 U/g solid	1 unit liberates 1 μmol fatty acid/hr at pH 7.0, 37°C	Powder; whole cell preparation		Nagase Lilipase B-4

SPECIFIC ACTIVITY	UNITS DEFINITION	PREPARATION FORM	ADDITIONAL ACTIVITIES	SUPPLIER CATALOG No.
Rhizopus niveus (optimum pH = 7.0, T = 30-45°C; stable pH 3-8)				
80,000 U/g	pH 7.0	Powder		Amano Lipase N
2.5 U/mg	1 unit liberates 1 µmol oleic acid/min at pH 8.0, 40°C with triolein as substrate	Powder		Fluka 62310
Wheat germ				
0.1 U/mg	1 unit liberates 1 µmol oleic acid/min at pH 8.0, 40°C with triolein as substrate			Fluka 62306
3000-8000 U/g agarose	1 unit hydrolyzes 1.0 µeq fatty acid from triglyceride/hr at pH 7.4, 37°C with triacetin as substrate; equivalent to 10 µL CO_2/30 min	Insoluble enzyme attached to 4% beaded agarose; suspension in 3.2 M $(NH_4)_2SO_4$ and 0.02% thimerosal		Sigma L2764
10 U/mg protein; inactive on olive oil	1 unit hydrolyzes 1.0 µeq fatty acid from triglyceride/hr at pH 7.4, 37°C with triacetin as substrate; equivalent to 10 µL CO_2/30 min	Lyophilized containing 95% protein	Acid phosphatase	Sigma L3001
Unspecified				
10 N.F. U/mg		Powder		Wako 120-01052

Phospholipase A$_2$ 3.1.1.4

REACTION CATALYZED
Phosphatidylcholine + H$_2$O ↔ 1-Acylglycerophosphocholine + a carboxylate

SYSTEMATIC NAME
Phosphatidylcholine 2-acylhydrolase

SYNONYMS
Ammodytoxin, β-bungarotoxin, lecithinase A, notexin Np, phosphatidase, phosphatidolipase, PLA$_2$

REACTANTS
Phosphatidylcholine, phosphatidylethanolamine, choline plasmalogen, phosphatides, H$_2$O, 1-Acylglycerophosphocholine, carboxylate, phosphatidic acid, phosphatidyl glycerol, 1,3-bis(3-sn-phosphatidyl)glycerol (cardiolipin), micellar diheptanoylphosphatidylcholine

APPLICATIONS
- Probe for structure-function relationship studies of biomembranes
- Determination of phospholipids in serum, with choline
- Partial hydrolysis of phospholipids for improved emulsification

NOTES
- Classified as a hydrolase, acting on ester bonds: carboxylic ester hydrolases
- A heat-stable, calcium-dependent enzyme catalyzing the hydrolysis of the 2-acyl bond of 3-n-phosphoglycerides; widely found in nature
- A selective inhibitor of acetyl choline release at the neuromuscular junction
- Removes the fatty acid attached to the 2-position of secondary substrates
- The main antigen in bee venom, causing hydrolysis of erythrocyte and mast cell membrane phospholipids, acting in synergy with melittin (LD$_{50}$ = 0.37 μg/g in mice [ip]). Bee venom enzyme prefers neutral zwitterionic phospholipids; pig pancreas enzyme is more active with anionic phospholipids
- Activated by Ca^{2+} (optimal activity requires ≥6 mM), Triton X-100 and non-ionic detergents.
- Inhibited by EDTA, Zn^{2+}, Ba^{2+} and Mn^{2+}
- Porcine pancreatic enzyme is irreversibly inactivated by adding 8 mM free Zn^{2+} or heat treatment (5 min at 90°C, 30 min at 80°C or 2 hrs at 70°C); reversibly inactivated at pH < 5 or limiting [Ca^{2+}]

SPECIFIC ACTIVITY	UNITS DEFINITION	PREPARATION FORM	ADDITIONAL ACTIVITIES	SUPPLIER CATALOG NO.
Agkistrodon halys (snake venom)				
11 U/mg	1 unit hydrolyzes 1 μmol 3-sn-phosphatidylcholine/min at pH 8.0, 37°C	Powder		Fluka 79475
Apis mellifera (bee venom) (MW = 14,000-18,500 Da)				
1500 U/mg prot	1 μmol PL/min	Lyophilized with no salts		BIOMOL SE-104
3000 U/mg prot; 2400 U/mg solid	1 unit converts 1 μmol lecithin from egg yolk to product/min at 40°C	Lyophilized	<0.1% U/mg proteases (at 37°C, azocoll as substrate, no EDTA)	Boehringer 1243063

3.1.1.4 Phospholipase A$_2$ continued

SPECIFIC ACTIVITY	UNITS DEFINITION	PREPARATION FORM	ADDITIONAL ACTIVITIES	SUPPLIER CATALOG No.
Apis mellifera (bee venom) continued				
1000 U/mg	1 unit liberates 1 µmol titratable fatty acid from soybean lecithin/min at pH 8.9, 25°C	≥98% purity; lyophilized		Cayman 60500
1000 U/mg	1 unit hydrolyzes 1 µmol 3-sn-phosphatidylcholine/min at pH 8.0, 37°C	Powder		Fluka 79478
≥850 U/mg protein	pH 8.9, 25°C using egg yolk phosphatidylcholine	≥96% purity by HPLC, amino acid analysis; powder		Latoxan L84 08
4000-10,000 U/g agarose	1 unit hydrolyzes 1.0 µmol L-α-phosphatidylcholine to L-α-lysophosphatidylcholine (egg yolk) and a fatty acid/min at pH 8.5, 37°C	Insoluble enzyme attached to beaded agarose; suspension in 2.0 M (NH$_4$)$_2$SO$_4$, pH 7.0		Sigma P1264
600-1800 U/mg protein	1 unit hydrolyzes 1.0 µmol L-α-phosphatidylcholine to L-α-lysophosphatidylcholine (soybean) and a fatty acid/min at pH 8.9, 25°C	Salt-free; lyophilized		Sigma P9279
Bovine pancreas				
7 U/mg	1 unit hydrolyzes 1 µmol 3-sn-phosphatidylcholine/min at pH 8.0, 37°C	Powder containing 20% protein		Fluka 79476
≥100 IU/mg	1 unit hydrolyzes 1.0 µmol L-α-phosphatidylcholine to L-α-lysophosphatidylcholine and fatty acid/min at pH 8.0, 40°C at a reaction time of 5-10 min	Lyophilized		ICN 151873
≥2.5 U/mg solid	1 unit hydrolyzes 1.0 µmol L-α-phosphatidylcholine to L-α-lysophosphatidylcholine and a fatty acid/min at pH 8.0, 37°C	Powder		Can Inova A-002
25-75 U/mg protein	1 unit hydrolyzes 1.0 µmol L-α-phosphatidylcholine to L-α-lysophosphatidylcholine (soybean) and a fatty acid/min at pH 8.0, 37°C	Lyophilized containing 20% protein; balance Tris buffer salts		Sigma P8913
Bungarus multicinctus (snake venom) (MW = 7,000 , 13,500 Da)				
		≥97% purity		Latoxan L81 16
Crotalus adamanteus (snake venom) (MW = 30,000 Da)				
200-400 U/mg protein	1 unit hydrolyzes 1.0 µmol L-α-phosphatidylcholine to L-α-lysophosphatidylcholine (soybean) and a fatty acid/min at pH 8.9, 25°C	Lyophilized		Sigma P0790

SPECIFIC ACTIVITY	UNITS DEFINITION	PREPARATION FORM	ADDITIONAL ACTIVITIES	SUPPLIER CATALOG No.
Crotalus adamanteus (snake venom) *continued*				
≥200 U/mg solid	1 unit liberates 1 μmol titratable fatty acid from soybean lecithin/min at pH 8.9, 25°C	Highly purified; lyophilized		Worthington LS05660 LS05662
Crotalus atrox (snake venom)				
300–700 U/mg protein	1 unit hydrolyzes 1.0 μmol L-α-phosphatidylcholine to L-α-lysophosphatidylcholine (soybean) and a fatty acid/min at pH 8.9, 25°C	Lyophilized		Sigma P3770
Crotalus durissus terrificus (snake venom)				
200 U/mg protein	1 unit hydrolyzes 1.0 μmol L-α-phosphatidylcholine to L-α-lysophosphatidylcholine (soybean) and a fatty acid/min at pH 8.9, 25°C	Lyophilized containing 50% protein; balance primarily Tris buffer salts		Sigma P5910
Laticauda semifasciata, isozymes				
150–600 U/mg protein	1 unit hydrolyzes 1.0 μmol L-α-phosphatidylcholine to L-α-lysophosphatidylcholine and a fatty acid/min at pH 8.9, 25°C	Lyophilized containing 50% protein; balance KPO_4 buffer salts		Sigma P7177
Naja mocambique mocambique (snake venom)				
1500 U/mg protein	1 unit hydrolyzes 1.0 μmol L-α-phosphatidylcholine to L-α-lysophosphatidylcholine (soybean) and a fatty acid/min at pH 8.9, 25°C	Lyophilized containing 80% protein		Sigma P4034
400 U/mg protein	1 unit hydrolyzes 1.0 μmol L-α-phosphatidylcholine to L-α-lysophosphatidylcholine (soybean) and a fatty acid/min at pH 8.9, 25°C	Lyophilized containing 90% protein		Sigma P7653
1500 U/mg protein	1 unit hydrolyzes 1.0 μmol L-α-phosphatidylcholine to L-α-lysophosphatidylcholine (soybean) and a fatty acid/min at pH 8.9, 25°C	Lyophilized containing 90% protein		Sigma P7778
Naja naja atra (snake venom) (MW = 13,500 Da)				
≥5000 U/mg protein	1 unit releases 1 μmol free fatty acids from egg yolk lipoprotein/min at pH 8.0, 38°C	>99% purity; lyophilized		Calbiochem 525150
1000 U/mg	1 unit hydrolyzes 1 μmol 3-sn-phosphatidylcholine/min at pH 8.0, 37°C			Fluka 79479

3.1.1.4 Phospholipase A$_2$ continued

SPECIFIC ACTIVITY	UNITS DEFINITION	PREPARATION FORM	ADDITIONAL ACTIVITIES	SUPPLIER CATALOG NO.
Naja naja atra (snake venom) continued				
500-2500 U/mg protein	1 unit hydrolyzes 1.0 μmol L-α-phosphatidylcholine to L-α-lysophosphatidylcholine (soybean) and a fatty acid/min at pH 8.9, 25°C	Lyophilized containing 70% protein; balance primarily citrate buffer salts		Sigma P6139
Notechis scutatus (snake venom) (MW = 13,574 Da)				
		93% purity by HPLC; lyophilized		Latoxan L81 04
Pancreas (optimum pH = 8.0-9.5, T = 40-50°C)				
3500 U/mL		Solution		Biocatalysts Lipomod 22L L022L
Porcine pancreas (optimum pH = 7.9-8.4, T = 50°C, MW = 14,000 Da; stable to 80°C: 40°C [pH 10], 40-50°C [pH 9], 40-65°C [pH 8], 40-75°C [pH 4])				
700 U/mg	1 unit converts 1 μmol lecithin from egg yolk to product/min at 40°C	Suspension 3.2 M (NH$_4$)$_2$SO$_4$	<0.01 U/mg protein trypsin (BAEE as substrate) <1 U/mg total proteolytic activity (Hb as substrate)	Boehringer 161454
300 U/mg protein	1 unit hydrolyzes 1 μmol L-α-phosphatidylcholine and fatty acid/min	Suspension in 3.2 M (NH$_4$)$_2$SO$_4$ containing 10 mg/mL protein		Calzyme 105B0300
700 U/mg protein	1 unit hydrolyzes 1 μmol 3-sn-phosphatidylcholine/min at pH 8.0, 37°C	Crystallized; suspension in 3.2 M (NH$_4$)$_2$SO$_4$, pH 5.5		Fluka 79482
35 U/g	1 unit hydrolyzes 1 μmol 3-sn-phosphatidylcholine/min at pH 8.0, 37°C	Lyophilized		Fluka 79483
≥5.0 U/mg solid	1 unit hydrolyzes 1.0 μmol L-α-phosphatidylcholine to L-α-lysophosphatidylcholine and a fatty acid/min at pH 8.0, 37°C	Powder		Can Inova A-001
10,000 IU/mL	1 IU produces 1 μmol free fatty acid from egg yolk/min at pH 8, 40°C	Food grade; purified; solution	No detectable side activities	Novo Nordisk Lecitase™ 10 L
600 U/mg protein	1 unit hydrolyzes 1.0 μmol L-α-phosphatidylcholine to L-α-lysophosphatidylcholine (soybean) and a fatty acid/min at pH 8.0, 37°C	Suspension in 3.2 M (NH$_4$)$_2$SO$_4$, pH 5.5		Sigma P6534
Streptomyces chromofuscus (optimum pH = 8.0, MW = 50,000 Da)				
50 U/mg solid	1 unit hydrolyzes 1 μmol phosphatidylcholine/min at pH 8.0, 37°C	Lyophilized	No detectable catalase, GO	Calbiochem 525200

Phospholipase A₂ continued 3.1.1.4

SPECIFIC ACTIVITY	UNITS DEFINITION	PREPARATION FORM	ADDITIONAL ACTIVITIES	SUPPLIER CATALOG NO.
Streptomyces violaceoruber (optimum pH = 7.3-8.3, pI = 7.51, MW = 15,000 Da [Sephadex G-100]; K_M = 5.0 mM [lecithin]; stable pH 6-9.5 [55°C, 10 min], T < 50°C [pH 8, 10 min, CaCl₂])				
3.0-10.0 U/mg solid	1 unit liberates 1 μmol fatty acid/min at pH 8.0, 37°C	Lyophilized containing 10% protein		Asahi T-31 [PLA₂]
20-80 U/mg solid	1 unit hydrolyzes 1.0 μmol L-α-phosphatidylcholine to L-α-lysophosphatidylcholine and a fatty acid/min at pH 8.0, 37°C	Lyophilized containing mannitol and Tris buffer		Sigma P8685
Trimeresurus flavoviridis (pI = 7.9, MW = 28,000 Da)				
≥1200 U/mg protein	1 unit releases 1 μmol free fatty acids from egg yolk lipoprotein/min at pH 8.0, 38°C	≥99% purity; lyophilized		Calbiochem 525150
Vipera ammodytes (snake venom) (MW = 13,775 Da)				
		Purified by gel filtration and ion exchange chromatography; lyophilized		Latoxan L81 05

Lysophospholipase 3.1.1.5

REACTION CATALYZED
    2-Lysophosphatidylcholine + H₂O ↔ Glycerophosphocholine + a carboxylate

SYSTEMATIC NAME
    2-Lysophosphatidylcholine acylhydrolase

SYNONYMS
    Lecithinase B, lysolecithinase, phospholipase B, LYPL

REACTANTS
    2-Lysophosphatidylcholine, H₂O, glycerophosphocholine, carboxylate

NOTES
- Classified as a hydrolase, acting on ester bonds: carboxylic ester hydrolases
- Activated by Ca^{2+}
- Inhibited by cationic detergents

SPECIFIC ACTIVITY	UNITS DEFINITION	PREPARATION FORM	ADDITIONAL ACTIVITIES	SUPPLIER CATALOG NO.
Aspergillus niger				
≥1000 U/g		Solution stabilized with K sorbate, pH 5.0-6.0		ABM-RP G-zyme™ G 999

3.1.1.5 Lysophospholipase continued

SPECIFIC ACTIVITY	UNITS DEFINITION	PREPARATION FORM	ADDITIONAL ACTIVITIES	SUPPLIER CATALOG No.
Vibrio species (optimum pH = 9.0-9.5; K_M = 0.67 mM [lysolecithin]; stable pH 6.5-9 [37°C, 1 hr], T < 60°C [pH 7.2, 10 min])				
	1 unit hydrolyzes 1 μmol lysolecithin/min at pH 8.0, 37°C	Lyophilized containing 60% protein		Asahi LYPL T-32
20-60 U/mg protein	1 unit produces 1.0 μmol glycerophosphorylcholine from egg yolk L-α-lysophosphatidylcholine/min at pH 8.0, 37°C	Lyophilized containing 85% protein; balance Tris buffer salt		ICN 156234
20-60 U/mg protein	1 unit produces 1.0 μmol glycerophosphorylcholine from egg yolk L-α-lysophosphatidylcholine/min at pH 8.0, 37°C	Lyophilized containing 85% protein; balance Tris buffer salts		Sigma P8914

3.1.1.6 Acetylesterase

REACTION CATALYZED
   An acetic ester + H_2O ↔ An alcohol + acetate

SYSTEMATIC NAME
   Acetic-ester acetylhydrolase

SYNONYMS
   C-Esterase (animal tissue), citrus acetylesterase, acetinase

REACTANTS
   Acetic ester, H_2O, alcohol, acetate, triacetin, glycerol, methyl butyrate, ethyl butyrate, ethyleneglycol diformate

NOTES
- Classified as a hydrolase, acting on ester bonds: carboxylic ester hydrolases
- Easily heat inactivated in comparison with pectinesterase (EC 3.1.1.11)

SPECIFIC ACTIVITY	UNITS DEFINITION	PREPARATION FORM	ADDITIONAL ACTIVITIES	SUPPLIER CATALOG No.
Orange peel (optimum pH = 5.5-6.5)				
5 U/mg protein	1 unit produces 1.0 μmol acetic acid from triacetin/min at pH 6.5, 30°C	Crystallized; suspension in 2.5 M $(NH_4)_2SO_4$ containing 95% protein and 0.1 M Na oxalate	Pectinesterase	Calzyme 235B0005
4-12 U/mg protein	1 unit produces 1.0 μmol acetic acid from triacetin/min at pH 6.5, 30°C	Suspension in 2.5 M $(NH_4)_2SO_4$ and 0.1 M Na oxalate, pH 6.5	Pectinesterase	Sigma A4530

Acetylcholinesterase 3.1.1.7

REACTION CATALYZED
Acetylcholine + H_2O ↔ Choline + acetate

SYSTEMATIC NAME
Acetylcholine acetylhydrolase

SYNONYMS
True cholinesterase, choline esterase I, cholinesterase, red cell cholinesterase, AChE

REACTANTS
Acetylcholine, acetic esters, H_2O, choline, acetate, acetylthiocholine, thiocholine, DTNB, 5-thio-2-nitrobenzoate, 2-nitrobenzoate-5-mercaptothiocholine

APPLICATIONS
- Determination of organophosphorous compounds (e.g., pesticides)
- Hydrolysis/condensation of carboxylic esters

NOTES
- Classified as a hydrolase, acting on ester bonds: carboxylic ester hydrolases
- Bound to cellular membranes; key enzyme with cholinergic neurotransmission
- Acts on a variety of acetic esters
- Catalyzes transacetylations also
- Activated by Mg^{2+}
- Inhibited by organophosphate compounds and most carbamates

SPECIFIC ACTIVITY	UNITS DEFINITION	PREPARATION FORM	ADDITIONAL ACTIVITIES	SUPPLIER CATALOG NO.
Bovine brain, membrane-bound				
550 U/mg protein	1 unit hydrolyzes 1 µmol acetylthiocholine/min at pH 8.0, 25°C	Solution containing 10 mM Tris-HCl buffer, 144 mM NaCl, 0.05% NaN₃, 0.1% Triton X-100, pH 7.4		Fluka 01028
Bovine erythrocytes				
4 U/mg protein; 0.5 U/mg solid	1 unit hydrolyzes 1 µmol acetylthiocholine iodide/min at pH 8.0, 25°C	Lyophilized		Biozyme ACHE1
8 U/mg protein; 2 U/mg solid	1 unit hydrolyzes 1 µmol acetylthiocholine iodide/min at pH 8.0, 25°C	Lyophilized		Biozyme ACHE2
8 U/mg protein	1 unit hydrolyzes 1 µmol acetylthiocholine iodide/min at pH 8.0, 25°C	Lyophilized containing 90% protein		Calzyme 122A0008
0.4 U/mg protein; 0.5 U/mg solid	1 unit hydrolyzes 1 µmol acetylthiocholine iodide/min at pH 8.0, 25°C	Lyophilized		ICN 190674

3.1.1.7 Acetylcholinesterase continued

SPECIFIC ACTIVITY	UNITS DEFINITION	PREPARATION FORM	ADDITIONAL ACTIVITIES	SUPPLIER CATALOG NO.
Bovine erythrocytes *continued*				
0.25-1.0 U/mg solid	1 unit hydrolyzes 1.0 μmol acetylcholine to choline and acetate/min at pH 8.0, 37°C	Lyophilized containing PO_4 buffer salts		Sigma C5021
2-4 U/mg				Wako 016-11211
Bovine erythrocytes, membrane-bound				
0.3 U/mg	1 unit hydrolyzes 1 μmol acetylthiocholine/min at pH 8.0, 25°C	Powder		Fluka 01025
400 U/mg protein	1 unit hydrolyzes 1 μmol acetylthiocholine/min at pH 8.0, 25°C	Solution containing 10 mM Tris-HCl buffer, 144 mM NaCl, 0.05% NaN_3, 0.1% Triton X-100, pH 7.4		Fluka 01026
Electrophorus electricus **(electric eel) (optimum pH = 7.0, T = 60°C, MW = 75,000 [monomer], 260,000 Da)**				
1000 U/mg solid	1 unit converts 1 μmol acetylthiocholine to product/min at 25°C	Salt-free; lyophilized		Boehringer 101885
2000 U/mg protein	1 unit converts 1 μmol acetylthiocholine iodide/min at pH 8.0, 25°C	Lyophilized containing 60% protein		Calzyme 144A2000
850 U/mg	1 unit hydrolyzes 1 μmol acetylcholine/min at pH 8.0, 37°C	Powder		Fluka 01022
300 U/mg	1 unit hydrolyzes 1 μmol acetylcholine/min at pH 8.0, 37°C	Lyophilized		Fluka 01023
2000 U/mg protein		Lyophilized containing 60% protein		ICN 153875
7500-15,000 U/g agarose; 1 mL gel yields 240-480 U	1 unit hydrolyzes 1.0 μmol acetylcholine to choline and acetate/min at pH 8.0, 37°C	Insoluble enzyme attached to beaded agarose; suspension in 2.0 M $(NH_4)_2SO_4$, pH 7.0		Sigma C2511
1000 U/mg protein	1 unit hydrolyzes 1.0 μmol acetylcholine to choline and acetate/min at pH 8.0, 37°C	Frozen solution containing 5 mg $(NH_4)_2SO_4$/mg protein		Sigma C2629
1000-2000 U/mg protein	1 unit hydrolyzes 1.0 μmol acetylcholine to choline and acetate/min at pH 8.0, 37°C	Lyophilized containing 60% protein; balance primarily Tris buffer salts		Sigma C2888
200-600 U/mg protein	1 unit hydrolyzes 1.0 μmol acetylcholine to choline and acetate/min at pH 8.0, 37°C	Lyophilized containing 60% protein; balance primarily Tris buffer salts		Sigma C3389
≥60 U/mg solid	1 unit hydrolyzes 1 μmol acetylcholine/min at pH 7.0, 25°C	Partially purified by $(NH_4)_2SO_4$ fractionation; dialyzed; lyophilized		Worthington LS02228 LS02229 LS02231

Acetylcholinesterase continued   3.1.1.7

SPECIFIC ACTIVITY	UNITS DEFINITION	PREPARATION FORM	ADDITIONAL ACTIVITIES	SUPPLIER CATALOG NO.
Electrophorus electricus (electric eel) continued				
≥1000 U/mg solid	1 unit hydrolyzes 1 μmol acetylcholine/min at pH 7.0, 25°C	Chromatographically purified; lyophilized		Worthington LS05310 LS05316 LS05314 LS05313
Equine serum				
115 U/mg solid		Lyophilized containing 90% protein		ICN 153874
Human Erythrocytes				
0.25-1.0 U/mg solid	1 unit hydrolyzes 1.0 μmol acetylcholine to choline and acetate/min at pH 8.0, 37°C	Lyophilized containing PO_4 buffer salts		Sigma C5400
Human, recombinant (MW = 260,000 Da)				
5000 U/mg protein	1 unit hydrolyzes 1.0 μmol acetylcholine to choline and acetate/min at pH 8.0, 37°C	Lyophilized containing PO_4 buffer salts		Sigma C1682

Cholinesterase   3.1.1.8

REACTION CATALYZED
    An acylcholine + H_2O ↔ Choline + a carboxylate

SYSTEMATIC NAME
    Acylcholine acylhydrolase

SYNONYMS
    Pseudocholinesterase, butyrylcholine esterase, non-specific cholinesterase, choline esterase II (unspecific), benzoylcholinesterase, serum cholinesterase, BC

REACTANTS
    Acylcholine, choline esters, H_2O, choline, carboxylate, butyrylcholine, benzoylcholine

APPLICATIONS
- Determination of organophosphorous compounds (e.g., pesticides), due to its strong inhibition by these compounds
- Serum control in diagnosis of liver disease, malignant tumors and bronchial asthma
- Hydrolysis/condensation of carboxylic esters

NOTES
- Classified as a hydrolase, acting on ester bonds: carboxylic ester hydrolase
- Found in mammalian blood plasma, liver, pancreas, interstitial mucosa and white matter of the central nervous system
- A tetrameric glycoprotein with 4 equal subunits

3.1.1.8 Cholinesterase continued

- Acts on a variety of choline esters; hydrolyzes butyrylcholine at 4 time the rate of acetylcholine
- D-β-methyl acetylcholine is not a substrate
- Activated by Mg^{2+} and Ca^{2+}
- Inhibited by diisopropyl phosphofluoridate, isopropylmethylphosphoryl fluoridate, physostigmine, carbamate derivatives, quaternary ammonium salts and numerous organophosphate esters

SPECIFIC ACTIVITY	UNITS DEFINITION	PREPARATION FORM	ADDITIONAL ACTIVITIES	SUPPLIER CATALOG No.
Equine serum (optimum pH = 6.0-8.5, pI = 2.9-3.0 and 4.36, MW = 440,000 Da [300,000 Da]; stable pH 3-9, T < 40°C)				
≥7 U/mg solid	1 unit hydrolyzes 1 µmol butyrylthiocholine iodide/min at pH 7.4, 25°C	Salt-free; lyophilized		Biozyme BCE1
200 U/mg protein; 50 U/mg solid	1 unit hydrolyzes 1 µmol butyrylthiocholine iodide/min at pH 7.4, 25°C	Highly purified; lyophilized		Biozyme BCE3
200 U/mg protein; ≥15 U/mg solid	1 unit hydrolyzes 1 µmol butyrylthiocholine iodide/min at pH 7.4, 25°C	Lyophilized		Biozyme BCE3S
50-100 U/mg solid	1 unit hydrolyzes 1 µmol butyrylthiocholine iodide/min at pH 7.4, 25°C	Lyophilized containing 95% protein		Calzyme 123A0050
400-500 U/mg solid	1 unit hydrolyzes 1 µmol butyrylthiocholine iodide/min at pH 7.4, 25°C	Lyophilized containing 95% protein		Calzyme 123A0400
250 U/mg	1 unit hydrolyzes 1 µmol butyrylcholine/min at pH 8.0, 37°C	Powder		Fluka 20776
10 U/mg	1 unit hydrolyzes 1 µmol butyrylcholine/min at pH 8.0, 37°C	Lyophilized		Fluka 20777
4-10 U/mg solid	1 unit hydrolyzes 1.0 µmol butyrylthiocholine/min at pH 7.4, 20°C	Lyophilized		ICN 100451
200 U/mg protein; 50-60 U/mg solid		Lyophilized		ICN 150544
50 U/mg solid		Lyophilized containing 50% protein		ICN 153878
20 U/mg solid	1 unit hydrolyzes 1 µmol butyrylthiocholine/min at pH 7.4, 20°C	Lyophilized		ICN 190310
>3.0 U/mg protein; >3.0 U/mg solid	1 unit hydrolyzes 1 µmol butyrylthiocholine iodide/min at pH 7.4, 37°C	Grade 2; lyophilized	<0.05% LDH, GOT, GPT, CK, ALP, γ-GT	Randox BC 910L

SPECIFIC ACTIVITY	UNITS DEFINITION	PREPARATION FORM	ADDITIONAL ACTIVITIES	SUPPLIER CATALOG NO.
Equine serum *continued*				
>10.0 U/mg protein; >10.0 U/mg solid	1 unit hydrolyzes 1 µmol butyrylthiocholine iodide/min at pH 7.4, 37°C	Grade 1; lyophilized	<0.05% LDH, GOT, GPT, CK, ALP, γ-GT	Randox BC 911L
1000 U/mg protein	1 unit hydrolyzes 1.0 µmol butyrylcholine to choline and butyrate/min at pH 8.0, 37°C	Highly purified; lyophilized containing 30% protein; balance buffer salts		Sigma C1057
500 U/mg protein	1 unit hydrolyzes 1.0 µmol butyrylcholine to choline and butyrate/min at pH 8.0, 37°C	Highly purified; lyophilized containing 50% protein; balance buffer salts		Sigma C4290
10-20 U/mg protein	1 unit hydrolyzes 1.0 µmol butyrylcholine to choline and butyrate/min at pH 8.0, 37°C	Lyophilized containing 70% protein; balance buffer salts		Sigma C7512
10 U/mg protein		Lyophilized		Wako 031-13821
≥5 U/mg solid	1 unit hydrolyzes 1 µmol acetylcholine/min at pH 7.4, 25°C	Chromatographically purified; lyophilized		Worthington LS01618 LS01622 LS01626
≥4 U/mg solid	1 unit hydrolyzes 1 µmol acetylcholine/min at pH 7.4, 25°C	Lyophilized		Worthington LS01628 LS01632 LS01636
Human				
				Vital 17034
Human serum				
3-9 U/mg protein	1 unit hydrolyzes 1.0 µmol butyrylcholine to choline and butyrate/min at pH 8.0, 37°C	Crude; powder containing 70% protein; balance primarily salts		Sigma C5386

3.1.1.11 Pectinesterase

REACTION CATALYZED
 Pectin + n H_2O ↔ n Methanol + pectate

SYSTEMATIC NAME
 Pectin pectylhydrolase

SYNONYMS
 Pectin demethoxylase, pectin methoxylase, pectin methylesterase

REACTANTS
 Pectin, H_2O, methanol, pectate

NOTES
- Classified as a hydrolase, acting on ester bonds: carboxylic ester hydrolase

SPECIFIC ACTIVITY	UNITS DEFINITION	PREPARATION FORM	ADDITIONAL ACTIVITIES	SUPPLIER CATALOG NO.
Orange peel				
50-100 U/mg protein	1 unit releases 1.0 µeq acid from pectin/min at pH 7.5, 30°C	Suspension in 3.2 M $(NH_4)_2SO_4$ and 0.1 M NaCl, pH 7.0	May contain pectinase and acetylesterase	Sigma P0764
50-350 U/mg protein	1 unit releases 1.0 µeq acid from pectin/min at pH 7.5, 30°C	Lyophilized containing 50% protein; balance primarily $(NH_4)_2SO_4$ and NaCl buffer salts	May contain pectinase and acetylesterase	Sigma P1889
350-500 U/mg protein	1 unit releases 1.0 µeq acid from pectin/min at pH 7.5, 30°C	Lyophilized containing 35% protein; balance primarily $(NH_4)_2SO_4$ and NaCl buffer salts		Sigma P5400
Tomato				
150-250 U/mg protein	1 unit releases 1.0 µeq acid from pectin/min at pH 7.5, 30°C	Lyophilized containing 40% protein		Sigma P6763

3.1.1.13 Sterol Esterase

REACTION CATALYZED
 A steryl ester + H_2O ↔ A sterol + a fatty acid

SYSTEMATIC NAME
 Steryl-ester acylhydrolase

SYNONYMS
 Cholesterol esterase, cholesteryl ester synthase, triterpenol esterase, CE

REACTANTS
 Steryl ester, H_2O, sterol, fatty acid

APPLICATIONS
- Synthesis of optically active alcohols and carboxylic acids via ester hydrolysis, esterification or transesterification
- Determination of cholesterol and cholesterol esters in serum and plasma, with cholesterol oxidase or peroxidase

Sterol Esterase continued — 3.1.1.13

NOTES

- Classified as a hydrolase, acting on ester bonds: carboxylic ester hydrolase
- Found in pancreas, intestine, liver and kidney
- A group of enzymes of broad specificity
- Act on esters of sterols and long-chain fatty acids; may also esterify sterols
- Aggregates to the active hexamer form in the presence of bile salts
- Activated by bile salts and Triton X-100; bile salts protect it from proteolytic degradation in the intestine
- Stabilized by Mg^{2+}, sodium cholate, BSA
- Inactivated by proteolytic enzymes
- Inhibited by Hg^{2+}, Ag^+ and ionic detergents

SPECIFIC ACTIVITY	UNITS DEFINITION	PREPARATION FORM	ADDITIONAL ACTIVITIES	SUPPLIER CATALOG No.
Bovine pancreas (optimum pH = 6.5-8.0; relative activity: 25°C = 1.0, 30°C = 1.6, 37°C = 3.0)				
>25 U/mg protein; >10 U/mg solid	1 unit forms 1.0 μmol cholesterol/min at pH 7.0, 37°C	Lyophilized	<0.01% GO, uricase <0.05% trypsin, chymotrypsin	Genzyme 1081
800-1600 U/g protein	1 unit hydrolyzes 1.0 μmol cholesteryl oleate to cholesterol and oleic acid/min at pH 7.0, 37°C in the presence of taurocholate	Partially purified; lyophilized containing 50% protein		Sigma C3766
20,000-50,000 U/g protein	1 unit hydrolyzes 1.0 μmol cholesteryl oleate to cholesterol and oleic acid/min at pH 7.0, 37°C in the presence of taurocholate	Lyophilized containing 15% protein		Sigma C5921
Candida rugosa (formerly C. cylindraccae) (optimum pH = 7.0-8.0, T = 25°C)				
26 U/mg at 25°C; 55 U/mg at 37°C	1 unit converts 1 μmol cholesterol oleate to product/min	Suspension 3.2 M $(NH_4)_2SO_4$ and BSA as stabilizer, pH 6	<0.001% GO, GK <0.005% NADH oxidase, HK, ATPase, uricase	Boehringer 161772
26 U/mg at 25°C; 55 U/mg at 37°C	1 unit converts 1 μmol cholesterol oleate to product/min	Solution containing 3 M NaCl and BSA as stabilizer, pH 6	<0.001% GO, GK <0.005% NADH oxidase, HK, ATPase, uricase	Boehringer 393916
25-100 U/mg protein	1 unit hydrolyzes 1 μmol cholesterol ester/min at pH 6.7, 37°C in the presence of Na cholate	Lyophilized containing 90% protein		Calzyme 148A0025
≥20 U/mg protein	1 unit liberates 1 μmol cholesterol from cholesterol oleate/min at pH 7.5, 25°C	Lyophilized		Seikagaku 100308-1 100308-2
Microbial (optimum pH = 6.0-8.0, T = 37°C)				
≥100 U/mg	1 unit forms 1 μmol H_2O_2/min	Lyophilized		AMRESCO 0669

SPECIFIC ACTIVITY	UNITS DEFINITION	PREPARATION FORM	ADDITIONAL ACTIVITIES	SUPPLIER CATALOG NO.
Microbial *continued*				
>6 U/mg protein	1 unit releases 1 μmol free cholesterol/min at pH 7.0, 37°C using cholesterol linoleate as substrate	Lyophilized		GDS Tech C-020
40 U/mg protein	1 unit releases 1 μmol free cholesterol/min at pH 6.5, 37°C	Lyophilized	<0.05% GO, uricase	Randox CE 948L
Pancreas (optimum pH = 7.0, T = 37°C)				
5-7 IU/mg powder	1 unit decomposes 1 μmol cholesteryl linoleate/min at 37°C; assay based on the increase in A_{500} as 4-aminoantipyrine is coupled to phenol to form quinoneimine dye	Lyophilized	<0.003% L-AOD <0.1% GO, uricase <0.6% A/U CE protease activity	Beckman 683970
>4 U/mg protein	1 unit releases 1 μmol free cholesterol/min at pH 7.0, 37°C using cholesterol acetate as substrate	Lyophilized		GDS Tech (D)C-040
Porcine pancreas (optimum pH = 7-9, T = 40°C, MW = 30,000 Da; stable pH 2.5-7.5, T < 55°C)				
15 U/mg	1 unit forms 1 μmol H_2O_2/min	Lyophilized		AMRESCO 0748
45 U/mg protein; 35 U/mg solid	1 unit produces 1 μmol cholesterol/min at 37°C	Highly purified; lyophilized	<0.001% uricase <0.5% chymotrypsin, trypsin	Biozyme CE2
25 U/mg protein	1 unit hydrolyzes 1 μmol cholesterol ester/min at pH 6.7, 37°C in the presence of Na cholate	Lyophilized containing KPO_4 buffer, stabilizer, 90% protein		Calzyme 065A0025
>10 U/mg prot; >6 U/mg solid	1 unit produces 1 μmol cholesterol from cholesterol acetate/min at pH 7.0, 37°C	Lyophilized containing stabilizers and buffer salts	No detectable trypsin, chymotrypsin	Diagnostic C-040
20 U/mg	1 unit liberates 1 μmol cholesterol/min at pH 7.0, 37°C with cholesterol acetate as substrate	Lyophilized	<0.01% uricase, GO <1% chymotrypsin <5% trypsin	Fluka 26745
15 U/mg protein	1 unit hydrolyzes 1.0 μmol cholesterol from cholesteryl acetate/min at pH 7.0, 37°C	Lyophilized		ICN 150670
25 U/mg protein	1 unit hydrolyzes 1.0 μmol cholesterol from cholesteryl acetate/min at pH 7.0, 37°C	Lyophilized		ICN 150671
>15 U/mg protein	1 unit releases 1 μmol free cholesterol/min at pH 7.0, 37°C	Lyophilized	<0.005% GO, uricase <0.01% chymotrypsin, trypsin	Randox CE 949L
20 U/mg solid	1 unit decomposes 1 μmol cholesterol ester/min at pH 7.0, 37°C	Lyophilized	<0.004% L-AO <0.1% GO, uricase <0.5 absorbance units/unit CE activity	Scripps C0414

SPECIFIC ACTIVITY	UNITS DEFINITION	PREPARATION FORM	ADDITIONAL ACTIVITIES	SUPPLIER CATALOG NO.
Porcine pancreas *continued*				
800-1600 U/g protein	1 unit hydrolyzes 1.0 μmol cholesteryl oleate to cholesterol and oleic acid/min at pH 7.0, 37°C in the presence of taurocholate	Lyophilized containing 70% protein; balance primarily KPO$_4$ buffer salts		Sigma C9530
≥300 U/g solid	1 unit hydrolyzes 1 μmol cholesterol ester/min at pH 7.0, 37°C	Chromatographically purified; lyophilized	<0.05% GO, *L*-AO, uricase <0.1% GalO	Worthington LS04103 LS04104 LS04105
Pseudomonas fluorescence				
25 U/mg solid; 100 U/mg enzyme protein	1 unit converts 1 μmol cholesterol oleate to product/min at 25°C	Lyophilized	<0.001% GO, GK <0.005% NADH oxidase, HK, ATPase, uricase	Boehringer 691941
12,000 U/g protein	1 unit hydrolyzes 1.0 μmol cholesteryl oleate to cholesterol and oleic acid/min at pH 7.0, 37°C in the presence of taurocholate	Lyophilized containing 20% protein; balance primarily KPO$_4$ and Triton X-100		Sigma C9281
5000 U/g protein	1 unit hydrolyzes 1.0 μmol cholesteryl oleate to cholesterol and oleic acid/min at pH 7.0, 37°C in the presence of taurocholate	Lyophilized containing 20% protein; balance primarily KPO$_4$ and Triton X-100		Sigma C9406
***Pseudomonas* species** (optimum pH = 6.0-9.0, pI = 5.95, T = 40°C, MW = 300,000 Da, 29,500 Da [SDS-PAGE], 31,000 Da [Sephadex G-100]; K_M = 54 μM [linoleate], 66 μM [oleate], 37 μM [linolenate], 0.15 mM [palmitate], 0.12 mM [myristate], 23 μM [stearate]; stable pH 5.0-9.0 [25°C, 24 hr], T < 55°C [pH 7.5, 10 min] and < 50°C [pH 7.0, 10 min])				
≥10 U/mg	1 unit decomposes 1 μmol cholesterol ester/min at pH 7.0, 37°C	Powder		Amano Cholesterol esterase
100 U/mg	1 unit forms 1 μmol H$_2$O$_2$/min	Lyophilized		AMRESCO 0913
100-250 U/mg solid	1 unit liberates 1 μmol cholesterol/min at pH 6.8, 37°C	Lyophilized containing 100% protein		Asahi CE T-18
>100 U/mg solid	1 unit hydrolyzes 1.0 μmol cholesterol from cholesteryl acetate/min at pH 7.0, 37°C	Lyophilized		ICN 105439
>150 U/mg protein; >20 U/mg solid	1 unit produces 1 μmol cholesterol/min at pH 6.5, 37°C	Lyophilized	<0.05% GO, uricase, GK, HK	Randox CE 950L

3.1.1.13 Sterol Esterase continued

SPECIFIC ACTIVITY	UNITS DEFINITION	PREPARATION FORM	ADDITIONAL ACTIVITIES	SUPPLIER CATALOG NO.
Pseudomonas species continued				
200,000-400,000 U/g protein	1 unit hydrolyzes 1.0 μmol cholesteryl oleate to cholesterol and oleic acid/min at pH 7.0, 37°C in the presence of taurocholate	Lyophilized containing 70% protein		Sigma C1403
≥100 U/mg solid	1 unit forms 1 μmol H_2O_2 (0.5 μmol quinoneimine dye)/min at pH 7.0, 37°C	Lyophilized containing 40% BSA, Mg^{2+}, Na cholate as stabilizers	<0.01% catalase	Toyobo COE-311
10-30 U/mg		Lyophilized		Wako 037-11221

3.1.1.20 Tannase

REACTION CATALYZED
    Digallate + H_2O ↔ 2 Gallate

SYSTEMATIC NAME
    Tannin acylhydrolase

REACTANTS
    Digallate, H_2O, gallate

APPLICATIONS
- Clarifier and yield enhancer for tea

NOTES
- Classified as a hydrolase, acting on ester bonds: carboxylic ester hydrolase
- Also hydrolyzes ester links in other tannins

SPECIFIC ACTIVITY	UNITS DEFINITION	PREPARATION FORM	ADDITIONAL ACTIVITIES	SUPPLIER CATALOG NO.
Aspergillus oryzae (optimum pH = 5.0-6.0, pI = 4.0, T = 30-40°C, MW = 200,000 Da)				
		Food grade		EnzymeDev Enzeco[R] Tannase
≥30 U/mg		Powder		Wako 207-11621 203-11623
Unspecified				
20,000 U/g	1 unit causes a ΔA_{310} of 1 absorbance unit/min using a 0.004% tannic acid solution in 0.02 M acetate buffer, pH 9.7	Powder	Trace amylase, protease	ICN 152094

Lipoprotein Lipase 3.1.1.34

REACTION CATALYZED
Triacylglycerol + H_2O ↔ Diacylglycerol + a carboxylate

SYSTEMATIC NAME
Triacylglycero-protein acylhydrolase

SYNONYMS
Clearing factor lipase, diglyceride lipase, diacylglycerol lipase

REACTANTS
Triacylglycerol, diacylglycerol, H_2O, diacylglycerol, carboxylate

APPLICATIONS
- Determination of triglycerides, with GPO, GK and POD

NOTES
- Classified as a hydrolase, acting on ester bonds: carboxylic ester hydrolase
- Hydrolyzes triacylglycerols in chylomicrons and low-density lipoproteins
- Detergent is required for hydrolysis of triglycerides
- Stabilized by BSA and cholic acid
- Inhibited by Hg^{2+}, Ag^+ and ionic detergents

SPECIFIC ACTIVITY	UNITS DEFINITION	PREPARATION FORM	ADDITIONAL ACTIVITIES	SUPPLIER CATALOG NO.
Alcaligenes species strain 679 (optimum pH = 7-8.5, T = 37-40°C)				
≥7000 U/mg protein	1 unit liberates 1 µmol fatty acid/min at pH 8.0, 37°C	Lyophilized		Seikagaku 100908-1
Bovine milk				
2000-6000 U/mg protein	1 unit releases 1.0 nmol *p*-nitrophenol/min at pH 7.2, 37°C using *p*-nitrophenyl butyrate as substrate	Suspension in 3.8 M $(NH_4)_2SO_4$ and 0.02 M Tris-HCl, pH 8		Sigma L2254
Chromobacterium viscosum (optimum pH = 3.0-10.0)				
1300 U/mg	1 unit liberates 1 µmol oleic acid/min at pH 8.0, 40°C with triolein as substrate	Lyophilized		Fluka 62333
>2500 U/mg protein; >2000 U/mg solid	1 unit forms 1.0 µmol fatty acid/min at pH 6.0-7.0, 37°C	Lyophilized	<0.01% NADH oxidase, ChO, catalase	Genzyme 1461
Microbial (optimum pH = 5.0-8.0; relative activity: 25°C = 1.0, 30°C = 1.16, 37°C = 1.51)				
>2000 U/mg protein; >1000 U/mg solid	1 unit forms 1.0 µmol fatty acid/min at pH 6.0-7.0, 37°C	Lyophilized		Genzyme 1471

3.1.1.34 Lipoprotein Lipase continued

SPECIFIC ACTIVITY	UNITS DEFINITION	PREPARATION FORM	ADDITIONAL ACTIVITIES	SUPPLIER CATALOG NO.
Pseudomonas species (optimum pH = 6.0-9.0, pI = 4.3 and 5.9-6.0, T = 45-50°C, MW = 31,000 and 134,000 Da; stable pH 7.0-9.0 [25°C, 20 hr], T < 55°C [pH 7.0, 10 min])				
≥2000 U/mg	1 unit produces 1 μmol fatty acid/min at pH 7.0, 37°C	Powder		Amano LPL-200S
≥800 U/mg	1 unit produces 1 μmol fatty acid/min at pH 7.0, 37°C	Powder		Amano LPL-80
2100 U/mg	1 unit liberates 1 μmol oleic acid/min at pH 8.0, 40°C with triolein as substrate	Lyophilized		Fluka 62335
>2000 U/mg protein; >1500 U/mg solid	1 unit forms 1.0 μmol fatty acid/min at pH 6.0-7.0, 37°C	Lyophilized		Genzyme 1481
50 U/mg protein	1 unit produces 1 μmol glycerol from triglyceride/min at pH 7.0, 37°C in the presence of BSA	Lyophilized containing 50% BSA and enzyme protein; balance primarily Na cholate, $MgCl_2$, sucrose, KPO_4 buffer		ICN 190072
>100 U/mg solid	1 unit hydrolyzes 1 μmol cholesterol oleate/min at pH 7.5, 25°C	Lyophilized	<0.001% NADH oxidase, GO <0.005% ATPase, HK, GK, uricase	Randox LP 952L
≥20 U/mg solid	1 unit forms 1 μmol glycerol (0.5 μmol quinoneimine dye)/min at 37°C	Lyophilized containing 80% BSA, Mg^{2+}, Na cholate as stabilizers	<0.001% phosphatase, NADH oxidase <0.002% ChO <0.02% catalase	Toyobo LPL-311
≥10 U/mg solid	1 unit forms 1 μmol glycerol (0.5 μmol quinoneimine dye)/min at pH 7.0, 37°C	Lyophilized containing 60% Mg^{2+}, Na-cholate, BSA as stabilizers		Toyobo LPL-701
25-35 U/mg		Lyophilized		Wako 129-02801

3.1.2.6 Hydroxyacylglutathione Hydrolase

REACTION CATALYZED
  S-(2-Hydroxyacyl)glutathione + H_2O ↔ Glutathione + a 2-hydroxy carboyxlate

SYSTEMATIC NAME
  S-(2-Hydroxyacyl)glutathione hydrolase

SYNONYMS
  Glyoxalase II

REACTANTS
  S-(2-Hydroxyacyl)glutathione, S-acetoacetylglutathione, H_2O, glutathione, 2-hydroxy carboyxlate

NOTES
- Classified as a hydrolase, acting on ester bonds: thiolester hydrolase
- Hydrolyzes S-acetoacetylglutathione more slowly

SPECIFIC ACTIVITY	UNITS DEFINITION	PREPARATION FORM	ADDITIONAL ACTIVITIES	SUPPLIER CATALOG NO.
Bovine liver				
10 U/mg protein	1 unit hydrolyzes 1.0 μmol S-lactoyl-glutathione/min at pH 7.4, 25°C	Lyophilized containing 80% protein; balance primarily PO_4-citrate buffer salts	<0.2% glyoxalase I	Sigma G0131

Alkaline Phosphatase   3.1.3.1

REACTION CATALYZED
An orthophosphoric monoester + H_2O ↔ An alcohol + orthophosphate

SYSTEMATIC NAME
Orthophosphoric-monoester phosphohydrolase (alkaline optimum)

SYNONYMS
Alkaline phosphomonoesterase, phosphomonoesterase, glycerophosphatase, ALP, AP

REACTANTS
Orthophosphoric monoester, H_2O, alcohol, orthophosphate, p-nitrophenylphosphate, p-nitrophenol, o-carboxy phenyl phosphate, salicylic acid

APPLICATIONS
- Removal of 5'-phosphate from DNA and RNA fragments prior to labeling the 5'-end with polynucleotide kinase
- Removes phosphate from both 5'-termini, preventing self-annealing of linearized cloning vehicle DNA
- Indicator enzyme in highly sensitive immunoassays with LCEC detection of phenol and when coupled to antibodies in ELISA systems
- After activation, serves as a reporter enzyme for chemiluminescent and other detection systems
- Used without preactivation for labeling Ig, IgFab and IgF(ab')$_2$ fragments from rabbit, mouse, sheep and goat
- Liposome immunoassay, with coupled liposomes containing enzyme
- Nucleic acid-based diagnostic assays
- Determination of dexamethasone phosphate
- Detection of DNA on membranes with enzyme-labeled probes

NOTES
- Classified as a hydrolase, acting on ester bonds: phosphoric monoester hydrolase
- Wide specificity and alkaline pH optimum
- Also catalyzes transphosphorylations; some enzymes hydrolyze pyrophosphate. Does not cleave poly(U)
- Human placental enzyme is a zinc protein
- *E. coli* enzyme is a nonspecific phosphomonoesterase containing Zn^{2+} and Mg^{2+}
- Bacterial enzyme removes 5'-phosphates from DNA, RNA and ribo- and deoxyribonucleotide triphosphates

Alkaline Phosphatase *continued*

- Arctic shrimp enzyme has high activity and is heat labile
- Microbial enzyme is stable up to 80°C in the presence of Mg^{2+} and Zn^{2+}. Its high activity, heat stability and low nonspecific binding are likely due to lack of post translational modification, prevalent with less-stable calf intestinal enzyme
- Calf intestinal enzyme is inactivated by heat and contains 2 molecules of Zn^{2+}. Activity is 1.5-2X higher in diethanolamine buffer, pH 9.8. Treatment with proteinase K and extraction with phenol gives complete inactivation
- Bovine intestinal enzyme is a phosphomonoesterase removing 3'- and 5'-phosphates from DNA and RNA
- Higher incubation temperatures are recommended for dephosphorylation of blunt or recessed termini of nucleic acids
- Activated by $CaCl_2$, Na^+, Mg^{2+}, Zn^{2+} and alcohol (Tris) compounds
- Inhibited by Ca^{2+} chelating agents, low Zn^{2+} (0.1 mM), orthophosphate and potassium-glutamate buffer
- Heat stability/Inactivation:
    Antarctic bacterium — heating with chelating agent to 60°C for 30 min, 65°C for 15 min or 75°C for 10 min, followed by two phenol extractions; claimed to be the only bacterial enzyme easily heat-killed
    arctic shrimp — completely and irreversibly inactivated at 65°C for 15 min, pH 8.0-8.5
    bacterial — extraction with phenol or phenol chloroform
    bovine intestine — heat inactivated
    calf intestine — irreversibly inactivated at 60°C for 30 min with chelator, 65°C for 15 min with 2 phenol extractions, 75°C for 10 min or phenol extraction alone
    calf intestine mucosa — >95% by heating to 75°C for 10 min in 5 mM EDTA, pH 8.0
    E. coli strain A19 and C75 — heating with chelating agent at 100°C; recovers at room temperature
    E. coli strain C4 — retains activity at 80°C
    E. coli strain C90 — active at 65°C for 1 hr; inactivated by phenol extraction
    microbial — stable up to 80°C with Mg^{2+} and Zn^{2+}

SPECIFIC ACTIVITY	UNITS DEFINITION	PREPARATION FORM	ADDITIONAL ACTIVITIES	SUPPLIER CATALOG NO.
Antarctic bacterium (70% activity after 30°C, 2 hr; inactivated at 65°C, 15 min)				
195 U/mg	1 MBU (Molecular Biology Unit) dephosphorylates 1 μg Hind III-digested pUC19 DNA/hr at pH 7.8, 30°C	Solution containing 50% glycerol, 25 mM Tris-HCl, 0.1 M NaCl, 0.1 mM EDTA, 5 mM $CaCl_2$, 0.01% Triton X-100, pH 7.5	No detectable exo- and endonucleolytic DNase, RNase, protease	Epicentre HK™ H92025 H92050 H92100

SPECIFIC ACTIVITY	UNITS DEFINITION	PREPARATION FORM	ADDITIONAL ACTIVITIES	SUPPLIER CATALOG NO.
Bacterial				
	1 unit hydrolyzes 1 μmol PNPP/min at pH 8, 20°C; 1 DNA unit dephosphorylates 1 pmol 5'-phosphorylated ends EcoR I fragments from λ DNA/hr at pH 7.5, 60°C	Molecular Biology Grade; 10 mM Tris-HCl, 10 mM MgCl$_2$, 0.5 mM ZnCl$_2$, 50% glycerol, pH 8	No detectable nickase, endonuclease, exonuclease	Oncor 120091 120092
Bovine intestinal mucosa				
2000 U/mg protein	1 unit hydrolyzes 1 μmol PNPP/min at pH 9.8, 25°C	Solution containing 3 M NaCl, 1 mM MgCl$_2$, 0.1 mM ZnCl$_2$, 30 mM TEA, pH 7.6		Fluka 79385
20-30 U/μL and 1 U/μL	1 unit hydrolyzes 1 μmol PNPP/min at pH 9.8, 37°C	500 mM Tris-HCl, 1 mM EDTA, pH 8.5 and dilution buffer		Life Technol 18009-019 18009-027
Bovine kidney				
0.5 U/mg solid	1 unit hydrolyzes 1 μmol PNPP/min at pH 9.6, 25°C	Lyophilized	<0.05 U/mg GOT	Biozyme ALPK1
40 U/mg solid	1 unit liberates 1 μmol p-nitrophenol from PNPP/min at pH 9.8, 37°C in DEA buffer	Lyophilized containing 90% protein	<0.005% γGT <0.02% GOT	Calzyme 057A0040
750 U/mg solid	1 unit liberates 1 μmol p-nitrophenol from PNPP/min at pH 9.8, 37°C in DEA buffer	Lyophilized containing 90% protein	<0.005% γGT <0.02% GOT	Calzyme 057A0040
Bovine liver				
100 U/mg protein	1 unit liberates 1 μmol p-nitrophenol from PNPP/min at pH 9.8, 37°C in DEA buffer	Lyophilized containing 95% protein		Calzyme 119A0100
20 U/mg	1 unit hydrolyzes 1 μmol PNPP/min at pH 9.8, 25°C	Powder		Fluka 79395
Calf				
				Vital 3038
Calf intestinal mucosa (optimum pH = 9.8, 2 subunits of MW = 68,000 Da each)				
2500 U/mg protein	1 unit hydrolyzes 1 μmol PNPP/min at pH 9.8, 37°C	Suspension in 3.2 M (NH$_4$)$_2$SO$_4$, 1 mM MgCl$_2$, 0.1 mM ZnCl$_2$, pH 7.0	<10% PPase	Fluka 79387
2700 U/mg protein	1 unit hydrolyzes 1 μmol PNPP/min at pH 9.8, 37°C	Solution containing 30 mM TEA, 3 M NaCl, 1 mM MgCl$_2$, 0.1 mM ZnCl$_2$, pH 7.6	<0.001% ADA, guanase, nucleoside phosphorylase <0.005% PDE <0.1% 5'-AMP deaminase <5% PPase	Fluka 79389
2500 U/mg protein	1 unit hydrolyzes 1 μmol PNPP/min at pH 9.8, 37°C	30 mM TEA buffer, 3 M NaCl, 1 mM MgCl$_2$, 0.1 mM ZnCl$_2$, pH 7.6		Fluka 79390

3.1.3.1 Alkaline Phosphatase continued

SPECIFIC ACTIVITY	UNITS DEFINITION	PREPARATION FORM	ADDITIONAL ACTIVITIES	SUPPLIER CATALOG No.
Calf intestinal mucosa continued				
2000 U/mg protein	1 unit hydrolyzes 1 μmol PNPP/min at pH 9.8, 37°C	30 mM TEA buffer, 3 M NaCl, 1 mM MgCl$_2$, 1 mM ZnCl$_2$, pH 7.6	0.001% PDE	Fluka 79391
1 U/mg protein	1 unit hydrolyzes 1 μmol PNPP/min at pH 9.8, 37°C	Lyophilized		Fluka 79392
1000 U/mg protein	1 unit hydrolyzes 1 μmol PNPP/min at pH 9.8, 37°C	Suspension in 3.2 M (NH$_4$)$_2$SO$_4$ solution, 0.001 mM MgCl$_2$, 0.0001 mM ZnCl$_2$, pH 7.0		Fluka 79393
3.5 U/mg agarose	1 unit hydrolyzes 1 μmol PNPP/min at pH 9.8, 37°C	Suspension in 2 M (NH$_4$)$_2$SO$_4$, 1 mM MgCl$_2$, 0.1 mM ZnCl$_2$, pH 7.0; 30 mg agarose/mL packed gel		Fluka 79394
>3000 U/mg	1 unit hydrolyzes 1 μmol PNPP/min at 37°C	30 mM TEA-HCl, 3 M NaCl, 1 mM MgCl$_2$, 0.1 mM ZnCl$_2$, pH 7.6	No detectable endonuclease, nickase, RNase	MINOTECH 201
1 U/μL	1 unit hydrolyzes 1 μmol PNPP/min at pH 9.8 in DEA	30 mM TEA, 3 mM NaCl, 1 mM MgCl$_2$, 0.1 mM ZnCl$_2$, pH 7.6	No detectable endonuclease, RNase, nickase	NBL Gene 021304
10,000 U/mL	1 unit hydrolyzes 1 μmol p-nitrophenyl-PO$_4$ to p-nitrophenol/min at 37°C in a 1 mL volume	50 mM KCl, 10 mM Tris-HCl, 0.1 mM ZnCl$_2$, 1 mM MgCl$_2$, 50% glycerol, pH 8.2	No detectable exonuclease, endonuclease, RNase	NE Biolabs 290S 290L
900-3000 U/mg; 1000-2000 U/mL	1 unit hydrolyzes 1 μmol PNPP/min at pH 9.8, 37°C	Solution containing 10 mM Tris-HCl, 50 mM KCl, 1 mM MgCl$_2$, 0.1 mM ZnCl$_2$, 50% glycerol, pH 8.0	No detectable DNase, nickase, RNase	Pharmacia 27-0620-01
1 U/μL	1 unit hydrolyzes 1 μmol PNPP/min at pH 9.8, 37°C	Cloning Quality; 10X Reaction buffer: 0.5 M Tris-HCl, 1 mM ZnCl$_2$, 10 mM MgCl$_2$, 10 mM spermidine, pH 9.3	<1% DNase <3% RNase ≥90% supercoiled plasmid	Promega M1821 M1822
Calf intestine (optimum pH = 8 [low concentration] or 10 [high concentration], pI = 5.7, T = 37-40°C, MW = 100,000-140,000 Da [69,000/subunit]; K_M = 1.7 mM; stable pH 8.5-10.3 [25°C, 20 hr], T < 40°C [pH 9.5, 30 min])				
20-30 U/μL	1 unit liberates 1 μmol p-nitrophenol/min at pH 9.8, 37°C with PNPP as substrate	10 mM Tris-HCl, 50 M KCl, 1 mM MgCl$_2$, 0.1 mM ZnCl$_2$, 50% glycerol, pH 8.0	No detectable non-specific nuclease, endonuclease, RNase	Amersham E 2250Y
≥1500 U/mg	1 unit converts 1 μmol PNPP/min at 37°C	Suspension in (NH$_4$)$_2$SO$_4$		AMRESCO 0210
≥3000 U/mg	1 unit converts 1 μmol PNPP/min at 37°C	Glycerol solution		AMRESCO E552
5 U/mg solid	1 unit hydrolyzes 1 μmol PNPP/min at pH 9.6, 25°C	Lyophilized		Biozyme ALPI 2
20 U/mg solid	1 unit hydrolyzes 1 μmol PNPP/min at pH 9.6, 25°C	Lyophilized		Biozyme ALPI 3
≥900 U/mg protein	1 unit hydrolyzes 1 μmol PNPP/min at pH 9.6, 25°C	50% glycerol solution containing 0.005 M Tris-HCl, 0.005 M MgCl$_2$, 0.0001 M ZnCl$_2$, pH 7.0		Biozyme ALPI 6G

Alkaline Phosphatase continued

SPECIFIC ACTIVITY	UNITS DEFINITION	PREPARATION FORM	ADDITIONAL ACTIVITIES	SUPPLIER CATALOG No.
Calf intestine continued				
2000 U/mg at 37°C (diethanolamine buffer); 400 U/mg at 25°C (Gly buffer)	1 unit converts 1 μmol PNPP to product/min	Suspension in 3.2 M $(NH_4)_2SO_4$, 1 mM $MnCl_2$, 0.1 mM $ZnCl_2$, pH 7	<0.0004% PDE <0.004% ADA, A-5'-monophosphate deaminase	Boehringer 108138 108146
140 U/mg at 37°C (DEA buffer); 35 U/mg at 25°C (Gly buffer)	1 unit converts 1 μmol PNPP to product/min	Suspension in 3.2 M $(NH_4)_2SO_4$, 1 mM $MnCl_2$, 0.1 mM $ZnCl_2$, BSA as stabilizer, pH 7		Boehringer 108154 108162
2000 U/mg; 20,000 U/mL	1 unit hydrolyzes 1 μmol PNPP/min at 37°C	30 mM TEA, 3 M NaCl, 1 mM $MgCl_2$, 0.1 mM $ZnCl_2$, pH 7.6		Boehringer 1097075
1200 U/mg protein	1 unit converts 1 μmol PNPP to product/min at pH 9.8, 25°C	>95% purity by HPLC; lyophilized		Boehringer 1464752
100 U/mg solid or 2000 U/mg protein at 37°C (DEA buffer); 400 U/mg at 25°C (Gly buffer)	1 unit converts 1 μmol PNPP to product/min	Lyophilized containing Ficoll 70, Mg^{2+}, Zn^{2+} as stabilizers	<0.0004% PDE <0.004% ADA, A-5'-monophosphate deaminase	Boehringer 405612
3000 U/mg protein	1 unit converts 1 μmol PNPP to product/min at 37°C DEA as buffer	EIA grade; 3 M NaCl, 1 mM $MgCl_2$, 100 mM $ZnCl_2$, 30 μM TEA, 8-16 moles/mole enzyme amino groups, pH 7.6		Boehringer 567744 567752
2000 U/mg; 1000 U/mL	1 unit hydrolyzes 1 μmol PNPP/min at 37°C	30 mM TEA, 3 M NaCl, 1 mM $MgCl_2$, 0.1 mM $ZnCl_2$, pH 7.6		Boehringer 713023
>2000 U/mg protein	1 unit hydrolyzes 1 μmol p-nitrophenylphosphate at pH 9.8, 37°C	Molecular biology grade; solution containing 30 mM TEA buffer, 3 M NaCl, 1 mM $MgCl_2$, 100 μM $ZnCl_2$, pH 7.6	No detectable RNase, DNase	Calbiochem 524575
3000-6000 U/mg protein	1 unit liberates 1 μmol p-nitrophenol from PNPP/min at pH 9.8, 37°C in DEA buffer	Conjugation grade; lyophilized		Calzyme 140A4500
3000-6000 U/mg protein	1 unit liberates 1 μmol p-nitrophenol from PNPP/min at pH 9.8, 37°C in DEA buffer	Conjugation grade; suspension in $(NH_4)_2SO_4$		Calzyme 140B4500
3000-6000 U/mg protein	1 unit liberates 1 μmol p-nitrophenol from PNPP/min at pH 9.8, 37°C in DEA buffer	Conjugation grade; 50% glycerol solution		Calzyme 140C4500
3000-6000 U/mg protein	1 unit liberates 1 μmol p-nitrophenol from PNPP/min at pH 9.8, 37°C in DEA buffer	Conjugation grade; in 3 M NaCl/TEA buffer		Calzyme 140E4500

3.1.3.1 Alkaline Phosphatase continued

SPECIFIC ACTIVITY	UNITS DEFINITION	PREPARATION FORM	ADDITIONAL ACTIVITIES	SUPPLIER CATALOG No.
Calf intestine *continued*				
3000-6000 U/mg protein	1 unit liberates 1 μmol *p*-nitrophenol from PNPP/min at pH 9.8, 37°C in DEA buffer	Molecular biology grade; lyophilized	No detectable endonuclease, exonuclease, RNase	Calzyme 235A4500
3000-6000 U/mg protein	1 unit liberates 1 μmol *p*-nitrophenol from PNPP/min at pH 9.8, 37°C in DEA buffer	Molecular biology grade; suspension in $(NH_4)_2SO_4$	No detectable endonuclease, exonuclease, RNase	Calzyme 235B4500
3000-6000 U/mg protein	1 unit liberates 1 μmol *p*-nitrophenol from PNPP/min at pH 9.8, 37°C in DEA buffer	Molecular biology grade; 50% glycerol solution	No detectable endonuclease, exonuclease, RNase	Calzyme 235C4500
3000-6000 U/mg protein	1 unit liberates 1 μmol *p*-nitrophenol from PNPP/min at pH 9.8, 37°C in DEA buffer	Molecular biology grade; in 3 M NaCl/TEA buffer	No detectable endonuclease, exonuclease, RNase	Calzyme 235E4500
	1 unit hydrolyzes 1 nmol *p*-nitrophenylphosphate/min at 37°C in a buffer of 1.0 M DEA, 10 mM *p*-nitrophenylphosphate, 0.25 mM MgCl$_2$, pH 9.8	10 mM Tris-HCl, 0.05 M NaCl, 50% glycerol, pH 8.0	No detectable endonuclease, RNase	CHIMERx 1025-01 1025-02
500 and 1000 U	1 unit hydrolyzes 1 μmol *p*-nitrophenylphosphate/min at pH 10.2, 37°C	500 mM Tris-HCl, pH 8.0		CLONTECH 8410-1 8410-2
≥3000 U/mg	PNPP as substrate, DEA as buffer at 37°C	3 M NaCl, 1 mM MgCl$_2$, 0.1 mM ZnCl$_2$, 30 mM TEA, pH 7.6		Crystal
1-2 U/μL; 10-20 U/μL	1 unit hydrolyzes 1 μmol 4-nitrophenylphosphate/min at 37°C	20 mM Tris-HCl, 1 mM MgCl$_2$, 0.1 mM ZnCl$_2$, 50% glycerol, pH 8.0	No detectable endo- and exodeoxyribonucleases, RNase	Fermentas EF0341 EF0342 EF0343
>1400 U/mg protein; >10,000 U/mL	1 unit hydrolyzes 1.0 μmol PNPP/min at pH 9.6, 25°C	50% glycerol solution		Genzyme 1903
500-600 U/mg protein	1 unit hydrolyzes 1 μmol PNPP/min at pH 9.6, 25°C	Purified; suspension in 3.2 M $(NH_4)_2SO_4$, 0.0001 M MgCl$_2$, 0.001 M ZnCl$_2$, pH 7.0		ICN 150032
1600 U/mg protein	1 unit hydrolyzes 1 μmol PNPP/min at pH 9.6, 25°C	Purified; suspension in 70% $(NH_4)_2SO_4$, pH 7.5		ICN 150034
800-900 U/mg protein	1 unit hydrolyzes 1 μmol PNPP/min at pH 9.6, 25°C	Highly purified; suspension in 70% $(NH_4)_2SO_4$, 0.001 M MgCl$_2$, 0.001 M ZnCl$_2$, pH 7.0		ICN 150096
>1300 U/mg protein	1 unit hydrolyzes 1 μmol PNPP/min at pH 9.6, 25°C	Highly purified; salt-free; solution containing 50% glycerol, 0.5 mM MgCl$_2$, 0.05 mM ZnCl$_2$, 10-20 mg/mL protein		ICN 150273

SPECIFIC ACTIVITY	UNITS DEFINITION	PREPARATION FORM	ADDITIONAL ACTIVITIES	SUPPLIER CATALOG NO.
Calf intestine *continued*				
1500 U/mg	1 unit liberates 1 μmol p-nitrophenol/min at pH 10.15, 37°C in Tris buffer	Lyophilized		ICN 159347
1.0 U/mg	1 unit hydrolyzes 1 μmol PNPP/min at pH 9.6, 25°C	Conjugation grade; lyophilized		ICN 190680
1000 U/mg protein	1 unit liberates 1 μmol p-nitrophenol from p-nitrophenylphosphate/min at 37°C	Labeling grade; lyophilized containing 30% carbohydrate as stabilizer	No detectable ADA, adenosine 5'-monophosphate deaminase, PDE	ICN 364841
	1 unit hydrolyzes 1 μmol PNPP/min at pH 8, 20°C; 1 DNA unit dephosphorylates 1 pmol 5'-phosphorylated ends of EcoR I fragments from λ DNA/hr	Molecular Biology Grade; 10 mM Tris-HCl, 1 mM MgCl$_2$, 50 mM KCl, 0.15 mM ZnCl$_2$, 50% glycerol, pH 8	No detectable nickase, endonuclease, exonuclease, RNase	Oncor 120221 120222
≥500 IU/mg protein	1 IU transforms 1 μmol substrate/min under standard IUB conditions at 25°C	Solution containing 3 M NaCl		OYC
1500-1700 U/mg (Gly); 5000-5500 U/mg (DEA)	1 unit hydrolyzes 1 μmol PNPP/min at pH 9.6, 25°C	Solution containing 50% glycerol, 10 mM Tris, 5 mM MgCl$_2$, 0.1 mM ZnCl$_2$, 10-25 mg/mL protein, pH 7.5		Scripps A0424
1200-1600 U/mg (Gly); 4000-5300 U/mg (DEA)	1 unit hydrolyzes 1 μmol PNPP/min at pH 9.6, 25°C	Solution containing 30 mM TEA, 3.0 M NaCl, 1 mM MgCl$_2$, 0.1 mM ZnCl$_2$, 6-12 mg/mL protein, pH 7.6		Scripps A0524
4000-5000 U/mg	1 unit hydrolyzes 1 μmol PNPP/min at pH 9.6, 25°C	Molecular Biology Grade; solution containing 50% glycerol, 10 mM Tris, 50 mM KCl, 1 mM MgCl$_2$, 0.1 mM ZnCl$_2$, 1.5-2.5 mg/mL protein, pH 8.0	No detectable endonuclease, exonuclease, RNase	Scripps A0625
5000 U/mL	1 unit hydrolyzes 1.0 μmol PNPP/min at pH 10.4, 37°C in Gly buffer	Solution containing 30 mM TEA buffer, 3 M NaCl, 1 mM MgCl$_2$, 0.1 mM ZnCl$_2$, pH 7.6	No detectable DNase, RNase	Sigma P8048
10,000-50,000 U/mL	1 unit hydrolyzes 1 μmol PNPP/min at pH 9.6, 25°C	10 mM Tris-HCl, 0.2 mM ZnCl$_2$, 5 mM MgCl$_2$, 50% glycerol, pH 8.3	No detectable endonuclease, single- and ds-DNase and RNase	Stratagene 600015 600016
1000 U and 5000 U	1 unit produces 1 μmol p-nitrophenol/min at pH 9.8, 37°C with p-nitrophenylphosphate as substrate	Solution containing 10 mM Tris-HCl, 1 mM MgCl$_2$, 0.1 mM ZnCl$_2$, 10 mM DTT, 50 mM KCl, 50% glycerol, pH 8.0	No detectable nuclease	TaKaRa 2250
≥2500 U/mg protein	1 unit forms 1 μmol p-nitrophenol/min at pH 10.25, 37°C	50% glycerol solution	No detectable DNase, RNase by gel electrophoresis <0.0001% ADA <0.001% PDE	Toyobo LPP-209
2000 DEA U/mg		Solution		Wako 016-14631

3.1.3.1 Alkaline Phosphatase continued

SPECIFIC ACTIVITY	UNITS DEFINITION	PREPARATION FORM	ADDITIONAL ACTIVITIES	SUPPLIER CATALOG No.
Calf intestine continued				
≥3000 U/mg protein	1 unit hydrolyzes 1 μmol PNPP/min at pH 9.8, 37°C	EIA grade; chromatographically purified; solution containing 3 M NaCl, 1 mM MgCl$_2$, 0.1 mM ZnCl$_2$, 0.3 M TEA, pH 7.6		Worthington LS04228 LS04230 LS04232 LS04234
Calf intestine or mucosa (MW = 140,000 Da)				
≥1600 U/mg protein; 4800 DEA U/mg protein	1 unit hydrolyzes 1 μmol PNPP/min at pH 9.6, 25°C (Gly buffer); 1 U = 3 DEA U	>90% purity by molecular exclusion chromatography; 50% glycerol solution containing 0.005 M Tris-HCl, 0.005 M MgCl$_2$, 0.0001 M ZnCl$_2$, 4-6.5% carbohydrate, 10-20 mg/mL protein, pH 7.0	No detectable Bovine IgG	Biozyme ALPI 10G
≥1800 U/mg protein; 5400 DEA U/mg protein	1 unit hydrolyzes 1 μmol PNPP/min at pH 9.6, 25°C (Gly buffer); 1 U = 3 DEA U	>90% purity by molecular exclusion chromatography; 50% glycerol solution containing 0.005 M Tris-HCl, 0.005 M MgCl$_2$, 0.0001 M ZnCl$_2$, 4-6.5% carbohydrate, 10-20 mg/mL protein, pH 7.0	No detectable Bovine IgG	Biozyme ALPI 11G
≥2000 U/mg protein; 6000 DEA U/mg protein	1 unit hydrolyzes 1 μmol PNPP/min at pH 9.6, 25°C (Gly buffer); 1 U = 3 DEA U	>90% purity by molecular exclusion chromatography; 50% glycerol solution containing 0.005 M Tris-HCl, 0.005 M MgCl$_2$, 0.0001 M ZnCl$_2$, 4-6.5% carbohydrate, 10-20 mg/mL protein, pH 7.0	No detectable Bovine IgG	Biozyme ALPI 12G
1400 U/mg protein; 4200 DEA U/mg protein	1 unit hydrolyzes 1 μmol PNPP/min at pH 9.6, 25°C (Gly buffer); 1 U = 3 DEA U	>90% purity by molecular exclusion chromatography; 50% glycerol solution containing 0.005 M Tris-HCl, 0.005 M MgCl$_2$, 0.0001 M ZnCl$_2$, 4-6.5% carbohydrate, 10-20 mg/mL protein, pH 7.0	No detectable Bovine IgG	Biozyme ALPI 8G
≥2600 DEA U/mg protein	1 DEA unit hydrolyzes 1 μmol PNPP/min at pH 9.8, 37°C (DEA buffer)	>90% purity by molecular exclusion chromatography; solution containing 0.03 M TEA, 3 M NaCl, 0.005 M MgCl$_2$, 0.0002 M ZnCl$_2$, 4-6.5% carbohydrate, 10-20 mg/mL protein, pH 7.6	No detectable Bovine IgG	Biozyme ALPI 8T
1000 U/mg protein; 3000 DEA U/mg protein	1 unit hydrolyzes 1 μmol PNPP/min at pH 9.6, 25°C (Gly buffer); 1 U = 3 DEA U	>90% purity by molecular exclusion chromatography; 50% glycerol solution containing 0.005 M Tris-HCl, 0.005 M MgCl$_2$, 0.0001 M ZnCl$_2$, 4-6.5% carbohydrate, 10-20 mg/mL protein, pH 7.0	No detectable Bovine IgG	Biozyme ALPI XG

SPECIFIC ACTIVITY	UNITS DEFINITION	PREPARATION FORM	ADDITIONAL ACTIVITIES	SUPPLIER CATALOG No.
Calf intestine or mucosa continued				
≥2000 DEA U/mg protein	1 DEA unit hydrolyzes 1 μmol PNPP/min at pH 9.8, 37°C (DEA buffer)	>90% purity by molecular exclusion chromatography; solution containing 0.03 M TEA, 3 M NaCl, 0.005 M MgCl$_2$, 0.0002 M ZnCl$_2$, 4-6.5% carbohydrate, 10-20 mg/mL protein, pH 7.6	No detectable Bovine IgG	Biozyme ALPI XT
Chicken intestine (optimum pH = 8.8, MW = 110,000 Da)				
35 U/mg	1 unit liberates 1 μmol p-nitrophenol from PNPP/min at pH 9.8, 37°C in DEA buffer	Lyophilized containing 95% protein		Calzyme 143A0035
6-8 U/mg solid	1 unit hydrolyzes 1 μmol PNPP/min at pH 10.4, 25°C	Partially purified; lyophilized		Elastin A123
30-40 U/mg protein	1 unit hydrolyzes 1 μmol PNPP/min at pH 10.4, 25°C	Chromatographically purified; salt-free; lyophilized		Elastin AC223
0.9-2.2 U/mg dry weight	1 unit hydrolyzes 1 μmol o-carboxyphenyl PO$_4$/min at pH 8.8, 25°C	Salt-free; lyophilized		ICN 100180
≥0.9 U/mg solid	1 unit hydrolyzes 1 μmol o-carboxyphenyl PO$_4$/min at pH 8.8, 25°C	Partially purified; lyophilized		Worthington LS03172 LS03171 LS03170 LS03174
Escherichia coli (optimum pH = 9.0, pI = 4.5, T = 37°C, MW = 80,000 Da [Sephadex G-200]; stable pH 8.5-10 [37°C, 1 hr], T < 45°C [pH 9.0, 10 min])				
40-70 U/mg solid	1 unit liberates 1 μmol p-nitrophenol/min at pH 9, 37°C	Lyophilized containing 100% protein		Asahi ALP T-08
≥500 U/mL	1 unit hydrolyzes 1 μmol PNPP/min in Tris-HCl at pH 8.0, 25°C	Solution containing 50 mM Tris-HCl and 50% glycerol, pH 7.5	No detectable nonspecific endonuclease and RNase	Calbiochem 524545
60 U/mg protein	1 unit hydrolyzes 1 μmol PNPP/min at pH 9.8, 37°C	Solution containing 50% glycerol, 5 mM Tris HCl, 0.5 mM MgCl$_2$, 0.5 mM ZnCl$_2$, pH 7.4	No detectable DNases, RNases, nickases	Fluka 79386
120 U/mg protein	1 unit hydrolyzes 1 μmol PNPP/min at pH 9.8, 37°C	Aqueous suspension in 2.5 M (NH$_4$)$_2$SO$_4$		Fluka 79388
>40 U/mg protein; >40 U/mg solid	1 unit hydrolyzes 1.0 μmol PNPP/min at pH 9.0, 37°C	Lyophilized		Genzyme 6041
10 U/mg	1 unit hydrolyzes 1 μmol PNPP/min	Partially purified; suspension in (NH$_4$)$_2$SO$_4$		ICN 100174
30-40 U/mg; 120-240 U/mL	1 unit hydrolyzes 1 μmol PNPP/min at pH 8.0, 25°C	Suspension in 65% saturated (NH$_4$)$_2$SO$_4$	No detectable DNase, nickase, RNase	Pharmacia 27-0598-02

SPECIFIC ACTIVITY	UNITS DEFINITION	PREPARATION FORM	ADDITIONAL ACTIVITIES	SUPPLIER CATALOG No.
Escherichia coli continued				
>40 U/mg	Tris, pH 8.0, 25°C	Suspension in 80% $(NH_4)_2SO_4$		ProZyme AP20 000 AP20 010 AP20 100
40-70 U/mg		Lyophilized		Wako 012-10691
≥10 U/mg protein	1 unit hydrolyzes 1 μmol PNPP/min at pH 8.0, 25°C	Partially purified salt fraction; suspension in 2.6 M $(NH_4)_2SO_4$, pH 8.0		Worthington LS04081 LS04082
≥20 U/mg protein	1 unit hydrolyzes 1 μmol PNPP/min at pH 8.0, 25°C	Chromatographically purified; suspension in 2.6 M $(NH_4)_2SO_4$, pH 8.0		Worthington LS05129 LS05130 LS05131
≥30 U/mg protein	1 unit hydrolyzes 1 μmol PNPP/min at pH 8.0, 25°C	Chromatographically purified; suspension in 2.6 M $(NH_4)_2SO_4$, pH 8.0	No detectable PDE when assayed at 0.1 mg/mL with bis(*p*-nitrophenyl) phosphate substrate <0.0002% RNase by weight as RNase A	Worthington LS06130 LS06124 LS06123 LS06122
Escherichia coli strain A19 (RNase-negative) (optimum pH = 8.0)				
0.2-0.5 U/μL	1 unit liberates 1 μmol *p*-nitrophenol/min at pH 8.0, 25°C with PNPP as substrate	10 mM Tris-HCl, 100 mM KCl, 1 mM $MgSO_4$, 50% glycerol, pH 8.0	No detectable non-specific nuclease, endonuclease, RNase	Amersham E 2110Y
20 U and 100 U	1 unit produces 1 μmol *p*-nitrophenol/min at pH 8.0, 25°C with PNPP as substrate	Solution containing 10 mM Tris-HCl, 1 mM $MgSO_4$, 100 mM KCl, 50% glycerol, pH 8.0	No detectable nuclease	TaKaRa 2110
Escherichia coli strain C4				
200 U/mL	1 unit hydrolyzes 1 μmol PNPP/min at pH 8.0, 37°C	10 mM Tris-HCl, 1 mM $MgCl_2$, 50 mM NaCl, 100 μg/mL BSA, 50% glycerol, pH 7.5	No detectable endonuclease, exonuclease, RNase	AGS Heidelb F00695S F00695M F00695L
0.05-0.2 U/μL; 20-50 U/mg protein	1 unit hydrolyzes 1 μmol PNPP/min at 37°C	10 mM Tris-HCl, 1 mM $MgCl_2$, 50 mM NaCl, 50% glycerol, pH 7.5	No detectable endo-, exodeoxyribonucleases, RNase	Fermentas EF0261 EF0262
Escherichia coli strain C75 (optimum pH = 8.0)				
0.2-0.5 U/μL	1 unit liberates 1 μmol *p*-nitrophenol/min at pH 8.0, 25°C	10 mM Tris-HCl, 100 mM KCl, 1 mM $MgSO_4$, 50% glycerol, pH 8.0	No detectable non-specific nuclease, endonuclease, RNase Contains little DNase	Amersham E 2120Y

Alkaline Phosphatase continued

SPECIFIC ACTIVITY	UNITS DEFINITION	PREPARATION FORM	ADDITIONAL ACTIVITIES	SUPPLIER CATALOG NO.
Escherichia coli strain C75 *continued*				
0.45 U/µL; 50 U and 250 U	1 unit produces 1 µmol *p*-nitrophenol/min at pH 8.0, 25°C with PNPP as substrate	Solution containing 10 mM Tris-HCl, 1 mM MgSO$_4$, 100 mM KCl, 50% glycerol, pH 8.0	No detectable nuclease	TaKaRa 2120A
Escherichia coli strain C90				
100-250 U/µL	1 unit hydrolyzes 1 nmol ATP/30 min at pH 8.0, 37°C	100 mM Tris-HCl, pH 8.0		Life Technol 18011-015 18011-049
Escherichia coli strain K12SW1033/pKI-5				
50 and 250 U	1 unit hydrolyzes 1 µmol PNPP/min at pH 8.0, 25°C	500 mM Tris-HCl, 10 mM MgCl$_2$, pH 8.0		CLONTECH 8409-1 8409-2
Human bone				
1-10 U/mg solid	1 unit liberates 1 µmol *p*-nitrophenol from PNPP/min at pH 9.8, 37°C in DEA buffer	Lyophilized		Calzyme 124A0001
Human intestine				
1-10 U/mg solid	1 unit liberates 1 µmol *p*-nitrophenol from PNPP/min at pH 9.8, 37°C in DEA buffer	Lyophilized		Calzyme 155A0001
Human kidney				
1-10 U/mg solid	1 unit liberates 1 µmol *p*-nitrophenol from PNPP/min at pH 9.8, 37°C in DEA buffer	Lyophilized		Calzyme 179A0001
Human liver				
1-10 U/mg solid	1 unit liberates 1 µmol *p*-nitrophenol from PNPP/min at pH 9.8, 37°C in DEA buffer	Lyophilized		Calzyme 178A0001
Human placenta				
	1 unit liberates 1 µmol *p*-nitrophenol from PNPP/min at pH 9.8, 37°C in DEA buffer	Lyophilized containing 90% protein		Calzyme 058A0100
	1 unit liberates 1 µmol *p*-nitrophenol from PNPP/min at pH 9.8, 37°C in DEA buffer	Lyophilized containing 90% protein		Calzyme 058A0750

3.1.3.1 Alkaline Phosphatase continued

SPECIFIC ACTIVITY	UNITS DEFINITION	PREPARATION FORM	ADDITIONAL ACTIVITIES	SUPPLIER CATALOG No.
Microbial (MW = 99,000 Da)				
> 700 U/mg	1 unit hydrolyzes 1 μmol PNPP/min at pH 9.8, 37°C	Single-band by SDS-PAGE; suspension in 80% $(NH_4)_2SO_4$		ProZyme DE12 000 DE12 010 DE12 100
***Pandalus borealis* (arctic shrimp)**				
5-10 U/μL	1 unit hydrolyzes 1 μmol PNPP/min at pH 9.6, 37°C with Gly/NaOH buffer	Homogeneous; 25 mM Tris-HCl, 1 mM $MgCl_2$, 0.1 mM $ZnCl_2$, 50% glycerol, pH 7.7	No detectable endonuclease, exonuclease, RNase	Amersham E 70092Y E 70092Z E 70092X
Porcine kidney				
300 U/mg solid	1 unit liberates 1 μmol *p*-nitrophenol from PNPP/min at pH 9.8, 37°C in DEA buffer	Lyophilized containing 90% protein		Calzyme 181A0300

3.1.3.2 Acid Phosphatase

REACTION CATALYZED
    An orthophosphoric monoester + H_2O ↔ An alcohol + orthophosphate

SYSTEMATIC NAME
    Orthophosphoric-monoester phosphohydrolase (acid optimum)

SYNONYMS
    Acid phosphomonoesterase, phosphomonoesterase, glycerophosphatase, prostatic acid phosphatase, PAP

REACTANTS
    Orthophosphoric monoester, H_2O, alcohol, orthophosphate, *o*-carboxyphenyl phosphate, salicylic acid

APPLICATIONS
- Dephosphorylation of phosphoproteins (e.g., β-casein, pepsinogen, ovalbumin)

NOTES
- Immunogen for antisera and calibrator to determine levels
- Associated with Gaucher's disease
- Monitoring the progression of prostate cancer
- Classified as a hydrolase, acting on ester bonds: phosphoric monoester hydrolase
- Name is associated with non-specific phosphomonoesterases with optimum activity in the range pH 4-6
- Isozymes EI, EII and EIII have pH optima of 5.5, 4.5 and 4.0, respectively; they are similar in size (approx. 55,000 Da)
- Distribution in nature includes plants, animals and microorganisms. Commercial sources include potatoes, wheat germ, milk and bovine prostate gland

- Wide specificity, hydrolyzing a variety of phosphomonoesters and phosphoproteins, but not phosphodiesters
- Also catalyzes transphosphorylations
- Unlike that from most other sources, potato enzyme is active at pH 7 and requires no activators, cofactors or reducing agents for maximum activity

SPECIFIC ACTIVITY	UNITS DEFINITION	PREPARATION FORM	ADDITIONAL ACTIVITIES	SUPPLIER CATALOG NO.
Bovine milk				
0.5-1.0 U/mg protein	1 unit hydrolyzes 1.0 μmol PNPP/min at pH 4.8, 37°C	Lyophilized containing 35% protein; balance primarily Na citrate buffer salt		Sigma P1267
Bovine prostate gland				
10 U/g solid	1 unit hydrolyzes 1.0 μmol PNPP/min at pH 4.8, 37°C; 60 Sigma U = 1 μM U	Partially purified; lyophilized	20 U/g total solid acid phosphatase	Sigma P6409
Bovine semen				
100-300 U/g solid	1 unit hydrolyzes 1.0 μmol PNPP/min at pH 4.8, 37°C; 60 Sigma U = 1 μM U	Crude; lyophilized	3 U/g total solid acid phosphatase	Sigma P2897
2 U/g solid	1 unit hydrolyzes 1.0 μmol PNPP/min at pH 4.8, 37°C; 60 Sigma U = 1 μM U	Crude; lyophilized	3 U/g total solid acid phosphatase	Sigma P3147
Human prostate gland (optimum pH = 4-6)				
10-20 U/mg solid	1 unit liberates 1 μmol p-nitrophenyl/min at pH 4.8, 37°C	Lyophilized containing 95% protein		Calzyme 050A0020
20 U/mg		Lyophilized containing 85% protein		ICN 153872
Human semen				
10-25 U/mg; 1.0 mg/mL		Antigen grade; >95% purity by SDS-PAGE; solution containing 10 mM Tris, 300 mM NaCl, 0.05% NaN$_3$, pH 8.0	No detectable HIV I, HIV II, HCV Ab, HBsAg	Cortex CP1016
22 U/mg prostatic acid phosphatase	1 unit hydrolyzes 1 μmol PNPP/min at pH 4.8, 37°C		3 U/mg tartrate insensitive phosphatase	Fluka 79423
		>98% purity; lyophilized	No detectable HGsAg, HCV, HIV-1 Ab	Scripps P0514
		>50% purity; solution containing 10 mM Tris, 150 mM NaCl, 0.1% NaN$_3$, 1-5 mg/mL protein, pH 7.2	No detectable HGsAg, HCV, HIV-1 Ab	Scripps P0523
		>98% purity; solution containing 10 mM Tris, 150 mM NaCl, 0.1% NaN$_3$, 1-5 mg/mL protein, pH 7.2	No detectable HGsAg, HCV, HIV-1 Ab	Scripps P0524

3.1.3.2 Acid Phosphatase continued

SPECIFIC ACTIVITY	UNITS DEFINITION	PREPARATION FORM	ADDITIONAL ACTIVITIES	SUPPLIER CATALOG NO.
Human semen continued				
200 U/mg protein	1 unit hydrolyzes 1.0 μmol PNPP/min at pH 4.8, 37°C; 60 Sigma U = 1 μM U	Affinity chromatographically purified; lyophilized containing 8% protein; balance primarily buffer salts, pH 4.8	≥220 U/mg protein total acid phosphatase	Sigma P1649
40 U/mg protein	1 unit hydrolyzes 1.0 μmol PNPP/min at pH 4.8, 37C; 60 Sigma U = 1 μM U	Crude; lyophilized containing 20% protein; balance primarily buffer salts, pH 4.8	50 U/mg total acid phosphatase	Sigma P1774
Human, unspecified				
				Vital 16970
Potato				
60 U/mg protein	1 unit converts 1 μmol PNPP to product/min at 25°C	Suspension in 3.2 M $(NH_4)_2SO_4$ and BSA as stabilizer, pH 6		Boehringer 108197
2 U/mg solid	1 unit converts 1 μmol PNPP to product/min at 25°C	Lyophilized		Boehringer 108219 108227
60 U/mg; 6 U/mg total protein	1 unit hydrolyzes 1 μmol PNPP/min at pH 4.8, 25°C	Crystallized; suspension in 3.2 M $(NH_4)_2SO_4$ and 1% BSA as stabilizer		Calbiochem 524528
3 U/mg protein	1 unit hydrolyzes 1 μmol PNPP/min at pH 4.8, 25°C	Suspension in 3.2 M $(NH_4)_2SO_4$ and 1% BSA, pH 6.6		Fluka 79397
3-10 U/mg solid	1 unit hydrolyzes 1.0 μmol PNPP/min at pH 4.8, 37°C	Lyophilized	5 U/mg solid apyrase (ATPase) activity	Sigma P1146
0.5-1.0 U/mg solid	1 unit hydrolyzes 1.0 μmol PNPP/min at pH 4.8, 37°C	Lyophilized		Sigma P3752
60 U/mg protein	1 unit hydrolyzes 1.0 μmol PNPP/min at pH 4.8, 37°C	Suspension in 3.2 M $(NH_4)_2SO_4$ and BSA stabilizer, pH 6.0		Sigma P6760
Potato, sweet				
20-60 U/mg protein	1 unit hydrolyzes 1.0 μmol PNPP/min at pH 4.8, 37°C	Suspension in 1.8 M $(NH_4)_2SO_4$ and 10 mM $MgCl_2$, pH 5.3	Apyrase (ATPase) activity < acid phosphatase activity	Sigma P1435
Potato, white				
200 U/mg protein	1 unit hydrolyzes 1.0 μmol PNPP/min at pH 4.8, 37°C	Suspension in 1.8 M $(NH_4)_2SO_4$ and 10 mM $MgCl_2$, pH 5.5	Apyrase (ATPase) activity: <5 % of acid phosphatase activity	Sigma P0157
Wheat germ (optimum pH = 4-6, MW =50,000-60,000 Da)				
1 U/mg solid	1 unit liberates 1 μmol p-nitrophenyl/min at pH 4.8, 37°C	Lyophilized containing 90% protein		Calzyme 145A0001
0.15 U/mg	1 unit hydrolyzes 1 μmol PNPP/min at pH 4.8, 25°C	Salt-free; lyophilized		Fluka 79410
1 U/mg solid		Lyophilized containing 90% protein		ICN 153876

Acid Phosphatase continued

3.1.3.2

SPECIFIC ACTIVITY	UNITS DEFINITION	PREPARATION FORM	ADDITIONAL ACTIVITIES	SUPPLIER CATALOG NO.
Wheat germ *continued*				
0.4 U/mg solid	1 unit hydrolyzes 1.0 μmol PNPP/min at pH 4.8, 37°C		Lipase	Sigma P3627
≥15 U/mg solid	1 unit hydrolyzes 1 μmol o-carboxy-(or PNPP)/min at pH 5.0, 25°C	Partially purified; lyophilized	Lipase	Worthington LS01141 LS01144

5'-Nucleotidase

3.1.3.5

REACTION CATALYZED

A 5'-ribonucleotide + H_2O ↔ A ribonucleoside + orthophosphate

SYSTEMATIC NAME

5'-Ribonucleotide phosphohydrolase

REACTANTS

5'-Ribonucleotide, H_2O, ribonucleoside, orthophosphate

NOTES

- Classified as a hydrolase, acting on ester bonds: phosphoric monoester hydrolase
- Wide specificity for 5'-nucleotides

SPECIFIC ACTIVITY	UNITS DEFINITION	PREPARATION FORM	ADDITIONAL ACTIVITIES	SUPPLIER CATALOG NO.
Crotalus adamanteus **(snake venom)**				
100 U/mg	1 unit hydrolyzes 1 μmol adenosine-5'-monophosphate/min at pH 9.0, 37°C	Powder containing 35% protein and 65% Gly buffer salt	<0.1% AP, PPase, nucleotide pyrophosphatase, PDE, protease	Fluka 74685
500-800 U/g agarose; 1 mL yields 15-25 U	1 unit hydrolyzes 1.0 μmol inorganic phosphorous from adenosine 3'-monophosphate/min at pH 7.5, 37°C	Suspension in 0.5 M NaCl		Sigma N3264
200-500 U/mg protein	1 unit hydrolyzes 1.0 μmol inorganic phosphorous from adenosine 3'-monophosphate/min at pH 7.5, 37°C	Partially purified; lyophilized containing 35% protein; balance primarily Gly buffer salts		Sigma N4005
Crotalus atrox **(snake venom)**				
30 U/mg	1 unit hydrolyzes 1 μmol adenosine-5'-monophosphate/min at pH 9.0, 37°C	Powder containing 90% Gly buffer salt and 10% protein	<0.05% AP, PPase <0.1% PDE, protease	Fluka 74692

3.1.3.5 5'-Nucleotidase continued

SPECIFIC ACTIVITY	UNITS DEFINITION	PREPARATION FORM	ADDITIONAL ACTIVITIES	SUPPLIER CATALOG NO.
Crotalus atrox (snake venom) *continued*				
200-500 U/mg protein	1 unit hydrolyzes 1.0 μmol inorganic phosphorous from adenosine 3'-monophosphate/min at pH 7.5, 37°C	Partially purified; lyophilized containing 10% protein; balance primarily Gly buffer salts		Sigma N5880

3.1.3.6 3'-Nucleotidase

REACTION CATALYZED
    A 3'-ribonucleotide + H_2O ↔ A ribonucleoside + orthophosphate

SYSTEMATIC NAME
    3'-Ribonucleotide phosphohydrolase

REACTANTS
    3'-Ribonucleotide, H_2O, ribonucleoside, orthophosphate

NOTES
- Classified as a hydrolase, acting on ester bonds: phosphoric monoester hydrolase
- Wide specificity for 3'-nucleotides

SPECIFIC ACTIVITY	UNITS DEFINITION	PREPARATION FORM	ADDITIONAL ACTIVITIES	SUPPLIER CATALOG NO.
Rye grass				
40-120 U/mg protein	1 unit hydrolyzes 1.0 μmol inorganic phosphorous from adenosine 3'-monophosphate/min at pH 7.5, 37°C	Lyophilized containing 25% protein; balance primarily Na citrate buffer salts		Sigma N7008

3-Phytase 3.1.3.8

REACTION CATALYZED

myo-Inositol hexakisphosphate + H_2O ↔ D-myo-Inositol 1,2,4,5,6-pentakisphosphate + orthophosphate

SYSTEMATIC NAME

myo-Inositol hexakisphosphate 3-phosphohydrolase

SYNONYMS

Phytase

REACTANTS

myo-Inositol hexakisphosphate, H_2O, D-myo-inositol 1,2,4,5,6-pentakisphosphate, orthophosphate

APPLICATIONS

- In food applications, removes plant seed phytins which bind essential trace minerals and proteins, reducing their bioavailability
- Soy bean processing
- Baking
- Feed improvement
- Waste treatment

NOTES

- Classified as a hydrolase, acting on ester bonds: phosphoric monoester hydrolase

SPECIFIC ACTIVITY	UNITS DEFINITION	PREPARATION FORM	ADDITIONAL ACTIVITIES	SUPPLIER CATALOG NO.
Aspergillus ficuum				
3-5 U/mg	1 unit liberates 1 μmol P_i from phytic acid/min at pH 2.5, 37°C	Lyophilized		Fluka 80170
1-5 U/mg solid (as phosphatase activity)	1 unit liberates 1.0 μmol P_i from 4.2×10^{-2} M Mg-K phytate/min at pH 2.5, 37°C	Lyophilized	May contain glucosidases and phosphatases	Sigma P9792
Aspergillus niger (optimum pH = 4.0-5.0)				
≥5000 FTU/g	1 FTU liberates 1 μmol inorganic phosphorus from an excess of Na phytate/min at pH 5.5, 37°C	Powder		BASF Natuphos^R 5.000
≥5000 FTU/g	1 FTU liberates 1 μmol inorganic phosphorus from an excess of Na phytate/min at pH 5.5, 37°C	Solution		BASF Natuphos^R 5.000 L
		Food grade		EnzymeDev Finase^R phytase
		Feed grade		EnzymeDev Finase^R S-40 Phytase

3.1.3.9 Glucose-6-Phosphatase

REACTION CATALYZED
: D-Glucose 6-phosphate + H_2O ↔ D-Glucose + orthophosphate

SYSTEMATIC NAME
: D-Glucose-6-phosphate phosphohydrolase

REACTANTS
: D-Glucose 6-phosphate, carbamoyl phosphate, hexose phosphates, pyrophosphate, phosphoenolpyruvate, nucleoside di- and triphosphates, H_2O, D-glucose, D-mannose, 3-methyl-D-glucose, 2-deoxy-D-glucose, orthophosphate

NOTES
- Classified as a hydrolase, acting on ester bonds: phosphoric monoester hydrolase
- Wide distribution in animal tissues
- Also catalyzes transphosphorylations

SPECIFIC ACTIVITY	UNITS DEFINITION	PREPARATION FORM	ADDITIONAL ACTIVITIES	SUPPLIER CATALOG NO.
Rabbit liver				
0.05-0.1 U/mg protein	1 unit releases 1.0 µmol P_i from Glc6P/min at pH 6.5, 37°C	Crude; microsomal preparation containing 30% protein; balance primarily sucrose		sigma G5758

3.1.3.11 Fructose-Bisphosphatase

REACTION CATALYZED
: D-Fructose 1,6-bisphosphate + H_2O ↔ D-Fructose 6-phosphate + orthophosphate

SYSTEMATIC NAME
: D-Fructose 1,6-bisphosphate 1-phosphohydrolase

SYNONYMS
: Hexosediphosphatase, D-fructose-1,6-diphosphatase

REACTANTS
: D-Fructose 1,6-bisphosphate, sedoheptulose 1,7-bisphosphate, H_2O, D-fructose 6-phosphate, orthophosphate

NOTES
- Classified as a hydrolase, acting on ester bonds: phosphoric monoester hydrolase
- Animal enzyme also acts on sedoheptulose 1,7-bisphosphate

Fructose-Bisphosphatase continued — 3.1.3.11

SPECIFIC ACTIVITY	UNITS DEFINITION	PREPARATION FORM	ADDITIONAL ACTIVITIES	SUPPLIER CATALOG NO.
Rabbit liver				
3-8 U/mg protein	1 unit converts 1.0 μmol fructose 1,6-diphosphate to F6P and P_i/min at pH 9.5, 25°C	Lyophilized containing 85% protein; balance primarily Na malonate and MgOAc	1% PPase	Sigma F5252
4-6 U/mg protein	1 unit converts 1.0 μmol fructose 1,6-diphosphate to F6P and P_i/min at pH 9.5, 25°C	Suspension in 3.2 M $(NH_4)_2SO_4$, 2 mM Na malonate, 1 mM MgOAc, 0.5 mM EDTA, pH 6	1% PPase	Sigma F9503
Rabbit muscle				
4 U/mg protein	1 unit converts 1.0 μmol fructose 1,6-diphosphate to F6P and inorganic phosphorus/min at pH 9.5, 25°C	Suspension in 3.2 M $(NH_4)_2SO_4$, pH 6		ICN 158194
4 U/mg protein	1 unit converts 1.0 μmol fructose 1,6-diphosphate to F6P and P_i/min at pH 9.5, 25°C	Suspension in 3.2 M $(NH_4)_2SO_4$, pH 6		Sigma F3003
Yeast, *Torula*				
10-30 U/mg protein	1 unit converts 1.0 μmol fructose 1,6-diphosphate to F6P and P_i/min at pH 9.5, 25°C	Suspension in 3.2 M $(NH_4)_2SO_4$, 5 mM MgOAc, 0.1 mM EDTA, pH 6		Sigma F0254

Phosphoprotein Phosphatase — 3.1.3.16

REACTION CATALYZED
A phosphoprotein + H_2O ↔ A protein + orthophosphate

SYSTEMATIC NAME
Phosphoprotein phosphohydrolase

SYNONYMS
Calcineurin, protein phosphatase-1, protein phosphatase-2A, protein phosphatase-$2A_1$, protein phosphatase-$2A_2$, protein phosphatase-2B, protein phosphatase-2C, λ-protein phosphatase, PP

REACTANTS
Phosphoprotein, phenolic phosphates, phosphamides, H_2O, protein, orthophosphate

APPLICATIONS
- Dephosphorylation of phosphoserine/threonine proteins and peptides in solution, in immunoprecipitates and on blots

NOTES
- Classified as a hydrolase, acting on ester bonds: phosphoric monoester hydrolase
- A group of enzymes removing the serine- or threonine-bound phosphate group from a wide range of phosphoproteins. Spleen enzyme also works on phenolic phosphates and phosphamides

- Enzymes in this group can be Mn^{2+}-dependent (*E. coli*) or -independent (bovine kidney)
- Includes enzymes phosphorylated under the action of a kinase like protein-tyrosine-phosphatase (EC 3.1.3.48)
- The key signaling enzyme in T-lymphocyte activation
- Calcineurin is a Ca^{2+}/calmodulin-dependent protein phosphatase identified as the physiological target of immunosuppressants cyclosporin A, cyclophilin and FK-506
- Holoenzyme has a catalytic A subunit with specificity for phosphoserine/threonine, and a regulatory B subunit that binds Ca^{2+}
- Nearly inactive in the absence of activators
- Generally inhibited by divalent cations, vanadate, protein phosphatase inhibitor 2 and okadaic acid

SPECIFIC ACTIVITY	UNITS DEFINITION	PREPARATION FORM	ADDITIONAL ACTIVITIES	SUPPLIER CATALOG No.
Bovine brain (MW = 61,000 [A subunit] and 19,200 [B subunit] Da)				
4.3 kU/mg protein	1 unit causes 50% inhibition of activated PDE at pH 7.5, 30°C in the presence of excess calmodulin and 100 μM Ca^{2+}	Lyophilized		BIOMOL SE-112
>300 U/mg	1 unit converts 1 μmol PNPP to *p*-nitrophenol and P_i/min at 30°C	>90% purity by SDS-PAGE; solution	No detectable protease	Boehringer 1636740
>300 mU/mg protein	1 unit converts 1 PNPP to *p*-nitrophenol and P_i/min at pH 7.0, 30°C	>90% Purity; solution containing 20 mM Tris-HCl, 100 mM NaCl, 1 mM imidazole, 1 mM MgOAc, 10 mM β-MSH, 2% trehalose, pH 7.5	No detectable protease	Calbiochem 539565 PP-2B
		≥95% purity		Chemicon AG640
Bovine kidney (subunit MW = 36,000 [catalytic], 55,000 and 60,000 [regulatory] Da)				
1000 U/mg	1 unit hydrolyzes 1 nmol PO_4 from phosphorylase/min at pH 7.0, 30°C	>95% by SDS-PAGE; 50 μg/mL in 25 mM Tris, 10% glycerol, 0.1 mM EDTA, 14 mM β-MSH, 1 mM benzamidine, 0.1 mM PMSF, pH 7.0		BIOMOL SE-119
750 U/mg protein	1 unit hydrolyzes 1 nmol PO_4 from phosphorylase/min at pH 7.0, 30°C	>95% by SDS-PAGE; 50 μg/mL in 25 mM Tris, 10% glycerol, 0.1 mM EDTA, 14 mM β-MSH, 1 mM benzamidine, 0.1 mM PMSF, pH 7.0		BIOMOL SE-120
750-2000 U/mg	1 unit dephosphorylates 1 nmol ^{32}P-labeled phosphorylase/min at pH 7.0, 30°C	≥95% Purity; 50 mM Tris-HCl, 50% glycerol, 0.1 mM PMSF, 1 mM benzamidine, 14 mM β-MSH, pH 7.0		Calbiochem 539508 PP-2A$_1$

SPECIFIC ACTIVITY	UNITS DEFINITION	PREPARATION FORM	ADDITIONAL ACTIVITIES	SUPPLIER CATALOG NO.
Bovine kidney *continued*				
2000 U/mg	1 unit dephosphorylates ^{32}P-labeled phosphorylase at 1 nmol/min at pH 7.0, 30°C	25 mM Tris-HCl, 50% glycerol, 0.1 mM EDTA, 1 mM benzamidine, 0.1 mM PMSF, 14 mM β-MSH, pH 7.0		Calbiochem 539510 PP-2A$_2$
***Escherichia coli*, carrying bacteriophage λ IOF221**				
400,000 U/mL	1 unit hydrolyzes 1 nmol PNPP (50 mM)/min at 30°C in λ-PPase buffer with 2.0 mM MnCl$_2$ and 100 μg/mL BSA	MW = 25,000 Da; >95% homogeneous; 50 mM Tris-HCl, 250 mM NaCl, 2 mM MnCl$_2$, 1 mM Na$_2$EDTA, 5 mM DTT, 50% glycerol, pH 7.0	No detectable protease, kinase, DNase, RNase	NE Biolabs 753S 753L
500 U/mL	1 unit hydrolyzes 1 nmol PNPP (50 mM)/min at 30°C in PP1 buffer with 0.2 mM MnCl$_2$, 5 mM caffeine, 100 μg/mL BSA	MW = 37,500 Da; >95% homogeneous; 50 mM imidazole, 250 mM NaCl, 2.0 mM MnCl$_2$, 1.0 mM Na$_2$EDTA, 5 mM DTT, 0.025% Tween 20, 50% glycerol, pH 7.0	No detectable protease, kinase, DNase, RNase, tyrosine phosphatase	NE Biolabs 754S 754L
Human, recombinant carrying the γ-isoform catalytic subunit, expressed in *E. coli*				
2 U/mg protein	1 unit converts 1 μmol PNPP to *p*-nitrophenol and P$_i$/min at 37°C	>95% purity; solution containing 50% glycerol, 0.03% BRIJ, 0.1 mM EGTA, 0.1% β-MSH, 200 mM NaCl, 25 mM TEA-HCl, pH 7.5	<0.1% protease	Calbiochem 539555 PP-1
Human, recombinant from *E. coli* (MW = 37,000 Da)				
2 U/mg; 5 U/mg (in the presence of MnCl$_2$)	1 unit converts 1 μmol PNPP to *p*-nitrophenol and P$_i$/min at 37°C; 1 pNPP U = 10 phosphorylase a U	>95% purity by SDS-PAGE; solution	No detectable protease	Boehringer 1636758
Lambda, recombinant in *E. coli* (MW = 25,000 Da)				
300,000 U/mg protein	1 unit hydrolyzes 1 nmol PNPP/min at pH 7.0, 30°C	Solution containing 250 mM NaCl, 50 mM Tris-HCl, 5 mM DTT, 2 mM MnCl$_2$, 1 mM EDTA, 50% glycerol, pH 7.0	No detectable DNase, kinase, protease, RNase	Calbiochem 539514
***N. crassa*, recombinant**				
		≥95% purity		Chemicon AG645
Rabbit skeletal muscle, recombinant (MW = 37,500 Da)				
500 U/mL	1 unit hydrolyzes 1 pmol PNPP/min at 30°C	>95% purity by SDS-PAGE; 500 U/mL in 50 mM imidazole, 0.25 M NaCl, 1 mM EDTA, 5 mM DTT, 0.25% Tween 20, 50% glycerol, pH 7.4	No detectable protease, kinase, tyrosine phosphatase	BIOMOL SE-129
Recombinant, unspecified				
		Calcineurin B subunit; ≥95% purity		Chemicon AG646
		Calcineurin Aα subunit; ≥95% purity		Chemicon AG647

3.1.3.16 Phosphoprotein Phosphatase continued

SPECIFIC ACTIVITY	UNITS DEFINITION	PREPARATION FORM	ADDITIONAL ACTIVITIES	SUPPLIER CATALOG NO.
Recombinant *continued*				
		Calcineurin AβB subunit; ≥95% purity		Chemicon AG648
		Calcineurin Aβ subunit; ≥95% purity		Chemicon AG651

3.1.3.26 6-Phytase

REACTION CATALYZED
  myo-Inositol hexakisphosphate + H_2O ↔ 1L-myo-Inositol 1,2,3,4,5-pentakisphosphate + orthophosphate

SYSTEMATIC NAME
  myo-Inositol-hexakisphosphate 6-phosphohydrolase

SYNONYMS
  Phytase, phytate 6-phosphatase

REACTANTS
  myo-Inositol hexakisphosphate, H_2O, 1L-myo-Inositol 1,2,3,4,5-pentakisphosphate, orthophosphate

APPLICATIONS
- Hydrolyzing phytic acid in animal feeds

NOTES
- Classified as a hydrolase, acting on ester bonds: phosphoric monoester hydrolase
- Phytic acid is the principal form of phosphorous in many seeds. It binds minerals, reducing their bioavailability; binds proteins limiting their digestibility; and binds enzymes, limiting their activity

SPECIFIC ACTIVITY	UNITS DEFINITION	PREPARATION FORM	ADDITIONAL ACTIVITIES	SUPPLIER CATALOG NO.
Aspergillus niger **(optimum pH = 5.5-7, T = 50-65°C; stable pH 2-6 [30°C, 30 min])**				
≥100 U/g	1 unit causes 1 μmol phosphorus/min at pH 6.0, 37°C	Powder	Cellulase, pectinase, xylanase	Amano Phytase "Amano"
Wheat				
0.04-0.07 U/mg	1 unit liberates 1 μmol P_i from phytic acid/min at pH 5.15, 55°C	Powder		Fluka 80172
0.015 U/mg solid	1 unit liberates 1.0 μmol P_i from 1.5×10^{-3} M phytate/min at pH 5.15, 55°C	Crude		Sigma P1259

Protein-Tyrosine-Phosphatase 3.1.3.48

REACTION CATALYZED
Protein tyrosine phosphate + H₂O ↔ Protein tyrosine + orthophosphate

SYSTEMATIC NAME
Protein-tyrosine-phosphate phosphohydrolase

SYNONYMS
Phosphotyrosine phosphatase

REACTANTS
Protein tyrosine phosphate, H₂O, protein tyrosine, orthophosphate

NOTES
- Classified as a hydrolase, acting on ester bonds: phosphoric monoester hydrolase
- Dephosphorylates o-phosphotyrosine groups in phosphoproteins, e.g., the products of protein-tyrosine kinase (EC 2.7.1.112)
- Inhibited by vanadate (0.1 mM) and molybdate (0.01 mM)

SPECIFIC ACTIVITY	UNITS DEFINITION	PREPARATION FORM	ADDITIONAL ACTIVITIES	SUPPLIER CATALOG NO.
Human CD34 cytoplasmic domain, recombinant expressed in yeast (MW = 95,000 Da)				
20,000 U/mg	1 unit hydrolyzes 1 nmol PNPP/min at pH 7.0, 30°C	>90% purity by SDS-PAGE; 50 mM HEPES, 0.15 M NaCl, 4 mM DTT, 0.0035% BRIJ 35, pH 7	No detectable protease	BIOMOL SE-135
Human, recombinant expressed in *E. coli* (MW = 38,000–40,000 Da)				
5 kU/mg	1 unit hydrolyzes 1 nmol PNPP/min at pH 7.0, 30°C	>95% purity by SDS-PAGE; 5000 U/mL in 50 mM imidazole, 0.25 M NaCl, 2.5 mM EDTA, 5 mM DTT, 50% glycerol, pH 7.0	No detectable protease, DNase, RNase, Ser/Thr phosphatase	BIOMOL SE-113
15 kU/mg	1 unit releases 1 nmol PNPP/min at pH 7, 30°C	>95% purity; 10 kU/mL in 50 mM imidazole, 0.25 M NaCl, 2.5 mM EDTA, 5 mM DTT, 50% glycerol, pH 7.0	No detectable protease, DNase, RNase, Ser/Thr phosphatase	BIOMOL SE-114
5000 U/mg protein	1 unit hydrolyzes 1 nmol PNPP/min at pH 7.0, 30°C	>95% purity; solution containing 250 mM NaCl, 50 mM imidazole, 5 mM DTT, 2.5 mM EDTA, 50% glycerol, pH 7.0	No detectable DNase, kinase, protease, RNase, Ser/Thr phosphatase	Calbiochem 539731
15,000 U/mg protein	1 unit hydrolyzes 1 nmol PNPP/min at pH 7.8, 30°C	>95% purity; solution containing 250 mM NaCl, 50 mM imidazole, 5 mM DTT, 2.5 mM EDTA, 50% glycerol, pH 7.0	No detectable DNase, kinase, protease, RNase, Ser/Thr phosphatase	Calbiochem 539732
5000 U/mL	1 unit hydrolyzes 1 nmol PNPP (50 mM)/min at 30°C in LAR buffer with 100 µg/mL BSA	>95% homogeneous; 50 mM imidazole, 250 mM NaCl, 2.5 mM Na₂EDTA, 5 mM DTT, 50% glycerol, pH 7.0	No detectable protease, kinase, DNase, RNase, Ser/Thr phosphatase	NE Biolabs 750S 750L
10,000 U/mL	1 unit hydrolyzes 1 nmol PNPP (50 mM)/min at 30°C in TC PTP buffer with 100 µg/mL BSA	>95% homogeneous; 50 mM imidazole, 250 mM NaCl, 2.5 mM Na₂EDTA, 5 mM DTT, 50% glycerol, pH 7.0	No detectable protease, kinase, DNase, RNase, Ser/Thr phosphatase	NE Biolabs 752S 752L

3.1.3.48 Protein-Tyrosine-Phosphatase continued

SPECIFIC ACTIVITY	UNITS DEFINITION	PREPARATION FORM	ADDITIONAL ACTIVITIES	SUPPLIER CATALOG NO.
Yersinia enterocolitica 34 kDa fragment, recombinant expressed in *E. coli*				
>20 U/mg protein	1 unit hydrolyzes 1 µmol PNPP/min at pH 7.5, 37°C	>95% purity; solution containing 100 mM NaOAc, 100 mM NaCl, 0.1% β-MSH, 50% glycerol, pH 5.5, sealed in glass vials under inert gas	No detectable protease	Calbiochem 539446
Yersinia enterocolitica, recombinant expressed in *E. coli* (MW = 34,000-51,000 Da)				
500 kU/mg	1 unit hydrolyzes 1 nmol PNPP/min at pH 7.0, 30°C	>95% purity; 50 kU/mL in 100 mM NaOAc, 100 mM NaCl, 2.5 mM EDTA, 5 mM DTT, 50% glycerol, pH 5.7	No detectable protease, DNase, RNase, Ser/Thr phosphatase	BIOMOL SE-115
>20 U/mg	1 unit hydrolyzes 1 µmol PNPP/min at pH 7.5, 37°C	>95% by SDS-PAGE; solution filled under N_2	No detectable protease	Boehringer 1500775
50,000 U/mL	1 unit hydrolyzes 1 nmol PNPP (50 mM)/min at 30°C in YOP buffer with 100 µg/mL BSA	>95% homogeneous; 100 mM NaOAc, 100 mM NaCl, 2.5 mM Na_2EDTA, 5 mM DTT, 50% glycerol, pH 5.7	No detectable protease, kinase, DNase, RNase, Ser/Thr phosphatase	NE Biolabs 751S 751L

3.1.4.1 Phosphodiesterase I

REACTION CATALYZED
    Hydrolytically removes 5'-nucleotides successively from the 3'-hydroxy termini of 3'-hydroxy-terminated oligonucleotides

SYSTEMATIC NAME
    Oligonucleate 5'-nucleotidohydrolase

SYNONYMS
    5'-Exonuclease, 5'-phosphodiesterase, PDE

REACTANTS
    5'-Mononucleotides, 3'-OH-terminated ribo- and deoxyribo-oligonucleotides, *p*-nitrophenyl thymidine-5'-phosphate

APPLICATIONS
- Structural and sequential DNA analysis

NOTES
- Classified as a hydrolase, acting on ester bonds: phosphoric diester hydrolase
- Low activity toward polynucleotides
- Bovine brain enzyme is calmodulin-dependent
- Inhibited by 3'-phosphate substrate terminus; reducing agents like glutathione, cysteine and ascorbate; and EDTA (5 mM)
- Partially inhibited by ATP, ADP and AMP

SPECIFIC ACTIVITY	UNITS DEFINITION	PREPARATION FORM	ADDITIONAL ACTIVITIES	SUPPLIER CATALOG No.
Bothrops atrox (snake venom)				
Sufficient activity to hydrolyze minimum of 0.01 μmol/mg solid	1 unit hydrolyzes 1.0 μmol bis(PNPP)/min at pH 8.8, 37°C	Crude; dried venom		Sigma P4631
Bovine brain				
		≥95% purity		Chemicon AG641
Bovine intestinal mucosa				
0.5-1.0 U/mg protein; <20 U/mg non-specific PO_4 activity at pH 10.4, 37°C using PNPP	1 unit hydrolyzes 1.0 μmol bis(PNPP)/min at pH 8.8, 37°C	Suspension in 3.2 M $(NH_4)_2SO_4$, pH 7.0	Pyrophosphatase, nucleotide and 5′-nucleotidase	Sigma P6903
Crotalus adamanteus (snake venom) (optimum pH = 9.8-10.4, MW = 115,000 Da)				
20-40 U/mg solid; 100 U/vial	1 unit hydrolyzes 1 μmol p-nitrophenyl thymidine-5-phosphate/min at pH 8.9, 25°C	Lyophilized		ICN 100978
≥20 U/mg dry weight	1 unit hydrolyzes 1 μmol p-nitrophenyl thymidine 5′-phosphate/min at pH 8.9, 25°C	Partially purified; lyophilized	Treated to remove 5′-nucleotidase	Pharmacia 27-0821-01
Sufficient activity to hydrolyze minimum of 0.01 μmol/mg solid	1 unit hydrolyzes 1.0 μmol bis(PNPP)/min at pH 8.8, 37°C	Crude; dried venom		Sigma P3134
0.2-0.4 U/mg solid	1 unit hydrolyzes 1.0 μmol bis(PNPP)/min at pH 8.8, 37°C	Lyophilized containing 35% Tris buffer salts	5′-nucleotidase activity: <1% PDE activity	Sigma P6877
≥20 U/mg solid	1 unit hydrolyzes 1 μmol p-nitrophenyl-thymidine-5′-phosphate/min at pH 8.9, 25°C	Purified; lyophilized	Inactivated 5′-nucleotidase	Worthington LS03926 LS03928
Crotalus atrox (snake venom)				
Hydrolyzes ≥ 0.01 μmol/mg solid	1 unit hydrolyzes 1.0 μmol bis(PNPP)/min at pH 8.8, 37°C	Crude; dried venom		Sigma P4506
0.2 U/mg protein	1 unit hydrolyzes 1.0 μmol bis(PNPP)/min at pH 8.8, 37°C	Lyophilized containing 65% protein; balance primarily Tris buffer salts		Sigma P6761

3.1.4.1 Phosphodiesterase I continued

SPECIFIC ACTIVITY	UNITS DEFINITION	PREPARATION FORM	ADDITIONAL ACTIVITIES	SUPPLIER CATALOG No.
Crotalus durissus terrificus (snake venom)				
6 U/mg protein	1 unit hydrolyzes 1 μmol bis(PNPP)/min at pH 8.9, 25°C	Solution containing 50% glycerol and 5 mM Tris HCl, pH 7.5	30 U/mg nucleotide pyrophosphatase	Fluka 79426
1-3 U/mg protein	1 unit hydrolyzes 1.0 μmol bis(PNPP)/min at pH 8.9, 25°C	50% glycerol solution, pH 6	Nucleotide pyrophosphatase	Sigma P7027
Penicillium species (optimum pH = 4.5-6.0, T = 65-85°C; stable pH 4.0-7.0 [50°C, 30 min], T > 20-60°C [pH 5.0, 15 min])				
≥13,000 U/mg	pH 4.8	Powder		Amano RP-1

3.1.4.2 Glycerophosphocholine Phosphodiesterase

REACTION CATALYZED

sn-Glycero-3-phosphocholine + H_2O ↔ Choline + sn-glycerol 3-phosphate

SYSTEMATIC NAME

sn-Glycero-3-phosphocholine glycerophosphohydrolase

SYNONYMS

Glycerophosphorylcholine phosphodiesterase, GPCP

REACTANTS

sn-Glycero-3-phosphocholine, sn-glycero-3-phosphoethanolamine, H_2O, choline, sn-glycerol 3-phosphate

NOTES

- Classified as a hydrolase, acting on ester bonds: phosphoric monoester hydrolase
- Activated by Ca^{2+}
- Inhibited by Zn^{2+} and EDTA

SPECIFIC ACTIVITY	UNITS DEFINITION	PREPARATION FORM	ADDITIONAL ACTIVITIES	SUPPLIER CATALOG No.
Gliocladium roseum (optimum pH = 8.5-9.5, pI = 3.75; K_M = 0.1 mM [glycerophosphorylcholine]; stable pH 7-9 [37°C, 1 hr], T < 50°C [pH 8, 10 min])				
1-2 U/mg solid	1 unit produces 1 μmol choline from glycerophosphorylcholine/min at pH 8.0, 37°C	Lyophilized containing 70% protein		Asahi GPCP T-33
Mold				
4-10 U/mg protein	1 unit produces 1.0 μmol choline from L-α-glycerophosphorylcholine/min at pH 8.0, 37°°C	Lyophilized containing 50% protein; balance primarily Tris buffer salt		ICN 157242

SPECIFIC ACTIVITY	UNITS DEFINITION	PREPARATION FORM	ADDITIONAL ACTIVITIES	SUPPLIER CATALOG NO.
Mold *continued*				
5-20 U/mg protein	1 unit produces 1.0 μmol choline from L-α-glycerophosphorylcholine/min at pH 8.0, 37°C	Lyophilized containing 40% protein; balance primarily Tris buffer salt		Sigma G1642

Phospholipase C   3.1.4.3

REACTION CATALYZED
    A phosphatidylcholine + H_2O ↔ 1,2-diacylglycerol + choline phosphate

SYSTEMATIC NAME
    Phosphatidylcholine cholinephosphohydrolase

SYNONYMS
    Clostridium oedematiens β- and γ-toxins, *Clostridium welchii* α-toxin, lecithinase C, lipophosphodiesterase I, PLC

REACTANTS
    Phosphatidylcholine, sphingomyelin, lecithin, phosphatide, diglyceride, phosphatidylinositol, H_2O, 1,2-diacylglycerol, choline phosphate

APPLICATIONS
- Determination of lecithins
- Preparation of C_6-NBd-diacylglycerol

NOTES
- Classified as a hydrolase, acting on ester bonds: phosphoric monoester hydrolase
- Hydrolyzes the link between glycerol and phosphate in lecithin and other phosphatides
- Consists of 3 major molecular forms, distinguishable by isoelectric points 5.2, 5.3 and 5.5
- A net positive charge on the phospholipid substrate is required for enzyme action
- Bacterial enzyme is a zinc protein that also acts on sphingomyelin and phosphatidylinositol
- Seminal plasma enzyme does not act on phosphatidylinositol
- Inactive toward lysophosphatidylcholine, lysophosphatidylethanolamine, phosphatidylglycerol, phosphatidylinositol, sphingomyelin and phosphatidic acid
- Hydrolyzes 10% of the total phospholipid of bovine brain myelin
- Activated by Ca^{2+} and sodium deoxycholate
- Inhibited by phosphonate analogs of glycerophosphatides, basic proteins (e.g., lysozyme), cytochrome c, ferricyanide, bleomycin, polymixin B and Tris-HCl buffer
- Inactivation by EDTA or phenanthroline is reversed by adding Zn^{2+} but not Ca^{2+}

3.1.4.3 Phospholipase C continued

SPECIFIC ACTIVITY	UNITS DEFINITION	PREPARATION FORM	ADDITIONAL ACTIVITIES	SUPPLIER CATALOG No.
Bacillus cereus (optimum pH = 6.6-9.0, pI = 7.0, MW = 20,000-25,000 Da [Sephadex G-100]; K_M = 20 mM [phosphatidylcholine]; stable pH 6.5-9, T < 60°C [pH 7.5, 10 min])				
30-100 U/mg solid	1 unit liberates 1 μmol choline PO$_4$/min at 37°C	Lyophilized containing 30% protein		Asahi PLC T-11
800 U/mg	1 unit converts 1 μmol lecithin to product/min at 37°C	Suspension 3.2 M (NH$_4$)$_2$SO$_4$	Sphingomyelinase	Boehringer 108502
2000 U/mg	1 unit converts 1 μmol lecithin to product/min at 37°C	Suspension 3.2 M (NH$_4$)$_2$SO$_4$	No detectable lipolytic activities <0.05% sphingomyelinase	Boehringer 691950
12 U/mg	1 unit hydrolyzes 1 μmol 3-*sn*-phosphatidylcholine/min at pH 7.5, 37°C	Powder containing 10% protein		Fluka 79477
800 U/mg protein	1 unit hydrolyzes 1 μmol L-α-phosphatidylcholine/min at pH 7.5, 37°C	Crystallized; suspension in 3.2 M (NH$_4$)$_2$SO$_4$, pH 6		Fluka 79484
>75 U/mg protein; >30 U/mg solid	1 unit liberates 1.0 μmol choline phosphate/min at 37°C	Lyophilized		Genzyme 1631
100-200 U/mg protein	1 unit liberates 1 μmol choline PO$_4$/min at pH 7.5, 37°C			ICN 151873
400 U/mg protein	1 unit liberates 1.0 μmol water soluble organic phosphorus from L-α-phosphatidylcholine (egg yolk or soybean)/min at pH 7.3, 37°C	Suspension in 3.2 M (NH$_4$)$_2$SO$_4$, pH 7.5		Sigma P4014
100-200 U/mg protein	1 unit liberates 1.0 μmol water soluble organic phosphorus from L-α-phosphatidylcholine (soybean)/min at pH 7.3, 37°C	Lyophilized containing 10% protein; balance primarily Tris buffer salts and ZnSO$_4$		Sigma P6135
500-2000 U/mg protein	1 unit liberates 1.0 μmol water soluble organic phosphorus from L-α-phosphatidylcholine (egg yolk)/min at pH 7.3, 37°C	Suspension in 3.2 M (NH$_4$)$_2$SO$_4$, pH 6		Sigma P7147
500-2000 U/mg protein	1 unit liberates 1.0 μmol water soluble organic phosphorus from L-α-phosphatidylcholine (egg yolk)/min at pH 7.3, 37°C	Lyophilized containing 10% protein; balance PO$_4$ buffer salts and NaCl	<20 U/mg protein sphingomyelinase	Sigma P9439
Clostridium perfringens (*C. welchii*) (optimum pH = 7.0-7.6, MW = 46,000 Da)				
40 U/mg	1 unit hydrolyzes 1 μmol 3-*sn*-phosphatidylcholine/min at pH 7.5, 37°C	Powder		Fluka 79485
4 U/mg	1 unit hydrolyzes 1 μmol 3-*sn*-phosphatidylcholine/min at pH 7.5, 37°C	Lyophilized		Fluka 79487

SPECIFIC ACTIVITY	UNITS DEFINITION	PREPARATION FORM	ADDITIONAL ACTIVITIES	SUPPLIER CATALOG No.
Clostridium perfringens (C. welchii) continued				
1-2 U/mg	1 unit liberates 1 μmol choline PO$_4$ from egg lecithin/min at pH 7.3, 37°C	Partially purified; salt-free; lyophilized		ICN 100976
50-150 U/mg protein	1 unit liberates 1.0 μmol water soluble organic phosphorus from L-α-phosphatidylcholine (egg yolk)/min at pH 7.3, 37°C	Partially purified; lyophilized in buffered salts		Sigma P1392
175-350 U/mg protein	1 unit liberates 1.0 μmol water soluble organic phosphorus from L-α-phosphatidylcholine (egg yolk)/min at pH 7.3, 37°C	Chromatographically purified; lyophilized in buffered salts		Sigma P4039
10-20 U/mg protein	1 unit liberates 1.0 μmol water soluble organic phosphorus from L-α-phosphatidylcholine (egg yolk)/min at pH 7.3, 37°C	Lyophilized		Sigma P7633
10-20 U/mg protein	1 unit liberates 1.0 μmol water soluble organic phosphorus from L-α-phosphatidylcholine (egg yolk)/min at pH 7.3, 37°C	Aseptically filled; lyophilized		Sigma P9185
$\geq$1 U/mg solid	1 unit liberates 1 μmol titratable fatty acid from soybean lecithin/min at pH 7.4, 25°C	Partially purified; dialyzed; lyophilized		Worthington LS04874 LS04875 LS04879
$\geq$60 U/mg protein	1 unit liberates 1 μmol titratable fatty acid from soybean lecithin/min at pH 7.4, 25°C	Chromatographically purified; 50% glycerol solution		Worthington LS05638 LS05639 LS05637

3.1.4.4 Phospholipase D

REACTION CATALYZED
A phosphatidylcholine + H_2O ↔ Choline + a phosphatidate

SYSTEMATIC NAME
Phosphatidylcholine phosphatidohydrolase

SYNONYMS
Choline phosphatase, lecithinase D, lipoPDE II, PLD

REACTANTS
Phosphatidylcholine, phosphatidyl esters, H_2O, choline, phosphatidate, lecithin

APPLICATIONS
- Determination of choline phospholipids and lecithins in biological fluids, with choline oxidase

NOTES
- Classified as a hydrolase, acting on ester bonds: phosphoric monoester hydrolase
- Important in signal transduction
- Broad specificity toward glycerophospholipids
- Relative reaction rates: lysophosphatidylcholine (100%), phosphatidylcholine (87%), sphingosylphosphorylcholine (22%) and plasmalogens (0%)
- Activated by Ca^{2+}, Triton X-100 and adekatol
- Inhibited by EDTA

SPECIFIC ACTIVITY	UNITS DEFINITION	PREPARATION FORM	ADDITIONAL ACTIVITIES	SUPPLIER CATALOG NO.
Cabbage				
300-800 U/mg protein	1 unit liberates 1.0 µmol choline from L-α-phosphatidylcholine (egg yolk)/hr at pH 5.6, 30°C	Lyophilized containing 40% protein		ICN 156235
1000-3000 U/mg protein	1 unit liberates 1.0 µmol choline from L-α-phosphatidylcholine (egg yolk)/hr at pH 5.6, 30°C	Lyophilized containing 30% protein; balance primarily Na EGTA		Sigma P0282
150-300 U/mg solid	1 unit liberates 1.0 µmol choline from L-α-phosphatidylcholine (egg yolk)/hr at pH 5.6, 30°C	Lyophilized containing 40% protein		Sigma P7758
300-500 U/mg protein	1 unit liberates 1.0 µmol choline from L-α-phosphatidylcholine (egg yolk)/hr at pH 5.6, 30°C	Lyophilized containing 25% protein		Sigma P8398
Cabbage, white				
0.3 U/mg	1 unit hydrolyzes 1 µmol L-α-phosphatidylcholine/min at pH 5.6, 37°C	Lyophilized		Fluka 79488
Peanut				
60-120 U/mg protein	1 unit liberates 1.0 µmol choline from L-α-phosphatidylcholine (egg yolk)/hr at pH 5.6, 30°C	Partially purified; lyophilized containing 30% protein; balance primarily buffer salts		Sigma P0515

SPECIFIC ACTIVITY	UNITS DEFINITION	PREPARATION FORM	ADDITIONAL ACTIVITIES	SUPPLIER CATALOG NO.
Peanut *continued*				
300-700 U/mg protein	1 unit liberates 1.0 μmol choline from L-α-phosphatidylcholine (egg yolk)/hr at pH 5.6, 30°C	Chromatographically purified; lyophilized containing 35% protein; balance primarily buffer salts		Sigma P0640
Streptomyces chromofuscus (optimum pH = 7.0-8.5, pI = 5.1, MW = 50,000 [Sephadex G-100], 57,000 [SDS-PAGE] Da; K_M = 1.4 m*M* [phosphatidylcholine]; stable pH 6-10 [0.1% BSA, 37°C, 1 hr], T < 60°C [pH 8, 10 min])				
30-90 U/mg solid	1 unit hydrolyzes 1 μmol phosphatidylcholine to phosphatidic acid and choline/min at 37°C	Lyophilized containing 70% protein	No detectable catalase, GO	Asahi PLD T-07
2650 U/mg solid	1 unit releases 1 μmol choline from egg yolk L-α-phosphatidylcholine/hr at pH 8.0, 30°C	Lyophilized	No detectable GO, catalase	BIOMOL SE-107
20 U/mg solid	1 unit converts 1 μmol egg lecithin to product/min at 37°C	Lyophilized		Boehringer 430331
20 U/mg	1 unit hydrolyzes 1 μmol 3-*sn*-phosphatidylcholine/min at pH 8.0, 37°C	Lyophilized		Fluka 79486
>40 U/mg protein; >30 U/mg solid	1 unit hydrolyzes 1.0 μmol phosphatidylcholine/min at pH 8.0, 37°C	Lyophilized	No detectable GO <0.3% U/mg catalase	Genzyme 1641
500-3000 U/mg solid	1 unit liberates 1.0 μmol choline from L-α-phosphatidylcholine (egg yolk)/hr at pH 5.6, 30°C	Lyophilized containing 50% protein		Sigma P8023
Streptomyces **species** (optimum pH = 5.5, pI = 4.2, MW = 46,000 Da [gel filtration]; stable pH 4.2-8.5 [0.05% BSA, 37°C, 1 hr])				
100-200 U/mg solid	1 unit hydrolyzes 1 μmol phosphatidylcholine to phosphatidic acid and choline/min at 37°C	Lyophilized containing 40% protein		Asahi PLDP T-39
>250 U/mg protein; >100 U/mg solid	1 unit hydrolyzes 1.0 μmol phosphatidylcholine/min at pH 5.5, 37°C	Lyophilized		Genzyme 1731
150-600 U/mg solid	1 unit liberates 1.0 μmol choline from L-α-phosphatidylcholine (egg yolk)/hr at pH 5.6, 30°C	Lyophilized		Sigma P4912

3.1.4.10 1-Phosphatidylinositol Phosphodiesterase

REACTION CATALYZED
1-Phosphatidyl-D-*myo*-inositol ↔ D-*myo*-Inositol 1,2-cyclic phosphate + diacylglycerol

SYSTEMATIC NAME
1-Phosphatidyl-D-*myo*-inositol inositolphosphohydrolase (cyclic-phosphate-forming)

SYNONYMS
Monophosphatidylinositol phosphodiesterase, phosphatidylinositol phospholipase C

REACTANTS
1-Phosphatidyl-D-*myo*-inositol, D-*myo*-inositol 1,2-cyclic phosphate, inositol-1-phosphate, diacylglycerol, lysophosphatidylinositol

APPLICATIONS
- Release of phosphatidylinositol-anchored membrane proteins
- Role of phosphatidylinositol derivatives in signal transduction

NOTES
- Classified as a hydrolase, acting on ester bonds: phosphoric monoester hydrolase
- Widely found in vertebrates, plants, protozoa, yeast and bacteria
- The animal enzyme hydrolyzes the cyclic phosphate to inositol-1-phosphate
- Does not hydrolyze phosphatidyl-choline, -ethanolamine, -glycerol, -serine, phosphatidic acid or sphingomyelin

SPECIFIC ACTIVITY	UNITS DEFINITION	PREPARATION FORM	ADDITIONAL ACTIVITIES	SUPPLIER CATALOG NO.
Bacillus cereus				
600 U/mg protein	1 unit liberates 4 units glycosyl-PI-anchored acetylcholinesterase/min at pH 7.4, 37°C	>95% purity by SDS-PAGE; 50 mM TEA, 10 mM EDTA, 10 mM NaN$_3$, pH 7.5	No detectable protease <0.2% phospholipase C <2% sphingomyelinase	BIOMOL SE-106
600 U/mg	1 unit liberates 2 units glycosyl-phosphatidylinositol-anchored acetylcholinesterase/min at pH 7.4, 37°C	Solution	No detectable protease <0.02% phospholipase C (phosphatidylcholine as substrate) <2% sphingomyelinase	Boehringer 1143069
500 U/mg protein	1 unit liberates 4 units glycosyl-PI-anchored acetylcholinesterase/min at pH 7.4, 37°C with soybean phosphatidylinositol as substrate	50 mM TEA, 10 mM EDTA, 10 mM NaN$_3$, pH 7.5	No detectable protease <0.02% phospholipase C <2% sphingomyelinase	Calbiochem 524640
600 U/mg protein	1 unit hydrolyzes 1 μmol 3-*sn*-phosphatidylcholine/min at pH 7.5, 37°C	Solution containing 50% glycerol, 10 mM Tris HCl, 10 mM EDTA, pH 8.0	<0.1% phospholipase C <2% sphingomyelinase	Fluka 79489
3000 U/mg protein	1 unit liberates 1 unit acetylcholinesterase from a membrane-bound crude preparation/min at pH 7.4, 30°C	Lyophilized containing PO$_4$ buffer salts, EDTA, stabilizer	<2 U/mg protein phospholipase C <10 U/mg protein sphingomyelinase	Sigma P5542

1-Phosphatidylinositol Phosphodiesterase continued — 3.1.4.10

SPECIFIC ACTIVITY	UNITS DEFINITION	PREPARATION FORM	ADDITIONAL ACTIVITIES	SUPPLIER CATALOG NO.
Bacillus cereus continued				
3000 U/mg protein	1 unit liberates 1 unit acetylcholinesterase from a membrane-bound crude preparation/min at pH 7.4, 30°C	Solution containing 60% glycerol, 10 m*M* Tris-HCl, 10 m*M* EDTA, pH 8.0	<1 U/mg protein phospholipase C <40 U/mg protein sphingomyelinase	Sigma P8804
Bacillus thuringiensis (optimum pH = 7.0-8.5)				
	1 unit cleaves 1.0 μmol phosphatidylinositol into *myo*-inositol phosphate ester/min at pH 7.5, 37°C determined after conversion to P_i			ICN 152354

Sphingomyelin Phosphodiesterase — 3.1.4.12

REACTION CATALYZED
 Sphingomyelin + H_2O ↔ *N*-Acylsphingosine + choline phosphate

SYSTEMATIC NAME
 Sphingomyelin cholinephosphohydrolase

SYNONYMS
 Neutral sphingomyelinase, sphingomyelinase

REACTANTS
 Sphingomyelin, phosphatidylcholine, H_2O, *N*-acylsphingosine, choline phosphate

APPLICATIONS
- Applied exogenously to initiate the signaling pathway

NOTES
- Classified as a hydrolase, acting on ester bonds: phosphoric monoester hydrolase
- Very little activity on phosphatidylcholine
- Activated by Mg^{2+}, Mn^{2+} and non-ionic detergents
- Stabilized by Mg^{2+}
- Inhibited by EDTA

SPECIFIC ACTIVITY	UNITS DEFINITION	PREPARATION FORM	ADDITIONAL ACTIVITIES	SUPPLIER CATALOG NO.
Bacillus cereus				
100-300 U/mg protein	1 unit hydrolyzes 1.0 μmol TNPAL-sphingomyelin/min at pH 7.4, 37°C	Lyophilized containing 10% protein; balance KPO_4 buffer salts and stabilizer		Sigma S7651
100-300 U/mg protein	1 unit hydrolyzes 1.0 μmol TNPAL-sphingomyelin/min at pH 7.4, 37°C	Solution containing 50% glycerol and 50 m*M* Tris-HCl, pH 7.5	<1% phospholipase C	Sigma S9396

3.1.4.12 Sphingomyelin Phosphodiesterase continued

SPECIFIC ACTIVITY	UNITS DEFINITION	PREPARATION FORM	ADDITIONAL ACTIVITIES	SUPPLIER CATALOG No.
Human placenta				
150 U/mg protein	1 unit hydrolyzes 1.0 nmol sphingomyelin to N-acylsphingosine and choline phosphate/hr at pH 5.0, 37°C	Solution containing 50% glycerol, 25 mM KPO_4, 0.1% Triton X-100, 0.05 mM PMSF, pH 4.5	<0.5% phospholipase C	Sigma S5383
Staphylococcus aureus				
225 U/mg protein	1 unit hydrolyzes 1 μmol TNPAL-sphingomyelin/min at pH 7.4, 37°C	0.37 mg/mL protein in 0.25 M PO_4, 50% glycerol, pH 7.5		BIOMOL SE-108
100-200 U/mg protein	1 unit hydrolyzes 1.0 μmol TNPAL-sphingomyelin/min at pH 7.4, 37°C	Solution containing 50% glycerol and 0.25 M PO_4 buffer, pH 7.5		ICN 195507
100-300 U/mg protein	1 unit hydrolyzes 1.0 μmol TNPAL-sphingomyelin/min at pH 7.4, 37°C	Solution containing 50% glycerol and 0.25 M Tris-HCl, pH 7.5		Sigma S8633
***Streptomyces* species** (optimum pH = 7.0-8.0, pI = 8.6, MW = 36,000 [SDS-PAGE], 42,000 [Sephadex G-100] Da; K_M = 1.42 mM [sphingomyelin]; stable pH 6-9 [37°C, 1 hr], T < 40°C, [pH 7.2, 10 min])				
20-100 U/mg solid	1 unit hydrolyzes 1 μmol sphingomyelin/min at pH 8.0, 37°C	Lyophilized containing 10% protein		Asahi SPC T-30
	1 unit hydrolyzes 1.0 μmol TNPAL-sphingomyelin/min at pH 7.4, 37°C	Lyophilized containing 30% protein		ICN 156646
500-100 U/mg protein	1 unit hydrolyzes 1.0 μmol TNPAL-sphingomyelin/min at pH 7.4, 37°C	Lyophilized containing 30% protein		Sigma S8889

3.1.4.17 3′,5′-Cyclic-Nucleotide Phosphodiesterase

REACTION CATALYZED
Nucleoside 3′,5′-cyclic phosphate + H_2O ↔ Nucleoside 5′-phosphate

SYSTEMATIC NAME
3′,5′-Cyclic-nucleotide 5′-nucleotidohydrolase

SYNONYMS
Phosphodiesterase, PDE

REACTANTS
Nucleoside 3′,5′-cyclic phosphate; 3′,5′-cyclic AMP; 3′,5′-cyclic dAMP; 3′,5′-cyclic IMP; 3′,5′-cyclic GMP; 3′,5′-cyclic CMP; H_2O, nucleoside 5′-phosphate

APPLICATIONS
- Determination of calmodulin
- Activator-insensitive enzymes (stimulated only 20% by Ca^{2+} and protein activator) are useful in Ca^{2+}-free systems

NOTES
- Classified as a hydrolase, acting on ester bonds: phosphoric monoester hydrolase
- Activated by Ca^{2+} and calmodulin (protein activator). Activator-sensitive enzymes can be stimulated 200% by activators

3',5'-Cyclic-Nucleotide Phosphodiesterase continued   3.1.4.17

SPECIFIC ACTIVITY	UNITS DEFINITION	PREPARATION FORM	ADDITIONAL ACTIVITIES	SUPPLIER CATALOG NO.
Bovine brain				
1-5 U/mg protein; 15-30 U/mg with Ca^{2+} and calmodulin	1 unit hydrolyzes 1 μmol 3',5'-cAMP to 5'-AMP/min at pH 7.5, 30°C	Affinity purified; lyophilized containing 5% protein, Tris-HCl, lactose as stabilizer	<0.1% 2',3'-cyclic nucleotide, 5'-nucleotidase, AP <0.2% 5'-ATPase <0.8% PPase	BIOMOL SE-103
1 U/mg	1 unit hydrolyzes 1 μmol 3',5'-cAMP to 5'-AMP/min at pH 7.5, 30°C, in the presence of 0.03 mM Ca^{2+} and a saturating level of the activator, calmodulin from bovine brain	Lyophilized; affinity purified	No detectable calmodulin	Fluka 79429
1-5 U/mg protein without added activator; 15-30 U/mg protein in the presence of 0.03 mM Ca^{2+} and saturating level of the activator (10 U/mL)	1 unit hydrolyzes 1.0 μmol 3',5'-cAMP to 5'-AMP/min at pH 7.5, 30°C	Activator deficient; affinity purified; lyophilized containing 5% protein; balance primarily Tris-HCl buffer salts and lactose as stabilizer	<0.1% 2',3'-cyclic nucleotide PDE, 5'-nucleotidase and AP <0.2% 5'-ATPase <0.8% PPase	Sigma P9529
Bovine heart				
0.2 U/mg protein	1 unit hydrolyzes 1.0 μmol 3',5'-cAMP to 5'-AMP/min at pH 7.5, 30°C	Crude; lyophilized containing 80% protein; balance primarily buffer salts as imidazole and Mg sulfate	No detectable 5'-nucleotidase PPase, 5'-ATPase, 2',3'-cyclic nucleotide PDE	Sigma P0134
0.08 U/mg protein	1 unit hydrolyzes 1.0 μmol 3',5'-cAMP to 5'-AMP/min at pH 7.5, 30°C	Activator-insensitive; crude; lyophilized containing 75% protein; balance primarily buffer salts as imidazole and $MgSO_4$	No detectable 5'-nucleotidase PPase, 5'-ATPase, 2',3'-cyclic nucleotide PDE	Sigma P0395
0.02-0.05 U/mg protein without added activator; 0.06-0.2 U/mg protein in the presence of 0.01 mM Ca^{2+} and saturating level of the activator (10 U)	1 unit hydrolyzes 1.0 μmol 3',5'-cAMP to 5'-AMP/min at pH 7.5, 30°C	Activator-sensitive; crude; lyophilized containing 75% protein; balance primarily buffer salts as imidazole and $MgSO_4$	<0.005 U 2',3'-cyclic nucleotide PDE, AP, 5'-ATPase <0.01 U 5'-nucleotidase <0.5 U/mg protein PPase	Sigma P0520

3.1.4.17 3',5'-Cyclic-Nucleotide Phosphodiesterase continued

SPECIFIC ACTIVITY	UNITS DEFINITION	PREPARATION FORM	ADDITIONAL ACTIVITIES	SUPPLIER CATALOG No.
Porcine brain				
1-5 U/mg protein without added activator; 15-30 U/mg protein in the presence of 0.03 mM Ca^{2+} and saturating level of the activator (10 U/mL)	1 unit hydrolyzes 1.0 μmol 3',5'-cAMP to 5'-AMP/min at pH 7.5, 30°C	Activator-deficient; affinity purified; lyophilized containing 5% protein; balance primarily buffer salts as Tris-HCl and lactose as stabilizer	<0.1% 2',3'-cyclic nucleotide PDE, AP, 5'-ATPase, PPase <0.5% 5'-nucleotidase	Sigma P1790

3.1.4.37 2',3'-Cyclic-Nucleotide 3'-Phosphodiesterase

REACTION CATALYZED
  Nucleoside 2',3'-cyclic phosphate + H$_2$O ↔ Nucleoside 2'-phosphate

SYSTEMATIC NAME
  Nucleoside 2',3'-cyclic phosphate 2'-nucleotidohydrolase

SYNONYMS
  Cyclic-CMP phosphodiesterase, CNPase

REACTANTS
  Nucleoside 2',3'-cyclic phosphate, H$_2$O, nucleoside 2'-phosphate, 2',3'-cyclic AMP, 2',3'-cyclic CMP, 2',3'-cyclic UMP

NOTES
- Classified as a hydrolase, acting on ester bonds: phosphoric monoester hydrolase
- Brain enzyme acts on 2',3'-cyclic AMP more rapidly than on UMP or CMP derivatives
- Liver enzyme acts on 2',3'-cyclic CMO more rapidly than on purine derivatives. It slowly hydrolyzes the corresponding 3',5'-cyclic phosphates (cyclic-CMP phosphodiesterase)

SPECIFIC ACTIVITY	UNITS DEFINITION	PREPARATION FORM	ADDITIONAL ACTIVITIES	SUPPLIER CATALOG No.
Bovine brain				
30-100 U/mg protein	1 unit converts 1.0 μmol 2',3'-cyclic NADP to NADP/min at pH 6.0, 25°C	Purified; solution containing 60% glycerol, 50 mM MES, 100 mM NaCl, pH 6.5	<1% AP, acid phosphatase, 2'-nucleotidase, 3'-nucleotidase, 5'-nucleotidase, PDE, 3',5'-cyclic nucleotide	Sigma P6274

Arylsulfatase 3.1.6.1

REACTION CATALYZED
A phenol sulfate + H_2O ↔ A phenol + sulfate

SYSTEMATIC NAME
Aryl-sulfate sulfohydrolase

SYNONYMS
Sulfatase, ARS

REACTANTS
Phenol sulfate, H_2O, phenol, sulfate

NOTES
- Classified as a hydrolase, acting on ester bonds: sulfuric ester hydrolase
- A group of enzymes with similar specificities

SPECIFIC ACTIVITY	UNITS DEFINITION	PREPARATION FORM	ADDITIONAL ACTIVITIES	SUPPLIER CATALOG NO.
Abalone entrails				
30-80 U/mg solid	1 unit hydrolyzes 1.0 µmol p-nitrocatechol sulfate/hr at pH 5.0, 37°C	Lyophilized	400-800 Sigma U/mg solid β-glucuronidase	Sigma S9629
20-40 U/mg solid	1 unit hydrolyzes 1.0 µmol p-nitrocatechol sulfate/hr at pH 5.0, 37°C	Lyophilized	<3.0 Sigma U/mg solid β-glucuronidase	Sigma S9754
Aerobacter aerogenes				
5 U/mg protein	1 unit hydrolyzes 1 µmol 4-nitrophenylsulfate/min at pH 7.1, 37°C	Aqueous solution containing 50% glycerol, 0.01 M Tris, pH 7.5	<0.1% β-glucuronidase	Fluka 86120
2-5 U/mg protein	1 unit hydrolyzes 1.0 µmol p-nitrophenyl sulfate/min at pH 7.1, 37°C	Partially purified; solution containing 50% glycerol and 0.01 M Tris, pH 7.5	No detectable β-glucuronidase at pH 7	Sigma S1629
Helix pomatia (edible snail)				
5 U/mg (4-nitrophenylsulfate substrate, 25°C); 1 U/mg (phenolphthalein disulfate substrate, 38°C); 290,000 Roy U/mg (2-hydroxy-5-nitrophenylsulfate substrate, 38°C)	1 Roy unit releases 1 µg 2-OH-5-nitrophenol from 2-OH-5-nitrophenylsulfate/hr at 38°C	Suspension in 3.2 M $(NH_4)_2SO_4$, pH 6	<2% β-glucuronidase (versus ARS activity with 4-nitrophenylsulfate)	Boehringer 102890
0.3 U/mg	1 unit hydrolyzes 1 µmol 4-nitrocatechol sulfate/min at pH 5.0, 37°C	Powder		Fluka 86123
15-40 U/mg solid	1 unit hydrolyzes 1.0 µmol p-nitrocatechol sulfate/hr at pH 5.0, 37°C	Partially purified; powder	300 U/mg β-glucuronidase (pH 5.0)	ICN 156719

3.1.6.1 Arylsulfatase continued

SPECIFIC ACTIVITY	UNITS DEFINITION	PREPARATION FORM	ADDITIONAL ACTIVITIES	SUPPLIER CATALOG No.
Helix pomatia (edible snail) *continued*				
2000-5000 U/mL	1 unit hydrolyzes 1.0 μmol *p*-nitro-catechol sulfate/hr at pH 5.0, 37°C during a 30 min assay	Crude; solution	100,000 U/mL β-glucuronidase (pH 5.0)	ICN 156720
15-40 U/mg solid	1 unit hydrolyzes 1.0 μmol *p*-nitro-catechol sulfate/hr at pH 5.0, 37°C	Lyophilized	400-600 Sigma U/mg solid β-glucuronidase	Sigma S3009
15-40 U/mg solid	1 unit hydrolyzes 1.0 μmol *p*-nitro-catechol sulfate/hr at pH 5.0, 37°C	Partially purified; powder	300 U/mg solid β-glucuronidase	Sigma S9626
2000-5000 U/mL	1 unit hydrolyzes 1.0 μmol *p*-nitro-catechol sulfate/hr at pH 5.0, 37°C	Crude; solution	100,000 Sigma U/mL β-glucuronidase at pH 5.0	Sigma S9751
Patella vulgata (limpet)				
10-25 U/mg solid	1 unit hydrolyzes 1.0 μmol *p*-nitro-catechol sulfate/hr at pH 5.0, 37°C	Salt-free; lyophilized	1000-2000 Sigma U/mg solid β-glucuronidase	Sigma S8504
5-15 U/mg solid	1 unit hydrolyzes 1.0 μmol *p*-nitro-catechol sulfate/hr at pH 5.0, 37°C	Salt-free; lyophilized	<2 Sigma U/mg solid β-glucuronidase	Sigma S8629

3.1.6.9 Chondro-4-Sulfatase

REACTION CATALYZED
   4-Deoxy-β-D-gluc-4-enuronosyl-(1,3)-*N*-acetyl-D-galactosamine 4-sulfate + H_2O ↔ 4-Deoxy-β-D-gluc-4-enuronosyl-(1,3)-*N*-acetyl-D-galactosamine + sulfate

SYSTEMATIC NAME
   4-Deoxy-β-D-gluc-4-enuronosyl-(1,3)-*N*-acetyl-D-galactosamine-4-sulfate 4-sulfohydrolase

SYNONYMS
   C4S

REACTANTS
   4-Deoxy-β-D-gluc-4-enuronosyl-(1,3)-*N*-acetyl-D-galactosamine 4-sulfate, H_2O, 4-deoxy-β-D-gluc-4-enuronosyl-(1,3)-*N*-acetyl-D-galactosamine, sulfate

NOTES
- Classified as a hydrolase, acting on ester bonds: sulfuric ester hydrolase
- Hydrolyzes saturated and unsaturated disaccharide 4-sulfates
- Does not act on higher oligosaccharides or any 6-sulfates
- Acts also on the saturated analog

Chondro-4-Sulfatase continued

3.1.6.9

SPECIFIC ACTIVITY	UNITS DEFINITION	PREPARATION FORM	ADDITIONAL ACTIVITIES	SUPPLIER CATALOG NO.
Proteus vulgaris (optimum pH = 7.5, T = 37-38°C)				
1.6 U/vial	1 unit liberates 1 μmol inorganic sulfate/min at pH 7.5, 37°C	Highly purified; lyophilized	<0.000016 U/vial CHase <0.000008 U/vial chondro-6-sulfatase	ICN 190336
1.6 U/vial; 4 vials	1 unit liberates 1 μmol inorganic sulfate/min at pH 7.5, 37°C	Highly purified; lyophilized	<0.000016 U/vial CHase <0.000008 U/vial chondro-6-sulfatase	ICN 320231
1.6 U/vial	1 unit forms 1 μmol H_2SO_4/min at pH 7.5, 37°C	Lyophilized		Seikagaku 100350-1
10 U/mg protein	1 unit liberates 1.0 μmol inorganic sulfate from 4-deoxy-β-D-gluc-4-enuronosyl-(1-3)-N-Ac-D-galactosamine 4-sulfate/min at pH 7.5, 37°C	Lyophilized containing 1-2% protein; balance primarily sucrose as stabilizer		Sigma C2655

Chondro-6-Sulfatase

3.1.6.10

REACTION CATALYZED

4-Deoxy-β-D-gluc-4-enuronosyl-(1,3)-N-acetyl-D-galactosamine 6-sulfate + H_2O ↔ 4-Deoxy-β-D-gluc-4-enuronosyl-(1,3)-N-acetyl-D-galactosamine + sulfate

SYSTEMATIC NAME

4-Deoxy-β-D-gluc-4-enuronosyl-(1,3)-N-acetyl-D-galactosamine-6-sulfate 6-sulfohydrolase

SYNONYMS

C6S

REACTANTS

4-Deoxy-β-D-gluc-4-enuronosyl-(1,3)-N-acetyl-D-galactosamine 6-sulfate, N-acetyl-D-galactosamine 4,6-disulfate, H_2O, 4-deoxy-β-D-gluc-4-enuronosyl-(1,3)-N-acetyl-D-galactosamine, sulfate

NOTES

- Classified as a hydrolase, acting on ester bonds: sulfuric ester hydrolase
- High specificity for saturated and unsaturated disaccharide 6-sulfates
- Does not act on higher oligosaccharides or any 4-sulfates
- Acts also on the saturated analog and N-acetyl-D-galactosamine 4,6-disulfate
- Activated by acetate ion

3.1.6.10 Chondro-6-Sulfatase continued

SPECIFIC ACTIVITY	UNITS DEFINITION	PREPARATION FORM	ADDITIONAL ACTIVITIES	SUPPLIER CATALOG No.
Proteus vulgaris (optimum pH = 6.2 [hyaluronic acid], 7.5 [Tris-HCl], 8.0 [chondroitin sulfate], T = 37-38°C)				
2.5 U/vial	1 unit liberates 1 μmol product/min at pH 7.4, 37°C		<0.000012 U/vial CHase <0.005 U/vial chondro-4-sulfatase	ICN 190337
2.5 U/vial; 4 vials	1 unit liberates 1 μmol product/min at pH 7.4, 37°C	Highly purified; lyophilized	<0.000012 U/vial CHase <0.005 U/vial chondro-4-sulfatase	ICN 320241
2.5 U/vial	1 unit forms 1 μmol H_2SO_4/min at pH 7.5, 37°C	Lyophilized		Seikagaku 100335-1
3-10 U/mg protein	1 unit liberates 1.0 μmol inorganic sulfate from 4-deoxy-β-D-gluc-4-enuronosyl-(1-3)-N-Ac-D-galactosamine 6-sulfate/min at pH 7.5, 37°C	Lyophilized containing 1-3% protein; balance primarily sucrose as stabilizer		Sigma C3030

3.1.11.1 Exodeoxyribonuclease I

REACTION CATALYZED
  Exonucleolytic cleavage in the 3'- to 5'-direction to yield 5'-phosphomononucleotides

SYSTEMATIC NAME
  Exodeoxyribonuclease I

SYNONYMS
  Exonuclease I

REACTANTS
  Single-stranded DNA, 5'-phosphomononucleotide

APPLICATIONS
- Eliminating residual ssDNA with 3'-termini
- Measuring endonucleolytic cleavage
- Measuring DNA helicase activity

NOTES
- Classified as a hydrolase, acting on ester bonds: exodeoxyribonuclease producing 5'-phosphomonoesters
- Preference for single-stranded DNA
- Requires Mg^{2+} (10 *mM* is optimal) and a free 3'-OH terminus
- Activities include:
  double-strand specific 3'-to-5' exonuclease
  DNA 3'-phosphatase
  Endonuclease at DNA apurinic sites
  RNase H
- *E. coli* enzyme hydrolyzes glucosylated DNA
- Similar enzymes: mammalian DNase III, exonuclease IV, and T_2- and T_4-induced exodeoxyribonucleases

Exodeoxyribonuclease I *continued* — 3.1.11.1

SPECIFIC ACTIVITY	UNITS DEFINITION	PREPARATION FORM	ADDITIONAL ACTIVITIES	SUPPLIER CATALOG No.
Escherichia coli SK 4258, overproducing strain of *E. coli* K12 (MW = 55,000 Da)				
50-200 U/µL	1 unit releases 10 nmol acid-soluble nucleotide from denatured DNA/30 min at 37°C	20 m*M* Tris-HCl, 0.5 m*M* EDTA, 5 m*M* β-MSH, 50% glycerol, pH 7.5	No detectable endonuclease, RNase, ds-exonuclease	Amersham E 70073Z
Escherichia coli, sbcB gene (inactivated at 80°C, 15 min)				
180,000 U/mg	1 unit releases 10 nmol acid-soluble nucleotides from calf thymus DNA/30 min at pH 7.8, 37°C	Solution containing 50% glycerol, 50 m*M* Tris-HCl, 0.1 *M* NaCl, 0.1 m*M* EDTA, 1 m*M* DTT, 0.1% Triton X-100, pH 7.5	No detectable RNase, endonuclease, ds-exonuclease	Epicentre X40501K X40505K

Exodeoxyribonuclease III — 3.1.11.2

REACTION CATALYZED
  Stepwise exonucleolytic cleavage in the 3'- to 5'-direction to yield 5'-phosphomononucleotides

SYSTEMATIC NAME
  Exodeoxyribonuclease III

SYNONYMS
  Exonuclease III

REACTANTS
  Ds-DNA, 5'-phosphomononucleotide

APPLICATIONS
- Preparing strand-specific radioactive probes (conjunction with Klenow)
- Preparing ssDNA templates for dideoxy sequencing
- Construction unidirectional deletion mutants
- Site-directed mutagenesis
- DNA footprinting

NOTES
- Removing nucleotides from the 3'-side of DNA nicks to produce gaps
- Classified as a hydrolase, acting on ester bonds: exodeoxyribonuclease producing 5'-phosphomonoesters
- Preference for double-stranded DNA
- Endonucleolytic phosphodiester bond hydrolysis at apurinic or apyrimidinic sites on DNA
- Preferentially cleaves 3'-recessed but not 3'-overhanging DNA ends
- Similar to *Haemophilus influenzae* exonuclease
- Activity is strongly dependent on temperature, salt concentration and DNA:enzyme ratio
- Activated by Mg^{2+} and sulfhydryl reagents
- Reversibly inhibited by EDTA

3.1.11.2 Exodeoxyribonuclease III continued

SPECIFIC ACTIVITY	UNITS DEFINITION	PREPARATION FORM	ADDITIONAL ACTIVITIES	SUPPLIER CATALOG NO.
Escherichia coli				
	1 unit produces 1 nmol acid-soluble radioactivity/30 min at 37°C	>95% purity by SDS-PAGE; 25 mM Tris-HCl, 0.05 mM DTT, 50% glycerol, pH 8.0	No detectable endonuclease	CHIMERx 1140-01 1140-02
	1 unit produces 1 nmol acid-soluble radioactivity/30 min at 37°C	Solution containing 25 mM Tris-HCl buffer, 0.05 mM DTT, 50% glycerol, pH 8.0		ICN 151109
140,000-500,000 U/mg protein; 100,000-250,000 U/mL	1 unit produces 1 nmol acid-soluble product from [^{3}H]DNA duplex/30 min at pH 8.0, 37°C	Solution containing 25 mM Tris-HCl, 66 mM KCl, 1 mM DTT, 50% glycerol, pH 8.0	No detectable nickase activity	Pharmacia 27-0874-01
Escherichia coli BE 257/pSGR 3 (optimum pH = 8.0, MW = 28,000 Da)				
50-200 U/µL	1 unit releases 1 nmol nucleotides to acid-soluble form/30 min at 37°C	>90% purity by SDS-PAGE; 50 mM Tris-HCl, 100 mM KCl, 1 mM DTT, 50% glycerol, pH 8.0	No detectable endonuclease	Amersham E 70023Y E 70023Z
200,000 U/mg; 100,000-200,000 U/mL	1 unit releases 1 nmol acid-soluble nucleotides from sonicated calf thymus DNA/30 min at 37°C	20 mM Tris-HCl, 66 mM KCl, 0.1 mM EDTA, 1 mM β-MSH, 200 µg/mL BSA, 50% glycerol, pH 8.0		Boehringer 779709 779717
5000 U and 25,000 U	1 unit produces 1 nmol acid-soluble DNA fragments/30 min at pH 8.0, 37°C with restriction-digested calf thymus DNA as substrate	Single band purity in SDS-PAGE; solution containing 25 mM Tris-HCl, 0.5 mM DTT, 50 mM KCl, 50% glycerol, pH 8.0	No detectable nuclease	TaKaRa 2170
Escherichia coli K-12, BE257/pSGR3 (Inactivated at 70°C, 20 min)				
100,000 U/mL	1 unit produces 1 nmol acid-insoluble total nucleotide/30 min at 37°C	Purified; 200 mM KCl, 5 mM KPO$_4$, 0.05 mM EDTA, 5 mM β-MSH, 200 µg/mL BSA, 50% glycerol, pH 6.5	No detectable endonuclease, contaminating exonuclease	NE Biolabs 206S 206L
Escherichia coli SR80, expressing the exo III gene on a plasmid				
50-100 U/µL	1 unit hydrolyzes 1 nmol ATP/30 min at pH 8.0, 37°C			Life Technol 18013-011 18013-037
Escherichia coli, overproducer				
50,000 U/mL	1 unit produces 1 nmol acid-soluble nucleotides from sonicated calf thymus DNA/30 min at pH 7.6, 37°C	20 mM Tris-HCl, 0.1 mM EDTA, 4 mM β-MSH, 200 µg/mL BSA, 50% glycerol, pH 8.0	No detectable deoxyribonuclease	AGS Heidelb A00730S A00730M
100-400 U/µL	1 unit releases 1 nmol acid-soluble reaction products from *E. coli* [^{3}H]-DNA/30 min at 37°C	50 mM Tris-HCl, 1 mM DTT, 50 mM KCl, 50% glycerol, pH 8.0	No detectable endodeoxyribonucleases	Fermentas EN0191 EN0192

Exodeoxyribonuclease III continued — 3.1.11.2

SPECIFIC ACTIVITY	UNITS DEFINITION	PREPARATION FORM	ADDITIONAL ACTIVITIES	SUPPLIER CATALOG NO.
Escherichia coli, overproducer continued				
	1 unit releases 1 nmol acid-soluble nucleotides from sonicated salmon sperm DNA/30 min at 37°C	50 mM Tris-HCl, 50 mM KCl, 1 mM DTT, 50% glycerol, pH 8.0	RNase H, 3'-phosphatase, specific apurinic and apyrimidic endonuclease	Oncor 120211 120212
Escherichia coli, recombinant				
150–200 U/μL	1 unit produces 1 nmol acid-soluble nucleotides/30 min at pH 7.6, 37°C	≥90% purity by SDS gel	≥90% supercoiled plasmid	Promega M1811 M1812
Escherichia coli, recombinant overproducer				
20,000–100,000 U/mL	1 unit degrades 20 nucleotides/pmol of 5' overhang ends/min at 37°C	20 mM Tris-HCl, 1 mM β-MSH, 50% glycerol, pH 8.0	No detectable endonuclease, 3'-exonuclease	Stratagene 600041 600042
Unspecified (inactivated at 65°C, 15 min)				
250,000 U/mg	1 unit releases 1 nmol acid-soluble nucleotides from calf thymus DNA/30 min at pH 7.8, 37°C	Solution containing 50% glycerol, 50 mM Tris-HCl, 0.1 M NaCl, 0.1 mM EDTA, 1 mM DTT, 0.1% Triton X-100, pH 7.5	No detectable exogenous RNase, endonuclease, ss-exonuclease	Epicentre EX4405K EX4425K

Exodeoxyribonuclease (λ-Induced) — 3.1.11.3

REACTION CATALYZED
Stepwise, nonprogressive exonucleolytic cleavage in the 5'- to 3'-direction to yield 5'-phosphomononucleotides

SYNONYMS
λ exonuclease

REACTANTS
Double-stranded DNA, 5'-phosphomononucleotides

APPLICATIONS
- Producing ssDNA templates for sequencing via the chain termination method
- Enriching large DNA fragments obtained through digestion of genomic DNA
- Exonuclease cycling assay
- Producing ssDNA primer for cDNA extensions

NOTES
- Classified as a hydrolase, acting on ester bonds: exodeoxyribonuclease producing 5'-phosphomonoesters
- Prefers blunt-ended DNA with a 5'-phosphate group; does not attack single-strand breaks
- Similar to T_4, T_5 and T_7 exonucleases, and mammalian DNase IV

3.1.11.3 Exodeoxyribonuclease (λ-Induced) continued

SPECIFIC ACTIVITY	UNITS DEFINITION	PREPARATION FORM	ADDITIONAL ACTIVITIES	SUPPLIER CATALOG NO.
Escherichia coli HMS7/pJL23/pGP6-1				
50-500 U/μL	1 unit releases 1.0 nmol acid-soluble nucleotide/15 min at 37°C	>95% purity by SDS-PAGE; 50 mM KPO$_4$, 1 mM EDTA, 1 mM DTT, 50% glycerol, pH 6.5		Amersham E 70025Y E 70025Z
Escherichia coli, λ lysogen SG5519				
1-10 U/μL	1 unit produces 10 nmol acid-soluble deoxyribonucleotide from dsDNA/30 min at pH 9.4, 37°C			Life Technol 28023-018
Unspecified				
5000-10,000 U/mL	1 unit produces 10 nmol acid-soluble product from 100 ng internally [^{32}P]-labeled 800-base pair PCR product containing a single 5'-terminal phosphate/30 min at pH 9.3, 37°C	Solution containing 10 mM Tris-HCl, 10 mM β-MSH, 200 μg/mL BSA, 50% glycerol, pH 7.6		Pharmacia 27-0865-01

3.1.11.5 Exodeoxyribonuclease V

REACTION CATALYZED
    Exonucleolytic cleavage (in presence of ATP) in either 5'- to 3'- or 3'- to 5'-direction to yield 5'-phosphooligonucleotides

SYSTEMATIC NAME
    Exodeoxyribonuclease V, deoxyribonuclease (ATP-dependent)

SYNONYMS
    Exonuclease V

REACTANTS
    Double-stranded DNA, 5'-phosphooligonucleotides

NOTES
- Classified as a hydrolase, acting on ester bonds: exodeoxyribonuclease producing 5'-phosphomonoesters
- Preference for double-stranded DNA; does not attack closed circular supercoiled or nicked circular dsDNA
- Possesses DNA-dependent ATPase activity; required ATP for activity
- Acts endonucleolytically on single-stranded circular DNA
- Similar to *Haemophilus influenzae* ATP-dependent DNase

Exodeoxyribonuclease V continued — 3.1.11.5

SPECIFIC ACTIVITY	UNITS DEFINITION	PREPARATION FORM	ADDITIONAL ACTIVITIES	SUPPLIER CATALOG NO.
Micrococcus luteus				
1000-5000 U/mL	1 unit produces 10 nmol acid-soluble nucleotides from 20 nmol E. coli DNA/30 min at pH 9.4, 37°C	50% glycerol solution containing 20 mM Tris-HCl, 10 mM β-MSH, 0.1 mM EDTA, 500 μg/mL BSA, pH 7.5		ICN 157556
2000-10,000 U/mL	1 unit produces 1.0 nmol acid-soluble nucleotides from 20 nmol E. coli DNA/30 min at pH 9.4, 37°C	Solution containing 50% glycerol, 20 mM Tris HCl, 10 mM β-MSH, 0.1 mM EDTA, 500 mg/mL BSA, pH 7.5		Sigma D3150
Micrococcus luteus ATCC 4698				
5-10 U/μL	1 unit produces 10 nmol acid-soluble nucleotides from 20 nmol E. coli [³H] DNA/30 min at 37°C	20 mM Tris-HCl, 0.1 mM EDTA, 10 mM β-MSH, 500 μg/mL BSA, 50% glycerol, pH 7.5	No detectable ds-exonuclease	Amersham E 70040Y E 70040Z
200 and 1000 U	1 unit produces 10 nmol acid-soluble nucleotides from E. coli H-DNA/30 min at pH 9.4, 37°C			CLONTECH 8407-1 8407-2
Unspecified (Inactivated at 70°C, 15 min)				
	1 unit converts 1 nmol deoxynucleotides in linear dsDNA into acid-soluble form/30 min at pH 37°C	Solution containing 50% glycerol, 25 mM Tris-HCl, 0.1 M NaCl, 0.1 mM EDTA, 1 mM DTT, 0.1% Triton X-100, pH 7.5	No detectable protease, RNase, ds-specific deoxyriboendonuclease	Epicentre Plasmid-Safe™ E3101K E3105K E3110K

Exodeoxyribonuclease VII — 3.1.11.6

REACTION CATALYZED

Exonucleolytic cleavage in either 5'- to 3'- or 3'- to 5'-direction to yield 5'-phosphomononucleotides

REACTANTS

Single-stranded DNA, 5'-phosphomononucleotides

APPLICATIONS
- Mapping intron positions in genomic DNA
- Excising segments of DNA inserted into plasmid vectors by the poly(dA-dT) tailing method
- Polishing overhangs produced by restriction endonucleases
- Mapping 5'- and 3'-ends of gene transcripts by nuclease protection assays

NOTES
- Classified as a hydrolase, acting on ester bonds: exodeoxyribonuclease producing 5'-phosphomonoesters
- The only bi-directional E. coli exonuclease with single-stranded DNA specificity
- No requirement for divalent cations; fully active with EDTA

3.1.11.6 Exodeoxyribonuclease VII continued

- Products of limit digestion are dimers to dodecamers
- Similar to *Micrococcus luteus* exonuclease

SPECIFIC ACTIVITY	UNITS DEFINITION	PREPARATION FORM	ADDITIONAL ACTIVITIES	SUPPLIER CATALOG NO.
Escherichia coli HMS137				
1-8 U/μL	1 unit produces 1 nmol acid-soluble oligodeoxyribonucleotide from denatured DNA/30 min at pH 7.9, 37°C			Life Technol 18015-016
Escherichia coli, two plasmid system				
10 U/μL	1 unit converts 1 nmol nucleotide to acid-soluble form/30 min at 37°C	>95% purity by SDS-PAGE; 50 mM Tris-HCl, 200 mM NaCl, 0.5 mM EDTA, 10 mM DTT, 50% glycerol, pH 8.0	No detectable RNase, ss- and ds-endonuclease, ds-exonuclease	Amersham E 70082Y E 70082Z

3.1.15.1 Venom Exonuclease

REACTION CATALYZED
  Exonucleolytic cleavage in the 3′- to 5′-direction to yield 5′-phosphomononucleotides

SYNONYMS
  Venom phosphodiesterase, phosphodiesterase, PDE

REACTANTS
  Single-stranded RNA, single-stranded DNA, 5′-phosphomononucleotides

APPLICATIONS
- Sequencing studies of RNA and DNA
- Converting supercoiled DNA to open/circular DNA

NOTES
- Classified as a hydrolase, acting on ester bonds: exonuclease active with either ribo- or deoxyribonucleic acids and producing 5′-phosphomonoesters
- Attacks high molecular weight, double-stranded DNA
- Similar to hog kidney phosphodiesterase and *Lactobacillus* exonuclease

SPECIFIC ACTIVITY	UNITS DEFINITION	PREPARATION FORM	ADDITIONAL ACTIVITIES	SUPPLIER CATALOG NO.
Crotalus durissus (snake venom)				
1.5 U/mg	1 unit converts 1 μmol bis(PNPP) to product/min at 25°C	50% glycerol solution, pH 6.0		Boehringer 108260

Spleen Exonuclease 3.1.16.1

REACTION CATALYZED
   Exonucleolytic cleavage in the 5'- to 3'-direction to yield 3'-phosphomononucleotides

SYSTEMATIC NAME
   Spleen exonuclease

SYNONYMS
   3'-Exonuclease, phosphodiesterase II, spleen phosphodiesterase

REACTANTS
   Single-stranded substrate, 3'-phosphomononucleotide

APPLICATIONS
- Sequencing studies on RNA and DNA oligonucleotides
- Characterization of polynucleotide chain length and base composition, and identity of terminal nucleotides

NOTES
- Classified as a hydrolase, acting on ester bonds: exonuclease active with either ribo- or deoxyribonucleic acids and producing other than 5'-phosphomonoesters
- Preference for single-stranded substrate
- Sensitive to substrate secondary structure
- Similar to *Lactobacillus acidophilus* nuclease, *Bacillus subtilis* nuclease, salmon testis nuclease

SPECIFIC ACTIVITY	UNITS DEFINITION	PREPARATION FORM	ADDITIONAL ACTIVITIES	SUPPLIER CATALOG NO.
Bovine spleen (optimum pH = 5.5 [succinate and phosphate buffer], 6-7 [acetate buffer])				
10-15 U/vial	1 unit causes a ΔA_{260} of 0.2 at pH 6.5, 37°C with an RNA substrate	Lyophilized containing 0.001 M Na pyrophosphate		ICN 100977
10-30 U/mg protein; 10-20 U/vial	1 unit produces acid-soluble nucleotides causing an ΔA_{260} of 16/30 min at pH 6.5, 37°C with RNA-Core as substrate; 2 mL reaction mixture	Lyophilized	≤2 U/vial 5'-nucleotidase	Sigma P6752
10-20 U/mg protein	1 unit produces acid-soluble nucleotides causing an ΔA_{260} of 16/30 min at pH 6.5, 37°C with RNA-Core as substrate; 2 mL reaction mixture	Lyophilized	5'-nucleotidase <1% PDE	Sigma P9041
≥10 U/vial	1 unit increases A_{260} by 0.200/30 min at pH 6.5, 37°C with an RNA substrate	Lyophilized from 0.001 M Na pyrophosphate, pH 6.9, alumina gel eluate		Worthington LS03603

SPECIFIC ACTIVITY	UNITS DEFINITION	PREPARATION FORM	ADDITIONAL ACTIVITIES	SUPPLIER CATALOG NO.
Calf spleen (optimum pH = 7.0)				
2 U/mg	1 unit converts 1 μmol thymidine-3′-nitrophenyl-phosphate to product/min at 25°C	Suspension in 3.2 M $(NH_4)_2SO_4$, pH 6.0		Boehringer 108251
≥2 U/mg	1 unit hydrolyzes 1.0 μmol UDP-glucose/min at pH 7.0, 25°C	Suspension in 3.2 M $(NH_4)_2SO_4$, pH 6.0		Calbiochem 524710
2 U/mg protein	1 unit hydrolyzes 1 μmol thymidine-3′-nitrophenyl phosphate/min at pH 6.0, 25°C	Suspension in 3.2 M $(NH_4)_2SO_4$ solution, pH 6	<0.2% phosphatase <0.5% 5′-nucleotidase <1% ADA	Fluka 79428
	1 unit produces acid-soluble nucleotides causing an ΔA_{260} of 16/30 min at pH 6.5, 37°C with RNA-Core as substrate; 2 mL reaction mixture	Suspension in 3.2 M $(NH_4)_2SO_4$, pH 6		Sigma P6897

Deoxyribonuclease I — 3.1.21.1

REACTION CATALYZED
Endonucleolytic cleavage of phosphodiester bonds to 5′-phosphodinucleotide and 5′-phosphooligonucleotide end-products

SYSTEMATIC NAME
Deoxyribonuclease I

SYNONYMS
Pancreatic DNase, DNase I, thymonuclease

REACTANTS
Double-stranded DNA, 5′-phosphodinucleotides, 5′-phosphooligonucleotides

APPLICATIONS
- Making a DNA library for shotgun sequencing (with Mn^{2+})
- Standardizing DNA and DNA methyl green
- Random cleavage of DNA clones
- Radioactive labeling of DNA by nick translation
- Removing DNA after *in vitro* transcription
- Nick translation with DNA polymerase I
- Removing DNA from RNA and protein preparations
- DNase footprinting in the study of protein-DNA binding specificity

NOTES
- Debriding agent to lyse fibrin and liquefy material, facilitating removal of necrotic debris from wounds
- Classified as a hydrolase, acting on ester bonds: endodeoxyribonuclease producing 5′-phosphomonoesters
- Preference for double-stranded DNA, but acts on single- and double-stranded DNA, chromatin and RNA/DNA hybrids
- Produces random, independent nicks in dsDNA with Mg^{2+}; with Mn^{2+}, both strands are cleaved at the same site to yield blunt-end fragments
- Bivalent cations are required for maximal activity
- Bovine pancreas produces 4 DNase I (A, B, C, D), differing in carbohydrate side chain or polypeptide component
- Activated by bivalent metal ions
- Stabilized by Ca^{2+}
- Reversibly inhibited by EDTA and SDS; irreversibly inhibited at 80°C, 10 min
- Similar to streptococcal DNase (streptodornase), T_4 endonuclease II, T_7 endonuclease II, *E. coli* endonuclease I, calf thymus 'nicking' enzyme, colicin E_2 and E_3

SPECIFIC ACTIVITY	UNITS DEFINITION	PREPARATION FORM	ADDITIONAL ACTIVITIES	SUPPLIER CATALOG NO.
Bovine pancreas (optimum pH = 7.8, T = 25°C, MW = 31,000 Da)				
50,000 U/mL	1 unit increases A_{260} by 0.001/min at pH 7.4, 25°C using calf thymus DNA as substrate	Highly purified by affinity chromatography; 10 m*M* Tris-HCl, 0.5 m*M* $CaCl_2$, 50 m*M* KCl, 100 μg/mL BSA, 50% glycerol, pH 7.4	No detectable RNase	AGS Heidelb A00720S A00720M
100,000 U/mg; 2 U/μL	1 unit completely degrades 1 μg DNA/10 min at 37°C in a 50 μL volume; 1 unit = 0.7 Kunitz unit	Single band of 4224 U visible by SDS-PAGE; 20 m*M* HEPES, 10 m*M* $CaCl_2$, 10 m*M* $MgCl_2$, 1 m*M* DTT, 50% glycerol, pH 7.5	No detectable RNase, contaminating proteins	Ambion 2222 2224

3.1.21.1 Deoxyribonuclease I continued

SPECIFIC ACTIVITY	UNITS DEFINITION	PREPARATION FORM	ADDITIONAL ACTIVITIES	SUPPLIER CATALOG No.
Bovine pancreas *continued*				
70 U/μL	1 unit increases A_{260} by 0.001/min at pH 5.0, 25°C with calf thymus DNA as substrate (Kunitz unit)	20 mM NaOAc, 150 mM NaCl, 50% glycerol, pH 5.0	No detectable RNase	Amersham E 2210A
3500 U/mg dry weight	1 unit increases A_{260} by 0.001/min at pH 5.0, 25°C with calf thymus DNA as substrate (Kunitz unit)	20 mM NaOAc, 150 mM NaCl, 50% glycerol, pH 5.0	No detectable RNase	Amersham E 2211Y
1800 Kunitz U	1 Kunitz unit produces a ΔA_{260} of 0.001/min at pH 5.0, 25°C using a DNA substrate	Lyophilized		AMRESCO 0649
3000 U/mg solid	1 unit increases E_{260} by 0.001/min at pH 5.0, 25°C; 1 U = 33 Dornase units	Salt-free; lyophilized		Biozyme DNP3
3000 U/mg solid	1 Kunitz unit increases absorbance by 0.001/min at 25°C with DNA as substrate	Lyophilized		Boehringer 104132
2000 U/mg solid	1 Kunitz unit increases absorbance by 0.001/min at 25°C with DNA as substrate	Lyophilized		Boehringer 104159
60,000 Dornase U/mg solid	1 unit causes a decrease of 1.0 relative viscosity unit in a solution of highly polymerized DNA/10 min at 30°C from initial relative viscosity of 4.0	Lyophilized		Calbiochem 260912
3000 U/mg protein	1 unit increases A_{260} by 0.001/min/mL at pH 5.00, 25°C	Salt-free; lyophilized containing 90% protein		Calzyme 073A3000
	1 MBU (Molecular Biology Unit) completely digests 1 μg pUC19 DNA to oligonucleotides/10 min at pH 7.8, 37°C	Solution containing 50% glycerol, 10 mM Tris-HCl, 10 mM $CaCl_2$, 10 mM $MgCl_2$, pH 7.5	No detectable RNase	Epicentre PK08DN D9902K D9905K D9910K
1800 U/mg	1 unit increases the A_{260} by 0.001/min at pH 5.0, 25°C with calf thymus DNA as substrate	Powder	<0.02% RNase <0.5% trypsin	Fluka 31130
2000 U/mg	1 unit increases the A_{260} by 0.001/min at pH 5.0, 25°C with calf thymus DNA as substrate	Sterile-filtered; endotoxin tested; lyophilized	<0.01% chymotrypsin, protease, RNase	Fluka 31131
2400 U/mg protein	1 unit increases the A_{260} by 0.001/min at pH 5.0, 25°C with calf thymus DNA as substrate	Powder		Fluka 31132
2000 U/mg	1 unit increases the A_{260} by 0.001/min at pH 5.0, 25°C with calf thymus DNA as substrate	Lyophilized	<0.0005% RNase <0.005% caseinase <0.01% chymotrypsin <0.5% trypsin	Fluka 31133

SPECIFIC ACTIVITY	UNITS DEFINITION	PREPARATION FORM	ADDITIONAL ACTIVITIES	SUPPLIER CATALOG NO.
Bovine pancreas *continued*				
2500 U/mg	1 unit increases the A_{260} by 0.001/min at 25°C with calf thymus DNA as substrate	Powder	<0.01% protease, RNase	Fluka 31134
3000 U/mg	1 unit increases the A_{260} by 0.001/min at pH 5.0, 25°C with calf thymus DNA as substrate	Lyophilized		Fluka 31135
1,000,000 Dornase U/vial	1 Kunitz unit = 46 Dornase units	Salt-free; non-pyrogenic; sterile; lyophilized		ICN 100574
2000-2600 Kunitz U/mg solid		Lyophilized containing Gly stabilizer		ICN 100575
65,000 Dornase U/mg		Lyophilized		ICN 100579
50,000-150,000 Dornase U/mg solid	1 Dornase unit decreases relative viscosity by 1.0 unit in a solution of highly polymerized DNA from the initial relative viscosity of 4.0/10 min at 30°C (95,000 Dornase units = 3000 Kunitz units)	Lyophilized		ICN 190062
3000 CU/mg	50 Kunitz Units = 3000 Christensen Units (CU)	Crude; powder containing mannitol excipient		Intergen 7050-00
60,000 DU/mg	1 Kunitz Unit = 35 Dornase Units	Purified; powder containing mannitol excipient		Intergen 7051-00
100,000 DU/mg	1 Kunitz Unit = 35 Dornase Units	Highly purified; salt-free; powder		Intergen 7053-00
5-15 U/µL	1 unit increases A_{260} of a high-molecular-weight DNA by 0.001/min/mL of reaction mixture at pH 5.0, 25°C			Life Technol 18047-019
>10,000 U/mg; 1 U/µL	1 unit increases A_{260} of a high-molecular-weight DNA by 0.001/min/mL of reaction mixture at pH 5.0, 25°C	Amplification grade; 200 mM Tris-HCl, 20 mM MgCl$_2$, 500 mM KCl, pH 8.4 and 25 mM EDTA, pH 8.0	No detectable RNase	Life Technol 18068-015
2000 Kunitz U/mg solid	1 unit increases A_{260} by 0.001/min at 25°C	Lyophilized		Oncor 120011
	1 unit completely hydrolyzes 1 µg λ DNA/min at 25°C	100 mM Tris-HCl, 50 mM KCl, 0.5 mM CaCl$_2$, 0.1 mg/mL BSA, 50% glycerol, pH 7.4	No detectable RNase	Oncor 120021 120022
≥1500 Kunitz U/mg	1 unit increases A_{260} by 0.001/min/mL at pH 5.0, 25°C with DNA as substrate	Chromatographically purified; lyophilized	<0.001% RNase	Pharmacia 27-0512-01
≥100,000 U/mg protein; 5000-10,000 U/mL	1 unit completely degrades 1 µg pBR322/10 min at 37°C in 20 µL assay buffer; 1 unit = 0.3 Kunitz unit	Molecular biology grade; homogeneous purity; solution containing 10 mM Tris-HCl, 10 mM CaCl$_2$, 10 mM MgCl$_2$, 50% glycerol, pH 7.5	No detectable RNase	Pharmacia 27-0514-01 27-0514-02

SPECIFIC ACTIVITY	UNITS DEFINITION	PREPARATION FORM	ADDITIONAL ACTIVITIES	SUPPLIER CATALOG NO.
Bovine pancreas *continued*				
≥1500 Kunitz U/mg	1 unit increases A_{260} by 0.001/min/mL at pH 5.0, 25°C with DNA as substrate	Salt-free; crystallized; lyophilized		Pharmacia 27-0516-01
1 U/μL	1 unit completely degrades 1 μg DNA at 100/10 min at pH 7.9, 37°C		No detectable RNase <1% RNase	Promega M6101 M6102
2000 Kunitz U DNase I/vial	1 Kunitz unit produces a ΔA_{260} of 0.001/min/mL at pH 5.0, 25°C using DNA, Type I or III as substrate	Total protein 1 mg		Sigma D4263
2000-3000 Kunitz U/mg protein	1 Kunitz unit produces a ΔA_{260} of 0.001/min/mL at pH 5.0, 25°C using DNA, Type I or III as substrate	Chromatographically purified; lyophilized containing 90% protein; balance primarily $CaCl_2$	<0.005% protease <0.01% chymotrypsin, RNase	Sigma D4513
2500 Kunitz U/mg protein	1 Kunitz unit produces a ΔA_{260} of 0.001/min/mL at pH 5.0, 25°C using DNA, Type I or III as substrate	Chromatographically purified; lyophilized containing 90% protein; balance primarily $CaCl_2$	<0.005% protease <0.01% chymotrypsin, RNase	Sigma D4527
1500-2500 Kunitz U/mg protein	1 Kunitz unit produces a ΔA_{260} of 0.001/min/mL at pH 5.0, 25°C using DNA, Type I or III as substrate	Chromatographically purified; lyophilized containing 90% protein; balance primarily $CaCl_2$	<0.02% RNase <0.05% protease <0.5% chymotrypsin	Sigma D5025
400-600 Kunitz U/mg protein	1 Kunitz unit produces a ΔA_{260} of 0.001/min/mL at pH 5.0, 25°C using DNA, Type I or III as substrate	Chromatographically purified; lyophilized containing 90% protein; balance primarily $CaCl_2$	<0.02% RNase	Sigma DN-25
1500 Kunitz U/mg protein	1 Kunitz unit produces a ΔA_{260} of 0.001/min/mL at pH 5.0, 25°C using DNA, Type I or III as substrate	Chromatographically purified; lyophilized containing 80% protein; balance primarily Gly	<0.01% RNase	Sigma DN-EP
10,000 U/mL	1 unit digests 1 μg DNA/10 min at 37°C in a 25 μL reaction volume	10 mM Tris-HCl, 10 mM $CaCl_2$, 10 mM $MgCl_2$, 50% glycerol, pH 7.5	No detectable RNase	Stratagene 600031 600032
30,000 U and 150,000 U	1 unit increases A_{260} by 0.001/min at pH 5.0, 25°C with calf thymus DNA as substrate (Kunitz unit)	Solution containing 20 mM NaOAc, 150 mM NaCl, 50% glycerol, pH 5.0	No detectable RNase	TaKaRa 2210
10 mg and 50 mg	1 unit increases A_{260} by 0.001/min at pH 5.0, 25°C with calf thymus DNA as substrate (Kunitz unit)	Lyophilized	No detectable RNase	TaKaRa 2210L
≥20,000 U/g solid; ≥800 U DNase/mL gel	1 unit increases A_{260} by 0.001/min/mL when acting upon highly polymerized DNA at pH 5.0, 25°C	Chromatographically purified; immobilized, bound to CNBr-activated Sepharose^R 4B; 50% suspension in 0.1 M acetate buffer and 0.02% NaN_3, pH 4.5		Worthington LS00135 LS00136

SPECIFIC ACTIVITY	UNITS DEFINITION	PREPARATION FORM	ADDITIONAL ACTIVITIES	SUPPLIER CATALOG NO.
Bovine pancreas *continued*				
≥2000 U/mg solid	1 unit increases A_{260} by 0.001/min/mL when acting upon highly polymerized DNA at pH 5.0, 25°C	Chromatographically prepared; 1X crystallized; lyophilized containing Gly stabilizer		Worthington LS02004 LS02006 LS02007 LS02009
1,000,000 Dornase U/vial (11 mg)	1 unit increases A_{260} by 0.001/min/mL when acting upon highly polymerized DNA at pH 5.0, 25°C; 1 Kunitz U = 45.5 Dornase U	Sterile-filtered; lyophilized		Worthington LS02058 LS02060
≥1400 U/mg solid	1 unit increases A_{260} by 0.001/min/mL when acting upon highly polymerized DNA at pH 5.0, 25°C	Purified; precrystallized; lyophilized		Worthington LS02138 LS02139 LS02141
2000 U/vial	1 unit increases A_{260} by 0.001/min/mL when acting upon highly polymerized DNA at pH 5.0, 25°C	Lyophilized		Worthington LS02173 LS02172
≥2000 U/mg solid	1 unit increases A_{260} by 0.001/min/mL when acting upon highly polymerized DNA at pH 5.0, 25°C	Chromatographically prepared; lyophilized containing Gly stabilizer	<0.0005% RNase by weight as RNase A	Worthington LS06330 LS06328 LS06332
≥10,000 U/vial	1 unit increases A_{260} by 0.001/min/mL when acting upon highly polymerized DNA at pH 5.0, 25°C	Molecular biology grade; chromatographically purified; lyophilized containing 2 mg Gly and 5 µmol Ca^{2+}/vial	No detectable RNase, protease	Worthington LS06333
2000 U/mL	1 unit increases A_{260} by 0.001/min/mL when acting upon highly polymerized DNA at pH 5.0, 25°C	Molecular biology grade; chromatographically purified; solution containing 50% glycerol and 1 mM $CaCl_2$	No detectable RNase, protease	Worthington LS06342 LS06344

3.1.22.1 Deoxyribonuclease II

REACTION CATALYZED
: Endonucleolytic cleavage of native and denatured DNA to 3'-phosphomononucleotide and 3'-phosphooligonucleotide end-products

SYSTEMATIC NAME
: Deoxyribonuclease II

SYNONYMS
: DNase II, pancreatic DNase II, acid DNase

REACTANTS
: Double-stranded DNA, 3'-phosphomononucleotides, 3'-phosphooligonucleotides, p-nitrophenylphosphodiesters

APPLICATIONS
- Mapping DNase-sensitive regions in eukaryotic DNA
- Radioactive labeling by nick translation
- Constructing plasmids
- Isolating DNA-free RNA produced by in vitro SP6 or T7 RNA polymerase systems

NOTES
- Classified as a hydrolase, acting on ester bonds: endodeoxyribonuclease producing other than 5'-phosphomonoesters
- Widely distributed in animal cells, localized in lysosomes
- Preference for double-stranded DNA
- Activated by bivalent cations
- Also acts on p-nitrophenylphosphodiesters at pH 5.6-5.9
- Strongly inhibited by iodoacetic acid, N-bromosuccinimide and H_2O_2; inhibited by sulfate; not strongly inhibited by diisopropylfluorophosphate (DFP)
- Similar to crab testis DNase, snail DNase, salmon testis DNase, liver acid DNase, human gastric mucosa and cervix acid DNases
- Differs from pancreatic DNase (EC 3.1.21.1) in pH optimum

SPECIFIC ACTIVITY	UNITS DEFINITION	PREPARATION FORM	ADDITIONAL ACTIVITIES	SUPPLIER CATALOG NO.
Bovine pancreas				
20,000-50,000 U/mL	1 Kunitz unit increases absorbance by 0.001/min	20 mM Tris-HCl, 1 mM DTT, 100 µg/mL BSA, 50 mM NaCl, 50% glycerol, pH 7.6	No detectable RNase	Boehringer 776785
Bovine spleen				
200 U/mg solid	1 unit increases A_{260} by 0.001/min at pH 4.6, 25°C	Salt-free; lyophilized		Biozyme DNS2
200 U/mg protein	1 unit increases A_{260} by 0.001/min/mL at pH 5.00, 25°C	Salt-free; lyophilized containing 90% protein		Calzyme 150A0200
250 U/mg	1 unit increases the A_{260} by 0.001/min at pH 4.6, 25°C with calf thymus DNA as substrate	Salt-free; lyophilized		Fluka 31150
250-300 Kunitz U/mg protein	1 Kunitz unit produces a ΔA_{260} of 0.001/min/mL at pH 4.6, 25°C (Mg ion concentration = 0.83 mM)	Salt-free; lyophilized	No detectable RNase	ICN 157555

Deoxyribonuclease II continued 3.1.22.1

SPECIFIC ACTIVITY	UNITS DEFINITION	PREPARATION FORM	ADDITIONAL ACTIVITIES	SUPPLIER CATALOG NO.
Bovine spleen continued				
3000-8000 Kunitz U/mg protein	1 Kunitz unit produces a ΔA_{260} of 0.001/min/mL at pH 4.6, 25°C	Lyophilized containing 85% protein	<0.02% RNase	Sigma D5150
400-1000 Kunitz U/mg protein	1 Kunitz unit produces a ΔA_{260} of 0.001/min/mL at pH 4.6, 25°C	Salt-free; lyophilized	<0.1% RNase	Sigma D8764
200-600 Kunitz U/mg protein	1 Kunitz unit produces a ΔA_{260} of 0.001/min/mL at pH 4.6, 25°C	Salt-free; lyophilized	<0.1% RNase	Sigma DN-II-B
Porcine spleen (optimum pH = 4.5-5.0, MW = 38,000 Da)				
		Lyophilized containing 70% protein; balance primarily NaCl	No detectable RNase	ICN 190370
2000-5000 Kunitz U/mg protein	1 Kunitz unit produces a ΔA_{260} of 0.001/min/mL at pH 4.6, 25°C	Lyophilized containing 70% protein; balance primarily NaCl	<0.05% RNase	Sigma D4138
12,000-20,000 Kunitz U/mg protein	1 Kunitz unit produces a ΔA_{260} of 0.001/min/mL at pH 4.6, 25°C	Lyophilized containing 70% protein		Sigma D5275
1500 Kunitz U/vial	1 Kunitz unit produces a ΔA_{260} of 0.001/min/mL at pH 4.6, 25°C	70% protein; balance primarily NaCl		Sigma D9784
≥800 U/mg solid	1 unit increases A_{260} by 0.002/min/mL when acting upon highly polymerized DNA at pH 4.6, 25°C	Chromatographically purified; dialyzed; lyophilized		Worthington LS02425 LS02427
≥2000 U/mg protein	1 unit increases A_{260} by 0.002/min/mL when acting upon highly polymerized DNA at pH 4.6, 25°C	Chromatographically prepared on DEAE cellulose and CM Sephadex^R; dialyzed; lyophilized		Worthington LS05410 LS05411
≥12,000 U/mg protein	1 unit increases A_{260} by 0.002/min/mL when acting upon highly polymerized DNA at pH 4.6, 25°C	Chromatographically prepared on DEAE cellulose and CM Sephadex^R; dialyzed; 50% glycerol solution		Worthington LS05416 LS05418 LS05420

Deoxyribonuclease (pyrimidine dimer)

REACTION CATALYZED
  Endonucleolytic cleavage near pyrimidine dimers to give products with 5′-phosphate

SYNONYMS
  Endodeoxyribonuclease (pyrimidine dimer)

REACTANTS
  Pyrimidine dimers

NOTES
- Classified as a hydrolase, acting on ester bonds: site-specific endodeoxyribonuclease specific for altered bases
- Initiates the process of repairing UV-damaged DNA by catalyzing the excision of pyrimidine dimers formed from either strand of DNA
- Acts on damaged DNA, 5′ from the damaged site
- Binds to DNA through electrostatic forces, then diffuses along DNA by a sliding mechanism until it reaches a pyrimidine dimer
- Similar to T4 endonuclease V, *E. coli* endonucleases III and V, and correndonuclease II

SPECIFIC ACTIVITY	UNITS DEFINITION	PREPARATION FORM	ADDITIONAL ACTIVITIES	SUPPLIER CATALOG NO.
T4 Endonuclease V, denV gene expressed in *E. coli*				
≥1000 U/mg protein	1 unit changes 10 ng UV-irradiated plasmid DNA from a supercoiled conformation to a relaxed-nicked conformation/30 min at 37°C as shown by agarose gel electrophoresis	Chromatographically purified; solution containing 0.01 M PO$_4$ buffer, 0.01 M EDTA, 0.001 M DTT, 20% glycerol, pH 6.5		Worthington LM01460 LM01462 LM01464

Calf Thymus Ribonuclease H

REACTION CATALYZED
Endonucleolytic cleavage of RNA-DNA hybrids to 5'-phosphomonoesters and single-stranded DNA

SYSTEMATIC NAME
Calf thymus ribonuclease H

SYNONYMS
Endoribonuclease H (calf thymus), RNase H

REACTANTS
RNA-DNA hybrids, 5'-phosphomonoesters

APPLICATIONS
- RNA mapping studies
- *In vivo* RNA-primed initiation of DNA synthesis
- Detection of DNA-RNA hybrids
- Specific fragmentation of RNA after hybridization with oligodeoxynucleotides
- Studies of *in vitro* polyadenylation reaction products
- Removes mRNA during second strand cDNA synthesis and cloning
- Removes poly(A) tails on mRNA after hybridization with oligo(dT)

NOTES
- Classified as a hydrolase, acting on ester bonds: endoribonuclease producing 5'-phosphomonoesters
- Acts on RNA-DNA hybrids; does not act on RNA-RNA, DNA-DNA or single-stranded RNA or DNA
- Similar to enzymes from *E. coli*, chicken embryo, human KB cells, rat liver, *Ustilago maydis*, human leukemic cells, *Saccharomyces cerevisiae* (H2) and *Tetrahymena pyriformis*
- Activated by Mg^{2+} and Mn^{2+}

SPECIFIC ACTIVITY	UNITS DEFINITION	PREPARATION FORM	ADDITIONAL ACTIVITIES	SUPPLIER CATALOG NO.
Escherichia coli				
	1 unit produces 1 nmol acid-soluble ribonucleotides from [³H]-poly(A)-poly(dT)/20 min at 37°C	20 m*M* Tris-HCl, 300 m*M* KCl, 20 m*M* MgOAc, 0.1 m*M* DTT, 7 m*M* EDTA, 50% glycerol, 200 µg/mL BSA, pH 7.5	No detectable DNase, endonuclease, RNase III, nonspecific RNase	CHIMERX 1330-01 1330-02
	1 unit produces 1 nmol total acid-soluble ribonucleotides from [³H]-poly(A)-poly(dT)/20 min at 37°C	Solution containing 25 m*M* HEPES buffer, 50 m*M* KCl, 1 m*M* DTT, 50% glycerol, pH 8.0	No detectable DNase, endonuclease, RNase III, non-specific RNase	ICN 152025
800-1200 U/mL	1 unit produces 1 nmol acid-soluble nucleotide/20 min at pH 7.5, 37°C with poly(rA)-poly(dT) as substrate	Solution containing 20 m*M* Tris-HCl, 300 m*M* KCl, 20 mM MgOAc, 7 m*M* EDTA, 0.1 m*M* DTT, 200 µg/mL BSA, 50% glycerol, pH 7.5	No detectable RNase, DNase	Pharmacia 27-0894-01 27-0894-02

SPECIFIC ACTIVITY	UNITS DEFINITION	PREPARATION FORM	ADDITIONAL ACTIVITIES	SUPPLIER CATALOG No.
Escherichia coli, carrying a plasmid containing the *Thermus thermophilus* RNase H gene (*Tth* RNase H)				
	1 unit solubilizes 1 nmol [^{3}H]-polyadenylic acid from a poly rA:poly dT substrate/10 min at pH 8.3, 65°C	50 mM Tris-HCl, 0.1 mM EDTA, 0.1% Triton X-100, 100 mM NaCl, 1 mM DTT, 50% glycerol, pH 8.0		Adv Biotech AB-0556 AB-0556b
Escherichia coli RNase H gene, expressed in *E. coli*				
1-5 U/μL	1 unit hydrolyzes 1 μmol RNA in [^{3}H]-labeled poly(A)-poly(dT) to acid-soluble material/20 min at 37°C	20 mM Tris-HCl, 0.1 M KCl, 10 mM MgCl$_2$, 0.1 mM EDTA, 0.1 mM DTT, 50% glycerol, pH 7.9	No detectable non-specific endonuclease, exonuclease, RNase	Amersham E 70054Y E 70054Z
1-4 U/μL	1 unit hydrolyzes 1 μmol RNA in [^{3}H]-labeled poly(A)-poly(dT) to acid-soluble material/20 min at pH 7.5, 37°C			Life Technol 18021-014 18021-071
1000-4000 U/mL	1 unit produces 1.0 nmol acid-soluble products from RNA-DNA hybrid/20 min at pH 7.5, 37°C	Solution containing 50% glycerol, 20 mM Tris-HCl, 100 mM KCl, 10 mM MgCl$_2$, 0.1 mM EDTA, 0.1 mM DTT, 0.05 mg/mL BSA, pH 7.5		Sigma R6501
Escherichia coli strain H560 polA1				
1000 U/mL	1 unit produces 1 nmol acid-soluble ribonucleotides from [^{3}H]-poly(A)-poly(dT)/20 min at 37°C	25 mM Tris-HCl, 50 mM KCl, 1 mM DTT, 0.1 mM EDTA, 50% glycerol, pH 8.0		Boehringer 786349 786357
Escherichia coli strain HB101 containing rnh plasmid (pkHII) and regulator plasmid (pNT203) (optimum pH = 8.0, MW = 21,000 Da)				
60 U/μL; 1000 U and 5000 U	1 unit produces 1 nmol acid-soluble ^{3}H/20 min at pH 7.0, 30°C with poly(rA)-poly(dT) as substrate	Highly purified; solution containing 25 mM Tris-HCl, 5 mM β-MSH, 0.5 mM EDTA, 30 mM NaCl, 50% glycerol, pH 7.5	No detectable nuclease	TaKaRa 2150A
Escherichia coli strain MRE-600				
1-5 U/μL	1 unit produces 1 nmol acid-soluble ribonucleotides from a [^{3}H]-poly(A)-poly(dT) hybrid/20 min at pH 7.8, 37°C	20 mM Tris-HCl, 100 mM NaCl, 1 mM EDTA, 50% glycerol, pH 7.5	No detectable nuclease, RNase, nickase	AGS Heidelb A00760S A00760M
1-7 U/μL	1 unit forms 1 nmol acid-soluble products/20 min at 37°C	25 mM HEPES-KOH, 50 mM KCl, 1 mM DTT, 60% glycerol, pH 8.0	No detectable endo- and exodeoxyribonucleases, RNase	Fermentas EN0201 EN0202
Escherichia coli, recombinant (optimum T = 37°C)				
	1 unit acid solubilizes 1 nmol [^{3}H]-polyadenylic acid/20 min at pH 7.5, 37°C in the presence of an equimolar concentration of polythymidylic acid	Solution containing 50% glycerol, 50 mM Tris-HCl, 0.1 M NaCl, 0.1 mM EDTA, 1 mM DTT, 0.1% Triton X-100, pH 7.5	No detectable DNA exo- and endonuclease, non-RNase H RNase, phosphatase	Epicentre R06100 R06250 R06500
0.5-2 U/μL	1 unit produces 1 nmol acid-soluble ribonucleotides from radiolabeled poly(rA)-poly(dT)/20 min at pH 7.8, 37°C	≥90% supercoiled plasmid	No detectable RNase H <1% DNase <3% RNase	Promega M4281 M4282

Calf Thymus Ribonuclease H continued — 3.1.26.4

SPECIFIC ACTIVITY	UNITS DEFINITION	PREPARATION FORM	ADDITIONAL ACTIVITIES	SUPPLIER CATALOG No.
Escherichia coli, recombinant *continued*				
1-2 U/mL	1 unit hydrolyzes 1 nmol RNA in [^{32}P]-labeled poly(A)-poly(dT) to acid-soluble products/20 min at 37°C	Solution containing 50% glycerol, 20 mM Tris-HCl, 0.1 M KCl, 0.1 mM DTT, 10 mM MgCl$_2$, 0.1 mM EDTA, pH 8.0		Seikagaku 120250-1 120250-2 120250-3
Unspecified ($T_{1/2}$ = several hrs at 70°C, 30 min at 95°C)				
	1 unit acid solubilizes 1 nmol [^{3}H]-polyadenylic acid/20 min at pH 7.5, 37°C in the presence of an equimolar concentration of polythymidylic acid	Solution containing 50% glycerol, 50 mM Tris-HCl, 0.1 M NaCl, 0.1 mM EDTA, 1 mM DTT, 0.1% Triton X-100, pH 7.5	No detectable DNA exo- and endonuclease, non-RNase H RNase, phosphatase	Epicentre Hydridase™ H39050 H39100 H39250

Ribonuclease T$_2$ — 3.1.27.1

REACTION CATALYZED
  Two-stage endonucleolytic cleavage to 3'-phosphomononucleotides and 3'-phosphooligonucleotides with 2',3'-cyclic phosphate intermediates
SYSTEMATIC NAME
  Ribonuclease T$_2$
SYNONYMS
  Ribonuclease II
REACTANTS
  3'-Phosphomononucleotides, 3'-phosphooligonucleotides

APPLICATIONS
- 3' terminal analysis of RNA
- RNase protection assay

NOTES
- Classified as a hydrolase, acting on ester bonds: endoribonuclease producing other than 5'-phosphomonoesters
- Similar to plant RNase, *E. coli* RNase I, RNase N$_2$, microbial RNase II

SPECIFIC ACTIVITY	UNITS DEFINITION	PREPARATION FORM	ADDITIONAL ACTIVITIES	SUPPLIER CATALOG No.
Aspergillus oryzae				
10-40 U/µL	1 unit hydrolyzes 1 A$_{260}$ unit of yeast RNA to acid-soluble material/15 min at pH 4.5, 37°C			Life Technol 18031-013

3.1.27.1 Ribonuclease T₂ continued

SPECIFIC ACTIVITY	UNITS DEFINITION	PREPARATION FORM	ADDITIONAL ACTIVITIES	SUPPLIER CATALOG No.
Aspergillus oryzae continued				
1000-3000 U/mg protein	1 unit produces acid-soluble oligonucleotides from polyadenylic acid (5′) equivalent to a ΔA_{260} of 1.0/15 min at pH 4.5, 37°C in a reaction volume of 1.0 mL	Electrophoretically purified; lyophilized	No detectable RNase T1, DNase, 5′-nucleotidase, PDE	Sigma R3751
1000 U/mg protein	1 unit produces acid-soluble oligonucleotides from polyadenylic acid (5′) equivalent to a ΔA_{260} of 1.0/15 min at pH 4.5, 37°C in a reaction volume of 1.0 mL	Electrophoretically purified; lyophilized containing 35% protein; balance primarily Na citrate	No detectable RNase T1, DNase, 5′-nucleotidase, PDE	Sigma R4376
	1 unit produces acid-soluble oligonucleotides from polyadenylic acid (5′) equivalent to a ΔA_{260} of 1.0/15 min at pH 4.5, 37°C in a reaction volume of 1.0 mL	Lyophilized containing 25% protein and 75% Na citrate	No detectable RNase T1, DNase, 5′-nucleotidase, PDE	Sigma R8376

3.1.27.3 Ribonuclease T₁

REACTION CATALYZED
  Two-stage endonucleolytic cleavage to 3′-phosphomononucleotides and 3′-phosphooligonucleotides ending in Gp with 2′,3′-cyclic phosphate intermediates

SYSTEMATIC NAME
  Ribonuclease T1

SYNONYMS
  Guanyloribonuclease, *Aspergillus oryzae* ribonuclease, *Ustilago sphaerogena* ribonuclease, RNase

REACTANTS
  Single-stranded RNA, 3′-phosphomononucleotides and 3′-phosphooligonucleotides ending in Gp

APPLICATIONS
- Removal of poly(A) sequences from mRNA in the presence of oligo(dT)
- Oligodeoxyribonucleotide-directed cleavage of RNA
- RNA sequence analysis
- Removal of mRNA during single-strand cDNA synthesis
- RNA fingerprinting
- Preparation of nucleoside 2′,3′-cyclic phosphates
- Synthesis of oligonucleotides

NOTES

- Classified as a hydrolase, acting on ester bonds: endoribonuclease producing other than 5'-phosphomonoesters
- A highly specific endoribonuclease
- Cleaves RNA or deaminated RNA between guanosine (or inosine) 3'-phosphate residues and the 5'-OH of adjacent nucleotides, forming the corresponding intermediate 2',3'-cyclic phosphates
- Preference for single-stranded RNA
- Activated by histidine and EDTA (chelation of inhibitory cation contaminants)
- Inhibited by heavy metals (Ag^+, Zn^{2+}, Cu^{2+} and Hg^{2+})
- Similar to *N. crassa* RNase N_1 and N_2, *Chalaropsis* RNase, *B. subtilis* RNase, microbial RNase I

SPECIFIC ACTIVITY	UNITS DEFINITION	PREPARATION FORM	ADDITIONAL ACTIVITIES	SUPPLIER CATALOG NO.
***Aspergillus oryzae* (optimum pH = 7.5, MW = 11,000 Da)**				
650,000 U/mg; 1.54 mg/mL		10 m*M* $NaPO_4$, 3 m*M* ammonium carbonate, 50% glycerol, pH 7	No detectable DNase	5'→3' 5305-122368
400,000 U/mg protein	1 unit increases OD_{260} by 1.0 at pH 7.5, 37°C	Highly purified; essentially pure by electrophoresis; suspension in 2.8 *M* $(NH_4)_2SO_4$		Amersham E 78021Y
1,000,000 U/mL and 5,000,000 U/mL	1 Egami unit releases sufficient acid-soluble oligonucleotides increasing A_{260} by 1.0	Suspension in 3.2 *M* $(NH_4)_2SO_4$, pH 6		Boehringer 109193 109207
≥200 U/mg solid	1 unit increases the A_{260} of a 0.3% yeast RNA solution by 1.0 unit/15 min at pH 7.5, 37°C	Lyophilized	<0.02% PME, PDE, DNase	Calbiochem 556785
380,000 U/mg protein	1 unit produces soluble oligonucleotides from yeast RNA/15 min at pH 7.5, 37°C in a 1 mL reaction volume, changing A_{260} by 1.0 according to the conditions of T. Egami et al.	Suspension in 3.2 *M* $(NH_4)_2SO_4$, pH 6		Fluka 83840
300,000 U/mg	1 unit produces acid-soluble oligonucleotides causing a ΔA_{260} of 1.0 at pH 7.5, 37°C	Highly purified; suspension in 0.70 saturated $(NH_4)_2SO_4$		ICN 101079
900-3000 U/µL	1 unit hydrolyzes 1 A_{260} unit of yeast RNA to acid-soluble material/15 min at pH 7.5, 37°C			Life Technol 18030-015
		Sequencing grade; lyophilized		Pharmacia 27-0991-01

SPECIFIC ACTIVITY	UNITS DEFINITION	PREPARATION FORM	ADDITIONAL ACTIVITIES	SUPPLIER CATALOG NO.
Aspergillus oryzae continued				
300,000-600,000 U/mg protein	1 unit produces acid-soluble oligonucleotides from yeast RNA equivalent to a ΔA_{260} of 1.0/15 min at pH 7.5, 37°C in a reaction volume of 1.0 mL	Suspension in 3.2 M $(NH_4)_2SO_4$ and 0.02 M Tris-HCl, pH 6		Sigma R1003
300,000-600,000 U/mg protein	1 unit produces acid-soluble oligonucleotides from yeast RNA equivalent to a ΔA_{260} of 1.0/15 min at pH 7.5, 37°C in a reaction volume of 1.0 mL	Lyophilized containing 70% protein; balance KPO_4 buffer salts		Sigma R7384
300,000-600,000 U/mg protein	1 unit produces acid-soluble oligonucleotides from yeast RNA equivalent to a ΔA_{260} of 1.0/15 min at pH 7.5, 37°C in a reaction volume of 1.0 mL	Suspension in 2.7 M $(NH_4)_2SO_4$ and 0.02 M Tris-HCl, pH 6		Sigma R8251
$\geq$300,000 U/mg protein	1 unit increases A_{260} by 1.0 at pH 7.5, 37°C	Highly purified; solution containing 2.8 M $(NH_4)_2SO_4$		Worthington LS03518 LS03519 LS03520
Aspergillus oryzae RNase T1 gene, expressed in *E. coli* (optimum pH = 7.5; active at extreme temperature and in denaturing solvents)				
650,000 U/mg; 1000 U/μL	25 units causes a ΔA_{260} of 0.01/min at pH 7.5, RT	Highly purified; 10 mM $NaPO_4$, 3 mM, ammonium carbonate, 50% glycerol, pH 7	No detectable non-specific endonuclease, exonuclease	Ambion 2280 2282
Ustilago sphaerogena				
5000-10,000 U/mg protein	1 unit produces acid-soluble oligonucleotides from yeast RNA equivalent to a ΔA_{260} of 1.0/15 min at pH 7.5, 37°C in a reaction volume of 1.0 mL	Suspension in 3.2 M $(NH_4)_2SO_4$		Sigma R8378

Ribonuclease U$_2$

REACTION CATALYZED
Two-stage endonucleolytic cleavage to 3′-phosphomononucleotides and 3′-phosphooligonucleotides ending in Ap or Gp with 2′,3′-cyclic phosphate intermediates

SYSTEMATIC NAME
Ribonuclease U$_2$

SYNONYMS
Ribonuclease PhyM, RNase

REACTANTS
3′-phosphomononucleotides and 3′-phosphooligonucleotides ending in Ap or Gp

NOTES
- Classified as a hydrolase, acting on ester bonds: endoribonuclease producing other than 5′-phosphomonoesters
- Similar to RNase U$_3$, *Pleospora* RNase, *Trichoderma koningi* RNase III

SPECIFIC ACTIVITY	UNITS DEFINITION	PREPARATION FORM	ADDITIONAL ACTIVITIES	SUPPLIER CATALOG No.
Physarum polycephalum				
	1 unit produces a uniform partial digestion of 3 µg RNA	Lyophilized		Amersham E 71537Y
		Sequencing grade; lyophilized		Pharmacia 27-0994-01
Ustilago sphaerogena				
	1 unit produces a uniform partial digestion of 3 µg RNA	Lyophilized		Amersham E 71538Y
		Sequencing grade; lyophilized		Pharmacia 27-0992-01
200 U/vial (polyadenylic acid (5′) substrate)	1 unit produces acid-soluble oligonucleotides from yeast RNA equivalent to a ΔA_{260} of 1.0/15 min at pH 4.5, 37°C in a reaction volume of 1.0 mL	Electrophoretically purified; lyophilized	No detectable DNase, 5′-nucleotidase, PDE	Sigma R8126

Pancreatic Ribonuclease

3.1.27.5

REACTION CATALYZED
Endonucleolytic cleavage to 3'-phosphomononucleotides and 3'-phosphooligonucleotides ending in Cp or Up, with 2',3'-cyclic phosphate intermediates

SYSTEMATIC NAME
Pancreatic ribonuclease

SYNONYMS
Ribonuclease, ribonuclease I, ribonuclease IA, RNase, ribonuclease A, ribonuclease B, endoribonuclease I

REACTANTS
Single-stranded RNA, 3'-phosphomononucleotides and 3'-phosphooligonucleotides ending in Cp or Up

APPLICATIONS
- Preparation of RNA-free DNA
- Mapping or quantitation of RNA by selective cleavage of single-stranded regions
- Purifying DNA plasmids
- Genome DNA isolation
- Mismatch detection
- RNase protection assays
- Molecular weight marker

NOTES
- Classified as a hydrolase, acting on ester bonds: endoribonuclease producing other than 5'-phosphomonoesters
- Hydrolyzes the phosphodiester bond between the 5'-ribose of a nucleotide and the phosphate group attached to the 3'-ribose of an adjacent pyrimidine nucleotide, forming a 2',3'-cyclic phosphate; this is then hydrolyzed to the corresponding 3'-nucleoside phosphate
- Ribonuclease B is a carbohydrate derivative of RNase A, containing 5 residues of mannose and 2 residues of N-acetylglucosamine per molecule
- Preference for single-stranded RNA
- Inhibited by heavy metal ions
- Competitively inhibited by DNA (denatured > native)
- Irreversibly inactivated by 0.1% SDS; 20% inhibited by 0.3 M NaCl
- Similar to venom RNase, *Thiobacillus thioparus* RNase, *Xenopus laevis* RNase, *Rhizopus oligosporus* RNase, ribonuclease M

SPECIFIC ACTIVITY	UNITS DEFINITION	PREPARATION FORM	ADDITIONAL ACTIVITIES	SUPPLIER CATALOG NO.
Aspergillus clavatus				
≥5000 U/mg solid	1 unit produces acid-soluble oligonucleotides equivalent to a ΔA_{260} of 1.0/30 min at pH 7.5, 37°C in a 1.5 mL reaction volume with yeast RNA as substrate	Lyophilized containing 50% protein; balance primarily PO_4 buffer salts		sigma R7003
Bacillus cereus				
	1 unit produces a uniform partial digestion of 3 μg RNA under standard RNA sequencing conditions in the absence of urea	Lyophilized		Amersham E 71539Y

SPECIFIC ACTIVITY	UNITS DEFINITION	PREPARATION FORM	ADDITIONAL ACTIVITIES	SUPPLIER CATALOG No.
Bacillus cereus continued				
		Sequencing grade; lyophilized		Pharmacia 27-0993-01
Bovine pancreas (optimum pH = 7.0-7.5, T = 65°C, MW = 13,700 [RNase A] and 14,700 [RNase B] Da)				
1140 U/mg; 1 mg/mL	1 unit increases A_{286} by 0.0146 absorbance units/min with 1 mM cCMP in a 1 mL volume; 1 Kunitz unit = 7.5 cCMP hydrolysis units	RNase A; affinity purified; 20 mM NaCl, 10 mM HEPES, 0.1% Triton X-100, 50% glycerol, pH 7.2	No detectable non-specific endonuclease, exonuclease	Ambion 2270 2271
700 U/mg; 1 mg/mL	1 unit increases A_{286} by 0.0146 absorbance units/min with 1 mM cCMP in a 1 mL volume; 1 Kunitz unit = 7.5 cCMP hydrolysis units	RNase A; RPA grade (molecular biology grade); 20 mM NaCl, 10 mM HEPES, 0.1% Triton X-100, 50% glycerol, pH 7.2	No detectable non-specific endonuclease, exonuclease, contaminating DNase and nicking activities	Ambion 2272
1800 U/mg	1 unit increases A_{260} by 1.0/min at pH 5.0, 25°C when yeast ribosomal RNA is hydrolyzed to acid soluble oligonucleotides; 1 Kunitz unit = 50 Amersham units	RNase A; chromatographically purified; 0.05 M NaOAc, 0.3 mM EDTA, 50% glycerol	No detectable protease, DNase	Amersham E 70194Y E 70194Z
≥90 U/mg	1 unit hydrolyzes RNA at a rate such that k (velocity constant) equals unity (Kunitz units) at pH 5.0, 25°C	RNase A; chromatographically purified; salt-free; lyophilized	No detectable protease	Amersham E 78020Y
70 Kunitz U/mg	1 Kunitz unit causes a maximum decrease at A_{300} in 1 mL 0.05% solution of a highly polymerized yeast tRNA/min at pH 5.0, 25°C using a DNA substrate	Lyophilized; solution		AMRESCO 0675 E345 E726 E218 E866 E106
70 U/mg solid	1 unit hydrolyzes RNA at a rate such that k (velocity constant) equals unity (Kunitz units) at pH 5.0, 25°C	Salt-free; crystallized; lyophilized containing 70% RNase A		Biozyme RN1
70 U/mg solid	1 unit hydrolyzes RNA at a rate such that k (velocity constant) equals unity (Kunitz units) at pH 5.0, 25°C	Chromatographically prepared; salt-free; lyophilized containing 70% RNase A	No detectable protease	Biozyme RN2
90 U/mg solid	1 unit hydrolyzes RNA at a rate such that k (velocity constant) equals unity (Kunitz units) at pH 5.0, 25°C	Chromatographically homogeneous; salt-free; lyophilized	No detectable DNase, protease	Biozyme RN3
40 U/mg solid	1 Kunitz unit decreases A_0 to A_1/min; A_0 to A_1 corresponds to the total conversions where A_1 is the final absorbance	Crude mixture of RNases; powder		Boehringer 109126 109134

SPECIFIC ACTIVITY	UNITS DEFINITION	PREPARATION FORM	ADDITIONAL ACTIVITIES	SUPPLIER CATALOG NO.
Bovine pancreas *continued*				
50 U/mg solid	1 Kunitz unit decreases A_0 to A_1/min; A_0 to A_1 corresponds to the total conversions where A_1 is the final absorbance	RNase A; powder		Boehringer 109142 109169
≥30 U/mg protein; 5000 U/mL	1 unit decreases A_{260} equivalent to a total conversion of RNA to oligonucleotides/min at 25°C	10 mM Tris-HCl, 5 mM $CaCl_2$, 50% glycerol, pH 7.0	No detectable RNase	Boehringer 1119915
≥30 U/mg; 10 mg/mL	1 Kunitz unit decreases A_0 to A_1/min; A_0 to A_1 corresponds to the total conversions where A_1 is the final absorbance	10 mM Tris-HCl, 5 mM $CaCl_2$, 50% glycerol, pH 7.0	No detectable DNase	Boehringer 1579681
70 U/mg dry weight	1 unit hydrolyzes RNA to yield a first-order velocity constant $k = 1$ (Kunitz units) at pH 5, 25°C	RNase A; salt-free; lyophilized	No detectable proteases	Calbiochem 55674
85 U/mg solid	1 unit hydrolyzes RNA to yield a first-order velocity constant $k = 1$ (Kunitz units) at pH 5, 25°C	RNase A; >95% purity; salt-free; lyophilized	No detectable proteases	Calbiochem 556746
3000 U/mg	1 unit increases A_{260} by 1.0 at pH 5.0, 37°C	95% purity; lyophilized		Calzyme 205A3000
>80 Kunitz U/mg solid		Chromatographically purified; salt-free; lyophilized		Crystal
100 U/mg protein	1 unit hydrolyzes the RNA at a rate constant $k=1$ at pH 5.0, 25°C (Kunitz units)	RNase A; solution containing 0.2 M $NaPO_4$, pH 6.4		Fluka 83830
90 U/mg	1 unit hydrolyzes the RNA at a rate constant $k=1$ at pH 5.0, 25°C (Kunitz units)	RNase A; powder containing 90% protein	<0.0001% protease 4% RNase B >90% RNase A	Fluka 83831
100 U/mg	1 unit hydrolyzes the RNA at a rate constant $k=1$ at pH 5.0, 25°C (Kunitz units)	RNase A; powder	<0.0001% protease	Fluka 83832
80 U/mg	1 unit hydrolyzes the RNA at a rate constant $k=1$ at pH 5.0, 25°C (Kunitz units)	RNase A; powder	<0.0001% protease >90% RNase A	Fluka 83833
70 Kunitz U/mg	1 unit hydrolyzes the RNA at a rate constant $k=1$ at pH 5.0, 25°C (Kunitz units)	RNase A; salt-free; 4X crystallized	No detectable protease	Fluka 83834
70 U/mg solid	1 unit hydrolyzes RNA at a rate such that the velocity constant (k) equals 1 at pH 5.0, 25°C	Salt-free; 5X crystallized; lyophilized	No detectable protease	ICN 101075

SPECIFIC ACTIVITY	UNITS DEFINITION	PREPARATION FORM	ADDITIONAL ACTIVITIES	SUPPLIER CATALOG NO.
Bovine pancreas *continued*				
50 Kunitz U/mg		RNase A; aggregate-free; lyophilized	No detectable PO_4, protease	ICN 101076
50 Kunitz U/mg		RNase B; salt-free; crystallized RNase	No detectable protease	ICN 101084
100 Kunitz U/mg		RNase B; lyophilized	No detectable PO_4, protease	ICN 104907
90 U/mg solid	1 unit hydrolyzes RNA at a rate such that the velocity constant (k) equals 1 at pH 5.0, 25°C	RNase A; chromatographically homogeneous; salt-free; lyophilized	No detectable protease	ICN 152024
50 U/mg	1 unit decreases absorbance from A_0 to A_1/min at pH 5.0, 25°C	Lyophilized		NBL Gene 021432 021434
≥40 Kunitz U/mg	1 unit hydrolyzes yeast RNA yielding a first-order velocity constant of 1.0 at pH 5.0, 25°C	RNase 1A; chromatographically purified; lyophilized	<0.5% RNase B	Pharmacia 27-0323-01
≥40 Kunitz U/mg	1 unit hydrolyzes yeast RNA yielding a first-order velocity constant of 1.0 at pH 5.0, 25°C	RNase A & B; chromatographically purified; lyophilized		Pharmacia 27-0330-01 27-0330-02
150-175 U/g solid	1 unit produces acid-soluble oligonucleotides equivalent to ΔA_{260} of 1.0/30 min at pH 7.5, 37°C	RNase A; insoluble enzyme attached to polyacrylamide; 30% borate buffer salts	RNase B	Sigma R1626
400-600 U/g agarose; 1 mL gel yields 12-20 U	1 unit produces acid-soluble oligonucleotides equivalent to ΔA_{260} of 1.0/30 min at pH 7.5, 37°C	RNase A; insoluble enzyme attached to beaded agarose; suspension in 2.0 M $(NH_4)_2SO_4$, pH 7.0	RNase B	Sigma R4001
50-100 Kunitz U/mg protein	1 unit produces acid-soluble oligonucleotides equivalent to ΔA_{260} of 1.0/30 min at pH 7.5, 37°C	RNase A; 60% purity by SDS-PAGE; chromatographically purified; salt-free; salt fractionated; 5X crystallized	RNase B No detectable protease	Sigma R4875
50-100 Kunitz U/mg solid	1 unit produces acid-soluble oligonucleotides equivalent to ΔA_{260} of 1.0/30 min at pH 7.5, 37°C	RNase A; 90% purity by SDS-PAGE; chromatographically purified; salt-free containing 90% protein	RNase B No detectable protease	Sigma R5000
100 Kunitz U/mg protein	1 unit produces acid-soluble oligonucleotides equivalent to ΔA_{260} of 1.0/30 min at pH 7.5, 37°C	RNase A; 95% purity by SDS-PAGE; chromatographically purified; salt-free	RNase B No detectable protease	Sigma R5125
100 Kunitz U/mg protein	1 unit produces acid-soluble oligonucleotides equivalent to ΔA_{260} of 1.0/30 min at pH 7.5, 37°C	RNase A; 95% purity by SDS-PAGE; chromatographically purified; solution containing 0.2 M $NaPO_4$ buffer, pH 6.4	RNase B No detectable protease	Sigma R5250
100 Kunitz U/mg protein	1 unit produces acid-soluble oligonucleotides equivalent to ΔA_{260} of 1.0/30 min at pH 7.5, 37°C	RNase A; chromatographically purified; salt-free; lyophilized	RNase B No detectable protease	Sigma R5500

SPECIFIC ACTIVITY	UNITS DEFINITION	PREPARATION FORM	ADDITIONAL ACTIVITIES	SUPPLIER CATALOG NO.
Bovine pancreas *continued*				
50-100 Kunitz U/mg protein	1 unit produces acid-soluble oligonucleotides equivalent to ΔA_{260} of 1.0/30 min at pH 7.5, 37°C	RNase A; chromatographically purified; salt-free; salt-fractionated	RNase B No detectable protease	Sigma R5503
50-100 Kunitz U/mg protein	1 unit produces acid-soluble oligonucleotides equivalent to ΔA_{260} of 1.0/30 min at pH 7.5, 37°C	RNase B; salt-free	No detectable protease	Sigma R5750
	1 unit produces acid-soluble oligonucleotides equivalent to ΔA_{260} of 1.0/30 min at pH 7.5, 37°C	RNase B; salt-free	No detectable protease	Sigma R5875
500-2000 U/g solid	1 unit produces acid-soluble oligonucleotides equivalent to ΔA_{260} of 1.0/30 min at pH 5.0, 37°C	RNase A; insoluble enzyme attached through covalent, non-charged bond to macroporous acrylic beads	RNase B	Sigma R7005
50-100 Kunitz U/mg protein	1 unit produces acid-soluble oligonucleotides equivalent to ΔA_{260} of 1.0/30 min at pH 7.5, 37°C	RNase B; 90% purity by SDS-PAGE; affinity chromatographed; salt-free	No detectable protease	Sigma R7884
≥2000 U/mg protein	1 unit increases A_{260} by 1.0 at pH 5.0, 37°C when yeast ribosomal RNA is hydrolyzed to acid soluble oligonucleotides; 1 Kunitz U = 50 U	RNase A; molecular biology grade; 50% glycerol solution	No detectable DNase, protease	Worthington LS02131 LS02132 LS02130
≥2500 U/mg solid	1 unit increases A_{260} by 1.0 at pH 5.0, 37°C when yeast ribosomal RNA is hydrolyzed to acid soluble oliognucleotides; 1 Kunitz U = 50 U	RNase A; 2X Crystallized from $(NH_4)_2SO_4$; 1X crystallized from EtOH; lyophilized		Worthington LS03431 LS03433 LS03435
≥3000 U/mg solid	1 unit increases A_{260} by 1.0 at pH 5.0, 37°C when yeast ribosomal RNA is hydrolyzed to acid soluble oliognucleotides; 1 Kunitz U = 50 U	RNase A; chromatographically purified; monophoretic on gel electrophoresis; lyophilized		Worthington LS05649 LS05650 LS05655
≥3000 U/mg protein	1 unit increases A_{260} by 1.0 at pH 5.0, 37°C when yeast ribosomal RNA is hydrolyzed to acid soluble oliognucleotides; 1 Kunitz U = 50 U	RNase A; chromatographically purified; monophoretic on gel electrophoresis; aggregate-free; solution containing 0.1 M PO_4 buffer and 0.1% phenol as preservative, pH 7.4		Worthington LS05677 LS05679 LS05680 LS05681
≥1000 U/mg solid	1 unit increases A_{260} by 1.0 at pH 5.0, 37°C when yeast ribosomal RNA is hydrolyzed to acid soluble oliognucleotides; 1 Kunitz U = 50 U	RNase B; partially purified chromatographically; dialyzed; lyophilized	50% RNase A	Worthington LS05708 LS05710 LS05715
Chicken liver				
50-200 U/mL	1 unit releases sufficient acid-soluble oligonucleotides increasing A_{260} by 1.0 decomposing 40 μg poly(C) as substrate			Boehringer 779725

Pancreatic Ribonuclease continued — 3.1.27.5

SPECIFIC ACTIVITY	UNITS DEFINITION	PREPARATION FORM	ADDITIONAL ACTIVITIES	SUPPLIER CATALOG NO.
Chicken liver *continued*				
≥50 U/mL	1 unit produces acid-soluble oligonucleotides equivalent to a ΔA_{260} of 1.0/15 min at pH 6.5, 37°C in a 1.1 mL reaction volume with polycytidylic acid as substrate	Solution containing 50% glycerol and 10 mM KPO$_4$, pH 6.0		Sigma R5133
Escherichia coli RNase I gene, expressed in E. coli (MW = 27,000 Da)				
1,000,000 U/mg; 1000 U/μL	50% degradation of [^{32}P]-labeled *in vitro* transcript mixed with 2 μg yeast RNA/30 min at 37°C as determined by TCA precipitation	RNase 1; >99% purity by SDS-PAGE; 100 mM NaCl, 10 mM Tris-HCl, 50% glycerol, pH 8.0	No detectable non-specific endonuclease, exonuclease	Ambion 2294 2295
Escherichia coli, recombinant (MW = 27,000 Da)				
5-10 U/μL	1 unit completely degrades 2 μg *E. coli* 5S RNA at 100 ng/sec at pH 7.5, 37°C in a 20 μL reaction volume	RNase 1; ≥80% purified by SDS gel; 10X reaction buffer: 100 mM Tris-HCl, 50 mM EDTA, 2 M NaOAc, pH 7.5; ≥90% supercoiled plasmid	<1% DNase	Promega RNase ONE M4261 M4262
Pancreas				
110 U/mg		RNase A; 10 mg/mL in 50 mM Tris-HCl, 200 mM NaCl, 3 mM KCl, 50% glycerol, pH 7.5	No detectable DNase	5'→3' 5305-888777

Ribonuclease V — 3.1.27.8

REACTION CATALYZED
Hydrolysis of poly(A), forming oligoribonucleotides and ultimately 3'-AMP

SYNONYMS
Endoribonuclease V, ribonuclease V1

REACTANTS
Poly(A), poly(U), oligoribonucleotides, 3'-AMP

NOTES
- Classified as a hydrolase, acting on ester bonds: endoribonuclease producing other than 5'-phosphomonoesters
- Cleaves RNA in double-stranded regions; cleaves single-stranded RNA at 1/10th the rate of double-stranded RNA
- Also hydrolyzes poly(U)

SPECIFIC ACTIVITY	UNITS DEFINITION	PREPARATION FORM	ADDITIONAL ACTIVITIES	SUPPLIER CATALOG No.
Cobra venom				
	1 unit produces 1 µg acid-soluble product/min at 37°C	0.2 M Tris succinate, 0.25 M KCl, 50% glycerol, pH 7.5		Amersham E 71536Y
		Solution		Pharmacia 27-0927-01

3.1.30.1 Aspergillus Nuclease S$_1$

REACTION CATALYZED
  Endonucleolytic cleavage to 5'-phosphomononucleotide and 5'-phosphooligonucleotide end-products

SYSTEMATIC NAME
  Aspergillus nuclease S$_1$

SYNONYMS
  Endonuclease S$_1$ (*Aspergillus*), single-stranded-nucleate endonuclease, deoxyribonuclease S$_1$, nuclease P1, Bal31 nuclease

REACTANTS
  Single-stranded, 5'-phosphomononucleotide, 5'-phosphooligonucleotide

APPLICATIONS
- Cleaves mismatches in double-stranded DNA
- Removes single-stranded protruding ends
- Nicks supercoiled plasmid DNA
- Opens hairpin loop structures formed during cDNA synthesis
- Large-scale production of 5'-mononucleotides
- Detection of 5'-terminal nucleotides of RNA and DNA in nucleotide sequence analysis
- Mapping restriction sites in DNA
- Making nested deletions of linear DNA
- Removing nucleotides from termini of double-stranded DNA to leave blunt ends
- Progressive shortening of double-stranded DNA fragments at both termini

NOTES
- Classified as a hydrolase, acting on ester bonds: endonuclease active with either ribo- or deoxyribonucleic acids and producing 5'-phosphomonoesters
- Specifically hydrolyzes both terminal and internal phosphodiester bonds of single-stranded DNA and RNA
- Activated by Mg^{2+}, Ca^{2+} and Zn^{2+} (some)
- Inhibited by dATP, pyrophosphates, SDS and chelating agents like EDTA and citric acid
- Nuclease S$_1$ from *Aspergillus oryzae*: rate of hydrolysis of ssDNA is 75,000X faster than dsDNA, and 5X faster than RNA
- Similar to *N. crassa* nuclease, mung bean nuclease, *Penicillium citrinum* nuclease P1

SPECIFIC ACTIVITY	UNITS DEFINITION	PREPARATION FORM	ADDITIONAL ACTIVITIES	SUPPLIER CATALOG NO.
Alteromonas espejiana				
1000-5000 U/mL	1 unit releases 600 Bp from 2 μg linearized pUR222 DNA/10 min at 30°C	20 mM Tris-HCl, 100 mM NaCl, 5 mM CaCl$_2$, 5 mM MgCl$_2$, 1 mM EDTA, 65% glycerol, pH 7.5		Boehringer 724793 724807
0.5-1.5 U/μL	1 unit produces 1 μg acid-soluble nucleotide from denatured DNA/min at pH 8.1, 30°C			Life Technol 18019-018 18019-059
1-3 U/μL	1 unit removes 200 base pairs from each end of 33 μg linear duplex DNA/10 min at pH 8.0, 30°C	5X Reaction buffer; 100 mM Tris-HCl, 60 mM CaCl$_2$, 60 mM MgCl$_2$, 5 mM EDTA, 3 M NaCl, pH 8.0		Promega M4111 M4112
Alteromonas espejiana Bal31 (optimum pH = 8.0)				
1-15 U/μL	1 unit removes 200 base pairs from each end of linearized pBR322 DNA/10 min at 30°C at a DNA concentration of 50 μg/mL	2 bands of equal intensity corresponding to MW of 109,000 and 85,000 seen on 10% SDS-polyacrylamide gel; 100 mM NaCl, 5 mM CaCl$_2$, 5 mM MgCl$_2$, 20 mM Tris-HCl, 1 mM EDTA, 50% glycerol, pH 8.0		Amersham E 70011Y
2500 U/mL	1 unit produces 1 nmol acid-soluble material from *E. coli* [^{3}H]-DNA/10 min at pH 8.1, 30°C in a 50 μL reaction	20 mM Tris-HCl, 5 mM CaCl$_2$, 100 mM NaCl, 1 mM EDTA, 5 mM MgCl$_2$, 50% glycerol, pH 8.1	No detectable nuclease	AGS Heidelb F00690S F00690M
1-5 U/μL	1 unit releases 600 base pairs from each end of linearized ds-pBR322 DNA/10 min at 30°C	20 mM Tris-HCl, 100 mM NaCl, 5 mM CaCl$_2$, 5 mM MgCl$_2$, 1 mM EDTA, 50% glycerol, pH 8.1	No detectable endodeoxyribonucleases	Fermentas EN0171 EN0172
1-3 U/μL	1 unit removes 400 base pairs from each end of linear dsDNA/10 min at pH 8.0, 30°C	10 mM Tris-HCl, 50 mM NaCl, 2 mM CaCl$_2$, 2 mM MgCl$_2$, 0.2 mM EDTA, 200 μg/mL BSA, 50% glycerol, pH 8.0	No detectable dsDNA endonucleases	NBL Gene 021001 021011
1000 U/mL	1 unit removes 200 base pairs from each end of linearized ds-φX174 DNA (650 μg/mL)/10 min at 30°C in 50 μL reaction mixture	Purified; 50 mM NaCl, 10 mM Tris-HCl, 0.25 mM EDTA, 1.5 mM CaCl$_2$, 1.5 mM MgCl$_2$, 200 μg/mL BSA, 50% glycerol, pH 7.4	No detectable ds-endonuclease	NE Biolabs 213S 213L
50 U and 250 U	1 unit produces 1 μg nucleotides from heat-denatured calf thymus DNA/min at pH 8.0, 30°C	Solution containing 20 mM Tris-HCl, 5 mM MgCl$_2$, 5 mM CaCl$_2$, 1 mM EDTA, 100 mM NaCl, 50% glycerol, pH 8.0		TaKaRa 2510
Aspergillus oryzae (optimum pH = 4.0-4.6, MW = 32,000-34,000 Da; 50% activity at pH 4.9)				
25-200 U/μL	1 unit releases 1 μg acid-soluble deoxynucleotides from denatured DNA/min at pH 4.5, 37°C	50 mM NaCl, 20 mM Tris-HCl, 0.1 mM ZnCl$_2$, 50% glycerol, pH 7.5	No detectable nuclease	AGS Heidelb A00770S A00770M
170,000 U/mg; 250 U/μL	1 unit produces 1 μg acid-soluble material/min at pH 4.5, 37°C	50 mM NaCl, 20 mM Tris-HCl, 0.1 mM ZnCl$_2$, 50% glycerol, pH 7.5	No detectable non-specific endonuclease, exonuclease	Ambion 2244 2245

SPECIFIC ACTIVITY	UNITS DEFINITION	PREPARATION FORM	ADDITIONAL ACTIVITIES	SUPPLIER CATALOG NO.
Aspergillus oryzae continued				
100-200 U/μL	1 unit converts 1 μg heat-denatured DNA to acid-soluble form/min at pH 4.6, 37°C	10 mM NaOAc, 150 mM NaCl, 0.05 mM ZnSO$_4$, 50% glycerol, pH 4.6	No detectable ds-exonuclease	Amersham E 2410Y
>50,000 U/mg; 400,000 U/mL	1 Vogt unit releases 1 μg acid-soluble deoxynucleotides from denatured DNA/min at 37°C	20 mM Tris-HCl, 50 mM NaCl, 0.1 mM ZnCl$_2$, 50% glycerol, pH 7.5		Boehringer 818330 818348
	1 unit produces 1 μg acid-soluble mononucleotide from denatured DNA/min at 37°C	20 mM Tris-HCl, 50 mM NaCl, 0.1 mM ZnCl$_2$, 50% glycerol, pH 7.5	No detectable ds-DNase	CHIMERx 1210-01 1210-02
20-100 U/μL	1 unit produces 1 μg acid-soluble deoxyribonucleotides/min at 37°C	20 mM Tris-HCl, 50 mM NaCl, 0.1 mM ZnCl$_2$, 50% glycerol, pH 7.5		Fermentas EN0321 EN0322
400,000 U/mg protein	1 unit releases 1 μg of acid-soluble deoxynucleotides from ss DNA/min at pH 4.6, 37°C	Solution	<0.0001% RNase A <0.1% DNase I <0.2% ds DNase	Fluka 74677
100,000-400,000 U/mg protein	1 unit causes 1.0 μg nucleic acid to become soluble in perchloric acid/min at pH 4.6, 37°C	Solution containing 50% glycerol, NaCl, NaOAc, ZnSO$_4$, pH 4.6	No detectable DNase I <0.1% native DNA	ICN 195353
400-1500 U/μL	1 unit hydrolyzes 1 μg denatured DNA to acid-soluble material/min at pH 4.6, 37°C	300 mM NaOAc, 10 mM ZnOAc, 50% glycerol, pH 4.6 and dilution buffer and 3 M NaCl		Life Technol 18000-016 18000-024
100-1000 U/μL	1 unit releases 1 μg acid-soluble dNMPs from 600 μg alkaline denatured herring sperm DNA/min at pH 4.6, 37°C	20 mM Tris-HCl, 50 mM NaCl, 0.1 mM ZnCl$_2$, 50% glycerol, pH 7.5	No detectable endonuclease	NBL Gene 021508 021510
≥100,000 U/mg protein	1 unit produces 1 μg acid-soluble nucleotide/min at 37°C	20 mM Tris-HCl, 50 mM NaCl, 0.1 mM ZnCl$_2$, 50% glycerol, pH 7.5		Oncor 120081 120082
100,000-600,000 U/mg protein; 200-500 U/μL	1 unit produces 1 μg acid-soluble nucleotides/min at pH 4.6, 37°C using heat-denatured DNA as substrate	Solution containing 20 mM Tris-HCl, 50 mM NaCl, 0.1 mM ZnCl$_2$, 50% glycerol, pH 7.5	No detectable DNase activity	Pharmacia 27-0920-01
200,000-600,000 U/mg protein	1 unit perchloric acid solubilizes 1.0 μg ss nucleic acid/min at pH 4.6, 37°C	Solution containing 50% glycerol, 250 mM NaCl, 50 mM ZnSO$_4$, 10 mM NaOAc, pH 4.6	<0.1% DNase I May contain up to 100 Kunitz U/mg protein RNase A	Sigma N7385
500,000 U/mg; 20,000 U and 100,000 U	1 unit converts 1 μg heat-denatured DNA into acid-soluble form/min at pH 4.6, 37°C	Solution containing 10 mM NaOAc, 150 mM NaCl, 0.05 mM ZnSO$_4$, 50% glycerol, pH 4.6	No detectable exonuclease	TaKaRa 2410
≥100,000 U/mL	1 unit renders acid-soluble 1 μg denatured DNA/min at pH 4.6, 37°C based on the release of acid soluble deoxyoligonucleotides from DNA	Chromatographically purified; solution containing 30 mM NaOAc, 50 mM NaCl, 1 mM ZnCl$_2$, 50% glycerol, pH 4.6		Worthington LS08029 LS08030 LS08031

SPECIFIC ACTIVITY	UNITS DEFINITION	PREPARATION FORM	ADDITIONAL ACTIVITIES	SUPPLIER CATALOG No.
Fungal α-amylase powder				
20-100 U/μL	1 unit produces 1 μg acid-soluble material/min at pH 4.6, 37°C	10X Reaction buffer; 0.5 M NaOAc, 2.8 M NaCl, 45 mM ZnSO$_4$, pH 4.5		Promega M5761 M5762
Mung bean (*Phaseolus aureus*) (optimum pH = 5.0)				
50 U/μL	1 unit releases 1 μg acid-soluble mononucleotide from denatured DNA/min at 37°C	Solution containing 10 mM NaOAc, 1 mM Cys, 50% glycerol, 100 μM ZnOAc, pH 5.0		Calbiochem 475907
	1 unit produces 1 μg acid-soluble material/min at 37°C using denatured calf thymus DNA	10 mM Tris-HCl, 0.1 mM ZnOAc, 50% glycerol, pH 7.5	No detectable ds-DNase	CHIMERx 1190-01 1190-02
1,500,000 U/mg	1 unit converts 1 μg heat-denatured calf thymus DNA to acid-soluble form/min at pH 4.6, 37°C	Solution containing 50% glycerol, 10 mM Tris-HCl, 50 mM NaCl, 0.01% Triton X-100, pH 7.5	No detectable acid phosphatase, AP; Ds-exo- or endonuclease <0.05% of activity on ssDNA	Epicentre M8202K M8205K M8210K
15 U/μL	1 unit converts 1 mg heat-denatured [^{32}P]-ssDNA into acid-soluble counts/min at 37°C using ss salmon sperm DNA as substrate	10 mM NaOAc, 0.1 mM ZnOAc, 1 mM Cys, 0.001% Triton X-100, 50% glycerol, pH 5.0	No detectable endonuclease, exonuclease	NBL Gene 022007 022008
10,000,000 U/mg protein	1 unit produces 1 μg acid soluble material/min at 37°C; 1 unit = 0.004 Kowalski unit	10 mM NaOAc, 1 mM Cys, 0.1 mM ZnOAc, 50% glycerol, pH 5.0		Oncor 120071 120072
Mung bean sprouts (*Phaseolus aureus*)				
30 U/μL	1 unit converts 1.0 μg heat-denatured calf thymus DNA into acid soluble form/min at pH 5.0, 37°C	10 mM Tris-HCl, 0.1 mM ZnOAc, 50% glycerol, pH 7.5	No detectable exonuclease	Amersham E 2420Y
50,000 U/mL	1 unit produces 1 μg acid-soluble material/min at 37°C using ss DNA as substrate	10 mM Tris-HCl, 50 mM NaCl, 0.1 mM ZnOAc, 0.01% Triton X-100, 50% glycerol, pH 7.5		Boehringer 1134485
20,000-40,000 U/mg protein	1 unit causes 1.0 μg denatured DNA (calf-thymus) to become acid-soluble /min at pH 5.0, 37°C	Lyophilized		ICN 195350
	1 unit produces 1 μg acid-soluble product/min at 37°C	Solution containing 0.01 M NaOAc, 0.1 mM ZnOAc, 1 mM Cys, 0.001% Triton X-100, 50% glycerol, pH 5.0		ICN 800707
25-50 U/μL	1 unit hydrolyzes 1 μg denatured DNA to acid-soluble material/min at pH 5.0, 37°C	100 mM NaOAc, 1 mM ZnOAc, 10 mM L-Cys, 500 mM NaCl, 50% glycerol, pH 5.0		Life Technol 18041-012

SPECIFIC ACTIVITY	UNITS DEFINITION	PREPARATION FORM	ADDITIONAL ACTIVITIES	SUPPLIER CATALOG NO.
Mung bean sprouts (*Phaseolus aureus*) continued				
10,000 U/mL	1 unit produces 1 μg acid-soluble total nucleotide/min at 37°C	10 mM NaOAc, 0.1 mM ZnOAc, 1 mM Cys, 0.001% Triton-X 100, 50% glycerol, pH 5.0	No detectable ds-exonuclease	NE Biolabs 250S 250L
750,000-1,000,000 U/mg protein; 100,000-150,000 U/mL	1 unit produces 1 μg acid-soluble material/min at pH 4.6, 37°C using heat-denatured DNA as substrate; 1 unit = 0.004 Kowalski unit	Molecular biology grade; homogeneous purity; solution containing 10 mM NaOAc, 0.1 mM ZnOAc, 1 mM Cys, 50% glycerol, pH 5.0	No detectable DNase activity	Pharmacia 27-0912-01 27-0912-02
50-100 U/μL	1 unit produces 1 μg acid-soluble nucleotides/min at pH 4.6, 37°C	10X Reaction buffer: 300 mM NaOAc, 500 mM NaCl, 10 mM $ZnCl_2$, 10 mM spermidine, pH 5		Promega M4311 M4312
50,000 U/mg protein	1 unit acid solubilizes 1.0 μg heat-denatured calf thymus DNA/min at pH 5.0, 37°C assuming 1 μg has A_{260} = 0.033 for mixed nucleotides	Lyophilized containing Na succinate, NaCl, ZnOAc		Sigma N6510
70,000-130,000 U/mL	1 unit produces 1 μg acid-soluble material/min at 37°C using ss salmon sperm DNA as substrate	10 mM NaOAc, 0.1 mM ZnOAc, 1 mM Cys HCl, 0.005% Triton X-100, 50% glycerol, pH 5.0	No detectable ds-nuclease	Stratagene 600052 600053
2000 U and 10,000 U	1 unit converts 1 μg heat-denatured calf thymus DNA into acid soluble form/min at pH 5.0, 37°C	Solution containing 10 mM Tris-HCl, 0.1 mM ZnOAc, 50% glycerol, pH 7.5	No detectable exonuclease	TaKaRa 2420
Neurospora crassa				
400-800 U/mg protein	1 unit produces acid-soluble oligonucleotides equivalent to a ΔA_{260} of 1.0/30 min at pH 8.0, 37°C	Suspension in 3.2 M $(NH_4)_2SO_4$, pH 6	Digests RNA also	Sigma E4253
Pencillium citrinum (optimum pH = 5.3 [RNA, heat-denatured DNA], 7.2 [3'-AMP], T = 70°C)				
200 U/mg protein (RNA); 1000 U/mg protein (A-3'-MP)	1 unit liberates 1.0 μmol acid-soluble nucleotides from RNA/min at pH 5.3, 37°C; 1 unit hydrolyzes 1.0 μmol orthophosphate from A-3'-MP/min at pH 7.2, 37°C	Lyophilized		ICN 195352
>400 U/mg protein	1 unit hydrolyzes 1 μmol 3'-AMP/min at pH 7.2, 37°C	Lyophilized		Amersham E 70018Y
300 U/mg PDE; 1000 U/mg 3'-PME	1 PDE unit hydrolyzes 1 μmol equivalent of phosphodiester linkages in yeast RNA/min at 37°C; 1 3'-PME unit forms 1 μmol P_i from 3'-AMP/min at 37°C	Lyophilized		Boehringer 236225
>400 U/mg solid	1 unit hydrolyzes 1 μmol 3'-AMP/min at pH 7.2, 37°C	Lyophilized		Calbiochem 493866

SPECIFIC ACTIVITY	UNITS DEFINITION	PREPARATION FORM	ADDITIONAL ACTIVITIES	SUPPLIER CATALOG NO.
Pencillium citrinum continued				
1200 U/mg 3'-PME activity	1 unit forms 1 μmol P_i from adenosine-3'-monophosphate/min at pH 7.2, 37°C	Lyophilized	220 U/mg 3'→5'-PDE activity where 1 unit catalyzes the hydrolysis of 1 μmol of phosphodiester linkages in yeast RNA/min at pH 5.3, 37°C	Fluka 74676
20-40 U/μL	1 unit hydrolyzes 1 A_{260} unit of yeast RNA to acid soluble material/15 min at pH 5.3, 37°C			Life Technol 18040-014
≥600 U/mg protein	1 unit hydrolyzes 1 μmol 3'-AMP/min at pH 7.2, 37°C	Lyophilized		Pharmacia 27-0852-01
400 U/mg product	1 unit produces 1.0 μmol acid-soluble nucleotides from RNA/min at pH 5.3, 37°C	Lyophilized		Seikagaku 120065-1 120065-2 120065-3
200 U/mg protein	1 unit liberates 1.0 μmol acid-soluble nucleotides from RNA/min at pH 5.3, 37°C		1000 U/mg protein 3'-AMP (3'-nucleotidase)	Sigma N8630
1200 U/mg 3'-nucleotidase activity; 400 U/mg PDE activity				Wako 147-04881 143-04883

Serratia marcescens Nuclease 3.1.30.2

REACTION CATALYZED
    Endonucleolytic cleavage to 5'-phosphomononucleotide and 5'-phosphooligonucleotide end-products

SYSTEMATIC NAME
    Serratia marcenscens nuclease

SYNONYMS
    Endonuclease (*Serratia marcescens*), nuclease S1

REACTANTS
    Double- or single-stranded, 5'-phosphomononucleotide, 5'-phosphooligonucleotide

NOTES
- Classified as a hydrolase, acting on ester bonds: endonuclease active with either ribo- or deoxyribonucleic acids and producing 5'-phosphomonoesters

3.1.30.2 Serratia marcescens Nuclease continued

- Hydrolyses double- or single-stranded substrate
- Similar to silkworm nuclease, potato nuclease, *Azotobacter* nuclease

SPECIFIC ACTIVITY	UNITS DEFINITION	PREPARATION FORM	ADDITIONAL ACTIVITIES	SUPPLIER CATALOG NO.
Escherichia coli				
10,000 U/mg	1 unit produces 1 μmol acid-soluble polynucleotides from native DNA (from calf thymus)/min at pH 8.0, 37°C	Solution containing 50% glycerol, 20 mM Tris HCl, 2 mM MgCl$_2$, 20 mM NaCl, pH 8.0	No detectable proteases >99% (GE) Proteases not detected	Fluka 74671
7000 U/mL	1 unit produces 1 μmol acid-soluble polynucleotides from native DNA (from calf thymus)/min at pH 8.0, 37°C	Solution containing 50% glycerol, 20 mM Tris HCl, 2 mM MgCl$_2$, 20 mM NaCl, pH 8.0	No detectable proteases >90% (GE)	Fluka 74673
Serratia marcescens, recombinant expressed in *E. coli*				
600–1200 U/μL	1 unit produces acid-soluble oligonucleotides equivalent to a ΔA_{260} of 1.0/30 min at pH 8.0, 37°C in 2.625 mL reaction volume using native or heat-denatured DNA substrate	≥99% SDS-PAGE; 50% glycerol, 20 mM Tris HCl, 2 mM MgCl$_2$, 20 mM NaCl, pH 8.0	No detectable protease	Sigma E8263

3.1.31.1 Micrococcal Nuclease

REACTION CATALYZED
    Endonucleolytic cleavage of DNA and RNA to 3'-phosphomononucleotide and 3'-phosphooligonucleotide end-products

SYSTEMATIC NAME
    Micrococcal nuclease

SYNONYMS
    Micrococcal endonuclease, nuclease S7

REACTANTS
    DNA, RNA, 3'-phosphomononucleotide, 3'-phosphooligonucleotide

APPLICATIONS
- Probe for bound and distorted DNA in *lac* transcription and repression complexes
- Removes endogenous RNA from *in vitro* translation systems
- Hydrolysis of nucleic acids in crude cell-free extracts
- Chromatin structural studies
- Probe for drug binding sites on DNA

Micrococcal Nuclease continued

3.1.31.1

NOTES

- Classified as a hydrolase, acting on ester bonds: endonuclease active with either ribo- or deoxyribonucleic acids and producing other than 5'-phosphomonoesters
- Demonstrates both exo- and endo-5'-phosphodiesterase activities
- Preferential endohydrolysis of double- or single-stranded DNA and RNA at sites rich in adenylate or uridylate, and deoxyadenylate or thymidylate
- Activated by Ca^{2+}
- Inhibited by EDTA, EGTA, 5'-deoxyribonucleotides and 5'-ribonucleotides
- Similar to *Chlamydomonas* nuclease, spleen phosphodiesterase, spleen endonuclease

SPECIFIC ACTIVITY	UNITS DEFINITION	PREPARATION FORM	ADDITIONAL ACTIVITIES	SUPPLIER CATALOG NO.
Staphylococcus aureus (MW = 16,800 Da)				
≥6000 U/mg protein	1 unit corresponds to a change in OD_{260} of 1.0 at pH 8.0, 37°C	Chromatographically and electrophoretically homogeneous; lyophilized		Amersham E 70196Y E 70196Z
300,000 U/mL	1 unit increases A_{260} by 1.0 OD/30 min at pH 8.8, 37°C	Pure protein by SDS gel electrophoresis; 50 mM NaCl, 2 mM Tris-HCl, 50% glycerol, pH 6.8	No detectable nuclease	ACS Heidelb
15,000 U/mg solid	1 unit increases A_{260} by 1.0	Lyophilized		Boehringer 107921
>15,000 U/mg solid	1 unit releases sufficient amounts of acid-soluble oligonucleotides increasing A_{260} by 1.0/30 min at 37°C	Lyophilized		Calbiochem 492899
100-300 U/μL	1 unit releases 1.0 A_{260} unit of acid-soluble products/30 min at 37°C	>95% purity by SDS-PAGE; 20 mM HEPES-KOH, 50 mM NaCl, 50% glycerol, pH 7.6		Fermentas EN0181 EN0182
100 U/mg	1 unit produces 1 μmol acid-soluble polynucleotides from native DNA (from calf thymus)/min at pH 8.8, 37°C	Powder		Fluka 74669
70 U/mg	1 unit produces 1 μmol acid-soluble polynucleotides from native DNA (from calf thymus)/min at pH 8.8, 37°C	Lyophilized		Fluka 74674
≥6000 U/mg protein	1 unit produces 1 A_{260} unit acid-soluble material/30 min at pH 8.8, 37°C with DNA as substrate	Lyophilized		Pharmacia 27-0584-01
100-200 μM U/mg protein	1 unit produces 1.0 μmol acid-soluble polynucleotides from native DNA/min at pH 8.8, 37°C based on E_{m260} = 10,000 for mixed nucleotides	60% protein; balance primarily Na citrate		Sigma N3755

3.1.31.1 Micrococcal Nuclease continued

SPECIFIC ACTIVITY	UNITS DEFINITION	PREPARATION FORM	ADDITIONAL ACTIVITIES	SUPPLIER CATALOG NO.
Staphylococcus aureus, Foggi strain				
10-200 μM U/mg protein	1 unit produces 1.0 μmol acid-soluble polynucleotides from native DNA/min at pH 8.8, 37°C based on E_{m260} = 10,000 for mixed nucleotides			Sigma N5386
≥6000 U/mg protein	1 unit changes A_{260} by 1.0 at pH 8.0, 37°C	Homogeneous chromatographically and electrophoretically by SDS-PAGE; lyophilized		Worthington LS04797 LS04798 LS04796

3.2.1.1 α-Amylase

REACTION CATALYZED
Endohydrolysis of 1,4-α-D-glucosidic linkages in polysaccharides containing three or more 1,4-α-linked D-glucose units

SYSTEMATIC NAME
1,4-α-D-Glucan glucanohydrolase

SYNONYMS
Glycogenase, diastase, fungal α-amylase, bacterial α-amylase

REACTANTS
1,4-α-D-Glucosidic linkages, starch, glycogen, polysaccharides, oligosaccharides, maltose, maltotriose, dextrins

APPLICATIONS
- General uses include starch and sugar syrups, brewing and distilling, textiles, baking, animal feeds, wine and fruit juices, industrial cleaning, paper and detergents
- Aids in starch dextrinization and liquefaction to high maltose and high conversion syrups by hydrolyzing amylose and amylopectin chains in a random manner
- Breaks down starch in cane juice
- Assists adjunct liquefaction in brewing and thins starch in distilling mashes
- Removes starch sizing from fabrics without damaging textile fibers, in preparation for subsequent dyeing and finishing
- Improves baking quality and fermentation, crumb softness, frozen doughs, and is a bromate replacement in the treatment of flour
- Diagnostic indicator of acute pancreatitis
- Determination of fats in foods
- Hydrolysis/condensation of glycosidic bonds

α-Amylase continued 3.2.1.1

NOTES
- Classified as a hydrolase: a glycosidase hydrolyzing o-glycosyl compounds
- Found in nearly all plants, animals and microorganisms, but vary markedly even from tissue to tissue within a species
- Acts randomly on internal bonds of starch, glycogen and related polysaccharides and oligosaccharides
- Reducing groups are liberated in the α-configuration
- Elevated serum levels reportedly occur in mumps, renal disease and abdominal disorders like cholecystitis
- Some enzyme varieties are quite heat stable, performing at temperatures up to 90°C
- Activated and stabilized by Ca^{2+}
- Requires Cl^- for activity
- Inhibited by urea and other amides

SPECIFIC ACTIVITY	UNITS DEFINITION	PREPARATION FORM	ADDITIONAL ACTIVITIES	SUPPLIER CATALOG No.
Aspergillus (optimum pH = 4-5)				
400 FAU/g				Danisco GRINDAMYL™ 9201
400 FAU/g				Danisco GRINDAMYL™ 9401 TAB
1000 FAU/g				Danisco GRINDAMYL™ A 1000
750 FAU/g				Danisco GRINDAMYL™ BR 58
1000 FAU/g				Danisco GRINDAMYL™ FD 11
200 FAU/g				Danisco GRINDAMYL™ S 100
Aspergillus niger (optimum pH = 3.0-6.0)				
200 Px/g		Powder containing gypsum		ABM-RP DP294A
		Food grade		EnzymeDev Multifresh[R]
		Food grade		EnzymeDev pHIozyme™

3.2.1.1 α-Amylase continued

SPECIFIC ACTIVITY	UNITS DEFINITION	PREPARATION FORM	ADDITIONAL ACTIVITIES	SUPPLIER CATALOG No.
Aspergillus niger continued				
300 U/mg	1 unit liberates 1 µmol maltose/min at pH 6.0, 25°C	Powder		Fluka 10060
Aspergillus oryzae (optimum pH = 4.5-6, T = 55-50°C; stable pH 5-7 [50°C, 4 hr])				
100 Px/mL	1 unit is measured by its dextrinization effect on soluble starch solution (modified SKB); 1 Px = 250 SKB	Solution stabilized with sorbitol, pH 5.0-6.0	Glucoamylase, proteinase	ABM-RP Amylozyme 100L
1.8 Px/g	1 unit is measured by its dextrinization effect on soluble starch solution (modified SKB); 1 Px = 250 SKB	Powder containing starch	Glucoamylase, proteinase	ABM-RP Amylozyme B200
20 Px/g	1 unit is measured by its dextrinization effect on soluble starch solution (modified SKB); 1 Px = 250 SKB	Powder containing starch	Glucoamylase, proteinase	ABM-RP Amylozyme B2000
20 Px/g	1 unit is measured by its dextrinization effect on soluble starch solution (modified SKB); 1 Px = 250 SKB	Powder containing gypsum		ABM-RP Amylozyme B2000G
1.8 Px/g	1 unit is measured by its dextrinization effect on soluble starch solution (modified SKB); 1 Px = 250 SKB	Powder containing gypsum		ABM-RP Amylozyme B200G
2.6 Px/g	1 unit is measured by its dextrinization effect on soluble starch solution (modified SKB); 1 Px = 250 SKB	Powder containing gypsum		ABM-RP Amylozyme B250G
3.6 Px/g	1 unit is measured by its dextrinization effect on soluble starch solution (modified SKB); 1 Px = 250 SKB	Powder	Glucoamylase, proteinase	ABM-RP Amylozyme B350
3.6 Px/g	1 unit is measured by its dextrinization effect on soluble starch solution (modified SKB); 1 Px = 250 SKB	Powder containing gypsum		ABM-RP Amylozyme B350G
40 Px/g	1 unit is measured by its dextrinization effect on soluble starch solution (modified SKB); 1 Px = 250 SKB	Powder containing gypsum		ABM-RP Amylozyme B4000G
40 Px/g	1 unit is measured by its dextrinization effect on soluble starch solution (modified SKB); 1 Px = 250 SKB	Powder	Glucoamylase, proteinase	ABM-RP Amylozyme C10
400 Px/g	1 unit is measured by its dextrinization effect on soluble starch solution (modified SKB); 1 Px = 250 SKB	Powder	Glucoamylase, proteinase	ABM-RP Amylozyme C100
200 Px/g	1 unit is measured by its dextrinization effect on soluble starch solution (modified SKB); 1 Px = 250 SKB	Powder containing dextrose	Glucoamylase, proteinase	ABM-RP Amylozyme C50

α-Amylase continued　　3.2.1.1

SPECIFIC ACTIVITY	UNITS DEFINITION	PREPARATION FORM	ADDITIONAL ACTIVITIES	SUPPLIER CATALOG No.
Aspergillus oryzae continued				
		Food grade		EnzymeDev Co-Adjuzyme^R
25 U/mg	1 unit liberates 1 μmol maltose/min at pH 6.9, 25°C	Powder		Fluka 10065
>1300 U/g amylase activity	1 unit liberates 1 μmol maltose/min at pH 6.9, 25°C with starch as substrate	Powder	Small secondary activities of GAM and pectinase	Fluka 33470
40 U/mg	1 unit liberates 1 μmol maltose/min at pH 6.0, 25°C with starch as substrate	Powder		Fluka 86247
1.5 U/mg	1 unit liberates 1 μmol maltose/min at pH 6.0, 25°C with starch as substrate	Powder		Fluka 86250
50-100 U/mg protein	1 unit reduces by half the soluble starch amount in incubation/4 min			ICN 150375
180 FAU/g	1 FAU breaks down 5.26 g starch/hr at pH 4.7, 37°C	Food grade; powder standardized with wheat flour		Novo Nordisk Fungamyl^R 180 S
800 FAU/g	1 FAU breaks down 5.26 g starch/hr at pH 4.7, 37°C	Food grade; solution		Novo Nordisk Fungamyl^R 800 L
35,000 FAU/g	1 FAU breaks down 5.26 g starch/hr at pH 4.7, 37°C	Food grade; granulate at 300 micron particle size		Novo Nordisk Fungamyl^R MG 35.00
45 AZ		Food grade; powder		Rohm Veron^R AV
96 AZ		Food grade; powder		Rohm Veron^R FD super
48 AZ		Food grade; powder		Rohm Veron^R GX
44 AZ		Food grade; powder		Rohm Veron^R HE
1353 AZ		Food grade; powder		Rohm Veron^R M3
48 AZ		Food grade; powder		Rohm Veron^R ST
96 AZ		Food grade; powder		Rohm Veron^R SX
100-200 U/mg protein	1 unit liberates 1.0 mg maltose from starch/3 min at pH 6.9, 20°C	Crude		Sigma A0273

3.2.1.1 α-Amylase continued

SPECIFIC ACTIVITY	UNITS DEFINITION	PREPARATION FORM	ADDITIONAL ACTIVITIES	SUPPLIER CATALOG NO.
Aspergillus species (optimum pH = 3.5-6.8)				
		Food grade		EnzymeDev Enzeco[R] Fungal Alpha Amylase
Aspergillus/Bacillus (optimum pH = 4.5-6.5)				
170 BAU/g				Danisco GRINDAMYL™ B TAB
200 FAU/g				Danisco GRINDAMYL™ MAX-LIFE 25
Bacillus amyloliquefaciens (optimum pH = 6.3 [5-8], T = 75°C, MW = 97,000 Da; Inactivated at 85°C, pH 4.0, 15 min)				
9 X/g		Powder containing starch		ABM-RP Bacterase CF9
9.0 X/g		Powder containing starch		ABM-RP Bacterase CF9
15 X/g 30 X/g 60 X/g 85 X/g	1 unit is measured by its dextrinization effect on soluble starch solution (modified SKB); 1X = 66 SKB at pH 4.7, 1X = 100 SKB at pH 6.0	Solution stabilized with salt, pH 5.5-7.5		ABM-RP Nervanase 180 Nervanase 360 Nervanase 720 Nervanase 1080
3 X/g 9 X/g 50 X/g 100 X/g 200 X/g	1 unit is measured by its dextrinization effect on soluble starch solution (modified SKB); 1X = 66 SKB at pH 4.7, 1X = 100 SKB at pH 6.0	Powder containing maltodextrin		ABM-RP Nervanase 3X Nervanase 10X Nervanase 50X Nervanase 100X Nervanase 200X
1800 U/mg protein	1 unit converts 1 μmol soluble starch to product/min at pH 5.5, 25°C	Lyophilized containing CaOAc and NaOAc as stabilizers		Boehringer 161764
120 KNU/g	1 KNU breaks down 5.26 g starch/hr at pH 5.6, 37°C	Food grade; solution active 70-90°C		Novo Nordisk BAN 120 L
240 KNU/g	1 KNU breaks down 5.26 g starch/hr at pH 5.6, 37°C	Food grade; solution active 70-90°C		Novo Nordisk BAN 240 L
480 KNU/g	1 KNU breaks down 5.26 g starch/hr at pH 5.6, 37°C	Food grade; solution active 70-90°C		Novo Nordisk BAN 480 L
800 KNU/g	1 KNU breaks down 5.26 g starch/hr at pH 5.6, 37°C	Food grade; granulate at 300 micron particle size; active 70-90°C		Novo Nordisk BAN 800 MG
1000 U/mg	1 unit liberates 1 μmol maltose/min at pH 6.0, 25°C	Lyophilized powder	90% β-amylase <0.01% protease	Fluka 10068

α-Amylase continued — 3.2.1.1

SPECIFIC ACTIVITY	UNITS DEFINITION	PREPARATION FORM	ADDITIONAL ACTIVITIES	SUPPLIER CATALOG No.
Bacillus licheniformis (optimum pH = 7.5 [5.0-9.5], T = 50-80°C; stable to 90°C; inactivated pH < 4.5, T > 100°C)				
2 U/mg	1 unit liberates 1 μmol maltose/min at pH 6.0, 25°C	Powder		Fluka 10067
900 BGNP U	Activity is determined against an internal standard using the synthetic starch substrate Blocked *p*-nitrophenylmaltoheptoaside (BGNP)	Solution		Genencor Desize DT
≥20,000 LU/mL	1 unit is the measure of the digestion time required to produce a color change with an iodine solution indicating a definite stage of dextrinization of starch substrate	Food grade; solution, pH 6.0-7.5		Genencor Spezyme^R AA20
500-1000 U/mg protein	1 unit liberates 1.0 mg maltose from starch/3 min at pH 6.9, 20°C	Solution containing 15% NaCl		ICN 190151
120 KNU/g	1 KNU breaks down 5.26 g starch/hr at pH 5.6, 37°C	Food grade; solution		Novo Nordisk Termamyl^R 120 L
500-1000 U/mg protein	1 unit liberates 1.0 mg maltose from starch/3 min at pH 6.9, 20°C	Aqueous containing 15% NaCl and 25% sucrose		Sigma A3403
500-1500 U/mg protein	1 unit liberates 1.0 mg maltose from starch/3 min at pH 6.9, 20°C	95% purity by SDS-PAGE; lyophilized containing 70% protein; balance primarily KPO$_4$		Sigma A4551
Bacillus species (MW = 50,000-55,000 Da [SDS-PAGE])				
2500 U/g dry agarose, 1 mL gel yields 80 U	1 unit liberates 1.0 mg maltose from starch/3 min at pH 6.9, 20°C	Insoluble enzyme attached to beaded agarose; suspension in 2.0 M (NH$_4$)$_2$SO$_4$, pH 7.0		Sigma A0909
175-300 U/g solid	1 unit liberates 1.0 μmol maltose from starch/min at pH 6.9, 30°C	Insoluble enzyme attached to polyacrylamide containing 30% borate buffer salts		Sigma A5386
1500-3000 U/mg protein	1 unit liberates 1.0 mg maltose from starch/3 min at pH 6.9, 20°C	4X Crystallized; lyophilized		Sigma A6380
400 U/mg protein	1 unit liberates 1.0 mg maltose from starch/3 min at pH 6.9, 20°C	Contains starch as an extender		Sigma A6814
Bacillus stearothermophilus (optimum pH = 5.5 [5.0-7.0]; thermostable to 110°C)				
16X/g 40X/g		Food grade; solution stabilized with salt, pH 5.0-7.0		ABM-RP Nervanase BT Nervanase BT2
		Feed grade		EnzymeDev Enzeco^R Amylase TS

3.2.1.1 α-Amylase continued

SPECIFIC ACTIVITY	UNITS DEFINITION	PREPARATION FORM	ADDITIONAL ACTIVITIES	SUPPLIER CATALOG No.
Bacillus stearothermophilus continued				
		Food grade		EnzymeDev Enzeco^R Bacterial Amylase-TS
		Food grade		EnzymeDev G-Zyme^R G995 Thermostable Alpha-Amylase
Bacillus subtilis (optimum pH = 6, T = 30-70°C; stable pH 5-10 [70°C, 1 hr], T < 80°C [starch substrate]; inactivated by boiling [5 min] or heating [85°C, pH 4.0, 15 min])				
≥360,000 U/g	1 unit decreases Iodine color (A_{660}) by 1%/min at pH 6.0, 40°C	Food grade; powder		Amano Bacterial Amylase
45,000 RAU/g				BASF Vevozyme
>1800 AU/mg solid	1 AU measures the starch-liquefying ability of amylase	Lyophilized containing 20% CaOAc and 10% NaOAc		Calbiochem 171568
		Food grade		EnzymeDev Cookerzyme^R
		Food grade		EnzymeDev Enzeco^R
		Food grade		EnzymeDev Fresh-N^R
		Food grade		EnzymeDev Megadex™ Amylase/ Transferase
380 U/mg	1 unit liberates 1 μmol maltose/min at pH 6.0, 25°C	Powder		Fluka 10069
50 U/mg	1 unit liberates 1 μmol maltose/min at pH 6.0, 25°C	Powder		Fluka 10070
230 BGNP/mL 920 BGNP/mL 2300 BGNP/mL	Activity is determined against an internal standard using the synthetic starch substrate Blocked *p*-nitrophenylmaltoheptoaside (BGNP)	Solution		Genencor Desize GC2X Desize GC8X Desize GC20X
165,000 bacterial amylase U/g	1 unit dextrinizes 1 mg starch/min at pH 6.6, 30°C			ICN 100447
		Premeasured units		ICN 104838

α-Amylase continued — 3.2.1.1

SPECIFIC ACTIVITY	UNITS DEFINITION	PREPARATION FORM	ADDITIONAL ACTIVITIES	SUPPLIER CATALOG No.
Bacillus subtilis continued				
50-100 U/mg solid	1 unit liberates 1.0 mg maltose from starch/3 min at pH 6.9, 20°C			ICN 190152
10,000 DUN/g solid	1 DUN reduces the blue value of starch-iodine complex by 1%/min at pH 6.0, 40°C	Purified; powder optimum T = 65-80°C		Nagase Speedase PN-4
120 KNU/g		Solution		Novo Nordisk Aquazym^R 120 L
240 KNU/g		Concentrated solution		Novo Nordisk Aquazym^R 240 L
800 KNU/g		Granulate at 300 micron particle size		Novo Nordisk Aquazym^R 800 MG
600 U/mg	1 unit forms 1 μmol reducing sugar (as glucose) from soluble starch/min at pH 6.0, 40°C	Lyophilized		Seikagaku 100200-1
Bacillus subtilis strain MN-385 (optimum pH = 6-8, T = 100°C, MW = 54,000 Da; stable pH 5-11 [50°C, 1 hr]; thermostable < 100°C)				
>14,000 DUN/g	1 DUN reduces the blue value of starch-iodine complex by 1%/min at pH 6.0, 40°C	Purified; solution		Nagase Speedase HK
>7000 DUN/g	1 DUN reduces the blue value of starch-iodine complex by 1%/min at pH 6.0, 40°C	Purified; solution		Nagase Speedase HS
				SpecialtyEnz
Fungal				
				SpecialtyEnz
Human pancreas (MW = 54,000 Da)				
380 RBB starch U/mg protein	1 RBB starch unit releases soluble products from insoluble RBB starch with A_{595} equal to that of 100 mM CuSO$_4$ solution/10 min at pH 7.0, 37°C	≥95% purity; lyophilized containing 50 mM NaOAc and 5 mM Ca^{2+}, pH 5.5	No detectable HBsAg, anti-HCV, anti-HBc, anti-HIV	ART 16-19-010000
380 RBB starch U/mg protein	1 RBB unit releases soluble products from insoluble RBB-starch with A_{595} equal to half that of 100 mM copper sulfate/10 min at pH 7.0, 37°C with amylopectin azure suspension as substrate	>95% purity; lyophilized from buffer containing 50 mM NaOAc and 5 mM CaCl$_2$, pH 5.5	No detectable HBsAg, anti-HIV	Calbiochem 171532

α-Amylase continued

SPECIFIC ACTIVITY	UNITS DEFINITION	PREPARATION FORM	ADDITIONAL ACTIVITIES	SUPPLIER CATALOG NO.
Human pancreas continued				
≥50 U/mg protein	1 unit liberates 1 μmol maltose from soluble starch/min at pH 6.9, 25°C	Lyophilized containing 90% protein		Calzyme 059A0050
900-950 U/mg protein	1 unit liberates 1 μmol reducing groups equivalent to maltose/min at pH 6.9, 25°C with soluble starch as substrate	3X Crystallized; lyophilized containing NaCl, CaCl$_2$, Tris buffer, pH 7.5		Elastin HA732
330-400 RBB starch U/mg protein; amylopectin azure as substrate	1 RBB starch unit releases soluble products from insoluble RBB starch having A$_{595}$ equal to 0.1 M copper sulfate solution/10 min at pH 7.0, 37°C	>95% purity by SDS-PAGE; lyophilized containing 50 mM NaOAc, 5 mM Ca^{2+}, pH 5.5		ICN 191339
Human saliva				
100 U/mg	1 unit liberates 1 μmol maltose/min at pH 6.0, 25°C	Powder		Fluka 10092
1000 U/mg protein	1 unit liberates 1.0 mg maltose from starch/3 min at pH 6.9, 20°C	Lyophilized containing 10% protein; balance primarily (NH$_4$)$_2$SO$_4$ and Na citrate		Sigma A0521
200-600 U/mg protein	1 unit liberates 1.0 mg maltose from starch/3 min at pH 6.9, 20°C	Crude; lyophilized containing 10% protein; balance primarily (NH$_4$)$_2$SO$_4$ and Na citrate		Sigma A1031
Human tissue				
>150 IU/mg protein	Sigma amylase defined substrate, 37°C	40% purity by SDS-PAGE; solution containing >0.2 mg/mL protein and Tris buffer, pH 7.0-8.0	No detectable HBsAg, HCV, HIV	Aalto 1005
Malt, barley (optimum pH = 3-8, T = 40-60°C; inactivated pH < 3 or > 8, T < 45°C or > 55°C)				
1000 U/g; 4000 U/g; 12,000 U/g	1 unit produces reducing sugar equivalent to 10 mg dextrose/30 min at pH 5.0, 40°C	Pharmaceutical grade; powder	α- and β-amylase, protease	Amano Biozyme M
1000 Lintner		Analytical grade		ICN 101538
	1 g digests 50 g starch in less than 30 min	1X N.F.; 60 mesh powder		ICN 101539
1-3 U/mg solid	1 unit liberates 1.0 mg maltose from starch/3 min at pH 6.9, 20°C		1-3 U/mg solid β-amylase	Sigma A2771
Porcine pancreas (optimum pH = 7.0-8.0, T = 40°C, MW = 50,000-54,000 Da; inactivated at pH < 3.0 or > 10.0, 65°C)				
25 USP U/mg		Powder		Am Labs
1000 U/mg	1 unit converts 1 μmol soluble starch to product/min at 25°C	Crystallized; suspension in 3.2 M (NH$_4$)$_2$SO$_4$ and BSA as stabilizer		Boehringer 102806 102814

α-Amylase continued

SPECIFIC ACTIVITY	UNITS DEFINITION	PREPARATION FORM	ADDITIONAL ACTIVITIES	SUPPLIER CATALOG NO.
Porcine pancreas *continued*				
≥100 U/mg protein	1 unit liberates 1 μmol maltose from soluble starch/min at pH 6.9, 25°C	Lyophilized containing 90% protein		Calzyme 146A0100
1500-1800 U/mg protein	1 unit liberates 1 μmol reducing groups equivalent to maltose/min at pH 6.9, 25°C with soluble starch as substrate	3X Crystallized; suspension in 0.5 saturated NaCl, 3 mM $CaCl_2$, 0.01% NaN_3 as preservative, pH 7.0	<0.01% trypsin	Elastin A663
1500-1800 U/mg protein; 1200-1450 U/mg solid	1 unit liberates 1 μmol reducing groups equivalent to maltose/min at pH 6.9, 25°C with soluble starch as substrate	3X Crystallized; dialyzed; lyophilized containing 15% sucrose as stabilizer, 0.5% salts	<0.01% trypsin	Elastin AL783
20 U/mg	1 unit liberates 1 μmol of maltose/min at pH 6.0, 25°C	Powder		Fluka 10094
1000 U/mg protein	1 unit liberates 1 μmol maltose/min at pH 6.0, 25°C	Suspension in 2.9 M NaCl containing 3 mM $CaCl_2$		Fluka 10095
700-1400 U/mg protein	1 unit liberates 1 μmol reducing group (as maltose)/min at pH 6.9, 25°C	Crystallized; suspension in 0.003 M $CaCl_2$ covered with toluene, stabilized with PMSF, pH 6.0		ICN 191239
500-1000 U/mg protein	1 unit liberates 1.0 mg maltose from starch/3 min at pH 6.9, 20°C	DFP-treated; crystallized; suspension in 3.2 M $(NH_4)_2SO_4$, pH 6.1		Sigma A2643
10-30 U/mg solid	1 unit liberates 1.0 mg maltose from starch/3 min at pH 6.9, 20°C		2-10 U/mg solid β-amylase	Sigma A3176
700-1400 U/mg protein	1 unit liberates 1.0 mg maltose from starch/3 min at pH 6.9, 20°C	PMSF-treated; 2X crystallized; suspension in 2.9 M NaCl containing 3 mM $CaCl_2$		Sigma A4268
700-1400 U/mg protein	1 unit liberates 1.0 mg maltose from starch/3 min at pH 6.9, 20°C	DFP-treated; 2X crystallized; suspension in 2.9 M NaCl containing 3 mM $CaCl_2$		Sigma A6255
≥700 U/mg protein	1 unit liberates 1 μmol reducing groups (calculated as maltose) from soluble starch/min at pH 6.9, 25°C	PMSF-treated; chromatographically prepared; suspension in 0.5 saturated NaCl and 0.003 M $CaCl_2$		Worthington LS01013 LS01015 LS01017
Porcine pancreas, PMSF treated				
200 U/mg protein	1 unit liberates 1 μmol maltose/min at pH 6.0, 25°C	2X Crystallized; suspension in 2.9 M NaCl containing 3 mM $CaCl_2$		Fluka 10090
Trichoderma longibrachiatum (T. reesei) (optimum pH = 4.5-5.5)				
		Food grade		EnzymeDev Asperzyme[R]

β-Amylase

REACTION CATALYZED
Exohydrolysis of 1,4-α-D-glucosidic linkages in polysaccharides to remove successive maltose units from non-reducing ends of the chains

SYSTEMATIC NAME
1,4-α-D-Glucan maltohydrolase

SYNONYMS
Saccharogen amylase, glycogenase, fungal amylase

REACTANTS
1,4-α-D-Glucosidic linkages, starch, glycogen, polysaccharides, oligosaccharides, β-maltose, 3,5-dinitrosalycilic acid

APPLICATIONS
- Saccharification of liquefied starch to produce high maltose syrups, very high maltose syrups and high conversion syrups
- Brewing and distilling industry: increases fermentability of brewing wort, especially when starchy adjuncts are added to the mash
- Molecular weight marker for gel chromatography
- Structural studies of starch and glycogen

NOTES
- Classified as a hydrolase: a glycosidase hydrolyzing o-glycosyl compounds
- A tetrameric enzyme with molecular symmetry and possessing 4 binding sites; tryptophan residues are involved at the active site
- Acts on starch, glycogen and related polysaccharides and oligosaccharides producing β-maltose by inversion
- Does not hydrolyze maltotriose or α-1,6-bonds in amylopectin; the shortest normal saccharide attacked is maltotetraose
- Unable to bypass branch links in branched polysaccharides like amylopectin and glycogen, hydrolysis is incomplete and the enzyme produces macromolecular limit dextrins
- Soybean enzyme is more stable than those from wheat and barley
- Inhibited by heavy metal ions, PMCB, iodoacetamide, ascorbate and urea; sulfhydryl sensitive; competitively inhibited by cyclohexamylose

SPECIFIC ACTIVITY	UNITS DEFINITION	PREPARATION FORM	ADDITIONAL ACTIVITIES	SUPPLIER CATALOG NO.
Aspergillus oryzae (optimum pH = 5.0, T = 55°C; stable pH 4.0-10, heat stable)				
≥20,000 U/g; ≥38,000 SKB/g	1 unit produces reducing sugar equivalent to 10 mg dextrose/30 min at pH 5.0, 40°C	Solution with pH 6.3, 54° Brix		Amano Biozyme-L
≥57,000 U/g; ≥100,000 SKB/g	1 unit produces reducing sugar equivalent to 10 mg dextrose/30 min at pH 5.0, 40°C	Powder		Amano Biozyme-S
Aspergillus species (optimum pH = 5.5, T = 60°C)				
		Powder		Amano Biozyme-F
40,000 U/g	SKB units	Powder		Biocatalysts A011P

SPECIFIC ACTIVITY	UNITS DEFINITION	PREPARATION FORM	ADDITIONAL ACTIVITIES	SUPPLIER CATALOG NO.
Aspergillus species continued				
40,000 U/g	SKB units	Powder		Biocatalysts Depol 243P D243P
5000 U/g	SKB units	Powder		Biocatalysts Depol 367P D267P
Hordeum distichon L (Barley) (optimum pH = 5.3; inactivated at 70°C [70 min] or 75°C [15 min])				
≥1425 L/mL		Solution stabilizes with sorbitol, pH 5.0-7.0		ABM-RP Beta-Amylase 1500L
20-80 U/mg protein	1 unit liberates 1 μmol maltose from soluble starch/min at pH 4.8, 37°C	Crystallized; suspension in 2.3 M $(NH_4)_2SO_4$ containing 10 mg/mL protein		Calzyme 234B0020
15 U/mg	1 unit liberates 1 μmol maltose/min at pH 4.8, 25°C	Crude; powder		Fluka 10100
≥1500 DP/mL	1 unit is the amount of enzyme contained in 0.1 mL of 5% solution producing sufficient reducing sugars to reduce 5 mL Fehling's solution/hr at 20°C	Food grade; solution	No detectable α-amylase	Genencor Spezyme^R BBA 1500
2000 Lintner U	100 Lintner units are 0.1 mL of a 5% infusion acting on starch substrate producing sufficient reducing sugars completely reducing 5 mL Fehling's solution; 2.6 Lintner U = 1 IU	Standardized with Na_2SO_4 to a diastatic power of 2000 Lintner units		ICN 160058
20-80 U/mg protein	1 unit liberates 1.0 mg maltose from starch/3 min at pH 4.8, 20°C	Crude		Sigma A7130
Potato, sweet (optimum pH = 4.0-5.0, MW = 206,000 Da)				
500-1000 U/mg protein	1 unit liberates 1 μmol maltose from soluble starch/min at pH 4.8, 37°C	Crystallized; suspension in 2.3 M $(NH_4)_2SO_4$ containing 10 mg/mL protein		Calzyme 234B0500
		Powder	>40% protein	Fluka 10112
750-1000 U/mg protein	1 unit liberates 1.0 mg maltose from starch/3 min at pH 4.8, 20°C	Crystallized suspension in 2.3 M $(NH_4)_2SO_4$		Sigma A7005
≥500 U/mg protein	1 unit liberates 1 μmol β-maltose/min at pH 4.8, 25°C	Crystallized; suspension in 2.4 M $(NH_4)_2SO_4$, pH 3.0		Worthington LS01202 LS01205
Soybean (optimum pH = 5.5, T = 60-65°C; stable pH 4-7 [55°C, 30 min], pH 4-8 [40°C, 2hr], 30-60°C [pH 5.5, 30 min])				
15,000 AUN/g solid	1 AUN liberates 100 μg reducing sugar (expressed as dextrose equivalent)/min at pH 5.5, 40°C	Purified; powder	No detectable α-amylase	Nagase β-Amylase #1500

3.2.1.3 Glucan 1,4-α-Glucosidase

REACTION CATALYZED
Hydrolysis of terminal 1,4-linked α-D-glucose residues successively from non-reducing ends of the chains with release of β-D-glucose

SYSTEMATIC NAME
1,4-α-D-Glucan glucohydrolase

SYNONYMS
Glucoamylase, amyloglucosidase, γ-amylase, lysosomal α-glucosidase, acid maltase, exo-1,4-α-glucosidase, glucozyme, AMG, GAM

REACTANTS
Terminal 1,4-linked α-D-glucose residues, polysaccharides, β-D-glucose, soluble starch, amylopectin, glycogen, α- or β-limit dextrin, amylose, maltooligosaccharides, panose

APPLICATIONS
- General uses include starch syrups, brewing and distilling, baking, animal feed, wine and fruit juices, industrial cleaning, pharmaceuticals and cosmetics
- Saccharification of liquefied starch to produce syrup and crystalline dextrose, isomerization to high fructose corn syrup (HFCS) and crystalline fructose, and fermentation to alcohol
- Brewing low calorie beers
- Whole grain hydrolysis for alcohol
- Increasing fruit juice yields
- Hydrolysis/condensation of glycosidic bonds
- Synthesis of hetero-oligosaccharides by the reverse reaction
- Determination of glycogen in whole yeast cells
- Determination of α-amylase activity

NOTES
- Classified as a hydrolase: a glycosidase hydrolyzing o-glycosyl compounds
- Acts on polysaccharides more rapidly than on oligosaccharides
- Most forms rapidly hydrolyze 1,6-α-D-glucosidic bonds when the next bond in sequence is 1,4-
- Some forms hydrolyze 1,6- and 1,3-α-D-glucosidic bonds in other polysaccharides
- β-D-Acetylglucosaminidase (included with EC 3.2.1.52) from mammalian intestine catalyzes similar reactions

SPECIFIC ACTIVITY	UNITS DEFINITION	PREPARATION FORM	ADDITIONAL ACTIVITIES	SUPPLIER CATALOG NO.
Aspergillus niger (optimum pH = 4.5, T = 50-55°C, MW = 97,000 Da; stable pH 2-7, T < 50°C; inactivation at 80°C/5 min, 75°C/40 min)				
200 U/mL	1 unit forms 1 mg dextrose from hydrolyzed starch/hr at pH 4.3, 60°C	Solution stabilized with K sorbate, pH 4.0-6.0	No detectable transglucosidase Contains aciduric α-holo-amylase, acid proteinase	ABM-RP Ambazyme LE200
300 U/mL	1 unit forms 1 mg dextrose from hydrolyzed starch/hr at pH 4.3, 60°C	Solution stabilized with K sorbate, pH 4.0-6.0	No detectable transglucosidase Contains aciduric α-holo-amylase, acid proteinase	ABM-RP Ambazyme LE300
90 U/mL	1 unit forms 1 mg dextrose from hydrolyzed starch/hr at pH 4.3, 60°C	Solution stabilized with K sorbate, pH 4.0-6.0	No detectable transglucosidase Contains aciduric α-holo-amylase, acid proteinase	ABM-RP Ambazyme LE90

SPECIFIC ACTIVITY	UNITS DEFINITION	PREPARATION FORM	ADDITIONAL ACTIVITIES	SUPPLIER CATALOG NO.
Aspergillus niger continued				
50 U/g		Powder containing starch		ABM-RP Ambazyme PC50
≥300 U/mL		Solution stabilized with K sorbate, pH 5.0-6.0		ABM-RP G-zyme^R G 990 G-zyme G 990 ZU
225 AG/mL; 3000 U/mL (Amano)		Solution	No detectable transglucosidase	Amano GNL-3000
14 U/mg	1 unit converts 1 μmol glycogen to product/min at 25°C	Suspension in 3.2 M $(NH_4)_2SO_4$ and BSA as stabilizer, pH 6		Boehringer 102857
6 U/mg solid	1 unit converts 1 μmol glycogen to product/min at 25°C	Special quality for starch determination; lyophilized containing Na citrate stabilizer and <1 mg/mg solid glucose	<0.001 U/mg solid β-glucanase (azobarleyglucan as substrate) <0.005% PGI <0.05% β-fructosidase, α- and β-Gal, 6PGDH	Boehringer 1202332 1202367
6 U/mg solid	1 unit converts 1 μmol glycogen to product/min at 25°C	Lyophilized containing <1 μg/mg solid glucose	<0.005% PGI	Boehringer 208469
		Food grade		EnzymeDev Adjuzyme^R
		Food grade		EnzymeDev Enzeco^R Glucoamylase AN
		Food grade		EnzymeDev Enzeco^R Glucoamylase Liquid
		Food grade		EnzymeDev G-Zyme^R G990 Glucoamylase
70 U/mg	1 unit liberates 1 μmol glucose/min at pH 4.8, 60°C	Powder	<0.1% protease, transglucosidase	Fluka 10113
120 U/mg	1 unit liberates 1 μmol glucose/min at pH 4.8, 60°C	Lyophilized		Fluka 10115
≥300 SGU/mL ≥300 SGU/mL ≥400 SGU/mL	1 unit liberates 1 g reducing sugars (calculated as glucose) from a soluble starch substrate/hr	Food grade; solution, pH 3.8-4.5	No detectable transglucosidase	Genencor Spezyme^R GA 300, 300 W, 400

3.2.1.3 Glucan 1,4-α-Glucosidase continued

SPECIFIC ACTIVITY	UNITS DEFINITION	PREPARATION FORM	ADDITIONAL ACTIVITIES	SUPPLIER CATALOG NO.
Aspergillus niger continued				
>20 U/mg protein; >10 U/mg solid	1 unit releases 1.0 μmol glucose/min at 37°C	Lyophilized	<0.0001% amylase	Genzyme 1161
4200 FLS/mg	1 FLS liberates 10 mg reducing sugar (expressed as dextrose equivalent)/min at pH 4.5, 40°C	Purified; solution		Nagase XL-4
300 AGU/g	1 AGU hydrolyzes 1 μmol maltose/min at pH 4.3, 25°C	Food grade; GRAS; solution	No detectable transglucosidase	Novo Nordisk AMG 300 L
300 AGU/g	1 AGU hydrolyzes 1 μmol maltose/min at pH 4.3, 25°C	Food grade; GRAS; granulate at 300 micron particle size	No detectable transglucosidase	Novo Nordisk AMG 300 MG
400 AGU/g	1 AGU hydrolyzes 1 μmol maltose/min at pH 4.3, 25°C	Food grade; GRAS; solution	No detectable transglucosidase	Novo Nordisk AMG 400 L
5000-8000 U/mL	1 unit liberates 1.0 mg glucose from starch/3 min at pH 4.5, 55°C	Solution containing 1 M glucose and 0.5% Na benzoate as preservative		Sigma A3042
40-80 U/mg protein	1 unit liberates 1.0 mg glucose from starch/3 min at pH 4.5, 55°C	Suspension in 3.2 M $(NH_4)_2SO_4$ and BSA, pH 6.0		Sigma A3514
30-60 U/mg protein	1 unit liberates 1.0 mg glucose from starch/3 min at pH 4.5, 55°C	Lyophilized containing <0.02% glucose		Sigma A7420
		Immobilized granulate in food grade salt buffer solution		UOP Aldomax GA100
Aspergillus species (optimum pH = 3.5-5.5, T = 50-75°C)				
300 U/g		Solution		Biocatalysts D339P
3000 AGU/g				Danisco AMYLASE 3L-HT
Rhizopus delemar (optimum pH = 5, T = 60°C; stable pH 4.5, T < 50°C [pH 5.5, 20 min])				
10,000 AUN/g solid	1 AUN liberates 100 μg reducing sugar (expressed as dextrose equivalent)/min at pH 4.5, 40°C	Powder		Nagase Glucozyme
20,000 AUN/g solid	1 AUN liberates 100 μg reducing sugar (expressed as dextrose equivalent)/min at pH 4.5, 40°C	Powder		Nagase Glucozyme 20,000
Rhizopus niveus (optimum pH = 4.5-5.0, T = 55-60°C; stable pH 4-7)				
≥12,000 U/g	1 unit produces reducing sugar equivalent to 10 mg dextrose/30 min at pH 4.5, 40°C	Powder	β-glucanase, protease, macerating activity	Amano Gluczyme 12
30 U/mg solid	1 unit produces 10 mg glucose from a buffered 1% starch solution/30 min at 40°C	95% homogeneous; lyophilized	No detectable α-amylase	ICN 320182

Glucan 1,4-α-Glucosidase continued   3.2.1.3

SPECIFIC ACTIVITY	UNITS DEFINITION	PREPARATION FORM	ADDITIONAL ACTIVITIES	SUPPLIER CATALOG NO.
Rhizopus niveus continued				
30 U/mg	1 unit forms 10 mg glucose from soluble starch/30 min at pH 4.5, 40°C	Lyophilized		Seikagaku 100580-1
Rhizopus oryzae (optimum pH = 3.5-5.0)				
		Food grade		EnzymeDev Enzeco[R] Glucoamylase
Rhizopus species (optimum pH = 4.5-5.0, T = 60°C, MW = 70,000 Da; K_M = 0.11 mM [maltose], 0.36 mM [maltotriose], 0.25 mM [maltotetraose], 0.16 mM [maltopentaose]; stable pH 4.0-8.5 [25°C, 20 hr], T < 45°C [pH 5.5, 10 min])				
≥30 U/mg solid	1 unit forms 10 mg glucose/30 min at pH 4.5, 40°C	Lyophilized		ICN 153488
5000 U/g solid	1 unit liberates 1.0 mg glucose from starch/3 min at pH 4.5, 55°C	Contains 35% protein; balance primarily diatomaceous earth, starch, sugar		Sigma A7255
≥30 U/mg solid	1 unit forms 10 mg glucose/30 min at pH 4.5, 40°C	Salt-free; lyophilized		Toyobo GLA-111
≥30 U/mg		Lyophilized		Wako 073-02851

Cellulase   3.2.1.4

REACTION CATALYZED
    Endohydrolysis of 1,4-β-*D*-glucosidic linkages in cellulose, lichenin and cereal β-*D*-glucans

SYSTEMATIC NAME
    1,4-(1,3;1,4)-β-*D*-Glucan 4-glucanohydrolase

SYNONYMS
    Endo-1,4-β-glucanase, β-glucanase

REACTANTS
    1,4-β-*D*-glucosidic linkages, cellulose, lichenin, β-*D*-glucan

APPLICATIONS
- Restores a smooth surface to cotton fibers by removing protruding microfibrils created by washing and wearing. Softens, improves dyeability and printing and eliminates fabric cling, all without weight loss, strength loss or lint generation
- Brightens colors, softens and removes particulate soil in laundering cotton and mixed fabrics
- Imparts a "stone-washed" look to denims
- Used with toweling, sheeting, home furnishings, shirting, high-speed finishing of garments
- Digestion of non-starch carbohydrates (e.g., lignins, pectins) cross-linked with each other and with proteins, starch and lipids
- In brewing, used with non-malt adjuncts like barley, wheat and sorghum

- Fruit and vegetable processing
- Separating protoplasts from plant tissue
- Improves starch/gluten/fiber separation in corn wet milling
- Improves starch/gluten separation in wheat starch milling
- Deep litter pig pen waste treatment
- Breaks down cellulosics for production of fermentable sugars
- Composting
- Silage production
- Vanilla extraction
- Alcohol production
- Saccharification of pretreated agricultural waste and waste material utilization
- Improves vegetable and fruit matter extractions (e.g., onion, carrot, orange, vanilla, brewing cereal grains)
- Hydrolysis/condensation of carboxylic ester bonds

NOTES

- Viscosity and functionality changes
- Bioconversions

- Classified as a hydrolase: a glycosidase hydrolyzing o-glycosyl compounds
- A family of enzymes acting together to hydrolyze cellulose
- Widely distributed in nature, with the highest sources being fungal and microbial organisms
- Will also hydrolyze 1,4-linkages in β-D-glucans also containing 1,3-linkages
- Activated by nonionic detergents like Triton X-100
- *Trichoderma reesei* cellulase is multienzymatic, contains at least 3 physically and chemically distinct components, and all components are indispensable in the conversion of cellulose to glucose

SPECIFIC ACTIVITY	UNITS DEFINITION	PREPARATION FORM	ADDITIONAL ACTIVITIES	SUPPLIER CATALOG NO.
Aspergillus niger (optimum pH = 4.5, T = 50-60°C; stable pH 3.0-5.0, T < 40°C [pH 4.5, 30 min])				
30,000 U/g; 60,000 U/g; 90,000 U/g	100 units produce reducing sugar equivalent to 1 mg dextrose/min at pH 4.5, 40°C	Pharmaceutical grade; powder	C_1-ase (filter paper degrading activity), β-glucosidase	Amano Cellulase AP 30000 Cellulase AP 60000 Cellulase AP 90000
50,000 CMC-ase U/mL	100 units produce reducing sugar equivalent to 1 mg dextrose/min at pH 4.5, 40°C	Food grade; solution		Amano Cellulase L
		Food grade		EnzymeDev Enzeco[R] Cellulase CRX
		Food grade		EnzymeDev Enzeco[R] Cellulase FG

SPECIFIC ACTIVITY	UNITS DEFINITION	PREPARATION FORM	ADDITIONAL ACTIVITIES	SUPPLIER CATALOG NO.
Aspergillus niger continued				
0.04 U/mg	1 unit hydrolyzes 1 μmol OBR-hydroxyethylcellulose/min at pH 4.8, 30°C	Powder		Fluka 22174
0.5 U/mg	1 unit hydrolyzes 1 μmol OBR-hydroxyethylcellulose/min at pH 4.8, 30°C	Powder		Fluka 22178
0.1 U/mg	1 unit hydrolyzes 1 μmol OBR-hydroxyethylcellulose/min at pH 4.8, 30°C	Powder	exo- and endo-1,4-β-glucanase-activities	Fluka 22180
20,000 cmc U/g	100 units produce reducing sugar equivalent to 1 mg glucose/min at pH 5.0, 40°C			ICN 101308
0.5-1.0 U/mg solid	1 unit liberates 1.0 μmol glucose from cellulose/hr at pH 5.0, 37°C			ICN 150583
1000 CUN/g solid	1 CUN liberates 100 μmol reducing sugar (expressed as dextrose equivalent)/min at pH 4.5, 40°C	Powder	200 U/g α-amylase 280 U/g polygalacturonase 570 U/g GAM 1000 U/g cellulase 1970 U/g β-1,3 glucanase 2000 U/g protease	Nagase Cellulase
0.3 U/mg solid	1 unit liberates 1.0 μmol glucose from cellulose/hr at pH 5.0, 37°C (2 hr incubation time)			Sigma C1184
	1 unit liberates 1.0 μmol glucose from cellulose/hr at pH 5.0, 37°C (2 hr incubation time)	γ-Irradiated		Sigma C1424
Aspergillus species (optimum pH = 4.0-5.0)				
1200 U/g cellulase		Food grade; solution	5 U/g β-glucosidase 800 U/g pectinase	Biocatalysts Depol 40L D040L
10,000 BGU/g				Danisco GLUCANASE GV 5L
Humicola insolens (optimum pH = 6.5, T = 50°C; stable 30-40°C [pH 9.3, 1 hr], 30-50°C [pH 7.0, 1 hr])				
0.02 U/mg	1 unit hydrolyzes 1 μmol OBR-hydroxyethylcellulose/min at pH 4.8, 30°C	Powder		Fluka 22175
				Novo Nordisk Celluzyme[R]

3.2.1.4 Cellulase continued

SPECIFIC ACTIVITY	UNITS DEFINITION	PREPARATION FORM	ADDITIONAL ACTIVITIES	SUPPLIER CATALOG No.
Penicillium funicullosum				
≥2000 U/g		Solution stabilized with Na benzoate, pH 3.7-4.2		ABM-RP Cellulase 2000L
≥4000 U/g		Solution stabilized with Na benzoate, pH 3.7-4.2		ABM-RP Cellulase 4000L
10,000 U/g CMCase/5000 U/g β-glucanase; 20,000 U/g CMCase/10,000 U/g β-glucanase; 30,000 U/g CMCase/15,000 U/g β-glucanase	1 cellulase unit produces 1 μmol reducing groups from Hercules 4M6F CMC substrate/min	Powder containing maltodextrin	Cellobiase, pentosanase	ABM-RP Cellulase CPD10 Cellulase CPD20 Cellulase CPD30
5-10 U/mg solid	1 unit liberates 1.0 μmol glucose from cellulose/hr at pH 5.0, 37°C (2 hr incubation time)	Powder		Sigma C0901
Trichoderma longibrachiatum (T. reesei) (optimum pH = 5; stable pH 4-7 [25°C, 16 hr], T < 50°C [pH 4.8, 30 min]; inactivated at 80°C [pH > 7.0, 10 min])				
≥20,000 U/g		Powder containing maltodextrin		ABM-RP Cellulase C
		Food grade		EnzymeDev Econase[R] Bake P
		Feed grade		EnzymeDev Econase[R] Sil Series
				EnzymeDev Ecostone[R] L Plus
1 U/mg	1 unit hydrolyzes 1 μmol OBR-hydroxyethylcellulose/min at pH 4.8, 30°C	Powder		Fluka 22173
0.1 U/mg	1 unit hydrolyzes 1 μmol OBR-hydroxyethylcellulose/min at pH 4.8, 30°C	Solution		Fluka 22176
0.3 U/mg	1 unit hydrolyzes 1 μmol OBR-hydroxyethylcellulose/min at pH 4.8, 30°C	Powder		Fluka 22177
100 GCU/mL 2500 IU/mL		Solution, pH 5.0		Genencor Primafast[R] 100 Primafast[R] 101

SPECIFIC ACTIVITY	UNITS DEFINITION	PREPARATION FORM	ADDITIONAL ACTIVITIES	SUPPLIER CATALOG No.
Trichoderma longibrachiatum (T. reesei) continued				
≥1800 IU/mL	Activity spectrophotometrically measures the release of soluble fragments dyed with Remazol Brilliant Blue against an internal standard	Solution, pH 5.0		Genencor Primafast[R] RFW
CMC 2500 IU/mL	1 unit liberates 1 µmol reducing sugars (as glucose equivalents)/min at pH 4.8, 50°C	Food grade; solution		Genencor Spezyme[R] CE
≥90 GCU/mL	Measured the amount of glucose released during incubation with a specified filter paper/hr at 50°C	Food grade; solution		Genencor Spezyme CP
≥140 FPU/mL	1 FPU produces 1 µmol reducing carbohydrate/min	Solution, pH 4.5		Iogen
8000 CMC U/g	1 unit liberates 1 mmol reducing sugars (as glucose)/min at pH 4.8, 50°C	Powder	Multi-component preparation	Karlan 2017
1500 NCU/g	1 NCU degrades CMC to reducing carbohydrates with a reduction power corresponding to 1 µmol glucose/min at pH 4.8, 40°C	Food grade; solution		Novo Nordisk Celluclast[R] 1.5 L
		Solution		Primalco Econase[R] Envo L
		Powder		Primalco Econase[R] Envo P
		Solution		Primalco Econase[R] Sil L
		Powder		Primalco Econase[R] Sil P
		On carrier		Primalco Ecopen
		Solution		Primalco Ecostone[R] L Plus
		Solution		Primalco Ecostone[R] L-20
		Powder		Primalco Ecostone[R] P
		Solution		Primalco Ecostone[R] S 15
		Solution		Primalco Ecostone[R] S 50

3.2.1.4 Cellulase continued

SPECIFIC ACTIVITY	UNITS DEFINITION	PREPARATION FORM	ADDITIONAL ACTIVITIES	SUPPLIER CATALOG NO.
Trichoderma longibrachiatum (T. reesei) continued				
		Solution		Primalco Ecostone[R] Softline
		Powder		Primalco Ecostone[R] SP-10
		Solution		Primalco Good Earth
1404 CU		Food grade; solution		Rohm Rohalase[R] 7069
1404 CU		Food grade; solution		Rohm Rohament[R] CL
1-5 U/mg solid	1 unit liberates 1.0 μmol glucose from cellulose/hr at pH 5.0, 37°C (2 hr incubation time)	Lyophilized		Sigma C8546
≥45 U/mg solid	1 unit liberates 0.01 mg glucose from microcrystalline cellulose/hr at pH 5.0, 37°C	Chromatographically purified; lyophilized	Exoglucanase, endoglucanase	Worthington LS02598 LS02601 LS02603 LS02600
≥25 U/mg solid	1 unit liberates 0.01 mg glucose from microcrystalline cellulose/hr at pH 5.0, 37°C	Partially purified; lyophilized		Worthington LS02610 LS02611 LS02609
Trichoderma species (optimum pH = 4.6, T = 60°C)				
1500 U/g		Solution		Biocatalysts C013L
3000 U/g solid		Powder		Biocatalysts C013P
Trichoderma viride (optimum pH = 4.0-5.0, T = 50°C, MW = 30,000-42,000-76,000 Da; stable pH 4.0-6.0, T < 40°C [pH 6.0, 1 hr])				
40,000 U/g; 60,000 U/g; 100,000 U/g		Pharmaceutical grade; powder	C_1-ase (filter paper degrading activity), β-glucanase, CMCase (a CMC-hydrolyzing activity)	Amano Cellulase TAP 40000, 60000, 100000
0.5 U/mg	1 unit converts 1 μmol CMC to product/min at 37°C	Lyophilized		Boehringer 238104
1 U/mg	1 unit hydrolyzes 1 μmol OBR-hydroxyethylcellulose/min at pH 4.8, 30°C	Powder		Fluka 22179

Cellulase continued — 3.2.1.4

SPECIFIC ACTIVITY	UNITS DEFINITION	PREPARATION FORM	ADDITIONAL ACTIVITIES	SUPPLIER CATALOG No.
Trichoderma viride continued				
1-2 U/mg solid	1 unit liberates 1.0 µmol glucose from cellulose/hr at pH 5.0, 37°C			ICN 150584
>16,000 U/g filter paper decomposing activity	1 unit measures the decomposing activity on filter paper at pH 4.0, 40°C	Powder	Amylase, hemicellulase, protease	ICN 152337
			Xylanase, hemicellulase	ICN 320961
≥16,000 U/g	Unit is based on filter paper decomposing activity (modified Sotoyama method)	Powder	3X the xylanase activity of cellulase R10 (#2020)	Karlan 2019
10,000 U/g	Unit is based on filter paper decomposing activity (modified Sotoyama method)	Powder	High level of hemicellulase	Karlan 2020
≥5,000 U/g	Unit is based on filter paper decomposing activity (modified Sotoyama method)	Powder		Karlan 2023
1600 FDUN/mg	500 FDUN/mg decomposes 1 sheet filter paper (1cm^2 TOYO #51) in 5 mL cellulase solution/hr at pH 4.0, 40°C with L-tube agitation	Solution		Nagase Cellulizer
0.3-1.0 U/mg solid	1 unit liberates 1.0 µmol glucose from cellulose/hr at pH 5.0, 37°C (2 hr incubation time)	Crude; lyophilized		Sigma C0898
3-10 U/mg solid	1 unit liberates 1.0 µmol glucose from cellulose/hr at pH 5.0, 37°C	Plant cell tested; crude; powder containing 50% protein; balance primarily lactose and glucose		Sigma C1794
3-10 U/mg solid	1 unit liberates 1.0 µmol glucose from cellulose/hr at pH 5.0, 37°C (2 hr incubation time)	Crude; powder containing 50% protein		Sigma C9422
940-1060 U/mg				Wako 039-15821 037-15822
Unspecified				
				SpecialtyEnz Cellulase

Endo-1,3(4)-β-Glucanase — 3.2.1.6

REACTION CATALYZED
Endohydrolysis of 1,3- or 1,4-linkages in β-D-glucans when the glucose residue whose reducing group is involved in the linkage to be hydrolyzed is itself substituted at C-3

SYSTEMATIC NAME
1,3-(1,3;1,4)-β-D-Glucan 3(4)-glucanohydrolase

SYNONYMS
Endo-1,3-β-glucanase, laminarinase, β-1,3-glucanase, funcelase, endoglucanase, β-glucanase

REACTANTS
1,3- or 1,4-Linkages in β-D-glucans, laminarin, lichenin, cereal D-glucans

APPLICATIONS
- General uses include starch syrups, animal feed and silage, wine and fruit juices, and paper
- In brewing, for quicker run-off, increased extract yield and brewing capacity, and improved wort
- Preparation of protoplasts from fibrous plants
- Dissolving cell walls of yeasts, filamentous fungi and mushrooms

NOTES
- Classified as a hydrolase: a glycosidase hydrolyzing o-glycosyl compounds
- Hydrolyzes cellulose, lichenan and cereal β-glucans to oligosaccharides (3-5 glucose units) and minor amounts of disaccharides

SPECIFIC ACTIVITY	UNITS DEFINITION	PREPARATION FORM	ADDITIONAL ACTIVITIES	SUPPLIER CATALOG NO.
Aspergillus niger (optimum pH = 4-5.5, T = 60°C)				
1 U/mg	1 unit releases 1 µmol reducing sugar equivalents (as glucose)/min at pH 5.0, 55°C using β-D-glucan as substrate	Powder		Fluka 49101
200 FBG/g	1 FBG unit degrades barley β-glucan to reducing carbohydrates with a reduction power corresponding to 1 µmol glucose/min at pH 5.0, 30°C	Food grade; solution		Novo Nordisk Finizym[R] 200 L
Bacillus subtilis (optimum pH = 7.0, T = 50-60°C; stable 30-50°C [pH 5.0, 45 min])				
7 U/mg protein	1 unit releases 1 µmol reducing sugar equivalents (as glucose)/min at pH 6.0, 55°C using β-D-glucan as substrate	Solution		Fluka 49104
1 U/mg	1 unit releases 1 µmol reducing sugar equivalents (as glucose)/min at pH 6.0, 55°C using β-D-glucan as substrate	Powder		Fluka 49106
200 BGU/g	1 BGU degrades barley β-glucan to reducing carbohydrates with a reduction power corresponding to 1 µmol glucose/min at pH 7.5, 30°C	Food grade; purified; solution	α-amylase	Novo Nordisk Cereflo[R] 200 L

SPECIFIC ACTIVITY	UNITS DEFINITION	PREPARATION FORM	ADDITIONAL ACTIVITIES	SUPPLIER CATALOG NO.
Bacillus subtilis and Trichoderma longibrachiatum (T. reesei) (optimum pH = 3.5-6.0)				
		Food grade		EnzymeDev Enzeco[R] Beta Glucanase Series
Geosmithia emersonii (optimum pH = 4.5, T = 80°C; stable pH 2.0-6.5)				
200 U/g	1 unit gives 1 mg maltose equivalent/min/g enzyme at pH 5.0, 50°C	Solution stabilized with Na benzoate, pH 3.3-4.2		ABM-RP Beta-Glucanase 150L
750 U/g	1 unit gives 1 mg maltose equivalent/min/g enzyme at pH 5.0, 50°C	Solution stabilized with Na benzoate, pH 3.3-4.2		ABM-RP Beta-Glucanase 750L
Geosmithia emersonii and Penicillium funiculosum (optimum pH = 4.5, T = 80°C; stable pH 2.0-6.5)				
1250 U/g		Solution stabilized with Na benzoate, pH 3.3-4.2		ABM-RP Beta-Glucanase 1000L
250 U/g		Solution stabilized with Na benzoate, pH 3.3-4.2		ABM-RP Beta-Glucanase 200L
250 U/g		Solution stabilized with Na benzoate, pH 3.3-4.2		ABM-RP Beta-Glucanase TCB
Mollusk				
25-125 U/g solid	1 unit liberates 1.0 mg reducing sugar (measured as glucose) from laminarin/min at pH 5.0, 37°C	Lyophilized containing 60% protein	Cellulase and α-amylase	Sigma L5144
Penicillium species				
5-10 U/mg protein	1 unit liberates 1.0 mg reducing sugar (measured as glucose) from laminarin/min at pH 5.0, 37°C	Lyophilized containing 70% protein; balance primarily acetate buffer salts	Cellulase and α-amylase	Sigma L9259
Trichoderma longibrachiatum (T. reesei) (optimum pH = 4.5-7.0, T = 65°C)				
		Feed grade		EnzymeDev Econase[R] Barley
		Feed grade		EnzymeDev Econase[R] BG, BGP
		Food grade		EnzymeDev Econase[R] Malt

3.2.1.6 Endo-1,3(4)-β-Glucanase continued

SPECIFIC ACTIVITY	UNITS DEFINITION	PREPARATION FORM	ADDITIONAL ACTIVITIES	SUPPLIER CATALOG NO.
Trichoderma longibrachiatum (T. reesei) continued				
850-1050 Azo-BBG U/mL	Activity is measured by the amount of Azo-barley-β-glucan hydrolyzed by endo-1,3:1,4-β-D-glucanase	Food grade; solution, pH 4.8-5.2	Protease, lipase	Genencor Laminex[R] BG
2250 BGL U/mL	1 unit liberates 1 μmol reducing sugars (as glucose equivalents)/min at pH 5.0, 30°C	Food grade; solution, pH 4.7-5.3	Amylase, protease, lipase	Genencor Multifect[R] B
		On carrier		Primalco Econase[R] Barley F
		Solution		Primalco Econase[R] Barley L
		Powder		Primalco Econase[R] Barley P
Trichoderma species (optimum pH = 4.8, T = 50-75°C)				
15,000 U/g		Solution		Biocatalysts G015L
2600 U/g		Solution		Biocatalysts G151L
100-400 U/g solid	1 unit liberates 1.0 mg reducing sugar (measured as glucose) from laminarin/min at pH 5.0, 37°C		Chitinase, cellulase and α-amylase	Sigma L5272
Trichoderma viride (optimum pH = 4.0-5.0, T = 50-60°C)				
≥15,000 U/g		Powder	High level of chitinase	Karlan Funcelase 3920

Endo-1,4-β-Xylanase　　3.2.1.8

REACTION CATALYZED
Endohydrolysis of 1,4-β-D-xylosidic linkages in xylans

SYSTEMATIC NAME
1,4-β-D-Xylan xylanohydrolase

SYNONYMS
Xylanase, pentosanase, endoxylanase

REACTANTS
1,4-β-D-Xylosidic linkages, xylans

APPLICATIONS
- In baked goods and brewing to treat whole grain, grain mashes and doughs to reduce viscosity for processing
- Improve bleachability of kraft pulp
- Upgrading wheat-based animal feeds by breaking down insoluble fibers to improve digestibility
- Production of wheat, rye and grain-based snack and bakery foods and ready-to-eat cereals
- Extraction of plant materials like tea and coffee

NOTES
- Classified as a hydrolase: a glycosidase hydrolyzing o-glycosyl compounds

SPECIFIC ACTIVITY	UNITS DEFINITION	PREPARATION FORM	ADDITIONAL ACTIVITIES	SUPPLIER CATALOG NO.
Aureobasidium pullulans				
100-300 U/mg protein	1 unit liberates 1.0 μmol reducing sugar (measured as xylose) from xylan/min at pH 4.5, 30°C	Lyophilized containing 30% protein; balance sorbitol and NaOAc buffer salts	<0.01% β-glucosidase, β-xylosidase <0.5% cellulase	Sigma X4001
Trichoderma longibrachiatum (T. reesei) (optimum pH = 4.5, T = 55°C; stable pH 3.0-7.0)				
1200 U/g pentosanase		Powder containing wheat flour	<0.1 Px/g α-amylase	ABM-RP Amylozyme PAF
75,000 U/g		Powder containing maltodextrin		ABM-RP Cellulase X
5000 U/g		Powder containing wheat flour		ABM-RP Cellulase XL5000P
		Food grade		EnzymeDev Econase^R HC 200 Econase^R HCP 4000
		Feed grade		EnzymeDev Econase^R Rye

3.2.1.8 Endo-1,4-β-Xylanase continued

SPECIFIC ACTIVITY	UNITS DEFINITION	PREPARATION FORM	ADDITIONAL ACTIVITIES	SUPPLIER CATALOG No.
Trichoderma longibrachiatum (T. reesei) continued				
		Feed grade		EnzymeDev Econase[R] Wheat
		Feed grade		EnzymeDev Econase[R] Wheat Plus
		Food grade		EnzymeDev Enzeco[R] Xylanase
		Food grade		EnzymeDev Enzeco[R] Xylanase
≥4000 GXU/mL	Activity is determined from the amount of dyed substrate released as measured spectrophotometrically/10 min at pH 4.5, 30°C using xylanase standard	Food grade; solution, pH 4.4-4.8	No detectable protease, lipase, amylase Contains cellulase	Genencor GC 140
≥445 XAU/mL	Activity is determined from the amount of dyed substrate released as measured at A_{590}/10 min at pH 4.5, 40°C using an endo-xylanase standard	Food grade; solution, pH 4.8-5.2	No detectable protease, lipase, amylase	Genencor Multifect XL Enzyme
≥200,000 BXU/mL	Determined on birch xylan substrate at pH 5.3, 50°C	Solution containing 0.35% Na benzoate as preservative		Primalco Econase[R] HC 200
		Powder		Primalco Econase[R] HCP 4000
		On carrier		Primalco Econase[R] Wheat F Plus
		Solution		Primalco Econase[R] Wheat L Plus
		Powder		Primalco Econase[R] Wheat P Plus
≥200,000 BXU/mL	Determined on birch xylan substrate at pH 5.3, 50°C	Solution containing 0.35% Na benzoate as preservative		Primalco Ecopulp[R] X-200

Endo-1,4-β-Xylanase continued

SPECIFIC ACTIVITY	UNITS DEFINITION	PREPARATION FORM	ADDITIONAL ACTIVITIES	SUPPLIER CATALOG NO.
Trichoderma species (optimum pH = 4-5.5, T = 30-60°C)				
1300 U/g pentosanase		Powder	Trace amylase	Biocatalysts Depol 222P D222P
11,000 U/g xylanase		Powder		Biocatalysts Depol 333P D333P
4000 U/g glucanase; 450 U/g cellulase; 400 U/g xylanase		Powder	450 U/g cellulase 400 U/g xylanase	Biocatalysts H334P
Trichoderma viride				
100-300 U/mg protein	1 unit liberates 1.0 μmol reducing sugar (measured as xylose) from xylan/min at pH 4.5, 30°C	Lyophilized containing 50% protein; balance sorbitol and NaOAc buffer salts	<0.002% b-xylosidase <0.01% β-glucosidase <0.2% cellulase	Sigma X3876

Oligo-1,6-Glucosidase 3.2.1.10

REACTION CATALYZED
   Hydrolysis of 1,6-α-D-glucosidic linkages in isomaltose and dextrins produced from starch and glycogen by α-amylase

SYSTEMATIC NAME
   Dextrin 6-α-D-glucanohydrolase

SYNONYMS
   Limit dextrinase, isomaltase, sucrase-isomaltase

REACTANTS
   1,6-α-D-Glucosidic linkages, isomaltose, dextrins, palatinose

NOTES
- Classified as a hydrolase: a glycosidase hydrolyzing o-glycosyl compounds
- Intestinal mucosa enzyme is a single polypeptide chain also exhibiting sucrose α-glucosidase (EC 3.2.1.48) activity

SPECIFIC ACTIVITY	UNITS DEFINITION	PREPARATION FORM	ADDITIONAL ACTIVITIES	SUPPLIER CATALOG NO.
Yeast, bakers				
3-5 U/mg protein	1 unit converts 1.0 μmol isomaltose to 2.0 μmol D-glucose/min at pH 6.8, 25°C	Lyophilized containing 10% protein; balance primarily Ficoll, 1% KPO$_4$, 0.5% EDTA as stabilizer	<0.5 U/mg protein maltase 15-35 U/mg protein α-glucosidase	Sigma I1256

3.2.1.11 Dextranase

REACTION CATALYZED
: Endohydrolysis of 1,6-α-D-glucosidic linkages in dextran

SYSTEMATIC NAME
: 1,6-α-D-Glucan 6-glucanohydrolase

REACTANTS
: 1,6-α-D-glucosidic linkages, dextran

APPLICATIONS
- Improve purification of sugar cane by eliminating filtration and crystallization problems

NOTES
- Classified as a hydrolase: a glycosidase hydrolyzing o-glycosyl compounds
- Activated by Co^{2+}, Mn^{2+} and Cu^{2+}
- Inhibited by Hg^{2+}, Ag^{+}, N-bromosuccinimide and I_2

SPECIFIC ACTIVITY	UNITS DEFINITION	PREPARATION FORM	ADDITIONAL ACTIVITIES	SUPPLIER CATALOG NO.
Chaetomium gracile				
1700 U/mg	1 unit produces 1 μmol glucose (as reducing sugars)/min at pH 5.1, 40°C	Suspension containing 80% $(NH_4)_2SO_4$	No detectable amylase, cellulase, hemicellulase, protease	ICN 321601
Leuconostoc mesenteroides (optimum pH = 5.5, T = 60°C; stable to 70-80°C with sucrose)				
≥30,000 U/mL	1 unit produces reducing sugar equivalent to 1 μmol Na thiosulfate/min at pH 4.5, 37°C	Solution	No detectable invertase, protease, cellulase, pectinase	Amano Dextranase L
Paecilomyces lilacinus				
60 U/mg	1 unit liberates 1 μmol isomaltose/min at pH 5.5, 50°C with dextran as substrate	Lyophilized		Fluka 31402
Penicillium species (optimum pH = 5.0-7.0, MW = 41,000 Da [GPC]; stable at pH 6.0 up to 40°C)				
250-500 U/mg protein		Salt-free; lyophilized containing 65% protein		ICN 190097
250-500 U/mg protein	1 unit liberates 1.0 μmol isomaltose (measured as maltose)/min at pH 6.0, 37°C using dextran as substrate	Salt-free; lyophilized containing 65% protein		Sigma D1508
100-200 U/mg protein	1 unit liberates 1.0 μmol isomaltose (measured as maltose)/min at pH 6.0, 37°C using dextran as substrate	Partially purified; lyophilized containing 25% protein		Sigma D4668
3-12 U/mg solid	1 unit liberates 1.0 μmol isomaltose (measured as maltose)/min at pH 6.0, 37°C using dextran as substrate	Lyophilized containing 20% protein		Sigma D4793
10-25 U/mg solid	1 unit liberates 1.0 μmol isomaltose (measured as maltose)/min at pH 6.0, 37°C using dextran as substrate	Crude; lyophilized		Sigma D5884

Dextranase continued 3.2.1.11

SPECIFIC ACTIVITY	UNITS DEFINITION	PREPARATION FORM	ADDITIONAL ACTIVITIES	SUPPLIER CATALOG NO.
Penicillium species *continued*				
400-800 U/mg protein	1 unit liberates 1.0 μmol isomaltose (measured as maltose)/min at pH 6.0, 37°C using dextran as substrate	Salt-free; lyophilized containing 35% protein		Sigma D8144
≥150 U/mg solid	1 unit releases 1 μmol isomaltose from dextran/min at pH 6.0, 37°C	Chromatographically purified; lyophilized		Worthington LS04315 LS04316
≥15 U/mg solid	1 unit releases 1 μmol isomaltose from dextran/min at pH 6.0, 37°C	Partially purified		Worthington LS04318 LS04317

Chitinase 3.2.1.14

REACTION CATALYZED

Random hydrolysis of *N*-acetyl-β-D-glucosaminide 1,4-β-linkages in chitin and chitodextrins

SYSTEMATIC NAME

Poly(1,4-(*N*-acetyl-β-D-glucosaminide)) glycanohydrolase

SYNONYMS

Chitodextrinase, 1,4-β-poly-*N*-acetylglucosaminidase, poly-β-glucosaminidase

REACTANTS

N-Acetyl-β-D-glucosaminide 1,4-β-linkages, chitin, chitodextrins

APPLICATIONS

- *In vitro* degradation of isolated pathogen cell walls
- Induction of the hydrolase defense reaction against pathogens
- Suppression of powdery mildew pathogen
- Degradation and hydrolysis of chitosan

NOTES

- Classified as a hydrolase: a glycosidase hydrolyzing *o*-glycosyl compounds
- Some enzymes also exhibit lysozyme (EC 3.2.1.17) activity

SPECIFIC ACTIVITY	UNITS DEFINITION	PREPARATION FORM	ADDITIONAL ACTIVITIES	SUPPLIER CATALOG NO.
Aeromonas hydrophila (optimum pH = 5.2, T = 50°C)				
600 U/mg	1 unit decreases A_{610} by 1%/min at pH 5.2, 37°C using colloidal chitin suspension	Lyophilized		Seikagaku 100289-1 100289-2

3.2.1.14 Chitinase continued

SPECIFIC ACTIVITY	UNITS DEFINITION	PREPARATION FORM	ADDITIONAL ACTIVITIES	SUPPLIER CATALOG NO.
Bacillus species				
40-70 U/mg		Lyophilized		Wako 038-14311
Bacillus species PI-7S (optimum pH = 7.8-8.6, 5.0 [soluble chitosan], 6.2 [glycol chitosan])				
40-60 mU/mg	1 unit liberates 1 μmol reducing sugar from N-Ac-chitosan/min at pH 6.8, 37°C	Lyophilized		Seikagaku 100292-1 100292-2
0.15-0.35 U/mg	1 unit liberates 1 μmol reducing sugar from chitosan/min at pH 6.0, 37°C	Lyophilized		Seikagaku 100295-1
Serratia marcescens				
10-30 U/g solid	1 unit liberates 1.0 mg N-Ac-D-glucosamine from chitin/hr at pH 6.0, 25°C in a 2 step reaction	Lyophilized containing 15% protein; balance primarily PO_4 buffer salts		Sigma C1650
400-1200 U/g solid	1 unit liberates 1.0 mg N-Ac-D-glucosamine from chitin/hr at pH 6.0, 25°C in a 2 step reaction	Lyophilized containing 30% protein; balance primarily PO_4 buffer salts		Sigma C7809
Streptomyces griseus				
0.1 U/mg	1 unit releases soluble chitooligosaccharides containing 1 μmol of N-Ac-glucosamine/min at pH 4.4, 37°C with colloidal chitin as substrate	Powder		Fluka 22725
500-2000 U/g solid	1 unit liberates 1.0 mg N-Ac-D-glucosamine from chitin/hr at pH 6.0, 25°C in a 2 step reaction	Crude; lyophilized containing 60% protein		Sigma C1525
200-600 U/g solid	1 unit liberates 1.0 mg N-Ac-D-glucosamine from chitin/hr at pH 6.0, 25°C in a 2 step reaction	Crude; lyophilized containing 20% protein		Sigma C6137

Polygalacturonase 3.2.1.15

REACTION CATALYZED
Random hydrolysis of 1,4-α-D-galactosiduronic linkages in pectate and other galacturonans

SYSTEMATIC NAME
Poly(1,4-α-D-galacturonide) glycanohydrolase

SYNONYMS
Pectin depolymerase, pectinase, hemicellulase

REACTANTS
1,4-α-D-Galactosiduronic linkages, pectin, galacturonans

APPLICATIONS
- Processing low-acid fruits with high methyl substitution on the pectin backbone
- Clarification
- Improved saccharification of plant solids
- Increased filtration rate
- Reducing viscosity
- Preventing gel formation
- Fruit juice and wine production
- Plant maceration and cell isolation, often with cellulase
- Processing apples, lemons, cranberries, oranges, cherries, grapes and tomatoes
- Increasing fruit juice yield

NOTES
- Classified as a hydrolase: a glycosidase hydrolyzing o-glycosyl compounds

SPECIFIC ACTIVITY	UNITS DEFINITION	PREPARATION FORM	ADDITIONAL ACTIVITIES	SUPPLIER CATALOG NO.
Aspergillus japonicus				
300-1500 U/mg protein	1 unit releases 1.0 μmol reducing sugar measured as D-galacturonic acid from polygalacturonic acid/min at pH 5.0, 30°C	Purified from pectolyase; suspension in 3.2 M $(NH_4)_2SO_4$ and 1 mM PMSF, pH 4.5	Pectin lyase; No detectable cellulase, hemicellulase, glycosaminidase	Sigma P3304
200-600 U/mg protein	1 unit releases 1.0 μmol reducing sugar (measured as D-galacturonic acid) from polygalacturonic acid/min at pH 5.0, 30°C	Sterile; solution containing 50% glycerol, 40 mM NaOAc buffer, 1 mM PMSF, pH 4.5	No detectable cellulase and hemicellulase; < 20 U/mg protein pectin lyase (pH 6.0)	Sigma P5079
Aspergillus niger (optimum pH = 4.0, T = 60°C; stable to 50°C [pH 3.5, 30 min])				
		Powder		Amano Pectinase G
		Powder		Amano Pectinase P
		Solution		Amano Pectinase PL

SPECIFIC ACTIVITY	UNITS DEFINITION	PREPARATION FORM	ADDITIONAL ACTIVITIES	SUPPLIER CATALOG No.
Aspergillus niger continued				
		Food grade		EnzymeDev Enzeco[R] Pectinase Concentrate
1 U/mL	1 unit produces 1 μmol proton/min at pH 7.0, 30°C with pectin from citrus peel as substrate	Solution		Fluka 76284
≥6600 APPV/mL	Pectin degradation is monitored by reduction of viscosity of a standardized apple pectin solution at pH 3.8, 22°C as measured using a capillary viscometer	Food grade; solution, pH 4.7-5.0		Genencor Multifect[R] PL Enzyme
3-9 U/mg protein	1 unit liberates 1 μmol galacturonic acid from polygalacturonic acid/min at pH 4.0, 25°C	40% glycerol solution		ICN 156058
26,000 PG/mL	Measured by reduction in viscosity of pectic acid solution at pH 3.5, 20°C	Food grade; solution, pH 4.5 optimum pH 3.5, T = 35°C	Pectolytic and hemicellulolytic; disintegrates plant cell walls	Novo Pectinex™ Ultra SP-L
5000 FDU$_{55°C}$/g 1000 FDU$_{20°C}$/g 1000 FDU$_{20°C}$/mL	Measured by depectinization of apple juice	Microgranulates or solution, pH 4.5 optimum pH = 3.5-4.1, T = 50°C	No detectable polyphenoloxydase	Novo Ultrazym[R] 100G, G, L
36,600 PGU		Food grade; solution		Rohm Rohament[R] MAX
9600 PGU		Food grade; solution		Rohm Rohament[R] PL
16,700 PGU		Food grade; solution		Rohm Rohapect[R] D5L
25,060 PGU		Food grade; solution		Rohm Rohapect[R] MA plus
5-20 U/mg protein	1 unit liberates 1.0 μmol galacturonic acid from polygalacturonic acid/min at pH 4.0, 25°C	Plant cell tested; solution containing KCl and sorbitol		Sigma P2829
500-2000 U/mg protein	1 unit releases 1.0 μmol reducing sugar measured as D-galacturonic acid from polygalacturonic acid/min at pH 5.0, 30°C	Purified from pectinase; solution containing 50% glycerol and 40 mM NaOAc buffer, pH 4.5	No detectable cellulase, hemicellulase, glycosaminidase < 2 U/mg protein pectin lyase	Sigma P3429
5-20 U/mg protein	1 unit liberates 1.0 μmol galacturonic acid from polygalacturonic acid/min at pH 4.0, 25°C	Solution containing KCl and sorbitol		Sigma P9179

Polygalacturonase continued

3.2.1.15

SPECIFIC ACTIVITY	UNITS DEFINITION	PREPARATION FORM	ADDITIONAL ACTIVITIES	SUPPLIER CATALOG No.
Aspergillus species (optimum pH = 3.5-4.8, T = 60°C)				
1150 PGU/g pectinase		Solution		Biocatalysts M263L
2400 PGU/g pectinase		Solution		Biocatalysts P062L
1000 PGU/g pectinase		Solution		Biocatalysts P162L
700 PGU/g pectinase		Solution		Biocatalysts P444L
				Danisco PEKTOLASE™ 3 PA
				Danisco PEKTOLASE™ CA
				Danisco PEKTOLASE™ LB
Fungal species				
	1 g reduces viscosity of 1500 g pectin by 50%/15 min at pH 3.5, 40°C	Technical powder; partially purified; concentrate		ICN 102588
Rhizopus species				
0.005 U/mg	1 unit produces 1 μmol proton/min at pH 7.0, 30°C with pectin from citrus peel as substrate	Powder		Fluka 76285
0.02 U/mg	1 unit produces 1 μmol proton/min at pH 7.0, 30°C with pectin from citrus peel as substrate	Powder		Fluka 76287
350-400 U/g solid	1 unit liberates 1 μmol galacturonic acid from polygalacturonic acid/min at pH 4.0, 25°C	Crude; powder		ICN 151803
400-800 U/g solid	1 unit liberates 1.0 μmol galacturonic acid from polygalacturonic acid/min at pH 4.0, 25°C	Crude		Sigma P2401
400-800 U/g solid	1 unit liberates 1.0 μmol galacturonic acid from polygalacturonic acid/min at pH 4.0, 25°C	Crude; powder		Sigma P4300
Unspecified				
				SpecialtyEnz Pectinase

Lysozyme

REACTION CATALYZED
Hydrolysis of 1,4-β-linkages between N-acetyl-muramic acid and N-acetyl-D-glucosamine residues in peptidoglycans, and between N-acetyl-D-glucosamine residues in chitodextrins

SYSTEMATIC NAME
Peptidoglycan N-acetylmuramoylhydrolase

SYNONYMS
Muramidase, N-acetylmuramidase

REACTANTS
Peptidoglycan, chitodextrins, mucopolysaccharides, mucopolypeptides

APPLICATIONS
- Hydrolysis of bacterial cell walls, mucopolysaccharides, mucopolypeptides and chitin
- Prevents late blowing in semi-hard cheeses due to *Cl. tyrobutyricum*
- Diagnostic tool for acute and chronic myelocytic leukemia, and lymphocytic leukemia
- Cell wall lysis in nucleic acid preparation
- Protein substrate for protease specificity determination
- Antibacterial agent
- Protein purification from inclusion bodies
- Break down of cell wall and outer membranes in plasmid preparation, with EDTA

NOTES
- Classified as a hydrolase: a glycosidase hydrolyzing o-glycosyl compounds
- Present in plasma in specific granules of polymorphonuclear leukocytes; takes part in antibacterial activity. Urinary levels increase significantly in renal disorders
- Enhanced by EDTA
- Inhibited by SDS, alcohols, N-acetyl-D-glucosamine, oxidizing agents and fatty acid; imidazole and indole derivatives form inhibitory charged transfer complexes

SPECIFIC ACTIVITY	UNITS DEFINITION	PREPARATION FORM	ADDITIONAL ACTIVITIES	SUPPLIER CATALOG NO.
Bovine, recombinant expressed in *Pichia pastoris*				
20,000 U/mg solid	1 unit decreases E_{450} by 0.001/min at pH 6.24, 25°C with *Micrococcus lysodeikticus* as substrate	Salt-free; albumin-free; lyophilized in the chloride form		Biozyme LZ3

Lysozyme continued 3.2.1.17

SPECIFIC ACTIVITY	UNITS DEFINITION	PREPARATION FORM	ADDITIONAL ACTIVITIES	SUPPLIER CATALOG No.
Bovine, recombinant *continued*				
25,000-60,000 U/mg solid	1 unit produces a ΔA_{450} of 0.001/min at pH 5.0, 25°C; suspension of *Micrococcus lysodeikticus* as substrate in 1.0 mL reaction mixture, 1 cm light path			Sigma L9772
Chicken egg white (optimum pH = 9.2, pI = 10.7, T = 25-35°C, MW = 14,400 Da; stable pH 4-5 [ambient T, days])				
>25,000 U/mg	1 unit decreases A_{450} by 0.001/min	Salt-free; lyophilized	No detectable ovalbumin, salmonella	ACS Heidelb F01060
≥20,000 U/mg	1 unit decreases A_{450} by 0.001/min at 25°C	3X Crystallized; lyophilized		AMRESCO 0663
50,000 Shugar U/mg	1 unit decreases absorbance by 0.001/min at 25°C with *Micrococcus luteus* as substrate	Crystallized	<1 U/mg proteases (37°C, azocoll as substrate)	Boehringer 107255 1243004 837059 1585657
50,000 U/mg	1 unit decreases A_{450} by 0.001/min at pH 6.24, 25°C using *Micrococcus lysodeikticus* as substrate	Salt-free; lyophilized containing 95% protein		Calzyme 097A50000
		Food grade		EnzymeDev Enzeco[R] Lysozyme
100,000 U/mg	1 unit decreases A_{450} by 0.001/min at pH 7.0, 25°C with *Micrococcus luteus* as substrate	Salt-free; dialyzed; lyophilized		Fluka 62970
70,000 U/mg	1 unit decreases A_{450} by 0.001/min at pH 7.0, 25°C with *Micrococcus luteus* as substrate	Crystallized		Fluka 62971
1000 MCG/mg	Activity is based on the rate of lysis of a suspension of *Micrococcus lysodeikticus* measured at A_{450} at pH 6.6, 37°C	Powder		Genencor Multifect[R] Lysozyme
20,000-25,000 U/mg protein	1 unit decreases A_{450} by 0.001/min at pH 6.24, 25°C using *Micrococcus lysodeikticus* as substrate	Salt-free; albumin-free; 3X crystallized; lyophilized		ICN 100831
>9000 U/mg protein	1 unit decreases A_{450} by 0.001/min at pH 6.24, 25°C using *Micrococcus lysodeikticus* as substrate	Salt-free; 2X crystallized; lyophilized		ICN 100834
≥22,000 Shugar U/mg	1 unit decreases A_{450} by 0.001/min at pH 6.2, 25°C	Dialyzed; powder		Can Inova
≥22,000 Shugar U/mg	1 unit decreases A_{450} by 0.001/min at pH 6.2, 25°C	Dialyzed; granulate		Can Inova

3.2.1.17 Lysozyme continued

SPECIFIC ACTIVITY	UNITS DEFINITION	PREPARATION FORM	ADDITIONAL ACTIVITIES	SUPPLIER CATALOG No.
Chicken egg white continued				
≥22,000 Shugar U/mg	1 unit decreases A_{450} by 0.001/min at pH 6.2, 25°C	Dialyzed; crystallized; powder		Can Inova
		Molecular Biology Grade; crystallized		Oncor 130172
	1 unit decreases A_{450} by 0.001/min at pH 6.24, 25°C using *Micrococcus luteus* as substrate	Lyophilized		Pharmacia 27-0267-01 27-0267-02
≥50,000 U/mg	1 unit decreases A_{450} by 0.001/min at pH 6.2, 35°C with *M. luteus* cells substrate	Lyophilized		Seikagaku 100940-1 100940-2
		Lyophilized		Seikagaku 130942-1
5,000-10,000 U/mg solid	1 unit produces a ΔA_{450} of 0.001/min at pH 6.24, 25°C; suspension of *Micrococcus lysodeikticus* as substrate in 2.6 mL reaction mixture, 1 cm light path	Insoluble enzyme attached to cross-linked beaded agarose; lyophilized containing lactose stabilizer		Sigma L1129
50,000 U/mg protein	1 unit produces a ΔA_{450} of 0.001/min at pH 6.24, 25°C; suspension of *Micrococcus lysodeikticus* as substrate in 2.6 mL reaction mixture, 1 cm light path	3X Crystallized; 90% protein, NaOAc, NaCl buffer salts		Sigma L2879
50,000 U/mg protein	1 unit produces a ΔA_{450} of 0.001/min at pH 6.24, 25°C; suspension of *Micrococcus lysodeikticus* as substrate in 2.6 mL reaction mixture, 1 cm light path	3X Crystallized; dialyzed; lyophilized containing 95% protein; balance primarily NaOAc and NaCl buffer salts		Sigma L6876
60,000 U/mg protein	1 unit produces a ΔA_{450} of 0.001/min at pH 6.24, 25°C; suspension of *Micrococcus lysodeikticus* as substrate in 2.6 mL reaction mixture, 1 cm light path	3X Crystallized; 85% protein, NaOAc, NaCl buffer salts		Sigma L7001
50,000 U/mg protein	1 unit produces a ΔA_{450} of 0.001/min at pH 6.24, 25°C; suspension of *Micrococcus lysodeikticus* as substrate in 2.6 mL reaction mixture (1cm light path)	3X Crystallized; dialyzed; lyophilized containing 95% protein; balance primarily NaOAc and NaCl buffer salts		Sigma L7651
50,000 U/mg protein	1 unit produces a ΔA_{450} of 0.001/min at pH 6.24, 25°C; suspension of *Micrococcus lysodeikticus* as substrate in 2.6 mL reaction mixture, 1 cm light path	3X Crystallized; dialyzed; aseptic; lyophilized containing 95% protein; balance primarily NaOAc and NaCl buffer salts		Sigma L7773
				Vital 17238

Lysozyme continued 3.2.1.17

SPECIFIC ACTIVITY	UNITS DEFINITION	PREPARATION FORM	ADDITIONAL ACTIVITIES	SUPPLIER CATALOG No.
Chicken egg white continued				
≥0.8 mg/mL		Powder		Wako 126-02671 122-02673
≥5000 U/mg solid	1 unit decreases turbidity of a cell suspension of *Micrococcus lysodeikticus* at A_{450} by 0.001/min at pH 7.0, 25°C	2X Crystallized; lyophilized containing NaCl and acetate		Worthington LS02880 LS02881 LS02883
≥8000 U/mg solid	1 unit decreases turbidity of a cell suspension of *Micrococcus lysodeikticus* at A_{450} by 0.001/min at pH 7.0, 25°C	2X Crystallized; dialyzed; lyophilized		Worthington LS02931 LS02933 LS02934
Human				
100,000-200,000 U/mg protein	1 unit produces a ΔA_{450} of 0.001/min at pH 6.24, 25°C; suspension of *Micrococcus lysodeikticus* as substrate in 2.6 mL reaction mixture, 1 cm light path	Lyophilized containing 10% protein; balance $NaPO_4$ and NaCl		Sigma L6394
Human milk (MW = 17,000 Da)				
30,000 Shugar U/mg	1 unit digests powdered cells of *Micrococcus lysodeikticus*, decreasing absorbance by 0.001/min at pH 7.0, 37°C	≥95% purity; lyophilized containing 50 mM NaOAc, 100 mM NaCl, pH 6.0	No detectable HBsAg, anti-HCV, anti-HBc, anti-HIV	ART 16-14-122519
Human neutrophil (MW = 17,000 Da)				
≥30,000 Shugar U/mg protein	1 unit digests powdered cells of *Micrococcus lysodeikticus* decreasing absorbance by 0.001/min at pH 7.0, 37°C	>98% purity by SDS-PAGE; lyophilized containing 0.05 M NaOAc buffer and 0.1 M NaCl, pH 6.0	No detectable HIV Ab, HBsAg	Cortex CP3026
30,000 Shugar U/mg protein	1 unit digests powdered cells of *Micrococcus lysodeikticus* decreasing absorbency by 0.001/min at pH 7.0, 37°C	>98% purity by SDS-PAGE; salt-free; lyophilized		ICN 191335
100,000 U/mg protein	1 unit produces a ΔA_{450} of 0.001/min at pH 6.24, 25°C; suspension of *Micrococcus lysodeikticus* as substrate in 2.6 mL reaction mixture, 1 cm light path	95% by SDS-PAGE; lyophilized containing 50 mM NaOAc and 100 mM NaCl, pH 6.0		Sigma L8402
Human plasma				
500,000 U/mg solid	1 unit produces a ΔA_{450} of 0.001/min at pH 7.4, 30°C; suspension of *Micrococcus lysodeikticus* as substrate in 1.0 mL reaction mixture, 1 cm light path			Sigma L2026
Human purulent sputum (leucocyte) (optimum pH = 5.5 [cell walls], 6.5 [whole cells]; MW = 20,300 Da)				
>2000 U/mg protein	1 unit lyses 1 μg *Streptococcus salivarius* IFO 3350 cells/min at pH 7.0, 37°C	Highly purified; lyophilized	<0.5% protease	ICN 320941

3.2.1.17 Lysozyme continued

SPECIFIC ACTIVITY	UNITS DEFINITION	PREPARATION FORM	ADDITIONAL ACTIVITIES	SUPPLIER CATALOG NO.
Human, recombinant expressed in *Pichia pastoris*				
>80,000 U/mg protein	1 unit decreases A_{450} of a *Micrococcus lysodeikticus* suspension by 0.001/min at pH 7.5, 25°C	>98% purity; salt-free; lyophilized	No detectable elastase, MPO, cathepsin G	Elastin SL754
***Streptomyces globisporus* (optimum pH = 5.5, T = 50°C)**				
2000 U/mg protein	1 unit lyses 1 μg *Streptococcus salivarius* IFO 3350 heat cells/min at pH 7.0, 37°C	Lyophilized		Seikagaku 100095-1
80,000 U/mg protein	1 unit produces a ΔA_{450} of 0.001/min at pH 6.24, 25°C; suspension of *Micrococcus lysodeikticus* as substrate in 2.6 mL reaction mixture, 1 cm light path	Crystallized; dialyzed; lyophilized containing 95% protein; balance primarily NaOAc and NaCl buffer salts		Sigma L6255
Turkey egg white (optimum pH = 6.5-7.5)				
2,000,000 U/mg	1 unit decreases A_{350} by 0.001/min at pH 7.5, 25°C with a 0.5 mg/mL suspension of lyophilized *E. coli* K802 cells in 50 mM Tris-HCl	Solution containing 50% glycerol, 50 mM Tris-HCl, 0.1 M NaCl, 0.1 mM EDTA, 1 mM DTT, 0.1% Triton X-100, pH 7.5	No detectable exonuclease, endonuclease (by agarose gel electrophoresis)	Epicentre Ready-Lyse™ R1802M R1804M R1810M
Unspecified (optimum pH = 6.0-7.0)				
20,000 U/mg	1 unit decreases A_{450} by 0.001/min at pH 6.24, 25°C using the lysis of *M. luteus* cells	Dialyzed; albumin-free (by electrophoresis); 3X crystallized; lyophilized		ICN 363234

3.2.1.18 Exo-α-Sialidase

REACTION CATALYZED
Hydrolysis of α-2,3-, α-2,6- and α-2,8-glycosidic linkages (at a decreasing rate, respectively) of terminal sialic residues in oligosaccharides, glycoproteins, glycolipids, colominic acid and synthetic substrates

SYSTEMATIC NAME
Acylneuraminyl hydrolase

SYNONYMS
Neuraminidase, sialidase

REACTANTS
Oligosaccharides, glycoproteins, glycolipids, colominic acid, mucin

APPLICATIONS
- Cell surface probe of glycoconjugate distribution
- Interest in using neuraminidase inhibitors as possible anti-viral and anti-bacterial agents

Notes

- Classified as a hydrolase: a glycosidase hydrolyzing o-glycosyl compounds
- Removes N-acetyl neuraminic acid from a variety of glycoproteins
- Does not act on 4-o-acetylated sialic acids; little activity against α-2,8-sialyl linkages
- In the absence of detergents and Ca^{2+}, the enzyme hydrolyzes N-acetylneuraminyl moiety of polysialogangliosides to produce monosialoganglioside Gm1. With detergent, Gm1 is further desialylated to asialoganglioside Ga1
- Has a 260-fold rate preference for α-2,3-sialyl linkages over α-2,6-
- Stabilized by BSA and EDTA
- Inhibited by PCMB, Hg^{2+} and Ag^+

SPECIFIC ACTIVITY	UNITS DEFINITION	PREPARATION FORM	ADDITIONAL ACTIVITIES	SUPPLIER CATALOG No.
Arthrobacter ureafaciens (optimum pH = 4.3-4.5 [colominic acid], 4.5-7.0 ([bovine submaxillary mucin], 5.0-5.5 [NANA-lactose]; MW = 51,000 [Band 1], 39000 [Band 2] Da)				
25 U/mg total protein at 25°C; 35 U/mg total protein at 37°C	1 unit converts 1 μmol N-Ac-neuraminosyl-D-lactose to product/min at pH 5.0	Solution containing 10 mM $NaPO_4$, 0.1% Micr-O-protect, 0.25 mg/mL BSA, pH 7	No detectable proteases (resorufin-labeled casein as substrate)	Boehringer 269611
>75 U/mg protein	1 unit liberates 1 μmol N-Ac-neuraminic acid/min at pH 5.0, 37°C	Lyophilized from buffer containing 10 mM $NaPO_4$, pH 7.0	No detectable aldolases, glycosidases, NANA, proteases	Calbiochem 480714
20 U/mg protein	1 unit releases 1 μmol N-Ac-neuraminic acid/min at pH 4.5, 37°C with Neu5Acα(2-3,6)Galβ(1-4)Glc as substrate	5 mM PO_4 buffer, pH 7.0	<0.01% protease, phospholipase C, NANA, α-glucosidase, β-glucosidase, α-Gal, β-Gal, α-mannosidase, α-fucosidase	Fluka 72207
>60 U/mg protein (NAN-lactose); >25 U/mg protein (bovine submaxillary mucin); >20 U/mg protein (colominic acid)	1 unit liberates 1.0 μmol N-Ac-neuraminic acid (NANA)/min at pH 5.0, 37°C using either NAN-lactose, bovine submaxillary mucin or colominic acid as substrate	Lyophilized containing Na-KPO_4 salts		ICN 153846

SPECIFIC ACTIVITY	UNITS DEFINITION	PREPARATION FORM	ADDITIONAL ACTIVITIES	SUPPLIER CATALOG No.
Arthrobacter ureafaciens continued				
>80 U/mg protein (NAN-lactose); >20 U/mg protein (bovine submaxillary mucin); >40 U/mg protein (colominic acid); >60 U/mg protein (bovine brain ganglioside)	1 unit liberates 1.0 μmol N-Ac-neuraminic acid (NANA)/min at pH 5.0, 37°C using either NAN-lactose, bovine submaxillary mucin, colominic acid or bovine brain ganglioside as substrate	Lyophilized		ICN 153855
>80 U/mg protein (NAN-lactose); >25 U/mg protein (bovine submaxillary mucin); >20 U/mg protein (colominic acid)	1 unit liberates 1.0 μmol N-Ac-neuraminic acid/min at pH 5.0, 37°C using NAN-lactose, bovine submaxillary mucin or colominic acid as substrate	Chromatographically purified; lyophilized containing NaPO$_4$ salts, pH 7	No detectable protease, phospholipase C, NANA, glycosidases	Nacalai 242-29 SP
>80 U/mg protein (NAN-lactose); >25 U/mg protein (bovine submaxillary mucin); >20 U/mg protein (colominic acid); >60 U/mg protein (bovine brain ganglioside)	1 unit liberates 1 μmol N-Ac-neuraminic acid/min at pH 5.0, 37°C using NAN-lactose, bovine submaxillary mucin, colominic acid or bovine brain ganglioside as substrate	Highly purified; lyophilized	No detectable protease, NANA, glycosidases	Nacalai 242-38
10 U/mL (NAN-lactose substrate)	1 unit liberates 1.0 μmol N-Ac-neuraminic acid/min at pH 5.0, 37°C using NAN-lactose or bovine submaxillary mucin	Suspension in 10 mM NaPO$_4$, 0.025% BSA, 0.05% NaN$_3$, pH 7.0		Sigma N3642

Exo-α-Sialidase continued 3.2.1.18

SPECIFIC ACTIVITY	UNITS DEFINITION	PREPARATION FORM	ADDITIONAL ACTIVITIES	SUPPLIER CATALOG NO.
Clostridium perfringens (optimum pH = 5.0-5.1, little activity at pH 4.0 or above pH 8.0; T = 50°C, MW = 60,000 Da)				
15 U/mg protein	1 unit releases 1 μmol sialic acid from bovine submaxillary mucin/min at pH 5.0, 37°C	Lyophilized		AMRESCO 1B1178
1.5 U/mg solid at 25°C; 3 U/mg solid at 37°C	1 unit converts 1 μmol N-Ac-neuraminosyl-D-lactose to product/min at pH 5.0	Lyophilized containing 5 μmol (1.86 μg) EDTA/vial		Boehringer 107590
60 U/mg protein	1 unit converts 1 μmol N-Ac-neuraminosyl-D-lactose to product/min at pH 5.0, 25°C	Lyophilized	<0.3% protease (resorufin-labeled casein as substrate)	Boehringer 1585886 1585894
160 U/mg	1 unit releases 1 μmol N-Ac-neuraminic acid/min at pH 4.5, 37°C with Neu5Acα(2-3,6)Galβ(1-4)Glc as substrate	Powder	<0.01% protease <0.1% NANA	Fluka 72199
1 U/mg	1 unit releases 1 μmol N-Ac-neuraminic acid/min at pH 4.5, 37°C with Neu5Acα(2-3,6)Galβ(1-4)Glc as substrate	Powder	<0.05% protease <3% NANA	Fluka 72201
3 U/mg	1 unit releases 1 μmol N-Ac-neuraminic acid/min at pH 4.5, 37°C with Neu5Acα(2-3,6)Galβ(1-4)Glc as substrate	Lyophilized	<0.01% NANA <0.02% protease	Fluka 72202
30 U/g dried material	1 unit releases 1 μmol N-Ac-neuraminic acid/min at pH 4.5, 37°C with Neu5Acα(2-3,6)Galβ(1-4)Glc as substrate	15% aqueous suspension in 2 M $(NH_4)_2SO_4$, pH 7.0		Fluka 72203
15 U/mg	1 unit releases 1 μmol N-Ac-neuraminic acid/min at pH 4.5, 37°C with Neu5Acα(2-3,6)Galβ(1-4)Glc as substrate	Lyophilized	<0.01% protease, NANA	Fluka 72204
200 U/mg	1 unit releases 1 μmol N-Ac-neuraminic acid/min at pH 4.5, 37°C with Neu5Acα(2-3,6)Galβ(1-4)Glc as substrate	Lyophilized	<0.001% protease, NANA	Fluka 72211
0.5-1.5 U/mg	1 unit releases 1 μmol sialic acid/min at pH 5.0, 37°C	Lyophilized		ICN 100872
150-400 U/mg protein (NAN-lactose substrate)	1 unit liberates 1.0 μmol N-Ac-neuraminic acid/min at pH 5.0, 37°C using NAN-lactose or bovine submaxillary mucin	Affinity chromatographed; dialyzed; lyophilized containing 85% protein		Sigma N2133
0.5-6 U/mg solid (NAN-lactose substrate); 0.1-3 U/mg solid (mucin substrate)	1 unit liberates 1.0 μmol N-Ac-neuraminic acid/min at pH 5.0, 37°C using NAN-lactose or bovine submaxillary mucin	Salt-fractionated; dialyzed; lyophilized	May contain protease and NANA	Sigma N2876

SPECIFIC ACTIVITY	UNITS DEFINITION	PREPARATION FORM	ADDITIONAL ACTIVITIES	SUPPLIER CATALOG No.
Clostridium perfringens continued				
6-10 U/mg protein (NAN-lactose substrate); 2-5 U/mg protein (mucin substrate)	1 unit liberates 1.0 μmol N-Ac-neuraminic acid/min at pH 5.0, 37°C using NAN-lactose or bovine submaxillary mucin	Dialyzed; lyophilized containing >95% protein	May contain protease and NANA	Sigma N3001
20-30 U/g agarose (NAN-lactose substrate); 1 mL gel yields 0.6-1.0 U	1 unit liberates 1.0 μmol N-Ac-neuraminic acid/min at pH 5.0, 37°C using NAN-lactose or bovine submaxillary mucin	Insoluble enzyme attached to beaded agarose; suspension in 2.0 M $(NH_4)_2SO_4$, pH 7.0		Sigma N4883
20-30 U/g agarose (NAN-lactose substrate); 1 mL gel yields 0.6-1.0 U	1 unit liberates 1.0 μmol N-Ac-neuraminic acid/min at pH 5.0, 37°C using NAN-lactose or bovine submaxillary mucin	Insoluble enzyme attached to beaded agarose; suspension in 2.0 M $(NH_4)_2SO_4$, pH 7.0		Sigma N5254
0.5-6 U/mg solid (NAN-lactose substrate); 0.1-3 U/mg solid (mucin substrate)	1 unit liberates 1.0 μmol N-Ac-neuraminic acid/min at pH 5.0, 37°C using NAN-lactose or bovine submaxillary mucin	Aseptic; salt-fractionated; dialyzed; lyophilized	May contain protease and NANA	Sigma N5505
10-20 U/mg protein (NAN-lactose substrate); 4 U/mg protein (mucin substrate)	1 unit liberates 1.0 μmol N-Ac-neuraminic acid/min at pH 5.0, 37°C using NAN-lactose or bovine submaxillary mucin	Dialyzed; lyophilized containing 90% protein	May contain protease and NANA	Sigma N5631
≥10 U/mg protein	1 unit releases 1 μmol sialic acid/min at pH 5.0, 37°C	Chromatographically purified; lyophilized containing sucrose stabilizer	Proteolytic activity <0.1% relative to that of 2X crystallized trypsin	Worthington LS04761 LS04762 LS04760
≥5 U/mg solid	1 unit releases 1 μmol sialic acid/min at pH 5.0, 37°C	Partially purified; lyophilized		Worthington LS04779 LS04780 LS04777
Clostridium perfringens, overexpressed in E. coli (MW = 43,000 Da)				
50,000 U/mL	1 unit cleaves >95% terminal αNeu5Ac from 1 nmol Neu5Acα2-3Galβ1-3GlcNAcβ1-3Galβ1-4Glc-AMC/hr at pH 4.5, 37°C in a 10 μL reaction	20 mM Tris-HCl, 50 mM NaCl, 5 mM Na_2EDTA, pH 7.5	No detectable exoglycosidase, protease	NE Biolabs 720S 720L

SPECIFIC ACTIVITY	UNITS DEFINITION	PREPARATION FORM	ADDITIONAL ACTIVITIES	SUPPLIER CATALOG NO.
Newcastle disease virus, Hitchner strain B1 (optimum pH = 5.0–6.0, MW = 74,000 Da)				
>5 U/mg protein	1 unit converts 1 μmol N-Ac-neuraminosyl-α-2,3-lactose to product/min at pH 5.5, 37°C	Lyophilized	No detectable endoglycosidases F and H, α- and β-NAGlc, α– and β-NAG, α- and β-mannosidase, α- and β-Gal, α- and β-fucosidase, α- and β-glucosidase, proteases	Boehringer 1521845
0.005 U/vial	1 unit hydrolyzes 1.0 μmol 2'-N(4-methylumbelliferyl)-α-N-Ac-neuraminic acid/min at pH 5.5, 37°C	Lyophilized containing PO₄ buffer salt		Sigma N5146
Salmonella typhimurium LT2, overexpressed in E. coli (optimum pH = 5.5–7.0, MW = 41,300 Da)				
50,000 U/mL	1 unit cleaves >95% terminal αNeu5Ac from 1 nmol Neu5Acα2-3Galβ1-3GlcNAcβ1-3Galβ1-4Glc-AMC/hr at pH 6.0, 37°C in a 10 μL reaction	20 mM Tris-HCl, 50 mM NaCl, 5 mM Na₂EDTA, 100 μg/mL BSA, pH 7.5	No detectable exoglycosidase, protease	NE Biolabs 728S 728L
	1 unit hydrolyzes 1 μmol 3'sialyllactose/min at pH 5.5, 37°C	Lyophilized containing 5 μmol KPO₄, 10 μmol NaCl, 0.02 mg NaN₃, pH 6.8		PanVera TAK 4455
Salmonella typhimurium, recombinant				
500 U/mg protein (NAN-lactose substrate)	1 unit liberates 1.0 μmol N-Ac-neuraminic acid/min at pH 5.0, 37°C using NAN-lactose or bovine submaxillary mucin	Chromatographically purified; dialyzed; lyophilized		Sigma N7771
Streptococcus species (optimum pH = 6.0–6.5, pI = 7.7, T = 37–50°C, MW = 85,000 Da; K_M = 0.33 mM [sialyllactose]; stable pH 4.0–9.0 [25°C, 16 hr], T < 50°C [pH 6.5, 10 min])				
50 U/mg	1 unit hydrolyzes 1 μmol sialyllactose/min at pH 6.5, 37°C			ICN 151738
50 U/mg	1 unit liberates 1 μmol N-Ac-neuraminic acid (NANA) from N-Ac-neuraminyl-(2,3)-lactose/min at pH 6.5, 37°C	Highly purified; lyophilized		ICN 321381
≥50 U/mg	1 unit liberates 1 μmol N-Ac-neuraminic acid from Ac-neuraminyl-(2-->3)-lactose/min at pH 6.5, 37°C	Lyophilized		Seikagaku 120050-1
≥50 U/mg	1 unit liberates 1 μmol N-Ac-neuraminic acid from Ac-neuraminyl-(2-->3)-lactose/min at pH 6.5, 37°C	Sterilized by membrane filtration		Seikagaku 120052-1
1 U/vial	1 unit liberates 1.0 μmol N-Ac-neuraminic acid/min at pH 6.5, 37°C using NAN-lactose or bovine submaxillary mucin	Lyophilized containing citrate-PO₄ buffer salts		Sigma N5271

SPECIFIC ACTIVITY	UNITS DEFINITION	PREPARATION FORM	ADDITIONAL ACTIVITIES	SUPPLIER CATALOG NO.
Streptococcus species continued				
≥50 U/mg solid	1 unit hydrolyzes 1 μmol sialyllactose (oxidizes 1 μmol NADH)/min at pH 6.5, 37°C	Lyophilized containing 60% BSA and EDTA as stabilizers	<0.001% NADH oxidase <0.1% catalase	Toyobo NRH-301
Vibrio cholerae (optimum pH = 5.5-6.2, MW = 95,000 Da)				
40 U/mg protein; 20 U/mg total protein	1 unit converts 1 μmol N-Ac-neuraminosyl-D-lactose to product/min at pH 5.5, 37°C	Solution containing 50 mM NaOAc, 154 mM NaCl, 9 mM CaCl$_2$, 0.1% Micr-O-protect, 25 mg/L HSA, pH 5.5	No detectable proteases	Boehringer 1080725
1 U/mL	1 unit releases 1 μmol N-Ac-neuraminic acid from human acid α$_1$-glycoprotein/min at pH 5.5, 37°C	Preservative-free; solution containing 4 mM CaCl$_2$, 50 mM NaOAc, 154 mM NaCl, pH 5.5		Calbiochem 480717
2 U/mg	1 unit releases 1 μmol N-Ac-neuraminic acid/min at pH 4.5, 37°C with Neu5Acα(2-3,6)Galβ(1-4)Glc as substrate	Sterile-filtered; aqueous solution containing 0.15 M NaCl and 4 mM CaCl$_2$, pH 5.5		Fluka 72197
0.5-1.0 IU/mL	1 unit liberates 1.0 mmol N-Ac-neuraminic acid from human acid α$_1$-glycoprotein/min at pH 5.5, 37°C	Suspension in 0.05 M NaOAc, pH 5.5		Life Technol 17050-014
45-135 U/g agarose (NAN-lactose substrate); 1 mL gel yields 1.5-4.5 U	1 unit liberates 1.0 μmol N-Ac-neuraminic acid/min at pH 5.0, 37°C using NAN-lactose or bovine submaxillary mucin	Insoluble enzyme attached to beaded agarose; lyophilized containing lactose stabilizer		Sigma N0621
8-24 U/mg protein (NAN-lactose substrate)	1 unit liberates 1.0 μmol N-Ac-neuraminic acid/min at pH 5.0, 37°C using NAN-lactose or bovine submaxillary mucin	Chromatographed; sterile-filtered; aqueous solution containing 0.15 M NaCl and 4 mM CaCl$_2$, pH 5.5	May contain protease and NANA	Sigma N6514
1.3 U/mg protein (NAN-lactose substrate)	1 unit liberates 1.0 μmol N-Ac-neuraminic acid/min at pH 5.0, 37°C using NAN-lactose or bovine submaxillary mucin	Chromatographed; sterile-filtered; aqueous solution containing 0.15 M NaCl and 4 mM CaCl$_2$, pH 5.5	May contain protease and NANA	Sigma N7885

α-Glucosidase — 3.2.1.20

REACTION CATALYZED
Hydrolysis of terminal, non-reducing 1,4-linked α-D-glucose residues with release of α-D-glucose

SYSTEMATIC NAME
α-D-Glucoside glucohydrolase

SYNONYMS
Glucoinvertase, glucosidosucrase, maltase, maltase-glucoamylase, αGluD, AGH

REACTANTS
1,4-Linked α-D-glucose residues, α-D-glucose, sucrose, maltose

APPLICATIONS
- Determination of α-amylase
- Determination of maltose in brewing

NOTES
- Classified as a hydrolase: a glycosidase hydrolyzing o-glycosyl compounds
- A group of enzymes that:
  - primarily exohydrolyze 1,4-α-glucosidic linkages;
  - hydrolyze oligosaccharides rapidly;
  - hydrolyze polysaccharides relatively slowly or not at all
- Intestinal enzyme also acts on polysaccharides by catalyzing the reactions of glucan 1,4-α-glucosidase (EC 3.2.1.3) and slow hydrolysis of 1,6-α-D-glucose links
- β-D-glucosides (like cellobiose) are not substrates
- Inhibited by Ag^+, Hg^{2+}, PCMB, MA, histidine, amines (including Tris) and sulfhydryl reagents
- Stabilized by BSA, glutathione (reduced), glycerol and sodium glutamate

SPECIFIC ACTIVITY	UNITS DEFINITION	PREPARATION FORM	ADDITIONAL ACTIVITIES	SUPPLIER CATALOG NO.
Bacillus stearothermophilus (optimum pH = 6.0-7.0, MW = 50,000 Da; K_M = 0.73 mM [PNPG], 1.3 mM [maltose], 2.4 mM [phenyl-α-glucopyranoside]; stable pH 5.0-11.0, T < 60°C)				
50 U/mg protein (p-nitrophenyl α-D-glucoside as substrate)	1 unit liberates 1.0 μmol D-glucose from p-nitrophenyl α-D-glucoside/min at pH 6.8, 37°C; 1 unit converts 1.0 μmol maltose to 2.0 μmol of D-glucose/min at pH 6.0, 25°C using maltose as substrate	Lyophilized containing KPO₄ buffer salts	Substantially free of β-glucosidase, α- and β-Gal	Sigma G3651
Microbial (optimum pH = 6.0-7.0, pI = 5.2, T = 60°C, MW = 65,000 Da [gel filtration, SDS-PAGE]; K_M = 0.63 mM [p-nitrophenyl-α-D-glucopyranoside]; stable pH 5.0-9.0, T < 60°C [pH 7.0, 15 min])				
≥20 U/mg solid	1 unit forms 1 μmol PNP/min at pH 7.0, 37°C	Lyophilized containing BSA as stabilizer		Toyobo AGH-211
Rice				
40-80 U/mg protein	1 unit converts 1.0 μmol maltose to 2.0 μmol of D-glucose/min at pH 4.0, 37°C using maltose as substrate	Suspension in 2.8 M (NH₄)₂SO₄ solution	Substantially free of β-glucosidase, α- and β-Gal	Sigma G9259

α-Glucosidase continued

SPECIFIC ACTIVITY	UNITS DEFINITION	PREPARATION FORM	ADDITIONAL ACTIVITIES	SUPPLIER CATALOG NO.
Saccharomyces cerevisiae, recombinant overproducer (optimum pH = 6.0-6.5, MW = 68,500 Da)				
>70 U/mg protein (4-nitrophenyl-α-D-glucopyranoside as substrate); 100 U/mg at 25°C or 220 U/mg at 37°C (maltose as substrate)	1 unit cleaves 1 μmol substrate/min at pH 6.8	Lyophilized containing 50 mM KPO$_4$ and 70% lactose, pH 7.15	No detectable proteases <0.01% α- and β-Gal, β-glucosidase <0.02% 6PGDH <0.1% PGI <1% HK	Boehringer 1630385
Saccharomyces species (optimum pH = 6.8-7.2, pI = 5.7, T = 40-42°C, MW = 52,000 Da; K_M = 0.36 mM [*p*-nitrophenyl-α-glucoside], 8.0 mM [phenyl-α-glucoside], 2.8 mM [phenyl-α-maltoside], 14 mM [maltose], 6.2 mM [maltotriose], 23 mM [sucrose]; stable pH 6.0-7.5 [25°C, 20 hr], T < 35°C [pH 7.0, 15 min])				
≥80 U/mg solid	1 unit forms 1 μmol PNP/min at pH 7.0, 37°C	Lyophilized containing 30-40% BSA		ICN 153487
≥80 U/mg solid	1 unit forms 1 μmol PNP/min at pH 7.0, 37°C	Lyophilized containing 30-40% BSA, glutathione, glycerol, Na Glu as stabilizers	<0.001% β-glucosidase, α- and β-Gal, AlcDH, NADPH oxidase <0.05% 6PGluDH <0.5% PGlcI, PGlcM	Toyobo AGH-201
≥70 U/mg				Wako 076-02841
Yeast (optimum pH = 7.0-7.5, MW = 68,500 Da; relative activity: 25°C = 1.0, 30°C = 1.5, 37°C = 2.6)				
150 U/mg protein; ≥100 U/mg solid	1 unit hydrolyzes 1μmol *p*-nitrophenyl α-D-glucopyranoside/min at 37°C	Salt-free; lyophilized		Biozyme AGLY3
65 U/mg	1 unit hydrolyzes 1 μmol *p*-nitrophenyl-α-D-glucopyranoside/min at pH 6.8, 37°C	Lyophilized	<0.01% α-amylase, β-glucosidase, α-Gal, β-Gal	Fluka 63412
>180 U/mg protein; >120 U/mg solid	1 unit hydrolyzes 1.0 μmol *p*NPG/min at pH 6.8, 37°C	Lyophilized	<0.00002% α-amylase <0.0001% α- and β-Gal, β-glucosidase	Genzyme 1231
≥50 IU/mg protein	1 IU transforms 1 μmol substrate/min under standard IUB conditions at 25°C	Lyophilized	<0.01% α- and β-Gal, β-glucosidase	OYC
85 U/mg protein (*p*-nitrophenyl α-D-glucoside as substrate); 40 U/mg protein (maltose as substrate)	1 unit liberates 1.0 μmol D-glucose from *p*-nitrophenyl α-D-glucoside/min at pH 6.8, 37°C; 1 unit converts 1.0 μmol maltose to 2.0 μmol of D-glucose/min at pH 6.0, 25°C using maltose as substrate	Suspension in 3.2 M (NH$_4$)$_2$SO$_4$ solution, pH 6; 40 U/mg total protein	No detectable β-glucosidase, α- and β-Gal <0.5 U/mg protein α-amylase	Sigma G7256

α-Glucosidase continued

SPECIFIC ACTIVITY	UNITS DEFINITION	PREPARATION FORM	ADDITIONAL ACTIVITIES	SUPPLIER CATALOG No.
Yeast continued				
85 U/mg protein (p-nitrophenyl α-D-glucoside as substrate); 40 U/mg protein (maltose as substrate)	1 unit liberates 1.0 μmol D-glucose from p-nitrophenyl α-D-glucoside/min at pH 6.8, 37°C; 1 unit converts 1.0 μmol maltose to 2.0 μmol of D-glucose/min at pH 6.0, 25°C using maltose as substrate	Solution containing 50% glycerol and BSA	No detectable β-glucosidase, α- and β-Gal <0.5 U/mg protein α-amylase	Sigma G8889
≥45 U/mg solid	1 unit hydrolyzes 1 μmol p-nitrophenyl-D-glucopyranoside at pH 6.8, 37°C	Chromatographically purified; dialyzed; lyophilized		Worthington LS09888 LS09890 LS09886
Yeast, bakers				
40 U/mg	1 unit releases 1 μmol glucose/min at pH 6.0, 25°C	Suspension in $(NH_4)_2SO_4$ containing 5 mg/mL protein	0.01% α- and β-Gal, β-glucosidase	Calzyme 108B0040
28 U/mg	1 unit releases 1 μmol glucose/min at pH 6.0, 25°C	Lyophilized containing 30% BSA as stabilizer		Calzyme 109A0028
>6 U/mg	1 unit releases 1 μmol glucose/min at pH 6.0, 25°C	Lyophilized containing 90% protein	0.01% α- and β-Gal, β-glucosidase	Calzyme 121A0006
9 U/mg protein (p-nitrophenyl α-D-glucoside as substrate); 3 U/mg protein (maltose as substrate)	1 unit liberates 1.0 μmol D-glucose from p-nitrophenyl α-D-glucoside/min at pH 6.8, 37°C; 1 unit converts 1.0 μmol maltose to 2.0 μmol of D-glucose/min at pH 6.0, 25°C using maltose as substrate	Partially purified; powder containing 50% protein; balance primarily PO_4 buffer salts and EDTA	No detectable β-glucosidase, α- and β-Gal	Sigma G5003
10-35 U/mg protein	1 unit converts 1.0 μmol maltose to 2.0 μmol D-glucose/min at pH 6.0, 25°C	Lyophilized containing 10% protein; balance primarily Ficoll, 1% K PO_4, 0.5% EDTA as stabilizer	<0.5 U/mg protein isomaltase 35-85 U/mg protein α-glucosidase	Sigma M3145
Yeast, brewers				
10 U/mg protein (p-nitrophenyl α-D-glucoside as substrate); 4 U/mg protein (maltose as substrate)	1 unit liberates 1.0 μmol D-glucose from p-nitrophenyl α-D-glucoside/min at pH 6.8, 37°C; 1 unit converts 1.0 μmol maltose to 2.0 μmol of D-glucose/min at pH 6.0, 25°C using maltose as substrate	Partially purified; powder containing 75% protein; balance primarily PO_4 buffer salts and EDTA	No detectable β-glucosidase, α- and β-Gal <0.05 U/mg protein α-amylase	Sigma G4634

3.2.1.20 α-Glucosidase continued

SPECIFIC ACTIVITY	UNITS DEFINITION	PREPARATION FORM	ADDITIONAL ACTIVITIES	SUPPLIER CATALOG NO.
Yeast, brewers continued				
50-100 U/mg protein (p-nitrophenyl α-D-glucoside as substrate); 15-40 U/mg protein (maltose as substrate)	1 unit liberates 1.0 µmol D-glucose from p-nitrophenyl α-D-glucoside/min at pH 6.8, 37°C; 1 unit converts 1.0 µmol maltose to 2.0 µmol of D-glucose/min at pH 6.0, 25°C using maltose as substrate	Purified; powder containing 30% protein; balance primarily PO_4 buffer salts and EDTA	No detectable β-glucosidase, α- and β-Gal <0.05 U/mg protein α-amylase	Sigma G6136

3.2.1.21 β-Glucosidase

REACTION CATALYZED
   Hydrolysis of terminal, non-reducing β-D-glucose residues with release of β-D-glucose

SYSTEMATIC NAME
   β-D-Glucoside glucohydrolase

SYNONYMS
   Gentiobiase, cellobiase, emulsin

REACTANTS
   Terminal, non-reducing β-D-glucose residues, β-D-glucosides, β-D-galactosides, α-L-arabinosides, β-D-xylosides, β-D-fucosides, β-D-glucose, salicen, saligenen, p-nitrophenyl-β-D-glucopyranoside, 2,4-dichlorophenyl-β-D-glucopyranoside

APPLICATIONS
- Hydrolysis of cellulose for fuel ethanol
- Synthesis of alkyl-β-D-glucosides in organic media
- Determination of α-amylase
- Carbohydrate structure resolution

NOTES
- Classified as a hydrolase: a glycosidase hydrolyzing o-glycosyl compounds
- Two active components with molecular weights of 117,000 and 66,500 Da
- Found in many seeds, yeasts, molds and bacteria
- Wide specificity for β-D-glucosides
- Some enzymes also hydrolyze β-D-galactosides, α-L-arabinosides, β-D-xylosides and β-D-fucosides
- Inhibited by $HgCl_2$, other heavy metal ions, sulfhydryl binding compounds and polyols
- Stabilized by BSA and reduced glutathione

SPECIFIC ACTIVITY	UNITS DEFINITION	PREPARATION FORM	ADDITIONAL ACTIVITIES	SUPPLIER CATALOG NO.
Amygdalae dulces (sweet almond) (optimum pH = 2.0-9.5, 4.4 [butyrylglucoside in acetate], 5.2-6.0 [*p*-nitrophenylglycoside], pI = 7.3, T = 50-55°C, 2 subunits with MW = 66,500 and 117,000 Da; stable at pH 6.0-9.0 [25°C, 64 hr] and < 50°C [pH 7.3, 1 hr])				
≥1000 U/mg solid	1 unit liberates 1 µg glucose/min at 35°C	Salt-free; lyophilized		Biozyme GSA2
6000 U/mg protein; ≥2500 U/mg solid	1 unit liberates 1 µg glucose/min at 35°C	Salt-free; lyophilized	<0.0005% amylase	Biozyme GSA3
20 U/mg protein; 2 U/mg solid	1 unit converts 1 µmol salicin (28 *mM*) to product/min at 25°C	Lyophilized	<0.05% α-Gal, α-glucosidase <0.15% β-Gal	Boehringer 105422
10-60 U/mg protein	1 unit forms 1 µmol PNP/min at pH 5.0, 37°C	Salt-free; lyophilized containing 90% protein	Traces of α- and β-Gal <0.0001% amylase	Calzyme 079A0060
30 U/mg	1 unit hydrolyzes 1 µmol glucose/min at pH 5.0, 37°C with salicin as substrate	Powder		Fluka 49289
6 U/mg	1 unit hydrolyzes 1 µmol glucose/min at pH 5.0, 35°C with salicin as substrate	Salt-free; lyophilized		Fluka 49290
2500 U/mg solid	1 unit liberates 1.0 µmol glucose from salicin at pH 5.0, 37°C	Salt-free; lyophilized		ICN 100348
4-12 U/mg solid	1 unit liberates 1.0 µmol glucose from salicin/min at pH 5.0, 37°C	Salt-free; lyophilized		Sigma G0395
20-40 U/mg solid	1 unit liberates 1.0 µmol glucose from salicin/min at pH 5.0, 37°C	Chromatographically purified; salt-free; lyophilized		Sigma G4511
≥10 U/mg solid	1 unit forms 1 µmol PNP/min at pH 5.0, 37°C	Lyophilized containing 50% BSA and glutathione as stabilizers	<0.0005% α-amylase	Toyobo BGH-201
≥2.5 U/mg solid	1 unit releases 1 µmol glucose from salicin/min at pH 5.0, 37°C	Partially purified; lyophilized		Worthington LS02198 LS02201 LS02204
Caldocellum saccharolyticum, recombinant expressed in *E. coli* (thermostable)				
300-1000 U/g solid	1 unit liberates 1.0 µmol glucose from salicin/min at pH 5.0, 37°C	Lyophilized		Sigma G6906
Phaseolus lunatus (optimum pH = 5.5, T = 60°C)				
0.05-0.15 U/mg		Lyophilized		Wako 122-03251
Trichoderma species (optimum pH = 4.8, T = 50-75°C)				
10 U/g		Solution		Biocatalysts G016L

3.2.1.22 α-Galactosidase

REACTION CATALYZED
Hydrolysis of terminal, non-reducing α-D-galactose residues in α-D-galactosides, including galactose oligosaccharides, galactomannans and galactolipids

SYSTEMATIC NAME
α-D-Galactoside galactohydrolase

SYNONYMS
Melibiase, αGal

REACTANTS
Terminal, non-reducing α-D-galactose residues, α-D-galactosides, galactose oligosaccharides, galactomannans, galactolipids, α-D-fucosides, α-D-galactose

NOTES
- Classified as a hydrolase: a glycosidase hydrolyzing o-glycosyl compounds

SPECIFIC ACTIVITY	UNITS DEFINITION	PREPARATION FORM	ADDITIONAL ACTIVITIES	SUPPLIER CATALOG NO.
Aspergillus niger				
50-150 U/mg protein	1 unit hydrolyzes 1.0 μmol o-nitrophenyl α-D-galactoside to ONP and D-galactose/min at pH 4.0, 25°C	Suspension in 3.2 M $(NH_4)_2SO_4$ and 50 mM NaOAc, pH 5.5	<0.5% β-xylosidase, β-Gal <1% β-NAGlc	Sigma G4408
Coffee beans, green (optimum pH = 7.2, MW = 329,000 Da)				
10 U/mg protein	1 unit converts 1 μmol 4-nitrophenyl-α-D-galactoside to product/min at 25°C	Suspension in 3.2 M $(NH_4)_2SO_4$, pH 6	<0.005% β-NAGlc, α- and β-glucosidase, α-mannosidase (4-nitrophenylglucoside substrates) <0.2% β-Gal (lactose substrate)	Boehringer 105023
10 U/mg protein	1 unit hydrolyzes 1.0 μmol p-nitrophenyl α-D-galactoside to PNP and D-galactose/min at pH 6.5, 25°C	Suspension in 3.2 M $(NH_4)_2SO_4$ and BSA, pH 6.0		Sigma G8507
Escherichia coli				
20-40 U/mg protein	1 unit hydrolyzes 1.0 μmol p-nitrophenyl α-D-galactoside to PNP and D-galactose/min at pH 6.5, 25°C	Lyophilized containing 20% protein		Sigma G6762
Mortierella vinacea (optimum pH = 5.5-6.0, T = 40°C)				
200 U/mg	1 unit releases 1 μmol PNP from p-nitrophenyl-α-D-galactoside/min at pH 5.9, 40°C	Lyophilized		Seikagaku 100560-1
Xanthomonas manihotis				
1000 U/mL	1 unit cleaves >95% terminal α-D-galactose from 1 nmol Galα1-3Galβ1-4Gal-AMC/hr at pH 6.0, 37°C in a 10 μL reaction	20 mM Tris-HCl, 50 mM NaCl, 0.1 mM Na_2EDTA, 0.02% NaN_3, pH 7.5	No detectable exoglycosidase, protease	NE Biolabs 725S 725L

β-Galactosidase 3.2.1.23

REACTION CATALYZED
Hydrolysis of terminal, non-reducing β-D-galactose residues in β-D-galactosides

SYSTEMATIC NAME
β-D-Galactoside galactohydrolase

SYNONYMS
β-1,3-Galactosidase, lactase, βGal, GAH

REACTANTS
β-D-Galactosides, α-L-arabinosides, β-D-fucosides, β-D-glucosides, β-D-galactose, o-nitrophenyl-β-D-galactoside, o-nitrophenyl-β-D-galactopyranoside (ONPG), galactose, H_2O

APPLICATIONS
- Determination of lactose
- Synthesis of β-galactosyldipeptides and β-1,3-digalactosylserine
- Highly sensitive enzyme-linked immunoassay of human IgG with 4-methylumbelliferyl-β-D-galactoside
- Synthesis of glucose and galactose from lactose, with some production of galactobiose and -triose
- Starch hydrolysis
- Lactose reduction and flavor modification in dairy applications
- Reporter enzyme in immunoassays
- Fluorimetric assays
- Colorimetric assays, with ONPG

NOTES
- Classified as a hydrolase: a glycosidase hydrolyzing o-glycosyl compounds
- A tetramer of four identical subunits, each with an active site which may be independently active; probably involves a galactosyl-enzyme complex intermediate
- Widespread in animals, plants and microorganisms
- Some enzymes hydrolyze α-L-arabinosides
- Some animal enzymes also hydrolyze β-D-fucosides and β-D-glucosides
- Rich in glycine and leucine; poor in tryptophan, histidine and sulfur-containing amino acids
- Not found in eukaryotes
- Can be coupled to other protein via its -SH groups
- 100X kinetic preference for β-1,3-links over β-1,6-; 500X preference for β-1,3-links over β-1,4
- Activated by monovalent cations, and some alcohols, including 2-mercaptoethanol
- Protected against heat inactivation by 5′-phosphorylribose 1-pyrophosphate in the presence of β-mercaptoethanol
- Inhibited by Ag^+, Hg^{2+}, PCMB, MIA, O-phenanthroline, α-galactosides, chelating agents, organomercurics and heavy metals
- Inactivated by Mg^{2+}; *E. coli* enzyme stabilized by Mg^{2+}
- Powder stabilized by sugars and sugar alcohol

3.2.1.23 β-Galactosidase continued

SPECIFIC ACTIVITY	UNITS DEFINITION	PREPARATION FORM	ADDITIONAL ACTIVITIES	SUPPLIER CATALOG No.
Aspergillus niger (optimum pH = 4.5-6.5)				
20-40 U/mg protein	1 unit hydrolyzes 1.0 μmol o-nitrophenyl β-D-galactoside to ONP and D-galactose/min at pH 4.0, 25°C	Suspension in 3.5 M $(NH_4)_2SO_4$ and 50 mM NaOAc, pH 5.2	<1% β-NAGlc, β-xylosidase <2% α-Gal	Sigma G3522
		Food grade		EnzymeDev Enzeco[R] Fungal Lactase
		Food grade		EnzymeDev Enzeco[R] Fungal Lactase
4 U/mg solid, lactose substrate	1 unit hydrolyzes 1.0 μmol o-nitrophenyl β-D-galactoside to ONP and D-galactose/min at pH 4.5, 30°C	Stabilized with starch		Sigma G7138
Aspergillus species (optimum pH = 4.5-5.0, pI = 4.2, T = 55-60°C, MW = 105,000 Da; K_M = 1.3 mM [o-nitrophenyl-β-D-galactoside], 0.69 mM [p-nitrophenyl-β-D-galactoside], 5.4 mM [phenyl-β-D-galactoside], 64 mM [lactose]; stable pH 3.5-8.5 [25°C, 20 hr], T < 45°C [pH 5.0, 30 min])				
65,000 U/g		Powder		Biocatalysts L017P
180-190 U/mg protein	1 unit forms 1 μmol ONP/min at pH 5.0, 37°C	Lyophilized containing 15% mannitol		ICN 153489
180-190 U/mg protein; ≥100 U/mg solid	1 unit forms 1 μmol ONP/min at pH 5.0, 37°C	Lyophilized containing 15% mannitol	No detectable α-Gal, α- and β-glucosidase, α- and β-mannosidase, proteinase	Toyobo GAH-211 219
180-190 U/mg protein; ≥2000 U/mL	1 unit forms 1 μmol ONP/min at pH 5.0, 37°C	50% glycerol solution	No detectable α-Gal, α- and β-glucosidase, α- and β-mannosidase, proteinase	Toyobo GAH-211 219
Bacterial				
≥600 IU/mg protein	1 IU transforms 1 μmol substrate/min under standard IUB conditions at 25°C	Lyophilized	<0.001% α-Gal, α- and β-glucosidase	OYC
Bovine liver				
0.15 U/mg protein	1 unit hydrolyzes 1.0 μmol o-nitrophenyl β-D-galactoside to ONP and D-galactose/min at pH 7.3, 37°C	Lyophilized containing 95% protein and 5% buffer salts		Sigma G1875
Bovine testes (optimum pH = 4.3, MW = 67,000 Da)				
3 U/mg protein	1 unit converts 1 μmol 4-nitrophenyl-β-D-galactopyranoside to product/min at pH 4.3, 37°C	Solution containing 20 mM Tris-HCl, 100 mM KCl, 1 mM EDTA, 0.1% Micr-O-protect, pH 7.5	No detectable proteases <0.01% α-mannosidase, α-glucosidase, sialidase <0.05% α-L-fucosidase, β-glucosidase <1% β-NAH	Boehringer 903345

β-Galactosidase continued — 3.2.1.23

SPECIFIC ACTIVITY	UNITS DEFINITION	PREPARATION FORM	ADDITIONAL ACTIVITIES	SUPPLIER CATALOG NO.
Bovine testes continued				
1-3 U/mg protein	1 unit hydrolyzes 1.0 μmol o-nitrophenyl β-D-galactoside to ONP and D-galactose/min at pH 4.4, 25°C	Suspension in 3.2 M $(NH_4)_2SO_4$, pH 5	<0.1% α-Gal, α- and β-glucosidase, α-fucosidase, α-mannosidase <1% β-NAGlc	Sigma G4142
Canavalia ensiformis (Jack bean) (optimum pH = 3.5, T = 37°C)				
0.2 U/mg protein	1 unit releases 1 μmol PNP from p-nitrophenyl-β-D-galactoside/min at pH 3.5, 37°C	Lyophilized		Seikagaku 100570-1
10-25 U/mg protein	1 unit hydrolyzes 1.0 μmol o-nitrophenyl β-D-galactoside to ONP and D-galactose/min at pH 3.5, 25°C	Suspension in 3.0 M $(NH_4)_2SO_4$ and 25 mM Na citrate, pH 5.5		Sigma G0884
Diplococcus pneumoniae (optimum pH = 6.0-6.5, MW = 190,000 Da)				
25 U/mg protein; 1 U/mg total protein	1 unit converts 1 μmol 4-nitrophenyl-β-D-galactoside to product/min at pH 6.0, 37°C	Solution containing 20 mM Na cacodylate, 0.1% Micr-O-protect, 1 g/L BSA, pH 6.0	No detectable α- and β-mannosidase, α- and β-glucosidase, α-Gal, α-fucosidase, sialidase, proteases <0.01% endo-β-NAGlc D <0.05% β-NAGlc <0.2% endo-α-NAG	Boehringer 1088718
0.5 U/mg total protein	1 unit hydrolyzes 1.0 μmol o-nitrophenyl β-D-galactoside to ONP and D-galactose/min at pH 6.0, 37°C	Solution containing 20 mM Na cacodylate, 10 mM NaN_3, 1 mg/mL BSA, pH 6.0	No detectable neuraminidase or protease <0.01% endo-β-NAGlc D <0.05% NAGlc, α-Gal, α- and β-glucosidase, α-fucosidase, α- and β-mannosidase <0.2% endo-α-NAG	Sigma G0149
Escherichia coli (optimum pH = 6-8, pI = 4.61, T = 50-55°C, MW = 540,000 Da [4 subunits of 135,000 Da each]; K_M = 0.3 mM [o-nitrophenyl-β-D-galactoside], 67 μM [p-nitrophenyl-β-D-galactoside], 0.23 mM [phenyl-β-D-galactoside], 2.5 mM [lactose]; stable pH 6.5-8.5 [25°C, 20 hr], T < 50°C [pH 7.3, 15 min])				
450 U/mg	1 unit hydrolyzes 1 μmol 2-nitrophenyl-β-D-galactopyranoside/min at pH 7.8, 37°C			Fluka 48274
600 U/mg protein	1 unit hydrolyzes 1 μmol 2-nitrophenyl-β-D-galactopyranoside/min at pH 7.8, 37°C	Lyophilized containing PO_4 buffer, sucrose as stabilizers and 28% protein		Fluka 48275
300 U/mg protein	1 unit oxidizes 1 μmol Glc6P/min at pH 7.6, 25°C	Suspension in 3.2 M $(NH_4)_2SO_4$ with BSA as stabilizer, pH 6.0	<0.001% α-Gal, β-fructosidase	Fluka 48276
≥50 U/mg	1 unit hydrolyzes 1 μmol o-nitrophenyl-β-D-galactopyranoside/min at pH 7.5, 25°C	Partially purified; lyophilized		ICN 104939

SPECIFIC ACTIVITY	UNITS DEFINITION	PREPARATION FORM	ADDITIONAL ACTIVITIES	SUPPLIER CATALOG NO.
Escherichia coli continued				
300-400 U/mg protein	1 unit hydrolyzes 1 μmol o-nitrophenyl-β-D-galactopyranoside/min at pH 7.5, 25°C	Chromatographically purified; suspension in 1.6 M $(NH_4)_2SO_4$		ICN 150039
>750 U/mg; >20 mg/mL	1 unit hydrolyzes 1 μmol ONPG to ONP and D-galactose/min at pH 7.3, 37°C	Single-band by SDS-PAGE; solution containing 50% glycerol and 25 mM $NaPO_4$, pH 7.0		ProZyme BG10 000 BG10 010 BG10 100
700-1000 U/mg	1 unit forms 1 μmol ONP/min at 37°C	Lyophilized		Scripps G0114
700-1000 U/mg	1 unit forms 1 μmol ONP/min at 37°C	Solution containing 40% glycerol, 30 mM $NaPO_4$, 1-5 mg/mL protein, pH 7.0		Scripps G0124
250-600 U/mg protein	1 unit hydrolyzes 1.0 μmol o-nitrophenyl β-D-galactoside to ONP and D-galactose/min at pH 7.3, 37°C	Purified; suspension in 1.7 M $(NH_4)_2SO_4$, 10 mM Tris buffer salts, 10 mM $MgCl_2$, pH 7.3		Sigma G2513
600-1200 U/mg protein	1 unit hydrolyzes 1.0 μmol o-nitrophenyl β-D-galactoside to ONP and D-galactose/min at pH 7.3, 37°C	Lyophilized containing 80% protein; balance primarily Tris buffer salts and $MgCl_2$; no BSA		Sigma G5635
250-500 U/mg protein	1 unit hydrolyzes 1.0 μmol o-nitrophenyl β-D-galactoside to ONP and D-galactose/min at pH 7.3, 37°C	Partially purified; lyophilized containing 95% protein; balance primarily Tris buffer salts and $MgCl_2$		Sigma G6008
600-1200 U/mg protein	1 unit hydrolyzes 1.0 μmol o-nitrophenyl β-D-galactoside to ONP and D-galactose/min at pH 7.3, 37°C	Purified; suspension in 1.7 M $(NH_4)_2SO_4$, 10 mM Tris buffer salts, 10 mM $MgCl_2$, pH 7.3		Sigma G6512
≥700 U/mg protein; ≥500 U/mg solid	1 unit forms 1 μmol ONP/min at pH 7.3, 37°C	Lyophilized	No detectable α-Gal, α- and β-glucosidase, α- and β-mannosidase, proteinase	Toyobo GAH-201 209
≥700 U/mg protein; ≥10,500 U/mL	1 unit forms 1 μmol ONP/min at pH 7.3, 37°C	Suspension in 60% saturated $(NH_4)_2SO_4$	No detectable α-Gal, α- and β-glucosidase, α- and β-mannosidase, proteinase	Toyobo GAH-201 209
10-20 U/L		60% $(NH_4)_2SO_4$ suspension		Wako 077-02371
≥50 U/mg solid	1 unit hydrolyzes 1 μmol o-nitrophenyl-β-D-galactopyranoside/min at pH 7.5, 25°C	Partially purified; powder		Worthington LS04090 LS04093
≥300 U/mg protein	1 unit hydrolyzes 1 μmol o-nitrophenyl-β-D-galactopyranoside/min at pH 7.5, 25°C	Chromatographically purified; suspension in 1.6 M $(NH_4)_2SO_4$		Worthington LS04099 LS04100 LS04102

SPECIFIC ACTIVITY	UNITS DEFINITION	PREPARATION FORM	ADDITIONAL ACTIVITIES	SUPPLIER CATALOG NO.
Escherichia coli, recombinant overproducer (optimum pH = 7.0, MW = 540,000 Da)				
300 U/mg (2-nitrophenyl-β-D-galactoside substrate at 37°C); 30 U/mg (lactose substrate at 25°C)	1 unit converts 1 μmol substrate to product/min at specified temperature	Suspension in 3.2 M (NH$_4$)$_2$SO$_4$, pH 6	<0.001% β-fructosidase, α-Gal, GDH, α-glucosidase, NADH oxidase	Boehringer 105031 634395
600 U/mg protein	1 unit converts 1 μmol 2-nitrophenyl-β-D-galactoside to product/min at 37°C	EIA-Grade; lyophilized containing PO$_4$ buffer and sucrose stabilizers, ≥12 free SH-groups/enzyme molecule	≥12 moles/mole free thiol groups <3% aggregated β-Gal (HPLC)	Boehringer 567779 745731
>300 U	1 unit hydrolyzes 1 μmol o-nitrophenyl-β-D-galactoside/min at pH 7.0, 28°C	Purity by SDS-PAGE; lyophilized containing 5 mM NaPO$_4$ buffer, pH 7.2		Life Technol 15531-015
	1 unit hydrolyzes 1.0 μmol o-nitrophenyl β-D-galactoside to ONP and D-galactose/min	Lyophilized containing PO$_4$ buffer and sucrose as stabilizers		Sigma G3153
Kluyveromyces fragilis (optimum pH = 6.5, T = 45°C; stable pH 6.5 [40°C, 1 hr; 20-30°C, 6 hr] and 40°C [pH 6.5-7, 1 hr; pH 6.5-7.5, 3 hr; pH 6.5-8, 5 hr])				
5 U/mg	1 unit hydrolyzes 1 μmol 2-nitrophenyl-β-D-galactopyranoside/min at pH 7.8, 37°C	Powder		Fluka 48277
3000 LAU/mL	1 LAU releases 1 μmol glucose/min at pH 6.5, 37°C	Food grade; highly purified; solution containing glycerol		Novo Nordisk Lactozym^R 3000 L
Saccharomyces fragilis				
4-12 U/mg protein	1 unit hydrolyzes 1.0 μmol o-nitrophenyl β-D-galactoside to ONP and D-galactose/min at pH 7.2, 37°C	Partially purified; 50% glycerol solution		Sigma G3782
Streptococcus strain 6646k (optimum pH = 5.5, T = 40°C)				
20 U/mg protein; 0.1 U/vial	1 unit releases 1 μmol PNP from p-nitrophenyl-β-D-galactoside/min at pH 5.5, 37°C	Lyophilized		Seikagaku 100573-1
Xanthomonas manihotis				
1000 U/mL	1 unit cleaves >95% terminal β-D-galactose from 1 nmol Galβ1-3GlcNAcβ1-3Galβ1-4Glc-AMC/hr at pH 4.5, 37°C in a 10 μL reaction	20 mM Tris-HCl, 50 mM NaCl, 0.1 mM Na$_2$EDTA, 0.02% NaN$_3$, pH 7.5	No detectable exoglycosidase, protease	NE Biolabs 726S 726L

α-Mannosidase

3.2.1.24

REACTION CATALYZED
Hydrolysis of terminal, non-reducing α-D-mannose residues in α-D-mannosides

SYSTEMATIC NAME
α-D-Mannoside mannohydrolase

REACTANTS
α-D-Mannosides, α-D-lyxosides, heptopyranosides, α-D-mannose

APPLICATIONS
- Study of sequence and anomeric configuration of sugars in various complex carbohydrates

NOTES
- Classified as a hydrolase: a glycosidase hydrolyzing o-glycosyl compounds
- Also hydrolyzes α-D-lyxosides and heptopyranosides with the same configuration as mannose at C-2, C-3 and C-4

SPECIFIC ACTIVITY	UNITS DEFINITION	PREPARATION FORM	ADDITIONAL ACTIVITIES	SUPPLIER CATALOG NO.
Almond				
15-30 U/mg protein	1 unit hydrolyzes 1.0 μmol p-nitrophenyl α-D-mannoside to PNP and D-mannose/min at pH 4.5, 25°C	Solution containing 10 mM KPO_4 buffer and 0.1% NaN_3, pH 6.0	0.2% β-glucosidase	Sigma M1266
Canavalia ensiformis (Jack bean) (optimum pH = 3.5-4.5, T = 37°C, MW = 190,000 Da)				
10 U/mg	1 unit converts 1 μmol 4-nitrophenyl-α-D-mannoside to product/min at 25°C	Suspension in 3.2 M $(NH_4)_2SO_4$, pH 6	<1% NAGlc (4-nitrophenyl glycoside as substrate)	Boehringer 107379
				ICN 321251
5 U/mg protein	1 unit releases 1 μmol PNP from p-nitrophenyl-α-D-mannoside/min at pH 4.5, 37°C	Lyophilized		Seikagaku 100962-1
20 U/mg protein	1 unit hydrolyzes 1.0 μmol p-nitrophenyl α-D-mannoside to PNP and D-mannose/min at pH 4.5, 25°C	Suspension in 3.0 M $(NH_4)_2SO_4$ and 0.1 mM ZnOAc, pH 7.5		Sigma M7257
10 U/mg		Solution		Wako 135-10751
T. cornutus				
4.3 U/mg	1 unit releases 1 μmol PNP or phenol/min at pH 4.0, 37°C	Lyophilized	<1% glycosidase	ICN 321261 321262

β-Mannosidase 3.2.1.25

REACTION CATALYZED
Hydrolysis of terminal, non-reducing β-D-mannose residues in β-D-mannosides

SYSTEMATIC NAME
β-D-Mannoside mannohydrolase

SYNONYMS
Mannanase, mannase

REACTANTS
β-D-Mannosides, β-D-mannose

NOTES
- Classified as a hydrolase: a glycosidase hydrolyzing o-glycosyl compounds

SPECIFIC ACTIVITY	UNITS DEFINITION	PREPARATION FORM	ADDITIONAL ACTIVITIES	SUPPLIER CATALOG NO.
Achatina fulica (Snail) (optimum pH = 4.5, T = 37°C)				
5 U/mg protein	1 unit releases 1 μmol PNP from p-nitrophenyl-β-D-mannopyranoside/min at pH 4.5, 37°C	Lyophilized		Seikagaku 100963-1
5-30 U/mL	1 unit hydrolyzes 1.0 μmol p-nitrophenyl β-D-mannopyranoside to PNP and D-mannopyranoside/min at pH 4.0, 25°C	Acetone powder; suspension in 3.0 M $(NH_4)_2SO_4$ and 10 mM NaOAc, pH 4	1% β-NAGlc, α-mannosidase	Sigma M9400
Xanthomonas holcicola				
1000 U/mL	1 unit cleaves >95% of the terminal β-D-mannose from Man-β-1,4-Man- β-1,4-Man-7-amino-4-methylcoumarin/hr at pH 5.5, 37°C	20 mM Tris-HCl, 50 mM NaCl, 5 mM EDTA, 0.02% NaN_3, 100 μg/mL BSA, pH 7.5	No detectable exoglycosidase, protease	NE Biolabs 732S 732L

β-Fructofuranosidase 3.2.1.26

REACTION CATALYZED
Hydrolysis of terminal non-reducing β-D-fructofuranoside residues in β-D-fructofuranosides

SYSTEMATIC NAME
β-D-Fructofuranoside fructohydrolase

SYNONYMS
Invertase, saccharase

REACTANTS
β-D-fructofuranosides, sucrose, β-D-fructofuranoside residues, raffinose

APPLICATIONS
- Production of invert sugar (glucose and fructose) from sucrose
- Determination of sucrose
- Structural investigations of carbohydrates containing β-D-fructofuranoside links

β-Fructofuranosidase continued

3.2.1.26

NOTES
- Classified as a hydrolase: a glycosidase hydrolyzing o-glycosyl compounds
- A glycoprotein containing 50% carbohydrate
- Also catalyzes fructotransferase reactions
- Not active against inulin or melezitose

SPECIFIC ACTIVITY	UNITS DEFINITION	PREPARATION FORM	ADDITIONAL ACTIVITIES	SUPPLIER CATALOG No.
Candida species (optimum pH = 3.5-4.0, T = 60-70°C, MW = 260,000 Da; K_M = 15 mM [Fru]; stable at pH 4.0-6.0 [50°C, 10 min], T < 60°C [pH 4.5, 10 min])				
≥100 U/mg solid	1 unit forms 1 mg reducing sugars (glucose)/3 min at pH 4.5, 20°C; 1 unit hydrolyzes 1 mmol saccharose/min at 20°C	Lyophilized containing 70% KH_2PO_4 as stabilizer		Toyobo IVH-101
Candida utilis (optimum pH = 3.5-4.0, T = 60-70°C)				
100 U/mg	1 unit hydrolyzes 1.0 μmol sucrose to invert sugar/min at pH 4.5, 55°C		No detectable melibiase	ICN 151345
≥100 U/mg solid	1 unit forms 1 mg reducing sugars (as glucose)/3 min; 1 IU hydrolyzes 1 μmol saccharose/min	Contains 70% stabilizers		ICN 321701
	1 unit hydrolyzes 1.0 μmol sucrose to invert sugar/min at pH 4.5, 55°C	Chromatographically purified	No detectable lactose and α-Gal (melibiase)	Sigma I4753
≥100 U/mg	1 unit forms 1 mg reducing sugar (as glucose) from saccharose/3 min at pH 4.5, 37°C	Lyophilized		Seikagaku 100770-1
Saccharomyces cerevisiae (optimum pH = 5.5, T = 50°C)				
5 bed volumes/hr	pH 5.5 at 50°C	Immobilized granulate in food grade salt buffer solution		UOP Hexomax™ INV-100
Yeast (optimum pH = 4.6 [5.0-5.5], pI = 3.8, T = 60-70°C, MW = 270,000 Da)				
300 U/mg solid	1 unit converts 1 μmol sucrose to product/min at 25°C	Lyophilized	<0.001% α-Gal (melibiase) <0.01% amyloglucosidase, β-Gal, α- and β-glucosidase	Boehringer 104914
Standardized to a 3.0/mL k value	1 WIA unit hydrolyzes 1.0 μmol sucrose/min at pH 4.5, 55°C; 1.0 k value equals 50 Wallerstein Invertase Activity Units where k value is the unimolecular reaction velocity constant as defined in A.O.A.C	Purified; solution		ICN 102057
4 U/mL		Concentrate		Wako 095-02112

β-Fructofuranosidase continued   3.2.1.26

SPECIFIC ACTIVITY	UNITS DEFINITION	PREPARATION FORM	ADDITIONAL ACTIVITIES	SUPPLIER CATALOG NO.
Yeast, bakers				
400 U/mg	1 unit hydrolyzes 1 μmol saccharose/min at pH 4.65, 25°C			Fluka 57628
100 U/mg	1 unit hydrolyzes 1 μmol saccharose/min at pH 4.65, 25°C	Powder containing up to 40% glucose	170 Summer-U/mg	Fluka 57629
400 U/mg solid	1 unit hydrolyzes 1.0 μmol sucrose to invert sugar/min at pH 4.5, 55°C		No detectable lactose and α-Gal (melibiase)	Sigma I4504
30-50 U/mg solid	1 unit hydrolyzes 1.0 μmol sucrose to invert sugar/min at pH 4.5, 55°C	Practical	No detectable lactose and α-Gal (melibiase)	Sigma I9253

α,α-Trehalase   3.2.1.28

REACTION CATALYZED
    α,α-Trehalose + H_2O ↔ 2 D-Glucose

SYSTEMATIC NAME
    α,α-Trehalose glucohydrolase

REACTANTS
    α,α-Trehalose, H_2O, D-glucose

NOTES
- Classified as a hydrolase: a glycosidase hydrolyzing o-glycosyl compounds

SPECIFIC ACTIVITY	UNITS DEFINITION	PREPARATION FORM	ADDITIONAL ACTIVITIES	SUPPLIER CATALOG NO.
Porcine kidney				
0.5-5 U/mg protein	1 unit converts 1.0 μmol trehalose to 2.0 μmol glucose/min at pH 5.7, 37°C; liberated glucose determined at pH 7.5	Solution containing 50% glycerol, 1% Triton X-100, 25 mM KPO_4, pH 6.5	≤1% α- and β-Gal, invertase, α- and β-glucosidase, amylase	Sigma T8778

3.2.1.31 β-Glucuronidase

REACTION CATALYZED
A β-D-glucuronoside + H_2O ↔ An alcohol + D-glucuronate

SYSTEMATIC NAME
β-D-Glucuronoside glucuronosohydrolase

SYNONYMS
GRS

REACTANTS
β-D-Glucuronoside, H_2O, alcohol, D-glucuronate, phenolphthalein glucuronide

APPLICATIONS
- Hydrolysis of steroid glucuronides

NOTES
- Classified as a hydrolase: a glycosidase hydrolyzing o-glycosyl compounds
- Found in tissue extracts of mammals and vertebrates, digestive juices of snails, mollusks, locusts, bacteria and plants. A structural protein of the endoplasmic reticulum
- A glycoprotein containing sialic acid
- Does not hydrolyze α- or β-glucosides
- Strongly and reversibly inhibited by various organic peroxides

SPECIFIC ACTIVITY	UNITS DEFINITION	PREPARATION FORM	ADDITIONAL ACTIVITIES	SUPPLIER CATALOG NO.
Abalone				
5 U/mg protein	1 unit forms 1 μmol PNP/min at pH 4.5, 37°C	$(NH_4)_2SO_4$ solution containing 10 mg/mL protein		Calzyme 188B0005
Abalone entrails				
400,000-800,000 U/g solid	1 unit liberates 1.0 μg phenolphthalein from phenolphthalein glucuronide/hr at pH 3.8, 37°C	Lyophilized	10,000-50,000 U/g solid sulfatase	Sigma G0258
Ampullaria (optimum pH = 4.2, T = 60°C, MW = 320,000 Da; stable 60°C [3 hr])				
20,000-40,000 Fishman U/mL		Purified; muggy liquid in $(NH_4)_2SO_4$		Wako 078-03141
Bovine liver (optimum pH = 4.4, MW = 290,000 Da)				
0.2 U/mg	1 unit liberates 1 μmol phenol-phthalein from phenolphthalein-glucuronide/min at pH 5.0, 37°C	Powder		Fluka 49309
0.04 U/mg	1 unit liberates 1 μmol phenol-phthalein from phenolphthalein-glucuronide/min at pH 4.66, 37°C	Powder	750 U/mg acc. to Fishman	Fluka 49310
250 U/g solid	1 unit cleaves 1 μmol phenolphthalein glucuronide/min at pH 4.5, 37°C; 1 U = 19,000 Fishman Units	Lyophilized		ICN 100352
500,000 U/g solid	1 unit liberates 1.0 μg phenolphthalein from phenolphthalein glucuronide/hr at pH 5.0, 37°C	Contains 15% salts		Sigma G0251

β-Glucuronidase continued 3.2.1.31

SPECIFIC ACTIVITY	UNITS DEFINITION	PREPARATION FORM	ADDITIONAL ACTIVITIES	SUPPLIER CATALOG NO.
Bovine liver *continued*				
3,000,000 U/g solid	1 unit liberates 1.0 μg phenolphthalein from phenolphthalein glucuronide/hr at pH 5.0, 37°C	Contains 50% salts		Sigma G0376
10,000,000 U/g solid	1 unit liberates 1.0 μg phenolphthalein from phenolphthalein glucuronide/hr at pH 5.0, 37°C	Contains 50% salts		Sigma G0501
200,000-400,000 U/g agarose; 1 mL yields 10,000 U	1 unit liberates 1.0 μg phenolphthalein from phenolphthalein glucuronide/hr at pH 5.0, 37°C	Insoluble enzyme attached to beaded agarose; suspension in 2.0 M $(NH_4)_2SO_4$, pH 7.0		Sigma G4506
5000 U/mL	1 unit liberates 1.0 μmol phenolphthalein from phenolphthalein glucuronide/hr at pH 5.0, 37°C	Solution containing 50% glycerol and acetate buffer, pH 5.0, 25°C		Sigma G4882
≥40 U/g solid	1 unit cleaves 1 μmol phenolphthalein glucuronide/min at pH 4.5, 37°C; 1 U = 19,000 Fishman U	Partially purified; lyophilized		Worthington LS02342 LS02344 LS02346 LS02348
Chlamys opercularis (scallop)				
5000-10,000 U/g solid	1 unit liberates 1.0 μg phenolphthalein from phenolphthalein glucuronide/hr at pH 3.8, 37°C	Crude; lyophilized	800 U/g solid sulfatase	Sigma G8635
Escherichia coli				
1.5 U/mg	1 unit liberates 1 μmol 4-nitrophenol/min at pH 6.5, 25°C with 4-nitrophenyl-glucuronide as substrate	Powder	30% (SDS-PAGE)	Fluka 49312
4 U/mg	1 unit liberates 1 μmol 4-nitrophenol/min at pH 6.5, 25°C with 4-nitrophenyl-glucuronide as substrate	Lyophilized	25% (SDS-PAGE)	Fluka 49313
0.04 U/mg	1 unit liberates 1 μmol 4-nitrophenol/min at pH 6.5, 25°C with 4-nitrophenyl-glucuronide as substrate	Powder	80% (SDS-PAGE)	Fluka 49314
5,000,000-20,000,000 U/g protein; 1000 U/vial	1 unit liberates 1.0 μg phenolphthalein from phenolphthalein glucuronide/hr at pH 6.8, 37°C	Purified; lyophilized containing 25% protein, PO_4 buffer, BSA as stabilizer		Sigma G5897
1,000,000-5,000,000 U/g protein	1 unit liberates 1.0 μg phenolphthalein from phenolphthalein glucuronide/hr at pH 6.8, 37°C	Lyophilized containing 60% protein, buffer salts, stabilizer		Sigma G7396

SPECIFIC ACTIVITY	UNITS DEFINITION	PREPARATION FORM	ADDITIONAL ACTIVITIES	SUPPLIER CATALOG NO.
Escherichia coli continued				
5,000,000-20,000,000 U/g protein	1 unit liberates 1.0 μg phenolphthalein from phenolphthalein glucuronide/hr at pH 6.8, 37°C	Purified; lyophilized containing 25% protein, buffer salts, stabilizer		Sigma G7646
5,000,000-20,000,000 U/g protein	1 unit liberates 1.0 μg phenolphthalein from phenolphthalein glucuronide/hr at pH 6.8, 37°C	Purified; solution containing 50% glycerol		Sigma G7771
20,000,000-60,000,000 U/g protein	1 unit liberates 1.0 μg phenolphthalein from phenolphthalein glucuronide/hr at pH 6.8, 37°C	Highly purified; lyophilized containing 30% protein, buffer salts, stabilizer		Sigma G7896
5,000,000-20,000,000 U/g protein; 1000 U/vial	1 unit liberates 1.0 μg phenolphthalein from phenolphthalein glucuronide/hr at pH 6.8, 37°C	Purified; lyophilized containing 25% protein		Sigma G8271
5,000,000-20,000,000 U/g protein; 1000 U/vial	1 unit liberates 1.0 μg phenolphthalein from phenolphthalein glucuronide/hr at pH 6.8, 37°C	Purified; lyophilized containing 25% protein and phosphate buffer		Sigma G8396
Escherichia coli K12 (RNase negative) (optimum pH = 6.0-6.5, MW = 220,000 Da)				
10 U/mg at 25°C; 20 U/mg at 37°C	1 unit converts 1 μmol 4-nitrophenyl-β-D-glucuronide to product/min at pH 7	50% glycerol solution, pH 6		Boehringer 127680 1585665
10 U/mg protein	1 unit liberates 1 μmol 4-nitrophenol/min at pH 6.5, 25°C with 4-nitrophenyl-glucuronide as substrate	50% glycerol solution, pH 6.5; stabilized with streptomycin		Fluka 49315
Helix aspersa (snail)				
250,000-500,000 U/g solid	1 unit liberates 1.0 μg phenolphthalein from phenolphthalein glucuronide/hr at pH 5.0, 37°C	Partially purified; powder	5000 U/g solid sulfatase	Sigma G4259
Helix pomatia (snail)				
5 U/mg protein	1 unit forms 1 μmol PNP/min at pH 4.5, 37°C	$(NH_4)_2SO_4$ solution containing 10 mg/mL protein		Calzyme 189B0005
3 U/mL	1 unit liberates 1 μmol phenolphthalein/min at pH 5.0, 37°C with phenolphthalein-glucuronide as substrate	Solution		Fluka 49283
3 U/mL	1 unit liberates 1 μmol phenolphthalein/min at pH 5.0, 37°C with phenolphthalein-glucuronide as substrate	Solution		Fluka 49284

SPECIFIC ACTIVITY	UNITS DEFINITION	PREPARATION FORM	ADDITIONAL ACTIVITIES	SUPPLIER CATALOG NO.
Helix pomatia (snail) continued				
9 U/mg	1 unit liberates 1 μmol phenolphthalein/min at pH 5.0, 37°C with phenolphthalein-glucuronide as substrate	Powder	<0.5% sulfatase	Fluka 49285
2000 U/mL	1 unit liberates 1 μmol phenolphthalein/min at pH 5.0, 37°C with phenolphthalein-glucuronide as substrate	Crude; solution	<0.5% sulfatase	Fluka 49286
17 U/mg	1 unit liberates 1 μmol phenolphthalein/min at pH 5.0, 37°C with phenolphthalein-glucuronide as substrate	Lyophilized	<0.2% acid phosphatase <0.5% sulfatase	Fluka 49287
300,000-400,000 U/g solid; 15,000-40,000 U/g solid (sulfatase activity)	1 unit liberates 1.0 μg phenolphthalein from phenolphthalein glucuronide/hr at pH 5.0, 37°C; 1 unit sulfatase hydrolyzes 1.0 μmol p-nitrocatechol sulfate/hr at pH 5.0, 37°C	Partially purified powder		ICN 152283
100,000 U/mL; 1000-5000 U/mL (sulfatase activity)	1 unit liberates 1.0 μg phenolphthalein from phenolphthalein glucuronide/hr at pH 5.0, 37°C; 1 unit sulfatase hydrolyzes 1.0 μmol p-nitrocatechol sulfate/hr at pH 5.0, 37°C	Crude; solution		ICN 152284
300,000-400,000 U/g solid	1 unit liberates 1.0 μg phenolphthalein from phenolphthalein glucuronide/hr at pH 5.0, 37°C	Partially purified; powder	15,000-40,000 U/g solid sulfatase	Sigma G0751
100,000 U/mL	1 unit liberates 1.0 μg phenolphthalein from phenolphthalein glucuronide/hr at pH 5.0, 37°C	Crude; solution; agglutinin removed	1000 U/mL sulfatase	Sigma G0762
400,000-600,000 U/g solid	1 unit liberates 1.0 μg phenolphthalein from phenolphthalein glucuronide/hr at pH 5.0, 37°C	Chromatographically purified; solution	15,000-40,000 U/g solid sulfatase	Sigma G1512
100,000 U/mL	1 unit liberates 1.0 μg phenolphthalein from phenolphthalein glucuronide/hr at pH 5.0, 37°C	Crude; solution	5000 U/mL sulfatase	Sigma G7017 G0876
100,000 U/mL	1 unit liberates 1.0 μg phenolphthalein from phenolphthalein glucuronide/hr at pH 5.0, 37°C	Sterile-filtered; crude; solution	1000-5000 U/mL sulfatase	Sigma G7770
100,000 U/mL	1 unit liberates 1.0 μg phenolphthalein from phenolphthalein glucuronide/hr at pH 5.0, 37°C	Crude; solution	1000 U/mL sulfatase	Sigma G8885

3.2.1.31 β-Glucuronidase continued

SPECIFIC ACTIVITY	UNITS DEFINITION	PREPARATION FORM	ADDITIONAL ACTIVITIES	SUPPLIER CATALOG No.
Patella vulgata (limpet)				
30 U/mg	1 unit liberates 1 μmol phenolphthalein/min at pH 5.0, 37°C with phenolphthalein-glucuronide as substrate	Powder	<2% sulfatase	Fluka 49325
1,000,000- 3,000,000 U/g solid	1 unit liberates 1.0 μg phenolphthalein from phenolphthalein glucuronide/hr at pH 3.8, 37°C	Lyophilized	Sulfatase (inhibited by 0.1 M PO_4)	Sigma G8132

3.2.1.32 Xylan Endo-1,3-β-Xylosidase

REACTION CATALYZED
   Random hydrolysis of 1,3-β-D-xylosidic linkages in 1,3-β-D-xylans

SYSTEMATIC NAME
   1,3-β-D-Xylan xylanohydrolase

SYNONYMS
   Xylanase, endo-1,3-β-xylanase

REACTANTS
   1,3-β-D-Xylans, 1,3-β-D-xylosidic linkages

APPLICATIONS
- Improving extractions
- Changing viscosity and functionality
- Products for starch processing
- Bioconversions

NOTES
- Classified as a hydrolase: a glycosidase hydrolyzing *o*-glycosyl compounds

SPECIFIC ACTIVITY	UNITS DEFINITION	PREPARATION FORM	ADDITIONAL ACTIVITIES	SUPPLIER CATALOG No.
Trichoderma longibrachiatum (T. reesei) (optimum pH = 4.5-6.0)				
3780 mUxyl.H/mg		Food grade; solution		Rohm Rohalase[R] 7118
		Food grade		EnzymeDev Enzeco[R] Hemicellulase
Trichoderma viride				
2.5 U/mg	1 unit liberates 1 μmol remazol brilliant blue R at pH 5.4, 30°C with remazol brilliant blue R-xylan as substrate	Powder		Fluka 95595

Hyaluronoglucosaminidase 3.2.1.35

REACTION CATALYZED
> Random hydrolysis of 1,4-linkages between N-acetyl-β-D-glucosamine and D-glucuronate residues in hyaluronate

SYSTEMATIC NAME
> Hyaluronate 4-glycanohydrolase

SYNONYMS
> Hyaluronidase

REACTANTS
> Hyaluronate, chondroitin, chondroitin 4-sulfate, chondroitin 6-sulfate, dermatan, glycosaminoglycan

APPLICATIONS
- Cell surface research
- Drug and pharmaceutical processing
- Tissue and cell dissociation
- Medical research

NOTES
- Classified as a hydrolase: a glycosidase hydrolyzing o-glycosyl compounds
- A glycoprotein containing 5% mannose and 2.7% glucosamine
- Also hydrolyzes 1,4-β-D-glycosidic linkages between N-acetyl-galactosamine or N-acetylgalactosamine sulfate and glucuronic acid in chondroitin, chondroitin 4- and 6-sulfates and dermatan
- Inhibited by Fe^{2+}, Fe^{3+}, Mn^{2+} and Cu^{2+}

SPECIFIC ACTIVITY	UNITS DEFINITION	PREPARATION FORM	ADDITIONAL ACTIVITIES	SUPPLIER CATALOG NO.
Bovine testes (optimum pH = 4.5-6, MW = 61,000 Da)				
≥300 U/mg solid	USP XXII-NF XVII	Lyophilized		Adv Biofact AY22-1
≥3000 U/mg solid	USP XXII-NF XVII	Lyophilized		Adv Biofact AY22-2
≥500 U/mg solid	1 unit causes the same turbidity reduction as the 'International Unit' (IU) as compared with the International Standard	Salt-free; lyophilized		Biozyme HYB3
≥1500 U/mg solid	1 unit causes the same turbidity reduction as the 'International Unit' (IU) as compared with the International Standard	Salt-free; lyophilized		Biozyme HYB4
≥2500 U/mg solid	1 unit causes the same turbidity reduction as the 'International Unit' (IU) as compared with the International Standard	Salt-free; lyophilized		Biozyme HYB6

3.2.1.35 Hyaluronoglucosaminidase continued

SPECIFIC ACTIVITY	UNITS DEFINITION	PREPARATION FORM	ADDITIONAL ACTIVITIES	SUPPLIER CATALOG NO.
Bovine testes *continued*				
2500 U/mg solid	1 unit causes the same turbidity reduction as the 'International Unit' (IU) as compared with the International Standard	Salt-free; lyophilized		Calbiochem 385931
0.5 U/mg	1 unit releases 1 μmol *N*-Ac-D-glucosamine end groups/min at pH 4.0, 37°C with hyaluronic acid K salt as substrate	Powder		Fluka 53717
0.1 U/mg	1 unit releases 1 μmol *N*-Ac-D-glucosamine end groups/min at pH 4.0, 37°C with hyaluronic acid K salt as substrate	Powder		Fluka 53718
0.1 U/mg	1 unit releases 1 μmol *N*-Ac-D-glucosamine end groups/min at pH 4.0, 37°C with hyaluronic acid K salt as substrate	Powder		Fluka 53719
0.02 U/mg	1 unit releases 1 μmol *N*-Ac-D-glucosamine end groups/min at pH 4.0, 37°C with hyaluronic acid K salt as substrate	Salt-free; lyophilized		Fluka 53720
≥300 USP U/mg		Lyophilized		ICN 100740
2500 U/mg solid	1 unit causes the same turbidity reduction as 1.0 unit of International Standard preparation	Salt-free; lyophilized		ICN 151276
≥2000 IU/mg solid		Salt-free; lyophilized		ICN 151277
300-500 U/mg solid	Assayed per USP XXII-NF XVII combined edition	Lyophilized		Sigma H3506
3000-15,000 U/mg solid	Assayed per USP XXII-NF XVII combined edition	Chromatographically purified; dialyzed; lyophilized		Sigma H3631
300 U/mg solid	Assayed per USP XXII-NF XVII combined edition	Sterile-filtered; lyophilized		Sigma H3757
750-1500 U/mg solid	Assayed per USP XXII-NF XVII combined edition	Lyophilized containing 90% protein; balance primarily buffer salts		Sigma H3884
≥300 USP/NF U/mg solid	1 unit corresponds to USP/NF XIII Unit and is referenced to a standard NF Hyaluronidase	Lyophilized		Worthington LS02594 LS02592 LS02591

Hyaluronoglucosaminidase continued

SPECIFIC ACTIVITY	UNITS DEFINITION	PREPARATION FORM	ADDITIONAL ACTIVITIES	SUPPLIER CATALOG No.
Bovine testes *continued*				
≥3000 USP/NF U/mg solid	1 unit corresponds to USP/NF XIII Unit and is referenced to a standard NF Hyaluronidase	Chromatographically purified; dialyzed; lyophilized	No detectable albumin-like proteins by gel electrophoresis	Worthington LS05477 LS05475 LS05474 LS05479
Ovine testes (optimum pH = 4.5-6.0, MW = 55,000 Da)				
≥300 U/mg solid	1 unit causes the same turbidity reduction as the 'International Unit' (IU) as compared with the International Standard	Salt-free; lyophilized		Biozyme HYO2
≥500 U/mg solid	1 unit causes the same turbidity reduction as the 'International Unit' (IU) as compared with the International Standard	Salt-free; lyophilized		Biozyme HYO3
≥1000 U/mg solid	1 unit causes the same turbidity reduction as the 'International Unit' (IU) as compared with the International Standard	Salt-free; lyophilized		Biozyme HYO4
≥1500 U/mg solid	1 unit causes the same turbidity reduction as the 'International Unit' (IU) as compared with the International Standard	Salt-free; lyophilized		Biozyme HYO5
5000 U/mg solid	1 unit causes the same turbidity reduction as the 'International Unit' (IU) as compared with the International Standard	Salt-free; lyophilized		Biozyme HYO6
5000 U/mg solid	1 unit causes the same turbidity reduction as the 'International Unit' (IU) as compared with the International Standard	Salt-free; lyophilized		Calbiochem 38594
25 U/mg	1 unit releases 1 μmol N-Ac-D-glucosamine end groups/min at pH 4.0, 37°C with hyaluronic acid K salt as substrate	Lyophilized		Fluka 53708
0.02 U/mg	1 unit releases 1 μmol N-Ac-D-glucosamine end groups/min at pH 4.0, 37°C with hyaluronic acid K salt as substrate	Salt-free; lyophilized		Fluka 53710

SPECIFIC ACTIVITY	UNITS DEFINITION	PREPARATION FORM	ADDITIONAL ACTIVITIES	SUPPLIER CATALOG NO.
Ovine testes *continued*				
0.03 U/mg	1 unit releases 1 μmol N-Ac-D-glucosamine end groups/min at pH 4.0, 37°C with hyaluronic acid K salt as substrate	Salt-free; lyophilized		Fluka 53712
≥5000 IU/mg solid		Salt-free; lyophilized		ICN 151271
500 IU/mg solid		Salt-free; lyophilized		ICN 151272
300 U/mg solid	1 unit causes the same turbidity reduction as 1.0 unit of International Standard preparation			ICN 151273
≥1500 IU/mg solid		Salt-free; lyophilized		ICN 151274
≥1000 IU/mg solid		Salt-free; lyophilized		ICN 151275
300 U/mg solid	Assayed per USP XXII-NF XVII combined edition	Lyophilized containing lactose		Sigma H2126
500 U/mg solid	Assayed per USP XXII-NF XVII combined edition	Lyophilized containing 20-50% lactose		Sigma H2251
1500 U/mg solid	Assayed per USP XXII-NF XVII combined edition	Lyophilized		Sigma H6254
Streptococcus hyalurolyticus				
2000 U/mg solid	1 unit changes OD in a solution of hyaluronic acid compared to a hyaluronidase International Standard	Salt-free; lyophilized	No detectable protease, lysozyme	ICN 151270

Hyaluronoglucuronidase 3.2.1.36

REACTION CATALYZED
Random hydrolysis of 1,3-linkages between β-D-glucuronate and N-acetyl-D-glucosamine residues in hyaluronate

SYSTEMATIC NAME
Hyaluronate 3-glycanohydrolase

SYNONYMS
Hyaluronidase

SUBSTRATES
Hyaluronate, chondroitin sulfates A and C

APPLICATIONS
- Facilitates absorption in intradermal administration of large fluid volumes

NOTES
- Classified as a hydrolase: a glycosidase hydrolyzing o-glycosyl compounds
- Most concentrated in bovine and ovine testes; also produced by bacteria

SPECIFIC ACTIVITY	UNITS DEFINITION	PREPARATION FORM	ADDITIONAL ACTIVITIES	SUPPLIER CATALOG NO.
Bovine testes (optimum pH = 7.0, T = 37°C, MW = 55,000 Da)				
≥300 U/mg protein	1 unit liberates 1 μmol N-Ac-glucosamine/min at pH 4.0, 37°C	Lyophilized containing 90% protein		Calzyme 090A0300
300 U/mg protein				Wako 080-06201
Leeches				
	1 unit produces color equivalent to 1 μg glucuronic acid from hyaluronic acid/hr using 3,5-dinitrosalicylic acid to develop color			Sigma H0140
Ovine testes				
≥1000 U/mg protein	1 unit liberates 1 μmol N-Ac-glucosamine/min at pH 4.0, 37°C	Lyophilized containing 90% protein		Calzyme 091A1000

3.2.1.37 Xylan 1,4-β-Xylosidase

REACTION CATALYZED
  Hydrolysis of 1,4-β-D-xylans to remove successive D-xylose residues from the non-reducing termini

SYSTEMATIC NAME
  1,4-β-D-Xylan xylohydrolase

SYNONYMS
  Xylobiase, β-xylosidase, exo-1,4-β-xylosidase

REACTANTS
  1,4-β-D-Xylans, xylobiose, D-xylose

NOTES
- Classified as a hydrolase: a glycosidase hydrolyzing o-glycosyl compounds
- Sheep liver enzyme shows additional exoglycosidase activities

SPECIFIC ACTIVITY	UNITS DEFINITION	PREPARATION FORM	ADDITIONAL ACTIVITIES	SUPPLIER CATALOG NO.
Aspergillus niger				
5-10 U/mg protein	1 unit hydrolyzes 1.0 μmol o-nitrophenyl β-D-xyloside to ONP and D-xylose/min at pH 5.0, 25°C	Partially purified; suspension in 3.5 M $(NH_4)_2SO_4$ and 50 mM NaOAc, pH 5.2	<3% β-NAGlc, α- and β-Gal	Sigma X3501

3.2.1.39 Glucan Endo-1,3-β-D-Glucosidase

REACTION CATALYZED
  Hydrolysis of 1,3-β-D-glucosidic linkages in 1,3-β-D-glucans

SYSTEMATIC NAME
  1,3-β-D-Glucan glucanohydrolase

SYNONYMS
  Endo-1,3-β-glucanase, laminarinase, fungal pectinase, glucanase

REACTANTS
  1,3-β-D-Glucans, laminarin, paramylon, pachyman

APPLICATIONS
- Improves wine filtration by decomposing carbohydrate-containing colloids

NOTES
- Classified as a hydrolase: a glycosidase hydrolyzing o-glycosyl compounds
- Very limited action on mixed-link (1,3-1,4)-β-D-glucans

Glucan Endo-1,3-β-D-Glucosidase continued

3.2.1.39

SPECIFIC ACTIVITY	UNITS DEFINITION	PREPARATION FORM	ADDITIONAL ACTIVITIES	SUPPLIER CATALOG NO.
Helix pomatia (Snail)				
0.3 U/mg protein	1 unit hydrolyzes 1 μmol OBR-hydroxyethylcellulose/min at pH 4.8, 30°C using β-D-glucan as substrate	Powder		Fluka 49103
Penicillium funiculosum				
≥500 U/g laminarinase		Solution stabilized with KCl, pH 5.0-7.0		ABM-RP Pectinase CPT
Trichoderma species (optimum pH = 5.5, T = 50°C)				
12,000 U/g glucanase		Powder		Biocatalysts Depol 39P D039P

α-L-Rhamnosidase

3.2.1.40

REACTION CATALYZED

Hydrolysis of terminal non-reducing α-L-rhamnose residues in α-L-rhamnosides

SYSTEMATIC NAME

α-L-Rhamnoside rhamnohydrolase

REACTANTS

α-L-Rhamnosides, α-L-rhamnose

NOTES

- Classified as a hydrolase: a glycosidase hydrolyzing o-glycosyl compounds

See Chapter 5. Multiple Enzyme Preparations

α-Dextrin Endo-1,6-α-Glucosidase

REACTION CATALYZED

Hydrolysis of 1,6-α-D-glucosidic linkages in pullulan, amylopectin and glycogen, and in the α- and β-amylase limit dextrins of amylopectin and glycogen

SYSTEMATIC NAME

α-Dextrin 6-glucanohydrolase

SYNONYMS

Limit dextrinase, debranching enzyme, amylopectin 6-glucanohydrolase, pullulanase

REACTANTS

Pullulan, amylopectin, glycogen, limit dextrins, maltose, maltodextrins

APPLICATIONS

- Continuous hydrolysis of soluble starch for the production of maltose
- Debranching starch after liquefaction

NOTES

- Classified as a hydrolase: a glycosidase hydrolyzing o-glycosyl compounds
- Requires a minimum of two glucose units in the carbohydrate side chain
- Waxy starch substrate produces maltodextrins free of D-glucose

SPECIFIC ACTIVITY	UNITS DEFINITION	PREPARATION FORM	ADDITIONAL ACTIVITIES	SUPPLIER CATALOG NO.
Bacillus acidopullulyticus (optimum pH = 5, T = 60°C; heat stable; inactivated by heating to 85°C/5 min or 80°C/40 min)				
200 PUN/g	1 PUN hydrolyzes pullulan, liberating reducing carbohydrate with reducing power equivalent to 1 μmol glucose/min at pH 5.0, 40°C	Food grade; solution		Novo Nordisk Promozyme^R 200 L
600 PUN/g	1 PUN hydrolyzes pullulan, liberating reducing carbohydrate with reducing power equivalent to 1 μmol glucose/min at pH 5.0, 40°C	Food grade; solution		Novo Nordisk Promozyme^R 600 L
Bacillus species				
1 U/mg	1 unit releases 1 μmol maltotriose/min at pH 5.0, 37°C using pullulan as substrate	Powder		Fluka 82577
Enterobacter aerogenes				
10-30 U/mg protein	1 unit liberates 1.0 μmol maltotriose (as glucose) from pullulan/min at pH 5.0, 25°C	Lyophilized containing 10% protein; balance KPO$_4$ buffer salts and stabilizer		Sigma P1067
10-30 U/mg protein	1 unit liberates 1.0 μmol maltotriose (as glucose) from pullulan/min at pH 5.0, 25°C	Suspension in 3.2 M (NH$_4$)$_2$SO$_4$, pH 6.2		Sigma P5420

α-Dextrin Endo-1,6-α-Glucosidase continued

3.2.1.41

SPECIFIC ACTIVITY	UNITS DEFINITION	PREPARATION FORM	ADDITIONAL ACTIVITIES	SUPPLIER CATALOG NO.
Klebsiella planticola (optimum pH = 5.2, T = 55°C; stable pH 4.5-7.5 [50°C, 60 hr], T < 60°C [pH 5, 1 hr])				
750 U/g	1 unit produces 1 mg maltose equivalent from pullulan/min at pH 5.0, 50°C	Solution stabilized with salt, pH 5.5-7.5		ABM-RP Pulluzyme 750 L
Klebsiella pneumoniae (Aerobacter aerogenes) (optimum pH = 5.5-6.0, T = 45-55°C, MW = 143,000 Da)				
2000 U/g	1 unit liberates 1.0 μmol maltotriose from pullulan/min at pH 6.0, 30°C	Partially purified; powder in diatomaceous earth		Hayashibara EN201
40 U/mg protein	1 unit liberates 1.0 μmol maltotriose from pullulan/min at pH 6.0, 30°C	Purified; crystallized; suspension in 3.2 M $(NH_4)_2SO_4$, pH 6.0		Hayashibara EN202
400 U/10 mg	1 unit releases 1 μmol maltotriose from pullulan/min at pH 5.0, 30°C	Crystallized; suspension in 2.7 M $(NH_4)_2SO_4$		ICN 321721
1000 U/0.5 g	1 unit releases 1 μmol maltotriose from pullulan/min at pH 5.0, 30°C	Crude; partially purified; powder containing diatomaceous earth		ICN 321731
2 U/mg	1 unit releases 1 μmol maltotriose from pullulan/min at pH 5.0, 30°C	Crude; powder		Seikagaku 120160-1
40 U/mg	1 unit releases 1 μmol maltotriose from pullulan/min at pH 5.0, 30°C	Crystallized; suspension in 2.7 M $(NH_4)_2SO_4$		Seikagaku 120162-1
40 U/mg protein		Suspension		Wako 165-15651

Glucosylceramidase

3.2.1.45

REACTION CATALYZED

D-Glucosyl-N-acylsphingosine + H_2O ↔ D-Glucose + N-acylsphingosine

SYSTEMATIC NAME

D-Glucosyl-N-acylsphingosine glucohydrolase

SYNONYMS

Psychosine hydrolase, glucosphingosine glucosylhydrolase, ceramide glycanase

REACTANTS

D-Glucosyl-N-acylsphingosine, glucosylsphingosine, H_2O, D-glucose, N-acylsphingosine, oligosaccharide glycosylceramides, gangliosides

APPLICATIONS

- Separates the oligosaccharide moiety from ceramide to facilitate complex glycolipid analysis

NOTES

- Classified as a hydrolase: a glycosidase hydrolyzing o-glycosyl compounds
- Hydrolyzes the Glcβ1→1Cer linkage
- Does not attack galactosylceramides

3.2.1.45 Glucosylceramidase continued

SPECIFIC ACTIVITY	UNITS DEFINITION	PREPARATION FORM	ADDITIONAL ACTIVITIES	SUPPLIER CATALOG NO.
Leeches (optimum pH = 4.0-5.0)				
14 mU/mg protein	1 unit converts 1 μmol monosialoganglioside GM$_1$ to product/min at 37°C	Solution containing 50 mM NaOAc and 10 mM NaN$_3$, pH 6	No detectable β-glucosidase, β-Gal, α- and β-mannosidase, α-fucosidase, α-NAGlc <0.1% sialidase <5% β-hexosaminidase <10% α-Gal	Boehringer 1143085
0.01 U/mg protein	1 unit hydrolyzes 1.0 μmol monosialoganglioside GM$_1$/min at pH 5.0, 37°C	Solution containing 50 mM NaOAc and 10 mM NaN$_3$, pH 6.0		Sigma C2557

3.2.1.49 α-N-Acetylglucosaminidase

REACTION CATALYZED
  Hydrolysis of terminal non-reducing N-acetyl-D-galactosamine residues in N-acetyl-α-D-galactosaminides

SYSTEMATIC NAME
  α-N-acetyl-D-galactosaminide N-acetylgalactosaminohydrolase, NAG

SYNONYMS
  α-N-Acetylgalactosaminidase

REACTANTS
  N-acetyl-α-D-galactosaminides

NOTES
- Classified as a hydrolase: a glycosidase hydrolyzing o-glycosyl compounds
- Splits N-acetylgalactosaminyl groups from O-3 of Ser and Thr

SPECIFIC ACTIVITY	UNITS DEFINITION	PREPARATION FORM	ADDITIONAL ACTIVITIES	SUPPLIER CATALOG NO.
Acremonium species (optimum pH = 4.0-4.5, T = 55°C)				
120 U/mg protein	1 unit releases 1 μmol PNP from p-nitrophenyl-α-D-N-Ac-galactosaminide/min at pH 4.5, 37°C	Solution containing 20% glycerol and 10 mM KPO$_4$ buffer, pH 7.0		Seikagaku 100086-1
Chicken liver				
5-20 U/mg protein	1 unit releases 1.0 μmol PNP from p-nitrophenyl-N-Ac-α-D-galactosaminide/min at pH 3.65, 37°C	Lyophilized containing <10% protein, trehalose, citrate, KPO$_4$		Sigma A9763

α-N-Acetylglucosaminidase continued — 3.2.1.49

SPECIFIC ACTIVITY	UNITS DEFINITION	PREPARATION FORM	ADDITIONAL ACTIVITIES	SUPPLIER CATALOG NO.
Squid liver (optimum pH = 2.7-3.5, MW = 500,000 Da)				
≥20 U/mg protein		Frozen		Wako 019-12421 015-12423

α-L-Fucosidase — 3.2.1.51

REACTION CATALYZED

An α-L-fucoside + H_2O ↔ An alcohol + L-fucose

SYSTEMATIC NAME

α-L-Fucoside fucohydrolase

REACTANTS

α-L-Fucoside, H_2O, alcohol, L-fucose

NOTES
- Classified as a hydrolase: a glycosidase hydrolyzing o-glycosyl compounds

SPECIFIC ACTIVITY	UNITS DEFINITION	PREPARATION FORM	ADDITIONAL ACTIVITIES	SUPPLIER CATALOG NO.
Bovine epididymis				
2-3 U/mg protein	1 unit hydrolyzes 1.0 μmol p-nitrophenyl α-L-fucoside to PNP and L-fucose/min at pH 6.5, 25°C	Suspension in 2.5 M $(NH_4)_2SO_4$, pH 5.8	<1.5% β-NAGlc	Sigma F7753
Bovine kidney (optimum pH = 5.0, MW = 217,000 Da [mammalian liver])				
2 U/mg	1 unit converts 1 μmol 4-nitrophenyl-α-L-fucoside to product/min at 25°C	Suspension in 3.2 M $(NH_4)_2SO_4$, pH 6	<0.2% α-mannosidase <5% β-NAGlc (at pH 4.5 in the presence of 0.2 M acetate, activity decreases to 0.3%)	Boehringer 104949
5-15 U/mg protein	1 unit hydrolyzes 1.0 μmol p-nitrophenyl α-L-fucoside to PNP and L-fucose/min at pH 5.5, 25°C	Suspension in 3.2 M $(NH_4)_2SO_4$, 10 mM $NaPO_4$, 10 mM citrate, pH 6.0	<0.1% α-mannosidase, β-Gal <0.2% β-NAGlc	Sigma F5884
2 U/mg protein		Suspension		Wako 061-02971
Fusarium oxysporium (optimum pH = 4.5-6.0, T = 60°C)				
110 U/mg	1 unit releases 1 μmol PNP from p-nitrophenyl-α-L-fucoside/min at pH 4.5, 37°C	Lyophilized		Seikagaku 100532-1

3.2.1.51 α-L-Fucosidase continued

SPECIFIC ACTIVITY	UNITS DEFINITION	PREPARATION FORM	ADDITIONAL ACTIVITIES	SUPPLIER CATALOG NO.
Fusarium oxysporium continued				
2 U/vial	1 unit hydrolyzes 1.0 μmol p-nitrophenyl α-L-fucoside to PNP and L-fucose/min	Lyophilized		Sigma F8768
Human placenta				
2-4 U/mg protein	1 unit hydrolyzes 1.0 μmol p-nitrophenyl α-L-fucoside to PNP and L-fucose/min at pH 5.6, 37°C	Suspension in 80% $(NH_4)_2SO_4$, 0.1 M $NaPO_4$/citrate buffer, pH 6.0		Sigma F6151

3.2.1.52 β-N-Acetylhexosaminidase

REACTION CATALYZED
  Hydrolysis of terminal, non-reducing N-acetyl-D-hexosamine residues in N-acetyl-β-D-hexosaminides

SYSTEMATIC NAME
  β-N-Acetyl-D-hexosaminide N-acetylhexosaminohydrolase

SYNONYMS
  Hexosaminidase, NAGlc, NAH

REACTANTS
  N-Acetyl-β-D-hexosaminides, N-acetylglucosides, N-acetylgalactosides, N-acetyl-D-hexosamine

APPLICATIONS
- Characterizing glycoconjugate lectin acceptors

NOTES
- Classified as a hydrolase: a glycosidase hydrolyzing o-glycosyl compounds

SPECIFIC ACTIVITY	UNITS DEFINITION	PREPARATION FORM	ADDITIONAL ACTIVITIES	SUPPLIER CATALOG NO.
Ascidiacea (optimum pH = 4.0, MW = 330,000 Da)				
≥170 U/mg protein		Frozen		Wako 012-12411
Bovine kidney (optimum pH = 4.5, MW = 150,000-160,000 Da [rat])				
4 U/mg	1 unit converts 1 μmol 4-nitrophenyl-N-Ac-β-D-glucosaminide to product/min at 25°C	Suspension in 3.2 M $(NH_4)_2SO_4$	<0.1% α-mannosidase <2% α-fucosidase (4-nitrophenyl-glycosides as substrate)	Boehringer 1017098

SPECIFIC ACTIVITY	UNITS DEFINITION	PREPARATION FORM	ADDITIONAL ACTIVITIES	SUPPLIER CATALOG NO.
Diplococcus pneumoniae (optimum pH = 5.0, MW = 180,000 Da)				
200 U/mg protein; 2 U/mg total protein	1 unit converts 1 μmol 4-nitrophenyl-N-Ac-β-D-glucosaminide to product/min at pH 4.8, 37°C	Solution containing 50 mM NaPO$_4$, 0.1% Micr-O-protect, 0.5 g/L BSA, pH 7.0	No detectable α- and β-Gal, α- and β-mannosidase, α- and β-glucosidase, α- and β-fucosidase, sialidase, proteases <0.02% endo-α-NAG <0.5% endo-β-NAGlc D	Boehringer 1088700
Jack bean (optimum pH = 5.5-8.5 [37°C, 24 hr] and 7 [55°C, 100 min])				
50 U/mg protein	1 unit hydrolyzes 1 μmol 4-nitrophenyl-N-acety-β-D-glucosaminide/min at pH 5.0, 25°C	Crystallized; suspension in 2.5 M (NH$_4$)$_2$SO$_4$, pH 7.0	<0.05% α-glucosidase, α-NAGlc, β-Gal, α-mannosidase, α-L-fucosidase	Fluka 01146
40 U/mg protein	1 unit releases 1 μmol PNP from p-nitrophenyl-β-D-N-Ac-glucosaminide/min at pH 5.0, 37°C	Lyophilized		Seikagaku 100094-1
Penicillium oxalicum (optimum pH = 3.0-4.5, T = 55°C)				
190 U/mg protein	1 unit releases 1 μmol PNP from p-nitrophenyl-β-D-N-Ac-galactosaminide/min at pH 4.5, 37°C	Solution containing 30% glycerol and 10 mM KPO$_4$ buffer, pH 7.5		Seikagaku 100096-1
Streptomyces plicatus, overexpressed in *E. coli* (MW = 100,000 Da)				
5000 U/mL	1 unit cleaves >95% β-D-N-Ac-galactosamine from 1 nmol GalNAcβ1-4Galβ1-4Glc-AMC/hr at pH 4.5, 37°C in a 10 μL reaction	20 mM Tris-HCl, 50 mM NaCl, 5 mM Na$_2$EDTA, pH 7.5	No detectable exoglycosidase, protease	NE Biolabs 721S 721L
T. cornutus				
50 U/mg	1 unit releases 1 μmol PNP or phenol/min at pH 4.0, 37°C	Lyophilized	50 U/mg β-NAGlc 10 U/mg β-NAG	ICN 321241

3.2.1.61 Mycodextranase

REACTION CATALYZED
Endohydrolysis of 1,4-α-D-glucosidic linkages in α-D-glucans containing both 1,3- and 1,4-bonds

SYSTEMATIC NAME
1,3-1,4-α-D-Glucan 4-glucanohydrolase

REACTANTS
α-D-Glucans, 1,4-α-D-glucosidic linkages, nigerose, 4-α-D-nigerosylglucose

NOTES
- Classified as a hydrolase: a glycosidase hydrolyzing o-glycosyl compounds
- No hydrolysis of α-D-glucans containing only 1,3- or 1,4- bonds

SPECIFIC ACTIVITY	UNITS DEFINITION	PREPARATION FORM	ADDITIONAL ACTIVITIES	SUPPLIER CATALOG NO.
Penicillium funiculosum				
50-100 U/mg protein	1 unit liberates 1.0 μmol reducing sugar (as glucose) from nigeran/min at pH 4.5, 37°C	Highly purified; lyophilized containing 50% protein; balance NaOAc and NaCl	<0.1% α-glucosidase <0.5% laminarinase 1% α-amylase	Sigma M3648

3.2.1.63 1,2-α-L-Fucosidase

REACTION CATALYZED
Methyl-2-α-L-fucopyranosyl-β-D-galactoside + H_2O ↔ L-Fucose + methyl β-D-galactoside

SYSTEMATIC NAME
2-α-L-Fucopyranosyl-β-D-galactoside fucohydrolase

SYNONYMS
Almond emulsin fucosidase II

REACTANTS
Methyl-2-α-L-fucopyranosyl-β-D-galactoside, H_2O, L-fucose, methyl β-D-galactoside

NOTES
- Classified as a hydrolase: a glycosidase hydrolyzing o-glycosyl compounds
- Highly specific for non-reducing terminal L-fucose residues linked to D-galactose residues by 1,2-α-linkage
- Not active against p-nitrophenyl-α-L-fucopyranoside
- Inhibited by Ag^+, Hg^{2+} and Cu^{2+}
- Not identical with 1,3-α-L-fucosidase (EC 3.2.1.111)

1,2-α-L-Fucosidase continued

3.2.1.63

SPECIFIC ACTIVITY	UNITS DEFINITION	PREPARATION FORM	ADDITIONAL ACTIVITIES	SUPPLIER CATALOG NO.
Arthrobacter oxidans (optimum pH = 8.5, T = 37°C, MW = 43,000 Da)				
	1 unit liberates 1 μmol PNP from p-nitrophenyl-α-L-fucopyranoside/min at pH 8.5, 30°C	Lyophilized		PanVera TAK 4451
Bacillus species K40T (optimum pH = 5.5-7.0; pH = 6.5, T = 37, 55°C, 30 min)				
60 U/mg	1 unit releases 1 μmol L-fucose from porcine gastric mucin/min at pH 6.5, 37°C	Salt- and stabilizer-free; lyophilized		Seikagaku 100534-1
Xanthomonas manihotis				
1000 U/mL	1 unit cleaves >95% terminal α-L-fuose from 1 nmol fucα1-2Galβ1-4Glc-AMC/hr at pH 6.0, 37°C in a 10 μL reaction	20 mM Tris-HCl, 50 mM NaCl, 0.1 mM Na$_2$EDTA, 5 mM CaCl$_2$, 0.02% NaN$_3$, 100 μg/mL BSA, pH 7.5	No detectable exoglycosidase, protease	NE Biolabs 724S 724L

Isoamylase

3.2.1.68

REACTION CATALYZED
  Hydrolysis of 1,6-α-D-glucosidic branch linkages in glycogen, amylopectin and their β-limit dextrins
SYSTEMATIC NAME
  Glycogen 6-glucanohydrolase
SYNONYMS
  Debranching enzyme
REACTANTS
  Glycogen, amylopectin, β-limit dextrins
APPLICATIONS
- Structural determination of polysaccharides

NOTES
- Classified as a hydrolase: a glycosidase hydrolyzing o-glycosyl compounds
- Distinguished from α-dextrin endo-1,6-α-glucosidase (EC 3.2.1.41) by the inability of isoamylase to attack pullulan and by limited action on α-limit dextrins. Completely hydrolyzes glycogen, in contrast to limited action by α-dextrin glucanohydrolase
- 1,6-Linkage hydrolyzed only if at a branch point

SPECIFIC ACTIVITY	UNITS DEFINITION	PREPARATION FORM	ADDITIONAL ACTIVITIES	SUPPLIER CATALOG NO.
Pseudomonas amyloderamosa (optimum pH = 3.0-4.0, T = 52°C)				
59,000 U/mg protein		Purified; crystallized; suspension in 0.5 saturated (NH$_4$)$_2$SO$_4$, pH 4.5		Hayashibara EN102

3.2.1.68 Isoamylase continued

SPECIFIC ACTIVITY	UNITS DEFINITION	PREPARATION FORM	ADDITIONAL ACTIVITIES	SUPPLIER CATALOG NO.
Pseudomonas amyloderamosa continued				
10,000-40,000 U/mg protein	1 unit increases A_{616} by 0.1/hr using rice starch as substrate	Suspension in 2.0 M $(NH_4)_2SO_4$		ICN 190106
59,000 U/mg protein				ICN 321711
59,000 U/mg protein	1 unit increases A_{610} by 0.1/hr	Suspension in 0.5 saturated $(NH_4)_2SO_4$		Seikagaku 100780-1
2,000,000-10,000,000 U/mg protein	1 unit increases A_{610} 0.1/hr using rice starch as substrate	Suspension 2.0 M $(NH_4)_2SO_4$		sigma I2758

3.2.1.70 Glucan 1,6-α-Glucosidase

REACTION CATALYZED

Hydrolysis of successive glucose residues from 1,6-α-D-glucans and derived oligosaccharides

SYSTEMATIC NAME

1,6-α-D-Glucan glucohydrolase

SYNONYMS

Exo-1,6-β-glucosidase, glucodextranase

REACTANTS

1,6-α-D-Glucans, dextrans, isomaltosaccharides, isomaltose (very slowly), glucose, β-D-glucose (inversion)

NOTES

- Classified as a hydrolase: a glycosidase hydrolyzing o-glycosyl compounds
- Some enzymes hydrolyze 1,3-α-D-glucosidic bonds in dextrans

SPECIFIC ACTIVITY	UNITS DEFINITION	PREPARATION FORM	ADDITIONAL ACTIVITIES	SUPPLIER CATALOG NO.
Arthrobacter globiformis (optimum pH = 6.0, pI = 4.31, T = 45°C, MW = 120,000 Da [SDS-PAGE])				
1-3 U/mg		Lyophilized		Wako 070-02741

Glucan 1,4-β-Glucosidase 3.2.1.74

REACTION CATALYZED
Hydrolysis of 1,4-linkages in 1,4-β-D-glucans to remove successive glucose units

SYSTEMATIC NAME
1,4-β-D-Glucan glucohydrolase

SYNONYMS
Exo-1,4-β-glucosidase

REACTANTS
1,4-β-D-Glucans, cellobiose, glucose

NOTES
- Classified as a hydrolase: a glycosidase hydrolyzing o-glycosyl compounds
- Acts on 1,4-β-D-glucans and related oligosaccharides
- Cellobiose is very slowly hydrolyzed

See Chapter 5. Multiple Enzyme Preparations

Mannan Endo-1,4-β-Mannosidase 3.2.1.78

REACTION CATALYZED
Random hydrolysis of 1,4-β-D-mannosidic linkages in mannans, galactomannans and glucomannans

SYSTEMATIC NAME
1,4-β-D-Mannan mannanohydrolase

SYNONYMS
Endo-1,4-β-mannanase, hemicellulase

REACTANTS
Mannans, galactomannans, glucomannans, gums

APPLICATIONS
- Preparation of plant protoplasts
- Reducing viscosity of coffee extract in the manufacturing of instant coffee

NOTES
- Classified as a hydrolase: a glycosidase hydrolyzing o-glycosyl compounds
- Hydrolyzes hexose and pentose polymers (gums), and galactomannans in seeds, coffee beans and locust beans

SPECIFIC ACTIVITY	UNITS DEFINITION	PREPARATION FORM	ADDITIONAL ACTIVITIES	SUPPLIER CATALOG NO.
Aspergillus niger (optimum pH = 4.5, T = 40-50°C; stable pH 3.0-10.0)				
≥2000 U/g		Powder containing maltodextrin		ABM-RP Cellulase AC

SPECIFIC ACTIVITY	UNITS DEFINITION	PREPARATION FORM	ADDITIONAL ACTIVITIES	SUPPLIER CATALOG No.
Aspergillus niger continued				
90,000 U/g	100 unit produces reducing sugar equivalent to 1 mg xylose/min at pH 4.5, 40°C	Pharmaceutical grade; powder		Amano Hemi-Cellulase
		Powder	Pentosanase, hexosanase, xylanase, fucosidase, mannosidase, hemicellulase, anthocyanase, pectinase and protease	Karlan 4525
1,500,000 VHCU/g	pH 5.0, 30°C based on locust bean gum	Food grade; solution containing 0.2% Na benzoate and 0.1% K sorbate optimum T = 60-70°C		Novo Nordisk Gamanase™ 1.5 L
0.01-0.1 U/mg solid (β-galactose dehydrogenase system and locust bean gum substrate)	1 unit liberates 1.0 μmol D-galactose from hemicellulose/hr at pH 5.5, 37°C in a 2 hr assay	Plant cell culture tested; powder stabilized with lactose	Cellulase	Sigma H0771
0.01-0.1 U/mg solid	1 unit liberates 1.0 μmol D-galactose from hemicellulose/hr at pH 5.5, 37°C using β-galactose dehydrogenase system and locust bean gum as substrate	Crude; powder containing lactose stabilizer	Cellulase	Sigma H2125
≥5 U/mg solid	1 unit liberates 1.0 μmol D-galactose from hemicellulose/hr at pH 5.5, 37°C; 2 hr assay	Powder	Cellulase	Sigma H7649
Aspergillus species (optimum pH = 4-5)				
20,000 GPU/g				Danisco GRINDAMYL™ H 121
0.005-0.05 U/mg solid; locust bean gum as substrate in a β-galactose dehydrogenase system	1 unit liberates 1.0 μmol D-galactose from hemicellulose/hr at pH 5.5, 37°C	Powder		ICN 151230
Pseudomonas species (optimum pH = 7-8, T = 30-40°C)				
			β-1,4-mannase β-1,3-xylanase porphylanase	Wako 015-13763 019-13761

Agarase 3.2.1.81

REACTION CATALYZED
Hydrolysis of 1,3-β-D-galactosidic linkages in agarose, giving the tetramer as the predominant product

SYSTEMATIC NAME
Agarose 3-glycanohydrolase

SYNONYMS
β-Agarase 1

REACTANTS
Agarose, porphyran, neoagarooligosaccharides

APPLICATIONS
- Recovering DNA, RNA and high molecular weight DNA fragments embedded in low melting temperature agarose after pulsed-field electrophoresis
- Solubilizing gel slices in preparative isolation of intact yeast artificial chromosomes
- Recovering nucleic acids

NOTES
- Classified as a hydrolase: a glycosidase hydrolyzing o-glycosyl compounds
- Unique in its ability to hydrolyze the agarose core into neoagarobiose oligosaccharides
- Carbohydrate products of agarase digestion do not interfere with subsequent restriction endonuclease digestion, ligation or transforming DNA
- Stabilized by agarose and BSA

SPECIFIC ACTIVITY	UNITS DEFINITION	PREPARATION FORM	ADDITIONAL ACTIVITIES	SUPPLIER CATALOG No.
Flavobacterium				
	1 unit completely degrades 200 μL molten 1% agarose/15 min at 42-47°C in 1X TBE	Cloning Qualified; ≥90% purity by SDS gel	<1% DNase <3% RNase ≥90% supercoiled plasmid	Promega AgarACE™ M1741 M1742
Mung bean				
15 U/μL	1 unit converts 1 μg heat-denatured [^{32}P]-ssDNA into acid-soluble counts/min at 37°C using ss-salmon sperm DNA as substrate	10 mM NaOAc, 0.1 mM ZnOAc, 1 mM Cys, 0.001% Triton X-100, 50% glycerol, pH 5.0	No detectable endonuclease, exonuclease	NBL Gene Gelase™ 022007 022008
Pseudomonas atlantica (optimum pH = 6.0 [5.0-8.5], T = 40-42°C; denatured at 95°C/2 min)				
0.1-1 U/μL	1 unit digests 200 μg low melting point agarose to neoagaro oligosaccharides/2 hr at 40°C	50 mM Bis Tris-HCl, 10 mM EDTA, 50% glycerol, pH 6.5		Amersham E 73430Y E 73430Z

SPECIFIC ACTIVITY	UNITS DEFINITION	PREPARATION FORM	ADDITIONAL ACTIVITIES	SUPPLIER CATALOG No.
Pseudomonas atlantica continued				
0.5-1.0 U/µL	1 unit digests 100 µL molten low-melting point agarose to neoagaro-oligosaccharides/hr at 45°C	Solution		Boehringer 1417215 1417223
6 U/mg	1 unit liberates 1 µmol reducing sugar (measured as D-galactose)/min at pH 6.0, 40°C with agar as substrate	Powder		Fluka 05053
	1 unit solubilizes 500 mg molten 1% LMP agarose/hr at 40-42°C	Purified to a single band on SDS-PAGE; solution containing BSA and 10X reaction buffer concentrate	No detectable DNase, RNase, phosphatase	ICN 158824
1 U/µL	1 unit digests 1 µL 1% LMP agarose/hr at pH 6.5, 40°C	100 mM Bis-Tris-HCl, 10 mM EDTA, pH 6.5	No detectable RNase, DNase	Life Technol 10195-014
1000-3000 U/mg solid	1 unit produces 1.0 µg reducing sugar (measured as D-galactose) from agar/min at pH 6.0, 40°C	Lyophilized containing BSA and PO$_4$ buffer salts		Sigma A6306
≥10 U/mg protein	1 unit solubilizes 500 mg molten 1% agarose/hr at 40-42°C; the Worthington unit is based upon digestion of 500 mg agarose/hr	Extensively purified to a single band on SDS-PAGE; solution containing 10 mM NaPO$_4$ and BSA as stabilizer, pH 6.1	No detectable DNase, RNase, phosphatase	Worthington LM01323 LM01326 LM01320
≥10 U/mg protein	1 unit solubilizes 500 mg molten 1% agarose/hr at 40-42°C; the Worthington unit is based upon digestion of 500 mg agarose/hr	Extensively purified to a single band on SDS-PAGE; lyophilized containing BSA as stabilizer	No detectable DNase, RNase, phosphatase	Worthington LM01510 LM01512 LM01514
Unspecified (optimum pH = 6.5; stable 45-50°C, several hr; inactivated at 95°C/2 min or 65°C/15 min)				
3000 U/mg	1 unit digests 600 mg molten 1% LMP-agarose in GELase Buffer/hr at pH 6.0, 45°C	Solution containing 50% glycerol, 50 mM Tris-HCl, 0.1 M NaCl, 0.1 mM EDTA, 1 mM DTT, 0.1% Triton X-100, pH 7.5	No detectable DNA endo- and exonuclease, phosphatase, RNase	Epicentre GELase™ G09050 G09100 G09200 G31050 G31100 G31200
1000 U/mL	1 unit digests 200 µL molten 1% SeaPlaque GTF agarose to nonprecipitable neoagarooligosaccharides/hr at 40°C	50 mM Bis Tris-HCl, 1.0 mM Na$_2$EDTA, 50% glycerol, pH 6.5	No detectable DNase, RNase	NE Biolabs 392S 392L

Cellulose 1,4-β-Cellobiosidase 3.2.1.91

REACTION CATALYZED
Hydrolysis of 1,4-β-D-glucosidic linkages in cellulose and cellotetraose, releasing cellobiose from the non-reducing ends of the chains

SYSTEMATIC NAME
1,4-β-D-Glucan cellobiohydrolase

SYNONYMS
Exo-cellobiohydrolase

REACTANTS
Cellulose, cellotetraose, cellobiose

NOTES
- Classified as a hydrolase: a glycosidase hydrolyzing o-glycosyl compounds

See Chapter 5. Multiple Enzyme Preparations

Mannosyl-Glycoprotein Endo-β-N-Acetylglucosamidase 3.2.1.96

REACTION CATALYZED
Endohydrolysis of the N,N'-diacetylchitobiosyl unit in high-mannose glycopeptides and glycoproteins containing the -[Man(GlcNAc)$_2$]Asn- structure

SYSTEMATIC NAME
Glycopeptide-D-mannosyl-N^4-(N-acetyl-D-glucosaminyl)$_2$-asparagine 1,4-N-acetyl-β-glucosaminohydrolase

SYNONYMS
N,N'-Diacetylchitobiosyl β-N-acetylglucosaminidase, endo-β-N-acetylglucosaminidase, endoglycosidase D, endoglycosidase F, endoglycosidase F1, endoglycosidase F2, endoglycosidase H, PNGase F

REACTANTS
Mannose glycopeptides, mannose glycoproteins

APPLICATIONS
- Studies of glycoconjugates with complex sugar chains; this enzyme offers possibilities not available with exoglycosidases
- Removal of carbohydrate residues from proteins

NOTES
- Classified as a hydrolase: a glycosidase hydrolyzing o-glycosyl compounds
- One N-acetyl-D-glucosamine residue remains attached to the protein; the rest of the oligosaccharide is released intact
- Will not cleave intact hybrids linked to the mannose core, or complex oligosaccharides

Mannosyl-Glycoprotein Endo-β-N-Acetylglucosamidase continued

SPECIFIC ACTIVITY	UNITS DEFINITION	PREPARATION FORM	ADDITIONAL ACTIVITIES	SUPPLIER CATALOG NO.
Diplococcus pneumoniae (optimum pH = 6.5, MW = 280,000 Da)				
20 U/mg protein	1 unit converts 1 μmol dansyl-Asn(GlcNAc)$_2$(Man)$_5$ to product/min at pH 6.5, 37°C	Lyophilized containing BSA as stabilizer	No detectable α- and β-Gal, α- and β-mannosidase, α-L-fucosidase, proteases <1% β-NAGlc	Boehringer 752991
20 U/mg protein		Lyophilized containing BSA	No detectable α-L-fucosidase, α- and β-D-Gal, α- and β-mannosidase, protease <1% β-NAGlc	ICN 190446
Escherichia coli, transfected with Endo H gene				
45 U/mg protein	1 unit hydrolyzes 1.0 μmol [^{3}H]-dansyl-Asn(GlcNAc)$_2$/min at pH 5.0, 37°C	Lyophilized	No detectable protease	ICN 190447
45 U/mg protein	1 unit hydrolyzes 1 μmol [^{3}H]-dansyl-Asn(GlcNAc)$_2$Man$_5$/min at 37°C as determined by ascending paper chromatography	Lyophilized	No detectable protease	ICN 321311 321312
Flavobacterium meningosepticum (optimum pH = 5.0-7.0, MW = 32,000-32,700 Da)				
1000 U/mg protein	1 unit hydrolyzes 1 μmol dansyl-Asn(GlcNAc)$_2$(Man)$_5$/hr at pH 5, 37°C	20 mM KPO$_4$, 50 mM EDTA, 0.05% Micr-O-protect, pH 7.2	No detectable β-Gal, β-glucosidase, α- and β-mannosidase, β-NAH, α-L-fucosidase, sialidase, proteases	Boehringer 1636197
1000 U/mg protein	1 unit hydrolyzes 1 μmol dansylated-fibrinogen glycopeptide/min at pH 4.5, 37°C	5 mM NaOAc, 50 mM NaCl, 50% glycerol, 0.05% Micr-O-protect, pH 4.5	No detectable α- and β-Gal, β-glucosidase, α- and β-mannosidase, α- and β-NAH, α-L-fucosidase, sialidase, proteases	Boehringer 1694413
1400 U/mg protein	1 unit hydrolyzes 1 μmol dansyl-Asn(GlcNAc)$_2$(Man)$_5$/hr at pH 5, 37°C	Solution containing 20 mM KPO$_4$, 50 mM EDTA, 0.1% Micr-O-protect, pH 7.2	No detectable β-Gal, β-glucosidase, α- and β-mannosidase, β-NAH, α-L-fucosidase, sialidase, proteases <0.1% N-glycosidase F	Boehringer 903329
400 U/mg protein	1 unit hydrolyzes 1 μmol dansyl-Asn(GlcNAc)$_2$(Man)$_5$/hr at pH 5, 37°C	Solution containing 20 mM PBS, 50 mM EDTA, 0.05% NaN$_3$, pH 7.2	<160 U/vial N-glycosidase F	Calbiochem 324703
1400 U/mg protein	1 unit hydrolyzes 1 μmol dansyl-Asn(GlcNAc)$_2$(Man)$_5$/hr at pH 5, 37°C	Solution containing 2 mM KPO$_4$, 50 mM EDTA, 0.05% NaN$_3$, pH 7.2	No detectable β-NAH, α-L-fucosidase, β-Gal, β-glucosidase, α-mannosidase, β-mannosidase, protease <0.1% N-glycosidase F	Calbiochem 324706
	1 unit hydrolyzes 1 μmol substrate/min at 37°C	Lyophilized		ICN 151032

SPECIFIC ACTIVITY	UNITS DEFINITION	PREPARATION FORM	ADDITIONAL ACTIVITIES	SUPPLIER CATALOG No.
Flavobacterium meningosepticum continued				
	1 unit hydrolyzes 1 μmol Dansyl-Asn-(GlcNAc)$_2$(Man)$_5$/hr at pH 5.0, 37°C	Solution containing 20 mM KPO$_4$, 50 mM EDTA, 0.05% NaN$_3$, pH 7.2		Sigma E1262
	1 unit hydrolyzes 1 μmol Dansyl-Leu-Met-Gly-Glu-Asp-Arg/min at pH 4.5, 37°C	Solution containing 10 mM NaOAc buffer and 0.1 M NaCl, pH 4.5		Sigma E2266
1400 U/mg protein		Suspension		Wako 055-05491
Flavobacterium species (optimum pH = 6.0, T = 50°C)				
190 U/mg	1 unit releases 1 μmol DNS-Asn-GlcNAc from DNS-ovalbumin-glycopeptide/min at pH 6.0, 37°C	Lyophilized		Seikagaku 100468-1
Streptococcus pneumoniae (optimum pH = 6.5, T = 37°C)				
20 U/mg protein	1 unit hydrolyzes 1 μmol (Man)$_5$(GlcNAc)$_2$Asn-(^{14}C)-acetyl/min at pH 6.5, 37°C	Homogeneous by SDS-PAGE; lyophilized containing BSA as stabilizer	No detectable α-L-fucosidase, α- and β-D-Gal, protease < 1% β-NAGlc	ICN 321281 321282
20 U/mg protein	1 unit hydrolyzes 1 μmol (Man)$_5$(GlcNAc)$_2$Asn-[^{14}C]-acetyl/min at pH 6.5, 37°C at saturated substrate concentration	Lyophilized		Seikagaku 100460-1 100460-2
Streptomyces griseus (optimum pH = 6.0, T = 37°C)				
30 U/mg protein	1 unit hydrolyzes 1 μmol (Man)$_5$(GlcNAc)$_2$Asn-[^{14}C]-acetyl/min at pH 5.0, 37°C at saturated substrate concentration	Lyophilized		Seikagaku 100465-1 100465-2
Streptomyces lividans, recombinant (optimum pH = 5.5-6.0, T = 37°C)				
40 U/mg protein	1 unit hydrolyzes 1 μmol dansyl-Asn(GlcNAc)$_2$(Man)$_5$/hr at pH 5, 37°C	Lyophilized	No detectable β-NAH, α-L-fucosidase, β-Gal, α- and β-glucosidase, α-mannosidase, protease	Calbiochem 324704
Streptomyces plicatus, overexpressed in *E. coli* (optimum pH = 5.5, MW = 29,000 Da, 70,000 Da)				
40 U/mg protein	1 unit converts 1 μmol dansyl-Asn(GlcNAc)$_2$(Man)$_5$ to product/min at pH 5.5, 37°C	Solution containing 50 mM NaPO$_4$, 25 mM EDTA, 0.1% Micr-O-protect, pH 7	No detectable β-Gal, α- and β-glucosidase, α-mannosidase, β-NAH, α-L-fucosidase, proteases	Boehringer 1088726 1643053 1088734
400,000 U/mg; 500,000 U/mL	1 unit removes >95% carbohydrate from 10 μg denatured RNase B/hr at 37°C; 10 NEB unit = 1 IUB unit	Purified; 20 mM Tris-HCl, 50 mM NaCl, 5 mM Na$_2$EDTA, pH 7.5	No detectable exoglycosidase, proteolytic activity	NE Biolabs 702S 702L
165,000 U/mg; 1,000,000 U/mL	1 unit removes >95% carbohydrate from 10 μg denatured RNase B/hr at 37°C; 10 NEB unit = 1 IUB unit	Purified; 20 mM Tris-HCl, 50 mM NaCl, 5 mM Na$_2$EDTA, pH 7.5	No detectable exoglycosidase, proteolytic activity	NE Biolabs 703S 703L

3.2.1.96 Mannosyl-Glycoprotein Endo-β-N-Acetylglucosamidase continued

SPECIFIC ACTIVITY	UNITS DEFINITION	PREPARATION FORM	ADDITIONAL ACTIVITIES	SUPPLIER CATALOG NO.
Streptomyces plicatus, recombinant from *Streptomyces lividans*				
15 U/mg protein	1 unit converts 1 mmol resorufin-labeled N-glycopeptide to product/min	Solution containing 10 mM PO_4 buffer, pH 7	No detectable exoglycosidase, proteases	Boehringer 100117 100119
Xanthomonas manihotis				
1000 U/mL	1 unit cleaves >95% of β-D-N-Ac-glucosamine from GlcNAc-β-1,4-GlcNAc-AMC/hr at pH 4.5, 37°C	20 mM Tris-HCl, 50 mM NaCl, 0.1 mM EDTA, 0.02% NaN_3, pH 7.5	No detectable exoglycosidase, protease	NE Biolabs 722S 722L

3.2.1.97 Glycopeptide α-N-Acetylgalactosaminidase

REACTION CATALYZED

Hydrolysis of terminal D-galactosyl-N-acetyl-α-D-galactosaminidic residues from a variety of glycopeptides and glycoproteins

SYSTEMATIC NAME

D-Galactosyl-N-acetyl-α-D-galactosamine D-galactosyl-N-acetyl-galactosaminohydrolase

REACTANTS

Glycopeptides, glycoproteins, serine aglycone, threonine aglycone

APPLICATIONS

- Cleaves the o-glycosidic link between the disaccharide Galβ1→3GalNAc and a serine or threonine residue in a polypeptide

NOTES

- Classified as a hydrolase: a glycosidase hydrolyzing o-glycosyl compounds
- BSA in some preparations may interfere with analysis of glycoproteins by SDS-PAGE or standard proteins sequencing methods

SPECIFIC ACTIVITY	UNITS DEFINITION	PREPARATION FORM	ADDITIONAL ACTIVITIES	SUPPLIER CATALOG NO.
Alcarigenes species (optimum pH = 4.5, 6.0-7.6, T = 40°C, MW = 160,000 Da)				
3 U/mg	1 unit releases 1 μmol Gal-GalNAc from asialofetuin/min at pH 4.5, 37°C	Solution		Seikagaku 100453-1
Diplococcus pneumoniae (optimum pH = 6.0-7.6, MW = 160,000 Da)				
	1 unit releases 1 μmol Gal-GalNAc from native asialofetuin/min at pH 6.0, 37°C	Solution containing 25% glycerol, 200 mM NaCl, 10 mM Na cacodylate, pH 6.0		Sigma E2391
10 U/mg protein				Wako 059-05651

Glycopeptide α-N-Acetylgalactosaminidase continued

3.2.1.97

SPECIFIC ACTIVITY	UNITS DEFINITION	PREPARATION FORM	ADDITIONAL ACTIVITIES	SUPPLIER CATALOG No.
Diplococcus pneumoniae continued				
>10 U/mg protein	1 unit converts 1 μmol asialofetuin to product/min at pH 6.0, 37°C	Solution containing 15 mM Na cacodylate, 10 mM NaN$_3$, 0.1 g/L BSA, 25% glycerol, 0.15 M NaCl, pH 6.0	No detectable α-Gal, α-NAGlc, α-NAG, α-L-fucosidase, α-mannosidase, endoglycosidase D, proteases <0.05% β-Gal <0.2% sialidase, β-NAGlc	Boehringer 1012142 1012169
>10 U/mg protein	1 unit converts 1 μmol asialofetuin to product/min at pH 6.0, 37°C	Solution containing 15 mM Na cacodylate, 10 mM NaN$_3$, 25% glycerol, 0.15 M NaCl, pH 6.0	No detectable α-Gal, α-NAGlc, α-NAG, α-L-fucosidase, α-mannosidase, endoglycosidase D, proteases <0.05% β-Gal <0.2% sialidase, β-NAGlc, α-NAGlc	Boehringer 1347101 1643061

3.2.1.99   Arabinan Endo-1,5-α-L-Arabinosidase

REACTION CATALYZED
    Endohydrolysis of 1,5-α-arabinofuranosidic linkages in 1,5-arabinans

SYSTEMATIC NAME
    1,5-α-L-Arabinan 1,5-α-L-arabinanohydrolase

SYNONYMS
    Endo-1,5-α-L-arabinanase, arabinase

REACTANTS
    1,5-arabinans, beet arabinan (slowly)

NOTES
- Classified as a hydrolase: a glycosidase hydrolyzing o-glycosyl compounds

---

See Chapter 5. Multiple Enzyme Preparations

3.2.1.103 Keratan-Sulfate Endo-1,4-β-Galactosidase

REACTION CATALYZED

Endohydrolysis of 1,4-β-D-galactosidic linkages in keratan sulfate

SYSTEMATIC NAME

Keratan-sulfate 1,4-β-D-galactanohydrolase

SYNONYMS

Endo-β-galactosidase, keratanase, βGal

REACTANTS

Keratan sulfate, blood group substances, non-sulfated oligosaccharides, milk oligosaccharides, glycosphingolipids

NOTES

- Classified as a hydrolase: a glycosidase hydrolyzing o-glycosyl compounds
- Hydrolyzes the 1,4-β-D-galactosyl linkages adjacent to 1,3-α-D-N-acetyl-glucosaminyl residues
- Acts on some non-sulfated oligosaccharides
- Only acts on blood group substances when the 1,2-linked fucosyl residues have been removed
- Activity decreases with repeated freeze-thaw

SPECIFIC ACTIVITY	UNITS DEFINITION	PREPARATION FORM	ADDITIONAL ACTIVITIES	SUPPLIER CATALOG NO.
Bacillus species KS 36 (optimum pH = 6.0, T = 37°C)				
0.1 U/vial	1 unit releases 1 μmol reducing group (as galactose) from bovine cornea keratan sulfate/hr at pH 7.4, 37°C	Lyophilized		Seikagaku 100812-1
Bacteriodes fragilis (optimum pH = 5.8, T = 55°C, MW = 32,000 Da)				
5 U/mg protein	1 unit releases 1 μmol reducing sugar from bovine corneal keratin sulfate/min at pH 5.8, 37°C	200 m*M* NaCl, 50 m*M* NaOAc, 25% glycerol, 10 m*M* NaN$_3$, 200 mg/mL BSA, pH 5.8	<0.1% α- and β-Gals, α- and βNAGlc, α-NAG, α-L-fucosidase, keratin sulfatase, α-mannosidase, proteases, sialidase	Calbiochem 324701
140-160 U/mg	1 unit releases 1 μmol reducing power as galactose/min at pH 5.8, 37°C using bovine corneal keratan sulfate as substrate			ICN 151031
150 U/mg protein	1 unit converts 1 μmol keratan sulfate to product/min at pH 5.8, 37°C	Solution containing 50 m*M* NaOAc, 20 m*M* NaCl, 0.2 mg/mL BSA, 10 m*M* NaN$_3$, 25% glycerol, pH 5.8		Boehringer 982954
5 U/mg protein		Suspension		Wako 050-05561

SPECIFIC ACTIVITY	UNITS DEFINITION	PREPARATION FORM	ADDITIONAL ACTIVITIES	SUPPLIER CATALOG No.
Escherichia freundii (optimum pH = 5-6, T = 50°C)				
3.0 U/mg protein	1 unit hydrolyzes 1 µmol β-galactosidic linkage of keratan sulfate liberating reducing groups corresponding to 1 µmol galactose/min at pH 5.8, 37°C when reducing sugars (as galactose) are determined by Park & Johnson method	Lyophilized	No detectable α-L-fucosidase, α- and β-D-Gal, α- and β-D-glucosidase	ICN 190445
3.0 U/mg protein	1 unit hydrolyzes 1 µmol β-galactosidic linkage of keratan sulfate liberating reducing groups corresponding to 1 µmol galactose/min at pH 5.8, 37°C when reducing sugars (as galactose) are determined by Park & Johnson method	Lyophilized containing <2.5 µmol KPO$_4$ buffer, pH 7.0	No detectable α-L-fucosidase, α- and β-D-Gal, α- and β-D-glucosidase, α- and β-NAG, α- and β-NAGlc, α- and β-mannosidase, β-xylosidase, neuraminidase, chondroitinase, heparinase	ICN 321321
3.0 U/mg protein; 0.1 U/vial	1 unit liberates 1 µmol galactose from bovine cornea keratan sulfate/min at pH 5.8, 37°C	Lyophilized		Seikagaku 100455-1
Pseudomonas species (optimum pH = 7.4, T = 37°C)				
20 U/mg protein	1 unit releases 1 µmol reducing group (as galactose) from bovine cornea keratan sulfate/hr at pH 7.4, 37°C	Lyophilized		Seikagaku 100810-1
	1 unit liberates 1.0 µmol reducing sugar (measured as galactose) from keratan sulfate/hr at pH 7.4, 37°C	Lyophilized		Sigma K2876
Pseudomonas species IFO-13309				
20 U/mg solid	1 unit reduces 1 µmol reducing sugar from keratan sulfate/hr at pH 7.4, 37°C	Lyophilized containing <0.05 µmol Tris-HCl/10 U vial	No detectable protease, sulfatase, hyaluronidase, chondroitinase, glycosidase, α- and β-D-Gal, α- and β-NAGlc, α- and β-mannosidase, β-mannosaminidase, β-xylosidase	ICN 320321

3.2.1.108 Lactase

REACTION CATALYZED
  Lactose + H_2O ↔ D-Glucose + D-galactose
SYSTEMATIC NAME
  Lactose galactohydrolase
REACTANTS
  Lactose, H_2O, D-glucose, D-galactose

NOTES
- Classified as a hydrolase: a glycosidase hydrolyzing o-glycosyl compounds
- Intestinal mucosal enzyme is isolated as a complex also exhibiting glycosylceramidase (EC 3.2.1.62) activity
- Hydrolysis products glucose and galactose are sweeter than lactose

SPECIFIC ACTIVITY	UNITS DEFINITION	PREPARATION FORM	ADDITIONAL ACTIVITIES	SUPPLIER CATALOG No.
Aspergillus oryzae (optimum pH = 4.5-4.8 [o-nitrophenyl-β-D-galactopyranoside] and 4.8 [lactose], T = 50-60°C; stable pH 4-8 [37°C, 1 hr], T < 55°C [pH 4.5-6.5, 30 min])				
14,000 U/g; 25,000 U/g	1 unit liberates 1 μmol ONP/min at pH 4.5, 37°C	Food grade; powder		Amano Lactase 14000, 25000
>5000 U/g	1 unit hydrolyzes 1.0 μmol o-nitrophenyl-β-D-galactoside to ONP/min at pH 4.5, 30°C; lactose is hydrolyzed at the same rate			ICN 100780

3.2.1.111 1,3-α-L-Fucosidase

REACTION CATALYZED
  Hydrolysis of 1,3-linkages between α-L-fucose and N-acetylglucosamine residues in glycoproteins
SYSTEMATIC NAME
  3-α-L-Fucosyl-N-acetylglucosaminyl-glycoprotein fucohydrolase
SYNONYMS
  Almond emulsin fucosidase I

REACTANTS
  Glycoproteins, lacto-N-fucopentaose II, lacto-N-fucopentaose III
NOTES
- Classified as a hydrolase: a glycosidase hydrolyzing o-glycosyl compounds
- p-Nitrophenyl-α-L-fucopyranoside is not a substrate
- Not identical with 1,2-α-L-fucosidase (EC 3.2.1.63)

SPECIFIC ACTIVITY	UNITS DEFINITION	PREPARATION FORM	ADDITIONAL ACTIVITIES	SUPPLIER CATALOG No.
Almond meal				
1 vial = 20 μU	1 unit liberates 1.0 μmol fucose from lacto-N-fucopentaose II/min at pH 5.0, 37°C	Lyophilized containing NaOAc and BSA		sigma F8899
***Streptomyces* species (optimum pH = 6.0-6.5, MW = 40,000 [gel filtration], 55,000 [SDS-PAGE] Da)**				
300 U/mg protein	1 unit releases 1 μmol L-fucose from lacto-N-fucopentaose III/min at 37°C	Solution containing 50 mM KPO_4, 100 mM NaCl, 0.02% NaN_3, 0.1% Brij 58, pH 6.0	No detectable α- and β-Gal, α-mannosidase, β-NAH, β-xylosidase, proteases	Boehringer 1630377
***Streptomyces* species 142 (optimum pH = 6.0, MW = 40,000 Da)**				
	1 unit produces 1 μmol L-fucose from 2 μM PA-lacto-N-fucopentaose III/min at pH 6.0, 37°C	50 mM KPO_4 buffer solution containing 0.1 M NaCl, 0.02% NaN_3, 0.1% Brij 58		PanVera TAK 4453
Xanthomonas manihotis				
1000 U/mL	1 unit cleaves >95% terminal α-L-fucose from 1 nmol Galβ1-4(fuc1-3)GlcNAcβ1-3Galβ1-4Glc-AMC/hr at pH 6.0, 37°C in a 10 μL reaction	20 mM Tris-HCl, 50 mM NaCl, 0.1 mM Na_2EDTA, 0.02% NaN_3, pH 7.5	No detectable exoglycosidase, protease	NE Biolabs 723S 723L

Mannosyl-Oligosaccharide 1,2-α-Mannosidase

3.2.1.113

REACTION CATALYZED
> Hydrolysis of the terminal 1,2-linked α-D-mannose residues in the mannosyl-oligosaccharide $Man_9(GlcNAc)_2$

SYSTEMATIC NAME
> 1,2-α-Mannosyl-oligosaccharide α-D-mannohydrolase

SYNONYMS
> Mannosidase 1A, mannosidase 1B

REACTANTS
> $Man_9(GlcNAc)_2$, α-D-mannose

NOTES
- Classified as a hydrolase: a glycosidase hydrolyzing o-glycosyl compounds
- Involved in glycoprotein synthesis
- p-Nitrophenyl-α-D-mannopyranoside is not a substrate

3.2.1.113 Mannosyl-Oligosaccharide 1,2-α-Mannosidase continued

SPECIFIC ACTIVITY	UNITS DEFINITION	PREPARATION FORM	ADDITIONAL ACTIVITIES	SUPPLIER CATALOG NO.
Xanthomonas manihotis				
1000 U/mL	1 unit cleaves >95% terminal α-D-mannose from 1 nmol Manα1-3Manβ1-4GlcNAc-AMC/hr at pH 6.0, 37°C in a 10 μL reaction	20 mM Tris-HCl, 50 mM NaCl, 0.02% NaN$_3$, 0.1 mM Na$_2$EDTA, pH 7.5	No detectable exoglycosidase, protease	NE Biolabs 729S 729L

3.2.1.114 Mannosyl-Oligosaccharide 1,3-1,6-α-Mannosidase

REACTION CATALYZED
Hydrolysis of the terminal 1,3- and 1,6-linked α-D-mannose residues in the mannosyl-oligosaccharide Man$_5$(GlcNAc)$_3$

SYSTEMATIC NAME
1,3-(1,6-)Mannosyl-oligosaccharide α-D-mannohydrolase

SYNONYMS
Mannosidase II, α-1,6-mannosidase

REACTANTS
Man$_5$(GlcNAc)$_3$, α-D-mannose

NOTES
- Classified as a hydrolase: a glycosidase hydrolyzing o-glycosyl compounds
- Involved in glycoprotein synthesis
- p-Nitrophenyl-α-D-mannopyranoside is not a substrate

SPECIFIC ACTIVITY	UNITS DEFINITION	PREPARATION FORM	ADDITIONAL ACTIVITIES	SUPPLIER CATALOG NO.
Xanthomonas manihotis				
1000 U/mL	1 unit cleaves >95% terminal α-D-mannose from 1 nmol Manα1-6Manα1-6(Manα1-3)Man-AMC/hr at pH 4.5, 37°C in a 10 μL reaction	20 mM Tris-HCl, 50 mM NaCl, 0.1 mM Na$_2$EDTA, 0.02% NaN$_3$, 100 μg/mL BSA, pH 7.5	No detectable exoglycosidase, protease	NE Biolabs 727S 727L

Endoglycosylceramidase 3.2.1.123

REACTION CATALYZED
    Oligoglycosylglucosylceramide + H_2O ↔
    Oligoglycosylglucose + ceramide

SYSTEMATIC NAME
    Oligoglycosylglucosylceramide glycohydrolase

SYNONYMS
    Endoglycoceramidase

REACTANTS
    Oligoglycosylglucosylceramide, glycosphingolipids, H_2O, oligosaccharides, ceramides

NOTES
- Classified as a hydrolase: a glycosidase hydrolyzing o-glycosyl compounds
- A *Rhodococcus* sp. enzyme which degrades various acidic and neutral glycosphingolipids by cleaving a glucosyl bond
- Does not act on monoglycosylceramides, glycoglycerolipids and glycoproteins
- Inhibited by Hg^{2+}, Zn^{2+} and Cu^{2+}

SPECIFIC ACTIVITY	UNITS DEFINITION	PREPARATION FORM	ADDITIONAL ACTIVITIES	SUPPLIER CATALOG NO.
***Rhodococcus* species**				
0.3 U/mg	1 unit liberates 1 μmol reducing sugar (as glucose) from bovine brain ganglioside mixture at pH 6.0, 37°C	Solution containing 0.01 M NaOAc, 0.003% BSA as stabilizer, pH 6.0		Fluka 45171
***Rhodococcus* species G-74-2 (optimum pH = 5-6 [globoside or G_{m1}], 4-5 [neogalatriaoxylceramide])**				
5 mU	1 unit releases 1 μmol reducing end as glucose from glycolipids/min at pH 6.0, 37°C	Lyophilized	No detectable protease, glycanase, α-NAG, α-NAGlc, α- and β-Gal, β-NAH, α-L-fucosidase, α-mannosidase, α- and β-glucosidase, neuraminidase	PanVera TAK 4457
5 mU	1 unit releases 1 μmol reducing end as glucose from glycolipids/min at pH 6.0, 37°C	Lyophilized	No detectable protease, α-NAG, α-NAGlc, α- and β-Gal, β-NAH, α-L-fucosidase, α- and β-mannosidase, α- and β-glucosidase, neuraminidase; Contains sphingomyelinase activity	TaKaRa 4457

3.2.1.132 Chitosanase

REACTION CATALYZED
Endohydrolysis of β-1,4-linkages between N-acetyl-D-glucosamine and D-glucosamine residues in a partly acetylated chitosan

SYSTEMATIC NAME
Chitosan N-acetylglucosaminohydrolase

REACTANTS
Chitosan

NOTES
- Classified as a hydrolase: a glycosidase hydrolyzing o-glycosyl compounds
- Acts only on polymers with 30-60% acetylation

SPECIFIC ACTIVITY	UNITS DEFINITION	PREPARATION FORM	ADDITIONAL ACTIVITIES	SUPPLIER CATALOG No.
Bacillus punilus BN-262 (optimum pH = 6.0-7.0, pI = 9.3, T = 30-55°C, MW = 31,000 Da)				
640-960 U/g				Wako 033-12943 037-12941

3.2.2.5 NAD⁺ Nucleosidase

REACTION CATALYZED
$NAD^+ + H_2O \leftrightarrow$ Nicotinamide + ADPribose + H^+

SYSTEMATIC NAME
NAD^+ glycohydrolase

SYNONYMS
NADase, DPNase, DPN hydrolase, ADP-ribosyl cyclase

REACTANTS
NAD^+, H_2O, nicotinamide, ADPribose

APPLICATIONS
- Widely used in NAD, NADH, NADP and NADPH research

NOTES
- Classified as a hydrolase: a glycosidase hydrolyzing N-glycosyl compounds
- Found primarily in animal tissue
- *N. crassa* enzyme differs from animal enzyme in that it does not form analogs; not inhibited by nicotinamide > 0.1 M
- Some animal enzymes also transfer ADPribose residues

NAD$^+$ Nucleosidase continued 3.2.2.5

SPECIFIC ACTIVITY	UNITS DEFINITION	PREPARATION FORM	ADDITIONAL ACTIVITIES	SUPPLIER CATALOG NO.
Apylsia californica				
1000-3000 U/mg protein; 100 µg protein/vial	1 unit produces 1 µmol cyclic ADP ribose from β-NAD$^+$/5 min at pH 7.0, 25°C	Lyophilized containing 10% protein; balance trehalose, KPO$_4$, DTT		Sigma A8950
Neurospora crassa (optimum pH = 3.0-9.0)				
3.0 U/mg protein	1 unit hydrolyzes 1.0 µmol β-NAD to nicotinamide and ADP-ribose/min at pH 7.3, 37°C	Crude; lyophilized containing 20% protein; balance PO$_4$ buffer salts, pH 7		Sigma N5263
0.5-3.0 U/mg protein	1 unit hydrolyzes 1.0 µmol β-NAD to nicotinamide and ADP-ribose/min at pH 7.3, 37°C	Crude; lyophilized containing 20% protein; balance PO$_4$ buffer salts, pH 7		Sigma N9629
≥0.05 U/vial	1 unit cleaves 1 µmol NAD/min at pH 7.5, 37°C	Partially purified; lyophilized		Worthington LS04461 LS04463
Porcine brain				
	1 unit hydrolyzes 1.0 µmol β-NAD to nicotinamide and ADP-ribose/min at pH 7.3, 37°C	Purified; acetone-dried; powder		Sigma N9879

Deoxyribopyrimidine Endonucleosidase 3.2.2.17

REACTION CATALYZED
Cleaves the *N*-glycosidic bond between the 5'-pyrimidine residue in cyclobutadipyrimidine (in DNA) and the corresponding deoxy-*D*-ribose residue

SYSTEMATIC NAME
Deoxy-*D*-cyclobutadipyrimidine polynucleotidodeoxyribohydrolase

SYNONYMS
Pyrimidine dimer DNA-glycosylase, uracil-DNA glycosylase, UNG

REACTANTS
Cyclobutadipyrimidine

APPLICATIONS
- Increases efficiency of site-directed mutagenesis
- Probe for protein-DNA interaction studies
- For rapid/efficient cloning of polymerase chain reaction products
- Eliminates PCR carry-over contamination from previous DNA reactions; results in fewer false positive results for cloning PCR fragments
- Production of highly-labeled oligonucleotide probes

3.2.2.17 Deoxyribopyrimidine Endonucleosidase continued

NOTES

- Classified as a hydrolase: a glycosidase hydrolyzing N-glycosyl compounds
- Releases uracil from uracil-containing DNA
- Will act on single- or double-stranded DNA, but not oligomers with fewer than 6 bases
- RNA and normal dT-DNA are not substrates
- Inhibited by high ionic strength (>200 mM)
- Forms abasic sites in DNA under alkaline and heat conditions

SPECIFIC ACTIVITY	UNITS DEFINITION	PREPARATION FORM	ADDITIONAL ACTIVITIES	SUPPLIER CATALOG NO.
Escherichia coli (optimum pH = 8.0; 95% inactivated at 95°C, 10 min)				
1000–10,000 U/mL	1 unit releases 60 pmol uracil from double-stranded uracil-containing DNA/30 min at 37°C in a 50 μL reaction containing 0.2 μg DNA	50 mM NaCl, 20 mM Tris-HCl, 1 mM EDTA, 1 mM DTT, 100 μg/mL BSA, 50% glycerol, pH 8.0		NE Biolabs 280S 280L
Escherichia coli strain K12 (inactivated at 95°C/10 min; activity partially restored at T < 55°C)				
1000 U/mL	1 unit completely degrades 1 μg purified single-stranded uracil-containing DNA/hr at 37°C	50 mM K HEPES buffer, 0.3 M NaCl, 1 mM EDTA, 1 mM DTT, 100 μg/mL BSA, 50% glycerol, pH 8.0		Boehringer 1269062 1444646
1 U/μL	1 unit degrades 1 μg single-stranded uracil-containing DNA/hr at 37°C	30 mM HEPES-KOH, 150 mM NaCl, 1 mM EDTA, 1 mM DTT, 0.05% Tween 20, 50% glycerol, pH 7.5	No detectable endo- and exodeoxyribonucleases, RNase	Fermentas EN0361 EN0362
Escherichia coli ung gene, expressed on a plasmid in *E. coli*				
1 U/μL	1 unit releases 1 μmol free uracil from [^{3}H]-poly(dU)/hr at pH 8.3, 37°C			Life Technol 18054-015
Escherichia coli, overexpressed from a two plasmid system				
1 U/μL	1 unit releases 1 nmol [^{3}H]-uracil from dU containing DNA into acid soluble counts/hr at 37°C	30 mM Tris-HCl, 150 mM NaCl, 1 mM EDTA, 1 mM DTT, 50% glycerol, pH 7.5	No detectable ss- and ds-exo- and endonuclease	Amersham E 71960Y

DNA-3-Methyladenine Glycosidase I   3.2.2.20

REACTION CATALYZED
    Hydrolysis of alkylated DNA, releasing 3-methyladenine

SYSTEMATIC NAME
    Alkylated-DNA glycohydrolase (releasing methyladenine and methylguanine)

REACTANTS
    Alkylated-DNA, 3-methyladenine

NOTES
- Classified as a hydrolase: a glycosidase hydrolyzing N-glycosyl compounds
- Removes alkylated bases from DNA in *E. coli*

See Chapter 5. Multiple Enzyme Preparations

DNA-3-Methyladenine Glycosidase II   3.2.2.21

REACTION CATALYZED
    Hydrolysis of alkylated DNA, releasing 3-methyladenine, 3-methylguanine, 7-methyladenine and 7-methylguanine

SYSTEMATIC NAME
    Alkylated-DNA glycohydrolase (releasing methyladenine and methylguanine)

REACTANTS
    Alkylated-DNA, 3-methyladenine, 3-methylguanine, 7-methyladenine, 7-methylguanine

NOTES
- Classified as a hydrolase: a glycosidase hydrolyzing N-glycosyl compounds
- Removes alkylated bases from DNA in *E. coli*

See Chapter 5. Multiple Enzyme Preparations

3.2.3.1 Thioglucosidase

REACTION CATALYZED
A thioglucoside + H_2O ↔ A thiol + a sugar

SYSTEMATIC NAME
Thioglucoside glucohydrolase

SYNONYMS
Myrosinase, sinigrinase

REACTANTS
Thioglucosides, thioglycosides, H_2O, thiols, sugars

NOTES
- Classified as a hydrolase: a glycosidase hydrolyzing S-glycosyl compounds
- Wide specificity

SPECIFIC ACTIVITY	UNITS DEFINITION	PREPARATION FORM	ADDITIONAL ACTIVITIES	SUPPLIER CATALOG NO.
Sinapis alba (white mustard seed)				
200 U/g		Powder		Biocatalysts M044P
200 U/g solid	1 unit produces 1.0 μmol glucose from sinigrin/min at pH 6.0, 25°C			ICN 156878
200 U/g solid	1 unit produces 1.0 μmol glucose from sinigrin/min at pH 6.0, 25°C			Sigma T4528

3.3.1.1 Adenosylhomocysteinase

REACTION CATALYZED
S-Adenosyl-L-homocysteine + H_2O ↔ Adenosine + L-homocysteine

SYSTEMATIC NAME
S-Adenosyl-L-homocysteine hydrolase

REACTANTS
S-Adenosyl-L-homocysteine, H_2O, adenosine, L-homocysteine

NOTES
- Classified as a hydrolase, acting on ether bonds: thioether hydrolase

SPECIFIC ACTIVITY	UNITS DEFINITION	PREPARATION FORM	ADDITIONAL ACTIVITIES	SUPPLIER CATALOG NO.
Rabbit erythrocytes				
3-10 U/mg protein	1 unit hydrolyzes 1.0 nmol S-adenosyl-L-homoCys to adenosine and L-homoCys/min at pH 7.2, 37°C	Solution containing 25 mM Tris, 1 mM DTT 1 mM EDTA, 20% glycerol, pH 7.4	<0.1 U/mg protein ADA	Sigma A3291

Leucyl Aminopeptidase 3.4.11.1

REACTION CATALYZED
Release of an N-terminal amino acid, Xaa↓Xbb-

SYNONYMS
Leucine aminopeptidase, leucyl peptidase, peptidase S, cytosol aminopeptidase, aminopeptidase, LAP

REACTANTS
Proteins, peptides, amino acid amides, amino acid methyl esters, leucine, proline, arylamides, leucinamide

APPLICATIONS
- Resolution of γ-methyl and γ-fluoroglutamic acids
- Cleavage of deferric form of albomycins
- Debittering
- Determination of L-peptides and amino acid amides containing N-terminal leucine or proline
- Serum control in protein sequence studies
- QC sera for diagnosis of liver and bile ducts

NOTES
- Classified as a hydrolase, acting on peptide bonds: aminopeptidase
- An exopeptidase consisting of 4 subunits, each having 1 zinc atom
- Isolated from pig kidney and cattle lens
- Xaa is preferably Leu, is not Arg or Lys, but may be other amino acids including Pro
- Xbb may be Pro
- Rates on arylamides are very low
- Activated by heavy metal ions (Mg^{2+} and Mn^{2+})
- Inhibited by Cd^{2+}, Cu^{2+}, Hg^{2+}, Pb^{2+}, alcohols, glycerol, p-chloromercuribenzoate and n-butanol
- Rapid inactivation by orthophenanthroline, bipyridyl, cupferron, sodium diethyldithiocarbamide, sodium sulfide and sodium cyanide
- Serum levels are elevated in obstructive jaundice, liver cirrhosis, liver carcinoma and latter stages of pregnancy

SPECIFIC ACTIVITY	UNITS DEFINITION	PREPARATION FORM	ADDITIONAL ACTIVITIES	SUPPLIER CATALOG NO.
Aspergillus oryzae (optimum pH = 6.0-8.0, T = 60°C)				
650 U/g	1 unit liberates 1 μmol Leu from Leu paranitroanilide/min at pH 7.0 using 2.0 mM Leu paranitroanilide	Powder	Endopeptidase	Imperial Debitrase DBS 50
Lactococcus lactis (optimum pH = 5.5-8.0, T = 35°C)				
32 U/g	1 unit liberates 1 μmol Leu from Leu paranitroanilide/min at pH 7.0 using 2.0 mM Leu paranitroanilide	Powder	Mixture of aminopeptidases	Imperial Accelase DB
Porcine kidney (optimum pH = 7.0-7.3, 9.1 [Mn^{2+}-activated enzyme], pI = 4.0-5.0, MW = 245,500-280,000 Da; stable at pH 8.0-8.5 [ambient T, Mg^{2+}, days], T < 65°C)				
≥100 U/mg protein	1 unit hydrolyzes 1 μmol L-leucinamide/min at pH 8.5, 25°C	Purified; suspension in 70% (NH4)2SO4, pH 8.7		Biozyme LAP1
200 U/mg protein	1 unit hydrolyzes 1 μmol L-leucinamide/min at pH 8.5, 25°C	Highly purified; suspension in 70% (NH4)2SO4, pH 8.7		Biozyme LAP2

3.4.11.1 Leucyl Aminopeptidase continued

SPECIFIC ACTIVITY	UNITS DEFINITION	PREPARATION FORM	ADDITIONAL ACTIVITIES	SUPPLIER CATALOG NO.
Porcine kidney continued				
>100 U/mg protein	1 unit hydrolyzes 1 μmol L-leucinamide/min at pH 8.5, 25°C	Suspension in 70% $(NH_4)_2SO_4$ containing 90% protein, Tris buffer, $MgCl_2$, pH 8		Calzyme 065B0100
10 U/mg	1 unit hydrolyzes 1 μmol L-leucinamide/min at pH 8.5, 25°C	Lyophilized		Fluka 61857
18 U/mg	1 unit hydrolyzes 1 μmol L-leucinamide/min at pH 8.5, 25°C	Powder		Fluka 61858
100 U/mg protein	1 unit hydrolyzes 1 μmol L-leucinamide/min at pH 8.5, 25°C	Crystallized; suspension in 3.8 M $(NH_4)_2SO_4$		Fluka 61860
5 U/mg	1 unit hydrolyzes 1 μmol L-leucinamide/min at pH 8.5, 25°C	Lyophilized		Fluka 61861
>100 U/mg protein		Suspension in 70% $(NH_4)_2SO_4$, Tris buffer, $MgCl_2$, pH 8; containing 90% protein		ICN 153884
>5 U/mg protein; >5 U/mg solid	1 unit oxidizes 1 μmol L-Leu-p-nitroanilide to L-Leu and p-nitroanilide/min at pH 7.2, 37°C	Lyophilized	<0.1% γGT, GPT, GOT <0.5% ALP, LDH	Randox LA 851L
Activated to ≥100 U/mg protein	1 unit hydrolyzes 1 μmol L-leucinamide/min at pH 8.5, 25°C	Chromatographically purified; suspension in 3.0 M $(NH_4)_2SO_4$, 0.1 M Tris, 0.005 M $MgCl_2$, pH 8.0	No detectable trypsin, chymotrypsin	Worthington LS05484 LS05485 LS05487
Porcine kidney cytosol				
80-200 U/mg protein	1 unit hydrolyzes 1.0 μmol L-leucinamide to L-Leu and NH_3/min at pH 8.5, 25°C	Lyophilized containing 80% protein; balance Tris buffer, $MgCl_2$, $(NH_4)_2SO_4$	No detectable trypsin and chymotrypsin	Sigma L1503
100-300 U/mg protein	1 unit hydrolyzes 1.0 μmol L-leucinamide to ≥Leu and NH_3/min at pH 8.5, 25°C	Chromatographically purified; suspension in 2.9 M $(NH_4)_2SO_4$, 0.1 M Tris, 5 mM $MgCl_2$, pH 8.0	No detectable trypsin and chymotrypsin	Sigma L9876
Porcine kidney microsomes				
20 U/mg protein	1 unit hydrolyzes 1 μmol L-leucinamide/min at pH 8.5, 25°C	Lyophilized containing 90% protein		Calzyme 152A0020
20 U/mg solid		Lyophilized containing 50% protein		ICN 153885
	1 unit hydrolyzes 1.0 μmol L-Leu-p-nitroanilide to L-Leu and p-nitroaniline/min at pH 7.2, 37°C	Lyophilized containing 50% protein; balance primarily PO_4 buffer salts, pH 7.1		Sigma L0632
10-40 U/mg protein	1 unit hydrolyzes 1.0 μmol L-Leu-p-nitroanilide to L-Leu and p-nitroaniline/min at pH 7.2, 37°C	Suspension in 3.5 M $(NH_4)_2SO_4$ and 10 mM $MgCl_2$, pH 7.7		Sigma L5006
15-25 U/mg protein	1 unit hydrolyzes 1.0 μmol L-Leu-p-nitroanilide to L-Leu and p-nitroaniline/min at pH 7.2, 37°C	Lyophilized containing 50% protein; balance primarily PO_4 buffer salts, pH 7.0		Sigma L6007

SPECIFIC ACTIVITY	UNITS DEFINITION	PREPARATION FORM	ADDITIONAL ACTIVITIES	SUPPLIER CATALOG NO.
Porcine kidney microsomes *continued*				
50 μg protein/vial		Sequencing grade; lyophilized		Sigma L9776

Membrane Alanyl Aminopeptidase 3.4.11.2

REACTION CATALYZED
Release of an N-terminal amino acid, Xaa↓Xbb-, from a peptide, amide or arylamide

SYNONYMS
Microsomal aminopeptidase, aminopeptidase M, aminopeptidase N, particle-bound aminopeptidase, amino-oligopeptidase, alanine aminopeptidase, membrane aminopeptidase I, pseudo leucine aminopeptidase, peptidase E, arylamidase, amino acid arylamidase

REACTANTS
Peptide, amide, arylamide, alanine, proline

APPLICATIONS
- Protein sequencing
- Identification of chemically-modified amino acid residues in proteins
- Two-stage assay of neutral endopeptidase and thermolysin

NOTES
- Classified as a hydrolase, acting on peptide bonds: aminopeptidase
- A zinc enzyme
- Not activated by heavy metal ions
- Xaa is preferably Ala, Pro (slow action) or most amino acids
- When a terminal hydrophobic residue is followed by a prolyl residue, the two may be released as an intact Xaa-Pro dipeptide
- Activity against alanine-β-naphthylamide is reportedly 2X that with leucine-β-naphthylamide
- No activity against L-asparagine, L-glutamate, β-alanine amide or γ-butyramide
- Amides from secondary amino groups are hydrolyzed slowly
- Inhibited by leucinethiol, 2,2'-bipyridine, 1,10-phenanthroline, acetone, alcohol and guanidine (0.5 M)

3.4.11.2 Membrane Alanyl Aminopeptidase continued

SPECIFIC ACTIVITY	UNITS DEFINITION	PREPARATION FORM	ADDITIONAL ACTIVITIES	SUPPLIER CATALOG NO.
Human placenta				
4-10 U/mg protein	1 unit hydrolyzes 1.0 μmol L-Ala β-naphthylamide to L-Ala and β-naphthylamine/min at pH 7.0, 37°C	Suspension in 3.2 M $(NH_4)_2SO_4$, 1 mM $NaPO_4$, pH 7.0		Sigma A9901
Porcine kidney (optimum pH = 7.0-7.5, T = 37°C, MW = 280,000 Da [2 identical subunits])				
4 U/mg at 25°C; 11 U/mg at 37°C	1 unit converts 1 μmol Leu-4-nitranilide to product/min	Suspension in $(NH_4)_2SO_4$		Boehringer 102768
>10 U/mg protein	1 unit hydrolyzes 1 μmol Leu-pNA/min at pH 7.2, 37°C (2.75 U/mg at 37°C = 1 U/mg at 25°C)	Crystallized; suspension in 3.2 M $(NH_4)_2SO_4$, 10 mM $MgCl_2$, 10 mM Tris, pH 7.5		Calbiochem 164598
Streptomyces griseus				
200-600 U/mg protein	1 unit hydrolyzes 1.0 μmol L-Leu-p-nitroanilide to L-Leu and p-nitroaniline/min at pH 8.0, 25°C; 3.0 mM substrate concentration	Lyophilized containing 50% protein; balance primarily CaOAc	<1 U/g endopeptidase (μmol Tyr equivalent/min release from casein)	Sigma A9934

3.4.11.5 Prolyl Aminopeptidase

REACTION CATALYZED
  Release of N-terminal proline from a peptide

SYNONYMS
  Proline aminopeptidase, Pro-X aminopeptidase, cytosol aminopeptidase V, proline iminopeptidase

REACTANTS
  Peptides, polyproline, prolyl-2-naphthylamide, proline

NOTES
- Classified as a hydrolase, acting on peptide bonds: aminopeptidase
- Requires Mn^{2+}
- Present in the cytosol of mammalian and microbial cells
- The mammalian enzyme is not specific for prolyl bonds. It is possibly identical with leucyl aminopeptidase (EC 3.4.11.1)
- Only the bacterial enzyme hydrolyzes both polyproline and prolyl-2-naphthylamide
- Stabilized by EDTA and 2-mercaptoethanol
- Inhibited by PCMB, Hg^{2+}, Zn^{2+} and Cu^{2+}

Prolyl Aminopeptidase continued

3.4.11.5

SPECIFIC ACTIVITY	UNITS DEFINITION	PREPARATION FORM	ADDITIONAL ACTIVITIES	SUPPLIER CATALOG NO.
Bacillus coagulans (optimum pH = 7.3, pI = 4.6, T = 45-50°C, MW = 40,000 Da [gel filtration, monomer]; K_M = 4.1 mM [Pro-Ala], 0.26 mM [Pro-2Nnap]; stable pH 6.5-7.5 [30°C, 15 min], T < 45°C [pH 7.0, 30 min])				
100 U/mL	1 unit hydrolyzes 1.0 μmol proline *p*-nitroanilide/min at pH 8.0, 30°C	30% glycerol solution		Sigma P4919
≥100 U/mL	1 unit forms 1 μmol *p*-nitroaniline/min at pH 7.0, 30°C	Suspension in 4.6 M $(NH_4)_2SO_4$ containing EDTA and β-MSH as stabilizers	<0.1% peptidase	Toyobo PIP-209

Bacterial Leucyl Aminopeptidase

3.4.11.10

REACTION CATALYZED
    Release of an N-terminal amino acid, preferentially leucine; not glutamic or aspartic acids

SYNONYMS
    Aeromonas proteolytica aminopeptidase, aminopeptidase

REACTANTS
    Peptides, leucine, amino acid

APPLICATIONS
- Removing N-terminal methionine residues from recombinant proteins expressed in bacteria

NOTES
- Classified as a hydrolase, acting on peptide bonds: aminopeptidase
- A zinc enzyme
- Isolated from *Aeromonas proteolytica*, *E. coli* and *Staphyloccus thermophilus*

SPECIFIC ACTIVITY	UNITS DEFINITION	PREPARATION FORM	ADDITIONAL ACTIVITIES	SUPPLIER CATALOG NO.
Aeromonas proteolytica				
50-150 U/mg protein	1 unit hydrolyzes 1.0 μmol L-Leu *p*-nitroanilide to L-Leu and *p*-nitroaniline/min at pH 8.0, 25°C	Lyophilized containing 40% protein; balance tricine buffer, $ZnCl_2$ and stabilizer, pH 8.0	No detectable endopeptidase	Sigma A8200

X-Pro Dipeptidase

REACTION CATALYZED
Hydrolysis of Xaa-Pro dipeptides

SYNONYMS
Prolidase, imidodipeptidase, proline dipeptidase, peptidase D, γ-peptidase

REACTANTS
Dipeptides, aminoacylhydroxyproline analogs

APPLICATIONS
- Protein sequencing studies

NOTES
- Classified as a hydrolase, acting on peptide bonds: dipeptidase
- Activated by Mn^{2+}
- Possibly thiol dependent
- Does not act on Pro-Pro
- Cytosolic from most animal tissues

SPECIFIC ACTIVITY	UNITS DEFINITION	PREPARATION FORM	ADDITIONAL ACTIVITIES	SUPPLIER CATALOG NO.
Lactococcus lactis				
15 U/g	1 unit hydrolyzes 1 μmol glycyl-L-proline/min at pH 8.0, 37°C	Powder		Fluka 81706
Porcine kidney				
150 U/mg protein; 40 U/mg solid	1 unit hydrolyzes 1 μmol glycyl-L-proline/min at pH 8.0, 40°C	Salt-free; lyophilized		Biozyme PR2F
30 U/mg	1 unit hydrolyzes 1 μmol glycyl-L-proline/min at pH 8.0, 37°C	Powder		Fluka 81708
200 U/mg protein	1 unit hydrolyzes 1.0 μmol glycyl-L-proline/min at pH 8.0, 37°C	Suspension in 2.7 M $(NH_4)_2SO_4$, pH 8.0		ICN 156376
100-200 U/mg protein	1 unit hydrolyzes 1.0 μmol Gly-Pro/min at pH 8.0, 37°C	Salt-free; lyophilized containing 25% protein		Sigma P6675

Dipeptidyl-Peptidase I 3.4.14.1

REACTION CATALYZED
Release of an N-terminal dipeptide, Xaa-Xbb↓Xcc-, except when Xaa is Arg or Lys, or Xbb or Xcc is Pro

SYNONYMS
Dipeptidyl aminopeptidase I, dipeptidyl transferase, cathepsin C

REACTANTS
Oligopeptides, dipeptides

APPLICATIONS
- Transpeptidation reactions
- Processing fusion proteins (e.g., cytokines, human growth hormone), where peptide cleavage stops before an X-Pro sequence
- Generation of authentic proteins from precursors in microbes
- Sequencing tryptic peptides
- Polymerizing dipeptide amides

NOTES
- Classified as a hydrolase, acting on peptide bonds: dipeptidyl-peptidase and tripeptidyl-peptidase
- A Cl^--dependent, lysosomal cysteine-type peptidase
- Maximally active at acidic pH
- Polymerizes dipeptide amides, arylamides and esters at neutral pH
- Inhibited by iodoacetate and formaldehyde

SPECIFIC ACTIVITY	UNITS DEFINITION	PREPARATION FORM	ADDITIONAL ACTIVITIES	SUPPLIER CATALOG NO.
Bovine spleen (optimum pH = 4.0-6.0 [hydrolytic], 7.0-8.0 [transferase/transamidase], MW = 210,000 Da)				
		Sequencing grade; highly purified; solution; not stabilized with albumin		Boehringer 1559621 1559630
≥4 U/mg protein	1 unit hydrolyzes 1 μmol Gly-Phe-pNA/min at pH 4.5, 37°C	Solution containing 50% glycerol, $NaPO_4$, 150 mM NaCl, 4 mM cysteamine, pH 6.7-7.0		Calbiochem 219395
3 U/mg	Gly-Phe-4-nitranilide as substrate at 37°C	Analytical grade; 50% glycerol solution		ICN 191164
10 U/mg protein	1 unit produces 1.0 μmol Gly-Phe-NHOH from Gly-Phe-NH_2 and hydroxylamine/min at pH 6.8, 37°C	Sulfate-free; lyophilized containing 30% protein; balance Na citrate and NaCl		Sigma C8511
Turkey liver				
≥4 U/mg protein	1 unit hydrolyzes 1 μmol Gly-Phe-pNA/min at pH 4.5, 37°C	Solution containing 50% glycerol, $NaPO_4$, 150 mM NaCl, 4 mM cysteamine, pH 6.7-7.0		Calbiochem 219397

3.4.14.2 Dipeptidyl-Peptidase II

REACTION CATALYZED
Release of an N-terminal dipeptide, Xaa-Xbb\

SYNONYMS
Dipeptidyl aminopeptidase II, dipeptidyl arylamidase II, carboxytripeptidase

REACTANTS
Tripeptides, oligopeptides, dipeptides

NOTES
- Classified as a hydrolase, acting on peptide bonds: dipeptidyl-peptidase and tripeptidyl-peptidase
- A serine-type, lysosomal peptidase
- Maximally active at acidic pH
- Preferred substrates are tripeptides, where Xbb is Ala or Pro

See Chapter 5. Multiple Enzyme Preparations

3.4.14.5 Dipeptidyl-Peptidase IV

REACTION CATALYZED
Release of an N-terminal dipeptide, Xaa-Xbb\Xcc, from a polypeptide.

SYNONYMS
Dipeptidyl aminopeptidase IV, Xaa-Pro-dipeptidyl-aminopeptidase, Gly-Pro naphthylamidase, postproline dipeptidyl aminopeptidase IV

REACTANTS
Dipeptides, casomorphine, substance P

APPLICATIONS
- Debittering

NOTES
- Classified as a hydrolase, acting on peptide bonds: dipeptidyl-peptidase and tripeptidyl-peptidase
- A serine-type peptidase
- Release occurs preferentially when Xbb is Pro and Xcc is neither Pro nor hydroxyproline
- Membrane-bound in mammals (lymphatic tissue, blood and bone marrow) and flavobacteria
- Binding of enzyme by the V-3 loop of CD4 results in HIV entry into lymphocytes

SPECIFIC ACTIVITY	UNITS DEFINITION	PREPARATION FORM	ADDITIONAL ACTIVITIES	SUPPLIER CATALOG NO.
Human placenta				
710 mU/mg	1 mU hydrolyzes 1 μmol Ala-Pro-7-amino-4-trifluoromethyl coumarin/min at pH 7.8, 30°C	50% glycerol, 10 mM Tris, 1 mM EDTA, 0.02% NaN₃, pH 7.8	No detectable HBsAg, HIV Ab	Calbiochem 317624

SPECIFIC ACTIVITY	UNITS DEFINITION	PREPARATION FORM	ADDITIONAL ACTIVITIES	SUPPLIER CATALOG No.
Lactococcus lactis (optimum pH = 6.0-8.0, T = 35°C)				
220 U/g	1 unit liberates 1 μmol Leu from Leu paranitroanilide/min at pH 7.0 using 2.0 mM Leu paranitroanilide	Powder	LAP	Imperial Debitrase DBP 20

Peptidyl-Dipeptidase A   3.4.15.1

REACTION CATALYZED
Release of a C-terminal dipeptide, -Xaa\Xbb-Xcc, when neither Xaa or Xbb is Pro

SYNONYMS
Angiotensin converting enzyme, kininase II, dipeptidyl carboxypeptidase I, peptidase P, carboxycathepsin, dipeptide hydrolase

REACTANTS
Peptides, angiotensin I, bradykinin, dipeptides, proangiotensin

NOTES
- Classified as a hydrolase, acting on peptide bonds: dipeptidyl-dipeptidase
- A zinc metalloglycoprotein
- Cl^--dependent
- Active at neutral pH
- Generally membrane-bound
- Only single dipeptides are released from angiotensin I and bradykinin because of lack of activity toward Pro bonds
- May have endopeptidase activity on some substrates

SPECIFIC ACTIVITY	UNITS DEFINITION	PREPARATION FORM	ADDITIONAL ACTIVITIES	SUPPLIER CATALOG No.
Rabbit lung				
3 U/mg	1 unit produces 1 μmol hippuric acid from hippuryl-His-Leu/min at pH 8.3, 37°C			Fluka 10386
2-4 U/mg protein	1 unit produces 1.0 μmol hippuric acid from hippuryl-His-Leu/min at pH 8.3, 37°C in 50 mM HEPES and 300 mM NaCl	Lyophilized containing traces of NaCl		ICN 152742
2-4 U/mg protein	1 unit produces 1.0 μmol hippuric acid from hippuryl-His-Leu-/min at pH 8.3, 37°C	Lyophilized containing traces of NaCl		Sigma A6778
10 U/mg		Lyophilized		Wako 016-13874, 012-13871 018-13873

3.4.16.1 Serine-Type Carboxypeptidase

REACTION CATALYZED
Release of a C-terminal amino acid

SYNONYMS
Carboxypeptidase C (yeast), carboxypeptidase P (*Penicillium*), carboxypeptidase S_1 (*Penicillium*), carboxypeptidase S_2 (*Penicillium*), carboxypeptidase Y (yeast), lysosomal carboxypeptidase A (higher animals), serine carboxypeptidase II (germinating wheat), carboxypeptidase W (wheat), cathepsin A (higher animals), phaseolin (beans), protease C (yeast), CPaseY

REACTANTS
Peptides, amino acids

APPLICATIONS
- Sequence analysis and limited hydrolysis of peptides and proteins, especially with carboxypeptidases A and B
- Sequencing polypeptides under denaturing conditions

NOTES
- Classified as a hydrolase, acting on peptide bonds: serine-type carboxypeptidase
- An acid and serine glycoprotein
- A carboxypeptidase of broad amino acid specificity, including proline and amidated amino acid residues
- Optimum pH is 4.5-6.0
- High catalysis rate with penultimate/terminal amino acid aromatic or aliphatic side chains
- Release of glycine and aspartate is retarded
- Dipeptides are completely resistant to cleavage
- Probably identical to lysosomal tyrosine carboxypeptidase, now included within this group
- Not homologous with serine-type endopeptidases
- Active under denaturing conditions of urea and SDS
- Inhibited by diisopropyl fluorophosphate, PMSF, APCK, 4-hydroxymercuribenzoate, aprotinin
- Sensitive to thiol-blocking reagents

SPECIFIC ACTIVITY	UNITS DEFINITION	PREPARATION FORM	ADDITIONAL ACTIVITIES	SUPPLIER CATALOG NO.
Wheat (optimum pH = 4.0 [Z-Glu-Tyr], T = 30°C)				
150 U/mg protein	1 unit hydrolyzes 1.0 µmol L-Ala from Z-Phe-Ala/min at pH 4.0, 30°C	Lyophilized	No detectable aminopeptidase, endopeptidase	Calbiochem 217363
150 U/mg protein	1 unit liberates 1.0 µmol L-Ala from CBZ-L-Phe-L-Ala/min at pH 4.0, 30°C	Highly purified; lyophilized	No detectable endopeptidase, aminopeptidase or other proteolytic activities	ICN 190095
				ICN 321271
≥150 U/mg protein	1 unit liberates 1 µmol L-Ala from Z-Phe-Ala/min at pH 4.0, 30°C	Lyophilized		Seikagaku 100250-1 100250-2

SPECIFIC ACTIVITY	UNITS DEFINITION	PREPARATION FORM	ADDITIONAL ACTIVITIES	SUPPLIER CATALOG NO.
Yeast (optimum pH = 5.5 [acidic amino acids], 7.0 [basic amino acids], pI = 3.6, MW = 61,000 Da)				
20-100 U/mg solid	1 unit hydrolyzes 1.0 μmol N-CBZ-Phe-Ala to N-CBZ-L-Phe and L-Ala/min at pH 4.0, 30°C			Sigma C6527
		Sequencing grade; free of impurities by SDS-PAGE; lyophilized containing <10 pmol each amino acid		Boehringer 1420348 1111914
20 U/mg solid	1 unit converts 1 μmol Z-Phe-Ala to product/min at 37°C	Lyophilized		Boehringer 238139
130 U/mg protein	1 unit hydrolyzes 1 μmol CBZ-Phe-Ala/min at pH 6.7, 25°C	Lyophilized from buffer containing 50 mM Na citrate, pH 5.0		Calbiochem 217369
400 U/mg protein	1 unit hydrolyzes 1 μmol Z-Phe-Ala/min at 25°C	Excision grade; >90% purity; lyophilized from buffer containing 50 mM Na citrate, pH 6.0		Calbiochem 217372
	1 IU transforms 1 μmol substrate/min under standard IUB conditions at 25°C	Lyophilized		OYC
20 μg/vial	1 unit hydrolyzes 1.0 μmol N-CBZ-Phe-Ala to N-CBZ-L-Phe and L-Ala/min at pH 6.75, 25°C	Sequencing grade; lyophilized		Sigma C4046
≥50 U/mg protein	1 unit hydrolyzes 1 μmol benzyl-oxycarbonyl-L-Phe-L-Leu/min at pH 6.5, 25°C	Highly purified; lyophilized		Worthington LS09070 LS09068 LS09071
Yeast, bakers				
50 U/mg protein	1 unit hydrolyzes 1 μmol substrate/min at pH 6.5, 25°C	Lyophilized containing 95% protein		Calzyme 184A0050
20 U/mg	1 unit hydrolyzes 1 μmol Z-L-phenyl-alanyl-L-Ala/min at pH 6.75, 25°C	Powder containing 20% protein	<1% amidase <80% esterase	Fluka 21943
12 U/mg	1 unit hydrolyzes 1 μmol Z-L-phenyl-alanyl-L-Ala/min at pH 6.75, 25°C	Lyophilized containing 10% protein; stabilized in 90% Na citrate		Fluka 21945
100 U/mg protein	1 unit liberates 1.0 μmol L-Ala from CBZ-L-Phe-L-Ala/min at pH 6.75, 25°C	Lyophilized containing citrate buffer, pH 5		ICN 150563
100 U/mg protein	1 unit hydrolyzes 1.0 μmol N-CBZ-Phe-Ala to N-CBZ-L-Phe and L-Ala/min at pH 6.75, 25°C based on E/m_{230} = 191.5	Lyophilized containing 20% protein; balance citrate buffer, pH 5	Amidase and esterase	Sigma C3888

3.4.17.1 Carboxypeptidase A

REACTION CATALYZED
Release of a C-terminal amino acid

SYNONYMS
Carboxypolypeptidase, pancreatic carboxypeptidase A, tissue carboxypeptidase A

REACTANTS
Peptides, amino acids, hippuryl-L-phenylalanine

APPLICATIONS
- Hydrolysis/condensation of amide bonds
- Sequence analysis of proteins by successive cleavage of unsubstituted amino acids from the C-terminus
- Downstream processing of fusion proteins that may precipitate at high salt concentration
- Resolution of racemic amino acids

NOTES
- Classified as a hydrolase, acting on peptide bonds: metallocarboxypeptidase
- A metallo-exoprotease; 1 zinc atom per molecule, likely situated near the active site
- Formed from procarboxypeptidase A
- Sources include cattle, pig and dogfish pancreas, mast cells and skeletal muscle
- C-terminal aromatic or branched side chain L-amino acids are preferentially cleaved
- Little or no action with -Asp, -Glu, -Arg, -Lys or -Pro
- Inactivation to apoenzyme occurs with removal of zinc; activity is fully restored by adding zinc to metal-free enzyme
- Inactivated by freezing and lyophilizing
- Inhibited by cysteine, sulfides, cyanide, chelating agents, metal ions and anions; strongly inhibited by chelator 1,10-phenanthroline
- Ochratoxin A is a competitive inhibitor
- Treatment with PMSF or DFP eliminates contaminating trypsin and chymotrypsin

SPECIFIC ACTIVITY	UNITS DEFINITION	PREPARATION FORM	ADDITIONAL ACTIVITIES	SUPPLIER CATALOG No.
Bovine pancreas (optimum pH = 7-8, pI = 6.0, MW = 35,250 Da)				
35 U/mg	1 unit converts 1 µmol hippuryl-L-Phe to product/min at 25°C	Aqueous suspension	<0.1% chymotrypsin, trypsin	Boehringer 103225
		Salt-free; lyophilized containing <200 pmol/10 µg protein each free amino acid	<0.1% chymotrypsin, trypsin	Boehringer 1418017
>40 U/mg protein	1 unit hydrolyzes 1 µmol hippuryl-L-Phe/min at pH 7.5, 25°C	Suspension in water saturated with toluene, pH 7.0	<0.1% trypsin, chymotrypsin	Calbiochem 217285
50 U/mg protein	1 unit hydrolyzes 1 µmol hippuryl-L-Phe/min at pH 7.5, 25°C	Lyophilized containing 95% protein		Calzyme 050A0050

Carboxypeptidase A continued

SPECIFIC ACTIVITY	UNITS DEFINITION	PREPARATION FORM	ADDITIONAL ACTIVITIES	SUPPLIER CATALOG No.
Bovine pancreas *continued*				
50 U/mg protein	1 unit hydrolyzes 1 μmol hippuryl-L-Phe/min at pH 7.5, 25°C	Crystallized; suspension containing toluene as preservative and 10-30 mg/mL protein		Calzyme 050C0050
55 U/mg protein	1 unit hydrolyzes 1 μmol hippuryl-L-Phe/min at pH 7.5, 25°C	Suspension containing additional N-terminal heptapeptide		Fluka 21937
70 U/mg protein	1 unit hydrolyzes 1 μmol hippuryl-L-Phe/min at pH 7.5, 25°C	Aqueous suspension containing 10% toluene as stabilizer		Fluka 21938
60 U/mg protein	1 unit hydrolyzes 1 μmol hippuryl-L-Phe/min at pH 7.5, 25°C	PMSF-treated; aqueous suspension containing 10% toluene as stabilizer	<0.001% trypsin, chymotrypsin	Fluka 21940
35-50 U/mg protein	1 unit hydrolyzes 1 μmol hippuryl-L-Phe/min at pH 7.5, 25°C	Aqueous suspension containing toluene		ICN 100403
35-50 U/mg protein	1 unit hydrolyzes 1 μmol hippuryl-L-Phe/min at pH 7.5, 25°C	PMSF-treated; aqueous suspension containing toluene		ICN 100407
35-50 U/mg protein	1 unit hydrolyzes 1 μmol hippuryl-L-Phe/min at pH 7.5, 25°C	DFP-treated; aqueous suspension containing toluene		ICN 150562
50 U/mg protein	1 unit hydrolyzes 1 μmol hippuryl-L-Phe/min at pH 7.5, 25°C	Contains additional N-terminal heptapeptide; crystallized; aqueous suspension containing toluene		Sigma C0261
50 U/mg protein	1 unit hydrolyzes 1 μmol hippuryl-L-Phe/min at pH 7.5, 25°C	2X Crystallized; aqueous suspension in toluene		Sigma C0386
200-400 U/g agarose 1 mL gel yields 6-12 U	1 unit hydrolyzes 1 μmol hippuryl-L-Phe/min at pH 7.5, 30°C	Insoluble enzyme attached to beaded agarose; suspension in 2.0 M $(NH_4)_2SO_4$, pH 7		Sigma C1261
50 U/mg protein	1 unit hydrolyzes 1 μmol hippuryl-L-Phe/min at pH 7.5, 25°C	PMSF-treated; dialyzed; recrystallized; aqueous suspension in toluene		Sigma C6393
50 U/mg protein	1 unit hydrolyzes 1 μmol hippuryl-L-Phe/min at pH 7.5, 25°C	DFP-treated; dialyzed; recrystallized; aqueous suspension in toluene		Sigma C6510
50 U/mg protein	1 unit hydrolyzes 1 μmol hippuryl-L-Phe/min at pH 7.5, 25°C	PMSF-treated; dialyzed; recrystallized; aqueous suspension in toluene		Sigma C9268
50 U/mg protein	1 unit hydrolyzes 1 μmol hippuryl-L-Phe/min at pH 7.5, 25°C	DFP-treated; dialyzed; recrystallized; aqueous suspension in toluene		Sigma C9762
50 U/mg				Wako 030-15111
≥35 U/mg protein	1 unit hydrolyzes 1 μmol hippuryl-L-Phe/min at pH 7.5, 25°C	2X Crystallized; suspension in toluene	<1% free amino acids	Worthington LS01670 LS01672 LS01674

3.4.17.1 Carboxypeptidase A continued

SPECIFIC ACTIVITY	UNITS DEFINITION	PREPARATION FORM	ADDITIONAL ACTIVITIES	SUPPLIER CATALOG No.
Bovine pancreas *continued*				
≥35 U/mg protein	1 unit hydrolyzes 1 μmol hippuryl-L-Phe/min at pH 7.5, 25°C	PMSF-treated; crystallized; suspension in toluene	<0.02% trypsin, chymotrypsin <2% carboxypeptidase B	Worthington LS01692 LS01694 LS01690

3.4.17.2 Carboxypeptidase B

REACTION CATALYZED
    Preferential release of C-terminal lysine, ornithine or arginine amino acids

SYNONYMS
    Protaminase, pancreatic carboxypeptidase B, tissue carboxypeptidase B

REACTANTS
    Peptides, L-lysine, L-ornithine, L-arginine, hippuryl-L-arginine

APPLICATIONS
- Sequence analysis by successive cleavage of basic amino acids from the C-terminus of proteins

NOTES
- Classified as a hydrolase, acting on peptide bonds: metallocarboxypeptidase
- A metallo-protease, containing 1 gram atom of zinc per mol
- Formed from procarboxypeptidase B
- Sources include cattle, pig and dogfish pancreas, skin fibroblasts and adrenal medulla
- Minimal activity toward carboxypeptidase A substrates; C-terminal His is not acted on
- Inhibited by EDTA, zinc chelators (e.g., 1,10-phenanthroline) and heavy metals; competitively inhibited by basic amino acids

SPECIFIC ACTIVITY	UNITS DEFINITION	PREPARATION FORM	ADDITIONAL ACTIVITIES	SUPPLIER CATALOG No.
Porcine pancreas (optimum pH = 7.0-9.0, MW = 34,500 Da)				
150 U/mg	1 unit converts 1 μmol hippuryl-L-Arg to product/min at 25°C	DFP-treated; solution	<0.7 mU/mg trypsin <2% carboxypeptidase A <5 mU/mg chymotrypsin	Boehringer 103233
>100 U/mg protein	1 unit hydrolyzes 1 μmol Hip-Arg/min at pH 7.7, 25°C	Lyophilized	No detectable chymotrypsin, trypsin <1.0% carboxypeptidase A	Calbiochem 217356
50 U/mg protein	1 unit hydrolyzes 1 μmol hippuryl-L-Arg/min at pH 7.65, 25°C	Lyophilized containing 80% protein		Calzyme 050A0200

SPECIFIC ACTIVITY	UNITS DEFINITION	PREPARATION FORM	ADDITIONAL ACTIVITIES	SUPPLIER CATALOG NO.
Porcine pancreas *continued*				
170 U/mg protein	1 unit hydrolyzes 1 µmol hippuryl-L-Arg/min at pH 7.65, 25°C	Frozen solution containing 0.1 M NaCl and 10-20 mg/mL protein		Calzyme 050H0200
>125 U/mg protein	1 unit hydrolyzes 1.0 µmol hippuryl-L-Arg/min at pH 7.7, 25°C	Chromatographically purified; salt-free; lyophilized containing 25 mM Tris-HCl and 0.1 M NaCl	<0.1% carboxypeptidase A	Elastin CB276
200 U/mg protein	1 unit hydrolyzes 1 µmol hippuryl-L-Arg/min at pH 7.65, 25°C	Solution containing 0.1 M NaCl	<1% carboxypeptidase A, chymotrypsin, trypsin	Fluka 21941
130 U/mg protein	1 unit hydrolyzes 1 µmol hippuryl-L-Arg/min at pH 7.65, 25°C	DFP-treated; solution containing 0.1 M NaCl	<0.1% carboxypeptidase A, chymotrypsin, trypsin	Fluka 21942
40 U/mg protein	1 unit hydrolyzes 1 µmol hippuryl-L-Arg/min at pH 7.65, 25°C	Purified; solution containing 0.1 M NaCl	No detectable carboxypeptidase A, chymotrypsin, trypsin	ICN 160026
150-250 U/mg protein	1 unit hydrolyzes 1 µmol hippuryl-L-Arg/min at pH 7.65, 25°C	Chromatographically purified; frozen solution containing 0.1 M NaCl	No detectable trypsin and carboxypeptidase A. May contain trace chymotrypsin	Sigma C7011
100-200 U/mg protein	1 unit hydrolyzes 1 µmol hippuryl-L-Arg/min at pH 7.65, 25°C	DFP-treated; frozen solution containing 0.1 M NaCl	Treated with DFP to eliminate trypsin and chymotrypsin	Sigma C7261
100 U/mg protein	1 unit hydrolyzes 1 µmol hippuryl-L-Arg/min at pH 7.65, 25°C	Frozen solution containing 0.1 M NaCl	May contain trypsin, chymotrypsin and carboxypeptidase A	Sigma C7386
≥70 U/mg protein	1 unit hydrolyzes 1 µmol hippuryl-L-Arg/min at pH 7.65, 25°C	PMSF-treated; solution containing 0.1 M NaCl	<0.02% trypsin, chymotrypsin	Worthington LS01722 LS01724 LS01720
≥170 U/mg protein	1 unit hydrolyzes 1 µmol hippuryl-L-Arg/min at pH 7.65, 25°C	Chromatographically prepared; solution containing 0.1 M NaCl	<0.1% trypsin, chymotrypsin <2% carboxypeptidase A	Worthington LS05305 LS05301 LS05304 LS05302

3.4.17.16 Membrane Pro-X Carboxypeptidase

REACTION CATALYZED
Release of a C-terminal residue other than proline, by preferential cleavage of a prolyl bond

SYNONYMS
Carboxypeptidase P, microsomal carboxypeptidase

REACTANTS
Peptides, amino acids, C-terminal

APPLICATIONS
- Protein structure
- Sequence analysis

NOTES
- Classified as a hydrolase, acting on peptide bonds: metallocarboxypeptidase
- One of the renal brush border exopeptidases
- A serine carboxypeptidase
- Release of serine and glycine is considerably retarded
- Inhibited by DFP, iodoacetic acid, PMCB and EDTA

SPECIFIC ACTIVITY	UNITS DEFINITION	PREPARATION FORM	ADDITIONAL ACTIVITIES	SUPPLIER CATALOG NO.
Penicillium janthinellum (optimum pH = 3.7-5.2, MW = 51,000 Da)				
		Sequencing grade; free of impurities by SDS-PAGE; lyophilized containing <10 pmol each amino acid		Boehringer 1420321 1111906
		Excision grade; >90% purity; lyophilized containing 50 mM Na citrate, pH 5.0	No detectable protease	Calbiochem 217365
18 U/mg	1 unit hydrolyzes 1 μmol N-carbobenzoxy-L-Glu/min at pH 3.7, 30°C	Lyophilized		Fluka 21946
40-70 U/mg protein	1 unit hydrolyzes 1.0 μmol N-CBZ-Glu-Tyr to N-CBZ-L-Glu and L-Tyr/min at pH 3.7, 30°C	Lyophilized containing Na citrate		Sigma C5396
Unspecified (optimum pH = 3.7 [cbz-Glu-Tyr], 4.5 [cbz-Gly-Pro-Leu-Glu], 5.2 [cyz-Gly-Lys], MW = 51,000 Da)				
40 U/mg protein	1 unit produces 1 μmol L-Tyr/min at 37°C	Lyophilized	No detectable protease	PanVera TAK 7304

Acylaminoacyl-Peptidase — 3.4.19.1

REACTION CATALYZED
Cleavage of an *N*-acetyl or *N*-formyl amino acid from the N-terminus of a polypeptide

SYNONYMS
Acylamino-acid-releasing enzyme, *N*-acylpeptide hydrolase, *N*-formylmethionine (fMet) aminopeptidase

REACTANTS
Polypeptide, *N*-acetyl or *N*-formyl amino acid, *N*-acetylalanine, glycine

APPLICATIONS
- Deblocking peptides for subsequent N-terminus sequence analysis by Edman degradation

NOTES
- Classified as a hydrolase, acting on peptide bonds: omega peptidase
- Active at neutral pH
- Inhibited by DFP, PCMB
- Human erythrocyte enzyme is relatively specific for removal of *N*-acetylalanine from peptides. It displays dipeptidyl-peptidase (EC 3.4.14.1-5) activity on glycyl-peptides, possibly as a result of misrecognition of the glycyl residue as an uncharged *N*-acyl group
- Preference for the sequence -acyl-X-QQ-Y is X = serine, alanine, methionine; Y is unspecified
- For sequence analysis, the protease:protein ratio of 1:100 or 1:10 is recommended for 2-18 hrs, 25°C

SPECIFIC ACTIVITY	UNITS DEFINITION	PREPARATION FORM	ADDITIONAL ACTIVITIES	SUPPLIER CATALOG NO.
Equine liver (optimum pH = 7.5-9.0, MW = 80,000 Da)				
		Sequencing grade; free of impurities; lyophilized		Boehringer 1370502
1 U/mg protein		>90% Purity; lyophilized	No detectable proteases	Calbiochem 114875
Unspecified (optimum pH = 7.2-7.6, MW = 75,000 [SDS-PAGE], 360,000 [GPC] Da)				
13.0 U/mg protein	1 unit produces 1 μmol Ala/min at 37°C	Homogeneous by PAGE; lyophilized containing sucrose		PanVera TAK 7301

3.4.19.3 Pyroglutamyl-Peptidase I

REACTION CATALYZED
Release of an N-terminal pyroglutamyl group from a polypeptide, provided the next residue is not proline

SYSTEMATIC NAME
Pyroglutamyl-peptidase

SYNONYMS
5-Oxoprolyl-peptidase, pyrrolidone-carboxylate peptidase, pyroglutamyl aminopeptidase

REACTANTS
Polypeptide, pyroglutamyl group, N-terminal

APPLICATIONS
- Deblocking proteins and peptides containing N-terminus pyroglutamyl residues, enabling subsequent Edman degradation

NOTES
- Classified as a hydrolase, acting on peptide bonds: omega peptidase
- A cysteine protease occurring in mammalian tissues, plants and microorganisms
- Activated by sulfhydryl reagents like DTE and DTT
- Stabilized by EDTA and 2-mercaptoethanol
- Inhibited by thiol-blocking reagents, PCMB and heavy metal ions

SPECIFIC ACTIVITY	UNITS DEFINITION	PREPARATION FORM	ADDITIONAL ACTIVITIES	SUPPLIER CATALOG NO.
Bacillus amyloliquefaciens (optimum pH = 7.0-8.0, T = 50°C, MW = 72,000 Da [gel filtration]; 3 subunits of 24,000 Da/mol enzyme]; K_M = 2.5 mM [PCA-L-Ala], 1.3 mM [PCA-β-naphthylamide]; stable pH 7.0-8.5 [30°C, 30 min], T < 40°C [pH 7.8, 30 min])				
20,000 U/mL	1 unit hydrolyzes 1.0 nmol L-pyro-Glu β-naphthylamide to L-pyroGlu and β-naphthyl-amine/min at pH 8.0, 37°C	Suspension in (NH4)2SO4		Sigma P4669
≥20 U/mL	1 unit forms 1 μmol β-naphthyl-amine/min at pH 8.0, 30°C	50% glycerol solution containing EDTA and β-MSH as stabilizers	<0.1% peptidase	Toyobo PGP-209
Bacillus amyloliquefaciens, recombinant expressed in *E. coli*				
7.0 U/mg protein	1 unit hydrolyzes 1.0 μmol Pyr-pNA/min at pH 8.0, 37°C	50% glycerol		Calbiochem 545125
Calf liver (optimum pH = 8.0, MW = 70,000-80,000 Da)				
		Sequencing grade; lyophilized	No detectable protease	Boehringer 1420445 1179861
100-400 U/mg protein	1 unit hydrolyzes 1.0 nmol L-pyro-Glu β-naphthylamide to L-pyroGlu and β-naphthyl-amine/min at pH 8.0, 37°C	Lyophilized containing 15% protein, 60% sucrose, 10% EDTA; balance NaPO4		Sigma P9419
	1 unit hydrolyzes 1.0 nmol L-pyro-Glu β-naphthylamide to L-pyroGlu and β-naphthyl-amine/min at pH 8.0, 37°C	Lyophilized containing 5% protein; balance sucrose and buffer salts		Sigma P9516

SPECIFIC ACTIVITY	UNITS DEFINITION	PREPARATION FORM	ADDITIONAL ACTIVITIES	SUPPLIER CATALOG NO.
Unspecified (optimum pH = 6.0–9.0, MW = 60,000–70,000 Da [GPC])				
5.0 U/mg protein; 0.5 U/mL	1 unit produces 1 µmol p-nitroaniline/min at 37°C	10 mM Tris-HCl, 5 mM DTT, 1 mM EDTA, 50% glycerol, pH 7.4	No detectable protease	PanVera TAK 7321

3.4.19.9 γ-Glu-X Carboxypeptidase

REACTION CATALYZED
Cleavage of a γ-glutamyl bond to release an unsubstituted C-terminal amino acid

SYNONYMS
Conjugase, folate conjugase, pteroyl-poly-γ-glutamate hydrolase, carboxypeptidase G, γ-glutamyl hydrolase, lysosomal γ-glutamyl carboxypeptidase

REACTANTS
Peptides, pteroyl-poly-γ-glutamate, glutamic acid, pteroyl-α-glutamate, C-terminal amino acid

NOTES
- Classified as a hydrolase, acting on peptide bonds: omega peptidase
- A lysosomal, thiol-dependent carboxypeptidase
- Progressively removes γ-glutamyl residues at acid pH from pteroyl-poly-γ-glutamate to yield pteroyl-α-glutamate (folic acid) and free glutamic acid
- More than one γ-glutamyl residue may be released from a poly-γ-glutamate chain at a time, possibly due to a free α-carboxyl group at each γ-glutamyl linkage
- Highly specific for the γ-glutamyl bond but not for the C-terminal amino acid leaving group
- Action on γ-glutamyl bonds is independent of an N-terminal pteroyl residue

SPECIFIC ACTIVITY	UNITS DEFINITION	PREPARATION FORM	ADDITIONAL ACTIVITIES	SUPPLIER CATALOG NO.
Pseudomonas species				
	1 unit hydrolyzes 1.0 µmol L-Glu from (+)amethopterin/min at pH 7.3, 30°C	Affinity chromatographed; lyophilized containing 10% protein; balance primarily PO$_4$ buffer salts		Sigma C4053
2–5 U/mg protein	1 unit hydrolyzes 1.0 µmol L-Glu from (+)amethopterin/min at pH 7.3, 30°C	Chromatographically purified; lyophilized containing 70% protein; balance primarily NaOAc		Sigma C9658

Chymotrypsin

REACTION CATALYZED

Preferential cleavage: Tyr↓, Trp↓, Phe↓, Leu↓

SYNONYMS

Chymotrypsin A, chymotrypsin B

REACTANTS

Polypeptides, tyrosine, tryptophan, phenylalanine, leucine, benzoyl-L-tyrosine ethyl ester (BTEE), acetyl-L-tyrosine ethyl ester (ATEE)

APPLICATIONS

- Peptide synthesis
- Oligosaccharide synthesis on a chymotrypsin-sensitive polymer
- Peptide mapping
- Peptide fingerprinting
- Sequence analysis
- Anti-inflammatory agent, reducing soft-tissue inflammation and edema

NOTES

- Classified as a hydrolase, acting on peptide bonds: serine endopeptidase
- Chymotrypsin A is formed from cattle and pig chymotrypsinogen A. Several isozymes are possible, depending on the number of bonds cleaved in the precursor
- Chymotrypsin B is formed from chymotrypsinogen B and is homologous with chymotrypsin A
- Widely distributed in nature
- Cleaves Leu↓, Ala↓, Asp↓ and Glu↓ at lower rates
- Also acts on amides and esters of accepted amino acids
- Inhibited by aprotinin, DFP, PMSF, phenothiazine-N-carbonylchloride, TPCK, ZPCK, α-macroglobulin, α1-antitrypsin, soybean trypsin inhibitor and chymostatin. *Not inhibited* by APMSF
- Elevated serum levels have been reported in patients with cystic fibrosis

SPECIFIC ACTIVITY	UNITS DEFINITION	PREPARATION FORM	ADDITIONAL ACTIVITIES	SUPPLIER CATALOG NO.
Bovine pancreas (optimum pH = 7.0-9.0, pI = 8.1-8.6, T = 37°C, MW = 25,000 Da; stable pH 3.0 [days])				
≥40 U/mg protein	1 unit hydrolyzes 1.0 μmol benzoyl-L-Tyr ethyl ester (BTEE)/min at pH 7.8, 25°C	Salt-free; lyophilized	Acts as esterase and amidase	Adv Biofact CH15-1
1000 U/mg	1 unit increases A_{237} by 0.0075/min at 25°C	Amorphous powder		AMRESCO 0164 E208
90 U/mg solid	1 unit converts 1 μmol acetyl-L-Tyr ethyl ester to product/min at 25°C	Salt-free; lyophilized from activated, crystallized chymotrypsinogen A		Boehringer 103306 103314

SPECIFIC ACTIVITY	UNITS DEFINITION	PREPARATION FORM	ADDITIONAL ACTIVITIES	SUPPLIER CATALOG NO.
Bovine pancreas *continued*				
		Sequencing grade; salt-free; lyophilized	Free of impurities that might interfere with the specific cleavage or in the separation of peptides using reversed-phase HPLC $<0.07\%$ trypsin	Boehringer 1418467 1334131
$\geq$1000 U/mg solid	1 unit decreases A_{237} by 0.0075/min at pH 7.0, 25°C with N-Ac-L-Tyr ethyl ester (ATEE) as substrate	Solid	$\geq$25 U/mg trypsin	Calbiochem 230832
1000 U/mg solid	1 unit hydrolyzes 1 µmol benzoyl-L-Tyr ethyl ester (BTEE)/min at pH 7.8, 25°C	Lyophilized containing 95% protein	Trypsin	Calzyme 066A1000
4-8 mg enzyme/mL gel; 40-100 U/mL	1 unit hydrolyzes 1 µmol ATEE/min at pH 8.0, 30°C	3X Crystallized; attached to beaded agarose in 0.02 M NaOAc and 0.01% NaN$_3$, pH 5		Elastin A424
4-8 mg enzyme/mL gel; 40-100 U/mL	1 unit hydrolyzes 1 µmol ATEE/min at pH 8.0, 30°C	3X Crystallized; attached to beaded agarose in 0.02 M NaOAc and 0.01% NaN$_3$, pH 5		Elastin AD424
4-8 mg enzyme/mL gel; 40-100 U/mL	1 unit hydrolyzes 1 µmol ATEE/min at pH 8.0, 30°C	3X Crystallized; attached to beaded agarose in 0.02 M NaOAc and 0.01% NaN$_3$, pH 5		Elastin AE424
65 U/mg	1 unit hydrolyzes 1 µmol Suc-(Ala)$_2$-Pro-Phe-4-NA/min at pH 7.8, 25°C		Free of low MW peptide fragments (GPC)	Fluka 27267
60 U/mg	1 unit hydrolyzes 1 µmol Suc-(Ala)$_2$-Pro-Phe-4-NA/min at pH 7.8, 25°C	Lyophilized	1200 ATEE- U/mg	Fluka 27270
50 U/mg	1 unit hydrolyzes 1 µmol Suc-(Ala)$_2$-Pro-Phe-4-NA/min at pH 7.8, 25°C	Powder		Fluka 27272
60 U/mg	1 unit hydrolyzes 1 µmol Suc-(Ala)$_2$-Pro-Phe-4-NA/min at pH 7.8, 25°C	TLCK-treated; lyophilized	$<0.001\%$ trypsin	Fluka 27280
40-50 U/mg protein	1 unit hydrolyzes 1.0 µmol N-benzoyl-L-Tyr ethyl ester (BTEE)/min at pH 7.8, 25°C	Salt-free; dialyzed; lyophilized; activation product of 3X crystallized zymogen		ICN 100461
45 U/mg solid	1 unit hydrolyzes 1.0 µmol N-benzoyl-L-Tyr ethyl ester (BTEE)/min at pH 7.8, 25°C	Salt-free; EtOH precipitate		ICN 100465
20-30 U/mg solid	1 unit hydrolyzes 1.0 µmol N-benzoyl-L-Tyr ethyl ester (BTEE)/min at pH 7.8, 25°C	Salt-free; crystallized; lyophilized		ICN 100468
45 U/mg		Salt-free; 2X crystallized from 3X crystallized chymotrypsinogen; lyophilized		ICN 100475
40-50 U/mg protein	1 unit hydrolyzes 1.0 µmol N-benzoyl-L-Tyr ethyl ester (BTEE)/min at pH 7.8, 25°C	Salt-free; 3X crystallized; lyophilized		ICN 100478

Chymotrypsin continued

SPECIFIC ACTIVITY	UNITS DEFINITION	PREPARATION FORM	ADDITIONAL ACTIVITIES	SUPPLIER CATALOG No.
Bovine pancreas continued				
35 U/mg protein	1 unit hydrolyzes 1.0 μmol N-benzoyl-L-Tyr ethyl ester (BTEE)/min at pH 7.8, 25°C	Salt-free; 1X crystallized; dialyzed; lyophilized		ICN 152272
≥1000 U/mg	1 unit decreases A_{237} by 0.0075/min at 25°C	Crystallized; powder	Trypsin <2.5% of chymotrypsin	Can Inova C-001
≥750 U/mg	1 unit decreases A_{237} by 0.0075/min at 25°C	Powder	≥750 U/mg trypsin	Can Inova C-011
≥1000 USP U/mg		USP grade; powder	<25 USP U trypsin/1000 USP U chymotrypsin	Intergen 7008-00
0.1 U/mg protein	1 unit hydrolyzes 1.0 μmol BTEE/min at pH 7.8, 25°C	DFP-inactivated; 3X crystallized; dialyzed; lyophilized		Sigma C1012
40-60 U/mg protein	1 unit hydrolyzes 1.0 μmol BTEE/min at pH 7.8, 25°C	TLCK-inactivated; salt-free; dialyzed; lyophilized	TLCK reportedly inactivates trypsin usually present without affecting chymotrypsin activity	Sigma C3142
40-60 U/mg protein	1 unit hydrolyzes 1.0 μmol BTEE/min at pH 7.8, 25°C	Salt-free; 3X crystallized from 4X crystallized chymotrypsinogen; dialyzed; lyophilized		Sigma C4129
20-50 U/mg solid	1 unit hydrolyzes 1.0 μmol BTEE/min at pH 7.8, 25°C	Salt-free; crystallized; lyophilized		Sigma C4629
25-50 U/mg solid	1 unit hydrolyzes 1.0 μmol BTEE/min at pH 7.8, 25°C	Salt-free; 2X crystallized; lyophilized		Sigma C4754
400-600 U/g solid	1 unit hydrolyzes 1.0 μmol ATEE/min at pH 8.0, 30°C	Insoluble enzyme attached to carboxymethyl cellulose containing 30% borate buffer salts		Sigma C7260
40-60 U/mg protein	1 unit hydrolyzes 1.0 μmol BTEE/min at pH 7.8, 25°C	Salt-free; 3X crystallized; lyophilized	No detectable autolysis products, low MW contaminants	Sigma C7762
30 U/mg protein	1 unit hydrolyzes 1.0 μmol BTEE/min at pH 7.8, 25°C	Biotin-labeled (4 mol biotin/mol protein) and TLCK-inactivated; salt-free; powder		Sigma C9021
2000-3500 U/g of agarose; 1 mL gel yields 65-120 U	1 unit hydrolyzes 1.0 μmol ATEE/min at pH 8.0, 30°C	Insoluble enzyme attached to agarose; lyophilized containing lactose as stabilizer		Sigma C9134
40-80 U/mg solid	1 unit hydrolyzes 1.0 μmol BTEE/min at pH 7.8, 25°C	Salt-free		Sigma C9381
100 mg/vial	1 unit hydrolyzes 1.0 μmol BTEE/min at pH 7.8, 25°C	Sterile-filtered; free of autolysis products and low MW contaminants		Sigma CHY-100S
50 mg/vial	1 unit hydrolyzes 1.0 μmol BTEE/min at pH 7.8, 25°C	Sterile-filtered; free of autolysis products and low MW contaminants		Sigma CHY-50S
5 mg/vial	1 unit hydrolyzes 1.0 μmol BTEE/min at pH 7.8, 25°C	Sterile-filtered; free of autolysis products and low MW contaminants		Sigma CHY-5S

Chymotrypsin continued 3.4.21.1

SPECIFIC ACTIVITY	UNITS DEFINITION	PREPARATION FORM	ADDITIONAL ACTIVITIES	SUPPLIER CATALOG No.
Bovine pancreas *continued*				
≥1000 USP chymotrypsin U/mg solid		Crystallized		Wako 036-10691 032-10693
≥35 U/mg protein	1 unit hydrolyzes 1 μmol BTEE/min at pH 7.8, 25°C	1X Crystallized; dialyzed against 1 mM HCl; lyophilized containing traces of Cl⁻		Worthington LS01333 LS01334 LS01332
≥45 U/mg protein	1 unit hydrolyzes 1 μmol BTEE/min at pH 7.8, 25°C	TLCK-treated; 3X crystallized; dialyzed against 1 mM HCl; lyophilized containing traces of Cl⁻		Worthington LS01430 LS01432 LS01434 LS01438
≥45 U/mg protein	1 unit hydrolyzes 1 μmol BTEE/min at pH 7.8, 25°C	3X Crystallized; dialyzed against 1 mM HCl; lyophilized containing traces of Cl⁻		Worthington LS01448 LS01450 LS01451 LS01453
≥45 U/mg protein	1 unit hydrolyzes 1 μmol BTEE/min at pH 7.8, 25°C	Dialyzed against 1 mM HCl; lyophilized containing traces of Cl⁻		Worthington LS01475 LS01479 LS01477
Human pancreas (MW = 25,000 Da)				
40-70 U/mg protein	1 unit hydrolyzes 1 μmol Suc-Ala-Ala-Pro-Phe-pNA/min at pH 8.0, 25°C	≥95% purity; salt-free; lyophilized	No detectable HBsAg, anti-HCV, anti-HBc, anti-HIV No detectable trypsin, trypsinogen by immunodiffusion against antiserum to trypsin	ART 16-19-030820
>400 U/mg protein	1 unit hydrolyzes 1 μmol Suc-Ala-Ala-Pro-Phe-pNA/min at pH 8.0, 25°C	>95% purity; lyophilized; salt-free	No detectable trypsin, trypsinogen by immunodiffusion against antiserum to trypsin No detectable HBsAg, HIV Ab	Calbiochem 230900
85% by burst titrant analysis		>95% purity by SDS-PAGE; salt-free; lyophilized containing 2 mM HCl	No detectable HIV Ab, HBsAg	Cortex CP3011
75-95% active sites by burst titrant analysis		>99% purity by SDS-PAGE; lyophilized containing 1 mM HCl		ICN 191338
10 BTEE U/vial	1 unit hydrolyzes 1.0 μmol BTEE/min at pH 7.8, 25°C	Lyophilized		Sigma C8946

3.4.21.1 Chymotrypsin continued

SPECIFIC ACTIVITY	UNITS DEFINITION	PREPARATION FORM	ADDITIONAL ACTIVITIES	SUPPLIER CATALOG NO.
Unspecified				
	1 unit hydrolyzes 1.0 μmol BTEE/min at pH 7.8, 25°C	Sequencing grade		Sigma C6423
				SpecialtyEnz

3.4.21.4 Trypsin

REACTION CATALYZED
    Preferential cleavage: Arg↓, Lys↓

SYNONYMS
    Cocoonase (insects), alkaline protease, alkalophilic protease

REACTANTS
    Polypeptides, amides, esters, peptide bonds, *p*-toluene-sulfonyl-L-arginine methyl ester (TAME)

APPLICATIONS
- Disintegrates primary tissues in cultures
- Desiccation of tryptic enzyme is used in tissue culture for obtaining uniform cell suspensions
- Removes monolayers of cells from glass and plastic
- Treatment for erythrocytes in Rha and various hemagglutination procedures
- Sample preparation for flow cytometric DNA analysis
- Environmental monitoring
- Improved recovery in cell harvesting from microcarriers
- Protein structure elucidation
- Tryptic mapping
- Fingerprinting and sequence analysis
- Translocation studies
- Generating glycopeptides from purified glycoproteins
- Isolation of intact, detergent-free phycobilisomes
- Cleaving fusion proteins
- Preparing proteins and F_{ab} fragments from IgM antibodies
- Isolating cells from a wide variety of tissues/organs, often with collagenase, hyaluronidase, etc.
- Subculturing cells
- Reducing cell density in tissue culture.

NOTES
- Classified as a hydrolase, acting on peptide bonds: serine endopeptidase
- Found in animals and some bacteria
- Trypsinogen is the inactive precursor of trypsin. It is secreted by exocrine cells of the pancreas and released into the lumen of the small intestine. Trypsinogen is converted to trypsin by enterokinase and trypsin
- Hydrolyzes peptides, amides and esters involving carboxyl groups of L-arginine and L-lysine
- Cattle β-trypsin is a single polypeptide chain formed from trypsinogen by cleavage of one

- peptide bond. Further peptide bond cleavages produce α- and other iso-forms
- Isolated as multiple cationic and anionic trypsins from the pancreas of many vertebrates and from lower species
- Forms complexes with α_2-macroglobulin
- Diphenyl carbamyl chloride (DCC) treatment reduces levels of the chymotrypsin usually present

- Stabilized by Ca^{2+}
- Inhibited by pancreatic-, soybean-, lima bean- and egg white-trypsin inhibitors; DFP, aprotinin, Ag^+, benzamidine and EDTA
- Elevated serum levels are found in patients with cystic fibrosis

SPECIFIC ACTIVITY	UNITS DEFINITION	PREPARATION FORM	ADDITIONAL ACTIVITIES	SUPPLIER CATALOG No.
Bovine pancreas (optimum pH = 7.0-9.0, T = 37°C, MW = 23,800 Da)				
10,000-13,000 BAEE U/mg protein	1 BAEE unit changes A_{253} by 0.001/min at pH 7.6, 25°C with N-benzoyl-L-Arg ethyl ester as substrate	Salt-free; lyophilized	Acts as esterase and amidase	Adv Biofact TY14-1
2500 U/mg solid	1 unit increases A_{253} by 0.003/min at 25°C	Amorphous powder		AMRESCO 0785
		Sequencing grade; salt-free	Free of impurities that might interfere with the specific cleavage or in the separation of peptides using reversed-phase HPLC	Boehringer 1418025 1418033
		Sequencing grade; salt-free; lyophilized	Free of impurities that might interfere with the specific cleavage or in the separation of peptides using reversed-phase HPLC <0.05% chymotrypsin	Boehringer 1418475 1047841
2500 USP U/mg solid	1 unit digests BAEE increasing A_{253} by 0.003/min at pH 7.6, 25°C	Powder	No detectable salmonella <50 NF U/mg chymotrypsin	Calbiochem 6502
80 U/mg protein		Excision grade; salt-free; lyophilized	<0.02% α–chymotrypsin	Calbiochem 650211
3000 U/mg solid	1 unit increases A_{253} by 0.003/min at pH 7.6, 25°C using BAEE as substrate	Lyophilized containing 90-95% protein		Calzyme 112A3000
8-12 mg enzyme/mL gel; 800-1400 U/mL	1 unit hydrolyzes 1 μmol BAEE/min at pH 7.6, 25°C	2X Crystallized; attached to beaded agarose in 0.02 M NaOAc and 0.01% NaN_3, pH 5		Elastin A432

SPECIFIC ACTIVITY	UNITS DEFINITION	PREPARATION FORM	ADDITIONAL ACTIVITIES	SUPPLIER CATALOG NO.
Bovine pancreas *continued*				
8-12 mg enzyme/mL gel; 800-1400 U/mL	1 unit hydrolyzes 1 μmol BAEE/min at pH 7.6, 25°C	2X Crystallized; attached to beaded agarose in 0.02 M NaOAc and 0.01% NaN$_3$, pH 5		Elastin AD432
8-12 mg enzyme/mL gel; 800-1400 U/mL	1 unit hydrolyzes 1 μmol BAEE/min at pH 7.6, 25°C	2X Crystallized; attached to beaded agarose in 0.02 M NaOAc and 0.01% NaN$_3$, pH 5		Elastin AE432
8000 U/mg	1 unit increases A$_{253}$ by 0.001/min at pH 7.6, 25°C with *N*-benzoyl-L-Arg ethyl ester as substrate			Fluka 93608
9000 U/mg	1 unit increases A$_{253}$ by 0.001/min at pH 7.6, 25°C with *N*-benzoyl-L-Arg ethyl ester as substrate	Salt-free; crystallized; lyophilized	<0.2% chymotrypsin	Fluka 93610
8300 U/mg	1 unit increases A$_{253}$ by 0.001/min at pH 7.6, 25°C with *N*-benzoyl-L-Arg ethyl ester as substrate	DPCC-treated; lyophilized	<0.001% chymotrypsin	Fluka 93611
7500 U/mg	1 unit increases A$_{253}$ by 0.001/min at pH 7.6, 25°C with *N*-benzoyl-L-Arg ethyl ester as substrate	Salt-free; lyophilized		Fluka 93612
7500 U/mg	1 unit increases A$_{253}$ by 0.001/min at pH 7.6, 25°C with *N*-benzoyl-L-Arg ethyl ester as substrate	TPCK-treated; crystallized		Fluka 93630
2500 NF U; 150 TAME U/mg		Acetylated; salt-free; 1X crystallized; lyophilized		ICN 101171
3000 NF U; 180 TAME U/mg		Salt-free; 2X crystallized; lyophilized	3.5% chymotrypsin	ICN 101179
>3000 NF U/mg		Salt-free; sterile; 3X crystallized; lyophilized	0.3% chymotrypsin	ICN 101192
		DCC-treated; crystallized		ICN 104922
1 mg enzyme/mL gel		Trypsin immobilized to agarose; suspension in acetic acid, pH 3.2		ICN 191324
≥3000 U/mg	1 unit increases A$_{253}$ by 0.003/min at 25°C	Crystallized; powder	<50 U/mg chymotrypsin	Can Inova T-001
≥2500 USP U/mg		USP grade; highly purified; lyophilized	No detectable porcine parvo virus, protease ≥50 USP U/mg chymotrypsin	Intergen 7004-00
≥2500 USP U/mg	1 USP unit causes a change in absorbance of 0.003/min at pH 7.6, 25°C	USP grade; crystallized; lyophilized		Life Technol 17068-032 17068-024

Trypsin continued

SPECIFIC ACTIVITY	UNITS DEFINITION	PREPARATION FORM	ADDITIONAL ACTIVITIES	SUPPLIER CATALOG No.
Bovine pancreas *continued*				
6000-9000 BAEE U/mg protein	1 BAEE unit yields ΔA_{253} of 0.001/min at pH 7.6, 25°C with BAEE as substrate; 1 BAEE U = 320 ATEE U	DPCC-treated; salt-free; dialyzed; lyophilized	<0.1 BTEE U/mg protein chymotrypsin (DPCC reduces chymotrypsin usually present)	Sigma T1005
50-100 U/mL packed gel	1 unit hydrolyzes 1.0 μmol BAEE/min at pH 8.0, 30°C	Insoluble enzyme attached to cross-linked beaded agarose; suspension in 10 mM acetic acid, pH 3.2		Sigma T1763
75-100 U/mL packed gel	1 unit hydrolyzes 1.0 μmol BAEE/min at pH 8.0, 30°C	TPCK-treated; insoluble enzyme attached to beaded agarose; suspension in 10 mM acetic acid and 0.01% thimerosal, pH 3.2		Sigma T4019
5000 BAEE U/mg protein	1 BAEE unit yields ΔA_{253} of 0.001/min at pH 7.6, 25°C with BAEE as substrate; 1 BAEE U = 320 ATEE U	Biotin-labeled; DPCC-treated; salt-free; powder	3 mol biotin/mol protein (DPCC reduces chymotrypsin usually present)	Sigma T6640
10,000 BAEE U/mg protein	1 BAEE unit yields ΔA_{253} of 0.001/min at pH 7.6, 25°C with BAEE as substrate; 1 BAEE U = 320 ATEE U	Acetylated		Sigma T6763
10,000 BAEE U/mg protein	1 BAEE unit yields ΔA_{253} of 0.001/min at pH 7.6, 25°C with BAEE as substrate	EtOH precipitate (denser than a lyophilized product)	<4 BTEE U/mg protein chymotrypsin	Sigma T8003
10,000-13,000 BAEE U/mg protein	1 BAEE unit yields ΔA_{253} of 0.001/min at pH 7.6, 25°C with BAEE as substrate; 1 BAEE U = 320 ATEE U	Salt-free; dialyzed; lyophilized	<4 BTEE U/mg protein chymotrypsin	Sigma T8253
100-200 U/g solid	1 unit hydrolyzes 1.0 μmol BAEE/min at pH 8.0, 30°C	Insoluble enzyme attached to polyacrylamide containing 20% $CaCl_2$ salt		Sigma T8386
10,000-13,000 BAEE U/mg protein	1 BAEE unit yields ΔA_{253} of 0.001/min at pH 7.6, 25°C with BAEE as substrate; 1 BAEE U = 320 ATEE U	TPCK-treated; salt-free; dialyzed; lyophilized	<0.1 BTEE U/mg protein chymotrypsin (TPCK reduces chymotrypsin usually present)	Sigma T8642
100 μg trypsin/vial	1 BAEE unit yields ΔA_{253} of 0.001/min at pH 7.6, 25°C with BAEE as substrate; 1 BAEE U = 320 ATEE U	Sequencing grade; lyophilized		Sigma T8658
100 μg/vial		Sequencing grade; lyophilized		Sigma T8658
1 g solid = 2 mL packed volume	1 BAEE unit yields ΔA_{253} of 0.001/min at pH 7.6, 25°C with BAEE as substrate	TPCK-treated; insoluble enzyme attached to DITC controlled pore glass 80-120 mesh; 700 Angstrom average pore size		Sigma T8899
10,000 BAEE U/mg protein	1 BAEE unit yields ΔA_{253} of 0.001/min at pH 7.6, 25°C with BAEE as substrate; 1 BAEE U = 320 ATEE U	Salt-free; dialyzed; lyophilized	<4 BTEE U/mg protein chymotrypsin	Sigma T8918

SPECIFIC ACTIVITY	UNITS DEFINITION	PREPARATION FORM	ADDITIONAL ACTIVITIES	SUPPLIER CATALOG NO.
Bovine pancreas *continued*				
≥25 U trypsin/mL of settled gel	1 unit hydrolyzes 1 µmol TAME/hr at pH 8.2, 25°C in the presence of 0.01 M Ca^{2+}	Immobilized, bound to Sepharose^R 4B; 2X crystallized; dialyzed; suspension in 0.1 M NaOAc buffer, 0.04% EDTA, 0.02% NaN_3, pH 4.5		Worthington LS000104 LS000100 LS000102
≥150 U/mg protein	1 unit hydrolyzes 1 µmol TAME/hr at pH 8.2, 25°C in the presence of 0.01 M Ca^{2+}	1X Crystallized; dialyzed against 1 mM HCl; lyophilized		Worthington LS003665 LS003667 LS003670
≥180 U/mg protein	1 unit hydrolyzes 1 µmol TAME/hr at pH 8.2, 25°C in the presence of 0.01 M Ca^{2+}	2X Crystallized; dialyzed against 1 mM HCl; lyophilized		Worthington LS003702 LS003703 LS003704 LS003706
≥180 U/mg protein	1 unit hydrolyzes 1 µmol TAME/hr at pH 8.2, 25°C in the presence of 0.01 M Ca^{2+}	3X Crystallized; dialyzed against 1 mM HCl; lyophilized		Worthington LS003708 LS003707 LS003709
≥180 U/mg protein	1 unit hydrolyzes 1 µmol TAME/hr at pH 8.2, 25°C in the presence of 0.01 M Ca^{2+}	TPCK-treated; 3X crystallized; lyophilized		Worthington LS003740 LS003741 LS003744 LS003742
≥180 U/mg protein; ≥100 mg/vial	1 unit hydrolyzes 1 µmol TAME/hr at pH 8.2, 25°C in the presence of 0.01 M Ca^{2+}	TPCK-treated; sterile-filtered; γ-irradiated; 3X crystallized; lyophilized	No detectable virus, mycoplasma	Worthington LS003750 LS003752
≥180 U/mg protein	1 unit hydrolyzes 1 µmol TAME/hr at pH 8.2, 25°C in the presence of 0.01 M Ca^{2+}	Sterile-filtered; γ-irradiated; 2X crystallized; lyophilized	No detectable virus, mycoplasma	Worthington LS004454 LS004452 LS004458
≥180 U/mg protein; 50 mg/vial	1 unit hydrolyzes 1 µmol TAME/hr at pH 8.2, 25°C in the presence of 0.01 M Ca^{2+}	Sterile-filtered; 3X crystallized; lyophilized		Worthington LS03736 LS03734
Gadus morhua **(Atlantic cod)**				
5,000-10,000 BAEE U/mg protein; 5 mg protein/vial	1 BAEE unit yields ΔA_{253} of 0.001/min at pH 7.6, 25°C with BAEE as substrate; 1 BAEE U = 320 ATEE U	Affinity chromatographically purified; lyophilized containing 50% protein; balance primarily Tris buffer and Ca^{2+} salts		Sigma T9906

Trypsin continued

3.4.21.4

SPECIFIC ACTIVITY	UNITS DEFINITION	PREPARATION FORM	ADDITIONAL ACTIVITIES	SUPPLIER CATALOG NO.
Human				
				Vital 17325
Human pancreas (optimum pH = 8.0-9.0, MW = 22,000 Da)				
2.5 U/mg protein	1 unit hydrolyzes 1 µmol N-benzyl-DL-Arg-pNA/min at pH 7.8, 25°C	≥95% purity; salt-free; lyophilized	No detectable HBsAg, anti-HCV, anti-HBc, anti-HIV	ART 16-19-032000
2.5 U/mg protein	1 unit hydrolyzes 1 µmol N-benzyl-DL-Arg-pNA/min at pH 7.8, 25°C in 200 mM Tris-HCl, CaCl$_2$	>95% purity; iodination grade; salt-free; lyophilized	No detectable chymotrypsin and its zymogen, HBsAg, HIV Ab	Calbiochem 650275
25-30 U/mg protein	1 unit hydrolyzes 1 µmol N-benzoyl-Phe-Val-Arg-pNA/min at pH 8.3, 25°C	Chromatographically purified; lyophilized containing no more than 2% NaCl	<0.1% chymotrypsin	Elastin TR127
5-15 U/mg protein		Frozen in 1 mM HCl		ICN 191340
2-3 U/mg		>95% purity by SDS-PAGE; lyophilized containing 2 mM HCl	No detectable HGsAg, HCV, HIV-1 Ab, chymotrypsin	Scripps T0614
1000 BAEE U/vial	1 BAEE unit yields ΔA_{253} of 0.001/min at pH 7.6, 25°C with BAEE as substrate; 1 BAEE U = 320 ATEE U	Salt-free; lyophilized		Sigma T6424
Human, purulent sputum (leucocyte)				
45-50 U/mg protein	1 unit hydrolyzes 1 µmol N-benzoyl-Phe-Val-Arg-pNA/min at pH 8.3, 25°C	Chromatographically purified; salt-free; lyophilized	No detectable cathepsin G, other leucocyte enzymes of sputum	Elastin ST51
Porcine				
		Solution containing 2.5 g/L trypsin in Hanks' Balanced Salt Solution (without CaCl$_2$, MgCl$_2$, MgSO$_4$)	No detectable porcine parvo virus, mycoplasma	Life Technol KaryoMAX 15050-065 15050-057
		Solution	No detectable porcine parvo virus, mycoplasma	Life Technol 10585-024
		Lyophilized containing 2.5 g/L trypsin and 0.85 g/L NaCl	No detectable porcine parvo virus, mycoplasma	Life Technol 15055-015
		Solution containing 25 g/L trypsin and 8.5 g/L NaCl	No detectable porcine parvo virus, mycoplasma	Life Technol 15090-038 15090-046
		EDTA-treated; lyophilized containing 0.5 g/L trypsin, 0.2 g/L EDTA, 0.85 g/L NaCl	No detectable porcine parvo virus, mycoplasma	Life Technol 15305-014
		EDTA-treated; solution containing 5.0 g/L trypsin, 2.0 g/L EDTA, 8.5 g/L NaCl	No detectable porcine parvo virus, mycoplasma	Life Technol 15400-039 15400-054

SPECIFIC ACTIVITY	UNITS DEFINITION	PREPARATION FORM	ADDITIONAL ACTIVITIES	SUPPLIER CATALOG NO.
Porcine *continued*				
		EDTA-treated; lyophilized containing 5.0 g/L trypsin, 2.0 g/L EDTA, 8.5 g/L NaCl	No detectable porcine parvo virus, mycoplasma	Life Technol 15405-012
1:300	1 g trypsin digests 250 g casein substrate/10 min at pH 7.6, 25°C		No detectable porcine mycoplasma	Life Technol 17073-016 17073-024 17073-032 17073-040
1:250	1 g trypsin digests 250 g casein substrate/10 min at pH 7.6, 25°C	Embryonic stem cell-qualified; powder	No detectable porcine parvo virus, mycoplasma	Life Technol 18270-018
		Lyophilized containing 25 g/L trypsin and 8.5 g/L NaCl	No detectable porcine parvo virus, mycoplasma	Life Technol 25095-019
		EDTA-treated; solution containing 2.5 g/L trypsin and 0.38 g/L EDTA in Hanks' Balanced Salt Solution (without $CaCl_2$, $MgCl_2$, $MgSO_4$)	No detectable porcine parvo virus, mycoplasma	Life Technol 25200-056 25200-072
		EDTA-treated; solution containing 0.5 g/L trypsin and 0.2 g/L EDTA in Hanks' Balanced Salt Solution (without $CaCl_2$, $MgCl_2$, $MgSO_4$)	No detectable porcine parvo virus, mycoplasma	Life Technol 25300-047 25300-062
1:250	1 g trypsin digests 250 g casein substrate/10 min at pH 7.6, 25°C	Lyophilized	No detectable porcine parvo virus, mycoplasma	Life Technol 27250-042 27250-018 27250-026 27250-034
Porcine pancreas (optimum pH = 7.4-8.2, pI = 10.8, T = 40-50°C, MW = 23,400 Da)				
1:75 USP U/mg 1:125 USP U/mg 1:150 USP U/mg 1:200 USP U/mg		Powder		Am Labs
2600 U/mg solid	1 unit increases A_{253} by 0.003/min at 25°C	Amorphous powder		AMRESCO 0458
Trypsin 250 USP	USP	Powder		Biocatalysts T069P
Trypsin 250 USP	USP	Powder	No detectable virus	Biocatalysts T071P
5000 NF U/mg solid	1 unit increases A_{253} by 0.003/min at pH 7.6, 25°C using *N*-α-benzoyl-L-Arg ethyl ester as substrate	Salt-free; 2X crystallized; lyophilized	<0.25% chymotrypsin	Biozyme TRY1

SPECIFIC ACTIVITY	UNITS DEFINITION	PREPARATION FORM	ADDITIONAL ACTIVITIES	SUPPLIER CATALOG NO.
Porcine pancreas *continued*				
5 U/mg; 50 U/mL	1 unit converts 1 μmol Chromozym^R Try to product/min at 25°C	Solution containing EDTA	Proteases	Boehringer 1074474
150 U/mL (Chromozym^R Try as substrate); 40 U/mg solid (BAEE as substrate)	1 unit converts 1 μmol substrate to product/min at 25°C	Solution (2.5%)	Proteases	Boehringer 210234
5000 U/mg solid	1 unit increases A_{253} by 0.003/min at pH 7.6, 25°C using BAEE as substrate	Lyophilized containing 90-95% protein		Calzyme 113A5000
1:200-1:300	1 g digests 250 g casein under NF test conditions for casein digestive power in pancreatin	Sterile-filtered; powder	No detectable mycoplasma	Difco 0152-13-1 0152-15-9 0152-17-7 0152-05-1
1:200-1:300		Sterile-filtered; powder		Difco 0153-60-2 0153-61-1
90 U/mg	1 unit increases A_{253} by 0.001/min at pH 7.6, 25°C with *N*-benzoyl-L-Arg ethyl ester as substrate	Powder		Fluka 93613
16,000 U/mg	1 unit increases A_{253} by 0.001/min at pH 7.6, 25°C with *N*-benzoyl-L-Arg ethyl ester as substrate		<0.2% chymotrypsin	Fluka 93614
1500 U/mg	1 unit increases A_{253} by 0.001/min at pH 7.6, 25°C with *N*-benzoyl-L-Arg ethyl ester as substrate	Powder	100% chymotrypsin	Fluka 93615
		Powder	≥250,000 USP U/g protease	ICN 103139
		Powder	≥300,000 USP U/g protease	ICN 103140
1000-5000 BAEE U/mg solid	1 BAEE unit produces a ΔA_{253} of 0.001/min at pH 7.6, 25°C using *N*-α-benzoyl-L-Arg ethyl ester as substrate; 1 ATEE unit produces a ΔA_{237} of 0.001/min at pH 7.0, 25°C		500-1000 ATEE U/mg chymotrypsin	ICN 150213
>250 U/mg		γ-Irradiated; virus-free; powder		ICN 153571
75,000-125,000 BAEE U/mL		Sterile-filtered; solution		ICN 190046

SPECIFIC ACTIVITY	UNITS DEFINITION	PREPARATION FORM	ADDITIONAL ACTIVITIES	SUPPLIER CATALOG NO.
Porcine pancreas *continued*				
2400-2800 U/mg	1 unit increases A_{253} by 0.003/min at 25°C	Powder	400-600 U/mg chymotrypsin	Can Inova T-006
≥1300 U/mg	1 unit increases A_{253} by 0.003/min at 25°C	Powder	≥500 U/mg chymotrypsin	Can Inova T-013
≥250 U/mg	1 unit increases A_{253} by 0.003/min at 25°C	Powder	≥75 U/mg chymotrypsin	Can Inova T-250
≥225 USP U/mg		Partially purified; lyophilized	No detectable porcine parvo virus ≥75 USP U/mg chymotrypsin ≥250 U/mg protease	Intergen 7001-00 7001-70 7001-80
≥300 USP U/mg		Purified; lyophilized	No detectable porcine parvo virus ≥90 USP U/mg chymotrypsin ≥300 U/mg protease	Intergen 7002-00
≥500 USP U/mg		Purified; lyophilized	No detectable porcine parvo virus ≥150 USP U/mg chymotrypsin ≥400 U/mg protease	Intergen 7003-00
≥225 USP U/mg		Sterile-filtered; 1X solution containing Hanks' Balanced Salt Solution (Ca^{2+} and Mg^{2+}-free)	No detectable porcine parvo virus Contains chymotrypsin, elastase, protease	Intergen 7010-80 7010-90
≥225 USP U/mg		EDTA-treated; sterile-filtered; 1X solution containing EDTA and Hanks' Balanced Salt Solution (Ca^{2+} and Mg^{2+}-free)	No detectable porcine parvo virus Contains chymotrypsin, elastase, protease	Intergen 7011-80 7011-90
≥225 USP U/mg		EDTA-treated; sterile-filtered; 10X solution containing EDTA and 0.9% NaCl	No detectable porcine parvo virus Contains chymotrypsin, elastase, protease	Intergen 7021-65 7021-80
≥28,000 USP U/mL		Sterile-filtered; 40X solution, pH 4.0	No detectable porcine parvo virus Contains chymotrypsin, elastase	Intergen Enzar-T^R 7000-65 7000-80
13,000-20,000 BAEE U/mg protein	1 BAEE unit yields ΔA_{253} of 0.001/min at pH 7.6, 25°C with BAEE as substrate; 1 BAEE U = 320 ATEE U	Crystallized; dialyzed; lyophilized		Sigma T0134
1 mg trypsin/tablet	1 BAEE unit yields ΔA_{253} of 0.001/min at pH 7.6, 25°C with BAEE as substrate; 1 BAEE U = 320 ATEE U	Tablets containing 1 mg trypsin with buffer salts		Sigma T7168

SPECIFIC ACTIVITY	UNITS DEFINITION	PREPARATION FORM	ADDITIONAL ACTIVITIES	SUPPLIER CATALOG No.
Porcine pancreas *continued*				
1000-2000 BAEE U/mg solid	1 BAEE unit yields ΔA_{253} of 0.001/min at pH 7.6, 25°C with BAEE as substrate; 1 BAEE U = 320 ATEE U	Crude	<10 BTEE U/mg solid chymotrypsin	Sigma T7409
13,000-20,000 BAEE U/mg protein	1 BAEE unit yields ΔA_{253} of 0.001/min at pH 7.6, 25°C with BAEE as substrate; 1 BAEE U = 320 ATEE U	Crystallized; dialyzed; lyophilized		Sigma T7418
1000-2000 BAEE U/mg solid	1 BAEE unit yields ΔA_{253} of 0.001/min at pH 7.6, 25°C with BAEE as substrate; 1 BAEE U = 320 ATEE U	Crude	1000-2000 ATEE U/mg solid chymotrypsin (1 ATEE U = ΔA_{237} of 0.001/min in 3 mL at pH 7.0, 25°C)	Sigma T8128
250-350 USP U/mg				Wako 204-12231 202-12232 206-12235
4000-5000 USP U/mg; 12,000-15,000 BAEE U/mg		Crystallized		Wako 207-09891 203-09893 205-09892
Rat pancreas				
15-20 U/mg protein	1 unit hydrolyzes 1 μmol N-benzoyl-Phe-Val-Arg-pNA/min at pH 8.3, 25°C	Chromatographically purified; salt-free; lyophilized	<0.1% chymotrypsin	Elastin R747
Streptomyces griseus				
15-20 U/mg solid	1 unit hydrolyzes casein producing 1.0 μmol Tyr as peptide/min at pH 11.0, 30°C	Lyophilized		ICN 150209
***Streptomyces* species** (optimum pH = 12, pI = 8.7, T = 60°C, MW = 50,000 Da; stable pH 5.0-11.5 [25°C, 24 hr], T < 50°C [pH 8.3, 15 min])				
$\geq$20 U/mg solid	1 unit increases OD_{275} corresponding to 1 μmol Tyr/min at pH 11.0, 30°C	Lyophilized containing Ca^{2+} as stabilizer		Toyobo ALP-101
Unspecified				
				SpecialtyEnz

3.4.21.5 Thrombin

REACTION CATALYZED
  Selective cleavage of Arg↓Gly bonds in fibrinogen to form fibrin and release fibrinopeptides A and B

SYNONYMS
  Fibrinogenase

REACTANTS
  Fibrinogen, fibrin, fibrinopeptides A and B

APPLICATIONS
- Topically reduces small vessel bleeding in surgery
- Clots blood in clinical diagnostic tests
- Cleaves specific peptides from recombinant fusion proteins
- Determination of fibrinogen
- Stimulates cell proliferation in cell culture
- Defibrinates whole blood and plasma

NOTES
- Classified as a hydrolase, acting on peptide bonds: serine endopeptidase
- Formed in plasma during clotting process from prothrombin by activation with Factor Xa, Factor Va and phospholipid
- Converts soluble plasma protein fibrinogen to an insoluble fibrin clot
- Activated by thromboplastin and $CaCl_2$
- Inhibited by DFP, TLCK, PMSF, APMSF, benzamidine, antithrombin III-heparin, hirudin, α_1-antitrypsin and α_2-macroglobulin
- More selective than trypsin and plasmin
- A homologue of chymotrypsin

SPECIFIC ACTIVITY	UNITS DEFINITION	PREPARATION FORM	ADDITIONAL ACTIVITIES	SUPPLIER CATALOG NO.
Agkistrodon contortrix				
0.2 U/vial	1 unit hydrolyzes 1.0 μmol N-benzoyl-Pro-Phe-Arg p-nitroanilide/min at pH 8.4, 37°C	Lyophilized containing 10% protein, 88% lactose, 2% thimerosal		Sigma A8766
Bovine				
100 NIH U/mg		Reagent grade; highly purified		ICN 101141
>2000 NIH/mg protein		High purity grade containing 3% moisture	No detectable plasminogen, plasmin	ICN 154163
				ICN 820361 820362
			No detectable plasminogen	ICN 820371
		Purified		ICN 820381
		Non-pyrogenic		ICN 820401, 820402

SPECIFIC ACTIVITY	UNITS DEFINITION	PREPARATION FORM	ADDITIONAL ACTIVITIES	SUPPLIER CATALOG No.
Bovine *continued*				
				Vital 3037
Bovine plasma (MW = 33,580 Da)				
1800–2000 NIH U/mg protein	Units are determined by comparison with standard curve prepared by using the Bureau of Biologics standard thrombin	Lyophilized from buffer containing 200 mM NaCl, 50 mM Na citrate, 0.1% PEG, pH 6.5		Calbiochem 604980
> 70 NIH U/mg solid	Units are determined by comparison with standard curve prepared by using the Bureau of Biologics standard thrombin	Lyophilized from buffer containing 50 mM Na citrate, 0.1% PEG, pH 6.5		Calbiochem 605157
≥1000 NIH U/mg protein	Units are determined by comparison with standard curve prepared by using the Bureau of Biologics standard thrombin	Non-sterile; lyophilized containing $CaCl_2$, PEG, NaCl	No detectable plasminogen	Calbiochem 605160
1800–2200 NIH U/mg	1 unit is equal to 1 NIH unit using a Standard Clotting Assay	Homogeneous by SDS-PAGE; frozen in concentrated form		EnzymeRes BTIIa
0.2 U/mg	1 unit hydrolyzes 1 μmol Tos-Gly-Pro-Arg-pNA.AcOH (= Chromozym™TH)/min at pH 8.4, 37°C	Lyophilized containing 0.3% protein, 40% NaCl, 59.7% Na citrate	<0.1% plasmin, plasminogen	Fluka 89224
6 U/mg	1 unit hydrolyzes 1 μmol Tos-Gly-Pro-Arg-pNA.AcOH (= Chromozym™TH)/min at pH 8.4, 37°C	Lyophilized containing 50% protein		Fluka 89225
1 U/mg	1 unit hydrolyzes 1 μmol Tos-Gly-Pro-Arg-pNA.AcOH (= Chromozym™TH)/min at pH 8.4, 37°C	Powder		Fluka 89228
7 U/mg protein	1 unit hydrolyzes 1 μmol Tos-Gly-Pro-Arg-pNA.AcOH (= Chromozym™TH)/min at pH 8.4, 37°C	Frozen solution of a crude preparation in 0.05 M PO_4 buffer, pH 7.0		Fluka 89229
>1000 U/mg		Purified; powder containing mannitol and NaCl		Intergen 7030-00 7030-10
		Powder		Intergen 7035-00 7035-10
>300 U/mg; >1 mg/mL	1 unit releases 1 μmol p-nitroaniline/min at pH 8.3, 37°C	5 mM MES and 0.5 M NaCl, pH 6.0		ProZyme CP20 005 CP20 025 CP20 100

SPECIFIC ACTIVITY	UNITS DEFINITION	PREPARATION FORM	ADDITIONAL ACTIVITIES	SUPPLIER CATALOG No.
Bovine plasma *continued*				
≥1000 NIH U/mg protein		High potency; lyophilized containing 50% protein, $CaCl_2$, NaCl, polyethylene glycol	No detectable plasminogen	SPL 1240
≥75 NIH U/mg protein		Powder containing 50% protein, $CaCl_2$, NaCl, Gly		SPL 1250
175-350 NIH U/mg protein	Activity expressed in NIH units obtained by direct comparison to an NIH Thrombin Reference Standard; 0.2 mL diluted plasma (1:1 with saline) used as substrate with 0.1 mL thrombin sample (stabilized in 1% buffered albumin solution)	Lyophilized containing 25% protein and 75% Na citrate, 1% buffered albumin solution, pH 5.8		Sigma T4265
50-100 NIH U/mg protein	Activity expressed in NIH units obtained by direct comparison to an NIH Thrombin Reference Standard; 0.2 mL diluted plasma (1:1 with saline) used as substrate with 0.1 mL thrombin sample (stabilized in 1% buffered albumin solution)	Lyophilized containing 50% protein; balance primarily NaCl and Tris-HCl buffer salts, 1% buffered albumin solution, pH 7.0		Sigma T4648
600 NIH U/mg protein	Activity expressed in NIH units obtained by direct comparison to an NIH Thrombin Reference Standard; 0.2 mL diluted plasma (1:1 with saline) used as substrate with 0.1 mL thrombin sample (stabilized in 1% buffered albumin solution)	Lyophilized; when reconstituted, contains stated activity in 0.15 M NaCl, 0.05 M Na citrate, 1% buffered albumin solution, pH 6.5	No detectable other known clotting factors (non-activated and activated), plasminogen, plasmin	Sigma T6634
3000 NIH U/mg protein	Activity expressed in NIH units obtained by direct comparison to an NIH Thrombin Reference Standard; 0.2 mL diluted plasma (1:1 with saline) used as substrate with 0.1 mL thrombin sample (stabilized in 1% buffered albumin solution)	Lyophilized; when reconstituted, contains stated activity in 0.15 M NaCl, 0.05 M Na citrate, 1% buffered albumin solution, pH 6.5	No detectable other known clotting factors (non-activated and activated), plasminogen, plasmin	Sigma T6759
1500-2500 NIH U/mg protein	Activity expressed in NIH units obtained by direct comparison to an NIH Thrombin Reference Standard; 0.2 mL diluted plasma (1:1 with saline) used as substrate with 0.1 mL thrombin sample (stabilized in 1% buffered albumin solution)	Lyophilized; when reconstituted, contains stated activity in 0.15 M NaCl, 0.05 M Na citrate, 1% buffered albumin solution, pH 6.5	No detectable other known clotting factors (non-activated and activated), plasminogen, plasmin	Sigma T7513

SPECIFIC ACTIVITY	UNITS DEFINITION	PREPARATION FORM	ADDITIONAL ACTIVITIES	SUPPLIER CATALOG No.
Bovine plasma *continued*				
150 NIH U/mg protein	Activity expressed in NIH units obtained by direct comparison to an NIH Thrombin Reference Standard; 0.2 mL diluted plasma (1:1 with saline) used as substrate with 0.1 mL thrombin sample (stabilized in 1% buffered albumin solution)	Crude; frozen solution containing 1000 NIH U/mL in 0.05 M PO$_4$ buffer, 1% buffered albumin solution, pH 7.0		Sigma T9000
2000 NIH U/mg protein; ≥10 NIH U/vial	Activity expressed in NIH units obtained by direct comparison to an NIH Thrombin Reference Standard; 0.2 mL diluted plasma (1:1 with saline) used as substrate with 0.1 mL thrombin sample (stabilized in 1% buffered albumin solution)	Lyophilized; when reconstituted, contains stated activity in 0.15 M NaCl, 0.05 M Na citrate, 1% buffered albumin solution, pH 6.5	No detectable other known clotting factors (non-activated and activated), plasminogen, plasmin	Sigma T9010
1500-2500 NIH U/mg protein	Activity expressed in NIH units obtained by direct comparison to an NIH Thrombin Reference Standard; 0.2 mL diluted plasma (1:1 with saline) used as substrate with 0.1 mL thrombin sample (stabilized in 1% buffered albumin solution)	Cell culture tested; when reconstituted, vial contains stated activity in 0.15 M NaCl and 0.05 M Na citrate, 1% buffered albumin solution, pH 6.5	No detectable other known clotting factors (non-activated and activated), plasminogen, plasmin	Sigma T9549
1000 NIH U/mg protein	Activity expressed in NIH units obtained by direct comparison to an NIH Thrombin Reference Standard; 0.2 mL diluted plasma (1:1 with saline) used as substrate with 0.1 mL thrombin sample (stabilized in 1% buffered albumin solution)	Lyophilized; when reconstituted, contains stated activity in 0.15 M NaCl, 0.05 M Na citrate, 1% buffered albumin solution, pH 6.5	No detectable other known clotting factors (non-activated and activated), plasminogen, plasmin	Sigma T2514
Human				
				Vital 17288
Human plasma (optimum pH = 8.2-9.0, MW = 33,600-37,000 Da)				
120 U/mg enzyme protein	1 unit converts 1 μmol Chromozym^R TH to product/min at 25°C	Lyophilized	<3% factor Xa	Boehringer 602400
≥1000 NIH U/mg protein	Units are determined by comparison with standard curve prepared by using the Bureau of Biologics standard thrombin	>95% purity; lyophilized from buffer containing 50 mM Na citrate buffer, 0.2 M NaCl, 0.1% PEG-8000, pH 7.4	No detectable HBsAg, HIV Ab	Calbiochem 605190

SPECIFIC ACTIVITY	UNITS DEFINITION	PREPARATION FORM	ADDITIONAL ACTIVITIES	SUPPLIER CATALOG NO.
Human plasma *continued*				
2800 NIH U/mg protein	Units are determined by comparison with standard curve prepared by using the Bureau of Biologics standard thrombin	Lyophilized from buffer containing 50 mM Na citrate buffer, 0.2 M NaCl, 0.1% PEG-8000, pH 6.5	No detectable HBsAg, HIV Ab	Calbiochem 605195
250 μg thrombin/mL gel; 1000 U/mg before attachment of matrix	Units are determined by comparison with standard curve prepared by using the Bureau of Biologics standard thrombin	Immobilized on agarose gel in 50 mM KPO$_4$, pH 7.5; Matrix: cross-linked 6% beaded agarose without spacer; preserved with 0.02% NaN$_3$	No detectable HBsAg, HIV Ab	Calbiochem 605204
≥2500 NIH U/mg protein	Units are determined by comparison with standard curve prepared by using the Bureau of Biologics standard thrombin	Citrate-free; solution containing 20 mM Tris-HCl and 100 mM NaCl, pH 7.4	No detectable HBsAg, HIV Ab	Calbiochem 605206
<50 U/mg	1 unit is equal to 1 NIH unit using a Standard Clotting Assay	γ-Thrombin; single-band purity by SDS-PAGE; frozen in concentrated form	No detectable HIV Ab, HsAg	EnzymeRes HGT
2700-3300 NIH U/mg	1 unit is equal to 1 NIH unit using a Standard Clotting Assay	γ-Thrombin; homogeneous by SDS-PAGE; frozen in concentrated form	No detectable HIV Ab, HsAg	EnzymeRes HT1002a
2 U/mg	1 unit hydrolyzes 1 μmol Tos-Gly-Pro-Arg-pNA.AcOH (= Chromozym™TH)/min at pH 8.4, 37°C	Powder	No detectable hepatitis B surface antigen, HIV	Fluka 89219
0.3 U/mg	1 unit hydrolyzes 1 μmol Tos-Gly-Pro-Arg-pNA.AcOH (= Chromozym™TH)/min at pH 8.4, 37°C	Lyophilized containing 0.2% protein, 40% NaCl, 59.8% Na citrate	<0.1% plasmin, plasminogen No detectable hepatitis B virus, HIV	Fluka 89222
0.5 U/mg	1 unit hydrolyzes 1 μmol Tos-Gly-Pro-Arg-pNA.AcOH (= Chromozym™TH)/min at pH 8.4, 37°C	Lyophilized containing 0.5% protein, 40% NaCl, 59.5% Na citrate	<0.1% plasmin, plasminogen No detectable hepatitis B virus, HIV	Fluka 89223
4000 NIH U/mg protein; 5000 NIH U/mL for ≥1000 U, 500 NIH U/mL for smaller sizes	Activity expressed in NIH units obtained by direct comparison to an NIH Thrombin Reference Standard; 0.2 mL diluted plasma (1:1 with saline) used as substrate with 0.1 mL thrombin sample (stabilized in 1% buffered albumin solution)	Highly purified; frozen solution containing 0.15 M NaCl, 0.05 M Na citrate, 1% buffered albumin solution, pH 6.5		Sigma T3010

SPECIFIC ACTIVITY	UNITS DEFINITION	PREPARATION FORM	ADDITIONAL ACTIVITIES	SUPPLIER CATALOG NO.
Human plasma *continued*				
1000 NIH U/mg protein	Activity expressed in NIH units obtained by direct comparison to an NIH Thrombin Reference Standard; 0.2 mL diluted plasma (1:1 with saline) used as substrate with 0.1 mL thrombin sample (stabilized in 1% buffered albumin solution)	Sterile; cell culture tested; lyophilized containing 1% buffered albumin solution	No detectable hepatitis virus and HIV Ab	Sigma T4393
2000 NIH U/mg protein	Activity expressed in NIH units obtained by direct comparison to an NIH Thrombin Reference Standard; 0.2 mL diluted plasma (1:1 with saline) used as substrate with 0.1 mL thrombin sample (stabilized in 1% buffered albumin solution)	Lyophilized; when reconstituted, contains stated activity in 0.15 M NaCl, 0.05 M Na citrate, 1% buffered albumin solution, pH 6.5	No detectable other known clotting factors (non-activated and activated), plasminogen, plasmin	Sigma T6884
1000 NIH U/mg protein	Activity expressed in NIH units obtained by direct comparison to an NIH Thrombin Reference Standard; 0.2 mL diluted plasma (1:1 with saline) used as substrate with 0.1 mL thrombin sample (stabilized in 1% buffered albumin solution)	Lyophilized; when reconstituted, contains stated activity in 0.15 M NaCl, 0.05 M Na citrate, 1% buffered albumin solution, pH 6.5	No detectable other known clotting factors (non-activated and activated), plasminogen, plasmin	Sigma T7009
3000 NIH U/mg protein; ≥10 NIH U/vial	Activity expressed in NIH units obtained by direct comparison to an NIH Thrombin Reference Standard; 0.2 mL diluted plasma (1:1 with saline) used as substrate with 0.1 mL thrombin sample (stabilized in 1% buffered albumin solution)	Lyophilized; when reconstituted, contains stated activity in 0.15 M NaCl, 0.05 M Na citrate, 1% buffered albumin solution, pH 6.5	No detectable other known clotting factors (non-activated and activated), plasminogen, plasmin	Sigma T8885
1000 NIH U/mg protein; ≥10 NIH U/vial	Activity expressed in NIH units obtained by direct comparison to an NIH Thrombin Reference Standard; 0.2 mL diluted plasma (1:1 with saline) used as substrate with 0.1 mL thrombin sample (stabilized in 1% buffered albumin solution)	Lyophilized; when reconstituted, contains stated activity in 0.15 M NaCl, 0.05 M Na citrate, 1% buffered albumin solution, pH 6.5	No detectable other known clotting factors (non-activated and activated), plasminogen, plasmin	Sigma T9135

SPECIFIC ACTIVITY	UNITS DEFINITION	PREPARATION FORM	ADDITIONAL ACTIVITIES	SUPPLIER CATALOG NO.
Murine plasma				
1000 NIH U/mg protein	Activity expressed in NIH units obtained by direct comparison to an NIH Thrombin Reference Standard; 0.2 mL diluted plasma (1:1 with saline) used as substrate with 0.1 mL thrombin sample (stabilized in 1% buffered albumin solution)	Lyophilized; when reconstituted, contains stated activity in 0.15 M NaCl, 0.05 M Na citrate, 1% buffered albumin solution, pH 6.5		Sigma T8397
Rat plasma				
1000 NIH U/mg protein	Activity expressed in NIH units obtained by direct comparison to an NIH Thrombin Reference Standard; 0.2 mL diluted plasma (1:1 with saline) used as substrate with 0.1 mL thrombin sample (stabilized in 1% buffered albumin solution)	Lyophilized; when reconstituted, contains stated activity in 0.15 M NaCl, 0.05 M Na citrate, 1% buffered albumin solution, pH 6.5		Sigma T5772
Russell's viper venom				
20 U/mg protein	1 unit increases Factor V activity in 0.1 mL fresh normal human plasma by at least 2X after incubation/3 min at 37°C	Lyophilized containing albumin and NaCl stabilizer		Sigma F3380

3.4.21.6 Coagulation Factor Xa

REACTION CATALYZED
Selective cleavage of Arg↓Thr and then Arg↓Ile bonds in prothrombin to form thrombin

SYNONYMS
    Thrombokinase, prothrombase, prothrombinase

REACTANTS
    Prothrombin, thrombin

APPLICATIONS
- Processing fusion proteins at definite cleavage sites
- Determination of prothrombin and heparin
- Coagulation studies

NOTES
- Classified as a hydrolase, acting on peptide bonds: serine endopeptidase
- Formed from proenzyme Factor X by limited proteolysis
- Acetylation of the γ-amino group blocks nonspecific cleavage
- A homologue of chymotrypsin.
- Inhibited by APMSF, antithrombin III-heparin, DFP, PMSF and soybean trypsin inhibitor
- Similar in specificity to scutelarin (EC 3.4.21.60)

Coagulation Factor Xa

3.4.21.6

SPECIFIC ACTIVITY	UNITS DEFINITION	PREPARATION FORM	ADDITIONAL ACTIVITIES	SUPPLIER CATALOG No.
Bovine plasma (optimum pH = 8.3, MW = 16,000 and 27,000 Da [2 disulfide-linked chains])				
		Special quality for cleavage of fusion proteins; highly purified by SDS-PAGE; lyophilized		Boehringer 1179888 1585924 1179896
≥210 IU/mg	CBS 31.39 chromogenic substrate	Frozen in concentrated form		EnzymeRes BFXA
0.5-2.0 mg/mL	1 unit cleaves 50 μg test substrate to 95% completion/6 hr or less	20 mM HEPES, 500 mM NaCl, 2 mM $CaCl_2$, 50% glycerol, pH 8.0		NE Biolabs 800-10S 800-10L
>125 U/mg; >1 mg/mL	1 unit releases 1 μmol p-nitroaniline/min at pH 8.3, 37°C	5 mM MES, 0.5 M NaCl, 1 mM benzamidine, pH 6.0		ProZyme CP10 005 CP10 025 CP10 100
≥500 U/mg protein	1 unit activated Factor X in the absence of Russell's viper venom (RVV) gives the same substrate clotting time as 1 unit of plasma Factor X in the presence of RVV; lipid present in both systems	Lyophilized containing 8% protein; balance Na citrate and sucrose	No detectable thrombin or plasmin	Sigma F2027
Bovine serum				
400 nkat/mg; 200 μg/vial	1 nkat = 60 nmol/min	>95% purity by reverse phase HPLC; lyophilized	<10% Factor X	ICN 153579
Human plasma (optimum pH = 8.3, MW = 44,000-46,000 Da)				
1.3 U/mg total protein (Chromozym[R] X as substrate); 0.8 U/mg total protein (Bz-Ile-Glu-Gly-Arg-4-nitranilide as substrate)	1 unit converts 1 μmol substrate to product/min at 25°C	Suspension in $(NH_4)_2SO_4$ and BSA stabilizer	<5% thrombin	Boehringer 602388
125 U/mg protein	1 unit hydrolyzes 1 μmol Chromozym X™/min at 25°C	Complete activation observed on 10% SDS-PAGE; lyophilized from buffer containing 20 mM Tris-HCl and 100 mM NaCl, pH 7.4	No detectable HBsAg, HIV Ab	Calbiochem 233526
≥180 IU/mg	CBS 31.39 chromogenic substrate	Frozen in concentrated form	No detectable HIV Ab, HsAg	EnzymeRes HFXa 1011

3.4.21.7 Plasmin

REACTION CATALYZED
Preferential cleavage is Lys↓ > Arg↓, converting fibrin into soluble products

SYNONYMS
Fibrinase, fibrinolysin

REACTANTS
Fibrin

APPLICATIONS
- Coagulation research
- Degradation of extracellular matrix protein
- Protein structure and sequence analysis

NOTES
- Classified as a hydrolase, acting on peptide bonds: serine endopeptidase
- A serine protease with higher selectivity than trypsin
- Multiple forms of active enzyme are produced via proteolysis of plasminogen
- Inhibited by DFP, PMSF, soybean trypsin inhibitor, antithrombin III-heparin, APMSF, aprotinin, leupeptin, TLCK, α_1-antitrypsin, α_2-macroglobulin and α_2-antiplasmin
- A homolog of chymotrypsin

SPECIFIC ACTIVITY	UNITS DEFINITION	PREPARATION FORM	ADDITIONAL ACTIVITIES	SUPPLIER CATALOG No.
Bovine plasma (optimum pH = 8.9, MW = 85,000 Da)				
2 U/mg	1 unit converts 1 μmol Chromozym^R PL to product/min at 25°C	Suspension in $(NH_4)_2SO_4$		Boehringer 602370
2-4 U/mg protein	1 unit produces a ΔA_{275} of 1.0 from α-casein/20 min at pH 7.5, 37°C when measuring perchloric acid soluble products in a volume of 5.0 mL	Lyophilized containing 5% protein; balance primarily KCl, lysine, pH 2.5		Sigma P7911
Human				
				Vital 17261
Human plasma (optimum pH = 8.9, MW = 76,500-85,000 Da)				
	1 unit hydrolyzes 1 μmol tosyl-Gly-Pro-Lys-pNA/min at pH 7.8, 25°C; 1 U = 1.25 CU	Frozen in 100 mM $NaPO_4$, 1 mM 6-aminohexanoic acid, 25% glycerol, pH 7.3	No detectable HBsAg, anti-HCV, anti-HBc, anti-HIV	ART 16-16-161213
8 U/mg	1 unit converts 1 μmol Chromozym^R PL to product/min at 25°C	Suspension in $(NH_4)_2SO_4$		Boehringer 602361
>90%	Active site titration PNPGB	Purified; frozen in concentrated form	No detectable HIV Ab, HsAg	EnzymeRes HPlas
0.3 U/mg	1 unit hydrolyzes 1 μmol Tos-Gly-Pro-Lys-4-NA.AcOH(=Chromozym™ PL)/min at pH 8.2, 25°C	Lyophilized	No detectable Hepatitis B virus, HIV	Fluka 80955

Plasmin continued — 3.4.21.7

SPECIFIC ACTIVITY	UNITS DEFINITION	PREPARATION FORM	ADDITIONAL ACTIVITIES	SUPPLIER CATALOG NO.
Human plasma continued				
3-6 U/mg protein	1 unit produces a ΔA_{275} of 1.0 from α-casein/20 min at pH 7.5, 37°C when measuring perchloric acid soluble products in a volume of 5.0 mL	Lyophilized containing 5% protein; balance primarily KCl, lysine, pH 2.5		Sigma P4895
Human serum (MW = 78,000 Da)				
≥8 U/mg protein	1 unit hydrolyzes 1 µmol tosyl-Gly-Pro-Lys-pNA/min at pH 7.8, 25°C	Solution containing 100 mM NaPO$_4$, 1 mM ε-aminohexanoic acid, 25% glycerol, pH 7.3	No detectable HBsAg, HIV Ab	Calbiochem 527621
10 U/mg protein	1 unit hydrolyzes 1 µmol tosyl-Gly-Pro-Lys-pNA/min at pH 7.8, 25°C	EACA- and lysine-free; lyophilized from 10 mg D-mannitol, 10 mg NaCl and small amount of buffer salts	No detectable HBsAg, HIV Ab	Calbiochem 527624
Porcine blood				
3-5 U/mg protein	1 unit produces a ΔA_{275} of 1.0 from α-casein/20 min at pH 7.5, 37°C when measuring perchloric acid soluble products in a volume of 5.0 mL			Sigma P8644

Enteropeptidase — 3.4.21.9

REACTION CATALYZED
    Activation of trypsinogen by selective cleavage of -(Asp)$_4$-Lys^6↓Ile- bond

SYNONYMS
    Enterokinase

REACTANTS
    Polypeptides, trypsinogen

APPLICATIONS
- Cleaves recombinant fusion proteins containing the recognition sequence
- Protein modification and amino acid sequence determination

NOTES
- Classified as a hydrolase, acting on peptide bonds: serine endopeptidase
- Converts proenzyme trypsinogen to active trypsin
- Activated by Ca^{2+}
- Inhibited by NaCl, TPCK, TLCK, SDS and guanidine HCl
- Not inhibited by protein inhibitors of trypsin
- A serine protease homologue of chymotrypsin

3.4.21.9 Enteropeptidase continued

SPECIFIC ACTIVITY	UNITS DEFINITION	PREPARATION FORM	ADDITIONAL ACTIVITIES	SUPPLIER CATALOG NO.
Bovine intestine				
0.5-2 U/mg solid	1 unit produces 1.0 nmol trypsin from trypsinogen/min at pH 5.6, 25°C	Salt-free; lyophilized	<5% free trypsin, aminopeptidase	Sigma E5510
Bovine, recombinant expressed in *E. coli* (MW = 26,300 Da)				
250 U/mL	1 unit cleaves 50 μg of an MBP fusion protein with a fusion joint of Asp-Asp-Asp-Asp-Lys to 95% completion/8 hr at pH 7.4, 23°C	Bovine light chain; >90% purity; solution containing 200 mM NaCl, 20 mM Tris-HCl, 2 mM CaCl$_2$, 50% glycerol, pH 7.2		Calbiochem 324792
100-400 U/mL	1 unit cleaves 50 μg fusion protein containing a fusion joint of Asp-Asp-Asp-Asp-Lys to 95% completion/8 hr at 23°C	Bovine light chain; 20 mM Tris-HCl, 200 mM NaCl, 2 mM CaCl$_2$, 50% glycerol, pH 7.0		NE Biolabs 800-70S 800-70L
Calf intestine (optimum pH = 8.0, MW = 150,000 Da)				
		Special quality for cleavage of fusion proteins; highly purified; lyophilized		Boehringer 1334115 1351311
1000 U/mg solid	1 unit activates 1 μmol trypsinogen/min at pH 5.6, 25°C	60% purity; lyophilized		Calzyme 186A1000
1000 U/mg protein	1 unit produces 1.0 nmol trypsin from trypsinogen/min at pH 5.6, 25°C	Highly purified; solution containing 50% glycerol, pH 4.5		Sigma E0766
Calf intestine or mucosa				
>1000 U/mg protein; ≥400 U/mg solid	1 unit activates 1 nmol trypsinogen/hr at pH 5.6, 25°C	Salt-free; lyophilized	No detectable chymotrypsin <1% trypsin	Biozyme EK 2B
≥100 U/μg protein; ≥300 U/μL	1 unit activates 1 nmol trypsinogen/hr at pH 5.6, 25°C	Highly purified; 50% glycerol solution, pH 4.5	<10 πg trypsin/μg enterokinase (azocoll as substrate)	Biozyme EK 3
Porcine intestine				
5 U/mg solid	1 unit activates 0.065 mg trypsinogen/hr at pH 5.8	Lyophilized		ICN 160060
2 U/mg solid	1 unit produces 1.0 nmol trypsin from trypsinogen/min at pH 5.6, 25°C	Salt-free; lyophilized	<1% aminopeptidase <5% free trypsin	Sigma E0632
100-5000 U/mg protein	1 unit produces 1.0 nmol trypsin from trypsinogen/min at pH 5.6, 25°C	Chromatographically purified; lyophilized containing 50% protein; balance primarily NaPO$_4$ buffer salts	<0.75% aminopeptidase <1% free trypsin	Sigma E0885

α-Lytic Endopeptidase 3.4.21.12

REACTION CATALYZED
Preferential cleavage of Ala-↓- and Val-↓- in bacterial cell walls, elastin and other proteins

SYNONYMS
Myxobacter α-lytic proteinase

REACTANTS
Bacterial cell walls, elastin, proteins

NOTES
- Classified as a hydrolase, acting on peptide bonds: serine endopeptidase
- From the myxobacterium *Lysobacter enzymogenes*
- A serine protease homologue of chymotrypsin

See Chapter 5. Multiple Enzyme Preparations

Glutamyl Endopeptidase 3.4.21.19

REACTION CATALYZED
Preferential cleavage: Glu↓, Asp↓ (phosphate buffer, pH 7.8); Glu↓ (ammonium bicarbonate buffer, pH 7.8; ammonium acetate buffer, pH 4.0)

SYNONYMS
V8 Proteinase, endoproteinase Glu-C, Staphylococcal serine proteinase, Glu-C

REACTANTS
Peptide bonds, casein

APPLICATIONS
- Peptide mapping basic proteins by proteolysis

NOTES
- Classified as a hydrolase, acting on peptide bonds: serine endopeptidase
- A serine protease from *Staphylococcus aureus* strain V8
- Active in the pH range 4.0-9.5; precipitates around pH 4.0
- Inhibited by DFP, α1-macroglobulin, 3,4-dichloroisocoumarin, TLCK, Cl$^-$, F$^-$, NO$_3^-$ and acetate

SPECIFIC ACTIVITY	UNITS DEFINITION	PREPARATION FORM	ADDITIONAL ACTIVITIES	SUPPLIER CATALOG NO.
Staphylococcus aureus				
15 U/mg	1 unit hydrolyzes 1 μmol carbobenzoxy-Phe-Leu-Glu-4-nitroanilide/min at pH 7.8, 25°C	Lyophilized		Fluka 45174

SPECIFIC ACTIVITY	UNITS DEFINITION	PREPARATION FORM	ADDITIONAL ACTIVITIES	SUPPLIER CATALOG No.
Staphylococcus aureus, strain V8 (optimum pH = 4.0 and 7.8, MW = 27,000 Da; autolysis at T > 40°C)				
		Sequencing grade; salt-free; lyophilized	Free of impurities that might interfere with the specific cleavage or in the separation of peptides using reverse-phase HPLC	Boehringer 1420399 1047817
20 U/mg solid (Z-Phe-Leu-Glu-4-nitranilide as substrate, 25°C); 500 U/mg solid (casein as substrate, 37°C)	1 unit converts 1 μmol substrate to product/min	Sequencing grade; lyophilized		Boehringer 791156
20 U/mg solid	1 unit hydrolyzes 1 μmol Z-Phe-Leu-Glu-pNA/min at pH 7.8, 25°C	Lyophilized	No detectable proteases	Calbiochem 324713
800 U/mg	1 unit hydrolyzes 1 μmol N-t-BOC-L-Glu-α-phenyl ester/min at pH 7.8, 37°C			Fluka 45172
500 U/mg	1 unit hydrolyzes casein changing A_{280} by 0.001/min at pH 7.8, 37°C	Lyophilized		ICN 151972
500 U/mg	1 unit changes A_{280} by 0.001/min at pH 7.8, 37°C using casein as substrate	Homogeneous by SDS-PAGE; lyophilized		ICN 399001
500-1000 U/mg solid	1 unit hydrolyzes 1.0 μmol N-t-BOC-L-Glu α-phenyl ester/min/min at pH 7.8, 37°C	Chromatographically purified; lyophilized		Sigma P2922
250-500 U/g solid; 1 mL gel yields 50-100 U	1 unit hydrolyzes 1.0 μmol N-t-BOC-L-Glu α-phenyl ester/min at pH 7.8, 37°C ; 1 U = 0.004 casein digestion unit	Insoluble enzyme attached to 4% cross-linked beaded agarose; lyophilized containing lactose stabilizer		Sigma P6552
500-1000 U/mg solid	1 unit hydrolyzes 1.0 μmol N-t-BOC-L-Glu α-phenyl ester/min at pH 7.8, 37°C			Sigma P8400
≥20 U/mg protein		Sequencing grade; lyophilized		Wako 050-05941
20 U/mg		Lyophilized		Wako 164-13982
≥500 U/mg solid	1 unit liberates 0.001 A_{280} acid soluble fragments from casein/min at pH 7.8, 37°C	Highly purified; lyophilized		Worthington LS03608 LS03605 LS03606

Cathepsin G — 3.4.21.20

REACTION CATALYZED
Preferential cleavage: Leu↓, Tyr↓, Phe↓, Met↓, Trp↓, Gln↓, Asn↓

REACTANTS
Polypeptides, collagen, proteoglycan, angiotensin II, angiotensin I, angiotensinogen

NOTES
- Classified as a hydrolase, acting on peptide bonds: serine endopeptidase
- A serine protease with specificity similar to chymotrypsin C (EC 3.4.21.2)
- From azurophil granules of polymorphonuclear leukocytes
- Implicated in connective tissue diseases such as emphysema and rheumatoid arthritis
- Inhibited by chymostatins, DFP, Z-Phe-CH$_2$Br

SPECIFIC ACTIVITY	UNITS DEFINITION	PREPARATION FORM	ADDITIONAL ACTIVITIES	SUPPLIER CATALOG NO.
Human leukocyte				
60 U/mg protein	1 unit releases 1.0 nmol p-nitroaniline/sec from N-succinyl-Ala-Ala-Pro-Phe p-nitroanilide at pH 7.5, 37°C	Lyophilized containing 0.5 M pyridinium acetate, pH 5.3		Sigma C4428
Human neutrophil (optimum pH = 7.5, MW = 23,500 Da)				
2-4 U/mg protein	1 unit hydrolyzes 1 μmol succinyl-Ala-Ala-Pro-Phe-pNA (1 mM)/min at pH 7.5, 25°C	≥95% purity; salt-free; lyophilized	No detectable HBsAg, anti-HCV, anti-HBc, anti-HIV	ART 16-14-030107
≥2 U/mg protein	1 unit hydrolyzes 1 μmol succinyl-Ala-Ala-Pro-Phe-pNA/min at pH 7.5, 25°C	≥95% purity; salt-free; solid	No detectable HBsAg, HIV Ab	Calbiochem 219373
2-4 U/mg protein	1 unit digests 1 μmol succinyl-Ala-Ala-Pro-Phe-pNA/min at pH 7.5, 25°C	>95% purity by SDS-PAGE; salt-free; lyophilized	No detectable HIV Ab, HBsAg	Cortex CP3010
160,000 U/mg protein		>98% purity by SDS-PAGE		ICN 191344
Human purulent sputum (leucocyte) (MW = 23,500 Da)				
4000-4800 U/mg protein; 30-38 U/mg solid	1 unit hydrolyzes 1 nmol succinyl-Ala-Ala-Pro-Phe-pNA/min at pH 8.3, 25°C	>95% purity by chromatography; lyophilized containing 1% Tris buffer and 97% (NH$_4$)$_2$SO$_4$ as stabilizer, pH 8	No detectable elastase, MPO, lysozyme	Elastin SG45

Coagulation Factor VIIa

REACTION CATALYZED
Selective cleavage of Arg-↓-Ile bond in Factor X to form Factor Xa

SYNONYMS
Human Factor VIIa

REACTANTS
Factor X, Factor Xa

NOTES
- Classified as a hydrolase, acting on peptide bonds: serine endopeptidase
- A serine protease formed from the precursor Factor VII
- Active Factor VIIa consists of 2 chains linked by a sulfhydryl bond, formed when the Arg^{152}-Ile^{153} peptide bond in the single-chain precursor is cleaved
- Activated by Ca^{2+}
- Cattle enzyme is more readily inhibited by diisopropyl fluorophosphate than human enzyme

SPECIFIC ACTIVITY	UNITS DEFINITION	PREPARATION FORM	ADDITIONAL ACTIVITIES	SUPPLIER CATALOG NO.
Human plasma (MW = 48,000–50,000 Da)				
45,000 U/mg solid	Clotting assay	>90% Activation observed on 10% SDS-PAGE; lyophilized from buffer containing 20 mM Tris-HCl and 100 mM NaCl, pH 7.4	No detectable HBsAg, HIV Ab	Calbiochem 219370
>60,000 U/mg	Clotting assay	Frozen in concentrated form	No detectable HIV Ab, HsAg	EnzymeRes HFVIIa
Human, recombinant from yeast				
>20 clot U/µg using 1-stage clotting assay; 1 mg solid/vial		>95% purity by reverse phase HPLC; lyophilized containing buffer	<5% Factor VII (reduced SDS-PAGE)	ICN 153577

Coagulation Factor IXa 3.4.21.22

REACTION CATALYZED
    Selective cleavage of Arg-↓-Ile bond in Factor X to form Factor Xa

SYNONYMS
    Activated Christmas Factor

REACTANTS
    Peptide bonds

NOTES
- Classified as a hydrolase, acting on peptide bonds: serine endopeptidase
- One of the γ-carboxyglutamic acid-containing blood coagulation factors
- Formed by activation of Factor IX with Factor XIa
- A serine protease homologue of chymotrypsin

SPECIFIC ACTIVITY	UNITS DEFINITION	PREPARATION FORM	ADDITIONAL ACTIVITIES	SUPPLIER CATALOG NO.
Bovine plasma				
10,000 U/mg	Clotting assay	Factor IXaα; MW = 55,400 Da; frozen in concentrated form		EnzymeRes BFIXa 1070
25,000 U/mg	Clotting assay	Factor IXaβ; MW = 43,900 Da; frozen in concentrated form		EnzymeRes BFIXa 1080
Human plasma				
10,000 U/mg	Clotting assay	Factor IXaα; MW = 58,700 Da; frozen in concentrated form	No detectable HIV Ab, HsAg	EnzymeRes HFIXa 1070
25,000 U/mg	Clotting assay	Factor IXaβ; MW = 45,000 Da; frozen in concentrated form	No detectable HIV Ab, HsAg	EnzymeRes HFIXa 1080

Cucumisin 3.4.21.25

REACTION CATALYZED
    Hydrolysis of proteins with broad specificity

REACTANTS
    Polypeptides

NOTES
- Classified as a hydrolase, acting on peptide bonds: serine endopeptidase
- From the sarcocarp of the musk melon (*Cucumis melo*)
- A serine protease homologue of subtilisin (EC 3.4.21.62)
- Other plant serine endopeptidases, less well characterized and without specific EC numbers: euphorbain (*Euphorbia cerifera*), solanain (*Solanum elaeagnifolium* [horse nettle]), hurain (*Hura crepitans*), tabernamontanain (*Tabernamontana grandiflora*)

3.4.21.25 Cucumisin continued

SPECIFIC ACTIVITY	UNITS DEFINITION	PREPARATION FORM	ADDITIONAL ACTIVITIES	SUPPLIER CATALOG No.
Sarcocarp of melon fruit				
10-30 U/mg solid	1 unit hydrolyzes casein to produce peptides equivalent to 1.0 μmol Tyr/min at pH 10.0, 37°C	Lyophilized containing sucrose		sigma C1932

3.4.21.26 Prolyl Oligopeptidase

REACTION CATALYZED
  Hydrolysis of Pro↓ >> Ala↓ in oligopeptides
SYNONYMS
  Post-proline cleaving enzyme, proline-specific endopeptidase
REACTANTS
  Oligopeptides

NOTES
- Classified as a hydrolase, acting on peptide bonds: serine endopeptidase
- Commonly activated by thiol compounds
- Generally cytosolic, found in vertebrates, plants and *Flavobacterium*
- Inhibited by DFP, 3,4-dichloroisocoumarin and Z-Gly-Pro-CH$_2$Cl
- A serine protease not homologous with chymotrypsin or subtilisin

SPECIFIC ACTIVITY	UNITS DEFINITION	PREPARATION FORM	ADDITIONAL ACTIVITIES	SUPPLIER CATALOG No.
***Flavobacterium meningosepticum* (optimum pH = 7.0, T = 40°C)**				
35 U/mg	1 unit produces 1 μmol *p*-nitroaniline from carbobenzoxy-Gly-Pro-*p*-nitroaniline/min at pH 7.0, 30°C	Lyophilized	<0.1% aminopeptidase, trypsin	ICN 320821
≥35 U/mg	1 unit produces 1 μmol *p*-nitroaniline from Z-Gly-Pro-*p*NA/min at pH 7.0, 30°C	Lyophilized		Seikagaku 120125-1
***Flavobacterium* species (optimum pH = 7.5, pI = 9.1, T = 40°C, MW = 78,000 Da [monomer]; K_M = 25 μM [Z-Gly-Pro-MCA], 0.14 mM [Z-Gly-Pro-2NNap]; stable pH 5.0-9.5, T < 40°C)**				
≥5 U/mg solid	1 unit forms 1 μmol *p*-nitroaniline/min at pH 7.0, 30°C	Lyophilized	<0.1% LAP, trypsin-like activity	Toyobo PSP-001

Coagulation Factor XIa

3.4.21.27

REACTION CATALYZED
: Selective cleavage of Arg-↓-Ala and Arg-↓-Val bonds in Factor IX to form Factor IXa

REACTANTS
: Peptide bonds, Factor IX, Factor IXa

NOTES
- Classified as a hydrolase, acting on peptide bonds: serine endopeptidase
- One of the γ-carboxyglutamic acid-containing blood coagulation factors
- Formed by activation of Factor XI with Factor XIIa.
- A serine protease homologue of chymotrypsin

SPECIFIC ACTIVITY	UNITS DEFINITION	PREPARATION FORM	ADDITIONAL ACTIVITIES	SUPPLIER CATALOG NO.
Bovine plasma (a doublet on SDS-PAGE)				
		Frozen in concentrated form		EnzymeRes BXIa 11a
Human plasma (MW = 160,000 Da)				
	Chromogenic assay S-2366	Frozen in concentrated form	No detectable HIV Ab, HsAg	EnzymeRes HXIa 1111

Plasma Kallikrein

3.4.21.34

REACTION CATALYZED
: Selective cleavage of some Arg↓ and Lys↓ bonds, including Lys↓Arg and Arg↓Ser in (human) kininigen to release bradykinin

SYNONYMS
: Serum kallikrein, kininogenin

REACTANTS
: Kininigen, bradykinin, Factor VII, Factor IX, Factor XI, Factor XII, plasminogen, prorenin, hexokinase

NOTES
- Classified as a hydrolase, acting on peptide bonds: serine endopeptidase
- Formed by activation of plasma prokallikrein (Fletcher factor) with Factor XIIa
- Activates coagulation Factors XII, VII and plasminogen
- Selective for Arg > Lys in P1, in small molecule substrates
- Present in plasma as the inactive precursor prokallikrein
- A serine protease homologue of chymotrypsin

SPECIFIC ACTIVITY	UNITS DEFINITION	PREPARATION FORM	ADDITIONAL ACTIVITIES	SUPPLIER CATALOG NO.
Human plasma (MW = 86,000 Da)				
15 U/mg	1 unit hydrolyzes 1 μmol D-Pro-Phe-Arg-pNA/min at pH 8.0, 25°C	≥95% purity; frozen in 10 mM Tris-HCl and 100 mM NaCl, pH 8.5	No detectable HBsAg, anti-HCV, anti-HBc, anti-HIV	ART 16-16-110112
≥15 U/mg protein	1 unit hydrolyzes 1 μmol D-Pro-Phe-Arg-pNA/min at pH 8.0, 25°C	Frozen in 100 mM Tris-HCl and 100 mM NaCl, pH 8.5	No detectable HBsAg, HIV Ab	Calbiochem 420307
2.5-15 μM	Chromogenic assay S-2302	Purified; frozen in concentrated form	No detectable HIV Ab, HsAg	EnzymeRes HPKa 1303
1-5 U/mg protein	1 unit hydrolyzes 1.0 μmol BAEE to Nα-benzyl-L-Arg and EtOH/min at pH 8.7, 25°C	Lyophilized		Sigma K1004
5-15 U/mg protein	1 unit hydrolyzes 1.0 μmol BAEE to Nα-benzyl-L-Arg and EtOH/min at pH 8.7, 25°C	Solution containing 50 mM PO_4 buffer and 50% glycerol, pH 7.5		Sigma K3126
Human urine (MW = 50,000 Da)				
0.5 U/mg protein	1 unit hydrolyzes 1.0 μmol BAEE/min at pH 8.0, 35°C	In 250 mM NaCl, 25 mM Tris-HCl and 50% glycerol, pH 8.0		Calbiochem 420313

3.4.21.35 Tissue Kallikrein

REACTION CATALYZED
Preferential cleavage of Arg↓ bonds in small molecule substrates. Highly selective hydrolysis of Met↓ or Leu↓ releases kallidin from kininogen. Rat enzyme cleaves two Arg↓ bonds to liberate bradykinin directly from autologous kininogens

SYNONYMS
Glandular kallikrein, pancreatic kallikrein, submandibular kallikrein, submaxillary kallikrein, kidney kallikrein, urinary kallikrein, γ-seminoprotein, endoproteinase Arg-C, arginylendopeptidase, submaxillary protease D, Arg-C

REACTANTS
Kininogens, kallidin, bradykinin

APPLICATIONS
- Liberates vasoactive peptides from inactive kininogen precursors
- Protein structure and sequence studies

Tissue Kallikrein continued   3.4.21.35

NOTES

- Classified as a hydrolase, acting on peptide bonds: serine endopeptidase
- Formed by activation of tissue prokallikrein with trypsin
- A serine protease homologue of chymotrypsin
- Inhibited by DFP, α_2-macroglobulin, TLCK, Hg^{2+}, Cu^{2+}, Zn^{2+} and PCMB
- *Not inhibited* by EDTA or hydroxyquinoline.
- Related enzymes include: mouse γ-renin (EC 3.4.21.40), submandibular proteinase A, epidermal growth-factor-binding protein, nerve growth factor γ-subunit, rat tonin, submaxillary proteinases A and B, T-kininogenase, kallikreins k7 and k8 and human prostate-specific antigen (γ-seminoprotein)

SPECIFIC ACTIVITY	UNITS DEFINITION	PREPARATION FORM	ADDITIONAL ACTIVITIES	SUPPLIER CATALOG NO.
Murine submaxillary gland (optimum pH = 8.0-9.0, MW = 21,300-30,000 Da)				
220 U/mg protein; 20 U/mg solid	1 unit converts 1 µmol N-tosyl-L-Arg-methyl ester/min at 25°C	Sequencing grade; lyophilized		Boehringer 269590
>200 U/mg protein	1 unit hydrolyzes 1 µmol N-Tos-Arg-methyl ester/min at pH 8.0, 25°C	Lyophilized		Calbiochem 324710
100 U/mg	1 unit hydrolyzes 1 µmol N-tosyl-L-Arg methylester HCl/min at pH 8.0, 37°C	Lyophilized		Fluka 45173
	1 unit produces 1 µmol p-nitroaniline/min at 37°C	TLCK-treated; TPCK-treated; homogeneous by PAGE; 5 mM NaPO$_4$ buffer solution containing 50% glycerol, pH 7.2	No detectable protease	PanVera TAK 7308
250-800 U/mg protein	1 unit hydrolyzes 1.0 µmol Nα-p-tosyl-L-Arg methyl ester/min at pH 8.0, 25°C	Lyophilized containing 5% protein; balance primarily NaCl and NaOAc		Sigma P5171
5 µg/vial		Sequencing grade; lyophilized		Sigma P6056
200-300 U/mg protein	1 unit hydrolyzes 1.0 µmol Nα-p-tosyl-L-Arg methyl ester/min at pH 8.0, 25°C	Lyophilized		Sigma P8402
Porcine pancreas (optimum pH = 8.2-8.7, MW = 26,200 Da)				
100 U/mg solid	1 unit hydrolyzes 1 µmol BAEE/min at pH 8.0, 25°C	Salt-free; lyophilized		Calbiochem 420306
50 U/mg protein	1 unit hydrolyzes 1 µmol Nα-benzoyl-L-Arg and EtOH/min at pH 8.7, 20°C	Lyophilized		ICN 190107
≥200 IU/mg		Powder		Intergen 8460-00
40 U/mg solid	1 unit hydrolyzes 1.0 µmol BAEE to Nα-benzyl-L-Arg and EtOH/min at pH 8.7, 25°C			Sigma K3627

3.4.21.35 Tissue Kallikrein continued

SPECIFIC ACTIVITY	UNITS DEFINITION	PREPARATION FORM	ADDITIONAL ACTIVITIES	SUPPLIER CATALOG NO.
Unspecified				
				ICN 190221

2.4.21.36 Pancreatic Elastase

REACTION CATALYZED
    Preferential cleavage: Ala↓

SYNONYMS
    Pancreatopeptidase E, pancreatic elastase I

REACTANTS
    Elastin, polypeptides, hemoglobin, fibrin, casein, albumin, denatured collagen, soy protein

APPLICATIONS
- Tissue dissociation, with collagenase, trypsin and chymotrypsin
- Protein sequence studies
- Solubilization of membrane proteins

NOTES
- Classified as a hydrolase, acting on peptide bonds: serine endopeptidase
- A serine protease consisting of a single peptide chain of 240 amino acid residues with 4 disulfide bridges
- Produced in the pancreas, occurs in leukocytes and serum; also found in bacteria
- Formed by activation of mammalian pancreatic proelastase by trypsin
- Unique among proteases in its ability to hydrolyze native elastin (not attacked by trypsin, chymotrypsin or pepsin)
- Also exhibits esterase and amidase activity
- May play a role in induction of atherosclerosis and acute hemorrhagic pancreatitis
- Inhibited by DFP, sulfonyl fluorides, p-dinitrophenyl diethyl phosphate, alkyl isocyanates, α_2-macroglobulin, leech inhibitor; NaCl and copper sulfate reduce elastolytic activity by 50%
- Soybean trypsin inhibitor and kallikrein inhibitor suppress proteolytic but not elastolytic activity
- Competitively inhibited by derivatives of dipeptides of alanine, valine, leucine and isoleucine
- A homologue of chymotrypsin

SPECIFIC ACTIVITY	UNITS DEFINITION	PREPARATION FORM	ADDITIONAL ACTIVITIES	SUPPLIER CATALOG No.
Murine pancreas (optimum pH = 8.8, MW = 28,000 Da)				
3.2–3.8 U/mg protein	1 unit hydrolyzes 1 μmol succinyl-(Ala)$_3$-pNA/min at pH 8.8, 37°C	Chromatographically purified; salt-free; lyophilized	No detectable trypsin, chymotrypsin, pancreatic enzymes	Elastin MS838
Porcine				
6–10 mg enzyme/mL gel; 3400–7500 U/mL	1 unit hydrolyzes 1 μmol succinyl-(Ala)$_3$-pNA/min at pH 8.0, 37°C	Chromatographically purified; attached to beaded agarose in 0.02 M NaOAc and 0.01% NaN$_3$, pH 5		Elastin A443
6–10 mg enzyme/mL gel; 3400–7500 U/mL	1 unit hydrolyzes 1 μmol succinyl-(Ala)$_3$-pNA/min at pH 8.0, 37°C	Chromatographically purified; attached to beaded agarose in 0.02 M NaOAc and 0.01% NaN$_3$, pH 5		Elastin AD443
6–10 mg enzyme/mL gel; 3400–7500 U/mL	1 unit hydrolyzes 1 μmol succinyl-(Ala)$_3$-pNA/min at pH 8.0, 37°C	Chromatographically purified; attached to beaded agarose in 0.02 M NaOAc and 0.01% NaN$_3$, pH 5		Elastin AE443
Porcine pancreas (optimum pH = 7.8–8.8, pI = 9.5, MW = 25,000 Da; stable pH 4–10.4)				
≥200 U/mg solid	1 unit hydrolyzes 1 μmol N-Ac-tri-L-Ala methyl ester/min at pH 8.5, 25°C	Chromatographically prepared; salt-free; lyophilized		Biozyme E2
≥120 U/mg protein	1 unit hydrolyzes 1 μmol N-Ac-tri-L-Ala methyl ester/min at pH 8.5, 25°C	Partially purified; suspension in 70% (NH$_4$)$_2$SO$_4$		Biozyme E3
≥3 U/mg protein	1 unit converts 1 μmol N-succinyl-L-alanyl-L-alanyl-L-Ala-4-nitroanilide to product/min	Lyophilized		Boehringer 100905 100907 100909
>150 U/mg protein	1 unit hydrolyzes 1 μmol N-Ac-(Ala)$_3$-methyl ester/min at pH 8.5, 25°C	Lyophilized	<0.5% chymotrypsin	Calbiochem 324689
5 U/mg protein	1 unit hydrolyzes 1 μmol succinyl Ala$_3$NA/min at pH 8.0, 25°C	Crystallized; aqueous suspension containing 90% protein		Calzyme 051D0005
10 U/mg protein	1 unit hydrolyzes 1 μmol SucAla$_3$NA/min at pH 8.0, 25°C	Salt-free; lyophilized containing 90% protein		Calzyme 075A0010
60–80 U/mg protein; >4 U/mg protein (Suc-(Ala)$_3$-pNA)	1 unit solubilizes 1 mg elastin/20 min at pH 8.8, 37°C with elastin-orcein as substrate; 1 unit hydrolyzes 1 μmol Suc-(Al)$_3$-pNA/min at pH 8.3, 25°C	2X Crystallized; aqueous suspension containing 0.01% NaN$_3$ as preservative	No detectable trypsin, chymotrypsin	Elastin E134
115–125 U/mg protein; >9 U/mg protein (Suc-(Ala)$_3$-pNA)	1 unit solubilizes 1 mg elastin/20 min at pH 8.8, 37°C with elastin-orcein as substrate; 1 unit hydrolyzes 1 μmol Suc-(Al)$_3$-pNA/min at pH 8.3, 25°C	Chromatographically purified; salt-free; 2X crystallized; lyophilized	No detectable trypsin, chymotrypsin, lipase, α-amylase, carboxypeptidase	Elastin EC134

SPECIFIC ACTIVITY	UNITS DEFINITION	PREPARATION FORM	ADDITIONAL ACTIVITIES	SUPPLIER CATALOG NO.
Porcine pancreas *continued*				
120-145 U/mg protein; >9 U/mg protein (Suc-(Ala)$_3$-pNA)	1 unit solubilizes 1 mg elastin/20 min at pH 8.8, 37°C with elastin-orcein as substrate; 1 unit hydrolyzes 1 μmol Suc-(Al)$_3$-pNA/min at pH 8.3, 25°C	Chromatographically purified; salt-free; 2X crystallized; suspension in 50% glycerol, 0.02 M NaOAc, 0.01% NaN$_3$, pH 6.0	No detectable trypsin, chymotrypsin, lipase, α-amylase, carboxypeptidase	Elastin EC439
70-95 U/mg protein, 65-85 U/mg solid; >4 U/mg protein (Suc-(Ala)$_3$-pNA)	1 unit solubilizes 1 mg elastin/20 min at pH 8.8, 37°C with elastin-orcein as substrate; 1 unit hydrolyzes 1 μmol Suc-(Al)$_3$-pNA/min at pH 8.3, 25°C	2X crystallized; lyophilized containing 90% protein and 10% Na$_2$CO$_3$	No detectable protease Contains euglobin proteins	Elastin EL357
120-145 U/mg protein; >9 U/mg protein (Suc-(Ala)$_3$-pNA)	1 unit solubilizes 1 mg elastin/20 min at pH 8.8, 37°C with elastin-orcein as substrate; 1 unit hydrolyzes 1 μmol Suc-(Al)$_3$-pNA/min at pH 8.3, 25°C	Chromatographically purified; salt-free; 2X crystallized; suspension in 0.02 M NaOAc, 0.01% NaN$_3$, pH 6.0	No detectable trypsin, chymotrypsin, lipase, α-amylase, carboxypeptidase	Elastin ES438
70-80 U/mg solid; >4 U/mg protein (Suc-(Ala)$_3$-pNA)	1 unit solubilizes 1 mg elastin/20 min at pH 8.8, 37°C with elastin-orcein as substrate; 1 unit hydrolyzes 1 μmol Suc-(Al)$_3$-pNA/min at pH 8.3, 25°C	2X crystallized; lyophilized containing 65% protein and 35% Na$_2$CO$_3$	No detectable trypsin, chymotrypsin, contaminating pancreatic enzymes, euglobin	Elastin ET947
3 U/mg	1 unit liberates 1 μmol 4-nitroaniline/min at pH 7.8, 25°C with succinyl-(L-Ala)$_3$-4-nitroanilide as substrate	Powder		Fluka 45122
2 U/mg	1 unit liberates 1 μmol 4-nitroaniline/min at pH 7.8, 25°C with succinyl-(L-Ala)$_3$-4-nitroanilide as substrate	Lyophilized		Fluka 45123
15 U/mg	1 unit liberates 1 μmol 4-nitroaniline/min at pH 7.8, 25°C with succinyl-(L-Ala)$_3$-4-nitroanilide as substrate	Powder		Fluka 45124
8 U/mg	1 unit liberates 1 μmol 4-nitroaniline/min at pH 7.8, 25°C with succinyl-(L-Ala)$_3$-4-nitroanilide as substrate	Lyophilized		Fluka 45125
4 U/mg protein	1 unit liberates 1 μmol 4-nitroaniline/min at pH 7.8, 25°C with succinyl-(L-Ala)$_3$-4-nitroanilide as substrate	Crystallized; suspension containing 0.01% thymol as preservative		Fluka 45127
>50 U/mg protein	1 unit solubilizes 1 mg elastin/20 min at pH 8.8, 37°C	2X Crystallized; aqueous suspension containing 0.01% NaN$_3$ as preservative		ICN 100617

SPECIFIC ACTIVITY	UNITS DEFINITION	PREPARATION FORM	ADDITIONAL ACTIVITIES	SUPPLIER CATALOG NO.
Porcine pancreas *continued*				
95-125 U/mg	1 unit solubilizes 1 mg elastin/20 min at pH 8.8, 37°C	Lyophilized		ICN 100619
60-120 U/mg protein	1 unit solubilizes 1 mg elastin/20 min at pH 8.8, 37°C	Chromatographically purified; lyophilized	>50 U/mg protein trypsin	Sigma E0127
60-120 U/mg protein	1 unit solubilizes 1 mg elastin/20 min at pH 8.8, 37°C	Affinity chromatographed to reduce trypsin	<50 U/mg protein trypsin	Sigma E0258
25-100 U/mg protein	1 unit solubilizes 1 mg elastin/20 min at pH 8.8, 37°C	2X Crystallized; aqueous suspension	Up to 50 BAEE U/mg protein trypsin	Sigma E1250
40-80 U/mg protein	1 unit solubilizes 1 mg elastin/20 min at pH 8.8, 37°C	2X Crystallized; lyophilized containing 70% protein; balance primarily Na_2CO_3	Up to 500 BAEE U/mg protein trypsin	Sigma E6883
60-120 U/mg protein	1 unit solubilizes 1 mg elastin/20 min at pH 8.8, 37°C	Cell culture tested; lyophilized containing <50 BAEE U/mg protein		Sigma E7885
150 U/mg protein		Lyophilized		Wako 058-05361
≥3 U/mg protein		2X Crystallized; suspension		Worthington LS02274 LS02279 LS02280 LS02276
≥3 U/mg protein	1 unit converts 1 μmol N-succinyl-trialynyl-p-nitroanilide/min at pH 8.0, 25°C	2X Crystallized; lyophilized		Worthington LS02290 LS02292 LS02294 LS02298
≥8 U/mg protein	1 unit converts 1 μmol N-succinyl-trialynyl-p-nitroanilide/min at pH 8.0, 25°C	Chromatographically prepared; lyophilized		Worthington LS06363 LS06365 LS06367
***Pseudomonas aeruginosa* (MW = 37,000 Da)**				
261 U/mg protein	1 unit hydrolyzes 1 μg insoluble elastin/hr at 37°C	Chromatographically purified; lyophilized		Elastin PE961
Rat pancreas (optimum pH = 8.8, MW = 26,400 Da)				
1.9-2.1 U/mg protein	1 unit hydrolyzes 1 μmol succinyl-(Ala)$_3$-pNA/min at pH 8.8, 37°C	Chromatographically purified; salt-free; lyophilized	No detectable trypsin, chymotrypsin	Elastin RE945

3.4.21.37 Leukocyte Elastase

REACTION CATALYZED
Preferential cleavage: Val↓ > Ala↓

SYNONYMS
Lysosomal elastase, neutrophil elastase

REACTANTS
Elastin, peptide bonds, collagen, proteoglycan, fibrin, hemoglobin, albumin

NOTES
- Classified as a hydrolase, acting on peptide bonds: serine endopeptidase
- Differs from pancreatic elastase (EC 3.4.21.36) in specificity toward synthetic substrates and inhibitor sensitivity
- A serine protease homologue of chymotrypsin
- Implicated in the development of pulmonary emphysema and rheumatoid arthritis

SPECIFIC ACTIVITY	UNITS DEFINITION	PREPARATION FORM	ADDITIONAL ACTIVITIES	SUPPLIER CATALOG NO.
Human leukocyte				
6 U/mg	1 unit releases 1 μmol 4-nitrophenol from BOC-Ala-4-nitrophenyl-ester/min at pH 6.5, 37°C	Lyophilized	No detectable HBsAg, HIV	Fluka 45121
50 U/mg protein	1 unit releases 1.0 nmol p-nitrophenol from N-t-BOC-L-Ala p-nitrophenyl ester/sec at pH 6.5, 37°C	Lyophilized containing 0.5 M pyridinium acetate, pH 5.3		Sigma E8140
Human neutrophil (white leukocytes) (optimum pH = 7.0–8.8, MW = 29,500 Da)				
20–22 U/mg protein	1 unit hydrolyzes 1 μmol MeO-Suc-Ala-Ala-Pro-Val-pNA/min at pH 8.0, 25°C	≥95% purity; salt-free; lyophilized	No detectable HBsAg, anti-HCV, anti-HBc, anti-HIV	ART 16-14-051200
>20 U/mg protein	1 unit hydrolyzes 1 μmol MeO-Suc-Ala-Ala-Pro-Val-pNA/min at pH 8.0, 25°C	>95% purity; salt-free; lyophilized	No detectable HBsAg, HIV Ab	Calbiochem 324681
20–22 U/mg protein	1 unit hydrolyzes 1 μmol MeO-Suc-Ala-Ala-Pro-Val-pNA/min at pH 8.0, 25°C	>95% purity by SDS-PAGE; salt-free; lyophilized	No detectable HIV Ab, HBsAg	Cortex CP3016
18–30 U/mg protein	1 unit hydrolyzes 1 μmol MeO-Suc-Ala-Ala-Pro-Val-pNA/min at pH 8.0, 25°C	>98% purity by SDS-PAGE	No detectable HBsAg, HIV Ab	ICN 191337
20–22 U/mg protein	1 unit hydrolyzes 1 μmol MeO-Suc-Ala-Ala-Pro-Val-pNA/min at pH 8.0, 25°C	≥98% purity by SDS-PAGE; lyophilized	No detectable HGsAg, HCV, HIV-1 Ab	Scripps E0214
Human purulent sputum (leucocyte) (MW = 29,500 Da)				
800–900 U/mg protein (Suc-Ala-Ala-Ala-pNA); 15,000–19,000 U/mg protein (MeO-Suc-Ala-Ala-Pro-Val-pNA)	1 unit hydrolyzes 1 nmol substrate/min at pH 7.5, 25°C	>95% purity by chromatography; salt-free; lyophilized	No detectable cathepsin G, MPO, lysozyme	Elastin SE563

Leukocyte Elastase continued

3.4.21.37

SPECIFIC ACTIVITY	UNITS DEFINITION	PREPARATION FORM	ADDITIONAL ACTIVITIES	SUPPLIER CATALOG NO.
Murine macrophage (MW = 53,000 Da)				
75 U/mg protein	1 unit degrades 1 µg elastin/hr at pH 7.8, 37°C	>95% purity by chromatography; lyophilized		Elastin ME325

Coagulation Factor XIIa

3.4.21.38

REACTION CATALYZED
Selective cleavage of Arg-↓-Ile bonds in Factor VII to form Factor VIIa and Factor XI to form Factor XIa

SYNONYMS
Hageman Factor (activated)

REACTANTS
Factor VII, Factor XI, plasminogen, plasma prokallikrein, Factor VIIa, Factor XIa

NOTES
- Classified as a hydrolase, acting on peptide bonds: serine endopeptidase
- Formed by activation of Factor XII with plasma kallikrein or Factor XIIa
- A serine protease homologue of chymotrypsin
- Initiates kinin generation, intrinsic fibrinolysis and renin-angiotensin activation
- Triggers the Contact Factor cascade by activating Factor XI to XIa

SPECIFIC ACTIVITY	UNITS DEFINITION	PREPARATION FORM	ADDITIONAL ACTIVITIES	SUPPLIER CATALOG NO.
Human plasma				
0.4 U/mg	1 unit hydrolyzes 1 µmol D-Pro-Phe-Arg-pNA/min at pH 8.0, 25°C	Factor β-XIIa; Lyophilized containing 10 mg D-mannitol MW = 30,000 Da		Calbiochem 233496
>50 U/mg	Clotting assay	Purified; frozen in concentrated form	No detectable HIV Ab, HsAg <5% β form	EnzymeRes HXII 1212a
		Factor α-XIIa; in 150 mM NaCl, 4 mM NaOAc, pH 5.3 MW = 80,000 Da	No detectable HBsAg, HIV Ab < 5% Factor β-XIIa	Calbiochem 233493

Lysyl Endopeptidase

REACTION CATALYZED
 Preferential cleavage: Lys↓, including -Lys↓Pro-

SYNONYMS
 Achromobacter proteinase I, Lys-C

REACTANTS
 Polypeptides

APPLICATIONS
- Sequence and structural studies
- Synthesis of Lys-X compounds

NOTES
- Classified as a hydrolase, acting on peptide bonds: serine endopeptidase
- A serine protease homologue of chymotrypsin from *Achromobacter lyticus*
- Retains complete activity after incubation in 4 M urea or in 0.1% SDS solution at 30°C, up to 6 hrs
- Hydrolyzes amide, ester and peptide bonds at the C-side of Lys
- Inhibited by DFP, TLCK, aprotinin and leupeptin
- *Not inhibited* by EDTA, PMSF or α_1-antitrypsin
- Similar enzymes include: endoproteinase Lys-C (*Lysobacter enzymogenes*) and Ps-1 (*Pseudomonas aeruginosa*)

SPECIFIC ACTIVITY	UNITS DEFINITION	PREPARATION FORM	ADDITIONAL ACTIVITIES	SUPPLIER CATALOG NO.
Achromobacter lyticus				
10 AU/vial 2 AU/vial	Amidase units	Sequencing grade; lyophilized		Wako 129-02541 125-02543
Achromobacter lyticus M497-1				
4 U/mg; 10 U/vial	1 unit forms 1 μmol *p*-nitroaniline/min at pH 9.5, 30°C	Salt-free; lyophilized		ICN 153898
Lysobacter enzymogenes (optimum pH = 8.5-8.8, MW = 33,000 [reduced], 30,000 [non-reduced] Da)				
		Sequencing grade; free of impurities by HPLC, SDS-PAGE using silver staining; lyophilized		Boehringer 1047825 1420429
150 U/mg protein	1 unit converts 1 μmol Chromozym^R PL to product/min at 25°C	Sequencing grade; lyophilized		Boehringer 476986
>150 U/mg protein	1 unit hydrolyzes 1 μmol Tos-Gly-Pro-Lys-*p*NA/min at pH 7.7, 25°C	Lyophilized from buffer containing 100 mM NH$_4$HCO$_3$		Calbiochem 324714
		>90% Purity; lyophilized from buffer containing 10 mM EDTA and 50 mM Tricine, pH 8.0	No other detectable proteases	Calbiochem 324715

Lysyl Endopeptidase continued

3.4.21.50

SPECIFIC ACTIVITY	UNITS DEFINITION	PREPARATION FORM	ADDITIONAL ACTIVITIES	SUPPLIER CATALOG NO.
Lysobacter enzymogenes continued				
30 U/mg	1 unit liberates 1 μmol 4-nitroanilide/min at 25°C with Chromozym® PL as substrate	Lyophilized		Fluka 45175

Tryptase

3.4.21.59

REACTION CATALYZED
  Preferential cleavage: Arg↓, Lys↓, but with more restricted specificity than trypsin

SYNONYMS
  Mast cell tryptase, mast cell protease II, skin tryptase, lung tryptase, pituitary tryptase

REACTANTS
  Peptide bonds

NOTES
- Classified as a hydrolase, acting on peptide bonds: serine endopeptidase
- A tetrameric molecule in mast cell granules with high affinity for heparin
- A serine protease homologue of chymotrypsin
- *Not inhibited* by α_1-proteinase inhibitor or α_2-macroglobulin
- Function is obscure; evidence exists to suggest both intracellular and extracellular roles

SPECIFIC ACTIVITY	UNITS DEFINITION	PREPARATION FORM	ADDITIONAL ACTIVITIES	SUPPLIER CATALOG NO.
Human				
				Vital 17330
Human lung (MW = 135,000 Da)				
≥500 mU/mg protein	1 unit hydrolyzes 1 μmol N-benzyl-DL-Arg-pNA/min at pH 8.0, 25°C	≥95% purity; lyophilized containing 50 mM NaOAc and 500 mM NaCl, pH 5.5	No detectable HBsAg, anti-HCV, anti-HBc, anti-HIV	ART 16-21-201825
5-10 U/mg protein	1 unit hydrolyzes 1 μmol N-benzyl-DL-Arg-pNA/min at pH 8.0, 37°C	>95% purity; 10 mM MES, 200 mM NaCl, pH 6.1	No detectable HIV Ab, HBsAg	Cortex CP3033
3 U/mg protein		>98% purity; lyophilized containing 50 mM NaOAc buffer with 0.5 M NaCl, pH 5.5		ICN 153575

Subtilisin

REACTION CATALYZED
: Hydrolysis of proteins with broad specificity for peptide bonds, and a preference for a large uncharged residue in P1

SYNONYMS
: Alcalase *Novo*, alkaline protease, bacterial proteinase *Novo*, nagarse, nagarse proteinase, subtilisin A, subtilisin B, subtilisin BPN′, subtilisin *Carlsberg*, subtilisin *Novo*, subtilopeptidase A, subtilopeptidase B, subtilopeptidase C

REACTANTS
: Polypeptides, peptide amides

APPLICATIONS
- Hydrolysis of a wide variety of proteins
- Increases solubility of proteinaceous material
- Meat tenderizer
- Cheese flavor developer
- Improves dough texture, flavor and color in cookies, cakes and pretzels
- Improves digestibility of animal feeds
- Reduces viscosity of fish press water, permitting the preparation of solutions having > 50% solids content without forming gels
- Hydrolyzes and solubilizes fabric protein stains (blood, grass, milk, gravy, egg, mucous, feces)
- Treatment for flour in the manufacture of baked goods
- Increases recovery of fish oils
- Reduces buildup of scale in evaporation tubes
- Degrades undesirable gelation buildup in soup stocks
- Transesterification and transpeptidation reactions
- Strips gelatin coatings from film product emulsions for silver recovery
- Silk degumming
- Leather

NOTES
- Classified as a hydrolase, acting on peptide bonds: serine endopeptidase
- A serine protease which evolved independently of chymotrypsin
- Contains no cysteine residues, although homologues do
- Activated by Ca^{2+}
- Resistant to alcohol and other solvents
- Inhibited by EDTA, phosphate buffers, triaminoacetic acid and glucoheptonate (>50°C), DFP, PMSF, soybean trypsin inhibitor, α_2-macroglobulin, indole and phenol
- Similar enzymes are produced by various *Bacillus subtilis* strains and other bacilli

SPECIFIC ACTIVITY	UNITS DEFINITION	PREPARATION FORM	ADDITIONAL ACTIVITIES	SUPPLIER CATALOG NO.
Aspergillus oryzae (optimum pH = 5-9, T = 50°C; inactivated at 75°C, 15-20 min)				
72,720 Azocasein U	Activity is expressed as the amount of soluble dye released from azocasein	Food grade; powder		Genencor Multifect^R P41

SPECIFIC ACTIVITY	UNITS DEFINITION	PREPARATION FORM	ADDITIONAL ACTIVITIES	SUPPLIER CATALOG No.
Bacillus amyloliquefaciens (optimum pH = 8.5, T = 50-55°C; inactivated at T > 60°C)				
7-9 U/mg solid	1 unit hydrolyzes casein producing the color equivalent to 1.0 µmol Tyr/min at pH 7.5, 37°C	4X Crystallized	No detectable DNase, RNase	ICN 101129
≥3000 GSU/mL	Activity determined spectrophotometrically against an internal standard using a synthetic substrate	Food grade; solution	Esterase	Genencor Multifect[R] P-3000 Enzyme
Bacillus lentus, expressed in another *Bacillus* (optimum pH = 10, T = 40°C; stable pH 5-11 [25°C, 24 hr]; inactivated T > 45°C [pH 10.0 10 min])				
≥20,000 GSU/kg ≥38,000 GSU/kg	Activity determined spectrophotometrically against an internal standard using a peptide substrate	Coated granulate with 0.3-0.85 mm particle size		Genencor Purafect[R] 2000G, 4000G with Enzoguard[R]
≥42,000 GSU/L	Activity determined spectrophotometrically against an internal standard using a peptide substrate	Solution, pH 5.8		Genencor Purafect[R] 4000L
Bacillus lentus, expressed in *Bacillus subtilis* (optimum pH = 9, T = 70°C; stable pH 6-12 [25°C, 2 hr], T < 40°C [pH 10.5, 2 hr]				
≥20,000 GxPU/kg ≥38,000 GxPU/kg	Activity determined spectrophotometrically against an internal standard using a peptide substrate	Coated granulate with 0.3-0.85 mm particle size		Genencor Purafect[R] OxP 2000G, 4000G with Enzoguard[R]
Bacillus licheniformis (optimum pH = 10.5, T = 60°C, MW = 27,300 Da; stable pH 5-11)				
150,000 Delft U/g	250,000 Delft units produce an OD_{275} of 1.000 for 1 g/L solution at pH 8.5, 40°C	Solution stabilized with glycerol, pH 6.0-7.0		ABM-RP Proteinase D
350,000 Delft U/g	250,000 Delft units produce an OD_{275} of 1.000 for 1 g/L solution at pH 8.5, 40°C	Microgranulate		ABM-RP Proteinase DS
8 U/mg	1 unit liberates 1 µmol folin-positive amino acids and peptides (as Tyr)/min at pH 7.5, 37°C	Powder	No detectable DNases and RNases	Fluka 82459
43 U/mg	1 unit hydrolyzes 1 µmol $N\alpha$-Tosyl-L-Arg methyl ester hydrochloride/min at pH 8.0, 30°C	Powder		Fluka 85967

SPECIFIC ACTIVITY	UNITS DEFINITION	PREPARATION FORM	ADDITIONAL ACTIVITIES	SUPPLIER CATALOG NO.
Bacillus licheniformis continued				
50 U/mg	1 unit liberates 1 μmol Folin-positive amino acids and peptides (calculated as Tyr)/min at pH 8.5, 25°C with hemoglobin as substrate			Fluka 85968
10,000 Azocasein U	Activity is expressed as the amount of soluble dye released from azocasein	Food grade; solution		Genencor Multifect^R P64
0.6 AU/g		Food grade; solution		Novo Nordisk Alcalase^R 0.6 L
2.4 AU/g		Food grade; solution		Novo Nordisk Alcalase^R 2.4 L
7-15 U/mg solid	1 unit hydrolyzes casein to produce color equivalent to 1.0 μmol (181 μg) Tyr/min at pH 7.5, 37°C	Crystallized; lyophilized	No detectable DNase, RNase	Sigma P5380
75-150 U/g agarose; 1 mL gel yields 2-3 U	1 unit hydrolyzes casein to produce color equivalent to 1.0 μmol (181 μg) Tyr/min at pH 7.5, 37°C	Insoluble enzyme attached to 4% cross-linked beaded agarose; lyophilized containing lactose stabilizer		Sigma P8790
Bacillus species (optimum pH = 8-11, T = 55-60°C)				
150 U/g proteinase		Solution		Biocatalysts Promod 298L P298L
100 U/g proteinase		Solution		Biocatalysts Promod 32L P032L
		Powder or solution	No detectable side activities	Novo Nordisk Esperase^R
		Granulate or solution	No detectable side activities	Novo Nordisk Savinase^R
Bacillus subtilis (optimum pH = 6-8 [10, with 0.6% casein], MW = 30,000 Da; stable pH 5-10 [30°C, 20 hr], T < 55°C [pH 9.0, 30 min])				
5 U/mg solid	1 unit liberates Folin-positive amino acids and peptides corresponding to 1 μmol Tyr/min at pH 8.0, 37°C with casein as substrate	Lyophilized		Boehringer 165905
15,700 Azocasein U	Activity is expressed as the amount of soluble dye released from azocasein	Food grade; powder		Genencor Multifect^R P53
≥3000 GSU/mL	Activity determined spectrophotometrically against an internal standard using a synthetic substrate	Solution		Genencor Protease 899

SPECIFIC ACTIVITY	UNITS DEFINITION	PREPARATION FORM	ADDITIONAL ACTIVITIES	SUPPLIER CATALOG NO.
Bacillus subtilis continued				
150,000 PUN/mL	1 PUN liberates the digestion product not precipitated with TCA giving the same Folin color as 1 µg Tyr/min at pH 7.5, 30°C	Purified solution		Nagase Bioprase APL-30
25,000 PUN/mL	1 PUN liberates the digestion product not precipitated with TCA giving the same Folin color as 1 µg Tyr/min at pH 7.5, 30°C	Purified solution		Nagase Bioprase XL-416
445 mU$_{Hb}$/mg	pH 7.5	Food grade; solution		Rohm Corolase^R 7089
757 mU$_{Hb}$/mg	pH 7.5	Food grade; powder		Rohm Veron^R MBS
27 mU$_{Hb}$/mg	pH 7.5	Food grade; powder		Rohm Veron^R P
49 mU$_{Hb}$/mg	pH 7.5	Food grade; powder		Rohm Veron^R W
Bacillus subtilis var. biotecus A				
20 U/mg	1 unit liberates 1 µmol folin-positive amino acids and peptides (as Tyr)/min at pH 7.5, 35.5°C with hemoglobin as substrate	Lyophilized		Fluka 82490
Bacterial				
3 U/mg	1 unit hydrolyzes 1 µmol N-carbobenzoxy-Gly-4-nitrophenylester/min at pH 6.0, 25°C	Powder	<0.1% RNase, DNase	Fluka 82518
10 U/mg	1 unit hydrolyzes 1 µmol N-carbobenzoxy-Gly-4-nitrophenylester/min at pH 6.0, 25°C	Powder		Fluka 82528
7-14 U/mg solid	1 unit hydrolyzes casein to produce color equivalent to 1.0 µmol (181 µg) Tyr/min at pH 7.5, 37°C	Crystallized; lyophilized	Substantially free of DNase, RNase	Sigma P4789
75-150 U/g agarose; 1 mL gel yields 3-6 U	1 unit hydrolyzes casein to produce color equivalent to 1.0 µmol (181 µg) Tyr/min at pH 7.5, 37°C	Coupled through an amide linkage to epichlorohydrin-activated 4% beaded agarose; suspension in 50% glycerol, 0.15 M NaCl, 0.01 M NaPO$_4$, 0.05 M phenylboronic acid, 0.02% NaN$_3$, pH 7.5		Sigma P8298

Endopeptidase K

REACTION CATALYZED
Hydrolysis of keratin and other proteins with subtilisin-like specificity

SYNONYMS
Tritirachium alkaline proteinase, proteinase K, PROK

REACTANTS
Keratin, peptide amides, polypeptides, peptide bonds, hemoglobin (denatured)

APPLICATIONS
- Removes proteins in the isolation of intact, native high molecular weight eukaryotic nucleic acids
- Specific modification of cell surface proteins and glycoproteins for analysis of membrane structure
- Rapid proteolytic inactivation of endogenous nucleases during isolation of mRNA and high molecular weight DNA
- Solubilizes membrane-bound enzymes, e.g., acetyl cholinesterase from *Torpedo marmorata*

NOTES
- Classified as a hydrolase, acting on peptide bonds: serine endopeptidase
- A highly active and stable serine protease with broad specificity, cleaving on the C-side of hydrophobic, aliphatic and aromatic amino acids
- From the mold *Tritirachium album limber*
- A homologue of subtilisin containing two disulfide bridges and one free Cys near the active site His
- Active against nitroanilides of amino acids with protected amino groups (excluding Arg)
- Inactivates mammalian DNase and RNase with SDS, urea or DTT
- Ca^{2+} dependent
- Inhibited by PMSF, DFP and Hg^{2+}
- *Not inactivated* by metal chelators, sulfhydryl reagents, PCMB, TLCK, TPCK or SDS(< 5 mg/mL)

SPECIFIC ACTIVITY	UNITS DEFINITION	PREPARATION FORM	ADDITIONAL ACTIVITIES	SUPPLIER CATALOG NO.
Tritirachium album (optimum pH = 7.5-12, pI = 8.9, T = 65°C, MW = 28,800 Da; most stable at pH 8.0)				
20 U/mg solid	1 unit liberates Folin-positive amino acids and peptides equivalent to 1 µmol Tyr/min at pH 7.5, 37°C using hemoglobin as substrate	Lyophilized	<30 Kunitz U DNase/mg solid <0.003 Kunitz U RNase/mg solid	Adv Biotech AB-0504 AB-0504b
20 U/mg solid	1 unit liberates Folin-positive amino acids and peptides corresponding to 1 µmol Tyr/min at 37°C using hemoglobin as substrate	Lyophilized	No detectable RNase, DNase	Amersham E 76230Y E 76230Z
32 U/mg	1 unit releases 1 µmol *p*-nitroaniline/min at 25°C	Lyophilized; Solution		AMRESCO 0706 E634 E195

SPECIFIC ACTIVITY	UNITS DEFINITION	PREPARATION FORM	ADDITIONAL ACTIVITIES	SUPPLIER CATALOG NO.
Tritirachium album continued				
30 U/mg protein	1 unit liberates Folin-positive amino acids and peptides corresponding to 1 μmol Tyr/min at pH 7.5, 37°C using hemoglobin as substrate	Chromatographically purified	No detectable RNase, DNase	AGS Heidelb F01080 F01081 F01082
30 U/mg; >600 U/mL (14-22 mg/mL) volume activity	1 unit liberates Folin-positive amino acids and peptides corresponding to 1 μmol Tyr/min at 37°C with Hb as substrate	Solution		Boehringer 1373196 1373200 1413783
20 U/mg solid	1 unit liberates Folin-positive amino acids and peptides corresponding to 1 μmol Tyr/min at 37°C with Hb as substrate	Lyophilized		Boehringer 161519 745723 1000144 1092766
>20 U/mg dry weight	1 unit liberates 1 μmol Tyr/min at 37°C	Lyophilized	<0.01 U/mg DNase, RNase	Calbiochem 539480
290 U/mg	1 unit hydrolyzes 1 μmol *N*-Ac-*L*-Tyr-ethylester (ATEE)/min at pH 9.0, 30°C	Powder		Fluka 82495
30 U/mg proteolytic activity (Anson assay)	1 unit liberates 1 μmol Folin-positive amino acids/min at pH 7.5, 35°C using hemoglobin as substrate	Chromatographically purified; >80% protein	<0.0005 U/mg RNase, DNase	ICN 193504
	1 unit releases the equivalent of 1 μmol Tyr in a standard Folin assay/min using hemoglobin as substrate	Lyophilized	No detectable endonuclease, exonuclease, RNase	NBL Gene 073002 073003
30 U/mg (mAnson units with Hb as substrate)		Molecular Biology Grade; lyophilized	No detectable nuclease	Oncor 130201 130202 130203
10-20 U/mg protein	1 unit hydrolyzes casein to produce color equivalent to 1.0 μmol Tyr/min at pH 7.5, 37°C	Lyophilized		Sigma P0390
10-20 U/mg protein	1 unit hydrolyzes casein to produce color equivalent to 1.0 μmol (181 μg) Tyr/min at pH 7.5, 37°C	Lyophilized containing 80% protein	No detectable nuclease activity after 200 μg/mL incubated in endonuclease (nickase), endonuclease-exonuclease (non-radioactive) or RNase (non-radioactive) assays supplemented with 0.2% SDS and 10 m*M* EDTA	Sigma P2308

SPECIFIC ACTIVITY	UNITS DEFINITION	PREPARATION FORM	ADDITIONAL ACTIVITIES	SUPPLIER CATALOG NO.
Tritirachium album continued				
10-20 U/mg protein	1 unit hydrolyzes casein to produce color equivalent to 1.0 μmol (181 μg) Tyr/min at pH 7.5, 37°C	Lyophilized containing 80% protein	No detectable nuclease activity after 200 μg/mL incubated in endonuclease (nickase), endonuclease-exonuclease (non-radioactive) or RNase (non-radioactive) assays supplemented with 0.2% SDS and 10 mM EDTA	Sigma P4914
10-20 U/mg protein	1 unit hydrolyzes casein to produce color equivalent to 1.0 μmol Tyr/min at pH 7.5, 37°C	Aseptic; lyophilized		Sigma P5056
140 U/mL	1 unit hydrolyzes casein to produce color equivalent to 1.0 μmol Tyr/min at pH 7.5, 37°C	Solution containing 40% glycerol, 10 mM Tris-HCl, 1 mM CaOAc, pH 7.5		Sigma P5568
10-20 U/mg protein	1 unit hydrolyzes casein to produce color equivalent to 1.0 μmol Tyr/min at pH 7.5, 37°C	Lyophilized containing 80% protein	<30 Kunitz U/mg solid DNase <0.003 Kunitz U/mg solid RNase	Sigma P6556
1-3 U/mg solid	1 unit hydrolyzes casein to produce color equivalent to 1.0 μmol Tyr/min at pH 7.5, 37°C	Crude; lyophilized containing 30% protein		Sigma P8044
15-22 U/mg				Wako 160-14001 166-14003
Tritirachium album limber (MW = 27,000 Da)				
≥20 U/mg solid	1 unit releases 1 μmol Folin-positive amino acids, measured as Tyr, at pH 7.5, 37°C using urea denatured Hb as substrate	Highly purified; lyophilized	<0.001% DNase, RNase	Worthington LS04220 LS04222 LS04224 LS04226
Unspecified				
20 mg/mL	1 unit produces 1 μmol Folin-positive amino acid/min at 37°C	Powder or solution		Life Technol 25530-015 25530-031 25530-049

Thermitase   3.4.21.66

REACTION CATALYZED
  Hydrolysis of proteins, including collagen

SYNONYMS
  Thermophilic *Streptomyces* serine proteinase

REACTANTS
  Peptides, collagen

APPLICATIONS
- Preparing DNA samples from blood or in agarose plugs for pulsed field gel electrophoresis
- Preparing crude cell lysates suitable for RT-PCR

NOTES
- Classified as a hydrolase, acting on peptide bonds: serine endopeptidase
- A serine protease homologue of subtilisin from *Thermoactinomyces vulgaris*
- Contains a single Cys near the active site His
- Inhibited by PMCB
- Ca^{2+} binding and high thermostability relative to subtilisin are due to an N-terminal extension of the polypeptide chain
- Similar in amino acid composition and properties to the thermostable enzyme from *Streptomyces rectus* var. *proteolyticus*

SPECIFIC ACTIVITY	UNITS DEFINITION	PREPARATION FORM	ADDITIONAL ACTIVITIES	SUPPLIER CATALOG NO.
***Thermus* species strain Rt41A (optimum T = 75°C)**				
0.3 U/μL	1 unit solubilizes 1 A_{420} unit from azocasein/hr at 75°C	Purity by SDS-PAGE		Life Technol 18061-010

Protein C (activated)   3.4.21.69

REACTION CATALYZED
  Degradation of blood coagulation Factors Va and VIIIa

REACTANTS
  Factor Va, Factor VIIIa

NOTES
- Classified as a hydrolase, acting on peptide bonds: serine endopeptidase
- One of the γ-carboxyglutamic acid-containing coagulation factors
- Formed by activation of protein C (the proenzyme that circulates in plasma) with the thrombin/thrombomodulin complex or snake venom serine endopeptidase
- Inhibits blood coagulation through selective inactivation of Coagulation Factors Va and VIIa
- Completely reduced by incubation with 2-mercaptoethanol

3.4.21.69 Protein C (activated) continued

SPECIFIC ACTIVITY	UNITS DEFINITION	PREPARATION FORM	ADDITIONAL ACTIVITIES	SUPPLIER CATALOG NO.
Human plasma (MW = 56,000 Da)				
20-40 μM	Baxter Protein C chromogenic substrate	Frozen in concentrated form	No detectable HIV Ab, HsAg	EnzymeRes APC

3.4.21.70 Pancreatic Endopeptidase E

REACTION CATALYZED
    Preferential cleavage of Ala-↓-

SYNONYMS
    Cholesterol-binding proteinase

REACTANTS
    Peptide bonds

APPLICATIONS
- Protein hydrolysis

NOTES
- Classified as a hydrolase, acting on peptide bonds: serine endopeptidase
- Does not hydrolyze elastin
- A serine protease homologue of chymotrypsin from pancreatic juice
- Distinguished from elastases by an acidic pI
- Binds cholesterol

SPECIFIC ACTIVITY	UNITS DEFINITION	PREPARATION FORM	ADDITIONAL ACTIVITIES	SUPPLIER CATALOG NO.
Pancreas (optimum pH = 6-8.5)				
209,220 LVE/g		Food grade; powder		Rohm Corolase^R PP

Pancreatic Elastase II 3.4.21.71

REACTION CATALYZED
  Preferential cleavage of Leu-↓-, Met-↓- and Phe-↓-
SYNONYMS
  Human pancreatic elastase
REACTANTS
  Elastin

NOTES
- Classified as a hydrolase, acting on peptide bonds: serine endopeptidase
- Formed by activation of mammalian pancreas proelastase II with trypsin
- Only one pancreatic elastase is typically expressed in a given species

---

See Chapter 5. Multiple Enzyme Preparations

IgA-Specific Serine Endopeptidase 3.4.21.72

REACTION CATALYZED
  Cleavage of immunoglobulin A at Pro-↓- bonds in the hinge region
SYNONYMS
  IgA protease
REACTANTS
  Peptide bonds
APPLICATIONS
- Cleavage of fusion protein if the amino acid sequence begins with X-Pro, where X = Thr, Ser or Ala

NOTES
- Classified as a hydrolase, acting on peptide bonds: serine endopeptidase
- No small molecule substrates are known
- Species variants differing slightly in specificity are secreted by the Gram negative bacteria *Neisseria gonorrhoeae* and *Haemophilus influenzae*
- A distant serine protease homologue of chymotrypsin
- Bacterial IgA-specific metalloendopeptidase (EC 3.4.24.13) has similar specificity

SPECIFIC ACTIVITY	UNITS DEFINITION	PREPARATION FORM	ADDITIONAL ACTIVITIES	SUPPLIER CATALOG NO.
Neisseria gonorrhoeae, from *E. coli*				
		Special quality for fusion of proteins; highly purified by SDS-PAGE; solution		Boehringer 1461265 1461273

μ-Plasminogen Activator

REACTION CATALYZED
Specific cleavage of Arg↓Val bond in plasminogen to form plasmin (fibrinolysin)

SYNONYMS
Urokinase, urinary plasminogen activator, cellular plasminogen activator

REACTANTS
Plasminogen, plasmin, fibrinolysin

NOTES
- Classified as a hydrolase, acting on peptide bonds: serine endopeptidase
- Formed from inactive precursor by plasmin or plasma kallikrein
- Differs in structure from t-plasminogen activator (EC 3.4.21.68)
- Does not bind to fibrin

SPECIFIC ACTIVITY	UNITS DEFINITION	PREPARATION FORM	ADDITIONAL ACTIVITIES	SUPPLIER CATALOG NO.
Human kidney cells				
12 U/g	1 unit activates that amount of plasminogen which hydrolyzes 1 μmol Tos-Gly-Pro-Lys-4-NA.AcOH (=Chromozym® PL)/min at pH 8.2, 25°C	Powder		Fluka 94420
0.3-1.2 U/mg protein	1 unit activates that amount of porcine plasminogen which produces a ΔA_{275} of 1.0 per mL/min at pH 7.5, 37°C when measuring perchloric acid soluble products from α-casein (1 cm light path)	Lyophilized containing 70% protein (90-95% is human albumin); balance mannitol and NaCl		Sigma U5004
20-30 U/mL	1 unit activates amount of porcine plasminogen which produces a ΔA_{275} of 1.0 per mL/min at pH 7.5, 37°C when measuring perchloric acid soluble products from α-casein (1 cm light path)	Solution containing 2% aqueous NaCl and 1 mg/mL protein		Sigma U8627
Human urine				
60,000 U/mg protein		High MW (54,000 Da); >95% purity; lyophilized from buffer containing 50 mM Tris-HCl, 100 mM NaCl, 200 mM mannitol, 0.1% PEG, 0.2% NaN₃, pH 7.5 pI = 8.5-8.9	<5% low MW urokinase	Calbiochem 672081
>50,000 U/mg protein	1 unit is equal to an international standard; activity is tested by the fibrinolytic method of Johnson, A.J., et al. 1969	Low MW (33,000 Da); >95% purity; lyophilized containing 1 mg KPO₄ and 10 mg D-mannitol, pH 7.7 pI = 8.9	<5% high MW urokinase	Calbiochem 672101

μ-Plasminogen Activator continued 3.4.21.73

Human urine continued

SPECIFIC ACTIVITY	UNITS DEFINITION	PREPARATION FORM	ADDITIONAL ACTIVITIES	SUPPLIER CATALOG No.
100,000 U/mg protein	1 unit is equal to an international standard; activity is tested by the fibrinolytic method of Johnson, A.J., et al. 1969 (1 unit = 1 CTA unit = 0.7 Plough units)	Lyophilized from buffer containing 50 mM Tris-HCl, 0.1 M NaCl, 20 mM mannitol, 0.1% PEG, 0.2% NaN$_3$		Calbiochem 672112
160,000 IU/mg protein; 3000 IU/vial		MW = 33,000 Da; >98% purity by SDS-PAGE; lyophilized containing 10 mg mannitol and 1 mg PO$_4$	<2% high MW form	Cortex CP4014
80,000 IU/mg; 3000 IU/vial		MW = 54,000 Da; >98% purity by SDS-PAGE; lyophilized containing 0.1 M NaCl, 50 mM Tris-HCl, 0.1% PEG, 20 mM mannitol, 0.2% NaN$_3$, pH 7.5	No detectable HIV I, HIV II, HCV Ab, HBsAg	Cortex CP4016
	1 unit activates that amount of porcine plasminogen which produces a ΔA_{275} of 1.0 per mL/min at pH 7.5, 37°C when measuring perchloric acid soluble products from α-casein (1 cm light path)	Lyophilized		Sigma U1131

Venombin A 3.4.21.74

REACTION CATALYZED
   Selective cleavage of Arg↓ bond in fibrinogen, to form fibrin and release fibrinopeptide A

SYNONYMS
   Ancrod (*Agkistrodon rhodostoma* [Malayan pit viper]), batroxobin and thrombocytin (*Bothrops atrox* [South American pit viper]), crotalase (*Crotalus adamanteus* [Eastern diamondback rattlesnake])

REACTANTS
   Peptide bonds

NOTES
- Classified as a hydrolase, acting on peptide bonds: serine endopeptidase
- The specificity of further fibrinogen degradation varies with enzyme biological source
- Does not require activation by Ca^{2+}
- A thrombin-like enzyme from venoms of snakes of the viper/rattlesnake group
- A serine protease homologue of chymotrypsin
- Activates platelets but lacks fibrinogen clotting activity
- Not to be confused with serotonin, also called thrombocytin

SPECIFIC ACTIVITY	UNITS DEFINITION	PREPARATION FORM	ADDITIONAL ACTIVITIES	SUPPLIER CATALOG NO.
Agkistrodon rhodostoma (Malayan pit viper venom)				
200 NIH U/mg protein	1 NIH unit equals 3.2 WHO units	Lyophilized		Sigma A5042
Bothrops atrox (venom)				
5 mU/mg solid	1 mU hydrolyzes 1.0 nmol Tos-Gly-Pro-Arg-pNA/min at pH 8.4, 37°C	Lyophilized		Calbiochem 605162
5000 U/mg solid	1 unit releases 1.0 nmol p-nitroaniline from N-p-tosyl-Gly-Pro-Arg p-nitroanilide/min at pH 8.4, 37°C	Lyophilized containing 40% protein; balance primarily Lys hydrochloride and NaCl		ICN 156910
5000 U/mg solid	1 unit releases 1.0 nmol p-nitroaniline from N-p-tosyl-Gly-Pro-Arg p-nitroanilide/min at pH 8.4, 37°C	Lyophilized containing 40% protein; balance primarily Lys-HCl and NaCl		Sigma T2643
Crotalus adamanteus (venom)				
100-300 U/mg protein	Activity expressed in NIH units obtained by direct comparison to NIH thrombin reference standard	After reconstitution with 1 mL water, vial contains indicated units in 0.1 M NaOAc, pH 7.0		Sigma C6397

3.4.22.1 Cathepsin B

REACTION CATALYZED
   Preferentially cleaves -Arg-Arg↓ bonds in small molecule substrates. Also shows peptidyl-dipeptidase activity, liberating C-terminal dipeptides

REACTANTS
   Polypeptides, dipeptides

NOTES
- Classified as a hydrolase, acting on peptide bonds: cysteine endopeptidase
- Located in lysosomes and involved in tissue degradation and restructuring
- Believed to be involved in intracellular digestion of extracellular protein taken up by endocytosis
- Different from cathepsin L (EC 3.4.22.15)
- A cysteine protease with broad specificity for peptide bonds

SPECIFIC ACTIVITY	UNITS DEFINITION	PREPARATION FORM	ADDITIONAL ACTIVITIES	SUPPLIER CATALOG NO.
Bovine spleen (optimum pH = 3-4)				
10 U/mg	1 unit hydrolyzes 1 µmol $N\alpha$-carbobenzoxy-L-Lys-4-nitrophenyl ester/min at pH 5.0, 25°C	Powder		Fluka 22131
18-35 U/mg protein	1 unit hydrolyzes 1 µmol $N\alpha$-CBZ-Lys p-nitrophenyl ester/min at pH 5.0, 25°C	Lyophilized containing 40% protein; balance primarily $NaPO_4$, NaCl, 6% EDTA as stabilizer		Sigma C6286
Human				
				Vital 17022
Human liver (optimum pH = 3.5-6.0, MW = 27,500 Da)				
50-200 U/mg protein	1 unit hydrolyzes 1 µmol Z-Arg-Arg-β-NA/min at pH 6.0, 40°C in the presence of DTT	≥95% purity; frozen in 50 mM NaOAc and 1 mM EDTA, pH 5.0	No detectable HBsAg, anti-HCV, anti-HBc, anti-HIV	ART 16-12-030102
50-200 U/mg protein	1 unit hydrolyzes 1 µmol Z-Arg-Arg-βNA (1-naphtylamine)/min at pH 6.0, 40°C	>95% purity; solution containing 20 mM NaOAc buffer, 100 mM NaCl, 1 mM EDTA, pH 5.0	No detectable HBsAg, HIV Ab	Calbiochem 219364
50 µL/vial; 50 µg/103 µL		>95% purity by SDS-PAGE; solution containing 20 mM NaOAc and 1 mM EDTA, pH 5.0	No detectable HIV Ab, HBsAg	Cortex CP3009
		>95% purity; solution containing 25 mM NaOAc buffer, 0.1 mM EDTA, 0.02% NaN_3; 50% glycerol, pH 5.1		ICN 191343
Human placenta				
3-12 U/mg protein	1 unit hydrolyzes 1 µmol $N\alpha$-CBZ-Lys p-nitrophenyl ester/min at pH 5.0, 25°C	Lyophilized containing 50% protein; balance PO_4 buffer salts		Sigma C0150

Papain — 3.4.22.2

REACTION CATALYZED

Hydrolysis of proteins with a preference for residues bearing a large hydrophobic side chain at the P2 position

SYNONYMS

Papaya peptidase I, calotropin (*Calotropis gigantea* [madar plant]), phytolacin (*Phytolacca americana* [pokeweed]), mexicanain (*Pileus mexicanus*)

REACTANTS

Polypeptides, benzoyl-L-arginine ethyl ester

APPLICATIONS

- General uses: animal feed, brewing and distilling, wine and fruit juice, industrial cleaning, dairy products, leather and wool, cosmetics and bath powders
- Solubilization of integral membrane proteins
- Production of glycopeptides from purified proteoglycans
- Release and determination of synthetic colors in foods
- Peptide mapping by limited proteolysis
- Solubilization of oat protein
- Preparation of Fab fragments of IgG antibodies
- Meat tenderizing and extracts; fish solubilization

NOTES

- Classified as a hydrolase, acting on peptide bonds: cysteine endopeptidase
- A cysteine protease from latex of the papaya (*Carica papaya*)
- Has wide specificity, degrading most protein substrates more extensively than pancreatic proteases; functions also as an esterase
- Hydrolyzes peptide, amide and ester bonds of amino acids and peptides; especially bonds involving Arg, Lys, Glu, His, Gly and Tyr Additional bonds are cleaved on prolonged incubation. Does not accept Val at P1'
- Activated by cysteine, sulfide, sulfite, heavy metal chelating agents like EDTA, and *N*-bromosuccinimide
- Inhibited by PMSF, TLCK, TPCK, α_2-macroglobulin, Hg^{2+} and other heavy metals, AEBSF, antipain, cystatin, E-64, leupeptin, sulfhydryl binding agents, carbonyl reagents and alkylating agents
- Inactivated by H_2O_2 generated by γ-irradiation of H_2O as a result of oxidation of the active SH group to sulfenic acid
- Stabilized by EDTA, cysteine and dimercaptoethanol
- Many plants contain cysteine endopeptidase homologues of papain with similar specificities

Papain continued

SPECIFIC ACTIVITY	UNITS DEFINITION	PREPARATION FORM	ADDITIONAL ACTIVITIES	SUPPLIER CATALOG No.
Carica papaya (optimum pH = 6.0-7.0, pI = 8.75, T = 55-70°C, MW = 23,000 Da; stable pH 6-7.5 [0.1% Cys, 40°C, 20 hr], relatively heat stable)				
150 XS/g 240 XS/g	1 unit produces 465 µg Tyr equivalent of aromatic nitrogen compounds soluble in 4.5% TCA/min at pH 6.5, 35°C	Powder		ABM-RP Scintillase 150D Scintillase 240D
150 XS/g 300 XS/g	1 unit produces 465 µg Tyr equivalent of aromatic nitrogen compounds soluble in 4.5% TCA/min at pH 6.5, 35°C	Solution stabilized with corn syrup, SO_2 added as a preservative pH 4.2-5.0	Esterase, thiolesterase, transamidase, transesterase	ABM-RP Scintillase CS 150L Scintillase CS 300L
100 TU/g proteinase	Tyr	Solution		Biocatalysts Promod 144L P144L
100-800 TU/g proteinase	Tyr	Powder		Biocatalysts Promod 144P P144P
30 U/mg	1 unit converts 1 µmol BAEE to product/min at 25°C	Crystallized; suspension	Esterase and transaminase	Boehringer 108014 1693379
30,000 USP U/mg		Crystallized		Calbiochem 5125
≥6000 U/mg		Powder		Difco 0253-17-5
6-10 mg enzyme/mL gel; 75-125 U/mL	1 unit hydrolyzes 1 µmol BAEE/min at pH 6.2, 25°C	2X Crystallized; attached to beaded agarose in 0.02 *M* NaOAc and 0.01% NaN_3, pH 5		Elastin A431
6-10 mg enzyme/mL gel; 75-125 U/mL	1 unit hydrolyzes 1 µmol BAEE/min at pH 6.2, 25°C	2X Crystallized; attached to beaded agarose in 0.02 *M* NaOAc and 0.01% NaN_3, pH 5		Elastin AD431
6-10 mg enzyme/mL gel; 75-125 U/mL	1 unit hydrolyzes 1 µmol BAEE/min at pH 6.2, 25°C	2X Crystallized; attached to beaded agarose in 0.02 *M* NaOAc and 0.01% NaN_3, pH 5		Elastin AE431
20-30 U/mg protein	1 unit hydrolyzes 1 µmol BAEE/min at pH 6.2, 25°C	2X Crystallized; suspension in 0.05 *M* NaOAc and 0.15 *M* NaCl, 0.01% NaN_3, pH 5		Elastin P66
15-30 U/mg protein	1 unit hydrolyzes 1 µmol BAEE/min at pH 6.2, 25°C	2X Crystallized; dialyzed; lyophilized		Elastin PL66
		Food grade		EnzymeDev Enzeco[R] Chillproof

SPECIFIC ACTIVITY	UNITS DEFINITION	PREPARATION FORM	ADDITIONAL ACTIVITIES	SUPPLIER CATALOG NO.
Carica papaya continued				
		Food grade		EnzymeDev Enzeco[R] Purified Papain
		Food grade		EnzymeDev Liquipanol[R] Liquid Papain
		Food grade		EnzymeDev Panol[R] Purified Papain
20 U/mg protein	1 unit hydrolyzes 1 μmol N-benzoyl-L-Arg ethyl ester/min at pH 6.2, 25°C	Crystallized; suspension		Fluka 76216
12 U/mg	1 unit hydrolyzes 1 μmol N-benzoyl-L-Arg ethyl ester/min at pH 6.2, 25°C	Powder		Fluka 76218
3 U/mg	1 unit hydrolyzes 1 μmol N-benzoyl-L-Arg ethyl ester/min at pH 6.2, 25°C	Powder		Fluka 76220
0.5 U/mg	1 unit hydrolyzes 1 μmol N-benzoyl-L-Arg ethyl ester/min at pH 6.2, 25°C	Powder		Fluka 76222
10-20 U/mg protein	1 unit hydrolyzes 1.0 mmol Nα-benzoyl-L-Arg ethyl ester (BAEE)/min at pH 6.2, 25°C	2X Crystallized; suspension in 0.05 M NaOAc		ICN 100921
15-40 U/mg protein	1 unit hydrolyzes 1.0 μmol Nα-benzoyl-L-Arg ethyl ester (BAEE)/min at pH 6.2, 25°C	2X Crystallized; suspension in 0.05 M NaOAc, pH 4.5		ICN 100924
$\geq$1750 USP U/mg		Technical grade; crude		ICN 102566
2 mg enzyme/mL gel		Purified; immobilized to agarose; suspension in 50 mM NaOAc, pH 4.5		ICN 191290
100,000 PaUN/g solid	1 PaUN liberates the digestion product not precipitated with TCA giving the same A_{275} as 1 μg Tyr/min at pH 7.5, 30°C	Powder		Nagase Papain
757 mU$_{Hb}$/mg	pH 7.5	Food grade; solution		Rohm Veron[R] L10
16-40 U/mg protein	1 unit hydrolyzes 1.0 μmol BAEE/min at pH 6.2, 25°C	2X Crystallized; suspension in 0.05 M NaOAc and 0.01% thymol, pH 4.5		Sigma P3125
1-2 U/mg solid	1 unit hydrolyzes 1.0 μmol BAEE/min at pH 6.2, 25°C	Crude; dried just as received from Africa; not cleaned or purified		Sigma P3250
1.5-3.5 U/mg solid	1 unit hydrolyzes 1.0 μmol BAEE/min at pH 6.2, 25°C	Crude; not stabilized with lactose or other adulterants		Sigma P3375

SPECIFIC ACTIVITY	UNITS DEFINITION	PREPARATION FORM	ADDITIONAL ACTIVITIES	SUPPLIER CATALOG NO.
Carica papaya continued				
3000-5000 U/g agarose; 1 mL gel yields 90-150 U	1 unit hydrolyzes 1.0 µmol BAEE/min at pH 7.0, 30°C	Insoluble enzyme attached to beaded agarose; lyophilized containing lactose stabilizer		Sigma P4406
10-20 U/mg protein	1 unit hydrolyzes 1.0 µmol BAEE/min at pH 6.2, 25°C	2X Crystallized; lyophilized containing 80% protein; balance primarily NaCl and NaOAc salts		Sigma P4762
10-20 U/mg protein	1 unit hydrolyzes 1.0 µmol BAEE/min at pH 6.2, 25°C	2X Crystallized; aseptically filled; lyophilized containing 80% protein; balance primarily NaCl and NaOAc salts		Sigma P5306
150-250 U/g solid	1 unit hydrolyzes 1.0 µmol BAEE/min at pH 7.0, 30°C	Insoluble enzyme attached to CMC containing 30% borate buffer salts		Sigma P8011
16-40 BAEE U/mg protein	1 unit hydrolyzes 1.0 µmol BAEE/min at pH 6.2, 25°C	Crystallized; suspension in 70% EtOH		Sigma P9886
		Crude		SpecialtyEnz
		Purified		SpecialtyEnz
3000 USP U/mg		Crude; powder		Wako 164-00172 166-00171
Activates on removal of mercury to ≥10 U/mg protein	1 unit hydrolyzes 1 µmol benzoyl-L-Arg ethyl ester/min at pH 6.2, 25°C after activation	Crystallized mercuripapain; suspension in 70% EtOH	No detectable free mercury	Worthington LS02487 LS02489 LS02491
Activates to ≥15 U/mg protein	1 unit hydrolyzes 1 µmol benzoyl-L-Arg ethyl ester/min at pH 6.2, 25°C after activation	2X Crystallized; lyophilized containing NaOAc		Worthington LS03118 LS03119 LS03120 LS03122
Activates to ≥20 U/mg protein	1 unit hydrolyzes 1 µmol benzoyl-L-Arg ethyl ester/min at pH 6.2, 25°C after activation	2X Crystallized; sterile-filtered; suspension in 0.05 M NaOAc, pH 4.5		Worthington LS03124 LS03126 LS03127 LS03128

3.4.22.3 Ficain

REACTION CATALYZED
  Hydrolysis of proteins with a preference for residues bearing a large hydrophobic side chain at the P2 position

SYNONYMS
  Ficin

REACTANTS
  Polypeptides, gelatin, collagen, milk protein, hemoglobin, elastin, soy protein, fibrin, fibrinogen, living ascaris

NOTES
- Classified as a hydrolase, acting on peptide bonds: cysteine endopeptidase
- The major proteolytic component of the latex of fig (*Ficus glabrata*)
- Hydrolyzes peptide, ester and amide bonds at the C-side of Gly, Ser, Thr, Met, Lys, Arg, Tyr, Ala, Asn and Val
- Does not accept Val at P1'
- Inhibited by DFP, TPCK, TLCK, α_2-macroglobulin, iodoacetic acid, mercuric chloride, E-64, NEM and cystatin
- A cysteine protease similar to papain
- The large genus *Ficus* contains cysteine endopeptidases with similar properties

SPECIFIC ACTIVITY	UNITS DEFINITION	PREPARATION FORM	ADDITIONAL ACTIVITIES	SUPPLIER CATALOG No.
Ficus carica (fig tree)				
3 U/mg	1 unit converts 1 μmol BAEE to product/min at 25°C	Suspension in NaOAc		Boehringer 1520822
1-2 U/mg protein	1 unit increases A_{400} by 1 unit/min	Lyophilized containing 90% protein		Calzyme 187A0001
750-1000 milk clotting U/g	1 unit liberates 1 mg Tyr from purified hemoglobin/10 min at 25°C			ICN 101688
500-1000 U/g of agarose; 1 mL gel yields 15-30 U	1 unit produces a ΔA_{280} of 1.0/min at 37°C when measuring TCA soluble products from casein in a 10 mL volume	Insoluble enzyme attached to beaded agarose; lyophilized containing lactose as stabilizer		Sigma F1634
0.25-0.75 U/mg solid	1 unit produces a ΔA_{280} of 1.0/min at 37°C when measuring TCA soluble products from casein in a 10 mL volume	Powder containing 90% protein		Sigma F3266
1.5-2.5 U/mg protein	1 unit produces a ΔA_{280} of 1.0/min at 37°C when measuring TCA soluble products from casein in a 10 mL volume	Suspension in 2.0 *M* NaCl and 0.03 *M* Cys, pH 5.0		Sigma F4125
1-2 U/mg protein	1 unit produces a ΔA_{280} of 1.0/min at 37°C when measuring TCA soluble products from casein in a 10 mL volume	Lyophilized containing 90% protein; balance primarily L-Cys and trace EDTA		Sigma F6008

Chymopapain

REACTION CATALYZED
Hydrolysis of proteins with a preference for residues bearing a large hydrophobic side chain at the P2 position

REACTANTS
Polypeptides, benzoyl-L-arginine ethyl ester (BAEE)

NOTES
- Classified as a hydrolase, acting on peptide bonds: cysteine endopeptidase
- The major endopeptidase from the latex of papaya (*Carica papaya*)
- Hydrolyzes a wide variety of substrates, similar to papain, but at slower rates
- Named because of its higher ratio of milk clotting-to-proteolytic activity when compared with papain; clotting capacities of the two enzymes are equal
- Does not accept Val at P1′
- A cysteine protease with multiple chromatographic forms

SPECIFIC ACTIVITY	UNITS DEFINITION	PREPARATION FORM	ADDITIONAL ACTIVITIES	SUPPLIER CATALOG NO.
Papaya latex				
0.5-2.0 U/mg protein	1 unit hydrolyzes 1.0 μmol BAEE to $N\alpha$-benzoyl-L-Arg/min at pH 6.2, 25°C	Chromatographically purified; lyophilized containing 75% protein; balance primarily L-Cys and EDTA		Sigma C8526
2-5 U/mg protein	1 unit hydrolyzes 1.0 μmol BAEE to $N\alpha$-benzoyl-L-Arg/min at pH 6.2, 25°C	Partially purified; powder containing 90% protein and 10% buffer salts as NaOAc and EDTA		Sigma C9007
Activates to ≥3 U/mg protein	1 unit hydrolyzes 1 μmol benzoyl-L-Arg ethyl ester/min at pH 6.2, 25°C after activation	Partially purified; lyophilized	Papain, lysozyme	Worthington LS03132 LS03130

Clostripain

REACTION CATALYZED
  Preferential cleavage: Arg↓, including Arg↓Pro, but not Lys-

SYNONYMS
  Clostridiopeptidase B, endoproteinase Arg-C

REACTANTS
  Peptide bonds, benzoyl-L-arginine ethyl ester (BAEE)

APPLICATIONS
- Peptide mapping, fingerprinting and sequence analysis
- Hydrolysis/condensation of amide bonds

NOTES
- Classified as a hydrolase, acting on peptide bonds: cysteine endopeptidase
- A two-chain cysteine protease hydrolyzing peptide, ester and amide bonds at the C-side of Arg
- Order of specificity for -Arg↓P_1'-: Leu>Ser>Phe.Val>Ala=Gly>Pro
- Activated by reducing agents like DTT, and Ca^{2+}
- Inhibited by EDTA, oxidizing agents, sulfhydryl reagents (e.g., TLCK), Co^{2+}, Cu^{2+} and Cd^{2+}; citrate, borate and Tris anions are less inhibitory
- Not a homologue of papain
- From the bacterium *Clostridium histolyticum*

SPECIFIC ACTIVITY	UNITS DEFINITION	PREPARATION FORM	ADDITIONAL ACTIVITIES	SUPPLIER CATALOG NO.
Clostridium histolyticum (optimum pH = 7.1-8.0, T = 37°C, MW = 50,000 Da)				
≥100 U/mg after activation with 2.5 m*M* DTT	1 unit hydrolyzes 1 μmol N-benzoyl-L-Arg ethyl ester (BAEE)/min at 25°C	Lyophilized	Acts as esterase	Adv Biofact CP21-1
		Sequencing grade; free of impurities; lyophilized containing activator and incubation buffer		Boehringer 1370529 1420364
50 U/mg dry weight	1 unit hydrolyzes 1 μmol BAEE/min at pH 7.6, 25°C in the presence of 2.5 μmol DTT	Lyophilized		Calbiochem 233185
100 U/mg	1 unit hydrolyzes 1 μmol N-benzoyl-L-Arg ethyl ester/min at pH 7.1, 25°C	Lyophilized		Fluka 27549
50 U/mg	1 unit hydrolyzes 1 μmol N-benzoyl-L-Arg ethyl ester/min at pH 7.1, 25°C	Lyophilized		Fluka 27550
50-150 U/mg solid	1 unit hydrolyzes 1 μmol Nα-benzoyl-L-Arg ethyl ester/min at pH 7.6, 25°C in the presence of 2.5 m*M* DTT	Salt-free; lyophilized		ICN 151457
100-300 U/mg protein	1 unit hydrolyzes 1.0 μmol BAEE/min at pH 7.6, 25°C in the presence of 2.5 m*M* DTT	Aseptically filled; lyophilized containing MOPS buffer, DTT, $CaCl_2$		Sigma C0799

Clostripain continued

SPECIFIC ACTIVITY	UNITS DEFINITION	PREPARATION FORM	ADDITIONAL ACTIVITIES	SUPPLIER CATALOG NO.
Clostridium histolyticum continued				
	1 unit hydrolyzes 1.0 μmol BAEE/min at pH 7.6, 25°C in the presence of 2.5 mM DTT	Purified; salt-free; lyophilized		Sigma C0888
100-300 U/mg protein	1 unit hydrolyzes 1.0 μmol BAEE/min at pH 7.6, 25°C in the presence of 2.5 mM DTT	Purified; lyophilized containing 80% protein; balance primarily NaCl		Sigma C7403
Activates to ≥50 U/mg solid	1 unit hydrolyzes 1.0 μmol BAEE/min at pH 7.6, 25°C in the presence of 2.5 mM DTT	Extensively purified; lyophilized		Worthington LS01641 LS01643 LS01646 LS01647

Actinidain 3.4.22.14

REACTION CATALYZED
  Hydrolysis of proteins with a preference for residues bearing a large hydrophobic side chain at the P2 position

SYNONYMS
  Actinidin, *Actinidia* anionic protease

REACTANTS
  Polypeptides

APPLICATIONS
- Structure/function relationships between cysteine proteases and inhibitors

NOTES
- Classified as a hydrolase, acting on peptide bonds: cysteine endopeptidase
- Does not accept Val at P1′
- Inhibited by E-64
- A cysteine protease similar to papain
- From the kiwi fruit or Chinese gooseberry (*Actinidia chinensis*)

SPECIFIC ACTIVITY	UNITS DEFINITION	PREPARATION FORM	ADDITIONAL ACTIVITIES	SUPPLIER CATALOG NO.
Actinidia chinensis (kiwi fruit) (MW = 23,700 Da)				
10 U/mg protein	1 unit hydrolyzes 1 μmol Z-Phe-Arg-AMC/min at pH 4.3, 25°C	Suspension in 100 mM NaPO$_4$ buffer, 1 mM EDTA, 50% saturated (NH$_4$)$_2$SO$_4$, pH 6.0		Calbiochem 113255

3.4.22.15 Cathepsin L

REACTION CATALYZED
 Hydrolysis of proteins with a preference for residues bearing a large hydrophobic side chain at the P2 position

REACTANTS
 Polypeptides

NOTES
- Classified as a hydrolase, acting on peptide bonds: cysteine endopeptidase
- Does not accept Val at P1'
- A lysosomal cysteine protease similar to papain
- Readily inhibited by Z-Phe-Phe-CHN$_2$ (a diazomethane inhibitor) and E-64 (an epoxide inhibitor)
- Distinguished from cathepsin B in its higher activity towards protein substrates, low activity on Z-Arg-Arg-NHMec and no peptidyl-dipeptidase activity

SPECIFIC ACTIVITY	UNITS DEFINITION	PREPARATION FORM	ADDITIONAL ACTIVITIES	SUPPLIER CATALOG NO.
Human				
				Vital 17027
Human liver (MW = 29,000 Da)				
≥1 U/mg protein	1 unit hydrolyzes 1 µmol Z-Phe-Arg-AFC/min at pH 5.5, 25°C	≥80% purity; frozen in 20 mM NaOAc, 1 mM EDTA, 200 mM NaCl, pH 5.0	No detectable HBsAg, anti-HCV, anti-HBc, anti-HIV	ART 16-12-030112

3.4.22.16 Cathepsin H

REACTION CATALYZED
 Hydrolysis of proteins through aminopeptidase (notably Arg↓ bond cleavage) and endopeptidase activities

SYNONYMS
 Cathepsin B$_3$

REACTANTS
 Polypeptides

NOTES
- Classified as a hydrolase, acting on peptide bonds: cysteine endopeptidase
- A cysteine protease functioning both as an aminopeptidase and an endopeptidase
- Catabolizes proteins in lysosomes of mammalian cells
- More basic than cathepsin B or L

Cathepsin H continued — 3.4.22.16

SPECIFIC ACTIVITY	UNITS DEFINITION	PREPARATION FORM	ADDITIONAL ACTIVITIES	SUPPLIER CATALOG No.
Human				
				Vital 1702
Human liver (MW = 28,000 Da)				
≥1 U/mg protein	1 unit hydrolyzes 1 μmol L-Arg-β-naphthylamide/min at pH 6.8, 40°C	≥90% purity; frozen in 50 mM NaOAc and 1 mM EDTA, pH 5.5	No detectable HBsAg, anti-HCV, anti-HBc, anti-HIV	ART 16-12-030108
				Calbiochem 219404

Calpain — 3.4.22.17

REACTION CATALYZED
  Preferential cleavage: Tyr↓, Met↓ or Arg↓, with Leu or Val as the P2 residue

SYNONYMS
  Ca^{2+}-activated neutral protease, calpain I, calpain II, m-calpain, μ-calpain, CANP, mCANP, μCANP

REACTANTS
  Polypeptides

APPLICATIONS
- Used in connection with dystrophy, cataract, inflammation and accumulation of memory

NOTES
- Classified as a hydrolase, acting on peptide bonds: cysteine endopeptidase
- An intracellular, non-lysosomal cysteine protease
- Of two primary types of calpain, one has high Ca^{2+} sensitivity in the micromolar range and is called μ-calpain, calpain I or μCANP
- The other has low Ca^{2+} sensitivity in the millimolar range and is called m-calpain, calpain II or mCANP
- A component of lens neutral proteinase
- Participant in the ATP release reaction of platelets stimulated with thrombin
- The Ca^{2+} requirement for proteolysis of nuclear matrix protein is dramatically decreased in the presence of DNA
- Inhibited by calpastatin, calpain inhibitor I, calpain inhibitor II and leupeptin

3.4.22.17 Calpain continued

SPECIFIC ACTIVITY	UNITS DEFINITION	PREPARATION FORM	ADDITIONAL ACTIVITIES	SUPPLIER CATALOG NO.
Porcine erythrocyte (optimum pH = 7.0-7.5, pI = 5.3, MW = 112,000 Da)				
120 U/mg protein	1 unit increases A_{750} by 1.0/30 min at pH 7.5, 30°C with casein as substrate	20 mM imidazole, 1 mM EDTA, 1 mM EGTA, 5 mM β-MSH, 30% glycerol, pH 6.8	<0.1% other peptidases	Calbiochem 208712
120 U/mg protein		Solution containing 20 mM imidazole buffer, 1 mM EDTA, 1 mM EGTA, 5 mM β-MSH, pH 6.8		Nacalai 070-43
Porcine kidney (optimum pH = 7.0-8.0, pI = 4.6, MW = 109,000 Da)				
200 U/mg protein	1 unit increases A_{750} by 1.0/30 min at pH 7.5, 30°C with casein as substrate	20 mM imidazole, 1 mM EDTA, 1 mM EGTA, 5 mM β-MSH, 30% glycerol, pH 6.8	<0.1% other peptidases	Calbiochem 208715
120 U/mg protein		Solution containing 20 mM imidozole buffer, 1 mM EDTA, 1 mM EGTA, 5 mM β-MSH, pH 6.8		Nacalai 070-44
Rabbit skeletal muscle				
15-40 U/mg protein	1 unit produces a ΔA_{280} of 0.5/30 min at pH 7.5, 30°C, measured as TCA soluble products using dimethylated casein as substrate in 1.8 mL volume, 1 cm light path	Lyophilized containing 5% protein; balance primarily lactose, DTT, Tris buffer salts		Sigma P4533

3.4.22.32 Stem Bromelain

REACTION CATALYZED

Strong preference for Z-Arg-Arg↓NHMec among small molecules along with broad protein cleavage specificity

SYNONYMS

Bromelain

REACTANTS

Z-Arg-Arg↓NHMec, polypeptides

APPLICATIONS

- Protein hydrolysis
- Resolution of β-hydroxy-α-amino acids by the action of protease on their *N*-acyl methyl esters

NOTES

- Classified as a hydrolase, acting on peptide bonds: cysteine endopeptidase
- Most abundant of the stem cysteine proteases from the pineapple plant (*Ananas comosus*)
- A basic protein, distinct from acidic pineapple fruit bromelain (EC 3.4.22.33)
- Activated by sulfhydryl reagents
- Inhibited by iodoacetic acid and mercurials, barely inhibited by chicken cystatin and very slowly inactivated by E-64 (an epoxide inhibitor)

Stem Bromelain continued

3.4.22.32

SPECIFIC ACTIVITY	UNITS DEFINITION	PREPARATION FORM	ADDITIONAL ACTIVITIES	SUPPLIER CATALOG NO.
Ananas comosus (pineapple stem) (optimum pH = 6.0, MW = 33,000 Da)				
>1000 U/g protein	1 unit hydrolyzes 1 μmol amino nitrogen from gelatin/20 min at pH 4.5, 45°C	Crystallized		Calbiochem 203761
10 U/mg protein	1 unit yields a rate of 1 absorption unit/min at 25°C when added to the reaction mixture	Homogeneous by re-chromatography and electrophoresis; lyophilized containing 98% protein		Calzyme 183A0010
		Food grade		EnzymeDev Enzeco[R] Bromelain
2 U/mg	1 unit releases 1 μmol 4-nitrophenol/min at pH 4.6, 25°C (Nα-carbobenzoxy-L-Lys-4-nitrophenyl ester as substrate)	Powder		Fluka 16990
5-15 U/mg protein	1 unit releases 1.0 μmol p-nitrophenol from Nα-CBZ-L-Lys p-nitrophenyl ester/min at pH 4.6, 25°C	Suspension in 3.2 M $(NH_4)_2SO_4$, pH 6		Sigma B0652
2-4 U/mg protein	1 unit hydrolyzes 1.0 mg amino nitrogen from gelatin/20 min at pH 4.5, 45°C	Contains 50% protein		Sigma B2252
10 U/mg protein	1 unit releases 1.0 μmol p-nitrophenol from Nα-CBZ-L-Lys p-nitrophenyl ester/min at pH 4.6, 25°C	Chromatographically purified; lyophilized containing 40% protein; balance mannitol and KPO_4 buffer salts		Sigma B5144

Fruit Bromelain

3.4.22.33

REACTION CATALYZED
    Broad specificity for hydrolysis of peptide bonds

SYNONYMS
    Pinguinain (*Bromelia pinguin*)

REACTANTS
    Peptide bonds, Bz-Phe-Val-Arg↓NHMec

NOTES
- Classified as a hydrolase, acting on peptide bonds: cysteine endopeptidase
- From the fruit of the pineapple plant (*Ananas comosus*)
- No action on Z-Arg-Arg-NHMec (see stem bromelain)
- Distinguished from a related enzyme with similar small molecule specificity in being scarcely inhibited by chicken cystatin

See Chapter 5. Multiple Enzyme Preparations

3.4.23.1 Pepsin A

REACTION CATALYZED
Preferential cleavage: hydrophobic and preferably aromatic residues in P1 and P1' positions

SYNONYMS
Pepsin

REACTANTS
Insulin, peptide bonds, hemoglobin, casein

APPLICATIONS
- Nonspecific hydrolysis of proteins and peptides in acidic media
- Subculturing mammary epithelial cells without toxic effects associated with trypsin
- Preparation of F(ab')$_2$ fragments of IgG antibodies
- In growth media, cleaves proteins to smaller molecules as nitrogen source
- Standard for serum pepsinogen determination
- Peptide synthesis
- Milk clotting
- Processing collagen to remove telio-proteins

NOTES
- Classified as a hydrolase, acting on peptide bonds: aspartic endopeptidase
- An acidic aspartyl protease, pepsin is the predominant endopeptidase in vertebrate gastric juice
- Formed from pepsinogen A by limited proteolysis below pH 5
- Cleaves proteins preferentially at peptide bonds involving the C-groups of aromatic amino acids and other hydrophobic amino acids, especially Phe and Leu
- Will not cleave bonds containing Val, Ala or Gly
- Cleavage points in the β chain of insulin: Phe$^1\downarrow$Val, Gln$^4\downarrow$His, Glu$^{13}\downarrow$Ala, Ala$^{14}\downarrow$Leu, Leu$^{15}\downarrow$Tyr, Tyr$^{16}\downarrow$Leu, Gly$^{23}\downarrow$Phe, Phe$^{24}\downarrow$Phe and Phe$^{25}\downarrow$Tyr
- Human enzyme occurs in five molecular forms
- Unphosphorylated pig pepsin A is called pepsin D
- Activated by pepsinogen
- Inhibited by aliphatic alcohols, substrate-like epoxides and pepstatin A

SPECIFIC ACTIVITY	UNITS DEFINITION	PREPARATION FORM	ADDITIONAL ACTIVITIES	SUPPLIER CATALOG No.
Porcine gastric mucosa (optimum pH = 1.0 [hemoglobin or casein], range = 1.5-2.0, T = 46-52°C, MW = 35,000 Da; inactivated at pH > 6.0, 70°C)				
1:3000 1:10,000 1:15,000	1:3000 digests 3000-3500X its weight of coagulated egg albumin using a milk clotting test in which the activity is compared to a USP standard	Powder		Am Labs
2500 U/mg solid; activity equivalent to 1:50,000 as defined by British Pharmacopoeia	1 unit increases A_{280} by 0.001/min at pH 2.0, 37°C	Salt-free; 2X crystallized; lyophilized		Biozyme PEP2
2500 U/mg	1 unit liberates sufficient acid-soluble product to increase A_{280} by 0.001/min at 37°C	Lyophilized		Boehringer 100911 100913
2500 U/mg solid	1 unit digests hemoglobin substrates to TCA-soluble products, measured as change in A_{280} by 0.001/min at pH 2.0, 37°C	Salt-free; lyophilized		Calbiochem 516360
3000 U/mg	1 unit renders TCA soluble 0.001 A_{280} nm/min at 37°C using denatured Hb as substrate	2X Crystallized; lyophilized containing 95% protein		Calzyme 099A3000
1:10,000	Digests 10,000 times its weight of coagulated egg albumin	Powder		Difco 0151-17-8
4.0-5.1 U/mg protein	1 unit increases A_{280} of TCA-soluble peptides from denatured hemoglobin by 1/min at 37°C	Chromatographically purified; salt-free; crystallized; lyophilized		Elastin CP719
2.0-2.5 U/mg solid	1 unit increases A_{280} of TCA-soluble peptides from denatured hemoglobin by 1/min at pH 2, 37°C	Crude; powder containing 1% NaCl		Elastin P715
100 U/mg	1 unit hydrolyzes 1 μmol acetyl-L-phenylalanyl-3,5-diiodo-L-Tyr/min at pH 2.0, 37°C	Powder		Fluka 77151
150 U/g	1 unit hydrolyzes 1 μmol acetyl-L-phenylalanyl-3,5-diiodo-L-Tyr/min at pH 2.0, 37°C	Salt-free; 2X crystallized; lyophilized	10 mAnson units/mg	Fluka 77152
50 U/g	1 unit hydrolyzes 1 μmol acetyl-L-phenylalanyl-3,5-diiodo-L-Tyr/min at pH 2.0, 37°C	Crystallized containing 2% ash and a 3-5% LOD	3 mAnson units/mg	Fluka 77160

3.4.23.1 Pepsin A continued

SPECIFIC ACTIVITY	UNITS DEFINITION	PREPARATION FORM	ADDITIONAL ACTIVITIES	SUPPLIER CATALOG NO.
Porcine gastric mucosa *continued*				
20 U/g	1 unit hydrolyzes 1 µmol acetyl-L-phenylalanyl-3,5-diiodo-L-Tyr/min at pH 2.0, 37°C	Crystallized containing 5% ash and a 5% LOD	0.8 mAnson-units/mg	Fluka 77163
3X USP/NF Std		Powder		ICN 102598
5X USP/NF Std		Purified; powder		ICN 102599
3-4 mg enzyme/mL gel		Pepsin immobilized to agarose; suspension in 50 mM NaOAc, pH 4.5		ICN 191291
2000-2400 U/mg protein	1 unit causes a ΔA_{280} of 0.001/min at pH 2.0, 37°C measured as TCA-soluble products using hemoglobin as substrate	Purified by chromatography; salt-free; lyophilized		ICN 195367
100-200 U/mg dry agarose	1 unit produces a ΔA_{280} of 0.001/min at pH 2.0, 37°C measured as TCA-soluble products with hemoglobin as substrate; light path = 1 cm	Insoluble enzyme attached to 4% cross-linked beaded agarose; lyophilized containing lactose and 20% agarose		Sigma P3286
3200-4500 U/mg protein	1 unit produces a ΔA_{280} of 0.001/min at pH 2.0, 37°C measured as TCA-soluble products with hemoglobin as substrate; light path = 1 cm	Chromatographically purified; salt-free; crystallized; lyophilized		Sigma P6887
800-2500 U/mg protein	1 unit produces a ΔA_{280} of 0.001/min at pH 2.0, 37°C measured as TCA-soluble products with hemoglobin as substrate; light path = 1 cm	Powder		Sigma P7000
2500-3500 U/mg protein	1 unit produces a ΔA_{280} of 0.001/min at pH 2.0, 37°C measured as TCA-soluble products with hemoglobin as substrate; light path = 1 cm	Crystallized; lyophilized		Sigma P7012
600-1000 U/mg protein	1 unit produces a ΔA_{280} of 0.001/min at pH 2.0, 37°C measured as TCA-soluble products with hemoglobin as substrate; light path = 1 cm	Powder		Sigma P7125
500-1000 U/mg		Crude; digestive powder		Wako 163-00642 169-05742 163-05745
≥2500 U/mg solid	1 unit releases 0.001 A_{280} as TCA-soluble hydrolysis products/min at 37°C using denatured Hb substrate; 1 unit X 0.0064 = FIP units	2X Crystallized pepsin from alcohol; lyophilized		Worthington LS03319 LS03317 LS03322

Chymosin

REACTION CATALYZED
Broad specificity similar to that of pepsin A. Clots milk by cleavage of a single Ser-Phe^{105}↓Met-Ala bond in the κ-chain of casein

SYNONYMS
Rennet, rennin

REACTANTS
Casein, peptide bonds

APPLICATIONS
- Aids proper setting of cottage cheese and other cultured cream products; provides coagulation strength for cutting in pH 4.6-4.75 and increases viscosity

NOTES
- Classified as a hydrolase, acting on peptide bonds: aspartic endopeptidase
- Neonatal gastric enzyme
- Found among mammals with postnatal immunoglobulin uptake
- High milk clotting, but weak general proteolytic activity
- An aspartyl protease formed from prochymosin
- Activated by prorennin A and B
- Use of the synonym rennin is discouraged to avoid confusion with renin (EC 3.4.23.15)

SPECIFIC ACTIVITY	UNITS DEFINITION	PREPARATION FORM	ADDITIONAL ACTIVITIES	SUPPLIER CATALOG No.
Aspergillus niger var. awamori (inactivated at 72°C/15 sec)				
		Kosher certified; 100% chymosin solution containing NaCl, Na benzoate as preservative; single or double strength, pH 5.5-5.9		ChrHansens Chymogen™
Calf stomach (optimum pH = 3.7 [hemoglobin], pI = 4.5, MW = 40,000 Da; most stable at pH 5.5-6.0; unstable pH > 6.5)				
20-30 U/mg protein	1 unit coagulates 10 mL milk/min at 30°C	Lyophilized containing 95% protein		Calzyme 220A0020
20 U/mg protein	1 unit coagulates 10 mL milk/min at 30°C	Crystallized; lyophilized containing 98% protein; balance NaCl		ICN 152019
20 U/mg protein	1 unit coagulates 10 mL milk/min at 30°C	Crystallized; lyophilized containing 40% protein; balance NaCl		Sigma R4877
50-100 U/mg protein	1 unit coagulates 10 mL milk/min at 30°C	Crystallized; lyophilized containing 90% protein; balance NaCl		Sigma R4879
20 U/mg protein	1 unit coagulates 10 mL milk/min at 30°C	Crystallized; lyophilized containing 98% protein; balance NaCl		Sigma R7751
Fermentation				
		Kosher certified (excluding Passover); solution containing NaCl, Ca^{2+} salts, propylene glycol, pH 5.55-5.65		ChrHansens Cottage Cheese Coagulator

3.4.23.4 Chymosin continued

SPECIFIC ACTIVITY	UNITS DEFINITION	PREPARATION FORM	ADDITIONAL ACTIVITIES	SUPPLIER CATALOG NO.
Mucor miehei (inactivated at 72-74°C [pH 5.6, 15 sec])				
		Kosher certified (including Passover); solution containing NaCl and Na benzoate pH 5.10-5.20; single or double strength		ChrHansens Hannilase[R] XL

3.4.23.5 Cathepsin D

REACTION CATALYZED
  Specificity similar to but narrower than pepsin A
  Does not cleave the $Gln^4 \downarrow His$ bond in the B chain of insulin

REACTANTS
  Insulin, peptide bonds

APPLICATIONS
- Immunogen for antisera production
- Tracer for iodination

NOTES
- Classified as a hydrolase, acting on peptide bonds: aspartic endopeptidase
- A major lysosomal aspartyl protease in mammalian cells
- A zymogen form has been identified
- An estrogen-regulated protein associated with tissue breakdown
- Cathepsins A and D work synergistically to hydrolyze protein substrates
- Associated with amyloid formation in Alzheimer's plaques
- High levels are positively correlated with breast cancer recurrence in node-negative and -positive disease
- Linked to lung tissue damage in smokers

SPECIFIC ACTIVITY	UNITS DEFINITION	PREPARATION FORM	ADDITIONAL ACTIVITIES	SUPPLIER CATALOG NO.
Bovine spleen				
10 U/mg protein	1 unit increases unit extinction by 0.001/min of digestion of the substrate at 37°C	Lyophilized containing 40% protein		Calzyme 135A0010

SPECIFIC ACTIVITY	UNITS DEFINITION	PREPARATION FORM	ADDITIONAL ACTIVITIES	SUPPLIER CATALOG No.
Bovine spleen *continued*				
0.02 U/mg	1 unit hydrolyzes 1 μmol Phe-Ala-Ala-Phe(4-NO_2)-Phe-Val-Leu-4-hydroxymethylpyridine ester/min at pH 3.5, 25°C	Powder		Fluka 22132
5-15 U/mg protein	1 unit increases A_{280} by 1.0/min/mL at pH 3.0, 37°C measured as TCA-soluble products using hemoglobin as substrate (1 cm light path)	Lyophilized containing 40% protein; balance primarily citrate buffer salts		Sigma C3138
Human liver (MW = 42,000 Da)				
≥300 U/mg	1 unit digests hemoglobin releasing peptides which are soluble in 10% TCA measured by a 1.0 increase at A_{280}/hr at 37°C using acid denatured hemoglobin as substrate	≥95% purity; lyophilized containing 2 m*M* $NaPO_4$, pH 6.5	No detectable HBsAg, anti-HCV, anti-HBc, anti-HIV	ART 16-12-030104
≥300 U/mg protein	1 unit digests acid-denatured hemoglobin-releasing peptides soluble in 10% TCA measured by increase in A_{280} by 1.0/hr	>98% purity; lyophilized from buffer containing 2 m*M* $NaPO_4$, pH 6.5	No detectable HBsAg, HIV Ab	Calbiochem 219401
300 U/mg	1 unit digests hemoglobin releasing peptides soluble in 10% TCA increasing A_{280} by 1.0/hr at pH 3.3 using acid denatured hemoglobin	>95% purity by SDS-PAGE; lyophilized containing 2 m*M* $NaPO_4$, pH 6.5	No detectable HIV Ab, HCV, HBsAg	Cortex CP3090
>300 U/mg	1 unit digests hemoglobin releasing peptides soluble in 10% TCA at pH 3.3, 37°C	>95% purity by SDS-PAGE; lyophilized containing 2 m*M* $NaPO_4$, pH 6.5	No detectable HGsAg, HCV, HIV-1 Ab	Scripps C2414
Human spleen				
120 U/mg protein	1 unit increases unit extinction by 0.001/min of digestion of the substrate at 37°C	Lyophilized containing 40% protein		Calzyme 135A0120

Renin

REACTION CATALYZED
Cleavage of Leu↓ bond in angiotensinogen to generate angiotensin I

SYNONYMS
Angiotensin-forming enzyme, angiotensinogenase

REACTANTS
Angiotensinogen, angiotensin

NOTES
- Classified as a hydrolase, acting on peptide bonds: aspartic endopeptidase
- An aspartyl protease formed from prorenin in plasma and kidney

SPECIFIC ACTIVITY	UNITS DEFINITION	PREPARATION FORM	ADDITIONAL ACTIVITIES	SUPPLIER CATALOG NO.
Human kidney				
0.4 U/mg	1 unit raises Hg by 30 mm in the majority of dogs	Lyophilized		Calzyme 165A0000
0.05-0.3 U/mg	1 unit increases Hg by 30 mm in the majority of dogs	Lyophilized	No detectable HGsAg, HCV, HIV-1 Ab	Scripps R0814
Porcine kidney				
0.01 U/mg	1 unit liberates 1 μmol angiotensin I from angiotensinogen/min at pH 6.0, 37°C	Lyophilized		Fluka 83555
5-15 U/mg protein	1 unit liberates 100 μg angiotensin I from angiotensinogen/hr at pH 6.0, 37°C	Lyophilized containing 50% protein; balance primarily pyrophosphate buffer salts		ICN 158045
0.5-5.0 U/mg protein	1 unit liberates 100 μg angiotensin I from angiotensinogen/hr at pH 6.0, 37°C	Lyophilized containing 35% protein; balance primarily pyrophosphate buffer salts		Sigma R2761

Reaction Catalyzed

Generally favors hydrophobic residues in P1 and P1′, but also accepts Lys in P1 leading to activation of trypsinogen. Does not clot milk

Synonyms

Awamorin and aspergillopepsin A (*Aspergillus awamori*), aspergillopepsin F (*Aspergillus foetidus*), proteinase B and proctase B (*Aspergillus niger*), trypsinogen kinase (*Aspergillus oryzae*), aspergillopeptidase A (*Aspergillus saitoi*), fungal proteinase, neutral protease, protease

Reactants

Peptide bonds

Applications

- Manufacture of proteins without bitterness and flavors from natural proteins
- Bio-polishing wool fabrics
- Protein hydrolysis
- Decomposition of soybean and rice protein
- Flour treatment and manufacture of baked goods
- Lactose reduction and flavor modification in dairy applications

Notes

- Classified as a hydrolase, acting on peptide bonds: aspartic endopeptidase
- An aspartyl protease found in a variety of *Aspergillus* species
- Produces shorter peptide lengths than conventional bacterial proteinases
- Activated by Cys

SPECIFIC ACTIVITY	UNITS DEFINITION	PREPARATION FORM	ADDITIONAL ACTIVITIES	SUPPLIER CATALOG NO.
Aspergillus oryzae (optimum pH = 7, T = 50°C; stable pH 6-9 [30°C, 1 hr] and T < 50°C [pH 7.0, 1 hr])				
3.5 k Tys/g (17 Xs/g) 20.0 k Tys/g	1 unit produces 1 µg Tyr equivalent of aromatic nitrogen compounds soluble in 4% TCA from casein/min; 1 k Tys produces 1 mg soluble Tyr equivalent/min	Powder containing starch	<0.3 X/g α-amylase	ABM-RP Panazyme 77A Panazyme 1000
≥20,000 U/g	1 unit produces amino acids equal to 100 mg L-Tyr/hr at pH 7.0 in 1 mL filtrate	Powder		Amano Protease 2A
≥5500 U/g	1 unit produces amino acids equal to 100 mg L-Tyr/hr at pH 3.0 in 1 mL filtrate	Powder	Aminopeptidase	Amano Protease M
		Food grade		EnzymeDev Enzeco[R] Fungal Protease
		Food grade		EnzymeDev Enzeco[R] Fungal Protease 180

3.4.23.18 Aspergillopepsin I continued

SPECIFIC ACTIVITY	UNITS DEFINITION	PREPARATION FORM	ADDITIONAL ACTIVITIES	SUPPLIER CATALOG NO.
Aspergillus oryzae continued				
		Food grade		EnzymeDev Enzeco[R] Fungal Protease Conc
50,000 PUN/g solid	1 PUN liberates the digestion product not precipitated with TCA giving the same Folin color as 1 μg Tyr/min at pH 7.5, 30°C	Powder		Nagase Denazyme AP
227 mU$_{Hb}$/mg	pH 5	Food grade; powder		Rohm Veron[R] PS
Aspergillus saitoi				
0.3 U/mg solid	1 unit hydrolyzes hemoglobin producing color equivalent to 1.0 μmol (181 μg) Tyr/min at pH 2.8, 37°C		No detectable AP	ICN 151973
Aspergillus sojae (optimum pH = 5-6)				
290 PU/mg	pH 7.0	Food grade; powder		Rohm Corolase[R] PN

3.4.23.19 Aspergillopepsin II

REACTION CATALYZED
  Preferential cleavage in the B chain of insulin: Asn^3↓Gln, Gly^{13}↓Ala, Tyr^{26}↓Thr

SYNONYMS
  Aspergillus niger var. *macrosporus* aspartic proteinase, proteinase A, proctase A

REACTANTS
  Insulin, peptide bonds

NOTES
- Classified as a hydrolase, acting on peptide bonds: aspartic endopeptidase
- Isolated from *Aspergillus niger* var. *macrosporus*
- An aspartyl protease distinct from aspergillopepsin I (EC 3.4.23.18) in specificity and insensitivity to pepstatin

SPECIFIC ACTIVITY	UNITS DEFINITION	PREPARATION FORM	ADDITIONAL ACTIVITIES	SUPPLIER CATALOG NO.
Aspergillus saitoi				
0.6 U/mg solid	1 unit hydrolyzes hemoglobin to produce color equivalent to 1.0 μmol Tyr/min at pH 2.8, 37°C	Crude; powder	Substantially AP free Many extraneous enzymes	Sigma P2143

Penicillopepsin 3.4.23.20

REACTION CATALYZED
Hydrolysis of proteins with broad specificity similar to that of pepsin A. Prefers hydrophobic residues at P1 and P1', but also cleaves $Gly^{20}\downarrow Glu$ in the β-chain of insulin

SYNONYMS
Peptidase A, *Penicillium janthinellum* aspartic proteinase

REACTANTS
Insulin β-chain

APPLICATIONS
- Manufacture of peptides without bitterness

NOTES
- Classified as a hydrolase, acting on peptide bonds: aspartic endopeptidase
- Homologue and similar in structure to pepsin A
- Clots milk
- An aspartyl protease which activates trypsinogen

SPECIFIC ACTIVITY	UNITS DEFINITION	PREPARATION FORM	ADDITIONAL ACTIVITIES	SUPPLIER CATALOG No.
Penicillium species (optimum pH = 6.0, T = 45°C)				
≥1800 U/g	1 unit produces amino acids equal to 100 mg L-Tyr/hr at pH 6.0 in 1 mL filtrate	Powder		Amano Protease B

3.4.23.21 Rhizopuspepsin

REACTION CATALYZED
Hydrolysis of proteins with broad specificity similar to that of pepsin A, preferring hydrophobic residues at P1 and P1'; cleaves insulin β chain $His^{10}\downarrow Leu$ and $Val^{12}\downarrow Glu$, but not $Gln^{4}\downarrow His$

SYNONYMS
Rhizopus aspartic proteinase, fungal protease

REACTANTS
Peptides

NOTES
- Classified as a hydrolase, acting on peptide bonds: aspartic endopeptidase
- Clots milk and activates trypsinogen
- From the zygomycete fungus *Rhizopus chinensis*
- An acid, aspartyl protease homologue of pepsin A
- Similar to an endopeptidase from *Rhizopus niveus*

3.4.23.21 Rhizopuspepsin continued

SPECIFIC ACTIVITY	UNITS DEFINITION	PREPARATION FORM	ADDITIONAL ACTIVITIES	SUPPLIER CATALOG No.
Rhizopus niveus (optimum pH = 3.0, T = 45°C)				
≥15,000 U/g	1 unit produces amino acids equal to 100 mg L-Tyr/hr at pH 3.0 in 1 mL filtrate	Powder		Amano Newlase II
Rhizopus species				
0.5 U/mg solid	pH 3.0, 37°C			ICN 151974
0.2-0.6 U/mg solid	1 unit hydrolyzes casein to produce color equivalent to 1.0 µmol (181 µg) Tyr/min at pH 3.0, 37°C			Sigma P5027

3.4.23.23 Mucorpepsin

REACTION CATALYZED
Hydrolysis of proteins, favoring hydrophobic residues at P1 and P1'; does not accept Lys at P1

SYNONYMS
Mucor rennin

REACTANTS
Proteins, casein

NOTES
- Classified as a hydrolase, acting on peptide bonds: aspartic endopeptidase
- Clots milk
- P1 specificity prevents activation of trypsinogen
- An aspartyl protease isolated from zygomycete fungi *Mucor pusillus* and *Mucor miehei*; species variants with 83% sequence identity and immunological crossreactivity

SPECIFIC ACTIVITY	UNITS DEFINITION	PREPARATION FORM	ADDITIONAL ACTIVITIES	SUPPLIER CATALOG No.
Mucor miehei				
0.1 U/mg	1 unit releases 1 µmol soluble macropeptides from casein/min at pH 6.5, 37°C	Lyophilized		Fluka 83553

Candidapepsin

3.4.23.24

REACTION CATALYZED
: Preferential cleavage at the carboxyl of hydrophobic amino acids, but fails to cleave Leu^{15}-Tyr, Tyr^{16}-Leu and Phe^{24}-Phe of insulin B chains. Activates trypsinogen and degrades keratin

SYNONYMS
: *Candida albicans* aspartic proteinase, proteinase A

REACTANTS
: Keratin, protein

APPLICATIONS
- Producing overlap peptides in sequencing studies

NOTES
- Classified as a hydrolase, acting on peptide bonds: aspartic endopeptidase
- An aspartyl protease from *Candida albicans*
- Inhibited by pepstatin
- Not inhibited by methyl 2-diazoacetamidohexanoate or 1,2-epoxy-3-(*p*-nitrophenoxy)propane

SPECIFIC ACTIVITY	UNITS DEFINITION	PREPARATION FORM	ADDITIONAL ACTIVITIES	SUPPLIER CATALOG NO.
Yeast, bakers				
15-50 U/mg protein	1 unit hydrolyzes 1 mg insulin chain β (oxidized)/min at pH 6.0, 25°C	Lyophilized containing 85% protein; balance primarily Na citrate, pH 5.0	No detectable carboxypeptidase Y	sigma P8892

Physaropepsin

3.4.23.27

REACTION CATALYZED
: Milk clotting activity. Preferential cleavage in the B chain of insulin: Gly^8↓Ser most rapid, followed by Leu^{11}↓Val, Cys(SO$_3$H)^{19}↓Gly and Phe^{24}↓Phe. Does not act on *N*-acetyl-Phe-Tyr(I)$_2$

SYNONYMS
: *Physarum* aspartic proteinase

REACTANTS
: Casein, insulin, peptide bonds

NOTES
- Classified as a hydrolase, acting on peptide bonds: aspartic endopeptidase
- An aspartyl protease from the slime mold *Physarum polycephalum*
- Not inhibited by pepstatin
- Blocked by methyl 2-diazoacetamidohexanoate
- Similar enzymes found in *Dictyostelium discoideum* and *Physarum flavicomum*

See Chapter 5. Multiple Enzyme Preparations

3.4.24.3 Microbial Collagenase

REACTION CATALYZED
Digestion of native collagen in the triple helical region at ↓Gly bonds. Preference with synthetic peptides: Gly at P3 and P1', Pro and Ala at P2 and P2', and hydroxyproline, Ala or Arg at P3'

SYNONYMS
Achromobacter iophagus collagenase, *Clostridium histolyticum* collagenase, clostridiopeptidase A, collagenase A, collagenase I, collagenolytic proteinase

REACTANTS
Collagen, peptide bonds

APPLICATIONS
- Determination of collagen
- Debriding agent
- Dissociation of tissues for establishment of primary cell cultures. Tissues include epithelial, lung, fat, adrenal, liver, bone, pancreas, mammary, thyroid, heart and salivary
- Dissolves connective tissues and exposes embedded cells without destroying cell membranes and other structures
- Preparation of endothelial cells from large vessels and adipocytes from epididymal fat pads of rats
- Isolation of rat pancreatic islets and hepatocytes

NOTES
- Classified as a hydrolase, acting on peptide bonds: metalloendopeptidase
- Two classes comprising six forms of metalloendopeptidases acting on native collagen have been isolated from *Clostridium histolyticum* medium:
    - class I forms includes α (68 kDa), β (115 kDa) and γ (79 kDa)
    - class II forms include δ (100 kDa), ε (110 kDa) and ξ (125 kDa)

  Classes I and II are immunologically crossreactive, but have significantly different sequences and specificities. Their actions on collagen are complementary
- Also act as peptidyl-tripeptidases
- Crude collagenases contain several collagenases, a sulfhydryl protease (clostripain [EC 3.4.22.8]), a trypsin-like enzyme and an aminopeptidase; very effective at breaking down intracellular matrices in tissue dissociation
- Enzyme variants have been purified from *Bacillus cereus*, *Empedobacter collagenolyticum*, *Pseudomonas marinoglutinosa*, *Vibrio* B-30, *Vibrio alginolyticus* and *Streptomyces* sp.
- Inhibited by EDTA, EGTA, Cys, His, DTT, 2-mercaptoethanol, Hg^{2+}, Pb^{2+}, Cd^{2+}, Cu^{2+}, Zn^{2+} and O-phenanthroline
- *Not inhibited* by DFP or serum

Microbial Collagenase continued

3.4.24.3

SPECIFIC ACTIVITY	UNITS DEFINITION	PREPARATION FORM	ADDITIONAL ACTIVITIES	SUPPLIER CATALOG No.
Achromobacter iophagus				
2000 U/mg solid	1 unit liberates 1.0 µmol Pz-Pro-Leu from Pz-Pro-Leu-Gly-Pro-D-Arg/15 min at pH 7.1, 37°C	Lyophilized		Sigma C1913
Clostridium histolyticum (optimum pH = 6.5-8.0, T = 30-37°C, MW = 56,000 Da)				
75-90 BTC U/mg	1 BTC unit liberates 1 nmol Leu equivalent from undenatured bovine tendon collagen/min at 37°C	Lyophilized		Adv Biofact C10-1
100-150 BTC U/mg	1 BTC unit liberates 1 nmol Leu equivalent from undenatured bovine tendon collagen/min at 37°C	Lyophilized		Adv Biofact C10-2
≥2000 BTC U/mg	1 BTC unit liberates 1 nmol Leu equivalent from undenatured bovine tendon collagen/min at 37°C	Lyophilized		Adv Biofact C10-3
160 U/mg	1 unit releases 1 µmol amino acid from collagen/5 hr at pH 7.5, 37°C	Powder		AMRESCO 0784
>0.15 U/mg	1 unit forms 1 µmol of product with 4-phenylazobenzyl-oxycarbonyl-Pro-Leu-Gly-Pro-D-Arg/min at 25°C (Wunsch units)	Crude	Trypsin, clostripain, neutral proteases Balanced ratio of enzyme activities	Boehringer 103578 1088785 1088793
>0.15 U/mg	1 unit forms 1 µmol of product with 4-phenylazobenzyl-oxycarbonyl-Pro-Leu-Gly-Pro-D-Arg/min at 25°C (Wunsch units)	Crude	Trypsin, clostripain, total proteases Balanced ratio of enzyme activities	Boehringer 1074032 1074059 1087789
>0.15 U/mg	1 unit forms 1 µmol of product with 4-phenylazobenzyl-oxycarbonyl-Pro-Leu-Gly-Pro-D-Arg/min at 25°C (Wunsch units)	Crude	Trypsin, clostripain, total proteases Normal to high collagenase and higher than average clostripain (>10 U/mg)	Boehringer 1088807 1088823 1088831
>0.15 U/mg	1 unit forms 1 µmol of product with 4-phenylazobenzyl-oxycarbonyl-Pro-Leu-Gly-Pro-D-Arg/min at 25°C (Wunsch units)	Crude	Trypsin, clostripain, total proteases Normal to high collagenase and very low tryptic activity (<0.1 U/mg [BAEE])	Boehringer 1088858 1088874 1088882
>1.5 U/mg	1 unit forms 1 µmol of product with 4-phenylazobenzyl-oxycarbonyl-Pro-Leu-Gly-Pro-D-Arg/min at 25°C (Wunsch units)	Crude	Trypsin, clostripain, total proteases	Boehringer 1213857 1249002 1213873

SPECIFIC ACTIVITY	UNITS DEFINITION	PREPARATION FORM	ADDITIONAL ACTIVITIES	SUPPLIER CATALOG NO.
Clostridium histolyticum continued				
>2000 collagenase U/mg protein	1 unit equals 1 nmol Leu equivalent release from collagen/min at pH 7.2, 37°C	Highly purified; lyophilized	<5 fluorescence U/mg protein proteolytic activity	Calbiochem 234134
80 collagenase U/mg solid	1 unit equals 1 nmol Leu equivalent release from collagen/min at pH 7.2, 37°C	Crude; salt-free; lyophilized	Collagenases, sulfhydryl protease, clostripain, trypsin-like proteolytic activities <12 fluorescence U/mg	Calbiochem 234153
100-150 collagenase U/mg solid	1 unit equals 1 nmol Leu equivalent release from collagen/min at pH 7.2, 37°C	Crude; salt-free; lyophilized	Higher clostripain and tryptic activity than type I 10-20 fluorescence U/mg protein	Calbiochem 234155
>130 U/mg protein	1 unit liberates ninhydrin color equivalent to 1 μmol L-Leu/5 hr at pH 7.5, 37°C	Crude containing 2% salts, Tris, $CaCl_2$, pH 7.5	<0.1 U/mg protein tryptic activity <0.5 U/mg protein clostripain 1-2 U/mg protein FALGPA 30-40 U/mg protein protease	Elastin CC203
1000-2000 U/mg protein	1 unit liberates ninhydrin color equivalent to 1 μmol L-Leu/5 hr at pH 7.5, 37°C with bovine Achilles tendon collagen as substrate	Chromatographically purified; salt-free; lyophilized	<0.1 U/mg protein clostripain <0.3 U/mg protein tryptic activity <1 U/mg protein protease 8-12 U/mg protein FALGPA	Elastin CL103
350-450 U/mg protein	1 unit liberates ninhydrin color equivalent to 1 μmol L-Leu/5 hr at pH 7.5, 37°C	Partially purified by chromatography; crude	<0.2 U/mg protein clostripain <0.5 U/mg protein tryptic activity 2-4 U/mg protein protease 3-5 U/mg protein FALGPA	Elastin CP303
0.15 U/mg	1 unit hydrolyzes 1 μmol PZ-Pro-Leu-Gly-Pro-D-Arg/min at pH 7.1, 25°C	Lyophilized		Fluka 27665
1 U/mg	1 unit hydrolyzes 1 μmol PZ-Pro-Leu-Gly-Pro-D-Arg/min at pH 7.1, 25°C	Powder	<5% caseinase, trypsin <50% clostripain	Fluka 27666
1 U/mg	1 unit hydrolyzes 1 μmol PZ-Pro-Leu-Gly-Pro-D-Arg/min at pH 7.1, 25°C	Crystallized	<0.1% caseinase <5% trypsin <35% clostripain	Fluka 27667
0.5 U/mg	1 unit hydrolyzes 1 μmol PZ-Pro-Leu-Gly-Pro-D-Arg/min at pH 7.1, 25°C	Powder	<10% trypsin <50% caseinase <100% clostripain	Fluka 27668
2 U/mg	1 unit hydrolyzes 1 μmol PZ-Pro-Leu-Gly-Pro-D-Arg/min at pH 7.1, 25°C	Crystallized	<2% trypsin <6% caseinase <35% clostripain	Fluka 27669

Microbial Collagenase continued — 3.4.24.3

SPECIFIC ACTIVITY	UNITS DEFINITION	PREPARATION FORM	ADDITIONAL ACTIVITIES	SUPPLIER CATALOG NO.
Clostridium histolyticum continued				
0.6 U/mg	1 unit hydrolyzes 1 μmol PZ-Pro-Leu-Gly-Pro-D-Arg/min at pH 7.1, 25°C	Powder	<2% trypsin <50% caseinase <100% clostripain	Fluka 27670
0.7 U/mg	1 unit hydrolyzes 1 μmol PZ-Pro-Leu-Gly-Pro-D-Arg/min at pH 7.1, 25°C	Powder	<2% trypsin <20% caseinase <50% clostripain	Fluka 27671
2 U/mg	1 unit hydrolyzes 1 μmol PZ-Pro-Leu-Gly-Pro-D-Arg/min at pH 7.1, 25°C	Lyophilized		Fluka 27680
5 U/mg	1 unit hydrolyzes 1 μmol PZ-Pro-Leu-Gly-Pro-D-Arg/min at pH 7.1, 25°C	Lyophilized		Fluka 27682
1 U/mg	1 unit hydrolyzes 1 μmol PZ-Pro-Leu-Gly-Pro-D-Arg/min at pH 7.1, 25°C	Sterile-filtered; lyophilized		Fluka 27683
1 U/mg	1 unit hydrolyzes 1 μmol PZ-Pro-Leu-Gly-Pro-D-Arg/min at pH 7.1, 25°C	Sterile-filtered; lyophilized		Fluka 27684
2 U/mg	1 unit hydrolyzes 1 μmol PZ-Pro-Leu-Gly-Pro-D-Arg/min at pH 7.1, 25°C	Sterile-filtered; lyophilized		Fluka 27685
125-220 U/mg solid	1 unit liberates peptides from collagen equivalent in ninhydrin color to 1.0 μmol L-Leu/5 hr at pH 7.4, 37°C in the presence of calcium ions	Salt-free; sterile-filtered; lyophilized	Clostripain, neutral protease, trypsin	ICN 100501
125-220 U/mg solid	1 unit liberates peptides from collagen equivalent in ninhydrin color to 1.0 μmol L-Leu/5 hr at pH 7.4, 37°C in the presence of calcium ions	Grade II; lyophilized	Clostripain, neutral protease, trypsin	ICN 100502
≥3500 U/g	1 unit liberates peptides from collagen equivalent in ninhydrin color to 3.0 μmol L-Leu/18 hr at pH 7.4, 37°C	Isolation grade; lyophilized	<300 U/g clostripain, trypsin, caseinase	ICN 150704
≥2000 U/g	1 unit liberates peptides from collagen equivalent in ninhydrin color to 3.0 μmol L-Leu/18 hr at pH 7.4, 37°C	Cell preparation grade; lyophilized	<500 U/g clostripain, trypsin, caseinase	ICN 150705
2000-3000 U/mL	1 unit liberates peptides from collagen equivalent in ninhydrin color to 3.0 μmol L-Leu/18 hr at pH 7.4, 37°C	Highly purified	No detectable non-specific protease	ICN 151459
125-220 U/mg solid	1 unit liberates peptides from collagen equivalent in ninhydrin color to 1.0 μmol L-Leu/5 hr at pH 7.4, 37°C in the presence of Ca^{2+}	Grade I; salt-free; lyophilized	Clostripain, neutral protease, trypsin	ICN 195109

3.4.24.3 Microbial Collagenase continued

SPECIFIC ACTIVITY	UNITS DEFINITION	PREPARATION FORM	ADDITIONAL ACTIVITIES	SUPPLIER CATALOG NO.
Clostridium histolyticum continued				
125-200 U/mg	1 unit liberates 1 mmol amino acid from collagen, as L-Leu/18 hr at 37°C	Lyophilized		Life Technol 17018-011 17018-029 17018-037
		Type I		Life Technol 17100-025 17100-017 17100-033
		Type II		Life Technol 17101-023 17101-015 17101-031
		Type III		Life Technol 17102-021 17102-013 17102-039
		Type IV		Life Technol 17104-027 17104-019 17104-035
		Hepatocyte qualified		Life Technol 17130-011 17130-029 17130-037
2.5 U/mg (1000 Mandle U/mg)	1 unit decreases specific viscosity by 1.0/min at pH 7.5, 30°C; 1 Mandle U increases A_{570} by 1.0/18 hr at pH 7.4, 37°C	Lyophilized		Seikagaku 100370-1
≥1200 collagen digestion U/mg solid; 2-5 FALGPA hydrolysis U/mg solid	1 Collagen Digestion Unit liberates peptides from collagen equivalent in ninhydrin color to 1.0 μmol Leu/5 hr at pH 7.4, 37°C in presence of Ca^{2+}; 1 FALGPA H Unit hydrolyzes 1.0 μmol furylacryloyl-Leu-Gly-Pro-Ala/min at pH 7.5, 25°C with Ca^{2+}	Crude	Clostripain, neutral protease	Sigma 9407

Microbial Collagenase continued

SPECIFIC ACTIVITY	UNITS DEFINITION	PREPARATION FORM	ADDITIONAL ACTIVITIES	SUPPLIER CATALOG NO.
Clostridium histolyticum continued				
0.5-1.0 FALGPA hydrolysis U/mg solid; 125 collagen digestion U/mg solid	1 unit hydrolyzes 1.0 µmol FALGPA (furylacryloyl-Leu-Gly-Pro-Ala)/min at pH 7.5, 25°C in the presence of Ca^{2+}; 1 unit liberates peptides from collagen equivalent in ninhydrin color to 1.0 µmol Leu/5 hr at pH 7.4, 37°C in the presence of Ca^{2+}		Clostripain, neutral protease and trypsin	Sigma C0130
2-10 FALGPA hydrolysis U/mg solid; 400 collagen digestion U/mg solid	1 unit hydrolyzes 1.0 µmol FALGPA (furylacryloyl-Leu-Gly-Pro-Ala)/min at pH 7.5, 25°C in the presence of Ca^{2+}; 1 unit liberates peptides from collagen equivalent in ninhydrin color to 1.0 µmol Leu/5 hr at pH 7.4, 37°C in the presence of Ca^{2+}	Chromatographically purified; lyophilized containing 90% protein	<1 neutral protease U/mg protein May contain clostripain	Sigma C0255
4-12 FALGPA hydrolysis U/mg solid; 1000-3000 collagen digestion U/mg solid	1 unit hydrolyzes 1.0 µmol FALGPA (furylacryloyl-Leu-Gly-Pro-Ala)/min at pH 7.5, 25°C in the presence of Ca^{2+}; 1 unit liberates peptides from collagen equivalent in ninhydrin color to 1.0 µmol Leu/5 hr at pH 7.4, 37°C in the presence of Ca^{2+}	Chromatographically purified; lyophilized containing 95% protein; balance primarily $CaCl_2$	<1 neutral protease, clostripain U/mg protein	Sigma C0773
0.5-2.0 FALGPA hydrolysis U/mg solid; 125 collagen digestion U/mg solid	1 unit hydrolyzes 1.0 µmol FALGPA (furylacryloyl-Leu-Gly-Pro-Ala)/min at pH 7.5, 25°C in the presence of Ca^{2+}; 1 unit liberates peptides from collagen equivalent in ninhydrin color to 1.0 µmol Leu/5 hr at pH 7.4, 37°C in the presence of Ca^{2+}			Sigma C2139
≥125 collagen digestion U/mg solid; 0.5-2.0 FALGPA hydrolysis U/mg solid	1 Collagen Digestion Unit liberates peptides from collagen equivalent in ninhydrin color to 1.0 µmol Leu/5 hr at pH 7.4, 37°C in presence of Ca^{2+}; 1 FALGPA H Unit hydrolyzes 1.0 µmol furylacryloyl-Leu-Gly-Pro-Ala/min at pH 7.5, 25°C with Ca^{2+}	Crude; cell culture tested	Clostripain, neutral protease	Sigma C2674

SPECIFIC ACTIVITY	UNITS DEFINITION	PREPARATION FORM	ADDITIONAL ACTIVITIES	SUPPLIER CATALOG NO.
Clostridium histolyticum continued				
≥125 collagen digestion U/mg solid; 0.5-2.0 FALGPA hydrolysis U/mg solid	1 Collagen Digestion Unit liberates peptides from collagen equivalent in ninhydrin color to 1.0 µmol Leu/5 hr at pH 7.4, 37°C in presence of Ca^{2+}; 1 FALGPA H Unit hydrolyzes 1.0 µmol furylacryloyl-Leu-Gly-Pro-Ala/min at pH 7.5, 25°C with Ca^{2+}	Crude; cell culture tested	Clostripain, neutral protease	Sigma C2674
1000-3000 collagen digestion U/mg solid; 4-12 FALGPA hydrolysis U/mg solid	1 Collagen Digestion Unit liberates peptides from collagen equivalent in ninhydrin color to 1.0 µmol Leu/5 hr at pH 7.4, 37°C in presence of Ca^{2+}; 1 FALGPA H Unit hydrolyzes 1.0 µmol furylacryloyl-Leu-Gly-Pro-Ala/min at pH 7.5, 25°C with Ca^{2+}	Highly purified; cell culture tested	<1 U/mg solid neutral protease, clostripain	Sigma C2799
0.5-2.0 FALGPA hydrolysis U/mg solid; 125 collagen digestion U/mg solid	1 unit hydrolyzes 1.0 µmol FALGPA (furylacryloyl-Leu-Gly-Pro-Ala)/min at pH 7.5, 25°C in the presence of Ca^{2+}; 1 unit liberates peptides from collagen equivalent in ninhydrin color to 1.0 µmol Leu/5 hr at pH 7.4, 37°C in the presence of Ca^{2+}			Sigma C5138
0.5-2.0 FALGPA hydrolysis U/mg solid; 125 collagen digestion U/mg solid	1 unit hydrolyzes 1.0 µmol FALGPA (furylacryloyl-Leu-Gly-Pro-Ala)/min at pH 7.5, 25°C in the presence of Ca^{2+}; 1 unit liberates peptides from collagen equivalent in ninhydrin color to 1.0 µmol Leu/5 hr at pH 7.4, 37°C in the presence of Ca^{2+}			Sigma C6885
2-5 FALGPA hydrolysis U/mg solid; 1200 collagen digestion U/mg solid	1 unit hydrolyzes 1.0 µmol FALGPA (furylacryloyl-Leu-Gly-Pro-Ala)/min at pH 7.5, 25°C in the presence of Ca^{2+}; 1 unit liberates peptides from collagen equivalent in ninhydrin color to 1.0 µmol Leu/5 hr at pH 7.4, 37°C in the presence of Ca^{2+}		Clostripain, nonspecific neutral protease and trypsin	Sigma C7657
1.8-2.2 FALGPA hydrolysis U/mg solid	1 unit hydrolyzes 1.0 µmol furylacryloyl-Leu-Gly-Pro-Ala/min at pH 7.5, 25°C in the presence of Ca^{2+}	Purified; lyophilized		Sigma C7926

SPECIFIC ACTIVITY	UNITS DEFINITION	PREPARATION FORM	ADDITIONAL ACTIVITIES	SUPPLIER CATALOG No.
Clostridium histolyticum continued				
1.1-1.5 FALGPA hydrolysis U/mg solid	1 unit hydrolyzes 1.0 µmol furylacryloyl-Leu-Gly-Pro-Ala/min at pH 7.5, 25°C in the presence of Ca^{2+}	Purified; lyophilized		Sigma C8051
0.5-0.9 FALGPA hydrolysis U/mg solid	1 unit hydrolyzes 1.0 µmol furylacryloyl-Leu-Gly-Pro-Ala/min at pH 7.5, 25°C in the presence of Ca^{2+}	Purified; lyophilized		Sigma C8176
0.1 FALGPA hydrolysis U/mg solid	1 unit hydrolyzes 1.0 µmol furylacryloyl-Leu-Gly-Pro-Ala/min at pH 7.5, 25°C in the presence of Ca^{2+}	Purified; lyophilized		Sigma C8301
1-3 FALGPA hydrolysis U/mg solid; 125 collagen digestion U/mg solid	1 unit hydrolyzes 1.0 µmol FALGPA (furylacryloyl-Leu-Gly-Pro-Ala)/min at pH 7.5, 25°C in the presence of Ca^{2+}; 1 unit liberates peptides from collagen equivalent in ninhydrin color to 1.0 µmol Leu/5 hr at pH 7.4, 37°C in the presence of Ca^{2+}			Sigma C9263
≥1200 collagen digestion U/mg solid; 2-5 FALGPA hydrolysis U/mg solid	1 Collagen Digestion Unit liberates peptides from collagen equivalent in ninhydrin color to 1.0 µmol Leu/5 hr at pH 7.4, 37°C in presence of Ca^{2+}; 1 FALGPA H Unit hydrolyzes 1.0 µmol furylacryloyl-Leu-Gly-Pro-Ala/min at pH 7.5, 25°C with Ca^{2+}	Crude; cell culture tested	Clostripain, neutral protease	Sigma C9407
0.5-2.0 FALGPA hydrolysis U/mg solid; 125 collagen digestion U/mg solid	1 unit hydrolyzes 1.0 µmol FALGPA (furylacryloyl-Leu-Gly-Pro-Ala)/min at pH 7.5, 25°C in the presence of Ca^{2+}; 1 unit liberates peptides from collagen equivalent in ninhydrin color to 1.0 µmol Leu/5 hr at pH 7.4, 37°C in the presence of Ca^{2+}		Clostripain, neutral protease and trypsin	Sigma C9891
≥1000 Mandl U/mg				Wako 034-13291
150-300 U/mg		Standardized		Wako 038-10531 034-10533 032-10534

SPECIFIC ACTIVITY	UNITS DEFINITION	PREPARATION FORM	ADDITIONAL ACTIVITIES	SUPPLIER CATALOG NO.
Clostridium histolyticum continued				
≥100 U/mg solid	1 unit liberates 1 μmol L-Leu equivalents from collagen/5 hr at pH 7.5, 37°C		Low proteolytic activities; Normal collagenase activity	Worthington LS04180 LS04182 LS04183 LS04185
≥160 U/mg solid	1 unit liberates 1 μmol L-Leu equivalents from collagen/5 hr at pH 7.5, 37°C		Low tryptic activity; Normal clostripain activity; High collagenase activity	Worthington LS04186 LS04188 LS04189 LS04191
≥125 U/mg solid	1 unit liberates 1 μmol L-Leu equivalents from collagen/5 hr at pH 7.5, 37°C	Crude		Worthington LS04194 LS04196 LS04197 LS04200
≥125 U/mg solid	1 unit liberates 1 μmol L-Leu equivalents from collagen/5 hr at pH 7.5, 37°C		High clostripain and tryptic activities	Worthington LS04194 LS04196 LS04197 LS04200
≥125 U/mg solid; 50 mg/vial	1 unit liberates 1 μmol L-Leu equivalents from collagen/5 hr at pH 7.5, 37°C	Sterile-filtered; lyophilized	High clostripain and tryptic activities	Worthington LS04202 LS04204
≥100 U/mg solid; 50 mg/vial	1 unit liberates 1 μmol L-Leu equivalents from collagen/5 hr at pH 7.5, 37°C	Sterile-filtered; lyophilized	Low proteolytic activities; Normal collagenase activity	Worthington LS04206 LS04208
≥160 U/mg solid; 50 mg/vial	1 unit liberates 1 μmol L-Leu equivalents from collagen/5 hr at pH 7.5, 37°C	Sterile-filtered; lyophilized	Low tryptic activity; Normal clostripain activity; High collagenase activity	Worthington LS04210 LS04212
≥125 U/mg solid; 50 mg/vial	1 unit liberates 1 μmol L-Leu equivalents from collagen/5 hr at pH 7.5, 37°C	Sterile-filtered; lyophilized		Worthington LS04214 LS04216
≥300 U/mg solid	1 unit liberates 1 μmol L-Leu equivalents from collagen/5 hr at pH 7.5, 37°C	Chromatographically purified; lyophilized	< 50 U/mg caseinase	Worthington LS05275 LS05273 LS05277

SPECIFIC ACTIVITY	UNITS DEFINITION	PREPARATION FORM	ADDITIONAL ACTIVITIES	SUPPLIER CATALOG No.
Clostridium histolyticum continued				
≥5 U/mg solid	1 unit liberates 1 μmol p-phenylazo-benzyloxycarbonyl-L-prolyl-L-Leu/min at pH 7.1, 37°C	Chromatographically purified; lyophilized	<0.2% clostripain (DTT-activated, BAEE) <3 U/mg caseinolytic activity (non-specific protease, see units definition)	Worthington LS05281 LS05280
Clostridium histolyticum strain F214; mixture				
300 Mandl U/mg solid	1 Mandl unit liberates 1 μmol amino acid determined by ninhydrin and expressed as Leu equivalents from native collagen/5 hr at pH 7.5, 37°C	Highly purified; mixture	Very low levels of caseinase No detectable pathogenic spores	ICN 362001
Paralithodes camtschatica (Kamchatka crab) hepatopancreas (optimum pH = 7.7, pI = 2.3, T = 37°C, MW = 70,000 Da)				
≥1000 Mandl U/mg solid	1 unit releases 1 μmol amino acid, expressed as Leu, from native collagen/5 hr at pH 7.5, 37°C	Lyophilized		Calbiochem 234133
120 Mandl U/mg	1 unit releases 1 μmol amino acid (expressed as Leu) from native collagen/5 hr at pH 7.5, 37°C	>99% purity; lyophilized		Calzyme 169A0120
0.5-1.5 U/mg solid	1 unit hydrolyzes casein to produce color equivalent to 1.0 μmol Tyr/min at pH 7.5, 37°C	Lyophilized	Mixture of proteases with broad specificity; active against insoluble collagen and protease substrates	Sigma C0299
Vibrio alginolyticus (Achromobacter iophagus)				
20 U/mg protein	1 unit forms 1 μmol of product with 4-phenylazobenzyl-oxycarbonyl-Pro-Leu-Gly-Pro-D-Arg/min at 25°C (Wunsch units)	Highly purified by affinity chromatography	<0.1 U/mg proteases (Hb as substrate)	Boehringer 602426

Leucolysin 3.4.24.6

REACTION CATALYZED
Cleavage of insulin B chain Phe^1↓Val, His^5↓Leu, Ala^{14}↓Leu, Gly^{20}↓Glu, Gly^{23}↓Phe and Phe^{24}↓Phe bonds and N-blocked dipeptides

SYNONYMS
Dispase, *Leucostoma* neutral proteinase, *Leucostoma* peptidase A

REACTANTS
Insulin B chain, N-blocked dipeptides

NOTES
- Classified as a hydrolase, acting on peptide bonds: metalloendopeptidase
- From the venom of the western cottonmouth moccasin snake, *Agkistrodon piscivorus leucostoma*

See Chapter 5. Multiple Enzyme Preparations

Neprilysin 3.4.24.11

REACTION CATALYZED
Preferential cleavage of polypeptides between hydrophobic residues, preferentially with Phe or Tyr at P1′

SYNONYMS
Endopeptidase 24.11, kidney-brush-border neutral peptidase, neutral endopeptidase

REACTANTS
Peptide bonds

NOTES
- Classified as a hydrolase, acting on peptide bonds: metalloendopeptidase
- A membrane-bound, zinc-containing glycoprotein
- Widely distributed in animal tissues, including brain, liver and lung; abundant in kidney brush border membrane
- Inhibited by phosphoramidon and thiorphan
- Common acute lymphoblastic leukemia antigen (CALLA)

SPECIFIC ACTIVITY	UNITS DEFINITION	PREPARATION FORM	ADDITIONAL ACTIVITIES	SUPPLIER CATALOG No.
Porcine kidney				
850-900 U/mg protein	1 unit liberates 1 μmol pNA/mg protein from Glutaryl-(Ala)$_2$-Phe-pNA/min at pH 7.4, 25°C	>98% purity by chromatography; lyophilized		Elastin NP47

Peptidyl-Lys Metalloendopeptidase 3.4.24.20

REACTION CATALYZED
Preferential cleavage: -Xaa↓Lys- (in which Xaa may be Pro)

SYNONYMS
Armillaria mellea neutral proteinase

REACTANTS
Peptide bonds

NOTES
- Classified as a hydrolase, acting on peptide bonds: metalloendopeptidase
- From the honey fungus *Armillaria mellea*
- Similar specificity is shown by *Myxobacter* AL-1 proteinase II

See Chapter 5. Multiple Enzyme Preparations

3.4.24.27 Thermolysin

REACTION CATALYZED
Preferential cleavage: ↓Leu > ↓Phe

SYNONYMS
Bacillus thermoproteolyticus neutral proteinase

REACTANTS
Peptide bonds, cytochrome c, insulin, tobacco mosaic virus protein

APPLICATIONS
- Probing the surface of intact pea chloroplasts
- Identify the location of disulfide bridges resistant to other proteases

NOTES
- Classified as a hydrolase, acting on peptide bonds: metalloendopeptidase
- A neutral, thermostable extracellular metalloendopeptidase
- Contains zinc and four calcium ions
- Species variants have been identified from *Micrococcus caseolyticus* and *Aspergillus oryzae*
- Inhibited by metal chelators (EDTA, 1,10-phenanthroline), mercuric chloride, α_2-macroglobulin, phospoamidates and phosphinates
- Not inhibited by aprotinin or other serine protease inhibitors
- Closely related but distinct enzymes are aeromonolysin (EC 3.4.24.25), aureolysin (EC 3.4.24.29), bacillolysin (EC 3.4.24.28), mycolysin (EC 3.4.24.31) and pseudolysin (EC 3.4.24.26)

SPECIFIC ACTIVITY	UNITS DEFINITION	PREPARATION FORM	ADDITIONAL ACTIVITIES	SUPPLIER CATALOG NO.
Aspergillus (optimum pH = 4.5-7.5, T = 50-55°C)				
500 U/g amino-peptidase		Food grade; powder		Biocatalysts Flavorpro 192P F192P
70 U/g proteinase		Powder		Biocatalysts Promod 192P P192P
200 U/g proteinase		Powder		Biocatalysts Promod 194P P194P
120 U/g endo-proteinase; 100 U/g amino-peptidase		Food grade; powder	20 U/g carboxypeptidase	Biocatalysts Promod 215P P215P
400 U/g proteinase		Powder		Biocatalysts Promod 25P P025P
60 U/g peptidase		Powder		Biocatalysts Promod 279P P279P
5 U/g proteinase		Powder	20,000 U/g amylase	Biocatalysts Promod 280P P280P
Bacillus thermoproteolyticus (optimum pH = 7.0-9.0, MW = 37,500 Da; thermostable in Ca^{2+} at 4-80°C)				
35 U/mg solid	1 unit liberates Folin-positive amino acids and peptides corresponding to 1 μmol Tyr/min at 35°C with casein as substrate; PU U = specific activity X MW of Tyr (M_r = 181.19)	Lyophilized		Boehringer 161586
10,000 U/mg protein 7000 U/mg dry weight	1 unit liberates a Folin's color equivalent to 1 mg Tyr from milk casein/min at pH 7.2, 35°C	3X Crystallized; lyophilized containing 20% CaOAc and 10% NaOAc		Calbiochem 58656
40 U/mg	1 unit liberates 1 μmol Folin-positive amino acids and peptides (calculated as Tyr)/min at pH 7.2, 37°C with casein as substrate	Lyophilized containing CaOAc and NaOAc as stabilizers		Fluka 88303
7000 PU/mg protein		3X Crystallized; crystalline		ICN 321351

Thermolysin continued — 3.4.24.27

SPECIFIC ACTIVITY	UNITS DEFINITION	PREPARATION FORM	ADDITIONAL ACTIVITIES	SUPPLIER CATALOG NO.
Bacillus thermoproteolyticus continued				
≥7000 PU/mg	1 PU unit liberates non-proteinaceous digestion product from milk casein to give 1 μg Tyr/min at pH 7.2, 35°C	Lyophilized		Seikagaku 120360-1 120360-2
Bacillus thermoproteolyticus Rokko (optimum pH = 7.0–9.0; calcium form retains 50% activity after incubation at 80°C/1 hr)				
50-100 U/mg protein	1 unit hydrolyzes casein producing color equivalent to 1.0 μmol Tyr/min at pH 7.5, 37°C using casein as substrate	Crystallized; lyophilized containing 20% buffer salts		ICN 150210
50-100 U/mg protein	1 unit hydrolyzes casein to produce color equivalent to 1.0 μmol (181 μg) Tyr/min at pH 7.5, 37°C	Crystallized; lyophilized containing calcium and Na buffer salts		Sigma P1512
100-200 U/g agarose; 1 mL gel yields 2-3 U	1 unit hydrolyzes casein to produce color equivalent to 1.0 μmol (181 μg) Tyr/min at pH 7.5, 37°C	Insoluble enzyme attached to 4% cross-linked beaded agarose; lyophilized containing lactose stabilizer		Sigma P9040
50-100 U/mg protein	1 unit hydrolyzes casein to produce color equivalent to 1.0 μmol (181 μg) Tyr/min at pH 7.5, 37°C	Crystallized; lyophilized containing 20% Ca^{2+} and NaOAc buffer salts; cell culture tested		Sigma T7902
7000 PU/mg		Lyophilized		Wako 201-08331

Bacillolysin — 3.4.24.28

REACTION CATALYZED

Similar bond cleavage to thermolysin (EC 3.4.24.27):
↓Leu > ↓Phe

SYNONYMS

Alkaline protease, *Bacillus* metalloendopeptidase, *Bacillus subtilis* neutral proteinase, dispase, megateriopeptidase (*Bacillus megaterium*), neutral protease

REACTANTS

Peptide bonds, fibronectin, type IV collagen

APPLICATIONS

- General uses: textiles, animal feed, brewing and distilling, baking, silage, wine and fruit juice, industrial cleaning, dairy, leather and wool
- Dairy applications include lactose reduction and flavor modification
- Fortifies brewing malt proteases
- Softens wheat gluten in baking
- Upgrades feed vegetable and animal protein
- Rapidly and gently separates intact epidermis from dermis and intact epithelial sheets in culture from substratum

Bacillolysin continued

NOTES

- Prevents unwanted clumping of cells cultured in suspension
- Gently dissociates a wide variety of animal tissues and organs to release individual cells
- Silk degumming
- Classified as a hydrolase, acting on peptide bonds: metalloendopeptidase
- Non-specific bond cleavage
- Activated by divalent cations
- Stabilized by Ca^{2+}
- Inhibited by EDTA, EGTA, Hg^{2+} and other heavy metals
- *Not inhibited* by serum or barley protease inhibitor
- Species variants include *Bacillus subtilis*, *Bacillus amyloliquefaciens*, *Bacillus megaterium*, *Bacillus mesentericus*, *Bacillus cereus* and *Bacillus stearothermophilus*

SPECIFIC ACTIVITY	UNITS DEFINITION	PREPARATION FORM	ADDITIONAL ACTIVITIES	SUPPLIER CATALOG NO.
***Bacillus licheniformis* (optimum pH = 7.0-11.0)**				
		Industrial grade		EnzymeDev Enzeco^R Alkaline Protease
		Industrial grade		EnzymeDev Enzeco^R Alkaline Protease-L
		Food grade		EnzymeDev Enzeco^R Alkaline Protease-L FG
0.5 U/mg solid	1 unit hydrolyzes casein to produce color equivalent to 1.0 μmol (181 μg) Tyr/min at pH 7.5, 37°C	Dust-free preparation of "alkaline protease" encapsulated in an inert, water-soluble layer		Sigma P6670
***Bacillus polymyxa* (optimum pH = 8.5, MW = 35,900 Da)**				
>0.5 U/mg	1 unit converts 1 μmol casein to product/min at pH 7.5, 37°C	Crude		Boehringer 165859
>6 U/mg	1 unit converts 1 μmol casein to product/min at pH 7.5, 37°C	Crude		Boehringer 210455 1284908 241750
>2.4 U/mL	1 unit converts 1 μmol casein to product/min at pH 7.5, 37°C	Crude		Boehringer 295825

SPECIFIC ACTIVITY	UNITS DEFINITION	PREPARATION FORM	ADDITIONAL ACTIVITIES	SUPPLIER CATALOG NO.
Bacillus polymyxa continued				
>6 U/mg solid	1 unit liberates 1 μmol Tyr/min from casein at pH 7.5, 37°C	Lyophilized		Calbiochem 322120
5000 caseinolytic U/100 mL		Sterile-filtered; frozen containing Hanks' Balanced Salt Solution, pH 7.4	No detectable bacteria, fungi, mycoplasma	Collaborative 40235
				Life Technol 17105-032 17105-041
0.4 U/mg solid	1 unit hydrolyzes casein to produce color equivalent to 1.0 μmol Tyr/min at pH 7.5, 37°C	Crude; powder	Many extraneous enzymes	Sigma P5647
1.0 U/mg solid	1 unit hydrolyzes casein to produce color equivalent to 1.0 μmol Tyr/min at pH 7.5, 37°C	Crude; powder	Many extraneous enzymes	Sigma P6141
≥6 U/mg solid	1 unit releases Folin-positive amino acids equivalent to 1 μmol Tyr from casein/min at pH 7.5, 37°C	Chromatographically purified; lyophilized		Worthington LM02100 LM02104 LM02108
Bacillus species (optimum pH = 7-8, T = 30-60°C, MW = 35,000 Da [SDS-PAGE]; stable pH 6.0-9.0 [40°C, 30 min], T < 70°C [pH 7.5, 10 min])				
≥10,000 U/g	1 unit produces amino acids equal to 100 mg L-Tyr/hr at pH 7.0 in 1 mL filtrate	Powder optimum pH = 8.0, T = 70°C		Amano Protease S
200 U/g endo-proteinase		Food grade; powder	30 U/g glutaminase	Biocatalysts Flavorpro 373P F373P
90 U/g proteinase		Powder		Biocatalysts Promod 223P P223P
70 U/g proteinase		Solution		Biocatalysts Promod 24L P024L
140 U/g proteinase		Powder		Biocatalysts Promod 24P P024P
700 U/g proteinase		Powder		Biocatalysts Promod 278P P278P
80 U/g proteinase		Solution		Biocatalysts Promod 31L P031L

SPECIFIC ACTIVITY	UNITS DEFINITION	PREPARATION FORM	ADDITIONAL ACTIVITIES	SUPPLIER CATALOG NO.
Bacillus species continued				
		Industrial grade		EnzymeDev Enzeco[R] High Alkaline Protease
$\geq$1000 U/mg solid	1 unit increases OD_{275} corresponding to 1 mg Tyr/min at pH 7.2, 35°C	Powder or granulate containing Ca^{2+} as stabilizer	<0.02% protease	Toyobo NEP-201
0.5 Xs/g 18 XS/g 36 XS/g 72 XS/g 200 XS/g 600 XS/g	1 unit produces 200 µg Tyr equivalent of aromatic nitrogen compounds soluble in 4% TCA from casein/min at pH 6.5, 35°C	Powder containing corn starch		ABM-RP Proteinase 05 Proteinase 18 Proteinase 36 Proteinase 72 Proteinase 200 Proteinase CXS
Bacillus subtilis (optimum pH = 5.5-7.5, T = 45-55°C)				
200 XS/mL	1 unit produces 200 µg Tyr equivalent of aromatic nitrogen compounds soluble in 4% TCA from casein/min at pH 6.5, 35°C	Solution stabilized with salt and sorbitol, pH 6.0-7.0	Bacterial α-amylase, β-glucanase	ABM-RP Proteinase 200L
$\geq$10,000 U/g	1 unit produces amino acids equal to 100 mg L-Tyr/hr at pH 10.0 in 1 mL filtrate	Powder optimum pH = 10.0, T = 60c		Amano Proleather
$\geq$150,000 U/g	1 unit produces amino acids equal to 100 mg L-Tyr/hr at pH 7.0 in 1 mL filtrate	Powder optimum pH = 7.0, T = 55°C		Amano Protease N
		Food grade		EnzymeDev Enzeco[R] Neutral Bacterial Protease
		Food grade		EnzymeDev Enzeco[R] Neutral Bacterial Protease
1 U/mg	1 unit hydrolyzes 1 µmol $N\alpha$-Tosyl-L-Arg methyl ester hydrochloride/min at pH 8.0, 30°C	Powder		Fluka 82462
35 U/mg protein	1 unit liberates 1 µmol Folin-positive amino acids and peptides (as Tyr)/min at pH 7.5, 35.5°C with hemoglobin as substrate	Solution		Fluka 82464

Bacillolysin continued 3.4.24.28

SPECIFIC ACTIVITY	UNITS DEFINITION	PREPARATION FORM	ADDITIONAL ACTIVITIES	SUPPLIER CATALOG NO.
Bacillus subtilis continued				
0.5 AU/g		Food grade; solution	No detectable α-amylase Contains β-glucanase	Novo Nordisk Neutrase^R 0.5 L
1.5 AU/g		Food grade; granulate at 300 micron particle size		Novo Nordisk Neutrase^R 1.5 MG
4.5 AU/g		Food grade; powder at 150 micron particle size		Novo Nordisk Neutrase^R 4.5 BG

Aureolysin 3.4.24.29

REACTION CATALYZED
    Preferential cleavage: hydrophobic P1' residues. Activates *Staphylococcus aureus* glutamyl endopeptidase (EC 3.4.21.19)

SYNONYMS
    Staphylococcus aureus neutral proteinase

REACTANTS
    Insulin β chain, *Staphylococcus aureus* glutamyl endopeptidase (EC 3.4.21.19), peptide bonds

NOTES
- Classified as a hydrolase, acting on peptide bonds: metalloendopeptidase
- A metalloenzyme from *Staphylococcus aureus*
- Confused with staphylokinase, the non-enzymatic activator of plasminogen

---

See Chapter 5. Multiple Enzyme Preparations

Mycolysin

REACTION CATALYZED
Preferential cleavage: bonds with hydrophobic residues in P1'

SYNONYMS
Pronase component, Streptomyces griseus neutral proteinase

REACTANTS
Peptide bonds, mucins

APPLICATIONS
- Degrading proteins completely to amino acids without decomposition
- Total degradation of proteins during isolation of DNA and RNA, without phenol
- Production of glycopeptides from purified glycoproteins
- Isolation of living chondrocytes

NOTES
- Classified as a hydrolase, acting on peptide bonds: metalloendopeptidase
- Isolated from Streptomyces griseus, S. naraensis and S. cacaoi
- Unusually non-specific protease, capable of hydrolyzing almost all peptide bonds in proteins
- Specificity similar to thermolysin, but more sensitive to inhibition by mercaptoacetyl-Phe-Leu
- Inhibited by EDTA and DFP
- Little structural similarity to other bacterial metalloendopeptidases

SPECIFIC ACTIVITY	UNITS DEFINITION	PREPARATION FORM	ADDITIONAL ACTIVITIES	SUPPLIER CATALOG NO.
Streptomyces griseus (optimum pH = 7.5, T = 40-60°C)				
7000 U/g solid	1 unit liberates Folin-positive amino acids and peptides equivalent to 1 μmol Tyr/min at pH 7.5, 40°C with casein as substrate	Lyophilized		AMRESCO E629
7000 U/mg solid	1 unit liberates Folin-positive amino acids and peptides corresponding to 1 μmol Tyr/min at pH 7.5, 40°C with casein as substrate	Lyophilized containing 20% CaOAc	No detectable nuclease Mixture of endo- and exoproteinases with unspecific proteolytic activities	Boehringer 165921 1459643
45,000 PUK/g solid (proteolytic units)	1 unit liberates acid soluble material from a 2% casein solution, equivalent to 25 μg Tyr/min measured at A_{275} at pH 7.4, 40°C	Lyophilized containing 10% CaOAc as stabilizer		Calbiochem 53702 Pronase[R]
>70,000 PUK (proteolytic units)/g dry weight	1 unit liberates a digestion product equivalent to 25 μg Tyr/min at pH 7.5, 40°C	Nuclease-free; lyophilized	<0.01% U/mg nuclease	Calbiochem 537088 Pronase[R]

Mycolysin continued   3.4.24.31

SPECIFIC ACTIVITY	UNITS DEFINITION	PREPARATION FORM	ADDITIONAL ACTIVITIES	SUPPLIER CATALOG NO.
Streptomyces griseus continued				
6 U/mg	1 unit liberates 1 μmol Folin-positive amino acids and peptides (as Tyr)/min at pH 7.5, 40°C with casein as substrate	Lyophilized		Fluka 81748
5 U/mg	1 unit liberates 1 μmol Folin-positive amino acids and peptides (as Tyr)/min at pH 7.5, 40°C with casein as substrate	Powder containing 70% protein	<0.01% starch 20% CaOAc	Fluka 81750
1,000,000 PU/g	1 PU liberates Folin-positive amino acids and peptides equivalent to 1 μg Tyr/min at pH 7.5, 40°C using casein as substrate	Grade I		ICN Neutralase™ 150208
250,000 PU/g	1 PU liberates Folin-positive amino acids and peptides equivalent to 1 μg Tyr/min at pH 7.5, 40°C using casein as substrate	Grade II		ICN Neutralase™ 152341
125-175 U/g solid	1 unit hydrolyzes 1.0 μmol BAEE/min at pH 7.0, 30°C	Insoluble enzyme attached to CMC containing 30% borate buffer salts		Sigma P0387
15-25 U/mg solid	1 unit hydrolyzes casein to produce peptide equivalent to 1.0 μmol (181 μg) Tyr/min at pH 11.0, 30°C	Twice as active at pH 11.0, 30°C than at usual assay conditions of pH 7.5, 37°C		Sigma P0652
500-750 U/g agarose; 1 mL gel yields 15-25 U	1 unit hydrolyzes 1.0 μmol BAEE/min at pH 7.0, 30°C	Insoluble enzyme attached to 4% cross-linked beaded agarose; lyophilized containing lactose stabilizer		Sigma P4531
4 U/mg solid	1 unit hydrolyzes casein to produce color equivalent to 1.0 μmol (181 μg) Tyr/min at pH 7.5, 37°C	25% as active at pH 11.0, 30°C; containing CaOAc		Sigma P5147
4 U/mg solid	1 unit hydrolyzes casein to produce color equivalent to 1.0 μmol (181 μg) Tyr/min at pH 7.5, 37°C	Containing 25% CaOAc	No detectable nuclease activity after 500 μg/mL incubated in endonuclease (nickase), endonuclease-exonuclease (non-radioactive) or RNase (non-radioactive) assays supplemented with 0.2% SDS and 10 m*M* EDTA	Sigma P6911
4 U/mg solid	1 unit hydrolyzes casein to produce color equivalent to 1.0 μmol (181 μg) Tyr/min at pH 7.5, 37°C	Starch-free; embryo tested; containing 25% CaOAc		Sigma P8811

β-Lytic Metalloendopeptidase

3.4.24.32

REACTION CATALYZED
Cleavage: the insulin β chain at Gly^{23}↓Phe > Val^{18}↓Cys(SO$_3$H) and N-acetylmuramoyl↓Ala

SYNONYMS
Achromopeptidase, *Myxobacter* β-lytic proteinase

REACTANTS
Peptide bonds, insulin β chain, N-acetylmuramoyl-Ala

APPLICATIONS
- Preparation of protoplasts and spheroplasts
- Preparation of bacterial enzymes
- Isolation of bacterial proteins via cell wall lysis

NOTES
- Classified as a hydrolase, acting on peptide bonds: metalloendopeptidase
- From *Achromobacter lyticus* and *Lysobacter enzymogenes*
- Digests bacterial cell walls
- Not a homologue of thermolysin
- Heat stable, bacteriolytic enzyme active against both Gram-(+) and -(-) organisms
- *Myxobacter* AL-1 proteinase I has similar characteristics

SPECIFIC ACTIVITY	UNITS DEFINITION	PREPARATION FORM	ADDITIONAL ACTIVITIES	SUPPLIER CATALOG NO.
Achromobacter lyticus				
20,000-40,000 U/mg solid	1 unit produces a ΔA$_{600}$ of 0.001/min/mL at pH 8.0, 37°C using *Micrococcus lysodeikticus* as substrate (1 cm light path)	Partially purified; powder containing lactose	Collagenase	Sigma A3422
	1 unit produces a ΔA$_{600}$ of 0.001/min/mL at pH 8.0, 37°C using *Micrococcus lysodeikticus* as substrate (1 cm light path)	Crude; powder containing lactose	Collagenase	Sigma A3547
300-600 U/mg solid	1 unit produces a ΔA$_{600}$ of 0.001/min/mL at pH 8.0, 37°C using *Micrococcus lysodeikticus* as substrate (1 cm light path)	Crude; powder containing 5% protein; balance primarily salts and medium components	Collagenase	Sigma A7550
***Achromobacter lyticus* M497-1 (optimum pH = 7.5-8.5, MW = 20,000 Da; heat stability at pH 6.0, 10 min: < 40°C = 100%, 50°C = 90%, 60°C = 75%, 70°C = very low)**				
10,000-20,000 U/mg	1 unit decreases turbidity at the rate of 0.001 absorbance units/min in a 1 mL reaction solution	Lyophilized	No detectable DNase	ICN 153530
1000 U/mg	1 unit decreases turbidity at the rate of 0.001 absorbance units/min in a 1 mL reaction solution	Powder containing lactose as excipient	No detectable DNase	ICN 153531
10,000-20,000 U/mg		Purified; lyophilized		Wako 015-09951

SPECIFIC ACTIVITY	UNITS DEFINITION	PREPARATION FORM	ADDITIONAL ACTIVITIES	SUPPLIER CATALOG NO.
Achromobacter lyticus continued				
1000 U/mg		Crude; powder		Wako 986-10121

Peptidyl-Asp Metalloendopeptidase

3.4.24.33

REACTION CATALYZED
    Cleavage: Xaa↓Asp, Xaa↓Glu and Xaa↓cysteic acid

SYNONYMS
    Endoproteinase Asp-N

REACTANTS
    Peptide bonds

APPLICATIONS
- Useful in sequencing because of its limited specificity
- Cleavage of fusion proteins

NOTES
- Classified as a hydrolase, acting on peptide bonds: metalloendopeptidase
- A metalloenzyme
- Isolated from *Pseudomonas fragi*
- Does not cleave at reduced or *S*-acetylated cysteine residues
- Inhibited by aprotinin, DFP, leupeptin, TLCK, EDTA and α-phenanthroline

SPECIFIC ACTIVITY	UNITS DEFINITION	PREPARATION FORM	ADDITIONAL ACTIVITIES	SUPPLIER CATALOG NO.
Pseudomonas fragi **(MW = 27,000 Da)**				
25,000 U/mg protein		>90% purity; lyophilized from buffer containing 10 m*M* Tris-HCl, pH 7.5	No other detectable proteases	Calbiochem 324708
Pseudomonas fragi, **mutant strain (optimum pH = 7.0-8.0, MW = 27,000 Da)**				
		Sequencing grade; free of impurities by SDS-PAGE, silver staining; lyophilized		Boehringer 1420488 1054589
≥20 U/μg		Sequencing grade; lyophilized		Wako 056-05921

3.4.24.40 Serralysin

REACTION CATALYZED
Preferential cleavage: bonds with hydrophobic residues in P′

SYNONYMS
Escherichia freundii proteinase, *Pseudomonas aeruginosa* alkaline proteinase, *Serratia marcescens* extracellular proteinase

REACTANTS
Insulin B chain, peptide bonds

NOTES
- Classified as a hydrolase, acting on peptide bonds: metalloendopeptidase
- A 50 kDa extracellular endopeptidase
- Isolated from *Pseudomonas aeruginosa*, *Escherichia freundii*, *Serratia marcescens* and *Erwinia chrysanthemi*
- Broad specificity and some species variation in cleavage of the insulin B chain
- pH optimum for protein digestion is 9-10

SPECIFIC ACTIVITY	UNITS DEFINITION	PREPARATION FORM	ADDITIONAL ACTIVITIES	SUPPLIER CATALOG NO.
Serratia species				
4-8 U/mg solid	1 unit hydrolyzes casein to produce color equivalent to 1.0 μmol (181 μg) Tyr/min at pH 7.5, 37°C			Sigma P2789

3.5.1.1 Asparaginase

REACTION CATALYZED
L-Asparagine + H_2O ↔ L-Aspartate + NH_3

SYSTEMATIC NAME
L-Asparagine amidohydrolase

SYNONYMS
Asparaginase II

REACTANTS
L-Asparagine, H_2O, L-aspartate, NH_3

APPLICATIONS
- Treatment of acute lymphoblastic leukemia

Asparaginase continued 3.5.1.1

NOTES

- Classified as a hydrolase, acting on carbon-nitrogen bonds other than peptide bonds, in linear amides
- Present in many animal tissues, bacteria, plants and the serum of certain rodents; not present in humans
- EC-2 from *E. coli* differs from EC-1 by its broad pH profile and its higher substrate affinity; EC-1 is commonly removed during purification
- EC-2 and *Erwinia* enzymes possess antilymphoma activity
- Competitively inhibited by 5-diazo-4-oxo-L-norvaline (DONV)

SPECIFIC ACTIVITY	UNITS DEFINITION	PREPARATION FORM	ADDITIONAL ACTIVITIES	SUPPLIER CATALOG NO.
Erwinia chrysanthemi				
100-500 U/mg protein	1 unit liberates 1.0 µmol NH₃ from L-Asn/min at pH 8.6, 37°C	Lyophilized containing 40% protein; balance primarily α-lactose and NaCl		Sigma A2925
Escherichia coli (MW = 141,000 Da, tetrameric)				
80 U/mg	1 unit converts 1 µmol L-Asn to product/min at 25°C	50% glycerol solution, pH 6.5	2% side conversion of L-Gln	Boehringer 102903
100-200 U/mg	1 unit liberates 1 µmol NH₃ from L-Asn/min at pH 8.6, 37°C	Lyophilized		Fluka 11185
70-100 U/mg protein	1 unit liberates 1 µmol NH₃ from L-Asn/min at pH 8.6, 25°C	50% glycerol solution, pH 6.5		Fluka 11187
250-300 U/mg protein	1 unit liberates 1.0 µmol NH₃ nitrogen from L-Asn/min at pH 8.6, 37°C	Chromatographically purified; lyophilized		ICN 150400
50-150 U/mg protein	1 unit liberates 1.0 µmol NH₃ from L-Asn/min at pH 8.6, 37°C	50% glycerol solution, pH 6.5		Sigma A3684
100-300 U/mg protein	1 unit liberates 1.0 µmol NH₃ from L-Asn/min at pH 8.6, 37°C	Chromatographically purified; lyophilized containing 80% protein; balance NaCl		Sigma A3809
100 U/mg protein	1 unit liberates 1.0 µmol NH₃ from L-Asn/min at pH 8.6, 37°C	50% glycerol solution, pH 6.5		Sigma A4887
	1 unit liberates 1.0 µmol NH₃ from L-Asn/min at pH 8.6, 37°C	Insoluble enzyme attached to beaded agarose; suspension in 2.0 M NaCl, pH 8		Sigma A9636
≥150 U/mg protein	1 unit releases 1 µmol NH₃ from L-Asn/min at pH 8.6, 37°C	Chromatographically purified; dialyzed; lyophilized		Worthington LS05042 LS05044 LS05048

3.5.1.2 Glutaminase

REACTION CATALYZED
L-Glutamine + H_2O ↔ L-Glutamate + NH_3

SYSTEMATIC NAME
L-Glutamine amidohydrolase

REACTANTS
L-Glutamine, H_2O, L-glutamate, NH_3

APPLICATIONS
- Determination glutamine
- Hydrolysis of defatted soy beans

NOTES
- Classified as a hydrolase, acting on carbon-nitrogen bonds other than peptide bonds, in linear amides

SPECIFIC ACTIVITY	UNITS DEFINITION	PREPARATION FORM	ADDITIONAL ACTIVITIES	SUPPLIER CATALOG No.
Bacillus subtilis (optimum pH = 6-7, T = 60°C; stable pH 6-10 [30°C, 1 hr], T < 50°C [pH 7.0, 30 min])				
60 U/g solid	1 unit liberates 1 µmol L-Glu/min at pH 7.0, 37°C	Powder		Nagase
Escherichia coli				
25 U/mg	1 unit deaminates 1 µmol L-Gln/min at pH 4.9, 37°C	Powder containing K succinate stabilizer, 1% EDTA, <0.2% NH_3	<0.01% NADH-oxidase	Fluka 49440
5-15 U/mg protein	1 unit deaminates 1 µmol L-Gln/min at pH 4.9, 37°C	Lyophilized containing 50% protein; balance primarily buffer salts	1 µg free ammonium ions/U of glutaminase	Sigma G5382
500-1500 U/mg protein	1 unit deaminates 1 µmol L-Gln/min at pH 4.9, 37°C	Chromatographically purified; lyophilized containing 10% protein; balance K succinate and stabilizer		Sigma G5894
50-200 U/mg protein	1 unit deaminates 1 µmol L-Gln/min at pH 4.9, 37°C	Lyophilized containing 30% protein; balance K succinate and EDTA	<0.1 µg free ammonium ions/U of glutaminase 0.01% NADH oxidase	Sigma G8880
Porcine kidney				
	1 unit deaminates 1.0 µmol L-Gln/min	Suspension in 2.0 M KPO_4, 0.01 M borate, 1% ethylene glycol, pH 8.0		Sigma G8025

Amidase 3.5.1.4

REACTION CATALYZED
Monocarboxylic acid amide + H_2O ↔ Monocarboxylate + NH_3

SYSTEMATIC NAME
Acylamide amidohydrolase

SYNONYMS
Acylamidase, acylase

REACTANTS
Monocarboxylic acid amide, H_2O, monocarboxylate, NH_3

NOTES
- Classified as a hydrolase, acting on carbon-nitrogen bonds other than peptide bonds, in linear amides
- Assayed as hydroxamate transferase

SPECIFIC ACTIVITY	UNITS DEFINITION	PREPARATION FORM	ADDITIONAL ACTIVITIES	SUPPLIER CATALOG NO.
Pseudomonas aeruginosa, recombinant expressed in *E. coli*				
200-400 U/mg protein	1 unit converts 1.0 µmol acetamide and hydroxylamine to acethydroxamate and NH_3/min at pH 7.2, 37°C	Solution containing 50% glycerol, trace amounts of β-MSH and PO_4 buffer salts		Sigma A6691

Urease 3.5.1.5

REACTION CATALYZED
Urea + H_2O ↔ CO_2 + 2 NH_3

SYSTEMATIC NAME
Urea amidohydrolase

SYNONYMS
Acid urease

REACTANTS
Urea, H_2O, CO_2, NH_3

APPLICATIONS
- Determination of urea in biological fluids
- Conjugate preparation for enzyme immunoassays, for determination of antigens, antibodies and metabolites

NOTES
- Treatment of straw
- Removal of urea from alcoholic beverages, especially Japanese sake
- Classified as a hydrolase, acting on C-N bonds other than peptide bonds, in linear amides
- A nickel protein occurring in many bacteria, several yeast species and many higher plants
- Activated by inorganic phosphate
- Stabilized by EDTA, glutathione, succinate and BSA
- Inhibited by Na^+, NH_3, and heavy metal ions; suramin and thiourea are competitive

3.5.1.5 Urease continued

SPECIFIC ACTIVITY	UNITS DEFINITION	PREPARATION FORM	ADDITIONAL ACTIVITIES	SUPPLIER CATALOG NO.
Bacillus pasteurii (7X more active at pH 8.2 than 7.0)				
100,000-300,000 U/g solid	1 unit liberates 1.0 µmol NH$_3$ from urea/min at pH 8.2, 25°C	Partially purified; powder containing 40% protein; balance primarily buffer salts, EDTA, stabilizer; <0.01 mg/U total reducing substance as glucose; <0.05 mg/U free NH$_3$	0.01% NADH and NADPH oxidase	Sigma U7127
Canavalia ensiformis (jack bean) (optimum pH = 6.0-7.3, pI = 5.0-5.1, T = 60°C, MW = 480,000 Da; K_M = 10.5 mM [urea]; stable pH 6.0-9.0 [30°C, 17 hr], T < 50°C [pH 8.0, 60 min])				
≥250 µg/mg protein	1 unit liberates 1 µmol ammonium/min at 25°C	Lyophilized		AMRESCO 0272
300 IU/mg protein; 6 U/mg powder	1 unit converts 1 µmol urea/min at 37°C	Powder containing stabilizers and buffer salts; <0.0010 mmol/IU NH$_3$		Beckman 682041
500 IU/mg protein; 7000-9000 IU/mL	1 unit converts 1 µmol urea/min at 37°C	50% glycerol solution, free of particulates; <0.008 mmol/IU NH$_3$		Beckman 683659
≥80 Nessler U/mg solid; ≥30 Bergmeyer U/mg solid	1 unit liberates 1 µmol NH$_3$/min at pH 7.0, 25°C; 1 Bergmeyer unit hydrolyzes 1 µmol urea/min at pH 8.0, 25°C; 1 Bergmeyer U = 2.4 Nessler U	Salt-free; lyophilized		Biozyme URE1
≥220 Nessler U/mg solid; ≥90 Bergmeyer U/mg solid	1 unit liberates 1 µmol NH$_3$/min at pH 7.0, 25°C; 1 Bergmeyer unit hydrolyzes 1 µmol urea/min at pH 8.0, 25°C; 1 Bergmeyer U = 2.4 Nessler U	Salt-free; lyophilized		Biozyme URE2
1400 Nessler U/mg protein; 580 Bergmeyer U/mg protein; 450 Nessler U/mg solid	1 unit liberates 1 µmol NH$_3$/min at pH 7.0, 25°C; 1 Bergmeyer unit hydrolyzes 1 µmol urea/min at pH 8.0, 25°C; 1 Bergmeyer U = 2.4 Nessler U	Salt-free; lyophilized		Biozyme URE3
100 U/mg protein	1 unit converts 1 µmol urea to product/min at 25°C	Lyophilized containing 25 mg protein, 60 mg sucrose, 15 mg citrate	<0.01% arginase	Boehringer 127442 737348
250 U/mg protein; 45 U/mg solid	1 unit converts 1 µmol urea to product/min at 25°C	Lyophilized	<0.01% L-asparaginase, L-arginase	Boehringer 174882
	1 unit oxidizes 1 µmol NADH/min at pH 7.6, 25°C in a coupled reaction using GlDH; 1 Calzyme U = 2.8 NH$_3$ U	Lyophilized containing 65% protein	No detectable glucose, "free" NH$_3$	Calzyme 116A0030

SPECIFIC ACTIVITY	UNITS DEFINITION	PREPARATION FORM	ADDITIONAL ACTIVITIES	SUPPLIER CATALOG No.
Canavalia ensiformis (jack bean) *continued*				
	1 unit oxidizes 1 μmol NADH/min at pH 7.6, 25°C in a coupled reaction using GIDH; 1 Calzyme U = 2.8 NH$_3$ U	Lyophilized containing 90% protein	No detectable glucose, "free" NH$_3$	Calzyme 116A0100
	1 unit oxidizes 1 μmol NADH/min at pH 7.6, 25°C in a coupled reaction using GIDH; 1 Calzyme U = 2.8 NH$_3$ U	Single band protein by SDS electrophoresis; lyophilized containing 95% protein		Calzyme 116A0600
>250 U/mg protein; >150 U/mg solid	1 unit liberates 1 μmol NH$_3$/min at pH 7.3, 25°C	Highly purified; salt-free; lyophilized containing stabilizers, <0.0002 μmol/U NH$_3$		Diagnostic U-080
60 U/mg	1 unit hydrolyzes 1 μmol urea/min at pH 7.0, 25°C	Powder		Fluka 94278
1000 U/mg	1 unit hydrolyzes 1 μmol urea/min at pH 7.0, 25°C	Powder		Fluka 94279
1 U/mg	1 unit hydrolyzes 1 μmol urea/min at pH 8.0, 25°C	Powder		Fluka 94280
35 U/mg	1 unit hydrolyzes 1 μmol urea/min at pH 7.0, 25°C	Powder		Fluka 94282
100 U/mg	1 unit hydrolyzes 1 μmol urea/min at pH 8.0, 25°C	Lyophilized		Fluka 94285
>250 U/mg protein	1 unit forms 1 μmol NH$_3$/min at pH 7.3, 25°C	Lyophilized		GDS Tech U-080
>200 U/mg protein; >100 U/mg solid	1 unit liberates 1.0 μmol NH$_3$/min at pH 7.6, 25°C	Lyophilized containing <0.0002 μmol/U free NH$_3$		Genzyme 1651
3500-4500 U/g	1 unit liberates 1 μmol NH$_3$/min at pH 7.0, 25°C			ICN 103211
≥100,000 U/g	1 unit liberates 1 μmol NH$_3$/min at pH 7.0, 25°C			ICN 160048
>100 U/mg protein; >40 U/mg solid	1 unit hydrolyzes 1 μmol urea/min at pH 8.0, 25°C	Lyophilized containing <0.0002 μmol/U free NH$_3$		Randox UR 889L
600,000-1,200,000 U/g solid	1 unit liberates 1.0 μmol NH$_3$ from urea/min at pH 7.0, 25°C	Crystallized		Sigma U0251
400,000-800,000 U/g solid	1 unit liberates 1.0 μmol NH$_3$ from urea/min at pH 7.0, 25°C	Highly purified; powder containing trace β-MSH		Sigma U0376
17,000-34,000 U/g	1 unit liberates 1.0 μmol NH$_3$ from urea/min at pH 7.0, 25°C	Powder containing <0.05 μg/U free NH$_3$ and <0.5 mg/U total reducing substance as glucose		Sigma U1500

3.5.1.5 Urease continued

SPECIFIC ACTIVITY	UNITS DEFINITION	PREPARATION FORM	ADDITIONAL ACTIVITIES	SUPPLIER CATALOG NO.
Canavalia ensiformis (jack bean) *continued*				
17,000-34,000 U/g; 230 U/tablet	1 unit liberates 1.0 μmol NH$_3$ from urea/min at pH 7.0, 25°C	Tablets containing <0.05 μg/U free NH$_3$ and <0.5 mg/U total reducing substance as glucose		Sigma U1625
17,000-34,000 U/g; 910 U/tablet	1 unit liberates 1.0 μmol NH$_3$ from urea/min at pH 7.0, 25°C	Tablets containing <0.05 μg/U free NH$_3$ and <0.5 mg/U total reducing substance as glucose		Sigma U1750
17,000-34,000 U/g; 680 U/mL	1 unit liberates 1.0 μmol NH$_3$ from urea/min at pH 7.0, 25°C	Solution containing glycerol, <0.05 μg/U free NH$_3$, <0.5 mg/U total reducing substance as glucose		Sigma U1875
30,000-80,000 U/g solid	1 unit liberates 1.0 μmol NH$_3$ from urea/min at pH 7.0, 25°C	Purified; powder containing <0.02 μg/U free NH$_3$, <0.2 mg/U total reducing substance as glucose	Substantially reduced free amino acids	Sigma U2000
60,000-120,000 U/g solid	1 unit liberates 1.0 μmol NH$_3$ from urea/min at pH 7.0, 25°C	Purified; powder containing <0.01 μg/U free NH$_3$ and <0.1 mg/U total reducing substance as glucose	Substantially reduced free amino acids	Sigma U2125
50,000-80,000 U/g solid	1 unit liberates 1.0 μmol NH$_3$ from urea/min at pH 7.0, 25°C	Powder containing <0.03 μg/U free NH$_3$ and <0.3 mg/U total reducing substance as glucose		Sigma U4002
800-1200 U/g solid	1 unit liberates 1.0 μmol NH$_3$ from urea/min at pH 7.0, 25°C	Insoluble enzyme attached to polyacrylamide containing 30% borate buffer salts		Sigma U5376
30,000-55,000 U/g agarose	1 unit liberates 1.0 μmol NH$_3$ from urea/min at pH 7.0, 25°C	Attached to 4% cross-linked beaded agarose; lyophilized containing lactose stabilizer		Sigma U7878
6000-1200 U/g solid	1 unit liberates 1.0 μmol NH$_3$ from urea/min at pH 7.0, 25°C	Powder containing some insolubles		Sigma U8876
≥100 U/mg solid	1 unit forms 2 μmol NH$_3$/min at pH 8.0, 37°C	Lyophilized containing 60-80% EDTA, glutathione, succinate, BSA as stabilizers, <0.0005% μg/U NH$_3$	<0.02% asparaginase <0.002% arginase	Toyobo URH-201
≥45 U/mg solid	1 unit forms 2 μmol NH$_3$/min at pH 8.0, 37°C	Lyophilized containing 60-80% EDTA, glutathione, succinate, BSA as stabilizers, <0.0005% μg/U NH$_3$	<0.02% asparaginase <0.002% arginase	Toyobo URH-301
80-150 U/mg				Wako 210-00781
≥45 U/mg solid	1 unit liberates 1 μmol NH$_3$/min at pH 7.6, 25°C in a GlDH coupled system	Lyophilized	No detectable NH$_3$	Worthington LS03885 LS03886 LS03887 LS03889

Urease continued   3.5.1.5

SPECIFIC ACTIVITY	UNITS DEFINITION	PREPARATION FORM	ADDITIONAL ACTIVITIES	SUPPLIER CATALOG NO.
Lactobacillus fermentum (optimum pH = 3.0-4.0, T = 60°C; stable pH 4-8 [30°C, 30 min], T < 50°C [pH 4, 10 min])				
		Food grade		EnzymeDev Nagapsin^R Acid Urease
2000 U/g solid	1 unit liberates 1 μmol NH$_3$/min at pH 4.0, 37°C	Powder		Nagase Nagapsin
Methylophilus species (50% activity at pH 7.0, 25°C)				
500 U/mg protein	1 unit liberates 1.0 μmol NH$_3$ from urea/min at pH 8.0, 30°C in a coupled system with L-GlDH	Lyophilized containing Tris buffer salts		Sigma U9754
Microbial (optimum pH = 8.9, MW = 210,000 Da [100,000/subunit]; K_M = 0.36 mM [urea]; stable pH 6.6-9.6 [4°C, 24 hr], T < 50°C [pH 8.0, 15 min])				
> 250 U/mg protein	1 unit forms 2 μmol NH$_3$/min at 30°C	Lyophilized	<0.01% asparaginase, arginase, NADH oxidase	Unitika

Penicillin Amidase   3.5.1.11

REACTION CATALYZED
    Penicillin + H$_2$O ↔ A carboxylate + 6-aminopenicillanate

SYSTEMATIC NAME
    Penicillin amidohydrolase

SYNONYMS
    Penicillin acylase

REACTANTS
    Penicillin, H$_2$O, carboxylate, 6-aminopenicillanate (6-APA)

APPLICATIONS
- Commercial manufacture of 6-APA and 7-ADCA
- Resolution of secondary alcohols

NOTES
- Classified as a hydrolase, acting on carbon-nitrogen bonds other than peptide bonds, in linear amides

SPECIFIC ACTIVITY	UNITS DEFINITION	PREPARATION FORM	ADDITIONAL ACTIVITIES	SUPPLIER CATALOG NO.
Escherichia coli (optimum pH = 7.8-8.0)				
8-16 U/mg protein	1 unit hydrolyzes 1 μmol benzylpenicillin/min at pH 7.6, 37°C	Suspension in 0.1 M KPO$_4$ buffer, pH 7.5		Fluka 76427

3.5.1.11 Penicillin Amidase continued

SPECIFIC ACTIVITY	UNITS DEFINITION	PREPARATION FORM	ADDITIONAL ACTIVITIES	SUPPLIER CATALOG NO.
Escherichia coli continued				
100 U/g	1 unit cleaves 1 μmol penicillin G to 6-APA/min at pH 7.6, 37°C	Powder		Fluka 76429
280 IU/g solid	1 IU generates 1 μmol 6-APA/min at pH 7.8-8.0, 37°C using 4% penicillin G solution	Immobilized on polymer beads; suspension in PO_4 buffer and preservatives	No detectable β-lactamase	Hindustan
10 U/mg protein	1 unit hydrolyzes 1.0 μmol benzylpenicillin/min at pH 7.8, 37°C	Solution containing 0.1 M KPO_4, pH 7.5		Sigma P3319
60-120 U/g solid	1 unit hydrolyzes 1.0 μmol benzylpenicillin/min at pH 7.8, 37°C	Insoluble enzyme attached to macroporous oxirane acrylic beads; lyophilized containing 50% glucose stabilizer		Sigma P3942
Escherichia coli 5K (optimum pH = 7.5-8.1, pI = 6.35 and 6.75, T = 54°C, MW = 70,000 Da)				
15 U/mg protein	1 unit cleaves 1 μmol phenylacetic acid from penicillin G/min at pH 7.8, 37°C	Sterile; solution containing 100 mM $NaPO_4$, pH 7.5		Calbiochem 516329

3.5.1.13 Aryl-Acylamidase

REACTION CATALYZED
    Anilide + H_2O ↔ Carboxylate + aniline
SYSTEMATIC NAME
    Aryl-acylamide amidohydrolase
REACTANTS
    Anilide, H_2O, carboxylate, aniline

NOTES
- Classified as a hydrolase, acting on carbon-nitrogen bonds other than peptide bonds, in linear amides
- Also acts on 4-substituted anilides

SPECIFIC ACTIVITY	UNITS DEFINITION	PREPARATION FORM	ADDITIONAL ACTIVITIES	SUPPLIER CATALOG NO.
Microbial (optimum pH = 8.5, T = 30°C)				
>15 U/mg protein	1 unit converts 1 μmol *p*-nitro-acetanilide to PNP/min at pH 8.5, 30°C	Solution		GDS Tech AR-100
Pseudomonas fluorescens ATCC 39005				
2-8 U/mg protein	1 unit converts 1.0 μmol *N*-Ac-*p*-aminophenol (acetaminophen) to *p*-aminophenol/min at pH 9.0, 37°C	Lyophilized containing 5% protein; balance Trizma buffer salts and stabilizer		Sigma A4951

Aminoacylase 3.5.1.14

REACTION CATALYZED
An *N*-acyl-*L*-amino acid + H_2O ↔ A carboxylate + an *L*-amino acid

SYSTEMATIC NAME
N-acyl-*L*-amino acid amidohydrolase

SYNONYMS
Acylase I, benzamidase, dehydropeptidase II, hippuricase, histozyme

REACTANTS
N-acyl-*L*-amino acid, dehydropeptides, H_2O, carboxylate, *L*-amino acid

APPLICATIONS
- Resolving *D*- and *L*-amino acids
- Removal of acetyl groups in amino acids

NOTES
- Classified as a hydrolase, acting on carbon-nitrogen bonds other than peptide bonds, in linear amides
- Wide specificity

SPECIFIC ACTIVITY	UNITS DEFINITION	PREPARATION FORM	ADDITIONAL ACTIVITIES	SUPPLIER CATALOG NO.
Aspergillus melleus				
0.5-1.0 U/mg	1 unit hydrolyzes 1 μmol *N*-Ac-*L*-Met/min at pH 7.0, 25°C	Powder		Fluka 01818
	1 unit hydrolyzes 1.0 μmol *N*-Ac-*L*-Met/hr at pH 7.0, 25°C	Lyophilized		Sigma A2156
Aspergillus species (optimum pH = 7.0 and 8.0-8.5, T = 40-50°C and 60°C; stable pH 6-8, T < 40°C and < 60°C)				
30,000 U/g	1 unit produces 1 μmol *L*-Met/30 min at pH 8.0, 37°C with *N*-Ac-*DL*-Met as substrate	Purified; powder		Amano Acylase
Porcine kidney (optimum pH = 7.0, T = 25°C, m2 = 86,000 Da; stable at pH 7 [70°C], but irreversibly inactivated at pH < 5.0)				
≥1000 U/mg solid	1 unit hydrolyzes 1.0 μmol *N*-Ac-*L*-Met/hr at pH 7.0, 25°C	Salt-free; lyophilized		Biozyme AC1
≥2000 U/mg solid	1 unit hydrolyzes 1.0 μmol *N*-Ac-*L*-Met/hr at pH 7.0, 25°C	Salt-free; lyophilized		Biozyme AC2
≥4000 U/mg solid	1 unit hydrolyzes 1.0 μmol *N*-Ac-*L*-Met/hr at pH 7.0, 25°C	Chromatographically prepared; salt-free; lyophilized		Biozyme AC3
1000 U/mg protein	1 unit hydrolyzes 1.0 μmol *N*-Ac-*L*-Met/hr at pH 7.0, 25°C	Salt-free; lyophilized containing 90% protein		Calzyme 051A1000
5000 U/mg protein	1 unit hydrolyzes 1.0 μmol *N*-Ac-*L*-Met/hr at pH 7.0, 25°C	Salt-free; lyophilized containing 90% protein		Calzyme 051A5000
30 U/mg	1 unit hydrolyzes 1.0 μmol *N*-Ac-*L*-Met/hr at pH 7.0, 25°C	Powder		Fluka 01816
15 U/mg	1 unit hydrolyzes 1.0 μmol *N*-Ac-*L*-Met/hr at pH 7.0, 25°C	Salt-free; lyophilized		Fluka 01821

3.5.1.14 Aminoacylase continued

SPECIFIC ACTIVITY	UNITS DEFINITION	PREPARATION FORM	ADDITIONAL ACTIVITIES	SUPPLIER CATALOG NO.
Porcine kidney continued				
70-110 U/mg	1 unit hydrolyzes 1.0 µmol N-Ac-L-Met/hr at pH 7.0, 25°C	Powder		Fluka 01831
10-20 U/mg	1 unit hydrolyzes 1.0 µmol N-Ac-L-Met/hr at pH 7.0, 25°C	Powder		Fluka 01833
2000-3000 U/mg protein	1 unit hydrolyzes 1.0 µmol N-Ac-L-Met/hr at pH 7.0, 25°C	Lyophilized		ICN 100062
2000-3000 U/mg protein	1 unit hydrolyzes 1.0 µmol N-Ac-L-Met/hr at pH 7.0, 25°C	Lyophilized containing 75% protein		Sigma A3010
5000-10,000 U/mg protein	1 unit hydrolyzes 1.0 µmol N-Ac-L-Met/hr at pH 7.0, 25°C	>90% purity by HPLC; salt-free; lyophilized		Sigma A5810
5000-8000 U/mg protein	1 unit hydrolyzes 1.0 µmol N-Ac-L-Met/hr at pH 7.0, 25°C	Salt-free; lyophilized		Sigma A7264
500-1500 U/mg protein	1 unit hydrolyzes 1.0 µmol N-Ac-L-Met/hr at pH 7.0, 25°C	Salt-free; lyophilized		Sigma A8376
3500-5000 U/mg		Lyophilized		Wako 018-11151

3.5.1.15 Aspartoacylase

REACTION CATALYZED

N-Acyl-L-aspartate + H_2O ↔ Carboxylate + L-aspartate

SYSTEMATIC NAME

N-Acyl-L-aspartate amidohydrolase

SYNONYMS

Acylase II, aminoacylase II

REACTANTS

N-Acyl-L-aspartate, H_2O, carboxylate, L-aspartate

NOTES

- Classified as a hydrolase, acting on carbon-nitrogen bonds other than peptide bonds, in linear amides

SPECIFIC ACTIVITY	UNITS DEFINITION	PREPARATION FORM	ADDITIONAL ACTIVITIES	SUPPLIER CATALOG NO.
Porcine kidney				
1.5 U/mg solid	1 unit hydrolyzes 1 µmol N-Ac-L-Asp/hr at pH 7.0, 37°C	Salt-free; lyophilized		ICN 191229

Choloylglycine Hydrolase 3.5.1.24

REACTION CATALYZED

3α, 7α, 12α-Trihydroxy-5β-cholan-24-oylglycine + H$_2$O ↔ 3α, 7α, 12α-trihydroxy-5β-cholanate + glycine

SYSTEMATIC NAME

3α, 7α, 12α-Trihydroxy-5β-cholan-24-oylglycine amidohydrolase

SYNONYMS

Bile salt hydrolase, choloyltaurine hydrolase, glycocholase

REACTANTS

3α, 7α, 12α-Trihydroxy-5β-cholan-24-oylglycine, 3α, 12α-dihydroxy-5β-cholan-24-oylglycine, choloyltaurine, H$_2$O, 3α, 7α, 12α-trihydroxy-5β-cholanate, glycine

NOTES

- Classified as a hydrolase, acting on carbon-nitrogen bonds other than peptide bonds, in linear amides

SPECIFIC ACTIVITY	UNITS DEFINITION	PREPARATION FORM	ADDITIONAL ACTIVITIES	SUPPLIER CATALOG NO.
Clostridium perfringens (welchii)				
10-20 U/mg solid	1 unit hydrolyzes 1.0 μmol glycocholic acid to Gly and cholic acid/5 min at pH 5.6, 37°C	Crude; acetone powder		Sigma C3636
100-300 U/mg protein	1 unit hydrolyzes 1.0 μmol glycocholic acid to Gly and cholic acid/5 min at pH 5.6, 37°C	Partially purified; lyophilized containing 30% protein; balance primarily buffer salts and stabilizer		Sigma C4018

Peptide -N^4-(N-Acetyl-β-Glucosaminyl)Asparagine Amidase 3.5.1.52

REACTION CATALYZED

Hydrolysis of an N^4-(acetyl-β-glucosaminyl)-asparagine residue in which the glucosamine residue may be further glycosylated, to yield a (substituted) N-acetyl-β-D-glucosaminylamine and a peptide containing an aspartic residue

SYSTEMATIC NAME

N-Linked-glycopeptide-(N-acetyl-β-D-glucosaminyl)-L-asparagine amidohydrolase

SYNONYMS

Glycopeptidase, glycopeptidase A, glycopeptidase F, glycopeptide N-glycosidase, N-glycanase, N-glycosidase A, N-glycosidase F, N-oligosaccharide glycopeptidase, PNGase F

REACTANTS

N^4-(Acetyl-β-glucosaminyl)-asparagine residue, N-Acetyl-β-D-glucosaminylamine, aspartic peptide, ammonia, RNase B, human transferrin, α-1-acid glycoprotein, human IGM, fetuin ovomucoid, invertase

3.5.1.52 Peptide -N^4-(N-Acetyl-β-Glucosaminyl)Asparagine Amidase continued

APPLICATIONS
- Removes carbohydrate residues from protein
- Glycosylates protein

NOTES
- Classified as a hydrolase, acting on carbon-nitrogen bonds other than peptide bonds, in linear amides
- Hydrolyzes a wider range of glycopeptides than endoglycosidase D or H
- Does not act on (GlcNAc)Asn; requires more than two amino acid residues in the substrate
- Inhibited by SDS; reaction mix should contain NP40 to counteract SDS
- Distinct from endoglycosidase F

SPECIFIC ACTIVITY	UNITS DEFINITION	PREPARATION FORM	ADDITIONAL ACTIVITIES	SUPPLIER CATALOG NO.
Almond (optimum pH = 4.0-6.0, T = 37°C, MW = 52,500 Da)				
>0.5 U/mg protein	1 unit hydrolyzes 1 μmol ovalbumin glycopeptide (Glu-Glu-Lys-Tyr-Asn(CHO)-Leu-Thr-Ser-Val)/min at pH 5, 37°C	Solution containing 50 mM citrate/PO$_4$ buffer, 50% glycerol, pH 5.0	No detectable proteases <0.1% α- and β-Gal, β-glucosidase, α- and β-mannosidase, β-NAH, α-L-fucosidase, sialidase, β-xylosidase	Boehringer 1642995
15 U/mg	1 unit hydrolyzes 1 μmol ovalbumin glycopeptide/min at pH 5, 37°C	Lyophilized		ICN 321291
10-20 U/mg protein	1 unit hydrolyzes 1 μmol ovalbumin glycopeptide/min at pH 5.0, 37°C	Lyophilized		Seikagaku 100676-1
Flavobacterium meningosepticum (optimum pH = 7.0-9.0, MW = 36,000 Da)				
25,000 U/mg protein	1 unit hydrolyzes 1 nmol dansyl fetuin glycopeptide/min at pH 7.2, 37°C	Solution containing 50 mM NaPO$_4$, 12.5 mM EDTA, 50% glycerol, pH 7.2	No detectable β-Gal, β-glucosidase, α- and β-mannosidase, β-NAH, α-L-fucosidase, sialidase, proteases <0.02% endoglycosidase F	Boehringer 903337 913782
25,000 U/mg protein	1 unit hydrolyzes 1 μmol dansyl fetuin/min at pH 7.2, 37°C	20 mM KPO$_4$, 50 mM EDTA, 0.05% NaN$_3$, pH 7.2	No detectable β-NAH, α-fucosidase, β-Gal, β-glucosidase, α-mannosidase, β-mannosidase, protease, sialidase <0.02% endoglyosidase F	Calbiochem 362184
1,000,000 U/mL	1 unit removes >95% carbohydrate from 10 μg denatured RNase B/hr at 37°C; 500 NEB unit=1 IUB milliunit	Purified; 20 mM Tris-HCl, 50 mM NaCl, 1 mM Na$_2$EDTA, 50% glycerol, pH 7.5	No detectable contaminating exoglycosidase, proteolytic activity	NE Biolabs 701S 701L
	1 unit hydrolyzes 1.0 nmol [^{3}H]-dansyl-fetuin/min at pH 7.2, 37°C	Solution containing 50% glycerol, 100 mM NaPO$_4$, 25 mM EDTA, 5 mM NaN$_3$, pH 7.2		Sigma G8031
25,000 U/mg		Suspension		Wako 078-03521

Peptide -N^4-(N-Acetyl-β-Glucosaminyl)Asparagine Amidase *continued*   3.5.1.52

SPECIFIC ACTIVITY	UNITS DEFINITION	PREPARATION FORM	ADDITIONAL ACTIVITIES	SUPPLIER CATALOG NO.
Flavobacterium meningosepticum, recombinant expressed in *E. coli*				
	1 unit hydrolyzes 1.0 nmol [³H]-dansyl-fetuin/min at pH 7.2, 37°C	Solution containing 50% glycerol, 0.08 M NaPO₄, 1 mM EDTA, 0.15 M NaCl, pH 7.5		Sigma G4403
Flavobacterium meningosepticum, recombinant from *E. coli* (optimum pH = 7.0–8.0, MW = 35,500 Da)				
25,000 U/mg protein	1 unit hydrolyzes 1 nmol dansyl fetuin glycopeptide/min at pH 7.2, 37°C	Solution containing 50 mM NaPO₄, 12.5 mM EDTA, 50% glycerol, pH 7.2	No detectable endoglycosidase F, β-Gal, β-glucosidase, α- and β-mannosidase, β-NAH, α-L-fucosidase, sialidase, proteases	Boehringer 1365169 1365177 1643037
25,000 U/mg protein	1 unit hydrolyzes 1 nmol dansyl fetuin glycopeptide/min at pH 7.2, 37°C	Lyophilized	No detectable endoglycosidase F, β-Gal, β-glucosidase, α- and β-mannosidase, β-NAH, α-L-fucosidase, sialidase, proteases	Boehringer 1365185 1365193 1643045
Unspecified				
	1 unit hydrolyzes 1 nmol fetuin/min at pH 8.6, 37°C			ICN 151406

Allantoinase   3.5.2.5

REACTION CATALYZED
    Allantoin + H₂O ↔ Allantoate

SYSTEMATIC NAME
    Allantoin amidohydrolase

REACTANTS
    Allantoin, H₂O, allantoate

NOTES
- Classified as a hydrolase, acting on carbon-nitrogen bonds other than peptide bonds, in cyclic amides

SPECIFIC ACTIVITY	UNITS DEFINITION	PREPARATION FORM	ADDITIONAL ACTIVITIES	SUPPLIER CATALOG NO.
Peanut				
50-200 U/g protein	1 unit hydrolyzes 1.0 μmol allantoin to allantoate/min at pH 7.0, 25°C	Lyophilized containing 50% protein; balance primarily PO₄ buffer salts and EDTA	Allantoicase	Sigma A3040

3.5.2.6 β-Lactamase

REACTION CATALYZED

A β-Lactam + H_2O ↔ A substituted β-amino acid

SYSTEMATIC NAME

β-Lactamhydrolase

SYNONYMS

Cephalosporinase, penicillinase

REACTANTS

β-Lactam, H_2O, substituted β-amino acid, penicillins, cephalosporins, acetate, D-benzylpenicilloate

APPLICATIONS

- Determination of penicillin or cephalosporin levels in fluids
- Destroys penicillins and cephalosporins in body fluids and culture media
- Enzyme label in ELISA kits

NOTES

- Classified as a hydrolase, acting on carbon-nitrogen bonds other than peptide bonds, in cyclic amides
- A group of enzymes of varying specificity for β-lactam hydrolysis. Some act more rapidly on penicillins; some more rapidly on cephalosporins
- Inactivates the antimicrobial properties of penicillins and cephalosporins by splitting the β-lactam ring

SPECIFIC ACTIVITY	UNITS DEFINITION	PREPARATION FORM	ADDITIONAL ACTIVITIES	SUPPLIER CATALOG NO.
Bacillus cereus (optimum pH = 7.0)				
2000 Levy U/mL/min	1 Levy unit inactivates 59.3 IU Na penicillin G/hr at pH 7.0, 25°C	Highly purified; sterile; solution		Difco 0345-63-8
20,000 Levy U/mL/min	1 Levy unit inactivates 59.3 IU Na penicillin G/hr at pH 7.0, 25°C	Highly purified; sterile; solution		Difco 0346-63-7 0346-72-6 0346-73-5
13 U/mg	1 unit hydrolyzes 1 μmol benzyl-penicillin (β-lactamase I)/min at pH 7.0, 25°C	Lyophilized	3 U/mg β-lactamase II	Fluka 61305
>15 U/mg solid	1 unit hydrolyzes 1.0 μmol penicillin/min at pH 7.0, 25°C	Lyophilized		Genzyme 1541
>15 U/mg solid	1 unit hydrolyzes 1.0 μmol penicillin/min at pH 7.0, 25°C	Sterile product in vials		Genzyme 1545
2000-2500 IU/mg	1 IU hydrolyzes 1 μmol Na penicillin G/min at pH 7.8, 37°C	Lyophilized	No detectable cephalosporinase	Hindustan
1200 U/vial	1 unit inactivates 1 μmol (600 IU) penicillin/hr at pH 7.0, 30°C	Highly purified; lyophilized		ICN 104771

β-Lactamase continued

3.5.2.6

SPECIFIC ACTIVITY	UNITS DEFINITION	PREPARATION FORM	ADDITIONAL ACTIVITIES	SUPPLIER CATALOG No.
Bacillus cereus continued				
500 U β-Lactamase I and 50 U β-Lactamase II/ vial	1 unit hydrolyzes 1.0 μmol benzyl-penicillin and 1 μmol cephalosporin C/min at pH 7.0, 25°C in the presence of EDTA	Lyophilized		ICN 190034
1500-3000 U/mg protein benzylpenicillin as substrate; 10-30 U/mg protein cephaloridine as substrate	1 unit hydrolyzes 1.0 μmol substrate/min at pH 7.0, 25°C	Lyophilized containing 10% protein; balance PO_4 and citrate buffer salts		Sigma P0389
50-100 U/mg protein benzylpenicillin as substrate; 0.2-0.6 U/mg protein cephaloridine as substrate	1 unit hydrolyzes 1.0 μmol substrate/min at pH 7.0, 25°C	Lyophilized containing 40% protein; balance PO_4 and citrate buffers and zinc salts		Sigma P1284
30-70 U/mg protein benzylpenicillin as substrate; 0.2-0.6 U/mg protein cephaloridine as substrate; 10 mg protein/vial	1 unit hydrolyzes 1.0 μmol substrate/min at pH 7.0, 25°C	Crude; lyophilized containing 40% protein; balance buffer and zinc salts		Sigma P6018
Bacillus cereus 569/H (MW = 28,000 Da)				
>500 U/mg β-lactamase I; >50 U/mg β-lactamase II	1 unit hydrolyzes 1.0 μmol penicillin/min at 25°C	Lyophilized		ICN 191489
7000 U/mg	1 unit converts 1 μmol benzyl penicillin/min at pH 7.0, 30°C	Purified; Type I; lyophilized		ICN 191492
2000 U/mg	1 unit converts 1 μmol cephalosporin C/min at pH 7.0, 30°C	Purified; Type II; lyophilized		ICN 191493
Bacillus cereus 569/H9 (MW = 31,500 Da)				
≥500 U β-lactamase I; 50 + U β-lactamase II	1 unit hydrolyzes 1 μmol benzyl-penicillin and 1 μmol cephalosporin C/min at 25°C; 1 IU = 100 Pollack units	Sterile; lyophilized		Calbiochem 426205

3.5.2.6 β-Lactamase continued

SPECIFIC ACTIVITY	UNITS DEFINITION	PREPARATION FORM	ADDITIONAL ACTIVITIES	SUPPLIER CATALOG NO.
Bacillus cereus 569/H9 continued				
>500 U/vial β-lactamase I; >50 U/vial β-lactamase II	1 unit hydrolyzes 1.0 μmol penicillin (βI) or cephalosporin (βII)/min at 25°C	Lyophilized		Genzyme 1401
>4 U/mg protein; >2 U/mg solid	1 unit hydrolyzes 1.0 μmol cephalosporin (βII)/min at 25°C	Lyophilized		Genzyme 1431
>40 U/mg protein; >20 U/mg solid	1 unit hydrolyzes 1.0 μmol penicillin (βI)/min at 25°C	Lyophilized		Genzyme 1441
Enterobacter cloacae				
0.2-0.6 U/mg β-lactamase I, benzylpenicillin as substrate; 15-25 U/mg β-lactamase II, cephalosporin C as substrate	1 unit hydrolyzes 1 μmol indicated substrate/min at pH 7.0, 25°C	Powder		Fluka 61304
6-18 U/mg protein benzylpenicillin as substrate; 50-150 U/mg protein cephaloridine as substrate	1 unit hydrolyzes 1.0 μmol substrate/min at pH 7.0, 25°C	Chromatographically purified; lyophilized containing 10% protein; balance $NaPO_4$ and Na citrate buffer salts		Sigma P4399
0.2-0.6 U/mg protein benzylpenicillin as substrate; 2-6 U/mg protein cephaloridine as substrate	1 unit hydrolyzes 1.0 μmol substrate/min at pH 7.0, 25°C	Lyophilized containing 60% protein; balance $NaPO_4$ buffer salts		Sigma P4524
Enterobacter cloacae P99 (MW = 40,000 Da)				
750 U/mg	1 unit converts 1 μmol cephalosporin C/min at pH 7.0, 30°C	Purified		ICN 191490
Escherichia coli 205 RTem + (566)				
200 U/mg solid	1 unit converts 1 μmol penicillin G to product/min at 37°C	Salt-free; lyophilized		Boehringer 663441

β-Lactamase continued 3.5.2.6

SPECIFIC ACTIVITY	UNITS DEFINITION	PREPARATION FORM	ADDITIONAL ACTIVITIES	SUPPLIER CATALOG NO.
Escherichia coli 205 RTem + (566) *continued*				
100-300 U/mg solid benzylpenicillin as substrate; 20-40 U/mg solid cephaloridine as substrate	1 unit hydrolyzes 1.0 μmol substrate/min at pH 7.0, 25°C	Lyophilized		Sigma P3553
Escherichia coli RTem (MW = 24,000 Da)				
3300 U/mg	1 unit converts 1 μmol benzyl-penicillin/min at pH 7.0, 30°C	Purified; lyophilized		ICN 191491
Proteus vulgaris (optimum pH = 7.0-8.0 [30°C], pI = 8.7, T = 45°C [pH 6.9], MW = 28,000 Da)				
≥120 U/mg protein		Solution		Wako 121-03581 127-03583
Unspecified				
				ICN 191488

Creatininase 3.5.2.10

REACTION CATALYZED
  Creatinine + H_2O ↔ Creatine
SYSTEMATIC NAME
  Creatinine amidohydrolase
REACTANTS
  Creatinine, H_2O, creatine
APPLICATIONS
- Determination of creatinine and creatine in biological fluids, with sarcosine dehydrogenase or sarcosine oxidase and formaldehyde dehydrogenase

NOTES
- Classified as a hydrolase, acting on carbon-nitrogen bonds other than peptide bonds, in cyclic amides
- Inhibited by Co^{2+}, Ag^+, Hg^{2+}, *N*-bromosuccinimide and EDTA
- Stabilized by sucrose and BSA

SPECIFIC ACTIVITY	UNITS DEFINITION	PREPARATION FORM	ADDITIONAL ACTIVITIES	SUPPLIER CATALOG NO.
Alcaligenes species				
70 U/mg	1 unit converts 1 μmol creatinine to product/min at 25°C	Solution containing 50% glycerol, 10 mM TEA, BSA as stabilizer, pH 8	<0.01% ATPase, HK, fructokinase <0.1% creatinase	Boehringer 126942
70 U/mg protein	1 unit hydrolyzes 1.0 μmol creatinine to creatine/min at pH 8.0, 25°C	50% glycerol solution containing 10 mM TEA		ICN 150716
Flavobacterium species				
200 U/mg protein	1 unit hydrolyzes 1.0 μmol creatinine to creatine/min at pH 6.5, 37°C	Lyophilized		ICN 150717
150-400 U/mg solid	1 unit hydrolyzes 1.0 μmol creatinine to creatine/min at pH 6.5, 37°C	Lyophilized	No detectable creatinase, HK, urease, ATPase	Sigma C7399
Microbial (optimum pH = 7.5)				
>8 U/mg protein; >5 U/mg solid	1 unit hydrolyzes 1.0 μmol creatinine/min at pH 7.0, 37°C	Lyophilized		Genzyme 1181
Pseudomonas species (optimum pH = 6.5-7.5, pI = 4.7, T = 70°C, MW = 175,00 Da [54,000 Da by TSK gel], 8 subunits per mol enzyme; K_M = 32 mM [creatinine], 57 mM [creatine]; stable pH 7.5-9.0 [25°C, 16 hr], T < 70°C [pH 7.5, 30 min])				
5-30 U/mg solid	1 unit hydrolyzes 1 μmol creatinine to creatine/min at 37°C	Lyophilized containing 50% protein		Asahi [CRN] T-22
70-140 U/mg	1 unit hydrolyzes 1 μmol creatinine to creatine/min at pH 8.0, 25°C	Powder	<1% creatinase	Fluka 27915
100-300 U/mg protein	1 unit hydrolyzes 1.0 μmol creatinine to creatine/min at pH 8.0, 25°C	Lyophilized		ICN 150718
≥150 U/mg solid	1 unit forms 1 μmol creatinine-picrate/min at pH 7.5, 37°C	Lyophilized containing 70% stabilizers		ICN 153491
100-300 U/mg protein	1 unit hydrolyzes 1.0 μmol creatinine to creatine/min at pH 8.0, 25°C	Lyophilized containing 75% protein; balance primarily sucrose; 70% of total protein is BSA	No detectable HK, urease, ATPase May contain 1% creatinase	Sigma C3173
≥150 U/mg solid	1 unit forms 1 μmol orange dye/min at pH 7.5, 37°C	Lyophilized containing 70% BSA and sucrose as stabilizers	No detectable creatine amidinohydrolase, sarcosine dehydrogenase, formaldehyde dehydrogenase <0.05% NADH oxidase <2% catalase	Toyobo CNH-301

N-Methylhydantoinase (ATP-hydrolyzing) 3.5.2.14

REACTION CATALYZED
ATP + N-methylimidazolidine-2,4-dione + 2 H_2O ↔ ADP + orthophosphate + N-carbamoylsarcosine

SYSTEMATIC NAME
N-Methylimidazolidine-2,4-dione amidohydrolase (ATP-hydrolyzing)

SYNONYMS
N-Methylhydantoin amidohydrolase

REACTANTS
ATP, N-methylimidazolidine-2,4-dione, H_2O, ADP, orthophosphate, N-carbamoylsarcosine

APPLICATIONS
- Determination of creatinine in biological fluids, with creatinine deiminase and N-carbamoyl sarcosine amidohydrolase; needs no preincubation

NOTES
- Classified as a hydrolase, acting on carbon-nitrogen bonds other than peptide bonds, in cyclic amides

SPECIFIC ACTIVITY	UNITS DEFINITION	PREPARATION FORM	ADDITIONAL ACTIVITIES	SUPPLIER CATALOG NO.
Arthrobacter species, from E. coli				
1.8 U/mg protein; 0.6 U/mg solid	1 unit converts 1 μmol N-methylhydantoin to product/min at 25°C	Lyophilized	<0.01% creatinase, creatininase, uricase	Boehringer 1481487

3.5.3.1 Arginase

REACTION CATALYZED
  L-Arginine + H_2O ↔ L-Ornithine + urea

SYSTEMATIC NAME
  L-Arginine amidinohydrolase

SYNONYMS
  Arginine amidinase, canavanase

REACTANTS
  L-Arginine, α-N-substituted L-arginines, canavanine, H_2O, L-ornithine, urea, 2,3-butanedione

APPLICATIONS
- Determination of L-Arg in plasma and urine
- Preparation of L-ornithine

NOTES
- Classified as a hydrolase, acting on carbon-nitrogen bonds other than peptide bonds, in linear amidines
- Present in many tissues and organs, mammalian liver is the richest source
- EDTA dissociates the enzyme into subunits of 30,000 Da each
- Mn^{2+} restores activity and reassociates subunits to the native form of 12,000 Da
- Activated by Mn^{2+}, Ni^{2+} and Co^{2+}
- Inhibited by Hg^{2+}, Ag^+ and Zn^{2+}
- Activity requires free guanidino and carbonyl groups of arginine

SPECIFIC ACTIVITY	UNITS DEFINITION	PREPARATION FORM	ADDITIONAL ACTIVITIES	SUPPLIER CATALOG NO.
Bovine liver				
20 U/mg solid	1 unit hydrolyzes 1 μmol L-Arg/min at pH 9.5, 37°C	Salt-free; lyophilized		Biozyme ARG1
150 U/mg protein; 100 U/mg solid	1 unit hydrolyzes 1 μmol L-Arg/min at pH 9.5, 37°C	Salt-free; lyophilized		Biozyme ARG2
100 U/mg protein	1 unit hydrolyzes 1 μmol L-Arg/min at pH 9.5, 37°C	Lyophilized containing 95% protein		Calzyme 182A0100
100-200 U/mg	1 unit converts 1 μmol L-Arg to ornithine and urea/min at pH 9.5, 37°C	Powder		Fluka 10995
20-30 U/mg	1 unit converts 1.0 μmol L-Arg to ornithine and urea/min at pH 9.5, 37°C	Purified; lyophilized		ICN 100277
	1 unit converts 1.0 μmol L-Arg to ornithine and urea/min at pH 9.5, 37°C	Lyophilized containing 95% protein; balance $MnSO_4$ and Na maleate		Sigma A2137
150-250 U/mg protein	1 unit converts 1.0 μmol L-Arg to ornithine and urea/min at pH 9.5, 37°C	Lyophilized containing 85% protein; balance $MnSO_4$ and Na maleate		Sigma A8013
≥20 U/mg solid	1 unit releases 1 μmol urea from L-Arg/min at pH 9.5, 37°C	Lyophilized		Worthington LS01164 LS01160

Creatinase 3.5.3.3

REACTION CATALYZED
Creatine + H_2O ↔ Sarcosine + urea

SYSTEMATIC NAME
Creatine amidinohydrolase

REACTANTS
Creatine, H_2O, sarcosine, urea

APPLICATIONS
- Determination of creatine and creatinine with sarcosine dehydrogenase or sarcosine oxidase/formaldehyde dehydrogenase

NOTES
- Classified as a hydrolase, acting on carbon-nitrogen bonds other than peptide bonds, in linear amidines
- Activated by EDTA
- Inhibited by Cu^{2+}, Hg^{2+}, Ag^+, Mg^{2+}, Ca^{2+}, Co^{2+}, Zn^{2+} and Mn^{2+}
- Stabilized by sugars and EDTA

SPECIFIC ACTIVITY	UNITS DEFINITION	PREPARATION FORM	ADDITIONAL ACTIVITIES	SUPPLIER CATALOG NO.
Actinobacillus species (optimum pH = 8.0, pI = 4.6, T = 40°C, MW = 100,000 Da [2 subunits/mol enzyme]; K_M = 19 mM [creatine]; stable pH 5.5-9.0 [25°C, 16 hr]. T < 50°C [pH 7.5, 30 min])				
20-40 U/mg protein	1 unit hydrolyzes 1.0 μmol creatine to urea and sarcosine/min at pH 7.5, 37°C	Lyophilized containing 50% protein; balance primarily gluconate, KPO_4, EDTA		Sigma C2409
≥6 U/mg solid	1 unit forms 1 μmol yellow dye/min at pH 7.5, 37°C	Lyophilized containing 40% EDTA and sugars as stabilizers	<0.05% NADH oxidase <2% catalase	Toyobo CRH-211
Bacillus species (optimum pH = 7.5, pI = 4.9, MW = 74,000 Da [Sephadex G-150]; K_M = 22.2 mM; stable pH 6-9, T < 40°C [pH 7.5, 10 min])				
4-10 U/mg solid	1 unit hydrolyzes 1 μmol creatine to sarcosine/min at 37°C	Lyophilized containing 70% protein		Asahi [CR] T-23
>4.5 U/mg protein; >4 U/mg solid	1 unit hydrolyzes 1.0 μmol creatine/min at 37°C	Lyophilized		Genzyme 1131
Flavobacterium species				
10-15 U/mg	1 unit hydrolyzes 1 μmol creatine to urea and sarcosine/min at pH 7.5, 37°C	Powder		Fluka 27905
10-20 U/mg protein	1 unit hydrolyzes 1.0 μmol creatine to urea and sarcosine/min at pH 7.5, 37°C	Lyophilized containing 75% protein; balance PO_4 buffer and lactose		Sigma C7024
Pseudomonas species				
10 U/mg protein	1 unit hydrolyzes 1.0 μmol creatine to urea and sarcosine/min at pH 7.5, 37°C	Lyophilized containing 60% protein; balance primarily sucrose		Sigma C0398
Pseudomonas species, recombinant expressed in *E. coli*				
10-15 U/mg protein	1 unit hydrolyzes 1.0 μmol creatine to urea and sarcosine/min at pH 7.5, 37°C	Lyophilized containing 70% protein		Sigma C3921

3.5.3.15 Protein-Arginine Deiminase

REACTION CATALYZED
    Protein L-arginine + H_2O ↔ Protein L-citrulline + NH_3

SYSTEMATIC NAME
    Protein L-arginine iminohydrolase

SYNONYMS
    Peptidylarginine deiminase

REACTANTS
    Protein L-arginine, H_2O, protein L-citrulline, NH_3, N-acyl-L-arginine, L-arginine esters (slowly)

APPLICATIONS
- Modification of Arg residues of proteins and peptides in enzymology, protein chemistry and biology

NOTES
- Classified as a hydrolase, acting on carbon-nitrogen bonds other than peptide bonds, in linear amidines
- Arg residues function as ligand recognition sites in many proteins and peptides
- Activated by Ca^{2+}
- Inhibited by IAA, MIA and N-ethylmaleimide

SPECIFIC ACTIVITY	UNITS DEFINITION	PREPARATION FORM	ADDITIONAL ACTIVITIES	SUPPLIER CATALOG NO.
Unspecified (optimum pH = 7.2-7.6 [Bz-Arg-OEt]; MW = 130,000 [gel filtration], 83,000 [SDS-PAGE] Da)				
	1 unit produces 1 μmol Cit/hr at 37°C	Homogeneous on SDS-PAGE; 20 mM Tris-HCl buffer solution containing 10 mM β-MSH and 50% glycerol, pH 7.6		PanVera TAK 7309

Cytosine Deaminase 3.5.4.1

REACTION CATALYZED
 Cytosine + H_2O ↔ Uracil + NH_3
SYSTEMATIC NAME
 Cytosine aminohydrolase
REACTANTS
 Cytosine, 5-methylcytosine, H_2O, uracil, NH_3

NOTES
- Classified as a hydrolase, acting on carbon-nitrogen bonds other than peptide bonds, in cyclic amidines

SPECIFIC ACTIVITY	UNITS DEFINITION	PREPARATION FORM	ADDITIONAL ACTIVITIES	SUPPLIER CATALOG NO.
Yeast, bakers (MW = 20,000 Da)				
70 U/mg protein	37°C	Suspension in 80% $(NH_4)_2SO_4$ containing 10 mg/mL protein		Calzyme 180B0070

Guanine Deaminase 3.5.4.3

REACTION CATALYZED
 Guanine + H_2O ↔ Xanthine + NH_3
SYSTEMATIC NAME
 Guanine aminohydrolase
SYNONYMS
 Guanase, guanine aminase

REACTANTS
 Guanine, H_2O, xanthine, NH_3
NOTES
- Classified as a hydrolase, acting on carbon-nitrogen bonds other than peptide bonds, in cyclic amidines

SPECIFIC ACTIVITY	UNITS DEFINITION	PREPARATION FORM	ADDITIONAL ACTIVITIES	SUPPLIER CATALOG NO.
Rabbit liver				
0.06–0.20 U/mg protein	1 unit deaminates 1.0 µmol guanine to xanthine/min at pH 8.0, 25°C	Suspension in 3.2 M $(NH_4)_2SO_4$, pH 6		ICN 157282
0.06–0.20 U/mg protein	1 unit deaminates 1.0 µmol guanine to xanthine/min at pH 8.0, 25°C	Suspension in 3.2 M $(NH_4)_2SO_4$, pH 6		Sigma G5752

3.5.4.4 Adenosine Deaminase

REACTION CATALYZED
Adenosine + $H_2O \leftrightarrow$ Inosine + NH_3

SYSTEMATIC NAME
Adenosine aminohydrolase

SYNONYMS
ADA

REACTANTS
Adenosine, H_2O, inosine, NH_3

APPLICATIONS
- Determination of adenosine in biological fluids
- Determination of 5′-nucleotidase in blood
- Synthesis of inosine, deoxyinosine and dideoxyinosine
- Deamination of adenosine analogs to the corresponding inosine analogs
- Elevated serum levels are a diagnostic test for portal cirrhosis

NOTES
- Classified as a hydrolase, acting on carbon-nitrogen bonds other than peptide bonds, in cyclic amidines
- Highest activity is found in animal small intestine mucosa, appendix and spleen; localized in cell cytoplasm and nucleus
- Primary function is detoxification of pharmacologically active adenosine
- Erythrocyte deficiency is associated with severe combined immunodeficiency (SCID)
- Serum levels are elevated in primary liver disease and secondary neoplasia
- Inhibited by Ag^+, Hg^{2+}, Cu^{2+}, TPCK, DFP and sulfhydryl reagents

SPECIFIC ACTIVITY	UNITS DEFINITION	PREPARATION FORM	ADDITIONAL ACTIVITIES	SUPPLIER CATALOG NO.
Bovine spleen				
>40 U/A_{280}; >500 U/mL	1 unit deaminates 1 μmol adenosine to inosine/min at pH 7.4, 25°C	Suspension in 70% $(NH_4)_2SO_4$		Diagnostic A-020
>40U/A_{280}; >500 U/mL	1 unit deaminates 1 μmol adenosine to inosine/min at pH 7.4, 25°C	Solution containing 50% glycerol and 10 mM KPO_4		Diagnostic A-030
30-90 U/mg protein	1 unit deaminates 1.0 μmol adenosine to inosine/min at pH 7.5, 25°C	Suspension in 3.2 M $(NH_4)_2SO_4$ and 0.01 M KPO_4, pH 6.0	No detectable AP	Sigma A5773
60-130 U/mg protein	1 unit deaminates 1.0 μmol adenosine to inosine/min at pH 7.5, 25°C	Solution containing 50% glycerol and 50 mM KPO_4, <25 mg free ammonium ions/mg protein, pH 6.0	No detectable AP	Sigma A6648
Calf intestinal mucosa				
160 U/mg protein	1 unit deaminates 1 μmol adenosine to inosine/min at pH 7.5, 25°C	Solution containing 50% glycerol and 0.01 M KH_2PO_4, pH 6.0	<0.002% AP <0.003% guanase, nucleoside phosphorylase <0.004% 5′-AMP deaminase	Fluka 01898

SPECIFIC ACTIVITY	UNITS DEFINITION	PREPARATION FORM	ADDITIONAL ACTIVITIES	SUPPLIER CATALOG No.
Calf intestinal mucosa *continued*				
200 U/mg protein	1 unit deaminates 1.0 μmol adenosine to inosine/min at pH 7.5, 25°C	Suspension in 3.2 M $(NH_4)_2SO_4$, pH 6		Sigma A1030
160-200 U/mg protein	1 unit deaminates 1.0 μmol adenosine to inosine/min at pH 7.5, 25°C	Solution containing 50% glycerol and 0.01 M KPO_4, pH 6.0		Sigma A1155
150-250 U/mg protein	1 unit deaminates 1.0 μmol adenosine to inosine/min at pH 7.5, 25°C	Lyophilized containing 20% protein and 60% lactose; balance KPO_4, pH 7.6		Sigma A1280
1-5 U/mg protein	1 unit deaminates 1.0 μmol adenosine to inosine/min at pH 7.5, 25°C	Crude; powder	Possible phosphatase impurity may deaminate 5'-AMP at a rate 0.5% of primary activity	Sigma A9876
Calf intestine (optimum pH = 7.0-7.4)				
200 U/mg	1 unit converts 1 μmol adenosine to product/min at 25°C	Solution containing 50% glycerol and 10 mM KPO_4, pH 6	<0.01% AP, AMP-deaminase, guanase, nucleoside phosphorylase	Boehringer 102091 102105
200 U/mg	1 unit converts 1 μmol adenosine to product/min at 25°C	Suspension in 3.2 M $(NH_4)_2SO_4$, pH 6	<0.01% AP, AMP-deaminase, guanase, nucleoside phosphorylase	Boehringer 102113 102121
200 U/mg protein	1 unit deaminates 1.0 μmol adenosine to inosine/min at pH 7.4, 25°C	Lyophilized containing 10 mg/mL protein		Calzyme 231A0200
Calf spleen (optimum pH = 6.3, MW = 32,500-33,000 Da; stable pH 4-10, [≤65°C, 10 min])				
150-300 U/mg protein	1 unit deaminates 1.0 μmol adenosine to inosine/min at pH 7.5, 25°C	Solution containing 50% glycerol and 5 mM KPO_4, pH 6.0	<10 mg free ammonium ions/mg protein <0.01% AP, guanase <0.1% nucleoside phosphorylase	Sigma A5043
150-200 U/mg protein	1 unit deaminates 1.0 μmol adenosine to inosine/min at pH 7.5, 25°C	Suspension in 3.2 M $(NH_4)_2SO_4$ and 0.01 M KPO_4, pH 6.0	<0.01% 5'-AMP deaminase, AP, guanase <0.05% nucleoside phosphorylase	Sigma A5168
≥15 U/mg solid	1 unit converts 1 μmol adenosine to inosine/min at pH 7.4, 25°C	Chromatographically purified; dialyzed; lyophilized		Worthington LS09043 LS09044

3.5.4.6 AMP Deaminase

REACTION CATALYZED
  AMP + H_2O ↔ IMP + NH_3

SYSTEMATIC NAME
  AMP aminohydrolase

SYNONYMS
  Adenylic acid deaminase, AMP aminase, 5'-adenylic acid deaminase, deaminase

REACTANTS
  AMP, H_2O, IMP, NH_3

APPLICATIONS
- Tinned mushrooms
- Increases natural flavor of fruits, vegetables, fish and shellfish containing 5'-AMP

NOTES
- Classified as a hydrolase, acting on carbon-nitrogen bonds other than peptide bonds, in cyclic amidines
- Not specific for 5'-AMP; deaminates adenosine at 50% the rate of 5'-AMP

SPECIFIC ACTIVITY	UNITS DEFINITION	PREPARATION FORM	ADDITIONAL ACTIVITIES	SUPPLIER CATALOG NO.
Aspergillus species (optimum pH = 5.6; stable pH 4.0-7.0, heat stable)				
50,000 U/mg	10 units decrease A_{265} by 0.001/hr at pH 5.6, 37°C	Powder		Amano Deamizyme
0.1 U/mg solid	1 unit deaminates 1.0 μmol 5'-AMP to 5'-IMP/min at pH 6.5, 25°C	Lyophilized containing 20% protein; balance primarily diatomaceous earth		Sigma A1907
Rabbit muscle				
50-200 U/mg protein	1 unit deaminates 1.0 μmol 5'-AMP to 5'-IMP/min at pH 6.5, 25°C	Solution containing 66% glycerol, 0.33 M KCl, 1 mM β-MSH, pH 7.0		Sigma A8384

3.5.4.21 Creatinine Deaminase

REACTION CATALYZED
  Creatinine + H_2O ↔ N-Methylhydantoin + NH_3

SYSTEMATIC NAME
  Creatinine iminohydrolase

SYNONYMS
  Creatinine deiminase

REACTANTS
  Creatinine, H_2O, N-Methylhydantoin, NH_3

APPLICATIONS
- Determination of creatinine in serum or urine, with N-methylhydantoin hydrolase and N-carbamoylsarcosine amidohydrolase; needs no preincubation

Creatinine Deaminase continued — 3.5.4.21

NOTES
- Classified as a hydrolase, acting on carbon-nitrogen bonds other than peptide bonds, in cyclic amidines
- Inhibited by o-phenanthroline, monoiodoacetate, Ag^+ and Hg^{2+}
- Stabilized by mannitol

SPECIFIC ACTIVITY	UNITS DEFINITION	PREPARATION FORM	ADDITIONAL ACTIVITIES	SUPPLIER CATALOG NO.
Corynebacterium lilium				
150 U/mg protein; 45 U/mg solid	1 unit converts 1 µmol creatinine to product/min at pH 7.8, 25°C	Lyophilized containing <0.01 µg/U NH_3	<0.007% urease <0.01% creatininase, creatinase, uricase	Boehringer 1482548 1481452
10-25 U/mg	1 unit hydrolyzes 1 µmol creatinine to N-methylhydantoin and NH_3/min at pH 7.5, 37°C in a coupled system with L-GlDH	Powder		Fluka 27917
Microbial (optimum pH = 9.5, pI = 4.4, T = 65-75°C, MW = 260,000 Da [6 subunits/mol enzyme]; K_M = 3.5 mM; stable pH 7.0-11.0 [30°C, 20 hr], T < 65°C [pH 7.5, 1 hr])				
>10 U/mg protein	1 unit releases 1 µmol NH_3/min at pH 7.6, 37°C	Lyophilized optimum pH = 7.0-8.5, T = 37°C		GDS Tech C-100
≥10 U/mg solid	1 unit forms 1 µmol NH_3 (oxidation of 1 µmol NADPH)/min at pH 7.5, 37°C	Lyophilized containing 30% stabilizers		ICN 153490
25-50 U/mg protein	1 unit hydrolyzes 1.0 µmol creatinine to N-methylhydantoin and NH_3/min at pH 7.5, 37°C in a coupled system with L-GlDH	Lyophilized containing 45% protein; balance mannitol and KPO_4 buffer salts		Sigma C9409
≥10 U/mg solid	1 unit forms 1 µmol NH_3 (oxidation of 1 µmol NADPH)/min at pH 7.5, 37°C	Lyophilized containing 30% mannitol as stabilizer	<0.01% creatinine amidohydrolase, creatine amidinohydrolase, urease, NADH oxidase <0.01 µg/U NH_3	Toyobo CNI-301

3.6.1.1 Inorganic Pyrophosphatase

REACTION CATALYZED
Pyrophosphate + H_2O ↔ 2 Orthophosphate

SYSTEMATIC NAME
Pyrophosphate phosphohydrolase

SYNONYMS
Pyrophosphatase, thermophilic pyrophosphatase, PPase

REACTANTS
Pyrophosphate, H_2O, orthophosphate

APPLICATIONS
- Maintains forward direction of reactions generating pyrophosphate (e.g., DNA polymerization)
- DNA sequencing, when selective bond weakening is observed
- Removing CAP structure on mRNA, viral RNA and low molecular weight RNA, by cleaving the connecting 5'-5' triphosphate link
- ELISA enzyme label
- Ligation
- Radiolabeling RNA
- Oligo-labeling
- High temperature sequencing
- Amplification of long fragments
- Hybridization probe
- Terminal radiation of TAP-treated viral genome RNA
- 5'- and 3'-end mapping of mRNA

NOTES
- Classified as a hydrolase, acting on acid anhydrides in phosphorus-containing anhydrides
- Specificity varies with source and the activating metal ion
- Enzyme from some sources may be identical with alkaline phosphatase (EC 3.1.3.1) or glucose-6-phosphatase (EC 3.1.3.9)

SPECIFIC ACTIVITY	UNITS DEFINITION	PREPARATION FORM	ADDITIONAL ACTIVITIES	SUPPLIER CATALOG No.
Bacillus stearothermophilus				
15-25 U/mg protein	1 unit liberates 1.0 μmol inorganic orthophosphate/min at pH 9.0, 50°C	Lyophilized		Sigma I2891
Escherichia coli				
1000 U/mg protein	1 unit liberates 1.0 μmol inorganic orthophosphate/min at pH 9.0, 25°C	Lyophilized containing Tris buffer salts		Sigma I2267
Thermus thermophilus				
2500 U/mg	1 unit converts 1 μmol pyrophosphate into 2 nmol PO_4/min at pH 9.0, 75°C using inorganic pyrophosphate as substrate	20 mM Tris-HCl, 2 mM $MgCl_2$, 1 mM $(NH_4)_2SO_4$, pH 8.0	No detectable ATPase and phosphatase	Adv Immuno IPTt1
Tobacco				
≥1000 U/mL	1 unit releases 1.0 nmol inorganic phosphorus from ATP/30 min at pH 5.0, 37°C	10 mM Tris HCl, 10 mM NaCl, 0.1 mM EDTA, 1 mM DTT, 0.01% Triton X-100, 50% glycerol, pH 7.5		Sigma P0414

Inorganic Pyrophosphatase continued 3.6.1.1

SPECIFIC ACTIVITY	UNITS DEFINITION	PREPARATION FORM	ADDITIONAL ACTIVITIES	SUPPLIER CATALOG NO.
Tobacco continued				
20 U/μL				Wako 206-13271
Yeast				
0.4 U/μL	1 unit hydrolyzes 1 μmol pyrophosphate/min at 25°C	10 mM Tris-HCl, 0.1 mM EDTA, 50% glycerol, pH 7.5	No detectable contaminating ss- and ds-DNA endonuclease, exonuclease, phosphatase	Amersham E 70953Y E 70953Z
200 U/mg	1 unit converts 1 μmol inorganic pyrophosphate to product/min at 25°C	Suspension in 3.2 M $(NH_4)_2SO_4$, pH 6.0	<0.01% ATPase, phosphatases (pH 7.0 with 4-nitrophenyl phosphate as substrate)	Boehringer 108987
600-800 U/mg	1 unit liberates 1 μmol inorganic o-phosphate/min at pH 7.2, 20°C	Suspension in 0.013 M citrate buffer		ICN 100759
Yeast, bakers				
70 U/mg	1 unit liberates 1 μmol phosphate from inorganic pyrophosphate/min at pH 7.2, 25°C	Powder		Fluka 83205
500-1000 U/mg protein	1 unit liberates 1.0 μmol inorganic orthophosphate/min at pH 7.2, 25°C	Lyophilized containing 85% buffer salts		Sigma I1643
5-1500 U/mg protein	1 unit liberates 1.0 μmol inorganic orthophosphate/min at pH 7.2, 25°C	Purified by HPLC; salt-free; lyophilized		Sigma I1891
Unspecified				
	1 unit liberates 1.0 μmol inorganic orthophosphate/min at pH 7.5, 65°C	20 mM Tris-HCl, 1.0 mM DTT, 1.0 mM EDTA, 50 mM NaCl, 50% glycerol, pH 7.5	No detectable endonuclease, 3′- and 5′-exonuclease	CHIMERX 1267-01 1267-02
	1 unit hydrolyzes 1 nmol organic PO_4 from ATP/30 min at pH 6.0, 37°C	Solution containing 50% glycerol, 10 mM Tris-HCl, 0.1 M NaCl, 0.1 mM EDTA, 1 mM DTT, 0.01% Triton X-100, pH 7.5	No detectable contaminating DNA exo- and endonuclease, phosphatase, RNase (by electrophoresis)	Epicentre T19050 T19100 T19250 T19500

3.6.1.3 Adenosinetriphosphatase

REACTION CATALYZED
ATP + H_2O ↔ ADP + orthophosphate

SYSTEMATIC NAME
ATP phosphohydrolase

SYNONYMS
Adenylpyrophosphatase, ATP monophosphatase, triphosphatase, ATPase, adenosine 5′-triphosphatase

REACTANTS
ATP, H_2O, ADP, orthophosphate

APPLICATIONS
- Determination of digitonin and digoxin

NOTES
- Classified as a hydrolase, acting on acid anhydrides in phosphorus-containing anhydrides
- Responsible for coupled active transport of Na^+ and K^+ across the plasma membrane
- Some enzymes in this group require Ca^{2+}, Mg^{2+}, anions, H^+ or DNA
- Activated by Na^+ and K^+
- Inhibited by vanadate and ouabain

SPECIFIC ACTIVITY	UNITS DEFINITION	PREPARATION FORM	ADDITIONAL ACTIVITIES	SUPPLIER CATALOG NO.
Dog kidney				
0.1 U/mg	1 unit liberates 1 μmol P_i from ATP/min at pH 7.4, 37°C in the presence of Na^+, K^+, Mg^{2+}, but without ouabain	Powder containing 10% protein	Activity sensitive to ouabain	Fluka 02075
1-2 U/mg protein	1 unit liberates 1.0 μmol inorganic phosphorus from ATP/min at pH 7.4, 37°C in the presence of Na^+, K^+, Mg^{2+}	Lyophilized containing 10% protein; balance primarily sucrose		Sigma A0142
1-2 U/mg protein	1 unit liberates 1.0 μmol of inorganic phosphorus from ATP/min at pH 7.4, 37°C in the presence of Na^+, K^+, Mg^{2+}	Solution containing 50% glycerol, 0.02 M Tris, 0.26 M sucrose, pH 7.4		Sigma A7305
Porcine cerebral cortex				
0.3 U/mg protein	1 unit liberates 1.0 μmol inorganic phosphorus from ATP/min at pH 7.4, 37°C in the presence of Na^+, K^+, Mg^{2+}	Na iodide-extracted fraction; lyophilized containing 10% protein, 90% sucrose, 0.4% EDTA, 0.06% NaCl		Sigma A7510
Rabbit kidney				
0.3-1.0 U/mg protein	1 unit liberates 1.0 μmol inorganic phosphorus from ATP/min at pH 7.4, 37°C in the presence of Na^+, K^+, Mg^{2+}	Lyophilized containing 10% protein; balance primarily sucrose		Sigma A2414

Apyrase 3.6.1.5

REACTION CATALYZED
ATP + 2 H$_2$O ↔ AMP + 2 orthophosphate

SYSTEMATIC NAME
ATP diphosphohydrolase

SYNONYMS
Adenosine diphosphatase, ATP-diphosphatase, ADPase

REACTANTS
ATP, ADP, H$_2$O, AMP, orthophosphate

NOTES
- Classified as a hydrolase, acting on acid anhydrides in phosphorus-containing anhydrides
- Requires Ca^{2+}
- Also acts on ADP and other nucleoside triphosphates and diphosphates

SPECIFIC ACTIVITY	UNITS DEFINITION	PREPARATION FORM	ADDITIONAL ACTIVITIES	SUPPLIER CATALOG NO.
Solanum tuberosum (potato)				
100 U/mg ATPase	1 unit liberates 1 μmol P$_i$ from the PO$_4$ ester/min at pH 6.5, 30°C	Powder	<0.3% acid phosphatase <3% AMPase	Fluka 10827
3-10 U/mg protein ATPase; 1-2 U/mg protein ADPase	1 unit liberates 1.0 μmol P$_i$/min at pH 6.5, 30°C	Partially purified; lyophilized containing 60% protein; balance primarily NaOAc buffer salts, pH 5.5	<1.5 U/mg protein acid phosphatase	Sigma A6132
>200 U/mg protein ATPase; ADPase is variable	1 unit liberates 1.0 μmol P$_i$/min at pH 6.5, 30°C	Partially purified; lyophilized containing 40% protein; balance primarily K succinate buffer salts	Acid phosphatase <0.2% of ATPase	Sigma A6160
>200 U/mg protein ATPase; ADPase <15% of ATPase	1 unit liberates 1.0 μmol P$_i$/min at pH 6.5, 30°C	Partially purified; lyophilized containing 40% protein; balance primarily K succinate buffer salts	Acid phosphatase <0.2% of ATPase	Sigma A6410
>200 U/mg protein ATPase; ADPase >50% of ATPase	1 unit liberates 1.0 μmol P$_i$/min at pH 6.5, 30°C	Partially purified; lyophilized containing 40% protein; balance primarily K succinate buffer salts	Acid phosphatase <0.3% of ATPase	Sigma A6535
5-10 U/mg solid ATPase; 1-2 U/mg solid ADPase	1 unit liberates 1.0 μmol P$_i$/min at pH 6.5, 30°C	Lyophilized		Sigma A7521
40-100 U/mg protein ATPase; 10-30 U/mg protein ADPase	1 unit liberates 1.0 μmol P$_i$/min at pH 6.5, 30°C		5 U/mg protein acid phosphatase	Sigma A7646

3.6.1.5 Apyrase continued

SPECIFIC ACTIVITY	UNITS DEFINITION	PREPARATION FORM	ADDITIONAL ACTIVITIES	SUPPLIER CATALOG NO.
Solanum tuberosum (potato) *continued*				
3-10 U/mg protein ATPase	1 unit liberates 1.0 μmol P$_i$/min at pH 6.5, 30°C	Partially purified; lyophilized containing >95% protein	<5 U/mg protein acid phosphatase ADPase >50% of ATPase	Sigma A9149

Nucleotide Pyrophosphatase 3.6.1.9

REACTION CATALYZED
    A dinucleotide + H$_2$O ↔ 2 Mononucleotides

SYSTEMATIC NAME
    Dinucleotide nucleotidohydrolase

REACTANTS
    Dinucleotide, NAD$^+$, NADP$^+$, FAD, CoA, ATP, ADP, H$_2$O, mononucleotides

NOTES
- Classified as a hydrolase, acting on acid anhydrides in phosphorus-containing anhydrides

SPECIFIC ACTIVITY	UNITS DEFINITION	PREPARATION FORM	ADDITIONAL ACTIVITIES	SUPPLIER CATALOG NO.
Crotalus adamanteus venom				
4-8 U/mg protein	1 unit hydrolyzes 1.0 μmol β-NAD to NMN and AMP/min at pH 7.4, 37°C in the presence of Mg^{2+}	Lyophilized containing 35% Tris buffer salts	<0.002 U/mg protein AP <0.003 U/mg protein 5'-nucleotidase 0.2-0.4 U/mg protein phosphodiesterase	Sigma P7383
Crotalus atrox venom				
4-8 U/mg protein	1 unit hydrolyzes 1.0 μmol β-NAD to NMN and AMP/min at pH 7.4, 37°C in the presence of Mg^{2+}	Lyophilized containing 80% protein and 20% Tris buffer salts	0.2 U/mg protein phosphodiesterase	Sigma P9138

Myosin ATPase 3.6.1.32

REACTION CATALYZED
 ATP + H_2O ↔ ADP + orthophosphate
SYSTEMATIC NAME
 Myosin ATP phosphohydrolase (actin-translocating)
SYNONYMS
 Actomyosin
REACTANTS
 ATP, H_2O, ADP, orthophosphate

NOTES
- Classified as a hydrolase, acting on acid anhydrides in phosphorus-containing anhydrides
- Involved with muscle contraction
- In the absence of actin, myosin shows only low, Ca^{2+}-requiring ATPase activity

SPECIFIC ACTIVITY	UNITS DEFINITION	PREPARATION FORM	ADDITIONAL ACTIVITIES	SUPPLIER CATALOG NO.
Rabbit muscle				
0.2-0.4 U/mg protein	1 unit liberates 1.0 µmol inorganic phosphorus from ATP/min at pH 9.0, 25°C in the presence of Ca^{2+}	Solution containing 50% glycerol and 0.6 M KCl, pH 6.8		Sigma A3425
Rabbit skeletal muscle (2 heavy chains of 200,000 Da each and 4 light chains of 15,000-30,000 Da each)				
>20 mg/mL		Buffered glycerol solution		ProZyme MW20 000 MW20 010 MW20 100

2-Haloacid Dehalogenase 3.8.1.2

REACTION CATALYZED
 (S)-2-Haloacid + H_2O ↔ (R)-2-Hydroxyacid + halide
SYSTEMATIC NAME
 2-Haloacid halidohydrolase
REACTANTS
 (S)-2-Haloacid, H_2O, (R)-2-hydroxyacid, halide

NOTES
- Classified as a hydrolase, acting on halide bonds in C-halide compounds
- Acts on acid of short (C_2-C_4) chain lengths, inverting configuration at C-2

3.8.1.2 2-Haloacid Dehalogenase continued

SPECIFIC ACTIVITY	UNITS DEFINITION	PREPARATION FORM	ADDITIONAL ACTIVITIES	SUPPLIER CATALOG NO.
Pseudomonas putida				
40 U/mg protein	1 unit liberates 1.0 μmol Cl⁻ from DL-2-chloropropionate/min at pH 10.5, 30°C	Lyophilized containing 50% protein; balance KPO$_4$ buffer salts		ICN 157295
Pseudomonas species				
20-60 U/mg protein	1 unit liberates 1.0 μmol Cl⁻ from DL-2-chloropropionate/min at pH 9.5, 30°C	Lyophilized containing 40% protein; balance KPO$_4$ buffer salts		ICN 157294
20-60 U/mg protein	1 unit liberates 1.0 μmol Cl⁻ from DL-2-chloropropionate/min at pH 9.5, 30°C	Lyophilized containing 40% protein; balance KPO$_4$ buffer salts		Sigma H5138

4.1.1.1 Pyruvate Decarboxylase

REACTION CATALYZED

A 2-oxo acid ↔ An aldehyde + CO_2

SYSTEMATIC NAME

2-Oxo-acid carboxy-lyase

SYNONYMS

α-Carboxylase, α-ketoacid carboxylase, pyruvic decarboxylase

REACTANTS

2-Oxo acid, aldehyde, acyloin, CO_2

NOTES
- Classified as a lyase: carbon-carbon lyase: carboxy-lyase
- A thiamin-diphosphate protein

SPECIFIC ACTIVITY	UNITS DEFINITION	PREPARATION FORM	ADDITIONAL ACTIVITIES	SUPPLIER CATALOG NO.
Yeast, brewers				
5-20 U/mg protein	1 unit converts 1.0 μmol pyruvate to acetaldehyde/min at pH 6.0, 25°C	Suspension in 3.2 M $(NH_4)_2SO_4$, 5% glycerol, 5 mM KPO$_4$, 1 mM MgOAc, 0.5 mM EDTA, 25 μM cocarboxylase, pH 6.5		Sigma P6810

Oxalate Decarboxylase

4.1.1.2

REACTION CATALYZED
    Oxalate ↔ Formate + CO_2
SYSTEMATIC NAME
    Oxalate carboxy-lyase
REACTANTS
    Oxalate, formate, CO_2

NOTES
- Classified as a lyase: carbon-carbon lyase: carboxy-lyase

SPECIFIC ACTIVITY	UNITS DEFINITION	PREPARATION FORM	ADDITIONAL ACTIVITIES	SUPPLIER CATALOG No.
Aspergillus species				
20 U/mg protein	1 unit converts 1 μmol oxalate to product/min at pH 5.0, 37°C	Suspension in 3.2 M $(NH_4)_2SO_4$, pH 6		Boehringer 479586
20–60 U/mg solid	1 unit converts 1.0 μmol oxalate to formate and CO_2/min at pH 5.0, 37°C	Suspension in 3.2 M $(NH_4)_2SO_4$, pH 6.0		Sigma O8630
Collybia velutipes				
	1 unit converts 1.0 μmol oxalate to formate and CO_2/min	Partially purified; powder		Sigma O3500

Oxaloacetate Decarboxylase

4.1.1.3

REACTION CATALYZED
    Oxaloacetate ↔ Pyruvate + CO_2
SYSTEMATIC NAME
    Oxaloacetate carboxy-lyase
SYNONYMS
    Oxaloacetate β-decarboxylase
REACTANTS
    Oxaloacetate, pyruvate, CO_2
APPLICATIONS
- Determination of citrate

NOTES
- Classified as a lyase: carbon-carbon lyase: carboxy-lyase
- *Klebsiella aerogenes* enzyme is a biotinyl-protein. It requires Na^+ and acts as a sodium pump while linked to phospholipid vesicles
- Some animal enzymes require Mn^{2+}
- Activated by Mg^{2+} and Mn^{2+}
- Inhibited by SDS, sodium laurylbenzene and sulfonate

4.1.1.3 Oxaloacetate Decarboxylase continued

SPECIFIC ACTIVITY	UNITS DEFINITION	PREPARATION FORM	ADDITIONAL ACTIVITIES	SUPPLIER CATALOG NO.
Pseudomonas species (optimum pH = 7.5, pI = 5.16, T = 35-40°C, MW = 110,000 Da [Sephadex G-150]; K_M = 1.52 mM [oxaloacetate]; stable pH = 7.5-9 [50°C, 10 min], T < 50°C [pH 7.5, 10 min])				
100-350 U/mg solid	1 unit produces 1 μmol pyruvate/min at pH 8.0, 25°C	Lyophilized containing 90% protein	<0.005% GOT <1.0% catalase	Asahi T-14
200-300 U/mg	1 unit converts 1 μmol oxalacetate to pyruvate and CO_2/min at pH 8.0, 25°C	Lyophilized		Fluka 75657
100-350 U/mg solid	1 unit converts 1.0 μmol oxalacetate to pyruvate and CO_2/min at pH 8.0, 25°C			ICN 156007
100-350 U/mg solid	1 unit converts 1.0 μmol oxalacetate to pyruvate and CO_2/min at pH 8.0, 30°C			Sigma 04878

4.1.1.4 Acetoacetate Decarboxylase

REACTION CATALYZED
Acetoacetate + H^+ ↔ Acetone + CO_2

SYSTEMATIC NAME
Acetoacetate carboxy-lyase

REACTANTS
Acetoacetate, H^+, acetone, CO_2

NOTES
- Classified as a lyase: carbon-carbon lyase: carboxy-lyase

SPECIFIC ACTIVITY	UNITS DEFINITION	PREPARATION FORM	ADDITIONAL ACTIVITIES	SUPPLIER CATALOG NO.
Bacillus polymyxa (optimum pH = 5.9, MW = 280,000 Da)				
100-360 U/mg protein; 100-110 U/mL		In 10 mM phosphoric acid buffer with 0.3% NaN_3, pH 7.5		Wako 019-12921

Glutamate Decarboxylase 4.1.1.15

REACTION CATALYZED
  L-Glutamate ↔ 4-Aminobutanoate + CO_2

SYSTEMATIC NAME
  L-Glutamate 1-carboxy-lyase

SYNONYMS
  GAD

REACTANTS
  L-Glutamate, α-methyl glutamate, L-cysteate, 3-sulfino-L-alanine, 4-aminobutanoate, CO_2, L-arginine, L-lysine, γ-aminobutyrate

APPLICATIONS
- Determination of Glu and Asp

NOTES
- Classified as a lyase: carbon-carbon lyase: carboxy-lyase
- A 310,000 Da molecular weight hexamer of 50,000 Da subunits, each with 1 molecule of pyridoxal phosphate
- The following are neither substrates nor inhibitors: D-glutamate, D- and L-aspartate, and α-amino adipic and α-aminopimelic acids
- Pure L-glutamine is not a substrate
- Brain enzyme acts on all listed substrates
- E. coli enzyme is highly specific, with significant activity only on L-glutamic acid and α-methyl glutamic acid
- Inhibited by semicarbazide

SPECIFIC ACTIVITY	UNITS DEFINITION	PREPARATION FORM	ADDITIONAL ACTIVITIES	SUPPLIER CATALOG No.
Clostridium perfringens				
0.02-0.06 U/mg	1 unit liberates 1 μmol CO_2 from L-Glu/min at pH 5.0, 37°C	Powder	Aspartic acid decarboxylase and glutaminase	Fluka 49535
Clostridium perfringens (welchii)				
0.2 U/mg solid	1 unit forms 1.0 μmol CO_2 from L-Glu/min at pH 5.0, 37°C	Crude; acetone powder	Aspartic decarboxylase and glutaminase	Sigma G2251
0.8-1.5 U/mg solid	1 unit forms 1.0 μmol CO_2 from L-Glu/min at pH 5.0, 37°C	Partially purified; powder		Sigma G2376
Escherichia coli (optimum pH 4.0-4.3, MW = 300,000-310,000 Da)				
15-30 U/mg	1 unit liberates 1 μmol CO_2 from L-Glu/min at pH 5.0, 37°C	Powder		Fluka 49537
25-50 U/mg protein	1 unit releases 1.0 μmol CO_2 from L-Glu/min at pH 5.0, 37°C	Suspension in 4.2 M $(NH_4)_2SO_4$, 0.1 mM pyridoxal PO_4, 0.1 mM DTT		ICN 157218
25-50 U/mg protein	1 unit forms 1.0 μmol CO_2 from L-Glu/min at pH 5.0, 37°C	Suspension in 4.2 M $(NH_4)_2SO_4$, 0.1 mM pyridoxal PO_4, 0.1 mM DTT		Sigma G0894
0.45-0.75 U/mg solid	1 unit forms 1.0 μmol CO_2 from L-Glu/min at pH 5.0, 37°C	Crude; acetone powder	No detectable glutaminase	Sigma G2001

4.1.1.15 Glutamate Decarboxylase continued

SPECIFIC ACTIVITY	UNITS DEFINITION	PREPARATION FORM	ADDITIONAL ACTIVITIES	SUPPLIER CATALOG No.
Escherichia coli continued				
2.5-5.5 U/mg solid	1 unit forms 1.0 µmol CO_2 from L-Glu/min at pH 5.0, 37°C	Partially purified; powder		Sigma G2126
20-40 U/mg protein	1 unit forms 1.0 µmol CO_2 from L-Glu/min at pH 5.0, 37°C	Purified; lyophilized	Low in GOT and GPT <1% Gln, Arg, Lys decarboxylase	Sigma G3757
≥200 µL/hr mg		Powder		Wako 079-00511
≥10 U/mg	1 unit releases 1 µmol CO_2/min at pH 5.0, 37°C	Chromatographically purified; suspension in 2.8 M $(NH_4)_2SO_4$		Worthington LS004562 LS004561 LS004563

4.1.1.17 Ornithine Decarboxylase

REACTION CATALYZED
    L-Ornithine ↔ Putrescine + CO_2
SYSTEMATIC NAME
    L-Ornithine carboxy-lyase
REACTANTS
    L-Ornithine, putrescine, CO_2

NOTES
- Classified as a lyase: carbon-carbon lyase: carboxy-lyase
- A pyridoxal-phosphate protein

SPECIFIC ACTIVITY	UNITS DEFINITION	PREPARATION FORM	ADDITIONAL ACTIVITIES	SUPPLIER CATALOG No.
Escherichia coli				
0.2-0.5 U/mg	1 unit liberates 1 µmol CO_2 from L-ornithine/min at pH 5.2, 37°C	Lyophilized		Fluka 75435
0.5-1.5 U/mg protein	1 unit releases 1.0 µmol CO_2 from L-ornithine/min at pH 5.2, 37°C	Lyophilized containing 30% protein; balance primarily buffer salts		Sigma O1502
0.03-0.12 U/mg protein	1 unit releases 1.0 µmol CO_2 from L-ornithine/min at pH 5.2, 37°C	Partially purified		Sigma O3001

Lysine Decarboxylase 4.1.1.18

REACTION CATALYZED
    L-Lysine ↔ Cadaverine + CO_2

SYSTEMATIC NAME
    L-Lysine carboxy-lyase

REACTANTS
    L-Lysine, 5-hydroxy-L-lysine, cadaverine, CO_2, hydroxy-D-lysine

NOTES
- Classified as a lyase: carbon-carbon lyase: carboxy-lyase
- A pyridoxal-phosphate protein
- Not active against D-lysine, or acetyl- and methyl-derivatives of lysine

SPECIFIC ACTIVITY	UNITS DEFINITION	PREPARATION FORM	ADDITIONAL ACTIVITIES	SUPPLIER CATALOG NO.
Bacterium cadaveris				
>0.11 U/mg substance	1 unit decarboxylates 1 μmol L-Lys/min at pH 5.2, 37°C with linoleic as substrate	Crude; acetone extract containing <8% ash		Fluka 62890
1-3 U/mg protein	1 unit releases 1.0 μmol CO_2 from L-Lys/min at pH 6.0, 37°C	Partially purified; powder		sigma L0882
50-100 U/mg protein	1 unit releases 1.0 μmol CO_2 from L-Lys/min at pH 6.0, 37°C	Lyophilized containing 85% protein; balance primarily KPO_4 buffer salts and stabilizer		sigma L5509
0.08 U/mg solid	1 unit releases 1.0 μmol CO_2 from L-Lys/min at pH 6.0, 37°C	Crude; acetone powder from which soluble enzyme may be obtained by aqueous extraction; the cell suspension used directly, or the preparation clarified by centrifugation		sigma L6126
Escherichia coli				
90-180 U/mg protein	1 unit releases 1.0 μmol CO_2 from L-Lys/min at pH 6.0, 37°C	Lyophilized containing 20% protein; balance primarily buffer salts and stabilizer		sigma L2508
0.01 U/mg solid	1 unit releases 1.0 μmol CO_2 from L-Lys/min at pH 6.0, 37°C	Crude; acetone powder		sigma L6251

4.1.1.19 Arginine Decarboxylase

REACTION CATALYZED
    L-Arginine $\leftrightarrow$ Agmatine + CO_2

SYSTEMATIC NAME
    L-Arginine carboxy-lyase

REACTANTS
    L-Arginine, agmatine, CO_2

NOTES
- Classified as a lyase: carbon-carbon lyase: carboxy-lyase
- A pyridoxal-phosphate protein

SPECIFIC ACTIVITY	UNITS DEFINITION	PREPARATION FORM	ADDITIONAL ACTIVITIES	SUPPLIER CATALOG No.
Escherichia coli				
2-6 U/mg	1 unit liberates 1 μmol CO_2 from L-Arg/min at pH 5.2, 37°C	Powder		Fluka 11025
0.2 U/mg L-Arg cleaved to agmatine	1 unit releases 1.0 μmol CO_2 from Arg/min at pH 5.2, 37°C	Acetone powder		ICN 100279
5-15 U/mg protein	1 unit releases 1.0 μmol CO_2 from L-Arg/min at pH 5.2, 37°C	Lyophilized containing 50% protein		ICN 150385
	1 unit releases 1.0 μmol CO_2 from L-Arg/min at pH 5.2, 37°C	Partially purified; powder containing 5% protein; balance primarily succinate buffer salts and sucrose		Sigma A8134
Escherichia coli strain ATCC 10787				
5-15 U/mg protein	1 unit releases 1.0 μmol CO_2 from L-Arg/min at pH 5.2, 37°C	Lyophilized containing 50% protein		Sigma A5381

4.1.1.22 Histidine Decarboxylase

REACTION CATALYZED
    L-Histidine $\leftrightarrow$ Histamine + CO_2

SYSTEMATIC NAME
    L-Histidine carboxy-lyase

REACTANTS
    L-Histidine, histamine + CO_2

NOTES
- Classified as a lyase: carbon-carbon lyase: carboxy-lyase
- Animal enzyme is a pyridoxal-phosphate protein
- Bacterial enzyme has a pyruvoyl residue as prosthetic group

Histidine Decarboxylase continued

4.1.1.22

SPECIFIC ACTIVITY	UNITS DEFINITION	PREPARATION FORM	ADDITIONAL ACTIVITIES	SUPPLIER CATALOG NO.
Clostridium perfringens (welchii)				
0.01-0.03 U/mg solid	1 unit releases 1.0 µmol CO_2 from L-His/min at pH 4.5, 37°C	Crude; acetone powder		Sigma H8250
0.1 U/mg protein	1 unit releases 1.0 µmol CO_2 from L-His/min at pH 4.5, 37°C	Partially purified; powder		Sigma H8375
Escherichia coli				
2-5 U/mg protein	1 unit releases 1.0 µmol CO_2 from L-His/min at pH 4.5, 37°C	Purified; powder containing 20% protein; balance primarily succinate buffer salts		Sigma H6134
Lactobacillus 30a				
0.25-0.5 U/mg solid	1 unit releases 1.0 µmol CO_2 from L-His/min at pH 4.5, 37°C	Crude; acetone powder		Sigma H3266

Orotidine-5′-Phosphate Decarboxylase

4.1.1.23

REACTION CATALYZED
    Orotidine 5′-phosphate ↔ UMP + CO_2

SYSTEMATIC NAME
    Orotidine-5′-phosphate carboxy-lyase

SYNONYMS
    Orotidine-5′-monophosphate decarboxylase

REACTANTS
    Orotidine 5′-phosphate, UMP, CO_2

NOTES
- Classified as a lyase: carbon-carbon lyase: carboxy-lyase
- Enzyme from higher eukaryotes is identical with orotate phosphoribosyltransferase (EC 2.4.2.10)

SPECIFIC ACTIVITY	UNITS DEFINITION	PREPARATION FORM	ADDITIONAL ACTIVITIES	SUPPLIER CATALOG NO.
Yeast, bakers				
0.2-0.4 U/mg	1 unit converts 1 µmol orotidine 5′-monophosphate to uridine 5′-monophosphate and CO_2/min at pH 8.0, 30°C	Powder		Fluka 75495
30-65 U/mg protein	1 unit converts 1.0 µmol orotidine 5′-monophosphate to uridine 5′-monophosphate/hr at pH 8.0, 30°C	Lyophilized containing 40% protein; balance primarily citrate buffer salt	May contain up to 10% OMP pyrophosphorylase	Sigma O9251

4.1.1.25 — Tyrosine Decarboxylase

REACTION CATALYZED
L-Tyrosine ↔ Tyramine + CO_2

SYSTEMATIC NAME
L-Tyrosine carboxy-lyase

SYNONYMS
TD

REACTANTS
L-Tyrosine, 3-hydroxytyrosine, 3-hydroxyphenylalanine, tyramine, CO_2

APPLICATIONS
- Determination of pyridoxal 5′-phosphate
- Determination of tyrosine, phenylalanine and dihydroxyphenylalanine either manometrically or colorimetrically, using holoenzyme

NOTES
- Classified as a lyase: carbon-carbon lyase: carboxy-lyase
- A pyridoxal-phosphate protein
- Bacterial enzyme also acts on 3-hydroxytyrosine and more slowly on 3-hydroxyphenylalanine

SPECIFIC ACTIVITY	UNITS DEFINITION	PREPARATION FORM	ADDITIONAL ACTIVITIES	SUPPLIER CATALOG NO.
Streptococcus faecalis				
0.3-1.0 U/mg	1 unit liberates 1 μmol CO_2 from L-Tyr/min at pH 5.5, 37°C	Powder		Fluka 93904
1 U/mg	1 unit liberates 1 μmol CO_2 from L-Tyr/min at pH 5.5, 37°C	Powder		Fluka 93905
0.3-1.0 U/mg solid	1 unit liberates 1.0 μmol CO_2 from L-Tyr/min at pH 5.5, 37°C	Dried cells	L-phenylalanine decarboxylase	Sigma T4379
0.5 U/mg solid with excess pyridoxal 5-PO_4; <0.05 U/mg solid without	1 unit liberates 1.0 μmol CO_2 from L-Tyr/min at pH 5.5, 37°C	Dried cells grown on B_6-deficient medium		Sigma T4629
Streptococcus faecalis NCTC 6783				
≥0.1 U/mg solid	1 unit yields 1 μmol CO_2 from L-Tyr/min at pH 5.5, 37°C; APO enzyme activity is measured in the presence of excess pyridoxal PO_4	Dried cells		Worthington LS04966 LS04964
Activates to ≥0.1 U/mg solid with 5 μg pyridoxal PO_4	1 unit yields 1 μmol CO_2 from L-Tyr/min at pH 5.5, 37°C; APO enzyme activity is measured in the presence of excess pyridoxal PO_4	Dried cells grown on B_6-deficient medium		Worthington LS04968 LS04970 LS04973

Phosphoenolpyruvate Carboxylase

4.1.1.31

REACTION CATALYZED

Orthophosphate + oxaloacetate ↔ H_2O + phosphoenolpyruvate + CO_2

SYSTEMATIC NAME

Orthophosphate:oxaloacetate carboxy-lyase (phosphorylating)

SYNONYMS

PEPC

REACTANTS

Orthophosphate, oxaloacetate, H_2O, phosphoenolpyruvate, CO_2

APPLICATIONS

- Determination of bicarbonate and CO_2 in reagents and buffer solutions, with MDH
- Bicarbonate fixation
- Determination of blood CO_2 levels

NOTES

- Classified as a lyase: carbon-carbon lyase: carboxy-lyase
- Catalyzes the fixation of carbon dioxide with phosphoenolpyruvate to produce oxaloacetate and inorganic phosphate
- Found in most plants and animals
- Activated by acetyl CoA, fructose-1,6-diphosphate and peroxide-free dioxane
- Inhibited by L-aspartate, fumarate, L-malate and sulfate

SPECIFIC ACTIVITY	UNITS DEFINITION	PREPARATION FORM	ADDITIONAL ACTIVITIES	SUPPLIER CATALOG NO.
Escherichia coli (optimum pH = 8-9, MW = 400,000 Da; stable 6 hr in Mg^{2+} buffer)				
0.05 U/mg protein	1 unit forms 1.0 μmol oxaloacetate from PEP and CO_2/min at pH 8.5, 25°C	Crude; lyophilized containing 60% protein; balance primarily Tris buffer salts		Sigma P8079
≥0.3 U/mg solid	1 unit oxidizes 1 μmol NADH/min at pH 8.5, 25°C	Partially purified; lyophilized		Worthington LS03383 LS03381 LS03376
Microbial (optimum pH = 7.0)				
>15 U/mg protein; >7 U/mg solid	1 unit carboxylates 1.0 μmol PEP/min at pH 8.0, 25°C	Lyophilized	<0.02% NADH oxidase <0.05% LDH <0.5% PK	Genzyme 6671
Wheat				
5 U/mg	1 unit converts 1 μmol PEP to product/min at 25°C	Suspension in 3.2 M $(NH_4)_2SO_4$, pH 6	<0.02% NADH oxidase <0.5% MDH (decarboxylating, NADH-dependent) <1% LDH	Boehringer 165794

4.1.1.31 Phosphoenolpyruvate Carboxylase continued

SPECIFIC ACTIVITY	UNITS DEFINITION	PREPARATION FORM	ADDITIONAL ACTIVITIES	SUPPLIER CATALOG No.
Wheat germ				
5 U/mg protein; 1 U/mg material	1 unit forms 1 μmol oxalacetate from PEP and CO_2 at pH 8.0, 37°C	Suspension in $(NH_4)_2SO_4$ containing 10 mM PO_4 buffer, 1.0 mM Biotin, 5.0 mM DTT, 1.0 mM PMSF, 10 mg/mL protein, pH 7.0		Calzyme 154B0001
Zea maize				
1-4 U/mg protein	1 unit forms 1.0 μmol oxaloacetate from PEP and CO_2/min at pH 8.5, 25°C	Suspension in 2.4 M $(NH_4)_2SO_4$ containing 10 mM PO_4 buffer, 1 mM biotin, 5 mM DTT, 1 mM PMSF, pH 7.0	MDH and malic enzyme	Sigma P2023
Zea maize leaves				
≥7 U/mg protein; ≥1 U/mg solid	1 unit converts 1 μmol CO_2/min at pH 8.0, 30°C	Salt-free; lyophilized	<0.1% LDH	Biozyme PEPC2F
>20 U/mg protein; >5-12 U/mg solid	1 unit consumes 1 μmol bicarbonate/min at pH 8.0, 30°C as measured by a ΔA_{340}	Highly purified; lyophilized containing stabilizers and buffer salts	<0.02% NADH oxidase, LDH	Diagnostic P-050
1 U/mg	1 unit carboxylates 1 μmol PEP/min at pH 8.0, 25°C	Powder	<0.1% LDH	Fluka 79414
>8 U/mg protein; >2 U/mg solid	1 unit carboxylates 1.0 μmol PEP/min at pH 8.0, 30°C	Lyophilized	<0.03% NADH oxidase <0.16% LDH	Genzyme 1581
1-3 U/mg solid	1 unit converts 1 μmol CO_2/min at pH 8.0, 30°C	Lyophilized		ICN 153532
4-6 U/mg	1 unit converts 1 μmol CO_2/min at pH 8.5, 30°C	Lyophilized containing 50% protein		Scripps P0114
5 U/mg protein		Lyophilized		Wako 166-14721
Zea maize shoots				
5 U/mg protein; 1 U/mg material	1 unit forms 1 μmol oxalacetate from PEP and CO_2 at pH 8.0, 37°C	Suspension in $(NH_4)_2SO_4$ containing 10 mM PO_4 buffer, 1.0 mM Biotin, 5.0 mM DTT, 1.0 mM PMSF, 10 mg/mL protein, pH 7.0		Calzyme 138B0001
5 U/mg protein; 1 U/mg solid		Suspension in $(NH_4)_2SO_4$, 10 mM PO_4 buffer, 1.0 mM biotin, 5.0 mM DTT, 1.0 mM PMSF, pH 7.0		ICN 153888

Ribulose-Bisphosphate Carboxylase 4.1.1.39

REACTION CATALYZED
    D-Ribulose 1,5-bisphosphate + CO_2 ↔ 2 3-Phospho-D-glycerate

SYSTEMATIC NAME
    3-Phospho-D-glycerate carboxy-lyase (dimerizing)

SYNONYMS
    Ribulosebisphosphate carboxylase/oxygenase, rubisco, D-ribulose 1,5-diphosphate carboxylase

REACTANTS
    D-Ribulose 1,5-bisphosphate, CO_2, O_2, 3-Phospho-D-glycerate, 2-phosphoglycolate

NOTES
- Classified as a lyase: carbon-carbon lyase: carboxy-lyase
- A copper protein
- Forms 3-Phospho-D-glycerate and 2-phosphoglycolate when O_2 is utilized in place of CO_2

SPECIFIC ACTIVITY	UNITS DEFINITION	PREPARATION FORM	ADDITIONAL ACTIVITIES	SUPPLIER CATALOG NO.
Spinach				
0.01-0.1 U/mg solid	1 unit converts 1.0 μmol D-RuDP and CO_2 to 2.0 μmol D-3-phosphoglycerate/min at pH 7.8, 25°C	Partially purified; powder		Sigma R8000

Phenylalanine Decarboxylase 4.1.1.53

REACTION CATALYZED
    L-Phenylalanine ↔ Phenylethylamine + CO_2

SYSTEMATIC NAME
    L-Phenylalanine carboxy-lyase

REACTANTS
    L-Phenylalanine, tyrosine, aromatic amino acids, phenylethylamine, CO_2

NOTES
- Classified as a lyase: carbon-carbon lyase: carboxy-lyase
- A pyridoxal-phosphate protein

4.1.1.53 Phenylalanine Decarboxylase continued

SPECIFIC ACTIVITY	UNITS DEFINITION	PREPARATION FORM	ADDITIONAL ACTIVITIES	SUPPLIER CATALOG NO.
Streptococcus faecalis				
0.05-0.15 U/mg	1 unit liberates 1 μmol CO_2 from L-Phe/min at pH 5.5, 37°C	Powder		Fluka 78087
0.05-0.15 U/mg solid	1 unit liberates 1.0 μmol CO_2 from L-Phe/min at pH 5.5, 37°C	Dried	L-TD	Sigma P2626

4.1.2.10 Mandelonitrile Lyase

REACTION CATALYZED
    Mandelonitrile ↔ Cyanide + benzaldehyde

SYSTEMATIC NAME
    Mandelonitrile benzaldehyde-lyase

SYNONYMS
    Hydroxynitrile lyase

REACTANTS
    Mandelonitrile, cyanide, benzaldehyde

NOTES
- Classified as a lyase: carbon-carbon lyase: aldehyde-lyase
- A flavoprotein

SPECIFIC ACTIVITY	UNITS DEFINITION	PREPARATION FORM	ADDITIONAL ACTIVITIES	SUPPLIER CATALOG NO.
Almond				
100 U/mg protein	1 unit forms 1.0 μmol benzaldehyde and HCN from mandelonitrile/min at pH 5.4, 25°C	Suspension in 50 mM imidazole, 3.2 M $(NH_4)_2SO_4$, pH 6.0		ICN 155332
	1 unit forms 1.0 μmol benzaldehyde and HCN from mandelonitrile/min at pH 5.4, 25°C	Suspension in 3.2 M $(NH_4)_2SO_4$ and 50 mM imidazole, pH 6.0		Sigma M0646
80-240 U/mg protein	1 unit forms 1.0 μmol benzaldehyde and HCN from mandelonitrile/min at pH 5.4, 25°C	Suspension in 2.8 M $(NH_4)_2SO_4$ and 50 mM imidazole, pH 6.0		Sigma M6782

Hydroxymandelonitrile Lyase 4.1.2.11

REACTION CATALYZED
: 4-Hydroxymandelonitrile ↔ Cyanide + 4-hydroxybenzaldehyde

SYSTEMATIC NAME
: 4-Hydroxymandelonitrile hydroxybenzaldehyde-lyase

SYNONYMS
: Hydroxynitrile lyase

REACTANTS
: 4-Hydroxymandelonitrile, cyanide, 4-hydroxybenzaldehyde

NOTES
- Classified as a lyase: carbon-carbon lyase: aldehyde-lyase

SPECIFIC ACTIVITY	UNITS DEFINITION	PREPARATION FORM	ADDITIONAL ACTIVITIES	SUPPLIER CATALOG NO.
Sorghum seedlings				
2-4 U/mg protein	1 unit forms 1.0 μmol 4-hydroxybenzaldehyde and HCN from 4-hydroxymandelonitrile/min at pH 5.4, 25°C	Lyophilized containing 50% protein; balance primarily PO_4 and citrate buffer salts		sigma H7880

Fructose-Bisphosphate Aldolase 4.1.2.13

REACTION CATALYZED
: D-Fructose 1,6-bisphosphate ↔ Glycerone phosphate + D-glyceraldehyde 3-phosphate

SYSTEMATIC NAME
: D-Fructose 1,6-bisphosphate D-glyceraldehyde-3-phosphate-lyase

SYNONYMS
: Aldolase, fructose-1,6-bisphosphate triosephosphate-lyase

REACTANTS
: D-Fructose 1,6-bisphosphate, (3S,4R)-ketose 1-phosphates, glycerone phosphate, D-glyceraldehyde 3-phosphate, dihydroxyacetone phosphate, aldehyde

APPLICATIONS
- Phosphate group transfers
- Asymmetric C-C bond formation
- Synthesis of polyhydroxylated compounds
- Synthesis of rare and unnatural sugars
- Determination of metabolites in coupled reactions
- Determination of D-fructose-1,6,-diphosphate
- Native size standard in column calibration
- Preparation of dihydroxyacetone phosphate, D-glyceraldehyde-3-phosphate and condensation products from dihydroxyacetone phosphate and aldehydes

Fructose-Bisphosphate Aldolase continued

NOTES

- Classified as a lyase: carbon-carbon lyase: aldehyde-lyase
- Found in animals, plants and most microorganisms
- Yeast and bacterial enzymes are zinc proteins
- Animals produce 5 different isozymes which may be organ-specific, including: type A from muscle, type B from kidney and liver, and type C from brain
- Contains 4 active subunits
- Elevated serum levels are seen in carcinomas, muscular dystrophy, hepatitis and myocardial infarction
- Inhibited by heavy metals, especially Cu^{2+}, Zn^{2+} and Ag^+

SPECIFIC ACTIVITY	UNITS DEFINITION	PREPARATION FORM	ADDITIONAL ACTIVITIES	SUPPLIER CATALOG NO.
Rabbit muscle (optimum pH = 7.0, pI = 6.1, MW = 161,000 Da; irreversibly denatured at pH < 4.5)				
9 U/mg	1 unit converts 1 μmol fructose-1,6-phosphate to product/min at 25°C	Suspension in 3.2 M $(NH_4)_2SO_4$, pH 6	<0.01% GAPDH <0.03% GDH <0.1% TIM	Boehringer 102652
10 U/mg protein	1 unit converts 1 μmol fructose-1,6-diphosphate to DHAP and GAP/min at pH 7.6, 25°C	Crystallized; suspension in 2.5 M $(NH_4)_2SO_4$ containing 10 mg/mL protein, 0.01 M Tris, 0.001 M EDTA, pH 7.5		Calzyme 050A0010
15-20 U/mg	1 unit converts 1 μmol fructose-1,6-diphosphate to DHAP and GAP/min at pH 7.6, 25°C	Powder		Fluka 05518
9 U/mg protein	1 unit converts 1 μmol fructose-1,6-diphosphate to DHAP and GAP/min at pH 7.6, 25°C	Crystallized; suspension in 3.2 M $(NH_4)_2SO_4$, pH 6		Fluka 05520
15-20 U/mg protein	1 unit converts 1.0 μmol fructose-1,6-diphosphate to DHAP/min at pH 7.4, 25°C	Crystallized; suspension in 3.2 M $(NH_4)_2SO_4$ at pH 7.5		ICN 100168
10-20 U/g solid	1 unit converts 1.0 μmol fructose-1,6-diphosphate to DHAP and GAP/min at pH 7.4, 30°C	Insoluble enzyme attached to polyacrylamide containing 30% borate buffer salts		Sigma A1386
10-20 U/mg protein	1 unit converts 1.0 μmol fructose-1,6-diphosphate to DHAP and GAP/min at pH 7.4, 25°C	Crystallized; suspension in 2.5 M $(NH_4)_2SO_4$, 0.01 M Tris, 0.001 M EDTA, pH 7.5	<0.03% PK, LDH, G3PDH, α-GPDH 0.05% TPI 0.5% PGI	Sigma A1893
10 U/mg protein	1 unit converts 1.0 μmol fructose-1,6-diphosphate to DHAP and GAP/min at pH 7.4, 25°C	Sulfate-free; lyophilized containing 85% protein; balance primarily citrate buffer salts	<0.03% PK, LDH, G3PDH, α-GPDH 0.05% TPI 0.5% PGI	Sigma A7145

Fructose-Bisphosphate Aldolase continued   4.1.2.13

SPECIFIC ACTIVITY	UNITS DEFINITION	PREPARATION FORM	ADDITIONAL ACTIVITIES	SUPPLIER CATALOG No.
Rabbit muscle *continued*				
≥10 U/mg protein	1 unit changes absorbance by 1.00/min at pH 7.5, 25°C	2X Crystallized; suspension in 2.1 M $(NH_4)_2SO_4$, pH 7.8		Worthington LS01123 LS01125
≥10 U/mg protein	1 unit changes absorbance by 1.00/min at pH 7.5, 25°C	Lyophilized containing sucrose as stabilizer		Worthington LS01130 LS01128
Spinach				
0.3-1.0 U/mg protein	1 unit converts 1.0 μmol fructose 1,6-diphosphate to DHAP and GAP/min at pH 7.4, 25°C			Sigma A9329
Staphylococcus aureus				
10-20 U/mg	1 unit converts 1 μmol fructose-1,6-diphosphate to DHAP and GAP/min at pH 7.6, 25°C	Powder containing TPI	TPI	Fluka 05522
20 U/mg protein	1 unit converts 1.0 μmol fructose 1,6-diphosphate to DHAP and GAP/min at pH 7.4, 25°C	Chromatographically purified; lyophilized containing 65% protein; balance Tris buffer salts and stabilizers	TPI	Sigma A2548
Trout muscle				
7-15 U/mg protein	1 unit converts 1.0 μmol fructose 1,6-diphosphate to DHAP and GAP/min at pH 7.4, 25°C	Crystallized; suspension in 3.2 M $(NH_4)_2SO_4$, 0.01 M Tris, 0.001 M EDTA, pH 7.5	0.03% PGI, LDH, G3PDH, α-GPDH 0.3% PK, TPI	Sigma A7024
Yeast, bakers				
20-50 U/mg protein	1 unit converts 1.0 μmol fructose 1,6-diphosphate to DHAP and GAP/min at pH 7.4, 25°C	Suspension in 4.0 M $(NH_4)_2SO_4$, 10 mM 2-mercaptoethanol, 10 mM His, 0.1 mM $ZnCl_2$, pH 6.8	<0.01% α-GPDH 0.1% TPI	Sigma A9562

N-Acetylneuraminate Lyase

REACTION CATALYZED
: N-Acetylneuraminate ↔ N-Acetyl-D-mannosamine + pyruvate

SYSTEMATIC NAME
: N-Acetylneuraminate pyruvate-lyase

SYNONYMS
: N-Acetylneuraminic acid aldolase, NANA

REACTANTS
: N-Acetylneuraminate, N-glycoloylneuraminate, o-acetylated sialic acids, N-acetyl-D-mannosamine, pyruvate

APPLICATIONS
- Determination of sialic acid and N-acetylneuraminic acid, with related enzymes

NOTES
- Classified as a lyase: carbon-carbon lyase: oxo-acid-lyase
- Acts on o-acetylated sialic acids other than 4-o-acetylated derivatives
- Inhibited by Ag^+, Hg^{2+}, PCMB and SDS

SPECIFIC ACTIVITY	UNITS DEFINITION	PREPARATION FORM	ADDITIONAL ACTIVITIES	SUPPLIER CATALOG NO.
Clostridium perfringens				
≥30 U/mg protein; 15 U/mg solid	1 unit oxidizes 1 μmol NADH/min at pH 7.5, 37°C	Lyophilized		ICN 153493
Clostridium perfringens (welchii)				
0.5-5 U/mg protein	1 unit releases 1.0 μmol pyruvate from NANA/min at pH 7.2, 37°C	Lyophilized containing 60% protein and PO_4 buffer salts		Sigma A5884
Escherichia coli (optimum pH = 7.5-8.0, pI = 4.6, T = 70°C, MW = 98,000 Da)				
30 U/mg	1 unit converts 1 μmol N-Ac-neuraminic acid to product/min at 37°C	Solution containing PO_4 buffer		Boehringer 878758
15 U/mg protein	1 unit hydrolyzes 1 μmol N-Ac-neuraminic acid/min at pH 7.7, 37°C	50 mM Tris-HCl buffer, pH 8.0	<0.01% NADH oxidase, NADH dehydrogenase	Fluka 01402
≥20 U/mg protein; 5 U/mL	1 unit liberates 1.0 μmol N-Ac-mannosamine (or pyruvate)/min at pH 7.7, 37°C using NANA as substrate	Solution containing 50 mM Tris-HCl, pH 8.0	No detectable NADH oxidase	ICN 153852
>20 U/mg protein	1 unit liberates 1.0 μmol N-Ac-mannosamine (or pyruvate)/min at pH 7.7, 37°C using NANA as substrate	Purified; solution containing 50 mM Tris-HCl buffer, pH 8.0	No detectable NADH oxidase	Nacalai 006-28 SP
	1 unit releases 1.0 μmol pyruvate from NANA/min at pH 7, 37°C	Solution containing 0.04 M PO_4 buffer, pH 7		Sigma A0805
12-18 U/mg				Wako 012-11671

N-Acetylneuraminate Lyase continued — 4.1.3.3

SPECIFIC ACTIVITY	UNITS DEFINITION	PREPARATION FORM	ADDITIONAL ACTIVITIES	SUPPLIER CATALOG NO.
Microbial (optimum pH = 7.5–8.0, pI = 4.6, T = 70°C, MW = 98,000 Da [3 X 35,000 Da subunits]; K_M = 2.5 mM [N-acetylneuraminic acid]; stable pH 6.0–9.0 [10°C, 25 hr], T < 65°C [pH 7.5, 30 min])				
≥15 U/mg solid; ≥30 U/mg protein	1 unit oxidizes 1 μmol NADH/min at pH 7.5, 37°C	Lyophilized containing 30% EDTA and mannitol as stabilizers	<0.001% NADH oxidase <1% catalase	Toyobo NAL-301

[Citrate (pro-3S)-Lyase] — 4.1.3.6

REACTION CATALYZED
    Citrate ↔ Acetate + oxaloacetate

SYSTEMATIC NAME
    [Citrate oxaloacetate-lyase ((pro-3S)-CH_2COO^- →acetate)]

SYNONYMS
    Citrase, citratase, citritase, citridesmolase, citrate aldolase, citrate lyase, CL

REACTANTS
    Citrate, acetate, oxaloacetate

NOTES
- Classified as a lyase: carbon-carbon lyase: oxo-acid-lyase
- The enzyme can be dissociated into components, two of which are identical with citrate CoA-transferase (EC 2.8.3.10) and citryl-CoA lyase (EC 4.1.3.34)
- [Citrate-(pro-3S)-lyase] (EC 3.1.2.16) deacetylates and inactivates the enzyme

SPECIFIC ACTIVITY	UNITS DEFINITION	PREPARATION FORM	ADDITIONAL ACTIVITIES	SUPPLIER CATALOG NO.
Enterobacter aerogenes				
0.3 U/mg	1 unit converts 1 μmol citrate to oxalacetic acid/min at pH 7.6, 25°C	Powder		Fluka 27457
0.25 U/mg solid	1 unit converts 1.0 μmol citrate to oxalacetate/min at pH 7.6, 25°C	Lyophilized containing BSA, sucrose, $MgSO_4$, EDTA		Sigma C0897
Klebsiella pneumoniae (Aerobacter aerogenes)				
≥0.25 U/mg solid	1 unit converts 1 μmol citrate to product/min at 25°C	Lyophilized containing BSA, sucrose, $MgSO_4$, EDTA stabilizers	<0.05% ICDH (NAD specific), NADH oxidase	Boehringer 354074

4.1.3.7 Citrate (si)-Synthase

REACTION CATALYZED
Citrate + CoA ↔ Acetyl-CoA + H_2O + oxaloacetate

SYSTEMATIC NAME
Citrate oxaloacetate-lyase ((pro-3S)-CH_2COO^- → acetyl-CoA)

SYNONYMS
Condensing enzyme, citrate condensing enzyme, citrogenase, oxaloacetate transacetase, CS

REACTANTS
Citrate, CoA, acetyl-CoA, H_2O, oxaloacetate

NOTES
- Classified as a lyase: carbon-carbon lyase: oxo-acid-lyase

SPECIFIC ACTIVITY	UNITS DEFINITION	PREPARATION FORM	ADDITIONAL ACTIVITIES	SUPPLIER CATALOG No.
Chicken heart				
100-200 U/mg protein	1 unit forms 1.0 μmol citrate from oxalacetate and acetyl-CoA /min at pH 8.0, 37°C	Chromatographically purified; crystallized suspension in 2.2 M $(NH_4)_2SO_4$ solution, trace PO_4, pH 7.0	No detectable ICDH, aconitase	Sigma C6897
Pigeon breast muscle				
80-150 U/mg protein	1 unit forms 1.0 μmol citrate from oxalacetate and acetyl-CoA/min at pH 8.0, 37°C	Crystallized; suspension in 2.2 M $(NH_4)_2SO_4$, 6 mM PO_4, 0.5 mM citrate, pH 7.0	<0.01% ICDH, aconitase <0.1% MDH	Sigma C4140
Porcine heart				
110 U/mg	1 unit converts 1 μmol oxalacetic acid and acetyl-CoA to product/min at 25°C	Suspension in 3.2 M $(NH_4)_2SO_4$, pH 7	0.1% MDH	Boehringer 103373 103381
100-200 U/mg protein	1 unit forms 1.0 μmol citrate from oxalacetate and acetyl-CoA /min at pH 8.0, 37°C	Crystallized; suspension in 2.2 M $(NH_4)_2SO_4$, 6 mM PO_4, 0.5 mM citrate, pH 7.0	<0.01% ICDH, aconitase <0.1% MDH	Sigma C3260

Tryptophanase 4.1.99.1

REACTION CATALYZED
  L-Tryptophan + H_2O ↔ Indole + pyruvate + NH_3

SYSTEMATIC NAME
  L-Tryptophan indole-lyase (deaminating)

REACTANTS
  L-Tryptophan, H_2O, indole, pyruvate, NH_3

NOTES
- Classified as a lyase: carbon-carbon lyase: other carbon-carbon lyase
- Contains a pyridoxal 5'-phosphate prosthetic group which can be removed by dialysis to yield apotryptophanase
- Requires K^+
- Also catalyzes 2,3-elimination and β-replacement reactions of some indole-substituted tryptophan analogs of L-cysteine, L-serine and other 3-substituted amino acids

SPECIFIC ACTIVITY	UNITS DEFINITION	PREPARATION FORM	ADDITIONAL ACTIVITIES	SUPPLIER CATALOG No.
Escherichia coli				
	1 unit releases 15-40 μg indole from L-Trp/10 min at pH 8.3, 37°C	Crude		sigma T0754

Carbonate Dehydratase 4.2.1.1

REACTION CATALYZED
  H_2CO_3 ↔ CO_2 + H_2O

SYSTEMATIC NAME
  Carbonate hydro-lyase

SYNONYMS
  Carbonic anhydrase, carbonic anhydrase III, CAIII, CA

REACTANTS
  H_2CO_3, CO_2, H_2O, p-nitrophenol acetate, p-nitrophenol, acetate, alkyl pyruvates

APPLICATIONS
- Determination of CO_2 in blood
- Elimination of CO_2 in reagents for acidity testing, e.g., in wine
- Elevated myoglobin in the absence of CAIII suggests that myoglobin is a product of cardiac muscle injury
- Carboxy group transfers
- Reduction reactions

4.2.1.1 Carbonate Dehydratase continued

NOTES

- Classified as a lyase: carbon-oxygen lyase: hydro-lyase
- A zinc protein; the zinc may be replaced by cobalt
- Widespread in nature, occurring in animals, plants and bacteria
- Highly polymorphic in mammalian species, differing in enzymatic properties, amino acid sequences and inhibitor binding
- Facilitates transport of CO_2 and is involved in the transfer and accumulation of H^+ and HCO_3^- in animal respiration; facilitates photosynthetic fixation of CO_2 in autotroph chloroplasts
- Mammalian erythrocytes contain 2 forms with distinct catalytic activity; III is not present in cardiac muscle
- Possesses a very high turnover number
- Activated by Zn^{2+}
- Inhibited by monovalent anions and sulfonamides

SPECIFIC ACTIVITY	UNITS DEFINITION	PREPARATION FORM	ADDITIONAL ACTIVITIES	SUPPLIER CATALOG NO.
Bovine erythrocytes (MW = 30,000 Da)				
≥900 U/g solid	1 unit hydrolyzes 1 μmol p-nitrophenyl acetate/min at pH 7.6, 25°C	Salt-free; lyophilized		Biozyme CAB1
1500 U/g solid	1 unit hydrolyzes 1 μmol p-nitrophenyl acetate/min at pH 7.6, 25°C	Salt-free; lyophilized		Biozyme CAB2
3000 U/mg solid	1 unit hydrolyzes 1 μmol p-nitrophenyl acetate/min at pH 7.6, 25°C	Mixture of A & B isoforms; salt-free; lyophilized		Calbiochem 215755
2000 U/mg protein	1 unit hydrolyzes 1 μmol p-nitrophenyl acetate/min at pH 7.6, 25°C	Lyophilized containing 90% protein		Calzyme 147A2000
200,000 U/mg	1 unit catalyzes the hydration of 1 μmol CO_2/min at pH 8.3, 25°C	Powder		Fluka 21805
240,000 U/mg	1 unit catalyzes the hydration of 1 μmol CO_2/min	Powder		Fluka 21808
2000 U/mg		Lyophilized containing 90% protein		ICN 153879
2500 W-A U/mg protein	1 Wilbur-Anderson unit causes the pH of a 0.02 M Trizma buffer to drop from 8.3 to 6.3/min at 0°C	Dialyzed; lyophilized		Sigma C3934
≥3000 U/mg solid	Activity is determined by the time required for saturated CO_2 to lower the pH of 0.02 M Tris-HCl buffer from 8.3 to 6.3 at 0°C	Dialyzed; lyophilized		Worthington LS01260 LS01263 LS01265

SPECIFIC ACTIVITY	UNITS DEFINITION	PREPARATION FORM	ADDITIONAL ACTIVITIES	SUPPLIER CATALOG No.
Bovine erythrocytes, isozyme				
3000 W-A U/mg protein	1 Wilbur-Anderson unit causes the pH of a 0.02 M Trizma buffer to drop from 8.3 to 6.3/min at 0°C	Electophoretically purified; dialyzed; lyophilized		Sigma C2522
2000 W-A U/mg protein	1 Wilbur-Anderson unit causes the pH of a 0.02 M Trizma buffer to drop from 8.3 to 6.3/min at 0°C	Electophoretically purified; dialyzed; lyophilized		Sigma C3640
Human erythrocytes, isozyme				
100-500 W-A U/mg protein	1 Wilbur-Anderson unit causes the pH of a 0.02 M Trizma buffer to drop from 8.3 to 6.3/min at 0°C	Electophoretically purified; dialyzed; lyophilized		Sigma C4396
100-500 W-A U/mg protein	1 Wilbur-Anderson unit causes the pH of a 0.02 M Trizma buffer to drop from 8.3 to 6.3/min at 0°C	Dialyzed; lyophilized		Sigma C5290
3000-5000 W-A U/mg protein	1 Wilbur-Anderson unit causes the pH of a 0.02 M Trizma buffer to drop from 8.3 to 6.3/min at 0°C	Electophoretically purified; dialyzed; lyophilized containing 85% protein		Sigma C6165
Human liver (MW = 30,000 Da)				
2000 U/mg protein	1 unit hydrolyzes 1 μmol p-nitrophenyl acetate/min at pH 7.6, 25°C	Suspension in 3.6 M $(NH_4)_2SO_4$ containing 10 mg/mL protein		Calzyme 079B2000
Human tissue (carbonic anhydrase III)				
70-400 U/mg	Wilbur-Anderson colorimetric assay, 25°C	>75% purity by SDS-PAGE; solution containing >0.3 mg/mL protein, 10 mM $NaPO_4$ buffer, 0.1% NaN_3, pH 7.0-8.0	No detectable HBsAg, HCV, HIV	Aalto 2001
100-500 U/mg	Wilbur-Anderson colorimetric assay, 25°C	Immunopure; >95% purity by SDS-PAGE; solution containing >0.3 mg/mL protein, 10 mM $NaPO_4$ buffer, 0.1% NaN_3, pH 7.0-8.0	No detectable HBsAg, HCV, HIV	Aalto 2002

4.2.1.2 Fumarate Hydratase

REACTION CATALYZED
    (S)-Malate ↔ Fumarate + H_2O

SYSTEMATIC NAME
    (S)-Malate hydro-lyase

SYNONYMS
    Fumarase

REACTANTS
    (S)-Malate, fumarate, H_2O

NOTES
- Classified as a lyase: carbon-oxygen lyase: hydro-lyase

SPECIFIC ACTIVITY	UNITS DEFINITION	PREPARATION FORM	ADDITIONAL ACTIVITIES	SUPPLIER CATALOG No.
Chicken heart				
200-400 U/mg protein	1 unit converts 1.0 μmol L-malate to fumarate/min at pH 7.6, 25°C	Suspension in 3.2 M $(NH_4)_2SO_4$, pH 7.5		Sigma F4631
Porcine heart				
300-500 U/mg protein	1 unit converts 1.0 μmol L-malate to fumarate/min at pH 7.6, 25°C	Affinity purified; suspension in 3.2 M $(NH_4)_2SO_4$, 0.05 M KPO_4, 0.014 M β-MSH, pH 7.5	<0.02% MDH	Sigma F1757
Rabbit liver				
200-500 U/mg protein	1 unit converts 1.0 μmol L-malate to fumarate/min at pH 7.6, 25°C	Suspension in 3.2 M $(NH_4)_2SO_4$, pH 7.5		Sigma F4756

4.2.1.11 Phosphopyruvate Hydratase

REACTION CATALYZED
    2-Phospho-D-glycerate ↔ Phosphoenolpyruvate + H_2O

SYSTEMATIC NAME
    2-Phospho-D-glycerate hydro-lyase

SYNONYMS
    Enolase, 2-phosphoglycerate dehydratase, neuron specific enolase

REACTANTS
    2-Phospho-D-glycerate, 3-phospho-D-erythronate, phosphoenolpyruvate, H_2O

APPLICATIONS
- Neuron specific enzyme is useful in studying neuronal differentiation and is used to visualize the entire nervous and neuroendocrine system
- Monitoring disease states such as Alzheimer's, Huntington's Chorea, neuroblastoma, head trauma, neuroendocrine cancer and small cell carcinomas of the lung

Phosphopyruvate Hydratase continued — 4.2.1.11

NOTES

- Classified as a lyase: carbon-oxygen lyase: hydro-lyase
- Dimeric 'GG' is the most abundant isozyme and is found in adult neurons

SPECIFIC ACTIVITY	UNITS DEFINITION	PREPARATION FORM	ADDITIONAL ACTIVITIES	SUPPLIER CATALOG No.
Bovine brain (neuron specific) (MW = 90,000 Da)				
1.00 mg/mL		>95% purity by SDS-PAGE; sterile-filtered; solution containing 15 mM Tris, 0.2 M NaCl, 5 mM MgCl$_2$, 0.1 mM EDTA, 0.1% NaN$_3$, pH 7.9		Cortex CP4110
Human brain (neuron specific) (MW = 45,000 Da)				
45 U/mg	1 unit converts NSE to 1 mol 2-phospho-D-glycerate/min at 25°C; subunit composition is γ–γ	≥95% purity by chromatography and gel-filtration; 0.1 M Tris-PO$_4$ buffer, 5 mM MgSO$_4$, 0.2 M KCl, pH 7.0	No detectable HBsAg, HIV-1 Ab, HBC	Adv Immuno NSE1
45 U/mg protein	1 unit converts 1 μmol 2-phospho-D-glyceric acid to PEP/min at 25°C	>95% purity; lyophilized from buffer containing 10 mM Tris-PO$_4$ and 1 mM MgSO$_4$, pH 7.3	No detectable HBsAg, HIV	Calbiochem 480732
50-70 U/mg	1 unit forms 1 μmol PEP/min at 25°C	>95% purity by SDS-PAGE; lyophilized containing 10 mM Tris-PO$_4$, 1 mM MgSO$_4$, pH 7.3	No detectable HGsAg, HCV, HIV-1 Ab	Scripps N0214
Rabbit muscle (MW = 82,000 Da)				
40 U/mg	1 unit converts 1 μmol glycerate-2-phosphate to product/min at 25°C	Suspension in 3.2 mM (NH$_4$)$_2$SO$_4$ solution, pH 6	<0.02% PGM, PK	Boehringer 104647
40 U/mg protein	1 unit converts 1.0 μmol 2-phospho-glycerate to PEP/min at pH 7.4, 37°C	Suspension in 3.2 M (NH$_4$)$_2$SO$_4$, pH 6.0		Calzyme 196B0040
30 U/mg protein	1 unit converts 1.0 μmol 2-phosphoglycerate to PEP/min at pH 7.4, 25°C	Crystallized; suspension in 3.2 M (NH$_4$)$_2$SO$_4$		ICN 104877
25-35 U/mg protein	1 unit converts 1.0 μmol 2-phospho-glycerate to PEP/min at pH 7.4, 25°C	Crystallized; suspension in 2.8 M (NH$_4$)$_2$SO$_4$, 0.05 M imidazole, 0.001 M MgSO$_4$, pH 7.5	<0.05% LDH, Pk <0.07% PGM	Sigma E0379
Yeast				
40 U/mg protein		Highly purified; 50% glycerol solution	<0.02% PGM, PK	ICN 193498
≥40 IU/mg protein	1 IU transforms 1 μmol substrate/min under standard IUB conditions at 25°C	50% glycerol solution	<0.02% PK, PGM	OYC
Yeast, bakers				
50 U/mg protein	1 unit converts 1.0 μmol 2-phospho-glycerate to PEP/min at pH 7.4, 37°C	Lyophilized containing 90-95% protein		Calzyme 185A0050
60 U/mg protein	1 unit converts 1.0 μmol 2-phospho-glycerate to PEP/min at pH 7.4, 25°C	Lyophilized containing 70% protein; balance primarily Tris buffer salts		Sigma E6126

Enoyl-CoA Hydratase

4.2.1.17

REACTION CATALYZED
   (3S)-3-Hydroxyacyl-CoA ↔ Trans-2(or 3)-enoyl-CoA + H_2O

SYSTEMATIC NAME
   (3S)-3-Hydroxyacyl-CoA hydro-lyase

SYNONYMS
   Enoyl hydrase, unsaturated acyl-CoA hydratase, crotonase

REACTANTS
   (3S)-3-Hydroxyacyl-CoA, cis-2(or 3)-enoyl-CoA, H_2O, trans-2(or 3)-enoyl-CoA, H_2O, (3R)-3-hydroxyacyl-CoA

APPLICATIONS
- Analysis of enoyl CoA derivatives

NOTES
- Classified as a lyase: carbon-oxygen lyase: hydro-lyase
- Acts in the reverse direction
- Similar to long-chain-enoyl-CoA hydratase (EC 4.2.1.74)

SPECIFIC ACTIVITY	UNITS DEFINITION	PREPARATION FORM	ADDITIONAL ACTIVITIES	SUPPLIER CATALOG NO.
Bovine liver				
500-1000 U/mg protein	1 unit hydrates 1.0 µmol crotonoyl CoA to hydroxybutyryl CoA/min at pH 7.5, 25°C	Crystallized; lyophilized containing 50% protein; balance KPO_4, Na citrate, EDTA, pH 7.4		Sigma C6073

Porphobilinogen Synthase

4.2.1.24

REACTION CATALYZED
   2 5-Aminolevulinate ↔ Porphobilinogen + 2 H_2O

SYSTEMATIC NAME
   5-Aminolevulinate hydro-lyase (adding 5-aminolevulinate and cyclizing)

SYNONYMS
   Aminolevulinate dehydratase

REACTANTS
   5-Aminolevulinate, porphobilinogen, H_2O

NOTES
- Classified as a lyase: carbon-oxygen lyase: hydro-lyase
- The fungal enzyme is a metalloprotein
- The second enzyme in the tetrapeptide biosynthetic pathway

Porphobilinogen Synthase continued

4.2.1.24

SPECIFIC ACTIVITY	UNITS DEFINITION	PREPARATION FORM	ADDITIONAL ACTIVITIES	SUPPLIER CATALOG NO.
Bovine liver (optimum pH = 6.8)				
≥6 U/mg protein	1 unit forms 1 μmol porphobilinogen from 5'-aminolevulinic acid/hr/mg protein at pH 6.7, 37°C	Suspension in 2.5 M $(NH_4)_2SO_4$ containing 50 mM KPO_4 and 1 mg/mL protein		Calzyme 237B0006
	1 unit produces 1.0 μmol porphobilinogen from δ-aminolevulinic acid/hr at pH 6.5, 37°C	Lyophilized containing 50% protein; balance primarily KPO_4 and DTT		Sigma A0442
2-8 U/mg protein	1 unit produces 1.0 μmol porphobilinogen from δ-aminolevulinic acid/hr at pH 6.5, 37°C	Suspension in 2.5 M $(NH_4)_2SO_4$, 50 mM KPO_4, 0.1 mM DTT, pH 6.8		Sigma A0644

Hyaluronate Lyase

4.2.2.1

REACTION CATALYZED
  Hyaluronate ↔ n 3-(4-Deoxy-β-D-gluc-4-enuronosyl)-N-acetyl-D-glucosamine

SYSTEMATIC NAME
  Hyaluronate lyase

SYNONYMS
  Hyaluronidase, hyaluroglucosidase

REACTANTS
  Hyaluronate, chondroitin, 3-(4-deoxy-β-D-gluc-4-enuronosyl)-N-acetyl-D-glucosamine

NOTES
- Classified as a lyase: carbon-oxygen lyase acting on polysaccharides
- Found in animals and bacteria
- Activated by chitosan, poly-L-lysine and 1,10-diaminodecane
- Inhibited by high molecular weight polysaccharides and heavy metals (Mn^{2+} and Hg^{2+}), and potassium ferricyanide
- Distinct from other hyaluronidases, this enzyme is specific for hyaluronate and inactive against chondroitin and chondroitin sulfate

SPECIFIC ACTIVITY	UNITS DEFINITION	PREPARATION FORM	ADDITIONAL ACTIVITIES	SUPPLIER CATALOG NO.
Streptococcus dysgalactiae **(optimum pH = 5.8-6.6, T = 37°C)**				
0.5 U/vial	1 unit liberates 1 μmol unsaturated disaccharide from pig skin hyaluronic acid/min at 37°C	Lyophilized		Seikagaku 100741-1

4.2.2.1 Hyaluronate Lyase continued

SPECIFIC ACTIVITY	UNITS DEFINITION	PREPARATION FORM	ADDITIONAL ACTIVITIES	SUPPLIER CATALOG No.
Streptomyces hyalurolyticus (optimum pH = 6.0, T = 60-65°C)				
0.05 U/mg	1 unit releases 1 μmol *N*-Ac-*D*-glucosamine end groups/min at pH 4.0, 37°C with hyaluronic acid K salt as substrate	Film		Fluka 53725
2000 U/mg solid	1 unit changes OD in a solution of hyaluronic acid compared to a hyaluronidase International Standard	Salt-free; lyophilized	No detectable protease, lysozyme	ICN 151270
2000 U/mg; 100 U/vial		Homogeneous by molecular exclusion and DEAE cellulose chromatography; salt-free; lyophilized containing <0.1 μg/U sugar	No detectable protease, lysozyme	ICN 320421
2000 TRU/mg protein; ≥100 TRU/vial	1 turbidity reducing unit (TRU) decreases A_{660} by 50%/30 min at 60°C	Lyophilized		Seikagaku 100740-1
	Assayed per USP XXI-NF XVI combined edition	Lyophilized		Sigma H1136
>2000 TRU/A_{280} (TRU = turbidity reducing unit)	1 TRU hydrolyzes 50% substrate as seen by a 50% reduction in turbidity/30 min at pH 6.0, 60°C	Lyophilized	<0.05 U/ampoule protease	Calbiochem 389561

Pectate Lyase 4.2.2.2

REACTION CATALYZED
  Eliminative cleavage of pectate to give oligosaccharides with 4-deoxy-α-*D*-gluc-4-enuronosyl groups at their non-reducing ends

SYSTEMATIC NAME
  Poly(1,4-a-*D*-galacturonide) lyase

SYNONYMS
  Pectate transeliminase

REACTANTS
  Pectate, polygalacturonides, 4-deoxy-α-*D*-gluc-4-enuronosyl oligosaccharides

NOTES
- Classified as a lyase: carbon-oxygen lyase acting on polysaccharides
- Also acts on other polygalacturonides
- Does not act on pectin

See Chapter 5. Multiple Enzyme Preparations

Poly(β-D-Mannuronate) Lyase   4.2.2.3

REACTION CATALYZED
Eliminative cleavage of polysaccharides containing β-D-mannuronate residues to give oligosaccharides with 4-deoxy-α-L-erythro-hex-4-enopyranuronosyl groups at their ends

SYSTEMATIC NAME
Poly(β-D-1,4-mannuronide) lyase

SYNONYMS
Alginate lyase I

REACTANTS
β-D-Mannuronate residues, 4-deoxy-α-L-erythro-hex-4-enopyranuronosyl oligosaccharides

NOTES
- Classified as a lyase: carbon-oxygen lyase acting on polysaccharides

---

See Chapter 5. Multiple Enzyme Preparations

Chondroitin ABC Lyase   4.2.2.4

REACTION CATALYZED
Eliminative degradation of polysaccharides containing 1,4-β-D-hexosaminyl and 1,3-β-D-glucuronosyl or 1,3-α-L-iduronosyl linkages to disaccharides containing 4-deoxy-β-D-gluc-4-enuronosyl groups

SYSTEMATIC NAME
Chondroitin ABC lyase

SYNONYMS
Chondroitinase, chondroitin ABC eliminase

REACTANTS
1,4-β-D-Hexosaminyl-linked polysaccharides, 1,3-β-D-glucuronosyl-linked polysaccharides, 1,3-α-L-iduronosyl-linked polysaccharides, chondroitin 4-sulfate, chondroitin 6-sulfate, dermatan sulfate, hyaluronate (slowly), 4-deoxy-β-D-gluc-4-enuronosyl disaccharides

APPLICATIONS
- Differentiates between several chondroitin sulfates of cartilage; facilitates study relative to disease states
- Oligosaccharide mapping of heparin sulfate
- Preparation and analysis of proteoglycan core protein

NOTES
- Classified as a lyase: carbon-oxygen lyase acting on polysaccharides
- Removes polysaccharide side chains, like chondroitin A and B and dermatan sulfate, from connective tissue and proteoglycan
- Also hydrolyzes hyaluronic acid side chains slowly
- Not active against keratosulfate, heparin or heparin sulfate
- Activated by heparin and heparin sulfate

4.2.2.4 Chondroitin ABC Lyase continued

SPECIFIC ACTIVITY	UNITS DEFINITION	PREPARATION FORM	ADDITIONAL ACTIVITIES	SUPPLIER CATALOG NO.
Proteus vulgaris (optimum pH = 6.2 [hyaluronic acid], 8.0 [chondroitin sulfate], T = 37°C, MW = 120,000-145,000, 80,000 Da [SDS-PAGE])				
10 U/mg protein	1 unit converts 1 μmol chondroitin-6-sulfate to product/min at pH 8, 37°C	Lyophilized	No detectable protease	Boehringer 1080717
0.7 U/mg	1 unit forms 1 μmol unsaturated disaccharide from chondroitin-6-sulfate/min at pH 8.0, 37°C	Lyophilized	<0.05% C6S <0.1% C4S	Fluka 27038
5 U/vial	1 unit liberates 1 μmol product/min at pH 8.0, 37°C	Highly purified by salt fractionation, DEAE, phosphocellulose chromatography; lyophilized	Very low levels lysozyme, protease <0.005 U/vial C4S <0.000025 U/vial C6S	ICN 190334
5 U/vial; 4 vials	1 unit liberates 1 μmol product/min at pH 8.0, 37°C	Highly purified by salt fractionation, DEAE, phosphocellulose chromatography; lyophilized	Very low levels lysozyme, protease <0.005 U/vial C4S <0.000025 U/vial C6S	ICN 320211
1 U/vial; 4 vials	1 unit liberates 1 μmol product/min at pH 8.0, 37°C	Highly purified; lyophilized	No detectable protease	ICN 320301
5 U/vial	1 unit forms 1 μmol unsaturated disaccharide from chondroitin-6-sulfate/min at pH 8.0, 37°C	Lyophilized		Seikagaku 100330-1
110 U/mg protein	1 unit forms 1 μmol unsaturated disaccharide from chondroitin-6-sulfate/min at pH 8.0, 37°C	Lyophilized	Protease-free	Seikagaku 100332-1
0.2-2.0 U/mg solid	1 unit liberates 1.0 μmol [2-acetamido-2-deoxy-3-O-(β-D-gluc-4-ene-pyranosyluronic acid)]-4-O-sulfo-D-galactose from chondroitin sulfate A or 1.0 μmol of ["..."]-6-O-sulfo-D-galactose from chondroitin sulfate C/min at pH 8.0, 37°C	Lyophilized containing 35% protein; balance Tris buffer salts and BSA stabilizer	Also acts on chondroitin sulfate B	Sigma C2905
50-250 U/mg protein	1 unit liberates 1.0 μmol [2-acetamido-2-deoxy-3-O-(β-D-gluc-4-ene-pyranosyluronic acid)]-4-O-sulfo-D-galactose from chondroitin sulfate A or 1.0 μmol of ["..."]-6-O-sulfo-D-galactose from chondroitin sulfate C/min at pH 8.0, 37°C	Lyophilized containing 10% protein; balance KPO$_4$ buffer salts and stabilizer	Substantially free of protease	Sigma C3667

Chondroitin AC Lyase 4.2.2.5

REACTION CATALYZED
Eliminative degradation of polysaccharides containing 1,4-β-D-hexosaminyl and 1,3-β-D-glucuronosyl linkages to disaccharides containing 4-deoxy-β-D-gluc-4-enuronosyl groups

SYSTEMATIC NAME
Chondroitin AC lyase

SYNONYMS
Chondroitinase, chondroitin sulfate lyase, chondroitin AC eliminase

REACTANTS
1,4-β-D-Hexosaminyl-linked polysaccharides, 1,3-β-D-glucuronosyl-linked polysaccharides, chondroitin 4-sulfate, chondroitin 6-sulfate, hyaluronate (less well), 4-deoxy-β-D-gluc-4-enuronosyl disaccharides

APPLICATIONS
- Structural analyses of copolymers

NOTES
- Classified as a lyase: carbon-oxygen lyase acting on polysaccharides
- Inactive toward keratosulfate, heparin sulfate B, heparin and dermatan sulfate
- Completely inhibited by chondroitin sulfate B

SPECIFIC ACTIVITY	UNITS DEFINITION	PREPARATION FORM	ADDITIONAL ACTIVITIES	SUPPLIER CATALOG NO.
Arthrobacter				
5 U/vial; 4 vials	1 unit liberates 1 μmol product/min at pH 7.3, 37°C	Highly purified by salt fractionation, DEAE, phosphocellulose chromatography; lyophilized	No detectable GRS, protease	ICN 320221
Arthrobacter aurescens (optimum pH = 6.0, T = 37°C)				
1 U/mg	1 unit forms 1 μmol unsaturated disaccharide from chondroitin-6-sulfate/min at pH 8.0, 37°C	Powder		Fluka 27040
5 U/vial	1 unit liberates 1 μmol product/min at pH 7.3, 37°C	Highly purified; lyophilized		ICN 190335
5 U/vial	1 unit forms 1 μmol unsaturated disaccharide from chondroitin-6-sulfate/min at pH 6.0, 37°C	Lyophilized		Seikagaku 100335-1
	1 unit liberates 1.0 μmol of [2-acetamido-2-deoxy-3-O-(β-D-gluc-4-ene-pyranosyluronic acid)]-4-O-sulfo-D-galactose from chondroitin sulfate A or 1.0 μmol of [...]-6-O-sulfo-D-galactose from chondroitin sulfate C/min at pH 6.0, 37°C	BSA as stabilizer		Sigma C2262

4.2.2.5 Chondroitin AC Lyase *continued*

SPECIFIC ACTIVITY	UNITS DEFINITION	PREPARATION FORM	ADDITIONAL ACTIVITIES	SUPPLIER CATALOG NO.
Flavobacterium heparinum (optimum pH = 5.5 [chondroitin], 6.0 [hyaluronic acid], 7.3 [chondroitin sulfate]; T = 37°C)				
1 U/vial; 4 vials		Highly purified; lyophilized		ICN 320311
1.0 U/vial	1 unit forms 1 μmol unsaturated disaccharide from chondroitin-6-sulfate/min at pH 7.3, 37°C	Lyophilized		Seikagaku 100334-1
0.5–1.5 U/mg solid		Lyophilized containing 15% protein; balance KPO_4 salts and BSA as stabilizer	Also cleaves chondroitin sulfate C	Sigma C2780

4.2.2.7 Heparin Lyase

REACTION CATALYZED
Eliminative cleavage of polysaccharides containing 1,4-linked *D*-glucuronate or *L*-iduronate residues and 1,4-α-linked 2-sulfoamino-2-deoxy-6-sulfo-*D*-glucose residues to give oligosaccharides with terminal 4-deoxy-α-*D*-gluc-4-enuronosyl groups at their non-reducing ends

SYSTEMATIC NAME
Heparin lyase

SYNONYMS
Heparin eliminase, heparinase

REACTANTS
1,4-Linked *D*-glucuronate polysaccharides, 1,4-linked *L*-iduronate polysaccharides, 1,4-α-linked 2-sulfoamino-2-deoxy-6-sulfo-*D*-glucose polysaccharides, 4-deoxy-α-*D*-gluc-4-enuronosyl oligosaccharides

NOTES
- Classified as a lyase: carbon-oxygen lyase acting on polysaccharides
- Inactive toward heparin sulfate

SPECIFIC ACTIVITY	UNITS DEFINITION	PREPARATION FORM	ADDITIONAL ACTIVITIES	SUPPLIER CATALOG NO.
Flavobacterium heparinum (optimum pH = 7.0, T = 35°C)				
1 U/mg	1 unit forms 1 μmol δ-4,5-unsaturated hexuronate residues from heparin/min at pH 7.0, 37°C	Crystalline		Fluka 51534
0.4 U/mg	1 unit forms 1 μmol δ-4,5-unsaturated hexuronate residues from heparin/min at pH 7.0, 37°C	Lyophilized	<2% chondroitinase <5% heparitinase, hyaluronidase	Fluka 51539

Heparin Lyase continued   4.2.2.7

SPECIFIC ACTIVITY	UNITS DEFINITION	PREPARATION FORM	ADDITIONAL ACTIVITIES	SUPPLIER CATALOG NO.
Flavobacterium heparinum continued				
	1 IU splits typical heparin liberating UV-absorbing materials corresponding to 1 μmol Δ4,5-hexuronate residues/min at pH 7.0, 37°C	Highly purified; lyophilized		ICN 190103
≥1.5 IU/mg protein	1 IU splits typical heparin liberating UV-absorbing materials corresponding to 1 μmol Δ⁴-hexuronate residues/min as calculated with 5500 for the molar extinction coefficient	Lyophilized containing <1 mg/vial BSA as stabilizer	Chondroitinase	ICN 321411
1.5 U/mg protein	1 unit cleaves heparin by elimination, yielding UV-absorbing materials corresponding to 1 μmol Δ⁴-hexuronate residues/min	Lyophilized		Seikagaku 100700-3
200-600 U/mg protein (heparin Na)	1 unit forms 0.1 μmol unsaturated uronic acid/hr at pH 7.5, 25°C	Lyophilized containing 25% BSA stabilizer		Sigma H2519
100-300 U/mg solid	1 unit forms 1.0 μmol unsaturated uronic acid/hr at pH 7.0, 25°C	Lyophilized containing 25% BSA stabilizer		Sigma H6512

Heparitin-Sulfate Lyase   4.2.2.8

REACTION CATALYZED
: Apparent cleavage of linkages between *N*-Ac-D-glucosamine and uronate, eliminating sulfate and producing an unsaturated sugar

SYSTEMATIC NAME
: Heparin-sulfate lyase

SYNONYMS
: Heparin-sulfate eliminase, heparitinase

REACTANTS
: *N*-Acetyl-D-glucosamine uronate linkages, sulfate, unsaturated sugar

NOTES
- Classified as a lyase: carbon-oxygen lyase acting on polysaccharides
- Does not act on *N,o*-desulfated glucosamine, *N*-acetyl-*o*-sulfated glucosamine linkages or heparin

4.2.2.8 Heparitin-Sulfate Lyase continued

SPECIFIC ACTIVITY	UNITS DEFINITION	PREPARATION FORM	ADDITIONAL ACTIVITIES	SUPPLIER CATALOG NO.
Flavobacterium heparinum (optimum pH = 7.0, T = 43°C)				
0.3 U/mg	1 unit forms 1 μmol unsaturated uronic acid/min at pH 7.5, 25°C	Lyophilized containing 25% BSA as stabilizer		Fluka 51538
	1 IU splits typical heparitin sulfate liberating UV-absorbing material corresponding to 1 μmol Δ4,5-hexuronate residues/min at pH 7.0, 37°C	Highly purified; lyophilized		ICN 190102
≥1.5 IU/mg protein	1 IU splits typical heparin sulfate giving UV-absorbing materials corresponding to 1 μmol Δ^4-hexuronate residues/min as calculated with 5500 for the molar extinction coefficient	Lyophilized containing <1 mg/vial BSA as stabilizer	Chondroitinase	ICN 321511
1.5 U/mg protein	1 unit cleaves bovine kidney heparin sulfate by elimination, yielding UV-absorbing materials corresponding to 1 μmol Δ^4-hexuronate residues/min	Lyophilized		Seikagaku 100703-3
0.1 U/vial	1 unit cleaves bovine kidney heparin sulfate by elimination, yielding UV-absorbing materials corresponding to 1 μmol Δ^4-hexuronate residues/min	Lyophilized		Seikagaku 100704-1
200-600 U/mg protein (bovine kidney heparin sulfate)	1 unit forms 0.1 μmol unsaturated uronic acid/hr at pH 7.5, 25°C	Lyophilized containing 25% BSA stabilizer		Sigma H8891

4.2.2.10 Pectin Lyase

REACTION CATALYZED
 Eliminative cleavage of pectin to give oligosaccharides with terminal 4-deoxy-6-methyl-α-D-galact-4-enuronosyl groups

SYSTEMATIC NAME
 Poly(methoxy-L-galacturonide) lyase

SYNONYMS
 Pectinase, pectolyase

REACTANTS
 Pectin, 4-deoxy-6-methyl-α-D-galact-4-enuronosyl oligosaccharides, polygalacturonic acid, D-galacturonic acid

APPLICATIONS
- Generating good yields of viable protoplasts in corn, soybean, red beet, sunflower, tomato and citrus, with cellulase

Pectin Lyase continued 4.2.2.10

NOTES

- Classified as a lyase: carbon-oxygen lyase acting on polysaccharides
- A multicomponent preparation very effective in depolymerizing plant pectins with varying degrees of esterification. May contain substantial hemicellulase, cellulase, pectinesterase, xylanase, pectin lyase and polygalacturonase
- Does not act on de-esterified pectin

SPECIFIC ACTIVITY	UNITS DEFINITION	PREPARATION FORM	ADDITIONAL ACTIVITIES	SUPPLIER CATALOG NO.
Aspergillus japonicus				
250-750 U/mg protein	1 unit causes a ΔA_{235} of 1.0/min at pH 6.0, 40°C due to the release of unsaturated products from pectin	Basic enzyme; solution containing 50% glycerol, 50 m*M* NaOAc buffer, 1 m*M* PMSF, pH 4.5	No detectable cellulase, hemicellulase, glycosaminidase <20 U/mg protein polygalacturonase	Sigma P2679
50-150 U/mg protein	1 unit causes a ΔA_{235} of 1.0/min at pH 5.0, 40°C due to the release of unsaturated products from pectin	Acidic enzyme; solution containing 50% glycerol, 40 m*M* NaOAc buffer, 1 m*M* PMSF, pH 4.5	No detectable cellulase, hemicellulase, glycosaminidase <10 U/mg protein polygalacturonase	Sigma P2804
2-4 U/mg solid	1 unit liberates 1.0 μmol galacturonic acid from polygalacturonic acid/min at pH 5.5, 25°C	Plant cell tested; lyophilized containing 60% protein; balance primarily lactose	Endopolygalacturonase, endopectin lyase and maceration stimulating factor	Sigma P5936
Aspergillus niger				
50-150 U/mg protein	1 unit causes a ΔA_{235} of 1.0/min at pH 5.0, 40°C due to the release of unsaturated products from pectin	Acidic enzyme; solution containing 50% glycerol and 40 m*M* NaOAc buffer, pH 4.5	<1 U/mg protein cellulase, hemicellulase, glycosaminidase <2 U/mg protein polygalacturonase	Sigma P7052
≥20 U/mg solid	1 unit liberates 1 μmol *D*-galacturonic acid from polygalacturonic acid/min at pH 5.0, 37°C	Chromatographically purified; lyophilized	Hemicellulase, cellulase, pectinesterase, xylanase	Worthington LS04297 LS04298 LS04296

4.3.1.1 Aspartate Ammonia-Lyase

REACTION CATALYZED
  L-Aspartate ↔ Fumarate + NH_3
SYSTEMATIC NAME
  L-Aspartate ammonia-lyase
SYNONYMS
  Aspartase, fumaric aminase

REACTANTS
  L-Aspartate, fumarate, NH_3
NOTES
- Classified as a lyase: carbon-nitrogen lyase: ammonia-lyase

SPECIFIC ACTIVITY	UNITS DEFINITION	PREPARATION FORM	ADDITIONAL ACTIVITIES	SUPPLIER CATALOG NO.
Hafnia alvei (Bacterium cadaveris)				
4-6 U/mg protein	1 unit converts 1.0 μmol L-Asp to fumarate/min at pH 8.5, 30°C	Lyophilized containing 40% protein; balance Tris buffer salts and stabilizer		Sigma A8147

4.3.1.3 Histidine Ammonia-Lyase

REACTION CATALYZED
  L-Histidine ↔ Urocanate + NH_3
SYSTEMATIC NAME
  L-Histidine ammonia-lyase
SYNONYMS
  Histidase, histidinase, histidine α-deaminase
REACTANTS
  L-Histidine, urocanate, NH_3

NOTES
- Classified as a lyase: carbon-nitrogen lyase: ammonia-lyase
- Appears to be specific for L-His
- Inactivated with purification, thorough dialysis or aging; reactivated with addition of glutathione or sodium thioglycolate

SPECIFIC ACTIVITY	UNITS DEFINITION	PREPARATION FORM	ADDITIONAL ACTIVITIES	SUPPLIER CATALOG NO.
Porcine liver (optimum pH = 9.5)				
2000 U/mg protein	1 unit increases A_{277} by 0.001/min at 25°C	Lyophilized containing 95% protein		Calzyme 228A2000
Pseudomonas fluorescens				
10-20 U/mg		Lyophilized		ICN 100722

Histidine Ammonia-Lyase continued

4.3.1.3

SPECIFIC ACTIVITY	UNITS DEFINITION	PREPARATION FORM	ADDITIONAL ACTIVITIES	SUPPLIER CATALOG NO.
Pseudomonas fluorescens continued				
10-20 U/mg	1 unit deaminates 1.0 nmol L-His to urocanic acid/min at pH 9.0, 25°C	Dried cells		Sigma H7500
500-2000 U/mg protein	1 unit deaminates 1.0 nmol L-His to urocanic acid/min at pH 9.0, 25°C	Partially purified; preparation containing 40% protein; balance buffer salts and stabilizer		Sigma H8883

Phenylalanine Ammonia-Lyase

4.3.1.5

REACTION CATALYZED
    L-Phenylalanine ↔ trans-Cinnamate + NH_3
SYSTEMATIC NAME
    L-Phenylalanine ammonia-lyase
SYNONYMS
    Phenylalanine deaminase

REACTANTS
    L-Phenylalanine, L-tyrosine, trans-cinnamate, NH_3
NOTES
- Classified as a lyase: carbon-nitrogen lyase: ammonia-lyase
- May also act on L-tyrosine

SPECIFIC ACTIVITY	UNITS DEFINITION	PREPARATION FORM	ADDITIONAL ACTIVITIES	SUPPLIER CATALOG NO.
Potato				
0.001-0.01 U/mg protein L-Phe as substrate	1 unit deaminates 1.0 μmol L-Phe to trans-cinnamate and NH_3/min at pH 8.5, 30°C; 1 unit deaminates 1.0 μmol L-Tyr to p-coumarate and NH_3/min at pH 8.5, 30°C	Lyophilized containing 70% protein; balance K borate buffer salts, DTT, $MgCl_2$, EDTA		Sigma P7774
Rhodotorula glutinis				
1 U/mg protein	1 unit deaminates 1 μmol L-phenyl-alanine to trans-cinnamate/min at pH 8.5, 30°C	60% glycerol, 3 mM Tris-HCl, 0.5 M $(NH_4)_2SO_4$, pH 7.5		Fluka 78085
0.08-2.0 U/mg protein using L-Phe; 0.2-0.4 U/mg protein using L-Tyr	1 unit deaminates 1.0 μmol L-phenyl-alanine to trans-cinnamate and NH_3/min at pH 8.5, 30°C; 1 unit deaminates 1.0 μmol L-Tyr to p-coumarate and NH_3/min at pH 8.5, 30°C	Solution containing 60% glycerol, 3 mM Tris-HCl, 0.5 M $(NH_4)_2SO_4$, pH 7.5		Sigma P1016
0.2-0.8 U/mg protein L-Phe as substrate	1 unit deaminates 1.0 μmol L-Phe to trans-cinnamate and NH_3/min at pH 8.5, 30°C; 1 unit deaminates 1.0 μmol L-Tyr to p-coumarate and NH_3/min at pH 8.5, 30°C	Solution containing 60% glycerol, 3 mM Tris-HCl, 0.5 M $(NH_4)_2SO_4$, pH 7.5		Sigma P9519

4.3.2.1 Argininosuccinate Lyase

REACTION CATALYZED
  N-(L-Arginino)succinate ↔ Fumarate + L-arginine

SYSTEMATIC NAME
  N-(L-Arginino)succinate arginine-lyase

SYNONYMS
  Arginosuccinase

REACTANTS
  N-(L-Arginino)succinate, fumarate, L-arginine

NOTES
- Classified as a lyase: carbon-nitrogen lyase: amidine-lyase

SPECIFIC ACTIVITY	UNITS DEFINITION	PREPARATION FORM	ADDITIONAL ACTIVITIES	SUPPLIER CATALOG NO.
Bovine liver				
	1 unit forms 1.0 μmol each of L-Arg and fumarate from L-arginino-succinate/min at pH 7.5, 37°C	Lyophilized containing 95% protein; balance primarily buffer salts		Sigma A3647
Porcine kidney				
0.1 U/mg protein	1 unit forms 1.0 μmol each of L-Arg and fumarate from L-arginino-succinate/min at pH 7.5, 37°C	Crude; lyophilized containing 90% protein; balance primarily buffer salts		Sigma A9012

4.3.2.2 Adenylosuccinate Lyase

REACTION CATALYZED
  N^6-(1,2-Dicarboxyethyl)AMP ↔ Fumarate + AMP

SYSTEMATIC NAME
  N^6-(1,2-Dicarboxyethyl)AMP AMP-lyase

REACTANTS
  N^6-(1,2-Dicarboxyethyl)AMP, 1-(5-phosphoribosyl)-4-(N-succinocarboxamide)-5-aminoimidazole, fumarate, AMP

NOTES
- Classified as a lyase: carbon-nitrogen lyase: amidine-lyase

SPECIFIC ACTIVITY	UNITS DEFINITION	PREPARATION FORM	ADDITIONAL ACTIVITIES	SUPPLIER CATALOG NO.
Yeast				
0.2-0.5 U/mg protein	1 unit converts 1.0 μmol adenylosuccinic acid to fumaric acid and 5'-AMP/min at pH 7.0, 25°C	Partially purified; lyophilized containing 50% protein; balance primarily KPO₄, EDTA, pH 7.0	<0.02 U/mg protein fumarase	Sigma A4653

Lactoylglutathione Lyase   4.4.1.5

REACTION CATALYZED
   (R)-S-Lactoylglutathione ↔ Glutathione + methylglyoxal

SYSTEMATIC NAME
   (R)-S-Lactoylglutathione methylglyoxal-lyase (isomerizing)

SYNONYMS
   Methylglyoxalase, aldoketomutase, Ketone-aldehyde mutase, glyoxalase I

REACTANTS
   (R)-S-Lactoylglutathione, 3-phosphoglycerolglutathione, glutathione, methylglyoxal

NOTES
- Classified as a lyase: carbon-sulfur lyase

SPECIFIC ACTIVITY	UNITS DEFINITION	PREPARATION FORM	ADDITIONAL ACTIVITIES	SUPPLIER CATALOG NO.
Yeast				
600-800 U/mg protein	1 unit forms 1.0 μmol S-lactoylglutathione from methyl glyoxal and GSH/min at pH 6.6, 25°C	Lyophilized containing >95% protein; balance citrate buffer salts		Sigma G0256
400-800 U/mg protein	1 unit forms 1.0 μmol S-lactoylglutathione from methyl glyoxal and GSH/min at pH 6.6, 25°C	Solution containing 50% glycerol, 0.4 M $(NH_4)_2SO_4$, 0.002 M KPO_4, pH 6.5	<1% glyoxalase II	Sigma G4252

Methionine γ-Lyase   4.4.1.11

REACTION CATALYZED
   L-Methionine ↔ Methanethiol + NH_3 + 2-oxobutanoate

SYSTEMATIC NAME
   L-Methionine methanethiol-lyase (deaminating)

SYNONYMS
   L-Methionase

REACTANTS
   L-Methionine, methanethiol, NH_3, 2-oxobutanoate, homocysteine, S-methylcysteine

APPLICATIONS
- Determination of minute amounts of L-methionine and L-homocysteine

NOTES
- Classified as a lyase: carbon-sulfur lyase
- A pyridoxal-phosphate protein

SPECIFIC ACTIVITY	UNITS DEFINITION	PREPARATION FORM	ADDITIONAL ACTIVITIES	SUPPLIER CATALOG NO.
***Pseudomonas putida* (optimum pH = 8.0, T = 37°C, MW = 172,000 Da [gel filtration] and 43,000 Da [SDS-PAGE])**				
≥2 U/mg		Lyophilized		Wako 126-02671 122-02673

4.6.1.1 Adenylate Cyclase

REACTION CATALYZED
    ATP ↔ 3',5'-cyclic AMP + pyrophosphate

SYSTEMATIC NAME
    ATP pyrophosphate-lyase (cyclizing)

SYNONYMS
    Adenylylcyclase, adenyl cyclase, 3',5'-cyclic AMP synthetase

REACTANTS
    ATP, dATP, 3',5'-cyclic AMP, 3',5'-cyclic dAMP, pyrophosphate

NOTES
- Classified as a lyase: phosphorus-oxygen lyase
- Requires pyruvate
- Activated by NAD^+ in the presence of $NAD(P)^+$-arginine ADP-ribosyltransferase (EC 2.4.2.31)

SPECIFIC ACTIVITY	UNITS DEFINITION	PREPARATION FORM	ADDITIONAL ACTIVITIES	SUPPLIER CATALOG NO.
Escherichia coli				
5-25 U/mg protein	1 unit forms 1.0 nmol cyclic AMP from ATP/min at pH 7.0, 37°C	Solution containing 20% glycerol, 20 mM KPO_4, 10 mM DTT, 5 mM $MgCl_2$, 5 mM ATP, pH 7.8		Sigma A0951

Ribulose-Phosphate 3-Epimerase   5.1.3.1

REACTION CATALYZED
    D-Ribulose 5-phosphate ↔ D-Xylulose 5-phosphate

SYSTEMATIC NAME
    D-Ribulose-5-phosphate 3-epimerase

SYNONYMS
    Phosphoribulose epimerase, erythrose-4-phosphate isomerase

REACTANTS
    D-Ribulose 5-phosphate, D-erythrose 4-phosphate, D-xylulose 5-phosphate, D-erythrulose 4-phosphate, D-threose 4-phosphate

NOTES
- Classified as an isomerase: racemases and epimerases acting on carbohydrates and derivatives

SPECIFIC ACTIVITY	UNITS DEFINITION	PREPARATION FORM	ADDITIONAL ACTIVITIES	SUPPLIER CATALOG No.
Yeast, bakers				
50-100 U/mg protein	1 unit converts 1.0 μmol D-ribulose 5-phosphate to D-xylulose 5-phosphate/min at pH 7.7, 25°C in a coupled system	Sulfate-free; lyophilized containing 35% citrate buffer salts	<0.1% phosphoriboisomerase, ADH, transketolase, transaldolase	Sigma R3251

UDP-Glucose 4-Epimerase   5.1.3.2

REACTION CATALYZED
    UDPGlucose ↔ UDPGalactose

SYSTEMATIC NAME
    UDPGlucose 4-epimerase

SYNONYMS
    UDPGalactose 4-epimerase

REACTANTS
    UDPGlucose, UDP-2-deoxyglucose, UDPgalactose

NOTES
- Classified as an isomerase: racemases and epimerases acting on carbohydrates and derivatives
- Requires NAD^+

5.1.3.2 UDP-Glucose 4-Epimerase continued

SPECIFIC ACTIVITY	UNITS DEFINITION	PREPARATION FORM	ADDITIONAL ACTIVITIES	SUPPLIER CATALOG NO.
Yeast, galactose adapted				
10-20 U/mg protein	1 unit converts 1.0 μmol UDP-galactose to UDP-glucose/min at pH 8.8, 25°C	Lyophilized in vials containing 40% buffer salts	<0.2% UDPG-pyrophosphorylase, UDPG-DH, Gal1P uridyl transferase 0.4% galactokinase	Sigma U3251

5.1.3.3 Aldose 1-Epimerase

REACTION CATALYZED

α-D-Glucose ↔ β-D-Glucose

SYSTEMATIC NAME

Aldose 1-epimerase

SYNONYMS

Mutarotase, aldose mutarotase

REACTANTS

α-D-Glucose, L-arabinose, D-xylose, D-galactose, maltose, lactose, β-D-Glucose

APPLICATIONS

- Rapid determination of glucose, with glucose oxidase and peroxidase

NOTES

- Classified as an isomerase: racemases and epimerases acting on carbohydrates and derivatives

SPECIFIC ACTIVITY	UNITS DEFINITION	PREPARATION FORM	ADDITIONAL ACTIVITIES	SUPPLIER CATALOG NO.
Porcine kidney (optimum pH = 7.4; pI = 5.56, 5.29, 5.24, 5.04; T = 30°C, MW = 37,000-40,700 Da; stable pH 7.0-9.0 [5°C, 10 days], T < 50°C [pH 7.2, 10 min])				
≥70 U/mL	1 unit produces 1 μmol β-D-glucose/min at pH 7.2, 37°C	Solution		Amano Mutarotase
5000 U/mg protein	1 unit increases the rate of spontaneous mutarotation of α-D-glucose to β-D-glucose 1 mmol/min at pH 7.4, 25°C	Suspension in 80% (NH$_4$)$_2$SO$_4$, pH 8.0		Biozyme MUR1
≥1500 U/mg solid	1 unit increases the rate of spontaneous mutarotation of α-D-glucose to β-D-glucose 1 mmol/min at pH 7.4, 25°C	Salt-free; lyophilized		Biozyme MUR1F
5000 U/mg protein	1 unit increase spontaneous mutarotation of 1 μmol α-D-glucose to β-D-glucose/min at pH 7.4, 25°C; 1 rotation = 0.04 U when measured with GlcDH	Suspension in (NH$_4$)$_2$SO$_4$ containing 10 mg/mL protein		Calzyme 136B5000

SPECIFIC ACTIVITY	UNITS DEFINITION	PREPARATION FORM	ADDITIONAL ACTIVITIES	SUPPLIER CATALOG NO.
Porcine kidney *continued*				
5000 U/mg protein	1 unit increase spontaneous mutarotation of 1 µmol α-D-glucose to β-D-glucose/min at pH 7.4, 25°C; 1 rotation = 0.04 U when measured with GlcDH	Lyophilized containing 95% protein		Calzyme 137A5000
>4000 U/mg protein; >1200 U/mg solid	1 unit increases the rate of spontaneous mutarotation 1.0 µmol α-D-glucose to β-D-glucose/min at 25°C	Lyophilized		Genzyme 1531
>70 U/mL	1 unit increases the rate of spontaneous mutarotation 1.0 µmol α-D-glucose to β-D-glucose/min at pH 7.2, 37°C	Solution		Genzyme 1534
≥50 IU/mg protein	1 IU transforms 1 µmol substrate/min under standard IUB conditions at 25°C	Lyophilized; suspension in $(NH_4)_2SO_4$ containing 50% glycerol	<0.01% NADH oxidase, GO <0.1% LDH <1% catalase	OYC
2500-7500 U/mg protein	1 unit increases spontaneous mutarotation of α-D-glucose to β-D-glucose by 1.0 µmol/min at pH 7.4, 25°C	Lyophilized containing 80% protein; balance PO_4 buffer salts		Sigma M4286
2500-7500 U/mg protein	1 unit increases spontaneous mutarotation of α-D-glucose to β-D-glucose by 1.0 µmol/min at pH 7.4, 25°C			Sigma M5526
10,000 U/mL		Suspension		Wako 133-07501
Porcine Liver				
5000 U/mg	1 unit converts 1 µmol α-D-glucose to product/min at 25°C	Suspension in 3.2 M $(NH_4)_2SO_4$, pH 6		Boehringer 127264

5.2.1.8 Peptidylprolyl Isomerase

REACTION CATALYZED
    Peptidylproline (w=180) ↔ Peptidylproline (w=0)

SYSTEMATIC NAME
    Peptidylproline *cis-trans*-isomerase

SYNONYMS
    Cyclophilin

REACTANTS
    Peptidylproline (w=180), peptidylproline (w=0)

NOTES
- Classified as an isomerase: *cis-trans*-isomerase
- Catalyzes the *cis-trans* isomerization of X-Pro peptide bonds
- Accelerates protein folding
- Inhibited by cyclosporin A
- Cyclophilin and cyclosporin A bind calcineurin and modulate its phosphatase activity

SPECIFIC ACTIVITY	UNITS DEFINITION	PREPARATION FORM	ADDITIONAL ACTIVITIES	SUPPLIER CATALOG NO.
Human, recombinant expressed in *E. coli* (MW = 18,000 Da)				
		>90% purity by SDS-PAGE and HPLC; 0.5 mg/mL in 20 mM Tris-HCl, pH 7.8		BIOMOL SE-105
		>90% by HPLC, SDS-PAGE; solution		Boehringer 1532197 1532219

5.3.1.1 Triose-Phosphate Isomerase

REACTION CATALYZED
    D-Glyceraldehyde 3-phosphate ↔ Glycerone phosphate

SYSTEMATIC NAME
    D-Glyceraldehyde-3-phosphate ketol-isomerase

SYNONYMS
    Phosphotriose isomerase, triosephosphate mutase, TIM, TPI

REACTANTS
    D-Glyceraldehyde 3-phosphate, glycerone phosphate

NOTES
- Classified as an isomerase: intramolecular oxidoreductase interconverting aldoses and ketoses

SPECIFIC ACTIVITY	UNITS DEFINITION	PREPARATION FORM	ADDITIONAL ACTIVITIES	SUPPLIER CATALOG NO.
Chicken muscle				
10,000 U/mg protein	1 unit produces 1 μmol D-GAP to DHAP/min at pH 7.6, 25°C	Crystallized; suspension in 3.2 M $(NH_4)_2SO_4$ containing 95% protein, pH 6.0		Calzyme 075B10000
Dog muscle				
1000 U/mg protein	1 unit converts 1.0 μmol D-GAP to DHAP/min at pH 7.6, 25°C	Suspension in 3.2 M $(NH_4)_2SO_4$, pH 6.5	<0.01% PK, 3PGPK, α-GPDH, aldolase, GAPDH <0.02% LDH	Sigma T6635
Porcine muscle				
5000 U/mg protein	1 unit converts 1.0 μmol D-GAP to DHAP/min at pH 7.6, 25°C	Suspension in 3.2 M $(NH_4)_2SO_4$, pH 6.0	<0.001% PK, LDH, PGlcI, α-GPDH, aldolase, GAPDH <0.01% 3PGPK	Sigma T7401
5000 U/mg protein	1 unit converts 1.0 μmol D-GAP to DHAP/min at pH 7.6, 25°C	Sulfate-free; lyophilized containing 70% protein; balance primarily EDTA and borate buffer salts		Sigma T7526
Rabbit muscle (MW = 43,000 Da)				
7000 U/mg protein	1 unit converts 1 μmol of D-GAP to DHAP/min at pH 7.6, 25°C	Crystallized; suspension in 75% $(NH_4)_2SO_4$	<0.005% LDH, PK, G3PDH, GAPDH	Biozyme TIM1
5000 U/mg	1 unit converts 1 μmol GAP to product/min at 25°C	Suspension in 3.2 M $(NH_4)_2SO_4$, pH 6	<0.001% GAPDH <0.01% aldolase, GDH	Boehringer 109754 109762
5000 U/mg protein	1 unit produces 1 μmol D-GAP to DHAP/min at pH 7.6, 25°C	Crystallized; suspension in 3.2 M $(NH_4)_2SO_4$ containing 95% protein, pH 6.0		Calzyme 077B5000
5000 U/mg protein	1 unit converts 1.0 μmol D-GAP to DHAP/min at pH 7.6, 25°C	Crystallized; suspension in 3.2 M $(NH_4)_2SO_4$, pH 6.0	<0.001% PK, LDH, PGlcI, α-GPDH, aldolase, GAPDH <0.01% 3PGPK	Sigma T2391
5000 U/mg protein	1 unit converts 1.0 μmol D-GAP to DHAP/min at pH 7.6, 25°C	Sulfate-free; lyophilized containing 70% protein; balance primarily EDTA and borate buffer salts	<0.01% PK, LDH, PGlcI, α-GPDH, aldolase, GAPDH, 3PGPK	Sigma T6258
Yeast, bakers				
10,000 U/mg protein	1 unit converts 1.0 μmol D-GAP to DHAP/min at pH 7.6, 25°C	Crystallized; suspension in 2.7 M $(NH_4)_2SO_4$ and 0.5 mM EDTA, pH 6.5	<0.01% PK, LDH, 3PGPK, PGlcI, α-GPDH, aldolase, GAPDH	Sigma T2507

5.3.1.5 Xylose Isomerase

REACTION CATALYZED
D-Xylose ↔ D-Xylulose

SYSTEMATIC NAME
D-Xylose ketol-isomerase

SYNONYMS
Glucose isomerase

REACTANTS
D-Xylose, D-glucose, D-xylulose, D-fructose

APPLICATIONS
- Immobilized enzyme is critical to the industrial production of high fructose corn syrup from starch through isomerization of dextrose (glucose) to fructose

NOTES
- Classified as an isomerase: intramolecular oxidoreductase interconverting aldoses and ketoses
- Activated by Mg^{2+}, HSO_3^- and $NaHSO_3$

SPECIFIC ACTIVITY	UNITS DEFINITION	PREPARATION FORM	ADDITIONAL ACTIVITIES	SUPPLIER CATALOG NO.
Streptomyces murinus (optimum pH = 7.5-7.8, T = 55-60°C)				
350 IGIU/g	1 IGIU converts glucose to fructose at 1 μmol/min at pH 7.5, 60°C	Food grade; granulate at 0.3-1.0 mm particle size		Novo Nordisk Sweetzyme^R T
Streptomyces olivochromogenes (optimum pH = 7.8-8.6)				
		Food grade, soluble		EnzymeDev G-Zyme^R G993
Streptomyces rubiginosus (optimum pH = 7.6-7.8, T = 55-57°C)				
1100-1500 GIU/g	1 unit converts 1 μmol glucose to fructose/min	Food grade; immobilized particles at 0.3-1.0 mm particle size		Genencor Spezyme^R CIGI
3500 GIU/g	1 unit converts 1 μmol glucose to fructose/min	Highly purified; food grade; solution		Genencor Spezyme^R GI
1100-1400 GIU/g	1 unit converts 1 μmol glucose to fructose/min	Food grade; immobilized on DEAE-cellulose carrier in polystyrene and titanium dioxide matrix at 0.3-0.84 mm particle size		Genencor Spezyme^R IGI
Streptomyces species (optimum pH = 8.0)				
≥1150 IGIU/g		Immobilized granulate in food grade salt buffer solution		UOP Ketomax^R GI-100

Ribose-5-Phosphate Isomerase

REACTION CATALYZED
D-Ribose 5-phosphate ↔ D-Ribulose 5-phosphate

SYSTEMATIC NAME
D-Ribose 5-phosphate ketol-isomerase

SYNONYMS
Phosphopentoisomerase, phosphoriboisomerase

REACTANTS
D-Ribose 5-phosphate, D-ribose 5-diphosphate, D-Ribose 5-triphosphate, D-ribulose 5-phosphate

NOTES
- Classified as an isomerase: intramolecular oxidoreductase interconverting aldoses and ketoses

SPECIFIC ACTIVITY	UNITS DEFINITION	PREPARATION FORM	ADDITIONAL ACTIVITIES	SUPPLIER CATALOG NO.
Spinach				
1000-2000 U/mg protein	1 unit converts 1.0 μmol D-ribose 5-phosphate to D-ribulose 5-phosphate/min at pH 7.7, 30°C	Sulfate-free; lyophilized containing 85% protein; balance primarily citrate buffer salts		Sigma P1780
75-150 U/mg protein	1 unit converts 1.0 μmol D-ribose 5-phosphate to D-ribulose 5-phosphate/min at pH 7.7, 30°C	Partially purified; powder		Sigma P9752
Yeast, *Torula*				
200-600 U/mg protein	1 unit converts 1.0 μmol D-ribose 5-phosphate to D-ribulose 5-phosphate/min at pH 7.7, 30°C	Sulfate-free; lyophilized containing 70% protein; balance primarily PO_4 buffer salts		Sigma P7434

5.3.1.8 Mannose-6-Phosphate Isomerase

REACTION CATALYZED
D-Mannose 6-phosphate ↔ D-Fructose 6-phosphate

SYSTEMATIC NAME
D-Ribulose 5-phosphate ketol-isomerase

SYNONYMS
Phosphomannose isomerase, phosphohexomutase, phosphohexoisomerase, PMI

REACTANTS
D-Ribulose 5-phosphate, D-fructose 6-phosphate

NOTES
- Classified as an isomerase: intramolecular oxidoreductase interconverting aldoses and ketoses
- A zinc protein
- Isomerizes ketoses to aldoses

5.3.1.8 Mannose-6-Phosphate Isomerase continued

SPECIFIC ACTIVITY	UNITS DEFINITION	PREPARATION FORM	ADDITIONAL ACTIVITIES	SUPPLIER CATALOG NO.
Yeast				
60 U/mg	1 unit converts 1 μmol mannose-6-phosphate to product/min at 25°C	Suspension in 3.2 M $(NH_4)_2SO_4$, pH 7	<0.001% NADPH oxidase <0.01% GR, HK, 6PGDH	Boehringer 131229
Yeast, bakers				
50-100 U/mg protein	1 unit converts 1.0 μmol D-mannose 6-phosphate to D-F6P/ min at pH 7.6, 25°C in a coupled system	Suspension in 3.2 M $(NH_4)_2SO_4$ and 5 mM KPO_4, pH 7		sigma P5153

5.3.1.9 Glucose-6-Phosphate Isomerase

REACTION CATALYZED
  D-Glucose 6-phosphate ↔ D-Fructose 6-phosphate

SYSTEMATIC NAME
  D-Glucose 6-phosphate ketol-isomerase

SYNONYMS
  Phosphohexose isomerase, phosphohexomutase, oxoisomerase, hexosephoshate isomerase, phosphosaccharomutase, phosphoglucose isomerase, phosphohexose isomerase, PGIcI, PHI

REACTANTS
  D-Glucose 6-phosphate, D-Fructose 6-phosphate

APPLICATIONS
- Determination of D-sedoheptulose 7-phosphate in blood and animal tissue

NOTES
- Classified as an isomerase: intramolecular oxidoreductase interconverting aldoses and ketoses
- Isomerizes ketoses to aldoses
- Also anomerizes D-glucose 6-phosphate

SPECIFIC ACTIVITY	UNITS DEFINITION	PREPARATION FORM	ADDITIONAL ACTIVITIES	SUPPLIER CATALOG NO.
Bacillus stearothermophilus **(optimum pH = 9-10, pI = 4.2, MW = 200,000 Da [54,000/subunit]; K_M = 0.27 mM [F6P]; stable pH 6-10.5, T < 60°C)**				
300-1000 U/mg protein	1 unit converts 1.0 μmol D-F6P to D-Glc6P/min at pH 9.0, 30°C	Lyophilized containing 70% protein; balance Tris buffer salt		sigma P5538
>400 U/mg protein	1 unit forms 1 μmol Glc6P/min at 30°C	Lyophilized	<0.01% PFK, 6PGlcDH, PGlcM, GR, NADPH oxidase	Unitika
>400 U/mg protein	1 unit forms 1 μmol Glc6P/min at 30°C	50% glycerol solution	<0.01% PFK, 6PGlcDH, PGlcM, GR, NADPH oxidase	Unitika

Glucose-6-Phosphate Isomerase continued — 5.3.1.9

SPECIFIC ACTIVITY	UNITS DEFINITION	PREPARATION FORM	ADDITIONAL ACTIVITIES	SUPPLIER CATALOG NO.
Rabbit muscle				
400-600 U/mg protein	1 unit converts 1.0 µmol D-F6P to D-Glc6P/min at pH 7.4, 25°C	Crystallized; suspension in 2.3 M $(NH_4)_2SO_4$, 25 mM imidazole, 1 mM EDTA, 0.1% β-MSH, pH 8.2		Sigma P8391
5-15 U/mg solid	1 unit converts 1.0 µmol D-F6P to D-Glc6P/min at pH 7.4, 25°C	Crude		Sigma P9252
350-5500 U/mg protein	1 unit converts 1.0 µmol D-F6P to D-Glc6P/min at pH 7.4, 25°C	Sulfate-free; lyophilized containing 85% protein; balance primarily citrate buffer salts		Sigma P9544
Yeast				
350 U/mg	1 unit converts 1 µmol F6P to product/min at 25°C	Suspension in 3.2 M $(NH_4)_2SO_4$, pH 6	<0.01% F6PK, GR, 6PGDH, PGlcM <0.2% β-fructosidase	Boehringer 127396 128139 1693662
500 U/mg protein	1 unit converts 1.0 µmol D-F6P to D-Glc6P/min at pH 7.4, 25°C	Sulfate-free; lyophilized containing 95% protein; balance primarily citrate buffer salts		Sigma P9010
Yeast, bakers				
530 U/mg protein	1 unit converts 1.0 µmol D-F6P to D-Glc6P/min at pH 7.4, 25°C	Aqueous solution containing 2.6 M $(NH_4)_2SO_4$ and 0.1 mM Mg^{2+}, pH 7.0	<0.05% HK <0.5% TPI	Fluka 79460
400 U/mg protein	1 unit converts 1 µmol F6P to Glc6P/min at pH 7.4, 25°C	Crystallized; suspension in 3.2 M $(NH_4)_2SO_4$		ICN 100983
	1 unit converts 1.0 µmol D-F6P to D-Glc6P/min at pH 7.4, 30°C	Insoluble enzyme attached to polyacrylamide containing 30% borate buffer salts		Sigma P4135
500-800 U/mg protein	1 unit converts 1.0 µmol D-F6P to D-Glc6P/min at pH 7.4, 25°C	Crystallized; suspension in 2.6 M $(NH_4)_2SO_4$ and 0.1 mM MgOAc, pH 7.0		Sigma P5381

5.3.2.1 Phenylpyruvate Tautomerase

REACTION CATALYZED
: keto-Phenylpyruvate ↔ enol-Phenylpyruvate

SYSTEMATIC NAME
: Phenylpyruvate keto-enol-isomerase

SYNONYMS
: Tautomerase

REACTANTS
: keto-Phenylpyruvate, arylpyruvates, enol-Phenylpyruvate

NOTES
- Classified as an isomerase: intramolecular oxidoreductase interconverting keto- and enol- groups

SPECIFIC ACTIVITY	UNITS DEFINITION	PREPARATION FORM	ADDITIONAL ACTIVITIES	SUPPLIER CATALOG NO.
Bovine kidney				
1-4 U/mg protein	1 unit produces a first-order rate constant (k) of 1.0 at pH 6.2, 25°C using p-hydroxyphenylpyruvate(keto)	Aqueous solution containing 10 U/mL		sigma T6004
Porcine kidney				
0.05 U/mg protein	1 unit produces a first-order rate constant (k) of 1.0 at pH 6.2, 25°C using p-hydroxyphenylpyruvate(keto)	Aqueous solution		sigma T1000

5.3.4.1 Protein Disulfide-Isomerase

REACTION CATALYZED
: Catalyzes the formation and rearrangement of -S-S- bonds in proteins

SYSTEMATIC NAME
: Protein disulfide-isomerase

SYNONYMS
: S-S Rearrangase, thioredoxin, DsbA

REACTANTS
: Disulfide bonds

APPLICATIONS
- Returning reduced or unfolded proteins back to native form
- Facilitates formation of correct disulfide bonds by promoting rapid reshuffling of incorrect disulfide pairings
- Refolding recombinant protein expresses in *E. coli* with a tendency to fold improperly

Protein Disulfide-Isomerase continued — 5.3.4.1

NOTES
- Classified as an isomerase: intramolecular oxidoreductase transposing S-S bonds
- Requires reducing agents or partly-reduced enzyme
- The reaction depends on sulfhdryl-disulfide interchange

SPECIFIC ACTIVITY	UNITS DEFINITION	PREPARATION FORM	ADDITIONAL ACTIVITIES	SUPPLIER CATALOG NO.
Bovine liver (optimum pH = 7-9, MW = 107,000 Da [homodimer])				
300 U/mg protein	1 unit recovers 1 RNase A unit from reduced bovine RNase A at pH 7.5, 25°C	Homogeneous purity; lyophilized from buffer containing 200 μL 50 mM NAPO$_4$ buffer, pH 7.5		Calbiochem 539425
100-400 U/mg protein	1 unit reactivates 1 unit ribonuclease 2 from scrambled RNase/min at pH 7.5, 30°C	>95% by SDS-PAGE; lyophilized containing 10% protein; balance KPO$_4$ buffer salts and stabilizer		Sigma P3818
363 U/mg protein; 1 mg protein/vial	1 unit recovers 1 RNase A unit from reduced bovine RNase A at pH 7.5, 25°C; 1 RNase A unit hydrolyzes cCMP increasing A$_{284}$ by 0.001/min	Homogeneous on SDS-PAGE; lyophilized containing 50 mM NAPO$_4$ buffer, pH 7.5		TaKaRa 7318
Escherichia coli (MW = 21,118 Da)				
		>95% purity by SDS-PAGE; lyophilized	No detectable protease	Boehringer 1585339
Escherichia coli, recombinant (MW = 11,700 Da)				
> 4 mg/mL protein	1 unit protein results in ΔA$_{650}$ of 0.001/min	Overproducer; >95% purity by SDS-PAGE; 20 mM KPO$_4$, 1 mM DTT, 0.1 mM EDTA, 50% glycerol, pH 7.4	No detectable proteases	Fermentas EO0301 EO0302
	≥4.00 A$_{650}$ insulin reduction U/min/mg	≥98% purity by SDS-PAGE; salt-free; lyophilized		Promega Z7051 Z7052
Unspecified (optimum pH = 7-9, MW = 107,000 Da)				
		Homogeneous on SDS-PAGE; lyophilized		PanVera TAK 7318

5.3.99.3 Prostaglandin-E Synthase

REACTION CATALYZED
(5Z,13E)-(15S)-9α,11α-Epidioxy-15-hydroxyprosta-5,13-dienoate ↔ (5Z,13E)-(15S)-11α,15-Dihydroxy-9-oxyprosta-5,13-dienoate

SYSTEMATIC NAME
(5Z,13E)-(15S)-9α,11α-Epidioxy-15-hydroxyprosta-5,13-dienoate E-isomerase

SYNONYMS
Prostaglandin-H_2 E-isomerase, endoperoxide isomerase

REACTANTS
(5Z,13E)-(15S)-9α,11α-Epidioxy-15-hydroxyprosta-5,13-dienoate, prostaglandin H_2, (5Z,13E)-(15S)-11α,15-Dihydroxy-9-oxyprosta-5,13-dienoate

NOTES
- Classified as an isomerase: intramolecular oxidoreductase, other intramolecular oxidoreductases
- Opens the epidioxy bridge
- Requires glutathione

See Chapter 5. Multiple Enzyme Preparations

5.3.99.4 Prostaglandin-I Synthase

REACTION CATALYZED
(5Z,13E)-(15S)-9α,11α-Epidioxy-15-hydroxyprosta-5,13-dienoate ↔ (5Z,13E)-(15S)-6,9α-Epoxy-11α,15-dihydroxyprosta-5,13-dienoate

SYSTEMATIC NAME
(5Z,13E)-(15S)-9α,11α-Epidioxy-15-hydroxyprosta-5,13-dienoate 6-isomerase

SYNONYMS
Prostacyclin synthase

REACTANTS
(5Z,13E)-(15S)-9α,11α-Epidioxy-15-hydroxyprosta-5,13-dienoate, prostaglandin H_2, (5Z,13E)-(15S)-6,9α-epoxy-11α,15-dihydroxyprosta-5,13-dienoate, prostaglandin I_2, prostacyclin

NOTES
- Classified as an isomerase: intramolecular oxidoreductase, other intramolecular oxidoreductases
- Converts prostaglandin H_2 into prostaglandin I_2 (prostacyclin)
- A heme-thiolate protein

See Chapter 5. Multiple Enzyme Preparations

Thromboxane-A Synthase

REACTION CATALYZED

(5Z,13E)-(15S)-9α,11α-Epidioxy-15-hydroxyprosta-5,13-dienoate ↔ (5Z,13E)-(15S)-9α,11α-Epoxy-15-hydroxythromba-5,13-dienoate

SYSTEMATIC NAME

(5Z,13E)-(15S)-9α,11α-Epidioxy-15-hydroxyprosta-5,13-dienoate thromboxane-A_2-isomerase

SYNONYMS

Thromboxane synthase

REACTANTS

(5Z,13E)-(15S)-9α,11α-Epidioxy-15-hydroxyprosta-5,13-dienoate, prostaglandin H_2, (5Z,13E)-(15S)-9α,11α-epoxy-15-hydroxythromba-5,13-dienoate, thromboxane A_2

NOTES

- Classified as an isomerase: intramolecular oxidoreductase, other intramolecular oxidoreductases
- Converts prostaglandin H_2 into thromboxane A_2
- A heme-thiolate protein

SPECIFIC ACTIVITY	UNITS DEFINITION	PREPARATION FORM	ADDITIONAL ACTIVITIES	SUPPLIER CATALOG NO.
Human platelet				
Provided with each lot		Lyophilized; enriched, but not purified; not suitable for kinetic studies or use as an antigen	No detectable HTLV-III/HIV Ab, hepatitis	BIOMOL EP-024

5.4.2.1 Phosphoglycerate Mutase

REACTION CATALYZED
  2-Phospho-D-glycerate ↔ 3-Phospho-D-glycerate

SYSTEMATIC NAME
  D-Phosphoglycerate 2,3-phosphomutase

SYNONYMS
  Phosphoglycerate phosphomutase, phosphoglyceromutase, PGlyM

REACTANTS
  2-Phospho-D-glycerate, 3-phospho-D-glycerate

APPLICATIONS
- Colorimetric determination of 2,3-diphosphoglycerate

NOTES
- Classified as an isomerase: intramolecular transferase (mutase): phosphotransferase (phosphomutase)
- Mammal and yeast enzyme are phosphorylated by (2R)-2,3-bisphosphoglycerate, an intermediate in the reaction
- Dissociation of bisphosphate from the rabbit muscle enzyme is much slower than the overall isomerization
- Mammal, yeast and rabbit muscle enzymes also slowly catalyze the reactions of bisphosphoglycerate mutase (EC 5.4.2.4)
- Wheat, rice, insect and some fungal enzymes have maximal activity in the absence of 2,3-bisphosphoglycerate

SPECIFIC ACTIVITY	UNITS DEFINITION	PREPARATION FORM	ADDITIONAL ACTIVITIES	SUPPLIER CATALOG NO.
Rabbit muscle				
2 U/mg protein	1 unit converts 1.0 μmol 3-phosphoglycerate to 2-phosphoglycerate/min at pH 7.6, 25°C in the presence of 1.3 mM 2,3-diphosphoglycerate (cofactor); activity considerably lower without cofactor	Crystallized; suspension in 2.4 M (NH$_4$)$_2$SO$_4$, pH 6.5	2 U/mg protein 2,3-diphosphoglycerate phosphatase <0.01% enolase, LDH, PK, PGK <0.03% TPI	Sigma P8252 665-3

Phosphoglucomutase 5.4.2.2

REACTION CATALYZED
α-D-Glucose 1-phosphate ↔ α-D-Glucose 6-phosphate

SYSTEMATIC NAME
α-D-Glucose 1,6-phosphomutase

SYNONYMS
Glucose phosphomutase, PGlcM, PGM,

REACTANTS
α-D-Glucose 1-phosphate, α-D-hexose 1-phosphate (slow), α-D-ribose 1-phosphate (slow), α-D-glucose 6-phosphate, α-D-hexose 6-phosphate, α-D-ribose 5-phosphate

NOTES
- Classified as an isomerase: intramolecular transferase (mutase): phosphotransferase (phosphomutase)
- Although reversible, the formation of glucose-6-phosphate is favored
- The dissociation of bisphosphate from the enzyme complex is much slower than the overall isomerization
- The reaction intermediate, α-D-glucose 1,6-bisphosphate, is required for optimal activity. It is formed during transfer of a phosphate from enzyme to substrate. Also activated by divalent cations (e.g., Mg^{2+}) and glutathione
- Inhibited by high concentrations of chelating agents, gluconate-6-phosphate, nucleotides and acetate

SPECIFIC ACTIVITY	UNITS DEFINITION	PREPARATION FORM	ADDITIONAL ACTIVITIES	SUPPLIER CATALOG NO.
Chicken muscle				
100 U/mg protein	1 unit converts 1.0 μmol α-D-Glc1P to α-D-Glc6P/min at pH 7.4, 30°C	Crystallized; suspension in 3.2 M $(NH_4)_2SO_4$ and 0.01% EDTA, pH 6		Sigma P6156
Rabbit muscle (optimum pH = 7.5-8.0, MW = 65,000 Da)				
200 U/mg	1 unit converts 1 μmol Glc1P to product/min at 25°C	Suspension in 3.2 M $(NH_4)_2SO_4$, pH 5	<0.01% GR, HK, PGI	Boehringer 108375 108383
300 U/mg protein	1 unit converts 1.0 μmol α-D-Glc1P to α-D-Glc6P/min at pH 7.4, 30°C	Crystallized; suspension in 2.5 M $(NH_4)_2SO_4$ and 0.01% EDTA, pH 6.0	<0.01% PGlcI, LDH, MDH	Fluka 79440
100-300 U/mg protein	1 unit converts 1.0 μmol α-D-Glc1P to α-D-Glc6P/min at pH 7.4, 30°C	Crystallized; suspension in 2.5 M $(NH_4)_2SO_4$ and 0.01% EDTA, pH 6.0	<0.01% PK, PGlcI, LDH	Sigma P3397
≥200 U/mg protein	1 unit oxidizes 1 μmol NAD/min at pH 7.6, 30°C using Glc1P as substrate in a G6PDH coupled system	Chromatographically purified; suspension in $(NH_4)_2SO_4$		Worthington LS08336 LS08338 LS08340

5.4.99.2   Methylmalonyl-CoA Mutase

REACTION CATALYZED
    (R)-2-Methyl-3-oxopropanoyl-CoA ↔ Succinyl-CoA
SYSTEMATIC NAME
    (R)-2-Methyl-3-oxopropanoyl-CoA CoA-carbonylmutase
REACTANTS
    (R)-2-Methyl-3-oxopropanoyl-CoA, succinyl-CoA

NOTES
- Classified as an isomerase: intramolecular transferase (mutase), transferring other groups
- Requires a cobamide coenzyme

See Chapter 5. Multiple Enzyme Preparations

5.99.1.2   DNA Topoisomerase

REACTION CATALYZED
    ATP-independent breakage of single-stranded DNA, followed by passage and rejoining
SYSTEMATIC NAME
    DNA topoisomerase
SYNONYMS
    Type I DNA topoisomerase, untwisting enzyme, relaxing enzyme, nicking-closing enzyme, swivelase, ω-protein
REACTANTS
    Single-stranded DNA
APPLICATIONS
- Analysis of DNA conformation and topology
- DNA repair, drug resistance and cell proliferation studies
- *In vitro* transcription
- Chromatin reconstitution
- Production of DNA topoisomers by relaxing positively and negatively supercoiled DNA

NOTES
- Classified as an isomerase: other isomerases
- Converts one topological isomer of DNA into another via 4 distinct activities:
  - relaxes the super coil of supercoiled circular molecules;
  - winds and unwinds knots in single-stranded, circular DNA;
  - forms double-stranded circular DNA from 2 single-stranded circular DNAs complementary to each other;
  - joins 2 molecules when a nick exists in 1 of 2 double-stranded circular DNAs and separates them
- Eukaryotic enzyme is active without Mg^{2+}; prokaryotic is not
- Inhibited by monovalent cations K^+ and Na^+ at levels > 0.2 M

SPECIFIC ACTIVITY	UNITS DEFINITION	PREPARATION FORM	ADDITIONAL ACTIVITIES	SUPPLIER CATALOG No.
Calf thymus				
5 U/μL	1 unit completely relaxes 0.5 μg supercoiled pBR322 DNA/30 min at 37°C in 20 μL reaction mixture	20 mM KPO$_4$, 50 mM KCl, 0.05 mM EDTA, 5 mM β-MSH, 50% glycerol, pH 7.2	No detectable endonuclease	Amersham E 0501Y
25,000 U/mL	1 unit fully relaxes 0.5 μg of supercoiled pBR322 DNA/30 min at pH 8.0, 37°C in a 0.02 mL volume	20 mM Tris-HCl, 1 mM DTT, 50 mM KCl, 0.1 mM EDTA, 50% glycerol, pH 7.5	No detectable contaminating nuclease	AGS Heidelb F00710S
10-20 U/μL	1 unit completely relaxes 1.0 μg pBR322 DNA/30 min at 37°C	20 mM Tris-HCl, 50 mM KCl, 1 mM DTT, 0.5 mM EDTA, 60% glycerol, pH 7.5	No detectable endo- and exodeoxyribonucleases	Fermentas ET0211 ET0212
5-15 U/μL	1 unit converts 0.5 μg superhelical φX174 RF DNA to the relaxed state/30 min at 37°C	Solution containing 30 mM KPO$_4$, 5 mM DTT, 0.2 mM Na$_2$EDTA, 0.2 mg/mL BSA, 50% glycerol, pH 7.0		ICN 152311
5-15 U/μL	1 unit converts 5 μg superhelical φX174 RF DNA to a relaxed state/30 min at pH 7.5, 37°C			Life Technol 38042-016 38042-024
6 U/μL; 100 U and 500 U	1 unit completely relaxes 0.5 μg supercoiled pBR322 DNA/30 min at pH 8.0, 37°C in 20 μL reaction mixture	Solution containing 20 mM KPO$_4$, 50 mM KCl, 0.05 mM EDTA, 5 mM β-MSH, 50% glycerol, pH 7.2	No detectable endonuclease	TaKaRa 2240A
Wheat germ				
	1 unit converts 1 μg supercoiled closed circular (Form I) pUC19 plasmid DNA to relaxed closed circular form (Form II)/30 min at pH 7.9, 37°C	Solution containing 50% glycerol, 50 mM Tris-HCl, 500 mM NaCl, 1 mM EDTA, 1 mM DTT, 0.1% Triton X-100, pH 7.5	No detectable contaminating exonuclease, endonuclease, RNase	Epicentre W9905H W9901K W9903K
2-10 U/μL	1 unit converts 1 μg Form I to Form II pBR322 DNA/30 min at pH 7.5, 37°C		≥90% supercoiled plasmid	Promega M2851 M2852

DNA Topoisomerase (ATP-hydrolyzing)

REACTION CATALYZED
ATP-dependent breaking, passage and rejoining of double-stranded DNA

SYSTEMATIC NAME
DNA topoisomerase (ATP-hydrolyzing)

SYNONYMS
Type II DNA topoisomerase, DNA-gyrase

REACTANTS
Double-stranded DNA

APPLICATIONS
- Supercoiling closed circular duplex DNA in the presence of ATP
- Relaxing negatively supercoiled DNA in the absence of ATP
- Catenating and decatenating duplex DNA rings

NOTES
- Classified as an isomerase: other isomerases
- Can introduce negative superhelical turns into double-stranded circular DNA
- One unit has nicking-closing activity; another catalyzes super-twisting and hydrolysis of ATP
- Requires divalent cation and ATP or dATP

SPECIFIC ACTIVITY	UNITS DEFINITION	PREPARATION FORM	ADDITIONAL ACTIVITIES	SUPPLIER CATALOG No.
Drosophila melanogaster embryos, 6-12 hour old				
20 U/µL	1 unit fully relaxes 0.3 µg (5 nM) negatively supercoiled pBR322 plasmid DNA/15 min at 30°C	15 mM NaPO$_4$, 700 mM KCl, 0.1 mM EDTA, 0.5 mM DTT, 50% glycerol, pH 7.1	No detectable contaminating exonuclease, endonuclease, RNase, Topoisommerase I	Amersham E 73590Y E 73590Z
Micrococcus luteus				
5-15 U/µL	1 unit converts 0.5 µg relaxed pBR322 DNA to a supercoiled state/30 min at 37°C	Solution containing 10 mM Tris-HCl, 20 mM β-MSH, 1 mg/mL BSA, 20% glycerol, pH 7.5		ICN 151004
5-15 U/µL	1 unit converts 0.5 µg relaxed pBR322 DNA to a supercoiled form/30 min at pH 7.5, 37°C	Purity by SDS-PAGE		Life Technol 18043-018 18043-026

Histidine-tRNA Ligase

6.1.1.21

REACTION CATALYZED
ATP + L-histidine + tRNA^{His} ↔ AMP + pyrophosphate + L-histidyl-tRNA^{His}

SYSTEMATIC NAME
L-Histidine:tRNA^{His} ligase (AMP-forming)

SYNONYMS
Histidyl-tRNA synthetase, Jo-1 antigen

REACTANTS
ATP, L-histidine, tRNA^{His}, AMP, pyrophosphate, L-histidyl-tRNA^{His}

APPLICATIONS
- Defines a group of polymyositis patients with interstitial lung disease, arthritis and fevers

NOTES
- Classified as a ligase forming carbon-oxygen bonds: ligases forming aminoacyl-tRNA and related compounds
- Specific protein for patients with polymyositis

SPECIFIC ACTIVITY	UNITS DEFINITION	PREPARATION FORM	ADDITIONAL ACTIVITIES	SUPPLIER CATALOG NO.
Human tissue culture cells (MW = 53,000 Da [SDS-PAGE])				
	1 unit generates an absorbance of 1.2 when using the standard protocol for enzyme immunoassay	Affinity purified; solution containing 10 mM Tris buffer, 0.02% N$_3$ as preservative, 50% glycerol as stabilizer		ImmunoVis JO1-3000 JO1-3010 JO1-3020

Acetate-CoA Ligase

6.2.1.1

REACTION CATALYZED
ATP + acetate + CoA ↔ AMP + pyrophosphate + acetyl-CoA

SYSTEMATIC NAME
Acetate:CoA ligase (AMP-forming)

SYNONYMS
Acetyl-CoA synthetase, acetyl activating enzyme, acetate thiokinase, acyl-activating enzyme

REACTANTS
ATP, acetate, propanoate, propenoate, CoA, AMP, pyrophosphate, acetyl-CoA

APPLICATIONS
- Determination of acetate and ATP

NOTES
- Classified as a ligase forming carbon-sulfur bonds: acid-thiol ligase

6.2.1.1 Acetate-CoA Ligase continued

SPECIFIC ACTIVITY	UNITS DEFINITION	PREPARATION FORM	ADDITIONAL ACTIVITIES	SUPPLIER CATALOG NO.
Yeast				
3 U/mg protein	1 unit converts 1 μmol acetate, CoA and ATP to product/min at 37°C	Lyophilized containing 5 mg protein/20 mg solid, KPO$_4$, sucrose, BSA and GSH as stabilizers		Boehringer 161675
Yeast, bakers				
3-6 U/mg protein	1 unit forms 1.0 μmol S-acetyl coenzyme A from acetate, ATP, CoA/min at pH 7.5, 37°C	Lyophilized containing 25% protein, KPO$_4$, sucrose, GSH		Sigma A1765
2-5 U/mg protein	1 unit forms 1.0 μmol S-acetyl coenzyme A from acetate, ATP, CoA/min at pH 7.5, 37°C	Solution containing 50% glycerol, 1.0 M (NH$_4$)$_2$SO$_4$, 50 mM KPO$_4$, 1 mM DTT, pH 7.5		Sigma A5269

6.2.1.3 Long-Chain-Fatty-Acid-CoA Ligase

REACTION CATALYZED

ATP + a long-chain carboxylic acid + CoA ↔ AMP + pyrophosphate + an acyl-CoA

SYSTEMATIC NAME

Acid:CoA ligase (AMP-forming)

SYNONYMS

Acyl-CoA synthetase, fatty acid thiokinase (long chain), acyl-activating enzyme, palmitoyl-CoA synthase, lignoceroyl-CoA synthase, arachidonyl-CoA synthetase

REACTANTS

ATP, long-chain carboxylic acid, CoA, AMP, pyrophosphate, acyl-CoA

APPLICATIONS
- Determination of free fatty acids, with acyl-CoA oxidase; or adenylate kinase, pyruvate kinase and lactate dehydrogenase

NOTES
- Classified as a ligase forming carbon-sulfur bonds: acid-thiol ligase
- Acts on a wide range of long-chain saturated and unsaturated fatty acids
- Enzymes from different tissues show variation in specificity:
  - liver enzyme acts on C_6 to C_{20} acids;
  - brain enzyme shows high activity up to C_{24}
- Activated by Triton X-100
- Stabilized by ATP and sugars
- Inhibited by metal ions and ionic detergents

Long-Chain-Fatty-Acid-CoA Ligase continued

6.2.1.3

SPECIFIC ACTIVITY	UNITS DEFINITION	PREPARATION FORM	ADDITIONAL ACTIVITIES	SUPPLIER CATALOG NO.
Pseudomonas fragi (optimum pH = 7.7, pI = 5.2, MW = 38,000 Da [Sephadex G-150]; K_M = 11 μM [palmitate], 0.17 mM [ATP], 0.32 mM [CoA]; stable pH 6-8 [37°C, 2 hr], T < 50°C [pH 7.5, 10 min])				
2-8 U/mg solid	1 unit converts 1 μmol fatty acid to acyl-CoA/min at pH 7.5, 37°C	Lyophilized containing 60% protein		Asahi ACS T-16
1.5 U/mg solid	1 unit converts 1 μmol palmitic acid to product/min at 37°C	Lyophilized		Boehringer 1002406
Pseudomonas species (optimum pH = 8.5-8.8, T = 45°C, MW = 600,000 Da; K_M = 63 μM [palmitate], 0.34 mM [CoA]; stable pH 6.5-7.7 [25°C, 19 hr], T < 45°C [pH 7.5, 10 min])				
3 U/mg	1 unit forms 1 μmol AMP and oleoyl-CoA from ATP and oleate/min at pH 8.1, 25°C in the presence of CoA	Lyophilized		Fluka 01819
	1 unit forms 1 μmol oleyl-CoA (oxidation of 2 μmol NADH)/min at pH 7.5, 20°C	Lyophilized		ICN 190678
2 U/mg protein	1 unit forms 1.0 μmol AMP and oleoyl-CoA from ATP, CoA and oleate/min at pH 8.1, 25°C	Lyophilized containing 35% protein; balance ATP, sucrose, KPO_4 salts		Sigma A2777
≥0.8 U/mg solid	1 unit forms 1 μmol oleyl-CoA (2 μmol NADH)/min at pH 8.1, 25°C	Lyophilized containing 60% ATP and sugars as stabilizers	<0.5% NADH oxidase <0.9% ATPase <100% catalase	Toyobo ACS-301

Succinate-CoA Ligase (GDP-Forming)

6.2.1.4

REACTION CATALYZED
    GTP + succinate + CoA ↔ ADP + orthophosphate + succinyl-CoA

SYSTEMATIC NAME
    Succinate:CoA ligase (GDP-forming)

SYNONYMS
    Succinyl-CoA synthetase (GDP-forming), succinate thiokinase, SCS

REACTANTS
    GTP, ITP, succinate, itaconate, CoA, ADP, orthophosphate, succinyl-CoA

APPLICATIONS
- Determination of succinate

NOTES
- Classified as a ligase forming carbon-sulfur bonds: acid-thiol ligase

Succinate-CoA Ligase (GDP-Forming) continued   6.2.1.4

SPECIFIC ACTIVITY	UNITS DEFINITION	PREPARATION FORM	ADDITIONAL ACTIVITIES	SUPPLIER CATALOG NO.
Porcine heart				
10 U/mg at 25°C; 13 U/mg at 37°C	1 unit converts 1 µmol succinate to product/min	Suspension in 3.2 M $(NH_4)_2SO_4$ and BSA as stabilizer, pH 6	<0.05% ADH, GIDH	Boehringer 161543
10-50 U/mg protein	1 unit converts 1.0 µmol succinate to succinyl-CoA/min at pH 7.4, 30°C in the presence of CoA and GTP	Partially purified; lyophilized containing 25% protein; balance primarily Tris-citrate buffer salts	No detectable acyl-CoA lyase and GTPase	Sigma S4755

Long-Chain-Fatty-Acid-[Acyl-Carrier-Protein] Ligase   6.2.1.20

REACTION CATALYZED
  ATP + an acid + [acyl-carrier protein] ↔ AMP + pyrophosphate + acyl-[acyl-carrier protein]

SYSTEMATIC NAME
  Long-chain-fatty-acid:[acyl-carrier-protein] ligase (AMP-forming)

SYNONYMS
  Acyl-[acyl-carrier-protein] synthetase

REACTANTS
  ATP, an acid, [acyl-carrier protein], AMP, pyrophosphate, acyl-[acyl-carrier protein]

NOTES
- Classified as a ligase forming carbon-sulfur bonds: acid-thiol ligase
- Not identical with long-chain-fatty-acid-CoA ligase (EC 6.2.1.3)

SPECIFIC ACTIVITY	UNITS DEFINITION	PREPARATION FORM	ADDITIONAL ACTIVITIES	SUPPLIER CATALOG NO.
Escherichia coli				
0.1-0.5 U/mg protein	1 unit produces 1.0 nmol palmitoyl-acyl carrier protein from palmitic acid and acyl carrier protein/min at pH 8.0, 37°C	50% glycerol, 50 mM $MgCl_2$, 5 mM ATP, 2% Triton X-100, 0.02% NaN_3, pH 8.0		Sigma A4037

Glutamate-Ammonia Ligase 6.3.1.2

REACTION CATALYZED

ATP + L-glutamate + NH_3 ↔ ADP + orthophosphate + L-glutamine

SYSTEMATIC NAME

L-Glutamate:ammonia ligase (ADP-forming)

SYNONYMS

Glutamine synthetase, GS

REACTANTS

ATP, L-glutamate, NH_3, 4-methylene-L-glutamate (slower acting), ADP, orthophosphate, L-glutamine, D-glutamate (5-15% the rate of L-glutamate)

NOTES

- Classified as a ligase forming carbon-nitrogen bonds: acid-ammonia (or amide) ligase (amide synthetase)

SPECIFIC ACTIVITY	UNITS DEFINITION	PREPARATION FORM	ADDITIONAL ACTIVITIES	SUPPLIER CATALOG NO.
Bacillus stearothermophilus (optimum pH = 7.0, MW = 510,000 Da [43,000/subunit]; K_M = 1.2 mM [L-Glu], 1.3 mM [ATP], 80 μM [NH_4^+]; stable pH 6.5-8.0, T < 55°C)				
>10 U/mg protein	1 unit forms 1 μmol ADP/min at 30°C	Lyophilized	<0.01% ATPase, GlDH	Unitika
>10 U/mg protein	1 unit forms 1 μmol ADP/min at 30°C	50% glycerol solution	<0.01% ATPase, GlDH	Unitika
Escherichia coli				
400-2000 U/mg protein	1 unit converts 1.0 μmol L-glutamate to L-Gln/15 min at pH 7.1, 37°C	Affinity purified; lyophilized containing 5% protein; balance KPO_4, Na citrate, MgOAc buffer salts, DTT as preservative	<0.2% ATPase	Sigma G1270
Escherichia coli W, ATCC 9637				
100-300 U/mg protein	1 unit converts 1.0 μmol L-Glu to L-Gln/15 min at pH 7.1, 37°C	Lyophilized containing 10% protein; balance buffer salts and stabilizer; grown in medium containing glucose and ammonium chloride		Sigma G3144
Ovine brain				
100 U/mg protein	1 unit converts 1.0 μmol L-Glu to L-Gln/15 min at pH 7.1, 37°C	Lyophilized containing 5% protein; balance KPO_4, Na citrate, MgOAc buffer salts	No detectable ATPase, MK	Sigma G6632

6.3.4.6 Urea Carboxylase

REACTION CATALYZED
   ATP + urea + CO_2 ↔ ADP + orthophosphate + urea-1-carboxylate

SYSTEMATIC NAME
   Urea:carbon-dioxide ligase (ADP-forming)

SYNONYMS
   Urease (ATP-hydrolyzing), urea carboxylase (hydrolyzing), ATP-urea amidolyase

REACTANTS
   ATP, urea, CO_2, ADP, orthophosphate, urea-1-carboxylate

NOTES
- Classified as a ligase forming carbon-nitrogen bonds: other carbon-nitrogen ligases
- A biotinyl-protein
- Yeast enzyme also hydrolyzes urea to CO_2 and NH_3, similar to allophanate hydrolase (EC 3.5.1.54)
- Activated by Mg^{2+}, K^+ and HCO_3^-
- Stabilized by mannitol and EDTA
- Inhibited by PCMB, SDS, DAC and heavy metal ions

SPECIFIC ACTIVITY	UNITS DEFINITION	PREPARATION FORM	ADDITIONAL ACTIVITIES	SUPPLIER CATALOG NO.
Candida species (optimum pH = 7.5-8.0, pI = 4.9, MW = 160,000 Da; K_M = 0.17 mM [urea], 60 µM [ATP], 1 mM [HCO_3], 0.3 mM [K^+]; stable pH 7.5-8.5 [25°C, 16 hr], T < 35°C [pH 8.0, 10 min])				
2-6 U/mg protein	1 unit liberates 2 µmol NH_3 from 1 µmol urea/min at pH 8.0, 30°C in a coupled system with PK/LDH	Lyophilized containing 40% protein; balance primarily mannitol and EDTA		ICN 158242
2-6 U/mg protein	1 unit liberates 2 µmol NH_3 from 1 µmol urea/min at pH 8.0, 30°C in a coupled system with PK/LDH	Lyophilized containing 50% protein; balance primarily mannitol and EDTA		Sigma U4129
≥1 U/mg solid	1 unit oxidizes 1 µmol NADH/min at pH 8.0, 30°C	Lyophilized containing 20% mannitol and EDTA as stabilizers	<0.1% HK <0.3% GK <1% ATPase	Toyobo URL-301
Yeast (optimum pH = 7.0-7.5, pI = 4.9, T = 40-50°C, MW = 200,000 Da; K_M = 0.18 mM [urea], 78 µM [ATP], 4.1 mM [HCO_3], 7.2 mM [K^+]; stable pH 7.5-8.5 [25°C, 16 hr], T < 45°C [pH 7.3, 10 min])				
≥1 U/mg solid	1 unit oxidizes 1 µmol NADH/min at pH 7.3, 30°C	Lyophilized containing 20% mannitol and EDTA as stabilizers	<0.1% HK <0.3% GK <1% ATPase	Toyobo URL-311

NAD⁺ Synthase (Glutamine-Hydrolyzing) — 6.3.5.1

REACTION CATALYZED

ATP + deamido-NAD$^+$ + L-glutamine + H$_2$O ↔ AMP + pyrophosphate + NAD$^+$ + L-Glu

SYSTEMATIC NAME

Deamido-NAD$^+$:L-glutamine amido-ligase (AMP-forming)

SYNONYMS

NAD$^+$ synthetase (glutamine-hydrolyzing)

REACTANTS

ATP, deamido-NAD$^+$, L-glutamine, H$_2$O, NH$_3$, AMP, pyrophosphate, NAD$^+$, L-Glu

NOTES

- Classified as a ligase forming carbon-nitrogen bonds: carbon-nitrogen ligase with glutamine as amido-N-donor

SPECIFIC ACTIVITY	UNITS DEFINITION	PREPARATION FORM	ADDITIONAL ACTIVITIES	SUPPLIER CATALOG NO.
Escherichia coli				
0.1-1.0 U/mg protein	1 unit forms 1.0 μmol β-NAD from NAD/min at pH 8.5, 37°C	Suspension in 50% glycerol, 0.02 M Tris-HCl, 0.01 M GSH, 0.01 M MgCl$_2$, pH 7.0		Sigma N1772

Pyruvate Carboxylase — 6.4.1.1

REACTION CATALYZED

ATP + pyruvate + HCO$_3^-$ ↔ ADP + orthophosphate + oxaloacetate

SYSTEMATIC NAME

Pyruvate:carbon-dioxide ligase (ADP-forming)

SYNONYMS

Pyruvic carboxylase

REACTANTS

ATP, pyruvate, HCO$_3^-$, ADP, orthophosphate, oxaloacetate

NOTES

- Classified as a ligase forming carbon-carbon bonds
- A biotinyl-protein
- Enzyme from animal tissues contains manganese; that from yeast contains zinc
- Animal enzyme requires acetyl-CoA

SPECIFIC ACTIVITY	UNITS DEFINITION	PREPARATION FORM	ADDITIONAL ACTIVITIES	SUPPLIER CATALOG NO.
Bovine liver				
5-25 U/mg protein	1 unit converts 1.0 μmol pyruvate and CO$_2$ to oxalacetate/min at pH 7.8, 30°C	Solution containing 50% glycerol, 0.05 M Tris-HCl, 2 mM MgOAc, 1 mM EDTA, pH 7.4	<0.5% LDH	Sigma P7173

6.5.1.1 DNA Ligase (ATP)

REACTION CATALYZED
 ATP + (deoxyribonucleotide)$_n$ + (deoxyribonucleotide)$_m$ ↔ AMP + pyrophosphate + (deoxyribonucleotide)$_{n+m}$

SYSTEMATIC NAME
 Poly(deoxyribonucleotide):poly(deoxyribonucleotide) ligase (AMP-forming)

SYNONYMS
 Polydeoxyribonucleotide synthase (ATP), polynucleotide ligase, sealase, DNA repair enzyme, DNA joinase, T$_4$ DNA ligase, *Pfu* DNA ligase, *Tth* DNA ligase

REACTANTS
 ATP, deoxyribonucleotides, ribonucleotides, AMP, pyrophosphate, deoxyribonucleotides, ribonucleotides

APPLICATIONS
- Repairs single-stranded nicks in duplex DNA
- Joins both blunt-ended and cohesive-ended restriction fragments of duplex DNA
- Ligation of cohesive- or blunt-ended DNA fragments
- Adding linkers or adapters to blunt-ended DNA
- 2nd strand cDNA synthesis
- Cloning restriction fragments
- Joining RNA single-strands via bridging oligonucleotide adapters
- Repairing nicks in duplex DNA, RNA or DNA/RNA hybrids
- Ligase chain reaction (LCR)

NOTES
- Classified as a ligase forming phosphoric ester bonds
- Forms a phosphodiester bond at the site of a single-strand 3'-hydroxyl and 5'-phosphate break in duplex DNA
- Makes limited use of RNA substrates
- Single-stranded nucleic acids are not substrates
- Requires Mg^{2+} and ATP
- Polyethylene glycol stimulates blunt-ended DNA joining
- Ligation of blunt ends is inhibited by phosphate (25 *mM*) and salt (50 *mM*)
- Only T4 DNA ligase joins fragments with overlapping complementary single-stranded protrusions as well as blunt ends
- Unlike *E. coli* DNA ligase, T4 enzyme joins duplex DNA molecules at blunt ends
- *Pfu* enzyme stability permits higher melt temperatures, increasing reliability of LCR and eliminating the template-predenaturation step. Higher ligation specificity and lower background than *Tth* DNA ligase

DNA Ligase (ATP) continued — 6.5.1.1

SPECIFIC ACTIVITY	UNITS DEFINITION	PREPARATION FORM	ADDITIONAL ACTIVITIES	SUPPLIER CATALOG NO.
Escherichia coli				
	1 unit yields 50% ligation of *Hind* III fragments of λ DNA/30 min at 16°C in 20 μL assay mixture	Solution containing 50 mM KCl, 10 mM Tris-HCl, 10 mM (NH$_4$)$_2$SO$_4$, 1 mM DTT, 50% glycerol, pH 7.4		ICN 151006
Escherichia coli (lysogenic strain), overexpressing the T4 DNA ligase gene (MW = 68,000 Da)				
1 U/μL	1 unit exchanges 1 nmol [^{32}P] from pyrophosphate into Norit^R-adsorbable material/20 min at 37°C	60 mM KCl, 20 mM Tris-HCl, 1 mM EDTA, 1 mM DTT, 50% glycerol, pH 7.6	No detectable non-specific endonuclease, exonuclease, RNase	Ambion 2130 2132 2134
1-4 U/μL > 10 U/μL	1 Weiss unit exchanges 1 nmol [^{32}P]-labeled pyrophosphate to ATP into Norit^R-adsorbable material/20 min at 37°C	Under c/857 control; 10 mM KPO$_4$, 50 mM KCl, 10 mM β-MSH, 50% glycerol, pH 7.5	No detectable endonuclease	NBL Gene 020106 020103 020153
Escherichia coli 1100, lysogenic for NM989 (λ T4lig) (optimum pH = 7.6-7.8, MW = 62,000 Da)				
1000 U/mL	1 unit changes 1 nmol [^{32}P] from PP$_i$ into Norit^R-adsorbable material/20 min at pH 7.6, 37°C	10 mM Tris-HCl, 0.1 mM EDTA, 1 mM DTT, 50 mM KCl, 200 μg/mL BSA, 50% glycerol, pH 7.6	No detectable contaminating endo- and exonuclease	AGS Heidelb A00800S A00800M
3000 U/mg; 1000 and 5000 U/mL	1 unit converts 1 nmol [^{32}P] from pyrophosphate into Norit^R-adsorbable material/20 min at 37°C	20 mM Tris-HCl, 60 mM KCl, 5 mM DTT, 1 mM EDTA, 50% glycerol, pH 7.5		Boehringer 481220 716359
100 and 500 U	1 unit converts 1 nmol [^{32}P] from pyrophosphate into a Norit^R-adsorbable form/20 min at pH 7.6, 37°C	660 mM Tris-HCl, 100 mM DTT, 66 mM MgCl$_2$, pH 7.6		CLONTECH 8406-1 8406-2
0.5-2 U/μL and 4-6 U/μL	1 unit exchanges 1 nmol [^{32}P]-labeled pyrophosphate into ATP/20 min at pH 7.6, 37°C; 1 unit = 300 cohesive-end ligation units	250 mM Tris-HCl, 50 mM MgCl$_2$, 5 mM ATP, 5 mM DTT, 25% PEG-8000, pH 7.6		Life Technol 15224-017 15224-025 15224-090 15224-041
	1 Weiss unit exchanges 1 nmol [^{32}PP$_i$] into (γ,β ^{32}P) ATP/20 min at 37°C	10 mM Tris-HCl, 0.1 mM EDTA, 50 mM KCl, 1 mM β-MSH, 200 μg/mL BSA, 50% glycerol, pH 7.4	No detectable exonuclease, endonuclease	MINOTECH 203
400,000 U/mg; 25,000 U and 125,000 U	1 unit ligates > 90% of 6 μg λ DNA *Hind* III fragments/30 min at pH 7.6, 16°C in 20 μL mixture; 1 unit = 0.008 Weiss units by the ATP-PP$_i$ exchange reaction	Solution containing 10 mM Tris-HCl, 10 mM β-MSH, 0.1 mM EDTA, 50 mM KCl, 50% glycerol, pH 7.5	No detectable nuclease	TaKaRa 2011
≥2000 U/mg protein	1 unit exchanges 1 nmol ^{32}P from pyrophosphate into a Norit^R-adsorbable compound (Weiss unit)	Solution containing 10 mM Tris-HCl, 1 mM EDTA, 1 mM DTT, 5 mM KCl, 20% glycerol, pH 7.6	No detectable endonuclease, exonuclease, phosphatase	Worthington LM01023 LM01026 LM01020

6.5.1.1 DNA Ligase (ATP) continued

SPECIFIC ACTIVITY	UNITS DEFINITION	PREPARATION FORM	ADDITIONAL ACTIVITIES	SUPPLIER CATALOG NO.
Escherichia coli 1100/λT4 Lig $E_{am}W_{am}S_{am}$100 cl 857 nin 5				
	1 unit converts 1 nmol [^{32}P] from pyrophosphate into Norit^R-adsorbable material/20 min at 37°C	>95% purity by SDS-PAGE; >90% of transformed colonies are blue; 25 mM Tris-HCl, 100 mM NaCl, 0.1 mM EDTA, 1 mM DTT, 50% glycerol, pH 7.6	No detectable contaminating endonuclease, exonuclease, RNase	Amersham E 70005Y E 70005Z E 70005X E 70042X
Escherichia coli 594 (Su$^-$) bearing λ lysogen gt41op-11 lig$^+$S7				
10 U/μL	1 unit gives 50% ligation of *Hind* III-digested λ DNA/30 min at pH 8.3, 16°C in a 20 μL final volume	188 mM Tris-HCl, 46 mM MgCl$_2$, 906 mM KCl, 37.5 mM DTT, 1.5 mM β-NAD, 100 mM (NH$_4$)$_2$SO$_4$, pH 8.3		Life Technol 18052-019
Escherichia coli C600 pcl857 pPLc28 lig8 (inactivated at 65°C, 10 min)				
400,000 U/mL; 2,000,000 U/mL	1 NEB unit gives 50% ligation of *Hind* III fragments of λ DNA/30 min at 16°C in 20 μL reaction mixture and a 5′ DNA termini concentration of 0.12 μM (300 μg/mL); 1 Weiss unit = 67 cohesive-end ligation units	Purified; 50 mM KCl, 10 mM Tris-HCl, 0.1 mM EDTA, 1 mM DTT, 200 μg/mL BSA, 50% glycerol, pH 7.4	No detectable contaminating endonuclease, exonuclease	NE Biolabs 202S 202L 202CS 202CL
Escherichia coli, recombinant				
1-2 U/μL; 3-6 U/μL; 20-40 U/μL	1 unit converts 1 nmol [^{32}PP$_i$] into a Norit^R-adsorbable form/20 min at 37°C (Weiss unit); 1 Weiss unit = 67 cohesive-ended ligation units	Overproducer; 10 mM Tris-HCl, 0.1 mM EDTA, 1 mM DTT, 50 mM KCl, 50% glycerol, pH 7.5	No detectable endo-, exodeoxyribonuclease, RNase	Fermentas EL0011 EL0012 EL0013 EL0014 EL0015 EL0016 EL0017
1-3 U/μL and 10-20 U/μL	1 Weiss unit exchanges 1 nmol [^{32}P] from pyrophosphate to ATP into Norit^R-adsorbable material/20 min at pH 7.6, 37°C	Cloning Qualified; ≥90% purity by SDS gel; 10X reaction buffer: 300 mM Tris-HCl, 100 mM MgCl$_2$, 100 mM DTT, 5 mM ATP, pH 7.8	<1% DNase <3% RNase ≥90% supercoiled plasmid	Promega M1801 M1802 M1794
2000-5000 U/mL	4 unit concatemerizes 2 μg *Eco*R I 8-mer linkers into fragments >1353 Bp in length/2 hr at 23°C	Inhibited at 65°C, 10 min; 20 mM KPO$_4$, 50 mM KCl, 10 mM DTT, 50% glycerol, pH 7.5	No detectable endonuclease, exonuclease with both ss- and ds-DNA	Stratagene 600011 600012
Escherichia coli, recombinant from *Pyrococcus furiosus* (*Pfu*)				
4 U/μL	1 unit ligates to completion 0.5 μg nicked pBluescript^R DNA/15 min at 55°C	Can't be heat inactivated, must be removed by protein extraction; 50 mM Tris-HCl, 10 mM DTT, 1 mM EDTA, 0.1% NP40, 0.1% Tween 20, 50% glycerol, pH 7.5	No detectable endonuclease, exonuclease	Stratagene 600191

DNA Ligase (ATP) continued   6.5.1.1

SPECIFIC ACTIVITY	UNITS DEFINITION	PREPARATION FORM	ADDITIONAL ACTIVITIES	SUPPLIER CATALOG NO.
Escherichia coli, recombinant from *Thermus thermophilus* (Tth) (optimum T = 65-72°C; stable 37°C/1 wk, 65°C/48 hr, 95°C/>1 hr)				
	1 unit converts 0.5 μg nicked pBR322 to circular form/15 min at pH 8.3, 65°C	20 m*M* Tris-HCl, 100 m*M* KCl, 1 m*M* EDTA, 1 m*M* DTT, 0.1% Triton X-100, 50% glycerol, pH 8.0	No detectable 5'-exonuclease, endonuclease	Adv Biotech AB-0325 AB-0325b
1-5 U/μL	1 unit ligates to completion 0.5 μg nicked pBluescript^R DNA/15 min at 55°C	$T_{1/2}$ >30 min at 95°C; can't be heat inactivated, must be removed by protein extraction; 50 m*M* Tris-HCl, 10 m*M* DTT, 1 m*M* EDTA, 0.1% NP40, 0.1% Tween 20, 50% glycerol, pH 7.5	No detectable endonuclease, exonuclease	Stratagene 600193
T4				
1100 Weiss U/mg	1 Weiss unit catalyzes 1 nmol [^{32}P] (of pyrophosphate) into Norit-absorbable material/20 min at 37°C; 1 Weiss U ligase is equivalent to 1 ligation U which catalyzes ≥95% ligation of 1 μg λ/Hind III fragments/20 min at 16°C	10 m*M* Tris-HCl, 50 m*M* KCl, 0.1 m*M* EDTA, 1 m*M* DTT, 50% glycerol, pH 7.4	No detectable endonuclease	Adv Immuno T41
	1 unit converts 1 nmol [^{32}P]pyrophosphate into a Norit^R-adsorbable form/20 min at 37°C	95% purity by SDS-PAGE; 10 m*M* Tris-HCl, 1.0 m*M* DTT, 50 m*M* KCl, 50% glycerol, pH 7.5	No detectable endonuclease, exonuclease, phosphatase	CHIMERX 1060-01 1060-02
T4 infected *E. coli* (MW = 68,000 Da)				
	1 Weiss unit converts 1 nmol [^{32}P]-ATP/20 min at pH 7.6, 37°C	10 m*M* KPO$_4$, 1 m*M* DTT, 50 m*M* KCl, 50% glycerol, pH 7.6	No detectable 5'-exonuclease, endonuclease	Adv Biotech AB-0324 AB-0324b
	1 Weiss unit exchanges 1 nmol [^{32}P] from pyrophosphate into α,β-^{32}P-ATP/20 min at 37°C	50 m*M* KCl, 10 m*M* Tris-HCl, 0.1 m*M* EDTA, 1 m*M* DTT, 50% glycerol, pH 7.4		ICN 151005
	1 unit converts 1 nmol [^{32}P] from pyrophosphate into Norit^R-adsorbable form/20 min at 37°C	Solution containing 50 m*M* KCl, 10 m*M* Tris-HCl, 1 m*M* DTT, 50% glycerol, pH 7.5		ICN 152278
≥3000 Weiss U/mg protein	1 Weiss unit exchanges 1 nmol [^{32}PP$_i$] into Norit^R-adsorbable material/20 min at 37°C; 1 Weiss unit = 0.2 units with d(A-T) as substrate; 50% ligation yield of Hind III fragments of λ DNA obtained with 0.006 Weiss unit	10 m*M* Tris-HCl, 50 m*M* KCl, 1 m*M* DTT, 0.1 m*M* EDTA, pH 7.4		Oncor 120031 120032
T4, cloned				
3000 U/mg; 1 U/μL and 10 U/μL	1 unit converts 1 nmol pyrophosphate into Norit^R-adsorbable material/20 min at pH 7.8, 37°C	Solution containing 50% glycerol, 50 m*M* Tris-HCl, 0.1 *M* NaCl, 0.1 m*M* EDTA, 1 m*M* DTT, 0.1% Triton X-100, pH 7.5	No detectable exonuclease, endonuclease, RNase (by agarose gel electrophoresis)	Epicentre L0805H, L0810H L0820H, LH805H LH810H, LH820H

6.5.1.1 DNA Ligase (ATP) continued

SPECIFIC ACTIVITY	UNITS DEFINITION	PREPARATION FORM	ADDITIONAL ACTIVITIES	SUPPLIER CATALOG NO.
T4, cloned *continued*				
3000-5000 U/mg protein; 5000-8000 U/mL	1 ligation unit gives ≥90% ligation of λ DNA *Hind* III fragments/30 min at pH 7.6, 16°C; 1 ligation unit = 0.0073 Weiss unit	Molecular biology grade; homogeneous purity; solution containing 10 mM Tris-HCl, 50 mM KCl, 1 mM DTT, 0.1 mM EDTA, 50% glycerol, pH 7.4	No detectable contaminating DNase or nickase activity	Pharmacia 27-0870-03 27-0870-04

6.5.1.2 DNA Ligase (NAD$^+$)

REACTION CATALYZED

NAD$^+$ + (deoxyribonucleotide)$_n$ + (deoxyribonucleotide)$_m$ ↔ AMP + nicotinamide nucleotide + (deoxyribonucleotide)$_{n+m}$

SYSTEMATIC NAME

Poly(deoxyribonucleotide):poly(deoxyribonucleotide) ligase (AMP-forming, NMN-forming)

SYNONYMS

Polydeoxyribonucleotide synthase (NAD$^+$), polynucleotide ligase (NAD$^+$), DNA repair enzyme, DNA joinase, *Taq* DNA ligase

REACTANTS

NAD$^+$, deoxyribonucleotides, ribonucleotides, AMP, nicotinamide nucleotide, deoxyribonucleotides, ribonucleotides

NOTES

- Classified as a ligase forming phosphoric ester bonds
- Forms a phosphodiester bond at the site of a cohesive, single-stranded break in duplex DNA
- Condensation of a 5'-phosphoryl group with adjacent 3'-OH is coupled with hydrolysis of NAD$^+$
- Makes limited use of RNA substrate; cannot ligate 5'-P-DNA with 3'-OH-RNA
- Closes single-stranded nicks in double-stranded DNA
- Will ligate blunt-ended DNA in the presence of polyethylene glycol
- Requires NAD$^+$ cofactor
- *Taq* enzyme stability at elevated temperatures allows enhanced hybridization stringency
- *Escherichia coli* 594/l gt4 lop 11 lig$^+$ S7 enzyme requires longer overlapping sticky ends for ligation than T4 DNA ligase

SPECIFIC ACTIVITY	UNITS DEFINITION	PREPARATION FORM	ADDITIONAL ACTIVITIES	SUPPLIER CATALOG NO.
Escherichia coli				
	1 unit yields 50% ligation of *Hind* III fragments of λ DNA at 16°C in 20 μL assay mixture with DNA terminus concentration of 0.02 μM (50 μg/mL)	10 mM Tris-HCl, 1.0 mM DTT, 0.1 mM EDTA, 50 mM KCl, 10 mM $(NH_4)_2SO_4$, 50% glycerol, pH 7.4	No detectable endonuclease, exonuclease	CHIMERx 1065-01 1065-02
Escherichia coli 594/λ gt4 lop 11 lig⁺ S7				
1000 U	1 unit converts 100 nmol poly [d(A-T)] to an exonuclease III-resistant form/30 min at 30°C	20 mM Tris-HCl, 50 mM KCl, 10 mM $(NH_4)_2SO_4$, 0.1 mM EDTA, 50% glycerol, 0.2% Triton X-100, pH 7.8		Boehringer 862509 862517
Escherichia coli strain 594(su⁻), carrying the prophage λgt4 lop11 lig⁺Sam 7				
10-100 U/μL	1 unit gives 50% ligation of *Hind* III fragments of λ DNA/30 min at 16°C and a 5′ DNA termini concentration of 0.8 nM (7 μg/mL)	>95% purity by SDS-PAGE; 10 mM Tris-HCl, 50 mM KCl, 0.1 mM EDTA, 10 mM $(NH_4)_2SO_4$, 1 mM DTT, 50% glycerol, pH 7.4	No detectable contaminating endonuclease, exonuclease, RNase	Amersham E 70020Y E 70020Z
4000 U/mL	1 unit gives 50% ligation of *Hind* III fragments of λ DNA/30 min at 16°C in 20 μL reaction mixture and a 5′ DNA termini concentration of 0.12 μM (300 μg/mL)	Purified; 50 mM KCl, 10 mM Tris-HCl, 0.1 mM EDTA, 1 mM DTT, 200 μg/mL BSA, 50% glycerol, pH 7.4	No detectable contaminating endonuclease, exonuclease	NE Biolabs 205S 205L
Escherichia coli strain, containing the cloned ligase gene from *Thermus aquaticus* HB8 (optimum T = 45-65°C)				
40,000 U/mL	1 unit gives 50% ligation of the 12-base pair cohesive ends of 1 μg *Bst*E II-digested λ DNA/15 min at 45°C in a 50 μL reaction volume	Purified; 50 mM KCl, 10 mM Tris-HCl, 0.1 mM EDTA, 10 mM DTT, 200 μg/mL BSA, 50% glycerol, pH 7.4	No detectable contaminating ss-DNA exonuclease, endonuclease, RNase, phosphatase	NE Biolabs 208S 208L
Escherichia coli UT481, carrying a plasmid encoding the ligase (optimum pH = 7.5-8.0, MW = 77,000 Da)				
1000 U and 5000 U	1 unit ligates >90% of 6 μg λ DNA-*Hind* III fragments/30 min at pH 8.0, 16°C in a 20 μL mixture	Solution containing 10 mM KPO_4, 50 mM KCl, 1 mM DTT, 1 mM EDTA, 50% glycerol, pH 7.5	No detectable nuclease	TaKaRa 2160
Thermophilic bacterium (optimum pH = 7-8)				
8,000,000 U/mg	1 unit ligates 50% of the cos sites in 1 μg bacteriophage λ DNA/min at pH 8.3, 45°C	Solution containing 50% glycerol, 50 mM Tris-HCl, 0.1 M NaCl, 0.1 mM EDTA, 1 mM DTT, 0.1% Triton X-100, pH 7.5	No detectable DNase	Epicentre Ampligase[R] A00101, A30201 A0102K, A0110K A0125K, A32250 A32750, A3202K A3210K, A3225K

6.5.1.2 DNA Ligase (NAD⁺) continued

SPECIFIC ACTIVITY	UNITS DEFINITION	PREPARATION FORM	ADDITIONAL ACTIVITIES	SUPPLIER CATALOG NO.
Thermus aquaticus (Taq)				
	1 unit yields 50% ligation of cos sites in 0.4 µg of Sma I- and Sal I-digested bacteriophage λ DNA/min at 45°C in 20 µL reaction	20 mM Tris-HCl, 0.1% Brij-35, 50 mM KCl, 50% glycerol, pH 7.6	No detectable endonuclease, 3'- and 5'-exonuclease, nonspecific ss- and ds-DNase	CHIMERx 1070-01 1070-02

6.5.1.3 RNA Ligase (ATP)

REACTION CATALYZED

ATP + (ribonucleotide)$_n$ + (ribonucleotide)$_m$ ↔ AMP + pyrophosphate + (ribonucleotide)$_{n+m}$

SYSTEMATIC NAME

Poly(ribonucleotide):poly(ribonucleotide) ligase (AMP-forming)

SYNONYMS

Polyribonucleotide synthase (ATP), T4 RNA ligase

REACTANTS

ATP, ribonucleotide, AMP, pyrophosphate, ribonucleotide, single-stranded DNA and RNA, dinucleotide phosphates

APPLICATIONS
- Ligation of RNA fragments
- Synthesis of single-stranded RNA or DNA
- Synthesis of oligomers with base analogs or fluorescence-labeled nucleotides
- 3'-End labeling of single-stranded RNA or DNA
- Specific modifications of tRNA
- Cloning full-length cDNA
- Formation of RNA-DNA co-polymers
- Inter- or intramolecular joining of RNA and DNA
- Circularization of oligonucleotides and RNA
- Incorporation of unnatural amino acids into proteins

NOTES
- Classified as a ligase forming phosphoric ester bonds
- Converts linear RNA to a circular form by transfer of the 5'-phosphate to the 3'-hydroxyl terminus
- Energy from ATP hydrolysis drives the joining of 5'-P-oligonucleotide to 3'-OH-oligonucleotide
- The smallest substrates are pNp and NpNpN
- DNA-to-DNA ligation rate is very slow
- Does not recognize double-stranded nucleic acids
- Oligonucleotides with both 5'-P and 3'-OH ends can serve as simultaneous donor and acceptor, yielding either circular or multimeric products

RNA Ligase (ATP) continued

6.5.1.3

SPECIFIC ACTIVITY	UNITS DEFINITION	PREPARATION FORM	ADDITIONAL ACTIVITIES	SUPPLIER CATALOG NO.
Cloned				
1000-4500 U/mg; 3000-15,000 U/mL	1 unit forms 1 nmol phosphatase-resistant [^{32}P]P$_i$ from 5'-[^{32}P]oligo(rA)$_{11}$/30 min at pH 7.5, 37°C	Molecular biology grade; homogeneous purity; solution containing 25 mM HEPES-NaOH, 13 mM KCl, 1 mM DTT, 50% glycerol, pH 7.5	No detectable contaminating RNase, DNase, nickase	Pharmacia 27-0883-01 27-0883-02
Escherichia coli **B, infected with T4 am N82 (inhibited at T > 15°C)**				
50 U/μL	1 unit converts 1 pmol [5'-^{32}P]pCp into its acid insoluble form/10 min at 5°C with oligo (A)$_n$ as the substrate, labeling the 3' terminus of RNA	>90% purity by SDS-PAGE; 20 mM Tris-HCl, 50 mM NaCl, 0.1 mM EDTA, 5 mM β-MSH, 50% glycerol, pH 7.5	No detectable nuclease	Amersham E 2050Y
1000 U and 5000 U	1 unit converts 1 pmol [5'-^{32}P]pCp into acid insoluble form/10 min at pH 7.5, 5°C with oligo (A)$_n$ as substrate	>90% purity by SDS-PAGE; solution containing 20 mM Tris-HCl, 5 mM β-MSH, 0.1 mM EDTA, 50 mM NaCl, 50% glycerol, pH 7.5	No detectable nuclease	TaKaRa 2050
Escherichia coli **strain, containing the plasmid pRF-E35 (inactivated at 65°C/15 min or 100°C/2 min)**				
20,000 U/mL	1 unit converts 1 nmol 5'-phosphoryl termini in 5'-[^{32}P]rA$_{20}$ to phosphatase-resistant form/30 min at 37°C	Purified; 50 mM KCl, 10 mM Tris-HCl, 0.1 mM EDTA, 1 mM DTT, 50% glycerol, pH 7.4	No detectable contaminating ss-DNA exonuclease, endonuclease, RNase, phosphatase	NE Biolabs 204S 204L
***Escherichia coli*, overproducer**				
5-20 U/μL	1 unit converts 1 nmol [5'-^{32}P]-(rA)$_{12-18}$ to a phosphatase-resistant form/30 min at 37°C	10 mM Tris-HCl, 0.1 mM EDTA, 1 mM DTT, 50 mM KCl, 50% glycerol, pH 7.5	No detectable endo-, exodeoxyribonuclease, RNase	Fermentas EL0021 EL0022
T4				
	1 unit converts 1 nmol [5'-^{32}P]A$_{12-18}$ termini into alkaline phosphatase-resistant form/30 min at 37°C	>95% purity by SDS-PAGE; 20 mM Tris-HCl, 1.0 mM DTT, 0.1 mM EDTA, 50% glycerol, pH 7.4	No detectable endonuclease, 3'-exonuclease, 5'-exonuclease/5'-phosphatase, nonspecific ss- and ds-DNase, RNase	CHIMERx 1270-01 1270-02
4000 U/mg	1 unit converts 1 nmol 5'-phosphoryl termini to phosphatase resistant form/30 min at pH 7.8, 37°C in poly-prA$_{12-18}$	Solution containing 50% glycerol, 50 mM Tris-HCl, 0.1 M NaCl, 0.1 mM EDTA, 1 mM DTT, 0.1% Triton X-100, pH 7.5	No detectable contaminating RNase, endo- and exo- DNase, phosphatase	Epicentre LR5010 LR5025, LR5050
T4 infected *E. coli* (inactivated at 100°C/2 min)				
7000 U/mL	1 unit converts 1 nmol 5'-phosphoryl termini to a phosphatase resistant form/30 min at pH 7.5, 37°C	10 mM Tris-HCl, 0.1 mM EDTA, 1 mM DTT, 50 mM KCl, 50% glycerol, pH 7.5	No detectable contaminating endo- and exonuclease, RNase	AGS Heidelb F00820S F00820M

6.5.1.3 RNA Ligase (ATP) continued

SPECIFIC ACTIVITY	UNITS DEFINITION	PREPARATION FORM	ADDITIONAL ACTIVITIES	SUPPLIER CATALOG No.
T4 infected *E. coli* continued				
	1 unit converts 1 nmol 5'-phosphoryl termini in 5'-[^{32}P]rA$_{20}$ to a phosphatase resistant form/30 min at 37°C at a 5'-termini concentration of 10 μM	10 mM Tris-HCl, 50 M KCl, 0.1 mM EDTA, 1 mM DTT, 50% glycerol, pH 7.4	DNA exonuclease: 90% of enzyme-generated λ DNA fragments are T4 DNA-ligated and recut with *Hind* III; DNA endonuclease: 10% φX174 RFI DNA converted to RFII; RNase: 5% conversion of 23S/16S rRNA to smaller fragments	Adv Biotech AB-0326 AB-0326b
5000-15,000 U/mL	1 unit forms 1 nmol [5'-^{32}P]-rA$_{12-18}$ into a phosphatase-resistant form/30 min at 37°C	10 mM Tris-HCl, 1 mM DTT, 0.1 mM EDTA, 50 mM KCl, 50% glycerol, pH 7.5		Boehringer 1449478 1449486
>5000 U/mg; >5 U/μL	1 unit incorporates 1 pmol pCp into acid precipitable material/30 min at pH 7.5, 4°C using tRNA as acceptor	Purity by SDS-PAGE		Life Technol 18003-012 18003-038

Chapter 5. Multiple Enzyme Preparations

BIOLOGICAL SOURCE	SPECIFIC ACTIVITY	UNIT DEFINITION	PREPARATION FORM	SUPPLIER CATALOG NO.
α-Amylase (3.2.1.1), β-Amylase (3.2.1.2) Produces maltose, maltotriose and limit dextrins with minimal glucose formation. Saccharifies liquefied starch in the production of high maltose syrups, high conversion syrups and very high maltose syrups for the brewing and distilling industries.				
Barley and *Aspergillus oryzae*	≥1600 DP/mL β-amylase; 600 DU/mL α-amylase	1 unit β-amylase is the amount of enzyme contained in 0.1 mL of 5% solution producing sufficient reducing sugars to reduce 5 mL Fehling's solution/hr at 20°C; 1 unit α-amylase dextrinizes soluble starch at 1g/hr at 20°C in the presence of excess β-amylase	Food grade; solution optimum pH = 5.3	Genencor Spezyme[R] DBA
α-Amylase (3.2.1.1), β-Glucanase (3.2.1.6)				
Bacillus subtilis and *Geosmithia emersonii*	21.5 X/g α-amylase; ≥75 U/g β-1,4-glucanase		Solution stabilized with salt, pH 6.0-6.5	ABM-RP Alphalase 250
Bacillus subtilis and *Penicillium funiculosum*	21.5 X/g α-amylase; ≥100 U/g β-1,4-glucanase		Powder containing wheat flour	ABM-RP Alphalase 250D
α-Amylase (3.2.1.1), Glucoamylase (3.2.1.3) Used commercially in the wet milling of starch.				
Aspergillus oryzae			Food grade optimum pH = 4.5-5.5	EnzymeDev Brewers Fermex[R]
Aspergillus oryzae			Food grade optimum pH = 4.5-5.5	EnzymeDev Brewers Mylase[R]
α-Amylase (3.2.1.1), Pentosanase (3.2.1.8)				
Aspergillus oryzae and *Trichoderma longibrachiatum*	1300 U/g pentosanase; 4 Px/g α-amylase		Powder containing wheat flour optimum pH = 5.5, T = 30-60°C	ABM-RP Amylozyme PA
α-Amylase (3.2.1.1), Pullulanase (3.2.1.41)				
Aspergillus oryzae and *Aspergillus niger*	375 U/g pullulanase; 7 Px/g α-amylase		Solution stabilized with sorbitol, pH 6.0-7.5	ABM-RP Amylozyme FVL
β-Amylase (3.2.1.2), Pentosanase (3.2.1.8)				
Aspergillus species	3000 U/g pentosanase; 600 U/g amylase		Powder; also contains 70 U/g proteinase optimum pH = 5.5, T = 30-60°C	Biocatalysts Depol 239P D239P

Carbohydrase Mixtures

BIOLOGICAL SOURCE	SPECIFIC ACTIVITY	UNIT DEFINITION	PREPARATION FORM	SUPPLIER CATALOG NO.
β-Amylase (3.2.1.2), Pentosanase (3.2.1.8) *continued*				
Aspergillus species	240,000 U/g amylase; 10,000 U/g pentosanase		Powder optimum pH = 5.5, T = 30-60°C	Biocatalysts Depol 260P D260P
β-Amylase (3.2.1.2), Pullulanase (3.2.1.41)				
Hordeum distichon and *Aspergillus niger*	220 U/g pullulanase; 500 U/g β-amylase		Solution stabilized with sorbitol, pH 5.5-6.5	ABM-RP Amylozyme NBE Amylozyme PB
Arylsulfatase (3.1.6.1), β-Glucuronidase (3.2.1.31) **Digests cell walls of plants and fungi; hydrolyzes urinary steroid conjugates.**				
Helix pomatia	5.5 U/mL β-Glucuronidase (phenolphthalein-β-glucuronide as substrate, 38°C); 4.5 U/mL (4-nitrophenyl-β-D-glucuronide substrate, 37°C); 2.6 U/mL Arylsulfatase (phenolphthalein disulfate as substrate, 38°C); 14 U/mL (4-nitrophenyl sulfate substrate, 25°C)	β-Glucuronidase: equal to 100,000 Fishman U/mL; 1 Fishman unit releases 1 µg phenolphthalein from phenolphthalein-β-glucuronide/hr at 37°C Arylsulfatase: equal to 800,000 Roy U/mL; 1 Roy unit releases 1 µg 2-hydroxy-5-nitrophenyl sulfate/hr at 38°C	Aqueous solution	Boehringer 127698
Helix pomatia	≥5.5 U/mL β-glucuronidase; ≥1.5 U/mL aryl sulfatase	1 unit liberates 1 µmol phenolphthalein from phenolphthalein glucuronide and phenolphthalein disulfate/min at pH 4.5 and 6.2, respectively, 37°C	Solution	Calbiochem 34742

BIOLOGICAL SOURCE	SPECIFIC ACTIVITY	UNIT DEFINITION	PREPARATION FORM	SUPPLIER CATALOG No.
β-Glucanase (3.2.1.6), Xylanase (3.2.1.8) Hydrolyzes β-D-glycans and arabinoxylans in barley and wheat worts and syrups. Improves viscosity, filtration and processing of starch syrups used in brewing, distilling and baking. Improves the nutritional value of wheat and barley animal feeds.				
Geosmithia emersonii and *Trichoderma longibrachiatum*	200 U/g β-glucanase; 1000 U/g pentosanase	1 β-glucanase unit produces 1 mg maltose equivalent from barley β-glucan/min at pH 5.0, 50°C; 1 pentosanase unit produces 1 mg maltose equivalent from xylan at pH 5.5, 55°C	Solution stabilized with Na benzoate, pH 3.7-4.1	ABM-RP Wheatzyme
Penicillium funiculosum and *Trichoderma longibrachiatum*	900 U/g β-glucanase; 700 U/g pentosanase	1 β-glucanase unit produces 1 mg maltose equivalent from barley β-glucan/min at pH 5.0, 50°C; 1 pentosanase unit produces 1 mg maltose equivalent from xylan at pH 5.5, 55°C	Solution stabilized with Na benzoate, pH 3.7-4.1	ABM-RP Wheatzyme WS
Trichoderma longibrachiatum	6000 BGU/g; 8250 EXU/g	1 BGU releases 0.278 μmol reducing sugars (measured as glucose equivalents)/min at pH 3.5, 40°C with 0.5% β-glucans from barley as substrate; 1 EXU releases 1 μmol xylose from a 1% xylan solution/min at pH 3.5, 40°C	Powder	BASF Natugrain
Trichoderma longibrachiatum	2000 BGU/g; 2750 EXU/g	1 BGU releases 0.278 μmol reducing sugars (measured as glucose equivalents)/min at pH 3.5, 40°C with 0.5% β-glucans from barley as substrate; 1 EXU releases 1 μmol xylose from a 1% xylan solution/min at pH 3.5, 40°C	Solution	BASF Natugrain L33%
Trichoderma species	5500 U/g β-glucanase 3300 U/g hemicellulase		Powder; also contains 60 U/g amylase optimum pH = 4.0-6.0, T < 70°C	Biocatalysts Depol 112P D112P
Trichoderma species	7000 U/g glucanase 2000 U/g xylanase		Food grade; solution; also contains 5 U/g β-glucosidase optimum pH = 3.5-6.0, T = 50-65°C	Biocatalysts Depol 112L D112L
Unspecified			On carrier	Primalco Econase[R] Wheat F
Unspecified			Solution	Primalco Econase[R] Wheat L

Carbohydrase Mixtures

BIOLOGICAL SOURCE	SPECIFIC ACTIVITY	UNIT DEFINITION	PREPARATION FORM	SUPPLIER CATALOG NO.
β-Glucanase (3.2.1.6), Xylanase (3.2.1.8) *continued*				
Unspecified			Powder	Primalco Econase[R] Wheat P
β-Glucosidase (3.2.1.21), α-L-Rhamnosidase (3.2.1.40)				
Aspergillus niger	0.01 U/mg solid	1 unit liberates 1.0 µmol reducing sugar (as glucose) from hesperidin/min at pH 3.8, 40°C	Lyophilized containing 2% protein; balance primarily sucrose	ICN 157338
Aspergillus niger	0.01 U/mg solid	1 unit liberates 1.0 µmol reducing sugar (as glucose) from hesperidin/min at pH 3.8, 40°C	Lyophilized containing 2% protein; balance primarily sucrose	Sigma H8137
Pencillium species			Powder	Amano Hesperidinase
Penicillium species	0.2-0.4 U/mg solid	1 unit liberates 1.0 µmol reducing sugar (as glucose) from hesperidin/min at pH 3.8, 40°C	Lyophilized containing 2% protein; balance primarily sucrose	ICN 157339
Penicillium species		1 unit liberates 1.0 µmol reducing sugar (as glucose) from hesperidin/min at pH 3.8, 40°C	Lyophilized containing sorbitol	Sigma H8510
Cellulase (3.2.1.4), β-Glucanase (3.2.1.6) Used in the processing of starch syrups, brewing and distilling, animal feeds, silage, wine and fruit juices, and paper.				
Pencillium emersonii	750 U/g	1 unit gives 1 mg maltose equivalent/min/g enzyme at pH 5.0, 50°C	Powder optimum pH = 4.5, T = 80°C; stable pH 4.5-6.5 [1 hr], T < 70°C [1 hr]	ABM-RP Beta Glucanase 750D
Cellulase (3.2.1.4), β-Glucanase (3.2.1.6), Cellobiohydrolase (3.2.1.91) Modifies cellulosic materials in the finishing processes of fabrics. Improves fabric hand and appearance and permanently reduces pilling.				
Trichoderma longibrachiatum				EnzymeDev BioTouch[R] L
Trichoderma longibrachiatum	100 FPU/mL		Solution optimum pH = 4.5-5.5, T = 45-55°C	Primalco Biotouch[R]
Trichoderma longibrachiatum	15,000 ECU/mL		Highly concentrated solution	Primalco Ecostone[R] L
Cellulase (3.2.1.4), β-Glucanase (3.2.1.6), Xylanase (3.2.1.8) Digests plant material for production of alcohol, starch and gluten for brewing, baking and malting.				
Trichoderma longibrachiatum	15,000 ECU/mL	Determined on hydroxyethyl cellulose as substrate at pH 4.8, 50°C	Solution containing 0.35% Na benzoate as preservative; also contains protease, amyloglucosidase optimum T = 30-50°C	Primalco Econase[R] CE

Carbohydrase Mixtures

BIOLOGICAL SOURCE	SPECIFIC ACTIVITY	UNIT DEFINITION	PREPARATION FORM	SUPPLIER CATALOG No.
Cellulase (3.2.1.4), β-Glucanase (3.2.1.6), Xylanase (3.2.1.8) *continued*				
Trichoderma longibrachiatum			Powder	Primalco Econase^R CEP
Cellulase (3.2.1.4), β-Glucanase (3.2.1.6), Xylanase (3.2.1.8), Pectinase (3.2.1.15), Hemicellulase (3.2.1.78), Arabinase (3.2.1.99) Breaks down cell walls, liberates bound materials for extraction and degrades non-starch polysaccharides. Used in starch, alcohol, cereal and vegetable processing, improves starch availability in fermentation, reduces viscosity and improves yields.				
Aspergillus species	100 FBG/g		Solution; no detectable amylase, lipase optimum pH = 3.3-5.5	Novo Nordisk Viscozyme^R L
Cellulase (3.2.1.4), β-Glucosidase (3.2.1.21) Digests crystalline cellulose to high glucose levels. Used in the production of nectars, fruit pulps and sugars from cellulosics; aids in wort filtration in brewing; and increases digestibility of animal feeds.				
Penicillium funicullosum	≥1000 CMCase U/g ≥10,000 CMCase U/g	1 unit produces 1 μmol reducing groups from Hercules 4M6F CMC substrate/min	Solution or powder optimum pH = 5, T = 60°C; stable pH 3-7 [25°C, 18 hr], T < 50°C [pH 4.8, 30 min]	ABM-RP Cellulase CPT
Cellulase (3.2.1.4), β-Glucosidase (3.2.1.21), Glucan 1,4-β-glucosidase (3.2.1.74) Breaks down crystalline cellulose, cereal b-glucans and wheat pentosans.				
Penicillium funicullosum	400 U/g CMCase/200 U/g β-glucanase; 2000 U/g CMCase/1000 U/g β-glucanase; 4000 U/g CMCase/2000 U/g β-glucanase	1 cellulase unit produces 1 μmol reducing groups from Hercules 4M6F CMC substrate/min	Solution optimum pH = 5, T = 55°C; stable up to pH 7 [25°C, 18 hr], T < 50 [pH 4.8, 30 min]	ABM-RP Cellulase BG200 Cellulase BG1000 Cellulase BG2000
Cellulase (3.2.1.4), Cellobiohydrolase (3.2.1.91) Used in stone-washing denim, giving a faded appearance and soft hand without damaging the fabric; provides a mild chemical abrasion, partially removing dye. Used also in bioconversions to change viscosity and functionality, and to improve extractions.				
Trichoderma longibrachiatum			Food grade optimum pH = 4.5-6.0	EnzymeDev Econase^R CE, CEP
Trichoderma longibrachiatum			Industrial grade optimum pH = 4.5-7.0	EnzymeDev Ecostone^R L,P,S
Trichoderma longibrachiatum			Food grade optimum pH = 4.5-6.0	EnzymeDev Enzeco^R Cellulase TL
Unspecified	3500 IU/mL 2500 IU/mL 2000 GDU/mL	Activity determined against an internal standard	Solution, pH 4.9-5.3 optimum pH = 4.5-6.0	Genencor IndiAge^R 2XL IndiAge^R 44L IndiAge^R 66L

BIOLOGICAL SOURCE	SPECIFIC ACTIVITY	UNIT DEFINITION	PREPARATION FORM	SUPPLIER CATALOG No.
Cellulase (3.2.1.4), Cellobiohydrolase (3.2.1.91) *continued*				
Unspecified	3000 GDU/g		Coated granulate at 16-50 mesh particle size optimum pH = 4.5-6.0	Genencor IndiAge[R] 77G with Enzoguard[R]
Unspecified	3000 GDU/g		Coated granulate at 16-50 mesh particle size; Triton X-100 recommended for optimal activity optimum pH = 5.2-5.7	Genencor IndiAge[R] Euro[R] G with Enzoguard[R]
Unspecified	5000 GDU/mL		Solution, pH 4.9-5.3; non-ionic ethyxylated surfactant recommended for optimal activity optimum pH = 5.5	Genencor IndiAge[R] Euro[R]-L
Cellulase (3.2.1.4), Glucan 1,4-β-glucosidase (3.2.1.74), Cellobiohydrolase (3.2.1.91) **Used in bioconversions to change viscosity and functionality, and to improve extractions.**				
Trichoderma longibrachiatum			Food grade optimum pH = 4.5-6.0	EnzymeDev Econase[R] SII
Cellulase (3.2.1.4), Hemicellulase (3.2.1.78) **Modifies, hydrolyzes and degrades cellulose, glucans and xylans. Facilitates extractions, improves separations and reduces viscosity. Used as a processing aid in the production of animal feeds, baked goods, botanical extracts, fruits, paper, ready-to-eat cereals, starch, vegetables, flavor and oil extraction, soy and vegetable protein extraction. Used also for waste treatment and improving the functionality of dietary fiber.**				
Trichoderma longibrachiatum	≥180 GCU/g	Activity determined spectrophotometrically measuring the amount of glucose released during incubation with filter paper/hr at 50°C	Food grade; Kosher certified; powder; also contains β-glucans and arabinoxylans; no detectable protease, lipase, amylase	Genencor Multifect[R] Cellulase 300
Trichoderma longibrachiatum	≥2500 IU/mL 8000 IU/g	1 unit liberates 1 μmol reducing sugars (as glucose equivalents)/min at pH 4.8, 50°C	Food grade; solution (CL) or powder (CS), pH 4.9-5.3; also contains β-glucans and arabinoxylans; no detectable protease, lipase, amylase	Genencor Multifect[R] CL Enzyme Multifect[R] CS Enzyme
Trichoderma longibrachiatum	≥90 GCU/mL	Activity determined spectrophotometrically measuring the amount of glucose released during incubation with filter paper/hr at 50°C	Food grade; solution, pH 4.8-5.2	Genencor Multifect[R] GC Enzyme
Unspecified				SpecialtyEnz
Cellulase (3.2.1.4), Pectinase (3.2.1.15)				
Aspergillus species	500 U/g cellulase; 825 U/g pectinase		Food grade; solution optimum pH = 3.5-6.0, T = 55-65°C	Biocatalysts Depol 165L D165L

BIOLOGICAL SOURCE	SPECIFIC ACTIVITY	UNIT DEFINITION	PREPARATION FORM	SUPPLIER CATALOG NO.
Cellulase (3.2.1.4), Pectinase (3.2.1.15) *continued*				
Aspergillus species	700 PGU/g pectinase		Solution optimum pH = 4.8, T = 60°C	Biocatalysts M264L
Aspergillus species	1000 PGU/g pectinase		Solution optimum pH = 4.8, T = 60°C	Biocatalysts M265L
Aspergillus species	900 PGU/g pectinase		Solution optimum pH = 4.8, T = 60°C	Biocatalysts M282L
Cellulase (3.2.1.4), Pectinase (3.2.1.15), Hemicellulase (3.2.1.78) Used in the total degradation of plant cell walls, especially vegetable tissue.				
Aspergillus niger	300 SPS U/g	1 SPS unit releases carbohydrate soluble in 50% EtOH equivalent to 1.0 μmol galactose/min at pH 4.5, 50°C	Powder	ICN 152339
Cellulase (3.2.1.4), Xylanase (3.2.1.8) Improves the nutritive value of animal feed.				
Aspergillus species	12,000 FXU/g; 5000 BGU/g			Danisco GRINDAZYM™ GP 5000
Aspergillus species	12,000 FXU/g; 5000 BGU/g		optimum pH = 4.0-5.0	Danisco GRINDAZYM™ GPL 5000
Aspergillus species	10,000 FXU/g; 4000 BGU/g		optimum pH = 4.0-5.0	Danisco GRINDAZYM™ GV Feed
Aspergillus species	10,000 FXU/g; 4000 BGU/g		optimum pH = 4.0-5.0	Danisco GRINDAZYM™ GVL Feed
Aspergillus species	12,000 FXU/g; 5000 BGU/g		optimum pH = 4.0-5.0	Danisco GRINDAZYM™ PF
Aspergillus species	12,000 FXU/g; 5000 BGU/g		optimum pH = 4.0-5.0	Danisco GRINDAZYM™ PFL
Endoglycosidase F (3.2.1.96), *N*-Glycosidase F (3.5.1.52)				
Flavobacterium meningosepticum	600 U/mg protein	1 unit hydrolyzes 1 μmol dansyl-Asn(GlcNAc)$_2$(Man)$_5$/hr at pH 5, 37°C; 100-150 U/vial *N*-glycosidase F (nmol/min)	Solution containing 20 mM KPO$_4$, 50 mM EDTA, 0.05% NaN$_3$, pH 7.2; no detectable β-galactosidase, β-glucosidase, α- and β-mannosidase, β-*N*-*N*-Ac-hexosamini-dase, α-*L*-fucosidase, sialidase, proteases optimum pH = 5.0-7.0	Boehringer 878740

Carbohydrase Mixtures

BIOLOGICAL SOURCE	SPECIFIC ACTIVITY	UNIT DEFINITION	PREPARATION FORM	SUPPLIER CATALOG NO.
Endoglycosidase F (3.2.1.96), N-Glycosidase F (3.5.1.52) continued				
Flavobacterium meningosepticum	600 U/mg protein	1 unit hydrolyzes 1.0 μmol dansyl-Asn(GlcNAc)$_2$/hr at pH 5.0, 37°C	Solution containing 20 mM KPO$_4$, 50 mM EDTA, 0.05% NaN$_3$, pH 7.2	ICN 157920
Flavobacterium meningosepticum	600 U/mg protein; equal amounts of each enzyme	1 unit hydrolyzes 1.0 μmol dansyl-Asn-(GlcNAc)$_2$(Man)$_5$/hr at pH 5.0, 37°C	20 mM KPO$_4$, 50 mM EDTA, 0.05% NaN$_3$, pH 7.2	Sigma E8762
Glucoamylase (3.2.1.3), Pullulanase (3.2.1.41) Used in the starch syrup industry for saccharification of liquefied starch in the production of high dextrose syrups. Also used in brewing and distilling, baking, animal feed, wine and fruit, rice and textile industries.				
Aspergillus niger	200 U/g GAM 200 U/g pullulanase	1 amyloglucosidase unit forms 1 mg dextrose from hydrolyzed starch/hr at pH 4.3, 60°C; 1 pullulanase unit forms 1 mg maltose equivalent from pullulan/min at pH 5.0, 50°C	Solution stabilized with NaCl, pH 5.5-7.0; also contains aciduric α-holo-amylase, acid proteinase; no detectable transglucosidase optimum pH = 4, T = 60°C; stable pH 3.5-5 [60°C, 72 hr], T < 60°C [pH 4.3, 30 min]	ABM-RP Ambazyme P20
Aspergillus niger and *Bacillus acidopullulyticus*	225 AGU/mL glucoamylase 75 PUN/mL pullulanase	1 AGU hydrolyzes 1 μmol maltose/min; 1 PUN hydrolyzes pullulan, liberating reducing carbohydrate with reducing power equivalent to 1 μmol glucose/min	Food grade; solution optimum pH = 4.3	Novo Nordisk Dextrozyme™ 225/75 L
Glycosidases				
Charonia lampas		1 unit releases 1.0 μmol p-nitrophenyl from glycoside/min at pH 4.0, 37°C	Crude	Seikagaku 100682-1
Turbo cornutus			Lyophilized	ICN 321051 321052
Turbo cornutus		1 unit releases 1.0 μmol p-nitrophenyl from glycoside/min at pH 4.0, 37°C	Crude	Seikagaku 100680-1
Lactose synthase (2.4.1.22), β-N-Acetylglucosaminyl glycopeptide β-1,4-galactosyl-transferase (2.4.1.38)				
Human milk	5.5 U/mg protein (glucose as acceptor, with α-lactalbumin); 3.5 U/mg protein (N-Ac-glucosamine as acceptor)	1 unit transfers 1 μmol galactose from UDP-galactose to acceptor/min at pH 8.4, 37°C	Homogeneous by SDS-PAGE; lyophilized; no detectable proteases optimum pH = 7.5-8.5	Boehringer 1088696

Carbohydrase Mixtures

BIOLOGICAL SOURCE	SPECIFIC ACTIVITY	UNIT DEFINITION	PREPARATION FORM	SUPPLIER CATALOG NO.
Pectinesterase (3.1.1.11), Polygalacturonase (3.2.1.15) Degrades pectin without altering fruit juice flavor. Used also in liquefaction of apple pulp.				
Mold	0.01 U/mg	1 unit produces 1 μmol proton/min at pH 7.0, 30°C with pectin from citrus peel as substrate	Also contains low activities of cellulase, hemicellulase and protease; <20% ash optimum pH = 4.0-5.0, T = 50°C	Fluka 76290
Pectinase (3.2.1.115), Endopectin lyase (4.2.2.3) Used in the preparation of active protoplasts from a wide spectrum of higher plants and tissues, with cellulase.				
Aspergillus japonicus	4 U/mg	1 unit liberates 1 μmol galacturonic acid from polygalacturonic acid/min at pH 5.5, 25°C	Powder; also contains maceration stimulating factor; <40% lactose, >60% protein	Fluka 76305
Aspergillus japonicus	2-4 U/mg solid	1 unit liberates 1.0 μmol galacturonic acid from polygalacturonic acid/min at pH 5.5, 25°C	Lyophilized containing 60% protein; balance primarily lactose	ICN 151804
Aspergillus japonicus	100,000 maceration U/g		Highly purified; lyophilized	ICN 320951 320952
Aspergillus japonicus	100,000 maceration U/g	Unit is based on the volume of single cells released from potato tuber disks	Highly purified; powder optimum pH = 5.5	Karlan Pectolyase Y23 8006
Pectinase (3.2.1.115), Hemicellulase (3.2.1.78) Used in the treatment of fruit mashes and maceration of plant tissues; primarily apples and pears.				
Aspergillus niger	26,000 PG/mL		Food grade; sterile-filtered; aseptically bottled; solution, pH 4.5	Novo Nordisk Pectinex™ Ultra SP-L
Pectinesterase (3.1.1.11), Polygalacturonase (3.2.1.15), Arabinase (3.2.1.99), Pectin-transeliminase (4.2.2.2) Breaks down pectic substances in crushed grapes, musts and wine. Used for red and white grape maceration, viscosity reduction, clarification and depectinization.				
Aspergillus niger	5000 FDU/g		Food grade; purified; granulate; no detectable polyphenoloxidase optimum pH = 4.1, T = 50°C	Novo Nordisk Ultrazym^R 100 G
Aspergillus niger	1000 FDU/g		Food grade; purified; granulate; no detectable polyphenoloxidase optimum pH = 4.1, T = 50°C	Novo Nordisk Ultrazym^R G
Aspergillus niger	1000 FDU/mL		Food grade; sterile-filtered; aseptically bottled; solution, pH 4.5; no detectable polyphenoloxidase optimum pH = 4.1, T = 50°C	Novo Nordisk Ultrazym^R L

Carbohydrase Mixtures

BIOLOGICAL SOURCE	SPECIFIC ACTIVITY	UNIT DEFINITION	PREPARATION FORM	SUPPLIER CATALOG NO.
Pectinesterase (3.1.1.11), Polygalacturonase (3.2.1.15), Hemicellulase (3.2.1.78), Pectin-transeliminase (4.2.2.2) Degrades soluble and insoluble pectins with varying esterification. Used for apple and pear concentrates, processing citrus and tropical fruit, depectinizing fruit juice, processing berries and wine making.				
Aspergillus niger	1000 FDU/mL		Food grade; sterile-filtered; aseptically bottled; solution optimum pH = 4-5	Novo Nordisk Pectinex™ 1X L
Aspergillus niger	2000 FDU/mL		Food grade; sterile-filtered; aseptically bottled; solution optimum pH = 4-5	Novo Nordisk Pectinex™ 2X L
Aspergillus niger	3000 FDU/mL		Food grade; sterile-filtered; aseptically bottled; solution optimum pH = 4-5	Novo Nordisk Pectinex™ 3X L
Xylanase (3.2.1.8), Mannanase (3.2.1.25) Used in biobleaching.				
Trichoderma longibrachiatum			Solution	Primalco Ecopulp[R] XM

BIOLOGICAL SOURCE	SPECIFIC ACTIVITY	UNIT DEFINITION	PREPARATION FORM	SUPPLIER CATALOG NO.
Alkaline Serine Protease (3.4.21.62) and Neutral Metalloproteinase (3.4.24.28) Hydrolyzes a wide range of proteins to peptides preferentially in an exo manner, producing noticeably shorter chain length peptides than conventional bacterial proteinases. Special application in baking due to its very low fungal α-amylase activity and relatively low thermal stability.				
Aspergillus oryzae	35 k Tys/g	1 k Tys unit produces 1 mg soluble Tyr/min at pH 6.5, 35°C	Starch diluent optimum pH = 9, T = 55-60°C	ABM-RP Panazyme 77A
Aspergillus oryzae	20 k Tys/g	1 k Tys unit produces 1 mg soluble Tyr/min at pH 6.5, 35°C	Dextrose diluent	ABM-RP Panazyme 1000
Chymotrypsin (3.4.21.1), Trypsin (3.4.21.4) Used as an anti-inflammatory agent and for pre-treating foods to aid in digestion.				
Bovine and porcine	>280 USP U/mg chymotrypsin; >200 USP U/mg trypsin		USP grade; powder	Intergen Zymolean[R] I 7061-00
Bovine and porcine	>192 USP U/mg chymotrypsin; >1500 USP U/mg trypsin		USP grade; powder	Intergen Zymolean[R] XI 7060-00
Unspecified				SpecialtyEnz
Coagulation Factor VIIa (3.4.21.21), Coagulation Factor IXa (3.4.21.22) Coagulation Factor X activating enzyme.				
Russell's viper venom	25 U/mg protein	1 unit produces a clotting time of 14-22 sec for a 1:10 dilution of normal human plasma in the presence of rabbit brain cephalin (1:15 dilution) and bovine Factor VII and X deficient plasma using a one stage determination of Factor X	Lyophilized	Sigma F3630
Russell's viper venom	1000 U/mg protein	1 unit produces a clotting time of 14-22 sec for a 1:10 dilution of normal human plasma in the presence of rabbit brain cephalin (1:15 dilution) and bovine Factor VII and X deficient plasma using a one stage determination of Factor X	Lyophilized	Sigma F8754
Collagenase (3.4.24.3), Dispase non-specific protease (3.4.24.28)				
Achromobacter iophagus *Bacillus polymyxa*			Lyophilized	Sigma C3180

BIOLOGICAL SOURCE	SPECIFIC ACTIVITY	UNIT DEFINITION	PREPARATION FORM	SUPPLIER CATALOG NO.
Papain (3.4.22.2), Bromelain (3.4.22.32) **Used in protein digestion.**				
Papaya Pineapple stem			Food grade optimum pH = 5.0-7.0	EnzymeDev Enzeco[R] Dual Protease
Pepsin A (3.4.23.1), Chymosin (3.4.23.4) **Used in cheese production.**				
Bovine calf stomach (abomasum)	Clotting strength: 93.3-96.7 CHU/mL		≥80% chymosin solution containing NaCl, propylene glycol, Na benzoate, Na propionate as preservatives, pH 5.55-5.75	ChrHansens Calf Rennet Extract 921010906 921010901-4
Bovine calf stomach (abomasum)			≥85% chymosin solution containing NaCl, propylene glycol, Na benzoate, Na propionate as preservatives, pH 5.55-5.75	ChrHansens Calf Rennet Extract 921010954
Bovine calf stomach (abomasum)			≥90% chymosin solution containing NaCl, propylene glycol, Na benzoate, Na propionate as preservatives, pH 5.55-5.75	ChrHansens Calf Rennet Extract 921011004
Peptidase (3.4.11.1), Protease (3.4.X.X) **Used for debittering protein hydrolysates at low degrees of hydrolysis. Also for extensive hydrolysis of proteins resulting in flavor development. Inactivated at 85°C for 5 min or 120°C for 5 sec.**				
Aspergillus oryzae	1000 LAPU/g	1 LAP unit hydrolyzes 1 μmol L-Leu-*p*-nitroanilide/min	Food grade; GRAS; microgranulate on NaCl optimum pH = 5.0-7.5	Novo Nordisk Flavourzyme™ 1000 MG A
Aspergillus oryzae	1000 LAPU/g	1 LAP unit hydrolyzes 1 μmol L-Leu-*p*-nitroanilide/min	Food grade; GRAS; microgranulate on wheat grits	Novo Nordisk Flavourzyme™ 1000 MG B
Peptidase and Aminopeptidase (3.4.11.1)				
Porcine intestinal mucosa	50-100 U/g solid	1 unit liberates 1.0 μmol β-naphthylamine from L-Leu β-naphthylamide/min at pH 7.1, 37°C		Sigma P7500
Proteases (3.4.X.X) **Used for debittering protein hydrolysates, even at low degrees of hydrolysis. Inactivated at 90°C for 5 min.**				
Bacillus species	1.5 AU/g	Anson unit (AU) is based on proteolysis of denatured hemoglobin	Food grade; microgranulate at 250-450 micron particle size	Novo Nordisk Protamex™

BIOLOGICAL SOURCE	SPECIFIC ACTIVITY	UNIT DEFINITION	PREPARATION FORM	SUPPLIER CATALOG NO.
α-Amylase (3.2.1.1), Amyloglucosidase (3.2.1.3), β-Glucanase (3.2.1.6), Neutral proteinase (3.4.24.28)				
Bacillus subtilis *Penicillium funiculosum* *Aspergillus niger*	≥5 X/g α-amylase ≥5 Xs/g proteinase ≥5 U/g amyloglucosidase		Solution stabilized with salt, pH 6.0-6.5	ABM-RP Alphalase SVP
α-Amylase (3.2.1.1), Aspergillopepsin I (3.4.23.18) Used in brewing and malting, and in digestive aids.				
Aspergillus oryzae *Bacillus subtilis*			Food grade optimum pH = 4.0-7.0	EnzymeDev Enzeco[R] Mash-zyme Series
Aspergillus species			Food grade optimum pH = 4.0-7.0	EnzymeDev Enzopharm[R]
α-Amylase (3.2.1.1), β-Glucanase (3.2.1.6), Protease (3.4.24.28) Used in brewing when malt is replaced with barley, and for production of malt extract and barley syrups.				
Bacillus subtilis	80 KNU/g α-amylase 300 BGU/g β-glucanase 0.33 AU/g protease	1 KNU breaks down 5.26 g starch/hr at pH 5.6, 37°C; 1 BGU degrades barley β-glucan to reducing carbohydrates as 1 μmol glucose/min at pH 7.5, 30°C; 1 AU liberates TCA-soluble product giving the same color as 1 meq Tyr at pH 7.5, 25°C	Food grade; solution optimum pH = 5-7, T = 50°C (proteolysis, saccharification), 64°C (debranching)	Novo Nordisk Ceremix[R] 2X L
Bacillus *Trichoderma*	15,000 U/g amylase 3000 U/g β-glucanase 20 U/g proteinase		Solution optimum pH = 4.8	Biocatalysts Combizyme 108L C108L
α-Amylase (3.2.1.1), Neutral protease (3.4.24.28)				
Aspergillus oryzae *Bacillus subtilis*	≥35 Xs/g proteinase ≥2.75 Px/g α-amylase		Solution stabilized with sorbitol, pH 6.0-6.5	ABM-RP Amylozyme CC
Bacillus subtilis	≥28 X/g α-amylase ≥90 Xs/g proteinase		Solution stabilized with salt, pH 5.0-6.5	ABM-RP Alphalase AP
Bacillus subtilis	≥28 X/g α-amylase ≥45 Xs/g proteinase		Solution stabilized with salt, pH 5.0-6.5; also contains ≥200 U/g β(1,3)1,4-gucanase	ABM-RP Alphalase AP3
Bacillus subtilis	12 X/g α-amylase 32 Xs/g proteinase		Powder containing starch	ABM-RP Dox 1
Bacillus subtilis			Powder containing additional proteinase and buffer	ABM-RP Proteinase DC 20 Proteinase DC 25

BIOLOGICAL SOURCE	SPECIFIC ACTIVITY	UNIT DEFINITION	PREPARATION FORM	SUPPLIER CATALOG No.
α-Amylase (3.2.1.1), Neutral protease (3.4.24.28) *continued*				
Bacillus subtilis			Solution containing additional α-amylase, alkaline proteinase, amphoteric detergent; stabilized with salt, pH 5.0-7.5	ABM-RP Proteinase DC 40L
α-Amylase (3.2.1.1), Protease (3.4.24.28) **Used in animal feed and silage production.**				
Bacillus species Aspergillus species			Food/feed grade optimum pH = 3.5-7.5	EnzymeDev Multizyme[R] II
β-Amylase (3.2.1.2), Pentosanase (3.2.1.8, Proteinase (3.4.24.27)				
Aspergillus species	240,000 U/g amylase 10,000 U/g pentosanase 70 U/g proteinase		Powder optimum pH = 5.5, T = 30-60°C	Biocatalysts Combizyme 261P C261P
Aspergillus species	240,000 U/g amylase 10,000 U/g pentosanase		Powder optimum pH = 5.5, T = 30-60°C	Biocatalysts Combizyme 274P C274P
Aspergillus species	12,000 U/g amylase 10,000 U/g pentosanase 70 U/g proteinase		Powder	Biocatalysts Combizyme 275P C275P
Aspergillus species	90,000 U/g amylase 5000 U/g pentosanase 17 U/g proteinase		Powder optimum pH = 5.5, T = 30-60°C	Biocatalysts Depol 276P D276P
β-Glucanase (3.2.1.6), Pectinase (3.2.1.15), Acid protease (3.4.23.21) **Decomposed plant tissue (e.g., onion), softens kelp and decreases viscosity of barley moromi.**				
Rhizopus delemar	>100,000 APUN/g solid	1 APUN liberates the digestion product not precipitated with TCA giving the same Folin color as 1 μg Tyr/min at pH 3.0, 30°C	Powder optimum pH = 3.0, T = 55°C; stable pH 4-5 [30°C, 20 hr], T < 35°C [pH 4.5, 15 min]	Nagase XP-415

Carbohydrase and Protease Mixtures

BIOLOGICAL SOURCE	SPECIFIC ACTIVITY	UNIT DEFINITION	PREPARATION FORM	SUPPLIER CATALOG NO.
Cellulase (3.2.1.4), β-Glucanase (3.2.1.6), Xylanase (3.2.1.8), Pectinase (3.2.1.15), Hemicellulase (3.2.1.78), Polysaccharidase (3.2.1.87), Arabinase (3.2.1.99), Protease (3.4.23.18) High activity toward vegetable tissue. Primary application is total degradation of plant cell walls to yield low molecular weight carbohydrates.				
Aspergillus niger	>100,000 PGU/g >400 SPSU/g soy polysaccharidase >700 XU/g xylanase >800 FBG/g fungal β-glucanase >900 AU/g arabinase >3000 KVHCU/g hemicellulase >10,000 NCU/g cellulase >12,000 HUT/g protease >20,000 KPU/g pectinase	1 PGU liberates 1.0 μmol galacturonic acid from polygalacturonic acid/min at pH 4.5, 40°C	Lyophilized optimum pH = 3.5-5.5	ICN 151906
Cellulase (3.2.1.4), Chitinase (3.2.1.14), Laminarinase (3.2.1.6$_{39}$), Xylanase (3.2.1.8), Proteinase (3.4.X.X)				
Trichoderma harzianum	>1500 BGX U/g	1 BGX unit releases reducing carbohydrates equivalent to 1.0 μmol glucose/min at pH 4.4, 30°C	Powder	ICN 152338
Cellulase (3.2.1.4), Chitinase (3.2.1.14), Protease (3.4.X.X)				
Trichoderma harzianum			Powder containing 80% protein	Fluka 62815
Collagenase (3.4.24.3), Dispase non-specific protease (3.4.24.28) For preparation of cells from a wide variety of tissues and organs.				
Vibrio alginolyticus (Achromobacter iophagus) Bacillus polymyxa	>0.1 U/mg solid collagenase >0.8 U/mg solid dispase	1 collagenase unit forms 1 μmol of product with 4-phenylazobenzyl-oxycarbonyl-Pro-Leu-Gly-Pro-D-Arg/min at 25°C (Wunsch units); 1 dispase unit converts 1 μmol casein to product/min at pH 7.5, 37°C	Crude	Boehringer 269638 1097113
Pectinase (3.2.1.115), Protease (3.4.23.X) Used in fruit juice and wine production.				
Aspergillus niger	3788 mU/mg	pH 5	Food grade; solution optimum pH = 2.5	Rohm Rohapect^R VRSL

BIOLOGICAL SOURCE	SPECIFIC ACTIVITY	UNIT DEFINITION	PREPARATION FORM	SUPPLIER CATALOG NO.
Lipase (3.1.1.3), Amylase (3.2.1.1), Cellulase (3.2.1.4), Protease (3.4.24.28) **Used in waste treatment.**				
Unspecified, multiple			Industrial grade optimum pH = 3.0-7.0	EnzymeDev Enzeco^R Septizyme
Unspecified, multiple			Industrial grade	EnzymeDev Enzeco^R Septizyme Series

Carbohydrase, Lipase, Nuclease and Protease Mixtures

BIOLOGICAL SOURCE	SPECIFIC ACTIVITY	UNIT DEFINITION	PREPARATION FORM	SUPPLIER CATALOG NO.
Pancreatin: Elastase, Lipase, α-Amylase, Trypsin, Chymotrypsin, RNase, etc. Used in the treatment of pancreatic insufficiency, the enzymatic liberation of folic acid from its conjugated state, digestion of casein and rapeseed, cleaning reverse osmosis membranes, improving the efficiency of giemsa staining, *in vitro* diagnostics and in various food and industrial processes.				
Chicken pancreas, desiccated			Powder	Difco 0459-12-2
Pancreas	≥2 USP U/mg lipase ≥25 USP U/mg amylase ≥25 USP U/mg protease	Pancreatin USP converts ≥25X its weight of USP Potato Starch Reference Standard into soluble carbohydrates and ≥25X its weight of casein into proteases	USP; powder	Scientific Protein Labs 0310
Pancreas	≥8 USP U/mg lipase ≥100 USP U/mg amylase ≥100 USP U/mg protease	Pancreatin USP converts ≥100 times its weight of USP Potato Starch Reference Standard into soluble carbohydrates and ≥100 times its weight of casein into proteases	4X USP; powder	Scientific Protein Labs 0340
Pancreas	≥12 USP U/mg lipase ≥150 USP U/mg amylase ≥150 USP U/mg protease	Pancreatin USP converts ≥150 times its weight of USP Potato Starch Reference Standard into soluble carbohydrates and ≥150 times its weight of casein into proteases	6X USP; powder	Scientific Protein Labs 0360
Pancreas	≥16 USP U/mg lipase ≥200 USP U/mg amylase ≥200 USP U/mg protease	Pancreatin USP converts ≥200 times its weight of USP Potato Starch Reference Standard into soluble carbohydrates and ≥200 times its weight of casein into proteases	8X USP; powder	Scientific Protein Labs 0380
Pancreas	≥32 USP U/mg lipase ≥225 USP U/mg amylase ≥200 USP U/mg protease	Pancreatin USP converts ≥225 times its weight of USP Potato Starch Reference Standard into soluble carbohydrates and ≥225 times its weight of casein into proteases	Powder	Scientific Protein Labs 0510
Pancreas				SpecialtyEnz
Pancreas			4X	SpecialtyEnz
Pancreas			8X	SpecialtyEnz
Porcine pancreas	25 USP U/mg protease 25 USP U/mg amylase 2 USP U/mg lipase		USP powder	Am Labs 1X USP 4X USP 8X USP

Carbohydrase, Lipase, Nuclease and Protease Mixtures

BIOLOGICAL SOURCE	SPECIFIC ACTIVITY	UNIT DEFINITION	PREPARATION FORM	SUPPLIER CATALOG No.
Pancreatin: Elastase, Lipase, α-Amylase, Trypsin, Chymotrypsin, RNase, etc. *continued*				
Porcine pancreas	100 USP/mg amylase 100 USP/mg protease 8 USP/mg lipase	USP	4X NF; powder optimum pH = 7.0-8.2, T = 37-40°C	Biocatalysts P211P
Porcine pancreas	$\geq$8X USP specifications		80% soluble in 50 mM NaN$_3$, 100 mM NaCl, pH 5	Elastin P8X
Porcine pancreas	>100 U/mg amylase >8 U/mg lipase >100 U/mg protease	1 unit corresponds to 1 USP-unit		Fluka 76190
Porcine pancreas	Proteolytic and diastatic activities are 1 times that of NF Std.		Powder	ICN 102557
Porcine pancreas	3X USP (IX)		Powder	ICN 102558
Porcine pancreas	5X USP (IX)		Powder	ICN 102559
Porcine pancreas	4X USP		10X; lyophilized containing 25.0 g pancreatin USP and 8.5 g NaCl/L	Life Technol 15725-013
Porcine pancreas	4X USP		10X solution containing 25.0 g pancreatin USP and 8.5 g NaCl/L	Life Technol 25720-012
Porcine pancreas			10X solution containing 25.0 g pancreatin and 8.5 g NaCl/L	Life Technol KaryMAX 10590-024
Porcine pancreas	$\geq$3000 U/g starch saccharifying activity (pH 7.0); $\geq$750 U/g fat digesting activity (pH 7.0); $\geq$26,000 U/g proteolytic activity (pH 8.0)		Powder optimum pH = 8.0-10.0, T = 45°C	Wako 163-00142 167-00145
Unspecified			Vegetable mixture	SpecialtyEnz
Unspecified blend of purified enzymes for dissociation of pancreas and recovery of functional islets of Langerhans.				
Unspecified	0.5 g			Boehringer Liberase™ 1666720

BIOLOGICAL SOURCE	SPECIFIC ACTIVITY	UNIT DEFINITION	PREPARATION FORM	SUPPLIER CATALOG No.
DNA polymerase (2.7.7.7), Reverse transcriptase (2.7.7.49)				
Thermophilic bacteria		Transcriptase: 1 unit incorporates 4 nmol dTTP into acid insoluble material/30 min at pH 8.3, 45°C using oligo(dT)18-primed poly(A)n as template; DNA Polymerase: 1 unit incorporates 10 nmol dNTP into acid insoluble material/30 min at pH 8.3, 74°C	Solution containing 50% glycerol, 50 mM Tris-HCl, 0.1 M NaCl, 0.1 mM EDTA, 1 mM DTT, 0.5% Tween 20, 0.5% NP40, pH 7.5; no detectable DNA exo- and endonuclease, protease, RNase	Epicentre Retrotherm™ RT R19250 R19500 R1910H
DNA polymerase I (2.7.7.7), DNase I (3.1.21.1) Used to label DNA with radiolabeled or biotinylated nucleotides. Gives efficient nick translation of DNA.				
Bovine pancreas	0.4 U/μL	1 unit incorporates 10 nmol total deoxyribonucleotide into acid precipitable material/30 min at pH 7.5, 37°C using poly(dA-dT) as template primer in the absence of DNase I		Life Technol 18062-016
RNase A (3.1.27.5), Non-specific RNase (3.1.27.X) Used for degradation of RNA to small oligonucleotides.				
Aspergillus oryzae	10,000 U/mL (2 mg/mL RNase A, 4000 U/mL Aspergillus oryzae RNase)	20 units degrade the RNA in a 1.5 mL plasmid preparation	10 mM Tris-HCl, 15 mM NaCl, pH 7.5; no detectable DNase	Stratagene 400720

Miscellaneous Enzyme Mixtures

BIOLOGICAL SOURCE	SPECIFIC ACTIVITY	UNIT DEFINITION	PREPARATION FORM	SUPPLIER CATALOG NO.
Glucose oxidase (1.1.3.4), Catalase (1.11.1.6) Used to obtain stable, high quality dried egg production during desugaring; to remove glucose in egg white; for preventing the Maillard spoilage reaction between proteins and reducing sugars; for preventing oxidative deterioration in fruit juices prior to pasteurization by removing O_2 in the presence of excess glucose; for removing O_2 in canned food; for removing H_2O_2 in foods; for improving mayonnaise shelf life; in diagnostic kits; and for the determination of glucose in body fluids.				
Aspergillus niger	25,000-30,000 IU/g GOD 6000-10,000 Baker U/g catalase	1 unit liberates 1 μmol H_2O_2/min at 25°C	Crude; solution containing PO_4 buffer optimum pH = 5.0, T = 30°C	ABM-RP Glucox^R-C
Aspergillus niger	$\geq$1500 IU/mL GOD 400-500 Baker U/mL catalase	1 unit glucose oxidase liberates 1 μmol H_2O_2/min at 25°C	Solution containing 0.1 M PO_4 buffer, 10% propylene glycol, pH 5.0-7.0 optimum pH = 5.0, T = 30°C	ABM-RP Glucox^R-RD
Aspergillus niger	$\geq$1500 IU/mL GOD 400-500 Baker U/mL catalase	1 unit glucose oxidase liberates 1 μmol H_2O_2/min at 25°C	Solution containing PO_4 buffer, 30% propylene glycol, 0.2% methyl paraben as preservative, pH 5.0-7.0 optimum pH = 5.0, T = 30°C	ABM-RP Glucox^R-RF
Aspergillus niger	15,000 U/g GOD 200,000 U/g catalase	1 unit glucose oxidase produces 1 μmol of both gluconic acid H_2O_2/min at pH 7.0, 25°C; 1 unit catalase catalyzes 1 μmol H_2O_2 and produces 0.5 μmol of each H_2O and O_2/min at pH 7.0, 25°C	Food grade; GRAS; powder optimum pH = 6.5, T = 30-50°C	Amano Hidelase
Aspergillus niger	1500 U/g GOD 20,000 U/g catalase	1 unit glucose oxidase produces 1 μmol of both gluconic acid and H_2O_2/min at pH 7.0, 25°C; 1 unit catalase catalyzes 1 μmol H_2O_2 producing 0.5 μmol of both H_2O and O_2/min at pH 7.0, 25°C	Food grade; GRAS; powder optimum pH = 6.5, T = 30-50°C	Amano Hidelase 15-I
Aspergillus niger	1500 U/g GOD 60,000 U/g catalase	1 unit glucose oxidase produces 1 μmol of both gluconic acid and H_2O_2/min at pH 7.0, 25°C; 1 unit catalase catalyzes 1 μmol H_2O_2 producing 0.5 μmol of both water and oxygen/min at pH 7.0, 25°C	Food grade; GRAS; powder optimum pH = 6.5, T = 30-50°C	Amano Hidelase 15-II
Fungi, various		1 titrimetric unit GO oxidizes 3.0 mg glucose to gluconic acid/15 min at pH 5.1, 35°C; 1 Baker unit catalase decomposes 264 mg H_2O_2/hr at pH 7.0, 25°C		Genencor
Penicillium amagasakiense	>100,000 GOUN/g solid	1 unit oxidizes 1 μmol glucose/min at pH 5.8, 35 °C; 1 GOUN = 11.2 unit	Powder	Nagase Deoxin

Miscellaneous Enzyme Mixtures

BIOLOGICAL SOURCE	SPECIFIC ACTIVITY	UNIT DEFINITION	PREPARATION FORM	SUPPLIER CATALOG NO.
Glucose-6-phosphate dehydrogenase (1.1.1.49), Hexokinase (2.7.1.1)				
Yeast	340 U/mL HK (glucose and ATP as substrates, 25°C) 170 U/mL G6P-DH (glucose-6-P as substrate, 25°C)	1 unit converts 1 μmol substrate to product/min	Suspension in 3.2 M $(NH_4)_2SO_4$, pH 6; HK:G6PDH protein ratio is 2:1	Boehringer 127183 127825 737275
Yeast	500-1000 U/g HK 500-1000 U/g G-6-PDH 200-600 U/g coupled	1 unit produces 1.0 μmol NADPH, as glucose is phosphorylated by ATP/min at pH 7.4, 30°C in the coupled mode	Mixed enzymes simultaneously attached to beaded agarose; suspension in 2.0 M $(NH_4)_2SO_4$, pH 7.0	Sigma H2130
Glyceraldehyde-3-phosphate dehydrogenase (1.2.1.12), 3-Phosphoglyceric phosphokinase (2.7.2.3)				
Yeast	50-140 U/mg protein GAPDH 50-140 U/mg protein 3-PGK	1 unit converts 1.0 μmol 3-phosphoglycerate to D-GAP/min at pH 7.0, 25°C in a coupled system with 3-PGK	Sulfate-free; lyophilized containing 70% protein and 30% citrate buffer salts	Sigma G8505
Glycerol-3-phosphate dehydrogenase (1.1.1.8), Triose-phosphate isomerase (5.3.1.1)				
Rabbit muscle	310 U/mL GDH (c = 2 mg/mL) 1550 U/mL (c = 10 mg/mL) 900 U/mL TIM (c = 2 mg/mL) 4500 U/mL (c = 10 mg/mL)	1 GDH unit converts 1 μmol DAP to product/min at 25°C; 1 TIM unit converts 1 μmol GAP to product/min at 25°C	Crystallized; suspension in 3.2 M $(NH_4)_2SO_4$, pH 6; GDH:TIM is 10:1 by protein content	Boehringer 1246763 127787 737259
Rabbit muscle	75-200 U/mg protein α-GDH 750-2000 U/mg protein TPI	1 unit α-GDH converts 1.0 μmol DHAP to α-glycerophosphate/min at pH 7.4, 25°C; 1 unit TPI converts 1.0 μmol D-GAP to DHAP/min at pH 7.6, 25°C	Crystallized; suspension in 2.4 M $(NH_4)_2SO_4$, pH 6	Sigma G1881
Rabbit muscle	125-250 U/mg protein α-GDH	1 unit α-GDH converts 1.0 μmol DHAP to α-glycerophosphate/min at pH 7.4, 25°C; 1 unit TPI converts 1.0 μmol D-GAP to DHAP/min at pH 7.6, 25°C	Sulfate-free; lyophilized containing 60% protein; balance primarily citrate buffer salts and EDTA	Sigma G6755
3-Hydroxyacyl-CoA dehydrogenase (1.1.1.35), 3-Oxoacyl-CoA thiolase (2.3.1.16), 2-Enoylacyl-CoA hydratase (4.2.1.17) optimum pH = 7 [hydratase], 9 [dehydrogenase], 9.5 [thiolase]; pI = 4.9, MW = 67,000 [α-subunit] and 42,000 Da [β-subunit]; K_M [hydratase] = 83 mM [palmito-2-enoyl-CoA], K_M [dehydrogenase] = 0.4 mM [L-3-hydroxy-palmitoyl-CoA] and 0.8 mM [NAD], K_M [thiolase] = 2.5 mM [3-oxo-palmitoyl-CoA] and 7.1 mM [CoA]; stable pH 5-7 [37°C, 1 hr] and T < 50°C [pH 6.5, 10 min]				
Pseudomonas fragi	20-40 U/mg solid	1 unit produces 1 μmol reduced NAD/min at pH 9.0, 37°C	Lyophilized	Asahi HDT T-37

Miscellaneous Enzyme Mixtures

BIOLOGICAL SOURCE	SPECIFIC ACTIVITY	UNIT DEFINITION	PREPARATION FORM	SUPPLIER CATALOG NO.
3α-Hydroxysteroid dehydrogenase (1.1.1.50), 3(or 17)β-Hydroxysteroid dehydrogenase (1.1.1.51)				
Pseudomonas testosteroni (ATCC 11996)	0.05-0.1 U/mg solid α-enzyme (androsterone as substrate) 0.03-0.08 U/mg solid β-enzyme (testosterone as substrate)	1 unit oxidizes 1.0 μmol substrate/min at pH 8.9, 25°C in the presence of β-NAD	Crude; dried cells grown on a medium containing testosterone	Sigma H7127
Pseudomonas testosteroni (ATCC 11996)	0.5-2.0 U/mg protein (androsterone or testosterone as substrate)	1 unit oxidizes 1.0 μmol substrate/min at pH 8.9, 25°C in the presence of β-NAD		Sigma H8879
Pseudomonas testosteroni (ATCC 11996)	0.5-1.5 U/mg protein α-enzyme (androsterone as substrate) 3-5 U/mg protein β-enzyme (testosterone as substrate)	1 unit oxidizes 1.0 μmol substrate/min at pH 8.9, 25°C in the presence of β-NAD		Sigma H9004
Lactate dehydrogenase (1.1.1.27), Pyruvate kinase (2.7.1.40)				
Rabbit muscle	450 U/mL PK (ADP and PEP as substrates) 450 U/mL LDH (pyruvate as substrate)	1 unit converts 1 μmol substrate to product/min at 25°C	Crystallized; suspension in 3.2 M $(NH_4)_2SO_4$, pH 6; PK:LDH protein ratio 3:1	Boehringer 109096 109100 737291
Orotidine-5′-phosphate pyrophosphorylase (2.4.2.10), Orotidine-5′-phosphate decarboxylase (4.1.1.23) Preferred enzyme for the determination of PRPP and for preparation of UMP analogs from the corresponding orotic acid.				
Yeast	0.2-0.3 U/mg solid	1 unit phosphorylates 1.0 μmol orotic acid to 5′-OMP, which is then decarboxylated to 5′-UMP/hr at pH 8.0, 25°C in a PRPP system	Crude; powder containing 50% buffer salts	Sigma O6250
Peroxidase (1.11.1.7), Cyclooxygenase (1.14.99.1), Prostacyclin synthase (5.3.99.4)				
Bovine aorta	Provided with each lot		Lyophilized	BIOMOL EP-023
Peroxidase (1.11.1.7), Cyclooxygenase (1.14.99.1), Prostaglandin-H_2 E-isomerase (5.3.99.3)				
Ram seminal vesicles	Provided with each lot		Lyophilized	BIOMOL EP-012
Unspecified hydrolytic and oxidative activities for bakery applications Strengthens bakery dough.				
Aspergillus species				Danisco GRINDAMYL™ S 747

Part 3. Alphabetical Listing of Enzymes Sharing Common EC Classifications

Chapter 6. Restriction Endonucleases

Aat II

RECOGNITION SEQUENCE
5'...GACGT▼C...3'
3'...C▲TGCAG...5'

OPTIMUM REACTION CONDITIONS
pH 7.4-8.0 and 25°C

HEAT INACTIVATION
65°C for 20 min

METHYLATION EFFECTS
Does not cleave AGGm^{5}CCT, AGGCm^{4}CT, GACGTm^{5}C or GAm^{5}CGTC

NOTES
- EC 3.1.21.4
- A type II site specific deoxyribonuclease
- Cleaves pBR322 DNA 5-10X more efficiently than λ DNA
- A large group of enzymes recognizing specific short DNA sequences. They cleave either within the recognition site, or at a short specific distance from it
- Requires Mg^{2+} and BSA for full activity

CONCENTRATION or VOLUME ACTIVITY	UNIT DEFINITION	REACTION BUFFER (As Provided)	ADDITIONAL ACTIVITIES and PURITY	SUPPLIER CATALOG NO.
Acetobacter aceti				
3-20 U/μL	1 unit completely digests 1 μg λ DNA/hr at 37°C in the buffer provided	100 mM Tris-HCl, 100 mM MgCl$_2$, 10 mM DTT, 500 mM NaCl, pH 7.5	>90% DNA fragments ligated and recut	Amersham E 0200Y E 0200Z
10,000 U/mL	1 unit degrades 1 μg λ dam$^-$ dcm$^-$ DNA/hr at 37°C in a 0.05 mL mixture	33 mM Tris-acetate, 10 mM MgOAc, 66 mM KOAc, 100 μg/mL BSA, pH 7.9	No detectable contaminating nuclease; 90% ligated and 100% recut after 50-fold overdigestion	AGS Heidelb F00005S F00005M
1-5 U/μL	1 unit completely digests 1 μg substrate DNA/hr	PFGE-tested; 33 mM Tris-acetate, 10 mM MgOAc, 66 mM KOAc, 0.5 mM DTT, pH 7.9		Boehringer 775207 1531484
4-8 U/μL	1 unit hydrolyzes 1 μg DNA/hr at 37°C in a 50 μL reaction volume using λ phage DNA as substrate	33 mM Tris-acetate, 10 mM MgOAc, 66 mM KOAc, 0.1 mg/mL BSA, pH 7.9	>90% of the DNA fragments ligated and recut after 50-fold overdigestion	Fermentas ER0991 ER0992
15,000-25,000 U/mL		50 mM KCl, 10 mM Tris-HCl, 0.1 mM EDTA, 1 mM DTT, 200 μg/mL acetylated BSA, 50% glycerol, pH 7.4	No detectable *Aat* I	ICN 153798
2-5 U/μL	1 unit digests completely 1 μg λ DNA substrate/hr at 37°C	10 mM Tris-HCl, 50 mM NaCl, 7 mM MgCl$_2$, 7 mM β-MSH, 100 μg/mL BSA, pH 7.8		NBL Gene 016112 016113
4710,000-20,000 U/mL	1 unit completely digests 1 μg λ DNA/hr at pH 7.5, 37°C in a 50 μL total assay mixture	100 mM Tris-acetate, 100 mM MgOAc, 500 mM KOAc, pH 7.5		Pharmacia 27-0953-04 27-0953-05

Aat II continued

CONCENTRATION or VOLUME ACTIVITY	UNIT DEFINITION	REACTION BUFFER (As Provided)	ADDITIONAL ACTIVITIES and PURITY	SUPPLIER CATALOG No.
Acetobacter aceti continued				
3-5 U/µL	1 unit completely digests 1 µg λ DNA/hr at 37°C in a 50 mL reaction volume	Blue/White Cloning Qualified; 100 mM Tris-HCl, 70 mM MgCl$_2$, 500 mM KCl, 10 mM DTT, pH 7.5	90% ligation	Promega R6541 R6542 R6545
500-5000 U/mL	1 unit cleaves 95% λ DNA/hr in a 50 µL reaction mixture	200 mM Tris-acetate, 500 mM KOAc, 100 mM MgOAc, 10 mM DTT, pH 7.9	95% cleavage, 50% ligation, 90% recleavage, 50% endonuclease (nickase) No detectable endonuclease (overdigestion)	Sigma R3507
2000-5000 U/mL		50 mM KCl, 10 mM Tris-HCl, 0.1 mM EDTA, 1 mM DTT, 200 µg/mL BSA, 50% glycerol, pH 7.4		Stratagene 500010 500011
100 U and 500 U	1 unit completely digests 1 µg λ DNA/hr at 37°C in 50 µL of supplied buffer and BSA	Genome grade; 330 mM Tris-acetate, 100 mM MgOAc, 5 mM DTT, 660 mM KOAc, BSA, pH 7.9		TaKaRa 1112
				Toyobo
Acetobacter aceti, expressed in E. coli				
20,000 U/mL	1 unit completely digests 1 µg λ DNA/hr in a total reaction volume of 50 µL using the buffer provided	20 mM Tris-acetate, 10 mM MgOAc, 50 mM KOAc, 1 mM DTT, pH 7.9	>95% DNA fragments ligated and recut after 5-fold overdigestion	NE Biolabs 117S 117L

Acc I

RECOGNITION SEQUENCE

$5'...GT\blacktriangledown(^A_C)(^G_T)AC..3'$
$3'...CA(^T_G)(^C_A)\blacktriangle TG...5'$

OPTIMUM REACTION CONDITIONS
  37°C; activity is enhanced 5X at 55°C

HEAT INACTIVATION
  70°C for 15 min.; resistant at 65°C for 10 min.;
  E. coli enzyme is resistant to heat inactivation

METHYLATION EFFECTS
  Does not cleave $GT(^A_C)(^G_T)m^6AC$ or $GT(^A_C)(^G_T)Am^5C$

NOTES
- EC 3.1.21.4
- A type II site specific deoxyribonuclease
- A large group of enzymes recognizing specific short DNA sequences. They cleave either within the recognition site, or at a short specific distance from it
- Polylinkers in M13 and pUC cloning vectors contain a single *Acc* I site
- Requires Mg^{2+}

Acc I continued

Acinetobacter calcoaceticus

CONCENTRATION or VOLUME ACTIVITY	UNIT DEFINITION	REACTION BUFFER (As Provided)	ADDITIONAL ACTIVITIES and PURITY	SUPPLIER CATALOG No.
8-12 U/μL	1 unit completely digests 1 μg λ DNA/hr at 37°C in the buffer provided	100 mM Tris-HCl, 100 mM MgCl$_2$, 10 mM DTT, 500 mM NaCl, pH 7.5	>90% DNA fragments ligated and 100% recut	Amersham E 1001Y
5000 U/mL	1 unit degrades 1 μg λ dam⁻ dcm⁻ DNA/hr at 55°C in a 0.05 mL mixture	33 mM Tris-acetate, 10 mM MgOAc, 66 mM KOAc, 100 μg/mL BSA, pH 7.9	No detectable contaminating nuclease	AGS Heidelb A00010S A00010M
5 U/μL	1 unit completely digests 1 μg substrate DNA/hr	33 mM Tris-acetate, 10 mM MgOAc, 66 mM KOAc, 0.5 mM DTT, pH 7.9		Boehringer 728420 728438
	1 unit completely cleaves 1 μg λ DNA/hr at pH 7.5, 37°C in a reaction volume of 50 μL	10 mM Tris-HCl, 10 mM MgCl$_2$, 30 mM NaCl, 1.0 mM DTT, 100 μg/mL BSA, pH 7.5		CHIMERx 2000-01 2000-02
3000-10,000 U/mL		50 mM KCl, 10 mM Tris-HCl, 0.1 mM EDTA, 1 mM DTT, 200 μg/mL acetylated BSA, 50% glycerol, pH 7.5		ICN 150221
2-8 U/μL	1 unit completely digests 1 mg λ DNA/hr at 37°C	500 mM Tris-HCl, 100 mM MgCl$_2$, pH 8.0	≥33% ligation; 100% cleavage	Life Technol 15415-011 15415-029
3-10 U/μL	1 unit digests completely 1 μg λ DNA substrate/hr at 37°C	33 mM Tris-acetate, 66 mM KOAc, 10 mM MgOAc, 0.5 mM DTT, pH 8.2		NBL Gene 014111 014102
				Nippon Gene
2.5 U/μL	1 unit hydrolyzes 1 μg λ DNA to completion/hr in a 50 μL total reaction volume	330 mM Tris-acetate, 660 mM KOAc, 100 mM MgOAc, 5 mM DTT, 1 mg/mL BSA, pH 8		Oncor 110011 110012
2000-5000 U/mL	1 unit completely digests 1 μg λ DNA/hr at pH 7.5, 37°C in a 50 μL total assay mixture	100 mM Tris-acetate, 100 mM MgOAc, 500 mM KOAc, pH 7.5		Pharmacia 27-0936-01 27-0936-02
3-10 U/μL	1 unit completely digests 1 μg λ DNA/hr at 37°C in a 50 mL reaction volume	Blue/White Cloning Qualified; 500 mM Tris-HCl, 50 mM MgCl$_2$, pH 8.2	90% ligation	Promega R6411 R6412 R6415
1000-8000 U/mL	1 unit cleaves 98% pBR322 DNA/hr in a 50 μL reaction mixture	200 mM Tris-acetate, 500 mM KOAc, 100 mM MgOAc, 10 mM DTT, pH 7.9	100% cleavage, 95% ligation, 95% recleavage No detectable endonuclease (overdigestion)	Sigma R6142
2000-5000 U/mL		50 mM KCl, 10 mM Tris-HCl, 0.1 mM EDTA, 1 mM DTT, 200 μg/mL BSA, 50% glycerol, pH 7.4		Stratagene 500020

Acc I continued

CONCENTRATION or VOLUME ACTIVITY	UNIT DEFINITION	REACTION BUFFER (As Provided)	ADDITIONAL ACTIVITIES and PURITY	SUPPLIER CATALOG No.
Acinetobacter calcoaceticus continued				
100 U and 500 U	1 unit completely digests 1 µg λ DNA/hr at 37°C in 50 µL of supplied buffer	100 mM Tris-HCl, 100 mM MgCl$_2$, 10 mM DTT, 500 mM NaCl, pH 7.5		TaKaRa 1001
				Toyobo
E. coli strain that carries cloned gene from Acinetobacter calcoaceticus				
10,000 U/mL	1 unit completely digests 1 µg λ DNA/hr in a total reaction volume of 50 µL using the buffer provided	20 mM Tris-acetate, 10 mM MgOAc, 50 mM KOAc, 1 mM DTT, pH 7.9	>95% DNA fragments ligated and recut after 10-fold overdigestion	NE Biolabs 161S 161L

Acc II

RECOGNITION SEQUENCE
    5'...CG▼CG...3'
    3'...GC▲GC...5'
ISOSCHIZOMERS
    Bsh1236 I, Bsp50 I, BstU I, FnuD II, Mvn I, Tha I
OPTIMUM REACTION CONDITIONS
    37°C
HEAT INACTIVATION
    None; use ethanol precipitation
METHYLATION EFFECTS
    Does not cleave m^{5}CGCG

NOTES
- EC 3.1.21.4
- A type II site specific deoxyribonuclease
- A large group of enzymes recognizing specific short DNA sequences. They cleave either within the recognition site, or at a short specific distance from it
- Requires Mg^{2+}

CONCENTRATION or VOLUME ACTIVITY	UNIT DEFINITION	REACTION BUFFER (As Provided)	ADDITIONAL ACTIVITIES and PURITY	SUPPLIER CATALOG No.
Acinetobacter calcoaceticus				
8-12 U/µL	1 unit completely digests 1 µg λ DNA/hr at 37°C in the buffer provided	100 mM Tris-HCl, 100 mM MgCl$_2$, 10 mM DTT, 500 mM NaCl, pH 7.5	>90% DNA fragments ligated and 100% recut	Amersham E 1002Y E 1002Z
	1 unit completely cleaves 1 µg λ DNA/hr at pH 7.5, 37°C in a reaction volume of 50 µL	10 mM Tris-HCl, 10 mM MgCl$_2$, 1.0 mM DTT, 100 µg/mL BSA, pH 7.5		CHIMERX 2002-01 2002-02

Acc II continued

Acinetobacter calcoaceticus continued

CONCENTRATION or VOLUME ACTIVITY	UNIT DEFINITION	REACTION BUFFER (As Provided)	ADDITIONAL ACTIVITIES and PURITY	SUPPLIER CATALOG No.
				Nippon Gene
100 U and 500 U	1 unit completely digests 1 μg λ DNA/hr at 37°C in 50 μL of supplied buffer	100 mM Tris-HCl, 100 mM MgCl$_2$, 10 mM DTT, 500 mM NaCl, pH 7.5		TaKaRa 1002

Acc III

RECOGNITION SEQUENCE
  5'...T▼CCGGA...3'
  3'...AGGCC▲T...5'
ISOSCHIZOMERS
  BspE I, BspM II, Mro I
OPTIMUM REACTION CONDITIONS
  60°C; 10-65% activity at 37°C
HEAT INACTIVATION
  Resistant
METHYLATION EFFECTS
  Cleaves Tm^{5}CCGGA and TCm^{5}CGGA
  Does not cleave TCCGGm^{6}A
  Sensitive to overlapping dam methylation

NOTES
- EC 3.1.21.4
- A type II site specific deoxyribonuclease
- Compatible ends: Bac77 I, BspE I, Cfr9 I, Cfr19 I, Mro I, Xma I
- A large group of enzymes recognizing specific short DNA sequences. They cleave either within the recognition site, or at a short specific distance from it
- Requires Mg^{2+}

Acinetobacter calcoaceticus

CONCENTRATION or VOLUME ACTIVITY	UNIT DEFINITION	REACTION BUFFER (As Provided)	ADDITIONAL ACTIVITIES and PURITY	SUPPLIER CATALOG No.
8000-12,000 U/mL	1 unit completely cleaves 1 μg λ DNA/hr at pH 7.5, 60°C in a reaction volume of 50 μL	50 mM Tris-HCl, 10 mM MgCl$_2$, 100 mM NaCl, 1.0 mM DTT, 100 μg/mL BSA, pH 7.5		CHIMERX 2001-01 2001-02
		50 mM NaCl, 10 mM Tris-HCl, 0.1 mM EDTA, 1 mM DTT, 500 μg/mL acetylated BSA, 50% glycerol, pH 7.4		ICN 159381
				Nippon Gene

Acc III continued

CONCENTRATION or VOLUME ACTIVITY	UNIT DEFINITION	REACTION BUFFER (As Provided)	ADDITIONAL ACTIVITIES and PURITY	SUPPLIER CATALOG No.
Acinetobacter calcoaceticus continued				
10 U/μL	1 unit completely digests 1 μg λ DNA/hr at 65°C in a 50 mL reaction volume	100 mM Tris-HCl, 100 mM MgCl$_2$, 1 M NaCl, 10 mM DTT, pH 8.5	90% ligation	Promega R6581 R6582
1000-5000 U/mL		50 mM KCl, 10 mM Tris-HCl, 0.1 mM EDTA, 1 mM DTT, 200 μg/mL BSA, 50% glycerol, pH 7.4		Stratagene 500040
20 U and 100 U	1 unit completely digests 1 μg unmethylated λ DNA/hr at 60°C in 50 μL of supplied buffer	10 mM Tris-HCl, 7 mM MgCl$_2$, 7 mM β-MSH, 200 mM NaCl, pH 8.5	DNA fragments ligated and recut after 10-fold overdigestion	TaKaRa 1113

Acc16 I

RECOGNITION SEQUENCE
   5'...TGC▼GCA...3'

ISOSCHIZOMERS
   Fsp I, Avi II

NOTES
- EC 3.1.21.4
- A type II site specific deoxyribonuclease
- A large group of enzymes recognizing specific short DNA sequences. They cleave either within the recognition site, or at a short specific distance from it
- Require Mg^{2+}

CONCENTRATION or VOLUME ACTIVITY	UNIT DEFINITION	REACTION BUFFER (As Provided)	ADDITIONAL ACTIVITIES and PURITY	SUPPLIER CATALOG No.
Acinetobacter calcoaceticus 16				
10,000 U/mL	1 unit degrades 1 μg λ dam$^-$ dcm$^-$ DNA/hr at 37°C in a 0.05 mL mixture	33 mM Tris-acetate, 10 mM MgOAc, 66 mM KOAc, 100 μg/mL BSA, pH 7.9	No detectable contaminating nuclease; 95% ligated and 95% recut after 5-fold overdigestion	AGS Heidelb F00015S F00015M
				SibEnzyme

Acc65 I

RECOGNITION SEQUENCE
5′...G▼GTACC...3′
3′...CCATG▲G...5′

ISOSCHIZOMERS
*Asp*718 I

NEOSCHIZOMERS
Kpn I (GGTAC▼C)

OPTIMUM REACTION CONDITIONS
37°C

HEAT INACTIVATION
65°C for 20 min.

METHYLATION EFFECTS
Does not cleave GGTACm^{5}C
Blocked by overlapping *dcm* and CG methylation

NOTES
- EC 3.1.21.4
- A type II site specific deoxyribonuclease
- A large group of enzymes recognizing specific short DNA sequences. They cleave either within the recognition site, or at a short specific distance from it
- Produces DNA fragments with a 4-base 5′-extension; *Kpn* I does not
- Does not exhibit star activity like *Kpn* I
- Compatible ends: *Asp*718 I, *Spl* I
- Requires Mg^{2+}

CONCENTRATION or VOLUME ACTIVITY	UNIT DEFINITION	REACTION BUFFER (As Provided)	ADDITIONAL ACTIVITIES and PURITY	SUPPLIER CATALOG No.
Acinetobacter aceti 655				
12,000 U/mL	1 unit degrades 1 μg λ dam⁻ dcm⁻ DNA/hr at 37°C in a 0.05 mL mixture	50 m*M* Tris-HCl, 10 m*M* MgCl₂, 100 m*M* NaCl, 100 μg/mL BSA, pH 7.5	No detectable contaminating nuclease; 90% ligated and 100% recut after 50-fold overdigestion	ACS Heidelb F00077S F00077M
8–12 U/μL	1 unit hydrolyzes 1 μg DNA/hr at 37°C in a 50 μL reaction volume using λ phage DNA as substrate	50 m*M* Tris-HCl, 10 m*M* MgCl₂, 100 m*M* NaCl, pH 7.5	90% of the DNA fragments ligated and recut after 50-fold overdigestion	Fermentas ER0901 ER0902
500–15,000 U/mL		50 m*M* KCl, 10 m*M* Tris-HCl, 0.1 m*M* EDTA, 1 m*M* DTT, 200 μg/mL acetylated BSA, 50% glycerol, pH 7.4		ICN 159382
10,000 U/mL	1 unit hydrolyzes 1 μg DNA/hr at 37°C in a 50 μL reaction volume using λ phage DNA as substrate	50 m*M* Tris-HCl, 10 m*M* MgCl₂, 100 m*M* NaCl, pH 7.5	90% of the DNA fragments ligated and recut after 50-fold overdigestion	NE Biolabs ER0901S ER0901L
				SibEnzyme
Acinetobacter calcoaceticus				
10 U/μL	1 unit completely digests 1 μg Ad2 DNA/hr at 37°C in a 50 mL reaction volume	Blue/White Cloning Qualified; 60 m*M* Tris-HCl, 60 m*M* MgCl₂, 1.5 *M* NaCl, 10 m*M* DTT, pH 7.9	90% ligation	Promega R6921 R6922

Acc113 I

RECOGNITION SEQUENCE
5'...AGT▼ACT...3'
3'...TCA▲TGA...5'

ISOSCHIZOMERS
Sca I, Eco255 I

NOTES
- EC 3.1.21.4
- A type II site specific deoxyribonuclease
- A large group of enzymes recognizing specific short DNA sequences. They cleave either within the recognition site, or at a short specific distance from it
- Requires Mg^{2+}

CONCENTRATION or VOLUME ACTIVITY	UNIT DEFINITION	REACTION BUFFER (As Provided)	ADDITIONAL ACTIVITIES and PURITY	SUPPLIER CATALOG NO.
Acinetobacter calcoaceticus 113				
				SibEnzyme

AccB1 I

RECOGNITION SEQUENCE
5'...G▼GYRCC...3'
3'...CCRYG▲G...5'

ISOSCHIZOMERS
Ban I, BshN I, Eco64 I

NOTES
- EC 3.1.21.4
- A type II site specific deoxyribonuclease
- A large group of enzymes recognizing specific short DNA sequences. They cleave either within the recognition site, or at a short specific distance from it
- Requires Mg^{2+}

CONCENTRATION or VOLUME ACTIVITY	UNIT DEFINITION	REACTION BUFFER (As Provided)	ADDITIONAL ACTIVITIES and PURITY	SUPPLIER CATALOG NO.
Acinetobacter calcoaceticus B1				
				SibEnzyme

AccB2 I

RECOGNITION SEQUENCE
5'...RGCGC▼Y...3'
3'...Y▲CGCGR...5'

ISOSCHIZOMERS
Bsp143 II, BstH2 I, Hae II

NOTES
- EC 3.1.21.4
- A type II site specific deoxyribonuclease
- A large group of enzymes recognizing specific short DNA sequences. They cleave either within the recognition site, or at a short specific distance from it
- Requires Mg^{2+}

CONCENTRATION or VOLUME ACTIVITY	UNIT DEFINITION	REACTION BUFFER (As Provided)	ADDITIONAL ACTIVITIES and PURITY	SUPPLIER CATALOG NO.
Acinetobacter calcoaceticus B2				
				SibEnzyme

AccB7 I

RECOGNITION SEQUENCE
5'...CCANNNN▼NTGG...3'
3'...GGTN▲NNNNACC...5'

ISOSCHIZOMERS
PflM I

METHYLATION EFFECTS
Not sensitive to dcm methylation, unlike PflM I

NOTES
- EC 3.1.21.4
- A type II site specific deoxyribonuclease
- A large group of enzymes recognizing specific short DNA sequences. They cleave either within the recognition site, or at a short specific distance from it
- Exhibits star activity at low [salt] or high pH
- Requires Mg^{2+}

CONCENTRATION or VOLUME ACTIVITY	UNIT DEFINITION	REACTION BUFFER (As Provided)	ADDITIONAL ACTIVITIES and PURITY	SUPPLIER CATALOG No.
Acinetobacter calcoac B7				
8000-12,000 U/mL		300 mM NaCl, 10 mM Tris-HCl, 0.1 mM EDTA, 1 mM DTT, 500 µg/mL acetylated BSA, 50% glycerol, pH 7.4		ICN 159383
10 U/µL	1 unit completely digests 1 µg λ DNA/hr at 37°C in a 50 mL reaction volume	60 mM Tris-HCl, 60 mM MgCl₂, 1 M NaCl, 10 mM DTT, pH 7.5	90% ligation	Promega R7081 R7082
				SibEnzyme

AccBS I

RECOGNITION SEQUENCE
5'...CCGCTC(-3/-3)▼...3'

ISOSCHIZOMERS
BsrB I

NOTES
- A EC 3.1.21.4
- A type II site specific deoxyribonuclease
- A large group of enzymes recognizing specific short DNA sequences. They cleave either within the recognition site, or at a short specific distance from it
- Requires Mg^{2+}

AccBS I continued

CONCENTRATION or VOLUME ACTIVITY	UNIT DEFINITION	REACTION BUFFER (As Provided)	ADDITIONAL ACTIVITIES and PURITY	SUPPLIER CATALOG NO.
Acinetobacter calcoaceticus BS				
				SibEnzyme

Aci I

RECOGNITION SEQUENCE
    5'...C▼CGC...3'
    3'...GGC▲G...5'
OPTIMUM REACTION CONDITIONS
    37°C
HEAT INACTIVATION
    65°C for 20 min.

NOTES
- EC 3.1.21.4
- A type II site specific deoxyribonuclease
- A large group of enzymes recognizing specific short DNA sequences. They cleave either within the recognition site, or at a short specific distance from it
- Has a non-palindromic recognition site
- Requires Mg^{2+}

CONCENTRATION or VOLUME ACTIVITY	UNIT DEFINITION	REACTION BUFFER (As Provided)	ADDITIONAL ACTIVITIES and PURITY	SUPPLIER CATALOG NO.
Arthrobacter citreus				
5000-15,000 U/mL		100 mM NaCl, 10 mM Tris-HCl, 0.1 mM EDTA, 1 mM DTT, 200 µg/mL acetylated BSA, 50% glycerol, pH 7.4		ICN 159384
5000 U/mL	1 unit completely digests 1 µg λ DNA/hr in a total reaction volume of 50 µL using the buffer provided	50 mM Tris-HCl, 10 mM MgCl₂, 100 mM NaCl, 1 mM DTT, pH 7.9	>95% DNA fragments ligated and 50% (due to non-palindromic recognition site) recut after 10-fold overdigestion; remaining ligation products form Hpa II/Msp I sites	NE Biolabs 551S 551L

Acl I

RECOGNITION SEQUENCE
5'...AA▼CGTT...3'
3'...TTGC▲AA...5'

ISOSCHIZOMERS
Psp1406 I

NOTES
- EC 3.1.21.4
- A type II site specific deoxyribonuclease
- A large group of enzymes recognizing specific short DNA sequences. They cleave either within the recognition site, or at a short specific distance from it
- Requires Mg^{2+}

CONCENTRATION or VOLUME ACTIVITY	UNIT DEFINITION	REACTION BUFFER (As Provided)	ADDITIONAL ACTIVITIES and PURITY	SUPPLIER CATALOG No.
Acinetobacter calcoaceticus M4				
				sibEnzyme

AclN I

RECOGNITION SEQUENCE
5'...A▼CTAGT...3'
3'...TGATC▲A...5'

ISOSCHIZOMERS
Spe I

NOTES
- EC 3.1.21.4
- A type II site specific deoxyribonuclease
- A large group of enzymes recognizing specific short DNA sequences. They cleave either within the recognition site, or at a short specific distance from.
- Requires Mg^{2+}

CONCENTRATION or VOLUME ACTIVITY	UNIT DEFINITION	REACTION BUFFER (As Provided)	ADDITIONAL ACTIVITIES and PURITY	SUPPLIER CATALOG No.
Acinetobacter calcoaceticus N20				
				sibEnzyme

AcIW I

RECOGNITION SEQUENCE
   5'...GGATC(4/5)▼...3'

ISOSCHIZOMERS
   Alw I

NOTES
- EC 3.1.21.4
- A type II site specific deoxyribonuclease
- A large group of enzymes recognizing specific short DNA sequences. They cleave either within the recognition site, or at a short specific distance from it
- Requires Mg^{2+}

CONCENTRATION or VOLUME ACTIVITY	UNIT DEFINITION	REACTION BUFFER (As Provided)	ADDITIONAL ACTIVITIES and PURITY	SUPPLIER CATALOG NO.
Acinetobacter calcoaceticus W2131				
				SibEnzyme

Acs I

RECOGNITION SEQUENCE
   5'...(^A_G)▼AATT(^T_C)...3'
   3'...(^T_C)TTAA▲(^A_G)...5'

HEAT INACTIVATION
   65°C for 20 min.

NOTES
- EC 3.1.21.4
- A type II site specific deoxyribonuclease
- A large group of enzymes recognizing specific short DNA sequences. They cleave either within the recognition site, or at a short specific distance from it
- Requires Mg^{2+}

CONCENTRATION or VOLUME ACTIVITY	UNIT DEFINITION	REACTION BUFFER (As Provided)	ADDITIONAL ACTIVITIES and PURITY	SUPPLIER CATALOG NO.
Arthrobacter citreus				
10,000 U/mL	1 unit degrades 1 μg λ dam⁻ dcm⁻ DNA/hr at 50°C in a 0.05 mL mixture	10 m*M* Tris-HCl, 100 m*M* NaCl, 100 μg/mL BSA, pH 8.0	No detectable contaminating nuclease; 75% ligated and 95% recut after 20-fold overdigestion	ACS Heidelb F00017S F00017M

Acs I continued

CONCENTRATION or VOLUME ACTIVITY	UNIT DEFINITION	REACTION BUFFER (As Provided)	ADDITIONAL ACTIVITIES and PURITY	SUPPLIER CATALOG No.
Arthrobacter citreus continued				
				SibEnzyme
Arthrobacter citreus 310				
10 U/µL	1 unit completely digests 1 µg substrate DNA/hr	10 mM Tris-HCl, 5 mM MgCl$_2$, 100 mM NaCl, 1 mM β-MSH, pH 8.0		Boehringer 1526456 1526464

Acy I

RECOGNITION SEQUENCE

$$5'...G(^A_G)\blacktriangledown CG(^C_T)C...3'$$
$$3'...C(^T_C)GC\blacktriangle(^G_A)G...5'$$

ISOSCHIZOMERS

Aha II, Bbi II, BsaH I, Hin1 I

HEAT INACTIVATION

65°C for 20 min.

NOTES
- EC 3.1.21.4
- A type II site specific deoxyribonuclease

- A large group of enzymes recognizing specific short DNA sequences. They cleave either within the recognition site, or at a short specific distance from it
- Require Mg^{2+}
- Compatible ends: Aha II, Asu II, Ban III, Bbi II, BstB I, Cla I, Csp45 I, Hap II, Hin1 I, HinP1 I, Hpa II, Mae II, Msp I, Nar I, Nsp V, Sfu I, Taq I, TthHB8 I

CONCENTRATION or VOLUME ACTIVITY	UNIT DEFINITION	REACTION BUFFER (As Provided)	ADDITIONAL ACTIVITIES and PURITY	SUPPLIER CATALOG No.
Anabaena cylindrica				
5 U/µL	1 unit completely digests 1 µg substrate DNA/hr	10 mM Tris-HCl, 5 mM MgCl$_2$, 100 mM NaCl, 1 mM β-MSH, pH 8.0		Boehringer 1081314
3000-20,000 U/mL		50 mM KCl, 10 mM Tris-HCl, 0.1 mM EDTA, 7 mM β-MSH, 50% glycerol, pH 7.6		ICN 159385
				Nippon Gene
5-20 U/µL	1 unit completely digests 1 µg λ DNA/hr at 37°C in a 50 mL reaction volume	100 mM Tris-HCl, 100 mM MgCl$_2$, 1 M NaCl, 10 mM DTT, pH 8.5	90% ligation	Promega R6631 R6632

Afa I

RECOGNITION SEQUENCE
5'...GT▼AC...3'
3'...CA▲TG...5'

ISOSCHIZOMERS
Csp6 I, Rsa I

NOTES
- EC 3.1.21.4
- A type II site specific deoxyribonuclease
- A large group of enzymes recognizing specific short DNA sequences. They cleave either within the recognition site, or at a short specific distance from it
- Requires Mg^{2+}

CONCENTRATION or VOLUME ACTIVITY	UNIT DEFINITION	REACTION BUFFER (As Provided)	ADDITIONAL ACTIVITIES and PURITY	SUPPLIER CATALOG No.
Acidiphilium facilis 28H				
1000 U and 5000 U	1 unit completely digests 1 μg λ DNA/hr at 37°C in 50 μL of supplied buffer and BSA	330 mM Tris-acetate, 100 mM MgOAc, 5 mM DTT, 660 mM KOAc, BSA, pH 7.9	DNA fragments ligated and recut after 10-fold overdigestion	TaKaRa 1116

Afe I

RECOGNITION SEQUENCE
5'...AGC▼GCT...3'
3'...TCG▲CGA...5'

ISOSCHIZOMERS
Aor51H I, Eco47 III

NOTES
- EC 3.1.21.4
- A type II site specific deoxyribonuclease
- A large group of enzymes recognizing specific short DNA sequences. They cleave either within the recognition site, or at a short specific distance from it
- Require Mg^{2+}

CONCENTRATION or VOLUME ACTIVITY	UNIT DEFINITION	REACTION BUFFER (As Provided)	ADDITIONAL ACTIVITIES and PURITY	SUPPLIER CATALOG No.
Alcaligenes faecalis				
				SibEnzyme

Afl II

RECOGNITION SEQUENCE
5'...C▼TTAAG...3'
3'...GAATT▲C...5'

ISOSCHIZOMERS
Bfr I, Esp4 I

OPTIMUM REACTION CONDITIONS
37°C

HEAT INACTIVATION
65°C for 20 min.

NOTES
- EC 3.1.21.4
- A type II site specific deoxyribonuclease
- A large group of enzymes recognizing specific short DNA sequences. They cleave either within the recognition site, or at a short specific distance from it
- Used for generating large fragments in genomes with large G+C content
- Requires Mg^{2+} and BSA for full activity

CONCENTRATION or VOLUME ACTIVITY	UNIT DEFINITION	REACTION BUFFER (As Provided)	ADDITIONAL ACTIVITIES and PURITY	SUPPLIER CATALOG No.
Anabaena flos-aquae				
5000-15,000 U/mL		50 mM KCl, 10 mM Tris-HCl, 0.1 mM EDTA, 1 mM DTT, 200 µg/mL acetylated BSA, 50% glycerol, pH 7.4		ICN 159386
8-12 U/µL	1 unit completely digests 1 mg Ad2 DNA/hr at 37°C	500 mM Tris-HCl, 100 mM MgCl₂, pH 8.0	≥33% ligation; 100% cleavage	Life Technol 15466-014
100 U and 500 U	1 unit completely digests 1 µg λ DNA/hr at 37°C in 50 µL of supplied buffer and BSA	100 mM Tris-HCl, 100 mM MgCl₂, 10 mM DTT, 500 mM NaCl, BSA, pH 7.5		Nippon Gene TaKaRa 1003
8-12 U/µL	1 unit completely digests 1 µg λ DNA/hr at 37°C in the buffer provided	100 mM Tris-HCl, 100 mM MgCl₂, 10 mM DTT, 500 mM NaCl, BSA, pH 7.5	>95% DNA fragments ligated and 100% recut	Toyobo Amersham E 1003Y
Escherichia coli strain, carrying the cloned Afl II gene from Anabaena flos-aquae				
10,000 U/mL	1 unit completely digests 1 µg φX174 DNA/hr in a total reaction volume of 50 µL using the buffer provided	10 mM Tris-HCl, 10 mM MgCl₂, 50 mM NaCl, 1 mM DTT, pH 7.9	>95% DNA fragments ligated and recut after 5-fold overdigestion	NE Biolabs 520S 520L
2000-20,000 U/mL	1 unit completely cleaves 1.0 µg λ DNA/hr in a 50 µL reaction mixture	100 mM Tris-HCl, 500 mM NaCl, 100 mM MgCl₂, 10 mM DTT, BSA, pH 7.9	100% cleavage, 40% ligation, 90% recleavage, 5% endonuclease (nickase) No detectable endonuclease (overdigestion)	Sigma R2635

Afl III

RECOGNITION SEQUENCE
  5'...A▼CPuPyGT...3'
  3'...TGPyPuC▲A...5'
OPTIMUM REACTION CONDITIONS
  37°C
HEAT INACTIVATION
  Resistant
METHYLATION EFFECTS
  Does not cleave Am^{5}C(^A_G)(^C_T)GT

NOTES
- EC 3.1.21.4
- A type II site specific deoxyribonuclease
- A large group of enzymes recognizing specific short DNA sequences. They cleave either within the recognition site, or at a short specific distance from it
- One of the 4 recognition sequences contains protein initiation codon ATG
- Requires Mg^{2+} and BSA for full activity

CONCENTRATION or VOLUME ACTIVITY	UNIT DEFINITION	REACTION BUFFER (As Provided)	ADDITIONAL ACTIVITIES and PURITY	SUPPLIER CATALOG No.
Anabaena flos-aquae				
≥3 U/μL	1 unit completely digests 1 μg λ DNA/hr at 37°C in the buffer provided	500 mM Tris-HCl, 100 mM MgCl$_2$, 10 mM DTT, 1000 mM NaCl, pH 7.5	>90% DNA fragments ligated and >95% recut	Amersham E 0201Y E 0201Z
5 U/μL	1 unit completely digests 1 μg substrate DNA/hr	50 mM Tris-HCl, 10 mM MgCl$_2$, 100 mM NaCl, 1 mM DTE, pH 7.5		Boehringer 1209183
1000-10,000 U/mL		100 mM NaCl, 50 mM Tris-HCl, 10 mM MgCl$_2$, 0.1 mM EDTA, 1 mM DTT, 200 μg/mL acetylated BSA, 50% glycerol, pH 7.4		ICN 159387
2-8 U/μL	1 unit completely digests 1 mg λ DNA/hr at 37°C	500 mM Tris-HCl, 100 mM MgCl$_2$, 1 M NaCl, pH 8.0	≥85% ligation; 100% cleavage	Life Technol 15467-012
Escherichia coli strain, carrying the cloned Afl III gene from Anabaena flos-aquae				
5000 U/mL	1 unit completely digests 1 μg λ DNA/hr in a total reaction volume of 50 μL using the buffer provided	50 mM Tris-HCl, 10 mM MgCl$_2$, 100 mM NaCl, 1 mM DTT, pH 7.9; add 100 μg/mL BSA	>95% DNA fragments ligated and recut after 5-fold overdigestion	NE Biolabs 541S 541L

Age I

RECOGNITION SEQUENCE
  5'...A▼CCGGT...3'
  3'...TGGCC▲A...5'

OPTIMUM REACTION CONDITIONS
  25°C

HEAT INACTIVATION
  Resistant

METHYLATION EFFECTS
  Does not cleave Am^{5}CCGGT or ACm^{5}CGGT

NOTES
- EC 3.1.21.4
- A type II site specific deoxyribonuclease
- A large group of enzymes recognizing specific short DNA sequences. They cleave either within the recognition site, or at a short specific distance from it
- Requires Mg^{2+}

CONCENTRATION or VOLUME ACTIVITY	UNIT DEFINITION	REACTION BUFFER (As Provided)	ADDITIONAL ACTIVITIES and PURITY	SUPPLIER CATALOG No.
Agrobacterium gelatinovorum				
1000-5000 U/mL		50 mM KCl, 10 mM Tris-HCl, 0.1 mM EDTA, 1 mM DTT, 200 µg/mL acetylated BSA, 50% glycerol, pH 7.4		ICN 159388
2000 U/mL	1 unit completely digests 1 µg λ DNA/hr in a total reaction volume of 50 µL using the buffer provided	10 mM Bis Tris-Propane-HCl, 10 mM MgCl$_2$, 1 mM DTT, pH 7.0	>95% DNA fragments ligated and recut after 3-fold overdigestion	NE Biolabs 552S 552L
				Nippon Gene
10 U/µL	1 unit completely digests 1 µg λ DNA/hr at 37°C in a 50 mL reaction volume	100 mM Tris-HCl, 1.5 M KCl, 100 mM MgCl$_2$, pH 7.4	90% ligation	Promega R7251 R7252

Ahd I

RECOGNITION SEQUENCE
 5'...GACNNN▼NNGTC...3'
 3'...CTGNN▲NNNCAG...5'

OPTIMUM REACTION CONDITIONS
 37°C

HEAT INACTIVATION
 65°C for 20 min.

NOTES
- EC 3.1.21.4
- A type II site specific deoxyribonuclease
- A large group of enzymes recognizing specific short DNA sequences. They cleave either within the recognition site, or at a short specific distance from it
- Requires Mg^{2+}

CONCENTRATION or VOLUME ACTIVITY	UNIT DEFINITION	REACTION BUFFER (As Provided)	ADDITIONAL ACTIVITIES and PURITY	SUPPLIER CATALOG NO.
Aeromonas hydrophilia (NEB 724)				
4000 U/mL	1 unit completely digests 1 μg λ DNA/hr in a total reaction volume of 50 μL using the buffer provided	20 m*M* Tris-acetate, 10 m*M* MgOAc, 50 m*M* KOAc, 1 m*M* DTT, pH 7.9; add 100 μg/mL BSA	<5% DNA fragments ligated and recut after 20-fold overdigestion	NE Biolabs 584S 584L

Alu I

RECOGNITION SEQUENCE
 5'...AG▼CT...3'
 3'...TC▲GA...5'

OPTIMUM REACTION CONDITIONS
 pH 7.9 and 37°C

HEAT INACTIVATION
 65°C for 20 min.

METHYLATION EFFECTS
 Does not cleave m^6AGCT, AGm^4CT, AGm^5CT or $AGhm^5CT$

NOTES
- EC 3.1.21.4
- A type II site specific deoxyribonuclease
- A large group of enzymes recognizing specific short DNA sequences. They cleave either within the recognition site, or at a short specific distance from it
- Requires Mg^{2+} and BSA for full activity
- Compatible ends: any blunt end

Alu I continued

CONCENTRATION or VOLUME ACTIVITY	UNIT DEFINITION	REACTION BUFFER (As Provided)	ADDITIONAL ACTIVITIES and PURITY	SUPPLIER CATALOG No.
Arthrobacter luteus				
1000 U and 2500 U	1 unit completely digests 1 μg DNA substrate/hr at 37°C	10 mM Tris-HCl, 10 mM MgCl$_2$, 1 mM DTT, 0.1 mg/mL BSA, pH 7.5	>90% DNA fragments ligated and recut after 10-fold overdigestion	Adv Biotech AB-0400-a AB-0400-b
8-12 U/μL	1 unit completely digests 1 μg λ DNA/hr at 37°C in the buffer provided	100 mM Tris-HCl, 100 mM MgCl$_2$, 10 mM DTT, pH 7.5	95% DNA fragments ligated and 100% recut	Amersham E 1004Z
10,000 U/mL	1 unit degrades 1 μg λ dam$^-$ dcm$^-$ DNA/hr at 37°C in a 0.05 mL mixture	33 mM Tris-acetate, 10 mM MgOAc, 66 mM KOAc, 100 μg/mL BSA, pH 7.9	No detectable contaminating nuclease; 95% ligated and 95% recut after 50-fold overdigestion	AGS Heidelb A00040S A00040M
10 U/μL and 40 U/μL	1 unit completely digests 1 μg substrate DNA/hr	33 mM Tris-acetate, 10 mM MgOAc, 66 mM KOAc, 0.5 mM DTT, pH 7.9		Boehringer 239275 656267 1047582
	1 unit completely cleaves 1 μg λ DNA/hr at pH 7.5, 37°C in a reaction volume of 50 μL	10 mM Tris-HCl, 10 mM MgCl$_2$, 50 mM NaCl, 1.0 mM DTT, 100 μg/mL BSA, pH 7.5		CHIMERx 2010-01 2010-02
8-12 U/μL	1 unit hydrolyzes 1 μg DNA/hr at 37°C in a 50 μL reaction volume using λ phage DNA as substrate	33 mM Tris-acetate, 10 mM MgOAc, 66 mM KOAc, 0.1 mg/mL BSA, pH 7.9	90% of the DNA fragments ligated and recut after 50-fold overdigestion	Fermentas ER0011 ER0012
3000-10,000 U/mL		50 mM KCl, 10 mM Tris-HCl, 0.1 mM EDTA, 1 mM DTT, 200 μg/mL acetylated BSA, 50% glycerol, pH 7.5		ICN 150277
8-12 U/μL and 50 U/μL	1 unit completely digests 1 mg λ DNA/hr at 37°C	500 mM Tris-HCl, 100 mM MgCl$_2$, pH 8.0	≥95% ligation; 100% cleavage	Life Technol 45200-029 45200-045 45200-011
	10 units does not produce unspecific cleavage products with 1 μg λ DNA/16 hr at 37°C	10 mM Tris-HCl, 10 mM MgCl$_2$, 1 mM DTT, 100 μg/mL BSA, pH 7.9	>90% DNA fragments ligated and recut after 10-fold overdigestion	MINOTECH 101
3-13 U/μL	1 unit digests completely 1 μg λ DNA substrate/hr at 37°C	33 mM Tris-acetate, 66 mM KOAc, 10 mM MgOAc, 0.5 mM DTT, pH 8.2		NBL Gene 010502 010505
				Nippon Gene
5 U/μL and 10-20 U/μL	1 unit hydrolyzes 1 μg λ DNA to completion/hr in a 50 μL total reaction volume	100 mM Tris-HCl, 500 mM NaCl, 100 mM MgCl$_2$, 100 mM β-MSH, 1.0 mg/mL BSA, pH 8		Oncor 110021 110022
2000-5000 U/mL	1 unit completely digests 1 μg λ DNA/hr at pH 7.5, 37°C in a 50 μL total assay mixture	100 mM Tris-acetate, 100 mM MgOAc, 500 mM KOAc, pH 7.5	Compatible ends: blunt	Pharmacia 27-0884-01 27-0884-02

Alu I continued

CONCENTRATION or VOLUME ACTIVITY	UNIT DEFINITION	REACTION BUFFER (As Provided)	ADDITIONAL ACTIVITIES and PURITY	SUPPLIER CATALOG NO.
Arthrobacter luteus continued				
10 U/μL	1 unit completely digests 1 μg λ DNA/hr at 37°C in a 50 mL reaction volume	60 mM Tris-HCl, 60 mM MgCl$_2$, 500 mM NaCl, 10 mM DTT, pH 7.5	90% ligation	Promega R6281 R6282
				SibEnzyme
3000-10,000 U/mL	1 unit cleaves 95% λ DNA/hr in a 50 μL reaction mixture	100 mM Bis Tris-Propane-HCl, 100 mM MgCl$_2$, 10 mM DTT, pH 7.0	95% cleavage, 100% ligation, 85% recleavage. No detectable endonuclease (overdigestion)	Sigma R6885
2000-12,000 U/mL		100 mM KCl, 10 mM Tris-HCl, 0.1 mM EDTA, 1 mM DTT, 0.02% Triton X-100, 200 μg/mL BSA, 50% glycerol, pH 7.4		Stratagene 500100
500 U and 2500 U	1 unit completely digests 1 μg λ DNA/hr at 37°C in 50 μL of supplied buffer	100 mM Tris-HCl, 100 mM MgCl$_2$, 10 mM DTT, pH 7.5	DNA fragments ligated and recut after 10-fold overdigestion	TaKaRa 1004
				Toyobo
Escherichia coli strain, carrying the cloned gene from Arthrobacter luteus				
8000 U/mL	1 unit completely digests 1 μg λ DNA/hr in a total reaction volume of 50 μL using the buffer provided	10 mM Bis Tris-Propane-HCl, 10 mM MgCl$_2$, 1 mM DTT, pH 7.0	>95% DNA fragments ligated and recut after 10-fold overdigestion	NE Biolabs 137S 137L

Alw I

RECOGNITION SEQUENCE
5'...GGATC(N)$_4$▼...3'
3'...CCTAG(N)$_5$▲...5'

OPTIMUM REACTION CONDITIONS
37 °C

HEAT INACTIVATION
65°C for 20 min.

METHYLATION EFFECTS
Does not cleave GGm^{6}ATC or GGATm^{4}C
Blocked by overlapping *dam* methylation

NOTES
- EC 3.1.21.4
- A type II site specific deoxyribonuclease
- A large group of enzymes recognizing specific short DNA sequences. They cleave either within the recognition site, or at a short specific distance from it
- Produces DNA fragments that have a single-base 5' extension
- Requires Mg^{2+}

CONCENTRATION or VOLUME ACTIVITY	UNIT DEFINITION	REACTION BUFFER (As Provided)	ADDITIONAL ACTIVITIES and PURITY	SUPPLIER CATALOG No.
Acinetobacter iwoffi				
500-5000 U/mL		50 m*M* KCl, 10 m*M* Tris-HCl, 0.1 m*M* EDTA, 1 m*M* DTT, 200 μg/mL acetylated BSA, 50% glycerol, pH 7.4		ICN 159391
2000 U/mL	1 unit completely digests 1 μg λ DNA (dam⁻)/hr in a total reaction volume of 50 μL using the buffer provided	20 m*M* Tris-acetate, 10 m*M* MgOAc, 50 m*M* KOAc, 1 m*M* DTT, pH 7.9	50% DNA fragments ligated and 90% recut after 2-fold overdigestion	NE Biolabs 513S 513L
1000-10,000 U/mL	1 unit completely cleaves 1.0 μg λ DNA/hr in a 50 μL reaction mixture	200 m*M* Tris-acetate, 500 m*M* KOAc, 100 m*M* MgOAc, 10 m*M* DTT, pH 7.9	100% cleavage, 50% ligation, 100% recleavage, 50% endonuclease (nickase) No detectable endonuclease (overdigestion)	Sigma R2885

Alw21 I

RECOGNITION SEQUENCE
5'...G(^A_T)GC(^A_T)▼C...3'
3'...C▲(^T_A)CG(^T_A)G...5'

ISOSCHIZOMERS
HgiA I

HEAT INACTIVATION
65°C for 20 min.

NOTES
- EC 3.1.21.4
- A type II site specific deoxyribonuclease
- A large group of enzymes recognizing specific short DNA sequences. They cleave either within the recognition site, or at a short specific distance from it
- Requires Mg^{2+} and BSA for full activity

CONCENTRATION or VOLUME ACTIVITY	UNIT DEFINITION	REACTION BUFFER (As Provided)	ADDITIONAL ACTIVITIES and PURITY	SUPPLIER CATALOG No.
Acinetobacter iwoffi				
5000 U/mL	1 unit degrades 1 μg λ dam⁻ dcm⁻ DNA/hr at 37°C in a 0.05 mL mixture	10 m*M* Tris-HCl, 10 m*M* MgCl₂, 100 m*M* KCl, 100 μg/mL BSA, pH 8.5	No detectable contaminating nuclease; 90% ligated and 100% recut after 50-fold overdigestion	ACS Heidelb F00350S F00350M
8-12 U/μL	1 unit hydrolyzes 1 μg DNA/hr 37°C in a 50 μL reaction volume using λ phage DNA as substrate	50 m*M* Tris-HCl, 10 m*M* MgCl₂, 100 m*M* NaCl, 0.1 mg/mL BSA, pH 7.5	>90% of the DNA fragments ligated and recut after 50-fold overdigestion	Fermentas ER0021 ER0022

Alw26 I

RECOGNITION SEQUENCE
5'...GTCTC(N)$_1$▼...3'
3'...CAGAG(N)$_5$▲...5'

ISOSCHIZOMERS
*Bsm*A I

HEAT INACTIVATION
65°C for 20 min.

METHYLATION EFFECTS
Does not cleave GTm^{5}CTC or GAGm^{6}AC
Blocked by overlapping CG methylation

NOTES
- EC 3.1.21.4
- A type II site specific deoxyribonuclease
- A large group of enzymes recognizing specific short DNA sequences. They cleave either within the recognition site, or at a short specific distance from it
- Large excess of enzyme results in the appearance of star activity
- Requires Mg^{2+} and BSA for full activity

CONCENTRATION or VOLUME ACTIVITY	UNIT DEFINITION	REACTION BUFFER (As Provided)	ADDITIONAL ACTIVITIES and PURITY	SUPPLIER CATALOG NO.
Acinetobacter iwoffi RFL26				
10,000 U/mL	1 unit degrades 1 μg λ dam$^-$ dcm$^-$ DNA/hr at 37°C in a 0.05 mL mixture	33 m*M* Tris-acetate, 10 m*M* MgOAc, 66 m*M* KOAc, 100 μg/mL BSA, pH 7.9	No detectable contaminating nuclease; 90% ligated and 100% recut after 50-fold overdigestion	AGS Heidelb F00050S F00050M
8-12 U/μL	1 unit hydrolyzes 1 μg DNA/hr at 37°C in a 50 μL reaction volume using λ phage DNA as substrate	33 m*M* Tris-acetate, 10 m*M* MgOAc, 66 m*M* KOAc, 0.1 mg/mL BSA, pH 7.9	>90% of the DNA fragments ligated and recut after 50-fold overdigestion	Fermentas ER0031 ER0032
				NE Biolabs
8-12 U/μL	1 unit completely digests 1 μg λ DNA/hr at 37°C in a 50 mL reaction volume	100 m*M* Tris-HCl, 100 m*M* MgCl$_2$, 500 m*M* NaCl, 10 m*M* DTT, pH 7.9	90% ligation	Promega R6761 R6762 R6765

Alw44 I

RECOGNITION SEQUENCE
  5'...G▼TGCAC...3'
  3'...CACGT▲G...5'

ISOSCHIZOMERS
  ApaL I, Sno I

HEAT INACTIVATION
  65°C for 20 min.

METHYLATION EFFECTS
  Does not cleave GTGm^{5}CAC
  Does cleave GTGCm^{6}AC
  Blocked by overlapping CG methylation

NOTES
- EC 3.1.21.4
- A type II site specific deoxyribonuclease
- A large group of enzymes recognizing specific short DNA sequences. They cleave either within the recognition site, or at a short specific distance from it
- Requires Mg^{2+} and BSA for full activity
- Compatible ends: ApaL I, Sno I

CONCENTRATION or VOLUME ACTIVITY	UNIT DEFINITION	REACTION BUFFER (As Provided)	ADDITIONAL ACTIVITIES and PURITY	SUPPLIER CATALOG No.
Acinetobacter lwoffi RFL44				
10 U/μL and 40 U/μL	1 unit completely digests 1 μg substrate DNA/hr	33 mM Tris-acetate, 10 mM MgOAc, 66 mM KOAc, 0.5 mM DTT, pH 7.9		Boehringer 1450506 1450514 1450573
8-12 U/μL	1 unit hydrolyzes 1 μg DNA/hr at 37°C in a 50 μL reaction volume using λ phage DNA as substrate	33 mM Tris-acetate, 10 mM MgOAc, 66 mM KOAc, 0.1 mg/mL BSA, pH 7.9	>90% of the DNA fragments ligated and recut after 50-fold overdigestion	Fermentas ER0041 ER0042
8000-12,000 U/mL		50 mM KCl, 10 mM Tris-HCl, 0.1 mM EDTA, 1 mM DTT, 200 μg/mL acetylated BSA, 50% glycerol, pH 7.5		ICN 159390
				Nippon Gene
8-12 U/μL	1 unit completely digests 1 μg λ DNA/hr at 37°C in a 50 mL reaction volume	100 mM Tris-HCl, 100 mM MgCl$_2$, 500 mM NaCl, 10 mM DTT, pH 7.9	90% ligation	Promega R6771 R6772
				Toyobo

AlwN I

RECOGNITION SEQUENCE
  5'...CAGNNN▼CTG...3'
  3'...GTC▲NNNGAC...5'
OPTIMUM REACTION CONDITIONS
  37°C
HEAT INACTIVATION
  65°C for 20 min.
METHYLATION EFFECTS
  Blocked by overlapping *dcm* methylation

NOTES
- EC 3.1.21.4
- A type II site specific deoxyribonuclease
- A large group of enzymes recognizing specific short DNA sequences. They cleave either within the recognition site, or at a short specific distance from it
- Requires Mg^{2+}

CONCENTRATION or VOLUME ACTIVITY	UNIT DEFINITION	REACTION BUFFER (As Provided)	ADDITIONAL ACTIVITIES and PURITY	SUPPLIER CATALOG No.
Acinetobacter lwoffii N				
3-15 U/μL	1 unit completely digests 1 μg λ DNA/hr at 37°C in the buffer provided	330 m*M* Tris-acetate, 100 m*M* MgOAc, 5 m*M* DTT, 660 m*M* KOAc, pH 7.9	>95% DNA fragments ligated and 95% recut	Amersham E 0202Y E 0202Z
5000-15,000 U/mL		50 m*M* NaCl, 10 m*M* Tris-HCl, 0.1 m*M* EDTA, 1 m*M* DTT, 100 μg/mL acetylated BSA, 50% glycerol, pH 7.4		ICN 159392
8-12 U/μL	1 unit completely digests 1 mg λ DNA/hr at 37°C	500 m*M* Tris-HCl, 100 m*M* MgCl₂, 500 m*M* NaCl, pH 8.0	≥95% ligation; 100% cleavage	Life Technol 15483-019
10,000 U/mL	1 unit completely digests 1 μg λ DNA/hr in a total reaction volume of 50 μL using the buffer provided	20 m*M* Tris-acetate, 10 m*M* MgOAc, 50 m*M* KOAc, 1 m*M* DTT, pH 7.9	>95% DNA fragments ligated and recut after 10-fold overdigestion	NE Biolabs 514S 514L
2000-20,000 U/mL	1 unit completely cleaves 1.0 μg λ DNA/hr in a 50 μL reaction mixture	200 m*M* Tris-acetate, 500 m*M* KOAc, 100 m*M* MgOAc, 10 m*M* DTT, pH 7.9	100% cleavage, 100% ligation, 90% recleavage, 100% endonuclease (overdigestion)	Sigma R3010

Ama87 I

RECOGNITION SEQUENCE
  5'...C▼YCGRG...3'
  3'...GRGCY▲C...5'

ISOSCHIZOMERS
  Ava I, Bco I, BsoB I, Eco88 I

NOTES
- EC 3.1.21.4
- A type II site specific deoxyribonuclease
- A large group of enzymes recognizing specific short DNA sequences. They cleave either within the recognition site, or at a short specific distance from it
- Requires Mg^{2+}

CONCENTRATION or VOLUME ACTIVITY	UNIT DEFINITION	REACTION BUFFER (As Provided)	ADDITIONAL ACTIVITIES and PURITY	SUPPLIER CATALOG No.
Alteromonas macleodii 87				
10,000 U/mL	1 unit degrades 1 μg λ dam⁻ dcm⁻ DNA/hr at 37°C in a 0.05 mL mixture	10 mM Tris-HCl, 10 mM MgCl₂, 100 mM KCl, 100 μg/mL BSA, pH 8.5	No detectable contaminating nuclease; 95% ligated and recut after 20-fold overdigestion	AGS Heidelb F00085S F00085M
				SibEnzyme

Aoc I

RECOGNITION SEQUENCE
  5'...CC▼TNAGG...3'
  3'...GGANT▲CC...5'

ISOSCHIZOMERS
  Axy I, Bsu36 I, Cvn I, Eco81 I, Mst II, Sau I

HEAT INACTIVATION
  65°C for 20 min.

NOTES
- EC 3.1.21.4
- A type II site specific deoxyribonuclease
- A large group of enzymes recognizing specific short DNA sequences. They cleave either within the recognition site, or at a short specific distance from it
- Requires Mg^{2+}

CONCENTRATION or VOLUME ACTIVITY	UNIT DEFINITION	REACTION BUFFER (As Provided)	ADDITIONAL ACTIVITIES and PURITY	SUPPLIER CATALOG No.
Anabaena oscillarioides				
10 U/μL	1 unit completely digests 1 μg substrate DNA/hr	10 mM Tris-HCl, 10 mM MgCl₂, 1 mM DTE, pH 7.5		Boehringer 1581023, 1581031

Aor51H I

RECOGNITION SEQUENCE
5'...AGC▼GCT...3'
3'...TCG▲CGA...5'

ISOSCHIZOMERS
Eco47 III

OPTIMUM REACTION CONDITIONS
37°C

HEAT INACTIVATION
60°C for 15 min.

NOTES
- EC 3.1.21.4
- A type II site specific deoxyribonuclease
- A large group of enzymes recognizing specific short DNA sequences. They cleave either within the recognition site, or at a short specific distance from it
- Requires Mg^{2+}

CONCENTRATION or VOLUME ACTIVITY	UNIT DEFINITION	REACTION BUFFER (As Provided)	ADDITIONAL ACTIVITIES and PURITY	SUPPLIER CATALOG No.
Acidiphilium organovorum 51H				
200 U and 1000 U	1 unit completely cleaves 1 μg DNA/hr in 50 μL reaction mixture	100 mM Tris-HCl, 100 mM $MgCl_2$, 10 mM DTT, 500 mM NaCl, pH 7.5		PanVera TAK 1118
400 U and 2000 U	1 unit completely digests 1 μg λ DNA/hr at 37°C in 50 μL of supplied buffer	100 mM Tris-HCl, 100 mM $MgCl_2$, 10 mM DTT, 500 mM NaCl, pH 7.5		TaKaRa 1118
Acidophilum organotrophum 51H				
8-12 U/μL	1 unit completely digests 1 μg λ DNA/hr at 37°C in the buffer provided	100 mM Tris-HCl, 100 mM $MgCl_2$, 10 mM DTT, 500 mM NaCl, pH 7.5	90% DNA fragments ligated and 95% recut	Amersham E 1118Y

Apa I

RECOGNITION SEQUENCE
5'...GGGCC▼C...3'
3'...C▲CCGGG...5'

ISOSCHIZOMERS
*Bsp*120 I

OPTIMUM REACTION CONDITIONS
25°C; 33% activity loss at 37°C

HEAT INACTIVATION
65°C for 20 min.

METHYLATION EFFECTS
Does not cleave GGGm^{5}CCC or GGGCCm^{5}C
Blocked by overlapping *dcm* methylation

NOTES
- EC 3.1.21.4
- A type II site specific deoxyribonuclease
- A large group of enzymes recognizing specific short DNA sequences. They cleave either within the recognition site, or at a short specific distance from it
- Yields a 3' extension
- Inhibited by high salt concentrations (>50-80 m*M*)
- Requires Mg^{2+}, KCl and BSA for full activity

CONCENTRATION or VOLUME ACTIVITY	UNIT DEFINITION	REACTION BUFFER (As Provided)	ADDITIONAL ACTIVITIES and PURITY	SUPPLIER CATALOG NO.
Acetobacter pasteurianus				
10 U/μL and 40 U/μL	1 unit completely digests 1 μg substrate DNA/hr	33 m*M* Tris-acetate, 10 m*M* MgOAc, 66 m*M* KOAc, 0.5 m*M* DTT, pH 7.9		Boehringer 703745 703753 899208
8-12 U/μL	1 unit completely digests 1 mg Ad2 DNA/hr at 30°C	200 m*M* Tris-HCl, 50 m*M* MgCl$_2$, 500 *M* KCl, pH 7.4	≥95% ligation; 100% cleavage	Life Technol 15440-019 15440-035
10-20 U/μL	1 unit digests completely 1 μg Ad2 DNA substrate/hr at 37°C	6 m*M* Tris-HCl, 6 m*M* NaCl, 6 m*M* MgCl$_2$, 6 m*M* β-MSH, 100 μg/mL BSA, pH 7.5		NBL Gene 015706 015708
10 U/μL and 15-30 U/μL	1 unit hydrolyzes 1 μg λ DNA to completion/hr in a 50 μL total reaction volume	100 m*M* Tris-HCl, 100 m*M* NaCl, 100 m*M* MgCl$_2$, 100 m*M* β-MSH, 1.0 mg/mL BSA, pH 7.5		Nippon Gene Oncor 110033 110034
10,000-25,000 U/mL	1 unit completely digests 1 μg Ad2 DNA/hr at pH 7.5, 37°C in a 50 μL total assay mixture	100 m*M* Tris-acetate, 100 m*M* MgOAc, 500 m*M* KOAc, pH 7.5		Pharmacia 27-0949-01 27-0949-02
10 U/μL and 40-80 U/μL	1 unit completely digests 1 μg Ad2 DNA/hr at 37°C in a 50 mL reaction volume	Blue/White Cloning Qualified; 60 m*M* Tris-HCl, 60 m*M* MgCl$_2$, 60 m*M* NaCl, 10 m*M* DTT, pH 7.5	90% ligation	Promega R6361 R6362 R4364

Apa I continued

CONCENTRATION or VOLUME ACTIVITY	UNIT DEFINITION	REACTION BUFFER (As Provided)	ADDITIONAL ACTIVITIES and PURITY	SUPPLIER CATALOG No.
Acetobacter pasteurianus continued				SibEnzyme
5000-40,000 U/mL	1 unit completely cleaves 1.0 µg Ad2 DNA/hr in a 50 µL reaction mixture	200 mM Tris-acetate, 500 mM KOAc, 100 mM MgOAc, 10 mM DTT, pH 7.9	100% cleavage, 90% ligation, 90% recleavage, 50% endonuclease (nickase) No detectable endonuclease (overdigestion)	Sigma R4258
5000-20,000 U/mL		100 mM NaCl, 20 mM Tris-HCl, 0.1 mM EDTA, 10 mM β-MSH, 0.01% Triton X-100, 200 µg/mL BSA, 50% glycerol, pH 7.4		Stratagene 500130
10,000 U and 50,000 U	1 unit completely digests 1 µg λ DNA/hr at 37°C in 50 µL of supplied buffer; 1 unit for Ad2 = 3 units for λ DNA	100 mM Tris-HCl, 100 mM MgCl$_2$, 10 mM DTT, pH 7.5		TaKaRa 1005
				Toyobo
Acetobacter pasteurianus sub. pasteurianus				
8-12 U/µL	1 unit completely digests 1 µg λ DNA/hr at 37°C in the buffer provided	100 mM Tris-HCl, 100 mM MgCl$_2$, 10 mM DTT, pH 7.5	>95% DNA fragments ligated and 100% recut	Amersham E 1005Y
	1 unit completely cleaves 1 µg Ad2 DNA/hr at pH 7.5, 25°C in a reaction volume of 50 µL	10 mM Tris-HCl, 10 mM MgCl$_2$, 1.0 mM DTT, 100 µg/mL BSA, pH 7.5		CHIMERx 2015-01 2015-02
20,000-60,000 U/mL		50 mM KCl, 10 mM Tris-HCl, 0.1 mM EDTA, 1 mM DTT, 200 µg/mL acetylated BSA, 50% glycerol, pH 7.4		ICN 153799
Escherichia coli strain, carrying the cloned gene from Acetobacter pasteurianus sub. pasteurianus				
20,000 U/mL	1 unit completely digests 1 µg Ad2 DNA/hr in a total reaction volume of 50 µL using the buffer provided	20 mM Tris-acetate, 10 mM MgOAc, 50 mM KOAc, 1 mM DTT, pH 7.9; add 100 µg/mL BSA	>95% DNA fragments ligated and recut after 20-fold overdigestion	NE Biolabs 114S 114L

ApaL I

RECOGNITION SEQUENCE
5'...G▼TGCAC...3'
3'...CACGT▲G...5'

ISOSCHIZOMERS
*Alw*44 I, *Sno* I

OPTIMUM REACTION CONDITIONS
37°C

HEAT INACTIVATION
Resistant

METHYLATION EFFECTS
Cleaves GTGCm^{6}AC
Does not cleave GTGCAm^{5}C

NOTES
- EC 3.1.21.4
- A type II site specific deoxyribonuclease
- A large group of enzymes recognizing specific short DNA sequences. They cleave either within the recognition site, or at a short specific distance from it
- Reports of a single site in M13 DNA have been attributed to a sequencing error
- Requires Mg^{2+} and BSA is required for full activity
- Sensitive to [NaCl] >50 mM

CONCENTRATION or VOLUME ACTIVITY	UNIT DEFINITION	REACTION BUFFER (As Provided)	ADDITIONAL ACTIVITIES and PURITY	SUPPLIER CATALOG NO.
Acetobacter pasteurianus				
8-12 U/μL	1 unit completely digests 1 μg λ DNA/hr at 37°C in the buffer provided	100 mM Tris-HCl, 100 mM MgCl$_2$, 10 mM DTT, pH 7.5	>95% DNA fragments ligated and 100% recut	Amersham E 1006Y
10,000 U/mL	1 unit degrades 1 μg λ dam$^-$ dcm$^-$ DNA/hr at 37°C in a 0.05 mL mixture	10 mM Tris-HCl, 10 mM MgCl$_2$, 100 μg/mL BSA, pH 7.5	No detectable contaminating nuclease; 90% ligated and 100% recut after 50-fold overdigestion	AGS Heidelb F00070S F00070M
5000-20,000 U/mL		50 mM KCl, 10 mM Tris-HCl, 0.1 mM EDTA, 1 mM DTT, 200 μg/mL acetylated BSA, 50% glycerol, pH 7.4		ICN 153800
10,000 U/mL	1 unit completely digests 1 μg λ DNA (*Hind* III digest)/hr in a total reaction volume of 50 μL using the buffer provided	20 mM Tris-acetate, 10 mM MgOAc, 50 mM KOAc, 1 mM DTT, pH 7.9; add 100 μg/mL BSA	>95% DNA fragments ligated and recut after 100-fold overdigestion	NE Biolabs 507S 507L
2000-25,000 U/mL	1 unit completely cleaves 1.0 μg λ DNA/hr in a 50 μL reaction mixture	200 mM Tris-acetate, 500 mM KOAC, 100 mM MgOAC, 10 mM DTT, pH 7.9	100% cleavage, 90% ligation, 90% recleavage, 50% endonuclease (nickase) No detectable endonuclease (overdigestion)	Sigma R3135

CONCENTRATION or VOLUME ACTIVITY	UNIT DEFINITION	REACTION BUFFER (As Provided)	ADDITIONAL ACTIVITIES and PURITY	SUPPLIER CATALOG No.
***Acetobacter pasteurianus* continued**				
5000-15,000 U/mL		50 mM KCl, 10 mM Tris-HCl, 0.1 mM EDTA, 1 mM DTT, 200 µg/mL BSA, 50% glycerol, pH 7.5		Stratagene 500140
400 U and 2000 U	1 unit completely digests 1 µg λ DNA/hr at 37°C in 50 µL of supplied buffer	100 mM Tris-HCl, 100 mM MgCl$_2$, 10 mM DTT, pH 7.5		TaKaRa 1006

Apo I

RECOGNITION SEQUENCE
   5'...Pu▼AATTPy...3'
   3'...PyTTAA▲Pu...5'
OPTIMUM REACTION CONDITIONS
   50°C; 30% activity at 37°C
HEAT INACTIVATION
   Resistant

NOTES
- EC 3.1.21.4
- A type II site specific deoxyribonuclease
- A large group of enzymes recognizing specific short DNA sequences. They cleave either within the recognition site, or at a short specific distance from it
- Cleaves to leave a 5' AATT extension which can be ligated to DNA fragments generated by *Eco*R I digestion
- Conditions of low ionic strength, high enzyme concentration, glycerol concentration >5% or pH >8.0 may result in star activity
- Requires Mg^{2+} and BSA is required for full activity

Apo I continued

CONCENTRATION or VOLUME ACTIVITY	UNIT DEFINITION	REACTION BUFFER (As Provided)	ADDITIONAL ACTIVITIES and PURITY	SUPPLIER CATALOG No.
Acetobacter protophormiae				
2000-10,000 U/mL		100 mM NaCl, 10 mM Tris-HCl, 0.1 mM EDTA, 1 mM DTT, 200 µg/mL acetylated BSA, 50% glycerol, pH 7.4		ICN 159393
Escherichia coli strain, carrying the cloned gene from Arthrobacter protophormiae (NEB 760)				
4000 U/mL	1 unit completely digests 1 µg λ DNA/hr in a total reaction volume of 50 µL using the buffer provided	50 mM Tris-HCl, 10 mM MgCl$_2$, 100 mM NaCl, 1 mM DTT, pH 7.9; add 100 µg/mL BSA	>95% DNA fragments ligated and recut after 20-fold overdigestion	NE Biolabs 566S 566L

Asc I

RECOGNITION SEQUENCE
    5'...GG▼CGCGCC...3'
    3'...CCGCGC▲GG...5'
OPTIMUM REACTION CONDITIONS
    37°C
HEAT INACTIVATION
    65°C for 20 min.

NOTES
- EC 3.1.21.4
- A type II site specific deoxyribonuclease
- A large group of enzymes recognizing specific short DNA sequences. They cleave either within the recognition site, or at a short specific distance from it
- Recognizes octanucleotides
- Requires Mg^{2+}

CONCENTRATION or VOLUME ACTIVITY	UNIT DEFINITION	REACTION BUFFER (As Provided)	ADDITIONAL ACTIVITIES and PURITY	SUPPLIER CATALOG No.
Acetobacter species				
1000-5000 U/mL		50 mM KCl, 10 mM Tris-HCl, 0.1 mM EDTA, 1 mM DTT, 200 µg/mL acetylated BSA, 50% glycerol, pH 7.4		ICN 159394
Escherichia coli strain, carrying the cloned gene from Arthrobacter species (NEB 688)				
10,000 U/mL	1 unit completely digests 1 µg λ DNA/hr in a total reaction volume of 50 µL using the buffer provided	20 mM Tris-acetate, 10 mM MgOAc, 50 mM KOAc, 1 mM DTT, pH 7.9	>95% DNA fragments ligated and recut after 5-fold overdigestion	NE Biolabs 558S 558L

Ase I

RECOGNITION SEQUENCE
  5'...AT▼TAAT...3'
  3'...TAAT▲TA...5'
OPTIMUM REACTION CONDITIONS
  37°C
HEAT INACTIVATION
  65°C for 20 min.
METHYLATION EFFECTS
  Cleaves ATTm^{6}AAT

NOTES
- EC 3.1.21.4
- A type II site specific deoxyribonuclease
- A large group of enzymes recognizing specific short DNA sequences. They cleave either within the recognition site, or at a short specific distance from it
- Cleaves pBR322 and adenovirus-2 DNA at a rate 3 times that of λ DNA
- Conditions of low ionic strength, high enzyme concentration, glycerol concentration >5% or pH >8.0 may result in star activity
- Requires Mg^{2+}

CONCENTRATION or VOLUME ACTIVITY	UNIT DEFINITION	REACTION BUFFER (As Provided)	ADDITIONAL ACTIVITIES and PURITY	SUPPLIER CATALOG No.
Aquaspirillum serpens				
10 U/μL	1 unit completely digests 1 μg λ DNA/hr at 37°C in the buffer provided	500 mM Tris-HCl, 100 mM MgCl$_2$, 10 mM DTT, 1000 mM NaCl, pH 7.5	>95% DNA fragments ligated and 95% recut	Amersham E 0203Y E 0203Z
5000-50,000 U/mL		500 mM KCl, 10 mM Tris-HCl, 0.1 mM EDTA, 1 mM DTT, 200 μg/mL acetylated BSA, 50% glycerol, pH 7.4		ICN 159395
				Nippon Gene
				Pharmacia
				Toyobo
Escherichia coli strain, carrying the cloned gene from *Aquaspirillum serpens*				
10,000 U/mL; 50,000 U/mL	1 unit completely digests 1 μg λ DNA/hr in a total reaction volume of 50 μL using the buffer provided	50 mM Tris-HCl, 10 mM MgCl$_2$, 100 mM NaCl, 1 mM DTT, pH 7.9	95% DNA fragments ligated and recut after 10-fold overdigestion	NE Biolabs 526S 526L

AsiA I

Recognition Sequence
5'...A▼CCGGT...3'
3'...TGGCC▲A...5'

Isoschizomers
Age I, PinA I

Notes
- EC 3.1.21.4
- A type II site specific deoxyribonuclease
- A large group of enzymes recognizing specific short DNA sequences. They cleave either within the recognition site, or at a short specific distance from it
- Requires Mg^{2+}

CONCENTRATION or VOLUME ACTIVITY	UNIT DEFINITION	REACTION BUFFER (As Provided)	ADDITIONAL ACTIVITIES and PURITY	SUPPLIER CATALOG No.
Arthrobacter species A7359				
				SibEnzyme

Asn I

Recognition Sequence
5'...AT▼TAAT...3'
3'...TAAT▲TA...5'

Heat Inactivation
65°C for 20 min.

Notes
- EC 3.1.21.4
- A type II site specific deoxyribonuclease
- A large group of enzymes recognizing specific short DNA sequences. They cleave either within the recognition site, or at a short specific distance from it
- Requires Mg^{2+}

CONCENTRATION or VOLUME ACTIVITY	UNIT DEFINITION	REACTION BUFFER (As Provided)	ADDITIONAL ACTIVITIES and PURITY	SUPPLIER CATALOG No.
Arthrobacter species NCM				
10 U/μL	1 unit completely digests 1 μg substrate DNA/hr	10 mM Tris-HCl, 5 mM MgCl$_2$, 100 mM NaCl, 1 mM β-MSH, pH 8.0		Boehringer 1097008 1096982

Asn I continued

CONCENTRATION or VOLUME ACTIVITY	UNIT DEFINITION	REACTION BUFFER (As Provided)	ADDITIONAL ACTIVITIES and PURITY	SUPPLIER CATALOG NO.
Arthrobacter species NCM *continued*				
1000-10,000 U/mL	1 unit completely cleaves 1.0 μg λ DNA/hr in a 50 μL reaction mixture	500 m*M* Tris-HCl, 1 *M* NaCl, 100 m*M* MgCl$_2$, 10 m*M* DTT, pH 7.9	100% cleavage, 90% ligation, 100% recleavage No detectable endonuclease (overdigestion)	Sigma R1634

Asp I

RECOGNITION SEQUENCE
    5'...GACN▼NNGTC...3'
    3'...CTGNN▲NCAG...5'

ISOSCHIZOMERS
    *Tth*111 I

HEAT INACTIVATION
    65°C for 20 min.

NOTES
- EC 3.1.21.4
- A type II site specific deoxyribonuclease
- A large group of enzymes recognizing specific short DNA sequences. They cleave either within the recognition site, or at a short specific distance from it
- Requires Mg^{2+} and BSA is required for full activity

CONCENTRATION or VOLUME ACTIVITY	UNIT DEFINITION	REACTION BUFFER (As Provided)	ADDITIONAL ACTIVITIES and PURITY	SUPPLIER CATALOG NO.
Achromobacter species				
10 U/μL and 40 U/μL	1 unit completely digests 1 μg substrate DNA/hr	10 m*M* Tris-HCl, 5 m*M* MgCl$_2$, 100 m*M* NaCl, 1 m*M* β-MSH, pH 8.0		Boehringer 1131354 1131362 1207687
8000-12,000 U/mL	1 unit completely digests 1 μg λ DNA/hr at pH 8.0, 37°C in a 50 μL total assay mixture	10 m*M* Tris-HCl, 100 m*M* NaCl, 5 m*M* MgCl$_2$, 1 m*M* β-MSH, pH 8.0		Pharmacia 27-0839-01

Asp700 I

RECOGNITION SEQUENCE
5'...GAANN▼NNTTC...3'
3'...CTTNN▲NNAAG...5'

ISOSCHIZOMERS
Xmn I

HEAT INACTIVATION
65°C for 20 min.

METHYLATION EFFECTS
Cleaves $GAm^6A(N)_4TTC$ and $GAA(N)_4TTm^5C$
Does not cleave $Gm^6AA(N)4TTC$

NOTES
- EC 3.1.21.4
- A type II site specific deoxyribonuclease
- A large group of enzymes recognizing specific short DNA sequences. They cleave either within the recognition site, or at a short specific distance from it
- Requires Mg^{2+}

CONCENTRATION or VOLUME ACTIVITY	UNIT DEFINITION	REACTION BUFFER (As Provided)	ADDITIONAL ACTIVITIES and PURITY	SUPPLIER CATALOG No.
Acinetobacter species 700 (Achromobacter species 700)				
10 U/μL and 40 U/μL	1 unit completely digests 1 μg substrate DNA/hr	10 mM Tris-HCl, 5 mM MgCl₂, 100 mM NaCl, 1 mM β-MSH, pH 8.0		Boehringer 835277 835285 1097032

Asp718 I

RECOGNITION SEQUENCE
5'...G▼GTACC...3'
3'...CCATG▲G...5'

HEAT INACTIVATION
65°C for 20 min.

METHYLATION EFFECTS
Cleaves $GGTm^6Am^5CC$
Does not cleave $GGTACm^5C$ or $GGTAm^5Cm^5C$

NOTES
- EC 3.1.21.4
- A type II site specific deoxyribonuclease
- A large group of enzymes recognizing specific short DNA sequences. They cleave either within the recognition site, or at a short specific distance from it
- Requires Mg^{2+}

Asp718 I continued

CONCENTRATION or VOLUME ACTIVITY	UNIT DEFINITION	REACTION BUFFER (As Provided)	ADDITIONAL ACTIVITIES and PURITY	SUPPLIER CATALOG NO.
Acinetobacter species 718 (Achromobacter species 718)				
10 U/μL and 40 U/μL	1 unit completely digests 1 μg substrate DNA/hr	10 mM Tris-HCl, 5 mM MgCl$_2$, 100 mM NaCl, 1 mM β-MSH, pH 8.0		Boehringer 814245 814253 1175050
				Pharmacia

AspE I

RECOGNITION SEQUENCE
  5'...GACNNN▼NNGTC...3'
  3'...CTGNN▲NNNCAG...5'

HEAT INACTIVATION
  65°C for 20 min.

NOTES
- EC 3.1.21.4
- A type II site specific deoxyribonuclease
- A large group of enzymes recognizing specific short DNA sequences. They cleave either within the recognition site, or at a short specific distance from it
- Requires Mg^{2+}

CONCENTRATION or VOLUME ACTIVITY	UNIT DEFINITION	REACTION BUFFER (As Provided)	ADDITIONAL ACTIVITIES and PURITY	SUPPLIER CATALOG NO.
Auerobacterium species				
10 U/μL	1 unit completely digests 1 μg substrate DNA/hr	10 mM Tris-HCl, 10 mM MgCl$_2$, 1 mM DTE, pH 7.5		Boehringer 1428179

AspH I

RECOGNITION SEQUENCE
5'...G(^A_T)GC(^T_A)▼C...3'
3'...C▲(^T_A)CG(^A_T)G...5'

ISOSCHIZOMERS
  HgiA I

HEAT INACTIVATION
  65°C for 20 min.

NOTES
- EC 3.1.21.4
- A type II site specific deoxyribonuclease
- A large group of enzymes recognizing specific short DNA sequences. They cleave either within the recognition site, or at a short specific distance from it
- Requires Mg^{2+} and high [salt] for optimal activity

CONCENTRATION or VOLUME ACTIVITY	UNIT DEFINITION	REACTION BUFFER (As Provided)	ADDITIONAL ACTIVITIES and PURITY	SUPPLIER CATALOG NO.
Achromobacter species 773				
5 U/μL	1 unit completely digests 1 μg substrate DNA/hr	10 mM Tris-HCl, 5 mM MgCl₂, 100 mM NaCl, 1 mM β-MSH, pH 8.0		Boehringer 1081322 1081349
Herpetosiphon giganteus				
2000-10,000 U/mL		50 mM KCl, 10 mM Tris-HCl, 0.1 mM EDTA, 1 mM DTT, 200 μg/mL acetylated BSA, 50% glycerol, pH 7.4		ICN 153814

AspLE I

RECOGNITION SEQUENCE
5'...GCG▼C...3'
3'...C▲GCG...5'

ISOSCHIZOMERS
  Cfo I, Hha I, HinP1 I, Hin6 I, HspA I

NOTES
- EC 3.1.21.4
- A type II site specific deoxyribonuclease
- A large group of enzymes recognizing specific short DNA sequences. They cleave either within the recognition site, or at a short specific distance from it
- Requires Mg^{2+}

AspLE I *continued*

CONCENTRATION or VOLUME ACTIVITY	UNIT DEFINITION	REACTION BUFFER (As Provided)	ADDITIONAL ACTIVITIES and PURITY	SUPPLIER CATALOG NO.
Arthrobacter species LE3860				
				SibEnzyme

AspS9 I

RECOGNITION SEQUENCE
    5'...G▼GNCC...3'
    3'...CCNG▲G...5'

ISOSCHIZOMERS
    *Bsi*Z I, *Cfr*13 I, *Sau*96 I,

NOTES
- EC 3.1.21.4
- A type II site specific deoxyribonuclease
- A large group of enzymes recognizing specific short DNA sequences. They cleave either within the recognition site, or at a short specific distance from it
- Requires Mg^{2+}

CONCENTRATION or VOLUME ACTIVITY	UNIT DEFINITION	REACTION BUFFER (As Provided)	ADDITIONAL ACTIVITIES and PURITY	SUPPLIER CATALOG NO.
Arthrobacter species S9				
				SibEnzyme

Asu I

RECOGNITION SEQUENCE
  5'...G▼GNCC...3'
  3'...CCNG▲G...5'

ISOSCHIZOMERS
  AspS9 I, BsiZ I, Cfr13 I, Sau96 I

NOTES
- EC 3.1.21.4
- A type II site specific deoxyribonuclease
- A large group of enzymes recognizing specific short DNA sequences. They cleave either within the recognition site, or at a short specific distance from it
- Require Mg^{2+}

CONCENTRATION or VOLUME ACTIVITY	UNIT DEFINITION	REACTION BUFFER (As Provided)	ADDITIONAL ACTIVITIES and PURITY	SUPPLIER CATALOG No.
Anabaena subcylindrica				
				Sigma

AsuHP I

RECOGNITION SEQUENCE
  5'...GGTGA(8/7)...3'

ISOSCHIZOMERS
  Hph I

NOTES
- EC 3.1.21.4
- A type II site specific deoxyribonuclease
- A large group of enzymes recognizing specific short DNA sequences. They cleave either within the recognition site, or at a short specific distance from it
- Requires Mg^{2+}

CONCENTRATION or VOLUME ACTIVITY	UNIT DEFINITION	REACTION BUFFER (As Provided)	ADDITIONAL ACTIVITIES and PURITY	SUPPLIER CATALOG No.
Actinobacillus suis HP				
				SibEnzyme

AsuNH I

RECOGNITION SEQUENCE
5'...G▼CTAGC...3'
3'...CGATC▲G...5'

ISOSCHIZOMERS
Nhe I

NOTES
- EC 3.1.21.4
- A type II site specific deoxyribonuclease
- A large group of enzymes recognizing specific short DNA sequences. They cleave either within the recognition site, or at a short specific distance from it
- Requires Mg^{2+}

CONCENTRATION or VOLUME ACTIVITY	UNIT DEFINITION	REACTION BUFFER (As Provided)	ADDITIONAL ACTIVITIES and PURITY	SUPPLIER CATALOG NO.
Actinobacillus suis NH				
				SibEnzyme

Ats I

RECOGNITION SEQUENCE
5'...GACN▼NNGTC...3'
3'...CTGNN▲NCAG...5'

ISOSCHIZOMERS
Asp I, PflF I, Tth111 I

NOTES
- EC 3.1.21.4
- A type II site specific deoxyribonuclease
- A large group of enzymes recognizing specific short DNA sequences. They cleave either within the recognition site, or at a short specific distance from it
- Requires Mg^{2+}

CONCENTRATION or VOLUME ACTIVITY	UNIT DEFINITION	REACTION BUFFER (As Provided)	ADDITIONAL ACTIVITIES and PURITY	SUPPLIER CATALOG NO.
Aureobacterium testaceum 4842				
				SibEnzyme

Ava I

RECOGNITION SEQUENCE
  5'...C▼PyCGPuG...3'
  3'...GPuGCPy▲C...5'

ISOSCHIZOMERS
  Bco I, Eco88 I, Nsp III

OPTIMUM REACTION CONDITIONS
  pH 7.5 and 37°C; 2-fold greater activity at 45°C

HEAT INACTIVATION
  65°C for 20 min.; 70°C for 15 min.

METHYLATION EFFECTS
  Cleaves Cm^{5}CCGGG
  Does not cleave m^{5}C(^C_T)CG(^A_G)G, C(^C_T)m^{5}CG(^A_G)G or CTCGm^{6}AG

NOTES
- EC 3.1.21.4
- A type II site specific deoxyribonuclease
- A large group of enzymes recognizing specific short DNA sequences. They cleave either within the recognition site, or at a short specific distance from it
- Requires Mg^{2+}
- Compatible cohesive ends: Acc III, Age I, Bca77 I, BsaW I, BseA I, BsiM I, BspE I, BsrF I, Bsu23 I, Cfr10 I, Kpn2 I, Mro I, NgoA IV, NgoM I, PinA I, Sal I, SgrA I
- 75% glycerol alters specificity
- Optimum KCl or NaCl concentration is 60-80 mM

CONCENTRATION or VOLUME ACTIVITY	UNIT DEFINITION	REACTION BUFFER (As Provided)	ADDITIONAL ACTIVITIES and PURITY	SUPPLIER CATALOG No.
Anabaena variabilis				
8-12 U/μL	1 unit completely digests 1 μg λ DNA/hr at 37°C in the buffer provided	100 mM Tris-HCl, 100 mM MgCl₂, 10 mM DTT, 500 mM NaCl, pH 7.5	>90% DNA fragments ligated and 100% recut	Amersham E 1007Y
5 U/μL	1 unit completely digests 1 μg substrate DNA/hr	10 mM Tris-HCl, 5 mM MgCl₂, 100 mM NaCl, 1 mM β-MSH, pH 8.0		Boehringer 740721 740730
	1 unit completely cleaves 1 μg λ DNA/hr at pH 7.5, 37°C in a reaction volume of 50 μL	10 mM Tris-HCl, 10 mM MgCl₂, 50 mM NaCl, 1.0 mM DTT, 100 μg/mL BSA, pH 7.5		CHIMERx 2020-01 2020-02
2000-10,000 U/mL	1 unit completely digests 1 μg λ DNA/hr at 37°C in a 0.05 mL total volume	50 mM KCl, 10 mM Tris-HCl, 0.1 mM EDTA, 1 mM DTT, 200 μg/mL acetylated BSA, 50% glycerol, pH 7.4		ICN 197021
2-8 U/μL	1 unit completely digests 1 mg λ DNA/hr at 37°C	500 mM Tris-HCl, 100 mM MgCl₂, 500 mM NaCl, pH 8.0	≥95% ligation; 100% cleavage	Life Technol 25220-013 25220-054
4-12 U/μL	1 unit digests completely 1 μg λ DNA substrate/hr at 37°C	10 mM Tris-HCl, 100 mM NaCl, 5 mM MgCl₂, 1 mM β-MSH, pH 8.3		NBL Gene 010701 010703
				Nippon Gene

Ava I continued

CONCENTRATION or VOLUME ACTIVITY	UNIT DEFINITION	REACTION BUFFER (As Provided)	ADDITIONAL ACTIVITIES and PURITY	SUPPLIER CATALOG No.
Anabaena variabilis continued				
2.5 U/μL and 5-15 U/μL	1 unit hydrolyzes 1 μg λ DNA to completion/hr in a 50 μL total reaction volume	330 mM Tris-acetate, 660 mM KOAc, 100 mM MgOAc, 5 mM DTT, 1 mg/mL BSA, pH 8		Oncor 110041 110042
2000-10,000 U/mL	1 unit completely digests 1 μg λ DNA/hr at pH 7.5, 37°C in a 50 μL total assay mixture	100 mM Tris-acetate, 100 mM MgOAc, 500 mM KOAc, pH 7.5		Pharmacia 27-0962-01 27-0962-02
8-12 U/μL	1 unit completely digests 1 μg λ DNA/hr at 37°C in a 50 mL reaction volume	Blue/White Cloning Qualified; 60 mM Tris-HCl, 60 mM MgCl$_2$, 500 mM NaCl, 10 mM DTT, pH 7.5	90% ligation	Promega R6091 R6092 R6095
5000-20,000 U/mL	1 unit completely cleaves 1.0 μg λ DNA/hr in a 50 μL reaction mixture	200 mM Tris-acetate, 500 mM KOAc, 100 mM MgOAc, 10 mM DTT, pH 7.9	100% cleavage, 98% ligation, 95% recleavage, 50% endonuclease (nickase) No detectable endonuclease (overdigestion)	Sigma R3379
2000-10,000 U/mL		50 mM KCl, 10 mM Tris-HCl, 0.1 mM EDTA, 1 mM DTT, 200 μg/mL BSA, 50% glycerol, pH 7.4		Stratagene 500170
500 U and 2500 U	1 unit completely digests 1 μg λ DNA/hr at 37°C in 50 μL of supplied buffer	100 mM Tris-HCl, 100 mM MgCl$_2$, 10 mM DTT, 500 mM NaCl, pH 7.5	Purity is confirmed by overdigestion and ligation-recutting test	TaKaRa 1007
3-20 U/μL	1 unit completely digests 1 μg λ DNA/hr in 0.05 mL at 37°C	10 mM Tris-HCl, 10 mM MgCl$_2$, 50 mM NaCl, 1 mM DTT, pH 7.5	>90% ligation >95% recutting	Toyobo AVA-101 AVA-102
Escherichia coli strain, carrying the cloned gene from Anabaena variabilis				
10,000 U/mL; 50,000 U/mL	1 unit completely digests 1 μg λ DNA/hr in a total reaction volume of 50 μL using the buffer provided	20 mM Tris-acetate, 10 mM MgOAc, 50 mM KOAc, 1 mM DTT, pH 7.9	>95% DNA fragments ligated and recut after 100-fold overdigestion	NE Biolabs 152S 152L

Ava II

RECOGNITION SEQUENCE
5'...GG▼(^A_T)CC...3'
3'...CC(^T_A)G▲G...5'

ISOSCHIZOMERS
Eco47 I, NspH II, Sin I

OPTIMUM REACTION CONDITIONS
pH 7.4-7.9 and 37°C

HEAT INACTIVATION
65°C for 20 min; 70°C for 15 min

METHYLATION EFFECTS
Cleaves GG(^A_T)Cm^{4}C
Does not cleave GG(^A_T)m^{5}CC, GG(^A_T)Cm^{5}C or GG(^A_T)hm^{5}Chm^{5}C
Blocked by overlapping *dcm* methylation

NOTES
- EC 3.1.21.4
- A type II site specific deoxyribonuclease
- A large group of enzymes recognizing specific short DNA sequences. They cleave either within the recognition site, or at a short specific distance from it
- Requires Mg^{2+}
- Compatible cohesive ends: *Ppu*M I, *Rst* II

CONCENTRATION or VOLUME ACTIVITY	UNIT DEFINITION	REACTION BUFFER (As Provided)	ADDITIONAL ACTIVITIES and PURITY	SUPPLIER CATALOG No.
Anabaena variabilis				
8-12 U/μL	1 unit completely digests 1 μg λ DNA/hr at 37°C in the buffer provided	100 m*M* Tris-HCl, 100 m*M* MgCl$_2$, 10 m*M* DTT, 500 m*M* NaCl, pH 7.5	>95% DNA fragments ligated and 100% recut	Amersham E 1008Y
5 U/μL	1 unit completely digests 1 μg substrate DNA/hr	33 m*M* Tris-acetate, 10 m*M* MgOAc, 66 m*M* KOAc, 0.5 m*M* DTT, pH 7.9		Boehringer 740748 740756
	1 unit completely cleaves 1 μg λ DNA/hr at pH 7.5, 37°C in a reaction volume of 50 μL	10 m*M* Tris-HCl, 10 m*M* MgCl$_2$, 50 m*M* NaCl, 1.0 m*M* DTT, 100 μg/mL BSA, pH 7.5		CHIMERX 2030-01 2030-02
2000-20,000 U/mL	1 unit completely digests 1 μg λ DNA/hr at 37°C in a 0.05 mL total volume	50 m*M* KCl, 10 m*M* Tris-HCl, 0.1 m*M* EDTA, 1 m*M* DTT, 200 μg/mL acetylated BSA, 50% glycerol, pH 7.4		ICN 197011
4-6 U/μL	1 unit completely digests 1 mg λ DNA/hr at 37°C	500 m*M* Tris-HCl, 100 m*M* MgCl$_2$, 500 m*M* NaCl, pH 8.0	≥95% ligation; 100% cleavage	Life Technol 15221-013 15221-054
2.5 U/μL	1 unit hydrolyzes 1 μg λ DNA to completion/hr in a 50 μL total reaction volume	100 m*M* Tris-HCl, 100 m*M* NaCl, 100 m*M* MgCl$_2$, 100 m*M* β-MSH, 1.0 mg/mL BSA, pH 7.5		Nippon Gene Oncor 110051 110052

Ava II continued

CONCENTRATION or VOLUME ACTIVITY	UNIT DEFINITION	REACTION BUFFER (As Provided)	ADDITIONAL ACTIVITIES and PURITY	SUPPLIER CATALOG No.
Anabaena variabilis continued				
2000-10,000 U/mL	1 unit completely digests 1 μg λ DNA/hr at pH 7.5, 37°C in a 50 μL total assay mixture	100 mM Tris-acetate, 100 mM MgOAc, 500 mM KOAc, pH 7.5		Pharmacia 27-0964-01 27-0964-02
1-10 U/μL	1 unit completely digests 1 μg unmethylated λ DNA/hr at 37°C in a 50 mL reaction volume	100 mM Tris-HCl, 100 mM MgCl$_2$, 500 mM NaCl, 10 mM DTT, pH 7.9	90% ligation	Promega R6131 R6132 R6135
1000-12,000 U/mL	1 unit completely cleaves 1.0 μg λ DNA/hr in a 50 μL reaction mixture	200 mM Tris-acetate, 500 mM KOAc, 100 mM MgOAc, 10 mM DTT, pH 7.9	100% cleavage, 80% ligation, 95% recleavage. No detectable endonuclease (overdigestion)	Sigma R6004
1000-10,000 U/mL		50 mM KCl, 10 mM Tris-HCl, 0.1 mM EDTA, 1 mM DTT, 200 μg/mL BSA, 50% glycerol, pH 7.4		Stratagene 500180
100 U and 500 U	1 unit completely digests 1 μg λ DNA/hr at 37°C in 50 μL of supplied buffer	100 mM Tris-HCl, 100 mM MgCl$_2$, 10 mM DTT, 500 mM NaCl, pH 7.5	Purity is confirmed by overdigestion, ligation-recutting and cloning test	TaKaRa 1008
Escherichia coli strain, carrying the cloned gene from Anabaena variabilis				
10,000 U/mL; 50,000 U/mL	1 unit completely digests 1 μg λ DNA/hr in a total reaction volume of 50 μL using the buffer provided	20 mM Tris-acetate, 10 mM MgOAc, 50 mM KOAc, 1 mM DTT, pH 7.9	>95% DNA fragments ligated and recut after 10-fold overdigestion	NE Biolabs 153S 153L

Avi II

RECOGNITION SEQUENCE
5'...TGC▼GCA...3'
3'...ACG▲CGT...5'

ISOSCHIZOMERS
Aos I

OPTIMUM REACTION CONDITIONS
pH 7.5 and 37°C

HEAT INACTIVATION
65°C for 20 min.

METHYLATION EFFECTS
Cleaves m^{6}AGCGCT
Does not cleave AGm^{5}CGCT

NOTES
- EC 3.1.21.4
- A type II site specific deoxyribonuclease
- A large group of enzymes recognizing specific short DNA sequences. They cleave either within the recognition site, or at a short specific distance from it
- Requires Mg^{2+}

CONCENTRATION or VOLUME ACTIVITY	UNIT DEFINITION	REACTION BUFFER (As Provided)	ADDITIONAL ACTIVITIES and PURITY	SUPPLIER CATALOG NO.
Anabaena variabilis halle				
10 U/μL	1 unit completely digests 1 μg substrate DNA/hr	50 mM Tris-HCl, 10 mM MgCl$_2$, 100 mM NaCl, 1 mM DTE, pH 7.5		Boehringer 1481436 1481444

Avr II

RECOGNITION SEQUENCE
5'...C▼CTAGG...3'
3'...GGATC▲C...5'

OPTIMUM REACTION CONDITIONS
pH 7.9-8.0 and 37°C

HEAT INACTIVATION
Resistant

NOTES
- EC 3.1.21.4
- A type II site specific deoxyribonuclease
- A large group of enzymes recognizing specific short DNA sequences. They cleave either within the recognition site, or at a short specific distance from it
- Produces a 5'-CTAG extension which can be efficiently ligated to DNA fragments generated by *Nhe* I, *Spe* I or *Xba* I
- Requires Mg^{2+}

Avr II continued

CONCENTRATION or VOLUME ACTIVITY	UNIT DEFINITION	REACTION BUFFER (As Provided)	ADDITIONAL ACTIVITIES and PURITY	SUPPLIER CATALOG No.
Anabaena variabilis UW				
1000-10,000 U/mL		50 mM KCl, 10 mM Tris-HCl, 0.1 mM EDTA, 1 mM DTT, 200 µg/mL acetylated BSA, 50% glycerol, pH 8.0		ICN 159396
Escherichia coli strain, carrying the cloned gene from Anabaena variabilis UW				
2000-5000 U/mL	1 unit completely digests 1 µg λ DNA (Hind III digest)/hr in a total reaction volume of 50 µL using the buffer provided	10 mM Tris-HCl, 10 mM MgCl$_2$, 50 mM NaCl, 1 mM DTT, pH 7.9	>95% DNA fragments ligated and recut after 50-fold overdigestion	NE Biolabs 174S 174L

Axy I

RECOGNITION SEQUENCE
    5'...CC▼TNAGG...3'
    3'...GGANT▲CC...5'

ISOSCHIZOMERS
    Aoc I, Bse21 I, Bsu36 I, Cvn I, Eco81 I

NOTES
- EC 3.1.21.4
- A type II site specific deoxyribonuclease
- A large group of enzymes recognizing specific short DNA sequences. They cleave either within the recognition site, or at a short specific distance from it
- Requires Mg^{2+}

CONCENTRATION or VOLUME ACTIVITY	UNIT DEFINITION	REACTION BUFFER (As Provided)	ADDITIONAL ACTIVITIES and PURITY	SUPPLIER CATALOG No.
Acetobacter xylinus				
				Nippon Gene

Bal I

RECOGNITION SEQUENCE
5'...TGG▼CCA...3'
3'...ACC▲GGT...5'

ISOSCHIZOMERS
Msc I

OPTIMUM REACTION CONDITIONS
pH 7.4-8.5 and 37°C

HEAT INACTIVATION
60°C for 15 min.

METHYLATION EFFECTS
Does not cleave TGGm^{5}CCA or TGGCm^{5}CA
Blocked by overlapping *dcm* methylation

NOTES
- EC 3.1.21.4
- A type II site specific deoxyribonuclease
- A large group of enzymes recognizing specific short DNA sequences. They cleave either within the recognition site, or at a short specific distance from it.
- Requires Mg^{2+}
- Compatible cohesive ends: any blunt end
- Sensitive to NaCl >40 *mM*

CONCENTRATION or VOLUME ACTIVITY	UNIT DEFINITION	REACTION BUFFER (As Provided)	ADDITIONAL ACTIVITIES and PURITY	SUPPLIER CATALOG No.
Brevibacterium albidum				
2-5 U/μL	1 unit completely digests 1 μg λ DNA/hr at 37°C in the buffer provided	Bal I basal buffer	>90% DNA fragments ligated and 100% recut	Amersham E 1009Y
2-10 U/μL	1 unit completely digests 1 μg unmethylated λ DNA/hr at 37°C in a 50 mL reaction volume	500 *mM* Tris-HCl, 50 *mM* MgCl$_2$, pH 8.2	80% ligation	Nippon Gene Promega R6691 R6692 R6695
20 U and 100 U	1 unit completely digests 1 μg λ DNA/hr at 37°C in 50 μL of supplied buffer	20 *mM* Tris-HCl, 7 *mM* MgCl$_2$, 7 *mM* β-MSH, 0.01% BSA, pH 8.5	Purity is confirmed by overdigestion, ligation-recutting and cloning test	TaKaRa 1009
Brevibacterium albidum ATCC 15831				
2000-10,000 U/mL		50 *mM* KCl, 10 *mM* Tris-HCl, 0.1 *mM* EDTA, 1 *mM* DTT, 200 μg/mL acetylated BSA, 50% glycerol, pH 7.4		ICN 153801

BamH I

RECOGNITION SEQUENCE
  5'...G▼GATCC...3'
  3'...CCTAG▲G...5'

ISOSCHIZOMERS
  Bst I

OPTIMUM REACTION CONDITIONS
  pH 7.4-8.5 and 37°C

HEAT INACTIVATION
  Resistant

METHYLATION EFFECTS
  Cleaves GGATCm^{5}C, GGm^{6}ATCC, GGm^{6}ATCm^{5}C and GGATCm^{4}C
  Does not cleave GGATm^{4}CC, GGATm^{5}CC, GGAThm^{5}Chm^{5}C or GGAhm^{5}UCC
  Not blocked by *dam* methylation

NOTES
- EC 3.1.21.4
- A type II site specific deoxyribonuclease
- A large group of enzymes recognizing specific short DNA sequences. They cleave either within the recognition site, or at a short specific distance from it
- Conditions of low ionic strength, high enzyme concentration, glycerol >5% or pH >8.0 may result in star activity
- Requires Mg^{2+} and BSA for full activity
- Compatible cohesive ends: *Bcl* I, *Bgl* II, *BsiQ* I, *BstY* I, *Dpn* II, *Mbo* I, *Mfl* I, *Nde* II, *Sau3A* I, *Xho* II
- Optimal KCl is 100-160 m*M*

CONCENTRATION or VOLUME ACTIVITY	UNIT DEFINITION	REACTION BUFFER (As Provided)	ADDITIONAL ACTIVITIES and PURITY	SUPPLIER CATALOG NO.
Bacillus amyloliquefaciens				
10 U/μL; 40 U/μL	1 unit completely cleaves 1 μg substrate DNA/hr	50 m*M* NaCl, 10 m*M* Tris, 10 m*M* MgCl$_2$, 1 m*M* β-MSH, pH 7.8	90-98% cleavage-ligation-recleavage	Am Allied L,H-03-02500 L,H-03-10000 L,H-03-50000
10 U/μL and 40 U/μL	1 unit completely digests 1 μg substrate DNA/hr	10 m*M* Tris-HCl, 5 m*M* MgCl$_2$, 100 m*M* NaCl, 1 m*M* β-MSH, pH 8.0		Boehringer 220612 567604 656275 798975 1274031
2000-20,000 U/mL		50 m*M* KCl, 10 m*M* Tris-HCl, 0.1 m*M* EDTA, 1 m*M* DTT, 200 μg/mL acetylated BSA, 50% glycerol, pH 7.5		ICN 150421
				Nippon Gene

BamH I continued

CONCENTRATION or VOLUME ACTIVITY	UNIT DEFINITION	REACTION BUFFER (As Provided)	ADDITIONAL ACTIVITIES and PURITY	SUPPLIER CATALOG NO.
Bacillus amyloliquefaciens continued				
10,000-20,000 U/mL and 50,000-100,000 U/mL	1 unit completely digests 1 µg λ DNA/hr at pH 7.5, 37°C in a 50 µL total assay mixture	100 mM Tris-acetate, 100 mM MgOAc, 500 mM KOAc, pH 7.5		Pharmacia 27-0868-03 27-0868-04 27-0868-18
10,000 U and 50,000 U	1 unit completely digests 1 µg λ DNA/hr at 30°C in 50 µL of supplied buffer	200 mM Tris-HCl, 100 mM MgCl$_2$, 10 mM DTT, 1 M KCl, pH 8.5	Purity is confirmed by overdigestion, ligation-recutting and cloning test	SibEnzyme TaKaRa 1010
Bacillus amyloliquefaciens H				
7500 U and 15,000 U	1 unit completely digests 1 µg DNA substrate/hr at 37°C	10 mM Tris-HCl, 10 mM MgCl$_2$, 100 mM NaCl, 1 mM DTT, 0.1 mg/mL BSA, pH 7.5	>90% DNA fragments ligated and recut after 10-fold overdigestion	Adv Biotech AB-0207-a AB-0207-b
8-12 U/µL	1 unit completely digests 1 µg λ DNA/hr at 30°C in the buffer provided	200 mM Tris-HCl, 100 mM MgCl$_2$, 10 mM DTT, 1000 mM KCl, pH 8.5	>95% DNA fragments ligated and 100% recut	Amersham E 1010Y E 1010Z
>40 U/µL	1 unit completely digests 1 µg λ DNA/hr at 30°C in the buffer provided	200 mM Tris-HCl, 100 mM MgCl$_2$, 10 mM DTT, 1000 mM KCl, pH 8.5	>95% DNA fragments ligated and 100% recut	Amersham E 1010YH E 1010XH
12,000 U/mL	1 unit degrades 1 µg λ dam$^-$ dcm$^-$ DNA/hr at 37°C in a 0.05 mL mixture	10 mM Tris-HCl, 100 mM NaCl, 100 µg/mL BSA, pH 8.0	No detectable contaminating nuclease; 90% ligated and 100% recut after 50-fold overdigestion	AGS Heidelb B00110S B00110M
	1 unit completely cleaves 1 µg λ DNA/hr at pH 7.5, 37°C in a reaction volume of 50 µL	50 mM Tris-HCl, 10 mM MgCl$_2$, 100 mM NaCl, 1.0 mM DTT, 100 µg/mL BSA, pH 7.5		CHIMERx 2050-01 2050-02
8-12 U/µL; 40-60 U/µL	1 unit hydrolyzes 1 µg DNA/hr at 37°C in a 50 µL reaction volume using λ phage DNA as substrate	10 mM Tris-HCl, 5 mM MgCl2, 100 mM NaCl, 0.02% Triton X-100, pH 8.0, 37°C	>90% of the DNA fragments ligated and recut after 50-fold overdigestion	Fermentas ER0051 ER0052 ER0053
8-12 U/µL and 50 U/µL	1 unit completely digests 1 mg λ DNA/hr at 37°C	500 mM Tris-HCl, 100 mM MgCl$_2$, 1 M NaCl, pH 8.0	≥95% ligation; 100% cleavage	Life Technol 15201-023 15201-031 15201-064 15201-080 15201-049
	10 units does not produce unspecific cleavage products with 1 µg λ DNA/16 hr at 37°C	10 mM Tris-HCl, 10 mM MgCl$_2$, 150 mM NaCl, 1 mM DTT, 100 µg/mL BSA, pH 7.9	>90% DNA fragments ligated and recut after 10-fold overdigestion	MINOTECH 102

BamH I continued

CONCENTRATION or VOLUME ACTIVITY	UNIT DEFINITION	REACTION BUFFER (As Provided)	ADDITIONAL ACTIVITIES and PURITY	SUPPLIER CATALOG No.
Bacillus amyloliquefaciens H continued				
4-12 U/µL and >40 U/µL	1 unit digests completely 1 µg λ DNA substrate/hr at 37°C	10 mM Tris-HCl, 100 mM NaCl, 5 mM MgCl$_2$, 1 mM β-MSH, pH 8.3		NBL Gene 012704 012707 012757
10 U/µL and 20-150 U/µL	1 unit hydrolyzes 1 µg λ DNA to completion/hr in a 50 µL total reaction volume	330 mM Tris-acetate, 800 mM KOAc, 100 mM MgOAc, 1.0 mg/mL BSA, pH 8.0		Oncor 110064 110065 110063
10 U/µL and 40-80 U/µL	1 unit completely digests 1 µg λ DNA/hr at 37°C in a 50 mL reaction volume	Blue/White Cloning Qualified; 60 mM Tris-HCl, 60 mM MgCl$_2$, 1 M NaCl, 10 mM DTT, pH 7.5	90% ligation	Promega R6021 R6022 R6025 R6026 R4024
5000-20,000 U/mL	1 unit completely cleaves 1.0 µg λ DNA/hr in a 50 µL reaction mixture	Unique buffer + BSA	100% cleavage, 95% ligation, 100% recleavage, 30% endonuclease (nickase) No detectable endonuclease (overdigestion)	Sigma R0260
10,000-40,000 U/mL		50 mM KCl, 10 mM Tris-HCl, 0.1 mM EDTA, 1 mM DTT, 0.02% Triton X-100, 200 µg/mL BSA, 50% glycerol, pH 7.4		Stratagene 500220 500221 500229
Bacillus subtilis MT-2				
3-20 U/µL 40-50 U/µL	1 unit completely digests 1 µg λ DNA/hr in 0.05 mL at 37°C	50 mM Tris-HCl, 10 mM MgCl$_2$, 100 mM NaCl, 1 mM DTT, pH 7.5	>90% ligation >95% recutting	Toyobo BAH-102 BAH-104 BAH-105 BAH-155
Escherichia coli strain, carrying the cloned gene from Bacillus amyloliquefaciens H				
20,000 U/mL; 100,000 U/mL	1 unit completely digests 1 µg λ DNA/hr in a total reaction volume of 50 µL using the buffer provided	10 mM Tris-HCl, 10 mM MgCl$_2$, 150 mM NaCl, 1 mM DTT, pH 7.9; add 100 µg/mL BSA	>95% DNA fragments ligated and recut after 50-fold overdigestion	NE Biolabs 136S 136L

Ban I

RECOGNITION SEQUENCE
   5'...G▼GPyPuCC...3'
   3'...CCPuPyG▲G...5'

OPTIMUM REACTION CONDITIONS
   pH 7.4-8.2 and 50°C; 10-25% activity at 37°C

HEAT INACTIVATION
   65°C for 20 min.

METHYLATION EFFECTS
   Cleaves GGm^5CGCC and $GG(^C_T)(^A_G)Cm^4C$

NOTES
- EC 3.1.21.4
- A type II site specific deoxyribonuclease
- A large group of enzymes recognizing specific short DNA sequences. They cleave either within the recognition site, or at a short specific distance from it
- Requires Mg^{2+}
- Optimal salt is 0-30 mM; inhibited by KCl >30 mM and NaCl >80 mM
- Compatible cohesive ends: Asp718 I, HgiC I, HgiH I

CONCENTRATION or VOLUME ACTIVITY	UNIT DEFINITION	REACTION BUFFER (As Provided)	ADDITIONAL ACTIVITIES and PURITY	SUPPLIER CATALOG No.
Bacillus aneurinolyticus				
8-12 U/μL	1 unit completely digests 1 μg substrate DNA/hr	33 mM Tris-acetate, 10 mM MgOAc, 66 mM KOAc, 0.5 mM DTT, pH 7.9		Boehringer 775215 775223
8-12 U/μL	1 unit completely digests 1 μg λ DNA/hr at 50°C in a 50 mL reaction volume	500 mM Tris-HCl, 50 mM MgCl$_2$, pH 8.2	90% ligation	Pharmacia Promega R6891 R6892
5000-50,000 U/mL	1 unit cleaves 98% λ DNA/hr in a 50 μL reaction mixture	200 mM Tris-acetate, 500 mM KOAc, 100 mM MgOAc, 10 mM DTT, pH 7.9	≥98% cleavage, 90% ligation, 95% recleavage No detectable endonuclease (overdigestion)	Sigma R4518
1000-10,000 U/mL		50 mM KCl, 10 mM Tris-HCl, 0.1 mM EDTA, 1 mM DTT, 500 μg/mL BSA, 50% glycerol, pH 7.4		Stratagene 500231
Bacillus aneurinolyticus IAM 1077				
3-15 U/μL	1 unit completely digests 1 μg λ DNA/hr at 50°C in the buffer provided	100 mM Tris-HCl, 100 mM MgCl$_2$, 10 mM DTT, pH 7.5	>90% DNA fragments ligated and >95% recut	Amersham E 0204Y
40-50 U/μL	1 unit completely digests 1 μg λ DNA/hr at 50°C in the buffer provided	100 mM Tris-HCl, 100 mM MgCl$_2$, 10 mM DTT, pH 7.5	>90% DNA fragments ligated and >95% recut	Amersham E 0204YH

Ban I continued

CONCENTRATION or VOLUME ACTIVITY	UNIT DEFINITION	REACTION BUFFER (As Provided)	ADDITIONAL ACTIVITIES and PURITY	SUPPLIER CATALOG NO.
Bacillus aneurinolyticus IAM 1077 continued				
10,000-60,000 U/mL		50 mM KCl, 10 mM Tris-HCl, 0.1 mM EDTA, 1 mM DTT, 200 µg/mL acetylated BSA, 50% glycerol, pH 7.4	No detectable Ban II, Ban III	ICN 153802
3-20 U/µL 40-50 U/µL	1 unit completely digests 1 µg λ DNA/hr in 0.05 mL at 50°C	10 mM Tris-HCl, 10 mM MgCl₂, 1 mM DTT, pH 7.5	>90% ligation >95% recutting	Toyobo BAN-103 BAN-104 BAN-154
Escherichia coli strain, carrying the cloned gene from Bacillus aneurinolyticus				
20,000 U/mL	1 unit completely digests 1 µg λ DNA/hr in a total reaction volume of 50 µL using the buffer provided	20 mM Tris-acetate, 10 mM MgOAc, 50 mM KOAc, 1 mM DTT, pH 7.9	>95% DNA fragments ligated and recut after 10-fold overdigestion	NE Biolabs 118S 118L

Ban II

RECOGNITION SEQUENCE
    5'...GPuGCPy▼C...3'
    3'...C▲PyCGPuG...5'

OPTIMUM REACTION CONDITIONS
    pH 7.5-8.3 and 37°C

HEAT INACTIVATION
    65°C for 20 min.

METHYLATION EFFECTS
    Cleaves $G(^A_G)GC(^C_T)m^5C$
    Does not cleave $G(^A_G)Gm^5C(^C_T)C$

NOTES
- EC 3.1.21.4
- A type II site specific deoxyribonuclease
- A large group of enzymes recognizing specific short DNA sequences. They cleave either within the recognition site, or at a short specific distance from it
- Require Mg^{2+}
- Optimal salt (KCl or NaCl) is 0-100 mM
- pBR322 has recognition sites extremely near the Tetracycline resistance region

Ban II continued

CONCENTRATION or VOLUME ACTIVITY	UNIT DEFINITION	REACTION BUFFER (As Provided)	ADDITIONAL ACTIVITIES and PURITY	SUPPLIER CATALOG No.
Bacillus aneurinolyticus				
8-12 U/μL	1 unit completely digests 1 μg λ DNA/hr at 37°C in the buffer provided	500 mM Tris-HCl, 100 mM MgCl$_2$, 10 mM DTT, 1000 mM NaCl, pH 7.5	100% DNA fragments ligated and recut	Amersham E 1012Y
8-12 U/μL	1 unit completely digests 1 μg substrate DNA/hr	10 mM Tris-HCl, 5 mM MgCl$_2$, 100 mM NaCl, 1 mM β-MSH, pH 8.0		Boehringer 775240 836826
	1 unit completely cleaves 1 μg λ DNA/hr at pH 7.5, 37°C in a reaction volume of 50 μL	10 mM Tris-HCl, 10 mM MgCl$_2$, 1.0 mM DTT, 100 μg/mL BSA, pH 7.5		CHIMERx 2060-01 2060-02
8-12 U/μL	1 unit completely digests 1 mg λ DNA/hr at 37°C	500 mM Tris-HCl, 100 mM MgCl$_2$, pH 8.0	≥95% ligation; 100% cleavage	Life Technol 15465-016
4-12 U/μL	1 unit digests completely 1 μg λ DNA substrate/hr at 37°C	10 mM Tris-HCl, 100 mM NaCl, 5 mM MgCl$_2$, 1 mM β-MSH, pH 8.3		NBL Gene 016303 016306
8-12 U/μL	1 unit completely digests 1 μg λ DNA/hr at 37°C in a 50 mL reaction volume	60 mM Tris-HCl, 60 mM MgCl$_2$, 1 M NaCl, 10 mM DTT, pH 7.5	90% ligation	Pharmacia Promega R6561 R6562
10,000-50,000 U/mL	1 unit completely cleaves 1.0 μg λ DNA/hr in a 50 μL reaction mixture	200 mM Tris-acetate, 500 mM KOAc, 100 mM MgOAc, 10 mM DTT, pH 7.9	100% cleavage, 95% ligation, 100% recleavage, 50% endonuclease (nickase) No detectable endonuclease (overdigestion)	Sigma R2630
2000 U and 10,000 U	1 unit completely digests 1 μg λ DNA/hr at 37°C in 50 μL of supplied buffer	500 mM Tris-HCl, 100 mM MgCl$_2$, 10 mM DTT, 1 M NaCl, pH 7.5	Purity is confirmed by overdigestion, ligation-recutting and cloning test	TaKaRa 1012
Bacillus aneurinolyticus IAM 1077				
4000-20,000 U/mL		50 mM KCl, 10 mM Tris-HCl, 0.1 mM EDTA, 1 mM DTT, 200 μg/mL acetylated BSA, 50% glycerol, pH 7.4		ICN 153803
5000 U/mL	1 unit completely digests 1 μg λ DNA/hr in a total reaction volume of 50 μL using the buffer provided	20 mM Tris-acetate, 10 mM MgOAc, 50 mM KOAc, 1 mM DTT, pH 7.9	>95% DNA fragments ligated and recut after 10-fold overdigestion	NE Biolabs 119S 119L
3-20 U/μL 40-50 U/μL	1 unit completely digests 1 μg λ DNA/hr in 0.05 mL at 37°C	10 mM Tris-HCl, 10 mM MgCl$_2$, 50 mM NaCl, 1 mM DTT, pH 7.5	>90% ligation >95% recutting	Toyobo BAN-203 BAN-204 BAN-254

Ban III

RECOGNITION SEQUENCE
  5'...AT▼CGAT...3'
  3'...TAGC▲TA...5'
ISOSCHIZOMERS
  BspD I, Cla I
OPTIMUM REACTION CONDITIONS
  pH 7.5 and 37°C
HEAT INACTIVATION
  65°C for 5 min.

NOTES
- EC 3.1.21.4
- A type II site specific deoxyribonuclease
- A large group of enzymes recognizing specific short DNA sequences. They cleave either within the recognition site, or at a short specific distance from it
- Requires Mg^{2+}
- Optimal salt (KCl or NaCl) is 50-120 mM
- Inhibited by glycerol >3%

CONCENTRATION or VOLUME ACTIVITY	UNIT DEFINITION	REACTION BUFFER (As Provided)	ADDITIONAL ACTIVITIES and PURITY	SUPPLIER CATALOG NO.
Bacillus aneurinolyticus				
3000-20,000 U/mL		100 mM KCl, 10 mM Tris-HCl, 0.1 mM EDTA, 10 mM β-MSH, 500 µg/mL acetylated BSA, 50% glycerol, pH 7.5		ICN 159398
Escherichia coli JM109				
3-20 U/µL 40-50 U/µL	1 unit completely digests 1 µg λ DNA/hr in 0.05 mL at 37°C	10 mM Tris-HCl, 10 mM MgCl₂, 50 mM NaCl, 1 mM DTT, pH 7.5	>90% ligation >95% recutting	Toyobo BAN-303 BAN-304 BAN-354

Bbe I

RECOGNITION SEQUENCE
5'...GGCGC▼C...3'
3'...C▲CGCGG...5'

ISOSCHIZOMERS
Nar I

OPTIMUM REACTION CONDITIONS
pH 7.1 and 37°C

HEAT INACTIVATION
60°C for 15 min.

NOTES
- EC 3.1.21.4
- A type II site specific deoxyribonuclease
- A large group of enzymes recognizing specific short DNA sequences. They cleave either within the recognition site, or at a short specific distance from it
- Requires Mg^{2+}

CONCENTRATION or VOLUME ACTIVITY	UNIT DEFINITION	REACTION BUFFER (As Provided)	ADDITIONAL ACTIVITIES and PURITY	SUPPLIER CATALOG NO.
Bifidobacterium breve				
8-12 U/μL	1 unit completely digests 1 μg λ DNA/hr at 37°C in the buffer provided	*Bbe* I basal buffer	>95% DNA fragments ligated and 100% recut	Amersham E 1015Z
500 U and 2500 U	1 unit completely digests 1 μg pBR322 DNA/hr at 37°C in 50 μL of supplied buffer	5 m*M* Tris-HCl, 5 m*M* MgCl₂, 2 m*M* β-MSH, 0.01% BSA, pH 7.1	Purity is confirmed by overdigestion and ligation-recutting test	TaKaRa 1015

Bbi II

RECOGNITION SEQUENCE
5'...GPu▼CGPyC...3'
3'...CPyGC▲PuG...5'

ISOSCHIZOMERS
Acy I

OPTIMUM REACTION CONDITIONS
pH 7.5 and 37°C

NOTES
- EC 3.1.21.4
- A type II site specific deoxyribonuclease
- A large group of enzymes recognizing specific short DNA sequences. They cleave either within the recognition site, or at a short specific distance from it
- Requires Mg^{2+}

Bbi II continued

CONCENTRATION or VOLUME ACTIVITY	UNIT DEFINITION	REACTION BUFFER (As Provided)	ADDITIONAL ACTIVITIES and PURITY	SUPPLIER CATALOG NO.
Bifidobacterium bifidum				
30 U and 150 U	1 unit completely digests 1 μg λ DNA/hr at 37°C in 50 μL of supplied buffer and BSA	100 mM Tris-HCl, 100 mM MgCl$_2$, 10 mM DTT, BSA, pH 7.5	Purity is confirmed by overdigestion and ligation-recutting test	TaKaRa 1017

BbrP I

RECOGNITION SEQUENCE
    5'...CAC▼GTG...3'
    3'...GTG▲CAC...5'
ISOSCHIZOMERS
    PmaC I
OPTIMUM REACTION CONDITIONS
    pH 8.0 and 37°C
HEAT INACTIVATION
    65°C for 20 min.
METHYLATION EFFECTS
    Does not cleave m^{5}CAm^{5}CGTG

NOTES
- EC 3.1.21.4
- A type II site specific deoxyribonuclease
- A large group of enzymes recognizing specific short DNA sequences. They cleave either within the recognition site, or at a short specific distance from it
- Requires Mg^{2+}

CONCENTRATION or VOLUME ACTIVITY	UNIT DEFINITION	REACTION BUFFER (As Provided)	ADDITIONAL ACTIVITIES and PURITY	SUPPLIER CATALOG NO.
Bacillus brevis				
10 U/μL	1 unit completely digests 1 μg substrate DNA/hr	10 mM Tris-HCl, 5 mM MgCl$_2$, 100 mM NaCl, 1 mM β-MSH, pH 8.0		Boehringer 1168860
				Toyobo

Bbs I

RECOGNITION SEQUENCE
  5'...GAAGAC(N)$_2$▼...3'
  3'...CTTCTG(N)$_6$▲...5'
ISOSCHIZOMERS
  Bbv II
OPTIMUM REACTION CONDITIONS
  pH 7.5-7.9 and 37°C
HEAT INACTIVATION
  65°C for 20 min.
METHYLATION EFFECTS
  Cleaves GAAGAm^{5}C

NOTES
- EC 3.1.21.4
- A type II site specific deoxyribonuclease
- A large group of enzymes recognizing specific short DNA sequences. They cleave either within the recognition site, or at a short specific distance from it
- Requires Mg^{2+}

CONCENTRATION or VOLUME ACTIVITY	UNIT DEFINITION	REACTION BUFFER (As Provided)	ADDITIONAL ACTIVITIES and PURITY	SUPPLIER CATALOG No.
Bacillus laterosporus				
1000-10,000 U/mL		150 m*M* KCl, 10 m*M* Tris-HCl, 0.1 m*M* EDTA, 1 m*M* DTT, 200 μg/mL acetylated BSA, 50% glycerol, pH 7.5		ICN 159399
1000-5000 U/mL	1 unit completely digests 1 μg λ DNA/hr in a total reaction volume of 50 μL using the buffer provided	10 m*M* Tris-HCl, 10 m*M* MgCl$_2$, 50 m*M* NaCl, 1 m*M* DTT, pH 7.9	>95% DNA fragments ligated and recut after 10-fold overdigestion	NE Biolabs 539S 539L

Bbu I

RECOGNITION SEQUENCE
5'...GCATG▼C...3'
3'...C▲GTACG...5'

ISOSCHIZOMERS
Sph I

OPTIMUM REACTION CONDITIONS
pH 7.5 and 37°C

NOTES
- EC 3.1.21.4
- A type II site specific deoxyribonuclease
- A large group of enzymes recognizing specific short DNA sequences. They cleave either within the recognition site, or at a short specific distance from it
- Requires Mg^{2+}
- Optimal activity at low salt concentration; *SpH* I requires high salt
- Compatible cohesive ends: *Nla* III, *Nsp* I, *SpH* I

CONCENTRATION or VOLUME ACTIVITY	UNIT DEFINITION	REACTION BUFFER (As Provided)	ADDITIONAL ACTIVITIES and PURITY	SUPPLIER CATALOG No.
***Bacillus* species**				
8000-12,000 U/mL				ICN 159400
***Bacillus* species Bu17091**				
10 U/μL and 40-80 U/μL	1 unit completely digests 1 μg λ DNA/hr at 37°C in a 50 μL reaction volume	60 mM Tris-HCl, 60 mM MgCl$_2$, 60 mM NaCl, 10 mM DTT, pH 7.5	90% ligation	Promega R6621 R6622 R4624

Bbv I

RECOGNITION SEQUENCE
  5'...GCAGC(N)$_8$▼...3'
  3'...CGTCG(N)$_{12}$▲...5'

ISOSCHIZOMERS
  Bst71 I

OPTIMUM REACTION CONDITIONS
  pH 7.4-7.9 and 37°C

HEAT INACTIVATION
  65°C for 20 min.

NOTES
- EC 3.1.21.4
- A type II site specific deoxyribonuclease
- A large group of enzymes recognizing specific short DNA sequences. They cleave either within the recognition site, or at a short specific distance from it
- Requires Mg^{2+}

CONCENTRATION or VOLUME ACTIVITY	UNIT DEFINITION	REACTION BUFFER (As Provided)	ADDITIONAL ACTIVITIES and PURITY	SUPPLIER CATALOG NO.
Bacillus brevis ATCC 9999				
200-2000 U/mL		50 mM KCl, 10 mM Tris-HCl, 0.1 mM EDTA, 1 mM DTT, 200 µg/mL BSA, 50% glycerol, pH 7.4		ICN 153804
				SibEnzyme
Escherichia coli strain, carrying the cloned gene from Bacillus brevis (ATCC 9999)				
500-2000 U/mL	1 unit completely digests 1 µg pBR322 DNA/hr in a total reaction volume of 50 µL using the buffer provided	10 mM Tris-HCl, 10 mM $MgCl_2$, 50 mM NaCl, 1 mM DTT, pH 7.9	>95% DNA fragments ligated and recut after 2-fold overdigestion	NE Biolabs 173S 173L

Bbv12 I

RECOGNITION SEQUENCE
5'...GWGCW▼C...3'
3'...C▲WCGWG...5'

ISOSCHIZOMERS
Alw21 I, AspH I, BsiHKA I

NOTES
- EC 3.1.21.4
- A type II site specific deoxyribonuclease
- A large group of enzymes recognizing specific short DNA sequences. They cleave either within the recognition site, or at a short specific distance from it
- Requires Mg^{2+}

CONCENTRATION or VOLUME ACTIVITY	UNIT DEFINITION	REACTION BUFFER (As Provided)	ADDITIONAL ACTIVITIES and PURITY	SUPPLIER CATALOG NO.
Bacillus brevis 12				
				SibEnzyme

Bbv16 II

RECOGNITION SEQUENCE
5'...GAAGAC(2/6)...3'

ISOSCHIZOMERS
Bbs I, Bpi I, BpuA I

NOTES
- EC 3.1.21.4
- A type II site specific deoxyribonuclease
- A large group of enzymes recognizing specific short DNA sequences. They cleave either within the recognition site, or at a short specific distance from it
- Requires Mg^{2+}

CONCENTRATION or VOLUME ACTIVITY	UNIT DEFINITION	REACTION BUFFER (As Provided)	ADDITIONAL ACTIVITIES and PURITY	SUPPLIER CATALOG NO.
Bacillus brevis 16				
				SibEnzyme

Bcg I

RECOGNITION SEQUENCE

5'...▼$_{10}$(N)CGA(N)$_6$TGC(N)$_{12}$▼...3'
3'...▲$_{12}$(N)GCT(N)$_6$ACG(N)$_{10}$▲...5'

OPTIMUM REACTION CONDITIONS
pH 8.4 and 37°C

HEAT INACTIVATION
Resistant

NOTES

- EC 3.1.21.4
- A type II site specific deoxyribonuclease
- A large group of enzymes recognizing specific short DNA sequences. They cleave either within the recognition site, or at a short specific distance from it
- Cleaves DNA twice to excise its recognition site
- Generates a 32 base-pair fragment with 2-base 3' overhangs
- Requires S-adenosylmethionine and Mg^{2+} for optimal activity

CONCENTRATION or VOLUME ACTIVITY	UNIT DEFINITION	REACTION BUFFER (As Provided)	ADDITIONAL ACTIVITIES and PURITY	SUPPLIER CATALOG NO.
Escherichia coli strain, carrying the cloned gene from *Bacillus coagulans* (NEB 697)				
2000 U/mL	1 unit completely digests 1 μg λ DNA/hr in a total reaction volume of 50 μL using the buffer provided	10 m*M* Tris-HCl, 10 m*M* MgCl$_2$, 100 m*M* NaCl, 1 m*M* DTT, pH 8.4; add 20 μ*M* S-adenosylMet		NE Biolabs 545S 545L

Bcl I

RECOGNITION SEQUENCE
  5'...T▼GATCA...3'
  3'...ACTAG▲T...5'

ISOSCHIZOMERS
  BsiQ I, Fba I

OPTIMUM REACTION CONDITIONS
  pH 7.5-8.5 and 50°C; 50% activity at 37°C

HEAT INACTIVATION
  Resistant

METHYLATION EFFECTS
  Cleaves TGATm^{5}CA
  Does not cleave TGm^{6}ATCA or TGAThm^{5}CA
  Blocked by *dam* methylation

NOTES
- EC 3.1.21.4
- A type II site specific deoxyribonuclease
- A large group of enzymes recognizing specific short DNA sequences. They cleave either within the recognition site, or at a short specific distance from it
- Leaves a 5'-GATC extension which can be efficiently ligated to DNA fragments generated by BamH I, Bgl II, BstY I, Mbo I, and Sau3A I
- Large excess of enzyme results in the appearance of star activity
- Requires Mg^{2+} and BSA for full activity
- Optimal NaCl is 50-100 *mM*
- Compatible cohesive ends: any blunt end; BamH I, Bgl II, Bst I, BstY I, Dpn II, Mbo I, Mfl I, Nde II, Nla II, Sau3A I, Xho II

CONCENTRATION or VOLUME ACTIVITY	UNIT DEFINITION	REACTION BUFFER (As Provided)	ADDITIONAL ACTIVITIES and PURITY	SUPPLIER CATALOG No.
Bacillus caldolyticus (A. Atkinson)				
10,000 U/mL	1 unit degrades 1 μg λ dam⁻ dcm⁻ DNA/hr at 50°C in a 0.05 mL mixture	10 *mM* Tris-HCl, 10 *mM* MgCl₂, 100 *mM* KCl, 100 μg/mL BSA, pH 8.5	No detectable contaminating nuclease; 90% ligated and 100% recut after 50-fold overdigestion	AGS Heidelb F00135S F00135M
10 U/μL and 40 U/μL	1 unit completely digests 1 μg substrate DNA/hr	10 *mM* Tris-HCl, 10 *mM* MgCl₂, 50 *mM* NaCl, 1 *mM* DTE, pH 7.5		Boehringer 693952 693979 1097059
8-12 U/μL	1 unit hydrolyzes 1 μg DNA/hr at 37°C in a 50 μL reaction volume using λ phage DNA as substrate	10 *mM* Tris-HCl, 10 *mM* MgCl2, 50 *mM* NaCl, 0.1 mg/mL BSA, pH 7.5	>90% of the DNA fragments ligated and recut after 50-fold overdigestion	Fermentas ER0721 ER0722
3000-15,000 U/mL		50 *mM* KCl, 10 *mM* Tris-HCl, 0.1 *mM* EDTA, 1 *mM* DTT, 200 μg/mL BSA, 50% glycerol, pH 7.5		ICN 150426

Bcl I continued

CONCENTRATION or VOLUME ACTIVITY	UNIT DEFINITION	REACTION BUFFER (As Provided)	ADDITIONAL ACTIVITIES and PURITY	SUPPLIER CATALOG No.
Bacillus caldolyticus (A. Atkinson) continued				
8-12 U/μL	1 unit completely digests 1 mg Ad2 DNA/hr at 50°C	500 mM Tris-HCl, 100 mM $MgCl_2$, 500 mM NaCl, pH 8.0	≥95% ligation; 100% cleavage	Life Technol 15236-011
	10 units does not produce unspecific cleavage products with 1 μg λ DNA/16 hr at 50°C	10 mM Tris-HCl, 10 mM $MgCl_2$, 50 mM NaCl, 1 mM DTT, 100 μg/mL BSA, pH 7.9	>90% DNA fragments ligated and recut after 10-fold overdigestion	MINOTECH 103
4-12 U/μL	1 unit digests completely 1 μg λ DNA substrate/hr at 50°C	10 mM Tris-HCl, 50 mM NaCl, 10 mM $MgCl_2$, 1 mM DTT, pH 7.8		NBL Gene 014202 014205
10,000 U/mL	1 unit completely digests 1 μg λ DNA (dam⁻)/hr in a total reaction volume of 50 μL using the buffer provided	50 mM Tris-HCl, 10 mM $MgCl_2$, 100 mM NaCl, 1 mM DTT, pH 7.9	>95% DNA fragments ligated and recut after 40-fold overdigestion	NE Biolabs 160S 160L
2.5 U/μL	1 unit hydrolyzes 1 μg λ DNA to completion/hr in a 50 μL total reaction volume	100 mM Tris-HCl, 500 mM NaCl, 100 mM $MgCl_2$, 100 mM β-MSH, 1.0 mg/mL BSA, pH 8		Nippon Gene Oncor 110081 110082
2000-10,000 U/mL	1 unit completely digests 1 μg λ DNA/hr at pH 7.5, 50°C in a 50 μL total assay mixture	100 mM Tris-acetate, 100 mM MgOAc, 500 mM KOAc, pH 7.5		Pharmacia 27-0937-01 27-0937-02
10 U/μL and 40-80 U/μL	1 unit completely digests 1 μg unmethylated λ DNA/hr at 50°C in a 50 mL reaction volume	Genome qualified; 100 mM Tris-HCl, 100 mM $MgCl_2$, 500 mM NaCl, 10 mM DTT, pH 7.9	90% ligation	Promega R6651 R6652 R4654
5000-20,000 U/mL	1 unit completely cleaves 1.0 μg λ DNA/hr at 50°C in a 50 μL reaction mixture	500 mM Tris-HCl, 1 M NaCl, 100 mM $MgCl_2$, 10 mM DTT, pH 7.9	100% cleavage, 95% ligation, 100% recleavage, 60% endonuclease (nickase) No detectable endonuclease (overdigestion)	Sigma R8631
5000-20,000 U/mL		50 mM KCl, 10 mM Tris-HCl, 0.1 mM EDTA, 1 mM DTT, 200 μg/mL BSA, 50% glycerol, pH 7.4		Stratagene 500260
3-20 U/μL	1 unit completely digests 1 μg λ DNA/hr in 0.05 mL at 50°C	10 mM Tris-HCl, 10 mM $MgCl_2$, 50 mM NaCl, 1 mM DTT, pH 7.5	>90% ligation >95% recutting	Toyobo BCL-101 BCL-102

Bcn I

RECOGNITION SEQUENCE
  5'...CC(^G_C)▼GG...3'
  3'...GG▲(^C_G)CC...5'

ISOSCHIZOMERS
  Cau II, *Nci* I

OPTIMUM REACTION CONDITIONS
  pH 7.5-7.9 and 37°C

HEAT INACTIVATION
  Resistant; use ethanol precipitation

METHYLATION EFFECTS
  Cleaves Cm^{5}C(^C_G)GG and m^{5}CC(^C_G)GG
  Does not cleave Cm^{4}C(^C_G)GG
  Cleavage rate is slowed significantly by CG methylation at Cm^{5}CGGG

NOTES
- EC 3.1.21.4
- A type II site specific deoxyribonuclease
- A large group of enzymes recognizing specific short DNA sequences. They cleave either within the recognition site, or at a short specific distance from it
- Requires Mg^{2+} and BSA for full activity

CONCENTRATION or VOLUME ACTIVITY	UNIT DEFINITION	REACTION BUFFER (As Provided)	ADDITIONAL ACTIVITIES and PURITY	SUPPLIER CATALOG No.
Bacillus centrosporus				
500 U and 2500 U	1 unit completely digests 1 μg λ DNA/hr at 37°C in 50 μL of supplied buffer	500 mM Tris-HCl, 100 mM MgCl$_2$, 10 mM DTT, 1 M NaCl, pH 7.5	Purity is confirmed by overdigestion and ligation-recutting test	TaKaRa 1019
Bacillus centrosporus RFL1				
8-12 U/μL	1 unit completely digests 1 μg λ DNA/hr at 37°C in the buffer provided	500 mM Tris-HCl, 100 mM MgCl$_2$, 10 mM DTT, 1000 mM NaCl, pH 7.5	0% DNA fragments ligated and recut	Amersham E 1019Y
10,000 U/mL	1 unit degrades 1 μg λ dam$^-$ dcm$^-$ DNA/hr at 37°C in a 0.05 mL mixture	33 mM Tris-acetate, 10 mM MgOAc, 66 mM KOAc, 100 μg/mL BSA, pH 7.9	No detectable contaminating nuclease; 90% ligated and 100% recut after 50-fold overdigestion	AGS Heidelb F00420S F00420M
8-12 U/μL	1 unit hydrolyzes 1 μg DNA/hr at 37°C in a 50 μL reaction volume using λ phage DNA as substrate	10 mM Tris-HCl, 10 mM MgCl$_2$, 50 mM NaCl, 0.1 mg/mL BSA, pH 7.5	>80% of the DNA fragments ligated and 90% recut after 10-fold overdigestion	Fermentas ER0061 ER0062

Bco I

RECOGNITION SEQUENCE
    5'...C▼PyCGPuG...3'
    3'...GPuGCPy▲C...5'

ISOSCHIZOMERS
    Ava I

OPTIMUM REACTION CONDITIONS
    pH 7.5 and 65°C

HEAT INACTIVATION
    Resistant

NOTES
- EC 3.1.21.4
- A type II site specific deoxyribonuclease
- A large group of enzymes recognizing specific short DNA sequences. They cleave either within the recognition site, or at a short specific distance from it
- Requires Mg^{2+}

CONCENTRATION or VOLUME ACTIVITY	UNIT DEFINITION	REACTION BUFFER (As Provided)	ADDITIONAL ACTIVITIES and PURITY	SUPPLIER CATALOG No.
Bacillus coagulans				
2-5 U/µL	1 unit completely digests 1 µg λ DNA/hr at 65°C in the buffer provided	500 mM Tris-HCl, 100 mM $MgCl_2$, 10 mM DTT, 1000 mM NaCl, pH 7.5	>90% DNA fragments ligated and recut	Amersham E 1610Y E 1610Z
Bacillus coagulans SM1				
2500 U and 5000 U	1 unit completely digests 1 µg DNA substrate/hr at 65°C	10 mM Tris-HCl, 10 mM $MgCl_2$, 100 mM NaCl, 1 mM DTT, 0.1 mg/mL BSA, pH 7.5	>90% DNA fragments ligated and recut after 10-fold overdigestion	Adv Biotech AB-0208-a AB-0208-b

Bfa I

RECOGNITION SEQUENCE
  5'...C▼TAG...3'
  3'...GAT▲C...5'
ISOSCHIZOMERS
  Mae I
OPTIMUM REACTION CONDITIONS
  pH 7.4-7.9 and 37°C
HEAT INACTIVATION
  Resistant

NOTES
- EC 3.1.21.4
- A type II site specific deoxyribonuclease
- A large group of enzymes recognizing specific short DNA sequences. They cleave either within the recognition site, or at a short specific distance from it
- Requires Mg^{2+}

CONCENTRATION or VOLUME ACTIVITY	UNIT DEFINITION	REACTION BUFFER (As Provided)	ADDITIONAL ACTIVITIES and PURITY	SUPPLIER CATALOG NO.
Bacillus fragilis (NEB 688)				
5000 U/mL	1 unit completely digests 1 µg λ DNA/hr in a total reaction volume of 50 µL using the buffer provided	20 m*M* Tris-acetate, 10 m*M* MgOAc, 50 m*M* KOAc, 1 m*M* DTT, pH 7.9	>50% DNA fragments ligated and recut after 10-fold overdigestion	NE Biolabs 568S 568L
Bacteroides fragilis				
1000-10,000 U/mL		50 m*M* KCl, 10 m*M* Tris-HCl, 0.1 m*M* EDTA, 1 m*M* DTT, 200 µg/mL acetylated BSA, 50% glycerol, pH 7.4		ICN 159401

Bfm I

RECOGNITION SEQUENCE
5'...C▼TPuPyAG...3'
3'...GAPyPuT▲C...5'

ISOSCHIZOMERS
Sfe I

OPTIMUM REACTION CONDITIONS
pH 7.9 and 37°C

HEAT INACTIVATION
65°C for 20 min.

NOTES
- EC 3.1.21.4
- A type II site specific deoxyribonuclease
- A large group of enzymes recognizing specific short DNA sequences. They cleave either within the recognition site, or at a short specific distance from it
- Requires Mg^{2+} and BSA for full activity

CONCENTRATION or VOLUME ACTIVITY	UNIT DEFINITION	REACTION BUFFER (As Provided)	ADDITIONAL ACTIVITIES and PURITY	SUPPLIER CATALOG NO.
Bacillus firmus S8-336				
4-8 U/μL	1 unit hydrolyzes 1 μg DNA/hr at 37°C in a 50 μL reaction volume using λ phage DNA as substrate	33 mM Tris-acetate, 10 mM MgOAc, 66 mM KOAc, 0.1 mg/mL BSA, pH 7.9	>90% DNA fragments ligated and recut after 50-fold overdigestion	Fermentas ER1161 ER1162

Bfr I

RECOGNITION SEQUENCE
5'...C▼TTAAG...3'
3'...GAATT▲C...5'

ISOSCHIZOMERS
Afl II

OPTIMUM REACTION CONDITIONS
pH 7.5 and 37°C

HEAT INACTIVATION
65°C for 20 min.

METHYLATION EFFECTS
Does not cleave m^5CTTAAG, GCC(N)$_5$GGm^5C or GCm^4C(N)$_5$GGC

NOTES
- EC 3.1.21.4
- A type II site specific deoxyribonuclease
- A large group of enzymes recognizing specific short DNA sequences. They cleave either within the recognition site, or at a short specific distance from it
- Requires Mg^{2+}

Bfr I continued

CONCENTRATION or VOLUME ACTIVITY	UNIT DEFINITION	REACTION BUFFER (As Provided)	ADDITIONAL ACTIVITIES and PURITY	SUPPLIER CATALOG NO.
Bacteroides fragilis				
10 U/μL	1 unit completely digests 1 μg substrate DNA/hr	10 mM Tris-HCl, 10 mM MgCl$_2$, 50 mM NaCl, 1 mM DTE, pH 7.5		Boehringer 1198939 1198947
				Toyobo

Bgl I

RECOGNITION SEQUENCE
  5'...GCCNNNN▼NGGC...3'
  3'...CGGN▲NNNNCCG...5'
OPTIMUM REACTION CONDITIONS
  37°C; 2-fold greater activity at pH 8.0 than pH 7.4; activity further increased at pH 9.5
HEAT INACTIVATION
  65°C for 20 min.
METHYLATION EFFECTS
  Cleaves GCm^{5}C(N)$_5$GGC
  Does not cleave Gm^{5}CC(N)$_5$GGC, GCC(N)$_5$GGm^{5}C
  or GCm^{4}C(N)$_5$GCC
  Blocked by overlapping CG methylation

NOTES
- EC 3.1.21.4
- A type II site specific deoxyribonuclease
- A large group of enzymes recognizing specific short DNA sequences. They cleave either within the recognition site, or at a short specific distance from it
- Sticky ends produced by *Bgl* I cleavage can be used to reconstitute plasmid and phage genomes and to exchange wild-type and mutant DNA fragments
- Requires Mg^{2+}
- Optimal [KCl] = 100-160 mM and [NaCl] = 12-160 mM

CONCENTRATION or VOLUME ACTIVITY	UNIT DEFINITION	REACTION BUFFER (As Provided)	ADDITIONAL ACTIVITIES and PURITY	SUPPLIER CATALOG NO.
Bacillus globigii				
2500 U and 5000 U	1 unit completely digests 1 μg DNA substrate/hr at 37°C	10 mM Tris-HCl, 10 mM MgCl$_2$, 100 mM NaCl, 1 mM DTT, 0.1 mg/mL BSA, pH 7.5	>95% DNA fragments ligated and recut after 10-fold overdigestion	Adv Biotech AB-0401-a AB-0401-b

BgI I continued

CONCENTRATION or VOLUME ACTIVITY	UNIT DEFINITION	REACTION BUFFER (As Provided)	ADDITIONAL ACTIVITIES and PURITY	SUPPLIER CATALOG No.
Bacillus globigii continued				
10 U/µL; 40 U/µL	1 unit completely cleaves 1 µg substrate DNA/hr	100 mM NaCl, 50 mM Tris, 5 mM MgCl$_2$, 2 mM β-MSH, pH 7.8	90-98% cleavage-ligation-recleavage	Am Allied L-06-01000 H-06-01000 L-06-05000 H-06-05000
8-12 U/µL	1 unit completely digests 1 µg λ DNA/hr at 37°C in the buffer provided	500 mM Tris-HCl, 100 mM MgCl$_2$, 10 mM DTT, 1000 mM NaCl, pH 7.5	>90% DNA fragments ligated and recut	Amersham E 1020Y E 1020Z
8-12 U/µL	1 unit hydrolyzes 1 µg DNA/hr at 37°C in a 50 µL reaction volume using λ phage DNA as substrate	50 mM Tris-HCl, 10 mM MgCl$_2$, 100 mM NaCl, pH 7.5	>90% of the DNA fragments ligated and recut after 50-fold overdigestion	Fermentas ER0071 ER0072
8-12 U/µL	1 unit completely digests 1 mg λ DNA/hr at 37°C	500 mM Tris-HCl, 100 mM MgCl$_2$, 500 mM NaCl, pH 8.0	≥95% ligation; 100% cleavage	Life Technol 15219-017
	10 units does not produce unspecific cleavage products with 1 µg λ DNA/16 hr at 37°C	10 mM Tris-HCl, 10 mM MgCl$_2$, 150 mM KCl, 1 mM DTT, 100 µg/mL BSA, pH 7.8	>90% DNA fragments ligated and recut after 10-fold overdigestion	MINOTECH 104
4-12 U/µL	1 unit digests completely 1 µg λ DNA substrate/hr at 37°C	50 mM Tris-HCl, 100 mM NaCl, 10 mM MgCl$_2$, 1 mM DTT, pH 7.8		NBL Gene 013903 013906
10 U/µL and 15-100 U/µL	1 unit hydrolyzes 1 µg λ DNA to completion/hr in a 50 µL total reaction volume	100 mM Tris-HCl, 1 M NaCl, 100 mM MgCl$_2$, 100 mM β-MSH, 1.0 mg/mL BSA, pH 8		Nippon Gene Oncor 110091 110092
8000-15,000 U/mL	1 unit completely digests 1 µg λ DNA/hr at pH 7.5, 37°C in a 50 µL total assay mixture	100 mM Tris-acetate, 100 mM MgOAc, 500 mM KOAc, pH 7.5		Pharmacia 27-0944-01 27-0944-02
10 U/µL and 40-80 U/µL	1 unit completely digests 1 µg λ DNA/hr at 37°C in a 50 mL reaction volume	Genome qualified; 60 mM Tris-HCl, 60 mM MgCl$_2$, 1.5 M NaCl, 10 mM DTT, pH 7.9	90% ligation	Promega R6071 R6072 R4074
2000-20,000 U/mL	1 unit completely cleaves 1.0 µg λ DNA/hr in a 50 µL reaction mixture	Unique buffer + BSA	100% cleavage, 90% ligation, 90% recleavage, 50% endonuclease (nickase) No detectable endonuclease (overdigestion)	SibEnzyme Sigma R6753
5000-15,000 U/mL		50 mM KCl, 10 mM Tris-HCl, 0.1 mM EDTA, 1 mM DTT, 200 µg/mL BSA, 50% glycerol, pH 7.4		Stratagene 500270 500271 TaKaRa

BgI I continued

CONCENTRATION or VOLUME ACTIVITY	UNIT DEFINITION	REACTION BUFFER (As Provided)	ADDITIONAL ACTIVITIES and PURITY	SUPPLIER CATALOG NO.
Bacillus globigii continued				
3-20 U/μL 40-50 U/μL	1 unit completely digests 1 μg λ DNA/hr in 0.05 mL at 37°C	50 mM Tris-HCl, 10 mM MgCl$_2$, 100 mM NaCl, 1 mM DTT, pH 7.5	>90% ligation >95% recutting	Toyobo BGL-101 BGL-102 BGL-152
Bacillus globigii RUB 561				
10 U/μL and 40 U/μL	1 unit completely digests 1 μg substrate DNA/hr	50 mM Tris-HCl, 10 mM MgCl$_2$, 100 mM NaCl, 1 mM DTE, pH 7.5, 37°C		Boehringer 404101 621641 1047604
	1 unit completely cleaves 1 μg λ DNA/hr at pH 7.5, 37°C in a reaction volume of 50 μL	50 mM Tris-HCl, 10 mM MgCl$_2$, 100 mM NaCl, 1.0 mM DTT, 100 μg/mL BSA, pH 7.5		CHIMERx 2080-01 2080-02
2000-20,000 U/mL		200 mM KCl, 10 mM KPO$_4$, 0.1 mM EDTA, 10 mM β-MSH, 200 μg/mL BSA, 50% glycerol, pH 7.4		ICN 150463
Bacillus globigii strain, lacking BgI II				
10,000 U/mL	1 unit degrades 1 μg λ dam$^-$ dcm$^-$ DNA/hr at 37°C in a 0.05 mL mixture	10 mM Tris-HCl, 10 mM MgCl$_2$, 100 mM KCl, 100 μg/mL BSA, pH 8.5	No detectable contaminating nuclease; 90% ligated and 100% recut after 50-fold overdigestion	AGS Heidelb F00140S F00140M
Escherichia coli strain, carrying the cloned gene from Bacillus globigii				
10,000 U/mL	1 unit completely digests 1 μg λ DNA/hr in a total reaction volume of 50 μL using the buffer provided	50 mM Tris-HCl, 10 mM MgCl$_2$, 100 mM NaCl, 1 mM DTT, pH 7.9	>95% DNA fragments ligated and recut after 10-fold overdigestion	NE Biolabs 143S 143L

Bgl II

RECOGNITION SEQUENCE
5'...A▼GATCT...3'
3'...TCTAG▲A...5'

OPTIMUM REACTION CONDITIONS
37°C; 2-fold greater activity at pH 8.0 than 7.4

HEAT INACTIVATION
Resistant

METHYLATION EFFECTS
Cleaves AGm^{6}ATCT and AGAhm^{5}UChm^{5}U
Does not cleave AGATm^{5}CT or AGAThm^{5}CT
Not blocked by *dam* methylation

NOTES
- EC 3.1.21.4
- A type II site specific deoxyribonuclease
- A large group of enzymes recognizing specific short DNA sequences. They cleave either within the recognition site, or at a short specific distance from it
- Requires Mg^{2+}
- Optimal [KCl or NaCl] = 100-160 mM
- Compatible cohesive ends: *Bam*H I, *Bcl* I, *Bsi*Q I, *Bst* I, *Bst*Y I, *Dpn* II, *Mbo* I, *Mfl* I, *Nde* II, *Sau*3A I, *Xho* II

CONCENTRATION or VOLUME ACTIVITY	UNIT DEFINITION	REACTION BUFFER (As Provided)	ADDITIONAL ACTIVITIES and PURITY	SUPPLIER CATALOG No.
Bacillus globigii				
2000 U and 5000 U	1 unit completely digests 1 μg DNA substrate/hr at 37°C	10 mM Tris-HCl, 10 mM MgCl$_2$, 100 mM NaCl, 1 mM DTT, 0.1 mg/mL BSA, pH 7.5	>90% DNA fragments ligated and recut after 10-fold overdigestion	Adv Biotech AB-0209-a AB-0209-b
10 U/μL; 40 U/μL	1 unit completely cleaves 1 μg substrate DNA/hr	100 mM NaCl, 50 mM Tris, 5 mM MgCl$_2$, 2 mM β-MSH, pH 7.8	90-98% cleavage-ligation-recleavage	Am Allied L-07-00500 H-07-00500 L-07-02500 H-07-02500 L-07-05000 H-07-05000
>40 U/μL	1 unit completely digests 1 μg λ DNA/hr at 37°C in the buffer provided	500 mM Tris-HCl, 100 mM MgCl$_2$, 10 mM DTT, 1000 mM NaCl, pH 7.5	>95% DNA fragments ligated and 100% recut	Amersham E 1021XH
8-12 U/μL	1 unit completely digests 1 μg λ DNA/hr at 37°C in the buffer provided	500 mM Tris-HCl, 100 mM MgCl$_2$, 10 mM DTT, 1000 mM NaCl, pH 7.5	>95% DNA fragments ligated and 100% recut	Amersham E 1021Y E 1021Z
10 U/μL and 40 U/μL	1 unit completely digests 1 μg substrate DNA/hr	10 mM Tris-HCl, 10 mM MgCl$_2$, 50 mM NaCl, 1 mM DTE, pH 7.5		Boehringer 348767 567639 899224 1175068

BglII continued

CONCENTRATION or VOLUME ACTIVITY	UNIT DEFINITION	REACTION BUFFER (As Provided)	ADDITIONAL ACTIVITIES and PURITY	SUPPLIER CATALOG No.
Bacillus globigii continued				
	1 unit completely cleaves 1 μg λ DNA/hr at pH 7.5, 37°C in a reaction volume of 50 μL	10 mM Tris-HCl, 10 mM MgCl$_2$, 150 mM KCl, 1.0 mM DTT, 100 μg/mL BSA, pH 8.0		CHIMERx 2090-01 2090-02
8-12 U/μL	1 unit hydrolyzes 1 μg DNA/hr at 37°C in a 50 μL reaction volume using λ phage DNA as substrate	50 mM Tris-HCl, 10 mM MgCl$_2$, 100 mM NaCl, pH 7.5	>90% of the DNA fragments ligated and recut after 50-fold overdigestion	Fermentas ER0081 ER0082
8-12 U/μL and 50 U/μL	1 unit completely digests 1 mg λ DNA/hr at 37°C	500 mM Tris-HCl, 100 mM MgCl$_2$, 1 M NaCl, pH 8.0	≥95% ligation; 100% cleavage	Life Technol 15213-010 15213-028 15213-093 15213-036
	10 units does not produce unspecific cleavage products with 1 μg λ DNA/16 hr at 37°C	50 mM Tris-HCl, 10 mM MgCl$_2$, 10 mM NaCl, 1 mM DTT, 100 μg/mL BSA, pH 7.9	>90% DNA fragments ligated and recut after 10-fold overdigestion	MINOTECH 105
4-12 U/μL	1 unit digests completely 1 μg λ DNA substrate/hr at 37°C	20 mM Gly-NaOH, 200 mM NaCl, 10 mM MgCl$_2$, 7 mM β-MSH, pH 9.5		NBL Gene 010802 010805 010807
				Nippon Gene
10 U/μL and 20-100 U/μL	1 unit hydrolyzes 1 μg λ DNA to completion/hr in a 50 μL total reaction volume	100 mM Tris-HCl, 1 M NaCl, 100 mM MgCl$_2$, 100 mM β-MSH, 1.0 mg/mL BSA, pH 8		Oncor 110101 110102 110105
8000-15,000 U/mL and 50,000-80,000 U/mL	1 unit completely digests 1 μg λ DNA/hr at pH 7.5, 37°C in a 50 μL total assay mixture	100 mM Tris-acetate, 100 mM MgOAc, 500 mM KOAc, pH 7.5		Pharmacia 27-0946-01 27-0946-02 27-0946-18
10 U/μL and 40-80 U/μL	1 unit completely digests 1 μg λ DNA/hr at 37°C in a 50 mL reaction volume	60 mM Tris-HCl, 60 mM MgCl$_2$, 1.5 M NaCl, 10 mM DTT, pH 7.9	90% ligation	Promega R6081 R6082 R6085 R6086 R4084
				SibEnzyme
10,000-20,000 U/mL	1 unit cleaves 98% λ DNA/hr in a 50 μL reaction mixture	500 mM Tris-HCl, 1 M NaCl, 100 mM MgCl$_2$, 10 mM DTT, pH 7.9	98% cleavage, 95% ligation, 95% recleavage No detectable endonuclease (nickase), endonuclease (overdigestion)	Sigma R6377

Bgl II continued

CONCENTRATION or VOLUME ACTIVITY	UNIT DEFINITION	REACTION BUFFER (As Provided)	ADDITIONAL ACTIVITIES and PURITY	SUPPLIER CATALOG No.
Bacillus globigii continued				
4000-15,000 U/mL		200 mM KCl, 10 mM Tris-HCl, 0.1 mM EDTA, 10 mM β-MSH, 200 μg/mL BSA, 50% glycerol, pH 7.4		Stratagene 500280 500281
2000 U and 10,000 U	1 unit completely digests 1 μg λ DNA/hr at 37°C in 50 μL of supplied buffer	500 mM Tris-HCl, 100 mM MgCl$_2$, 10 mM DTT, 1 M NaCl, pH 7.5	Purity is confirmed by overdigestion, ligation-recutting and cloning test	TaKaRa 1021
3-20 U/μL 40-50 U/μL	1 unit completely digests 1 μg λ DNA/hr in 0.05 mL at 37°C	50 mM Tris-HCl, 10 mM MgCl$_2$, 100 mM NaCl, 1 mM DTT, pH 7.5	>90% ligation >95% recutting	Toyobo BGL-201 BGL-202 BGL-252
Bacillus globigii Rub 562				
3000-40,000 U/mL		10 mM MgCl$_2$, 10 mM Tris-HCl, 10 mM β-MSH, 100 μg/mL BSA, 50% glycerol, pH 7.4		ICN 150464
Bacillus globigii strain, lacking Bgl I				
10,000 U/mL	1 unit degrades 1 μg λ dam⁻ dcm⁻ DNA/hr at 37°C in a 0.05 mL mixture	10 mM Tris-HCl, 100 mM NaCl, 100 μg/mL BSA, pH 8.0	No detectable contaminating nuclease; 90% ligated and 100% recut after 50-fold overdigestion	AGS Heidelb A00150S A00150M
Escherichia coli strain, carrying the cloned gene from Bacillus globigii				
10,000 U/mL; 50,000 U/mL	1 unit completely digests 1 μg λ DNA/hr in a total reaction volume of 50 μL using the buffer provided	50 mM Tris-HCl, 10 mM MgCl$_2$, 100 mM NaCl, 1 mM DTT, pH 7.9	>95% DNA fragments ligated and recut after 10-fold overdigestion	NE Biolabs 144S 144L

Bln I

RECOGNITION SEQUENCE
5′...C▼CTAGG...3′
3′...GGATC▲C...5′

ISOSCHIZOMERS
Avr II

OPTIMUM REACTION CONDITIONS
pH 7.5-8.5 and 37°C

HEAT INACTIVATION
Resistant; use ethanol precipitation

NOTES
- EC 3.1.21.4
- A type II site specific deoxyribonuclease
- A large group of enzymes recognizing specific short DNA sequences. They cleave either within the recognition site, or at a short specific distance from it
- Requires Mg^{2+}

CONCENTRATION or VOLUME ACTIVITY	UNIT DEFINITION	REACTION BUFFER (As Provided)	ADDITIONAL ACTIVITIES and PURITY	SUPPLIER CATALOG No.
Brevibacterium linens				
8-12 U/μL	1 unit completely digests 1 μg λ DNA/hr at 37°C in the buffer provided	200 mM Tris-HCl, 100 mM MgCl$_2$, 10 mM DTT, 1000 mM KCl, pH 8.5	>90% DNA fragments ligated and 100% recut	Amersham E 1022Y E 1022Z
10 U/μL	1 unit completely digests 1 μg substrate DNA/hr	50 mM Tris-HCl, 10 mM MgCl$_2$, 100 mM NaCl, 1 mM DTE, pH 7.5		Boehringer 1558161 1558170
400 U and 2000 U	1 unit completely cleaves 1 μg DNA/hr in 50 μL reaction mixture	200 mM Tris-HCl, 100 mM MgCl$_2$, 10 mM DTT, 1 M KCl, pH 8.5		PanVera TAK 1022
400 U and 2000 U	1 unit completely digests 1 μg λ DNA/hr at 37°C in 50 μL of supplied buffer	Genomic grade; 200 mM Tris-HCl, 100 mM MgCl$_2$, 10 mM DTT, 1 mM KCl, pH 8.5	Purity is confirmed by overdigestion, ligation-recutting and genome DNA digestion test	TaKaRa 1022

Blp I

RECOGNITION SEQUENCE
   5'...GC▼TNAGC...3'
   3'...CGANT▲CG...5'

ISOSCHIZOMERS
   Bpu1102 I

OPTIMUM REACTION CONDITIONS
   pH 7.9 and 37°C

HEAT INACTIVATION
   Resistant

NOTES
- EC 3.1.21.4
- A type II site specific deoxyribonuclease
- A large group of enzymes recognizing specific short DNA sequences. They cleave either within the recognition site, or at a short specific distance from it
- Requires Mg^{2+}

CONCENTRATION or VOLUME ACTIVITY	UNIT DEFINITION	REACTION BUFFER (As Provided)	ADDITIONAL ACTIVITIES and PURITY	SUPPLIER CATALOG NO.
Bacillus lentus (NEB 819)				
10,000 U/mL	1 unit completely digests 1 μg λ DNA/hr in a total reaction volume of 50 μL using the buffer provided	20 m*M* Tris-acetate, 10 m*M* MgOAc, 50 m*M* KOAc, 1 m*M* DTT, pH 7.9	>80% DNA fragments ligated and >95% recut after 10-fold overdigestion	NE Biolabs 585S 585L

Bme18 I

RECOGNITION SEQUENCE
   5'...G▼GWCC...3'
   3'...CCWG▲G...5'

ISOSCHIZOMERS
   Ava II, Eco47 I, HgiE I, Sin I

NOTES
- EC 3.1.21.4
- A type II site specific deoxyribonuclease
- A large group of enzymes recognizing specific short DNA sequences. They cleave either within the recognition site, or at a short specific distance from it
- Requires Mg^{2+}

CONCENTRATION or VOLUME ACTIVITY	UNIT DEFINITION	REACTION BUFFER (As Provided)	ADDITIONAL ACTIVITIES and PURITY	SUPPLIER CATALOG NO.
Bacillus megaterium 18				
				SibEnzyme

Bmy I

RECOGNITION SEQUENCE

$5'...G(^G_{AT})GC(^C_{TA})\blacktriangledown C...3'$
$3'...C\blacktriangle(^C_{TA})CG(^G_{AT})G...5'$

ISOSCHIZOMERS
  *Bsp*1286 I

OPTIMUM REACTION CONDITIONS
  pH 7.9 and 37°C

HEAT INACTIVATION
  65°C for 20 min.

NOTES
- EC 3.1.21.4
- A type II site specific deoxyribonuclease
- A large group of enzymes recognizing specific short DNA sequences. They cleave either within the recognition site, or at a short specific distance from it
- Requires Mg^{2+}

CONCENTRATION or VOLUME ACTIVITY	UNIT DEFINITION	REACTION BUFFER (As Provided)	ADDITIONAL ACTIVITIES and PURITY	SUPPLIER CATALOG NO.
Bacillus mycoides				
10 U/μL	1 unit completely digests 1 μg substrate DNA/hr	33 m*M* Tris-acetate, 10 m*M* MgOAc, 66 m*M* KOAc, 0.5 m*M* DTT, pH 7.9		Boehringer 1269089

Bpi I

RECOGNITION SEQUENCE
  5'...GAAGAC(N)$_2$▼...3'
  3'...CTTCTG(N)$_6$▲...5'

ISOSCHIZOMERS
  Bbv II

OPTIMUM REACTION CONDITIONS
  pH 7.5-7.9 and 37°C

HEAT INACTIVATION
  65°C for 20 min.

NOTES
- EC 3.1.21.4
- A type II site specific deoxyribonuclease
- A large group of enzymes recognizing specific short DNA sequences. They cleave either within the recognition site, or at a short specific distance from it
- Requires Mg^{2+} and BSA for full activity

CONCENTRATION or VOLUME ACTIVITY	UNIT DEFINITION	REACTION BUFFER (As Provided)	ADDITIONAL ACTIVITIES and PURITY	SUPPLIER CATALOG No.
Bacillus pumilus Sw 4-3				
10,000 U/mL	1 unit degrades 1 μg λ dam⁻ dcm⁻ DNA/hr at 37°C in a 0.05 mL mixture	33 mM Tris-acetate, 10 mM MgOAc, 66 mM KOAc, 100 μg/mL BSA, pH 7.9	No detectable contaminating nuclease; 90% ligated and 100% recut after 50-fold overdigestion	AGS Heidelb F00153S F00153M
8-12 U/μL	1 unit hydrolyzes 1 μg DNA/hr at 37°C in a 50 μL reaction volume using λ phage DNA as substrate	10 mM Tris-HCl, 10 mM MgCl$_2$, 50 mM NaCl, 0.1 mg/mL BSA, pH 7.5	>90% of the DNA fragments ligated and recut after 50-fold overdigestion	Fermentas ER1011 ER1012

Bpm I

RECOGNITION SEQUENCE
5'...CTGGAG(N)$_{16}$▼...3'
3'...GACCTC(N)$_{14}$▲...5'

ISOSCHIZOMERS
Gsu I

OPTIMUM REACTION CONDITIONS
pH 7.4-7.9 and 37°C

HEAT INACTIVATION
65°C for 20 min.

METHYLATION EFFECTS
Blocked by overlapping *dcm* methylation

NOTES
- EC 3.1.21.4
- A type II site specific deoxyribonuclease
- A large group of enzymes recognizing specific short DNA sequences. They cleave either within the recognition site, or at a short specific distance from it
- Requires Mg^{2+} and BSA for full activity

CONCENTRATION or VOLUME ACTIVITY	UNIT DEFINITION	REACTION BUFFER (As Provided)	ADDITIONAL ACTIVITIES and PURITY	SUPPLIER CATALOG NO.
Bacillus pumilus				
1000-5000 U/mL		100 m*M* NaCl, 50 m*M* Tris-HCl, 10 m*M* MgCl$_2$, 1 m*M* DTT, 200 µg/mL acetylated BSA, 50% glycerol, pH 7.4		ICN 159402
Bacillus pumilus (NEB 711)				
2000 U/mL	1 unit completely digests 1 µg λ DNA/hr in a total reaction volume of 50 µL using the buffer provided	50 m*M* Tris-HCl, 10 m*M* MgCl$_2$, 100 m*M* NaCl, 1 m*M* DTT, pH 7.9; add 100 µg/mL BSA	>50% DNA fragments ligated and recut after 5-fold overdigestion	NE Biolabs 565S 565L

Bpu14 I

RECOGNITION SEQUENCE
5'...TT▼CGAA...3'
3'...AAGC▲TT...5'

ISOSCHIZOMERS
BsiC I, Bsp119 I, BstB I, Cbi I, Csp45 I, Lsp I, Nsp V, Sfu I

NOTES
- EC 3.1.21.4
- A type II site specific deoxyribonuclease
- A large group of enzymes recognizing specific short DNA sequences. They cleave either within the recognition site, or at a short specific distance from it
- Requires Mg^{2+}

CONCENTRATION or VOLUME ACTIVITY	UNIT DEFINITION	REACTION BUFFER (As Provided)	ADDITIONAL ACTIVITIES and PURITY	SUPPLIER CATALOG NO.
Bacillus pumilus 14				
				SibEnzyme

Bpu1102 I

RECOGNITION SEQUENCE
5'...GC▼TNAGC...3'
3'...CGANT▲CG...5'

ISOSCHIZOMERS
Cel II, Esp I

OPTIMUM REACTION CONDITIONS
pH 7.4-8.0 and 37°C

HEAT INACTIVATION
Resistant

NOTES
- EC 3.1.21.4
- A type II site specific deoxyribonuclease
- A large group of enzymes recognizing specific short DNA sequences. They cleave either within the recognition site, or at a short specific distance from it
- Requires Mg^{2+}

Bpu1102 I continued

CONCENTRATION or VOLUME ACTIVITY	UNIT DEFINITION	REACTION BUFFER (As Provided)	ADDITIONAL ACTIVITIES and PURITY	SUPPLIER CATALOG NO.
Bacillus pumilus				
5000 U/mL	1 unit degrades 1 μg λ dam⁻ dcm⁻ DNA/hr at 37°C in a 0.05 mL mixture	33 mM Tris-acetate, 10 mM MgOAc, 66 mM KOAc, 100 μg/mL BSA, pH 7.9	No detectable contaminating nuclease; 90% ligated and 100% recut after 50-fold overdigestion	AGS Heidelb F00313S F00313M
1000-20,000 U/mL		Reaction buffer		ICN 159403
100 U and 500 U	1 unit completely digests 1 μg λ DNA/hr at 37°C in 50 μL of supplied buffer and BSA	330 mM Tris-acetate, 100 mM MgOAc, 5 mM DTT, 660 mM KOAc, BSA, pH 7.9	Purity is confirmed by overdigestion and ligation-recutting test	TaKaRa 1023
Bacillus pumilus RFL1102				
8-12 U/μL	1 unit hydrolyzes 1 μg DNA/hr at 37°C in a 50 μL reaction volume using λ phage DNA as substrate	33 mM Tris-acetate, 10 mM MgOAc, 66 mM KOAc, pH 7.9	>80% of the DNA fragments ligated and >90% recut after 50-fold overdigestion	Fermentas ER0091 ER0092
4-6 U/μL	1 unit completely digests 1 mg λ DNA/hr at 37°C	500 mM Tris-HCl, 100 mM MgCl₂, 500 mM NaCl, pH 8.0	≥95% ligation; 100% cleavage	Life Technol 15486-012
5000-10,000 U/mL		50 mM KCl, 10 mM Tris-HCl, 0.1 mM EDTA, 1 mM DTT, 200 μg/mL BSA, 50% glycerol, pH 7.4		Stratagene 500286

BpuA I

RECOGNITION SEQUENCE
  5'...GAAGAC(N)$_2$▼...3'
  3'...CTTCTG(N)$_6$▲...5'
OPTIMUM REACTION CONDITIONS
  pH 8.0 and 37°C
HEAT INACTIVATION
  65°C for 20 min.

NOTES
- EC 3.1.21.4
- A type II site specific deoxyribonuclease
- A large group of enzymes recognizing specific short DNA sequences. They cleave either within the recognition site, or at a short specific distance from it
- Requires Mg^{2+}

BpuA I continued

CONCENTRATION or VOLUME ACTIVITY	UNIT DEFINITION	REACTION BUFFER (As Provided)	ADDITIONAL ACTIVITIES and PURITY	SUPPLIER CATALOG NO.
Bacillus pumilus				
10 U/μL	1 unit completely digests 1 μg substrate DNA/hr	10 mM Tris-HCl, 5 mM MgCl$_2$, 100 mM NaCl, 1 mM β-MSH, pH 8.0		Boehringer 1497944

Bsa I

RECOGNITION SEQUENCE
    5'...GGTCTC(N)$_1$▼...3'
    3'...CCAGAG(N)$_5$▲...5'

OPTIMUM REACTION CONDITIONS
    pH 7.4-7.9 and 55°C; 10% activity at 37°C

HEAT INACTIVATION
    65°C for 20 min.

METHYLATION EFFECTS
    Does not cleave GGTCTm^{5}C

NOTES
- EC 3.1.21.4
- A type II site specific deoxyribonuclease
- A large group of enzymes recognizing specific short DNA sequences. They cleave either within the recognition site, or at a short specific distance from it
- Requires Mg^{2+}

CONCENTRATION or VOLUME ACTIVITY	UNIT DEFINITION	REACTION BUFFER (As Provided)	ADDITIONAL ACTIVITIES and PURITY	SUPPLIER CATALOG NO.
Bacillus stearothermophilus				
1000-10,000 U/mL		100 mM NaCl, 10 mM Tris-HCl, 0.1 mM EDTA, 1 mM DTT, 200 μg/mL acetylated BSA, 50% glycerol, pH 7.4		ICN 159404
Bacillus stearothermophilus 6-55 (Z. Chen)				
5000 U/mL	1 unit completely digests 1 μg T7 DNA/hr in a total reaction volume of 50 μL using the buffer provided	20 mM Tris-acetate, 10 mM MgOAc, 50 mM KOAc, 1 mM DTT, pH 7.9	>95% DNA fragments ligated and recut after 10-fold overdigestion	NE Biolabs 535S 535L

Bsa29 I

RECOGNITION SEQUENCE
5'...AT▼CGAT...3'
3'...TAGC▲TA...5'

ISOSCHIZOMERS
Ban III, Bsc I, BseC I, BsiX I, Bsp106 I, BspD I, BspX I, Bsu15 I, Cla I

NOTES
- EC 3.1.21.4
- A type II site specific deoxyribonuclease
- A large group of enzymes recognizing specific short DNA sequences. They cleave either within the recognition site, or at a short specific distance from it
- Requires Mg^{2+}

CONCENTRATION or VOLUME ACTIVITY	UNIT DEFINITION	REACTION BUFFER (As Provided)	ADDITIONAL ACTIVITIES and PURITY	SUPPLIER CATALOG No.
Bacillus species 29				
				SibEnzyme

BsaA I

RECOGNITION SEQUENCE
5'...PyAC▼GTPu...3'
3'...PuTG▲CAPy...5'

OPTIMUM REACTION CONDITIONS
pH 7.4-7.9 and 37°C

HEAT INACTIVATION
Resistant

METHYLATION EFFECTS
Does not cleave $(^C_T)Am^5CGT(^A_G)$

NOTES
- EC 3.1.21.4
- A type II site specific deoxyribonuclease
- A large group of enzymes recognizing specific short DNA sequences. They cleave either within the recognition site, or at a short specific distance from it
- Requires Mg^{2+}

BsaA I continued

CONCENTRATION or VOLUME ACTIVITY	UNIT DEFINITION	REACTION BUFFER (As Provided)	ADDITIONAL ACTIVITIES and PURITY	SUPPLIER CATALOG NO.
Bacillus stearothermophilus A				
1000-5000 U/mL		50 mM KCl, 10 mM Tris-HCl, 0.1 mM EDTA, 1 mM DTT, 200 µg/mL acetylated BSA, 50% glycerol, pH 7.4		ICN 159405
Escherichia coli strain, carrying the cloned gene from *Bacillus stearothermophilus* A				
5000 U/mL	1 unit completely digests 1 µg λ DNA/hr in a total reaction volume of 50 µL using the buffer provided	50 mM Tris-HCl, 10 mM MgCl$_2$, 100 mM NaCl, 1 mM DTT, pH 7.9	> 95% DNA fragments ligated and recut after 10-fold overdigestion	NE Biolabs 531S 531L

BsaB I

RECOGNITION SEQUENCE

    5'...GATNN▼NNATC...3'
    3'...CTANN▲NNTAG...5'

ISOSCHIZOMERS

*Bsi*B I, *Mam* I

OPTIMUM REACTION CONDITIONS

pH 7.4-7.9 and 60°C; 20% activity at 37°C

HEAT INACTIVATION

Resistant

METHYLATION EFFECTS

Does not cleave GAT(N)$_4$ATm^{5}C
Blocked by overlapping *dam* methylation

NOTES

- EC 3.1.21.4
- A type II site specific deoxyribonuclease
- A large group of enzymes recognizing specific short DNA sequences. They cleave either within the recognition site, or at a short specific distance from it
- Requires Mg^{2+}

CONCENTRATION or VOLUME ACTIVITY	UNIT DEFINITION	REACTION BUFFER (As Provided)	ADDITIONAL ACTIVITIES and PURITY	SUPPLIER CATALOG NO.
Bacillus stearothermophilus B				
5000-20,000 U/mL		50 mM NaCl, 10 mM Tris-HCl, 0.1 mM EDTA, 1 mM DTT, 200 µg/mL acetylated BSA, 50% glycerol, pH 7.4		ICN 159406

BsaB I continued

CONCENTRATION or VOLUME ACTIVITY	UNIT DEFINITION	REACTION BUFFER (As Provided)	ADDITIONAL ACTIVITIES and PURITY	SUPPLIER CATALOG NO.
Bacillus stearothermophilus B (Z. Chen)				
20,000 U/mL	1 unit completely digests 1 μg λ DNA/hr in a total reaction volume of 50 μL using the buffer provided	10 mM Tris-HCl, 10 mM MgCl$_2$, 50 mM NaCl, 1 mM DTT, pH 7.9	>95% DNA fragments ligated and recut after 10-fold overdigestion	NE Biolabs 537S 537L

BsaH I

RECOGNITION SEQUENCE
    5'...GPu▼CGPyC...3'
    3'...CPyGC▲PuG...5'
ISOSCHIZOMERS
    Aha II
OPTIMUM REACTION CONDITIONS
    pH 7.5-7.9 and 37°C; 2-fold increase in activity at 60°C
HEAT INACTIVATION
    Resistant

NOTES
- EC 3.1.21.4
- A type II site specific deoxyribonuclease
- A large group of enzymes recognizing specific short DNA sequences. They cleave either within the recognition site, or at a short specific distance from it
- Requires Mg^{2+} and BSA for full activity

CONCENTRATION or VOLUME ACTIVITY	UNIT DEFINITION	REACTION BUFFER (As Provided)	ADDITIONAL ACTIVITIES and PURITY	SUPPLIER CATALOG NO.
Bacillus stearothermophilus				
2000-20,000 U/mL	1 unit cleaves 98% λ DNA/hr in a 50 μL reaction mixture	200 mM Tris-acetate, 500 mM KOAc, 100 mM MgOAc, 10 mM DTT, BSA, pH 7.9	98% cleavage, 100% ligation, 98% recleavage, 100% endonuclease (overdigestion)	Sigma R2760
Bacillus stearothermophilus (Z. Chen)				
10,000 U/mL	1 unit completely digests 1 μg λ DNA/hr in a total reaction volume of 50 μL using the buffer provided	20 mM Tris-acetate, 10 mM MgOAc, 50 mM KOAc, 1 mM DTT, pH 7.9; add 100 μg/mL BSA	>95% DNA fragments ligated and recut after 10-fold overdigestion	NE Biolabs 556S 556L
Herpetosiphon giganteus Hpa2				
2000-20,000 U/mL		50 mM KCl, 10 mM Tris-HCl, 0.1 mM EDTA, 1.0 mM DTT, 200 μg/mL BSA, 50% glycerol, pH 7.5		ICN 151256

BsaJ I

RECOGNITION SEQUENCE
5'...C▼CNNGG...3'
3'...GGNNC▲C...5'

ISOSCHIZOMERS
Sec I

OPTIMUM REACTION CONDITIONS
pH 7.4-7.9 and 60°C; 20% activity at 37°C

HEAT INACTIVATION
Resistant

NOTES
- EC 3.1.21.4
- A type II site specific deoxyribonuclease
- A large group of enzymes recognizing specific short DNA sequences. They cleave either within the recognition site, or at a short specific distance from it
- Requires Mg^{2+}

CONCENTRATION or VOLUME ACTIVITY	UNIT DEFINITION	REACTION BUFFER (As Provided)	ADDITIONAL ACTIVITIES and PURITY	SUPPLIER CATALOG NO.
Bacillus stearothermophilus J				
1000-10,000 U/mL		50 mM NaCl, 10 mM Tris-HCl, 0.1 mM EDTA, 1 mM DTT, 200 µg/mL acetylated BSA, 50% glycerol, pH 7.4		ICN 159408
Bacillus stearothermophilus J (Z. Chen)				
5000 U/mL	1 unit completely digests 1 µg λ DNA/hr in a total reaction volume of 50 µL using the buffer provided	10 mM Tris-HCl, 10 mM MgCl₂, 50 mM NaCl, 1 mM DTT, pH 7.9	>95% DNA fragments ligated and recut after 10-fold overdigestion	NE Biolabs 536S 536L

BsaM I

RECOGNITION SEQUENCE
5'...GAATGCN▼...3'
3'...CTTAC▲GN...5'

ISOSCHIZOMERS
Bsm I

OPTIMUM REACTION CONDITIONS
pH 7.2 and 65°C; <10% activity at 37°C

NOTES
- EC 3.1.21.4
- A type II site specific deoxyribonuclease
- A large group of enzymes recognizing specific short DNA sequences. They cleave either within the recognition site, or at a short specific distance from it
- Requires Mg^{2+}

BsaM I continued

CONCENTRATION or VOLUME ACTIVITY	UNIT DEFINITION	REACTION BUFFER (As Provided)	ADDITIONAL ACTIVITIES and PURITY	SUPPLIER CATALOG NO.
Bacillus stearothermophilus				
10 U/μL	1 unit completely digests 1 μg λ DNA/hr at 65°C in a 50 mL reaction volume	60 mM Tris-HCl, 60 mM MgCl$_2$, 1.5 M NaCl, 10 mM DTT, pH 7.2	90% ligation	Promega R6991 R6992

BsaO I

RECOGNITION SEQUENCE

5'...CG(^A_G)(^T_C)▼CG...3'
3'...GC▲(^T_C)(^A_G)GC...5'

ISOSCHIZOMERS

BsiE I, *Mcr* I (neoschizomer)

OPTIMUM REACTION CONDITIONS

pH 7.6 and 50°C; 50% activity at 37°C

NOTES
- EC 3.1.21.4
- A type II site specific deoxyribonuclease
- A large group of enzymes recognizing specific short DNA sequences. They cleave either within the recognition site, or at a short specific distance from it
- Requires Mg^{2+}

CONCENTRATION or VOLUME ACTIVITY	UNIT DEFINITION	REACTION BUFFER (As Provided)	ADDITIONAL ACTIVITIES and PURITY	SUPPLIER CATALOG NO.
Bacillus stearothermophilus				
10 U/μL	1 unit completely digests 1 μg λ DNA/hr at 50°C in a 50 mL reaction volume	100 mM Tris-HCl, 100 mM MgCl$_2$, 500 mM NaCl, 10 mM DTT, pH 7.6	90% ligation	Promega R6961 R6962

BsaW I

RECOGNITION SEQUENCE
5'...(^A_T)▼CCGG(^A_T)...3'
3'...(^T_A)GGCC▲(^T_A)...5'

ISOSCHIZOMERS
  Bet I

OPTIMUM REACTION CONDITIONS
  pH 7.6-7.9 and 60°C; 25% activity at 37°C

HEAT INACTIVATION
  Resistant

NOTES
- EC 3.1.21.4
- A type II site specific deoxyribonuclease
- A large group of enzymes recognizing specific short DNA sequences. They cleave either within the recognition site, or at a short specific distance from it
- Requires Mg^{2+} and BSA for full activity

CONCENTRATION or VOLUME ACTIVITY	UNIT DEFINITION	REACTION BUFFER (As Provided)	ADDITIONAL ACTIVITIES and PURITY	SUPPLIER CATALOG No.
Bacillus stearothermophilus W (Z. Chen)				
5000 U/mL	1 unit completely digests 1 μg λ DNA/hr in a total reaction volume of 50 μL using the buffer provided	10 mM Tris-HCl, 10 mM MgCl₂, 50 mM NaCl, 1 mM DTT, pH 7.9; add 100 μg/mL BSA	>95% DNA fragments ligated and recut after 10-fold overdigestion	NE Biolabs 567S 567L
Bacillus stearothermophilus WI				
1000-10,000 U/mL		100 mM NaCl, 50 mM Tris-HCl, 0.1 mM EDTA, 1 mM DTT, 200 μg/mL acetylated BSA, 50% glycerol, pH 7.6		ICN 159409

Bsc I

RECOGNITION SEQUENCE
5'...AT▼CGAT...3'
3'...TAGCT▲A...5'

ISOSCHIZOMERS
  None commercially available

OPTIMUM REACTION CONDITIONS
  pH 7.8 and 37°C

METHYLATION EFFECTS
  Does not cleave ATCGm^{6}AT
  Blocked by *dam* methylation

NOTES
- EC 3.1.21.4
- A type II site specific deoxyribonuclease
- A large group of enzymes recognizing specific short DNA sequences. They cleave either within the recognition site, or at a short specific distance from it
- Requires Mg^{2+}

Bsc I continued

CONCENTRATION or VOLUME ACTIVITY	UNIT DEFINITION	REACTION BUFFER (As Provided)	ADDITIONAL ACTIVITIES and PURITY	SUPPLIER CATALOG NO.
Bacillus species				
4-12 U/μL	1 unit digests completely 1 μg λ DNA substrate/hr at 37°C	50 mM Tris-HCl, 100 mM NaCl, 10 mM MgCl$_2$, 1 mM DTT, BSA, pH 7.8		NBL Gene 013304 013307

Bsc4 I

RECOGNITION SEQUENCE

5'...CCNNNNN▼NNGG...3'
3'...GGNN▲NNNNNCC...5'

ISOSCHIZOMERS

BsiY I, Bsl I

NOTES
- EC 3.1.21.4
- A type II site specific deoxyribonuclease
- A large group of enzymes recognizing specific short DNA sequences. They cleave either within the recognition site, or at a short specific distance from it
- Requires Mg^{2+}
- Exhibits star activity at low [salt]

CONCENTRATION or VOLUME ACTIVITY	UNIT DEFINITION	REACTION BUFFER (As Provided)	ADDITIONAL ACTIVITIES and PURITY	SUPPLIER CATALOG NO.
Bacillus schlegelii 4				
				SibEnzyme

Bsc91 I

RECOGNITION SEQUENCE
    5'...GAAGAC(N)₂▼...3'
    3'...CTTCTG(N)₆▲...5'
ISOSCHIZOMERS
    Bbs I
OPTIMUM REACTION CONDITIONS
    pH 7.5 and 37°C

NOTES
- A EC 3.1.21.4
- A type II site specific deoxyribonuclease
- A large group of enzymes recognizing specific short DNA sequences. They cleave either within the recognition site, or at a short specific distance from it
- Requires Mg^{2+}

CONCENTRATION or VOLUME ACTIVITY	UNIT DEFINITION	REACTION BUFFER (As Provided)	ADDITIONAL ACTIVITIES and PURITY	SUPPLIER CATALOG NO.
Bacillus species 91				
10 U/µL	1 unit completely digests 1 µg λ DNA/hr at 37°C in a 50 mL reaction volume	60 mM Tris-HCl, 60 mM MgCl₂, 1 M NaCl, 10 mM DTT, pH 7.5	90% ligation	Promega R7041 R7042

BscB I

RECOGNITION SEQUENCE
    5'...GGN▼NCC...3'
    3'...CCN▲NGG...5'
ISOSCHIZOMERS
    None commercially available
OPTIMUM REACTION CONDITIONS
    pH 7.5 and 55°C

NOTES
- EC 3.1.21.4
- A type II site specific deoxyribonuclease
- A large group of enzymes recognizing specific short DNA sequences. They cleave either within the recognition site, or at a short specific distance from it
- Requires Mg^{2+}

CONCENTRATION or VOLUME ACTIVITY	UNIT DEFINITION	REACTION BUFFER (As Provided)	ADDITIONAL ACTIVITIES and PURITY	SUPPLIER CATALOG NO.
Bacillus species A11				
500 U and 2000 U	1 unit completely digests 1 µg DNA substrate/hr at 55°C	10 mM Tris-HCl, 10 mM MgCl₂, 1 mM DTT, 0.1 mg/mL BSA, pH 7.5	>90% DNA fragments ligated and recut after 10-fold overdigestion	Adv Biotech AB-0210-a AB-0210-b

BscC I

RECOGNITION SEQUENCE
5'...GAATGCN▼...3'
3'...CTTAC▲GN...5'

OPTIMUM REACTION CONDITIONS
pH 7.5 and 65°C

NOTES
- EC 3.1.21.4
- A type II site specific deoxyribonuclease
- A large group of enzymes recognizing specific short DNA sequences. They cleave either within the recognition site, or at a short specific distance from it
- Requires Mg^{2+}

CONCENTRATION or VOLUME ACTIVITY	UNIT DEFINITION	REACTION BUFFER (As Provided)	ADDITIONAL ACTIVITIES and PURITY	SUPPLIER CATALOG NO.
Bacillus species 2G				
250 U and 1000 U	1 unit completely digests 1 μg DNA substrate/hr at 65°C	10 mM Tris-HCl, 10 mM MgCl₂, 100 mM NaCl, 1 mM DTT, 0.1 mg/mL BSA, pH 7.5	>90% DNA fragments ligated and recut after 2-fold overdigestion	Adv Biotech AB-0211-a AB-0211-b

BscF I

RECOGNITION SEQUENCE
5'...▼GATC...3'
3'...CTAG▲...5'

ISOSCHIZOMERS
Mbo I

OPTIMUM REACTION CONDITIONS
pH 7.5 and 55°C

NOTES
- EC 3.1.21.4
- A type II site specific deoxyribonuclease
- A large group of enzymes recognizing specific short DNA sequences. They cleave either within the recognition site, or at a short specific distance from it
- Requires Mg^{2+}

CONCENTRATION or VOLUME ACTIVITY	UNIT DEFINITION	REACTION BUFFER (As Provided)	ADDITIONAL ACTIVITIES and PURITY	SUPPLIER CATALOG NO.
Bacillus species				
250 U and 1000 U	1 unit completely digests 1 μg DNA substrate/hr at 55°C	10 mM Tris-HCl, 10 mM MgCl₂, 100 mM NaCl, 1 mM DTT, 0.1 mg/mL BSA, pH 7.5	>90% DNA fragments ligated and recut after 2-fold overdigestion	Adv Biotech AB-0316-a AB-0316-b

Bse1 I

RECOGNITION SEQUENCE
  5'...ACTGG(1/-1)▼...3'

ISOSCHIZOMERS
  BseN I, Bsr I, BsrS I

NOTES
- EC 3.1.21.4
- A type II site specific deoxyribonuclease
- A large group of enzymes recognizing specific short DNA sequences. They cleave either within the recognition site, or at a short specific distance from it
- Requires Mg^{2+}

CONCENTRATION or VOLUME ACTIVITY	UNIT DEFINITION	REACTION BUFFER (As Provided)	ADDITIONAL ACTIVITIES and PURITY	SUPPLIER CATALOG NO.
Bacillus stearothermophilus 1				
				SibEnzyme

Bse3D I

RECOGNITION SEQUENCE
  5'...GCAATG(2/0)▼...3'

ISOSCHIZOMERS
  BsrD I

NOTES
- EC 3.1.21.4
- A type II site specific deoxyribonuclease
- A large group of enzymes recognizing specific short DNA sequences. They cleave either within the recognition site, or at a short specific distance from it
- Requires Mg^{2+}

CONCENTRATION or VOLUME ACTIVITY	UNIT DEFINITION	REACTION BUFFER (As Provided)	ADDITIONAL ACTIVITIES and PURITY	SUPPLIER CATALOG NO.
Bacillus stearothermophilus 3D				
				SibEnzyme

Bse8 I

RECOGNITION SEQUENCE
5'...GATNN▼NNATC...3'
3'...CTANN▲NNTAG...5'

ISOSCHIZOMERS
BsaB I, Bsh1365 I, BsiB I, BsrBR I, Mam I

NOTES
- EC 3.1.21.4
- A type II site specific deoxyribonuclease
- A large group of enzymes recognizing specific short DNA sequences. They cleave either within the recognition site, or at a short specific distance from it
- Requires Mg^{2+}

CONCENTRATION or VOLUME ACTIVITY	UNIT DEFINITION	REACTION BUFFER (As Provided)	ADDITIONAL ACTIVITIES and PURITY	SUPPLIER CATALOG NO.
Bacillus species 8				
				SibEnzyme

Bse21 I

RECOGNITION SEQUENCE
5'...CC▼TNAGG...3'
3'...GGANT▲CC...5'

ISOSCHIZOMERS
Aoc I, Axy I, Bsu36 I, Cvn I, Eco81 I

NOTES
- EC 3.1.21.4
- A type II site specific deoxyribonuclease
- A large group of enzymes recognizing specific short DNA sequences. They cleave either within the recognition site, or at a short specific distance from it
- Requires Mg^{2+}

CONCENTRATION or VOLUME ACTIVITY	UNIT DEFINITION	REACTION BUFFER (As Provided)	ADDITIONAL ACTIVITIES and PURITY	SUPPLIER CATALOG NO.
Bacillus species 21				
				SibEnzyme

Bse118 I

RECOGNITION SEQUENCE
5'...R▼CCGGY...3'
3'...YGGCC▲R...5'

ISOSCHIZOMERS
BsrF I, BssA I, Cfr10 I

NOTES
- EC 3.1.21.4
- A type II site specific deoxyribonuclease
- A large group of enzymes recognizing specific short DNA sequences. They cleave either within the recognition site, or at a short specific distance from it
- Requires Mg^{2+}

CONCENTRATION or VOLUME ACTIVITY	UNIT DEFINITION	REACTION BUFFER (As Provided)	ADDITIONAL ACTIVITIES and PURITY	SUPPLIER CATALOG No.
Bacillus species 118				
				SibEnzyme

BseA I

RECOGNITION SEQUENCE
5'...T▼CCGGA...3'
3'...AGGCC▲T...5'

ISOSCHIZOMERS
BspM II

OPTIMUM REACTION CONDITIONS
pH 7.9-8.0 and 65°C

HEAT INACTIVATION
65°C for 20 min.

NOTES
- EC 3.1.21.4
- A type II site specific deoxyribonuclease
- A large group of enzymes recognizing specific short DNA sequences. They cleave either within the recognition site, or at a short specific distance from it
- Requires Mg^{2+}

CONCENTRATION or VOLUME ACTIVITY	UNIT DEFINITION	REACTION BUFFER (As Provided)	ADDITIONAL ACTIVITIES and PURITY	SUPPLIER CATALOG No.
Bacillus stearothermophilus				
8-12 U/μL	1 unit completely digests 1 μg substrate DNA/hr	10 mM Tris-HCl, 5 mM MgCl$_2$, 100 mM NaCl, 1 mM β-MSH, pH 8.0		Boehringer 1417169

BseA I continued

CONCENTRATION or VOLUME ACTIVITY	UNIT DEFINITION	REACTION BUFFER (As Provided)	ADDITIONAL ACTIVITIES and PURITY	SUPPLIER CATALOG NO.
Bacillus stearothermophilus continued				
	10 units does not produce unspecific cleavage products with 1 μg λ DNA/16 hr at 65°C	50 mM Tris-HCl, 10 mM MgCl$_2$, 100 mM NaCl, 1 mM DTT, 100 μg/mL BSA, 0.02% Triton X-100, pH 7.9	>90% DNA fragments ligated and recut after 10-fold overdigestion	MINOTECH 106

BseC I

RECOGNITION SEQUENCE
5'...AT▼CGAT...3'
3'...TAGC▲TA...5'

ISOSCHIZOMERS
Cla I

OPTIMUM REACTION CONDITIONS
pH 7.9 and 60°C

NOTES
- EC 3.1.21.4
- A type II site specific deoxyribonuclease
- A large group of enzymes recognizing specific short DNA sequences. They cleave either within the recognition site, or at a short specific distance from it
- Requires Mg^{2+}

CONCENTRATION or VOLUME ACTIVITY	UNIT DEFINITION	REACTION BUFFER (As Provided)	ADDITIONAL ACTIVITIES and PURITY	SUPPLIER CATALOG NO.
Bacillus stearothermophilus species				
384,600 U/mg	50 units does not produce unspecific cleavage products with 1 μg λ DNA/16 hr at 60°C	50 mM Tris-HCl, 10 mM MgCl$_2$, 100 mM NaCl, 1 mM DTT, 100 μg/mL BSA, pH 7.9	>90% DNA fragments ligated and recut after 20-fold overdigestion	MINOTECH 138

BseD I

RECOGNITION SEQUENCE
5'...C▼CNNGG...3'
3'...GGNNC▲C...5'

ISOSCHIZOMERS
Sec I

OPTIMUM REACTION CONDITIONS
pH 7.9 and 60°C; 10% activity at 37°C

HEAT INACTIVATION
65°C for 20 min.

NOTES
- EC 3.1.21.4
- A type II site specific deoxyribonuclease
- A large group of enzymes recognizing specific short DNA sequences. They cleave either within the recognition site, or at a short specific distance from it
- Requires Mg^{2+} and BSA for full activity

CONCENTRATION or VOLUME ACTIVITY	UNIT DEFINITION	REACTION BUFFER (As Provided)	ADDITIONAL ACTIVITIES and PURITY	SUPPLIER CATALOG No.
Bacillus stearothermophilus RFL1434				
8000 U/mL	1 unit degrades 1 µg λ dam⁻ dcm⁻ DNA/hr at 60°C in a 0.05 mL mixture	33 mM Tris-acetate, 10 mM MgOAc, 66 mM KOAc, 100 µg/mL BSA, pH 7.9	No detectable contaminating nuclease; 90% ligated and 100% recut after 50-fold overdigestion	ACS Heidelb F00156S F00156M
4-8 U/µL	1 unit hydrolyzes 1 µg DNA/hr at 37°C in a 50 µL reaction volume using λ phage DNA as substrate	33 mM Tris-acetate, 10 mM MgOAc, 66 mM KOAc, 0.1 mg/mL BSA, pH 7.9	>90% of the DNA fragments ligated and recut after 50-fold overdigestion	Fermentas ER1081 ER1082

BseN I

RECOGNITION SEQUENCE
5'...ACTGGN▼...3'
3'...TGAC▲CN...5'

ISOSCHIZOMERS
Bsr I

OPTIMUM REACTION CONDITIONS
pH 7.5-7.9 and 65°C; <10% activity at 37°C

HEAT INACTIVATION
Resistant

NOTES
- EC 3.1.21.4
- A type II site specific deoxyribonuclease
- A large group of enzymes recognizing specific short DNA sequences. They cleave either within the recognition site, or at a short specific distance from it
- Requires Mg^{2+} and BSA for full activity

CONCENTRATION or VOLUME ACTIVITY	UNIT DEFINITION	REACTION BUFFER (As Provided)	ADDITIONAL ACTIVITIES and PURITY	SUPPLIER CATALOG No.
Bacillus species N				
10,000 U/mL	1 unit degrades 1 μg λ dam⁻ dcm⁻ DNA/hr at 65°C in a 0.05 mL mixture	33 mM Tris-acetate, 10 mM MgOAc, 66 mM KOAc, 100 μg/mL BSA, pH 7.9	No detectable contaminating nuclease; 95% ligated and 95% recut after 10-fold overdigestion	AGS Heidelb F00152S F00152M
8-12 U/μL	1 unit hydrolyzes 1 μg DNA/hr at 37°C in a 50 μL reaction volume using λ phage DNA as substrate	10 mM Tris-HCl, 10 mM MgCl₂, 0.1 mg/mL BSA, pH 7.5	>90% of the DNA fragments ligated and recut after 50-fold overdigestion	Fermentas ER0881 ER0882

BseP I

RECOGNITION SEQUENCE
5'...G▼CGCGC...3'
3'...CGCGC▲G...5'

ISOSCHIZOMERS
BssHI I, Pau I

NOTES
- EC 3.1.21.4
- A type II site specific deoxyribonuclease
- A large group of enzymes recognizing specific short DNA sequences. They cleave either within the recognition site, or at a short specific distance from it
- Requires Mg^{2+}

BseP I continued

CONCENTRATION or VOLUME ACTIVITY	UNIT DEFINITION	REACTION BUFFER (As Provided)	ADDITIONAL ACTIVITIES and PURITY	SUPPLIER CATALOG No.
Bacillus stearothermophilus P6				
				SibEnzyme

BseR I

RECOGNITION SEQUENCE
    5'...GAGGAG(N)$_{10}$▼...3'
    3'...CTCCTC(N)$_8$▲...5'
OPTIMUM REACTION CONDITIONS
    pH 7.0 and 37°C
HEAT INACTIVATION
    65°C for 20 min.

NOTES
- EC 3.1.21.4
- A type II site specific deoxyribonuclease
- A large group of enzymes recognizing specific short DNA sequences. They cleave either within the recognition site, or at a short specific distance from it
- Require Mg^{2+}

CONCENTRATION or VOLUME ACTIVITY	UNIT DEFINITION	REACTION BUFFER (As Provided)	ADDITIONAL ACTIVITIES and PURITY	SUPPLIER CATALOG No.
Bacillus species R				
4000 U/mL	1 unit completely digests 1 μg λ DNA/hr in a total reaction volume of 50 μL using the buffer provided	10 m*M* Bis Tris-Propane-HCl, 10 m*M* MgCl$_2$, 1 m*M* DTT, pH 7.0	>95% DNA fragments ligated and recut after 4-fold overdigestion	NE Biolabs 581S 581L

Bsg I

RECOGNITION SEQUENCE
5'...GTGCAG(N)$_{16}$▼...3'
3'...CACGTC(N)$_{14}$▲...5'

OPTIMUM REACTION CONDITIONS
pH 7.9 and 37°C

HEAT INACTIVATION
65°C for 20 min.

NOTES
- EC 3.1.21.4
- A type II site specific deoxyribonuclease
- A large group of enzymes recognizing specific short DNA sequences. They cleave either within the recognition site, or at a short specific distance from it
- 25% activity without 80 μM S-adenosylmethionine
- Require Mg^{2+}

CONCENTRATION or VOLUME ACTIVITY	UNIT DEFINITION	REACTION BUFFER (As Provided)	ADDITIONAL ACTIVITIES and PURITY	SUPPLIER CATALOG No.
Bacillus sphaericus (NEB 581)				
2000 U/mL	1 unit completely digests 1 μg λ DNA/hr in a total reaction volume of 50 μL using the buffer provided	20 mM Tris-acetate, 10 mM MgAOc, 50 mM KOAc, 1 mM DTT, pH 7.9; add 80 μM S-adenosylmethionine	>95% DNA fragments ligated and 80% recut after 4-fold overdigestion	NE Biolabs 559S 559L

Bsh I

RECOGNITION SEQUENCE
5'...GG▼CC...3'
3'...CC▲GG...5'

ISOSCHIZOMERS
Hae III

OPTIMUM REACTION CONDITIONS
pH 7.5 and 37°C

NOTES
- EC 3.1.21.4
- A type II site specific deoxyribonuclease
- A large group of enzymes recognizing specific short DNA sequences. They cleave either within the recognition site, or at a short specific distance from it
- Requires Mg^{2+}

Bsh I continued

CONCENTRATION or VOLUME ACTIVITY	UNIT DEFINITION	REACTION BUFFER (As Provided)	ADDITIONAL ACTIVITIES and PURITY	SUPPLIER CATALOG No.
Bacillus sphaericus				
5000 U and 10,000 U	1 unit completely digests 1 μg DNA substrate/hr at 37°C	10 mM Tris-HCl, 10 mM MgCl$_2$, 50 mM NaCl, 1 mM DTT, 0.1 mg/mL BSA, pH 7.5	>90% DNA fragments ligated and recut after several-fold overdigestion	Adv Biotech AB-0212-a AB-0212-b

Bsh1236 I

RECOGNITION SEQUENCE
   5'...CG▼CG...3'
   3'...GC▲GC...5'
ISOSCHIZOMERS
   Acc II, *Bst*U I, *Fnu*D II, *Mvn* I, *Tha* I
OPTIMUM REACTION CONDITIONS
   pH 7.4-8.5 and 37°C
HEAT INACTIVATION
   65°C for 20 min.
METHYLATION EFFECTS
   Blocked by CG methylation

NOTES
- EC 3.1.21.4
- A type II site specific deoxyribonuclease
- A large group of enzymes recognizing specific short DNA sequences. They cleave either within the recognition site, or at a short specific distance from it
- Requires Mg^{2+} and BSA for full activity

CONCENTRATION or VOLUME ACTIVITY	UNIT DEFINITION	REACTION BUFFER (As Provided)	ADDITIONAL ACTIVITIES and PURITY	SUPPLIER CATALOG No.
Bacillus sphaericus				
3000-10,000 U/mL		50 mM KCl, 10 mM Tris-HCl, 0.1 mM EDTA, 1 mM DTT, 200 μg/mL acetylated BSA, 50% glycerol, pH 7.5		ICN 159410
Bacillus sphaericus RFL1236				
10,000 U/mL	1 unit degrades 1 μg λ dam⁻ dcm⁻ DNA/hr at 37°C in a 0.05 mL mixture	10 mM Tris-HCl, 10 mM MgCl$_2$, 100 mM KCl, 100 μg/mL BSA, pH 8.5	No detectable contaminating nuclease; 90% ligated and 100% recut after 50-fold overdigestion	AGS Heidelb F00320S F00320M
8-12 U/μL	1 unit hydrolyzes 1 μg DNA/hr at 37°C in a 50 μL reaction volume using λ phage DNA as substrate	10 mM Tris-HCl, 10 mM MgCl$_2$, 100 mM KCl, 0.1 mg/mL BSA, pH 8.5	>90% of the DNA fragments ligated and recut after 50-fold overdigestion	Fermentas ER0091 ER0091

Bsh1236 I continued

CONCENTRATION or VOLUME ACTIVITY	UNIT DEFINITION	REACTION BUFFER (As Provided)	ADDITIONAL ACTIVITIES and PURITY	SUPPLIER CATALOG NO.
Bacillus sphaericus RFL1236 continued				
3000-10,000 U/mL		50 mM KCl, 10 mM Tris-HCl, 0.1 mM EDTA, 1 mM DTT, 200 µg/mL BSA, 50% glycerol, pH 7.4		stratagene 500304

Bsh1285 I

RECOGNITION SEQUENCE
  5'...CGPuPy▼CG...3'
  3'...GC▲PyPuGC...5'
ISOSCHIZOMERS
  Mcr I
OPTIMUM REACTION CONDITIONS
  pH 7.5 and 37°C
HEAT INACTIVATION
  65°C for 20 min.
METHYLATION EFFECTS
  Blocked by CG methylation

NOTES
- EC 3.1.21.4
- A type II site specific deoxyribonuclease
- A large group of enzymes recognizing specific short DNA sequences. They cleave either within the recognition site, or at a short specific distance from it
- Requires Mg^{2+} and BSA for full activity

CONCENTRATION or VOLUME ACTIVITY	UNIT DEFINITION	REACTION BUFFER (As Provided)	ADDITIONAL ACTIVITIES and PURITY	SUPPLIER CATALOG NO.
Bacillus sphaericus RFL1285				
10,000 U/mL	1 unit degrades 1 µg λ dam⁻ dcm⁻ DNA/hr at 37°C in a 0.05 mL mixture	10 mM Tris-HCl, 10 mM $MgCl_2$, 50 mM NaCl, 100 µg/mL BSA, pH 7.5	No detectable contaminating nuclease; 90% ligated and 100% recut after 50-fold overdigestion	AGS Heidelb F00154S F00154M
8-12 U/µL	1 unit hydrolyzes 1 µg DNA/hr at 37°C in a 50 µL reaction volume using λ phage DNA as substrate	10 mM Tris-HCl, 10 mM $MgCl_2$, 50 mM NaCl, 0.1 mg/mL BSA, pH 7.5	>90% of the DNA fragments ligated and recut after 50-fold overdigestion	Fermentas ER0891 ER0892

Bsh1365 I

RECOGNITION SEQUENCE
  5'...GATNN▼NNATC...3'
  3'...CTANN▲NNTAG...5'

ISOSCHIZOMERS
  BsaB I

OPTIMUM REACTION CONDITIONS
  pH 7.5 and 37°C

HEAT INACTIVATION
  65°C for 20 min.

METHYLATION EFFECTS
  Does not cleave $Gm^6AT(N)_4ATC$ or $GAT(N)_4m^6ATC$
  Blocked by overlapping *dam* methylation

NOTES
- EC 3.1.21.4
- A type II site specific deoxyribonuclease
- A large group of enzymes recognizing specific short DNA sequences. They cleave either within the recognition site, or at a short specific distance from it
- Star activity is observed with >8-10-fold overdigestion of substrate with enzyme
- Requires Mg^{2+} and BSA for full activity

CONCENTRATION or VOLUME ACTIVITY	UNIT DEFINITION	REACTION BUFFER (As Provided)	ADDITIONAL ACTIVITIES and PURITY	SUPPLIER CATALOG NO.
Bacillus sphaericus RFL1365				
8000 U/mL	1 unit degrades 1 μg λ dam⁻ dcm⁻ DNA/hr at 37°C in a 0.05 mL mixture	10 mM Tris-HCl, 10 mM $MgCl_2$, 100 μg/mL BSA, pH 7.5	No detectable contaminating nuclease; 80% ligated and 100% recut after 10-fold overdigestion	AGS Heidelb F00158S F00158M
4-8 U/μL	1 unit hydrolyzes 1 μg DNA/hr at 37°C in a 50 μL reaction volume using λ phage DNA as substrate	10 mM Tris-HCl, 10 mM $MgCl_2$, 0.1 mg/mL BSA, pH 7.5	80% of the DNA fragments ligated and >90% recut after 10-fold overdigestion	Fermentas ER1041 ER1042

BshN I

RECOGNITION SEQUENCE
5'...G▼GPyPuCC...3'
3'...CCPuPyG▲G...5'

ISOSCHIZOMERS
HgiC I

OPTIMUM REACTION CONDITIONS
pH 7.5 and 37°C

HEAT INACTIVATION
65°C for 20 min.

METHYLATION EFFECTS
Cleavage rate is slowed significantly by overlapping dcm methylation
Blocked by overlapping CG methylation

NOTES
- EC 3.1.21.4
- A type II site specific deoxyribonuclease
- A large group of enzymes recognizing specific short DNA sequences. They cleave either within the recognition site, or at a short specific distance from it
- Requires Mg^{2+}

CONCENTRATION or VOLUME ACTIVITY	UNIT DEFINITION	REACTION BUFFER (As Provided)	ADDITIONAL ACTIVITIES and PURITY	SUPPLIER CATALOG NO.
Bacillus sphaericus Tk 4-5				
8-12 U/μL	1 unit hydrolyzes 1 μg DNA/hr at 37°C in a 50 μL reaction volume using λ phage DNA as substrate	50 mM Tris-HCl, 10 mM MgCl₂, 100 mM NaCl, pH 7.5	>90% of the DNA fragments ligated and recut after 50-fold overdigestion	Fermentas ER1001 ER1002

BsiB I

RECOGNITION SEQUENCE
5'...GATNN▼NNATC...3'
3'...CTANN▲NNTAG...5'

ISOSCHIZOMERS
BsaB I

OPTIMUM REACTION CONDITIONS
pH 7.5 and 55°C

NOTES
- EC 3.1.21.4
- A type II site specific deoxyribonuclease
- A large group of enzymes recognizing specific short DNA sequences. They cleave either within the recognition site, or at a short specific distance from it
- Requires Mg^{2+}

BsiB I continued

CONCENTRATION or VOLUME ACTIVITY	UNIT DEFINITION	REACTION BUFFER (As Provided)	ADDITIONAL ACTIVITIES and PURITY	SUPPLIER CATALOG No.
Bacillus species				
2000 U and 5000 U	1 unit completely digests 1 µg DNA substrate/hr at 55°C	10 mM Tris-HCl, 10 mM MgCl$_2$, 100 mM NaCl, 1 mM DTT, 0.1 mg/mL BSA, pH 7.5	>90% of the DNA fragments ligated and recut after 2-fold overdigestion	Adv Biotech AB-0213-a AB-0213-b

BsiC I

RECOGNITION SEQUENCE
  5'...TT▼CGAA...3'
  3'...AAGC▲TT...5'

ISOSCHIZOMERS
  Asu II

OPTIMUM REACTION CONDITIONS
  pH 7.5 and 60°C

NOTES
- EC 3.1.21.4
- A type II site specific deoxyribonuclease
- A large group of enzymes recognizing specific short DNA sequences. They cleave either within the recognition site, or at a short specific distance from it
- Requires Mg^{2+}

CONCENTRATION or VOLUME ACTIVITY	UNIT DEFINITION	REACTION BUFFER (As Provided)	ADDITIONAL ACTIVITIES and PURITY	SUPPLIER CATALOG No.
Bacillus species				
2500 U and 5000 U	1 unit completely digests 1 µg DNA substrate/hr at 60°C	10 mM Tris-HCl, 10 mM MgCl$_2$, 1 mM DTT, 0.1 mg/mL BSA, pH 7.5	>90% of the DNA fragments ligated and recut after 10-fold overdigestion	Adv Biotech AB-0214-a AB-0214-b

BsiE I

RECOGNITION SEQUENCE
5'...CGPuPy▼CG...3'
3'...GC▲PuPuGC...5'

OPTIMUM REACTION CONDITIONS
pH 7.4-7.9 and 60°C; 30% activity at 37°C

HEAT INACTIVATION
Resistant

NOTES
- EC 3.1.21.4
- A type II site specific deoxyribonuclease
- A large group of enzymes recognizing specific short DNA sequences. They cleave either within the recognition site, or at a short specific distance from it
- Requires Mg^{2+} and BSA for full activity

CONCENTRATION or VOLUME ACTIVITY	UNIT DEFINITION	REACTION BUFFER (As Provided)	ADDITIONAL ACTIVITIES and PURITY	SUPPLIER CATALOG No.
Bacillus stearothermophilus				
5000-15,000 U/mL		50 mM NaCl, 10 mM Tris-HCl, 0.1 mM EDTA, 1 mM DTT, 200 µg/mL acetylated BSA, 50% glycerol, pH 7.4		ICN 159411
Bacillus stearothermophilus (D. Clark)				
10,000 U/mL	1 unit completely digests 1 µg λ DNA/hr in a total reaction volume of 50 µL using the buffer provided	10 mM Tris-HCl, 10 mM MgCl₂, 50 mM NaCl, 1 mM DTT, pH 7.9; add 100 µg/mL BSA	>95% DNA fragments ligated and recut after 10-fold overdigestion	NE Biolabs 554S 554L

BsiHKA I

RECOGNITION SEQUENCE

$5'...G(^A_T)GC(^A_T)\blacktriangledown C...3'$
$3'...C\blacktriangle(^T_A)CG(^T_A)G...5'$

ISOSCHIZOMERS
HgiA I

OPTIMUM REACTION CONDITIONS
pH 7.6-7.9 and 65°C; minimal activity at 37°C

HEAT INACTIVATION
Resistant

NOTES
- EC 3.1.21.4
- A type II site specific deoxyribonuclease
- A large group of enzymes recognizing specific short DNA sequences. They cleave either within the recognition site, or at a short specific distance from it
- Requires Mg^{2+} and BSA for full activity

CONCENTRATION or VOLUME ACTIVITY	UNIT DEFINITION	REACTION BUFFER (As Provided)	ADDITIONAL ACTIVITIES and PURITY	SUPPLIER CATALOG No.
Bacillus stearothermophilus				
5000-15,000 U/mL		100 mM NaCl, 50 mM Tris-HCl, 0.1 mM EDTA, 1 mM DTT, 200 µg/mL acetylated BSA, 50% glycerol, pH 7.6		ICN 159412
Bacillus stearothermophilus (P. Shaw)				
10,000 U/mL; 50,000 U/mL	1 unit completely digests 1 µg λ DNA/hr in a total reaction volume of 50 µL using the buffer provided	50 mM Tris-HCl, 10 mM $MgCl_2$, 100 mM NaCl, 1 mM DTT, pH 7.9; add 100 µg/mL BSA	>95% DNA fragments ligated and recut after 10-fold overdigestion	NE Biolabs 570S 570L

BsiL I

RECOGNITION SEQUENCE

$5'...CC\blacktriangledown(^A_T)GG...3'$
$3'...GG(^T_A)\blacktriangle CC...5'$

ISOSCHIZOMERS
BstN I

OPTIMUM REACTION CONDITIONS
pH 7.5 and 60°C

NOTES
- EC 3.1.21.4
- A type II site specific deoxyribonuclease
- A large group of enzymes recognizing specific short DNA sequences. They cleave either within the recognition site, or at a short specific distance from it
- Requires Mg^{2+}

BsiL I continued

CONCENTRATION or VOLUME ACTIVITY	UNIT DEFINITION	REACTION BUFFER (As Provided)	ADDITIONAL ACTIVITIES and PURITY	SUPPLIER CATALOG NO.
Bacillus species				
2500 U and 5000 U	1 unit completely digests 1 μg DNA substrate/hr at 60°C	10 mM Tris-HCl, 10 mM MgCl$_2$, 1 mM DTT, 0.1 mg/mL BSA, pH 7.5	0% DNA fragments ligated and recut after 10-fold overdigestion	Adv Biotech AB-0215-a AB-0215-b

BsiM I

RECOGNITION SEQUENCE
  5'...T▼CCGGA...3'
  3'...AGGCC▲T...5'
ISOSCHIZOMERS
  BspM I, Mro I
OPTIMUM REACTION CONDITIONS
  pH 7.5 and 60°C
HEAT INACTIVATION
  Resistant

NOTES
- EC 3.1.21.4
- A type II site specific deoxyribonuclease
- A large group of enzymes recognizing specific short DNA sequences. They cleave either within the recognition site, or at a short specific distance from it
- Requires Mg^{2+}

CONCENTRATION or VOLUME ACTIVITY	UNIT DEFINITION	REACTION BUFFER (As Provided)	ADDITIONAL ACTIVITIES and PURITY	SUPPLIER CATALOG NO.
Bacillus species				
2000 U and 5000 U	1 unit completely digests 1 μg DNA substrate/hr at 60°C	10 mM Tris-HCl, 10 mM MgCl$_2$, 100 mM NaCl, 1 mM DTT, 0.1 mg/mL BSA, pH 7.5	>90% DNA fragments ligated and recut after 10-fold overdigestion	Adv Biotech AB-0216-a AB-0216-b
8-12 U/μL	1 unit completely digests 1 μg dam$^-$ λ DNA/hr at 55°C in the buffer provided	500 mM Tris-HCl, 100 mM MgCl$_2$, 10 mM DTT, 1000 mM NaCl, pH 7.5	>90% DNA fragments ligated and recut	Amersham E 1615Y

BsiQ I

RECOGNITION SEQUENCE
5'...T▼GATCA...3'
3'...ACTAG▲T...5'

ISOSCHIZOMERS
Bcl I

OPTIMUM REACTION CONDITIONS
pH 7.5 and 60°C

NOTES
- EC 3.1.21.4
- A type II site specific deoxyribonuclease
- A large group of enzymes recognizing specific short DNA sequences. They cleave either within the recognition site, or at a short specific distance from it
- Requires Mg^{2+}

CONCENTRATION or VOLUME ACTIVITY	UNIT DEFINITION	REACTION BUFFER (As Provided)	ADDITIONAL ACTIVITIES and PURITY	SUPPLIER CATALOG NO.
Bacillus species				
2500 U and 5000 U	1 unit completely digests 1 μg DNA substrate/hr at 60°C	10 mM Tris-HCl, 10 mM MgCl₂, 50 mM NaCl, 1 mM DTT, 0.1 mg/mL BSA, pH 7.5	>90% DNA fragments ligated and recut after 10-fold overdigestion	Adv Biotech AB-0217-a AB-0217-b

BsiS I

RECOGNITION SEQUENCE
5'...C▼CGG...3'
3'...GGC▲C...5'

ISOSCHIZOMERS
Hpa II

OPTIMUM REACTION CONDITIONS
pH 7.9 and 55°C

NOTES
- EC 3.1.21.4
- A type II site specific deoxyribonuclease
- A large group of enzymes recognizing specific short DNA sequences. They cleave either within the recognition site, or at a short specific distance from it
- Requires Mg^{2+}

*Bsi*S I continued

CONCENTRATION or VOLUME ACTIVITY	UNIT DEFINITION	REACTION BUFFER (As Provided)	ADDITIONAL ACTIVITIES and PURITY	SUPPLIER CATALOG NO.
Bacillus stearothermophilus				
	10 units does not produce unspecific cleavage products with 1 µg λ DNA/16 hr at 55°C	50 mM Tris-HCl, 10 mM MgCl$_2$, 100 mM NaCl, 1 mM DTT, 100 µg/mL BSA, pH 7.9	>90% DNA fragments ligated and recut after 10-fold overdigestion	MINOTECH 107

*Bsi*W I

RECOGNITION SEQUENCE
  5'...C▼GTACG...3'
  3'...GCATG▲C...5'

ISOSCHIZOMERS
  Spl I

OPTIMUM REACTION CONDITIONS
  pH 7.5-7.9 and 55°C; 50% activity at 37°C

HEAT INACTIVATION
  Resistant

NOTES
- EC 3.1.21.4
- A type II site specific deoxyribonuclease
- A large group of enzymes recognizing specific short DNA sequences. They cleave either within the recognition site, or at a short specific distance from it
- Requires Mg^{2+}

CONCENTRATION or VOLUME ACTIVITY	UNIT DEFINITION	REACTION BUFFER (As Provided)	ADDITIONAL ACTIVITIES and PURITY	SUPPLIER CATALOG NO.
Bacillus species				
10 U/µL	1 unit completely digests 1 µg substrate DNA/hr	50 mM Tris-HCl, 10 mM MgCl$_2$, 100 mM NaCl, 1 mM DTE, pH 7.5		Boehringer 1388959 1388967
				Toyobo
Bacillus stearothermophilus (D. Clark)				
12,000 U/mL	1 unit completely digests 1 µg λ DNA/hr in a total reaction volume of 50 µL using the buffer provided	50 mM Tris-HCl, 10 mM MgCl$_2$, 100 mM NaCl, 1 mM DTT, pH 7.9	>95% DNA fragments ligated and recut after 2-fold overdigestion	NE Biolabs 553S 553L

*Bsi*X I

RECOGNITION SEQUENCE
5'...AT▼CGAT...3'
3'...TAGC▲TA...5'
ISOSCHIZOMERS
Cla I
OPTIMUM REACTION CONDITIONS
pH 7.5 and 65°C

NOTES
- EC 3.1.21.4
- A type II site specific deoxyribonuclease
- A large group of enzymes recognizing specific short DNA sequences. They cleave either within the recognition site, or at a short specific distance from it
- Requires Mg^{2+}

CONCENTRATION or VOLUME ACTIVITY	UNIT DEFINITION	REACTION BUFFER (As Provided)	ADDITIONAL ACTIVITIES and PURITY	SUPPLIER CATALOG No.
Bacillus species				
2500 U and 5000 U	1 unit completely digests 1 μg DNA substrate/hr at 65°C	10 mM Tris-HCl, 10 mM $MgCl_2$, 50 mM NaCl, 1 mM DTT, 0.1 mg/mL BSA, pH 7.5	>90% DNA fragments ligated and recut after 10-fold overdigestion	Adv Biotech AB-0218-a AB-0218-b

*Bsi*Y I

RECOGNITION SEQUENCE
5'...CC(N)₅▼NNGG...3'
3'...GGNN▲(N)₅CC...5'
ISOSCHIZOMERS
Bsl I
OPTIMUM REACTION CONDITIONS
pH 7.5 and 55°C
HEAT INACTIVATION
65°C for 20 min.

NOTES
- EC 3.1.21.4
- A type II site specific deoxyribonuclease
- A large group of enzymes recognizing specific short DNA sequences. They cleave either within the recognition site, or at a short specific distance from it
- Requires Mg^{2+}

BsiY I continued

CONCENTRATION or VOLUME ACTIVITY	UNIT DEFINITION	REACTION BUFFER (As Provided)	ADDITIONAL ACTIVITIES and PURITY	SUPPLIER CATALOG No.
Bacillus species				
500 U and 1000 U	1 unit completely digests 1 μg DNA substrate/hr at 55°C	10 mM Tris-HCl, 10 mM MgCl$_2$, 50 mM NaCl, 1 mM DTT, 0.1 mg/mL BSA, pH 7.5	>90% DNA fragments ligated and recut after 10-fold overdigestion	Adv Biotech AB-0219-a AB-0219-b
10 U/μL	1 unit completely digests 1 μg substrate DNA/hr	10 mM Tris-HCl, 10 mM MgCl$_2$, 50 mM NaCl, 1 mM DTE, pH 7.5		Boehringer 1388916

BsiZ I

RECOGNITION SEQUENCE
  5'...G▼GNCC...3'
  3'...CCNG▲G...5'
ISOSCHIZOMERS
  Sau96 I
OPTIMUM REACTION CONDITIONS
  pH 7.5 and 60°C

NOTES
- EC 3.1.21.4
- A type II site specific deoxyribonuclease
- A large group of enzymes recognizing specific short DNA sequences. They cleave either within the recognition site, or at a short specific distance from it
- Requires Mg^{2+}

CONCENTRATION or VOLUME ACTIVITY	UNIT DEFINITION	REACTION BUFFER (As Provided)	ADDITIONAL ACTIVITIES and PURITY	SUPPLIER CATALOG No.
Bacillus species				
2500 U and 5000 U	1 unit completely digests 1 μg DNA substrate/hr 60°C	10 mM Tris-HCl, 10 mM MgCl$_2$, 100 mM NaCl, 1 mM DTT, 0.1 mg/mL BSA, pH 7.5	>90% DNA fragments ligated and recut after 10-fold overdigestion	Adv Biotech AB-0220-a AB-0220-b

Bsl I

RECOGNITION SEQUENCE
5'...CCNNNNN▼NNGG...3'
3'...GGNN▲NNNNNCC...5'

OPTIMUM REACTION CONDITIONS
pH 7.4-7.9 and 55°C; 30% activity at 37°C

HEAT INACTIVATION
Resistant

NOTES
- EC 3.1.21.4
- A type II site specific deoxyribonuclease
- A large group of enzymes recognizing specific short DNA sequences. They cleave either within the recognition site, or at a short specific distance from it
- Requires Mg^{2+}

CONCENTRATION or VOLUME ACTIVITY	UNIT DEFINITION	REACTION BUFFER (As Provided)	ADDITIONAL ACTIVITIES and PURITY	SUPPLIER CATALOG NO.
Bacillus **species**				
3000-20,000 U/mL		50 mM KCl, 10 mM Tris-HCl, 0.1 mM EDTA, 1 mM DTT, 200 μg/mL acetylated BSA, 50% glycerol, pH 7.4		ICN 159413
Bacillus **species (D. Cowan)**				
5000 U/mL	1 unit completely digests 1 μg λ DNA/hr in a total reaction volume of 50 μL using the buffer provided	50 mM Tris-HCl, 10 mM MgCl₂, 100 mM NaCl, 1 mM DTT, pH 7.9	>95% DNA fragments ligated and recut after 10-fold overdigestion	NE Biolabs 555S 555L

Bsm I

RECOGNITION SEQUENCE
  5'...GAATGCN▼...3'
  3'...CTTAC▲GN...5'

OPTIMUM REACTION CONDITIONS
  pH 7.5-8.5 and 65°C; 20% activity at 37°C

HEAT INACTIVATION
  Resistant

METHYLATION EFFECTS
  Cleaves GAATGm^{5}C
  Does not cleave Gm^{6}AATGC

NOTES
- EC 3.1.21.4
- A type II site specific deoxyribonuclease
- A large group of enzymes recognizing specific short DNA sequences. They cleave either within the recognition site, or at a short specific distance from it
- Recognizes a nonpalandromic sequence, giving it 2 possible top strand recognition sequences
- Requires Mg^{2+}

CONCENTRATION or VOLUME ACTIVITY	UNIT DEFINITION	REACTION BUFFER (As Provided)	ADDITIONAL ACTIVITIES and PURITY	SUPPLIER CATALOG NO.
Bacillus stearothermophilus				
5000 U/mL	1 unit degrades 1 μg λ dam$^-$ dcm$^-$ DNA/hr at 65°C in a 0.05 mL mixture	10 mM Tris-HCl, 10 mM MgCl$_2$, 100 mM KCl, 100 μg/mL BSA, pH 8.5	No detectable contaminating nuclease; 90% ligated and 100% recut after 10-fold overdigestion	AGS Heidelb A00155S A00155M
				Nippon Gene
				Toyobo
Bacillus stearothermophilus NUB36				
3-20 U/μL	1 unit completely digests 1 μg λ DNA/hr at 65°C in the buffer provided	100 mM Tris-HCl, 100 mM MgCl$_2$, 10 mM DTT, 500 mM NaCl, pH 7.5	>90% DNA fragments ligated and 95% recut	Amersham E 0205Y E 0205Z
10 U/μL	1 unit completely digests 1 μg substrate DNA/hr	50 mM Tris-HCl, 10 mM MgCl$_2$, 100 mM NaCl, 1 mM DTE, pH 7.5		Boehringer 1292307 1292315
4-6 U/μL	1 unit completely digests 1 mg λ DNA/hr at 65°C	500 mM Tris-HCl, 100 mM MgCl$_2$, 500 mM NaCl, pH 8.0	≥95% ligation; 100% cleavage	Life Technol 35405-018
4-12 U/μL	1 unit digests completely 1 μg λ DNA substrate/hr at 65°C	10 mM Tris-HCl, 50 mM NaCl, 10 mM MgCl$_2$, 1 mM DTT, pH 7.8		NBL Gene 016901 016903
2000-20,000 U/mL	1 unit completely cleaves 1.0 μg λ DNA/hr in a 50 μL reaction mixture	100 mM Tris-HCl, 500 mM NaCl, 100 mM MgCl$_2$, 10 mM DTT, pH 7.9	100% cleavage, 100% ligation, 100% recleavage, 5% endonuclease (nickase) No detectable endonuclease (overdigestion)	Sigma R3635

Bsm I continued

CONCENTRATION or VOLUME ACTIVITY	UNIT DEFINITION	REACTION BUFFER (As Provided)	ADDITIONAL ACTIVITIES and PURITY	SUPPLIER CATALOG NO.
Bacillus stearothermophilus NUB36 *continued*				
1000-10,000 U/mL		50 mM KCl, 10 mM Tris-HCl, 0.1 mM EDTA, 1 mM DTT, 200 μg/mL BSA, 50% glycerol, pH 7.4		Stratagene 500290
3-20 U/μL	1 unit completely digests 1 μg DNA/hr at 65°C	10 mM Tris-HCl, 10 mM MgCl$_2$, 50 mM NaCl, 1 mM DTT, pH 7.5	>90% ligation; >95% recut	Toyobo BSM-101 BSM-102
Bacillus stearothermophilus NUB36 (N. Welker)				
4000-20,000 U/mL		50 mM KCl, 10 mM Tris-HCl, 0.1 mM EDTA, 1 mM DTT, 200 μg/mL BSA, 50% glycerol, pH 7.4		ICN 153805
5000 U/mL	1 unit completely digests 1 μg λ DNA/hr in a total reaction volume of 50 μL using the buffer provided	10 mM Tris-HCl, 10 mM MgCl$_2$, 50 mM NaCl, 1 mM DTT, pH 7.9	>95% DNA fragments ligated and recut after 10-fold overdigestion	NE Biolabs 134S 134L

BsmA I

RECOGNITION SEQUENCE
  5'...GTCTC(N)$_1$▼...3'
  3'...CAGAG(N)$_5$▲...5'

ISOSCHIZOMERS
  *Alw*26 I

OPTIMUM REACTION CONDITIONS
  pH 7.4-8.4 and 55°C; 10% activity at 37°C, 30% activity at 65°C

HEAT INACTIVATION
  Resistant

METHYLATION EFFECTS
  Does not cleave GTCTm^{5}C

NOTES
- EC 3.1.21.4
- A type II site specific deoxyribonuclease
- A large group of enzymes recognizing specific short DNA sequences. They cleave either within the recognition site, or at a short specific distance from it
- Requires Mg^{2+} and BSA for full activity

BsmA I continued

CONCENTRATION or VOLUME ACTIVITY	UNIT DEFINITION	REACTION BUFFER (As Provided)	ADDITIONAL ACTIVITIES and PURITY	SUPPLIER CATALOG NO.
Bacillus stearothermophilus				
3000-20,000 U/mL		200 mM KCl, 10 mM Tris-HCl, 0.1 mM EDTA, 1 mM DTT, 500 µg/mL acetylated BSA, 50% glycerol, pH 7.4		ICN 159414
Bacillus stearothermophilus (NEB 481)				
5000 U/mL	1 unit completely digests 1 µg λ DNA/hr in a total reaction volume of 50 µL using the buffer provided	10 mM Tris-HCl, 10 mM MgCl$_2$, 100 mM NaCl, 10 mM DTT, pH 8.4; add 100 µg/mL BSA	>95% DNA fragments ligated and recut after 10-fold overdigestion	NE Biolabs 529S 529L

BsmB I

RECOGNITION SEQUENCE
  5'...CGTCTC(N)$_1$▼...3'
  3'...GCAGAG(N)$_5$▲...5'
ISOSCHIZOMERS
  Esp3 I
OPTIMUM REACTION CONDITIONS
  pH 7.9 and 55°C; 20% activity at 37°C
HEAT INACTIVATION
  Resistant

NOTES
- EC 3.1.21.4
- A type II site specific deoxyribonuclease
- A large group of enzymes recognizing specific short DNA sequences. They cleave either within the recognition site, or at a short specific distance from it
- Requires Mg^{2+}

CONCENTRATION or VOLUME ACTIVITY	UNIT DEFINITION	REACTION BUFFER (As Provided)	ADDITIONAL ACTIVITIES and PURITY	SUPPLIER CATALOG NO.
Bacillus stearothermophilus B61				
2000 U/mL	1 unit completely digests 1 µg λ DNA/hr in a total reaction volume of 50 µL using the buffer provided	50 mM Tris-HCl, 10 mM MgCl$_2$, 100 mM NaCl, 1 mM DTT, pH 7.9	>95% DNA fragments ligated and recut after 2-fold overdigestion	NE Biolabs 580S 580L

BsmF I

RECOGNITION SEQUENCE
   5'...GGGAC(N)$_{10}$▼...3'
   3'...CCCTG(N)$_{14}$▲...5'

ISOSCHIZOMERS
   Fin I

OPTIMUM REACTION CONDITIONS
   pH 7.4-7.9 and 65°C; 50% activity at 37°C

HEAT INACTIVATION
   Resistant

NOTES
- EC 3.1.21.4
- A type II site specific deoxyribonuclease
- A large group of enzymes recognizing specific short DNA sequences. They cleave either within the recognition site, or at a short specific distance from it
- Requires Mg^{2+}

CONCENTRATION or VOLUME ACTIVITY	UNIT DEFINITION	REACTION BUFFER (As Provided)	ADDITIONAL ACTIVITIES and PURITY	SUPPLIER CATALOG NO.
Bacillus stearothermophilus F				
2000-10,000 U/mL		50 mM NaCl, 10 mM Tris-HCl, 0.1 mM EDTA, 1 mM DTT, 200 µg/mL acetylated BSA, 50% glycerol, pH 7.4		ICN 159415
4000 U/mL	1 unit completely digests 1 µg λ DNA/hr in a total reaction volume of 50 µL using the buffer provided	20 mM Tris-acetate, 10 mM MgOAc, 50 mM KOAc, 1 mM DTT, pH 7.9	>95% DNA fragments ligated and recut after 10-fold overdigestion	NE Biolabs 572S 572L

BsoB I

RECOGNITION SEQUENCE
   5'...C▼YCGRG...3'
   3'...GRGCY▲C...5'

ISOSCHIZOMERS
   Ama87 I, Ava I, Bco I, Eco88 I

NOTES
- EC 3.1.21.4
- A type II site specific deoxyribonuclease
- A large group of enzymes recognizing specific short DNA sequences. They cleave either within the recognition site, or at a short specific distance from it
- Requires Mg^{2+}

BsoB I continued

CONCENTRATION or VOLUME ACTIVITY	UNIT DEFINITION	REACTION BUFFER (As Provided)	ADDITIONAL ACTIVITIES and PURITY	SUPPLIER CATALOG NO.
Bacillus stearothermophilus JN2091				
				NE Biolabs

BsoF I

RECOGNITION SEQUENCE
  5'...GC▼NGC...3'
  3'...CGN▲CG...5'

ISOSCHIZOMERS
  *Fnu*4H I

OPTIMUM REACTION CONDITIONS
  pH 7.9 and 55°C; <10% activity at 37°C

HEAT INACTIVATION
  65°C for 20 min.

NOTES
- EC 3.1.21.4
- A type II site specific deoxyribonuclease
- A large group of enzymes recognizing specific short DNA sequences. They cleave either within the recognition site, or at a short specific distance from it
- Leaves a difficult-to-ligate 1-base 5' extension
- >50% ligation is possible using high concentration T4 DNA ligase
- Requires Mg^{2+}

CONCENTRATION or VOLUME ACTIVITY	UNIT DEFINITION	REACTION BUFFER (As Provided)	ADDITIONAL ACTIVITIES and PURITY	SUPPLIER CATALOG NO.
Bacillus stearothermophilus (Z. Chen)				
5000 U/mL	1 unit completely digests 1 μg λ DNA/hr in a total reaction volume of 50 μL using the buffer provided	10 m*M* Tris-HCl, 10 m*M* MgCl₂, 50 m*M* NaCl, 1 m*M* DTT, pH 7.9	<10% DNA fragments ligated after 4-fold overdigestion	NE Biolabs 578S 578L

Bsp13 I

RECOGNITION SEQUENCE

5'...T▼CCGGA...3'
3'...AGGCC▲T...5'

ISOSCHIZOMERS

Acc III, BseA I, BsiM I, BspE I, Kpn2 I, Mro I

NOTES
- EC 3.1.21.4
- A type II site specific deoxyribonuclease
- A large group of enzymes recognizing specific short DNA sequences. They cleave either within the recognition site, or at a short specific distance from it
- Requires Mg^{2+}

CONCENTRATION or VOLUME ACTIVITY	UNIT DEFINITION	REACTION BUFFER (As Provided)	ADDITIONAL ACTIVITIES and PURITY	SUPPLIER CATALOG NO.
Bacillus species 13				
				sibEnzyme

Bsp19 I

RECOGNITION SEQUENCE

5'...C▼CATGG...3'
3'...GGTAC▲C...5'

ISOSCHIZOMERS

Nco I

NOTES
- EC 3.1.21.4
- A type II site specific deoxyribonuclease
- A large group of enzymes recognizing specific short DNA sequences. They cleave either within the recognition site, or at a short specific distance from it
- Requires Mg^{2+}

CONCENTRATION or VOLUME ACTIVITY	UNIT DEFINITION	REACTION BUFFER (As Provided)	ADDITIONAL ACTIVITIES and PURITY	SUPPLIER CATALOG NO.
Bacillus species 19				
				sibEnzyme

Bsp68 I

RECOGNITION SEQUENCE
5'...TCG▼CGA...3'
3'...AGC▲GCT...5'

ISOSCHIZOMERS
Nru I

OPTIMUM REACTION CONDITIONS
pH 7.5 and 37°C

HEAT INACTIVATION
65°C for 20 min.

METHYLATION EFFECTS
Blocked by CG methylation

NOTES
- EC 3.1.21.4
- A type II site specific deoxyribonuclease
- A large group of enzymes recognizing specific short DNA sequences. They cleave either within the recognition site, or at a short specific distance from it
- Large excess of the enzyme results in the appearance of star activity
- Requires Mg^{2+} and BSA for full activity

CONCENTRATION or VOLUME ACTIVITY	UNIT DEFINITION	REACTION BUFFER (As Provided)	ADDITIONAL ACTIVITIES and PURITY	SUPPLIER CATALOG NO.
Bacillus megaterium				
				AGS Heidelb
Bacillus megaterium RFL68				
8-12 U/μL	1 unit hydrolyzes 1 μg DNA/hr at 37°C in a 50 μL reaction volume using λ phage DNA as substrate	50 mM Tris-HCl, 10 mM MgCl₂, 100 mM NaCl, 0.1 mg/mL BSA, pH 7.5, 37°C	>80% of the DNA fragments ligated and >90% recut after 10-fold overdigestion	Fermentas ER0111 ER0112

Bsp106 I

RECOGNITION SEQUENCE
5'...AT▼CGAT...3'
3'...TAGC▲TA...5'

ISOSCHIZOMERS
Cla I, BspD I, Bsu15 I

OPTIMUM REACTION CONDITIONS
pH 7.4 and 37°C

HEAT INACTIVATION
65°C for 20 min.

METHYLATION EFFECTS
Does not cleave ATCGm^{5}AT

NOTES
- EC 3.1.21.4
- A type II site specific deoxyribonuclease
- A large group of enzymes recognizing specific short DNA sequences. They cleave either within the recognition site, or at a short specific distance from it
- Requires Mg^{2+}

CONCENTRATION or VOLUME ACTIVITY	UNIT DEFINITION	REACTION BUFFER (As Provided)	ADDITIONAL ACTIVITIES and PURITY	SUPPLIER CATALOG No.
Bacillus species 106				
5000-15,000 U/mL		50 mM KCl, 10 mM Tris-HCl, 0.1 mM EDTA, 10 mM β-MSH, 200 µg/mL BSA, 50% glycerol, pH 7.4		Stratagene 500160 500161

Bsp119 I

RECOGNITION SEQUENCE
5'...TT▼CGAA...3'
3'...AAGC▲TT...5'

ISOSCHIZOMERS
Asu II

OPTIMUM REACTION CONDITIONS
pH 7.9-8.0 and 37°C

HEAT INACTIVATION
Resistant

METHYLATION EFFECTS
Blocked by CG methylation

NOTES
- EC 3.1.21.4
- A type II site specific deoxyribonuclease
- A large group of enzymes recognizing specific short DNA sequences. They cleave either within the recognition site, or at a short specific distance from it
- Requires Mg^{2+}

Bsp119 I continued

CONCENTRATION or VOLUME ACTIVITY	UNIT DEFINITION	REACTION BUFFER (As Provided)	ADDITIONAL ACTIVITIES and PURITY	SUPPLIER CATALOG No.
Bacillus species RFL119				
10,000 U/mL	1 unit degrades 1 µg λ dam⁻ dcm⁻ DNA/hr at 37°C in a 0.05 mL mixture	10 mM Tris-HCl, 100 mM NaCl, 100 µg/mL BSA, pH 8.0	No detectable contaminating nuclease; 90% ligated and 100% recut after 50-fold overdigestion	AGS Heidelb F00080S F00080M
8-12 U/µL	1 unit hydrolyzes 1 µg DNA/hr at 37°C in a 50 µL reaction volume using λ phage DNA as substrate	33 mM Tris-acetate, 10 mM MgOAc, 66 mM KOAc, pH 7.9	>90% of the DNA fragments ligated and recut after 50-fold overdigestion	Fermentas ER0121 ER0122

Bsp120 I

RECOGNITION SEQUENCE
  5'...G▼GGCCC...3'
  3'...CCCGG▲G...5'
ISOSCHIZOMERS
  Apa I
OPTIMUM REACTION CONDITIONS
  pH 7.5-7.9 and 37°C
HEAT INACTIVATION
  Resistant
METHYLATION EFFECTS
  Blocked by overlapping *dcm* and CG methylation

NOTES
- EC 3.1.21.4
- A type II site specific deoxyribonuclease
- A large group of enzymes recognizing specific short DNA sequences. They cleave either within the recognition site, or at a short specific distance from it
- Produces DNA fragments that have a 4-base 5' extension; *Apa* I does not
- Requires Mg^{2+} and BSA for full activity

CONCENTRATION or VOLUME ACTIVITY	UNIT DEFINITION	REACTION BUFFER (As Provided)	ADDITIONAL ACTIVITIES and PURITY	SUPPLIER CATALOG No.
Bacillus species RFL120				
10,000 U/mL	1 unit degrades 1 µg λ dam⁻ dcm⁻ DNA EcoR I-fragments/hr at 37°C in a 0.05 mL mixture	33 mM Tris-acetate, 10 mM MgOAc, 66 mM KOAc, 100 µg/mL BSA, pH 7.9	No detectable contaminating nuclease; 90% ligated and 100% recut after 50-fold overdigestion	AGS Heidelb F00060S F00060M
8-12 U/µL; 40-60 U/µL	1 unit hydrolyzes 1 µg DNA/hr at 37°C in a 50 µL reaction volume using λ phage DNA as substrate	10 mM Tris-HCl, 10 mM MgCl₂, 0.1 mg/mL BSA, pH 7.5	>90% of the DNA fragments ligated and recut after 50-fold overdigestion	Fermentas ER0131 ER0132

Bsp120 I continued

CONCENTRATION or VOLUME ACTIVITY	UNIT DEFINITION	REACTION BUFFER (As Provided)	ADDITIONAL ACTIVITIES and PURITY	SUPPLIER CATALOG No.
Bacillus species RFL120 continued				
3000-20,000 U/mL				ICN 159416
8000 U/mL	1 unit hydrolyzes 1 µg DNA/hr at 37°C in a 50 µL reaction volume using λ phage DNA as substrate	10 mM Tris-HCl, 10 mM MgCl$_2$, 0.1 mg/mL BSA, pH 7.5	>90% of the DNA fragments ligated and recut after 50-fold overdigestion	NE Biolabs ER0131S ER0131L

Bsp143 I

RECOGNITION SEQUENCE
  5'...▼GATC...3'
  3'...CTAG▲...5'
ISOSCHIZOMERS
  Mbo I
OPTIMUM REACTION CONDITIONS
  pH 7.9-8.0 and 37°C
HEAT INACTIVATION
  65°C for 20 min.
METHYLATION EFFECTS
  Cleaves Gm^{6}ATC
  Not blocked by *dam* methylation; *Mbo* I is
  Blocked by overlapping CG methylation

NOTES
- EC 3.1.21.4
- A type II site specific deoxyribonuclease
- A large group of enzymes recognizing specific short DNA sequences. They cleave either within the recognition site, or at a short specific distance from it
- Large excess of enzyme results in the appearance of star activity
- Requires Mg^{2+}

CONCENTRATION or VOLUME ACTIVITY	UNIT DEFINITION	REACTION BUFFER (As Provided)	ADDITIONAL ACTIVITIES and PURITY	SUPPLIER CATALOG No.
Bacillus species RFL143				
10,000 U/mL	1 unit degrades 1 µg λ dam$^-$ dcm$^-$ DNA/hr at 37°C in a 0.05 mL mixture	10 mM Tris-HCl, 100 mM NaCl, 100 µg/mL BSA, pH 8.0	No detectable contaminating nuclease; 90% ligated and 100% recut after 50-fold overdigestion	AGS Heidelb
8-12 U/µL	1 unit hydrolyzes 1 µg DNA/hr at 37°C in a 50 µL reaction volume using λ phage DNA as substrate	33 mM Tris-acetate, 10 mM MgOAc, 66 mM KOAc, 0.02% Triton X-100, pH 7.9	>90% of the DNA fragments ligated and recut after 50-fold overdigestion	Fermentas ER0781 ER0782

Bsp143 II

RECOGNITION SEQUENCE
  5'...PuGCGC▼Py...3'
  3'...Py▲CGCGPu...5'

ISOSCHIZOMERS
  Hae II

OPTIMUM REACTION CONDITIONS
  pH 7.9 and 37°C

HEAT INACTIVATION
  65°C for 20 min.

METHYLATION EFFECTS
  Blocked by CG methylation

NOTES
- EC 3.1.21.4
- A type II site specific deoxyribonuclease
- A large group of enzymes recognizing specific short DNA sequences. They cleave either within the recognition site, or at a short specific distance from it
- Large excess of enzyme results in the appearance of star activity
- Requires Mg^{2+} and BSA for full activity

CONCENTRATION or VOLUME ACTIVITY	UNIT DEFINITION	REACTION BUFFER (As Provided)	ADDITIONAL ACTIVITIES and PURITY	SUPPLIER CATALOG NO.
Escherichia coli, carrying the cloned *Bsp*143 II gene from *Bacillus* species RFL143				
8-12 U/μL	1 unit hydrolyzes 1 μg DNA/hr at 37°C in a 50 μL reaction volume using λ phage DNA as substrate	33 m*M* Tris-acetate, 10 m*M* MgOAc, 66 m*M* KOAc, 0.1 mg/mL BSA, pH 7.9	>90% of the DNA fragments ligated and recut after 50-fold overdigestion	Fermentas ER0791 ER0792

Bsp1286 I

RECOGNITION SEQUENCE
$$5'...G(^G_{AT})GC(^C_{AT})\blacktriangledown C...3'$$
$$3'...C\blacktriangle(^C_{TA})CG(^G_{TA})G...5'$$

ISOSCHIZOMERS
  Aoc II, Nsp II, Sdu I

OPTIMUM REACTION CONDITIONS
  pH 7.4-7.9 and 30-37°C

HEAT INACTIVATION
  65°C for 20 min.

METHYLATION EFFECTS
  Cleaves $G(^G_{AT})GC(^C_{AT})m^5C$
  Does not cleave $G(^G_{AT})Gm^5C(^C_{AT})C$

NOTES
- EC 3.1.21.4
- A type II site specific deoxyribonuclease
- A large group of enzymes recognizing specific short DNA sequences. They cleave either within the recognition site, or at a short specific distance from it
- Recognizes six distinct sequences:
  GGGCCC
  GAGCCC (complement GGGCTC)
  GTGCCC (complement GGGCAC)
  GAGCAC
  GTGCAC
  GAGCTC (complement GTGCTC)
- Produces a difficult-to-ligate ambiguous extension
- Requires Mg^{2+} and BSA for full activity

CONCENTRATION or VOLUME ACTIVITY	UNIT DEFINITION	REACTION BUFFER (As Provided)	ADDITIONAL ACTIVITIES and PURITY	SUPPLIER CATALOG NO.
Bacillus sphaericus				
8-12 U/μL	1 unit completely digests 1 μg λ DNA/hr at 30°C in the buffer provided	100 mM Tris-HCl, 100 mM MgCl₂, 10 mM DTT, pH 7.5	>90% DNA fragments ligated and 100% recut	Amersham E 1024Y
3000-10,000 U/mL		50 mM KCl, 100 mM Tris-HCl, 0.1 mM EDTA, 1 mM DTT, 200 μg/mL BSA, 50% glycerol, pH 7.4		ICN 150527
500 U and 2500 U	1 unit completely digests 1 μg λ DNA/hr at 30°C in 50 μL of supplied buffer and BSA	100 mM Tris-HCl, 100 mM MgCl₂, 10 mM DTT, pH 7.5	Purity is confirmed by overdigestion and ligation-recutting test	Nippon Gene TaKaRa 1024
Bacillus sphaericus IAM 1286				
10,000 U/mL	1 unit degrades 1 μg λ dam⁻ dcm⁻ DNA/hr at 37°C in a 0.05 mL mixture	33 mM Tris-acetate, 10 mM MgOAc, 66 mM KOAc, 100 μg/mL BSA, pH 7.9	No detectable contaminating nuclease; >98% ligated and 100% recut after 50-fold overdigestion	AGS Heidelb A00160S A00160M

Bsp1286 I continued

CONCENTRATION or VOLUME ACTIVITY	UNIT DEFINITION	REACTION BUFFER (As Provided)	ADDITIONAL ACTIVITIES and PURITY	SUPPLIER CATALOG No.
Bacillus sphaericus IAM 1286 continued				
5000 U/mL	1 unit completely digests 1 µg λ DNA/hr in a total reaction volume of 50 µL using the buffer provided	10 mM Tris-HCl, 10 mM MgCl$_2$, 1 mM DTT, pH 7.9; add 100 µg/mL BSA	>95% DNA fragments ligated and recut after 10-fold overdigestion	NE Biolabs 120S 120L
10 U/µL	1 unit completely digests 1 µg λ DNA/hr at 37°C in a 50 mL reaction volume	60 mM Tris-HCl, 60 mM MgCl$_2$, 60 mM NaCl, 10 mM DTT, pH 7.5	85% ligation	Promega R6741 R6742

Bsp1407 I

RECOGNITION SEQUENCE
  5'...T▼GTACA...3'
  3'...ACATG▲T...5'
OPTIMUM REACTION CONDITIONS
  pH 7.9 and 37°C
HEAT INACTIVATION
  65°C for 20 min.

NOTES
- EC 3.1.21.4
- A type II site specific deoxyribonuclease
- A large group of enzymes recognizing specific short DNA sequences. They cleave either within the recognition site, or at a short specific distance from it
- Requires Mg^{2+}

CONCENTRATION or VOLUME ACTIVITY	UNIT DEFINITION	REACTION BUFFER (As Provided)	ADDITIONAL ACTIVITIES and PURITY	SUPPLIER CATALOG No.
Bacillus stearothermophilus RFL1407				
5000 U/mL	1 unit degrades 1 µg λ dam$^-$ dcm$^-$ DNA/hr at 37°C in a 0.05 mL mixture	33 mM Tris-acetate, 10 mM MgOAc, 66 mM KOAc, 100 µg/mL BSA, pH 7.9	No detectable contaminating nuclease; 90% ligated and 100% recut after 50-fold overdigestion	AGS Heidelb F00162S F00162M
8-12 U/µL	1 unit hydrolyzes 1 µg DNA/hr at 37°C in a 50 µL reaction volume using λ phage DNA as substrate	33 mM Tris-acetate, 10 mM MgOAc, 66 mM KOAc, 0.1 mg/mL BSA, pH 7.9	>90% of the DNA fragments ligated and recut after 50-fold overdigestion	Fermentas ER0931 ER0932

Bsp1720 I

RECOGNITION SEQUENCE
```
5'...GC▼TNAGC...3'
3'...CGANT▲CG...5'
```
ISOSCHIZOMERS
Blp I, Bpu1102 I, Cel II

NOTES
- EC 3.1.21.4
- A type II site specific deoxyribonuclease
- A large group of enzymes recognizing specific short DNA sequences. They cleave either within the recognition site, or at a short specific distance from it
- Requires Mg^{2+}

CONCENTRATION or VOLUME ACTIVITY	UNIT DEFINITION	REACTION BUFFER (As Provided)	ADDITIONAL ACTIVITIES and PURITY	SUPPLIER CATALOG No.
Bacillus species 1720				
				SibEnzyme

BspA2 I

RECOGNITION SEQUENCE
```
5'...C▼CTAGG...3'
3'...GGATC▲C...5'
```
ISOSCHIZOMERS
Avr II, Bln I

NOTES
- EC 3.1.21.4
- A type II site specific deoxyribonuclease
- A large group of enzymes recognizing specific short DNA sequences. They cleave either within the recognition site, or at a short specific distance from it
- Requires Mg^{2+}

CONCENTRATION or VOLUME ACTIVITY	UNIT DEFINITION	REACTION BUFFER (As Provided)	ADDITIONAL ACTIVITIES and PURITY	SUPPLIER CATALOG No.
Bacillus species A2				
				SibEnzyme

BspC I

RECOGNITION SEQUENCE
5'...CGAT▼CG...3'
3'...GC▲TAGC...5'

ISOSCHIZOMERS
Pvu I, Xml I, Xor II

OPTIMUM REACTION CONDITIONS
pH 7.4 and 37°C

HEAT INACTIVATION
65°C for 20 min.

NOTES
- EC 3.1.21.4
- A type II site specific deoxyribonuclease
- A large group of enzymes recognizing specific short DNA sequences. They cleave either within the recognition site, or at a short specific distance from it
- Superior to Pvu I for pulsed field gel electrophoresis
- Free of Pvu II contamination; Pvu I is not
- Requires Mg^{2+}

CONCENTRATION or VOLUME ACTIVITY	UNIT DEFINITION	REACTION BUFFER (As Provided)	ADDITIONAL ACTIVITIES and PURITY	SUPPLIER CATALOG NO.
Bacillus species CI				
2000-15,000 U/mL		50 mM KCl, 10 mM KPO₄, 0.1 mM EDTA, 10 mM β-MSH, 200 µg/mL BSA, 50% glycerol, pH 7.4		Stratagene 500302

BspD I

RECOGNITION SEQUENCE
5'...AT▼CGAT...3'
3'...TAGC▲TA...5'

ISOSCHIZOMERS
Cla I

OPTIMUM REACTION CONDITIONS
pH 7.4-7.9 and 37°C

HEAT INACTIVATION
Resistant

NOTES
- EC 3.1.21.4
- A type II site specific deoxyribonuclease
- A large group of enzymes recognizing specific short DNA sequences. They cleave either within the recognition site, or at a short specific distance from it
- Blocked by overlapping dam methylation
- Requires Mg^{2+}

BspD I continued

CONCENTRATION or VOLUME ACTIVITY	UNIT DEFINITION	REACTION BUFFER (As Provided)	ADDITIONAL ACTIVITIES and PURITY	SUPPLIER CATALOG No.
Bacillus species				
3000-15,000 U/mL		50 mM NaCl, 10 mM Tris-HCl, 0.1 mM EDTA, 1 mM DTT, 200 µg/mL acetylated BSA, 50% glycerol, pH 7.4		ICN 159417
Bacillus species (NEB 595)				
5000 U/mL	1 unit completely digests 1 µg λ DNA (dam⁻)/hr in a total reaction volume of 50 µL using the buffer provided	20 mM Tris-acetate, 10 mM MgOAc, 50 mM KOAc, 1 mM DTT, pH 7.9	>95% DNA fragments ligated and recut after 10-fold overdigestion	NE Biolabs 557S 557L

BspE I

RECOGNITION SEQUENCE
    5'...T▼CCGGA...3'
    3'...AGGCC▲T...5'
ISOSCHIZOMERS
    BspM II
OPTIMUM REACTION CONDITIONS
    pH 7.4-7.9 and 37°C
HEAT INACTIVATION
    Resistant
METHYLATION EFFECTS
    Blocked by overlapping dam methylation

NOTES
- EC 3.1.21.4
- A type II site specific deoxyribonuclease
- A large group of enzymes recognizing specific short DNA sequences. They cleave either within the recognition site, or at a short specific distance from it
- Requires Mg^{2+}

CONCENTRATION or VOLUME ACTIVITY	UNIT DEFINITION	REACTION BUFFER (As Provided)	ADDITIONAL ACTIVITIES and PURITY	SUPPLIER CATALOG No.
Bacillus species				
3000-20,000 U/mL		50 mM KCl, 10 mM Tris-HCl, 0.1 mM EDTA, 1 mM DTT, 200 µg/mL acetylated BSA, 50% glycerol, pH 7.4		ICN 159418
Escherichia coli strain, carrying the cloned gene from Bacillus species				
10,000 U/mL	1 unit completely digests 1 µg λ DNA/hr in a total reaction volume of 50 µL using the buffer provided	50 mM Tris-HCl, 10 mM MgCl₂, 100 mM NaCl, 1 mM DTT, pH 7.9	>95% DNA fragments ligated and recut after 10-fold overdigestion	NE Biolabs 540S 540L

*Bsp*H I

RECOGNITION SEQUENCE
5'...T▼CATGA...3'
3'...AGTAC▲T...5'

ISOSCHIZOMERS
*Rsp*X I

OPTIMUM REACTION CONDITIONS
pH 7.4-7.9 and 37°C

HEAT INACTIVATION
65°C for 20 min.

METHYLATION EFFECTS
Does not cleave TCm^{6}ATGA or TCATGm^{6}A
Blocked by overlapping *dam* methylation

NOTES
- EC 3.1.21.4
- A type II site specific deoxyribonuclease
- A large group of enzymes recognizing specific short DNA sequences. They cleave either within the recognition site, or at a short specific distance from it
- Requires Mg^{2+}

CONCENTRATION or VOLUME ACTIVITY	UNIT DEFINITION	REACTION BUFFER (As Provided)	ADDITIONAL ACTIVITIES and PURITY	SUPPLIER CATALOG NO.
Bacillus species H				
3-15 U/μL	1 unit completely digests 1 μg λ DNA/hr at 37°C in the buffer provided	330 m*M* Tris-acetate, 100 m*M* MgOAc, 5 m*M* DTT, 660 m*M* KOAc, pH 7.9	>95% DNA fragments ligated and 95% recut	Amersham E 0206Y E 0206Z
1000-10,000 U/mL		100 m*M* KCl, 10 m*M* Tris-HCl, 0.1 m*M* EDTA, 1 m*M* DTT, 200 μg/mL acetylated BSA, 50% glycerol, pH 7.4		ICN 159419
Escherichia coli strain, carrying the cloned gene from *Bacillus* species H (NEB 394)				
10,000 U/mL	1 unit completely digests 1 μg λ DNA/hr in a total reaction volume of 50 μL using the buffer provided	20 m*M* Tris-acetate, 10 m*M* MgOAc, 50 m*M* KOAc, 1 m*M* DTT, pH 7.9	>95% DNA fragments ligated and recut after 10-fold overdigestion	NE Biolabs 517S 517L

BspL I

RECOGNITION SEQUENCE
5'...GGN▼NCC...3'
3'...CCN▲NGG...5'

ISOSCHIZOMERS
Nla IV

OPTIMUM REACTION CONDITIONS
pH 7.9 and 37°C

HEAT INACTIVATION
65°C for 20 min.

METHYLATION EFFECTS

NOTES
Blocked by overlapping CG methylation

- EC 3.1.21.4
- A type II site specific deoxyribonuclease
- A large group of enzymes recognizing specific short DNA sequences. They cleave either within the recognition site, or at a short specific distance from it
- Requires Mg^{2+} and BSA for full activity

CONCENTRATION or VOLUME ACTIVITY	UNIT DEFINITION	REACTION BUFFER (As Provided)	ADDITIONAL ACTIVITIES and PURITY	SUPPLIER CATALOG No.
Bacillus species RJ3-212				
8-12 U/μL	1 unit hydrolyzes 1 μg DNA/hr at 37°C in a 50 μL reaction volume using λ phage DNA as substrate	33 m*M* Tris-acetate, 10 m*M* MgOAc, 66 m*M* KOAc, 0.1 mg/mL BSA, pH 7.9	>80% DNA fragments ligated and >90% recut after 50-fold overdigestion	Fermentas ER1151 ER1152

BspLU11 I

RECOGNITION SEQUENCE
5'...A▼CATGT...3'
3'...TGTAC▲A...5'
OPTIMUM REACTION CONDITIONS
pH 7.5 and 37°C
HEAT INACTIVATION
65°C for 20 min.

NOTES
- EC 3.1.21.4
- A type II site specific deoxyribonuclease
- A large group of enzymes recognizing specific short DNA sequences. They cleave either within the recognition site, or at a short specific distance from it
- Requires Mg^{2+}

CONCENTRATION or VOLUME ACTIVITY	UNIT DEFINITION	REACTION BUFFER (As Provided)	ADDITIONAL ACTIVITIES and PURITY	SUPPLIER CATALOG No.
Bacillus species				
8-12 U/μL	1 unit completely digests 1 μg substrate DNA/hr	50 mM Tris-HCl, 10 mM $MgCl_2$, 100 mM NaCl, 1 mM DTE, pH 7.5		Boehringer 1693743 1693751

BspM I

RECOGNITION SEQUENCE
5'...ACCTGC(N)$_4$▼...3'
3'...TGGACG(N)$_8$▲...5'
OPTIMUM REACTION CONDITIONS
pH 7.4-7.9 and 37°C
HEAT INACTIVATION
65°C for 20 min.
METHYLATION EFFECTS
Cleaves ACCTGm^{5}C

NOTES
- EC 3.1.21.4
- A type II site specific deoxyribonuclease
- A large group of enzymes recognizing specific short DNA sequences. They cleave either within the recognition site, or at a short specific distance from it
- Single BspM I sites in pBR322 and pUC18 and 19 are resistant to cleavage
- 100-fold overdigestion cleaves less than half of the DNA
- Other plasmid DNAs are also resistant to BspM I cleavage
- Requires Mg^{2+}

BspM I continued

CONCENTRATION or VOLUME ACTIVITY	UNIT DEFINITION	REACTION BUFFER (As Provided)	ADDITIONAL ACTIVITIES and PURITY	SUPPLIER CATALOG No.
Bacillus species M				
1000-5000 U/mL		150 mM KCl, 10 mM Tris-HCl, 10 mM MgCl$_2$, 0.1 mM EDTA, 1 mM DTT, 200 µg/mL acetylated BSA, 50% glycerol, pH 7.4		ICN 159420
Escherichia coli strain, carrying the cloned gene from Bacillus species M (NEB 356)				
2000 U/mL	1 unit completely digests 1 µg λ DNA/hr in a total reaction volume of 50 µL using the buffer provided	10 mM Tris-HCl, 10 mM MgCl$_2$, 150 mM NaCl, 1 mM DTT, pH 7.9	>95% DNA fragments ligated and recut after 5-fold overdigestion	NE Biolabs 502S 502L

BspT I

RECOGNITION SEQUENCE
    5'...C▼TTAAG...3'
    3'...GAATT▲C...5'
ISOSCHIZOMERS
    Afl II
OPTIMUM REACTION CONDITIONS
    pH 7.5-8.5 and 37°C
HEAT INACTIVATION
    65°C for 20 min.

NOTES
- EC 3.1.21.4
- A type II site specific deoxyribonuclease
- A large group of enzymes recognizing specific short DNA sequences. They cleave either within the recognition site, or at a short specific distance from it
- Requires Mg^{2+}

CONCENTRATION or VOLUME ACTIVITY	UNIT DEFINITION	REACTION BUFFER (As Provided)	ADDITIONAL ACTIVITIES and PURITY	SUPPLIER CATALOG No.
Bacillus species RFL1265I				
10,000 U/mL	1 unit degrades 1 µg λ dam$^-$ dcm$^-$ DNA/hr at 37°C in a 0.05 mL mixture	10 mM Tris-HCl, 10 mM MgCl$_2$, 100 mM KCl, 100 µg/mL BSA, pH 8.5	No detectable contaminating nuclease; 90% ligated and 100% recut after 50-fold overdigestion	AGS Heidelb F00166S F00166M
8-12 U/µL	1 unit hydrolyzes 1 µg DNA/hr at 37°C in a 50 µL reaction volume using λ phage DNA as substrate	50 mM Tris-HCl, 10 mM MgCl$_2$, 100 mM NaCl, pH 7.5	80% of the φX174 DNA fragments ligated and >90% recut after 50-fold overdigestion	Fermentas ER0831 ER0832

Bsr I

RECOGNITION SEQUENCE
  5'...ACTGGN▼...3'
  3'...TGAC▲CN...5'

OPTIMUM REACTION CONDITIONS
  pH 7.8 and 65°C; 20% activity at 37°C

HEAT INACTIVATION
  Resistant

NOTES
- EC 3.1.21.4
- A type II site specific deoxyribonuclease
- A large group of enzymes recognizing specific short DNA sequences. They cleave either within the recognition site, or at a short specific distance from it
- Requires Mg^{2+} and BSA for full activity

CONCENTRATION or VOLUME ACTIVITY	UNIT DEFINITION	REACTION BUFFER (As Provided)	ADDITIONAL ACTIVITIES and PURITY	SUPPLIER CATALOG NO.
Bacillus stearothermophilus (NEB 447)				
5000 U/mL	1 unit completely digests 1 µg φx174DNA/hr in a total reaction volume of 50 µL using the buffer provided	10 mM Tris-HCl, 10 mM $MgCl_2$, 150 mM KCl, 1 mM DTT, pH 7.8; add 100 µg/mL BSA	>95% DNA fragments ligated and recut after 4-fold overdigestion	NE Biolabs 527S 527L

BsrB I

RECOGNITION SEQUENCE
  5'...GAG▼CGG...3'
  3'...CTC▲GCC...5'

OPTIMUM REACTION CONDITIONS
  pH 7.9 and 37°C; fully active at 50°C

HEAT INACTIVATION
  Resistant

NOTES
- EC 3.1.21.4
- A type II site specific deoxyribonuclease
- A large group of enzymes recognizing specific short DNA sequences. They cleave either within the recognition site, or at a short specific distance from it
- Its recognition sequence is non-palindromic: when DNA is cut with BsrB I and then religated, only 50% of ligated sites regenerate BsrB I sites
- Requires Mg^{2+}

CONCENTRATION or VOLUME ACTIVITY	UNIT DEFINITION	REACTION BUFFER (As Provided)	ADDITIONAL ACTIVITIES and PURITY	SUPPLIER CATALOG NO.
Bacillus stearothermophilus cpw 193				
10,000 U/mL	1 unit completely digests 1 μg λ DNA/hr in a total reaction volume of 50 μL using the buffer provided	10 mM Tris-HCl, 10 mM $MgCl_2$, 50 mM NaCl, 1 mM DTT, pH 7.9	>90% DNA fragments ligated and 50% recut after 10-fold overdigestion	NE Biolabs 102S 102L

BsrBR I

RECOGNITION SEQUENCE
  5'...GATNN▼NNATC...3'
  3'...CTANN▲NNTAG...5'

OPTIMUM REACTION CONDITIONS
  pH 6.8 and 37°C

NOTES
- EC 3.1.21.4
- A type II site specific deoxyribonuclease
- A large group of enzymes recognizing specific short DNA sequences. They cleave either within the recognition site, or at a short specific distance from it
- Compatible cohesive ends: any blunt end
- Requires Mg^{2+}

BsrBR I continued

CONCENTRATION or VOLUME ACTIVITY	UNIT DEFINITION	REACTION BUFFER (As Provided)	ADDITIONAL ACTIVITIES and PURITY	SUPPLIER CATALOG NO.
Bacillus stearothermophilus				
10 U/μL	1 unit completely digests 1 μg λ DNA/hr at 37°C in a 50 mL reaction volume	900 mM Tris-HCl, 100 mM MgCl$_2$, 500 mM NaCl, pH 6.8	90% ligation	Promega R7121 R7122

BsrD I

RECOGNITION SEQUENCE
    5'...GCAATGNN▼...3'
    3'...CGTTAC▲NN...5'
OPTIMUM REACTION CONDITIONS
    pH 7.9 and 60°C
HEAT INACTIVATION
    Resistant

NOTES
- EC 3.1.21.4
- A type II site specific deoxyribonuclease
- A large group of enzymes recognizing specific short DNA sequences. They cleave either within the recognition site, or at a short specific distance from it
- Requires Mg^{2+} and BSA for full activity

CONCENTRATION or VOLUME ACTIVITY	UNIT DEFINITION	REACTION BUFFER (As Provided)	ADDITIONAL ACTIVITIES and PURITY	SUPPLIER CATALOG NO.
Bacillus stearothermophilus D70 (Z. Chen)				
2000 U/mL	1 unit completely digests 1 μg λ DNA/hr in a total reaction volume of 50 μL using the buffer provided	10 mM Tris-HCl, 10 mM MgCl$_2$, 50 mM NaCl, 1 mM DTT, pH 7.9; add 100 μg/mL BSA	>95% DNA fragments ligated and recut after 10-fold overdigestion	NE Biolabs 574S 574L

BsrF I

RECOGNITION SEQUENCE
　　5'...Pu▼CCGGPy...3'
　　3'...PyGGCC▲Pu...5'

ISOSCHIZOMERS
　　Cfr10 I

OPTIMUM REACTION CONDITIONS
　　pH 7.9 and 37°C

HEAT INACTIVATION
　　Resistant

NOTES
- EC 3.1.21.4
- A type II site specific deoxyribonuclease
- A large group of enzymes recognizing specific short DNA sequences. They cleave either within the recognition site, or at a short specific distance from it
- Requires Mg^{2+} and BSA for full activity

CONCENTRATION or VOLUME ACTIVITY	UNIT DEFINITION	REACTION BUFFER (As Provided)	ADDITIONAL ACTIVITIES and PURITY	SUPPLIER CATALOG NO.
Bacillus stearothermophilus (Z. Chen)				
3000 U/mL	1 unit completely digests 1 μg pBR322 DNA/hr in a total reaction volume of 50 μL using the buffer provided	10 m*M* Tris-HCl, 10 m*M* MgCl₂, 50 m*M* NaCl, 1 m*M* DTT, pH 7.9; add 100 μg/mL BSA	>95% DNA fragments ligated and recut after 20-fold overdigestion	NE Biolabs 562S 562L

BsrG I

RECOGNITION SEQUENCE
　　5'...T▼GTACA...3'
　　3'...ACATG▲T...5'

ISOSCHIZOMERS
　　Bsp1407 I

OPTIMUM REACTION CONDITIONS
　　pH 7.9 and 37°C; 2-fold greater activity at 60°C

HEAT INACTIVATION
　　Resistant

NOTES
- EC 3.1.21.4
- A type II site specific deoxyribonuclease
- A large group of enzymes recognizing specific short DNA sequences. They cleave either within the recognition site, or at a short specific distance from it
- Requires Mg^{2+} and BSA for full activity

BsrG I continued

CONCENTRATION or VOLUME ACTIVITY	UNIT DEFINITION	REACTION BUFFER (As Provided)	ADDITIONAL ACTIVITIES and PURITY	SUPPLIER CATALOG NO.
Bacillus stearothermophilus (Z. Chen)				
10,000 U/mL	1 unit completely digests 1 µg λ DNA/hr in a total reaction volume of 50 µL using the buffer provided	10 mM Tris-HCl, 10 mM MgCl$_2$, 50 mM NaCl, 1 mM DTT, pH 7.9; add 100 µg/mL BSA	>95% DNA fragments ligated and recut after 10-fold overdigestion	NE Biolabs 575S 575L

BsrS I

RECOGNITION SEQUENCE
    5'...ACTGG(1/-1)▼...3'

ISOSCHIZOMERS
    Bse1 I, BseN I, Bsr I

NOTES
- EC 3.1.21.4
- A type II site specific deoxyribonuclease
- A large group of enzymes recognizing specific short DNA sequences. They cleave either within the recognition site, or at a short specific distance from it
- Requires Mg^{2+}

CONCENTRATION or VOLUME ACTIVITY	UNIT DEFINITION	REACTION BUFFER (As Provided)	ADDITIONAL ACTIVITIES and PURITY	SUPPLIER CATALOG NO.
Bacillus stearothermophilus CPW19				
10 U/µL	1 unit completely digests 1 µg φ DNA/hr at 65°C in a 50 mL reaction volume	60 mM Tris-HCl, 1.5 M KCl, 60 mM MgCl$_2$, 1 mM DTT, pH 7.2	90% ligation	Promega R7241 R7242

BssA I

RECOGNITION SEQUENCE
  5'...R▼CCGGY...3'
  3'...YGGCC▲R...5'

ISOSCHIZOMERS
  Bse118 I, BsrF I, Cfr10 I

NOTES
- EC 3.1.21.4
- A type II site specific deoxyribonuclease
- A large group of enzymes recognizing specific short DNA sequences. They cleave either within the recognition site, or at a short specific distance from it
- Requires Mg^{2+}

CONCENTRATION or VOLUME ACTIVITY	UNIT DEFINITION	REACTION BUFFER (As Provided)	ADDITIONAL ACTIVITIES and PURITY	SUPPLIER CATALOG No.
Bacillus species				
	10 units does not produce unspecific cleavage products with 1 µg λ DNA/16 hr at 65°C	20 mM Tris-HCl, 3 mM MgCl₂, 100 mM KCl, 100 µg/mL BSA, 0.04% Triton X-100, pH 8.5	>90% DNA fragments ligated and recut after 10-fold overdigestion	MINOTECH 108

BssH II

RECOGNITION SEQUENCE
  5'...G▼CGCGC...3'
  3'...CGCGC▲G...5'

ISOSCHIZOMERS
  BseP I

OPTIMUM REACTION CONDITIONS
  pH 7.0-8.3 and 50°C; 75-100% activity at 37°C

HEAT INACTIVATION
  Resistant

METHYLATION EFFECTS
  Does not cleave Gm^{5}CGCGC

NOTES
- EC 3.1.21.4
- A type II site specific deoxyribonuclease
- A large group of enzymes recognizing specific short DNA sequences. They cleave either within the recognition site, or at a short specific distance from it
- GC-rich recognition sequence makes this enzyme useful for genomic DNA analysis; produces 100,000-sized fragments from mammalian genomic DNA
- Compatible cohesive ends: Asc I, Dsa I, Mlu I
- Requires Mg^{2+}

BssH II continued

CONCENTRATION or VOLUME ACTIVITY	UNIT DEFINITION	REACTION BUFFER (As Provided)	ADDITIONAL ACTIVITIES and PURITY	SUPPLIER CATALOG No.
Bacillus stearothermophilus				
5000 U/mL	1 unit degrades 1 μg λ dam⁻ dcm⁻ DNA/hr at 50°C in a 0.05 mL mixture	10 mM Tris-HCl, 100 mM NaCl, 100 μg/mL BSA, pH 8.0	No detectable contaminating nuclease; 100% ligated and recut after 50-fold overdigestion	AGS Heidelb A00170S A00170M
				Nippon Gene
2000-6000 U/mL	1 unit completely digests 1 μg λ DNA/hr at pH 7.5, 50°C in a 50 μL total assay mixture	100 mM Tris-acetate, 100 mM MgOAc, 500 mM KOAc, pH 7.5		Pharmacia 27-0887-01
1000-10,000 U/mL		200 mM KCl, 10 mM KPO₄, 0.1 mM EDTA, 1 mM DTT, 500 μg/mL BSA, 50% glycerol, pH 7.4		Stratagene 500310 500311
				Toyobo
Bacillus stearothermophilus strain H3				
3-20 U/μL	1 unit completely digests 1 μg λ DNA/hr at 50°C in the buffer provided	500 mM Tris-HCl, 100 mM MgCl₂, 10 mM DTT, 1000 mM NaCl, pH 7.5	>90% DNA fragments ligated and >98% recut	Amersham E 0207Y E 0207Z
10 U/μL	1 unit completely digests 1 μg substrate DNA/hr	33 mM Tris-acetate, 10 mM MgOAc, 66 mM KOAc, 0.5 mM DTT, pH 7.9		Boehringer 1168851
	1 unit completely cleaves 1 μg λ DNA/hr at pH 7.5, 50°C in a reaction volume of 50 μL	10 mM Tris-HCl, 10 mM MgCl₂, 50 mM NaCl, 1.0 mM DTT, 100 μg/mL BSA, pH 7.5		CHIMERX 2115-01 2115-02
4000-20,000 U/mL		50 mM KCl, 10 mM Tris-HCl, 0.1 mM EDTA, 1.0 mM DTT, 200 μg/mL BSA, 50% glycerol, pH 7.4		ICN 150528
8-12 U/μL	1 unit completely digests 1 mg λ DNA/hr at 50°C	500 mM Tris-HCl, 100 mM MgCl₂, 500 mM NaCl, pH 8.0	≥85% ligation; 100% cleavage	Life Technol 15468-010
6-12 U/μL	1 unit digests completely 1 μg λ DNA substrate/hr at 50°C	10 mM Tris-HCl, 100 mM NaCl, 5 mM MgCl₂, 1 mM β-MSH, pH 8.3		NBL Gene 016811 016802
10 U/μL	1 unit completely digests 1 μg λ DNA/hr at 50°C in a 50 mL reaction volume	Genome qualified; 900 mM Tris-HCl, 100 mM MgCl₂, 500 mM NaCl, pH 7.2	95% ligation	Promega R6831-2 R6835
300 U and 1500 U	1 unit completely digests 1 μg λ DNA/hr at 50°C in 50 μL of supplied buffer	Genomic grade; 100 mM Tris-HCl, 100 mM MgCl₂, 10 mM DTT, 500 mM NaCl, pH 7.5	Purity is confirmed by overdigestion, ligation-recutting and genome DNA digestion test	TaKaRa 1119
3-20 U/μL	1 unit completely digests 1 μg DNA/hr at 50°C	10 mM Tris-HCl, 10 mM MgCl₂, 50 mM NaCl, 1 mM DTT, pH 7.5	>90% ligation; >95% recut	Toyobo BSH-201 BSH-202

BssH II continued

CONCENTRATION or VOLUME ACTIVITY	UNIT DEFINITION	REACTION BUFFER (As Provided)	ADDITIONAL ACTIVITIES and PURITY	SUPPLIER CATALOG No.
Bacillus stearothermophilus strain H3 (N. Welker)				
4000 U/mL; 20,000 U/mL	1 unit completely digests 1 μg λ DNA/hr in a total reaction volume of 50 μL using the buffer provided	10 mM Bis Tris Propane-HCl, 10 mM MgCl$_2$, 100 mM NaCl, 1 mM DTT, pH 7.0	>95% DNA fragments ligated and recut after 20-fold overdigestion	NE Biolabs 199S 199L

BssK I

RECOGNITION SEQUENCE
    5'...▼CCNGG...3'
    3'...GGNCC▲...5'

ISOSCHIZOMERS
    *Msp*R9 I, *Scr*F I

NOTES
- EC 3.1.21.4
- A type II site specific deoxyribonuclease
- A large group of enzymes recognizing specific short DNA sequences. They cleave either within the recognition site, or at a short specific distance from it
- Requires Mg^{2+}

CONCENTRATION or VOLUME ACTIVITY	UNIT DEFINITION	REACTION BUFFER (As Provided)	ADDITIONAL ACTIVITIES and PURITY	SUPPLIER CATALOG No.
Bacillus stearothermophilus				
				NE Biolabs

BssS I

RECOGNITION SEQUENCE
  5'...C▼TCGTG...3'
  3'...GAGCA▲C...5'

OPTIMUM REACTION CONDITIONS
  pH 7.9 and 37°C; activity increases 50% at 60°C

HEAT INACTIVATION
  Resistant

NOTES
- EC 3.1.21.4
- A type II site specific deoxyribonuclease
- A large group of enzymes recognizing specific short DNA sequences. They cleave either within the recognition site, or at a short specific distance from it
- Low salt reaction buffers, >10 U/μg and incubation time >2 hrs may result in star activity
- Requires Mg^{2+}

CONCENTRATION or VOLUME ACTIVITY	UNIT DEFINITION	REACTION BUFFER (As Provided)	ADDITIONAL ACTIVITIES and PURITY	SUPPLIER CATALOG NO.
Bacillus stearothermophilus S719 (Z. Chen)				
2000 U/mL	1 unit completely digests 1 μg λ DNA/hr in a total reaction volume of 50 μL using the buffer provided	50 mM Tris-HCl, 10 mM MgCl₂, 100 mM NaCl, 1 mM DTT, pH 7.9	>95% DNA fragments ligated and recut after 5-fold overdigestion	NE Biolabs 587S 587L

BssT1 I

RECOGNITION SEQUENCE
  5'...C▼CWWGG...3'
  3'...GGWWC▲C...5'

ISOSCHIZOMERS
  *Eco*130 I, *Eco*T14 I, *Erh* I, *Sty* I

NOTES
- EC 3.1.21.4
- A type II site specific deoxyribonuclease
- A large group of enzymes recognizing specific short DNA sequences. They cleave either within the recognition site, or at a short specific distance from it
- Requires Mg^{2+}

CONCENTRATION or VOLUME ACTIVITY	UNIT DEFINITION	REACTION BUFFER (As Provided)	ADDITIONAL ACTIVITIES and PURITY	SUPPLIER CATALOG NO.
Bacillus stearothermophilus T1				
				SibEnzyme

Bst I

RECOGNITION SEQUENCE
5'...G▼GATCC...3'
3'...CCTAG▲G...5'

ISOSCHIZOMERS
BamH I

NOTES
- EC 3.1.21.4
- A type II site specific deoxyribonuclease
- A large group of enzymes recognizing specific short DNA sequences. They cleave either within the recognition site, or at a short specific distance from it
- Requires Mg^{2+}

CONCENTRATION or VOLUME ACTIVITY	UNIT DEFINITION	REACTION BUFFER (As Provided)	ADDITIONAL ACTIVITIES and PURITY	SUPPLIER CATALOG No.
Bacillus stearothermophilus 1503-4R				
				Pharmacia

Bst2B I

RECOGNITION SEQUENCE
5'...CACGAG(-5/-1)▼...3'

ISOSCHIZOMERS
BssS I

NOTES
- EC 3.1.21.4
- A type II site specific deoxyribonuclease
- A large group of enzymes recognizing specific short DNA sequences. They cleave either within the recognition site, or at a short specific distance from it
- Requires Mg^{2+}

CONCENTRATION or VOLUME ACTIVITY	UNIT DEFINITION	REACTION BUFFER (As Provided)	ADDITIONAL ACTIVITIES and PURITY	SUPPLIER CATALOG No.
Bacillus stearothermophilus 2B				
				SibEnzyme

Bst2U I

RECOGNITION SEQUENCE
5'...CC▼WGG...3'
3'...GGW▲CC...5'

ISOSCHIZOMERS
BsiL I, BstN I, BstO I, EcoR II, Mva I

NOTES
- EC 3.1.21.4
- A type II site specific deoxyribonuclease
- A large group of enzymes recognizing specific short DNA sequences. They cleave either within the recognition site, or at a short specific distance from it
- Requires Mg^{2+}

CONCENTRATION or VOLUME ACTIVITY	UNIT DEFINITION	REACTION BUFFER (As Provided)	ADDITIONAL ACTIVITIES and PURITY	SUPPLIER CATALOG No.
Bacillus stearothermophilus 2U				
				SibEnzyme

Bst71 I

RECOGNITION SEQUENCE
   5'...GCAGC(N)$_8$▼...3'
   3'...CGTCG(N)$_{12}$▲...5'

OPTIMUM REACTION CONDITIONS
   pH 7.6 and 50°C; 50-75% activity at 37°C

NOTES
- EC 3.1.21.4
- A type II site specific deoxyribonuclease
- A large group of enzymes recognizing specific short DNA sequences. They cleave either within the recognition site, or at a short specific distance from it
- Requires Mg^{2+}

CONCENTRATION or VOLUME ACTIVITY	UNIT DEFINITION	REACTION BUFFER (As Provided)	ADDITIONAL ACTIVITIES and PURITY	SUPPLIER CATALOG NO.
Bacillus stearothermophilus strain 71				
5-10 U/μL	1 unit completely digests 1 μg λ DNA/hr at 50°C in a 50 mL reaction volume	60 mM Tris-HCl, 60 mM MgCl$_2$, 1.5 M NaCl, 10 mM DTT, pH 7.6	95% ligation	Promega R6871 R6872

Bst98 I

RECOGNITION SEQUENCE
   5'...C▼TTAAG...3'
   3'...GAATT▲C...5'

OPTIMUM REACTION CONDITIONS
   pH 7.9 and 37°C; 2-fold greater activity at 60°C

NOTES
- EC 3.1.21.4
- A type II site specific deoxyribonuclease
- A large group of enzymes recognizing specific short DNA sequences. They cleave either within the recognition site, or at a short specific distance from it
- Requires Mg^{2+}

CONCENTRATION or VOLUME ACTIVITY	UNIT DEFINITION	REACTION BUFFER (As Provided)	ADDITIONAL ACTIVITIES and PURITY	SUPPLIER CATALOG NO.
Bacillus stearothermophilus				
8-12 U/μL	1 unit completely digests 1 μg Ad2 DNA/hr at 37°C in a 50 mL reaction volume	60 mM Tris-HCl, 60 mM MgCl$_2$, 1.5 M NaCl, 10 mM DTT, pH 7.9	90% ligation	Promega R7141 R7142

Bst1107 I

RECOGNITION SEQUENCE
5'...GTA▼TAC...3'
3'...CAT▲ATG...5'

ISOSCHIZOMERS
Sna I

OPTIMUM REACTION CONDITIONS
pH 7.5-8.5 and 37°C

HEAT INACTIVATION
65°C for 20 min.

METHYLATION EFFECTS
Does not cleave GTATAm^{5}C
Blocked by overlapping CG methylation

NOTES
- EC 3.1.21.4
- A type II site specific deoxyribonuclease
- A large group of enzymes recognizing specific short DNA sequences. They cleave either within the recognition site, or at a short specific distance from it
- Large excess of enzyme results in the appearance of star activity
- Requires Mg^{2+}

CONCENTRATION or VOLUME ACTIVITY	UNIT DEFINITION	REACTION BUFFER (As Provided)	ADDITIONAL ACTIVITIES and PURITY	SUPPLIER CATALOG NO.
***Bacillus* species**				
1000-15,000 U/mL		In reaction buffer		ICN 159421
Bacillus stearothermophilus				
100 U and 500 U	1 unit completely digests 1 μg λ DNA/hr at 37°C in 50 μL of supplied buffer	200 mM Tris-HCl, 100 mM MgCl₂, 10 mM DTT, 1 M KCl, pH 8.5	Purity is confirmed by overdigestion, ligation-recutting and cloning test	TaKaRa 1028
***Bacillus stearothermophilus* RFL1107**				
5000 U/mL	1 unit degrades 1 μg λ dam⁻ dcm⁻ DNA/hr at 37°C in a 0.05 mL mixture	10 mM Tris-HCl, 100 mM NaCl, 100 μg/mL BSA, pH 8.0	No detectable contaminating nuclease; 80% ligated and 90% recut after 10-fold overdigestion	AGS Heidelb F00595S F00595M
8-12 U/μL	1 unit completely digests 1 μg substrate DNA/hr	50 mM Tris-HCl, 10 mM MgCl₂, 100 mM NaCl, 1 mM DTE, pH 7.5		Boehringer 1378953
8-12 U/μL	1 unit hydrolyzes 1 μg DNA/hr at 37°C in a 50 μL reaction volume using λ phage DNA as substrate	10 mM Tris-HCl, 10 mM MgCl₂, 100 mM KCl, pH 8.5	>90% of the pBR322 DNA fragments ligated and recut after 50-fold overdigestion	Fermentas ER0701 ER0702
8000 U/mL	1 unit hydrolyzes 1 μg DNA/hr at 37°C in a 50 μL reaction volume using λ phage DNA as substrate	10 mM Tris-HCl, 10 mM MgCl₂, 100 mM KCl, pH 8.5	>90% of the DNA fragments ligated and recut after 50-fold overdigestion	NE Biolabs ER0701S ER0701L

BstB I

RECOGNITION SEQUENCE
    5'...TT▼CGAA...3'
    3'...AAGC▲TT...5'
ISOSCHIZOMERS
    Fsp II
OPTIMUM REACTION CONDITIONS
    pH 7.4-7.9 and 65°C; 10% activity at 37°C
HEAT INACTIVATION
    Resistant
METHYLATION EFFECTS
    Does not cleave TTCGm^{6}AA or TTm^{5}CGAA

NOTES
- EC 3.1.21.4
- A type II site specific deoxyribonuclease
- A large group of enzymes recognizing specific short DNA sequences. They cleave either within the recognition site, or at a short specific distance from it
- Requires Mg^{2+}

CONCENTRATION or VOLUME ACTIVITY	UNIT DEFINITION	REACTION BUFFER (As Provided)	ADDITIONAL ACTIVITIES and PURITY	SUPPLIER CATALOG NO.
Bacillus stearothermophilus				
5000-40,000 U/mL	1 unit completely cleaves 1.0 µg λ DNA/hr in a 50 µL reaction mixture	200 mM Tris-acetate, 500 mM KOAc, 100 mM MgOAc, 10 mM DTT, pH 7.9	100% cleavage, 90% ligation, 100% recleavage, 90% endonuclease (nickase) No detectable endonuclease (overdigestion)	Sigma R2260
Bacillus stearothermophilus B				
5000-25,000 U/mL		50 mM KCl, 10 mM Tris-HCl, 0.1 mM EDTA, 1 mM DTT, 200 µg/mL acetylated BSA, 50% glycerol, pH 7.4		ICN 159422
Bacillus stearothermophilus B (Z. Chen)				
20,000 U/mL	1 unit completely digests 1 µg λ DNA/hr in a total reaction volume of 50 µL using the buffer provided	20 mM Tris-acetate, 10 mM MgOAc, 50 mM KOAc, 1 mM DTT, pH 7.9	>95% DNA fragments ligated and recut after 10-fold overdigestion	NE Biolabs 519S 519L

BstBA I

RECOGNITION SEQUENCE
5'...YAC▼GTR...3'
3'...RTG▲CAY...5'

ISOSCHIZOMERS
BsaA I

NOTES
- EC 3.1.21.4
- A type II site specific deoxyribonuclease
- A large group of enzymes recognizing specific short DNA sequences. They cleave either within the recognition site, or at a short specific distance from it
- Requires Mg^{2+}

CONCENTRATION or VOLUME ACTIVITY	UNIT DEFINITION	REACTION BUFFER (As Provided)	ADDITIONAL ACTIVITIES and PURITY	SUPPLIER CATALOG No.
Bacillus stearothermophilus BA				
				SibEnzyme

BstD102 I

RECOGNITION SEQUENCE
5'...CCGCTC(-3/-3)▼...3'

ISOSCHIZOMERS
AccBS I, BsrB I

NOTES
- EC 3.1.21.4
- A type II site specific deoxyribonuclease
- A large group of enzymes recognizing specific short DNA sequences. They cleave either within the recognition site, or at a short specific distance from it
- Requires Mg^{2+}

CONCENTRATION or VOLUME ACTIVITY	UNIT DEFINITION	REACTION BUFFER (As Provided)	ADDITIONAL ACTIVITIES and PURITY	SUPPLIER CATALOG No.
Bacillus stearothermophilus D102				
				Pharmacia

BstDE I

RECOGNITION SEQUENCE
5'...C▼TNAG...3'
3'...GANT▲C...5'

ISOSCHIZOMERS
Dde I

NOTES
- EC 3.1.21.4
- A type II site specific deoxyribonuclease
- A large group of enzymes recognizing specific short DNA sequences. They cleave either within the recognition site, or at a short specific distance from it
- Requires Mg^{2+}

CONCENTRATION or VOLUME ACTIVITY	UNIT DEFINITION	REACTION BUFFER (As Provided)	ADDITIONAL ACTIVITIES and PURITY	SUPPLIER CATALOG No.
Bacillus stearothermophilus DE				
				SibEnzyme

BstDS I

RECOGNITION SEQUENCE
5'...C▼CRYGG...3'
3'...GGYRC▲C...5'

ISOSCHIZOMERS
Dsa I

NOTES
- EC 3.1.21.4
- A type II site specific deoxyribonuclease
- A large group of enzymes recognizing specific short DNA sequences. They cleave either within the recognition site, or at a short specific distance from it
- Requires Mg^{2+}

CONCENTRATION or VOLUME ACTIVITY	UNIT DEFINITION	REACTION BUFFER (As Provided)	ADDITIONAL ACTIVITIES and PURITY	SUPPLIER CATALOG No.
Bacillus stearothermophilus DS				
				SibEnzyme

BstE II

RECOGNITION SEQUENCE
5'...G▼GTNACC...3'
3'...CCANTG▲G...5'

ISOSCHIZOMERS
BstP I, Eco91 I, EcoO65 I

OPTIMUM REACTION CONDITIONS
pH 7.3-8.3 and 60°C; 10-50% activity at 37°C

HEAT INACTIVATION
Resistant

METHYLATION EFFECTS
Cleaves GGTNAm^{5}Cm^{5}C or GGTNACm^{4}C
Does not cleave GGTNAhm^{5}Chm^{5}C

NOTES
- EC 3.1.21.4
- A type II site specific deoxyribonuclease
- A large group of enzymes recognizing specific short DNA sequences. They cleave either within the recognition site, or at a short specific distance from it
- Compatible cohesive ends: Mae III
- Requires Mg^{2+}

BstE II continued

CONCENTRATION or VOLUME ACTIVITY	UNIT DEFINITION	REACTION BUFFER (As Provided)	ADDITIONAL ACTIVITIES and PURITY	SUPPLIER CATALOG NO.
Bacillus stearothermophilus				
10 U/μL; 40 U/μL	1 unit completely cleaves 1 μg substrate DNA/hr	60 mM KCl, 20 mM Tris, 0.75 mM MgCl$_2$, pH 7.5	90-98% cleavage-ligation-recleavage	Am Allied L-21-02000 H-21-02000 L-21-10000 H-21-10000
10,000 U/mL	1 unit degrades 1 μg λ dam$^-$ dcm$^-$ DNA/hr at 60°C in a 0.05 mL mixture	50 mM Tris-HCl, 10 mM MgCl$_2$, 100 mM NaCl, 100 μg/mL BSA, pH 7.5	No detectable contaminating nuclease; 90% ligated and 100% recut after 50-fold overdigestion	AGS Heidelb F00180S F00180M
10 U/μL	1 unit completely digests 1 μg substrate DNA/hr	10 mM Tris-HCl, 5 mM MgCl$_2$, 100 mM NaCl, 1 mM β-MSH, pH 8.0		Boehringer 404233 567612
5000-50,000 U/mL		50 mM KCl, 10 mM Tris-HCl, 0.1 mM EDTA, 1.0 mM DTT, 200 μg/mL acetylated BSA, 50% glycerol, pH 7.4		ICN 150529
				Nippon Gene
				Pharmacia
10 U/μL	1 unit completely digests 1 μg λ DNA/hr at 60°C in a 50 mL reaction volume	60 mM Tris-HCl, 60 mM MgCl$_2$, 1.5 M NaCl, 10 mM DTT, pH 7.3	90% ligation	Toyobo Promega R6641 R6642
Bacillus stearothermophilus ATCC 12980				
6-10 U/μL	1 unit completely digests 1 mg λ DNA/hr at 60°C	500 mM Tris-HCl, 100 mM MgCl$_2$, 500 mM NaCl, pH 8.0	≥95% ligation; 100% cleavage	Life Technol 15225-014
Bacillus stearothermophilus ET				
8-12 U/μL	1 unit completely digests 1 μg λ DNA substrate/hr at 60°C	10 mM Tris-HCl, 100 mM NaCl, 5 mM MgCl$_2$, 1 mM β-MSH, pH 8.3		NBL Gene 014403 014406
10,000 U/mL; 50,000 U/mL	1 unit completely digests 1 μg λ DNA/hr in a total reaction volume of 50 μL using the buffer provided	50 mM Tris-HCl, 10 mM MgCl$_2$, 100 mM NaCl, 1 mM DTT, pH 7.9	>95% DNA fragments ligated and recut after 20-fold overdigestion	NE Biolabs 162S 162L
8 U/μL	1 unit hydrolyzes 1 μg λ DNA to completion/hr in a 50 μL total reaction volume	500 mM Tris-HCl, 1 M NaCl, 100 mM MgCl$_2$, 10 mM DTE, 1.0 mg/mL BSA, pH 8		Oncor 110121 110122

BstE II continued

CONCENTRATION or VOLUME ACTIVITY	UNIT DEFINITION	REACTION BUFFER (As Provided)	ADDITIONAL ACTIVITIES and PURITY	SUPPLIER CATALOG NO.
Bacillus stearothermophilus ET continued				
4000-20,000 U/mL	1 unit completely cleaves 1.0 μg λ DNA/hr at 60°C in a 50 μL reaction mixture	Unique buffer	100% cleavage, 100% ligation, 100% recleavage, 75% endonuclease (nickase) No detectable endonuclease (overdigestion)	Sigma R4253
3-20 U/μL	1 unit completely digests 1 μg DNA/hr at 60°C	50 mM Tris-HCl, 10 mM MgCl$_2$, 100 mM NaCl, 1 mM DTT, pH 7.5	>90% ligation; >95% recut	Toyobo BST-203 BST-204

BstF5 I

RECOGNITION SEQUENCE
    5'...GGATG(2/0)▼...3'

ISOSCHIZOMERS
    Fok I

NOTES
- EC 3.1.21.4
- A type II site specific deoxyribonuclease
- A large group of enzymes recognizing specific short DNA sequences. They cleave either within the recognition site, or at a short specific distance from it
- Requires Mg^{2+}

CONCENTRATION or VOLUME ACTIVITY	UNIT DEFINITION	REACTION BUFFER (As Provided)	ADDITIONAL ACTIVITIES and PURITY	SUPPLIER CATALOG NO.
Bacillus stearothermophilus F5				
				SibEnzyme

BstH2 I

RECOGNITION SEQUENCE
5'...RGCGC▼Y...3'
3'...Y▲CGCGR...5'

ISOSCHIZOMERS
Bsp143 II, Hae II

NOTES
- EC 3.1.21.4
- A type II site specific deoxyribonuclease
- A large group of enzymes recognizing specific short DNA sequences. They cleave either within the recognition site, or at a short specific distance from it
- Requires Mg^{2+}

CONCENTRATION or VOLUME ACTIVITY	UNIT DEFINITION	REACTION BUFFER (As Provided)	ADDITIONAL ACTIVITIES and PURITY	SUPPLIER CATALOG No.
Bacillus stearothermophilus H2				
				SibEnzyme

BstHP I

RECOGNITION SEQUENCE
5'...GTT▼AAC...3'
3'...CAA▲TTG...5'

ISOSCHIZOMERS
Hpa I

NOTES
- EC 3.1.21.4
- A type II site specific deoxyribonuclease
- A large group of enzymes recognizing specific short DNA sequences. They cleave either within the recognition site, or at a short specific distance from it
- Requires Mg^{2+}

CONCENTRATION or VOLUME ACTIVITY	UNIT DEFINITION	REACTION BUFFER (As Provided)	ADDITIONAL ACTIVITIES and PURITY	SUPPLIER CATALOG No.
Bacillus stearothermophilus HP				
				SibEnzyme

BstMC I

RECOGNITION SEQUENCE
5'...CGRY▼CG...3'
3'...GC▲YRGC...5'

ISOSCHIZOMERS
BsaO I, Bsh1285 I, BsiE I

NOTES
- EC 3.1.21.4
- A type II site specific deoxyribonuclease
- A large group of enzymes recognizing specific short DNA sequences. They cleave either within the recognition site, or at a short specific distance from it
- Requires Mg^{2+}

CONCENTRATION or VOLUME ACTIVITY	UNIT DEFINITION	REACTION BUFFER (As Provided)	ADDITIONAL ACTIVITIES and PURITY	SUPPLIER CATALOG NO.
Bacillus stearothermophilus MC				
				SibEnzyme

BstN I

RECOGNITION SEQUENCE
5'...CC▼(^A_T)GG...3'
3'...GG(^T_A)▲CC...5'

ISOSCHIZOMERS
Apy I, Mva I, EcoR II (cuts at a different location)

OPTIMUM REACTION CONDITIONS
pH 7.4-7.9 and 60°C; 30% activity at 37°C

HEAT INACTIVATION
Resistant

METHYLATION EFFECTS
Cleaves $m^5CC(^A_T)GG$, $Cm^5C(^A_T)GG$, $Cm^4C(^A_T)GG$ and $m^5Cm^5C(^A_T)GG$
Does not cleave $hm^5Chm^5C(^A_T)GG$
Not blocked by *dcm* methylation

NOTES
- EC 3.1.21.4
- A type II site specific deoxyribonuclease
- A large group of enzymes recognizing specific short DNA sequences. They cleave either within the recognition site, or at a short specific distance from it
- BstN I-cut DNA is difficult to ligate with T4 DNA ligase. Ligation is enhanced in the presence of 15% PEG and by filling in the extensions with DNA polymerase I-large (Klenow) fragment
- Requires Mg^{2+} and BSA for full activity

BstN I continued

CONCENTRATION or VOLUME ACTIVITY	UNIT DEFINITION	REACTION BUFFER (As Provided)	ADDITIONAL ACTIVITIES and PURITY	SUPPLIER CATALOG No.
Bacillus stearothermophilus				
5000-20,000 U/mL		50 mM KCl, 10 mM Tris-HCl, 0.1 mM EDTA, 1.0 mM DTT, 200 µg/mL acetylated BSA, 50% glycerol, pH 7.5		ICN 150530
Bacillus stearothermophilus N				
	10 units does not produce unspecific cleavage products with 1 µg λ DNA/16 hr at 60°C	10 mM Tris-HCl, 10 mM MgCl₂, 50 mM NaCl, 1 mM DTT, 100 µg/mL BSA, pH 7.9	>90% DNA fragments ligated and recut after 10-fold overdigestion	MINOTECH 109
2000-20,000 U/mL	1 unit completely cleaves 1.0 µg λ DNA/hr in a 50 µL reaction mixture	100 mM Tris-HCl, 500 mM NaCl, 100 mM MgCl₂, 10 mM DTT, BSA, pH 7.9	100% cleavage, <0.5% ligation, 100% recleavage No detectable endonuclease (overdigestion)	Sigma R2759
4000-20,000 U/mL		50 mM KCl, 10 mM Tris-HCl, 0.1 mM EDTA, 1 mM DTT, 200 µg/mL BSA, 50% glycerol, pH 7.4		Stratagene 500330 500331
Bacillus stearothermophilus N (NEB 197)				
10,000 U/mL	1 unit completely digests 1 µg λ DNA/hr in a total reaction volume of 50 µL using the buffer provided	10 mM Tris-HCl, 10 mM MgCl₂, 50 mM NaCl, 1 mM DTT, pH 7.9; add 100 µg/mL BSA	None of the DNA fragments ligated after 2-fold overdigestion	NE Biolabs 168S 168L

BstNS I

RECOGNITION SEQUENCE
5'...RCATG▼Y...3'
3'...Y▲GTACR...5'

ISOSCHIZOMERS
Nsp I

NOTES
- EC 3.1.21.4
- A type II site specific deoxyribonuclease
- A large group of enzymes recognizing specific short DNA sequences. They cleave either within the recognition site, or at a short specific distance from it
- Requires Mg^{2+}

BstNS I continued

CONCENTRATION or VOLUME ACTIVITY	UNIT DEFINITION	REACTION BUFFER (As Provided)	ADDITIONAL ACTIVITIES and PURITY	SUPPLIER CATALOG NO.
Bacillus stearothermophilus 1161NS				sibEnzyme

BstO I

RECOGNITION SEQUENCE

$5'...CC\blacktriangledown(^A_T)GG...3'$
$3'...GG(^T_A)\blacktriangle CC...5'$

OPTIMUM REACTION CONDITIONS

pH 7.3 and 60°C; 25–50% activity at 37°C

METHYLATION EFFECTS

Not blocked by *dcm* methylation

NOTES
- EC 3.1.21.4
- A type II site specific deoxyribonuclease
- A large group of enzymes recognizing specific short DNA sequences. They cleave either within the recognition site, or at a short specific distance from it
- Produces an ambiguous single base extension which is difficult to ligate
- Requires Mg^{2+}

CONCENTRATION or VOLUME ACTIVITY	UNIT DEFINITION	REACTION BUFFER (As Provided)	ADDITIONAL ACTIVITIES and PURITY	SUPPLIER CATALOG NO.
Bacillus stearothermophilus				
10 U/μL	1 unit completely digests 1 μg λ DNA/hr at 60°C in a 50 mL reaction volume	100 mM Tris-HCl, 100 mM MgCl₂, 500 mM NaCl, 10 mM DTT, pH 7.3	0% ligation	Promega R6931 R6932

BstP I

RECOGNITION SEQUENCE
5'...G▼GTNACC...3'
3'...CCANTG▲G...5'

ISOSCHIZOMERS
BstE II

OPTIMUM REACTION CONDITIONS
pH 7.5 and 60°C

NOTES
- EC 3.1.21.4
- A type II site specific deoxyribonuclease
- A large group of enzymes recognizing specific short DNA sequences. They cleave either within the recognition site, or at a short specific distance from it
- Requires Mg^{2+}

CONCENTRATION or VOLUME ACTIVITY	UNIT DEFINITION	REACTION BUFFER (As Provided)	ADDITIONAL ACTIVITIES and PURITY	SUPPLIER CATALOG NO.
Bacillus stearothermophilus				
2000 U and 10,000 U	1 unit completely digests 1 μg λ DNA/hr at 60°C in 50 μL of supplied buffer	500 mM Tris-HCl, 100 mM $MgCl_2$, 10 mM DTT, 1 M NaCl, pH 7.5	Purity is confirmed by overdigestion, ligation-recutting and cloning test	TaKaRa 1025

N. BstSE Nickase

RECOGNITION SEQUENCE
5'...GAGTCNNNN▼NN...3'
3'...CTCAGNNNNNN...5'

NOTES
- EC 3.1.21.4
- A type II site specific deoxyribonuclease
- A large group of enzymes recognizing specific short DNA sequences. They cleave either within the recognition site, or at a short specific distance from it
- Requires Mg^{2+}

CONCENTRATION or VOLUME ACTIVITY	UNIT DEFINITION	REACTION BUFFER (As Provided)	ADDITIONAL ACTIVITIES and PURITY	SUPPLIER CATALOG NO.
Bacillus stearothermophilus SE-589				
5000 U/mL		10 mM Tris-HCl, 10 mM MgCl₂, 150 mM KCl, 1 mM DTT, pH 8.5	90% DNA fragments ligated and recut after 5-fold overdigestion	sibEnzyme E401 E402

BstSF I

RECOGNITION SEQUENCE
    5'...C▼TRYAG...3'
    3'...GAYRT▲C...5'

ISOSCHIZOMERS
    Bfm I, Sfc I

NOTES
- EC 3.1.21.4
- A type II site specific deoxyribonuclease
- A large group of enzymes recognizing specific short DNA sequences. They cleave either within the recognition site, or at a short specific distance from it
- Requires Mg^{2+}

CONCENTRATION or VOLUME ACTIVITY	UNIT DEFINITION	REACTION BUFFER (As Provided)	ADDITIONAL ACTIVITIES and PURITY	SUPPLIER CATALOG NO.
Bacillus stearothermophilus SF				
				sibEnzyme

BstSN I

RECOGNITION SEQUENCE
5'...TAC▼GTA...3'
3'...ATG▲CAT...5'

ISOSCHIZOMERS
Eco105 I, SnaB I

NOTES
- EC 3.1.21.4
- A type II site specific deoxyribonuclease
- A large group of enzymes recognizing specific short DNA sequences. They cleave either within the recognition site, or at a short specific distance from it
- Requires Mg^{2+}

CONCENTRATION or VOLUME ACTIVITY	UNIT DEFINITION	REACTION BUFFER (As Provided)	ADDITIONAL ACTIVITIES and PURITY	SUPPLIER CATALOG NO.
Bacillus stearothermophilus SN				
				SibEnzyme

BstU I

RECOGNITION SEQUENCE
5'...CG▼CG...3'
3'...GC▲GC...5'

ISOSCHIZOMERS
Acc II, FnuD II

OPTIMUM REACTION CONDITIONS
pH 7.4-7.9 and 60°C; 20% activity at 37°C

HEAT INACTIVATION
Resistant

METHYLATION EFFECTS
Does not cleave m^5CGCG

NOTES
- EC 3.1.21.4
- A type II site specific deoxyribonuclease
- A large group of enzymes recognizing specific short DNA sequences. They cleave either within the recognition site, or at a short specific distance from it
- Requires Mg^{2+}

BstU I continued

CONCENTRATION or VOLUME ACTIVITY	UNIT DEFINITION	REACTION BUFFER (As Provided)	ADDITIONAL ACTIVITIES and PURITY	SUPPLIER CATALOG NO.
Bacillus stearothermophilus U				
3000-20,000 U/mL		50 mM NaCl, 10 mM Tris-HCl, 0.1 mM EDTA, 1 mM DTT, 200 μg/mL acetylated BSA, 50% glycerol, pH 7.4		ICN 159423
5000-25,000 U/mL	1 unit completely cleaves 1.0 μg λ DNA/hr in a 50 μL reaction mixture	100 mM Tris-HCl, 500 mM NaCl, 100 mM MgCl$_2$, 10 mM DTT, pH 7.9	100% cleavage, >80% ligation, 99% recleavage No detectable endonuclease (overdigestion)	Sigma R4135
Bacillus stearothermophilus U (Z. Chen)				
10,000 U/mL	1 unit completely digests 1 μg λ DNA/hr in a total reaction volume of 50 μL using the buffer provided	10 mM Tris-HCl, 10 mM MgCl$_2$, 50 mM NaCl, 1 mM DTT, pH 7.9	>70% DNA fragments ligated and >95% recut after 10-fold overdigestion	NE Biolabs 518S 518L

BstX I

RECOGNITION SEQUENCE
  5'...CCANNNN▼NTGG...3'
  3'...GGTN▲NNNNNACC...5'

OPTIMUM REACTION CONDITIONS
  pH 7.5-8.0 and 55°C; 50-100% activity at 37°C; minimal activity at 65°C

HEAT INACTIVATION
  65°C for 20 min.

METHYLATION EFFECTS
  Cleaves Cm^{5}CA(N)$_6$TGG
  Does not cleave m^{5}CCA(N)$_6$TGG

NOTES
- EC 3.1.21.4
- A type II site specific deoxyribonuclease
- A large group of enzymes recognizing specific short DNA sequences. They cleave either within the recognition site, or at a short specific distance from it
- Large excess of enzyme results in the appearance of star activity
- Ligation of fragments is difficult due to the presence of 4 undefined nucleotides in the 3'-cohesive ends
- Requires Mg^{2+}
- Inhibited by >10% glycerol

BstX I continued

CONCENTRATION or VOLUME ACTIVITY	UNIT DEFINITION	REACTION BUFFER (As Provided)	ADDITIONAL ACTIVITIES and PURITY	SUPPLIER CATALOG NO.
Bacillus stearothermophilus				
10 U/µL; 40 U/µL	1 unit completely cleaves 1 µg substrate DNA/hr	100 mM NaCl, 50 mM Tris, 5 mM MgCl$_2$, 2 mM β-MSH, pH 7.8	90-98% cleavage-ligation-recleavage	Am Allied L-27-01000 H-27-01000 L-27-05000 H-27-05000
5000-15,000 U/mL		50 mM KCl, 10 mM Tris-HCl, 0.1 mM EDTA, 1.0 mM DTT, 200 µg/mL BSA, 50% glycerol, pH 7.5		ICN 150531
				Nippon Gene Toyobo
1000 U and 5000 U	1 unit completely digests 1 µg λ DNA/hr at 45°C in 50 µL of supplied buffer	500 mM Tris-HCl, 100 mM MgCl$_2$, 10 mM DTT, 1 M NaCl, pH 7.5	Purity is confirmed by overdigestion and ligation-recutting test	TaKaRa 1027
Bacillus stearothermophilus X				
10,000 U/mL	1 unit degrades 1 µg λ dam$^-$ dcm$^-$ DNA/hr at 45°C in a 0.05 mL mixture	10 mM Tris-HCl, 100 mM NaCl, 100 µg/mL BSA, pH 8.0	No detectable contaminating nuclease; 85% ligated and 95% recut after 10-fold overdigestion	AGS Heidelb A00200S A00200M
	1 unit completely cleaves 1 µg λ DNA/hr at pH 7.5, 50°C in a reaction volume of 50 µL	10 mM Tris-HCl, 10 mM MgCl$_2$, 50 mM NaCl, 1.0 mM DTT, 100 µg/mL BSA, pH 7.5		CHIMERx 2118-01 2118-02
8-12 U/µL	1 unit hydrolyzes 1 µg DNA/hr at 37°C in a 50 µL reaction volume using λ phage DNA as substrate	50 mM Tris-HCl, 10 mM MgCl$_2$, 100 mM NaCl, pH 7.5	90% of the DNA fragments ligated and recut after 50-fold overdigestion	Fermentas ER1021 ER1022
8-12 U/µL	1 unit completely digests 1 mg λ DNA/hr at 55°C	500 mM Tris-HCl, 100 mM MgCl$_2$, 500 mM NaCl, pH 8.0	≥85% ligation; 100% cleavage	Life Technol 15463-011 15463-029
10,000 U/mL	1 unit completely digests 1 µg λ DNA/hr in a total reaction volume of 50 µL using the buffer provided	50 mM Tris-HCl, 10 mM MgCl$_2$, 100 mM NaCl, 1 mM DTT, pH 7.9	>95% DNA fragments ligated and recut after 20-fold overdigestion	NE Biolabs 113S 113L
5000-15,000 U/mL	1 unit completely cleaves 1.0 µg λ DNA/hr at 50°C in a 50 µL reaction mixture	500 mM Tris-HCl, 1 M NaCl, 100 mM MgCl$_2$, 10 mM DTT, pH 7.9	100% cleavage, 100% ligation, 100% recleavage, <1% endonuclease (nickase) No detectable endonuclease (overdigestion)	Sigma R2884
3-20 U/µL	1 unit completely digests 1 µg DNA/hr at 50°C	50 mM Tris-HCl, 10 mM MgCl$_2$, 100 mM NaCl, 1 mM DTT, pH 7.5	>90% ligation; >95% recut	Toyobo BSX-101 BSX-102

BstX I continued

CONCENTRATION or VOLUME ACTIVITY	UNIT DEFINITION	REACTION BUFFER (As Provided)	ADDITIONAL ACTIVITIES and PURITY	SUPPLIER CATALOG No.
Bacillus stearothermophilus XI				
3-20 U/μL	1 unit completely digests 1 μg λ DNA/hr at 50°C in the buffer provided	500 mM Tris-HCl, 100 mM MgCl$_2$, 10 mM DTT, 1000 mM NaCl, pH 7.5	>90% DNA fragments ligated and >95% recut	Amersham E 0208Y E 0208Z
10 U/μL	1 unit completely digests 1 μg substrate DNA/hr	50 mM Tris-HCl, 10 mM MgCl$_2$, 100 mM NaCl, 1 mM DTE, pH 7.5		Boehringer 1117777 1117785
6-12 U/μL	1 unit digests completely 1 μg λ DNA substrate/hr at 45°C	50 mM Tris-HCl, 100 mM NaCl, 10 mM MgCl$_2$, 1 mM DTT, pH 7.8		NBL Gene 015201 015203
10 U/μL	1 unit hydrolyzes 1 μg λ DNA to completion/hr in a 50 μL total reaction volume	100 mM Tris-HCl, 1 M NaCl, 100 mM MgCl$_2$, 100 mM β-MSH, 1.0 mg/mL BSA, pH 8		Oncor 110551 110552
8-12 U/μL	1 unit completely digests 1 μg λ DNA/hr at 50°C in a 50 mL reaction volume	Blue/White Cloning Qualified; 60 mM Tris-HCl, 60 mM MgCl$_2$, 1.5 M NaCl, 10 mM DTT, pH 7.3	90% ligation	Promega R6471 R6472 R6475
2000-12,000 U/mL		50 mM KCl, 10 mM Tris-HCl, 0.1 mM EDTA, 1 mM DTT, 200 μg/mL BSA, 50% glycerol, pH 7.4		Stratagene 500340 500341

BstX2 I

RECOGNITION SEQUENCE
 5'...R▼GATCY...3'
 3'...YCTAG▲R...5'

ISOSCHIZOMERS
 BstY I, Mfl I, Xho II

NOTES
- EC 3.1.21.4
- A type II site specific deoxyribonuclease
- A large group of enzymes recognizing specific short DNA sequences. They cleave either within the recognition site, or at a short specific distance from it
- Requires Mg^{2+}

CONCENTRATION or VOLUME ACTIVITY	UNIT DEFINITION	REACTION BUFFER (As Provided)	ADDITIONAL ACTIVITIES and PURITY	SUPPLIER CATALOG NO.
Bacillus stearothermophilus X2				
				SibEnzyme

BstY I

RECOGNITION SEQUENCE
 5'...Pu▼GATCPy...3'
 3'...PyCTAG▲Pu...5'

ISOSCHIZOMERS
 Mfl I

OPTIMUM REACTION CONDITIONS
 pH 7.4-8.0 and 60°C; 30% activity at 37°C

HEAT INACTIVATION
 Resistant

METHYLATION EFFECTS
 Cleaves $PuGm^6ATCPy$ and $PuGATm^5CPy$
 Does not cleave $PuGATm^4CPy$
 Not blocked by *dam* methylation

NOTES
- EC 3.1.21.4
- A type II site specific deoxyribonuclease
- A large group of enzymes recognizing specific short DNA sequences. They cleave either within the recognition site, or at a short specific distance from it
- Requires Mg^{2+}

BstY I continued

CONCENTRATION or VOLUME ACTIVITY	UNIT DEFINITION	REACTION BUFFER (As Provided)	ADDITIONAL ACTIVITIES and PURITY	SUPPLIER CATALOG NO.
Bacillus stearothermophilus Y				
3000-15,000 U/mL		50 mM KCl, 10 mM Tris-HCl, 0.1 mM EDTA, 1 mM DTT, 200 µg/mL acetylated BSA, 50% glycerol, pH 7.4		ICN 159424
8-12 U/µL	1 unit completely digests 1 mg λ DNA/hr at 60°C	500 mM Tris-HCl, 100 mM MgCl₂, 500 mM NaCl, pH 8.0	≥85% ligation; 100% cleavage	Life Technol 15477-011
Bacillus stearothermophilus Y (Z. Chen)				
10,000 U/mL	1 unit completely digests 1 µg λ DNA/hr in a total reaction volume of 50 µL using the buffer provided	10 mM Tris-HCl, 10 mM MgCl₂, 1 mM DTT, pH 7.9; add 100 µg/mL BSA	>95% DNA fragments ligated and recut after 100-fold overdigestion	NE Biolabs 523S 523L

BstZ I

RECOGNITION SEQUENCE
  5'...C▼GGCCG...3'
  3'...GCCGG▲C...5'

OPTIMUM REACTION CONDITIONS
  pH 7.6 and 50°C; 50% activity at 37°C

NOTES
- EC 3.1.21.4
- A type II site specific deoxyribonuclease
- A large group of enzymes recognizing specific short DNA sequences. They cleave either within the recognition site, or at a short specific distance from it
- Compatible cohesive ends: Eae I, Eag I, Eco52 I, Not I, Xma III
- Requires Mg^{2+}

CONCENTRATION or VOLUME ACTIVITY	UNIT DEFINITION	REACTION BUFFER (As Provided)	ADDITIONAL ACTIVITIES and PURITY	SUPPLIER CATALOG NO.
Bacillus stearothermophilus				
10 U/µL	1 unit completely digests 1 µg λ DNA/hr at 50°C in a 50 mL reaction volume	Blue/White Cloning Qualified; genome qualified; 60 mM Tris-HCl, 60 mM MgCl₂, 1.5 M NaCl, 10 mM DTT, pH 7.6	90% ligation	Promega R6881 R6882

Bsu6 I

RECOGNITION SEQUENCE
    5'...CTCTTC(1/4)▼...3'

ISOSCHIZOMERS
    *Eam*1104 I, *Ear* I, *Ksp*632 I

NOTES
- EC 3.1.21.4
- A type II site specific deoxyribonuclease
- A large group of enzymes recognizing specific short DNA sequences. They cleave either within the recognition site, or at a short specific distance from it
- Requires Mg^{2+}

CONCENTRATION or VOLUME ACTIVITY	UNIT DEFINITION	REACTION BUFFER (As Provided)	ADDITIONAL ACTIVITIES and PURITY	SUPPLIER CATALOG No.
Bacillus subtilis 6v1				
				SibEnzyme

Bsu15 I

RECOGNITION SEQUENCE
    5'...AT▼CGAT...3'
    3'...TAGC▲TA...5'

ISOSCHIZOMERS
    Cla I

OPTIMUM REACTION CONDITIONS
    pH 7.9 and 37°C

HEAT INACTIVATION
    65°C for 20 min.

METHYLATION EFFECTS
    Does not cleave ATCGm^{6}AT
    Blocked by overlapping *dam* and CG methylation

NOTES
- EC 3.1.21.4
- A type II site specific deoxyribonuclease
- A large group of enzymes recognizing specific short DNA sequences. They cleave either within the recognition site, or at a short specific distance from it
- Requires Mg^{2+} and BSA for full activity

Bsu15 I continued

CONCENTRATION or VOLUME ACTIVITY	UNIT DEFINITION	REACTION BUFFER (As Provided)	ADDITIONAL ACTIVITIES and PURITY	SUPPLIER CATALOG NO.
Bacillus subtilis 15				
10,000 U/mL	1 unit degrades 1 µg λ dam⁻ dcm⁻ DNA/hr at 37°C in a 0.05 mL mixture	33 mM Tris-acetate, 10 mM MgOAc, 66 mM KOAc, 100 µg/mL BSA, pH 7.9	No detectable contaminating nuclease; 90% ligated and 100% recut after 50-fold overdigestion	AGS Heidelb F00230S F00230M
8-12 U/µL	1 unit hydrolyzes 1 µg DNA/hr at 37°C in a 50 µL reaction volume using λ phage DNA as substrate	33 mM Tris-acetate, 10 mM MgOAc, 66 mM KOAc, 0.1 mg/mL BSA, pH 7.9	>90% of the DNA fragments ligated and recut after 50-fold overdigestion	Fermentas ER0141 ER0142

Bsu36 I

RECOGNITION SEQUENCE
  5'...CC▼TNAGG...3'
  3'...GGANT▲CC...5'

ISOSCHIZOMERS
  Eco81 I

HEAT INACTIVATION
  Resistant

NOTES
- EC 3.1.21.4
- A type II site specific deoxyribonuclease
- A large group of enzymes recognizing specific short DNA sequences. They cleave either within the recognition site, or at a short specific distance from it
- Produces ambiguous extensions which are difficult to ligate
- Requires Mg^{2+}

CONCENTRATION or VOLUME ACTIVITY	UNIT DEFINITION	REACTION BUFFER (As Provided)	ADDITIONAL ACTIVITIES and PURITY	SUPPLIER CATALOG NO.
Bacillus subtilis				
10 U/µL	1 unit completely digests 1 µg λ DNA/hr at 37°C in a 50 mL reaction volume	60 mM Tris-HCl, 60 mM MgCl₂, 1 M NaCl, 10 mM DTT, pH 7.5	<10% ligation	Promega R6821 R6822
8000-12,000 U/mL		50 mM KCl, 10 mM Tris-HCl, 0.1 mM EDTA, 1 mM DTT, 200 µg/mL BSA, 50% glycerol, pH 7.4		Stratagene 500346

Bsu36 I continued

CONCENTRATION or VOLUME ACTIVITY	UNIT DEFINITION	REACTION BUFFER (As Provided)	ADDITIONAL ACTIVITIES and PURITY	SUPPLIER CATALOG No.
Bacillus subtilis 36 (B. Zhou)				
10,000 U/mL	1 unit completely digests 1 μg λ DNA (Hind III digest)/hr in a total reaction volume of 50 μL using the buffer provided	50 mM Tris-HCl, 10 mM MgCl$_2$, 100 mM NaCl, 1 mM DTT, pH 7.9; add 100 μg/mL BSA	50% DNA fragments ligated and recut after 2-fold overdigestion	NE Biolabs 524S 524L
Bacillus subtilis 36 I				
3000-20,000 U/mL		50 mM KCl, 10 mM Tris-HCl, 0.1 mM EDTA, 1 mM DTT, 200 μg/mL acetylated BSA, 50% glycerol, pH 7.5		ICN 159425
5000-50,000 U/mL	1 unit completely cleaves 1.0 μg λ DNA/hr in a 50 μL reaction mixture	Unique buffer + BSA	100% cleavage, 95% ligation, 100% recleavage, 50% endonuclease (nickase) No detectable endonuclease (overdigestion)	Sigma R4385

BsuR I

RECOGNITION SEQUENCE
    5'...GG▼CC...3'
    3'...CC▲GG...5'

OPTIMUM REACTION CONDITIONS
    37°C

HEAT INACTIVATION
    Resistant

METHYLATION EFFECTS
    Does not cleave GGm^{5}CC

NOTES
- EC 3.1.21.4
- A type II site specific deoxyribonuclease
- A large group of enzymes recognizing specific short DNA sequences. They cleave either within the recognition site, or at a short specific distance from it
- Requires Mg^{2+} and BSA for full activity

BsuR I continued

CONCENTRATION or VOLUME ACTIVITY	UNIT DEFINITION	REACTION BUFFER (As Provided)	ADDITIONAL ACTIVITIES and PURITY	SUPPLIER CATALOG NO.
Bacillus subtilis R				
10,000 U/mL	1 unit degrades 1 μg λ dam⁻ dcm⁻ DNA/hr at 37°C in a 0.05 mL mixture	33 mM Tris-acetate, 10 mM MgOAc, 66 mM KOAc, 100 μg/mL BSA, pH 7.9	No detectable contaminating nuclease; 90% ligated and 100% recut after 10-fold overdigestion	AGS Heidelb
8-12 U/μL; 40-60 U/μL	1 unit hydrolyzes 1 μg DNA/hr at 37°C in a 50 μL reaction volume using λ phage DNA as substrate	10 mM Tris-HCl, 10 mM MgCl$_2$, 100 mM KCl, 0.1 mg/mL BSA, pH 8.5	>90% of the DNA fragments ligated and recut after 50-fold overdigestion	Fermentas ER0151 ER0152 ER0153
				SibEnzyme

Cac8 I

RECOGNITION SEQUENCE
   5'...GCN▼NGC...3'
   3'...CGN▲NCG...5'
OPTIMUM REACTION CONDITIONS
   37°C
HEAT INACTIVATION
   65°C for 20 min.

NOTES
- EC 3.1.21.4
- A type II site specific deoxyribonuclease
- A large group of enzymes recognizing specific short DNA sequences. They cleave either within the recognition site, or at a short specific distance from it
- Requires Mg^{2+}

CONCENTRATION or VOLUME ACTIVITY	UNIT DEFINITION	REACTION BUFFER (As Provided)	ADDITIONAL ACTIVITIES and PURITY	SUPPLIER CATALOG NO.
Clostridium acetobutyliticum (NEB 846)				
2000 U/mL	1 unit completely digests 1 μg λ DNA/hr in a total reaction volume of 50 μL using the buffer provided	50 mM Tris-HCl, 10 mM MgCl$_2$, 100 mM NaCl, 1 mM DTT, pH 7.9	>95% DNA fragments ligated and recut after 5-fold overdigestion	NE Biolabs 579S 579L

Cbi I

RECOGNITION SEQUENCE
  5'...TT▼CGAA...3'
  3'...AAGC▲TT...5'

ISOSCHIZOMERS
  Bpu14 I, BsiC I, Bsp119 I, BstB I, Csp45 I, Lsp I, Nsp V, Sfu I

NOTES
- EC 3.1.21.4
- A type II site specific deoxyribonuclease
- A large group of enzymes recognizing specific short DNA sequences. They cleave either within the recognition site, or at a short specific distance from it
- Requires Mg^{2+}

CONCENTRATION or VOLUME ACTIVITY	UNIT DEFINITION	REACTION BUFFER (As Provided)	ADDITIONAL ACTIVITIES and PURITY	SUPPLIER CATALOG NO.
Clostridium bifermentans B-4				
				Nippon Gene

CciN I

RECOGNITION SEQUENCE
  5'...GC▼GGCCGC...3'
  3'...CGCCGG▲CG...5'

ISOSCHIZOMERS
  Not I

NOTES
- EC 3.1.21.4
- A type II site specific deoxyribonuclease
- A large group of enzymes recognizing specific short DNA sequences. They cleave either within the recognition site, or at a short specific distance from it
- Requires Mg^{2+}

CONCENTRATION or VOLUME ACTIVITY	UNIT DEFINITION	REACTION BUFFER (As Provided)	ADDITIONAL ACTIVITIES and PURITY	SUPPLIER CATALOG NO.
Curtobacterium citreum N				
				SibEnzyme

Cel II

RECOGNITION SEQUENCE
5′...GC▼TNAGC...3′
3′...CGANT▲CG...5′

ISOSCHIZOMERS
*Bpu*1102 I, *Esp* I

OPTIMUM REACTION CONDITIONS
37°C

HEAT INACTIVATION
Resistant; use phenol extraction

NOTES
- EC 3.1.21.4
- A type II site specific deoxyribonuclease
- A large group of enzymes recognizing specific short DNA sequences. They cleave either within the recognition site, or at a short specific distance from it
- Requires Mg^{2+}

CONCENTRATION or VOLUME ACTIVITY	UNIT DEFINITION	REACTION BUFFER (As Provided)	ADDITIONAL ACTIVITIES and PURITY	SUPPLIER CATALOG NO.
Coccochloris elabens				
2-5 U/μL	1 unit completely digests 1 μg λ DNA/hr at 37°C in the buffer provided	500 m*M* Tris-HCl, 100 m*M* $MgCl_2$, 10 m*M* DTT, 1000 m*M* NaCl, pH 7.5	>80% DNA fragments ligated and >90% recut	Amersham E 0312Y E 0312Z
3-10 U/μL	1 unit digests completely 1 μg λ DNA substrate/hr at 37°C	50 m*M* Tris-HCl, 100 m*M* NaCl, 10 m*M* $MgCl_2$, 1 m*M* DTT, pH 7.8		NBL Gene 017912 017913
Coccochloris elabens 17a				
10 U/μL	1 unit completely digests 1 μg substrate DNA/hr	50 m*M* Tris-HCl, 10 m*M* $MgCl_2$, 100 m*M* NaCl, 1 m*M* DTE, pH 7.5		Boehringer 1449397 1464256

Cfo I

RECOGNITION SEQUENCE
5'...GCG▼C...3'
3'...C▲GCG...5'

ISOSCHIZOMERS
Hha I

HEAT INACTIVATION
65°C for 20 min.

METHYLATION EFFECTS
Does not cleave Gm^5CGC or Ghm^5CGhm^5C

NOTES
- EC 3.1.21.4
- A type II site specific deoxyribonuclease
- A large group of enzymes recognizing specific short DNA sequences. They cleave either within the recognition site, or at a short specific distance from it
- Compatible cohesive ends: Hha I
- Requires Mg^{2+}

CONCENTRATION or VOLUME ACTIVITY	UNIT DEFINITION	REACTION BUFFER (As Provided)	ADDITIONAL ACTIVITIES and PURITY	SUPPLIER CATALOG NO.
Clostridium formicoaceticum				
10 U/μL	1 unit completely digests 1 μg substrate DNA/hr	10 m*M* Tris-HCl, 10 m*M* MgCl₂, 1 m*M* DTE, pH 7.5		Boehringer 688541 688550
8-12 U/μL	1 unit completely digests 1 mg λ DNA/hr at 37°C	500 m*M* Tris-HCl, 100 m*M* MgCl₂, pH 8.0	≥95% ligation; 100% cleavage	Life Technol 15237-019
10-20 U/μL	1 unit digests completely 1 μg λ DNA substrate/hr at 37°C	10 m*M* Tris-HCl, 10 m*M* MgCl₂, 1 m*M* DTT, pH 7.8		NBL Gene 012304 012307
10 U/μL	1 unit completely digests 1 μg λ DNA/hr at 37°C in a 50 mL reaction volume	60 m*M* Tris-HCl, 60 m*M* MgCl₂, 500 m*M* NaCl, 10 m*M* DTT, pH 7.5	90% ligation	Promega R6241 R6242
2000-20,000 U/mL	1 unit completely cleaves 1.0 μg λ DNA/hr in a 50 μL reaction mixture	200 m*M* Tris-acetate, 500 m*M* KOAc, 100 m*M* MgOAc, 10 m*M* DTT, pH 7.9	100% cleavage, 80% ligation, 100% recleavage No detectable endonuclease (overdigestion)	Sigma R1761
Clostridium formicoaceticum species				
8000-12,000 U/mL		50 m*M* NaCl, 10 m*M* Tris-HCl, 0.1 m*M* EDTA, 1 m*M* DTT, 500 μg/mL acetylated BSA, 50% glycerol, pH 7.4		ICN 159426

Cfr I

RECOGNITION SEQUENCE
5'...Py▼GGCCPu...3'
3'...PuCCGG▲Py...5'

ISOSCHIZOMERS
Eae I

OPTIMUM REACTION CONDITIONS
37°C

HEAT INACTIVATION
65°C for 20 min.

METHYLATION EFFECTS
Does not cleave PyGGm^{5}CCPu or PyGGCm^{5}CPu
Blocked by overlapping dcm and CG methylation

NOTES
- EC 3.1.21.4
- A type II site specific deoxyribonuclease
- A large group of enzymes recognizing specific short DNA sequences. They cleave either within the recognition site, or at a short specific distance from it
- Requires Mg^{2+} and BSA for full activity

CONCENTRATION or VOLUME ACTIVITY	UNIT DEFINITION	REACTION BUFFER (As Provided)	ADDITIONAL ACTIVITIES and PURITY	SUPPLIER CATALOG NO.
Escherichia coli, carrying the cloned gene from *Citrobacter freundii* RFL1				
8-12 U/μL	1 unit hydrolyzes 1 μg DNA/hr at 37°C in a 50 μL reaction volume using λ phage dcm$^-$ DNA as substrate	33 mM Tris-acetate, 10 mM MgOAc, 66 mM KOAc, 0.1 mg/mL BSA, pH 7.9	>90% DNA fragments ligated and recut after 50-fold overdigestion	Fermentas ER0161 ER0162

Cfr9 I

RECOGNITION SEQUENCE
  5'...C▼CCGGG...3'
  3'...GGGCC▲C...5'

ISOSCHIZOMERS
  Sma I, Xma I

OPTIMUM REACTION CONDITIONS
  pH 7.5 and 37°C

HEAT INACTIVATION
  65°C for 20 min.; 70°C for 5 min.

METHYLATION EFFECTS
  Cleaves CCm^{5}CGGG and Cm^{5}CCGGG
  Does not cleave Cm^{4}CCGGG, m^{4}CCCGGG, CCm^{4}CGGG or m^{5}CCCGGG

NOTES
- EC 3.1.21.4
- A type II site specific deoxyribonuclease
- A large group of enzymes recognizing specific short DNA sequences. They cleave either within the recognition site, or at a short specific distance from it
- Produces DNA fragments with a 4-base 5'-extension; Sma I makes blunt ends
- To achieve complete substrate digestion with Cfr9 I, DNA in the reaction buffer should be ≥50 µg/ml
- >10-15-fold overdigestion of substrate with enzyme results in the appearance of star activity
- Requires Mg^{2+}

CONCENTRATION or VOLUME ACTIVITY	UNIT DEFINITION	REACTION BUFFER (As Provided)	ADDITIONAL ACTIVITIES and PURITY	SUPPLIER CATALOG NO.
Citrobacter freundii RFL9				
3-20 U/µL	1 unit completely digests 1 µg DNA/hr at 37°C	Cfr9 I	>90% ligation; >95% recut	Toyobo CF9-101 CF9-102
Escherichia coli, carrying the cloned *Cfr9IR* gene from *Citrobacter freundii* RFL9				
4-8 U/µL	1 unit hydrolyzes 1 µg DNA/hr at 37°C in a 50 µL reaction volume using λ phage DNA as substrate	10 m*M* Tris-HCl, 5 m*M* MgCl$_2$, 200 m*M* Na glutamate, pH 7.2	>90% of the DNA fragments ligated and recut after 10-fold overdigestion	Fermentas ER0171 ER0172

Cfr10 I

RECOGNITION SEQUENCE
    5'...Pu▼CCGGPy...3'
    3'...PyGGCC▲Pu...5'

ISOSCHIZOMERS
    BsrF I

OPTIMUM REACTION CONDITIONS
    pH 8.5 and 37°C

HEAT INACTIVATION
    Resistant; use phenol extraction

METHYLATION EFFECTS
    Does not cleave Pum^{5}CCGGPy or PuCm^{5}CGGPy
    Blocked by CG methylation

NOTES
- EC 3.1.21.4
- A type II site specific deoxyribonuclease
- A large group of enzymes recognizing specific short DNA sequences. They cleave either within the recognition site, or at a short specific distance from it
- Large excess of enzyme results in the appearance of star activity
- Requires Mg^{2+}

CONCENTRATION or VOLUME ACTIVITY	UNIT DEFINITION	REACTION BUFFER (As Provided)	ADDITIONAL ACTIVITIES and PURITY	SUPPLIER CATALOG No.
Citrobacter freundii				
1-5 U/μL	1 unit completely digests 1 μg substrate DNA/hr	10 mM Tris-HCl, 5 mM MgCl$_2$, 100 mM NaCl, 1 mM β-MSH, pH 8.0		Boehringer 1088564
Citrobacter freundii RFL10				
2-6 U/μL	1 unit completely digests 1 μg λ DNA/hr at 37°C in the buffer provided	Cfr10 I basal buffer	>90% DNA fragments ligated and 100% recut	Amersham E 1120Y
4000 U/mL	1 unit degrades 1 μg λ dam⁻ dcm⁻ DNA/hr at 37°C in a 0.05 mL mixture	10 mM Tris-HCl, 100 mM NaCl, 100 μg/mL BSA, pH 8.0	No detectable contaminating nuclease; 90% ligated and 100% recut after 10-fold overdigestion	AGS Heidelb F00220S F00220M
50 U and 250 U	1 unit completely digests 1 μg λ DNA/hr at 37°C in 50 μL of supplied buffer	20 mM Tris-HCl, 3 mM MgSO$_4$, 100 mM KCl, 0.02% Triton X-100, pH 8.5	Purity is confirmed by overdigestion and ligation-recutting test	TaKaRa 1120
3-20 U/μL	1 unit completely digests 1 μg DNA/hr at 37°C	Cfr10 I	>90% ligation; >95% recut	Toyobo CF10-101 CF10-102
Escherichia coli, carrying the cloned *Cfr10IR* gene from *Citrobacter freundii* RFL10				
8-12 U/μL	1 unit hydrolyzes 1 μg DNA/hr at 37°C in a 50 μL reaction volume using λ phage DNA as substrate	10 mM Tris-HCl, 5 mM MgCl$_2$, 100 mM NaCl, 0.02% Triton X-100, pH 8.0	>90% of the DNA fragments ligated and recut after 50-fold overdigestion	Fermentas ER0181 ER0182

Cfr13 I

RECOGNITION SEQUENCE
5'...G▼GNCC...3'
3'...CCNG▲G...5'

ISOSCHIZOMERS
Asu I, Sau96 I

OPTIMUM REACTION CONDITIONS
pH 8.5 and 37°C

HEAT INACTIVATION
65°C for 20 min.

METHYLATION EFFECTS
Does not cleave GGNCm^{5}C or GGNm^{5}CC
Blocked by overlapping *dcm* and CG methylation

NOTES

- EC 3.1.21.4
- A type II site specific deoxyribonuclease
- A large group of enzymes recognizing specific short DNA sequences. They cleave either within the recognition site, or at a short specific distance from it
- Requires Mg^{2+} and BSA for full activity

CONCENTRATION or VOLUME ACTIVITY	UNIT DEFINITION	REACTION BUFFER (As Provided)	ADDITIONAL ACTIVITIES and PURITY	SUPPLIER CATALOG No.
Citrobacter freundii RFL13				
10,000 U/mL	1 unit degrades 1 μg λ dam$^-$ dcm$^-$ DNA/hr at 37°C in a 0.05 mL mixture	33 mM Tris-acetate, 10 mM MgOAc, 66 mM KOAc, 100 μg/mL BSA, pH 7.9	No detectable contaminating nuclease; 90% ligated and 100% recut after 50-fold overdigestion	AGS Heidelb F00580S F00580M
8-12 U/μL	1 unit hydrolyzes 1 μg DNA/hr at 37°C in a 50 μL reaction volume using λ phage DNA as substrate	33 mM Tris-acetate, 10 mM MgOAc, 66 mM KOAc, 0.1 mg/mL BSA, pH 7.9	>90% of the DNA fragments ligated and recut after 50-fold overdigestion	Fermentas ER0191 ER0192
500 U and 2500 U	1 unit completely digests 1 μg λ DNA/hr at 37°C in 50 μL of supplied buffer	200 mM Tris-HCl, 100 mM MgCl$_2$, 10 mM DTT, 1 M KCl, pH 8.5	Purity is confirmed by overdigestion and ligation-recutting test	TaKaRa 1031
3-20 U/μL	1 unit completely digests 1 μg DNA/hr at 37°C	10 mM Tris-HCl, 10 mM MgCl$_2$, 1 mM DTT, 50 mM NaCl, pH 7.5	>90% ligation; >95% recut	Toyobo CF13-101 CF13-102

Cfr42 I

RECOGNITION SEQUENCE
  5'...CCGC▼GG...3'
  3'...GG▲CGCC...5'

ISOSCHIZOMERS
  Sac II

OPTIMUM REACTION CONDITIONS
  pH 7.5 and 37°C

HEAT INACTIVATION
  65°C for 20 min.

METHYLATION EFFECTS
  Blocked by CG methylation

NOTES
- EC 3.1.21.4
- A type II site specific deoxyribonuclease
- A large group of enzymes recognizing specific short DNA sequences. They cleave either within the recognition site, or at a short specific distance from it
- Certain sites in λ and φX174 DNAs are difficult to cleave with both Cfr42 I and Sac II
- Requires Mg^{2+} and BSA for full activity

CONCENTRATION or VOLUME ACTIVITY	UNIT DEFINITION	REACTION BUFFER (As Provided)	ADDITIONAL ACTIVITIES and PURITY	SUPPLIER CATALOG No.
Citrobacter freundii RFL42				
10,000 U/mL	1 unit degrades 1 µg λ dam⁻ dcm⁻ DNA BcI I-fragments/hr at 37°C in a 0.05 mL mixture	10 mM Tris-HCl, 10 mM MgCl₂, 100 µg/mL BSA, pH 7.5	No detectable contaminating nuclease; 90% ligated and 100% recut after 50-fold overdigestion	AGS Heidelb F00550S F00550M
Escherichia coli, carrying the cloned Cfr42IR gene from Citrobacter freundii RFL42				
8-12 U/µL	1 unit hydrolyzes 1 µg DNA/hr at 37°C in a 50 µL reaction volume using λ phage DNA as substrate	10 mM Tris-HCl, 10 mM MgCl₂, 0.1 mg/mL BSA, pH 7.5	>90% of the DNA fragments ligated and recut after 50-fold overdigestion	Fermentas ER0201 ER0202

Cla I

RECOGNITION SEQUENCE
5'...AT▼CGAT...3'
3'...TAGC▲TA...5'

ISOSCHIZOMERS
Ban III, Bsc I, BsiX I, Bsp106 I, BspD I, BspX I, Bsu15 I

OPTIMUM REACTION CONDITIONS
37°C

HEAT INACTIVATION
65°C for 20 min.

METHYLATION EFFECTS
Does not cleave m^6ATCGAT, ATm^5CGAT or ATCGm^6AT
Blocked by overlapping *dam* methylation

NOTES
- EC 3.1.21.4
- A type II site specific deoxyribonuclease
- A large group of enzymes recognizing specific short DNA sequences. They cleave either within the recognition site, or at a short specific distance from it
- DNA from dam^+ *E. coli* is partially resistant to cleavage
- Compatible cohesive ends: Acc I, Aci I, Acy I, Aha II, Asu II, Ban III, Bbi II, BsaH I, BsiC I, BstB I, Csp45 I, Hap II, HgiD I, Hin1 I, HinP1 I, Hpa II, Mae II, Msp I, Nar I, Nsp V, Nsp7524 V, Nun II, Psp1406 I, Sfu I, Taq I, TthHB8 I
- Requires Mg^{2+} and BSA for full activity

CONCENTRATION or VOLUME ACTIVITY	UNIT DEFINITION	REACTION BUFFER (As Provided)	ADDITIONAL ACTIVITIES and PURITY	SUPPLIER CATALOG NO.
Caryophanon latum				
10 U/μL; 40 U/μL	1 unit completely cleaves 1 μg substrate DNA/hr	60 mM KCl, 20 mM Tris, 0.75 mM MgCl₂, pH 7.5	90-98% cleavage-ligation-recleavage	Am Allied L-32-00500 H-32-00500 L-32-02500 H-32-02500
8-12 U/μL	1 unit completely digests 1 mg λ DNA/hr at 37°C	500 mM Tris-HCl, 100 mM MgCl₂, pH 8.0	≥95% ligation; 100% cleavage	Life Technol 15416-050 15416-068
8000-15,000 U/mL	1 unit completely digests 1 μg λ DNA/hr at pH 7.5, 37°C in a 50 μL total assay mixture	100 mM Tris-acetate, 100 mM MgOAc, 500 mM KOAc, pH 7.5		Pharmacia 27-0935-01 27-0935-02
Caryophanon latum L (H. Mayer)				
4000-10,000 U/mL		50 mM KCl, 10 mM Tris-HCl, 0.1 mM EDTA, 1 mM DTT, 200 μg/mL BSA, 50% glycerol, pH 7.4		ICN 153806

Cla I continued

CONCENTRATION or VOLUME ACTIVITY	UNIT DEFINITION	REACTION BUFFER (As Provided)	ADDITIONAL ACTIVITIES and PURITY	SUPPLIER CATALOG NO.
Caryophanon latum L				
1000 U and 2000 U	1 unit completely digests 1 μg DNA substrate/hr at 37°C	10 mM Tris-HCl, 10 mM MgCl$_2$, 50 mM NaCl, 1 mM DTT, 0.1 mg/mL BSA, pH 7.5	>90% of the DNA fragments ligated and recut after 10-fold overdigestion	Adv Biotech AB-0225-a AB-0225-b
8-12 U/μL	1 unit completely digests 1 μg dam$^-$ λ DNA/hr at 37°C in the buffer provided	100 mM Tris-HCl, 100 mM MgCl$_2$, 10 mM DTT, 500 mM NaCl, pH 7.5	>95% DNA fragments ligated and 100% recut	Amersham E 1034Y E 1034Z
10 U/μL and 40 U/μL	1 unit completely digests 1 μg substrate DNA/hr	50 mM Tris-HCl, 10 mM MgCl$_2$, 100 mM NaCl, 1 mM DTE, pH 7.5		Boehringer 404217 656291 1092758 1274074
10 U/μL	1 unit completely digests 1 μg λ DNA/hr at 37°C in a 50 mL reaction volume	Blue/White Cloning Qualified; genome qualified; 100 mM Tris-HCl, 100 mM MgCl$_2$, 500 mM NaCl, 10 mM DTT, pH 7.9	90% ligation	Promega R6551 R6552 R6555
2000-10,000 U/mL	1 unit completely cleaves 1.0 μg λ DNA/hr in a 50 μL reaction mixture	200 mM Tris-acetate, 500 mM KOAc, 100 mM MgOAc, 10 mM DTT, pH 7.9	100% cleavage, 95% ligation, 95% recleavage, 25% endonuclease (nickase) No detectable endonuclease (overdigestion)	Sigma R7763
1000 U and 5000 U	1 unit completely digests 1 μg λ DNA/hr at 30°C in 50 μL of supplied buffer	Genomic grade; 100 mM Tris-HCl, 100 mM MgCl$_2$, 10 mM DTT, 500 mM NaCl, pH 7.5	Purity is confirmed by overdigestion, ligation-recutting, genome DNA digestion and cloning test	TaKaRa 1034
Escherichia coli strain, carrying the cloned gene from Caryophanon latum				
5000 U/mL	1 unit completely digests 1 μg λ DNA (dam$^-$)/hr in a total reaction volume of 50 μL using the buffer provided	20 mM Tris-acetate, 10 mM MgOAc, 50 mM KOAc, 1 mM DTT, pH 7.9; add 100 μg/mL BSA	>95% DNA fragments ligated and recut after 10-fold overdigestion	NE Biolabs 197S 197L

Cpo I

RECOGNITION SEQUENCE
5'...CG▼G(^A_T)CCG...3'
3'...GCC(^T_A)G▲GC...5'

ISOSCHIZOMERS
Csp I, *Rsr* II

OPTIMUM REACTION CONDITIONS
37°C

HEAT INACTIVATION
60°C for 15 min.

METHYLATION EFFECTS
Blocked by CG methylation

NOTES
- EC 3.1.21.4
- A type II site specific deoxyribonuclease
- A large group of enzymes recognizing specific short DNA sequences. They cleave either within the recognition site, or at a short specific distance from it
- Requires Mg^{2+} and BSA for full activity

CONCENTRATION or VOLUME ACTIVITY	UNIT DEFINITION	REACTION BUFFER (As Provided)	ADDITIONAL ACTIVITIES and PURITY	SUPPLIER CATALOG No.
Caseobacter polymorphus				
8-12 U/μL	1 unit completely digests 1 μg λ DNA/hr at 37°C in the buffer provided	Genomic grade; 200 mM Tris-HCl, 100 mM MgCl$_2$, 10 mM DTT, 1000 mM KCl, pH 8.5	>95% DNA fragments ligated and 100% recut	Amersham E 1035Y
4000 U/mL	1 unit degrades 1 μg λ dam$^-$ dcm$^-$ DNA/hr at 37°C in a 0.05 mL mixture	33 mM Tris-acetate, 10 mM MgOAc, 66 mM KOAc, 100 μg/mL BSA, pH 7.9	No detectable contaminating nuclease; 90% ligated and 100% recut after 50-fold overdigestion	AGS Heidelb F00535S F00535M
8-12 U/μL	1 unit hydrolyzes 1 μg DNA/hr at 37°C in a 50 μL reaction volume using λ phage DNA as substrate	33 mM Tris-acetate, 10 mM MgOAc, 66 mM KOAc, 0.1 mg/mL BSA, pH 7.9	>90% of the DNA fragments ligated and recut after 50-fold overdigestion	Fermentas ER0741 ER0742
400 U and 2000 U	1 unit completely cleaves 1 μg DNA/hr in 50 μL reaction mixture	200 mM Tris-HCl, 100 mM MgCl$_2$, 10 mM DTT, 1 M KCl, pH 8.5		PanVera TAK 1035
400 U and 2000 U	1 unit completely digests 1 μg λ DNA/hr at 30°C in 50 μL of supplied buffer	Genomic grade; 200 mM Tris-HCl, 100 mM MgCl$_2$, 10 mM DTT, 1 M KCl, pH 8.5	Purity is confirmed by overdigestion, ligation-recutting and genome DNA digestion test	TaKaRa 1035

Csp I

RECOGNITION SEQUENCE
5'...CG▼G(^A_T)CCG...3'
3'...GCC(^T_A)G▲GC...5'

ISOSCHIZOMERS
Cpo I, *Rsr* II

OPTIMUM REACTION CONDITIONS
30°C; 50-75% activity at 37°C

HEAT INACTIVATION
65°C for 20 min.

METHYLATION EFFECTS
Cleaves CGG(^A_T)Cm^{5}CG
Does not cleave CGG(^A_T)m^{5}CCG or m^{5}CGG(^A_T)CCG

NOTES
- EC 3.1.21.4
- A type II site specific deoxyribonuclease
- A large group of enzymes recognizing specific short DNA sequences. They cleave either within the recognition site, or at a short specific distance from it
- Superior to *Rsr* II for pulsed field gel electrophoresis
- Requires Mg^{2+}

CONCENTRATION or VOLUME ACTIVITY	UNIT DEFINITION	REACTION BUFFER (As Provided)	ADDITIONAL ACTIVITIES and PURITY	SUPPLIER CATALOG NO.
***Clostridium* species**				
10 U/μL	1 unit completely digests 1 μg λ DNA/hr at 30°C in a 50 mL reaction volume	Genome qualified; 100 m*M* Tris-HCl, 100 m*M* MgCl$_2$, 1.5 *M* KCl, pH 7.6	90% ligation	Promega R6671 R6672 R6675
2000-10,000 U/mL		50 m*M* KCl, 10 m*M* Tris-HCl, 0.1 m*M* EDTA, 1 m*M* DTT, 500 μg/mL BSA, 50% glycerol, pH 7.4		Stratagene 500384
***Corynebacterium* species**				
3-20 U/μL	1 unit completely digests 1 μg DNA/hr at 30°C	50 m*M* Tris-HCl, 10 m*M* MgCl$_2$, 100 m*M* NaCl, 1 m*M* DTT, pH 7.5	>95% ligation; >95% recut	Toyobo CSP-101 CSP-102

Csp6 I

RECOGNITION SEQUENCE
5'...G▼TAC...3'
3'...CAT▲G...5'

ISOSCHIZOMERS
Rsa I

OPTIMUM REACTION CONDITIONS
pH 7.5 and 37°C

HEAT INACTIVATION
65°C for 20 min.

NOTES
- EC 3.1.21.4
- A type II site specific deoxyribonuclease
- A large group of enzymes recognizing specific short DNA sequences. They cleave either within the recognition site, or at a short specific distance from it
- Produces DNA fragments with a 2-base 5'-extension; *Rsa* I does not
- Requires Mg^{2+}

CONCENTRATION or VOLUME ACTIVITY	UNIT DEFINITION	REACTION BUFFER (As Provided)	ADDITIONAL ACTIVITIES and PURITY	SUPPLIER CATALOG NO.
***Corynebacterium* species RFL6**				
10,000 U/mL	1 unit degrades 1 μg λ dam⁻ dcm⁻ DNA/hr at 37°C in a 0.05 mL mixture	10 m*M* Tris-HCl, 10 m*M* MgCl₂, 100 μg/mL BSA, pH 7.5	No detectable contaminating nuclease; 90% ligated and 100% recut after 50-fold overdigestion	AGS Heidelb F00240S F00240M
8-12 U/μL	1 unit hydrolyzes 1 μg DNA/hr at 37°C in a 50 μL reaction volume using λ phage DNA as substrate	10 m*M* Tris-HCl, 10 m*M* MgCl₂, pH 7.5	>90% of the DNA fragments ligated and recut after 50-fold overdigestion	Fermentas ER0211 ER0212
10,000 U/mL	1 unit hydrolyzes 1 μg DNA/hr at 37°C in a 50 μL reaction volume using λ phage DNA as substrate	10 m*M* Tris-HCl, 10 m*M* MgCl₂, pH 7.5	>90% of the DNA fragments ligated and recut after 50-fold overdigestion	NE Biolabs ER0211S ER0211L

Csp45 I

RECOGNITION SEQUENCE
5'...TT▼CGAA...3'
3'...AAGC▲TT...5'

OPTIMUM REACTION CONDITIONS
pH 7.5

HEAT INACTIVATION
65°C for >15 min.

NOTES
- EC 3.1.21.4
- A type II site specific deoxyribonuclease
- A large group of enzymes recognizing specific short DNA sequences. They cleave either within the recognition site, or at a short specific distance from it
- Compatible cohesive ends: *Acy* I, *Aha* II, *Asu* II, *Ban* III, *Bbi* II, *Bst*B I, *Cla* I, *Hap* II, *Hin*1 I, *Hin*P I, *Hpa* II, *Mae* II, *Msp* I, *Nar* I, *Nsp* V, *Sfu* I, *Taq* I, *Tth*HB8 I
- Requires Mg^{2+}

CONCENTRATION or VOLUME ACTIVITY	UNIT DEFINITION	REACTION BUFFER (As Provided)	ADDITIONAL ACTIVITIES and PURITY	SUPPLIER CATALOG No.
Clostridium sporogenes				
10 U/μL	1 unit completely digests 1 μg λ DNA/hr at 37°C in a 50 mL reaction volume	Blue/White Cloning Qualified; genome qualified; 60 mM Tris-HCl, 60 mM MgCl$_2$, 500 mM NaCl, 10 mM DTT, pH 7.5	90% ligation	Promega R6571 R6572
3-20 U/μL	1 unit completely digests 1 μg DNA/hr at 37°C	10 mM Tris-HCl, 10 mM MgCl$_2$, 50 mM NaCl, 1 mM DTT, pH 7.5	>90% ligation; >90% recut	Toyobo C45-101 C45-102

CspA I

RECOGNITION SEQUENCE
5'...A▼CCGGT...3'
3'...TGGCC▲A...5'

ISOSCHIZOMERS
Age I

OPTIMUM REACTION CONDITIONS
pH 7.1

NOTES
- EC 3.1.21.4
- A type II site specific deoxyribonuclease
- A large group of enzymes recognizing specific short DNA sequences. They cleave either within the recognition site, or at a short specific distance from it
- Requires Mg^{2+}

CONCENTRATION or VOLUME ACTIVITY	UNIT DEFINITION	REACTION BUFFER (As Provided)	ADDITIONAL ACTIVITIES and PURITY	SUPPLIER CATALOG No.
Corynebacterium species				
	30 units does not produce unspecific cleavage products with 1 µg λ DNA/16 hr at 37°C	20 mM Tris-acetate, 10 mM MgOAc, 50 mM KOAc, 1 mM DTT, 100 µg/mL BSA, pH 7.9	>90% DNA fragments ligated and recut after 10-fold overdigestion	MINOTECH 143

CviJ I

RECOGNITION SEQUENCE
5'...PuG▼CPy...3'
3'...PyC▲GPu...5'

NOTES
- EC 3.1.21.4
- A type II site specific deoxyribonuclease
- A large group of enzymes recognizing specific short DNA sequences. They cleave either within the recognition site, or at a short specific distance from it
- Requires Mg^{2+}

CONCENTRATION or VOLUME ACTIVITY	UNIT DEFINITION	REACTION BUFFER (As Provided)	ADDITIONAL ACTIVITIES and PURITY	SUPPLIER CATALOG No.
Chlorella virus IL-3A				
	1 unit completely cleaves 1 µg pUC19 DNA/hr at pH 7.5, 50°C in a reaction volume of 50 µL	20 mM glycylglycine-KOH, 10 mM MgOAc, 7.0 mM β-MSH, pH 8.5		CHIMERx 2125-01 2125-02

Cvn I

RECOGNITION SEQUENCE
5'...CC▼TNAGG...3'
3'...GGANT▲CC...5'

OPTIMUM REACTION CONDITIONS
pH 7.4

HEAT INACTIVATION
65°C for 10 min.

NOTES
- EC 3.1.21.4
- A type II site specific deoxyribonuclease
- A large group of enzymes recognizing specific short DNA sequences. They cleave either within the recognition site, or at a short specific distance from it
- Requires Mg^{2+}

CONCENTRATION or VOLUME ACTIVITY	UNIT DEFINITION	REACTION BUFFER (As Provided)	ADDITIONAL ACTIVITIES and PURITY	SUPPLIER CATALOG NO.
Chromatium vinosum				
6-10 U/μL	1 unit completely digests 1 mg λ DNA/hr at 37°C	200 mM Tris-HCl, 50 mM $MgCl_2$, 500 M KCl, pH 7.4	0% ligation	Life Technol 15428-014

Dde I

RECOGNITION SEQUENCE
5'...C▼TNAG...3'
3'...GANT▲C...5'

OPTIMUM REACTION CONDITIONS
pH 7.5 and 37°C

HEAT INACTIVATION
65°C for 20 min.

METHYLATION EFFECTS
Cleaves hm^{5}CTNAG
Does not cleave m^{5}CTNAG or hm^{5}CTNAG

NOTES
- EC 3.1.21.4
- A type II site specific deoxyribonuclease
- A large group of enzymes recognizing specific short DNA sequences. They cleave either within the recognition site, or at a short specific distance from it
- Cleaves single-stranded 100-fold slower than double-stranded DNA
- Compatible cohesive ends: Acc I, Axy I, Cel II, Cvn I, Esp I, Mst II, OxaN I, Sau I
- High salt or glycerol >5% alter specificity
- Requires Mg^{2+}

CONCENTRATION or VOLUME ACTIVITY	UNIT DEFINITION	REACTION BUFFER (As Provided)	ADDITIONAL ACTIVITIES and PURITY	SUPPLIER CATALOG NO.
Desulfovibrio desulfuricans				
3-20 U/µL	1 unit completely digests 1 µg λ DNA/hr at 37°C in the buffer provided	500 mM Tris-HCl, 100 mM MgCl$_2$, 10 mM DTT, 1000 mM NaCl, pH 7.5	>90% DNA fragments ligated and >95% recut	Amersham E 0209Y E 0209Z
6-16 U/µL	1 unit digests completely 1 µg λ DNA substrate/hr at 37°C	50 mM Tris-HCl, 100 mM NaCl, 10 mM MgCl$_2$, 1 mM DTT, BSA, pH 7.8		NBL Gene 012901 012903
10 U/µL and 15-30 U/µL	1 unit hydrolyzes 1 µg λ DNA to completion/hr in a 50 µL total reaction volume	100 mM Tris-HCl, 1 M NaCl, 100 mM MgCl$_2$, 100 mM β-MSH, 1.0 mg/mL BSA, pH 8		Oncor 110141 110142
2000-10,000 U/mL	1 unit completely digests 1 µg λ DNA/hr at pH 7.5, 37°C in a 50 µL total assay mixture	100 mM Tris-acetate, 100 mM MgOAc, 500 mM KOAc, pH 7.5		Pharmacia 27-0943-01 27-0943-02
10 U/µL	1 unit completely digests 1 µg λ DNA/hr at 37°C in a 50 mL reaction volume	60 mM Tris-HCl, 60 mM MgCl$_2$, 1.5 M NaCl, 10 mM DTT, pH 7.9	90% ligation	Promega R6291 R6292 R6295
2000-10,000 U/mL		50 mM KCl, 10 mM Tris-HCl, 0.1 mM EDTA, 1 mM DTT, 200 µg/mL BSA, 50% glycerol, pH 7.4		Stratagene 500390

Dde I continued

CONCENTRATION or VOLUME ACTIVITY	UNIT DEFINITION	REACTION BUFFER (As Provided)	ADDITIONAL ACTIVITIES and PURITY	SUPPLIER CATALOG No.
Desulfovibrio desulfuricans continued				
3-20 U/μL	1 unit completely digests 1 μg DNA/hr at 37°C	50 mM Tris-HCl, 10 mM $MgCl_2$, 100 mM NaCl, 1 mM DTT, pH 7.5	>80% ligation; >95% recut	Toyobo DDE-101 DDE-102
Desulfovibrio desulfuricans NCIB 8310				
2000-10,000 U/mL		50 mM KCl, 10 mM Tris-HCl, 0.1 mM EDTA, 1 mM DTT, 200 μg/mL BSA, 50% glycerol, pH 7.4		ICN 153807
Desulfovibrio desulfuricans NCIB Norway				
10 U/μL	1 unit completely digests 1 μg substrate DNA/hr	50 mM Tris-HCl, 10 mM $MgCl_2$, 100 mM NaCl, 1 mM DTE, pH 7.5		Boehringer 835293 835307
8-12 U/μL	1 unit completely digests 1 mg λ DNA/hr at 37°C	500 mM Tris-HCl, 100 mM $MgCl_2$, 500 mM NaCl, pH 8.0	≥33% ligation; 100% cleavage	Life Technol 15238-041 15238-058
Escherichia coli strain, carrying the cloned gene from Desulfovibrio desulfuricans				
10,000 U/mL	1 unit completely digests 1 μg λ DNA/hr in a total reaction volume of 50 μL using the buffer provided	50 mM Tris-HCl, 10 mM $MgCl_2$, 100 mM NaCl, 1 mM DTT, pH 7.9	>95% DNA fragments ligated and recut after 2-fold overdigestion	NE Biolabs 175S 175L

Dpn I

RECOGNITION SEQUENCE
  5'...GAm▼TC...3'
  3'...CT▲AmG...5'

ISOSCHIZOMERS
  Mbo I, Sau3A I

OPTIMUM REACTION CONDITIONS
  37°C

HEAT INACTIVATION
  Resistant

METHYLATION EFFECTS
  Cleaves Gm^6ATC, Gm^6ATm^5C and Gm^6ATm^4C
  Does not cleave GATC, $GATm^4C$ or $GATm^5C$

NOTES
- EC 3.1.21.4
- A type II site specific deoxyribonuclease
- A large group of enzymes recognizing specific short DNA sequences. They cleave either within the recognition site, or at a short specific distance from it
- Cleaves only when its recognition site is methylated
- DNA purified from a dam^+ strain is a substrate
- Shares the same recognition sequence as Mbo I and Sau3A I, but their cutting site and methylation sensitivities differ
- Compatible cohesive ends: any blunt end
- Requires Mg^{2+}

CONCENTRATION or VOLUME ACTIVITY	UNIT DEFINITION	REACTION BUFFER (As Provided)	ADDITIONAL ACTIVITIES and PURITY	SUPPLIER CATALOG NO.
Diplococcus pneumoniae				
8-12 U/μL	1 unit completely digests 1 μg pBR322 DNA/hr at 37°C in the buffer provided	500 mM Tris-HCl, 100 mM MgCl₂, 10 mM DTT, 1000 mM NaCl, pH 7.5	>30% DNA fragments ligated and >90% recut	Amersham E 0302Z
3-10 U/μL	1 unit completely digests 1 μg substrate DNA/hr	33 mM Tris-acetate, 10 mM MgOAc, 66 mM KOAc, 0.5 mM DTT, pH 7.9		Boehringer 742970 742988
2-8 U/μL	1 unit completely digests 1 mg pBR322 DNA/hr at 37°C	200 mM Tris-HCl, 50 mM MgCl₂, 500 M KCl, pH 7.4	≥10% ligation; 100% cleavage	Life Technol 15242-019
2-6 U/μL	1 unit digests completely 1 μg pBR322 DNA substrate/hr at 37°C	33 mM Tris-acetate, 66 mM KOAc, 10 mM MgOAc, 0.5 mM DTT, pH 8.2		NBL Gene 012411 012402
10 U/μL	1 unit completely digests 1 μg pBR322 DNA/hr at 37°C in a 50 mL reaction volume	60 mM Tris-HCl, 60 mM MgCl₂, 500 mM NaCl, 10 mM DTT, pH 7.5	80% ligation	Promega R6231 R6232
1000-10,000 U/mL	1 unit completely cleaves 1.0 μg methylated pBR322 DNA/hr in a 50 μL reaction mixture	200 mM Tris-acetate, 500 mM KOAc, 100 mM MgOAc, 10 mM DTT, pH 7.9	100% cleavage, 50% ligation, 80% recleavage, 95% endonuclease (nickase) No detectable endonuclease (overdigestion)	Sigma R8381

Dpn I continued

CONCENTRATION or VOLUME ACTIVITY	UNIT DEFINITION	REACTION BUFFER (As Provided)	ADDITIONAL ACTIVITIES and PURITY	SUPPLIER CATALOG NO.
Diplococcus pneumoniae strain 641 (S. Lacks)				
5000-20,000 U/mL		100 mM KCl, 10 mM Tris-HCl, 0.1 mM EDTA, 0.1 mM DTT, 200 µg/mL BSA, 50% glycerol, pH 7.4		ICN 153808
Escherichia coli strain, carrying a *Dpn I* overproducing plasmid (S. Lacks)				
20,000 U/mL	1 unit completely digests 1 µg pBR322 DNA (dam methylated)/hr in a total reaction volume of 50 µL using the buffer provided	20 mM Tris-acetate, 10 mM MgOAc, 50 mM KOAc, 1 mM DTT, pH 7.9	>70% DNA fragments ligated and >95% recut after 20-fold overdigestion	NE Biolabs 176S 176L

Dpn II

RECOGNITION SEQUENCE
  5'...▼GATC...3'
  3'...CTAG▲...5'
ISOSCHIZOMERS
  Mbo I
OPTIMUM REACTION CONDITIONS
  37°C
HEAT INACTIVATION
  65°C for 20 min.
METHYLATION EFFECTS
  Does not cleave Gm^{6}ATC
  Blocked by *dam* methylation

NOTES
- EC 3.1.21.4
- A type II site specific deoxyribonuclease
- A large group of enzymes recognizing specific short DNA sequences. They cleave either within the recognition site, or at a short specific distance from it
- Leaves a 5'-GATC extension which is efficiently ligated to DNA compatible cohesive ends generated by *Bam*H I, *Bcl* I, *Bgl* II, *Bst*Y I, *Mbo* I, *Sau*3A I
- Requires Mg^{2+}

CONCENTRATION or VOLUME ACTIVITY	UNIT DEFINITION	REACTION BUFFER (As Provided)	ADDITIONAL ACTIVITIES and PURITY	SUPPLIER CATALOG NO.
Escherichia coli				
5000-20,000 U/mL		200 mM NaCl, 10 mM Tris-HCl, 0.1 mM EDTA, 1 mM DTT, 200 µg/mL acetylated BSA, 50% glycerol, pH 7.4		ICN 159429

Dpn II continued

CONCENTRATION or VOLUME ACTIVITY	UNIT DEFINITION	REACTION BUFFER (As Provided)	ADDITIONAL ACTIVITIES and PURITY	SUPPLIER CATALOG No.
Escherichia coli strain, carrying a *Dpn* I overproducing plasmid (S. Lacks)				
10,000; 50,000 U/mL	1 unit completely digests 1 μg λ DNA (dam⁻)/hr in a total reaction volume of 50 μL using the buffer provided	50 mM Bis Tris-HCl, 10 mM MgCl$_2$, 100 mM NaCl, 1 mM DTT, pH 6.0	>95% DNA fragments ligated and recut after 100-fold overdigestion	NE Biolabs 543S 543L

Dra I

RECOGNITION SEQUENCE
  5'...TTT▼AAA...3'
  3'...AAA▲TTT...5'

ISOSCHIZOMERS
  Aha III

OPTIMUM REACTION CONDITIONS
  pH 7.5 and 37°C

HEAT INACTIVATION
  65°C for 20 min.

METHYLATION EFFECTS
  Cleaves TTTAm^{6}AA

NOTES
- EC 3.1.21.4
- A type II site specific deoxyribonuclease
- A large group of enzymes recognizing specific short DNA sequences. They cleave either within the recognition site, or at a short specific distance from it
- Compatible cohesive ends: any blunt end
- An infrequent cutter of genomic DNA with high GC content
- Requires Mg^{2+} and BSA for full activity

CONCENTRATION or VOLUME ACTIVITY	UNIT DEFINITION	REACTION BUFFER (As Provided)	ADDITIONAL ACTIVITIES and PURITY	SUPPLIER CATALOG No.
Deinococcus radiophilus				
8-12 U/μL	1 unit completely digests 1 μg λ DNA/hr at 37°C in the buffer provided	Genomic grade; 100 mM Tris-HCl, 100 mM MgCl$_2$, 10 mM DTT, 500 mM NaCl, pH 7.5	>90% DNA fragments ligated and 100% recut	Amersham E 1037Y E 1037Z
8000-10,000 U/mL	1 unit degrades 1 μg λ dam⁻ dcm⁻ DNA/hr at 37°C in a 0.05 mL mixture	10 mM Tris-HCl, 100 mM NaCl, 100 μg/mL BSA, pH 8.0	No detectable contaminating nuclease; 90% ligated and 100% recut after 10-50 fold overdigestion	AGS Heidelb F00250S F00250M
10 U/μL and 40 U/μL	1 unit completely digests 1 μg substrate DNA/hr	10 mM Tris-HCl, 10 mM MgCl$_2$, 50 mM NaCl, 1 mM DTE, pH 7.5		Boehringer 779695 827754 1175076

Dra I continued

CONCENTRATION or VOLUME ACTIVITY	UNIT DEFINITION	REACTION BUFFER (As Provided)	ADDITIONAL ACTIVITIES and PURITY	SUPPLIER CATALOG No.
Deinococcus radiophilus continued				
	1 unit completely cleaves 1 µg λ DNA/hr at pH 7.5, 37°C in a reaction volume of 50 µL	10 mM Tris-HCl, 10 mM MgCl$_2$, 50 mM NaCl, 1.0 mM DTT, 100 µg/mL BSA, pH 7.5		CHIMERX 2145-01 2145-02
8-12 U/µL	1 unit hydrolyzes 1 µg DNA/hr at 37°C in a 50 µL reaction volume using λ phage DNA as substrate	10 mM Tris-HCl, 10 mM MgCl$_2$, 0.1 mg/mL BSA, pH 7.5	90% of the DNA fragments ligated and >90% recut after 10-fold overdigestion	Fermentas ER0221 ER0222
6-12 U/µL	1 unit digests completely 1 µg λ DNA substrate/hr at 37°C	10 mM Tris-HCl, 50 mM NaCl, 10 mM MgCl$_2$, 1 mM DTT, pH 7.8		NBL Gene 010604 010607
				Nippon Gene
10 U/µL	1 unit hydrolyzes 1 µg λ DNA to completion/hr in a 50 µL total reaction volume	100 mM Tris-HCl, 500 mM NaCl, 100 mM MgCl$_2$, 100 mM β-MSH, 1.0 mg/mL BSA, pH 8		Oncor 110154 110155
10,000-25,000 U/mL	1 unit completely digests 1 µg λ DNA/hr at pH 7.5, 37°C in a 50 µL total assay mixture	100 mM Tris-acetate, 100 mM MgOAc, 500 mM KOAc, pH 7.5		Pharmacia 27-0941-01 27-0941-02
10 U/µL	1 unit completely digests 1 µg λ DNA/hr at 37°C in a 50 mL reaction volume	60 mM Tris-HCl, 60 mM MgCl$_2$, 500 mM NaCl, 10 mM DTT, pH 7.5	80% ligation	Promega R6271 R6272
				SibEnzyme
5000-15,000 U/mL	1 unit completely cleaves 1.0 µg λ DNA/hr in a 50 µL reaction mixture	200 mM Tris-acetate, 500 mM KOAc, 100 mM MgOAc, 10 mM DTT, pH 7.9	100% cleavage, 60% ligation, 90% recleavage. No detectable endonuclease (overdigestion)	Sigma R4381
10,000-40,000 U/mL		50 mM KCl, 10 mM Tris-HCl, 0.1 mM EDTA, 1 mM DTT, 200 µg/mL BSA, 50% glycerol, pH 7.4		Stratagene 500410
4000 U and 20,000 U	1 unit completely digests 1 µg λ DNA/hr at 37°C in 50 µL of supplied buffer	Genomic grade; 100 mM Tris-HCl, 100 mM MgCl$_2$, 10 mM DTT, 500 mM NaCl, pH 7.5	Purity is confirmed by overdigestion, ligation-recutting and genome DNA digestion test	TaKaRa 1037
3-20 U/µL	1 unit completely digests 1 µg DNA/hr at 37°C	10 mM Tris-HCl, 10 mM MgCl$_2$, 50 mM NaCl, 1 mM DTT, pH 7.5	>90% ligation; >95% recut	Toyobo DRA-103 DRA-104
Deinococcus radiophilus ATCC 27603				
5000-20,000 U/mL		50 mM KCl, 10 mM Tris-HCl, 0.1 mM EDTA, 1.0 mM DTT, 200 µg/mL BSA, 50% glycerol, pH 7.5		ICN 190516

Dra I continued

CONCENTRATION or VOLUME ACTIVITY	UNIT DEFINITION	REACTION BUFFER (As Provided)	ADDITIONAL ACTIVITIES and PURITY	SUPPLIER CATALOG NO.
Deinococcus radiophilus ATCC 27603 continued				
8-12 U/μL and 50 U/μL	1 unit completely digests 1 mg λ DNA/hr at 37°C	500 mM Tris-HCl, 100 mM MgCl$_2$, pH 8.0	≥33% ligation; 100% cleavage	Life Technol 25430-018 25430-067 25430-026
20,000 U/mL	1 unit completely digests 1 μg λ DNA/hr in a total reaction volume of 50 μL using the buffer provided	20 mM Tris-acetate, 10 mM MgOAc, 50 mM KOAc, 1 mM DTT, pH 7.9	>80% DNA fragments ligated and recut after 20-fold overdigestion	NE Biolabs 129S 129L

Dra II

RECOGNITION SEQUENCE
    5'...PuG▼GNCCPy...3'
    3'...PyCCNG▲GPu...5'
ISOSCHIZOMERS
    *Eco*O109 I, *Pss* I
OPTIMUM REACTION CONDITIONS
    37°C
HEAT INACTIVATION
    60°C for 20 min.
METHYLATION EFFECTS
    Does not cleave PuGGNCm^{5}CPy

NOTES
- EC 3.1.21.4
- A type II site specific deoxyribonuclease
- A large group of enzymes recognizing specific short DNA sequences. They cleave either within the recognition site, or at a short specific distance from it
- Compatible cohesive ends: *Sau*96 I
- Requires Mg^{2+}

CONCENTRATION or VOLUME ACTIVITY	UNIT DEFINITION	REACTION BUFFER (As Provided)	ADDITIONAL ACTIVITIES and PURITY	SUPPLIER CATALOG NO.
Deinococcus radiophilus				
1-5 U/μL	1 unit completely digests 1 μg λ DNA/hr at 37°C in the buffer provided	100 mM Tris-HCl, 100 mM MgCl$_2$, 10 mM DTT, pH 7.5	>90% DNA fragments ligated and 90% recut	Amersham E 0226Y E 0226Z
1-3 U/μL	1 unit completely digests 1 μg substrate DNA/hr	10 mM Tris-HCl, 10 mM MgCl$_2$, 1 mM DTE, pH 7.5		Boehringer 843504 843512

Dra II continued

CONCENTRATION or VOLUME ACTIVITY	UNIT DEFINITION	REACTION BUFFER (As Provided)	ADDITIONAL ACTIVITIES and PURITY	SUPPLIER CATALOG NO.
Deinococcus radiophilus continued				
1000-5000 U/mL		10 mM β-MSH, 20 mM Tris-HCl, 200 mM NaCl, 0.1 mM EDTA, 200 μg/mL acetylated BSA, 50% glycerol, pH 8.0		ICN 159430
3 U/μL	1 unit hydrolyzes 1 μg λ DNA to completion/hr in a 50 μL total reaction volume	100 mM Tris-HCl, 100 mM NaCl, 100 mM MgCl$_2$, 100 mM β-MSH, 1.0 mg/mL BSA, pH 7.5		Oncor 110571 110572
1000-5000 U/mL	1 unit completely cleaves 1.0 μg λ DNA/hr in a 50 μL reaction mixture		100% cleavage, 90% ligation, 90% recleavage, 50% endonuclease (nickase) No detectable endonuclease (overdigestion)	sigma R0639

Dra III

RECOGNITION SEQUENCE
  5'...CACNNN▼GTG...3'
  3'...GTG▲NNNCAC...5'

OPTIMUM REACTION CONDITIONS
  37°C

HEAT INACTIVATION
  65°C for 20 min.

NOTES
- EC 3.1.21.4
- A type II site specific deoxyribonuclease
- A large group of enzymes recognizing specific short DNA sequences. They cleave either within the recognition site, or at a short specific distance from it
- Conditions of low ionic strength, high enzyme concentration, glycerol concentration >5% or pH >8.0 may result in star activity
- Requires Mg^{2+}

CONCENTRATION or VOLUME ACTIVITY	UNIT DEFINITION	REACTION BUFFER (As Provided)	ADDITIONAL ACTIVITIES and PURITY	SUPPLIER CATALOG NO.
Deinococcus radiophilus				
1-5 U/μL	1 unit completely digests 1 μg λ DNA/hr at 37°C in the buffer provided	500 mM Tris-HCl, 100 mM MgCl$_2$, 10 mM DTT, 1000 mM NaCl, pH 7.5	>90% DNA fragments ligated and 90% recut	Amersham E 0210Y E 0210Z

Dra III continued

CONCENTRATION or VOLUME ACTIVITY	UNIT DEFINITION	REACTION BUFFER (As Provided)	ADDITIONAL ACTIVITIES and PURITY	SUPPLIER CATALOG No.
Deinococcus radiophilus continued				
1-5 U/μL	1 unit completely digests 1 μg substrate DNA/hr	50 mM Tris-HCl, 10 mM MgCl$_2$, 100 mM NaCl, 1 mM DTE, pH 7.5		Boehringer 843539 843547
10 U/μL	1 unit hydrolyzes 1 μg λ DNA to completion/hr in a 50 μL total reaction volume	500 mM Tris-HCl, 1 M NaCl, 100 mM MgCl$_2$, 10 mM DTE, 1.0 mg/mL BSA, pH 8		Oncor 110601 110602
1000-10,000 U/mL	1 unit completely cleaves 1.0 μg λ DNA/hr in a 50 μL reaction mixture	500 mM Tris-HCl, 1 M NaCl, 100 mM MgCl$_2$, 10 mM DTT, pH 7.9	100% cleavage, 100% ligation, 100% recleavage, 50% endonuclease (nickase) No detectable endonuclease (overdigestion)	Sigma R4017
2000-10,000 U/mL		300 mM NaCl, 20 mM Tris-HCl, 0.1 mM EDTA, 1 mM DTT, 200 μg/mL BSA, 50% glycerol, pH 7.4		Stratagene 500430
Deinococcus radiophilus ATCC 27603				
2000-15,000 U/mL		50 mM KCl, 10 mM Tris-HCl, 0.1 mM EDTA, 1 mM DTT, 200 μg/mL BSA, 50% glycerol, pH 7.4		ICN 153809
3000 U/mL	1 unit completely digests 1 μg λ DNA/hr in a total reaction volume of 50 μL using the buffer provided	50 mM Tris-HCl, 10 mM MgCl$_2$, 100 mM NaCl, 1 mM DTT, pH 7.9; add 100 μg/mL BSA	>90% DNA fragments ligated and recut after 10-fold overdigestion	NE Biolabs 510S 510L

Drd I

RECOGNITION SEQUENCE
5'...GACNNNN▼NNGTC...3'
3'...CTGNN▲NNNNCAG...5'

OPTIMUM REACTION CONDITIONS
37°C

HEAT INACTIVATION
65°C for 20 min.

NOTES
- EC 3.1.21.4
- A type II site specific deoxyribonuclease
- A large group of enzymes recognizing specific short DNA sequences. They cleave either within the recognition site, or at a short specific distance from it
- Requires Mg^{2+}

CONCENTRATION or VOLUME ACTIVITY	UNIT DEFINITION	REACTION BUFFER (As Provided)	ADDITIONAL ACTIVITIES and PURITY	SUPPLIER CATALOG No.
Deinococcus radiodurans				
1000-10,000 U/mL		50 mM KCl, 10 mM Tris-HCl, 0.1 mM EDTA, 1 mM DTT, 200 µg/mL acetylated BSA, 50% glycerol, pH 7.5		ICN 159431
Deinococcus radiodurans (NEB 479)				
10,000 U/mL	1 unit completely digests 1 µg λ DNA/hr in a total reaction volume of 50 µL using the buffer provided	20 mM Tris-acetate, 10 mM MgOAc, 50 mM KOAc, 1 mM DTT, pH 7.9	>95% DNA fragments ligated and recut after 5-fold overdigestion	NE Biolabs 530S 530L

Dsa I

RECOGNITION SEQUENCE
5'...C▼C(^A_G)(^C_T)GG...3'
3'...GG(^T_C)(^G_A)C▲C...5'

HEAT INACTIVATION
65°C for 20 min.

NOTES
- EC 3.1.21.4
- A type II site specific deoxyribonuclease
- A large group of enzymes recognizing specific short DNA sequences. They cleave either within the recognition site, or at a short specific distance from it
- Requires Mg^{2+}

Dsa I continued

CONCENTRATION or VOLUME ACTIVITY	UNIT DEFINITION	REACTION BUFFER (As Provided)	ADDITIONAL ACTIVITIES and PURITY	SUPPLIER CATALOG NO.
Dactylococcopsis salina				
1-3 U/μL	1 unit completely digests 1 μg substrate DNA/hr	50 mM Tris-HCl, 10 mM MgCl$_2$, 100 mM NaCl, 1 mM DTE, pH 7.5		Boehringer 1167073

DseD I

RECOGNITION SEQUENCE
    5'...GACNNNN▼NNGTC...3'
    3'...CTGNN▲NNNNCAG...5'
ISOSCHIZOMERS
    Drd I

NOTES
- EC 3.1.21.4
- A type II site specific deoxyribonuclease
- A large group of enzymes recognizing specific short DNA sequences. They cleave either within the recognition site, or at a short specific distance from it
- Requires Mg^{2+}

CONCENTRATION or VOLUME ACTIVITY	UNIT DEFINITION	REACTION BUFFER (As Provided)	ADDITIONAL ACTIVITIES and PURITY	SUPPLIER CATALOG NO.
Deinococcus species Dx				
				SibEnzyme

Eae I

RECOGNITION SEQUENCE
5'...Py▼GGCCPu...3'
3'...PuCCGG▲Py...5'

ISOSCHIZOMERS
Cfr I

OPTIMUM REACTION CONDITIONS
37°C

HEAT INACTIVATION
65°C for 20 min.

METHYLATION EFFECTS
Does not cleave PyGGm^{5}CCPu or PyGGCm^{5}CPu
Blocked by overlapping *dcm* methylation

NOTES
- EC 3.1.21.4
- A type II site specific deoxyribonuclease
- A large group of enzymes recognizing specific short DNA sequences. They cleave either within the recognition site, or at a short specific distance from it
- DNA from *dcm*$^+$ *E. coli* is only partially cleaved
- Requires Mg^{2+}

CONCENTRATION or VOLUME ACTIVITY	UNIT DEFINITION	REACTION BUFFER (As Provided)	ADDITIONAL ACTIVITIES and PURITY	SUPPLIER CATALOG NO.
Enterobacter aerogenes				
8-12 U/μL	1 unit completely digests 1 μg λ DNA/hr at 37°C in the buffer provided	500 mM Tris-HCl, 100 mM MgCl$_2$, 10 mM DTT, 1000 mM NaCl, pH 7.5	>95% DNA fragments ligated and 100% recut	Amersham E 1123Y
10,000 U/mL	1 unit degrades 1 μg λ dam$^-$ dcm$^-$ DNA/hr at 37°C in a 0.05 mL mixture	33 mM Tris-acetate, 10 mM MgOAc, 66 mM KOAc, 100 μg/mL BSA, pH 7.9	No detectable contaminating nuclease; 95% ligated and 90% recut after 12-fold overdigestion	ACS Heidelb F00260S F00260M
10 U/μL	1 unit completely digests 1 μg substrate DNA/hr	33 mM Tris-acetate, 10 mM MgOAc, 66 mM KOAc, 0.5 mM DTT, pH 7.9		Boehringer 1062557 1062565
1000-10,000 U/mL	1 unit cleaves 98% λ DNA/hr in a 50 μL reaction mixture	100 mM Bis Tris-Propane-HCl, 100 mM MgCl$_2$, 10 mM DTT, pH 7.0	98% cleavage, 100% ligation, 70% recleavage, 90% endonuclease (nickase) No detectable endonuclease (overdigestion)	Sigma R3759
200 U and 1000 U	1 unit completely digests 1 μg λ DNA/hr at 37°C in 50 μL of supplied buffer	100 mM Tris-HCl, 100 mM MgCl$_2$, 10 mM DTT, 500 mM NaCl, pH 7.5	Purity is confirmed by overdigestion and ligation-recutting test	TaKaRa 1123
Enterobacter aerogenes (N.L. Brown)				
3000 U/mL	1 unit completely digests 1 μg λ DNA/hr in a total reaction volume of 50 μL using the buffer provided	10 mM Bis Tris-Propane-HCl, 10 mM MgCl$_2$, 1 mM DTT, pH 7.0	>95% DNA fragments ligated and recut after 10-fold overdigestion	NE Biolabs 508S 508L

Eae I continued

CONCENTRATION or VOLUME ACTIVITY	UNIT DEFINITION	REACTION BUFFER (As Provided)	ADDITIONAL ACTIVITIES and PURITY	SUPPLIER CATALOG No.
Enterobacter aerogenes (P.R. Whitehead)				
1000-10,000 U/mL		50 mM KCl, 10 mM Tris-HCl, 0.1 mM EDTA, 1 mM DTT, 200 µg/mL BSA, 50% glycerol, pH 7.4		ICN 153810
Enterobacter aerogenes PW201				
3-10 U/µL	1 unit digests completely 1 µg λ DNA substrate/hr at 37°C	10 mM Tris-HCl, 10 mM MgCl$_2$, 10 mM β-MSH, 100 µg/mL BSA, pH 7.5		NBL Gene 015101 015103

Eag I

RECOGNITION SEQUENCE
    5'...C▼GGCCG...3'
    3'...GCCGG▲C...5'
ISOSCHIZOMERS
    *Eco*52 I
OPTIMUM REACTION CONDITIONS
    pH 7.9-9.0 at 25°C; 50% activity at pH 7.4
HEAT INACTIVATION
    65°C for 20 min.
METHYLATION EFFECTS
    Does not cleave CGGm^{5}CCG or m^{5}CGGCm^{5}CG

NOTES
- EC 3.1.21.4
- A type II site specific deoxyribonuclease
- A large group of enzymes recognizing specific short DNA sequences. They cleave either within the recognition site, or at a short specific distance from it
- Compatible ends: *Bsp*120 I, *Eae* I, *Not* I
- Its GC-rich recognition sequence makes this enzyme useful for genomic DNA analysis
- Requires Mg^{2+}

CONCENTRATION or VOLUME ACTIVITY	UNIT DEFINITION	REACTION BUFFER (As Provided)	ADDITIONAL ACTIVITIES and PURITY	SUPPLIER CATALOG No.
Enterobacter agglomerans				
3000-20,000 U/mL		500 mM NaCl, 10 mM Tris-HCl, 0.1 mM EDTA, 1 mM DTT, 200 µg/mL acetylated BSA, 50% glycerol, pH 8.2		ICN 159432
8000-12,000 U/mL	1 unit completely digests 1 µg λ DNA or Ad2 DNA/hr at pH, 37°C in a 50 µL total assay mixture	50 mM Tris-HCl, 100 mM NaCl, 10 mM MgCl$_2$, 1 mM DTT, pH 7.9		Pharmacia 27-0885-01

Eag I continued

CONCENTRATION or VOLUME ACTIVITY	UNIT DEFINITION	REACTION BUFFER (As Provided)	ADDITIONAL ACTIVITIES and PURITY	SUPPLIER CATALOG No.
Escherichia coli strain, carrying the cloned *Eag* I gene from *Enterobacter agglomerans*				
10,000 U/mL; 50,000 U/mL	1 unit completely digests 1 μg λ DNA (BstE II digest)/hr in a total reaction volume of 50 μL using the buffer provided	50 mM Tris-HCl, 10 mM MgCl$_2$, 100 mM NaCl, 1 mM DTT, pH 7.9	>95% DNA fragments ligated and recut after 100-fold overdigestion	NE Biolabs 505S 505L
5000-50,000 U/mL	1 unit cleaves 98% λ DNA/hr in a 50 μL reaction mixture	500 mM Tris-HCl, 1 M NaCl, 100 mM MgCl$_2$, 10 mM DTT, pH 7.9	100% cleavage, 100% ligation, 90% recleavage, 70% endonuclease (nickase) No detectable endonuclease (overdigestion)	Sigma R4510

Eam1104 I

RECOGNITION SEQUENCE
   5'...CTCTTCN▼NNN...3'
   3'...GAG▲AAGNNNN...5'

ISOSCHIZOMERS
   Ear I, *Ksp*632 I

OPTIMUM REACTION CONDITIONS
   37°C; heat sensitive

HEAT INACTIVATION
   65°C for 20 min.

NOTES
- EC 3.1.21.4
- A type II site specific deoxyribonuclease
- A large group of enzymes recognizing specific short DNA sequences. They cleave either within the recognition site, or at a short specific distance from it
- λ DNA contains slow cleavage sites for *Eam*1104 I and its isoschizomers
- Requires Mg^{2+} and BSA for full activity

CONCENTRATION or VOLUME ACTIVITY	UNIT DEFINITION	REACTION BUFFER (As Provided)	ADDITIONAL ACTIVITIES and PURITY	SUPPLIER CATALOG No.
Enterobacter amnigenus RFL1104				
10,000 U/mL	1 unit degrades 1 μg λ dam$^-$ dcm$^-$ DNA/hr at 37°C in a 0.05 mL mixture	33 mM Tris-acetate, 10 mM MgOAc, 66 mM KOAc, 100 μg/mL BSA, pH 7.9	No detectable contaminating nuclease; 90% ligated and 90% recut after 10-fold overdigestion	AGS Heidelb F00394S F00394M

Eam1104 I continued

CONCENTRATION or VOLUME ACTIVITY	UNIT DEFINITION	REACTION BUFFER (As Provided)	ADDITIONAL ACTIVITIES and PURITY	SUPPLIER CATALOG NO.
Enterobacter amnigenus RFL1104 continued				
4-8 U/μL	1 unit hydrolyzes 1 μg DNA/hr at 37°C in a 50 μL reaction volume using λ phage DNA as substrate	33 mM Tris-acetate, 10 mM MgOAc, 66 mM KOAc, 0.1 mg/mL BSA, pH 7.9	>90% of the DNA fragments ligated and recut after 50-fold overdigestion	Fermentas ER0231 ER0232
4000-8000 U/mL		100 mM KCl, 10 mM Tris-HCl, 0.1 mM EDTA, 1 mM DTT, 1 mM BSA, 50% glycerol, pH 7.4		Stratagene 500445 500446

Eam1105 I

RECOGNITION SEQUENCE
    5'...GACNNN▼NNGTC...3'
    3'...CTGNN▲NNNCAG...5'
OPTIMUM REACTION CONDITIONS
    37°C
HEAT INACTIVATION
    65°C for 20 min.
METHYLATION EFFECTS
    Cleaves GAm^{5}C(N)$_5$GTm^{5}C

NOTES
- EC 3.1.21.4
- A type II site specific deoxyribonuclease
- A large group of enzymes recognizing specific short DNA sequences. They cleave either within the recognition site, or at a short specific distance from it
- Large excess of enzyme results in the appearance of star activity
- Requires Mg^{2+} and BSA for full activity

CONCENTRATION or VOLUME ACTIVITY	UNIT DEFINITION	REACTION BUFFER (As Provided)	ADDITIONAL ACTIVITIES and PURITY	SUPPLIER CATALOG NO.
Enterobacter amnigenus RFL1105				
10,000 U/mL	1 unit degrades 1 μg λ dam⁻ dcm⁻ DNA/hr at 37°C in a 0.05 mL mixture	10 mM Tris-HCl, 100 mM NaCl, 100 μg/mL BSA, pH 8.0	No detectable contaminating nuclease; 90% ligated and 100% recut after 50-fold overdigestion	AGS Heidelb F00265S F00265M
4-8 U/μL	1 unit hydrolyzes 1 μg DNA/hr at 37°C in a 50 μL reaction volume using λ phage DNA as substrate	10 mM Tris-HCl, 5 mM MgCl$_2$, 100 mM NaCl, 0.1 mg/mL BSA, pH 7.5	>80% of the DNA fragments ligated and >90% recut after 10-fold overdigestion	Fermentas ER0241 ER0242
50 U and 250 U	1 unit completely digests 1 μg λ DNA/hr at 37°C in 50 μL of supplied buffer	200 mM Tris-HCl, 100 mM MgCl$_2$, 10 mM DTT, 1 M KCl, pH 8.5	Purity is confirmed by overdigestion and ligation-recutting test	TaKaRa 1124

Ear I

RECOGNITION SEQUENCE
　5'...CTCTTC(N)$_1$▼...3'
　3'...GAGAAG(N)$_4$▲...5'

OPTIMUM REACTION CONDITIONS
　37°C

HEAT INACTIVATION
　65°C for 20 min.

METHYLATION EFFECTS
　Does not cleave Gm^{6}AAGAG, GAAGm^{6}AG or m^{5}CTm^{5}CTTm^{5}C

NOTES
- EC 3.1.21.4
- A type II site specific deoxyribonuclease
- A large group of enzymes recognizing specific short DNA sequences. They cleave either within the recognition site, or at a short specific distance from it
- Requires Mg^{2+}

CONCENTRATION or VOLUME ACTIVITY	UNIT DEFINITION	REACTION BUFFER (As Provided)	ADDITIONAL ACTIVITIES and PURITY	SUPPLIER CATALOG No.
Enterobacter aerogenes				
1000-5000 U/mL		10 m*M* Tris-HCl, 0.1 m*M* EDTA, 1 m*M* DTT, 200 µg/mL acetylated BSA, 50% glycerol, pH 7.4		ICN 159433
***Enterobacter aerogenes* (NEB 450)**				
5000 U/mL	1 unit completely digests 1 µg λ DNA/hr in a total reaction volume of 50 µL using the buffer provided	10 m*M* Bis Tris-Propane-HCl, 10 m*M* MgCl$_2$, 1 m*M* DTT, pH 7.0	>95% DNA fragments ligated and recut after 2-fold overdigestion	NE Biolabs 528S 528L

Ecl136 II

RECOGNITION SEQUENCE
5'...GAG▼CTC...3'
3'...CTC▲GAG...5'

ISOSCHIZOMERS
Sac I

OPTIMUM REACTION CONDITIONS
37°C

HEAT INACTIVATION
65°C for 20 min.

METHYLATION EFFECTS
Does not cleave GAGCTm^{5}C
Blocked by overlapping CG methylation

NOTES
- EC 3.1.21.4
- A type II site specific deoxyribonuclease
- A large group of enzymes recognizing specific short DNA sequences. They cleave either within the recognition site, or at a short specific distance from it
- Produces DNA fragments with blunt ends; Sac I does not
- Requires Mg^{2+} and BSA for full activity

CONCENTRATION or VOLUME ACTIVITY	UNIT DEFINITION	REACTION BUFFER (As Provided)	ADDITIONAL ACTIVITIES and PURITY	SUPPLIER CATALOG No.
Enterobacter cloacae RFL136				
10,000 U/mL	1 unit degrades 1 µg λ dam⁻ dcm⁻ DNA Sma I-fragments/hr at 37°C in a 0.05 mL mixture	33 mM Tris-acetate, 10 mM MgOAc, 66 mM KOAc, 100 µg/mL BSA, pH 7.9	No detectable contaminating nuclease; 90% ligated and 100% recut after 50-fold overdigestion	AGS Heidelb F00540S F00540M
8-12 U/µL	1 unit hydrolyzes 1 µg DNA/hr at 37°C in a 50 µL reaction volume using λ phage DNA as substrate	10 mM Bis Tris-propane-HCl, 10 mM MgCl₂, 1 mM DTT, 0.1 mg/mL BSA, pH 6.5	>90% of the DNA fragments ligated and recut after 50-fold overdigestion	Fermentas ER0251 ER0252
12,000 U/mL	1 unit hydrolyzes 1 µg DNA/hr at 37°C in a 50 µL reaction volume using λ phage DNA as substrate	33 mM Tris-acetate, 10 mM MgOAc, 66 mM KOAc, 100 µg/mL BSA, pH 7.9	>90% of the DNA fragments ligated and recut after 50-fold overdigestion	NE Biolabs ER0251S ER0251L

Ecl HK I

RECOGNITION SEQUENCE
5'...GACNNN▼NNGTC...3'
3'...CTGNN▲NNNCAG...5'

NOTES
- EC 3.1.21.4
- A type II site specific deoxyribonuclease
- A large group of enzymes recognizing specific short DNA sequences. They cleave either within the recognition site, or at a short specific distance from it
- Produces ambiguous single base extensions which are difficult to ligate
- Requires Mg^{2+}

CONCENTRATION or VOLUME ACTIVITY	UNIT DEFINITION	REACTION BUFFER (As Provided)	ADDITIONAL ACTIVITIES and PURITY	SUPPLIER CATALOG No.
Enterobacter cloacae				
10 U/μL	1 unit completely digests 1 μg λ DNA/hr at 37°C in a 50 mL reaction volume	60 mM Tris-HCl, 60 mM MgCl$_2$, 1 M NaCl, 10 mM DTT, pH 7.5	10% ligation	Promega R7111 R7112

Ecl X I

RECOGNITION SEQUENCE
5'...C▼GGCCG...3'
3'...GCCGG▲C...5'

ISOSCHIZOMERS
Xma III

HEAT INACTIVATION
65°C for 20 min.

METHYLATION EFFECTS
Does not cleave m^{5}CGGCm^{5}CG or CGGm^{5}CCG

NOTES
- EC 3.1.21.4
- A type II site specific deoxyribonuclease
- A large group of enzymes recognizing specific short DNA sequences. They cleave either within the recognition site, or at a short specific distance from it
- Requires Mg^{2+}

EcIX I continued

CONCENTRATION or VOLUME ACTIVITY	UNIT DEFINITION	REACTION BUFFER (As Provided)	ADDITIONAL ACTIVITIES and PURITY	SUPPLIER CATALOG No.
Enterobacter cloacae 590				
10 U/μL and 40 U/μL	1 unit completely digests 1 μg substrate DNA/hr	10 mM Tris-HCl, 5 mM MgCl$_2$, 100 mM NaCl, 1 mM β-MSH, pH 8.0		Boehringer 1131389 1131397 1131427

Eco24 I

RECOGNITION SEQUENCE
   5'...GPuGCPy▼C...3'
   3'...C▲PyCGPuG...5'
ISOSCHIZOMERS
   HgiJ II
OPTIMUM REACTION CONDITIONS
   37°C
HEAT INACTIVATION
   65°C for 20 min.

NOTES
- EC 3.1.21.4
- A type II site specific deoxyribonuclease
- A large group of enzymes recognizing specific short DNA sequences. They cleave either within the recognition site, or at a short specific distance from it
- Requires Mg^{2+} and BSA for full activity

CONCENTRATION or VOLUME ACTIVITY	UNIT DEFINITION	REACTION BUFFER (As Provided)	ADDITIONAL ACTIVITIES and PURITY	SUPPLIER CATALOG No.
Escherichia coli RFL24				
10,000 U/mL	1 unit degrades 1 μg λ dam⁻ dcm⁻ DNA/hr at 37°C in a 0.05 mL mixture	33 mM Tris-acetate, 10 mM MgOAc, 66 mM KOAc, 100 μg/mL BSA, pH 7.9	No detectable contaminating nuclease; 90% ligated and 100% recut after 50-fold overdigestion	AGS Heidelb F00130S F00130M
8-12 U/μL	1 unit hydrolyzes 1 μg DNA/hr at 37°C in a 50 μL reaction volume using λ phage DNA as substrate	33 mM Tris-acetate, 10 mM MgOAc, 66 mM KOAc, 0.1 mg/mL BSA, pH 7.9	>90% of the DNA fragments ligated and recut after 50-fold overdigestion	Fermentas ER0281 ER0282

Eco31 I

RECOGNITION SEQUENCE
  5'...GGTCTC(N)$_1$▼...3'
  3'...CCAGAG(N)$_5$▲...5'

OPTIMUM REACTION CONDITIONS
  37°C

HEAT INACTIVATION
  65°C for 20 min.

METHYLATION EFFECTS
  Cleaves GAGACm^{5}C at least 50-fold slower than unmethylated site
  Does not cleave GGTm^{5}CTC or GAGm^{6}ACC
  Blocked by overlapping CG methylation

NOTES
- EC 3.1.21.4
- A type II site specific deoxyribonuclease
- A large group of enzymes recognizing specific short DNA sequences. They cleave either within the recognition site, or at a short specific distance from it
- Large excess of enzyme results in the appearance of star activity
- Requires Mg^{2+} and BSA for full activity

CONCENTRATION or VOLUME ACTIVITY	UNIT DEFINITION	REACTION BUFFER (As Provided)	ADDITIONAL ACTIVITIES and PURITY	SUPPLIER CATALOG No.
Escherichia coli RFL31				
5000 U/mL	1 unit degrades 1 µg λ dam$^-$ dcm$^-$ DNA/hr at 37°C in a 0.05 mL mixture	33 mM Tris-acetate, 10 mM MgOAc, 66 mM KOAc, 100 µg/mL BSA, pH 7.9	No detectable contaminating nuclease; 90% ligated and 90% recut after 50-fold overdigestion	AGS Heidelb F00270S F00270M
Escherichia coli, carrying the cloned *Eco31IR* gene from *E. coli* RFL31				
8-12 U/µL	1 unit hydrolyzes 1 µg DNA/hr at 37°C in a 50 µL reaction volume using λ phage DNA as substrate	10 mM Tris-HCl, 10 mM MgCl$_2$, 50 mM NaCl, 0.1 mg/mL BSA, pH 7.5	>90% of the DNA fragments ligated and recut after 50-fold overdigestion	Fermentas ER0291 ER0292

Eco32 I

RECOGNITION SEQUENCE
  5'...GAT▼ATC...3'
  3'...CTA▲TAG...5'

ISOSCHIZOMERS
  *Eco*R V

OPTIMUM REACTION CONDITIONS
  37°C

HEAT INACTIVATION
  65°C for 20 min.

NOTES
- EC 3.1.21.4
- A type II site specific deoxyribonuclease
- A large group of enzymes recognizing specific short DNA sequences. They cleave either within the recognition site, or at a short specific distance from it
- Requires Mg^{2+}

CONCENTRATION or VOLUME ACTIVITY	UNIT DEFINITION	REACTION BUFFER (As Provided)	ADDITIONAL ACTIVITIES and PURITY	SUPPLIER CATALOG NO.
Escherichia coli RFL32				
8-12 U/μL; 40-60 U/μL	1 unit hydrolyzes 1 μg DNA/hr at 37°C in a 50 μL reaction volume using λ phage DNA as substrate	10 m*M* Tris-HCl, 10 m*M* MgCl₂, 100 m*M* KCl, pH 8.5, 37 °C	>90% of the DNA fragments ligated and recut after 50-fold overdigestion	Fermentas ER0301 ER0302 ER0303

Eco47 I

RECOGNITION SEQUENCE
5'...G▼G(^A_T)CC...3'
3'...CC(^T_A)G▲G...5'

ISOSCHIZOMERS
Ava II

OPTIMUM REACTION CONDITIONS
pH 9.0 and 37°C

HEAT INACTIVATION
65°C for 20 min.

METHYLATION EFFECTS
Does not cleave GG(^A_T)Cm^{5}C
Blocked by overlapping *dcm* and CG methylation

NOTES
- EC 3.1.21.4
- A type II site specific deoxyribonuclease
- A large group of enzymes recognizing specific short DNA sequences. They cleave either within the recognition site, or at a short specific distance from it
- Requires Mg^{2+}

CONCENTRATION or VOLUME ACTIVITY	UNIT DEFINITION	REACTION BUFFER (As Provided)	ADDITIONAL ACTIVITIES and PURITY	SUPPLIER CATALOG NO.
Escherichia coli RFL47				
3000-20,000 U/mL		100 m*M* KCl, 10 m*M* Tris-HCl, 0.1 m*M* EDTA, 1 m*M* DTT, 200 μg/mL acetylated BSA, 50% glycerol, pH 7.4		ICN 159435
3-20 U/μL	1 unit completely digests 1 μg DNA/hr at 37°C	50 m*M* Tris-HCl, 10 m*M* MgCl$_2$, 100 m*M* NaCl, 1 m*M* DTT, pH 7.5	>90% ligation; >95% recut	Toyobo E47-101 E47-102
Escherichia coli, carrying the cloned *Eco47*IR gene from *E.coli* RFL47				
8-12 U/μL	1 unit hydrolyzes 1 μg DNA/hr at 37°C in a 50 μL reaction volume using λ phage DNA as substrate	10 m*M* Tris-HCl, 10 m*M* MgCl$_2$, 100 m*M* KCl, pH 8.5	>90% of the DNA fragments ligated and recut after 50-fold overdigestion	Fermentas ER0311 ER0312

Eco47 III

RECOGNITION SEQUENCE
5'...AGC▼GCT...3'
3'...TCG▲CGA...5'

ISOSCHIZOMERS
Aor51H I

OPTIMUM REACTION CONDITIONS
pH 8.5 and 37°C

HEAT INACTIVATION
65°C for 20 min.

METHYLATION EFFECTS
Cleaves m^{6}AGCGCT
Does not cleave AGm^{5}CGCT
Blocked by CG methylation

NOTES
- EC 3.1.21.4
- A type II site specific deoxyribonuclease
- A large group of enzymes recognizing specific short DNA sequences. They cleave either within the recognition site, or at a short specific distance from it
- High enzyme concentration in digestions results in star activity
- Compatible cohesive ends: any blunt end
- Requires Mg^{2+}

CONCENTRATION or VOLUME ACTIVITY	UNIT DEFINITION	REACTION BUFFER (As Provided)	ADDITIONAL ACTIVITIES and PURITY	SUPPLIER CATALOG NO.
Escherichia coli RFL 47				
5000 U/mL	1 unit degrades 1 μg λ dam$^-$ dcm$^-$ DNA/hr at 37°C in a 0.05 mL mixture	10 mM Tris-HCl, 10 mM MgCl$_2$, 100 mM KCl, 100 μg/mL BSA, pH 8.5	No detectable contaminating nuclease; 80% ligated and 90% recut after 10-fold overdigestion	AGS Heidelb F00280S F00280M
5 U/μL	1 unit completely digests 1 μg substrate DNA/hr	50 mM Tris-HCl, 10 mM MgCl$_2$, 100 mM NaCl, 1 mM DTE, pH 7.5		Boehringer 1167103
8-12 U/μL; 40-60 U/μL	1 unit hydrolyzes 1 μg DNA/hr at 37°C in a 50 μL reaction volume using λ phage DNA as substrate	50 mM Tris-HCl, 10 mM MgCl$_2$, 100 mM NaCl, pH 7.5	>80% of pBR322 DNA fragments ligated and >90% recut after 50-fold overdigestion	Fermentas ER0321 ER0322 ER0323
1000-10,000 U/mL		In reaction buffer		ICN 159436
4-6 U/μL	1 unit completely digests 1 mg λ DNA/hr at 37°C	500 mM Tris-HCl, 100 mM MgCl$_2$, 1 M NaCl, pH 8.0	≥65% ligation; 100% cleavage	Life Technol 15472-012
2-6 U/μL	1 unit digests completely 1 μg λ DNA substrate/hr at 37°C	50 mM Tris-HCl, 100 mM NaCl, 10 mM MgCl$_2$, 1 mM DTT, pH 7.8		NBL Gene 017811
8000 U/mL	1 unit hydrolyzes 1 μg DNA/hr at 37°C in a 50 μL reaction volume using λ phage DNA as substrate	50 mM Tris-HCl, 10 mM MgCl$_2$, 100 mM NaCl, pH 7.5	>80% of DNA fragments ligated and >90% recut after 50-fold overdigestion	NE Biolabs ER0321S ER0321L

Eco47 III continued

CONCENTRATION or VOLUME ACTIVITY	UNIT DEFINITION	REACTION BUFFER (As Provided)	ADDITIONAL ACTIVITIES and PURITY	SUPPLIER CATALOG NO.
Escherichia coli RFL 47 continued				
2-5 U/μL	1 unit completely digests 1 μg λ DNA/hr at 37°C in a 50 mL reaction volume	Genome qualified; 60 mM Tris-HCl, 60 mM $MgCl_2$, 1.5 M NaCl, 10 mM DTT, pH 7.9	80% ligation	Promega R6731 R6732
4000-6000 U/mL		100 mM KCl, 10 mM Tris-HCl, 1 mM EDTA, 1 mM DTT, 200 μg/mL BSA, 50% glycerol, pH 7.4		Stratagene 500454
3-20 U/μL	1 unit completely digests 1 μg DNA/hr at 37°C	50 mM Tris-HCl, 10 mM $MgCl_2$, 100 mM NaCl, 1 mM DTT, pH 7.5	>90% ligation; >90% recut	Toyobo E47-301 E47-302

Eco52 I

RECOGNITION SEQUENCE
  5'...C▼GGCCG...3'
  3'...GCCGG▲C...5'
ISOSCHIZOMERS
  Eag I, EclX I, Xma III
OPTIMUM REACTION CONDITIONS
  pH 9.0 and 37°C
HEAT INACTIVATION
  60°C for 15 min.
METHYLATION EFFECTS
  Blocked by CG methylation

NOTES
- EC 3.1.21.4
- A type II site specific deoxyribonuclease
- A large group of enzymes recognizing specific short DNA sequences. They cleave either within the recognition site, or at a short specific distance from it
- Compatible cohesive ends: BstZ I, Cfr I, Eae I, Eag I, EclX I, Not I, Xma III
- Requires Mg^{2+} and BSA for full activity

CONCENTRATION or VOLUME ACTIVITY	UNIT DEFINITION	REACTION BUFFER (As Provided)	ADDITIONAL ACTIVITIES and PURITY	SUPPLIER CATALOG NO.
Escherichia coli RFL52				
8-12 U/μL	1 unit completely digests 1 μg λ DNA/hr at 37°C in the buffer provided	Eco52 I basal buffer	>95% DNA fragments ligated and 100% recut	Amersham E 1039Z

Eco52 I continued

CONCENTRATION or VOLUME ACTIVITY	UNIT DEFINITION	REACTION BUFFER (As Provided)	ADDITIONAL ACTIVITIES and PURITY	SUPPLIER CATALOG NO.
Escherichia coli RFL52 continued				
10,000 U/mL	1 unit degrades 1 µg λ dam⁻ dcm⁻ DNA/hr at 37°C in a 0.05 mL mixture	10 mM Tris-HCl, 3 mM MgCl$_2$, 100 mM NaCl, 100 µg/mL BSA, pH 8.5	No detectable contaminating nuclease; 90% ligated and 90% recut after 50-fold overdigestion	AGS Heidelb F00680S F00680M
8-12 U/µL	1 unit hydrolyzes 1 µg DNA/hr at 37°C in a 50 µL reaction volume using λ phage DNA as substrate	10 mM Tris-HCl, 3 mM MgCl$_2$, 100 mM NaCl, 0.1 mg/mL BSA, pH 8.5	>90% of the DNA fragments ligated and recut after 50-fold overdigestion	Fermentas ER0331 ER0332
1000-5000 U/mL		50 mM NaCl, 10 mM Tris-HCl, 0.1 mM EDTA, 1 mM DTT, 500 µg/mL acetylated BSA, 50% glycerol, pH 7.4		ICN 159437
1-5 U/µL	1 unit completely digests 1 µg EcoR I λ DNA fragments/hr at 37°C in a 50 mL reaction volume	Blue/White Cloning Qualified; genome qualified; 100 mM Tris-HCl, 30 mM MgCl$_2$, 1 M NaCl, pH 9.0	90% ligation	Promega R6751 R6752
100 U and 500 U	1 unit completely digests 1 µg λ DNA/hr at 37°C in 50 µL of supplied buffer	10 mM Tris-HCl, 3 mM MgCl$_2$, 100 mM NaCl, 0.01% BSA, pH 8.9	Purity is confirmed by overdigestion, ligation-recutting and cloning test	TaKaRa 1039
3-20 U/µL	1 unit completely digests 1 µg DNA/hr at 37°C	50 mM Tris-HCl, 10 mM MgCl$_2$, 100 mM NaCl, 1 mM DTT, pH 7.5	>90% ligation; >95% recut	Toyobo E52-101 E52-102

Eco57 I

RECOGNITION SEQUENCE
5'...CTGAAG(N)$_{16}$▼...3'
3'...GACTTC(N)$_{14}$▲...5'

ISOSCHIZOMERS
Xma III

OPTIMUM REACTION CONDITIONS
37°C

HEAT INACTIVATION
65°C for 20 min.

METHYLATION EFFECTS
Does not cleave CTGAm^{6}AG or CTTCm^{6}AG

NOTES
- EC 3.1.21.4
- A type II site specific deoxyribonuclease
- A large group of enzymes recognizing specific short DNA sequences. They cleave either within the recognition site, or at a short specific distance from it
- DNA cleavage by this enzyme is never complete
- Mg^{2+} and BSA are required for full activity; stimulated 100-fold by 10 m*M* S-adenosylmethionine

CONCENTRATION or VOLUME ACTIVITY	UNIT DEFINITION	REACTION BUFFER (As Provided)	ADDITIONAL ACTIVITIES and PURITY	SUPPLIER CATALOG No.
Escherichia coli RFL57				
6000 U/mL	1 unit degrades 1 µg λ dam$^-$ dcm$^-$ DNA/hr at 37°C in a 0.05 mL mixture	33 m*M* Tris-acetate, 10 m*M* MgOAc, 66 m*M* KOAc, 100 µg/mL BSA, pH 7.9	90% ligated and 0% recut after 10-fold overdigestion	AGS Heidelb F00290S F00290M
1000-15,000 U/mL		In reaction buffer		ICN 159438 Sigma
Escherichia coli, carrying the cloned *Eco*57IR gene from *E.coli* RFL57				
4-8 U/µL	1 unit hydrolyzes 1 µg DNA/hr at 37°C in a 50 µL reaction volume using λ phage DNA as substrate	10 m*M* Tris-HCl, 10 m*M* MgCl$_2$, 0.1 mg/mL BSA, 0.01 m*M* SAM, pH 7.5	70% of the DNA fragments ligated and none recut after 2- to 10-fold overdigestion	Fermentas ER0341 ER0342
6000 U/mL	1 unit hydrolyzes 1 µg DNA/hr at 37°C in a 50 µL reaction volume using λ phage DNA as substrate	10 m*M* Tris-HCl, 10 m*M* MgCl$_2$, 0.1 mg/mL BSA, 0.01 m*M* SAM, pH 7.5	>90% of the DNA fragments ligated and none recut after 10-fold overdigestion	NE Biolabs ER0341S ER0341L

Eco64 I

RECOGNITION SEQUENCE
  5'...G▼GPyPuCC...3'
  3'...CCPuPyG▲G...5'

ISOSCHIZOMERS
  HgiC I, BshN I

OPTIMUM REACTION CONDITIONS
  37°C

HEAT INACTIVATION
  Resistant

METHYLATION EFFECTS
  Cleavage rate slows significantly by overlapping dcm methylation
  Blocked by overlapping dcm and CG methylation

NOTES
- EC 3.1.21.4
- A type II site specific deoxyribonuclease
- A large group of enzymes recognizing specific short DNA sequences. They cleave either within the recognition site, or at a short specific distance from it
- GGCGCC sites in λ and pBR322 DNAs are cleaved 100-fold more slowly than remaining sites
- Demonstrates marked site preference; BshN I does not
- Requires Mg^{2+} and BSA for full activity

CONCENTRATION or VOLUME ACTIVITY	UNIT DEFINITION	REACTION BUFFER (As Provided)	ADDITIONAL ACTIVITIES and PURITY	SUPPLIER CATALOG No.
Escherichia coli RFL64				
10,000 U/mL	1 unit degrades 1 μg λ dam⁻ dcm⁻ DNA/hr at 37°C in a 0.05 mL mixture	10 mM Tris-HCl, 10 mM $MgCl_2$, 100 mM KCl, 100 μg/mL BSA, pH 8.5	No detectable contaminating nuclease; 90% ligated and 100% recut after 25-fold overdigestion	AGS Heidelb F00120S F00120M
Escherichia coli, **carrying the cloned** *Eco64IR* **gene from** *E. coli* **RFL64**				
8-12 U/μL	1 unit hydrolyzes 1 μg DNA/hr at 37°C in a 50 μL reaction volume using λ phage DNA as substrate	10 mM Tris-HCl, 10 mM $MgCl_2$, 100 mM KCl, 0.1 mg/mL BSA, pH 8.5	>90% of the DNA fragments ligated and recut after 50-fold overdigestion	Fermentas ER0351 ER0352

Eco72 I

RECOGNITION SEQUENCE
5'...CAC▼GTG...3'
3'...GAG▲CAC...5'

ISOSCHIZOMERS
*Bbr*P I, *Pma*C I, *Pml* I

OPTIMUM REACTION CONDITIONS
37°C

HEAT INACTIVATION
65°C for 20 min.

METHYLATION EFFECTS
Blocked by CG methylation

NOTES
- EC 3.1.21.4
- A type II site specific deoxyribonuclease
- A large group of enzymes recognizing specific short DNA sequences. They cleave either within the recognition site, or at a short specific distance from it
- Low salt or excess enzyme digestion may result in star activity
- Compatible cohesive ends: any blunt end
- Requires Mg^{2+}

CONCENTRATION or VOLUME ACTIVITY	UNIT DEFINITION	REACTION BUFFER (As Provided)	ADDITIONAL ACTIVITIES and PURITY	SUPPLIER CATALOG NO.
Escherichia coli RFL72				
10,000 U/mL	1 unit degrades 1 µg λ dam⁻ dcm⁻ DNA/hr at 37°C in a 0.05 mL mixture	33 m*M* Tris-acetate, 10 m*M* MgOAc, 66 m*M* KOAc, 100 µg/mL BSA, pH 7.9	No detectable contaminating nuclease; 90% ligated and 100% recut after 50-fold overdigestion with pBR322-*Hind* III DNA fragments as substrate	AGS Heidelb F00490S F00490M
5000-12,000 U/mL		100 m*M* KCl, 10 m*M* Tris-HCl, 0.1 m*M* EDTA, 1 m*M* DTT, 200 µg/mL acetylated BSA, 50% glycerol, pH 7.4		ICN 159439
8-12 U/µL	1 unit completely digests 1 µg λ DNA/hr at 37°C in a 50 mL reaction volume	100 m*M* Tris-HCl, 100 m*M* MgCl₂, 1.5 *M* KCl, pH 7.4	90% ligation	Promega R6981 R6982
Escherichia coli, carrying the cloned *Eco*72IR gene from *E.coli* RFL72				
8-12 U/µL	1 unit hydrolyzes 1 µg DNA/hr at 37°C in a 50 µL reaction volume using λ phage DNA as substrate	33 m*M* Tris-acetate, 10 m*M* MgOAc, 66 m*M* KOAc, pH 7.9	>80% of DNA fragments ligated and >90% recut after 10-fold overdigestion	Fermentas ER0361 ER0362
Esherichia coli, cloned				
5000-12,000 U/mL		100 m*M* KCl, 10 m*M* Tris-HCl, 1 m*M* EDTA, 1 m*M* DTT, 200 µg/mL BSA, 50% glycerol, pH 7.4		Stratagene 500474 500475

Eco81 I

RECOGNITION SEQUENCE
5'...CC▼TNAGG...3'
3'...GGANT▲CC...5'

ISOSCHIZOMERS
*Bsu*36 I, *Cvn* I, *Sau* I

OPTIMUM RECTION CONDITIONS
pH 8.5 and 37°C

HEAT INACTIVATION
Resistant; 70°C for 15 min.

NOTES
- EC 3.1.21.4
- A type II site specific deoxyribonuclease
- A large group of enzymes recognizing specific short DNA sequences. They cleave either within the recognition site, or at a short specific distance from it
- Requires Mg^{2+}

CONCENTRATION or VOLUME ACTIVITY	UNIT DEFINITION	REACTION BUFFER (As Provided)	ADDITIONAL ACTIVITIES and PURITY	SUPPLIER CATALOG NO.
Escherichia coli RFL81				
8-12 U/μL	1 unit completely digests 1 μg λ DNA/hr at 37°C in the buffer provided	100 m*M* Tris-HCl, 100 m*M* MgCl$_2$, 10 m*M* DTT, 500 m*M* NaCl, pH 7.5	0% DNA fragments ligated and recut	Amersham E 1131Y E 1131Z
10,000 U/mL	1 unit degrades 1 μg λ dam⁻ dcm⁻ DNA/hr at 37°C in a 0.05 mL mixture	10 m*M* Tris-HCl, 10 m*M* MgCl$_2$, 100 μg/mL BSA, pH 7.5	No detectable contaminating nuclease; 90% ligated and >90% recut after incubation with pBR322-*Hind* III fragments	ACS Heidelb F00210S F00210M
8-12 U/μL	1 unit hydrolyzes 1 μg DNA/hr at 37°C in a 50 μL reaction volume using λ phage DNA as substrate	33 m*M* Tris-acetate, 10 m*M* MgOAc, 66 m*M* KOAc, pH 7.9	>80% of DNA fragments ligated and >90% recut after 10-fold overdigestion	Fermentas ER0371 ER0372
200 U and 1000 U	1 unit completely digests 1 μg λ DNA/hr at 37°C in 50 μL of supplied buffer	100 m*M* Tris-HCl, 100 m*M* MgCl$_2$, 10 m*M* DTT, 500 m*M* NaCl, pH 7.5	Purity is confirmed by overdigestion and ligation-recutting test	TaKaRa 1131
3-20 U/μL	1 unit completely digests 1 μg DNA/hr at 37°C	10 m*M* Tris-HCl, 10 m*M* MgCl$_2$, 1 m*M* DTT, pH 7.5		Toyobo E81-101 E81-102

Eco88 I

RECOGNITION SEQUENCE
    5'...C▼PyCGPuG...3'
    3'...GPuGCPy▲C...5'

ISOSCHIZOMERS
    Ava I

OPTIMUM REACTION CONDITIONS
    4°C; 80-100% activity at 37°C

HEAT INACTIVATION
    65°C for 20 min.

NOTES
- EC 3.1.21.4
- A type II site specific deoxyribonuclease
- A large group of enzymes recognizing specific short DNA sequences. They cleave either within the recognition site, or at a short specific distance from it
- Large excess of enzyme results in the appearance of star activity, as does incubation at 37°C
- Requires Mg^{2+}

CONCENTRATION or VOLUME ACTIVITY	UNIT DEFINITION	REACTION BUFFER (As Provided)	ADDITIONAL ACTIVITIES and PURITY	SUPPLIER CATALOG NO.
Escherichia coli, carrying the cloned *Eco*88IR gene from *E.coli* RFL88				
8-12 U/μL	1 unit hydrolyzes 1 μg DNA/hr at 37°C in a 50 μL reaction volume using λ phage DNA as substrate	10 mM Tris-HCl, 10 mM MgCl₂, pH 7.5	>90% of the DNA fragments ligated and recut after 50-fold overdigestion	Fermentas ER0381 ER0382

Eco91 I

RECOGNITION SEQUENCE
    5'...G▼GTNACC...3'
    3'...CCANTG▲G...5'

ISOSCHIZOMERS
    BstE II

OPTIMUM REACTION CONDITIONS
    37°C

HEAT INACTIVATION
    65°C for 20 min.

NOTES
- EC 3.1.21.4
- A type II site specific deoxyribonuclease
- A large group of enzymes recognizing specific short DNA sequences. They cleave either within the recognition site, or at a short specific distance from it
- Requires Mg^{2+} and BSA for full activity

Eco91 I continued

CONCENTRATION or VOLUME ACTIVITY	UNIT DEFINITION	REACTION BUFFER (As Provided)	ADDITIONAL ACTIVITIES and PURITY	SUPPLIER CATALOG No.
Escherichia coli RFL91				
8-12 U/μL	1 unit hydrolyzes 1 μg DNA/hr at 37°C in a 50 μL reaction volume using λ phage DNA as substrate	50 mM Tris-HCl, 10 mM MgCl$_2$, 100 mM NaCl, 0.1 mg/mL BSA, pH 7.5	>90% of the DNA fragments ligated and recut after 50-fold overdigestion	Fermentas ER0391 ER0392

Eco105 I

RECOGNITION SEQUENCE
 5'...TAC▼GTA...3'
 3'...ATG▲CAT...5'
ISOSCHIZOMERS
 *Sna*B I
OPTIMUM REACTION CONDITIONS
 pH 7.5
HEAT INACTIVATION
 65°C for 15 min.

NOTES
- EC 3.1.21.4
- A type II site specific deoxyribonuclease
- A large group of enzymes recognizing specific short DNA sequences. They cleave either within the recognition site, or at a short specific distance from it
- May exhibit star activity at high enzyme or glycerol concentration
- Requires Mg^{2+}

CONCENTRATION or VOLUME ACTIVITY	UNIT DEFINITION	REACTION BUFFER (As Provided)	ADDITIONAL ACTIVITIES and PURITY	SUPPLIER CATALOG No.
Escherichia coli RFL105				
5000 U/mL	1 unit degrades 1 μg λ dam⁻ dcm⁻ DNA/hr at 37°C in a 0.05 mL mixture	12.5 mM Tris-acetate, 50 mM K-glycinate, 5 mM MgOAc, 100 μg/mL BSA, pH 7.5	No detectable contaminating nuclease; 90% ligated and 90% recut after 2-fold overdigestion	AGS Heidelb F00600S F00600M
3-20 U/μL	1 unit completely digests 1 μg DNA/hr at 37°C	10 mM Tris-HCl, 10 mM MgCl$_2$, 1 mM DTT, pH 7.5	>90% ligation; >90% recut	Toyobo E15-101 E15-102
Escherichia coli, carrying the cloned *Eco*105IR gene from *E.coli* RFL105				
8-12 U/μL	1 unit hydrolyzes 1 μg DNA/hr at 37°C in a 50 μL reaction volume using λ phage DNA as substrate	10 mM Tris-HCl, 10 mM MgCl$_2$, 0.1 mg/mL BSA, pH 7.5	>70% of the phage fd DNA fragments ligated and recut after 2-fold overdigestion	Fermentas ER0401 ER0402

Eco130 I

RECOGNITION SEQUENCE
5'...C▼C(^A_T)(^A_T)GG...3'
3'...GG(^T_A)(^T_A)C▲C...5'

NOTES
- EC 3.1.21.4
- A type II site specific deoxyribonuclease
- A large group of enzymes recognizing specific short DNA sequences. They cleave either within the recognition site, or at a short specific distance from it
- Requires Mg^{2+}

CONCENTRATION or VOLUME ACTIVITY	UNIT DEFINITION	REACTION BUFFER (As Provided)	ADDITIONAL ACTIVITIES and PURITY	SUPPLIER CATALOG No.
Escherichia coli RFL130				
10,000 U/mL	1 unit degrades 1 μg λ dam⁻ dcm⁻ DNA/hr at 37°C in a 0.05 mL mixture	50 m*M* Tris-HCl, 10 m*M* MgCl$_2$, 100 m*M* NaCl, 100 μg/mL BSA, pH 7.5	No detectable contaminating nuclease; 90% ligated and 100% recut after 50-fold overdigestion	AGS Heidelb F00630S F00630M
8-12 U/μL	1 unit hydrolyzes 1 μg DNA/hr at 37°C in a 50 μL reaction volume using λ phage DNA as substrate	50 m*M* Tris-HCl, 10 m*M* MgCl$_2$, 100 m*M* NaCl, pH 7.5	>90% of the DNA fragments ligated and recut after 50-fold overdigestion	Fermentas ER0411 ER0412
1000-5000 U/mL		50 m*M* KCl, 10 m*M* Tris-HCl, 0.1 m*M* EDTA, 1 m*M* DTT, 200 μg/mL acetylated BSA, 50% glycerol, pH 7.4		ICN 159434

Eco147 I

RECOGNITION SEQUENCE
5'...AGG▼CCT...3'
3'...TCC▲GGA...5'

NOTES
- EC 3.1.21.4
- A type II site specific deoxyribonuclease
- A large group of enzymes recognizing specific short DNA sequences. They cleave either within the recognition site, or at a short specific distance from it
- Requires Mg^{2+}

Eco147 I continued

CONCENTRATION or VOLUME ACTIVITY	UNIT DEFINITION	REACTION BUFFER (As Provided)	ADDITIONAL ACTIVITIES and PURITY	SUPPLIER CATALOG NO.
Escherichia coli RFL147				
10,000 U/mL	1 unit degrades 1 µg λ dam⁻ dcm⁻ DNA/hr at 37°C in a 0.05 mL mixture	33 mM Tris-acetate, 10 mM MgOAc, 66 mM KOAc, 100 µg/mL BSA, pH 7.9	No detectable contaminating nuclease; 90% ligated and 100% recut after 50-fold overdigestion	AGS Heidelb F00620S F00620M
Escherichia coli, carrying the cloned *Eco147IR* gene from *E.coli* RFL147				
8-12 U/µL	1 unit hydrolyzes 1 µg DNA/hr at 37°C in a 50 µL reaction volume using λ phage DNA as substrate	10 mM Tris-HCl, 10 mM MgCl₂, pH 7.5	>90% of the DNA fragments ligated and recut after 50-fold overdigestion	Fermentas ER0421 ER0422

Eco255 I

RECOGNITION SEQUENCE
  5'...AGT▼ACT...3'
  3'...TCA▲TGA...5'

NOTES
- EC 3.1.21.4
- A type II site specific deoxyribonuclease
- A large group of enzymes recognizing specific short DNA sequences. They cleave either within the recognition site, or at a short specific distance from it
- Requires Mg^{2+}

CONCENTRATION or VOLUME ACTIVITY	UNIT DEFINITION	REACTION BUFFER (As Provided)	ADDITIONAL ACTIVITIES and PURITY	SUPPLIER CATALOG NO.
Escherichia coli, carrying the cloned *Eco255IR* gene from *E.coli* RFL255				
8-12 U/µL	1 unit hydrolyzes 1 µg DNA/hr at 37°C in a 50 µL reaction volume using λ phage DNA as substrate	10 mM Tris-HCl, 10 mM MgCl₂, 0.1 mg/mL BSA, pH 7.5	>80% of the DNA fragments ligated and >90% recut after 50-fold overdigestion	Fermentas ER1051 ER1052

EcoICR I

RECOGNITION SEQUENCE
5'...GAG▼CTC...3'
3'...CTC▲GAG...5'

NOTES
- EC 3.1.21.4
- A type II site specific deoxyribonuclease
- A large group of enzymes recognizing specific short DNA sequences. They cleave either within the recognition site, or at a short specific distance from it
- Compatible cohesive ends: any blunt end
- Requires Mg^{2+}

CONCENTRATION or VOLUME ACTIVITY	UNIT DEFINITION	REACTION BUFFER (As Provided)	ADDITIONAL ACTIVITIES and PURITY	SUPPLIER CATALOG No.
Escherichia coli ICR				
8-12 U/μL and 40-80 U/μL	1 unit completely digests 1 μg EcoR I λ DNA fragments/hr at 37°C in a 50 mL reaction volume	Blue/White Cloning Qualified; 60 m*M* Tris-HCl, 60 m*M* MgCl₂, 500 m*M* NaCl, 10 m*M* DTT, pH 7.5	90% ligation	Promega R6951 R6952 R6954

EcoN I

RECOGNITION SEQUENCE
5'...CCTNN▼NNNAGG...3'
3'...GGANNN▲NNTCC...5'

OPTIMUM REACTION CONDITIONS
37°C

HEAT INACTIVATION
65°C for 20 min.

NOTES
- EC 3.1.21.4
- A type II site specific deoxyribonuclease
- A large group of enzymes recognizing specific short DNA sequences. They cleave either within the recognition site, or at a short specific distance from it
- Produces DNA fragments with a single-base 5' extension which are more difficult to ligate than blunt-ended fragments. High concentration T4 DNA ligase can produce more efficient ligation
- Requires Mg^{2+}

EcoN I continued

CONCENTRATION or VOLUME ACTIVITY	UNIT DEFINITION	REACTION BUFFER (As Provided)	ADDITIONAL ACTIVITIES and PURITY	SUPPLIER CATALOG No.
Escherichia coli				
3-20 U/μL	1 unit completely digests 1 μg λ DNA/hr at 37°C in the buffer provided	330 mM Tris-acetate, 100 mM MgOAc, 5 mM DTT, 660 mM KOAc, pH 7.9	>50% DNA fragments ligated and 50% recut	Amersham E 0211Y E 0211Z
3000-20,000 U/mL		50 mM KCl, 10 mM Tris-HCl, 0.1 mM EDTA, 1 mM DTT, 200 μg/mL acetylated BSA, 50% glycerol, pH 7.5		ICN 159440
5000-25,000 U/mL	1 unit completely cleaves 1.0 μg λ DNA/hr in a 50 μL reaction mixture	200 mM Tris-acetate, 500 mM KOAc, 100 mM MgOAc, 10 mM DTT, pH 7.9	100% cleavage, 60% ligation, 95% recleavage, 70% endonuclease (nickase) No detectable endonuclease (overdigestion)	Sigma R4635
Escherichia coli (NEB 441)				
15,000 U/mL	1 unit completely digests 1 μg λ DNA/hr in a total reaction volume of 50 μL using the buffer provided	20 mM Tris-acetate, 10 mM MgOAc, 50 mM KOAc, 1 mM DTT, pH 7.9	>50% DNA fragments ligated and >95% recut after 10-fold overdigestion	NE Biolabs 521S 521L

EcoO65 I

RECOGNITION SEQUENCE
    5'...G▼GTNACC...3'
    3'...CCANTG▲G...5'

ISOSCHIZOMERS
    BstE II

OPTIMUM REACTION CONDITIONS
    37°C

HEAT INACTIVATION
    70°C for 15 min.

NOTES
- EC 3.1.21.4
- A type II site specific deoxyribonuclease
- A large group of enzymes recognizing specific short DNA sequences. They cleave either within the recognition site, or at a short specific distance from it
- Requires Mg^{2+}

EcoO65 I continued

CONCENTRATION or VOLUME ACTIVITY	UNIT DEFINITION	REACTION BUFFER (As Provided)	ADDITIONAL ACTIVITIES and PURITY	SUPPLIER CATALOG NO.
Escherichia coli K11a				
8-12 U/μL	1 unit completely digests 1 μg λ DNA/hr at 37°C in the buffer provided	500 mM Tris-HCl, 100 mM MgCl$_2$, 10 mM DTT, 1000 mM NaCl, BSA, pH 7.5	>95% DNA fragments ligated and 100% recut	Amersham E 1135Y
1000 U and 5000 U	1 unit completely digests 1 μg λ DNA/hr at 37°C in 50 μL of supplied buffer and BSA	500 mM Tris-HCl, 100 mM MgCl$_2$, 10 mM DTT, 1 M NaCl, BSA, pH 7.5	Purity is confirmed by overdigestion, ligation-recutting and cloning test	TaKaRa 1135

EcoO109 I

RECOGNITION SEQUENCE
  5'...PuG▼GNCCPy...3'
  3'...PyCCNG▲GPu...5'
ISOSCHIZOMERS
  Dra II
OPTIMUM REACTION CONDITIONS
  37°C
HEAT INACTIVATION
  65°C for 20 min.
METHYLATION EFFECTS
  Does not cleave PuGGNCm^{5}CPy
  Blocked by overlapping *dcm* methylation

NOTES
- EC 3.1.21.4
- A type II site specific deoxyribonuclease
- A large group of enzymes recognizing specific short DNA sequences. They cleave either within the recognition site, or at a short specific distance from it
- Requires Mg^{2+} and BSA for full activity

CONCENTRATION or VOLUME ACTIVITY	UNIT DEFINITION	REACTION BUFFER (As Provided)	ADDITIONAL ACTIVITIES and PURITY	SUPPLIER CATALOG NO.
Escherichia coli H709C				
8-12 U/μL	1 unit hydrolyzes 1 μg DNA/hr at 37°C in a 50 μL reaction volume using λ phage DNA as substrate	33 mM Tris-acetate, 10 mM MgOAc, 66 mM KOAc, 0.1 mg/mL BSA, pH 7.9	>90% of the DNA fragments ligated and recut after 50-fold overdigestion	Fermentas ER0261 ER0262
8-12 U/μL	1 unit completely digests 1 mg λ DNA/hr at 37°C	500 mM Tris-HCl, 100 mM MgCl$_2$, 500 mM NaCl, pH 8.0	≥95% ligation; 100% cleavage	Life Technol 15495-013

EcoO109 I continued

CONCENTRATION or VOLUME ACTIVITY	UNIT DEFINITION	REACTION BUFFER (As Provided)	ADDITIONAL ACTIVITIES and PURITY	SUPPLIER CATALOG No.
Escherichia coli H709 *continued*				
6-12 U/μL	1 unit digests completely 1 μg λ DNA substrate/hr at 37°C	10 mM Tris-HCl, 10 mM MgCl$_2$, 1 mM DTT, pH 7.8		NBL Gene 015604 015607
7000-10,000 U/mL		100 mM KCl, 10 mM KPO$_4$, 1 mM EDTA, 7 mM β-MSH, 200 μg/mL BSA, 50% glycerol, pH 7.5		Nippon Gene Stratagene 500476
2000 U and 10,000 U	1 unit completely digests 1 μg λ DNA/hr at 37°C in 50 μL of supplied buffer	100 mM Tris-HCl, 100 mM MgCl$_2$, 10 mM DTT, pH 7.5	Purity is confirmed by overdigestion and ligation-recutting test	TaKaRa 1043
Escherichia coli H709c (I.Orskov)				
5000-20,000 U/mL		60 mM NaCl, 20 mM Tris-HCl, 0.5 mM EDTA, 14 mM β-MSH, 50% glycerol, pH 8.2		ICN 153811
Escherichia coli strain, carrying the cloned gene from *E. coli* H709c				
20,000 U/mL	1 unit completely digests 1 μg λ DNA (*Hind* III digest)/hr in a total reaction volume of 50 μL using the buffer provided	20 mM Tris-acetate, 10 mM MgOAc, 50 mM KOAc, 1 mM DTT, pH 7.9; add 100 μg/mL BSA	>95% DNA fragments ligated and recut after 10-fold overdigestion	NE Biolabs 503S 503L

EcoR I

RECOGNITION SEQUENCE
  5'...G▼AATTC...3'
  3'...CTTAA▲G...5'

OPTIMUM REACTION CONDITIONS
  pH 7.5-7.9 and 37°C

HEAT INACTIVATION
  65°C for 20 min.

METHYLATION EFFECTS
  Cleaves GAATThm^{5}C and GAAhm^{5}Uhm^{5}UC
  Does not cleave Gm^{6}AATTC, GAm^{6}ATTC or GAATTm^{5}C

NOTES
- EC 3.1.21.4
- A type II site specific deoxyribonuclease
- A large group of enzymes recognizing specific short DNA sequences. They cleave either within the recognition site, or at a short specific distance from it
- Conditions of low ionic strength, high enzyme concentration, glycerol >5%, pH >8.0 or replacement of Mg^{2+} by Mn^{2+} may result in star activity; β-mercaptoethanol inhibits star activity
- Requires Mg^{2+}

CONCENTRATION or VOLUME ACTIVITY	UNIT DEFINITION	REACTION BUFFER (As Provided)	ADDITIONAL ACTIVITIES and PURITY	SUPPLIER CATALOG No.
Escherichia coli BS5				
10 U/μL and 40 U/μL	1 unit completely digests 1 μg substrate DNA/hr	50 mM Tris-HCl, 10 mM MgCl$_2$, 100 mM NaCl, 1 mM DTE, pH 7.5		Boehringer 703737 1175084 200310 606189
Escherichia coli RY13				
25,000 U and 50,000 U and 250,000 U	1 unit completely digests 1 μg DNA substrate/hr 37°C	10 mM Tris-HCl, 10 mM MgCl$_2$, 100 mM NaCl, 1 mM DTT, 0.1 mg/mL BSA, pH 7.5	>90% DNA fragments ligated and recut after 10-fold overdigestion	Adv Biotech AB-0226-a AB-0226-b AB-0226-c
10 U/μL; 40 U/μL	1 unit completely cleaves 1 μg substrate DNA/hr	100 mM NaCl, 50 mM Tris, 5 mM MgCl$_2$, 2 mM β-MSH, pH 7.8	90-98% cleavage-ligation-recleavage	Am Allied L-01-05000 H-01-05000 L-01-25000 H-01-25000 L-01-50000 H-01-50000
8-12 U/μL	1 unit completely digests 1 μg λ DNA/hr at 37°C in the buffer provided	500 mM Tris-HCl, 100 mM MgCl$_2$, 10 mM DTT, 1000 mM NaCl, pH 7.5	100% DNA fragments ligated and recut	Amersham E 1040Y E 1040Z

EcoR I continued

CONCENTRATION or VOLUME ACTIVITY	UNIT DEFINITION	REACTION BUFFER (As Provided)	ADDITIONAL ACTIVITIES and PURITY	SUPPLIER CATALOG No.
Escherichia coli RY13 continued				
>40 U/µL	1 unit completely digests 1 µg λ DNA/hr at 37°C in the buffer provided	500 mM Tris-HCl, 100 mM MgCl$_2$, 10 mM DTT, 1000 mM NaCl, pH 7.5	100% DNA fragments ligated and recut	Amersham E 1040ZH E 1040XH
12,000 U/mL	1 unit degrades 1 µg λ dam$^-$ dcm$^-$ DNA/hr at 37°C in a 0.05 mL mixture	50 mM Tris-HCl, 10 mM MgCl$_2$, 100 mM NaCl, 100 µg/mL BSA, pH 7.5	No detectable contaminating nuclease; 90% ligated and 100% recut after 50-fold overdigestion	AGS Heidelb A00300S A00300M
	1 unit completely cleaves 1 µg λ DNA/hr at pH 7.5, 37°C in a reaction volume of 50 µL	50 mM Tris-HCl, 10 mM MgCl$_2$, 100 mM NaCl, 1.0 mM DTT, 100 µg/mL BSA, pH 7.5		CHIMERx 2150-01 2150-02
8-12 U/µL; 40-60 U/µL	1 unit hydrolyzes 1 µg DNA/hr at 37°C in a 50 µL reaction volume using λ phage DNA as substrate	50 mM Tris-HCl, 10 mM MgCl$_2$, 100 mM NaCl, 0.02% Triton X-100, pH 7.5	>90% of the DNA fragments ligated and recut after 50-fold overdigestion	Fermentas ER0271 ER0272 ER0273
		400 mM NaCl, 5 mM Tris-KPO$_4$, 0.1 mM EDTA, 5 mM β-MSH, 0.15% Triton X-100, 200 µg/mL BSA, 50% glycerol, pH 7.4		ICN 151025
8-12 U/µL and 50 U/µL	1 unit completely digests 1 mg λ DNA/hr at 37°C	500 mM Tris-HCl, 100 mM MgCl$_2$, 1 M NaCl, pH 8.0	≥95% ligation; 100% cleavage	Life Technol 15202-013 15202-021 15202-120 15202-039
	10 units does not produce unspecific cleavage products with 1 µg λ DNA/16 hr at 37°C	100 mM Tris-HCl, 5 mM MgCl$_2$, 100 mM NaCl, 100 µg/mL BSA, 0.025% Triton X-100, pH 7.9	>90% DNA fragments ligated and recut after 10-fold overdigestion	MINOTECH 110
8-12 U/µL 14-15 U/µL >40 U/µL	1 unit digests completely 1 µg λ DNA substrate/hr at 37°C	50 mM Tris-HCl, 100 mM NaCl, 10 mM MgCl$_2$, 1 mM DTT, pH 7.8		NBL Gene 010106 010108 010109 010158 010159
10 U/µL and 20-200 U/µL	1 unit hydrolyzes 1 µg λ DNA to completion/hr in a 50 µL total reaction volume	100 mM Tris-HCl, 1 M NaCl, 100 mM MgCl$_2$, 100 mM β-MSH, 1.0 mg/mL BSA, pH 8		Nippon Gene Oncor 110166 110167 110168

EcoR I continued

CONCENTRATION or VOLUME ACTIVITY	UNIT DEFINITION	REACTION BUFFER (As Provided)	ADDITIONAL ACTIVITIES and PURITY	SUPPLIER CATALOG No.
Escherichia coli RY13 *continued*				
10,000–25,000 U/mL and 50,000–100,000 U/mL	1 unit completely digests 1 µg *Bst*E II pre-digested λ DNA/hr at pH 7.9, 37°C in a 50 µL total assay mixture	100 mM Tris-acetate, 100 mM MgOAc, 500 mM KOAc, pH 7.5	Compatible ends: *Apo* I, *Mun* I, *Tsp*509 I	Pharmacia 27-0854-03 27-0854-04 27-0854-18
8–12 U/µL and 40–80 U/µL	1 unit completely digests 1 µg λ DNA/hr at 37°C in a 50 mL reaction volume	Blue/White Cloning Qualified; 900 mM Tris-HCl, 100 mM MgCl$_2$, 500 mM NaCl, pH 7.5	90% ligation	Promega R6011 R6012 R6013 R4014 R6015
				SibEnzyme
10,000–25,000 U/mL	1 unit completely cleaves 1.0 µg λ DNA/hr in a 50 µL reaction mixture	Unique buffer + BSA	100% cleavage, 95% ligation, 95% recleavage, 50% endonuclease (nickase) No detectable endonuclease (overdigestion)	Sigma R2627
40,000–125,000 U/mL	1 unit completely cleaves 1.0 µg λ DNA/hr in a 50 µL reaction mixture	Unique buffer + BSA	100% cleavage, 95% ligation, 95% recleavage, 50% endonuclease (nickase) No detectable endonuclease (overdigestion)	Sigma R2881
10,000 U and 50,000 U	1 unit completely digests 1 µg λ DNA/hr at 37°C in 50 µL of supplied buffer	500 mM Tris-HCl, 100 mM MgCl$_2$, 10 mM DTT, 1 M NaCl, pH 7.5	Purity is confirmed by overdigestion, ligation-recutting and cloning test	TaKaRa 1040
3–20 U/µL and 40–50 U/µL	1 unit completely digests 1 µg DNA/hr at 37°C	50 mM Tris-HCl, 10 mM MgCl$_2$, 100 mM NaCl, 1 mM DTT, pH 7.5	>90% ligation; >95% recut	Toyobo ECO-101 ECO-102 ECO-103 ECO-153
Escherichia coli strain, carrying the cloned gene from *E. coli* RY13				
20,000 U/mL; 100,000 U/mL	1 unit completely digests 1 µg λ DNA/hr in a total reaction volume of 50 µL using the buffer provided	100 mM Tris-HCl, 10 mM MgCl$_2$, 50 mM NaCl, 0.025% Triton X-100, pH 7.5	>95% DNA fragments ligated and recut after 100-fold overdigestion	NE Biolabs 101S 101L 101XL
Esherichia coli, cloned				
10,000–120,000 U/mL		300 mM NaCl, 5 mM KPO$_4$, 0.1 mM EDTA, 5 mM β-MSH, 0.15% Triton X-100, 200 µg/mL BSA, 50% glycerol, pH 7.4		Stratagene 500480 500481 500489

EcoR II

RECOGNITION SEQUENCE
$$5'...CC(^A_T)GG\blacktriangledown...3'$$
$$3'...GG(^T_A)CC\blacktriangle...5'$$

NEOSCHIZOMERS
Apy I, BsiL I, BstN I, Mva I (different cleavage site), TspA I

OPTIMUM REACTION CONDITIONS
pH 7.9 and 37°C

HEAT INACTIVATION
65°C for 20 min.

METHYLATION EFFECTS
Cleaves $m^5CC(^A_T)GG$
Does not cleave $m^4CC(^A_T)GG$, $Cm^4C(^A_T)GG$ $Cm^5C(^A_T)GG$, CCm^6AGG or $hm^5Chm^5C(^A_T)GG$

NOTES
- EC 3.1.21.4
- A type II site specific deoxyribonuclease
- A large group of enzymes recognizing specific short DNA sequences. They cleave either within the recognition site, or at a short specific distance from it
- Activity is dependent on the coordinate presence of at least two EcoR II restriction sites per DNA molecule
- Requires Mg^{2+}

CONCENTRATION or VOLUME ACTIVITY	UNIT DEFINITION	REACTION BUFFER (As Provided)	ADDITIONAL ACTIVITIES and PURITY	SUPPLIER CATALOG NO.
Escherichia coli				
3000-20,000 U/mL		50 mM KCl, 10 mM Tris-HCl, 0.1 mM EDTA, 1 mM DTT, 200 μg/mL acetylated BSA, 50% glycerol, pH 7.5		ICN 159442
Escherichia coli MV1193				Nippon Gene
10 U/μL	1 unit completely digests 1 μg substrate DNA/hr	50 mM Tris-HCl, 10 mM MgCl₂, 100 mM NaCl, 1 mM DTE, pH 7.5		Boehringer 1427881
Escherichia coli R245				
2-8 U/μL	1 unit completely digests 1 mg Ad2 DNA/hr at 37°C	500 mM Tris-HCl, 60 mM MgCl₂, 500 mM NaCl, 500 mM KCl, pH 7.4	≥95% ligation; 100% cleavage	Life Technol 15203-011
1000-5000 U/mL	1 unit cleaves 98% λ DNA/hr in a 50 μL reaction mixture	200 mM Tris-acetate, 500 mM KOAc, 100 mM MgOAc, 10 mM DTT, pH 7.9	98% cleavage, 100% ligation, 90% recleavage, 100% endonuclease (overdigestion)	Sigma R1636
1000-10,000 U/mL		50 mM KCl, 10 mM Tris-HCl, 0.1 mM EDTA, 1 mM DTT, 200 μg/mL BSA, 50% glycerol, pH 7.4		Stratagene 500490 500491
Escherichia coli RY23				
3-20 U/μL	1 unit completely digests 1 μg DNA/hr at 37°C	10 mM Tris-HCl, 10 mM MgCl₂, 50 mM NaCl, 1 mM DTT, pH 7.5	>90% ligation; >95% recut	Toyobo ECO-201 ECO-202

EcoR V

RECOGNITION SEQUENCE
5'...GAT▼ATC...3'
3'...CTA▲TAG...5'

ISOSCHIZOMERS
*Eco*32 I

OPTIMUM REACTION CONDITIONS
pH 7.5-7.9 and 37°C

HEAT INACTIVATION
Resistant

METHYLATION EFFECTS
Cleaves GATATm^{5}C and GATAThm^{5}C
Does not cleave Gm^{6}ATATC or GATm^{6}ATC

NOTES
- EC 3.1.21.4
- A type II site specific deoxyribonuclease
- A large group of enzymes recognizing specific short DNA sequences. They cleave either within the recognition site, or at a short specific distance from it
- Compatible ends: blunt; leaves a ligatable blunt end in the tetracycline resistance gene of pBR322
- Can be used in conjunction with plasmid cloning vectors to add sticky ends to blunt-ended fragments
- DMSO and >5% glycerol results in star activity
- Requires Mg^{2+} and BSA for full activity

CONCENTRATION or VOLUME ACTIVITY	UNIT DEFINITION	REACTION BUFFER (As Provided)	ADDITIONAL ACTIVITIES and PURITY	SUPPLIER CATALOG NO.
Escherichia coli				
10 U/μL; 40 U/μL	1 unit completely cleaves 1 μg substrate DNA/hr	100 mM NaCl, 50 mM Tris, 5 mM MgCl$_2$, 2 mM β-MSH, pH 7.8	90-98% cleavage-ligation-recleavage	Am Allied L-19-02000 H-19-02000 L-19-10000 H-19-10000
2-12 U/μL	1 unit completely digests 1 μg λ DNA/hr at 37°C in the buffer provided	500 mM Tris-HCl, 100 mM MgCl$_2$, 10 mM DTT, 1000 mM NaCl, pH 7.5	>90% DNA fragments ligated and 100% recut	Amersham E 1042Y E 1042Z Nippon Gene
5000-15,000 U/mL	1 unit completely digests 1 μg λ DNA/hr at pH 7.5, 37°C in a 50 μL total assay mixture	100 mM Tris-acetate, 100 mM MgOAc, 500 mM KOAc, pH 7.5		Pharmacia 27-0934-01 27-0934-02 SibEnzyme
5000-20,000 U/mL	1 unit completely cleaves 1.0 μg λ DNA/hr in a 50 μL reaction mixture	100 mM Tris-HCl, 500 mM NaCl, 100 mM MgCl$_2$, 10 mM DTT, pH 7.9	100% cleavage, 95% ligation, 95% recleavage, 95% endonuclease (nickase) No detectable endonuclease (overdigestion)	Sigma R2756

EcoR V continued

CONCENTRATION or VOLUME ACTIVITY	UNIT DEFINITION	REACTION BUFFER (As Provided)	ADDITIONAL ACTIVITIES and PURITY	SUPPLIER CATALOG No.
Escherichia coli continued				
3000 U and 15,000 U	1 unit completely digests 1 µg λ DNA/hr at 37°C in 50 µL of supplied buffer	500 mM Tris-HCl, 100 mM MgCl$_2$, 10 mM DTT, 1 M NaCl, pH 7.5	Purity is confirmed by overdigestion and ligation-recutting test	TaKaRa 1042
Escherichia coli B946				
8-12 U/µL and 50 U/µL	1 unit completely digests 1 mg λ DNA/hr at 37°C	500 mM Tris-HCl, 100 mM MgCl$_2$, 500 mM NaCl, pH 8.0	≥65% ligation; 100% cleavage	Life Technol 15425-010 15425-036 15425-028
Escherichia coli J62plg74				
7500 U and 20,000 U	1 unit completely digests 1 µg DNA substrate/hr at 37°C	10 mM Tris-HCl, 10 mM MgCl$_2$, 100 mM NaCl, 1 mM DTT, 0.1 mg/mL BSA, pH 7.5	>90% DNA fragments ligated and recut after 2-fold overdigestion	Adv Biotech AB-0227-a AB-0227-b
10,000 U/mL	1 unit degrades 1 µg λ dam$^-$ dcm$^-$ DNA/hr at 37°C in a 0.05 mL mixture	10 mM Tris-HCl, 100 mM NaCl, 100 µg/mL BSA, pH 8.0	No detectable contaminating nuclease; 98% ligated and recut after 15-fold overdigestion	AGS Heidelb A00310S A00310M
10 U/µL and 40 U/µL	1 unit completely digests 1 µg substrate DNA/hr	10 mM Tris-HCl, 5 mM MgCl$_2$, 100 mM NaCl, 1 mM β-MSH, pH 8.0		Boehringer 667145 667153 1040197
	1 unit completely cleaves 1 µg λ DNA/hr at pH 7.5, 37°C in a reaction volume of 50 µL	50 mM Tris-HCl, 10 mM MgCl$_2$, 100 mM NaCl, 1.0 mM DTT, 100 µg/mL BSA, pH 7.5		CHIMERX 2150-01 2150-02
5000-20,000 U/mL		50 mM KCl, 10 mM Tris-HCl, 0.1 mM EDTA, 1.0 mM DTT, 200 µg/mL BSA, 50% glycerol, pH 7.4		ICN 151026
	10 units does not produce unspecific cleavage products with 1 µg λ DNA/16 hr at 37°C	10 mM Tris-HCl, 5 mM MgCl$_2$, 50 mM NaCl, 1 mM DTT, 100 µg/mL BSA, pH 7.9	>90% DNA fragments ligated and recut after 10-fold overdigestion	MINOTECH 111
6-12 U/µL	1 unit digests completely 1 µg λ DNA substrate/hr at 37°C	10 mM Tris-HCl, 100 mM NaCl, 5 mM MgCl$_2$, 1 mM β-MSH, pH 8.3		NBL Gene 011604 011607
10 U/µL	1 unit hydrolyzes 1 µg λ DNA to completion/hr in a 50 µL total reaction volume	100 mM Tris-HCl, 1 M NaCl, 100 mM MgCl$_2$, 100 mM β-MSH, 1.0 mg/mL BSA, pH 8		Oncor 110171 110172
10 U/µL and 40-80 U/µL	1 unit completely digests 1 µg λ DNA/hr at 37°C in a 50 mL reaction volume	Blue/White Cloning Qualified; 60 mM Tris-HCl, 60 mM MgCl$_2$, 1.5 M NaCl, 10 mM DTT, pH 7.9	90% ligation	Promega R6351 R6355 R4354

EcoR V continued

CONCENTRATION or VOLUME ACTIVITY	UNIT DEFINITION	REACTION BUFFER (As Provided)	ADDITIONAL ACTIVITIES and PURITY	SUPPLIER CATALOG No.
Escherichia coli J62plg74 continued				
3-20 U/µL and 40-50 U/µL	1 unit completely digests 1 µg DNA/hr at 37°C	50 mM Tris-HCl, 10 mM MgCl$_2$, 100 mM NaCl, 1 mM DTT, pH 7.5	>90% ligation; >95% recut	Toyobo ER5-101 ER5-102 ER5-152
Escherichia coli strain, carrying the cloned gene from the plasmid J62plg74				
20,000 U/mL; 100,000 U/mL	1 unit completely digests 1 µg λ DNA/hr in a total reaction volume of 50 µL using the buffer provided	10 mM Tris-HCl, 10 mM MgCl$_2$, 50 mM NaCl, 1 mM DTT, pH 7.9; add 100 µg/mL BSA	>95% DNA fragments ligated and recut after 10-fold overdigestion	NE Biolabs 195S 195L
Esherichia coli, cloned				
5000-20,000 U/mL		50 mM KCl, 10 mM Tris-HCl, 0.1 mM EDTA, 1 mM DTT, 200 µg/mL BSA, 50% glycerol, pH 7.4		Stratagene 500500

EcoT14 I

RECOGNITION SEQUENCE
    5'...C▼C(^A_T)(^A_T)GG...3'
    3'...GG(^T_A)(^T_A)C▲C...5'
ISOSCHIZOMERS
    Eco130 I, Sty I
OPTIMUM REACTION CONDITIONS
    37°C
HEAT INACTIVATION
    60°C for 15 min.

NOTES
- EC 3.1.21.4
- A type II site specific deoxyribonuclease
- A large group of enzymes recognizing specific short DNA sequences. They cleave either within the recognition site, or at a short specific distance from it
- Requires Mg^{2+}

CONCENTRATION or VOLUME ACTIVITY	UNIT DEFINITION	REACTION BUFFER (As Provided)	ADDITIONAL ACTIVITIES and PURITY	SUPPLIER CATALOG No.
Escherichia coli TB14				
8-12 U/µL	1 unit completely digests 1 µg λ DNA/hr at 37°C in the buffer provided	500 mM Tris-HCl, 100 mM MgCl$_2$, 10 mM DTT, 1000 mM NaCl, pH 7.5	100% DNA fragments ligated and recut	Amersham E 1038Y
3000 U and 15,000 U	1 unit completely digests 1 µg λ DNA/hr at 37°C in 50 µL of supplied buffer	500 mM Tris-HCl, 100 mM MgCl$_2$, 10 mM DTT, 1 M NaCl, pH 7.5	Purity is confirmed by overdigestion, ligation-recutting and cloning test	TaKaRa 1038

EcoT22 I

RECOGNITION SEQUENCE
5'...ATGCA▼T...3'
3'...TACGT▲A...5'

ISOSCHIZOMERS
Ava III, *Nsi* I

OPTIMUM REACTION CONDITIONS
pH 7.5 and 37°C

HEAT INACTIVATION
60°C for 15 min.; 100°C for 5 min.

NOTES
- EC 3.1.21.4
- A type II site specific deoxyribonuclease
- A large group of enzymes recognizing specific short DNA sequences. They cleave either within the recognition site, or at a short specific distance from it
- Low salt or β-mercaptoethanol results in star activity
- Requires Mg^{2+}

CONCENTRATION or VOLUME ACTIVITY	UNIT DEFINITION	REACTION BUFFER (As Provided)	ADDITIONAL ACTIVITIES and PURITY	SUPPLIER CATALOG NO.
Escherichia coli TB22				
3-20 U/μL and 40-50 U/μL	1 unit completely digests 1 μg DNA/hr at 37°C	50 m*M* Tris-HCl, 10 m*M* MgCl$_2$, 100 m*M* NaCl, 1 m*M* DTT, pH 7.5	>90% ligation; >90% recut	Toyobo E22-103 E22-104 E22-154
Escherichia coli WA921				
8-12 U/μL	1 unit completely digests 1 μg λ DNA/hr at 37°C in the buffer provided	500 m*M* Tris-HCl, 100 m*M* MgCl$_2$, 10 m*M* DTT, 1000 m*M* NaCl, pH 7.5	95% DNA fragments ligated and 100% recut	Amersham E 1125Y
2000 U and 10,000 U	1 unit completely digests 1 μg λ DNA/hr at 37°C in 50 μL of supplied buffer	500 m*M* Tris-HCl, 100 m*M* MgCl$_2$, 10 m*M* DTT, 1 *M* NaCl, pH 7.5	Purity is confirmed by overdigestion and ligation-recutting test	TaKaRa 1125

*Eco*T38 I

RECOGNITION SEQUENCE
    5'...GRGCY▼C...3'
    3'...C▲YCGRG...5'

ISOSCHIZOMERS
    Ban II, *Eco*24 I, *Fri*O I

NOTES
- EC 3.1.21.4
- A type II site specific deoxyribonuclease
- A large group of enzymes recognizing specific short DNA sequences. They cleave either within the recognition site, or at a short specific distance from it
- Requires Mg^{2+}

CONCENTRATION or VOLUME ACTIVITY	UNIT DEFINITION	REACTION BUFFER (As Provided)	ADDITIONAL ACTIVITIES and PURITY	SUPPLIER CATALOG No.
Escherichia coli TH38				
				Nippon Gene

Ege I

RECOGNITION SEQUENCE
    5'...GGC▼GCC...3'
    3'...CCG▲CGG...5'

ISOSCHIZOMERS
    Bbe I, *Ehe* I, *Kas* I, *Mly*113 I, *Nar* I

NOTES
- EC 3.1.21.4
- A type II site specific deoxyribonuclease
- A large group of enzymes recognizing specific short DNA sequences. They cleave either within the recognition site, or at a short specific distance from it
- Requires Mg^{2+}

CONCENTRATION or VOLUME ACTIVITY	UNIT DEFINITION	REACTION BUFFER (As Provided)	ADDITIONAL ACTIVITIES and PURITY	SUPPLIER CATALOG No.
Enterobacter gergoviae NA				
				SibEnzyme

Ehe I

RECOGNITION SEQUENCE
5'...GGC▼GCC...3'
3'...CCG▲CGG...5'

ISOSCHIZOMERS
Nar I

OPTIMUM REACTION CONDITIONS
pH 8.0 and 37°C

HEAT INACTIVATION
65°C for 20 min.; 70°C for 5 min.

METHYLATION EFFECTS
Does not cleave GGm^{5}CGCC, GGCGm^{5}CC or GGhm^{5}CGhm^{5}Chm^{5}C
Blocked by CG methylation

NOTES
- EC 3.1.21.4
- A type II site specific deoxyribonuclease
- A large group of enzymes recognizing specific short DNA sequences. They cleave either within the recognition site, or at a short specific distance from it
- Large excess of enzyme results in the appearance of star activity
- Completely digests λ and pBR322 DNAs and produces fragments with blunt ends; *Nar* I does not
- Requires Mg^{2+} and BSA for full activity

CONCENTRATION or VOLUME ACTIVITY	UNIT DEFINITION	REACTION BUFFER (As Provided)	ADDITIONAL ACTIVITIES and PURITY	SUPPLIER CATALOG NO.
Erwinia herbicola 9/5				
3-20 U/μL	1 unit completely digests 1 μg λ DNA/hr at 37°C in the buffer provided	100 m*M* Tris-HCl, 100 m*M* MgCl₂, 10 m*M* DTT, pH 7.5	>80% DNA fragments ligated and >90% recut	Amersham E 0227Y E 0227Z
10,000 U/mL	1 unit degrades 1 μg λ dam⁻ dcm⁻ *Pst* I DNA fragments/hr at 37°C in a 0.05 mL mixture	33 m*M* Tris-acetate, 10 m*M* MgOAc, 66 m*M* KOAc, 100 μg/mL BSA, pH 7.9	No detectable contaminating nuclease; 60% ligated and 90% recut after 50-fold overdigestion	AGS Heidelb F00410S F00410M
8-12 U/μL	1 unit hydrolyzes 1 μg DNA/hr at 37°C in a 50 μL reaction volume using λ phage DNA as substrate	33 m*M* Tris-acetate, 10 m*M* MgOAc, 66 m*M* KOAc, 0.1 mg/mL BSA, pH 7.9	>60% of the pBR322 DNA fragments ligated and >90% recut after 50-fold overdigestion	Fermentas ER0441 ER0442
				NE Biolabs
3-20 U/μL	1 unit completely digests 1 μg DNA/hr at 37°C	10 m*M* Tris-HCl, 10 m*M* MgCl₂, 1 m*M* DTT, pH 7.5	>80% ligation; >90% recut	Toyobo EHE-101 EHE-102

Erh I

RECOGNITION SEQUENCE
  5'...C▼CWWGG...3'
  3'...GGWWC▲C...5'
ISOSCHIZOMERS
  BssT1 I, Eco130 I, EcoT14 I, Sty I

NOTES
- EC 3.1.21.4
- A type II site specific deoxyribonuclease
- A large group of enzymes recognizing specific short DNA sequences. They cleave either within the recognition site, or at a short specific distance from it
- Requires Mg^{2+}

CONCENTRATION or VOLUME ACTIVITY	UNIT DEFINITION	REACTION BUFFER (As Provided)	ADDITIONAL ACTIVITIES and PURITY	SUPPLIER CATALOG NO.
Erwinia rhaponici B9				
				SibEnzyme

Esp3 I

RECOGNITION SEQUENCE
  5'...CGTCTC(N)$_1$▼...3'
  3'...GCAGAG(N)$_5$▲...5'
ISOSCHIZOMERS
  Bpu1102 I, Cel II
OPTIMUM REACTION CONDITIONS
  37°C
HEAT INACTIVATION
  65°C for 20 min.
METHYLATION EFFECTS
  Does not cleave m^{5}CGTCTC, CGTm^{5}CTC or GAGm^{6}ACG
  Blocked by CG methylation

NOTES
- EC 3.1.21.4
- A type II site specific deoxyribonuclease
- A large group of enzymes recognizing specific short DNA sequences. They cleave either within the recognition site, or at a short specific distance from it
- Sensitive to DTT; freshly made DTT must be added to the reaction buffer
- Requires Mg^{2+} and BSA for full activity

Esp3 I continued

CONCENTRATION or VOLUME ACTIVITY	UNIT DEFINITION	REACTION BUFFER (As Provided)	ADDITIONAL ACTIVITIES and PURITY	SUPPLIER CATALOG No.
Erwinia species RFL3				
4000 U/mL	1 unit degrades 1 μg λ dam⁻ dcm⁻ DNA/hr at 37°C in a 0.05 mL mixture	33 mM Tris-acetate, 10 mM MgOAc, 66 mM KOAc, 100 μg/mL BSA, pH 7.9	No detectable contaminating nuclease; 90% ligated and 100% recut after 50-fold overdigestion	AGS Heidelb F00315S F00315M
4-8 U/μL	1 unit hydrolyzes 1 μg DNA/hr at 37°C in a 50 μL reaction volume using λ phage DNA as substrate	33 mM Tris-acetate, 10 mM MgOAc, 66 mM KOAc, 0.1 mg/mL BSA, 1 mM DTT, pH 7.9	>90% of the DNA fragments ligated and recut after 50-fold overdigestion	Fermentas ER0451 ER0452

Esp1396 I

RECOGNITION SEQUENCE
    5'...CCANNNN▼NTGG...3'
    3'...GGTN▲NNNNACC...5'

ISOSCHIZOMERS
    *Pfl*M I

OPTIMUM REACTION CONDITIONS
    37°C

HEAT INACTIVATION
    65°C for 20 min.

METHYLATION EFFECTS
    Not blocked by *dcm* methylation; *Pfl*M I and its isoschizomer *Van*91 I are

NOTES
- EC 3.1.21.4
- A type II site specific deoxyribonuclease
- A large group of enzymes recognizing specific short DNA sequences. They cleave either within the recognition site, or at a short specific distance from it
- >10-fold overdigestion of substrate with enzyme results in the appearance of star activity
- Requires Mg^{2+} and BSA for full activity

CONCENTRATION or VOLUME ACTIVITY	UNIT DEFINITION	REACTION BUFFER (As Provided)	ADDITIONAL ACTIVITIES and PURITY	SUPPLIER CATALOG No.
Enterobacter species RFL1396				
8-12 U/μL	1 unit hydrolyzes 1 μg DNA/hr at 37°C in a 50 μL reaction volume using λ phage DNA as substrate	10 mM Tris-HCl, 10 mM MgCl₂, 50 mM NaCl, 0.1 mg/mL BSA, pH 7.5	>90% of the DNA fragments ligated and recut after 10-fold overdigestion	Fermentas ER0951 ER0952

Fau I

RECOGNITION SEQUENCE
5'...CCCGC(4/6)▼...3'

NOTES
- EC 3.1.21.4
- A type II site specific deoxyribonuclease
- A large group of enzymes recognizing specific short DNA sequences. They cleave either within the recognition site, or at a short specific distance from it
- Requires Mg^{2+}

CONCENTRATION or VOLUME ACTIVITY	UNIT DEFINITION	REACTION BUFFER (As Provided)	ADDITIONAL ACTIVITIES and PURITY	SUPPLIER CATALOG NO.
Flavobacterium aquatile				
				sibEnzyme

FauND I

RECOGNITION SEQUENCE
5'...CA▼TATG...3'
3'...GTAT▲AC...5'

ISOSCHIZOMERS
Nde I

NOTES
- EC 3.1.21.4
- A type II site specific deoxyribonuclease
- A large group of enzymes recognizing specific short DNA sequences. They cleave either within the recognition site, or at a short specific distance from it
- Requires Mg^{2+}

CONCENTRATION or VOLUME ACTIVITY	UNIT DEFINITION	REACTION BUFFER (As Provided)	ADDITIONAL ACTIVITIES and PURITY	SUPPLIER CATALOG NO.
Flavobacterium aquatile ND				
				sibEnzyme

Fba I

RECOGNITION SEQUENCE
5'...T▼GATCA...3'
3'...ACTAG▲T...5'

ISOSCHIZOMERS
Bcl I

OPTIMUM REACTION CONDITIONS
37°C

HEAT INACTIVATION
Resistant; use ethanol precipitation

NOTES
- EC 3.1.21.4
- A type II site specific deoxyribonuclease
- A large group of enzymes recognizing specific short DNA sequences. They cleave either within the recognition site, or at a short specific distance from it
- Requires Mg^{2+}

CONCENTRATION or VOLUME ACTIVITY	UNIT DEFINITION	REACTION BUFFER (As Provided)	ADDITIONAL ACTIVITIES and PURITY	SUPPLIER CATALOG NO.
Flavobacterium balustinum				
8-12 U/μL	1 unit completely digests 1 μg λ DNA (N6 methyladenine-free)/hr at 37°C in the buffer provided	200 mM Tris-HCl, 100 mM MgCl₂, 10 mM DTT, 1000 mM KCl, pH 8.5	100% DNA fragments ligated and recut	Amersham E 1045Y E 1045Z
500 U and 2500 U	1 unit completely digests 1 μg N^6-methyladenine free λ DNA/hr at 37°C in 50 μL of supplied buffer	200 mM Tris-HCl, 100 mM MgCl₂, 10 mM DTT, 1 M KCl, pH 8.5	Purity is confirmed by overdigestion, ligation-recutting and cloning test	TaKaRa 1045

Fnu4H I

RECOGNITION SEQUENCE
5'...GC▼NGC...3'
3'...CGN▲CG...5'

ISOSCHIZOMERS
BsoF I, Fsp4H I, Ita I

NOTES
- EC 3.1.21.4
- A type II site specific deoxyribonuclease
- A large group of enzymes recognizing specific short DNA sequences. They cleave either within the recognition site, or at a short specific distance from it
- Requires Mg^{2+}

Fnu4H I continued

CONCENTRATION or VOLUME ACTIVITY	UNIT DEFINITION	REACTION BUFFER (As Provided)	ADDITIONAL ACTIVITIES and PURITY	SUPPLIER CATALOG No.
Fusobacterium nucleatum 4H				
1000-5000 U/mL		50 mM NaCl, 10 mM Tris-HCl, 0.1 mM EDTA, 1 mM DTT, 200 µg/mL acetylated BSA, 50% glycerol, pH 7.4		ICN 159443
				NE Biolabs

FnuD II

RECOGNITION SEQUENCE
    5'...CG▼CG...3'
    3'...GC▲GC...5'

ISOSCHIZOMERS
    Acc II, Bsh1236 I, BstU I, Mvn I, Tha I

NOTES
- EC 3.1.21.4
- A type II site specific deoxyribonuclease
- A large group of enzymes recognizing specific short DNA sequences. They cleave either within the recognition site, or at a short specific distance from it
- Requires Mg^{2+}

CONCENTRATION or VOLUME ACTIVITY	UNIT DEFINITION	REACTION BUFFER (As Provided)	ADDITIONAL ACTIVITIES and PURITY	SUPPLIER CATALOG No.
Fusobacterium nucleatum D				
				NE Biolabs

Fok I

RECOGNITION SEQUENCE
5'...GGATG(N)$_9$▼...3'
3'...CCTAC(N)$_{13}$▲...5'

OPTIMUM REACTION CONDITIONS
37°C

HEAT INACTIVATION
65°C for 20 min.

METHYLATION EFFECTS
Cleaves CATm^{5}CC and CATCm^{5}C
Does not cleave GGm^{6}ATG, Cm^{6}ATCC or CATCm^{4}C

NOTES
- EC 3.1.21.4
- A type II site specific deoxyribonuclease
- A large group of enzymes recognizing specific short DNA sequences. They cleave either within the recognition site, or at a short specific distance from it
- Cleaves between virtually any two nucleotides by constructing a complementary oligonucleotide to the sequence being cleaved
- Overdigestions of >5 U/μg DNA and incubation times >2 hr. are not recommended
- Recognizes nonpalindromic sequences, having two possible top strand recognition sequences
- Produces ambiguous ends which are difficult to ligate
- Requires Mg^{2+}

CONCENTRATION or VOLUME ACTIVITY	UNIT DEFINITION	REACTION BUFFER (As Provided)	ADDITIONAL ACTIVITIES and PURITY	SUPPLIER CATALOG NO.
Escherichia coli strain, carrying the cloned gene from *Flavobacterium okeanokoites*				
4000 U/mL	1 unit completely digests 1 μg λ DNA/hr in a total reaction volume of 50 μL using the buffer provided	20 m*M* Tris-acetate, 10 m*M* MgOAc, 50 m*M* KOAc, 1 m*M* DTT, pH 7.9	>95% DNA fragments ligated and recut after 5-fold overdigestion	NE Biolabs 109S 109L
Escherichia coli UT481 carrying the plasmid encoding *Fok* I gene				
8-12 U/μL	1 unit completely digests 1 μg λ DNA/hr at 37°C in the buffer provided	100 m*M* Tris-HCl, 100 m*M* MgCl$_2$, 10 m*M* DTT, 500 m*M* NaCl, BSA, pH 7.5	>95% DNA fragments ligated and 100% recut	Amersham E 1046Y
1000 U and 5000 U	1 unit completely digests 1 μg λ DNA/hr at 37°C in 50 μL of supplied buffer and BSA	100 m*M* Tris-HCl, 100 m*M* MgCl$_2$, 10 m*M* DTT, 500 m*M* NaCl, pH 7.5	Purity is confirmed by overdigestion and ligation-recutting test	TaKaRa 1046
Flavobacterium okeanokoites				
1-5 U/μL	1 unit completely digests 1 μg substrate DNA/hr	10 m*M* Tris-HCl, 10 m*M* MgCl$_2$, 50 m*M* NaCl, 1 m*M* DTE, pH 7.5		Boehringer 1004816 1004824

Fok I continued

CONCENTRATION or VOLUME ACTIVITY	UNIT DEFINITION	REACTION BUFFER (As Provided)	ADDITIONAL ACTIVITIES and PURITY	SUPPLIER CATALOG No.
Flavobacterium okeanokoites continued				
6-14 U/μL	1 unit digests completely 1 μg λ DNA substrate/hr at 37°C	10 mM Tris-HCl, 50 mM NaCl, 10 mM $MgCl_2$, 6 mM β-MSH, 100 μg/mL BSA, pH 7.5		NBL Gene 017111 017102
2 U/μL	1 unit hydrolyzes 1 μg λ DNA to completion/hr in a 50 μL total reaction volume	100 mM Tris-HCl, 500 mM NaCl, 100 mM $MgCl_2$, 100 mM β-MSH, 1.0 mg/mL BSA, pH 8		Nippon Gene Oncor 110634 110635
2-10 U/μL	1 unit completely digests 1 μg λ DNA/hr at 37°C in a 50 mL reaction volume	60 mM Tris-HCl, 60 mM $MgCl_2$, 500 mM NaCl, 10 mM DTT, pH 7.5	90% ligation	Promega R6781 R6782
				SibEnzyme
2000-12,000 U/mL		50 mM KCl, 10 mM Tris-HCl, 0.1 mM EDTA, 1 mM DTT, 200 μg/mL BSA, 50% glycerol, pH 7.4		Sigma ICN 153812

FriO I

RECOGNITION SEQUENCE
    5'...GRGCY▼C...3'
    3'...C▲YCGRG...5'

ISOSCHIZOMERS
    Ban II, Eco24 I, EcoT38 I

NOTES
- EC 3.1.21.4
- A type II site specific deoxyribonuclease
- A large group of enzymes recognizing specific short DNA sequences. They cleave either within the recognition site, or at a short specific distance from it
- Requires Mg^{2+}

CONCENTRATION or VOLUME ACTIVITY	UNIT DEFINITION	REACTION BUFFER (As Provided)	ADDITIONAL ACTIVITIES and PURITY	SUPPLIER CATALOG No.
Flavobacterium species 09				
				SibEnzyme

Fse I

RECOGNITION SEQUENCE
  5'...GGCCGG▼CC...3'
  3'...CC▲GGCCGG...5'

OPTIMUM REACTION CONDITIONS
  65°C

HEAT INACTIVATION
  Resistant

METHYLATION EFFECTS
  Does not cleave GGm^{5}CCGGm^{5}CC, GGCm^{5}CGGCC or GGm^{5}CCGGCC

NOTES
- EC 3.1.21.4
- A type II site specific deoxyribonuclease
- A large group of enzymes recognizing specific short DNA sequences. They cleave either within the recognition site, or at a short specific distance from it
- Requires Mg^{2+} and BSA for full activity

CONCENTRATION or VOLUME ACTIVITY	UNIT DEFINITION	REACTION BUFFER (As Provided)	ADDITIONAL ACTIVITIES and PURITY	SUPPLIER CATALOG NO.
Escherichia coli strain, carrying the cloned gene from *Frankia* species				
4000 U/mL	1 unit completely digests 1 μg λ DNA/hr in a total reaction volume of 50 μL using the buffer provided	10 m*M* Bis Tris-Propane-HCl, 10 m*M* MgCl$_2$, 1 m*M* DTT, pH 7.0; add 100 μg/mL BSA	>95% DNA fragments ligated and recut after 4-fold overdigestion	NE Biolabs 588S 588L
Frankia species Eul 1B				
8-12 U/μL	1 unit completely digests 1 μg Ad2 DNA/hr at 30°C in the buffer provided	Genomic grade; 100 m*M* Tris-HCl, 100 m*M* MgCl$_2$, 10 m*M* DTT, 500 m*M* NaCl, pH 7.5	>95% DNA fragments ligated and 100% recut	Amersham E 1047Y
50 U and 250 U	1 unit completely cleaves 1 μg DNA/hr in 50 μL reaction mixture	100 m*M* Tris-HCl, 100 m*M* MgCl$_2$, 10 m*M* DTT, 500 m*M* NaCl, pH 7.5		PanVera TAK 1047
50 U and 250 U	1 unit completely digests 1 μg Ad2 DNA/hr at 30°C in 50 μL of supplied buffer	Genomic grade; 100 m*M* Tris-HCl, 100 m*M* MgCl$_2$, 10 m*M* DTT, 500 m*M* NaCl, pH 7.5	Purity is confirmed by overdigestion, ligation-recutting and genome DNA digestion test	TaKaRa 1047A

Fsp I

RECOGNITION SEQUENCE
5'...TGC▼GCA...3'
3'...ACG▲CGT...5'

OPTIMUM REACTION CONDITIONS
pH 7.9 and 37°C

HEAT INACTIVATION
65°C for 20 min.; 80°C for 5 min.

METHYLATION EFFECTS
Does not cleave TGm^{5}CGCA

NOTES
- EC 3.1.21.4
- A type II site specific deoxyribonuclease
- A large group of enzymes recognizing specific short DNA sequences. They cleave either within the recognition site, or at a short specific distance from it
- Requires Mg^{2+}

CONCENTRATION or VOLUME ACTIVITY	UNIT DEFINITION	REACTION BUFFER (As Provided)	ADDITIONAL ACTIVITIES and PURITY	SUPPLIER CATALOG NO.
Escherichia coli strain, carrying the cloned gene from *Fischerella* species				
5000 U/mL	1 unit completely digests 1 μg λ DNA/hr in a total reaction volume of 50 μL using the buffer provided	20 m*M* Tris-acetate, 10 m*M* MgOAc, 50 m*M* KOAc, 1 m*M* DTT, pH 7.9	>95% DNA fragments ligated and recut after 5-fold overdigestion	NE Biolabs 135S 135L
Escherichia coli UT481, carrying the cloned gene for *Fsp* I from *Fischerella*				
1-5 U/μL	1 unit completely digests 1 μg λ DNA/hr at 37°C in the buffer provided	330 m*M* Tris-acetate, 100 m*M* MgOAc, 5 m*M* DTT, 660 m*M* KOAc, pH 7.9	>95% DNA fragments ligated and 95% recut	Amersham E 0212Y E 0212Z
Fischerella species				
1000-5000 U/mL		50 m*M* KCl, 10 m*M* Tris-HCl, 0.1 m*M* EDTA, 1 m*M* DTT, 200 μg/mL acetylated BSA, 50% glycerol, pH 7.5		ICN 159445
8-12 U/μL	1 unit completely digests 1 mg λ DNA/hr at 37°C	500 m*M* Tris-HCl, 100 m*M* MgCl$_2$, pH 8.0	≥85% ligation; 100% cleavage	Life Technol 15474-018
1000-10,000 U/mL	1 unit completely cleaves 1.0 μg λ DNA/hr in a 50 μL reaction mixture	200 m*M* Tris-acetate, 500 m*M* KOAc, 100 m*M* MgOAc, 10 m*M* DTT, pH 7.9	100% cleavage, 95% ligation, 95% recleavage, 50% endonuclease (nickase) No detectable endonuclease (overdigestion)	Nippon Gene Sigma R6505
200 U and 1000 U	1 unit completely digests 1 μg λ DNA/hr at 37°C in 50 μL of supplied buffer and BSA	Genomic grade; 330 m*M* Tris-acetate, 100 m*M* MgOAc, 5 m*M* DTT, 660 m*M* KOAc, BSA, pH 7.9	Purity is confirmed by overdigestion, ligation-recutting and genome DNA digestion test	TaKaRa 1048
3-20 U/μL	1 unit completely digests 1 μg DNA/hr at 37°C	10 m*M* Tris-HCl, 10 m*M* MgCl$_2$, 50 m*M* NaCl, 1 m*M* DTT, pH 7.5	>80% ligation; >95% recut	Toyobo FSP-101 FSP-102

Fsp4H I

RECOGNITION SEQUENCE
5'...GC▼NGC...3'
3'...CGN▲CG...5'

ISOSCHIZOMERS
BsoF I, Fnu4H I, Ita I

NOTES
- EC 3.1.21.4
- A type II site specific deoxyribonuclease
- A large group of enzymes recognizing specific short DNA sequences. They cleave either within the recognition site, or at a short specific distance from it
- Requires Mg^{2+}

CONCENTRATION or VOLUME ACTIVITY	UNIT DEFINITION	REACTION BUFFER (As Provided)	ADDITIONAL ACTIVITIES and PURITY	SUPPLIER CATALOG NO.
Flavobacterium species 4H				
				SibEnzyme

Gsu I

RECOGNITION SEQUENCE
5′...CTGGAG(N)$_{16}$▼...3′
3′...GACCTC(N)$_{14}$▲...5′

ISOSCHIZOMERS
Bpm I

OPTIMUM REACTION CONDITIONS
30°C; 70% activity at 37°C

HEAT INACTIVATION
65°C for 20 min.

METHYLATION EFFECTS
Blocked by overlapping dcm methylation

NOTES
- EC 3.1.21.4
- A type II site specific deoxyribonuclease
- A large group of enzymes recognizing specific short DNA sequences. They cleave either within the recognition site, or at a short specific distance from it.
- 10 μM S-adenosylmethionine gives a two-fold increase in activity
- Requires Mg^{2+}

CONCENTRATION or VOLUME ACTIVITY	UNIT DEFINITION	REACTION BUFFER (As Provided)	ADDITIONAL ACTIVITIES and PURITY	SUPPLIER CATALOG NO.
Gluconobacter suboxydans H-15T				
2000 U/mL	1 unit degrades 1 μg λ dam$^-$ dcm$^-$ DNA/hr at 30°C in a 0.05 mL mixture	10 mM Tris-HCl, 10 mM MgCl$_2$, 100 μg/mL BSA, pH 7.5	No detectable contaminating nuclease; 80% ligated and 90% recut after 25-fold overdigestion	AGS Heidelb F00330S F00330M
1-3 U/μL	1 unit hydrolyzes 1 μg DNA/hr at 37°C in a 50 μL reaction volume using λ phage DNA as substrate	10 mM Tris-HCl, 10 mM MgCl$_2$, pH 7.5	90% of the DNA fragments ligated and >50% recut after 50-fold overdigestion	Fermentas ER0461 ER0462
500-10,000 U/mL	1 unit completely cleaves 1.0 μg λ DNA/hr in a 50 μL reaction mixture	200 mM Tris-acetate, 500 mM KOAc, 100 mM MgoAc, 10 mM DTT, BSA, pH 7.9	100% cleavage, 75% ligation, 95% recleavage No detectable endonuclease (overdigestion)	Sigma R3386

Hae II

RECOGNITION SEQUENCE
  5'...PuGCGC▼Py...3'
  3'...Py▲CGCGPu...5'

ISOSCHIZOMERS
  *Bsp*143 II

OPTIMUM REACTION CONDITIONS
  pH 7.5 and 37°C

HEAT INACTIVATION
  Resistant

METHYLATION EFFECTS
  Does not cleave PuGm^{5}CGCPy, PuGCGm^{5}CPy or PuGhm^{5}CGhm^{5}CPy

NOTES
- EC 3.1.21.4
- A type II site specific deoxyribonuclease
- A large group of enzymes recognizing specific short DNA sequences. They cleave either within the recognition site, or at a short specific distance from it
- Compatible cohesive ends: *Bbe* I
- Requires Mg^{2+} and BSA for full activity

CONCENTRATION or VOLUME ACTIVITY	UNIT DEFINITION	REACTION BUFFER (As Provided)	ADDITIONAL ACTIVITIES and PURITY	SUPPLIER CATALOG NO.
Escherichia coli strain, carrying the cloned gene from *Haemophilus aegyptius*				
10,000 U/mL	1 unit completely digests 1 µg λ DNA/hr in a total reaction volume of 50 µL using the buffer provided	20 m*M* Tris-acetate, 10 m*M* MgOAc, 50 m*M* KOAc, 1 m*M* DTT, pH 7.9; add 100 µg/mL BSA	>95% DNA fragments ligated and recut after 10-fold overdigestion	NE Biolabs 107S 107L
Haemophilus aegyptius				
8-12 U/µL	1 unit completely digests 1 µg λ DNA/hr at 37°C in the buffer provided	100 m*M* Tris-HCl, 100 m*M* MgCl$_2$, 10 m*M* DTT, 500 m*M* NaCl, pH 7.5	100% DNA fragments ligated and recut	Amersham E 1050Y
10,000 U/mL	1 unit degrades 1 µg λ dam$^-$ dcm$^-$ DNA/hr at 37°C in a 0.05 mL mixture	33 m*M* Tris-acetate, 10 m*M* MgOAc, 66 m*M* KOAc, 100 µg/mL BSA, pH 7.9	No detectable contaminating nuclease; 100% ligated and 95% recut after 10-fold overdigestion	AGS Heidelb A00335S A00335M
3-10 U/µL	1 unit completely digests 1 µg substrate DNA/hr	33 m*M* Tris-acetate, 10 m*M* MgOAc, 66 m*M* KOAc, 0.5 m*M* DTT, pH 7.9		Boehringer 693910 693928
8-12 U/µL	1 unit completely digests 1 mg λ DNA/hr at 37°C	500 m*M* Tris-HCl, 100 m*M* MgCl$_2$, 500 m*M* NaCl, pH 8.0	≥95% ligation; 100% cleavage	Life Technol 15204-019
3-10 U/µL	1 unit digests completely 1 µg λ DNA substrate/hr at 37°C	33 m*M* Tris-acetate, 66 m*M* KOAc, 10 m*M* MgOAc, 0.5 m*M* DTT, pH 8.2		NBL Gene 014011 014002
				Nippon Gene

Hae II continued

CONCENTRATION or VOLUME ACTIVITY	UNIT DEFINITION	REACTION BUFFER (As Provided)	ADDITIONAL ACTIVITIES and PURITY	SUPPLIER CATALOG No.
Haemophilus aegyptius continued				
5 U/μL	1 unit hydrolyzes 1 μg λ DNA to completion/hr in a 50 μL total reaction volume	330 mM Tris-acetate, 660 mM KOAc, 100 mM MgOAc, 5 mM DTT, 1 mg/mL BSA, pH 8		Oncor 110181 110182
2000-5000 U/mL	1 unit completely digests 1 μg λ DNA/hr at pH 7.5, 37°C in a 50 μL total assay mixture	100 mM Tris-acetate, 100 mM MgOAc, 500 mM KOAc, pH 7.5		Pharmacia 27-0940-01 27-0940-02
10 U/μL	1 unit completely digests 1 μg λ DNA/hr at 37°C in a 50 mL reaction volume	60 mM Tris-HCl, 60 mM MgCl$_2$, 500 mM NaCl, 10 mM DTT, pH 7.5	90% ligation	Promega R6661 R6662
2000-10,000 U/mL	1 unit completely cleaves 1.0 μg λ DNA/hr in a 50 μL reaction mixture	Unique buffer	100% cleavage, 100% ligation, 95% recleavage No detectable endonuclease (overdigestion)	Sigma R4257
100 U and 500 U	1 unit completely digests 1 μg λ DNA/hr at 37°C in 50 μL of supplied buffer	100 mM Tris-HCl, 100 mM MgCl$_2$, 10 mM DTT, 500 mM NaCl, pH 7.5	Purity is confirmed by overdigestion and ligation-recutting test	TaKaRa 1050
3-20 U/μL	1 unit completely digests 1 μg DNA/hr at 37°C	10 mM Tris-HCl, 10 mM MgCl$_2$, 50 mM NaCl, 1 mM DTT, pH 7.5	>90% ligation; >95% recut	Toyobo HAE-201 HAE-202
Haemophilus aegyptius ATCC 11116				
2000-15,000 U/mL		50 mM KCl, 10 mM Tris-HCl, 0.1 mM EDTA, 1.0 mM DTT, 200 μg/mL BSA, 50% glycerol, pH 7.4		ICN 151220
Obtained from a cloned source				
5000-40,000 U/mL		50 mM KCl, 10 mM Tris-HCl, 0.1 mM EDTA, 1 mM DTT, 500 μg/mL BSA, 50% glycerol, pH 7.4		Stratagene 500550

Hae III

RECOGNITION SEQUENCE
  5'...GG▼CC...3'
  3'...CC▲GG...5'
ISOSCHIZOMERS
  BssC I, BsuR I, Pal I
OPTIMUM REACTION CONDITIONS
  pH 7.9 and 37°C
HEAT INACTIVATION
  Resistant
METHYLATION EFFECTS
  Cleaves GGCm^{5}C
  Does not cleave GGm^{5}CC or GGhm^{5}Chm^{5}C
  Not blocked by overlapping *dcm* methylation

NOTES
- EC 3.1.21.4
- A type II site specific deoxyribonuclease
- A large group of enzymes recognizing specific short DNA sequences. They cleave either within the recognition site, or at a short specific distance from it
- Cleaves single-stranded DNA 10-times more slowly than double-stranded
- High enzyme or glycerol concentration may result in star activity
- Compatible cohesive ends: any blunt end
- Requires Mg^{2+}

CONCENTRATION or VOLUME ACTIVITY	UNIT DEFINITION	REACTION BUFFER (As Provided)	ADDITIONAL ACTIVITIES and PURITY	SUPPLIER CATALOG NO.
Escherichia coli strain, carrying the cloned gene from *Haemophilus aegyptius*				
10,000 U/mL; 50,000 U/mL	1 unit completely digests 1 μg λ DNA/hr in a total reaction volume of 50 μL using the buffer provided	10 m*M* Tris-HCl, 10 m*M* MgCl$_2$, 50 m*M* NaCl, 1 m*M* DTT, pH 7.9	>95% DNA fragments ligated and recut after 20-fold overdigestion	NE Biolabs 108S 108L
Haemophilus aegyptius				
5000 U and 10,000 U	1 unit completely digests 1 μg DNA substrate/hr at 37°C	10 m*M* Tris-HCl, 10 m*M* MgCl$_2$, 50 m*M* NaCl, 1 m*M* DTT, 0.1 mg/mL BSA, pH 7.5	>95% DNA fragments ligated and recut after 10-fold overdigestion	Adv Biotech AB-0318-a AB-0318-b
40 U/μL	1 unit completely cleaves 1 μg substrate DNA/hr	RFLP grade; 100 m*M* NaCl, 50 m*M* Tris, 5 m*M* MgCl$_2$, 2 m*M* β-MSH, pH 7.8	90-98% cleavage-ligation-recleavage	Am Allied H-12-10000
10 U/μL; 40 U/μL	1 unit completely cleaves 1 μg substrate DNA/hr	100 m*M* NaCl, 50 m*M* Tris, 5 m*M* MgCl$_2$, 2 m*M* β-MSH, pH 7.8	90-98% cleavage-ligation-recleavage	Am Allied L-11-02500 H-11-02500 L-11-10000 H-11-10000 L-11-15000 H-11-15000
8-12 U/μL	1 unit completely digests 1 μg λ DNA/hr at 37°C in the buffer provided	100 m*M* Tris-HCl, 100 m*M* MgCl$_2$, 10 m*M* DTT, 500 m*M* NaCl, pH 7.5	>95% DNA fragments ligated and 100% recut	Amersham E 1051Y E 1051Z

Hae III continued

CONCENTRATION or VOLUME ACTIVITY	UNIT DEFINITION	REACTION BUFFER (As Provided)	ADDITIONAL ACTIVITIES and PURITY	SUPPLIER CATALOG NO.
Haemophilus aegyptius continued				
10,000 U/mL	1 unit degrades 1 µg λ dam⁻ dcm⁻ DNA/hr at 37°C in a 0.05 mL mixture	10 mM Tris-HCl, 100 mM NaCl, 100 µg/mL BSA, pH 8.0	No detectable contaminating nuclease; 95% ligated and recut after 25-fold overdigestion	AGS Heidelb A00340S A00340M
10 U/µL and 40 U/µL	1 unit completely digests 1 µg substrate DNA/hr	10 mM Tris-HCl, 10 mM MgCl$_2$, 50 mM NaCl, 1 mM DTE, pH 7.5		Boehringer 693936 693944 1336029 1347691
	1 unit completely cleaves 1 µg λ DNA/hr at pH 7.5, 37°C in a reaction volume of 50 µL	10 mM Tris-HCl, 10 mM MgCl$_2$, 50 mM NaCl, 1.0 mM DTT, 100 µg/mL BSA, pH 7.5		CHIMERx 2150-01 2150-02
5000-20,000 U/mL		50 mM KCl, 10 mM Tris-HCl, 0.1 mM EDTA, 1.0 mM DTT, 200 µg/mL BSA, 50% glycerol, pH 7.5		ICN 151221
8-12 U/µL and 50 U/µL	1 unit completely digests 1 mg λ DNA/hr at 37°C	500 mM Tris-HCl, 100 mM MgCl$_2$, 500 mM NaCl, pH 8.0	≥95% ligation; 100% cleavage	Life Technol 15205-016 15205-081 15205-024
	10 units does not produce unspecific cleavage products with 1 µg λ DNA/16 hr at 37°C	20 mM Tris-acetate, 10 mM MgOAc, 50 mM KOAc, 1 mM DTT, 100 µg/mL BSA, pH 7.9	>90% DNA fragments ligated and recut after 10-fold overdigestion	MINOTECH 112
8-12 U/µL	1 unit digests completely 1 µg λ DNA substrate/hr at 37°C	10 mM Tris-HCl, 50 mM NaCl, 10 mM MgCl$_2$, 1 mM DTT, pH 7.8		NBL Gene 011404 011407
10 U/µL	1 unit hydrolyzes 1 µg λ DNA to completion/hr in a 50 µL total reaction volume	330 mM Tris-acetate, 660 mM KOAc, 100 mM MgOAc, 5 mM DTT, 1 mg/mL BSA, pH 8		Nippon Gene Oncor 110191 110192
5000-10,000 U/mL	1 unit completely digests 1 µg λ DNA/hr at pH 7.5, 37°C in a 50 µL total assay mixture	100 mM Tris-acetate, 100 mM MgOAc, 500 mM KOAc, pH 7.5		Pharmacia 27-0866-01 27-0866-02
10 U/µL and 40-80 U/µL	1 unit completely digests 1 µg λ DNA/hr at 37°C in a 50 mL reaction volume	100 mM Tris-HCl, 100 mM MgCl$_2$, 500 mM NaCl, 10 mM DTT, pH 7.9	90% ligation	Promega R6171 R6172 R6175 R4174
				SibEnzyme

Hae III continued

CONCENTRATION or VOLUME ACTIVITY	UNIT DEFINITION	REACTION BUFFER (As Provided)	ADDITIONAL ACTIVITIES and PURITY	SUPPLIER CATALOG No.
Haemophilus aegyptius continued				
5000-20,000 U/mL	1 unit completely cleaves 1.0 μg λ DNA/hr in a 50 μL reaction mixture	200 mM Tris-acetate, 500 mM KOAc, 100 mM MgOAc, 10 mM DTT, pH 7.9	100% cleavage, 100% ligation, 100% recleavage. No detectable endonuclease (overdigestion)	Sigma R5628
4000 U and 20,000 U	1 unit completely digests 1 μg λ DNA/hr at 37°C in 50 μL of supplied buffer	100 mM Tris-HCl, 100 mM MgCl$_2$, 10 mM DTT, 500 mM NaCl, pH 7.5	Purity is confirmed by overdigestion and ligation-recutting test	TaKaRa 1051
3-20 U/μL	1 unit completely digests 1 μg DNA/hr at 37°C	10 mM Tris-HCl, 10 mM MgCl$_2$, 50 mM NaCl, 1 mM DTT, pH 7.5	>90% ligation; >95% recut	Toyobo HAE-301 HAE-302

Hap II

RECOGNITION SEQUENCE
  5'...C▼CGG...3'
  3'...GGC▲C...5'

ISOSCHIZOMERS
  Hpa II

OPTIMUM REACTION CONDITIONS
  37°C

HEAT INACTIVATION
  70°C for 15 min.

METHYLATION EFFECTS
  Does not cleave Cm^5CGG

NOTES
- EC 3.1.21.4
- A type II site specific deoxyribonuclease
- A large group of enzymes recognizing specific short DNA sequences. They cleave either within the recognition site, or at a short specific distance from it
- Requires Mg^{2+}

CONCENTRATION or VOLUME ACTIVITY	UNIT DEFINITION	REACTION BUFFER (As Provided)	ADDITIONAL ACTIVITIES and PURITY	SUPPLIER CATALOG No.
Haemophilus aphrophilus				
8-12 U/μL	1 unit completely digests 1 μg λ DNA/hr at 37°C in the buffer provided	100 m*M* Tris-HCl, 100 m*M* $MgCl_2$, 10 m*M* DTT, pH 7.5	>90% DNA fragments ligated and 100% recut	Amersham E 1053Y E 1053Z
2000 U and 10,000 U	1 unit completely digests 1 μg λ DNA/hr at 37°C in 50 μL of supplied buffer	100 m*M* Tris-HCl, 100 m*M* $MgCl_2$, 10 m*M* DTT, pH 7.5	Purity is confirmed by overdigestion and ligation-recutting test	TaKaRa 1053

Hga I

RECOGNITION SEQUENCE
  5′...GACGC(N)$_5$▼...3′
  3′...CTGCG(N)$_{10}$▲...5′

OPTIMUM REACTION CONDITIONS
  37°C

HEAT INACTIVATION
  65°C for 20 min.

METHYLATION EFFECTS
  Does not cleave GAm^{5}CGC or GACGm^{5}C

NOTES
- EC 3.1.21.4
- A type II site specific deoxyribonuclease
- A large group of enzymes recognizing specific short DNA sequences. They cleave either within the recognition site, or at a short specific distance from it
- Cleaves single-stranded DNA slowly
- One of the few restriction endonucleases to produce a 5 base 5′-extension
- Requires Mg^{2+}

CONCENTRATION or VOLUME ACTIVITY	UNIT DEFINITION	REACTION BUFFER (As Provided)	ADDITIONAL ACTIVITIES and PURITY	SUPPLIER CATALOG NO.
Escherichia coli, carrying the cloned *Hga* I gene from *Haemophilus gallinarum*				
0.2-0.5 U/μL	1 unit completely digests 1 μg φX174 RFI DNA/hr at 37°C in the buffer provided	100 m*M* Tris-HCl, 100 m*M* MgCl$_2$, 10 m*M* DTT, pH 7.5	>95% DNA fragments ligated and 95% recut	Amersham E 0213Y E 0213Z
2000 U/mL	1 unit completely digests 1 μg φX174 DNA/hr in a total reaction volume of 50 μL using the buffer provided	10 m*M* Bis Tris-Propane-HCl, 10 m*M* MgCl$_2$, 1 m*M* DTT, pH 7.0	>95% DNA fragments ligated and recut after 5-fold overdigestion	NE Biolabs 154S 154L
Haemophilus gallinarum ATCC 14385				
500-2000 U/mL		50 m*M* KCl, 10 m*M* Tris-HCl, 0.1 m*M* EDTA, 1 m*M* DTT, 200 μg/mL BSA, 50% glycerol, pH 7.4		ICN 153813

HgiE I

RECOGNITION SEQUENCE
5'...G▼GWCC...3'
3'...CCWG▲G...5'

ISOSCHIZOMERS
Ava II, Bme18 I, Eco47 I, Sin I

NOTES
- EC 3.1.21.4
- A type II site specific deoxyribonuclease
- A large group of enzymes recognizing specific short DNA sequences. They cleave either within the recognition site, or at a short specific distance from it
- Requires Mg^{2+}

CONCENTRATION or VOLUME ACTIVITY	UNIT DEFINITION	REACTION BUFFER (As Provided)	ADDITIONAL ACTIVITIES and PURITY	SUPPLIER CATALOG NO.
Herpetesiphon giganteus Hpg24				
10,000 U/mL	1 unit degrades 1 μg λ dam⁻ dcm⁻ DNA/hr at 37°C in a 0.05 mL mixture	10 mM Tris-HCl, 10 mM MgCl₂, 100 mM KCl, 100 μg/mL BSA, pH 8.5	No detectable contaminating nuclease; 90% ligated and 100% recut after 10-fold overdigestion	AGS Heidelb F00345S F00345M

Hha I

RECOGNITION SEQUENCE
5'...GCG▼C...3'
3'...C▲GCG...5'

ISOSCHIZOMERS
Cfo I, Hin6 I, HinP1 I

OPTIMUM REACTION CONDITIONS
pH 8.0 and 37°C

HEAT INACTIVATION
65°C for 20 min.; 70°C for 15 min.

METHYLATION EFFECTS
Does not cleave Gm^5CGC, $GCGm^5C$ or Ghm^5CGhm^5C

NOTES
- EC 3.1.21.4
- A type II site specific deoxyribonuclease
- A large group of enzymes recognizing specific short DNA sequences. They cleave either within the recognition site, or at a short specific distance from it
- Produces a 3' extension, compared to 5' extension produced by isoschizomer HinP I
- Cleaves single-stranded DNA at half the rate of double-stranded
- >5% glycerol and high enzyme concentration result in star activity
- Compatible cohesive ends: Cfo I
- Requires Mg^{2+} and BSA for full activity

CONCENTRATION or VOLUME ACTIVITY	UNIT DEFINITION	REACTION BUFFER (As Provided)	ADDITIONAL ACTIVITIES and PURITY	SUPPLIER CATALOG NO.
Escherichia coli strain, carrying the cloned gene from Haemophilus haemolyticus				
20,000 U/mL	1 unit completely digests 1 μg λ DNA/hr in a total reaction volume of 50 μL using the buffer provided	20 mM Tris-acetate, 10 mM MgOAc, 50 mM KOAc, 1 mM DTT, pH 7.9; add 100 μg/mL BSA	>95% DNA fragments ligated and recut after 10-fold overdigestion	NE Biolabs 139S 139L
Haemophilus haemolyticus				
8-12 U/μL	1 unit completely digests 1 μg λ DNA/hr at 37°C in the buffer provided	100 mM Tris-HCl, 100 mM MgCl₂, 10 mM DTT, 500 mM NaCl, pH 7.5	>90% DNA fragments ligated and 100% recut	Amersham E 1056Y E 1056Z
8-12 U/μL	1 unit completely digests 1 mg λ DNA/hr at 37°C	500 mM Tris-HCl, 100 mM MgCl₂, 500 mM NaCl, pH 8.0	≥95% ligation; 95% cleavage	Life Technol 25212-010
				Nippon Gene
10 U/μL and 15-80 U/μL	1 unit hydrolyzes 1 μg λ DNA to completion/hr in a 50 μL total reaction volume	100 mM Tris-HCl, 500 mM NaCl, 100 mM MgCl₂, 100 mM β-MSH, 1.0 mg/mL BSA, pH 8		Oncor 110201 110202
5000-10,000 U/mL	1 unit completely digests 1 μg λ DNA/hr at pH 7.5, 37°C in a 50 μL total assay mixture	100 mM Tris-acetate, 100 mM MgOAc, 500 mM KOAc, pH 7.5		Pharmacia 27-0888-01 27-0888-02

Hha I continued

CONCENTRATION or VOLUME ACTIVITY	UNIT DEFINITION	REACTION BUFFER (As Provided)	ADDITIONAL ACTIVITIES and PURITY	SUPPLIER CATALOG NO.
Haemophilus haemolyticus continued				
10 U/μL	1 unit completely digests 1 μg λ DNA/hr at 37°C in a 50 mL reaction volume	100 mM Tris-HCl, 100 mM MgCl$_2$, 500 mM NaCl, 10 mM DTT, pH 7.9	90% ligation	Promega R6441 R6442
10,000-25,000 U/mL	1 unit completely cleaves 1.0 μg λ DNA/hr in a 50 μL reaction mixture	200 mM Tris-acetate, 500 mM KOAc, 100 mM MgOAc, 10 mM DTT, pH 7.9	100% cleavage, 100% ligation, 60% recleavage. No detectable endonuclease (overdigestion)	Sigma R1506
2000 U and 10,000 U	1 unit completely digests 1 μg λ DNA/hr at 37°C in 50 μL of supplied buffer	100 mM Tris-HCl, 100 mM MgCl$_2$, 10 mM DTT, 500 mM NaCl, pH 7.5	Purity is confirmed by overdigestion and ligation-recutting test	TaKaRa 1056
3-20 U/μL	1 unit completely digests 1 μg DNA/hr at 37°C	10 mM Tris-HCl, 10 mM MgCl$_2$, 50 mM NaCl, 1 mM DTT, pH 7.5	>90% ligation; >95% recut	Toyobo HHA-101 HHA-102
Haemophilus haemolyticus ATCC 10014				
10,000-25,000 U/mL		150 mM KCl, 5 mM KPO$_4$, 0.1 mM EDTA, 1.0 mM DTT, 200 μg/mL BSA, 50% glycerol, pH 7.4		ICN 151257

Hin1 I

RECOGNITION SEQUENCE
    5'...GPu▼CGPyC...3'
    3'...CPyGC▲PuC...5'

ISOSCHIZOMERS
    Acy I, Aha II, Bbi II, BsaH I

OPTIMUM REACTION CONDITIONS
    pH 8.5 and 37°C

HEAT INACTIVATION
    65°C for 20 min.

METHYLATION EFFECTS
    Cleavage rate slowed significantly by overlapping dcm methylation
    Blocked by dcm and CG methylation

NOTES
- EC 3.1.21.4
- A type II site specific deoxyribonuclease
- A large group of enzymes recognizing specific short DNA sequences. They cleave either within the recognition site, or at a short specific distance from it
- Requires Mg^{2+} and BSA for full activity

CONCENTRATION or VOLUME ACTIVITY	UNIT DEFINITION	REACTION BUFFER (As Provided)	ADDITIONAL ACTIVITIES and PURITY	SUPPLIER CATALOG NO.
Haemophilus influenzae RFL1				
10,000 U/mL	1 unit degrades 1 μg λ dam⁻ dcm⁻ DNA/hr at 37°C in a 0.05 mL mixture	10 mM Tris-HCl, 100 mM NaCl, 100 μg/mL BSA, pH 8.0	No detectable contaminating nuclease; 90% ligated and 100% recut after 50-fold overdigestion	AGS Heidelb F00030S F00030M
8-12 U/μL	1 unit hydrolyzes 1 μg DNA/hr at 37°C in a 50 μL reaction volume using λ phage DNA as substrate	10 mM Tris-HCl, 10 mM $MgCl_2$, 50 mM NaCl, 0.1 mg/mL BSA, pH 7.5	>90% of the DNA fragments ligated and recut after 50-fold overdigestion	Fermentas ER0471 ER0472
1000-8000 U/mL		100 mM NaCl, 10 mM KPO_4, 0.1 mM EDTA, 10 mM β-MSH, 200 μg/mL BSA, 50% glycerol, pH 7.4		Stratagene 500614
3-20 U/μL	1 unit completely digests 1 μg DNA/hr at 37°C	Hin 1 I	>90% ligation; >95% recut	Toyobo HIN-101 HIN-102

Hin2 I

RECOGNITION SEQUENCE
5'...C▼CGG...3'
3'...GGC▲C...5'

ISOSCHIZOMERS
Hpa II

OPTIMUM REACTION CONDITIONS
37°C

HEAT INACTIVATION
65°C for 20 min.

METHYLATION EFFECTS
Does not cleave Cm^{5}CGG
Blocked by CG methylation

NOTES
- EC 3.1.21.4
- A type II site specific deoxyribonuclease
- A large group of enzymes recognizing specific short DNA sequences. They cleave either within the recognition site, or at a short specific distance from it
- Requires Mg^{2+} and BSA for full activity

CONCENTRATION or VOLUME ACTIVITY	UNIT DEFINITION	REACTION BUFFER (As Provided)	ADDITIONAL ACTIVITIES and PURITY	SUPPLIER CATALOG NO.
Haemophilus influenzae RFL2				
8-12 U/μL	1 unit hydrolyzes 1 μg DNA/hr at 37°C in a 50 μL reaction volume using λ phage DNA as substrate	33 mM Tris-acetate, 10 mM MgOAc, 66 mM KOAc, 0.1 mg/mL BSA, pH 7.9	>90% DNA fragments ligated and recut after 50-fold overdigestion	Fermentas ER1111 ER1112

Hin6 I

RECOGNITION SEQUENCE
5'...G▼CGC...3'
3'...CGC▲G...5'

ISOSCHIZOMERS
Hha I

OPTIMUM REACTION CONDITIONS
37°C

HEAT INACTIVATION
65°C for 20 min.

METHYLATION EFFECTS
Blocked by CG methylation

NOTES
- EC 3.1.21.4
- A type II site specific deoxyribonuclease
- A large group of enzymes recognizing specific short DNA sequences. They cleave either within the recognition site, or at a short specific distance from it
- Produces DNA fragments with a 2-base 5'-extension; Hha I does not
- Requires Mg^{2+} and BSA for full activity

CONCENTRATION or VOLUME ACTIVITY	UNIT DEFINITION	REACTION BUFFER (As Provided)	ADDITIONAL ACTIVITIES and PURITY	SUPPLIER CATALOG NO.
Haemophilus influenzae RFL6				
10,000 U/mL	1 unit degrades 1 μg λ dam⁻ dcm⁻ DNA/hr at 37°C in a 0.05 mL mixture	10 mM Tris-HCl, 100 mM NaCl, 100 μg/mL BSA, 1 mM DTT, pH 8.0	No detectable contaminating nuclease; 90% ligated and 100% recut after 50-fold overdigestion	AGS Heidelb F00360S F00360M
8-12 U/μL	1 unit hydrolyzes 1 μg DNA/hr at 37°C in a 50 μL reaction volume using λ phage DNA as substrate	33 mM Tris-acetate, 10 mM MgOAc, 66 mM KOAc, 0.1 mg/mL BSA, pH 7.9	>90% of the DNA fragments ligated and recut after 50-fold overdigestion	Fermentas ER0481 ER0482

Hinc II

RECOGNITION SEQUENCE
  5'...GTPy▼PuAC...3'
  3'...CAPu▲PyTG...5'

ISOSCHIZOMERS
  *Hin*d II

OPTIMUM REACTION CONDITIONS
  pH 7.5-7.9 and 37°C

HEAT INACTIVATION
  65°C for 20 min.

METHYLATION EFFECTS
  Cleaves GTm^{5}CPuAC and GTPyPuAm^{5}C
  Does not cleave GTPyPum^{6}AC or GTPyPuAhm^{5}C
  Cleavage is slowed significantly by CG methylation at GTm^{5}CGAC

NOTES
- EC 3.1.21.4
- A type II site specific deoxyribonuclease
- A large group of enzymes recognizing specific short DNA sequences. They cleave either within the recognition site, or at a short specific distance from it
- Mn^{2+}, DMSO or >5% glycerol result in star activity
- Compatible cohesive ends: any blunt end
- Requires Mg^{2+} and BSA for full activity

CONCENTRATION or VOLUME ACTIVITY	UNIT DEFINITION	REACTION BUFFER (As Provided)	ADDITIONAL ACTIVITIES and PURITY	SUPPLIER CATALOG No.
Escherichia coli strain, carrying the cloned gene from *Haemophilus influenzae* Rc				
10,000 U/mL	1 unit completely digests 1 µg λ DNA/hr in a total reaction volume of 50 µL using the buffer provided	50 mM Tris-HCl, 10 mM MgCl$_2$, 100 mM NaCl, 1 mM DTT, pH 7.9; add 100 µg/mL BSA	>95% DNA fragments ligated and recut after 10-fold overdigestion	NE Biolabs 103S 103L
Haemophilus influenzae Rc				
10 U/µL; 40 U/µL	1 unit completely cleaves 1 µg substrate DNA/hr	60 mM KCl, 20 mM Tris, 0.75 mM MgCl$_2$, pH 7.5	90-98% cleavage-ligation-recleavage	Am Allied L-15-01000 H-15-01000 L-15-05000 H-15-05000
>40 U/µL	1 unit completely digests 1 µg λ DNA/hr at 37°C in the buffer provided	100 mM Tris-HCl, 100 mM MgCl$_2$, 10 mM DTT, 500 mM NaCl, pH 7.5	>95% DNA fragments ligated and 100% recut	Amersham E 1059XH
8-12 U/µL	1 unit completely digests 1 µg λ DNA/hr at 37°C in the buffer provided	100 mM Tris-HCl, 100 mM MgCl$_2$, 10 mM DTT, 500 mM NaCl, pH 7.5	>95% DNA fragments ligated and 100% recut	Amersham E 1059Y E 1059Z
10,000 U/mL	1 unit degrades 1 µg λ dam$^-$ dcm$^-$ DNA/hr at 37°C in a 0.05 mL mixture	10 mM Tris-HCl, 100 mM NaCl, 100 µg/mL BSA, pH 8.0	No detectable contaminating nuclease; 90% ligated and 100% recut after 50-fold overdigestion	AGS Heidelb F00370S F00370M

Hinc II continued

CONCENTRATION or VOLUME ACTIVITY	UNIT DEFINITION	REACTION BUFFER (As Provided)	ADDITIONAL ACTIVITIES and PURITY	SUPPLIER CATALOG No.
Haemophilus influenzae Rc continued				
	1 unit completely cleaves 1 µg λ DNA/hr at pH 7.5, 37°C in a reaction volume of 50 µL	10 mM Tris-HCl, 10 mM MgCl$_2$, 50 mM NaCl, 1.0 mM DTT, 100 µg/mL BSA, pH 7.5		CHIMERX 2200-01 2200-02
8-12 U/µL	1 unit hydrolyzes 1 µg DNA/hr at 37°C in a 50 µL reaction volume using λ phage DNA as substrate	33 mM Tris-acetate, 10 mM MgOAc, 66 mM KOAc, 0.1 mg/mL BSA, pH 7.9	>90% of the DNA fragments ligated and recut after 50-fold overdigestion	Fermentas ER0491 ER0492
5000-25,000 U/mL		50 mM KCl, 10 mM Tris-HCl, 0.1 mM EDTA, 1.0 mM DTT, 200 µg/mL BSA, 50% glycerol, pH 7.4		ICN 151258
8-12 U/µL and 50 U/µL	1 unit completely digests 1 mg λ DNA/hr at 37°C	200 mM Tris-HCl, 50 mM MgCl$_2$, 500 M KCl, pH 7.4	≥65% ligation; 100% cleavage	Life Technol 15206-014 15206-022 15206-048 15206-030
	10 units does not produce unspecific cleavage products with 1 µg λ DNA/16 hr at 37°C	20 mM Tris-acetate, 10 mM MgOAc, 50 mM KOAc, 1 mM DTT, 100 µg/mL BSA, pH 7.9	>90% DNA fragments ligated and recut after 10-fold overdigestion	MINOTECH 113
6-16 U/µL	1 unit digests completely 1 µg λ DNA substrate/hr at 37°C	Hinc II buffer		NBL Gene 011501 011503
				Nippon Gene
5 U/µL and 15-40 U/µL	1 unit hydrolyzes 1 µg λ DNA to completion/hr in a 50 µL total reaction volume	100 mM Tris-HCl, 1 M NaCl, 100 mM MgCl$_2$, 100 mM β-MSH, 1.0 mg/mL BSA, pH 8		Oncor 110211 110212
2000-10,000 U/mL and 50,000-80,000 U/mL	1 unit completely digests 1 µg λ DNA/hr at pH 7.5, 37°C in a 50 µL total assay mixture	100 mM Tris-acetate, 100 mM MgOAc, 500 mM KOAc, pH 7.5		Pharmacia 27-0858-01 27-0858-02 27-0858-18
10 U/µL and 40-80 U/µL	1 unit completely digests 1 µg λ DNA/hr at 37°C in a 50 mL reaction volume	Blue/White Cloning Qualified; 60 mM Tris-HCl, 60 mM MgCl$_2$, 500 mM NaCl, 10 mM DTT, pH 7.5	80% ligation	Promega R6031 R6032 R6035 R4034
5000-25,000 U/mL	1 unit completely cleaves 1.0 µg λ DNA/hr in a 50 µL reaction mixture	200 mM Tris-acetate, 500 mM KOAc, 100 mM MgOAc, 10 mM DTT, BSA, pH 7.9	100% cleavage, 90% ligation, 90% recleavage No detectable endonuclease (overdigestion)	Sigma R7511
3-20 U/µL	1 unit completely digests 1 µg DNA/hr at 37°C	50 mM Tris-HCl, 10 mM MgCl$_2$, 100 mM NaCl, 1 mM DTT, pH 7.5	>90% ligation; >95% recut	Toyobo HNC-203 HNC-204

Hinc II continued

CONCENTRATION or VOLUME ACTIVITY	UNIT DEFINITION	REACTION BUFFER (As Provided)	ADDITIONAL ACTIVITIES and PURITY	SUPPLIER CATALOG No.
Haemophilus influenzae Rc, cloned				
1000 U and 5000 U	1 unit completely digests 1 μg λ DNA/hr at 37°C in 50 μL of supplied buffer	100 mM Tris-HCl, 100 mM MgCl$_2$, 10 mM DTT, 500 mM NaCl, pH 7.5	Purity is confirmed by overdigestion and ligation-recutting test	TaKaRa 1059
Obtained from a cloned source				
5000-15,000 U/mL		50 mM NaCl, 20 mM Tris-HCl, 0.1 mM EDTA, 15 mM DTT, 0.1% Triton X-100, 200 μg/mL BSA, 50% glycerol, pH 7.4		Stratagene 500594

Hind II

RECOGNITION SEQUENCE

$$5'...GT(^T_C)\blacktriangledown(^A_G)AC...3'$$
$$3'...CA(^A_G)\blacktriangle(^T_C)TG...5'$$

ISOSCHIZOMERS
    Hinc II

HEAT INACTIVATION
    65°C for 20 min.

METHYLATION EFFECTS
    Does not cleave GTPyPum^{6}AC

NOTES
- EC 3.1.21.4
- A type II site specific deoxyribonuclease
- A large group of enzymes recognizing specific short DNA sequences. They cleave either within the recognition site, or at a short specific distance from it
- Requires Mg^{2+}

VOLUME ACTIVITY	UNIT DEFINITION	REACTION BUFFER (As Provided)	ADDITIONAL ACTIVITIES and PURITY	SUPPLIER CATALOG No.
Haemophilus influenzae Rd com^{-10}				
3-10 U/μL and 40 U/μL	1 unit completely digests 1 μg substrate DNA/hr	10 mM Tris-HCl, 10 mM MgCl$_2$, 50 mM NaCl, 1 mM DTE, pH 7.5		Boehringer 220540 567655 656305 1175092
2000-10,000 U/mL	1 unit completely digests 1 μg λ DNA/hr at 37°C in a 0.05 mL total volume			ICN 197020
				Stratagene

Hind III

RECOGNITION SEQUENCE
5'...A▼AGCTT...3'
3'...TTCGA▲A...5'

OPTIMUM REACTION CONDITIONS
pH 7.9-8.0 and 37°C

HEAT INACTIVATION
Varies

METHYLATION EFFECTS
Cleaves Am^6AGCTT and $AAGChm^5Uhm^5U$
Does not cleave $m^6AAGCTT$, $AAGm^5CTT$ or $AAGhm^5CTT$

NOTES
- EC 3.1.21.4
- A type II site specific deoxyribonuclease
- A large group of enzymes recognizing specific short DNA sequences. They cleave either within the recognition site, or at a short specific distance from it
- Mn^{2+} presence may result in star activity
- Compared with λ DNA substrate, a 2-fold and 10-fold excess is required to cut pBR322 and SV40, respectively
- Requires a duplex of >10 base pairs, Mg^{2+} and BSA for full activity

CONCENTRATION or VOLUME ACTIVITY	UNIT DEFINITION	REACTION BUFFER (As Provided)	ADDITIONAL ACTIVITIES and PURITY	SUPPLIER CATALOG NO.
Escherichia coli strain, carrying the cloned gene from Haemophilus influenzae Rd				
20,000 U/mL; 100,000 U/mL	1 unit completely digests 1 μg λ DNA/hr in a total reaction volume of 50 μL using the buffer provided	10 mM Tris-HCl, 10 mM MgCl₂, 50 mM NaCl, 1 mM DTT, pH 7.9	>95% DNA fragments ligated and recut after 100-fold overdigestion	NE Biolabs 104S 104L
Haemophilus influenzae				
10,000-100,000 U/mL		50 mM KCl, 10 mM Tris-HCl, 0.1 mM EDTA, 1.0 mM DTT, 200 μg/mL BSA, 50% glycerol, pH 7.4		ICN 151259
				Nippon Gene
10,000-20,000 U/mL and 50,000-80,000 U/mL	1 unit completely digests 1 μg λ DNA/hr at pH 7.5, 37°C in a 50 μL total assay mixture	100 mM Tris-acetate, 100 mM MgOAc, 500 mM KOAc, pH 7.5		Pharmacia 27-0860-01 27-0860-02 27-0860-18
				SibEnzyme
Haemophilus influenzae Rd				
8-12 U/μL	1 unit completely digests 1 μg λ DNA/hr at 37°C in the buffer provided	100 mM Tris-HCl, 100 mM MgCl₂, 10 mM DTT, 500 mM NaCl, pH 7.5	100% DNA fragments ligated and recut	Amersham E 1060Y E 1060Z
>40 U/μL	1 unit completely digests 1 μg λ DNA/hr at 37°C in the buffer provided	100 mM Tris-HCl, 100 mM MgCl₂, 10 mM DTT, 500 mM NaCl, pH 7.5	100% DNA fragments ligated and recut	Amersham E 1060YH E 1060XH

Hind III continued

CONCENTRATION or VOLUME ACTIVITY	UNIT DEFINITION	REACTION BUFFER (As Provided)	ADDITIONAL ACTIVITIES and PURITY	SUPPLIER CATALOG NO.
Haemophilus influenzae Rd *continued*				
12,000 U/mL	1 unit degrades 1 µg λ dam⁻ dcm⁻ DNA/hr at 37°C in a 0.05 mL mixture	10 mM Tris-HCl, 10 mM MgCl$_2$, 100 mM KCl, 100 µg/mL BSA, 1 mM DTT, pH 8.5	No detectable contaminating nuclease; 95% ligated and recut after 4-fold overdigestion	AGS Heidelb A00380S A00380M
	1 unit completely cleaves 1 µg λ DNA/hr at pH 7.5, 37°C in a reaction volume of 50 µL	10 mM Tris-HCl, 10 mM MgCl$_2$, 50 mM NaCl, 1.0 mM DTT, 100 µg/mL BSA, pH 7.5		CHIMERx 2220-01 2220-02
8-12 U/µL; 40-60 U/µL	1 unit hydrolyzes 1 µg DNA/hr at 37°C in a 50 µL reaction volume using λ phage DNA as substrate	10 mM Tris-HCl, 10 mM MgCl$_2$, 100 mM KCl, 0.1 mg/mL BSA, pH 8.5	>90% of the DNA fragments ligated and recut after 50-fold overdigestion	Fermentas ER0501 ER0502
8-12 U/µL and 50 U/µL	1 unit completely digests 1 mg λ DNA/hr at 37°C	500 mM Tris-HCl, 100 mM MgCl$_2$, 500 mM NaCl, pH 8.0	≥95% ligation; 100% cleavage	Life Technol 15207-012 15207-020 15207-053 15207-038
	10 units does not produce unspecific cleavage products with 1 µg λ DNA/16 hr at 37°C	10 mM Tris-HCl, 10 mM MgCl$_2$, 50 mM NaCl, 1 mM DTT, 100 µg/mL BSA, pH 7.9	>90% DNA fragments ligated and recut after 10-fold overdigestion	MINOTECH 114
10 U/µL and 20-200 U/µL	1 unit hydrolyzes 1 µg λ DNA to completion/hr in a 50 µL total reaction volume	100 mM Tris-HCl, 500 mM NaCl, 100 mM MgCl$_2$, 100 mM β-MSH, 1.0 mg/mL BSA, pH 8		Oncor 110226 110227 110228
10 U/µL and 40-80 U/µL	1 unit completely digests 1 µg λ DNA/hr at 37°C in a 50 mL reaction volume	Blue/White Cloning Qualified; 60 mM Tris-HCl, 60 mM MgCl$_2$, 1 M NaCl, 10 mM DTT, pH 7.5	90% ligation	Promega R6041 R6042 R6045 R4044
10,000-40,000 U/mL	1 unit completely cleaves 1.0 µg λ DNA/hr in a 50 µL reaction mixture	100 mM Tris-HCl, 500 mM NaCl, 100 mM MgCl$_2$, 10 mM DTT, pH 7.9	100% cleavage, 90% ligation, 90% recleavage, 40% endonuclease (nickase) No detectable endonuclease (overdigestion)	Sigma R1137
10,000 U and 50,000 U	1 unit completely digests 1 µg λ DNA/hr at 37°C in 50 µL of supplied buffer	100 mM Tris-HCl, 100 mM MgCl$_2$, 10 mM DTT, 500 mM NaCl, pH 7.5	Purity is confirmed by overdigestion, ligation-recutting and cloning test	TaKaRa 1060
3-20 U/µL and 40-50 U/µL	1 unit completely digests 1 µg DNA/hr at 37°C	10 mM Tris-HCl, 10 mM MgCl$_2$, 50 mM NaCl, 1 mM DTT, pH 7.5	>90% ligation; >95% recut	Toyobo HND-302 HND-304 HND-305 HND-355

Hind III continued

CONCENTRATION or VOLUME ACTIVITY	UNIT DEFINITION	REACTION BUFFER (As Provided)	ADDITIONAL ACTIVITIES and PURITY	SUPPLIER CATALOG NO.
Haemophilus influenzae Rd com^{-10}				
10,000 and 30,000 U	1 unit completely digests 1 µg DNA substrate/hr at 37°C	10 mM Tris-HCl, 10 mM MgCl$_2$, 50 mM NaCl, 1 mM DTT, 0.1 mg/mL BSA, pH 7.5	>95% DNA fragments ligated and recut after 10-fold overdigestion	Adv Biotech AB-0317-a AB-0317-b
10 U/µL; 40 U/µL	1 unit completely cleaves 1 µg substrate DNA/hr	50 mM NaCl, 10 mM Tris, 10 mM MgCl$_2$, 1 mM β-MSH, pH 7.8	90-98% cleavage-ligation-recleavage	Am Allied L-13-10000 H-13-10000 L-13-50000 H-13-50000
10 U/µL and 40 U/µL	1 unit completely digests 1 µg substrate DNA/hr	10 mM Tris-HCl, 5 mM MgCl$_2$, 100 mM NaCl, 1 mM β-MSH, pH 8.0		Boehringer 656313 656321 798983 1274040
6-12 U/µL 14-15 U/µL >40 U/µL	1 unit digests completely 1 µg λ DNA substrate/hr at 37°C	50 mM Tris-HCl, 50 mM NaCl, 10 mM MgCl$_2$, 1 mM DTT, pH 8.3		NBL Gene 011806 011808 011858
Obtained from a cloned source				
10,000-50,000 U/mL		250 mM KCl, 10 mM Tris-HCl, 0.1 mM EDTA, 1 mM DTT, 500 µg/mL BSA, 50% glycerol, pH 7.4		Stratagene 500600 500601 500609

Hinf I

RECOGNITION SEQUENCE
  5'...G▼ANTC...3'
  3'...CTNA▲G...5'

OPTIMUM REACTION CONDITIONS
  pH 8.0 and 37°C

HEAT INACTIVATION
  Varies

METHYLATION EFFECTS
  Cleaves GANTm^{5}C
  Does not cleave Gm^{6}ANTC or GANThm^{5}C
  Cleavage rate is slowed significantly by CG methylation at GANTm^{5}C

NOTES
- EC 3.1.21.4
- A type II site specific deoxyribonuclease
- A large group of enzymes recognizing specific short DNA sequences. They cleave either within the recognition site, or at a short specific distance from it
- Cuts single-stranded DNA 100-fold slower than double-stranded
- Mn^{2+}, high enzyme or glycerol concentration results in star activity
- Requires Mg^{2+} and BSA for full activity

CONCENTRATION or VOLUME ACTIVITY	UNIT DEFINITION	REACTION BUFFER (As Provided)	ADDITIONAL ACTIVITIES and PURITY	SUPPLIER CATALOG NO.
Escherichia coli strain, carrying the cloned gene from *Haemophilus influenzae* RF				
10,000 U/mL; 40,000 U/mL	1 unit completely digests 1 μg λ DNA/hr in a total reaction volume of 50 μL using the buffer provided	10 m*M* Tris-HCl, 10 m*M* MgCl$_2$, 50 m*M* NaCl, 1 m*M* DTT, pH 7.9	>95% DNA fragments ligated and recut after 100-fold overdigestion	NE Biolabs 155S 155L
Haemophilus influenzae RF				
1250 U and 2500 U	1 unit completely digests 1 μg DNA substrate/hr at 37°C	10 m*M* Tris-HCl, 10 m*M* MgCl$_2$, 50 m*M* NaCl, 1 m*M* DTT, 0.1 mg/mL BSA, pH 7.5	>90% DNA fragments ligated and recut after 10-fold overdigestion	Adv Biotech AB-0228-a AB-0228-b
10 U/μL; 40 U/μL	1 unit completely cleaves 1 μg substrate DNA/hr	100 m*M* NaCl, 50 m*M* Tris, 5 m*M* MgCl$_2$, 2 m*M* β-MSH, pH 7.8	90-98% cleavage-ligation-recleavage	Am Allied L-16-01000 H-16-01000 L-16-05000 H-16-05000
8-12 U/μL	1 unit completely digests 1 μg λ DNA/hr at 37°C in the buffer provided	500 m*M* Tris-HCl, 100 m*M* MgCl$_2$, 10 m*M* DTT, 1000 m*M* NaCl, pH 7.5	>90% DNA fragments ligated and 100% recut	Amersham E 1061Y E 1061Z
10,000 U/mL	1 unit degrades 1 μg λ dam$^-$ dcm$^-$ DNA/hr at 37°C in a 0.05 mL mixture	10 m*M* Tris-HCl, 10 m*M* MgCl$_2$, 100 m*M* KCl, 100 μg/mL BSA, pH 8.5	No detectable contaminating nuclease; 100% ligated and 95% recut after 50-fold overdigestion	AGS Heidelb A00383S A00383M

Hinf I continued

CONCENTRATION or VOLUME ACTIVITY	UNIT DEFINITION	REACTION BUFFER (As Provided)	ADDITIONAL ACTIVITIES and PURITY	SUPPLIER CATALOG NO.
Haemophilus influenzae RF continued				
10 U/μL and 40 U/μL	1 unit completely digests 1 μg substrate DNA/hr	50 mM Tris-HCl, 10 mM MgCl$_2$, 100 mM NaCl, 1 mM DTE, pH 7.5		Boehringer 779652 779679 1097067 1274082
	1 unit completely cleaves 1 μg λ DNA/hr at pH 7.5, 37°C in a reaction volume of 50 μL	10 mM Tris-HCl, 10 mM MgCl$_2$, 50 mM NaCl, 1.0 mM DTT, 100 μg/mL BSA, pH 7.5		CHIMERX 2240-01 2240-02
8-12 U/μL; 40-60 U/μL	1 unit hydrolyzes 1 μg DNA/hr at 37°C in a 50 μL reaction volume using λ phage DNA as substrate	10 mM Tris-HCl, 10 mM MgCl$_2$, 100 mM KCl, 0.1 mg/mL BSA, pH 8.5	>90% of the DNA fragments ligated and recut after 50-fold overdigestion	Fermentas ER0801 ER0802
8-12 U/μL	1 unit completely digests 1 mg λ DNA/hr at 37°C	500 mM Tris-HCl, 100 mM MgCl$_2$, 500 mM NaCl, pH 8.0	≥90% ligation; 100% cleavage	Life Technol 15223-019
8-12 U/μL >40 U/μL	1 unit digests completely 1 μg λ DNA substrate/hr at 37°C	50 mM Tris-HCl, 100 mM NaCl, 10 mM MgCl$_2$, 1 mM DTT, pH 7.8		NBL Gene 011903 011906 011956
				Nippon Gene
10 U/μL and 15-50 U/μL	1 unit hydrolyzes 1 μg λ DNA to completion/hr in a 50 μL total reaction volume	100 mM Tris-HCl, 1 M NaCl, 100 mM MgCl$_2$, 100 mM β-MSH, 1.0 mg/mL BSA, pH 8		Oncor 110231 110232
8000-15,000 U/mL	1 unit completely digests 1 μg λ DNA/hr at pH 7.5, 37°C in a 50 μL total assay mixture	100 mM Tris-acetate, 100 mM MgOAc, 500 mM KOAc, pH 7.5		Pharmacia 27-0968-01 27-0968-02
10 U/μL and 40-80 U/μL	1 unit completely digests 1 μg λ DNA/hr at 37°C in a 50 mL reaction volume	60 mM Tris-HCl, 60 mM MgCl$_2$, 500 mM NaCl, 10 mM DTT, pH 7.5	90% ligation	Promega R6201 R6202 R6205 R4204
				SibEnzyme
8000-20,000 U/mL		50 mM KCl, 10 mM Tris-HCl, 0.1 mM EDTA, 1 mM DTT, 200 μg/mL BSA, 50% glycerol, pH 7.4		Stratagene 500610
3000 U and 15,000 U	1 unit completely digests 1 μg λ DNA/hr at 37°C in 50 μL of supplied buffer	500 mM Tris-HCl, 100 mM MgCl$_2$, 10 mM DTT, 1 M NaCl, pH 7.5	Purity is confirmed by overdigestion and ligation-recutting test	TaKaRa 1061
3-20 U/μL	1 unit completely digests 1 μg DNA/hr at 37°C	50 mM Tris-HCl, 10 mM MgCl$_2$, 100 mM NaCl, 1 mM DTT, pH 7.5	>90% ligation; >95% recut	Toyobo HNF-101 HNF-102

Hinf I continued

CONCENTRATION or VOLUME ACTIVITY	UNIT DEFINITION	REACTION BUFFER (As Provided)	ADDITIONAL ACTIVITIES and PURITY	SUPPLIER CATALOG No.
Haemophilus influenzae serotype f				
2000-10,000 U/mL	1 unit completely cleaves 1.0 μg λ DNA/hr in a 50 μL reaction mixture	100 mM Tris-HCl, 500 mM NaCl, 100 mM MgCl$_2$, 10 mM DTT, pH 7.9	100% cleavage, 95% ligation, 100% recleavage No detectable endonuclease (overdigestion)	Sigma R6760

Hinf II

RECOGNITION SEQUENCE
5'...G▼ANTC...3'
3'...CTNA▲G...5'

NOTES
- EC 3.1.21.4
- A type II site specific deoxyribonuclease
- A large group of enzymes recognizing specific short DNA sequences. They cleave either within the recognition site, or at a short specific distance from it
- Requires Mg^{2+}

CONCENTRATION or VOLUME ACTIVITY	UNIT DEFINITION	REACTION BUFFER (As Provided)	ADDITIONAL ACTIVITIES and PURITY	SUPPLIER CATALOG No.
Haemophilus influenzae				
2000-10,000 U/mL		50 mM KCl, 10 mM Tris-HCl, 0.1 mM EDTA, 1.0 mM DTT, 200 μg/mL BSA, 50% glycerol, pH 7.4		ICN 151260

HinP1 I

RECOGNITION SEQUENCE
5'...G▼CGC...3'
3'...CGC▲G...5'

ISOSCHIZOMERS
Hha I

OPTIMUM REACTION CONDITIONS
37°C

HEAT INACTIVATION
65°C for 20 min.

METHYLATION EFFECTS
Does not cleave Gm^{5}CGC

NOTES
- EC 3.1.21.4
- A type II site specific deoxyribonuclease
- A large group of enzymes recognizing specific short DNA sequences. They cleave either within the recognition site, or at a short specific distance from it
- Produces a 5' extension, compared to 3' extension produced by isoschizomer *Hha* I
- Cleaves single-stranded DNA at half the rate of double-stranded
- The 5' extension can be efficiently ligated into the *Acc* I site of M13 and pUC cloning vectors
- Requires Mg^{2+}

CONCENTRATION or VOLUME ACTIVITY	UNIT DEFINITION	REACTION BUFFER (As Provided)	ADDITIONAL ACTIVITIES and PURITY	SUPPLIER CATALOG No.
Haemophilus influenzae				
5000-20,000 U/mL		50 m*M* KCl, 10 m*M* Tris-HCl, 0.1 m*M* EDTA, 1.0 m*M* DTT, 200 μg/mL BSA, 50% glycerol, pH 7.4		ICN 151261
Escherichia coli strain, carrying the cloned gene from Haemophilus influenzae P$_1$				
10,000 U/mL	1 unit completely digests 1 μg λ DNA/hr in a total reaction volume of 50 μL using the buffer provided	10 m*M* Tris-HCl, 10 m*M* MgCl$_2$, 50 m*M* NaCl, 1 m*M* DTT, pH 7.9	>95% DNA fragments ligated and recut after 5-fold overdigestion	NE Biolabs 124S 124L

Hpa I

RECOGNITION SEQUENCE
   5'...GTT▼AAC...3'
   3'...CAA▲TTG...5'

OPTIMUM REACTION CONDITIONS
   pH 7.0 and 37°C

HEAT INACTIVATION
   Resistant

METHYLATION EFFECTS
   Cleaves GTTAAm^{5}C
   Does not cleave GTTAm^{6}AC, GTTAhm^{5}C or Ghm^{5}Uhm^{5}UAAC
   Blocked by overlapping CG methylation

NOTES
- EC 3.1.21.4
- A type II site specific deoxyribonuclease
- A large group of enzymes recognizing specific short DNA sequences. They cleave either within the recognition site, or at a short specific distance from it
- Large excess of enzyme and >5% glycerol results in the appearance of star activity
- Compatible cohesive ends: any blunt end
- Requires Mg^{2+} and BSA for full activity

CONCENTRATION or VOLUME ACTIVITY	UNIT DEFINITION	REACTION BUFFER (As Provided)	ADDITIONAL ACTIVITIES and PURITY	SUPPLIER CATALOG No.
Escherichia coli strain, carrying the cloned gene from *Haemophilus parainfluenzae*				
5000 U/mL	1 unit completely digests 1 μg λ DNA/hr in a total reaction volume of 50 μL using the buffer provided	20 m*M* Tris-acetate, 10 m*M* MgOAc, 50 m*M* KOAc, 1 m*M* DTT, pH 7.9	>95% DNA fragments ligated and recut after 20-fold overdigestion	NE Biolabs 105S 105L
Haemophilus parainfluenzae				
500 U and 2000 U	1 unit completely digests 1 μg DNA substrate/hr at 37°C	10 m*M* Tris-acetate, 10 m*M* MgOAc, 50 m*M* KOAc, 1 m*M* DTT, 0.1 mg/mL BSA, pH 7.9	>95% DNA fragments ligated and recut after 10-fold overdigestion	Adv Biotech AB-0341-a AB-0341-b
8-12 U/μL	1 unit completely digests 1 μg λ DNA/hr at 37°C in the buffer provided	200 m*M* Tris-HCl, 100 m*M* MgCl$_2$, 10 m*M* DTT, 1000 m*M* KCl, pH 8.5	>95% DNA fragments ligated and 100% recut	Amersham E 1064Y
10,000 U/mL	1 unit degrades 1 μg λ dam$^-$ dcm$^-$ DNA/hr at 37°C in a 0.05 mL mixture	33 m*M* Tris-acetate, 10 m*M* MgOAc, 66 m*M* KOAc, 100 μg/mL BSA, pH 7.9	No detectable contaminating nuclease; 90% ligated and 100% recut after 20-fold overdigestion	AGS Heidelb F00384S F00384M
3-10 U/μL	1 unit completely digests 1 μg substrate DNA/hr	33 m*M* Tris-acetate, 10 m*M* MgOAc, 66 m*M* KOAc, 0.5 m*M* DTT, pH 7.9		Boehringer 380385 567647
8-12 U/μL	1 unit hydrolyzes 1 μg DNA/hr at 37°C in a 50 μL reaction volume using λ phage DNA as substrate	33 m*M* Tris-acetate, 10 m*M* MgOAc, 66 m*M* KOAc, 0.1 mg/mL BSA, pH 7.9	>90% of the DNA fragments ligated and recut after 50-fold overdigestion	Fermentas ER0841 ER0842

Hpa I continued

CONCENTRATION or VOLUME ACTIVITY	UNIT DEFINITION	REACTION BUFFER (As Provided)	ADDITIONAL ACTIVITIES and PURITY	SUPPLIER CATALOG No.
Haemophilus parainfluenzae continued				
2000-10,000 U/mL		10 mM Tris-HCl, 0.1 mM EDTA, 1.0 mM DTT, 200 µg/mL BSA, 50% glycerol, pH 7.5		ICN 151266
4-6 U/µL	1 unit completely digests 1 mg λ DNA/hr at 37°C	200 mM Tris-HCl, 50 mM MgCl$_2$, 500 M KCl, pH 7.4	≥33% ligation; 100% cleavage	Life Technol 15208-010 15208-028
3-7 U/µL	1 unit digests completely 1 µg λ DNA substrate/hr at 37°C	33 mM Tris-acetate, 66 mM KOAc, 10 mM MgOAc, 0.5 mM DTT, pH 8.2		NBL Gene 013111 013102
				Nippon Gene
5 U/µL	1 unit hydrolyzes 1 µg λ DNA to completion/hr in a 50 µL total reaction volume	330 mM Tris-acetate, 660 mM KOAc, 100 mM MgOAc, 5 mM DTT, 1 mg/mL BSA, pH 8		Oncor 110241 110242
1000-5000 U/mL	1 unit completely digests 1 µg λ DNA/hr at pH 7.5, 37°C in a 50 µL total assay mixture	100 mM Tris-acetate, 100 mM MgOAc, 500 mM KOAc, pH 7.5		Pharmacia 27-0862-01 27-0862-02
3-10 U/µL	1 unit completely digests 1 µg λ DNA/hr at 37°C in a 50 mL reaction volume	100 mM Tris-HCl, 70 mM MgCl$_2$, 500 mM KCl, 10 mM DTT, pH 7.5	80% ligation	Promega R6301 R6302 R6305
				SibEnzyme
2000-10,000 U/mL	1 unit completely cleaves 1.0 µg Ad2 DNA/hr in a 50 µL reaction mixture	200 mM Tris-acetate, 500 mM KOAc, 100 mM MgOAc, 10 mM DTT, pH 7.9	100% cleavage, 75% ligation, 80% recleavage, 50% endonuclease (nickase) No detectable endonuclease (overdigestion)	Sigma R8507
1000-5000 U/mL		50 mM KCl, 10 mM Tris-HCl, 0.1 mM EDTA, 1 mM DTT, 200 µg/mL BSA, 50% glycerol, pH 7.5		Stratagene 500630 500631
500 U and 2500 U	1 unit completely digests 1 µg λ DNA/hr at 37°C in 50 µL of supplied buffer	200 mM Tris-HCl, 100 mM MgCl$_2$, 10 mM DTT, 1 M KCl, pH 8.5	Purity is confirmed by overdigestion and ligation-recutting test	TaKaRa 1064
3-20 U/µL	1 unit completely digests 1 µg DNA/hr at 37°C	10 mM Tris-HCl, 10 mM MgCl$_2$, 50 mM NaCl, 1 mM DTT, pH 7.5	>90% ligation; >95% recut	Toyobo HPA-101 HPA-102
	1 unit completely cleaves 1 µg λ DNA/hr at pH 7.5, 37°C in a reaction volume of 50 µL	10 mM Tris-HCl, 10 mM MgCl$_2$, 50 mM KCl, 1.0 mM DTT, 100 µg/mL BSA, pH 7.5		CHIMERx 2250-01 2250-02

Hpa II

RECOGNITION SEQUENCE
5'...C▼CGG...3'
3'...GGC▲C...5'

ISOSCHIZOMERS
Hap II, Msp I

OPTIMUM REACTION CONDITIONS
pH 7.5 and 37°C

HEAT INACTIVATION
65°C for 20 min.

METHYLATION EFFECTS
Does not cleave m^4CCGG, m^5CCGG, Cm^4CGG, Cm^5CGG or hm^5Chm^5CGG
Blocked by CG methylation

NOTES
- EC 3.1.21.4
- A type II site specific deoxyribonuclease
- A large group of enzymes recognizing specific short DNA sequences. They cleave either within the recognition site, or at a short specific distance from it
- Inhibited by salt concentrations of >50 mM KCl
- Hpa II and Msp I exhibit different methylation sensitivity
- Compatible cohesive ends: Acy I, Aha II, Asu II, Ban III, Bbi II, BstB I, Cla I, Csp45 I, Hap II, HinP1 I, Hin1 I, Mae II, Msp I, Nar I, Nsp V, Sfu I, Taq I, TthH8 I
- Requires Mg^{2+}

CONCENTRATION or VOLUME ACTIVITY	UNIT DEFINITION	REACTION BUFFER (As Provided)	ADDITIONAL ACTIVITIES and PURITY	SUPPLIER CATALOG NO.
Escherichia coli strain, carrying the cloned gene from *Haemophilus parainfluenzae*				
6000 U/mL; 30,000 U/mL	1 unit completely digests 1 μg λ DNA/hr in a total reaction volume of 50 μL using the buffer provided	10 mM Bis Tris-Propane-HCl, 10 mM MgCl₂, 1 mM DTT, pH 7.0	>95% DNA fragments ligated and recut after 10-fold overdigestion	NE Biolabs 171S 171L
Haemophilus parainfluenzae				
10,000 U/mL	1 unit degrades 1 μg λ dam⁻ dcm⁻ DNA/hr at 37°C in a 0.05 mL mixture	10 mM Tris-HCl, 10 mM MgCl₂, 100 μg/mL BSA, pH 7.5	No detectable contaminating nuclease; 90% ligated and 100% recut after 50-fold overdigestion	AGS Heidelb F00385S F00385M
10 U/μL and 40 U/μL	1 unit completely digests 1 μg substrate DNA/hr	10 mM Tris-HCl, 10 mM MgCl₂, 1 mM DTE, pH 7.5		Boehringer 239291 656330 1207598
8-12 U/μL	1 unit hydrolyzes 1 μg DNA/hr at 37°C in a 50 μL reaction volume using λ phage DNA as substrate	10 mM Tris-HCl, 10 mM MgCl₂, pH 7.5	>90% of the DNA fragments ligated and recut after 50-fold overdigestion	Fermentas ER0511 ER0512
8-12 U/μL	1 unit completely digests 1 mg λ DNA/hr at 37°C	200 mM Tris-HCl, 100 mM MgCl₂, pH 7.4	≥95% ligated; 100% recut	Life Technol 15209-018 15209-067

Hpa II continued

CONCENTRATION or VOLUME ACTIVITY	UNIT DEFINITION	REACTION BUFFER (As Provided)	ADDITIONAL ACTIVITIES and PURITY	SUPPLIER CATALOG No.
Haemophilus parainfluenzae continued				
8-12 U/μL	1 unit digests completely 1 μg λ DNA substrate/hr at 37°C	10 mM Tris-HCl, 10 mM MgCl$_2$, 1 mM DTT, pH 7.8		NBL Gene 012603 012606
5 U/μL	1 unit hydrolyzes 1 μg λ DNA to completion/hr in a 50 μL total reaction volume	100 mM Tris-HCl, 100 mM NaCl, 100 mM MgCl$_2$, 100 mM β-MSH, 1.0 mg/mL BSA, pH 7.5		Oncor 110251 110252
5000-10,000 U/mL	1 unit completely digests 1 μg λ DNA/hr at pH 7.5, 37°C in a 50 μL total assay mixture	100 mM Tris-acetate, 100 mM MgOAc, 500 mM KOAc, pH 7.5		Pharmacia 27-0864-01 27-0864-02
10 U/μL	1 unit completely digests 1 μg λ DNA/hr at 37°C in a 50 mL reaction volume	60 mM Tris-HCl, 60 mM MgCl$_2$, 60 mM NaCl, 10 mM DTT, pH 7.5	90% ligation	Promega R6311 R6312 R6315
				SibEnzyme
2000-20,000 U/mL	1 unit completely cleaves 1.0 μg λ DNA/hr in a 50 μL reaction mixture	100 mM Bis Tris-Propane-HCl, 100 mM MgCl$_2$, 10 mM DTT, pH 7.0	100% cleavage, 90% ligation, 100% recleavage No detectable endonuclease (overdigestion)	Sigma R0629
1000-25,000 U/mL		50 mM KCl, 10 mM Tris-HCl, 0.1 mM EDTA, 1 mM DTT, 200 μg/mL BSA, 50% glycerol, pH 7.4		Stratagene 500640 500641
3-20 U/μL	1 unit completely digests 1 μg DNA/hr at 37°C	10 mM Tris-HCl, 10 mM MgCl$_2$, 1 mM DTT, pH 7.5	>90% ligation; >90% recut	Toyobo HPA-201 HPA-202
	1 unit completely cleaves 1 μg λ DNA/hr at pH 7.5, 37°C in a reaction volume of 50 μL	10 mM Tris-HCl, 10 mM MgCl$_2$, 1.0 mM DTT, 100 μg/mL BSA, pH 7.5		CHIMERx 2260-01 2260-02
2000-20,000 U/mL		50 mM KCl, 10 mM Tris-HCl, 0.1 mM EDTA, 1.0 mM DTT, 200 μg/mL BSA, 50% glycerol, pH 7.4		ICN 151267

Hph I

RECOGNITION SEQUENCE
  5'...GGTGA(N)$_8$▼...3'
  3'...CCACT(N)$_7$▲...5'

OPTIMUM REACTION CONDITIONS
  37°C; enhanced by low pH and high glycerol concentration

HEAT INACTIVATION
  65°C for 20 min.

METHYLATION EFFECTS
  Cleaves TCACm^{5}C
  Does not cleave Tm^{5}CACC, TCAm^{5}CC or GGTGm^{6}A
  Blocked by overlapping *dam* methylation

NOTES
- EC 3.1.21.4
- A type II site specific deoxyribonuclease
- A large group of enzymes recognizing specific short DNA sequences. They cleave either within the recognition site, or at a short specific distance from it
- May cleave at N_9/N_8, depending on the sequence between recognition and cleavage sites
- Incubation of >12 U for >4 hours on φX174 DNA results in additional cleavage products. Has not yet been shown for other DNAs
- Requires Mg^{2+} and BSA for full activity

CONCENTRATION or VOLUME ACTIVITY	UNIT DEFINITION	REACTION BUFFER (As Provided)	ADDITIONAL ACTIVITIES and PURITY	SUPPLIER CATALOG NO.
Escherichia coli strain, carrying the cloned gene from *Haemophilus parahaemolyticus* (C.A. Hutchison III)				
4000 U/mL; 6000 U/mL	1 unit completely digests 1 μg λ DNA/hr in a total reaction volume of 50 μL using the buffer provided	20 m*M* Tris-acetate, 10 m*M* MgOAc, 50 m*M* KOAc, 1 m*M* DTT, pH 7.9	>75% DNA fragments ligated and recut after 2-fold overdigestion	NE Biolabs 158S 158L
Haemophilus parahaemolyticus				
3-10 U/μL	1 unit completely digests 1 μg λ DNA/hr at 37°C in the buffer provided	330 m*M* Tris-acetate, 100 m*M* MgOAc, 5 m*M* DTT, 660 m*M* KOAc, pH 7.9	>75% DNA fragments ligated and 75% recut	Amersham E 0214Y E 0214Z
8-12 U/μL	1 unit hydrolyzes 1 μg dam$^-$ DNA/hr at 37°C in a 50 μL reaction volume using λ phage DNA as substrate	10 m*M* Tris-HCl, 10 m*M* MgCl$_2$, 0.1 mg/mL BSA, pH 7.5	>70% of the DNA fragments ligated and >90% recut after 50-fold overdigestion	Fermentas ER1101 ER1102
500-10,000 U/mL	1 unit completely cleaves 1.0 μg λ DNA/hr in a 50 μL reaction mixture		100% cleavage, 70% ligation, 100% recleavage No detectable endonuclease (overdigestion)	Sigma R5260
Haemophilus parahaemolyticus (C.A. Hutchison III)				
1000-10,000 U/mL	Unstable at 37°C	50 m*M* KCl, 10 m*M* Tris-HCl, 0.1 m*M* EDTA, 1 m*M* DTT, 200 μg/mL BSA, 50% glycerol, pH 7.4		ICN 153815

Hsp92 I

Recognition Sequence

$5'...G(^A_G)\blacktriangledown CG(^T_C)C...3'$
$3'...C(^T_C)GC\blacktriangle(^A_G)G...5'$

Notes
- EC 3.1.21.4
- A type II site specific deoxyribonuclease
- A large group of enzymes recognizing specific short DNA sequences. They cleave either within the recognition site, or at a short specific distance from it
- Requires Mg^{2+}

CONCENTRATION or VOLUME ACTIVITY	UNIT DEFINITION	REACTION BUFFER (As Provided)	ADDITIONAL ACTIVITIES and PURITY	SUPPLIER CATALOG No.
Haemophilus influenzae				
10 U/μL	1 unit completely digests 1 μg λ DNA/hr at 37°C in a 50 mL reaction volume	100 m*M* Tris-HCl, 100 m*M* MgCl₂, 1 *M* NaCl, 10 m*M* DTT, pH 8.5	90% ligation	Promega R7151 R7152

Hsp92 II

RECOGNITION SEQUENCE

5'...CATG▼...3'
3'...▲GTAC...5'

NOTES

- EC 3.1.21.4
- A type II site specific deoxyribonuclease
- A large group of enzymes recognizing specific short DNA sequences. They cleave either within the recognition site, or at a short specific distance from it
- Requires Mg^{2+}

CONCENTRATION or VOLUME ACTIVITY	UNIT DEFINITION	REACTION BUFFER (As Provided)	ADDITIONAL ACTIVITIES and PURITY	SUPPLIER CATALOG No.
Haemophilus influenzae 92				
10 U/µL	1 unit completely digests 1 µg λ DNA/hr at 37°C in a 50 mL reaction volume	100 mM Tris-HCl, 100 mM $MgCl_2$, 1.5 M KCl, pH 7.4	90% ligation	Promega R7161 R7162

HspA I

RECOGNITION SEQUENCE
5'...G▼CGC...3'
3'...CGC▲G...5'

NOTES
- EC 3.1.21.4
- A type II site specific deoxyribonuclease
- A large group of enzymes recognizing specific short DNA sequences. They cleave either within the recognition site, or at a short specific distance from it
- Requires Mg^{2+}

CONCENTRATION or VOLUME ACTIVITY	UNIT DEFINITION	REACTION BUFFER (As Provided)	ADDITIONAL ACTIVITIES and PURITY	SUPPLIER CATALOG No.
Haemophilus **species A1**				
10,000 U/mL		33 mM Tris-acetate, 10 mM MgOAc, 66 mM KOAc, 1 mM DTT, pH 7.9	95% DNA fragments ligated and recut after 10-fold overdigestion	SibEnzyme E069 E070

I-*Ceu* I

RECOGNITION SEQUENCE
5'...TAACTATAACGGTCCTAA▼GGTAGCGA...3'
3'...ATTGATATTGCCAG▲GATTCCATCGCT...5'

NOTES
- An intron-encoded endonuclease
- EC 3.1.21.4
- A type II site specific deoxyribonuclease
- A large group of enzymes recognizing specific short DNA sequences. They cleave either within the recognition site, or at a short specific distance from it
- No detectable cleavage of pBR322, λ, Ad2, T7 or fX174 DNA after incubation with 5 U enzyme for 16 hr at 37°C
- Requires Mg^{2+}

I-Ceu I continued

CONCENTRATION or VOLUME ACTIVITY	UNIT DEFINITION	REACTION BUFFER (As Provided)	ADDITIONAL ACTIVITIES and PURITY	SUPPLIER CATALOG NO.
From the chloroplast large rRNA gene of *Chlamydomonas*, cloned and overexpressed in *E. coli*				
500–2000 U/mL	1 unit cleaves 1 μg agarose-embedded *E. coli* chromosomal DNA to completion/3 hr at 37°C; 0.5 unit digests 1 μg DNA/3 hr at 37°C in a reaction volume of 50 μL	10 mM Tris-HCl, 10 mM MgCl$_2$, 1 mM DTT, pH 8.6; add 100 μg/mL BSA	>95% DNA fragments ligated and recut after 5-fold overdigestion	NE Biolabs 699S 699L

I-Cla I

RECOGNITION SEQUENCE

5'...AT▼CGAT...3'
3'...TAGC▲TA...5'

NOTES

- An intron-encoded endonuclease
- EC 3.1.21.4
- A type II site specific deoxyribonuclease

- A large group of enzymes recognizing specific short DNA sequences. They cleave either within the recognition site, or at a short specific distance from it
- Compatible cohesive ends: *Acy* I, *Asu* II, *Hin*P I, *Mae* II, *Msp* I, *Nar* I, *Sfu* I, *Taq* I
- Requires Mg^{2+}

CONCENTRATION or VOLUME ACTIVITY	UNIT DEFINITION	REACTION BUFFER (As Provided)	ADDITIONAL ACTIVITIES and PURITY	SUPPLIER CATALOG NO.
Bacillus sphaericus				
10 U/μL and 20–150 U/μL	1 unit hydrolyzes 1 μg λ DNA to completion/hr in a 50 μL total reaction volume	100 mM Tris-HCl, 500 mM NaCl, 100 mM MgCl$_2$, 100 mM β-MSH, 1.0 mg/mL BSA, pH 8		Oncor 110501 110502

I-Dmo I

RECOGNITION SEQUENCE

5'...GCCTT[GCCGGGTAAG▼TTCC]GGC...3'
3'...CG<u>AA</u>[CGGCCC▲ATTCAAGG]<u>CCG</u>...5'

NOTES

- An intron-encoded endonuclease
- EC 3.1.21.4
- A type II site specific deoxyribonuclease
- A large group of enzymes recognizing specific short DNA sequences. They cleave either within the recognition site, or at a short specific distance from it
- The 14 bracketed base pairs represent the minimal recognition sequence
- An absence of underlined nucleotides reduces cleavage, but does not abolish it
- Requires Mg^{2+}

CONCENTRATION or VOLUME ACTIVITY	UNIT DEFINITION	REACTION BUFFER (As Provided)	ADDITIONAL ACTIVITIES and PURITY	SUPPLIER CATALOG NO.
Desulfurococcus mobilis				
6-14 U/μL	1 unit digests completely 1 μg λ DNA substrate/hr at 65°C			NBL Gene 013201 013203

I-Ppo I

RECOGNITION SEQUENCE

5'...CTCTCTTAA▼GGTAGC...3'
3'...GAGAG▲AATTCCATCG...5'

NOTES

- An intron-encoded endonuclease
- EC 3.1.21.4
- A type II site specific deoxyribonuclease
- A large group of enzymes recognizing specific short DNA sequences. They cleave either within the recognition site, or at a short specific distance from it
- Recognizes rDNA repeats in *S. cerevisiae* chromosome 12, generating 600 Bp, 400 Bp and 9 kb fragments
- No detectable cleavage of pBR322, I, Ad2, T7 or fX174 DNA after incubation with 5 U enzyme for 16 hr at 37°C
- Incubation with >50 U/16 hr gives cleavage at degenerate sites
- Requires Mg^{2+}

CONCENTRATION or VOLUME ACTIVITY	UNIT DEFINITION	REACTION BUFFER (As Provided)	ADDITIONAL ACTIVITIES and PURITY	SUPPLIER CATALOG NO.
From the nuclear extrachromosomal ribosomal DNA of *Physarum polycephalum*, cloned and overexpressed in *E. coli*				
1000-10,000 U/mL	1 unit cleaves 1 μg agarose-embedded *Saccharomyces cerevisiae* chromosomal DNA in 0.1 mL to completion/3 hr at 37°C; 0.5 unit digests 1 μg DNA/3 hr at 37°C in a reaction volume of 50 μL	10 m*M* Tris-HCl, 10 m*M* $MgCl_2$, 1 m*M* DTT, pH 8.6; add 100 μg/mL BSA	>95% DNA fragments ligated and recut after 5-fold overdigestion	NE Biolabs 697S 697L
Physarum polycephalum				
100-200 U/μL	1 unit completely digests 1 μg I-*Ppo* I plasmid DNA/hr at 37°C in a 50 mL reaction volume		90% ligation	Promega R7031 R7032

I-Sce I

RECOGNITION SEQUENCE

5'...TAGGGATAA▼CAGGGTAAT...3'

APPLICATIONS
- A super rare-cutting endonuclease, for cloning and mapping
- Generates a 4-base 3'-hydroxyl overhang
- Tolerates single-base substitutions within the recognition sequence

NOTES
- EC 3.1.21.4
- An intron-encoded endonuclease
- A type II site specific deoxyribonuclease
- A large group of enzymes recognizing specific short DNA sequences. They cleave either within the recognition site, or at a short specific distance from it
- Requires Mg^{2+}

CONCENTRATION or VOLUME ACTIVITY	UNIT DEFINITION	REACTION BUFFER (As Provided)	ADDITIONAL ACTIVITIES and PURITY	SUPPLIER CATALOG No.
Escherichia coli TG1				
	1 unit cleaves 1 µg pSCM522x Hinf λ-DNA/hr at 37°C in a total volume of 25 µL incubation buffer		Composed of omega nuclease and omega transposase	Boehringer Meganuclease 1497235 1362399
Saccharomyces cerevisiae				
				Boehringer

Iso-Nae I

NOTES
- EC 3.1.21.4
- A type II site specific deoxyribonuclease
- A large group of enzymes recognizing specific short DNA sequences. They cleave either within the recognition site, or at a short specific distance from it
- Requires Mg^{2+}

CONCENTRATION or VOLUME ACTIVITY	UNIT DEFINITION	REACTION BUFFER (As Provided)	ADDITIONAL ACTIVITIES and PURITY	SUPPLIER CATALOG No.
Streptomyces species				
	50 units does not produce unspecific cleavage products with 1 μg λ DNA/16 hr at 37°C	10 mM Tris-HCl, 10 mM MgCl$_2$, 1 mM DTT, 100 μg/mL BSA, pH 7.9	>90% DNA fragments ligated and recut after 10-fold overdigestion	MINOTECH 142

Ita I

RECOGNITION SEQUENCE
5'...GC▼NGC...3'
3'...CGN▲CG...5'

HEAT INACTIVATION
65°C for 20 min.

NOTES
- EC 3.1.21.4
- A type II site specific deoxyribonuclease
- A large group of enzymes recognizing specific short DNA sequences. They cleave either within the recognition site, or at a short specific distance from it
- Require Mg^{2+}

CONCENTRATION or VOLUME ACTIVITY	UNIT DEFINITION	REACTION BUFFER (As Provided)	ADDITIONAL ACTIVITIES and PURITY	SUPPLIER CATALOG No.
Ilyobacter tartaricus				
10 U/μL	1 unit completely digests 1 μg substrate DNA/hr	50 mM Tris-HCl, 10 mM MgCl$_2$, 100 mM NaCl, 1 mM DTE, pH 7.5		Boehringer 1497979 1497987

Kas I

RECOGNITION SEQUENCE
5'...G▼GCGCC...3'
3'...CCGCG▲G...5'

ISOSCHIZOMERS
Nar I

OPTIMUM REACTION CONDITIONS
37°C

HEAT INACTIVATION
65°C for 20 min.

NOTES
- EC 3.1.21.4
- A type II site specific deoxyribonuclease
- A large group of enzymes recognizing specific short DNA sequences. They cleave either within the recognition site, or at a short specific distance from it
- Produces a 4-base 5' extension, compared to the 2-base 5' extension produced by isoschizomer *Nar* I
- Marked site preference
- 25-times more active on λ than on pBR322 DNA
- Requires Mg^{2+} and BSA for full activity

CONCENTRATION or VOLUME ACTIVITY	UNIT DEFINITION	REACTION BUFFER (As Provided)	ADDITIONAL ACTIVITIES and PURITY	SUPPLIER CATALOG NO.
Kluyvera ascorbata				
1000-5000 U/mL		300 m*M* KCl, 10 m*M* Tris-HCl, 10 m*M* MgCl$_2$, 0.1 m*M* EDTA, 1 m*M* DTT, 200 μg/mL acetylated BSA, 50% glycerol, pH 7.4		ICN 159448
Escherichia coli strain, carrying the cloned gene from Kluyvera ascorbata				
1000-5000 U/mL	1 unit completely digests 1 μg pBR322 DNA/hr in a total reaction volume of 50 μL using the buffer provided	10 m*M* Tris-HCl, 10 m*M* MgCl$_2$, 50 m*M* NaCl, 1 m*M* DTT, pH 7.9; add 100 μg/mL BSA	>95% DNA fragments ligated and recut after 10-fold overdigestion	NE Biolabs 544S 544L

Kpn I

RECOGNITION SEQUENCE
    5'...GGTAC▼C...3'
    3'...C▲CATGG...5'

ISOSCHIZOMERS
    *Acc*65 I, *Asp*718 I

OPTIMUM REACTION CONDITIONS
    pH 7.0-7.5 and 37°C

HEAT INACTIVATION
    Resistant

METHYLATION EFFECTS
    Cleaves GGTAm^{5}CC, GGTACm^{5}C, GGTAm^{5}Cm^{5}C and GGTm^{6}ACC
    Does not cleave GGTm^{6}Am^{5}CC, GGTAm^{4}CC or GGTACm^{4}C

NOTES
- EC 3.1.21.4
- A type II site specific deoxyribonuclease
- A large group of enzymes recognizing specific short DNA sequences. They cleave either within the recognition site, or at a short specific distance from it
- Conditions of high enzyme concentration, glycerol concentration >5% or pH >8.0 may result in star activity
- Inhibited by salt concentrations >20 m*M* salt
- Requires Mg^{2+} and BSA for full activity

CONCENTRATION or VOLUME ACTIVITY	UNIT DEFINITION	REACTION BUFFER (As Provided)	ADDITIONAL ACTIVITIES and PURITY	SUPPLIER CATALOG No.
Klebsiella pneumoniae				
10 U/μL; 40 U/μL	1 unit completely cleaves 1 μg substrate DNA/hr	60 m*M* KCl, 20 m*M* Tris, 0.75 m*M* MgCl₂, pH 7.5	90-98% cleavage-ligation-recleavage	Am Allied L-24-02000 H-24-02000 L-24-05000 H-24-05000 L-24-25000 H-24-25000
8-12 U/μL	1 unit completely digests 1 μg λ DNA/hr at 37°C in the buffer provided	100 m*M* Tris-HCl, 100 m*M* MgCl₂, 10 m*M* DTT, pH 7.5	>90% DNA fragments ligated and >95% recut	Amersham E 1068Y E 1068Z
8-12 U/μL and 50 U/μL	1 unit completely digests 1 mg Ad2 DNA/hr at 37°C	200 m*M* Tris-HCl, 50 m*M* MgCl₂, 500 *M* KCl, pH 7.4	≥95% ligation; 100% cleavage	Life Technol 15232-010 15232-036 15232-077 15232-028
10 U/μL	1 unit hydrolyzes 1 μg λ DNA to completion/hr in a 50 μL total reaction volume	100 m*M* Tris-HCl, 100 m*M* NaCl, 100 m*M* MgCl₂, 100 m*M* β-MSH, 1.0 mg/mL BSA, pH 7.5		Nippon Gene Oncor 110261 110262

Kpn I continued

CONCENTRATION or VOLUME ACTIVITY	UNIT DEFINITION	REACTION BUFFER (As Provided)	ADDITIONAL ACTIVITIES and PURITY	SUPPLIER CATALOG NO.
Klebsiella pneumoniae continued				
8000-15,000 U/mL and 50,000-80,000 U/mL	1 unit completely digests 1 μg Ad2 DNA/hr at pH 7.5, 37°C in a 50 μL total assay mixture	100 mM Tris-acetate, 100 mM MgOAc, 500 mM KOAc, pH 7.5		Pharmacia 27-0908-01 27-0908-02 27-0908-18
15,000 U/mL		Blue/white certified; 10 mM Tris-HCl, 10 mM MgCl$_2$, 1 mM DTT, pH 7.6	70% DNA fragments ligated and recut after 5-fold overdigestion	SibEnzyme E079 E080
10,000-60,000 U/mL	1 unit completely cleaves 1.0 μg λ DNA/hr in a 50 μL reaction mixture	100 mM Bis Tris-Propane-HCl, 100 mM MgCl$_2$, 10 mM DTT, pH 7.0	100% cleavage, 70% ligation, 60% recleavage, 20% endonuclease (nickase) No detectable endonuclease (overdigestion)	Sigma R1258
8000-20,000 U/mL		50 mM KCl, 10 mM Tris-HCl, 0.1 mM EDTA, 1 mM DTT, 200 μg/mL BSA, 50% glycerol, pH 7.4		Stratagene 500660 500661
5000 U and 25,000 U	1 unit completely digests 1 μg λ DNA/hr at 37°C in 50 μL of supplied buffer	100 mM Tris-HCl, 100 mM MgCl$_2$, 10 mM DTT, pH 7.5	Purity is confirmed by overdigestion, ligation-recutting and cloning test	TaKaRa 1068
Klebsiella pneumoniae OK8				
7500 U and 20,000 U	1 unit completely digests 1 μg DNA substrate/hr at 37°C	10 mM Tris-HCl, 10 mM MgCl$_2$, 1 mM DTT, 0.1 mg/mL BSA, pH 7.5	>95% DNA fragments ligated and recut after 10-fold overdigestion	Adv Biotech AB-0229-a AB-0229-b
10,000 U/mL	1 unit degrades 1 μg λ dam$^-$ dcm$^-$ DNA/hr at 37°C in a 0.05 mL mixture	10 mM Tris-HCl, 10 mM MgCl$_2$, 1 mM DTT, 100 μg/mL BSA, pH 7.5	No detectable contaminating nuclease; 90% ligated and 100% recut after 50-fold overdigestion	AGS Heidelb A00390S A00390M
10 U/μL and 40 U/μL	1 unit completely digests 1 μg substrate DNA/hr	10 mM Tris-HCl, 10 mM MgCl$_2$, 1 mM DTE, pH 7.5		Boehringer 899186 742945 742953 742961
	1 unit completely cleaves 1 μg Ad2 DNA/hr at pH 7.5, 37°C in a reaction volume of 50 μL	10 mM Tris-HCl, 10 mM MgCl$_2$, 1.0 mM DTT, 100 μg/mL BSA, pH 7.5		CHIMERx 2280-01 2280-02
8-12 U/μL; 40-60 U/μL	1 unit hydrolyzes 1 μg DNA/hr at 37°C in a 50 μL reaction volume using λ phage DNA as substrate	10 mM Tris-HCl, 10 mM MgCl$_2$, 0.1 mg/mL BSA, Triton X-100, pH 7.5	>90% of the DNA fragments ligated and recut after 50-fold overdigestion	Fermentas ER0521 ER0522 ER0523

Kpn I continued

CONCENTRATION or VOLUME ACTIVITY	UNIT DEFINITION	REACTION BUFFER (As Provided)	ADDITIONAL ACTIVITIES and PURITY	SUPPLIER CATALOG No.
Klebsiella pneumoniae OK8 continued				
2000-20,000 U/mL		50 mM KCl, 10 mM Tris-HCl, 0.1 mM EDTA, 1.0 mM DTT, 200 μg/mL BSA, 50% glycerol, pH 7.4		ICN 151405
	10 units does not produce unspecific cleavage products with 1 μg λ DNA/16 hr at 37°C	10 mM Tris-HCl, 10 mM MgCl$_2$, 100 μg/mL BSA, 0.01% Triton X-100, pH 7.0	>90% DNA fragments ligated and recut after 10-fold overdigestion	MINOTECH 115
6-12 U/μL	1 unit digests completely 1 μg λ DNA substrate/hr at 37°C	6 mM Tris-HCl, 6 mM NaCl, 6 mM MgCl$_2$, 6 mM β-MSH, 100 μg/mL BSA, pH 7.5		NBL Gene 012504 012507
10,000 U/mL	1 unit completely digests 1 μg Ad2 DNA/hr in a total reaction volume of 50 μL using the buffer provided	10 mM Bis Tris-Propane-HCl, 10 mM MgCl$_2$, 1 mM DTT, pH 7.0; add 100 μg/mL BSA	>95% DNA fragments ligated and recut after 20-fold overdigestion	NE Biolabs 142S 142L
8-12 U/μL and 40-80 U/μL	1 unit completely digests 1 μg Ad2 DNA/hr at 37°C in a 50 mL reaction volume	Blue/White Cloning Qualified; 100 mM Tris-HCl, 70 mM MgCl$_2$, 500 mM KCl, 10 mM DTT, pH 7.5	90% ligation	Promega R6341 R6342 R6345 R4344
3-20 U/μL and 40-50 U/μL	1 unit completely digests 1 μg DNA/hr at 37°C	10 mM Tris-HCl, 10 mM MgCl$_2$, 1 mM DTT, pH 7.5	>90% ligation; >95% recut	Toyobo KPN-101 KPN-102 KPN-152

Kpn2 I

RECOGNITION SEQUENCE
　　5'...T▼CCGGA...3'
　　3'...AGGCC▲T...5'

ISOSCHIZOMERS
　　*Bsp*M II

OPTIMUM REACTION CONDITIONS
　　55°C; 50% activity at 37°C

HEAT INACTIVATION
　　Resistant

METHYLATION EFFECTS
　　Cleaves TCCGGm^{6}A
　　Does not cleave Tm^{5}CCGGA or TCm^{5}CGGA
　　Blocked by CG methylation

NOTES
- EC 3.1.21.4
- A type II site specific deoxyribonuclease
- A large group of enzymes recognizing specific short DNA sequences. They cleave either within the recognition site, or at a short specific distance from it
- Requires Mg^{2+}

CONCENTRATION or VOLUME ACTIVITY	UNIT DEFINITION	REACTION BUFFER (As Provided)	ADDITIONAL ACTIVITIES and PURITY	SUPPLIER CATALOG NO.
Escherichia coli, carrying the cloned *Kpn*2IR gene from *Klebsiella pneumoniae* RFL2				
8-12 U/μL	1 unit hydrolyzes 1 μg DNA/hr at 37°C in a 50 μL reaction volume using λ phage DNA as substrate	33 m*M* Tris-acetate, 10 m*M* MgOAc, 66 m*M* KOAc, pH 7.9	>90% of the DNA fragments ligated and recut after 50-fold overdigestion	Fermentas ER0531 ER0532
Klebsiella pneumoniae RFL2				
2000 U/mL	1 unit degrades 1 μg λ dam$^-$ dcm$^-$ DNA/hr at 55°C in a 0.05 mL mixture	10 m*M* Tris-HCl, 100 m*M* NaCl, 100 μg/mL BSA, pH 8.0	No detectable contaminating nuclease; 90% ligated and 100% recut after 10-fold overdigestion	AGS Heidelb F00020S F00020M
8-12 U/μL	1 unit completely digests 1 mg λ DNA/hr at 55°C	200 m*M* Tris-HCl, 50 m*M* MgCl$_2$, 500 *M* KCl, pH 7.4	≥95% ligation; 100% cleavage	Life Technol 15490-014

Ksp I

RECOGNITION SEQUENCE
　　5′...CCGC▼GG...3′
　　3′...GG▲CGCC...5′
ISOSCHIZOMERS
　　Sac II, Sst II
HEAT INACTIVATION
　　65°C for 20 min.
METHYLATION EFFECTS
　　Does not cleave m^{5}CCGCGG or Cm^{5}CGCGG

NOTES
- EC 3.1.21.4
- A type II site specific deoxyribonuclease
- A large group of enzymes recognizing specific short DNA sequences. They cleave either within the recognition site, or at a short specific distance from it
- Requires Mg^{2+}

CONCENTRATION or VOLUME ACTIVITY	UNIT DEFINITION	REACTION BUFFER (As Provided)	ADDITIONAL ACTIVITIES and PURITY	SUPPLIER CATALOG NO.
Kluyvera species				
10 U/μL and 40 U/μL	1 unit completely digests 1 μg substrate DNA/hr	10 mM Tris-HCl, 10 mM MgCl$_2$, 1 mM DTE, pH 7.5		Boehringer 1117807 1117793 1207709

Ksp22 I

RECOGNITION SEQUENCE
　　5′...T▼GATCA...3′
　　3′...ACTAG▲T...5′
HEAT INACTIVATION
　　65°C for 20 min.
METHYLATION EFFECTS
　　Blocked by overlapping *dam* methylation

NOTES
- EC 3.1.21.4
- A type II site specific deoxyribonuclease
- A large group of enzymes recognizing specific short DNA sequences. They cleave either within the recognition site, or at a short specific distance from it
- Requires Mg^{2+}

Ksp22 I continued

CONCENTRATION or VOLUME ACTIVITY	UNIT DEFINITION	REACTION BUFFER (As Provided)	ADDITIONAL ACTIVITIES and PURITY	SUPPLIER CATALOG NO.
Kurthia species 22				
30,000 U/mL		10 mM Tris-HCl, 10 mM MgCl$_2$, 50 mM NaCl, 1 mM DTT, pH 7.6	90% DNA fragments ligated and recut after 10-fold overdigestion	SibEnzyme E081 E082

Ksp632 I

RECOGNITION SEQUENCE
  5'...CTCTTC(N)$_1$▼...3'
  3'...GAGAAG(N)$_4$▲...5'
ISOSCHIZOMERS
  Eam1104 I, Ear I
HEAT INACTIVATION
  65°C for 20 min.

NOTES
- EC 3.1.21.4
- A type II site specific deoxyribonuclease
- A large group of enzymes recognizing specific short DNA sequences. They cleave either within the recognition site, or at a short specific distance from it
- Requires Mg^{2+}

CONCENTRATION or VOLUME ACTIVITY	UNIT DEFINITION	REACTION BUFFER (As Provided)	ADDITIONAL ACTIVITIES and PURITY	SUPPLIER CATALOG NO.
Kluyvera species				
10 U/µL	1 unit completely digests 1 µg substrate DNA/hr	33 mM Tris-acetate, 10 mM MgOAc, 66 mM KOAc, 0.5 mM DTT, pH 7.9		Boehringer 1081276 1081284

Kzo9 I

RECOGNITION SEQUENCE
5'...▼GATC...3'
3'...CTAG▲...5'

HEAT INACTIVATION
65°C for 20 min.

METHYLATION EFFECTS
Not blocked by overlapping *dam* methylation

NOTES
- EC 3.1.21.4
- A type II site specific deoxyribonuclease
- A large group of enzymes recognizing specific short DNA sequences. They cleave either within the recognition site, or at a short specific distance from it
- Requires Mg^{2+}

CONCENTRATION or VOLUME ACTIVITY	UNIT DEFINITION	REACTION BUFFER (As Provided)	ADDITIONAL ACTIVITIES and PURITY	SUPPLIER CATALOG No.
Kurthia zopfii 9				
5000 U/mL		10 mM Tris-HCl, 10 mM MgCl₂, 50 mM NaCl, 1 mM DTT, pH 7.6	90% DNA fragments ligated and recut after 10-fold overdigestion	SibEnzyme E187 E188

Lsp I

RECOGNITION SEQUENCE
5'...TT▼CGAA...3'
3'...AAGC▲TT...5'

ISOSCHIZOMERS
Asu II, *Bst*B I

NOTES
- EC 3.1.21.4
- A type II site specific deoxyribonuclease
- A large group of enzymes recognizing specific short DNA sequences. They cleave either within the recognition site, or at a short specific distance from it
- Requires Mg^{2+}

CONCENTRATION or VOLUME ACTIVITY	UNIT DEFINITION	REACTION BUFFER (As Provided)	ADDITIONAL ACTIVITIES and PURITY	SUPPLIER CATALOG No.
Lactobacillus species				
5-12 U/μL	1 unit digests completely 1 μg λ DNA substrate/hr at 60°C	10 mM Tris-HCl, 50 mM NaCl, 10 mM MgCl$_2$, 6 mM β-MSH, 100 μg/mL BSA, pH 7.5		NBL Gene 015402 015405

Mae I

RECOGNITION SEQUENCE
5'...C▼TAG...3'
3'...GAT▲C...5'

ISOSCHIZOMERS
Bfa I, Rma I

HEAT INACTIVATION
65°C for 20 min.

NOTES
- EC 3.1.21.4
- A type II site specific deoxyribonuclease
- A large group of enzymes recognizing specific short DNA sequences. They cleave either within the recognition site, or at a short specific distance from it
- Requires Mg^{2+}

CONCENTRATION or VOLUME ACTIVITY	UNIT DEFINITION	REACTION BUFFER (As Provided)	ADDITIONAL ACTIVITIES and PURITY	SUPPLIER CATALOG No.
Methanococcus aeolicus PI-15/H				
1-5 U/μL	1 unit completely digests 1 μg substrate DNA/hr			Boehringer 822213 822221

Mae II

RECOGNITION SEQUENCE
    5'...A▼CGT...3'
    3'...TGC▲A...5'
ISOSCHIZOMERS
    Tai I (different cleavage position)
HEAT INACTIVATION
    65°C for 20 min.
METHYLATION EFFECTS
    Does not cleave Am⁵CGT

NOTES
- EC 3.1.21.4
- A type II site specific deoxyribonuclease
- A large group of enzymes recognizing specific short DNA sequences. They cleave either within the recognition site, or at a short specific distance from it
- Requires Mg^{2+}

CONCENTRATION or VOLUME ACTIVITY	UNIT DEFINITION	REACTION BUFFER (As Provided)	ADDITIONAL ACTIVITIES and PURITY	SUPPLIER CATALOG NO.
Methanococcus aeolicus				
				Pharmacia
Methanococcus aeolicus PI-15/H				
1-5 U/μL	1 unit completely digests 1 μg substrate DNA/hr			Boehringer 862495

Mae III

RECOGNITION SEQUENCE
    5'...▼GTNAC...3'
    3'...CANTG▲...5'
OPTIMUM REACTION CONDITIONS
    55°C
HEAT INACTIVATION
    85°C for 30 min.

NOTES
- EC 3.1.21.4
- A type II site specific deoxyribonuclease
- A large group of enzymes recognizing specific short DNA sequences. They cleave either within the recognition site, or at a short specific distance from it
- Requires Mg^{2+}

Mae III continued

CONCENTRATION or VOLUME ACTIVITY	UNIT DEFINITION	REACTION BUFFER (As Provided)	ADDITIONAL ACTIVITIES and PURITY	SUPPLIER CATALOG NO.
Methanococcus aeolicus				
1000-5000 U/mL	1 unit completely digests 1 μg λ DNA/hr at pH 8.2, 55°C in a 50 μL total assay mixture	20 mM Tris-HCl, 275 mM NaCl, 6 mM MgCl$_2$, 7 mM β-MSH, pH 8.2		Pharmacia 27-0863-01
Methanococcus aeolicus PI-15/H				
1-5 U/μL	1 unit completely digests 1 μg substrate DNA/hr			Boehringer 822230 822248

Mam I

RECOGNITION SEQUENCE
    5′...GATNN▼NNATC...3′
    3′...CTANN▲NNTAG...5′

NOTES
- EC 3.1.21.4
- A type II site specific deoxyribonuclease
- A large group of enzymes recognizing specific short DNA sequences. They cleave either within the recognition site, or at a short specific distance from it
- Requires Mg^{2+}

CONCENTRATION or VOLUME ACTIVITY	UNIT DEFINITION	REACTION BUFFER (As Provided)	ADDITIONAL ACTIVITIES and PURITY	SUPPLIER CATALOG NO.
Corynebacterium species (Microbacterium ammoniaphilum)				
10 U/μL	1 unit completely digests 1 μg substrate DNA/hr	50 mM Tris-HCl, 10 mM MgCl$_2$, 100 mM NaCl, 1 mM DTE, pH 7.5		Boehringer 1131281 1131290

Mbo I

RECOGNITION SEQUENCE

5'...▼GATC...3'
3'...CTAG▲...5'

ISOSCHIZOMERS

Dpn I, Sau3A I

HEAT INACTIVATION

65°C for 20 min.; 75°C for 15 min.

METHYLATION EFFECTS

Cleaves GATm^{4}C, GATm^{5}C and GAhm^{5}UC
Does not cleave Gm^{6}ATC or GAThm^{5}C
Blocked by dam methylation; Sau3A I is not

NOTES

- EC 3.1.21.4
- A type II site specific deoxyribonuclease
- A large group of enzymes recognizing specific short DNA sequences. They cleave either within the recognition site, or at a short specific distance from it
- DNA from dam$^+$ E. coli is partially resistant to cleavage
- Compatible cohesive ends: BamH I, Bcl I, Bgl II, Bst I, BstY I, Dpn II, Mfl I, Nde II, Sau3A I, Xho II
- Requires Mg^{2+}

CONCENTRATION or VOLUME ACTIVITY	UNIT DEFINITION	REACTION BUFFER (As Provided)	ADDITIONAL ACTIVITIES and PURITY	SUPPLIER CATALOG NO.
Escherichia coli strain, carrying the cloned gene from *Moraxella bovis*				
5000 U/mL; 25,000 U/mL	1 unit completely digests 1 μg λ DNA (dam⁻)/hr in a total reaction volume of 50 μL using the buffer provided	50 mM Tris-HCl, 10 mM MgCl$_2$, 100 mM NaCl, 1 mM DTT, pH 7.9	>95% DNA fragments ligated and recut after 100-fold overdigestion	NE Biolabs 147S 147L
Moraxella bovis				
500 U and 1000 U	1 unit completely digests 1 μg DNA substrate/hr at 37°C	10 mM Tris-HCl, 10 mM MgCl$_2$, 100 mM NaCl, 1 mM DTT, 0.1 mg/mL BSA, pH 7.5	>95% of the DNA fragments ligated and recut after 2-fold overdigestion	Adv Biotech AB-0399-a AB-0399-b
8-12 U/μL	1 unit completely digests 1 μg dam⁻ λ DNA/hr at 37°C in the buffer provided	200 mM Tris-HCl, 100 mM MgCl$_2$, 10 mM DTT, 1000 mM KCl, pH 8.5	>95% DNA fragments ligated and 100% recut	Amersham E 1069Y E 1069Z
	1 unit completely cleaves 1 μg unmethylated λ DNA/hr at pH 7.5, 37°C in a reaction volume of 50 μL	10 mM Tris-HCl, 10 mM MgCl$_2$, 50 mM NaCl, 1.0 mM DTT, 100 μg/mL BSA, pH 7.5		CHIMERX 2283-01 2283-02
8-12 U/μL	1 unit hydrolyzes 1 μg DNA/hr at 37°C in a 50 μL reaction volume using λ phage DNA as substrate	10 mM Tris-HCl, 10 mM MgCl$_2$, 100 mM KCl, pH 8.5	>90% of the DNA fragments ligated and recut after 50-fold overdigestion	Fermentas ER0811 ER0812
6-14 U/μL	1 unit digests completely 1 μg λ DNA substrate/hr at 37°C	10 mM Tris-HCl, 10 mM MgCl$_2$, 1 mM DTT, pH 7.8		MINOTECH NBL Gene 017001 017003

Mbo I continued

CONCENTRATION or VOLUME ACTIVITY	UNIT DEFINITION	REACTION BUFFER (As Provided)	ADDITIONAL ACTIVITIES and PURITY	SUPPLIER CATALOG No.
Moraxella bovis continued				
5 U/μL	1 unit hydrolyzes 1 μg λ DNA to completion/hr in a 50 μL total reaction volume	100 mM Tris-HCl, 500 mM NaCl, 100 mM MgCl$_2$, 100 mM β-MSH, 1.0 mg/mL BSA, pH 8		Oncor 110581 110582
5000-15,000 U/mL	1 unit cleaves ≥95% λ DNA/hr in a 50 μL reaction mixture	500 mM Tris-HCl, 1 M NaCl, 100 mM MgCl$_2$, 10 mM DTT, BSA, pH 7.9	100% cleavage, 70% ligation, 100% recleavage, 50% endonuclease (nickase) No detectable endonuclease (overdigestion)	Sigma R7881
800-4000 U/mL		50 mM KCl, 10 mM Tris-HCl, 0.1 mM EDTA, 1 mM DTT, 200 μg/mL BSA, 50% glycerol, pH 7.4		Stratagene 500670
1000 U and 5000 U	1 unit completely digests 1 μg N^6-methyladenine free λ DNA/hr at 37°C in 50 μL of supplied buffer	200 mM Tris-HCl, 100 mM MgCl$_2$, 10 mM DTT, 1 M KCl, pH 8.5	Purity is confirmed by overdigestion and ligation-recutting test	TaKaRa 1069
Moraxella bovis ATCC 10900				
5000-15,000 U/mL		50 mM KCl, 10 mM Tris-HCl, 0.1 mM EDTA, 1.0 mM DTT, 200 μg/mL BSA, 50% glycerol, pH 7.4		ICN 151590
8-12 U/μL	1 unit completely digests 1 μg unmethylated λ DNA/hr at 37°C in a 50 mL reaction volume	100 mM Tris-HCl, 100 mM MgCl$_2$, 500 mM NaCl, 10 mM DTT, pH 7.9	95% ligation	Promega R6711 R6712
Moraxella bovis M				
8-12 U/μL	1 unit completely digests 1 mg SV40 DNA/hr at 37°C	500 mM Tris-HCl, 100 mM MgCl$_2$, 500 mM NaCl, pH 8.0	≥95% ligation; 100% cleavage	Life Technol 15248-016 15248-073

Mbo II

RECOGNITION SEQUENCE
   5'...GAAGA(N)$_8$▼...3'
   3'...CTTCT(N)$_7$▲...5'

OPTIMUM REACTION CONDITIONS
   pH 7.5

HEAT INACTIVATION
   65°C for 20 min.; 70°C for 15 min.; 80°C for 5 min.

METHYLATION EFFECTS
   Cleaves Tm^{5}CTTm^{5}C and Gm^{6}AAGA
   Does not cleave GAAGm^{6}A or GAm^{6}AGA
   Blocked by overlapping *dam* methylation

NOTES
- EC 3.1.21.4
- A type II site specific deoxyribonuclease
- A large group of enzymes recognizing specific short DNA sequences. They cleave either within the recognition site, or at a short specific distance from it
- Produces an ambiguous single base extension which is difficult to ligate
- Digests >1 hr are not recommended
- Partially cleaves DNA from *dam*$^+$ E. coli
- Requires Mg^{2+}

CONCENTRATION or VOLUME ACTIVITY	UNIT DEFINITION	REACTION BUFFER (As Provided)	ADDITIONAL ACTIVITIES and PURITY	SUPPLIER CATALOG NO.
Escherichia coli strain, carrying the cloned gene from *Moraxella bovis*				
5000 U/mL	1 unit completely digests 1 µg λ DNA (dam⁻)/hr in a total reaction volume of 50 µL using the buffer provided	10 m*M* Tris-HCl, 10 m*M* MgCl$_2$, 50 m*M* NaCl, 1 m*M* DTT, pH 7.9	50% DNA fragments ligated and >90% recut after 2-fold overdigestion	NE Biolabs 148S 148L
Moraxella bovis				
8-12 U/µL	1 unit completely digests 1 µg dam⁻ λ DNA/hr at 37°C in the buffer provided	100 m*M* Tris-HCl, 100 m*M* MgCl$_2$, 10 m*M* DTT, pH 7.5	95% DNA fragments ligated and 100% recut	Amersham E 1145Y E 1145Z
5000 U/mL	1 unit degrades 1 µg λ dam⁻ dcm⁻ DNA/hr at 37°C in a 0.05 mL mixture	33 m*M* Tris-acetate, 10 m*M* MgOAc, 66 m*M* KOAc, 100 µg/mL BSA, pH 7.9	No detectable contaminating nuclease; 50-60% ligated and 80-90% recut after 5-fold overdigestion	AGS Heidelb A00395S A00395M
	1 unit completely cleaves 1 µg unmethylated λ DNA/hr at pH 7.5, 37°C in a reaction volume of 50 µL	10 m*M* Tris-HCl, 10 m*M* MgCl$_2$, 1.0 m*M* DTT, 100 µg/mL BSA, pH 7.5		CHIMERx 2284-01 2284-02
4-8 U/µL	1 unit hydrolyzes 1 µg DNA/hr at 37°C in a 50 µL reaction volume using λ phage DNA as substrate	10 m*M* Tris-HCl, 10 m*M* MgCl$_2$, 0.1 mg/mL BSA, pH 7.5	>60% of the DNA fragments ligated and >90% recut after 50-fold overdigestion	Fermentas ER0821 ER0822
				Nippon Gene
				Oncor

Mbo II continued

CONCENTRATION or VOLUME ACTIVITY	UNIT DEFINITION	REACTION BUFFER (As Provided)	ADDITIONAL ACTIVITIES and PURITY	SUPPLIER CATALOG NO.
Moraxella bovis continued				
2000-20,000 U/mL	1 unit completely cleaves 1.0 μg λ DNA/hr in a 50 μL reaction mixture	100 mM Tris-HCl, 500 mM NaCl, 100 mM MgCl$_2$, 10 mM DTT, BSA, pH 7.9	100% cleavage, 90% ligation, 75% recleavage. No detectable endonuclease (overdigestion) at 10X	Sigma R8884
400 U and 2000 U	1 unit completely digests 1 μg N^6-methyladenine free λ DNA/hr at 37°C in 50 μL of supplied buffer	100 mM Tris-HCl, 100 mM MgCl$_2$, 10 mM DTT, pH 7.5	Purity is confirmed by overdigestion and ligation-recutting test	TaKaRa 1145
3-20 U/μL	1 unit completely digests 1 μg DNA/hr at 37°C	10 mM Tris-HCl, 10 mM MgCl$_2$, 1 mM DTT, pH 7.5		Toyobo MBO-201 MBO-202
Moraxella bovis ATCC 10900				
2000-20,000 U/mL		50 mM KCl, 10 mM Tris-HCl, 0.1 mM EDTA, 1 mM DTT, 200 μg/mL BSA, 50% glycerol, pH 7.4	No detectable Mbo I	ICN 153816
8-12 U/μL	1 unit completely digests 1 μg unmethylated λ DNA/hr at 37°C in a 50 mL reaction volume	60 mM Tris-HCl, 60 mM MgCl$_2$, 500 mM NaCl, 10 mM DTT, pH 7.5	20% ligation	Promega R6723 R6724
Moraxella bovis M				
8-12 U/μL	1 unit completely digests 1 mg SV40 DNA/hr at 37°C	500 mM Tris-HCl, 100 mM MgCl$_2$, pH 8.0	≥10% ligation; 65% cleavage	Life Technol 15241-011

McrBC I

Notes
- EC 3.1.21.4
- A type II site specific deoxyribonuclease
- A large group of enzymes recognizing specific short DNA sequences. They cleave either within the recognition site, or at a short specific distance from it
- Requires Mg^{2+}

CONCENTRATION or VOLUME ACTIVITY	UNIT DEFINITION	REACTION BUFFER (As Provided)	ADDITIONAL ACTIVITIES and PURITY	SUPPLIER CATALOG NO.
Escherichia coli K-12				
				NE Biolabs

Mfe I

RECOGNITION SEQUENCE
5'...C▼AATTG...3'
3'...GTTAA▲C...5'

HEAT INACTIVATION
65°C for 20 min.

NOTES
- EC 3.1.21.4
- A type II site specific deoxyribonuclease
- A large group of enzymes recognizing specific short DNA sequences. They cleave either within the recognition site, or at a short specific distance from it
- Requires Mg^{2+}

CONCENTRATION or VOLUME ACTIVITY	UNIT DEFINITION	REACTION BUFFER (As Provided)	ADDITIONAL ACTIVITIES and PURITY	SUPPLIER CATALOG NO.
Escherichia coli strain, carrying the cloned gene from *Mycoplasma fermentans*				
10,000 U/mL	1 unit completely digests 1 μg λ DNA/hr in a total reaction volume of 50 μL using the buffer provided	10 m*M* Tris-HCl, 10 m*M* MgCl₂, 50 m*M* NaCl, 1 m*M* DTT, pH 7.9	>95% DNA fragments ligated and recut after 20-fold overdigestion	NE Biolabs 589S 589L

Mfl I

RECOGNITION SEQUENCE
  5'...Pu▼GATCPy...3'
  3'...PyCTAG▲Pu...5'

ISOSCHIZOMERS
  Xho II

HEAT INACTIVATION
  Resistant; use phenol extraction

NOTES
- EC 3.1.21.4
- A type II site specific deoxyribonuclease
- A large group of enzymes recognizing specific short DNA sequences. They cleave either within the recognition site, or at a short specific distance from it
- Requires Mg^{2+}

CONCENTRATION or VOLUME ACTIVITY	UNIT DEFINITION	REACTION BUFFER (As Provided)	ADDITIONAL ACTIVITIES and PURITY	SUPPLIER CATALOG No.
Microbacterium flavum				
8-12 U/μL	1 unit completely digests 1 μg DNA/hr at 37°C in the buffer provided	100 mM Tris-HCl, 100 mM MgCl₂, 10 mM DTT, pH 7.5	>95% DNA fragments ligated and 100% recut	Amersham E 1070Y
200 U and 1000 U	1 unit completely cleaves 1 μg DNA/hr in 50 μL reaction mixture	100 mM Tris-HCl, 100 mM MgCl₂, 10 mM DTT, pH 7.5		PanVera TAK 1070
500 U and 2500 U	1 unit completely digests 1 μg N^6-methyladenine free λ DNA/hr at 37°C in 50 μL of supplied buffer	100 mM Tris-HCl, 100 mM MgCl₂, 10 mM DTT, pH 7.5	Purity is confirmed by overdigestion and ligation-recutting test	TaKaRa 1070

Mlu I

RECOGNITION SEQUENCE
5'...A▼CGCGT...3'
3'...TGCGC▲A...5'

OPTIMUM REACTION CONDITIONS
pH 7.5

HEAT INACTIVATION
Resistant; use ethanol precipitation

METHYLATION EFFECTS
Cleaves m^{6}ACGCGT
Does not cleave Am^{5}CGCGT

NOTES
- EC 3.1.21.4
- A type II site specific deoxyribonuclease
- A large group of enzymes recognizing specific short DNA sequences. They cleave either within the recognition site, or at a short specific distance from it
- Partially cleaves DNA from *dam*$^+$ *E. coli*
- Glycerol >5% inhibits activity
- Compatible cohesive ends: *Asc* I, *Bss*H II, *Dsa* I
- Requires Mg^{2+}

CONCENTRATION or VOLUME ACTIVITY	UNIT DEFINITION	REACTION BUFFER (As Provided)	ADDITIONAL ACTIVITIES and PURITY	SUPPLIER CATALOG No.
Micrococcus luteus				
1-5 U/μL	1 unit completely digests 1 μg λ DNA/hr at 37°C in the buffer provided	Genomic grade; 500 m*M* Tris-HCl, 100 m*M* MgCl$_2$, 10 m*M* DTT, 1000 m*M* NaCl, pH 7.5	100% DNA fragments ligated and recut	Amersham E 1071Y E 1071Z
10,000 U/mL	1 unit degrades 1 μg λ dam$^-$ dcm$^-$ DNA/hr at 37°C in a 0.05 mL mixture	50 m*M* Tris-HCl, 10 m*M* MgCl$_2$, 100 m*M* NaCl, 100 μg/mL BSA, pH 7.5	No detectable contaminating nuclease; 90% ligated and 100% recut after 10-fold overdigestion	AGS Heidelb F00397S F00397M
10 U/μL and 40 U/μL	1 unit completely digests 1 μg substrate DNA/hr	50 m*M* Tris-HCl, 10 m*M* MgCl$_2$, 100 m*M* NaCl, 1 m*M* DTE, pH 7.5		Boehringer 909700 909718 1207601
	1 unit completely cleaves 1 μg λ DNA/hr at pH 7.5, 37°C in a reaction volume of 50 μL	10 m*M* Tris-HCl, 10 m*M* MgCl$_2$, 50 m*M* NaCl, 1.0 m*M* DTT, 100 μg/mL BSA, pH 7.5		CHIMERx 2287-01 2287-02
8-12 U/μL; 40-60 U/μL	1 unit hydrolyzes 1 μg DNA/hr at 37°C in a 50 μL reaction volume using λ phage DNA as substrate	10 m*M* Tris-HCl, 10 m*M* MgCl$_2$, 100 m*M* KCl, 0.1 mg/mL BSA, pH 8.5	>90% of the DNA fragments ligated and recut after 50-fold overdigestion	Fermentas ER0561 ER0562 ER0563
1000-10,000 U/mL		50 m*M* KCl, 10 m*M* Tris-HCl, 0.1 m*M* EDTA, 1.0 m*M* DTT, 200 μg/mL BSA, 50% glycerol, pH 7.4		ICN 151701

Mlu I continued

CONCENTRATION or VOLUME ACTIVITY	UNIT DEFINITION	REACTION BUFFER (As Provided)	ADDITIONAL ACTIVITIES and PURITY	SUPPLIER CATALOG No.
Micrococcus luteus continued				
6-14 U/μL	1 unit digests completely 1 μg λ DNA substrate/hr at 37°C	50 mM Tris-HCl, 100 mM NaCl, 10 mM MgCl$_2$, 1 mM DTT, pH 7.8		NBL Gene 015802 015805
				Nippon Gene
10 U/μL	1 unit hydrolyzes 1 μg λ DNA to completion/hr in a 50 μL total reaction volume	500 mM Tris-HCl, 1 M NaCl, 100 mM MgCl$_2$, 10 mM DTE, 1.0 mg/mL BSA, pH 8		Oncor 110281 110282
8000-15,000 U/mL	1 unit completely digests 1 μg λ DNA/hr at pH 7.5, 37°C in a 50 μL total assay mixture	100 mM Tris-acetate, 100 mM MgOAc, 500 mM KOAc, pH 7.5		Pharmacia 27-0945-01 27-0945-02
10 U/μL	1 unit completely digests 1 μg λ DNA/hr at 37°C in a 50 mL reaction volume	Blue/White Cloning Qualified; genome qualified; 60 mM Tris-HCl, 60 mM MgCl$_2$, 1.5 M NaCl, 10 mM DTT, pH 7.9	90% ligation	Promega R6381 R6382
20,000 U/mL		50 mM Tris-HCl, 10 mM MgCl$_2$, 100 mM NaCl, 1 mM DTT, pH 7.6	90% DNA fragments ligated and recut after 10-fold overdigestion	SibEnzyme E085 E086
5000-25,000 U/mL	1 unit completely cleaves 1.0 μg λ DNA/hr in a 50 μL reaction mixture	500 mM Tris-HCl, 1 M NaCl, 100 mM MgCl$_2$, 10 mM DTT, pH 7.9	100% cleavage, 90% ligation, 95% recleavage, 10% endonuclease (nickase) No detectable endonuclease (overdigestion)	Sigma R8257
8000-40,000 U/mL		50 mM KCl, 10 mM Tris-HCl, 0.1 mM EDTA, 1 mM DTT, 200 μg/mL BSA, 50% glycerol, pH 7.5		Stratagene 500690
1000 U and 5000 U	1 unit completely digests 1 mg λ DNA/hr at 37°C in 50 mL of supplied buffer	Genomic grade; 500 mM Tris-HCl, 100 mM MgCl2, 10 mM DTT, 1 M NaCl, pH 7.5		TaKaRa 1071
3-20 U/μL	1 unit completely digests 1 μg DNA/hr at 37°C	50 mM Tris-HCl, 10 mM MgCl$_2$, 100 mM NaCl, 1 mM DTT, pH 7.5	>90% ligation; >95% recut	Toyobo MLU-101 MLU-102
Micrococcus luteus IFO 12992				
8-12 U/μL and 50 U/μL	1 unit completely digests 1 mg λ DNA/hr at 37°C	500 mM Tris-HCl, 100 mM MgCl$_2$, 1 M NaCl, pH 8.0		Life Technol 15432-016 15432-024
10,000 U/mL	1 unit completely digests 1 μg λ DNA/hr in a total reaction volume of 50 μL using the buffer provided	50 mM Tris-HCl, 10 mM MgCl$_2$, 100 mM NaCl, 1 mM DTT, pH 7.9	>95% DNA fragments ligated and recut after 10-fold overdigestion	NE Biolabs 198S 198L

MluN I

RECOGNITION SEQUENCE
5'...TGG▼CCA...3'
3'...ACC▲GGT...5'

ISOSCHIZOMERS
Bal I

NOTES
- A large group of enzymes recognizing specific short DNA sequences. They cleave either within the recognition site, or at a short specific distance from it
- Requires Mg^{2+}

CONCENTRATION or VOLUME ACTIVITY	UNIT DEFINITION	REACTION BUFFER (As Provided)	ADDITIONAL ACTIVITIES and PURITY	SUPPLIER CATALOG No.
Micrococcus luteus NI				
10 U/µL and 40 U/µL	1 unit completely digests 1 µg substrate DNA/hr	33 mM Tris-acetate, 10 mM MgOAc, 66 mM KOAc, 0.5 mM DTT, pH 7.9		Boehringer 1526430 1526448 1526472

Mnl I

RECOGNITION SEQUENCE
5'...CCTC(N)$_7$▼...3'
3'...GGAG(N)$_6$▲...5'

HEAT INACTIVATION
65°C for 20 min.

METHYLATION EFFECTS
Does not cleave m^{5}CCTC or m^{5}Cm^{5}CTm^{5}C

NOTES
- EC 3.1.21.4
- A type II site specific deoxyribonuclease
- A large group of enzymes recognizing specific short DNA sequences. They cleave either within the recognition site, or at a short specific distance from it
- Cleaves single-stranded DNA at 50% the rate of double-stranded DNA
- Requires Mg^{2+}

CONCENTRATION or VOLUME ACTIVITY	UNIT DEFINITION	REACTION BUFFER (As Provided)	ADDITIONAL ACTIVITIES and PURITY	SUPPLIER CATALOG NO.
Escherichia coli, carrying the cloned *Mnl*IR gene from *Moraxella nonliquefaciens*				
8-12 U/μL	1 unit hydrolyzes 1 μg DNA/hr at 37°C in a 50 μL reaction volume using λ phage DNA as substrate	10 m*M* Tris-HCl, 10 m*M* MgCl$_2$, 50 m*M* NaCl, 0.1 mg/mL BSA at pH 7.5	90% of the DNA fragments ligated and recut after 50-fold overdigestion	Fermentas ER1071 ER1072
5000 U/mL	1 unit completely digests 1 μg λ DNA/hr in a total reaction volume of 50 μL using the buffer provided	10 m*M* Tris-HCl, 10 m*M* MgCl$_2$, 50 m*M* NaCl, 1 m*M* DTT, pH 7.9; add 100 μg/mL BSA	80% DNA fragments ligated and >90% recut after 2-fold overdigestion	NE Biolabs 163S 163L
Moraxella nonliquefaciens				
1-5 U/μL	1 unit completely digests 1 μg λ DNA/hr at 37°C in the buffer provided	100 m*M* Tris-HCl, 100 m*M* MgCl$_2$, 10 m*M* DTT, 500 m*M* NaCl, pH 7.5	>80% DNA fragments ligated and 90% recut	Amersham E 0215Y E 0215Z
10,000 U/mL	1 unit degrades 1 μg λ dam$^-$ dcm$^-$ DNA/hr at 37°C in a 0.05 mL mixture	33 m*M* Tris-acetate, 10 m*M* MgOAc, 66 m*M* KOAc, 100 μg/mL BSA, pH 7.9	No detectable contaminating nuclease; 90% ligated and 100% recut after 50-fold overdigestion	AGS Heidelb F00398S F00398M
	1 unit completely cleaves 1 μg λ DNA/hr at pH 7.5, 37°C in a reaction volume of 50 μL	10 m*M* Tris-HCl, 10 m*M* MgCl$_2$, 50 m*M* NaCl, 1.0 m*M* DTT, 100 μg/mL BSA, pH 7.5		CHIMERx 2289-01 2289-02
400-2000 U/mL		50 m*M* KCl, 10 m*M* Tris-HCl, 0.1 m*M* EDTA, 1.0 m*M* DTT, 50% glycerol, pH 7.4		ICN 151702
400-2000 U/mL		50 m*M* KCl, 10 m*M* Tris-HCl, 0.1 m*M* EDTA, 1 m*M* DTT, 200 μg/mL BSA, 50% glycerol, pH 7.4		Stratagene 500700 500701

Mph1103 I

RECOGNITION SEQUENCE
5'...ATGCA▼T...3'
3'...T▲ACGTA...5'

NOTES
- EC 3.1.21.4
- A type II site specific deoxyribonuclease
- A large group of enzymes recognizing specific short DNA sequences. They cleave either within the recognition site, or at a short specific distance from it
- Requires Mg^{2+}

CONCENTRATION or VOLUME ACTIVITY	UNIT DEFINITION	REACTION BUFFER (As Provided)	ADDITIONAL ACTIVITIES and PURITY	SUPPLIER CATALOG No.
Moraxella phenylpyruvica RFL1103				
8-12 U/μL	1 unit hydrolyzes 1 μg DNA/hr at 37°C in a 50 μL reaction volume using λ phage DNA as substrate	10 mM Tris-HCl, 10 mM $MgCl_2$, 100 mM KCl, 0.1 mg/mL BSA, pH 8.5	>90% of the DNA fragments ligated and recut after 50-fold overdigestion	Fermentas ER0731 ER0732

Mro I

RECOGNITION SEQUENCE
5'...T▼CCGGC...3'
3'...AGGCC▲T...5'

ISOSCHIZOMERS
Acc III, BspM II

OPTIMUM REACTION CONDITIONS
pH 8.0

HEAT INACTIVATION
100°C for 5 min.

NOTES
- EC 3.1.21.4
- A type II site specific deoxyribonuclease
- A large group of enzymes recognizing specific short DNA sequences. They cleave either within the recognition site, or at a short specific distance from it
- Almost digests pBR322; Acc III does not
- Requires Mg^{2+}

Mro I continued

CONCENTRATION or VOLUME ACTIVITY	UNIT DEFINITION	REACTION BUFFER (As Provided)	ADDITIONAL ACTIVITIES and PURITY	SUPPLIER CATALOG No.
Micrococcus roseus				
1-5 U/μL	1 unit completely digests 1 μg substrate DNA/hr	33 m*M* Tris-acetate, 10 m*M* MgOAc, 66 m*M* KOAc, 0.5 m*M* DTT, pH 7.9		Boehringer 1102982
3000-15,000 U/mL		10 m*M* Tris-HCl, 200 m*M* NaCl, 0.1 m*M* EDTA, 5 m*M* β-MSH, 200 μg/mL BSA, 50% glycerol, pH 7.5		ICN 159449
3-20 U/μL	1 unit completely digests 1 μg DNA/hr at 37°C	10 m*M* Tris-HCl, 10 m*M* MgCl$_2$, 1 m*M* DTT, pH 7.5	>90% ligation; >95% recut	Toyobo MRO-101 MRO-102

MroN I

RECOGNITION SEQUENCE
    5'...G▼CCGGC...3'
    3'...CGGCC▲G...5'

HEAT INACTIVATION
    Resistant

NOTES
- EC 3.1.21.4
- A type II site specific deoxyribonuclease
- A large group of enzymes recognizing specific short DNA sequences. They cleave either within the recognition site, or at a short specific distance from it
- Requires Mg^{2+}

CONCENTRATION or VOLUME ACTIVITY	UNIT DEFINITION	REACTION BUFFER (As Provided)	ADDITIONAL ACTIVITIES and PURITY	SUPPLIER CATALOG No.
Micrococcus roseus N				
15,000 U/mL		10 m*M* Tris-HCl, 10 m*M* MgCl$_2$, 1 m*M* DTT, pH 7.6	90% DNA fragments ligated and recut after 10-fold overdigestion	SibEnzyme E087 E088

MroX I

Recognition Sequence
5'...GAANN▼NNTTC...3'
3'...CTTNN▲NNAAG...5'

Heat Inactivation
65°C for 20 min.

Notes
- EC 3.1.21.4
- A type II site specific deoxyribonuclease
- A large group of enzymes recognizing specific short DNA sequences. They cleave either within the recognition site, or at a short specific distance from it
- Requires Mg^{2+}

CONCENTRATION or VOLUME ACTIVITY	UNIT DEFINITION	REACTION BUFFER (As Provided)	ADDITIONAL ACTIVITIES and PURITY	SUPPLIER CATALOG NO.
Micrococcus roseus X				
2000 U/mL		10 mM Tris-HCl, 10 mM $MgCl_2$, 100 mM NaCl, 1 mM DTT, pH 8.5	50% DNA fragments ligated and recut after 5-fold overdigestion	SibEnzyme E249 E250

Msc I

Recognition Sequence
5'...TGG▼CCA...3'
3'...ACC▲GGT...5'

Isoschizomers
Bal I

Optimum Reaction Conditions
pH 7.9

Heat Inactivation
65°C for 20 min., >100°C for 5 min.

Methylation Effects
Blocked by overlapping *dcm* methylation.

Notes
- EC 3.1.21.4
- A type II site specific deoxyribonuclease
- A large group of enzymes recognizing specific short DNA sequences. They cleave either within the recognition site, or at a short specific distance from it
- Partially cleaves DNA from dcm^+ *E. coli*
- Requires Mg^{2+}

Msc I continued

CONCENTRATION or VOLUME ACTIVITY	UNIT DEFINITION	REACTION BUFFER (As Provided)	ADDITIONAL ACTIVITIES and PURITY	SUPPLIER CATALOG NO.
Escherichia coli strain, carrying the cloned gene from *Micrococcus* species				
3000 U/mL; 15,000 U/mL	1 unit completely digests 1 μg λ DNA/hr in a total reaction volume of 50 μL using the buffer provided	20 mM Tris-acetate, 10 mM MgOAc, 50 mM KOAc, 1 mM DTT, pH 7.9	>95% DNA fragments ligated and recut after 15-fold overdigestion	NE Biolabs 534S 534L
Micrococcus species				
3000 U/mL	1 unit degrades 1 μg λ dam⁻ dcm⁻ DNA/hr at 37°C in a 0.05 mL mixture	10 mM Tris-HCl, 10 mM MgCl$_2$, 100 mM KCl, 100 μg/mL BSA, pH 8.5	No detectable contaminating nuclease; 95% ligated and recut after 15-fold overdigestion	AGS Heidelb A00100S A00100M
3000-15,000 U/mL		150 mM KCl, 10 mM Tris-HCl, 0.1 mM EDTA, 1 mM DTT, 200 μg/mL acetylated BSA, 50% glycerol, pH 7.5		ICN 159450
4-6 U/μL	1 unit completely digests 1 mg λ DNA/hr at 37°C	500 mM Tris-HCl, 100 mM MgCl$_2$, 500 mM NaCl, pH 8.0≥	3% ligation; 100% cleavage	Life Technol 15469-018
3000-15,000 U/mL	1 unit completely cleaves 1.0 μg λ DNA/hr in a 50 μL reaction mixture	200 mM Tris-acetate, 500 mM KOAc, 100 mM MgOAc, 10 mM DTT, pH 7.9	100% cleavage, 95% ligation, 100% recleavage, 50% endonuclease (nickase) No detectable endonuclease (overdigestion)	Sigma R0510
3-20 U/μL	1 unit completely digests 1 μg DNA/hr at 37°C	10 mM Tris-HCl, 10 mM MgCl$_2$, 50 mM NaCl, 1 mM DTT, pH 7.5	>95% ligation; >95% recut	Toyobo MSC-101 MSC-102

Mse I

RECOGNITION SEQUENCE
5'...T▼TAA...3'
3'...AAT▲T...5'

HEAT INACTIVATION
Resistant; 65°C for 20 min.

NOTES
- EC 3.1.21.4
- A type II site specific deoxyribonuclease
- A large group of enzymes recognizing specific short DNA sequences. They cleave either within the recognition site, or at a short specific distance from it
- Requires Mg^{2+}

Mse I continued

CONCENTRATION or VOLUME ACTIVITY	UNIT DEFINITION	REACTION BUFFER (As Provided)	ADDITIONAL ACTIVITIES and PURITY	SUPPLIER CATALOG NO.
Micrococcus species				
4000-20,000 U/mL		50 mM NaCl, 10 mM Tris-HCl, 0.1 mM EDTA, 1 mM DTT, 200 µg/mL acetylated BSA, 50% glycerol, pH 7.4		ICN 159451
4-6 U/µL	1 unit completely digests 1 mg λ DNA/hr at 37°C	500 mM Tris-HCl, 100 mM MgCl$_2$, pH 8.0	≥95% ligation; 100% cleavage	Life Technol 15494-016
Micrococcus species (NEB 446)				
4000 U/mL; 20,000 U/mL	1 unit completely digests 1 µg λ DNA/hr in a total reaction volume of 50 µL using the buffer provided	10 mM Tris-HCl, 10 mM MgCl$_2$, 50 mM NaCl, 1 mM DTT, pH 7.9; add 100 µg/mL BSA	95% DNA fragments ligated and recut after 5-fold overdigestion	NE Biolabs 525S 525L

Msl I

RECOGNITION SEQUENCE
    5'...CAPyNN▼NNPuTG...3'
    3'...GTPuNN▲NNPyAC...5'

HEAT INACTIVATION
    65°C for 20 min.

NOTES
- EC 3.1.21.4
- A type II site specific deoxyribonuclease
- A large group of enzymes recognizing specific short DNA sequences. They cleave either within the recognition site, or at a short specific distance from it
- Requires Mg^{2+}

CONCENTRATION or VOLUME ACTIVITY	UNIT DEFINITION	REACTION BUFFER (As Provided)	ADDITIONAL ACTIVITIES and PURITY	SUPPLIER CATALOG NO.
Moraxella osloensis				
4000-20,000 U/mL		100 mM NaCl, 10 mM Tris-HCl, 0.1 mM EDTA, 1 mM DTT, 200 µg/mL acetylated BSA, 50% glycerol, pH 7.6		ICN 159452
Moraxella osloensis (NEB 722)				
5000 U/mL	1 unit completely digests 1 µg λ DNA/hr in a total reaction volume of 50 µL using the buffer provided	10 mM Tris-HCl, 10 mM MgCl$_2$, 50 mM NaCl, 1 mM DTT, pH 7.9	>95% DNA fragments ligated and recut after 10-fold overdigestion	NE Biolabs 571S 571L

Msp I

RECOGNITION SEQUENCE
    5'...C▼CGG...3'
    3'...GGC▲C...5'
ISOSCHIZOMERS
    Hpa II
OPTIMUM REACTION CONDITIONS
    pH 7.5
HEAT INACTIVATION
    65°C for 20 min.
METHYLATION EFFECTS
    Cleaves Cm^4CGG, m^4CCGG and Cm^5CGG
    Slowly cleaves $GGCm^5CGG$
    Does not cleave m^5CCGG or hm^5Chm^5CGG

NOTES
- EC 3.1.21.4
- A type II site specific deoxyribonuclease
- A large group of enzymes recognizing specific short DNA sequences. They cleave either within the recognition site, or at a short specific distance from it
- Used with *Hpa* II to study CpG methylation patterns in eukaryotic DNA
- Compatible cohesive ends: any blunt end; *Acc* I, *Aci* I, *Acy* I, *Ban* III, *Bbi* II, *BsaH* I, *BsiC* I, *BsiX* I, *Bsp*106 I, *Bsp*D I, *Bst*B I, *Cla* I, *Csp*45 I, *Hgi*D I, *Hin*1 I, *Hin*P I, *Mae* II, *Nar* I, *Nsp* V, *Nsp*7524 V, *Psp*1406 I, *Sfu* I, *Taq* I
- Requires Mg^{2+}

CONCENTRATION or VOLUME ACTIVITY	UNIT DEFINITION	REACTION BUFFER (As Provided)	ADDITIONAL ACTIVITIES and PURITY	SUPPLIER CATALOG NO.
Escherichia coli strain, carrying the cloned gene from Moraxella species				
20,000 U/mL; 100,000 U/mL	1 unit completely digests 1 μg λ DNA/hr in a total reaction volume of 50 μL using the buffer provided	10 m*M* Tris-HCl, 10 m*M* MgCl₂, 50 m*M* NaCl, 1 m*M* DTT, pH 7.9	>95% DNA fragments ligated and recut after 75-fold overdigestion	NE Biolabs 106S 106L
Moraxella species				
10 U/μL; 40 U/μL	1 unit completely cleaves 1 μg substrate DNA/hr	50 m*M* NaCl, 10 m*M* Tris, 10 m*M* MgCl₂, 1 m*M* β-MSH, pH 7.8	90-98% cleavage-ligation-recleavage	Am Allied L-18-02000 H-18-02000 L-18-10000 H-18-10000
8-12 U/μL	1 unit completely digests 1 μg λ DNA/hr at 37°C in the buffer provided	100 m*M* Tris-HCl, 100 m*M* MgCl₂, 10 m*M* DTT, 500 m*M* NaCl, pH 7.5	>90% DNA fragments ligated and 100% recut	Amersham E 1150Y E 1150Z
>40 U/μL	1 unit completely digests 1 μg λ DNA/hr at 37°C in the buffer provided	100 m*M* Tris-HCl, 100 m*M* MgCl₂, 10 m*M* DTT, 500 m*M* NaCl, pH 7.5	>90% DNA fragments ligated and 100% recut	Amersham E 1150ZH

Msp I continued

CONCENTRATION or VOLUME ACTIVITY	UNIT DEFINITION	REACTION BUFFER (As Provided)	ADDITIONAL ACTIVITIES and PURITY	SUPPLIER CATALOG No.
Moraxella species continued				
10,000 U/mL	1 unit degrades 1 µg λ dam⁻ dcm⁻ DNA/hr at 37°C in a 0.05 mL mixture	33 mM Tris-acetate, 10 mM MgOAc, 66 mM KOAc, 100 µg/mL BSA, pH 7.9	No detectable contaminating nuclease; 95% ligated and recut after 50-fold overdigestion	AGS Heidelb F00400S F00400M
10 U/µL and 40 U/µL	1 unit completely digests 1 µg substrate DNA/hr	10 mM Tris-HCl, 10 mM MgCl₂, 1 mM DTE, pH 7.5		Boehringer 633518 633526 1047647 1274058
	1 unit completely cleaves 1 µg λ DNA/hr at pH 7.5, 37°C in a reaction volume of 50 µL	10 mM Tris-HCl, 10 mM MgCl₂, 50 mM NaCl, 1.0 mM DTT, 100 µg/mL BSA, pH 7.5		CHIMERx 2290-01 2290-02
8-12 U/µL; 40-60 U/µL	1 unit hydrolyzes 1 µg DNA/hr at 37°C in a 50 µL reaction volume using λ phage DNA as substrate	33 mM Tris-acetate, 10 mM MgOAc, 66 mM KOAc, 0.1 mg/mL BSA, pH 7.9	>90% of the DNA fragments ligated and recut after 50-fold overdigestion	Fermentas ER0541 ER0542 ER0543
3000-20,000 U/mL		50 mM KCl, 10 mM Tris-HCl, 0.1 mM EDTA, 1.0 mM DTT, 200 µg/mL BSA, 50% glycerol, pH 7.5		ICN 151717
8-12 U/µL and 50 U/µL	1 unit completely digests 1 mg λ DNA/hr at 37°C	500 mM Tris-HCl, 100 mM MgCl₂, pH 8.0	≥95% ligation; 100% cleavage	Life Technol 15419-013 15419-039 15419-021
6-14 U/µL	1 unit digests completely 1 µg λ DNA substrate/hr at 37°C	10 mM Tris-HCl, 10 mM MgCl₂, 1 mM DTT, pH 7.8		NBL Gene 013404 013407
10 U/µL	1 unit hydrolyzes 1 µg λ DNA to completion/hr in a 50 µL total reaction volume	100 mM Tris-HCl, 100 mM NaCl, 100 mM MgCl₂, 100 mM β-MSH, 1.0 mg/mL BSA, pH 7.5		Nippon Gene Oncor 110291 110292
7000-15,000 U/mL and 50,000-80,000 U/mL	1 unit completely digests 1 µg λ DNA/hr at pH 7.5, 37°C in a 50 µL total assay mixture	100 mM Tris-acetate, 100 mM MgOAc, 500 mM KOAc, pH 7.5		Pharmacia 27-0988-01 27-0988-02 27-0988-18
10 U/µL	1 unit completely digests 1 µg λ DNA/hr at 37°C in a 50 mL reaction volume	100 mM Tris-HCl, 100 mM MgCl₂, 500 mM NaCl, 10 mM DTT, pH 7.9	90% ligation	Promega R7021 R7022
15,000 U/mL		10 mM Tris-HCl, 10 mM MgCl₂, 1 mM DTT, pH 7.6	90% DNA fragments ligated and recut after 20-fold overdigestion	SibEnzyme E091 E092

CONCENTRATION or VOLUME ACTIVITY	UNIT DEFINITION	REACTION BUFFER (As Provided)	ADDITIONAL ACTIVITIES and PURITY	SUPPLIER CATALOG No.
Moraxella species continued				
5000-15,000 U/mL	1 unit completely cleaves 1.0 µg λ DNA/hr in a 50 µL reaction mixture	100 mM Tris-HCl, 500 mM NaCl, 100 mM MgCl$_2$, 10 mM DTT, pH 7.9	100% cleavage, 95% ligation, 85% recleavage No detectable endonuclease (overdigestion)	Sigma R4506
5000-30,000 U/mL		50 mM KCl, 10 mM Tris-HCl, 0.1 mM EDTA, 1 mM DTT, 200 µg/mL BSA, 50% glycerol, pH 7.4		Stratagene 500710 500711
2000 U and 10,000 U	1 unit completely digests 1 µg λ DNA/hr at 37°C in 50 µL of supplied buffer	100 mM Tris-HCl, 100 mM MgCl$_2$, 10 mM DTT, 500 mM NaCl, pH 7.5	Purity is confirmed by overdigestion and ligation-recutting test	TaKaRa 1150
3-20 U/µL	1 unit completely digests 1 µg DNA/hr at 37°C	10 mM Tris-HCl, 10 mM MgCl$_2$, 50 mM NaCl, 1 mM DTT, pH 7.5	>90% ligation; >95% recut	Toyobo MSP-101 MSP-102

Msp17 I

RECOGNITION SEQUENCE
  5'...GR▼CGYC...3'
  3'...CYGC▲RG...5'

HEAT INACTIVATION
  65°C for 20 min.

NOTES
- EC 3.1.21.4
- A type II site specific deoxyribonuclease
- A large group of enzymes recognizing specific short DNA sequences. They cleave either within the recognition site, or at a short specific distance from it
- Requires Mg^{2+}

CONCENTRATION or VOLUME ACTIVITY	UNIT DEFINITION	REACTION BUFFER (As Provided)	ADDITIONAL ACTIVITIES and PURITY	SUPPLIER CATALOG NO.
Moraxella species 17				
5000 U/mL		10 mM Tris-HCl, 10 mM MgCl$_2$, 100 mM NaCl, 1 mM DTT, pH 8.5	90% DNA fragments ligated and recut after 10-fold overdigestion	SibEnzyme E093 E094

MspA1 I

RECOGNITION SEQUENCE
  5'...C(^A_C)G▼C(^G_T)G...3'
  3'...G(^T_G)C▲G(^C_A)C...5'

HEAT INACTIVATION
  65°C for 20 min.

NOTES
- EC 3.1.21.4
- A type II site specific deoxyribonuclease
- A large group of enzymes recognizing specific short DNA sequences. They cleave either within the recognition site, or at a short specific distance from it
- Compatible cohesive ends: *Acy* I, *Aha* II, *Asu* II, *Ban* III, *Bbi* I, *Bst*B I, *Cla* I, *Csp*45 I, *Hap* II, *Hin*1 I, *Hin*P I, *Hpa* II, *Mae* II, *Nar* I, *Nsp* V, *Sfu* I, *Taq* I, *Tth*HB8 I
- Requires Mg^{2+}

MspA1 I continued

CONCENTRATION or VOLUME ACTIVITY	UNIT DEFINITION	REACTION BUFFER (As Provided)	ADDITIONAL ACTIVITIES and PURITY	SUPPLIER CATALOG NO.
Moraxella species				
10,000 U/mL	1 unit completely digests 1 µg λ DNA/hr in a total reaction volume of 50 µL using the buffer provided	20 mM Tris-acetate, 10 mM MgOAc, 50 mM KOAc, 1 mM DTT, pH 7.9; add 100 µg/mL BSA	>95% DNA fragments ligated and recut after 10-fold overdigestion	NE Biolabs 577S 577L
10 U/µL and 40-80 U/µL	1 unit completely digests 1 µg λ DNA/hr at 37°C in a 50 mL reaction volume	60 mM Tris-HCl, 60 mM MgCl$_2$, 500 mM NaCl, 10 mM DTT, pH 7.5	90% ligation	Promega R6041 R6042 R6045 R4404

MspC I

RECOGNITION SEQUENCE
5'...C▼TTAAG...3'
3'...GAATT▲C...5'

ISOSCHIZOMERS
Afl II

OPTIMUM REACTION CONDITIONS
pH 7.9

NOTES
- EC 3.1.21.4
- A type II site specific deoxyribonuclease
- A large group of enzymes recognizing specific short DNA sequences. They cleave either within the recognition site, or at a short specific distance from it
- Requires Mg^{2+}

CONCENTRATION or VOLUME ACTIVITY	UNIT DEFINITION	REACTION BUFFER (As Provided)	ADDITIONAL ACTIVITIES and PURITY	SUPPLIER CATALOG NO.
Micrococcus species				
	30 units does not produce unspecific cleavage products with 1 µg λ DNA/Hind III digest/16 hr at 37°C	10 mM Tris-HCl, 10 mM MgCl$_2$, 150 mM NaCl, 1 mM DTT, 100 µg/mL BSA, pH 7.9	>95% DNA fragments ligated and recut after 5-fold overdigestion	MINOTECH 136

*Msp*R9 I

RECOGNITION SEQUENCE
  5'...CC▼NGG...3'
  3'...GGN▲CC...5'

HEAT INACTIVATION
  Resistant

METHYLATION EFFECTS
  Blocked by overlapping *dcm* methylation

NOTES
- EC 3.1.21.4
- A type II site specific deoxyribonuclease
- A large group of enzymes recognizing specific short DNA sequences. They cleave either within the recognition site, or at a short specific distance from it
- Requires Mg^{2+}

CONCENTRATION or VOLUME ACTIVITY	UNIT DEFINITION	REACTION BUFFER (As Provided)	ADDITIONAL ACTIVITIES and PURITY	SUPPLIER CATALOG No.
Micrococcus species R9				
40,000 U/mL		50 m*M* Tris-HCl, 10 m*M* MgCl$_2$, 100 m*M* NaCl, 1 m*M* DTT, pH 7.6	0% DNA fragments ligated and recut after 2-fold overdigestion	SibEnzyme E175 E176

Mun I

RECOGNITION SEQUENCE
  5'...C▼AATTG...3'
  3'...GTTAA▲C...5'

HEAT INACTIVATION
  Partially resistant at 65°C for 10 min.

METHYLATION EFFECTS
  Does not cleave CAm^{6}ATTG

NOTES
- EC 3.1.21.4
- A type II site specific deoxyribonuclease
- A large group of enzymes recognizing specific short DNA sequences. They cleave either within the recognition site, or at a short specific distance from it
- The only restriction enzyme generating an overhang compatible with *Eco*R I
- Requires Mg^{2+}

Mun I continued

CONCENTRATION or VOLUME ACTIVITY	UNIT DEFINITION	REACTION BUFFER (As Provided)	ADDITIONAL ACTIVITIES and PURITY	SUPPLIER CATALOG No.
Escherichia coli, carrying the cloned MunIR gene from Mycoplasma unidentified				
8-12 U/µL; 40-60 U/µL	1 unit hydrolyzes 1 µg DNA/hr at 37°C in a 50 µL reaction volume using λ phage DNA as substrate	10 mM Tris-HCl, 10 mM MgCl$_2$, 50 mM NaCl, 0.1 mg/mL BSA at pH 7.5	>90% of the DNA fragments ligated and recut after 50-fold overdigestion	Fermentas ER0751 ER0752 ER0753
Escherichia coli, carrying the plasmid encoding Mun I gene				
100 U and 500 U	1 unit completely digests 1 µg λ DNA/hr at 30°C in 50 µL of supplied buffer and BSA	100 mM Tris-HCl, 100 mM MgCl$_2$, 10 mM DTT, 500 mM NaCl, BSA, pH 7.5	Purity is confirmed by overdigestion and ligation-recutting test	TaKaRa 1153
Mycoplasma species				
8000-10,000 U/mL		100 mM KCl, 10 mM Tris-HCl, 1 mM EDTA, 1 mM DTT, 200 µg/mL acetylated BSA, 50% glycerol, pH 7.4		ICN 159454
4-6 U/µL	1 unit completely digests 1 mg λ DNA/hr at 37°C	500 mM Tris-HCl, 100 mM MgCl$_2$, pH 8.0	≥85% ligation; 100% cleavage	Life Technol 15491-012
Mycoplasma unidentified				
5000 U/mL	1 unit degrades 1 µg λ dam$^-$ dcm$^-$ DNA/hr at 37°C in a 0.05 mL mixture	33 mM Tris-acetate, 10 mM MgOAc, 66 mM KOAc, 100 µg/mL BSA, pH 7.9	No detectable contaminating nuclease; 90% ligated and 100% recut after 50-fold overdigestion	AGS Heidelb F00396S F00396M
3-10 U/µL	1 unit completely digests 1 µg substrate DNA/hr	10 mM Tris-HCl, 10 mM MgCl$_2$, 50 mM NaCl, 1 mM DTE, pH 7.5		Boehringer 1441337
Obtained from a cloned source				
10,000-80,000 U/mL		100 mM KCl, 10 mM Tris-HCl, 1 mM EDTA, 1 mM DTT, 200 µg/mL BSA, 50% glycerol, pH 7.4		Stratagene 500720 500721

Mva I

RECOGNITION SEQUENCE
5'...CC▼(^A_T)GG...3'
3'...GG(^T_A)▲CC...5'

ISOSCHIZOMERS
*Eco*R II

OPTIMUM REACTION CONDITIONS
pH 8.5

HEAT INACTIVATION
65°C for >15 min.

METHYLATION EFFECTS
Cleaves Cm^{5}C(^A_T)GG and m^{5}CC(^A_T)GG
Does not cleave Cm^{4}C(^A_T)GG, CCm^{6}AGG, m^{4}CC(^A_T)GG or m^{5}Cm^{5}C(^A_T)GG
Not blocked by *dam* methylation; *Eco*R II is blocked

NOTES
- EC 3.1.21.4
- A type II site specific deoxyribonuclease
- A large group of enzymes recognizing specific short DNA sequences. They cleave either within the recognition site, or at a short specific distance from it
- Produces DNA fragments with a 1-base 5'-extension
- Requires Mg^{2+}

CONCENTRATION or VOLUME ACTIVITY	UNIT DEFINITION	REACTION BUFFER (As Provided)	ADDITIONAL ACTIVITIES and PURITY	SUPPLIER CATALOG No.
Escherichia coli, carrying the cloned *Mva*IR gene from *Micrococcus varians* RFL19				
8-12 U/μL	1 unit hydrolyzes 1 μg DNA/hr at 37°C in a 50 μL reaction volume using λ phage DNA as substrate	10 m*M* Tris-HCl, 10 m*M* MgCl$_2$, 100 m*M* KCl, 0.1 mg/mL BSA, pH 8.5	>90% of the DNA fragments ligated and recut after 10-fold overdigestion	Fermentas ER0551 ER0552
Micrococcus varians				
3-20 U/μL	1 unit completely digests 1 μg DNA/hr at 37°C			Toyobo MVA-101 MVA-102
Micrococcus varians RFL19				
8-12 U/μL	1 unit completely digests 1 μg λ DNA/hr at 37°C in the buffer provided	200 m*M* Tris-HCl, 100 m*M* MgCl$_2$, 10 m*M* DTT, 1000 m*M* KCl, pH 8.5	0% DNA fragments ligated and recut	Amersham E 1072Y
10,000 U/mL	1 unit degrades 1 μg λ dam$^-$ dcm$^-$ DNA/hr at 37°C in a 0.05 mL mixture	10 m*M* Tris-HCl, 10 m*M* MgCl$_2$, 100 m*M* KCl, 100 μg/mL BSA, pH 8.5	No detectable contaminating nuclease; 90% ligated and 100% recut after 50-fold overdigestion with pBR322-*Pst* λ DNA fragments as substrate	AGS Heidelb F00190S F00190M
8-12 U/μL	1 unit completely digests 1 μg substrate DNA/hr	50 m*M* Tris-HCl, 10 m*M* MgCl$_2$, 100 m*M* NaCl, 1 m*M* DTE, pH 7.5		Boehringer 1288067 1288075

Mva I continued

CONCENTRATION or VOLUME ACTIVITY	UNIT DEFINITION	REACTION BUFFER (As Provided)	ADDITIONAL ACTIVITIES and PURITY	SUPPLIER CATALOG NO.
Micrococcus varians RFL19 continued				
1000 U and 5000 U	1 unit completely digests 1 μg λ DNA/hr at 37°C in 50 μL of supplied buffer	200 mM Tris-HCl, 100 mM MgCl$_2$, 10 mM DTT, 1 M KCl, pH 8.5	Purity is confirmed by overdigestion and ligation-recutting test	TaKaRa 1072

Mva1269 I

RECOGNITION SEQUENCE
5'...GAATGCN▼...3'
3'...CTTAC▲GN...5'

NOTES
- EC 3.1.21.4
- A type II site specific deoxyribonuclease
- A large group of enzymes recognizing specific short DNA sequences. They cleave either within the recognition site, or at a short specific distance from it
- Requires Mg^{2+}

CONCENTRATION or VOLUME ACTIVITY	UNIT DEFINITION	REACTION BUFFER (As Provided)	ADDITIONAL ACTIVITIES and PURITY	SUPPLIER CATALOG NO.
Micrococcus varians RFL1269				
4-8 U/μL	1 unit hydrolyzes 1 μg DNA/hr at 37°C in a 50 μL reaction volume using λ phage DNA as substrate	10 mM Tris-HCl, 10 mM MgCl$_2$, 100 mM KCl, 0.1 mg/mL BSA, pH 8.5	>90% of the DNA fragments ligated and recut after 50-fold overdigestion	Fermentas ER0961 ER0962

Mvn I

RECOGNITION SEQUENCE
5'...CG▼CG...3'
3'...CG▲GC...5'

ISOSCHIZOMERS
*Fnu*D II

NOTES
- EC 3.1.21.4
- A type II site specific deoxyribonuclease
- A large group of enzymes recognizing specific short DNA sequences. They cleave either within the recognition site, or at a short specific distance from it
- Requires Mg^{2+}

CONCENTRATION or VOLUME ACTIVITY	UNIT DEFINITION	REACTION BUFFER (As Provided)	ADDITIONAL ACTIVITIES and PURITY	SUPPLIER CATALOG NO.
Methanococcus vannielii				
10 U/μL	1 unit completely digests 1 μg substrate DNA/hr	10 mM Tris-HCl, 10 mM $MgCl_2$, 50 mM NaCl, 1 mM DTE, pH 7.5		Boehringer 1062573 1062581

Mwo I

RECOGNITION SEQUENCE
5'...GC(N)$_5$▼(N)$_2$GC...3'
3'...CG(N)$_2$▲(N)$_5$CG...5'

OPTIMUM REACTION CONDITIONS
10% activity at 37°C

HEAT INACTIVATION
Resistant

NOTES
- EC 3.1.21.4
- A type II site specific deoxyribonuclease
- A large group of enzymes recognizing specific short DNA sequences. They cleave either within the recognition site, or at a short specific distance from it
- Requires Mg^{2+}

Mwo I continued

CONCENTRATION or VOLUME ACTIVITY	UNIT DEFINITION	REACTION BUFFER (As Provided)	ADDITIONAL ACTIVITIES and PURITY	SUPPLIER CATALOG No.
Escherichia coli strain, carrying the cloned gene from *Methanobacterium wolfeii*				
5000 U/mL	1 unit completely digests 1 µg λ DNA/hr in a total reaction volume of 50 µL using the buffer provided	50 m*M* Tris-HCl, 10 m*M* MgCl$_2$, 150 m*M* NaCl, 1 m*M* DTT, pH 7.9; add 100 µg/mL BSA	>95% DNA fragments ligated and recut after 10-fold overdigestion	NE Biolabs 573S 573L
Methanobacterium wolfeii				
4000-20,000 U/mL		100 m*M* NaCl, 50 m*M* Tris-HCl, 0.1 m*M* EDTA, 1 m*M* DTT, 200 µg/mL acetylated BSA, 50% glycerol, pH 7.6		ICN 159455

Nae I

RECOGNITION SEQUENCE
   5'...GCC▼GGC...3'
   3'...CGC▲CCG...5'
OPTIMUM REACTION CONDITIONS
   pH 8.0
HEAT INACTIVATION
   65°C for 20 min.
METHYLATION EFFECTS
   Does not cleave GCCGGm^{5}C, GCm^{5}CGGC or Gm^{5}CCGGC

NOTES
- EC 3.1.21.4
- A type II site specific deoxyribonuclease
- A large group of enzymes recognizing specific short DNA sequences. They cleave either within the recognition site, or at a short specific distance from it
- Demonstrates marked site preference:
   - Ad2 DNA contains a slow cut site
   - cleaves the single site in λ DNA 50-fold slower than Ad2 DNA
   - cuts pBR322 DNA very slowly
   *Ngo*M I has less site preference
- Compatible cohesive ends: any blunt end
- Requires Mg^{2+}

Nae I continued

CONCENTRATION or VOLUME ACTIVITY	UNIT DEFINITION	REACTION BUFFER (As Provided)	ADDITIONAL ACTIVITIES and PURITY	SUPPLIER CATALOG NO.
Escherichia coli strain, carrying the cloned gene from Nocardia aerocolonigenes				
10,000 U/mL	1 unit completely digests 1 μg Ad2 DNA/hr in a total reaction volume of 50 μL using the buffer provided	10 mM Bis Tris-Propane-HCl, 10 mM MgCl$_2$, 1 mM DTT, pH 7.0	>80% DNA fragments ligated and recut after 10-fold overdigestion	NE Biolabs 190S 190L
Nocardia aerocolonigenes				
3000 U/mL	1 unit degrades 1 μg λ dam⁻ dcm⁻ DNA/hr at 37°C in a 0.05 mL mixture	33 mM Tris-acetate, 10 mM MgOAc, 66 mM KOAc, 100 μg/mL BSA, pH 7.9	No detectable contaminating nuclease; 98% ligated and 90% recut after 15-fold overdigestion	AGS Heidelb A00405S A00405M
10 U/μL	1 unit completely digests 1 μg substrate DNA/hr	33 mM Tris-acetate, 10 mM MgOAc, 66 mM KOAc, 0.5 mM DTT, pH 7.9		Boehringer 786314 786322
8-12 U/μL	1 unit digests completely 1 μg λ DNA substrate/hr at 37°C	33 mM Tris-acetate, 66 mM KOAc, 10 mM MgOAc, 0.5 mM DTT, pH 8.2		NBL Gene 017201 017203
10 U/μL	1 unit completely digests 1 μg Ad2 DNA/hr at 37°C in a 50 mL reaction volume	60 mM Tris-HCl, 60 mM MgCl$_2$, 60 mM NaCl, 10 mM DTT, pH 7.5	90% ligation	Promega R7131 R7132
4 U/μL	1 unit completely digests 1 μg Ad2 DNA/hr at 37°C in a 50 mL reaction volume	60 mM Tris-HCl, 60 mM NaCl, 60 mM MgCl$_2$, 10 mM DTT, 3.5 μM 14-base pair oligonucleotide DNA duplex, pH 7.5	90% ligation	Promega Turbo™ Nae I R7231 R7232
4000-15,000 U/mL		50 mM KCl, 10 mM Tris-HCl, 0.1 mM EDTA, 1 mM DTT, 200 μg/mL BSA, 50% glycerol, pH 7.4		Stratagene 500730
500 U and 2500 U	1 unit completely digests 1 μg pBR322 DNA/hr at 37°C in 50 μL of supplied buffer	Genomic grade; 100 mM Tris-HCl, 100 mM MgCl$_2$, 10 mM DTT, pH 7.5	Purity is confirmed by overdigestion, ligation-recutting and genome DNA digestion test	TaKaRa 1155
3-20 U/μL	1 unit completely digests 1 μg DNA/hr at 37°C	10 mM Tris-HCl, 10 mM MgCl$_2$, 1 mM DTT, pH 7.5	>80% ligation; >90% recut	Toyobo NAE-101 NAE-102
Nocardia aerocolonigenes ATCC 23870				
4000-12,000 U/mL		50 mM KCl, 10 mM Tris-HCl, 0.1 mM EDTA, 1 mM DTT, 200 μg/mL BSA, 50% glycerol, pH 7.4		ICN 153819

Nar I

RECOGNITION SEQUENCE
  5'...GG▼CGCC...3'
  3'...CCGC▲GG...5'
OPTIMUM REACTION CONDITIONS
  pH 7.5; unstable at 37°C
HEAT INACTIVATION
  65°C for 20 min.
METHYLATION EFFECTS
  Does not cleave GGm^{5}CGCC, GGCGCm^{4}C, GGCGCm^{5}C or GGhm^{5}CGhm^{5}Chm^{5}C

NOTES
- EC 3.1.21.4
- A type II site specific deoxyribonuclease
- A large group of enzymes recognizing specific short DNA sequences. They cleave either within the recognition site, or at a short specific distance from it
- Exhibits marked site preference: the single site in λ DNA is cut 50-fold slower than sites in other DNA
- Sensitive to NaCl >80 mM
- Compatible cohesive ends: Acc I, Aci I, Acy I, Aha II, Asu II, Ban III, Bbi I, BsaH I, BsiC I, BsiX I, Bsp106 I, BspD I, BstB I, Cla I, Csp45 I, Hap II, HgiD I, Hin1 I, HinP I, Hpa II, Mae II, Msp I, Nsp7524 V, Nsp V, Psp1406 I, Sfu I, Taq I, TthHB8 I
- Requires Mg^{2+}

CONCENTRATION or VOLUME ACTIVITY	UNIT DEFINITION	REACTION BUFFER (As Provided)	ADDITIONAL ACTIVITIES and PURITY	SUPPLIER CATALOG NO.
Nocardia argentinensis				
10 U/μL	1 unit completely digests 1 μg substrate DNA/hr	33 mM Tris-acetate, 10 mM MgOAc, 66 mM KOAc, 0.5 mM DTT, pH 7.9		Boehringer 1103016 1103024
				Nippon Gene
10,000-20,000 U/mL	1 unit completely digests 1 μg Ad2 DNA/hr at pH 7.5, 37°C in a 50 μL total assay mixture	100 mM Tris-acetate, 100 mM MgOAc, 500 mM KOAc, pH 7.5		Pharmacia 27-0822-01 27-0822-02
8-12 U/μL	1 unit completely digests 1 μg Ad2 DNA/hr at 37°C in a 50 mL reaction volume	Genome qualified; 500 mM Tris-HCl, 50 mM MgCl$_2$, pH 8.2	90% ligation	Promega R6861 R6862
10 U/μL	1 unit completely digests 1 μg Ad2 DNA/hr at 37°C in a 50 mL reaction volume	60 mM Tris-HCl, 60 mM NaCl, 60 mM MgCl$_2$, 10 mM DTT, 2.4 μM 18-base pair oligonucleotide DNA duplex, pH 7.5	95% ligation	Promega Turbo™ Nar I R7261 R7262

Nar I continued

CONCENTRATION or VOLUME ACTIVITY	UNIT DEFINITION	REACTION BUFFER (As Provided)	ADDITIONAL ACTIVITIES and PURITY	SUPPLIER CATALOG No.
Nocardia argentinensis continued				
2000-20,000 U/mL	1 unit completely cleaves 1.0 µg Ad2 DNA/hr in a 50 µL reaction mixture	Unique buffer + BSA	100% cleavage, 99% ligation, 95% recleavage, 10% endonuclease (nickase) No detectable endonuclease (overdigestion)	Sigma R5259
2000-10,000 U/mL		50 mM KCl, 10 mM Tris-HCl, 0.1 mM EDTA, 1 mM DTT, 200 µg/mL BSA, 50% glycerol, pH 7.4		Stratagene 500740 500741
3-20 U/µL	1 unit completely digests 1 µg DNA/hr at 37°C	10 mM Tris-HCl, 10 mM MgCl$_2$, 1 mM DTT, pH 7.5	>90% ligation; >95% recut	Toyobo NAR-101 NAR-102
Nocardia argentinensis ATCC 31306				
2000-20,000 U/mL		50 mM KCl, 10 mM Tris-HCl, 0.1 mM EDTA, 1 mM DTT, 200 µg/mL BSA, 50% glycerol, pH 7.4		ICN 153820
6-10 U/µL	1 unit completely digests 1 mg Ad2 DNA/hr at 37°C	500 mM Tris-HCl, 100 mM MgCl$_2$, pH 8.0	≥95% ligation; 90% cleavage	Life Technol 15422-017
2000-5000 U/mL	1 unit completely digests 1 µg φX174 DNA/hr in a total reaction volume of 50 µL using the buffer provided	10 mM Bis Tris-Propane-HCl, 10 mM MgCl$_2$, 1 mM DTT, pH 7.0	>95% DNA fragments ligated and recut after 10-fold overdigestion	NE Biolabs 191S 191L

Nci I

RECOGNITION SEQUENCE
5'...CC▼(^C_G)GG...3'
3'...GG(^G_C)▲CC...5'

OPTIMUM REACTION CONDITIONS
pH 7.5

HEAT INACTIVATION
65°C for 20min.; 80°C for 5 min.

METHYLATION EFFECTS
Cleaves m^{5}CCSGG
Does not cleave CmCSGG

NOTES
- EC 3.1.21.4
- A type II site specific deoxyribonuclease
- A large group of enzymes recognizing specific short DNA sequences. They cleave either within the recognition site, or at a short specific distance from it
- Produces an ambiguous single base extension which is difficult to ligate; prior T4 kinase treatment improves ligation
- The only restriction enzyme known to cleave 3'-phosphorylated fragment ends
- Requires Mg^{2+}

CONCENTRATION or VOLUME ACTIVITY	UNIT DEFINITION	REACTION BUFFER (As Provided)	ADDITIONAL ACTIVITIES and PURITY	SUPPLIER CATALOG NO.
Neisseria cinerea				
1500 U and 5000 U	1 unit completely digests 1 μg DNA substrate/hr at 37°C	10 mM Tris-acetate, 10 mM MgOAc, 50 mM KOAc, 1 mM DTT, 0.1 mg/mL BSA, pH 7.9	>50% DNA fragments ligated and 90% recut after 2-fold overdigestion	Adv Biotech AB-0232-a AB-0232-b
10 U/μL	1 unit completely digests 1 μg substrate DNA/hr	10 mM Tris-HCl, 10 mM MgCl$_2$, 1 mM DTE, pH 7.5		Boehringer 659363 667412
2000-10,000 U/mL		50 mM KCl, 10 mM Tris-HCl, 0.1 mM EDTA, 1.0 mM DTT, 200 μg/mL BSA, 50% glycerol, pH 7.4		ICN 151733
8-12 U/μL	1 unit digests completely 1 μg λ DNA substrate/hr at 37°C	10 mM Tris-HCl, 10 mM MgCl$_2$, 1 mM DTT, pH 7.8		NBL Gene 014802 014805
				Nippon Gene
5000-20,000 U/mL	1 unit completely cleaves 1.0 μg λ DNA/hr in a 50 μL reaction mixture	200 mM Tris-acetate, 500 mM KOAc, 100 mM MgOAc, 10 mM DTT, pH 7.9	100% cleavage, 50% ligation, 95% recleavage No detectable endonuclease (overdigestion)	Sigma R5635
2000-20,000 U/mL		50 mM KCl, 10 mM Tris-HCl, 0.1 mM EDTA, 1 mM DTT, 200 μg/mL BSA, 50% glycerol, pH 7.4		Stratagene 500750

CONCENTRATION or VOLUME ACTIVITY	UNIT DEFINITION	REACTION BUFFER (As Provided)	ADDITIONAL ACTIVITIES and PURITY	SUPPLIER CATALOG No.
Neisseria cinerea continued				
3-20 U/μL	1 unit completely digests 1 μg DNA/hr at 37°C	10 mM Tris-HCl, 10 mM MgCl$_2$, 1 mM DTT, pH 7.5		Toyobo NCI-101 NCI-102
Neisseria cinerea ATCC 14685				
10 U/μL	1 unit completely digests 1 μg λ DNA/hr at 37°C in a 50 mL reaction volume	60 mM Tris-HCl, 60 mM MgCl$_2$, 500 mM NaCl, 10 mM DTT, pH 7.5	30% ligation	Promega R7061 R7062
Neisseria cinerea NRCC 31006				
6-10 U/μL	1 unit completely digests 1 mg λ DNA/hr at 37°C	200 mM Tris-HCl, 100 mM MgCl$_2$, pH 7.4	0% ligation	Life Technol 15411-010
20,000 U/mL	1 unit completely digests 1 μg λ DNA/hr in a total reaction volume of 50 μL using the buffer provided	20 mM Tris-acetate, 10 mM MgOAc, 50 mM KOAc, 1 mM DTT, pH 7.9	50% DNA fragments ligated and >95% recut after 2-fold overdigestion	NE Biolabs 196S 196L

Nco I

RECOGNITION SEQUENCE
5'...C▼CATGG...3'
3'...GGTAC▲C...5'

OPTIMUM REACTION CONDITIONS
pH 7.5–7.9

HEAT INACTIVATION
65°C for 20 min.; 70°C for 15 min.; 80°C for 5 min.

METHYLATION EFFECTS
Cleaves CCm^{6}ATGG
Does not cleave m^{4}CCATGG or m^{5}CCATGG
Inhibited by 5-methylcytosine

NOTES
- EC 3.1.21.4
- A type II site specific deoxyribonuclease
- A large group of enzymes recognizing specific short DNA sequences. They cleave either within the recognition site, or at a short specific distance from it
- High glycerol concentration or DMSO solution results in star activity
- Useful for cutting near translation initiation sites of cDNA and fusing cDNA to translation initiation sites of expression vectors like pTrc99A
- Compatible cohesive ends: *Afl* III, *Bsp*H I, *Rca* I
- Requires Mg^{2+}

CONCENTRATION or VOLUME ACTIVITY	UNIT DEFINITION	REACTION BUFFER (As Provided)	ADDITIONAL ACTIVITIES and PURITY	SUPPLIER CATALOG NO.
Escherichia coli strain, carrying the cloned gene from *Nocardia corallina*				
10,000 U/mL; 50,000 U/mL	1 unit completely digests 1 μg λ DNA/hr in a total reaction volume of 50 μL using the buffer provided	20 m*M* Tris-acetate, 10 m*M* MgOAc, 50 m*M* KOAc, 1 m*M* DTT, pH 7.9	>95% DNA fragments ligated and recut after 5-fold overdigestion	NE Biolabs 193S 193L
Nocardia corallina				
500 U and 2000 U	1 unit completely digests 1 μg DNA substrate/hr at 37°C	10 m*M* Tris-acetate, 10 m*M* MgOAc, 50 m*M* KOAc, 1 m*M* DTT, 0.1 mg/mL BSA, pH 7.9	>90% DNA fragments ligated and recut after 5-fold overdigestion	Adv Biotech AB-0233-a AB-0233-b
10 U/μL; 40 U/μL	1 unit completely cleaves 1 μg substrate DNA/hr	50 m*M* NaCl, 10 m*M* Tris, 10 m*M* MgCl$_2$, 1 m*M* β-MSH, pH 7.8	90-98% cleavage-ligation-recleavage	Am Allied L-14-00200 L-14-01000 H-14-01000 L-14-05000 H-14-05000
8-12 U/μL	1 unit completely digests 1 μg λ DNA/hr at 37°C in the buffer provided	*Nco* I basal buffer	>95% DNA fragments ligated and 100% recut	Amersham E 1160Y E 1160Z
10 U/μL and 40 U/μL	1 unit completely digests 1 μg substrate DNA/hr	50 m*M* Tris-HCl, 10 m*M* MgCl$_2$, 100 m*M* NaCl, 1 m*M* DTE, pH 7.5, 37°C		Boehringer 835315 835323 1047698

Nco I continued

CONCENTRATION or VOLUME ACTIVITY	UNIT DEFINITION	REACTION BUFFER (As Provided)	ADDITIONAL ACTIVITIES and PURITY	SUPPLIER CATALOG NO.
Nocardia corallina continued				
	1 unit completely cleaves 1 µg λ DNA/hr at pH 7.5, 37°C in a reaction volume of 50 µL	50 mM Tris-HCl, 10 mM MgCl$_2$, 100 mM NaCl, 1.0 mM DTT, 100 µg/mL BSA, pH 7.5		CHIMERX 2292-01 2292-02
8-12 U/µL	1 unit hydrolyzes 1 µg DNA/hr at 37°C in a 50 µL reaction volume using λ phage DNA as substrate	33 mM Tris-acetate, 10 mM MgOAc, 66 mM KOAc, 0.1 mg/mL BSA, pH 7.9	>90% of the DNA fragments ligated and recut after 50-fold overdigestion	Fermentas ER0571 ER0572
2000-20,000 U/mL		50 mM KCl, Tris-HCl, 0.1 mM EDTA, 1.0 mM DTT, 200 µg/mL BSA, 50% glycerol, pH 7.4		ICN 151734
	10 units does not produce unspecific cleavage products with 1 µg λ DNA/16 hr at 37°C	50 mM Tris-HCl, 10 mM MgCl$_2$, 100 mM NaCl, 1 mM DTT, 100 µg/mL BSA, pH 7.9	>90% DNA fragments ligated and recut after 10-fold overdigestion	MINOTECH 116
6-12 U/µL	1 unit digests completely 1 µg λ DNA substrate/hr at 37°C	50 mM Tris-HCl, 100 mM NaCl, 10 mM MgCl$_2$, 1 mM DTT, pH 7.8		NBL Gene 017311 017302
10 U/µL and 20-80 U/µL	1 unit hydrolyzes 1 µg λ DNA to completion/hr in a 50 µL total reaction volume	100 mM Tris-HCl, 1 M NaCl, 100 mM MgCl$_2$, 100 mM β-MSH, 1.0 mg/mL BSA, pH 8		Nippon Gene Oncor 110611 110612
2000-5000 U/mL and 50,000-80,000 U/mL	1 unit completely digests 1 µg λ DNA/hr at pH 7.5, 37°C in a 50 µL total assay mixture	100 mM Tris-acetate, 100 mM MgOAc, 500 mM KOAc, pH 7.5		Pharmacia 27-0971-01 27-0971-02 27-0971-18
10 U/µL	1 unit completely digests 1 µg λ DNA/hr at 37°C in a 50 mL reaction volume	Blue/White Cloning Qualified; 60 mM Tris-HCl, 60 mM MgCl$_2$, 1.5 M NaCl, 10 mM DTT, pH 7.9	90% ligation	Promega R6513 R6514 R6515
2000-20,000 U/mL	1 unit completely cleaves 1.0 µg λ DNA/hr in a 50 µL reaction mixture	200 mM Tris-acetate, 500 mM KOAc, 100 mM MgOAc, 10 mM DTT, pH 7.9, 25°C	100% cleavage, 100% ligation, 100% recleavage, 25% endonuclease (nickase) No detectable endonuclease (overdigestion)	Sigma R8761
1000-10,000 U/mL		50 mM KCl, 10 mM Tris-HCl, 0.1 mM EDTA, 1 mM DTT, 200 µg/mL BSA, 50% glycerol, pH 7.4		Stratagene 500760 500761
500 U and 2500 U	1 unit completely digests 1 µg λ DNA/hr at 37°C in 50 µL of supplied buffer and BSA	200 mM Tris-HCl, 100 mM MgCl$_2$, 10 mM DTT, 1 M KCl, BSA, pH 8.5	Purity is confirmed by overdigestion, ligation-recutting and cloning test	TaKaRa 1160

Nco I continued

CONCENTRATION or VOLUME ACTIVITY	UNIT DEFINITION	REACTION BUFFER (As Provided)	ADDITIONAL ACTIVITIES and PURITY	SUPPLIER CATALOG No.
Nocardia corallina continued				
3-20 U/µL	1 unit completely digests 1 µg DNA/hr at 37°C	50 mM Tris-HCl, 10 mM MgCl$_2$, 100 mM NaCl, 1 mM DTT, pH 7.5	>90% ligation; >95% recut	Toyobo NCO-101 NCO-102
Nocardia corallina ATCC 19070				
5000 U/mL	1 unit degrades 1 µg λ dam⁻ dcm⁻ DNA/hr at 37°C in a 0.05 mL mixture	10 mM Tris-HCl, 100 mM NaCl, 100 µg/mL BSA, pH 8.0	No detectable contaminating nuclease; 90-95% ligated and 95-100% recut after 10-50-fold overdigestion	AGS Heidelb A00430S A00430M
8-12 U/µL	1 unit completely digests 1 mg λ DNA/hr at 37°C	500 mM Tris-HCl, 100 mM MgCl$_2$, 1 M NaCl, pH 8.0	≥95% ligation; 100% cleavage	Life Technol 15421-019 15421-050

Nde I

RECOGNITION SEQUENCE
   5'...CA▼TATG...3'
   3'...GTAT▲AC...5'

OPTIMUM REACTION CONDITIONS
   $T_{1/2}$ = 15 min. at 37°C

HEAT INACTIVATION
   65°C for 20 min.

METHYLATION EFFECTS
   Cleaves m^5CATATG
   Blocked by N^6-methyladenine methylation

NOTES

- EC 3.1.21.4
- A type II site specific deoxyribonuclease
- A large group of enzymes recognizing specific short DNA sequences. They cleave either within the recognition site, or at a short specific distance from it
- Yields a single base with a 5'-extension which is difficult to ligate with T4 DNA ligase; ligation is improved by the addition of 15% PEG
- Compatible ends: Ase I, Asn I, Bfa I, Csp6 I, Mae I, Mse I, Tru9 I, Vsp I
- Sensitive to impurities in some DNA preparations: DNA purified by the standard miniprep process is cleaved at lower rates
- Requires Mg^{2+}

CONCENTRATION or VOLUME ACTIVITY	UNIT DEFINITION	REACTION BUFFER (As Provided)	ADDITIONAL ACTIVITIES and PURITY	SUPPLIER CATALOG NO.
Escherichia coli, carrying the cloned *Nde* I gene from *Neisseria denitrificans*				
3-20 U/μL	1 unit completely digests 1 μg λ DNA/hr at 37°C in the buffer provided	330 m*M* Tris-acetate, 100 m*M* MgOAc, 5 m*M* DTT, 660 m*M* KOAc, pH 7.9	>95% DNA fragments ligated and 95% recut	Amersham E 0216Y E 0216Z
20,000 U/mL	1 unit completely digests 1 μg λ DNA/hr in a total reaction volume of 50 μL using the buffer provided	20 m*M* Tris-acetate, 10 m*M* MgOAc, 50 m*M* KOAc, 1 m*M* DTT, pH 7.9	>95% DNA fragments ligated and recut after 20-fold overdigestion	NE Biolabs 111S 111L
Neisseria denitrificans				
8-12 U/μL	1 unit hydrolyzes 1 μg DNA/hr at 37°C in a 50 μL reaction volume using λ phage DNA as substrate	10 m*M* Tris-HCl, 10 m*M* MgCl₂, 100 m*M* KCl, 0.1 mg/mL BSA, pH 8.5	>90% of the DNA fragments ligated and recut after 50-fold overdigestion	Fermentas ER0581 ER0582
3000-20,000 U/mL		50 m*M* KCl, 10 m*M* Tris-HCl, 0.1 m*M* EDTA, 1.0 m*M* DTT, 200 μg/mL BSA, 50% glycerol, pH 7.4		ICN 151735
5 U/μL and 10-20 U/μL	1 unit hydrolyzes 1 μg λ DNA to completion/hr in a 50 μL total reaction volume	100 m*M* Tris-HCl, 1 *M* NaCl, 100 m*M* MgCl₂, 100 m*M* β-MSH, 1.0 mg/mL BSA, pH 8		Oncor 110511 110512

Nde I continued

CONCENTRATION or VOLUME ACTIVITY	UNIT DEFINITION	REACTION BUFFER (As Provided)	ADDITIONAL ACTIVITIES and PURITY	SUPPLIER CATALOG No.
Neisseria denitrificans continued				
5000-10,000 U/mL	1 unit completely digests 1 µg λ DNA/hr at pH 7.5, 37°C in a 50 µL total assay mixture	100 mM Tris-acetate, 100 mM MgOAc, 500 mM KOAc, pH 7.5		Pharmacia 27-0915-01 27-0915-02
10 U/µL	1 unit completely digests 1 µg λ DNA/hr at 37°C in a 50 mL reaction volume	60 mM Tris-HCl, 60 mM MgCl$_2$, 1.5 M NaCl, 10 mM DTT, pH 7.9	90% ligation	Promega R6801 R6802
3000-20,000 U/mL	1 unit completely cleaves 1.0 µg λ DNA/hr in a 50 µL reaction mixture	500 mM Tris-HCl, 1 M NaCl, 100 mM MgCl$_2$, 10 mM DTT, pH 7.9	100% cleavage, 70% ligation, 90% recleavage, 50% endonuclease (nickase) No detectable endonuclease (overdigestion)	Sigma R5509
3000-10,000 U/mL		50 mM KCl, 10 mM Tris-HCl, 0.1 mM EDTA, 1 mM DTT, 200 µg/mL BSA, 50% glycerol, pH 7.4		Stratagene 500770
400 U and 2000 U	1 unit completely digests 1 µg λ DNA/hr at 37°C in 50 µL of supplied buffer	500 mM Tris-HCl, 100 mM MgCl$_2$, 10 mM DTT, 1 M NaCl, pH 7.5	Purity is confirmed by overdigestion, ligation-recutting and cloning test	TaKaRa 1161
Neisseria denitrificans NRCC 31009				
10,000 U/mL	1 unit degrades 1 µg λ dam$^-$ dcm$^-$ DNA/hr at 37°C in a 0.05 mL mixture (extrapolated from a 15 min. test)	50 mM Tris-HCl, 10 mM MgCl$_2$, 100 mM NaCl, 100 µg/mL BSA, pH 7.5	No detectable contaminating nuclease; 95% ligated and recut after 10-fold overdigestion	AGS Heidelb B00440S B00440M
10 U/µL	1 unit completely digests 1 µg substrate DNA/hr	50 mM Tris-HCl, 10 mM MgCl$_2$, 100 mM NaCl, 1 mM DTE, pH 7.5		Boehringer 1040219 1040227
4-6 U/µL	1 unit completely digests 1 mg λ DNA/hr at 37°C	500 mM Tris-HCl, 100 mM MgCl$_2$, 500 mM NaCl, pH 8.0	≥33% ligation; 100% cleavage	Life Technol 15426-018 15426-026
6-12 U/µL	1 unit digests completely 1 µg λ DNA substrate/hr at 37°C	50 mM Tris-HCl, 100 mM NaCl, 10 mM MgCl$_2$, 1 mM DTT, pH 7.8		NBL Gene 017411 017402
5-20 U/µL	λ DNA substrate, 37°C	1 M NaCl, 500 mM Tris-HCl, 100 mM MgCl$_2$, 10 mM DTT, pH 7.5)	90% ligation and 100% recutting	Nippon Gene 319-01142 313-01145 311-01146

Nde II

RECOGNITION SEQUENCE

5'...▼GATC...3'
3'...CTAG▲...5'

ISOSCHIZOMERS

Mbo I, Sau3A I

HEAT INACTIVATION

65°C for 30 min.

METHYLATION EFFECTS

Does not cleave DNA when the A residue is N^6-methyladenine

Blocked by overlapping *dam* methylation; *Sau*3A I is not

NOTES

- EC 3.1.21.4
- A type II site specific deoxyribonuclease
- A large group of enzymes recognizing specific short DNA sequences. They cleave either within the recognition site, or at a short specific distance from it
- Compatible cohesive ends: *Bam*H I, *Bcl* I, *Bgl* II, *Xho* II
- Requires Mg^{2+}

CONCENTRATION or VOLUME ACTIVITY	UNIT DEFINITION	REACTION BUFFER (As Provided)	ADDITIONAL ACTIVITIES and PURITY	SUPPLIER CATALOG NO.
Neisseria denitrificans				
10,000 U/mL	1 unit degrades 1 μg λ dam⁻ dcm⁻ DNA/hr at 37°C in a 0.05 mL mixture	100 mM Tris-HCl, 150 mM NaCl, 10 mM MgCl₂, 100 μg/mL BSA, pH 7.6	No detectable contaminating nuclease; 90% ligated and 100% recut after 50-fold overdigestion	AGS Heidelb A00450S A00450M
10 U/μL and 15-40 U/μL	1 unit hydrolyzes 1 μg λ DNA to completion/hr in a 50 μL total reaction volume	1 M Tris-HCl, 1.5 M NaCl, 100 mM MgCl₂, pH 8		Oncor 110271 110272
Neisseria denitrificans NRCC 31009				
5 U/μL	1 unit completely digests 1 μg substrate DNA/hr			Boehringer 1040235 1040243
4-6 U/μL	1 unit completely digests 1 mg SV40 DNA/hr at 37°C	1 M Tris-HCl, 100 mM MgCl₂, 1.5 M NaCl, pH 7.6 and 10 mM DTT	≥95% ligation; 100% cleavage	Life Technol 15427-016
5-20 U/μL	λ (dam⁻) DNA substrate, 37°C	150 mM NaCl, 100 mM Tris-HCl, 10 mM MgCl₂, 1 mM DTT, pH 7.5	90% ligation and 100% recutting	Nippon Gene 317-01861 313-01863 311-01864

NgoA IV

RECOGNITION SEQUENCE

5'...G▼CCGGC...3'
3'...CGGCC▲G...5'

NOTES
- EC 3.1.21.4
- A type II site specific deoxyribonuclease
- A large group of enzymes recognizing specific short DNA sequences. They cleave either within the recognition site, or at a short specific distance from it
- Unlike *Nae* I and *Ngo*M I, does not show a marked site preference
- Requires Mg^{2+}

CONCENTRATION or VOLUME ACTIVITY	UNIT DEFINITION	REACTION BUFFER (As Provided)	ADDITIONAL ACTIVITIES and PURITY	SUPPLIER CATALOG NO.
Neisseria gonorrhea				
8-12 U/μL	1 unit completely digests 1 mg Ad2 DNA/hr at 37°C	200 mM Tris-acetate, 100 mM MgOAc, 500 mM KOAc, pH 7.9	≥95% ligation; 100% cleavage	Life Technol 15498-017

NgoM I

RECOGNITION SEQUENCE
5'...G▼CCGGC...3'
3'...CGGCC▲G...5'

NOTES
- EC 3.1.21.4
- A type II site specific deoxyribonuclease
- A large group of enzymes recognizing specific short DNA sequences. They cleave either within the recognition site, or at a short specific distance from it
- Requires Mg^{2+}

CONCENTRATION or VOLUME ACTIVITY	UNIT DEFINITION	REACTION BUFFER (As Provided)	ADDITIONAL ACTIVITIES and PURITY	SUPPLIER CATALOG NO.
Escherichia coli strain, carrying the cloned gene from *Neisseria gonorrhea*				
10,000 U/mL	1 unit completely digests 1 µg λ DNA/hr in a total reaction volume of 50 µL using the buffer provided	20 mM Tris-acetate, 10 mM MgOAc, 50 mM KOAc, 1 mM DTT, pH 7.9	>95% DNA fragments ligated and recut after 10-fold overdigestion	NE Biolabs 564S 564L
Neisseria gonorrhea				
8-12 U/µL	1 unit completely digests 1 µg Ad2 DNA/hr at 37°C in a 50 mL reaction volume	250 mM Tris-acetate, 1 M KOAc, 100 mM MgOAc, 10 mM DTT, pH 7.8	90% ligation	Promega R7171 R7172

Nhe I

RECOGNITION SEQUENCE
5'...G▼CTAGC...3'
3'...CGATC▲G...5'

OPTIMUM REACTION CONDITIONS
pH 8.0

HEAT INACTIVATION
65°C for 30 min.; 80°C for 5 min.

METHYLATION EFFECTS
Does not cleave GCTAGm^{5}C

NOTES
- EC 3.1.21.4
- A type II site specific deoxyribonuclease
- A large group of enzymes recognizing specific short DNA sequences. They cleave either within the recognition site, or at a short specific distance from it
- Inhibited by salt >100 mM and 3'-5-methylcytosine
- High glycerol and pH, low ion concentration and DMSO solution result in star activity
- Compatible cohesive ends: Avr II, Bln I, Spe I, Xba I
- Requires Mg^{2+}

CONCENTRATION or VOLUME ACTIVITY	UNIT DEFINITION	REACTION BUFFER (As Provided)	ADDITIONAL ACTIVITIES and PURITY	SUPPLIER CATALOG No.
Neisseria mucosa subsp. *heidelbergensis* ATCC 25999				
8-12 U/μL	1 unit digests completely 1 μg λ DNA substrate/hr at 37°C	10 mM Tris-HCl, 50 mM NaCl, 7 mM MgCl$_2$, 7 mM β-MSH, 100 μg/mL BSA, pH 7.8		NBL Gene 016501 016503
5000 U/mL	1 unit completely digests 1 μg λ DNA (Hind III digest)/hr in a total reaction volume of 50 μL using the buffer provided	10 mM Tris-HCl, 10 mM MgCl$_2$, 50 mM NaCl, 1 mM DTT, pH 7.9; add 100 μg/mL BSA	>95% DNA fragments ligated and recut after 50-fold overdigestion	NE Biolabs 131S 131L
5-20 U/μL	λ DNA substrate, 37°C	500 mM NaCl, 100 mM Tris-HCl, 100 mM MgCl$_2$, 10 mM DTT, pH 7.5 (10X buffer)	90% ligation and 100% recutting	Nippon Gene 314-00891 310-00893 318-00894
10 U/μL and 20-50 U/μL	1 unit hydrolyzes 1 μg λ DNA to completion/hr in a 50 μL total reaction volume	100 mM Tris-HCl, 100 mM NaCl, 100 mM MgCl$_2$, 100 mM β-MSH, 1.0 mg/mL BSA, pH 7.5		Oncor 110621 110622
5000-10,000 U/mL	1 unit completely digests 1 μg Ad2 DNA/hr at pH 7.5, 37°C in a 50 μL total assay mixture	100 mM Tris-acetate, 100 mM MgOAc, 500 mM KOAc, pH 7.5		Pharmacia 27-0974-01 27-0974-02

Nhe I continued

CONCENTRATION or VOLUME ACTIVITY	UNIT DEFINITION	REACTION BUFFER (As Provided)	ADDITIONAL ACTIVITIES and PURITY	SUPPLIER CATALOG No.
Neisseria mucosa subsp. heidelbergensis continued				
10 U/µL	1 unit completely digests 1 µg EcoR I λ DNA fragments/hr at 37°C in a 50 mL reaction volume	Genome qualified; 60 mM Tris-HCl, 60 mM MgCl$_2$, 500 mM NaCl, 10 mM DTT, pH 7.5	90% ligation	Promega R6501 R6502 R6505
2000-20,000 U/mL	1 unit completely cleaves 1.0 µg Hind III digest λ DNA/hr in a 50 µL reaction mixture	100 mM Tris-HCl, 500 mM NaCl, 100 mM MgCl$_2$, 10 mM DTT, pH 7.9	100% cleavage, 95% ligation, 100% recleavage, 50% endonuclease (nickase) No detectable endonuclease (overdigestion)	Sigma R5634
2000-15,000 U/mL		50 mM KCl, 10 mM Tris-HCl, 0.1 mM EDTA, 1 mM DTT, 200 µg/mL BSA, 50% glycerol, pH 7.4		Stratagene 500780
500 U and 2500 U	1 unit completely digests 1 µg λ DNA/hr at 37°C in 50 µL of supplied buffer	100 mM Tris-HCl, 100 mM MgCl$_2$, 10 mM DTT, 500 mM NaCl, pH 7.5	Purity is confirmed by overdigestion, ligation-recutting and genome DNA digestion test	TaKaRa 1162
3-20 U/µL	1 unit completely digests 1 µg DNA/hr at 37°C	10 mM Tris-HCl, 10 mM MgCl$_2$, 50 mM NaCl, 1 mM DTT, pH 7.5	>90% ligation; >95% recut	Toyobo NHE-101 NHE-102

Nla III

RECOGNITION SEQUENCE
5'...CATG▼...3'
3'...▲GTAC...5'

OPTIMUM REACTION CONDITIONS
pH 7.9

HEAT INACTIVATION
65°C for 20 min.; 70°C for 5 min.

NOTES
- EC 3.1.21.4
- A type II site specific deoxyribonuclease
- A large group of enzymes recognizing specific short DNA sequences. They cleave either within the recognition site, or at a short specific distance from it
- NaCl and KCl inhibit activity; ammonium sulfate does not
- Compatible cohesive ends: Sph I
- Requires Mg^{2+}

CONCENTRATION or VOLUME ACTIVITY	UNIT DEFINITION	REACTION BUFFER (As Provided)	ADDITIONAL ACTIVITIES and PURITY	SUPPLIER CATALOG NO.
Escherichia coli, carrying the cloned Nla III gene from Neisseria lactamica				
1-5 U/μL	1 unit completely digests 1 μg φX174 RFI DNA/hr at 37°C in the buffer provided	330 mM Tris-acetate, 100 mM MgOAc, 5 mM DTT, 660 mM KOAc, pH 7.9	>95% DNA fragments ligated and 95% recut	Amersham E 0217Y E 0217Z
5000 U/mL	1 unit completely digests 1 μg φX174 DNA/hr in a total reaction volume of 50 μL using the buffer provided	20 mM Tris-acetate, 10 mM MgOAc, 50 mM KOAc, 1 mM DTT, pH 7.9; add 100 μg/mL BSA	>95% DNA fragments ligated and recut after 15-fold overdigestion	NE Biolabs 125S 125L
1000-10,000 U/mL	1 unit completely cleaves 1.0 μg λ DNA/hr in a 50 μL reaction mixture	200 mM Tris-acetate, 500 mM KOAc, 100 mM MgOAc, 10 mM DTT, BSA, pH 7.9	100% cleavage, 100% ligation, 95% recleavage 100% endonuclease (overdigestion)	Sigma R5760
Neisseria lactamica				
4000-20,000 U/mL		200 mM KCl, 10 mM Tris-HCl, 0.1 mM EDTA, 1 mM DTT, 500 μg/mL acetylated BSA, 50% glycerol, pH 7.5		ICN 159456
3-20 U/μL	1 unit completely digests 1 μg DNA/hr at 37°C	10 mM Tris-HCl, 10 mM $MgCl_2$, 50 mM NaCl, 1 mM DTT, pH 7.5	>95% ligation; >95% recut	Toyobo NLA-301 NLA-302

Nla IV

RECOGNITION SEQUENCE
5'...GGN▼NCC...3'
3'...CCN▲NGG...5'

HEAT INACTIVATION
65°C for 20 min.

NOTES
- EC 3.1.21.4
- A type II site specific deoxyribonuclease
- A large group of enzymes recognizing specific short DNA sequences. They cleave either within the recognition site, or at a short specific distance from it
- NaCl and KCl inhibit activity
- Requires Mg^{2+}

CONCENTRATION or VOLUME ACTIVITY	UNIT DEFINITION	REACTION BUFFER (As Provided)	ADDITIONAL ACTIVITIES and PURITY	SUPPLIER CATALOG NO.
Escherichia coli, carrying the Nla IV gene from Neisseria lactamica				
1000 U/mL	1 unit completely digests 1 μg pBR322 DNA/hr in a total reaction volume of 50 μL using the buffer provided	20 m*M* Tris-acetate, 10 m*M* MgOAc, 50 m*M* KOAc, 1 m*M* DTT, pH 7.9; add 100 μg/mL BSA	>95% DNA fragments ligated and recut after 5-fold overdigestion	NE Biolabs 126S 126L
500-5000 U/mL	1 unit completely cleaves 1.0 μg λ DNA/hr in a 50 μL reaction mixture	200 m*M* Tris-acetate, 500 m*M* KOAc, 100 m*M* MgOAc, 10 m*M* DTT, BSA, pH 7.9	98% cleavage, 99% ligation, 100% recleavage No detectable endonuclease (overdigestion)	Sigma R5885
Neisseria lactamica				
4000-20,000 U/mL		150 m*M* KCl, 10 m*M* Tris-HCl, 0.1 m*M* EDTA, 1 m*M* DTT, 200 μg/mL acetylated BSA, 50% glycerol, pH 7.5		ICN 159457

Not I

RECOGNITION SEQUENCE
5'...GC▼GGCCGC...3'
3'...CGCCGG▲CG...5'

OPTIMUM REACTION CONDITIONS
pH 7.5-7.9

HEAT INACTIVATION
65°C for 10-30 min.

METHYLATION EFFECTS
Cleaves GCGGCCGm^{5}C
Does not cleave GCGGm^{5}CCGC or GCGGCm^{5}CGC
Does not cleave DNA when internal cytosine is methylated by M. *Xma* III

NOTES
- EC 3.1.21.4
- A type II site specific deoxyribonuclease
- A large group of enzymes recognizing specific short DNA sequences. They cleave either within the recognition site, or at a short specific distance from it
- *Not* I and *Sfi* I are the only two known 8-base recognizing restriction endonucleases
- Compatible cohesive ends: *Bsp*120 I, *Bst*Z I, *Eae* I, *Eag* I, *Ecl*X I, *Eco*52 I, *Xma* II
- Supercoiled plasmids may require 5-fold more enzyme than linear DNA for complete digestion
- Requires Mg^{2+}; some variants require BSA and Triton X-100
- Useful for generating larger DNA fragments to be analyzed by pulse field gel electrophoresis

CONCENTRATION or VOLUME ACTIVITY	UNIT DEFINITION	REACTION BUFFER (As Provided)	ADDITIONAL ACTIVITIES and PURITY	SUPPLIER CATALOG No.
Escherichia coli strain, carrying the cloned gene from *Nocardia otitidis-caviarum*				
10,000 U/mL; 50,000 U/mL	1 unit completely digests 1 μg Ad2 DNA/hr in a total reaction volume of 50 μL using the buffer provided	50 m*M* Tris-HCl, 10 m*M* MgCl$_2$, 100 m*M* NaCl, 1 m*M* DTT, pH 7.9; add 100 μg/mL BSA	>95% DNA fragments ligated and recut after 25-fold overdigestion	NE Biolabs 189S 189L
Nocardia otitidis-caviarum				
400 U and 1000 U	1 unit completely digests 1 μg DNA substrate/hr at 37°C	10 m*M* Tris-HCl, 10 m*M* MgCl$_2$, 100 m*M* NaCl, 1 m*M* DTT, 0.1 mg/mL BSA, pH 7.5	>90% DNA fragments ligated and recut after 5-fold overdigestion	Adv Biotech AB-0234-a AB-0234-b
10 U/μL; 40 U/μL	1 unit completely cleaves 1 μg substrate DNA/hr	100 m*M* NaCl, 50 m*M* Tris, 5 m*M* MgCl$_2$, 2 m*M* β-MSH, pH 7.8	90-98% cleavage-ligation-recleavage	Am Allied L-31-00200 L-31-01000 H-31-01000
8-12 U/μL	1 unit completely digests 1 μg Ad2 DNA/hr at 37°C in the buffer provided	Genomic grade; 500 m*M* Tris-HCl, 100 m*M* MgCl$_2$, 10 m*M* DTT, 1000 m*M* NaCl, BSA, Triton X-100, pH 7.5	>95% DNA fragments ligated and 100% recut	Amersham E 0304Y E 0304Z

Not I continued

CONCENTRATION or VOLUME ACTIVITY	UNIT DEFINITION	REACTION BUFFER (As Provided)	ADDITIONAL ACTIVITIES and PURITY	SUPPLIER CATALOG No.
Nocardia otitidis-caviarum continued				
10,000 U/mL	1 unit degrades 1 μg Ad2 DNA/hr at 37°C in a 0.05 mL mixture	50 mM Tris-HCl, 10 mM MgCl$_2$, 100 mM NaCl, 100 μg/mL BSA, pH 7.5	No detectable contaminating nuclease; 90% ligated and 95-100% recut after 4-fold overdigestion	AGS Heidelb F00460S F00460M
10 U/μL and 40 U/μL	1 unit completely digests 1 μg substrate DNA/hr	50 mM Tris-HCl, 10 mM MgCl$_2$, 100 mM NaCl, 1 mM DTE, pH 7.5		Boehringer 1014706 1014714 1037668
	1 unit completely cleaves 1 μg Ad2 DNA/hr at pH 7.5, 37°C in a reaction volume of 50 μL	50 mM Tris-HCl, 10 mM MgCl$_2$, 100 mM NaCl, 1.0 mM DTT, 100 μg/mL BSA, pH 7.5		CHIMERx 2295-01 2295-02
8-12 U/μL	1 unit hydrolyzes 1 μg DNA/hr at 37°C in a 50 μL reaction volume using λ phage DNA as substrate	50 mM Tris-HCl, 10 mM MgCl$_2$, 100 mM NaCl, 0.02% Triton X-100, pH 7.5	>90% of the pTZ19RJL2 DNA fragments ligated and recut after 50-fold overdigestion	Fermentas ER0591 ER0592
15 U/μL and 50 U/μL	1 unit completely digests 1 mg Ad2 DNA/hr at 37°C	500 mM Tris-HCl, 100 mM MgCl$_2$, 1 M NaCl, pH 8.0	≥95% ligation; 100% cleavage	Life Technol 15441-025 15441-017 15441-041
	10 units does not produce unspecific cleavage products with 1 μg Ad2 DNA/16 hr at 37°C	10 mM Tris-HCl, 10 mM MgCl$_2$, 150 mM NaCl, 100 μg/mL BSA, 0.1% Triton X-100, pH 7.9	>90% DNA fragments ligated and recut after several-fold overdigestion	MINOTECH 117
5-12 U/μL	1 unit digests completely 1 μg Ad2 DNA substrate/hr at 37°C	50 mM Tris-HCl, 100 mM NaCl, 10 mM MgCl$_2$, 1 mM DTT, Triton X-100, pH 7.8		NBL Gene 015301 015303
4-20 U/μL	Ad2 DNA substrate, 37°C	1 M NaCl, 500 mM Tris-HCl, 100 mM MgCl$_2$, 10 mM DTT, 0.01% Triton X-100, pH 7.5 (10X buffer)	90% ligation and 100% recutting	Nippon Gene 316-01415 312-01453 310-01454
5 U/μL	1 unit hydrolyzes 1 μg Ad2 DNA to completion/hr in a 50 μL total reaction volume	500 mM Tris-HCl, 1 M NaCl, 100 mM MgCl$_2$, 10 mM DTE, 1.0 mg/mL BSA, pH 8		Oncor 110491 110492
10,000-20,000 U/mL	1 unit completely digests 1 μg Ad2 DNA/hr at pH 7.5, 37°C in a 50 μL total assay mixture	100 mM Tris-acetate, 100 mM MgOAc, 500 mM KOAc, pH 7.5		Pharmacia 27-0976-01 27-0976-02
10 U/μL and 40-80 U/μL	1 unit completely digests 1 μg Ad2 DNA/hr at 37°C in a 50 mL reaction volume	Blue/White Cloning Qualified; genome qualified; 60 mM Tris-HCl, 60 mM MgCl$_2$, 1.5 M NaCl, 10 mM DTT, pH 7.9	90% ligation	Promega R6431 R6432 R6435 R4434

Not I continued

CONCENTRATION or VOLUME ACTIVITY	UNIT DEFINITION	REACTION BUFFER (As Provided)	ADDITIONAL ACTIVITIES and PURITY	SUPPLIER CATALOG No.
Nocardia otitidis-caviarum continued				
2000-30,000 U/mL	1 unit cleaves 95% Ad2 DNA/hr in a 50 µL reaction mixture	Unique buffer + BSA	95% cleavage, 85% ligation, 80% recleavage, 85% endonuclease (nickase) No detectable endonuclease (overdigestion)	Sigma R8506
10,000-50,000 U/mL		200 mM KCl, 10 mM Tris-HCl, 0.1 mM EDTA, 0.2% Triton X-100, 200 µg/mL BSA, 50% glycerol, pH 7.4		Stratagene 500800 500801
500 U and 2500 U	1 unit completely digests 1 µg Ad2 DNA/hr at 37°C in 50 µL of supplied buffer, BSA, Triton X-100	Genomic grade; 500 mM Tris-HCl, 100 mM MgCl$_2$, 10 mM DTT, 1 M NaCl, BSA, Triton X-100, pH 7.5	Purity is confirmed by overdigestion, ligation-recutting, genome DNA digestion and cloning test	TaKaRa 1166
3-20 U/µL	1 unit completely digests 1 µg DNA/hr at 37°C	50 mM Tris-HCl, 10 mM MgCl$_2$, 100 mM NaCl, 1 mM DTT, pH 7.5	>90% ligation; >95% recut	Toyobo NOT-103 NOT-104
Nocardia otitidis-caviarum ATCC 14630				
2000-20,000 U/mL		50 mM KCl, 10 mM Tris-HCl, 0.1 mM EDTA, 1 mM DTT, 200 µg/mL BSA, 50% glycerol, pH 7.4		ICN 153836
400 U and 2000 U	1 unit completely cleaves 1 µg DNA/hr in 50 µL reaction mixture	500 mM Tris-HCl, 100 mM MgCl$_2$, 10 mM DTT, 1 M NaCl, 0.01% BSA, 0.1% Triton X-100, pH 7.5		PanVera TAK 1166

Nru I

RECOGNITION SEQUENCE
  5'...TCG▼CGC...3'
  3'...AGC▲CGA...5'
OPTIMUM REACTION CONDITIONS
  pH 8.0
METHYLATION EFFECTS
  Cleaves TCGm^{5}CGA
  Does not cleave Tm^{5}CGCGA or TCGCGm^{6}A
  Blocked by overlapping *dam* methylation

NOTES
- EC 3.1.21.4
- A type II site specific deoxyribonuclease
- A large group of enzymes recognizing specific short DNA sequences. They cleave either within the recognition site, or at a short specific distance from it
- Compatible cohesive ends: any blunt end
- Partially cleaves DNA from *dam*$^+$ *E. coli*
- Requires Mg^{2+}; some variants require NaCl

CONCENTRATION or VOLUME ACTIVITY	UNIT DEFINITION	REACTION BUFFER (As Provided)	ADDITIONAL ACTIVITIES and PURITY	SUPPLIER CATALOG NO.
Nocardia rubra (Rhodococcus rhodochrous ATCC 15906)				
500 U and 2000 U	1 unit completely digests 1 µg DNA substrate/hr at 37°C	10 m*M* Tris-HCl, 10 m*M* MgCl$_2$, 50 m*M* NaCl, 1 m*M* DTT, 0.1 mg/mL BSA, pH 7.5	>90% DNA fragments ligated and recut after 2-fold overdigestion	Adv Biotech AB-0235-a AB-0235-b
8-12 U/µL	1 unit completely digests 1 µg λ DNA/hr at 37°C in the buffer provided	Genomic grade; Nru I basal buffer	95% DNA fragments ligated and recut	Amersham E 0305Y E 0305Z
8000 U/mL	1 unit degrades 1 µg λ dam$^-$ dcm$^-$ DNA/hr at 37°C in a 0.05 mL mixture	10 m*M* Tris-HCl, 100 m*M* NaCl, 100 µg/mL BSA, pH 8.0	No detectable contaminating nuclease; 20% ligated and 90% recut after 10-fold overdigestion	AGS Heidelb A00470S A00470M
10 U/µL and 40 U/µL	1 unit completely digests 1 µg substrate DNA/hr	10 m*M* Tris-HCl, 5 m*M* MgCl$_2$, 100 m*M* NaCl, 1 m*M* β-MSH, pH 8.0		Boehringer 776769 776777 1175106
	1 unit completely cleaves 1 µg λ DNA/hr at pH 7.5, 37°C in a reaction volume of 50 µL	50 m*M* Tris-HCl, 10 m*M* MgCl$_2$, 100 m*M* NaCl, 1.0 m*M* DTT, 100 µg/mL BSA, pH 7.5		CHIMERx 2297-01 2297-02
5000-50,000 U/mL		50 m*M* KCl, 10 m*M* Tris-HCl, 0.1 m*M* EDTA, 1.0 m*M* DTT, 200 µg/mL BSA, 50% glycerol, pH 7.4		ICN 151776
8-12 U/µL	1 unit completely digests 1 mg λ DNA/hr at 37°C	500 m*M* Tris-HCl, 100 m*M* MgCl$_2$, 500 m*M* NaCl, 500 m*M* KCl, pH 8.0	≥33% ligation; 100% cleavage	Life Technol 15423-015

Nru I continued

CONCENTRATION or VOLUME ACTIVITY	UNIT DEFINITION	REACTION BUFFER (As Provided)	ADDITIONAL ACTIVITIES and PURITY	SUPPLIER CATALOG NO.
Nocardia rubra (Rhodococcus rhodochrous ATCC 15906) continued				
	10 units does not produce unspecific cleavage products with 1 µg λ DNA/16 hr at 37°C	50 mM Tris-HCl, 10 mM MgCl$_2$, 100 mM KCl, 100 µg/mL BSA, pH 8.0	>90% DNA fragments ligated and recut after 10-fold overdigestion	MINOTECH 118
8-12 U/µL	1 unit digests completely 1 µg λ DNA substrate/hr at 37°C	10 mM Tris-HCl, 100 mM NaCl, 5 mM MgCl$_2$, 1 mM β-MSH, pH 8.3		NBL Gene 014501 014503
10,000 U/mL; 50,000 U/mL	1 unit completely digests 1 µg λ DNA/hr in a total reaction volume of 50 µL using the buffer provided	50 mM Tris-HCl, 10 mM MgCl$_2$, 100 mM KCl, pH 7.7	<20% DNA fragments ligated after 20-fold overdigestion	NE Biolabs 192S 192L
5-20 U/µL	λ DNA substrate, 37°C	150 mM NaCl, 6 mM Tris-HCl, 6 mM MgCl$_2$, 6 mM β-MSH, pH 7.5	90% ligation and 100% recutting	Nippon Gene 315-00522 319-00525 317-00526
5 U/µL	1 unit hydrolyzes 1 µg λ DNA to completion/hr in a 50 µL total reaction volume	500 mM Tris-HCl, 1 M NaCl, 100 mM MgCl$_2$, 10 mM DTE, 1.0 mg/mL BSA, pH 8		Oncor 110301 110302
5000-10,000 U/mL	1 unit completely digests 1 µg λ DNA/hr at pH 7.5, 37°C in a 50 µL total assay mixture	100 mM Tris-acetate, 100 mM MgOAc, 500 mM KOAc, pH 7.5		Pharmacia 27-0820-01 27-0820-02
10 U/µL	1 unit completely digests 1 µg λ DNA/hr at 37°C in a 50 mL reaction volume	Genome qualified; 100 mM Tris-HCl, 100 mM MgCl$_2$, 1.5 M KCl, pH 7.4	10% ligation	Promega R7091 R7092
5000 U/mL		10 mM Tris-HCl, 10 mM MgCl$_2$, 100 mM NaCl, 1 mM DTT, pH 8.5	20% DNA fragments ligated and recut after 5-fold overdigestion	SibEnzyme E099 E100
5000-50,000 U/mL	1 unit cleaves 98% Ad2 DNA/hr in a 50 µL reaction mixture	Unique buffer	100% cleavage, 10% ligation, 90% recleavage, 10% endonuclease (nickase) No detectable endonuclease (overdigestion)	Sigma R5759
2000-10,000 U/mL		50 mM KCl, 10 mM Tris-HCl, 0.1 mM EDTA, 1 mM DTT, 200 µg/mL BSA, 50% glycerol, pH 7.4		Stratagene 500810
1000 U and 5000 U	1 unit completely digests 1 µg λ DNA/hr at 37°C in 50 µL of supplied buffer	Genomic grade; 10 mM Tris-HCl, 7 mM MgCl$_2$, 7 mM β-MSH, 150 mM KCl, 0.01% BSA, pH 7.5	Purity is confirmed by overdigestion, ligation-recutting and genome DNA digestion test	TaKaRa 1168
3-20 U/µL	1 unit completely digests 1 µg DNA/hr at 37°C	Nru I	>80% ligation; >95% recut	Toyobo NRU-101 NRU-102

NruG I

RECOGNITION SEQUENCE
5'...GACNNN▼NNGTC...3'
3'...CTGNN▲NNNCAG...5'

HEAT INACTIVATION
65°C for 20 min.

NOTES
- EC 3.1.21.4
- A type II site specific deoxyribonuclease
- A large group of enzymes recognizing specific short DNA sequences. They cleave either within the recognition site, or at a short specific distance from it
- Requires Mg^{2+}

CONCENTRATION or VOLUME ACTIVITY	UNIT DEFINITION	REACTION BUFFER (As Provided)	ADDITIONAL ACTIVITIES and PURITY	SUPPLIER CATALOG NO.
Nocardia rugosa G				
3000 U/mL		10 mM Tris-HCl, 10 mM $MgCl_2$, 1 mM DTT, pH 7.6	5% DNA fragments ligated and recut after 5-fold overdigestion	SibEnzyme E193 E194

Nsi I

RECOGNITION SEQUENCE
 5'...ATGCA▼T...3'
 3'...T▲ACGTA...5'

ISOSCHIZOMERS
 Ava III, *Eco*T22 I

HEAT INACTIVATION
 65°C for 20 min.

METHYLATION EFFECTS
 Does not cleave ATGCm^{6}AT or ATGm^{5}CAT

NOTES
- EC 3.1.21.4
- A type II site specific deoxyribonuclease
- A large group of enzymes recognizing specific short DNA sequences. They cleave either within the recognition site, or at a short specific distance from it
- Compatible cohesive ends: *Pst* I
- Requires Mg^{2+}

CONCENTRATION or VOLUME ACTIVITY	UNIT DEFINITION	REACTION BUFFER (As Provided)	ADDITIONAL ACTIVITIES and PURITY	SUPPLIER CATALOG No.
Neisseria sicca				
10 U/µL; 40 U/µL	1 unit completely cleaves 1 µg substrate DNA/hr	100 mM NaCl, 50 mM Tris, 5 mM MgCl$_2$, 2 mM β-MSH, pH 7.8	90-98% cleavage-ligation-recleavage	Am Allied L-29-01000 H-29-01000 L-29-05000 H-29-05000
10,000 U/mL	1 unit degrades 1 µg λ dam$^-$ dcm$^-$ DNA/hr at 37°C in a 0.05 mL mixture	10 mM Tris-HCl, 10 mM MgCl$_2$, 100 mM KCl, 100 µg/mL BSA, pH 8.5	No detectable contaminating nuclease; 90% ligated and 100% recut after 10-fold overdigestion	AGS Heidelb A00480S A00480M
10 U/µL and 40 U/µL	1 unit completely digests 1 µg substrate DNA/hr	50 mM Tris-HCl, 10 mM MgCl$_2$, 100 mM NaCl, 1 mM DTE, pH 7.5		Boehringer 909831 909840 1207628
8-12 U/µL	1 unit completely digests 1 mg λ DNA/hr at 37°C	500 mM Tris-HCl, 100 mM MgCl$_2$, 1 M NaCl, pH 8.0	≥95% ligation; 100% cleavage	Life Technol 15434-012 15434-038
6-12 U/µL	1 unit digests completely 1 µg λ DNA substrate/hr at 37°C	50 mM Tris-HCl, 100 mM NaCl, 10 mM MgCl$_2$, 1 mM DTT, pH 7.8		NBL Gene 017511 017502
10 U/µL	1 unit completely digests 1 µg λ DNA/hr at 37°C in a 50 mL reaction volume	Blue/White Cloning Qualified; 60 mM Tris-HCl, 60 mM MgCl$_2$, 1.5 M NaCl, 10 mM DTT, pH 7.9	90% ligation	Promega R6531 R6532

Nsi I continued

CONCENTRATION or VOLUME ACTIVITY	UNIT DEFINITION	REACTION BUFFER (As Provided)	ADDITIONAL ACTIVITIES and PURITY	SUPPLIER CATALOG NO.
Neisseria sicca continued				
2000-20,000 U/mL	1 unit completely cleaves 1.0 μg λ DNA/hr in a 50 μL reaction mixture	Unique buffer	100% cleavage, 100% ligation, 100% recleavage, <1% endonuclease (nickase) No detectable endonuclease (overdigestion)	Sigma R5884
5000-20,000 U/mL		50 mM KCl, 10 mM Tris-HCl, 0.1 mM EDTA, 1 mM DTT, 200 μg/mL BSA, 50% glycerol, pH 7.4		Stratagene 500820 500821
Neisseria sicca ATCC 29256				
4000-20,000 U/mL		50 mM KCl, 10 mM Tris-HCl, 0.1 mM EDTA, 1 mM DTT, 200 μg/mL BSA, 50% glycerol, pH 7.4		ICN 153822
10,000 U/mL	1 unit completely digests 1 μg λ DNA/hr in a total reaction volume of 50 μL using the buffer provided	10 mM Tris-HCl, 10 mM MgCl$_2$, 100 mM NaCl, 1 mM DTT, pH 8.4	>95% DNA fragments ligated and recut after 20-fold overdigestion	NE Biolabs 127S 127L
5-20 U/μL	λ DNA substrate, 37°C	1 M NaCl, 500 mM Tris-HCl, 100 mM MgCl$_2$, 10 mM DTT, pH 7.5	90% ligation and 100% recutting	Nippon Gene 317-00901 315-00902 319-00905

Nsp I

RECOGNITION SEQUENCE
    5'...PuCATG▼Py...3'
    3'...Py▲GTACPu...5'

HEAT INACTIVATION
    65°C for 10 min.

NOTES
- EC 3.1.21.4
- A type II site specific deoxyribonuclease
- A large group of enzymes recognizing specific short DNA sequences. They cleave either within the recognition site, or at a short specific distance from it
- Requires Mg^{2+}

CONCENTRATION or VOLUME ACTIVITY	UNIT DEFINITION	REACTION BUFFER (As Provided)	ADDITIONAL ACTIVITIES and PURITY	SUPPLIER CATALOG No.
Nostoc species PCC 7524				Amersham
10 U/µL	1 unit completely digests 1 µg substrate DNA/hr	10 m*M* Tris-HCl, 10 m*M* MgCl₂, 50 m*M* NaCl, 1 m*M* DTE, pH 7.5		Boehringer 1131419
4-6 U/µL	1 unit completely digests 1 mg λ DNA/hr at 37°C	500 m*M* Tris-HCl, 100 m*M* MgCl₂, pH 8.0	≥85% ligation; 100% cleavage	Life Technol 15407-018
				TaKaRa

Nsp V

RECOGNITION SEQUENCE
5'...TT▼CGAA...3'
3'...AAGC▲TT...5'

ISOSCHIZOMERS
*Bst*B I, *Csp*45 I, *Nsp*7542 I

OPTIMUM REACTION CONDITIONS
pH 7.9

HEAT INACTIVATION
65°C for >15 min.

NOTES
- EC 3.1.21.4
- A type II site specific deoxyribonuclease
- A large group of enzymes recognizing specific short DNA sequences. They cleave either within the recognition site, or at a short specific distance from it
- Requires Mg^{2+}

CONCENTRATION or VOLUME ACTIVITY	UNIT DEFINITION	REACTION BUFFER (As Provided)	ADDITIONAL ACTIVITIES and PURITY	SUPPLIER CATALOG No.
Nostoc species ATCC 29411				
1-5 U/µL	λ DNA substrate, 37°C	100 m*M* Tris-HCl, 100 m*M* MgCl₂, 10 m*M* DTT, pH 7.5	95% ligation and 100% recutting	Nippon Gene 312-00912 316-00915 314-00916
Nostoc species PCC 7524				Amersham
8-12 U/µL	1 unit completely digests 1 mg λ DNA/hr at 50°C	500 m*M* Tris-HCl, 100 m*M* MgCl₂, pH 8.0	≥65% ligation; 100% cleavage	Life Technol 15462-013
				TaKaRa

Nsp V continued

CONCENTRATION or VOLUME ACTIVITY	UNIT DEFINITION	REACTION BUFFER (As Provided)	ADDITIONAL ACTIVITIES and PURITY	SUPPLIER CATALOG No.
Nostoc species PCC 7524 continued				
3-20 U/µL	1 unit completely digests 1 µg DNA/hr at 37°C	10 mM Tris-HCl, 10 mM MgCl$_2$, 1 mM DTT, pH 7.5	>90% ligation; >95% recut	Toyobo NSP-101 NSP-102

Nsp7524 I

RECOGNITION SEQUENCE
  5'...PuCATG▼Py...3'
  3'...Py▲GTACPu...5'

HEAT INACTIVATION
  60°C for 15 min.

NOTES
- EC 3.1.21.4
- A type II site specific deoxyribonuclease
- A large group of enzymes recognizing specific short DNA sequences. They cleave either within the recognition site, or at a short specific distance from it
- Requires Mg^{2+}

CONCENTRATION or VOLUME ACTIVITY	UNIT DEFINITION	REACTION BUFFER (As Provided)	ADDITIONAL ACTIVITIES and PURITY	SUPPLIER CATALOG No.
Nostoc species				
200 U and 1000 U	1 unit completely digests 1 µg λ DNA/hr at 37°C in 50 µL of supplied buffer, BSA, Triton X-100	200 mM Tris-HCl, 100 mM MgCl$_2$, 10 mM DTT, 1 M KCl, BSA, Triton X-100, pH 8.5	Purity is confirmed by overdigestion and ligation-recutting test	TaKaRa 1171
Nostoc species PCC7524				
8-12 U/µL	1 unit completely digests 1 µg λ DNA/hr at 37°C in the buffer provided	200 mM Tris-HCl, 100 mM MgCl$_2$, 10 mM DTT, 1000 mM KCl, BSA, Triton X-100, pH 8.5	>95% DNA fragments ligated and 100% recut	Amersham E 1171Y

Nsp7524 V

RECOGNITION SEQUENCE
5'...TT▼CGAA...3'
3'...AAGC▲TT...5'

ISOSCHIZOMERS
Asu II

HEAT INACTIVATION
70°C for 15 min.

NOTES
- EC 3.1.21.4
- A type II site specific deoxyribonuclease
- A large group of enzymes recognizing specific short DNA sequences. They cleave either within the recognition site, or at a short specific distance from it
- Requires Mg^{2+}

CONCENTRATION or VOLUME ACTIVITY	UNIT DEFINITION	REACTION BUFFER (As Provided)	ADDITIONAL ACTIVITIES and PURITY	SUPPLIER CATALOG No.
Nostoc species				
500 U and 2500 U	1 unit completely digests 1 µg λ DNA/hr at 37°C in 50 µL of supplied buffer	Genomic grade; 100 mM Tris-HCl, 100 mM $MgCl_2$, 10 mM DTT, pH 7.5	Purity is confirmed by overdigestion, ligation-recutting, genome DNA digestion and cloning test	TaKaRa 1175
Nostoc species PCC7524				
8-12 U/µL	1 unit completely digests 1 µg λ DNA/hr at 37°C in the buffer provided	Genomic grade; 100 mM Tris-HCl, 100 mM $MgCl_2$, 10 mM DTT, pH 7.5	95% DNA fragments ligated and 100% recut	Amersham E 1175Y

NspB II

RECOGNITION SEQUENCE
5'...C(^A_C)G▼C(^G_T)G...3'
3'...G(^T_G)C▲G(^C_A)C...5'

NOTES
- EC 3.1.21.4
- A type II site specific deoxyribonuclease
- A large group of enzymes recognizing specific short DNA sequences. They cleave either within the recognition site, or at a short specific distance from it
- Requires Mg^{2+}

NspB II continued

CONCENTRATION or VOLUME ACTIVITY	UNIT DEFINITION	REACTION BUFFER (As Provided)	ADDITIONAL ACTIVITIES and PURITY	SUPPLIER CATALOG No.
Nostoc species B				
3-10 U/µL	1 unit completely digests 1 µg λ DNA/hr at 37°C in the buffer provided	100 mM Tris-HCl, 100 mM MgCl$_2$, 10 mM DTT, pH 7.5	>80% DNA fragments ligated and >95% recut	Amersham E 0218Y E 0218Z

Pac I

RECOGNITION SEQUENCE
     5'...TTAAT▼TAA...3'
     3'...AAT▲TAATT...5'
OPTIMUM REACTION CONDITIONS
     pH 7.0-7.5
HEAT INACTIVATION
     65°C for 20 min.; 90°C for 5 min.

NOTES
- EC 3.1.21.4
- A type II site specific deoxyribonuclease
- A large group of enzymes recognizing specific short DNA sequences. They cleave either within the recognition site, or at a short specific distance from it
- Requires Mg^{2+}

CONCENTRATION or VOLUME ACTIVITY	UNIT DEFINITION	REACTION BUFFER (As Provided)	ADDITIONAL ACTIVITIES and PURITY	SUPPLIER CATALOG No.
Nostoc species B				
3-10 U/µL	1 unit completely digests 1 µg λ DNA/hr at 37°C in the buffer provided	100 mM Tris-HCl, 100 mM MgCl$_2$, 10 mM DTT, pH 7.5	>80% DNA fragments ligated and >95% recut	Amersham E 0218Y E 0218Z

Pae I

RECOGNITION SEQUENCE
5'...GCATG▼C...3'
3'...C▲GTACG...5'

NOTES
- EC 3.1.21.4
- A type II site specific deoxyribonuclease
- A large group of enzymes recognizing specific short DNA sequences. They cleave either within the recognition site, or at a short specific distance from it
- Requires Mg^{2+}

CONCENTRATION or VOLUME ACTIVITY	UNIT DEFINITION	REACTION BUFFER (As Provided)	ADDITIONAL ACTIVITIES and PURITY	SUPPLIER CATALOG NO.
Pseudomonas aeruginosa				
10,000 U/mL	1 unit degrades 1 μg λ dam⁻ dcm⁻ DNA/hr at 37°C in a 0.05 mL mixture	10 mM Tris-HCl, 10 mM MgCl₂, 1 mM DTT, 100 μg/mL BSA, pH 7.5	No detectable contaminating nuclease; 90% ligated and 100% recut after 50-fold overdigestion	AGS Heidelb
8-12 U/μL	1 unit hydrolyzes 1 μg DNA/hr at 37°C in a 50 μL reaction volume using λ phage DNA as substrate	10 mM Tris-HCl, 10 mM MgCl₂, 0.1 mg/mL BSA, pH 7.5	>90% of the DNA fragments ligated and recut after 50-fold overdigestion	Fermentas ER0601 ER0602

PaeR7 I

RECOGNITION SEQUENCE
5'...C▼TCGAG...3'
3'...GAGCT▲C...5'

ISOSCHIZOMERS
Xho I

HEAT INACTIVATION
Resistant

NOTES
- EC 3.1.21.4
- A type II site specific deoxyribonuclease
- A large group of enzymes recognizing specific short DNA sequences. They cleave either within the recognition site, or at a short specific distance from it
- Sequence CT▼CTCGAG is resistant to cleavage
- Requires Mg^{2+}

PaeR7 I continued

CONCENTRATION or VOLUME ACTIVITY	UNIT DEFINITION	REACTION BUFFER (As Provided)	ADDITIONAL ACTIVITIES and PURITY	SUPPLIER CATALOG No.
Pseudomonas aeruginosa				
2000-40,000 U/mL		50 mM KCl, 10 mM Tris-HCl, 0.1 mM EDTA, 1.0 mM DTT, 200 µg/mL BSA, 50% glycerol, pH 7.4		ICN 151797
Pseudomonas aeruginosa strain PA0303 pMG7 (R.V. Miller)				
20,000 U/mL	1 unit completely digests 1 µg λ DNA (Hind III digest)/hr in a total reaction volume of 50 µL using the buffer provided	20 mM Tris-acetate, 10 mM MgOAc, 50 mM KOAc, 1 mM DTT, pH 7.9	>95% DNA fragments ligated and recut after 10-fold overdigestion	NE Biolabs 177S 177L

Pal I

RECOGNITION SEQUENCE
5'...GG▼CC...3'
3'...CC▲GG...5'

NOTES
- EC 3.1.21.4
- A type II site specific deoxyribonuclease
- A large group of enzymes recognizing specific short DNA sequences. They cleave either within the recognition site, or at a short specific distance from it
- Requires Mg^{2+}

CONCENTRATION or VOLUME ACTIVITY	UNIT DEFINITION	REACTION BUFFER (As Provided)	ADDITIONAL ACTIVITIES and PURITY	SUPPLIER CATALOG No.
Providencia alcalifaciens				
5000-25,000 U/mL		50 mM KCl, 10 mM Tris-HCl, 0.1 mM EDTA, 1 mM DTT, 200 µg/mL BSA, 50% glycerol, pH 7.5		Pharmacia Stratagene 500850

Pau I

RECOGNITION SEQUENCE
  5'...G▼CGCGC...3'
  3'...CGCGC▲G...5'

OPTIMUM REACTION CONDITIONS
  20% activity at 37°C

METHYLATION EFFECTS
  Blocked by overlapping CG methylation

NOTES
- EC 3.1.21.4
- A type II site specific deoxyribonuclease
- A large group of enzymes recognizing specific short DNA sequences. They cleave either within the recognition site, or at a short specific distance from it
- Prototype is *BseP1* I
- Requires Mg^{2+}

CONCENTRATION or VOLUME ACTIVITY	UNIT DEFINITION	REACTION BUFFER (As Provided)	ADDITIONAL ACTIVITIES and PURITY	SUPPLIER CATALOG NO.
Paracoccus alcaliphilus ZVK3-3				
8-12 U/µL	1 unit hydrolyzes 1 µg DNA/hr at 37°C in a 50 µL reaction volume using λ phage DNA as substrate	10 mM Tris-HCl, 10 mM MgCl₂, 100 mM KCl, pH 8.5	>90% of the DNA fragments ligated and recut after 50-fold overdigestion	Fermentas ER1091 ER1092

*Pfl*23 II

RECOGNITION SEQUENCE
  5'...C▼GTACG...3'
  3'...GCATG▲C...5'

NOTES
- EC 3.1.21.4
- A type II site specific deoxyribonuclease
- A large group of enzymes recognizing specific short DNA sequences. They cleave either within the recognition site, or at a short specific distance from it
- Requires Mg^{2+}

Pfl23 II continued

Pseudomonas fluorescens RFL23

CONCENTRATION or VOLUME ACTIVITY	UNIT DEFINITION	REACTION BUFFER (As Provided)	ADDITIONAL ACTIVITIES and PURITY	SUPPLIER CATALOG No.
1500 U/mL	1 unit degrades 1 µg λ dam⁻ dcm⁻ DNA Hind III-fragments/hr at 37°C in a 0.05 mL mixture	33 mM Tris-acetate, 10 mM MgOAc, 66 mM KOAc, 100 µg/mL BSA, pH 7.9	No detectable contaminating nuclease; 90% ligated and >90% recut after 50-fold overdigestion	AGS Heidelb F00483S F00483M
1-3 U/µL	1 unit hydrolyzes 1 µg DNA/hr at 37°C in a 50 µL reaction volume using λ phage DNA as substrate	33 mM Tris-acetate, 10 mM MgOAc, 66 mM KOAc, 0.1 mg/mL BSA, pH 7.9	>90% of the DNA fragments ligated and recut after 10-fold overdigestion	Fermentas ER0851 ER0852

PflM I

RECOGNITION SEQUENCE
    5'...CCA(N)$_4$▼NTGG...3'
    3'...GGTN▲(N)$_4$ACC...5'
HEAT INACTIVATION
    65°C for 20 min.
METHYLATION EFFECTS
    Blocked by overlapping dcm methylation

NOTES
- EC 3.1.21.4
- A type II site specific deoxyribonuclease
- A large group of enzymes recognizing specific short DNA sequences. They cleave either within the recognition site, or at a short specific distance from it
- Partially cleaves DNA from dcm$^+$ E. coli
- λ DNA is cleaved at significantly lower rates than other substrates
- Requires Mg^{2+}

Pseudomonas fluorescens

CONCENTRATION or VOLUME ACTIVITY	UNIT DEFINITION	REACTION BUFFER (As Provided)	ADDITIONAL ACTIVITIES and PURITY	SUPPLIER CATALOG No.
3-20 U/µL	1 unit completely digests 1 µg λ DNA/hr at 37°C in the buffer provided	500 mM Tris-HCl, 100 mM MgCl$_2$, 10 mM DTT, 1000 mM NaCl, pH 7.5	>95% DNA fragments ligated and 95% recut	Amersham E 0219Y E 0219Z
4000-20,000 U/mL		50 mM KCl, 10 mM Tris-HCl, 0.1 mM EDTA, 1 mM DTT, 200 µg/mL acetylated BSA, 50% glycerol, pH 7.5		ICN 159459

*Pfl*M I continued

CONCENTRATION or VOLUME ACTIVITY	UNIT DEFINITION	REACTION BUFFER (As Provided)	ADDITIONAL ACTIVITIES and PURITY	SUPPLIER CATALOG No.
Pseudomonas fluorescens continued				
2000-10,000 U/mL	1 unit cleaves 98% λ DNA/hr in a 50 μL reaction mixture	500 mM Tris-HCl, 1 M NaCl, 100 mM MgCl$_2$, 10 mM DTT, BSA, pH 7.9	98% cleavage, 98% ligation, 99% recleavage; No detectable endonuclease (nickase) or endonuclease (overdigestion)	sigma R6010
Pseudomonas fluorescens (NEB 375)				
8000 U/mL	1 unit completely digests 1 μg λ DNA/hr in a total reaction volume of 50 μL using the buffer provided	50 mM Tris-HCl, 10 mM MgCl$_2$, 100 mM NaCl, 1 mM DTT, pH 7.9; add 100 μg/mL BSA	>95% DNA fragments ligated and recut after 10-fold overdigestion	NE Biolabs 509S 509L

PI-*Psp* I

RECOGNITION SEQUENCE
5'...TGGCAAACAGCTATTAT▼GGGTATTATGGG...3'
3'...ACCGTTTGTCGAT▲AATACCCATAATACCCA...5'

NOTES
- An intron-encoded endonuclease
- EC 3.1.21.4
- A type II site specific deoxyribonuclease
- A large group of enzymes recognizing specific short DNA sequences. They cleave either within the recognition site, or at a short specific distance from it
- 10 U of enzyme gives no detectable cleavage of pBR322, λ, Ad2, T7 or fX174 DNA after 16 hr at 65°C
- Requires Mg^{2+}

CONCENTRATION or VOLUME ACTIVITY	UNIT DEFINITION	REACTION BUFFER (As Provided)	ADDITIONAL ACTIVITIES and PURITY	SUPPLIER CATALOG NO.
Escherichia coli strain, expressing DNA polymerase from the thermophile, *Pyrococcus* species GB-D				
1000–10,000 U/mL	1 unit cleaves 1 μg agarose-embedded *E. coli* chromosomal DNA to completion/3 hr at 55°C; 0.5 unit digests 1 μg DNA/hr at 65°C in a reaction volume of 50 μL	10 mM Tris-HCl, 10 mM MgCl$_2$, 150 mM KCl, 1 mM DTT, pH 8.6; add 100 μg/mL BSA	>95% DNA fragments ligated and recut after 5-fold overdigestion	NE Biolabs 695S 695L

PI-Sce I

RECOGNITION SEQUENCE

5′...ATCTATGTCGGGTGC▼GGAGAAAGAGGTAATGAAATGGCA...3′

3′...TAGATACAGCC▲CACGCCTCTTTCTCCATTACTTTACCGT...5′

NOTES
- An intron-encoded endonuclease
- EC 3.1.21.4
- A type II site specific deoxyribonuclease
- A large group of enzymes recognizing specific short DNA sequences. They cleave either within the recognition site, or at a short specific distance from it
- 100 U of enzyme gives no cleavage of pBR322, λ, Ad2, T7 or fX174 after 16 hr at 37°C
- Requires Mg^{2+}

CONCENTRATION or VOLUME ACTIVITY	UNIT DEFINITION	REACTION BUFFER (As Provided)	ADDITIONAL ACTIVITIES and PURITY	SUPPLIER CATALOG NO.
Escherichia coli, expressing the modified VMA1 ATPase gene of the yeast, *Saccharomyces cerevisiae*				
1000–10,000 U/mL	1 unit cleaves 1 μg agarose-embedded *Saccharomyces cerevisiae* chromosomal DNA to completion/3 hr at 50°C; 6.0 units digests 1 μg DNA/hr at 37°C in a reaction volume of 50 μL	10 mM KPO$_4$, 10 mM MgCl$_2$, 100 mM KCl, 1 mM DTT, pH 8.6; add 100 μg/mL BSA	>95% DNA fragments ligated and recut after 5-fold overdigestion	NE Biolabs 696S 696L

PI-*Tli* I

RECOGNITION SEQUENCE

5'...GGTTCTTTATGCGGACAC▼TGACGGCTTTATG...3'
3'...CCAAGAAATACGCC▲TGTGACTGCCGAAATAC...5'

NOTES

- An intron-encoded endonuclease
- EC 3.1.21.4
- A type II site specific deoxyribonuclease
- A large group of enzymes recognizing specific short DNA sequences. They cleave either within the recognition site, or at a short specific distance from it
- 5 U of enzyme gives no cleavage of pBR322, λ, Ad2, T7 or fX174 DNA after 16 hr at 37°C
- 50 U of enzyme cleaves at degenerate sites
- Requires Mg^{2+}

CONCENTRATION or VOLUME ACTIVITY	UNIT DEFINITION	REACTION BUFFER (As Provided)	ADDITIONAL ACTIVITIES and PURITY	SUPPLIER CATALOG NO.
Escherichia coli, carrying the cloned second intervening sequence in the DNA polymerase gene isolated from *archaea* thermophile, *Thermococcus litorali*				
1000-5000 U/mL	1 unit cleaves 1 μg agarose-embedded *Saccharomyces cerevisiae* chromosomal DNA to completion/3 hr at 50°C; 1 unit digests 1 μg DNA/hr at 50°C in a reaction volume of 50 μL	50 mM Tris-HCl, 10 mM MgCl₂, 100 mM NaCl, 1 mM DTT, pH 8.6; add 100 μg/mL BSA	>75% DNA fragments ligated and recut after 2-fold overdigestion	NE Biolabs 698S 698L

PinA I

RECOGNITION SEQUENCE
5'...A▼CCGGT...3'
3'...TGGCC▲A...5'

ISOSCHIZOMERS
Age I

HEAT INACTIVATION
65°C for 10 min.

NOTES
- EC 3.1.21.4
- A type II site specific deoxyribonuclease
- A large group of enzymes recognizing specific short DNA sequences. They cleave either within the recognition site, or at a short specific distance from it
- Requires Mg^{2+}

CONCENTRATION or VOLUME ACTIVITY	UNIT DEFINITION	REACTION BUFFER (As Provided)	ADDITIONAL ACTIVITIES and PURITY	SUPPLIER CATALOG NO.
Pseudomonas inqualis				
10 U/μL	1 unit completely digests 1 μg substrate DNA/hr	10 mM Tris-HCl, 5 mM MgCl₂, 100 mM NaCl, 1 mM β-MSH, pH 8.0		Boehringer 1464841 1464850
4-6 U/μL	1 unit completely digests 1 mg λ DNA/hr at 37°C	200 mM Tris-HCl, 50 mM MgCl₂, 500 M KCl, pH 7.4	≥95% ligation; 100% cleavage	Life Technol 15408-016

Ple I

RECOGNITION SEQUENCE
5'...GAGTC(N)₄▼...3'
3'...CTCAG(N)₅▲...5'

HEAT INACTIVATION
65°C for 20 min.

NOTES
- EC 3.1.21.4
- A type II site specific deoxyribonuclease
- A large group of enzymes recognizing specific short DNA sequences. They cleave either within the recognition site, or at a short specific distance from it
- Requires Mg^{2+}

CONCENTRATION or VOLUME ACTIVITY	UNIT DEFINITION	REACTION BUFFER (As Provided)	ADDITIONAL ACTIVITIES and PURITY	SUPPLIER CATALOG NO.
Pseudomonas lemoignei (NEB 418)				
400-1000 U/mL	1 unit completely digests 1 µg λ DNA/hr in a total reaction volume of 50 µL using the buffer provided	20 mM Tris-acetate, 10 mM MgOAc, 50 mM KOAc, 1 mM DTT, pH 7.9	20% DNA fragments ligated using concentrated T4 DNA ligase after 2-fold overdigestion	NE Biolabs 515S 515L

Ple19 I

RECOGNITION SEQUENCE
    5'...CGAT▼CG...3'
    3'...GC▲TAGC...5'

HEAT INACTIVATION
    65°C for 20 min.

METHYLATION EFFECTS
    Not blocked by overlapping *dam* methylation

NOTES
- EC 3.1.21.4
- A type II site specific deoxyribonuclease
- A large group of enzymes recognizing specific short DNA sequences. They cleave either within the recognition site, or at a short specific distance from it
- Requires Mg^{2+}

CONCENTRATION or VOLUME ACTIVITY	UNIT DEFINITION	REACTION BUFFER (As Provided)	ADDITIONAL ACTIVITIES and PURITY	SUPPLIER CATALOG NO.
Pseudomonas lemoignei 19				
1000-5000 U/mL		33 mM Tris-acetate, 10 mM MgOAc, 66 mM KOAc, 1 mM DTT, pH 7.9	90% DNA fragments ligated and recut after 2-fold overdigestion	sibEnzyme E195 E196

PmaC I

RECOGNITION SEQUENCE
  5'...CAC▼GTG...3'
  3'...GTC▲CAC...5'

HEAT INACTIVATION
  60°C for 15 min.

NOTES
- EC 3.1.21.4
- A type II site specific deoxyribonuclease
- A large group of enzymes recognizing specific short DNA sequences. They cleave either within the recognition site, or at a short specific distance from it
- Requires Mg^{2+}

CONCENTRATION or VOLUME ACTIVITY	UNIT DEFINITION	REACTION BUFFER (As Provided)	ADDITIONAL ACTIVITIES and PURITY	SUPPLIER CATALOG NO.
Pseudomonas maltophila CB50P				
8-12 U/μL	1 unit completely digests 1 μg λ DNA/hr at 37°C in the buffer provided	100 mM Tris-HCl, 100 mM MgCl₂, 10 mM DTT, pH 7.5	100% DNA fragments ligated and recut	Amersham E 1177Y
500 U and 2500 U	1 unit completely digests 1 μg λ DNA/hr at 37°C in 50 μL of supplied buffer	100 mM Tris-HCl, 100 mM MgCl₂, 10 mM DTT, pH 7.5	Purity is confirmed by overdigestion and ligation-recutting test	TaKaRa 1177

Pme I

RECOGNITION SEQUENCE
  5'...GTTT▼AAAC...3'
  3'...CAAA▲TTTG...5'

HEAT INACTIVATION
  65°C for 20 min.

NOTES
- EC 3.1.21.4
- A type II site specific deoxyribonuclease
- A large group of enzymes recognizing specific short DNA sequences. They cleave either within the recognition site, or at a short specific distance from it
- Requires Mg^{2+}

Pme I continued

CONCENTRATION or VOLUME ACTIVITY	UNIT DEFINITION	REACTION BUFFER (As Provided)	ADDITIONAL ACTIVITIES and PURITY	SUPPLIER CATALOG NO.
Pseudomonas mendicina				
4000–20,000 U/mL		100 mM NaCl, 10 mM Tris-HCl, 0.1 mM EDTA, 1 mM DTT, 200 µg/mL acetylated BSA, 50% glycerol, pH 7.4		ICN 159460
				Pharmacia
Pseudomonas mendicina (NEB 698)				
4000 U/mL	1 unit completely digests 1 µg λ DNA/hr in a total reaction volume of 50 µL using the buffer provided	20 mM Tris-acetate, 10 mM MgOAc, 50 mM KOAc, 1 mM DTT, pH 7.9; add 100 µg/mL BSA	>80% DNA fragments ligated and recut after 5-fold overdigestion	NE Biolabs 560S 560L

Pme55 I

RECOGNITION SEQUENCE
    5'...AGG▼CCT...3'
    3'...TCC▲GGA...5'
HEAT INACTIVATION
    Resistant

NOTES
- EC 3.1.21.4
- A type II site specific deoxyribonuclease
- A large group of enzymes recognizing specific short DNA sequences. They cleave either within the recognition site, or at a short specific distance from it
- Requires Mg^{2+}

CONCENTRATION or VOLUME ACTIVITY	UNIT DEFINITION	REACTION BUFFER (As Provided)	ADDITIONAL ACTIVITIES and PURITY	SUPPLIER CATALOG NO.
Pseudomonas mendicina 55				
5000 U/mL		10 mM Tris-HCl, 10 mM $MgCl_2$, 1 mM DTT, pH 7.6	90% DNA fragments ligated and recut after 5-fold overdigestion	SibEnzyme E105 E106

Pml I

RECOGNITION SEQUENCE
5'...CAC▼GTG...3'
3'...GTG▲CAC...5'

HEAT INACTIVATION
65°C for 20 min.

NOTES
- EC 3.1.21.4
- A type II site specific deoxyribonuclease
- A large group of enzymes recognizing specific short DNA sequences. They cleave either within the recognition site, or at a short specific distance from it
- Requires Mg^{2+}

CONCENTRATION or VOLUME ACTIVITY	UNIT DEFINITION	REACTION BUFFER (As Provided)	ADDITIONAL ACTIVITIES and PURITY	SUPPLIER CATALOG NO.
Pseudomonas maltophila				
4000-20,000 U/mL		50 mM KCl, 10 mM Tris-HCl, 0.1 mM EDTA, 10 mM MgCl$_2$, 1.0 mM DTT, 200 µg/mL acetylated BSA, 50% glycerol, pH 7.4		ICN 159461
Pseudomonas maltophila (NEB 515)				
20,000 U/mL	1 unit completely digests 1 µg λ DNA (Hind III digest)/hr in a total reaction volume of 50 µL using the buffer provided	10 mM Bis Tris-Propane-HCl, 10 mM MgCl$_2$, 1 mM DTT, pH 7.0; add 100 µg/mL BSA	>95% DNA fragments ligated and recut after 10-fold overdigestion	NE Biolabs 532S 532L

Ppu10 I

RECOGNITION SEQUENCE
5'...A▼TGCAT...3'
3'...TACGT▲A...5'

ISOSCHIZOMERS
Ava III

HEAT INACTIVATION
Resistant

NOTES
- EC 3.1.21.4
- A type II site specific deoxyribonuclease
- A large group of enzymes recognizing specific short DNA sequences. They cleave either within the recognition site, or at a short specific distance from it
- Produces DNA fragments with a 4-base 5'-extension
- Requires Mg^{2+}

CONCENTRATION or VOLUME ACTIVITY	UNIT DEFINITION	REACTION BUFFER (As Provided)	ADDITIONAL ACTIVITIES and PURITY	SUPPLIER CATALOG NO.
Pseudomonas putida RFL10				
8000 U/mL	1 unit degrades 1 µg λ dam⁻ dcm⁻ DNA/hr at 37°C in a 0.05 mL mixture	33 mM Tris-acetate, 10 mM MgOAc, 66 mM KOAc, 100 µg/mL BSA, pH 7.9	No detectable contaminating nuclease; 90% ligated and >90% recut after 50-fold overdigestion	ACS Heidelb F00493S F00493M
8-12 U/µL	1 unit hydrolyzes 1 µg DNA/hr at 37°C in a 50 µL reaction volume using λ phage DNA as substrate	33 mM Tris-acetate, 10 mM MgOAc, 66 mM KOAc, 0.1 mg/mL BSA, pH 7.9	>90% of the DNA fragments ligated and recut after 50-fold overdigestion	Fermentas ER0861 ER0862
3000-20,000 U/mL		In reagent buffer		ICN 159462
10,000 U/mL	1 unit hydrolyzes 1 µg DNA/hr at 37°C in a 50 µL reaction volume using λ phage DNA as substrate	33 mM Tris-acetate, 10 mM MgOAc, 66 mM KOAc, 0.1 mg/mL BSA, pH 7.9	>90% of the DNA fragments ligated and recut after 50-fold overdigestion	NE Biolabs ER0861S ER0861L

*Ppu*M I

RECOGNITION SEQUENCE
5'...PuG▼G(^A_T)CCPy...3'
3'...PyCC(^T_A)G▲GPu...5'

OPTIMUM REACTION CONDITIONS
pH 7.9

HEAT INACTIVATION
Resistant; 90°C for 5 min.

METHYLATION EFFECTS
Blocked by overlapping *dcm* methylation

NOTES

- EC 3.1.21.4
- A type II site specific deoxyribonuclease
- A large group of enzymes recognizing specific short DNA sequences. They cleave either within the recognition site, or at a short specific distance from it
- Requires Mg^{2+}

CONCENTRATION or VOLUME ACTIVITY	UNIT DEFINITION	REACTION BUFFER (As Provided)	ADDITIONAL ACTIVITIES and PURITY	SUPPLIER CATALOG No.
Pseudomonas putida				
1-5 U/μL	1 unit completely digests 1 μg λ DNA (*Hind* III)/hr at 37°C in the buffer provided	330 m*M* Tris-acetate, 100 m*M* MgOAc, 5 m*M* DTT, 660 m*M* KOAc, pH 7.9	>95% DNA fragments ligated and 95% recut	Amersham E 0220Y E 0220Z
1000-5000 U/mL		100 m*M* NaCl, 10 m*M* Tris-HCl, 0.1 m*M* EDTA, 1 m*M* DTT, 200 μg/mL acetylated BSA, 50% glycerol, pH 7.4		ICN 159463
3-20 U/μL	1 unit completely digests 1 μg DNA/hr at 37°C	50 m*M* Tris-HCl, 10 m*M* MgCl₂, 100 m*M* NaCl, 1 m*M* DTT, pH 7.5	>95% ligation; >95% recut	Toyobo PPM-101 PPM-102
Pseudomonas putida (NEB 372)				
2000 U/mL	1 unit completely digests 1 μg λ DNA (*Hind* III digest)/hr in a total reaction volume of 50 μL using the buffer provided	20 m*M* Tris-acetate, 10 m*M* MgOAc, 50 m*M* KOAc, 1 m*M* DTT, pH 7.9	>95% DNA fragments ligated and recut after 10-fold overdigestion	NE Biolabs 506S 506L

PshA I

RECOGNITION SEQUENCE
5'...GACNN▼NNGTC...3'
3'...CTGNN▲NNCAG...5'

HEAT INACTIVATION
60°C for 15 min.

METHYLATION EFFECTS
Blocked by overlapping *dcm* methylation

NOTES
- EC 3.1.21.4
- A type II site specific deoxyribonuclease
- A large group of enzymes recognizing specific short DNA sequences. They cleave either within the recognition site, or at a short specific distance from it
- Requires Mg^{2+}

CONCENTRATION or VOLUME ACTIVITY	UNIT DEFINITION	REACTION BUFFER (As Provided)	ADDITIONAL ACTIVITIES and PURITY	SUPPLIER CATALOG NO.
Plesiomonas shigelloides				
8-12 U/μL	1 unit completely digests 1 μg λ DNA/hr at 37°C in the buffer provided	200 m*M* Tris-HCl, 100 m*M* MgCl₂, 10 m*M* DTT, 1000 m*M* KCl, pH 8.5	95% DNA fragments ligated and 100% recut	Amersham E 1074Y E 1074Z
				NE Biolabs
200 U and 1000 U	1 unit completely cleaves 1 μg DNA/hr in 50 μL reaction mixture	200 m*M* Tris-HCl, 100 m*M* MgCl₂, 10 m*M* DTT, 1 *M* KCl, pH 8.5		PanVera TAK 1074
200 U and 1000 U	1 unit completely digests 1 μg λ DNA/hr at 37°C in 50 μL of supplied buffer	200 m*M* Tris-HCl, 100 m*M* MgCl₂, 10 m*M* DTT, 1 *M* KCl, pH 8.5	Purity is confirmed by overdigestion and ligation-recutting test	TaKaRa 1074

PshB I

RECOGNITION SEQUENCE
5'...AT▼TAAT...3'
3'...TAAT▲TA...5'

NOTES
- EC 3.1.21.4
- A type II site specific deoxyribonuclease
- A large group of enzymes recognizing specific short DNA sequences. They cleave either within the recognition site, or at a short specific distance from it
- Requires Mg^{2+}

PshB I continued

CONCENTRATION or VOLUME ACTIVITY	UNIT DEFINITION	REACTION BUFFER (As Provided)	ADDITIONAL ACTIVITIES and PURITY	SUPPLIER CATALOG NO.
Plesiomonas shigelloides				
1000 U and 5000 U	1 unit completely digests 1 μg λ DNA/hr at 37°C in 50 μL of supplied buffer	Genomic grade; 10 mM Tris-HCl, 10 mM MgCl$_2$, 1 mM DTT, 50 mM NaCl, pH 8.5	Purity is confirmed by overdigestion, ligation-recutting, genome DNA digestion and cloning test	TaKaRa 1109

Psp5 II

RECOGNITION SEQUENCE
    5'...PuG▼G(^A_T)CCPy...3'
    3'...PyCC(^T_A)G▲GPu...5'
HEAT INACTIVATION
    Partially resistant at 65°C for 10 min.

NOTES
- EC 3.1.21.4
- A type II site specific deoxyribonuclease
- A large group of enzymes recognizing specific short DNA sequences. They cleave either within the recognition site, or at a short specific distance from it
- Requires Mg^{2+}

CONCENTRATION or VOLUME ACTIVITY	UNIT DEFINITION	REACTION BUFFER (As Provided)	ADDITIONAL ACTIVITIES and PURITY	SUPPLIER CATALOG NO.
Pseudomonas fluorescens RFL5				
5000 U/mL	1 unit degrades 1 μg λ dam$^-$ dcm$^-$ DNA Hind III-fragments/hr at 37°C in a 0.05 mL mixture	10 mM Tris-HCl, 10 mM MgCl$_2$, 100 mM KCl, 100 μg/mL BSA, 1 mM DTT, pH 8.5	No detectable contaminating nuclease; 90% ligated and 100% recut after 50-fold overdigestion	AGS Heidelb F00495S F00495M
4-8 U/μL	1 unit hydrolyzes 1 μg DNA/hr at 37°C in a 50 μL reaction volume using λ phage DNA as substrate	10 mM Tris-HCl, 10 mM MgCl$_2$, 50 mM NaCl, 0.1 mg/mL BSA, pH 7.5	>90% of the DNA fragments ligated and recut after 50-fold overdigestion	Fermentas ER0761 ER0762
4-6 U/μL	1 unit completely digests 1 mg λ DNA/hr at 37°C	500 mM Tris-HCl, 100 mM MgCl$_2$, 500 mM NaCl, pH 8.0	≥85% ligation; 100% cleavage	Life Technol 1549-010

Psp124B I

RECOGNITION SEQUENCE
  5'...GAGCT▼C...3'
  3'...C▲TCGAG...5'
HEAT INACTIVATION
  65°C for 20 min.

NOTES
- EC 3.1.21.4
- A type II site specific deoxyribonuclease
- A large group of enzymes recognizing specific short DNA sequences. They cleave either within the recognition site, or at a short specific distance from it
- Requires Mg^{2+}

CONCENTRATION or VOLUME ACTIVITY	UNIT DEFINITION	REACTION BUFFER (As Provided)	ADDITIONAL ACTIVITIES and PURITY	SUPPLIER CATALOG NO.
Pseudomonas species 124B				
15,000 U/mL		10 mM Tris-HCl, 10 mM $MgCl_2$, 50 mM NaCl, 1 mM DTT, pH 7.6	90% DNA fragments ligated and recut after 10-fold overdigestion	SibEnzyme E107 E108

Psp1406 I

RECOGNITION SEQUENCE
  5'...AA▼CGTT...3'
  3'...TTGC▲AA...5'
HEAT INACTIVATION
  65°C for 20 min.

NOTES
- EC 3.1.21.4
- A type II site specific deoxyribonuclease
- A large group of enzymes recognizing specific short DNA sequences. They cleave either within the recognition site, or at a short specific distance from it
- Requires Mg^{2+}

Psp1406 I continued

CONCENTRATION or VOLUME ACTIVITY	UNIT DEFINITION	REACTION BUFFER (As Provided)	ADDITIONAL ACTIVITIES and PURITY	SUPPLIER CATALOG NO.
Pseudomonas species RFL1406				
5000 U/mL	1 unit degrades 1 µg λ dam⁻ dcm⁻ DNA/hr at 37°C in a 0.05 mL mixture	33 mM Tris-acetate, 10 mM MgOAc, 66 mM KOAc, 100 µg/mL BSA, pH 7.9	No detectable contaminating nuclease; 90% ligated and 100% recut after 50-fold overdigestion	AGS Heidelb F00497S F00497M
3-10 U/µL	1 unit completely digests 1 µg substrate DNA/hr	10 mM Tris-HCl, 10 mM MgCl₂, 1 mM DTE, pH 7.5		Boehringer 1533860
8-12 U/µL	1 unit hydrolyzes 1 µg DNA/hr at 37°C in a 50 µL reaction volume using λ phage DNA as substrate	33 mM Tris-acetate, 10 mM MgOAc, 66 mM KOAc, 0.1 mg/mL BSA, pH 7.9	>90% of the DNA fragments ligated and recut after 50-fold overdigestion	Fermentas ER0941 ER0942
10,000 U/mL	1 unit hydrolyzes 1 µg DNA/hr at 37°C in a 50 µL reaction volume using λ phage DNA as substrate	33 mM Tris-acetate, 10 mM MgOAc, 66 mM KOAc, 0.1 mg/mL BSA, pH 7.9,	>90% of the DNA fragments ligated and recut after 50-fold overdigestion	NE Biolabs ER0941S ER0941L

PspA I

RECOGNITION SEQUENCE
    5'...C▼CCGGG...3'
    3'...GGGCC▲C...5'

NOTES
- EC 3.1.21.4
- A type II site specific deoxyribonuclease
- A large group of enzymes recognizing specific short DNA sequences. They cleave either within the recognition site, or at a short specific distance from it
- Requires Mg^{2+}

CONCENTRATION or VOLUME ACTIVITY	UNIT DEFINITION	REACTION BUFFER (As Provided)	ADDITIONAL ACTIVITIES and PURITY	SUPPLIER CATALOG NO.
Pseudomonas species				
2000-15,000 U/mL		50 mM KCl, 10 mM Tris-HCl, 0.1 mM EDTA, 1 mM DTT, 200 µg/mL BSA, 50% glycerol, pH 7.5		Stratagene 500853 500854

PspE I

RECOGNITION SEQUENCE
5'...G▼GTNACC...3'
3'...CCANTG▲G...5'

HEAT INACTIVATION
65°C for 20 min.

NOTES
- EC 3.1.21.4
- A type II site specific deoxyribonuclease
- A large group of enzymes recognizing specific short DNA sequences. They cleave either within the recognition site, or at a short specific distance from it
- Requires Mg^{2+}

CONCENTRATION or VOLUME ACTIVITY	UNIT DEFINITION	REACTION BUFFER (As Provided)	ADDITIONAL ACTIVITIES and PURITY	SUPPLIER CATALOG No.
Pseudomonas species E				
20,000 U/mL		10 mM Tris-HCl, 10 mM MgCl₂, 1 mM DTT, pH 7.6	90% DNA fragments ligated and recut after 10-fold overdigestion	SibEnzyme E169 E170

PspL I

RECOGNITION SEQUENCE
5'...C▼GTACG...3'
3'...GCATG▲C...5'

HEAT INACTIVATION
65°C for 20 min.

NOTES
- EC 3.1.21.4
- A type II site specific deoxyribonuclease
- A large group of enzymes recognizing specific short DNA sequences. They cleave either within the recognition site, or at a short specific distance from it
- Requires Mg^{2+}

CONCENTRATION or VOLUME ACTIVITY	UNIT DEFINITION	REACTION BUFFER (As Provided)	ADDITIONAL ACTIVITIES and PURITY	SUPPLIER CATALOG No.
Pseudomonas species L				
20,000 U/mL		33 mM Tris-acetate, 10 mM MgOAc, 66 mM KOAc, 1 mM DTT, pH 7.9	90% DNA fragments ligated and recut after 5-fold overdigestion	SibEnzyme E223 E224

PspN4 I

RECOGNITION SEQUENCE
5'...GGN▼NCC...3'
3'...CCN▲NGG...5'

HEAT INACTIVATION
65°C for 20 min.

NOTES
- EC 3.1.21.4
- A type II site specific deoxyribonuclease
- A large group of enzymes recognizing specific short DNA sequences. They cleave either within the recognition site, or at a short specific distance from it
- Requires Mg^{2+}

CONCENTRATION or VOLUME ACTIVITY	UNIT DEFINITION	REACTION BUFFER (As Provided)	ADDITIONAL ACTIVITIES and PURITY	SUPPLIER CATALOG No.
Pseudomonas species N4				
20,000 U/mL		33 mM Tris-acetate, 10 mM MgOAc, 66 mM KOAc, 1 mM DTT, pH 7.9	90% DNA fragments ligated and recut after 20-fold overdigestion	SibEnzyme E089 E090

PspOM I

RECOGNITION SEQUENCE
5'...G▼GGCCC...3'
3'...CCCGG▲G...5'

HEAT INACTIVATION
65°C for 20 min.

NOTES
- EC 3.1.21.4
- A type II site specific deoxyribonuclease
- A large group of enzymes recognizing specific short DNA sequences. They cleave either within the recognition site, or at a short specific distance from it
- Requires Mg^{2+}

CONCENTRATION or VOLUME ACTIVITY	UNIT DEFINITION	REACTION BUFFER (As Provided)	ADDITIONAL ACTIVITIES and PURITY	SUPPLIER CATALOG No.
Pseudomonas species OM2164				
40,000 U/mL		33 mM Tris-acetate, 10 mM MgOAc, 66 mM KOAc, 1 mM DTT, pH 7.9	90% DNA fragments ligated and recut after 10-fold overdigestion	SibEnzyme E215 E216

PspP I

RECOGNITION SEQUENCE
5'...G▼GNCC...3'
3'...CCNG▲G...5'

ISOSCHIZOMERS
AspS9 I, BsiZ I, Cfr13 I, Sau96 I, Stu I

OPTIMUM REACTION CONDITIONS
pH 7.9

NOTES
- EC 3.1.21.4
- A type II site specific deoxyribonuclease
- A large group of enzymes recognizing specific short DNA sequences. They cleave either within the recognition site, or at a short specific distance from it
- Requires Mg^{2+}

CONCENTRATION or VOLUME ACTIVITY	UNIT DEFINITION	REACTION BUFFER (As Provided)	ADDITIONAL ACTIVITIES and PURITY	SUPPLIER CATALOG No.
Psychrobacter species				
53,840 U/mg	30 units does not produce unspecific cleavage products with 1 µg λ DNA/16 hr at 25°C	10 mM Tris-HCl, 10 mM MgCl₂, 50 mM NaCl, 1 mM DTT, 100 µg/mL BSA, pH 7.9	>95% DNA fragments ligated and recut after 10-fold overdigestion	MINOTECH 139

PspPP I

RECOGNITION SEQUENCE
5'...RG▼GWCCY...3'
3'...YCCWG▲GR...5'

HEAT INACTIVATION
65°C for 20 min.

NOTES
- EC 3.1.21.4
- A type II site specific deoxyribonuclease
- A large group of enzymes recognizing specific short DNA sequences. They cleave either within the recognition site, or at a short specific distance from it
- Requires Mg^{2+}

CONCENTRATION or VOLUME ACTIVITY	UNIT DEFINITION	REACTION BUFFER (As Provided)	ADDITIONAL ACTIVITIES and PURITY	SUPPLIER CATALOG No.
Pseudomonas species PP				
5000 U/mL		33 mM Tris-acetate, 10 mM MgOAC, 66 mM KOAc, 1 mM DTT, pH 7.9	90% DNA fragments ligated and recut after 10-fold overdigestion	SibEnzyme E255 E256

Pst I

RECOGNITION SEQUENCE
  5'...CTGCA▼G...3'
  3'...G▲ACGTC...5'

OPTIMUM REACTION CONDITIONS
  pH 7.5-7.9

HEAT INACTIVATION
  70°C for 5 min.

METHYLATION EFFECTS
  Does not cleave CTGCm^{6}AG, m^{5}CTGCAG or Chm^{5}UGCAG

NOTES

- EC 3.1.21.4
- A type II site specific deoxyribonuclease
- A large group of enzymes recognizing specific short DNA sequences. They cleave either within the recognition site, or at a short specific distance from it
- Glycerol >12% or enzyme:DNA ratio >25 U/µg results in star activity
- Compatible cohesive ends: *Alw*2 I, *Asp*H I, *Ava* III, *Bmy* I, *Bsi*HKA I, *Bsp*1286 I, *Eco*T22 I, *Hgi*A I, *Nsi* I, *Sse*8387 I
- Requires Mg^{2+}

CONCENTRATION or VOLUME ACTIVITY	UNIT DEFINITION	REACTION BUFFER (As Provided)	ADDITIONAL ACTIVITIES and PURITY	SUPPLIER CATALOG NO.
Escherichia coli ED8654, carrying the cloned Pst I gene				
8-12 U/µL	1 unit completely digests 1 µg λ DNA/hr at 37°C in the buffer provided	500 m*M* Tris-HCl, 100 m*M* MgCl$_2$, 10 m*M* DTT, 1000 m*M* NaCl, pH 7.5	>95% DNA fragments ligated and 100% recut	Amersham E 1073Y E 1073Z
>40 U/µL	1 unit completely digests 1 µg λ DNA/hr at 37°C in the buffer provided	500 m*M* Tris-HCl, 100 m*M* MgCl$_2$, 10 m*M* DTT, 1000 m*M* NaCl, pH 7.5	>95% DNA fragments ligated and 100% recut	Amersham E 1073YH
10,000 U and 50,000 U	1 unit completely digests 1 µg λ DNA/hr at 37°C in 50 µL of supplied buffer	500 m*M* Tris-HCl, 100 m*M* MgCl$_2$, 10 m*M* DTT, 1 *M* NaCl, pH 7.5	Purity is confirmed by overdigestion, ligation-recutting and cloning test	TaKaRa 1073
Escherichia coli strain, carrying the cloned gene from *Providencia stuartii*				
6-12 U/µL >40 U/µL	1 unit digests completely 1 µg λ DNA substrate/hr at 37°C	50 m*M* Tris-HCl, 100 m*M* NaCl, 10 m*M* MgCl$_2$, 1 m*M* DTT, pH 7.8		NBL Gene 010304 010307 010357
20,000 U/mL; 100,000 U/mL	1 unit completely digests 1 µg λ DNA/hr in a total reaction volume of 50 µL using the buffer provided	50 m*M* Tris-HCl, 10 m*M* MgCl$_2$, 100 m*M* NaCl, 1 m*M* DTT, pH 7.9	>95% DNA fragments ligated and recut after 100-fold overdigestion	NE Biolabs 140S 140L

Pst I continued

CONCENTRATION or VOLUME ACTIVITY	UNIT DEFINITION	REACTION BUFFER (As Provided)	ADDITIONAL ACTIVITIES and PURITY	SUPPLIER CATALOG NO.
Escherichia coli, carrying a Pst I overproducing plasmid				
7500-40,000 U/mL	1 unit completely cleaves 1.0 μg λ DNA/hr in a 50 μL reaction mixture	500 mM Tris-HCl, 1 M NaCl, 100 mM MgCl₂, 10 mM DTT, pH 7.9	100% cleavage, 95% ligation, 90% recleavage. No detectable endonuclease (overdigestion)	Sigma R2007
Obtained from a cloned source				
10,000-50,000 U/mL		200 mM NaCl, 10 mM Tris-HCl, 0.1 mM EDTA, 0.15% Triton X-100, 200 μg/mL BSA, 50% glycerol, pH 7.4		Stratagene 500860 500861
Providencia stuarti				
8-12 U/μL; 40-60 U/μL	1 unit hydrolyzes 1 μg DNA/hr at 37°C in a 50 μL reaction volume using λ phage DNA as substrate	10 mM Tris-HCl, 10 mM MgCl₂, 100 mM KCl, 0.1 mg/mL BSA, pH 8.5	>90% of the DNA fragments ligated and recut after 50-fold overdigestion	Fermentas ER0611 ER0612 ER0613
5000-40,000 U/mL		100 mM NaCl, 10 mM Tris-HCl, 0.1 mM EDTA, 1 mM DTT, 0.15% Triton X-100, 50% glycerol, pH 7.5		ICN 153837
25,000 U and 50,000 U	1 unit completely digests 1 μg DNA substrate/hr at 37°C	10 mM Tris-HCl, 10 mM MgCl₂, 100 mM NaCl, 1 mM DTT, 0.1 mg/mL BSA, pH 7.5		Adv Biotech AB-0236-a AB-0236-b
40 U/μL	1 unit completely cleaves 1 μg substrate DNA/hr	RFLP grade; 100 mM NaCl, 50 mM Tris, 15 mM MgCl₂, 1 mM β-MSH, pH 7.8	90-98% cleavage-ligation-recleavage	Am Allied H-10-10000
10 U/μL; 40 U/μL	1 unit completely cleaves 1 μg substrate DNA/hr	100 mM NaCl, 50 mM Tris, 15 mM MgCl₂, 1 mM β-MSH, pH 7.8	90-98% cleavage-ligation-recleavage	Am Allied L-02-05000 H-02-05000 L-02-10000 H-02-10000 L-02-50000 H-02-50000
12,000 U/mL	1 unit degrades 1 μg λ dam⁻ dcm⁻ DNA/hr at 37°C in a 0.05 mL mixture	50 mM Tris-HCl, 10 mM MgCl₂, 100 mM NaCl, 100 μg/mL BSA, pH 7.5	No detectable contaminating nuclease; 90% ligated and 100% recut after 10-fold overdigestion	AGS Heidelb B00500S B00500M
10 U/μL and 40 U/μL	1 unit completely digests 1 μg substrate DNA/hr	50 mM Tris-HCl, 10 mM MgCl₂, 100 mM NaCl, 1 mM DTE, pH 7.5		Boehringer 621625 621633 798991 1274066
	1 unit completely cleaves 1 μg λ DNA/hr at pH 7.5, 37°C in a reaction volume of 50 μL	20 mM Tris-HCl, 10 mM MgCl₂, 50 mM (NH₄)₂SO₄, 1.0 mM DTT, 100 μg/mL BSA, pH 7.5		CHIMERx 2300-01 2300-02

Pst I continued

CONCENTRATION or VOLUME ACTIVITY	UNIT DEFINITION	REACTION BUFFER (As Provided)	ADDITIONAL ACTIVITIES and PURITY	SUPPLIER CATALOG NO.
Providencia stuartii continued				
8-12 U/μL and 50 U/μL	1 unit completely digests 1 mg λ DNA/hr at 37°C	500 mM Tris-HCl, 100 mM MgCl$_2$, 500 mM NaCl, pH 8.0	≥95% ligation; 100% cleavage	Life Technol 15215-015 15215-023 15215-049 15215-064 15215-031
	10 units does not produce unspecific cleavage products with 1 μg λ DNA/16 hr at 37°C	10 mM Tris-HCl, 10 mM MgCl$_2$, 100 mM NaCl, 1 mM DTT, 100 μg/mL BSA, pH 7.9	>90% DNA fragments ligated and recut after 10-fold overdigestion	MINOTECH 119
10 U/μL and 20-150 U/μL	1 unit hydrolyzes 1 μg λ DNA to completion/hr in a 50 μL total reaction volume	500 mM Tris-HCl, 1 M NaCl, 100 mM MgCl$_2$, 10 mM DTE, 1.0 mg/mL BSA, pH 8		Oncor 110336 110337 110338
10,000-15,000 U/mL and 50,000-80,000 U/mL	1 unit completely digests 1 μg λ DNA/hr at pH 7.5, 37°C in a 50 μL total assay mixture	100 mM Tris-acetate, 100 mM MgOAc, 500 mM KOAc, pH 7.5		Pharmacia 27-0886-03 27-0886-04 27-0886-18
10 U/μL and 40-80 U/μL	1 unit completely digests 1 μg λ DNA/hr at 37°C in a 50 mL reaction volume	Blue/White Cloning Qualified; 900 mM Tris-HCl, 100 mM MgCl$_2$, 500 mM NaCl, pH 7.5	90% ligation	Promega R6111 R6112 R6115 R4114
50,000 U/mL		50 mM Tris-HCl, 10 mM MgCl$_2$, 100 mM NaCl, 1 mM DTT, pH 7.6	95% DNA fragments ligated and recut after 50-fold overdigestion	SibEnzyme E109 E110
2000-10,000 U/mL	1 unit completely cleaves 1.0 μg λ DNA/hr in a 50 μL reaction mixture	100 mM Tris-HCl, 500 mM NaCl, 100 mM MgCl$_2$, 10 mM DTT, pH 7.9	100% cleavage, 100% ligation, 95% recleavage No detectable endonuclease (overdigestion)	Sigma R7002
Providencia stuartii 164				
5-20 U/μL	λ DNA substrate, 37°C	1 M NaCl, 500 mM Tris-HCl, 100 mM MgCl$_2$, 10 mM DTT, pH 7.5	95% ligation and 100% recutting	Nippon Gene 312-01171 314-01175 312-01176 318-01771
3-20 U/μL and 40-50 U/μL	1 unit completely digests 1 μg DNA/hr at 37°C	50 mM Tris-HCl, 10 mM MgCl$_2$, 100 mM NaCl, 1 mM DTT, pH 7.5	>90% ligation; >95% recut	Toyobo PST-105 PST-107 PST-108 PST-158

Pvu I

RECOGNITION SEQUENCE
5'...CGAT▼CG...3'
3'...GC▲TAGC...5'

OPTIMUM REACTION CONDITIONS
pH 7.5

HEAT INACTIVATION
100°C for 5 min.

NOTES
- EC 3.1.21.4
- A type II site specific deoxyribonuclease
- A large group of enzymes recognizing specific short DNA sequences. They cleave either within the recognition site, or at a short specific distance from it
- Requires Mg^{2+}

CONCENTRATION or VOLUME ACTIVITY	UNIT DEFINITION	REACTION BUFFER (As Provided)	ADDITIONAL ACTIVITIES and PURITY	SUPPLIER CATALOG NO.
Proteus vulgaris				
3-20 U/µL	1 unit completely digests 1 µg DNA/hr at 37°C	50 mM Tris-HCl, 10 mM MgCl$_2$, 100 mM NaCl, 1 mM DTT, pH 7.5	>90% ligation; >95% recut	Toyobo PVU-101 PVU-102

Pvu II

RECOGNITION SEQUENCE
5'...CAG▼CTG...3'
3'...GTC▲GAC...5'
OPTIMUM REACTION CONDITIONS
pH 7.5
HEAT INACTIVATION
Resistant

NOTES
- EC 3.1.21.4
- A type II site specific deoxyribonuclease
- A large group of enzymes recognizing specific short DNA sequences. They cleave either within the recognition site, or at a short specific distance from it
- High glycerol or enzyme concentration results in star activity
- Compatible cohesive ends: blunt ends
- Requires Mg^{2+}

CONCENTRATION or VOLUME ACTIVITY	UNIT DEFINITION	REACTION BUFFER (As Provided)	ADDITIONAL ACTIVITIES and PURITY	SUPPLIER CATALOG NO.
Proteus vulgaris				
3-20 U/µL and 40-50 U/µL	1 unit completely digests 1 µg DNA/hr at 37°C	10 m*M* Tris-HCl, 10 m*M* MgCl$_2$, 50 m*M* NaCl, 1 m*M* DTT, pH 7.5	>90% ligation; >95% recut	Toyobo PVU-203 PVU-204 PVU-254
Proteus vulgaris 84				
30,000 U/mL		10 m*M* Tris-HCl, 10 m*M* MgCl$_2$, 50 m*M* NaCl, 1 m*M* DTT, pH 7.6	90% DNA fragments ligated and recut after 10-fold overdigestion	SibEnzyme E111 E112

Rca I

RECOGNITION SEQUENCE
  5'...T▼CATGA...3'
  3'...AGTAC▲T...5'

ISOSCHIZOMERS
  *Bsp*H I

HEAT INACTIVATION
  65°C for 10 min.

NOTES
- EC 3.1.21.4
- A type II site specific deoxyribonuclease
- A large group of enzymes recognizing specific short DNA sequences. They cleave either within the recognition site, or at a short specific distance from it
- Requires Mg^{2+}

CONCENTRATION or VOLUME ACTIVITY	UNIT DEFINITION	REACTION BUFFER (As Provided)	ADDITIONAL ACTIVITIES and PURITY	SUPPLIER CATALOG No.
Rhodococcus capsulatum				
5 U/μL	1 unit completely digests 1 μg substrate DNA/hr	10 m*M* Tris-HCl, 5 m*M* $MgCl_2$, 100 m*M* NaCl, 1 m*M* β-MSH, pH 8.0		Boehringer 1467123 1467131
4-6 U/μL	1 unit completely digests 1 mg λ DNA/hr at 37°C	200 m*M* Tris-HCl, 50 m*M* $MgCl_2$, 500 M KCl, pH 7.4	≥95% ligation; 100% cleavage	Life Technol 15409-014

Rsa I

RECOGNITION SEQUENCE
  5'...GT▼AC...3'
  3'...CA▲TG...5'

ISOSCHIZOMERS
  Afa I

OPTIMUM REACTION CONDITIONS
  pH 7.9–8.0

HEAT INACTIVATION
  Partially resistant; 65°C for 5–20 min.

METHYLATION EFFECTS
  Cleaves GTAm^{5}C
  Does not cleave GTm^{6}AC or GTAm^{4}C
  Cleavage is slowed by GTAm^{5}C

NOTES
- EC 3.1.21.4
- A type II site specific deoxyribonuclease
- A large group of enzymes recognizing specific short DNA sequences. They cleave either within the recognition site, or at a short specific distance from it
- Single-stranded DNA is cleaved slowly
- Compatible cohesive ends: any blunt end
- Requires Mg^{2+}

CONCENTRATION or VOLUME ACTIVITY	UNIT DEFINITION	REACTION BUFFER (As Provided)	ADDITIONAL ACTIVITIES and PURITY	SUPPLIER CATALOG No.
Rhodopseudomonas sphaeroides (Rhodobacter sphaeroides)				
2000 U and 5000 U	1 unit completely digests 1 μg DNA substrate/hr at 37°C	10 m*M* Tris-HCl, 10 m*M* MgCl$_2$, 50 m*M* NaCl, 1 m*M* DTT, 0.1 mg/mL BSA, pH 7.5	>90% DNA fragments ligated and recut after 10-fold overdigestion	Adv Biotech AB-0238-a AB-0238-b
8–12 U/μL	1 unit completely digests 1 μg λ DNA/hr at 37°C in the buffer provided	100 m*M* Tris-HCl, 100 m*M* MgCl$_2$, 10 m*M* DTT, pH 7.5	>90% DNA fragments ligated and recut	Amersham E 0310Y E 0310Z
10,000 U/mL	1 unit degrades 1 μg λ dam$^-$ dcm$^-$ DNA/hr at 37°C in a 0.05 mL mixture	33 m*M* Tris-acetate, 10 m*M* MgOAc, 66 m*M* KOAc, 100 μg/mL BSA, pH 7.9	No detectable contaminating nuclease; 98% ligated and 90% recut after 20-fold overdigestion	AGS Heidelb A00520S A00520M
10 U/μL and 40 U/μL	1 unit completely digests 1 μg substrate DNA/hr	10 m*M* Tris-HCl, 10 m*M* MgCl$_2$, 1 m*M* DTE, pH 7.5		Boehringer 729124 729132 1047671
	1 unit completely cleaves 1 μg λ DNA/hr at pH 7.5, 37°C in a reaction volume of 50 μL	10 m*M* Tris-HCl, 10 m*M* MgCl$_2$, 50 m*M* NaCl, 1.0 m*M* DTT, 100 μg/mL BSA, pH 7.5		CHIMERX 2340-01 2340-02
8–12 U/μL	1 unit hydrolyzes 1 μg DNA/hr at 37°C in a 50 μL reaction volume using λ phage DNA as substrate	33 m*M* Tris-acetate, 10 m*M* MgOAc, 66 m*M* KOAc, 0.1 mg/mL BSA, pH 7.9	>90% DNA fragments ligated and recut after 50-fold overdigestion	Fermentas ER1121 ER1122

CONCENTRATION or VOLUME ACTIVITY	UNIT DEFINITION	REACTION BUFFER (As Provided)	ADDITIONAL ACTIVITIES and PURITY	SUPPLIER CATALOG No.
Rhodopseudomonas sphaeroides (Rhodobacter sphaeroides) continued				
5000-40,000 U/mL		50 mM KCl, 10 mM Tris-HCl, 0.1 mM EDTA, 1.0 mM DTT, 200 µg/mL BSA, 50% glycerol		ICN 152035
8-12 U/µL	1 unit completely digests 1 mg λ DNA/hr at 37°C	500 mM Tris-HCl, 100 mM MgCl$_2$, pH 8.0	≥65% ligation; 100% cleavage	Life Technol 15424-013 15424-039
	10 units does not produce unspecific cleavage products with 1 µg λ DNA/16 hr at 37°C	10 mM Tris-HCl, 10 mM MgCl$_2$, 50 mM NaCl, 1 mM DTT, 100 µg/mL BSA, pH 7.9	>90% DNA fragments ligated and recut after 10-fold overdigestion	MINOTECH 121
8-12 U/µL	1 unit digests completely 1 µg λ DNA substrate/hr at 37°C	10 mM Tris-HCl, 10 mM MgCl$_2$, 1 mM DTT, pH 7.8		NBL Gene 013503 013506
10,000 U/mL	1 unit completely digests 1 µg λ DNA/hr in a total reaction volume of 50 µL using the buffer provided	10 mM Bis Tris-Propane-HCl, 10 mM MgCl$_2$, 1 mM DTT, pH 7.0	>95% DNA fragments ligated and recut after 10-fold overdigestion	NE Biolabs 167S 167L
5-20 U/µL 30-100 U/µL	λ DNA substrate, 37°C	500 mM NaCl, 100 mM Tris-HCl, 100 mM MgCl$_2$, 10 mM DTT, pH 7.5 (10X buffer)	90% ligation and 100% recutting	Nippon Gene 318-00291 314-00293 312-00294 314-02111
5 U/µL	1 unit hydrolyzes 1 µg λ DNA to completion/hr in a 50 µL total reaction volume	100 mM Tris-HCl, 100 mM NaCl, 100 mM MgCl$_2$, 100 mM β–MSH, 1.0 mg/mL BSA, pH 7.5		Oncor 110361 110362
5000-10,000 U/mL and 50,000-80,000 U/mL	1 unit completely digests 1 µg λ DNA/hr at pH 7.5, 37°C in a 50 µL total assay mixture	100 mM Tris-acetate, 100 mM MgOAc, 500 mM KOAc, pH 7.5		Pharmacia 27-0907-01 27-0907-02 27-0907-18
10 U/µL and 40-80 U/µL	1 unit completely digests 1 µg λ DNA/hr at 37°C in a 50 mL reaction volume	100 mM Tris-HCl, 100 mM MgCl$_2$, 500 mM NaCl, 10 mM DTT, pH 7.9	90% ligation	Promega R6371 R6372 R4375
10,000 U/mL		10 mM Tris-HCl, 10 mM MgCl$_2$, 1 mM DTT, pH 7.6	90% DNA fragments ligated and recut after 10-fold overdigestion	SibEnzyme E113 E114
5000-40,000 U/mL	1 unit completely cleaves 1.0 µg λ DNA/hr in a 50 µL reaction mixture	Unique buffer	100% cleavage, 90% ligation, 95% recleavage No detectable endonuclease (overdigestion)	Sigma R4756

Rsa I continued

CONCENTRATION or VOLUME ACTIVITY	UNIT DEFINITION	REACTION BUFFER (As Provided)	ADDITIONAL ACTIVITIES and PURITY	SUPPLIER CATALOG No.
Rhodopseudomonas sphaeroides (Rhodobacter sphaeroides) continued				
8000-20,000 U/mL		50 mM KCl, 10 mM Tris-HCl, 0.1 mM EDTA, 1 mM DTT, 200 µg/mL BSA, 50% glycerol, pH 7.4		Stratagene 500890
3-20 U/µL and 40-50 U/µL	1 unit completely digests 1 µg DNA/hr at 37°C	10 mM Tris-HCl, 10 mM MgCl$_2$, 1 mM DTT, pH 7.5	>90% ligation; >95% recut	Toyobo RSA-101 RSA-102 RSA-152

Rsr II

RECOGNITION SEQUENCE

5'...CG▼G(^A_T)CCG...3'
3'...GCC(^T_A)G▲GC...5'

HEAT INACTIVATION

65°C for 10-20 min.

NOTES
- EC 3.1.21.4
- A type II site specific deoxyribonuclease
- A large group of enzymes recognizing specific short DNA sequences. They cleave either within the recognition site, or at a short specific distance from it
- Sensitive to NaCl >50 mM
- Requires Mg^{2+} and DTT for maximum activity in extended digestions

CONCENTRATION or VOLUME ACTIVITY	UNIT DEFINITION	REACTION BUFFER (As Provided)	ADDITIONAL ACTIVITIES and PURITY	SUPPLIER CATALOG No.
Rhodopseudomonas sphaeroides				
10 U/µL	1 unit completely digests 1 µg substrate DNA/hr	10 mM Tris-HCl, 10 mM MgCl$_2$, 1 mM DTE, pH 7.5		Boehringer 1292587 1292595
20 U/µL	1 unit cleaves 1 µg λ DNA/18 hr at 37°C	100 mM Tris-HCl, 80 mM MgCl$_2$, pH 8.2 and 10 mM DTT	≥33% ligation; 90% cleavage	Life Technol 15456-015

CONCENTRATION or VOLUME ACTIVITY	UNIT DEFINITION	REACTION BUFFER (As Provided)	ADDITIONAL ACTIVITIES and PURITY	SUPPLIER CATALOG NO.
Rhodopseudomonas sphaeroides continued				
500-5000 U/mL	1 unit completely cleaves 1.0 µg λ DNA/hr in a 50 µL reaction mixture	Unique buffer	98% cleavage, 100% ligation, 95% recleavage, 10% endonuclease (nickase) No detectable endonuclease (overdigestion)	Sigma R5132
Rhodopseudomonas sphaeroides (S. Kaplan)				
500-5000 U/mL		50 mM KCl, 10 mM Tris-HCl, 0.1 mM EDTA, 1 mM DTT, 200 µg/mL BSA, 50% glycerol, pH 7.4		ICN 153823
2000 U/mL	1 unit completely digests 1 µg λ DNA/hr in a total reaction volume of 50 µL using the buffer provided	20 mM Tris-acetate, 10 mM MgOAc, 50 mM KOAc, 1 mM DTT, pH 7.9	>95% DNA fragments ligated and recut after 50-fold overdigestion	NE Biolabs 501S 501L

Sac I

RECOGNITION SEQUENCE
5'...GAGCT▼C...3'
3'...C▲TCGAG...5'

ISOSCHIZOMERS
Sst I

OPTIMUM REACTION CONDITIONS
pH 7.5-7.9

HEAT INACTIVATION
60°C for 5-30 min.

METHYLATION EFFECTS
Cleaves Gm^6AGCTC and $GAGCTm^5C$
Does not cleave $GAGm^5CTC$

NOTES
- EC 3.1.21.4
- A type II site specific deoxyribonuclease
- A large group of enzymes recognizing specific short DNA sequences. They cleave either within the recognition site, or at a short specific distance from it
- Inhibited by salt >10 mM
- Sensitive to DNA contaminants
- High glycerol concentration or DMSO solution results in star activity
- Compatible cohesive ends: Sst I
- Requires Mg^{2+}

CONCENTRATION or VOLUME ACTIVITY	UNIT DEFINITION	REACTION BUFFER (As Provided)	ADDITIONAL ACTIVITIES and PURITY	SUPPLIER CATALOG NO.
Streptomyces achromogenes				
1750 U and 5000 U	1 unit completely digests 1 μg DNA substrate/hr at 37°C	10 mM Tris-HCl, 10 mM MgCl₂, 1 mM DTT, 0.1 mg/mL BSA, pH 7.5	>95% DNA fragments ligated and recut after 10-fold overdigestion	Adv Biotech AB-0397-a AB-0397-b
10 U/μL; 40 U/μL	1 unit completely cleaves 1 μg substrate DNA/hr	60 mM KCl, 20 mM Tris, 0.75 mM MgCl₂, pH 7.5	90-98% cleavage-ligation-recleavage	Am Allied L-25-01000 H-25-01000 L-25-05000 H-25-05000
8-12 U/μL	1 unit completely digests 1 μg λ DNA/hr at 37°C in the buffer provided	100 mM Tris-HCl, 100 mM MgCl₂, 10 mM DTT, pH 7.5	>95% DNA fragments ligated and 100% recut	Amersham E 1078Y E 1078Z
	1 unit completely digests 1 μg λ DNA/hr at 37°C in the buffer provided	100 mM Tris-HCl, 100 mM MgCl₂, 10 mM DTT, pH 7.5	>95% DNA fragments ligated and 100% recut	Amersham E 1078YH
10 U/μL and 40 U/μL	1 unit completely digests 1 μg substrate DNA/hr	33 mM Tris-acetate, 10 mM MgOAc, 66 mM KOAc, 0.5 mM DTT, pH 7.9		Boehringer 669792 669806 1047655
	1 unit completely cleaves 1 μg Ad2 DNA/hr at pH 7.5, 37°C in a reaction volume of 50 μL	10 mM Tris-HCl, 10 mM MgCl₂, 1.0 mM DTT, 100 μg/mL BSA, pH 7.5		CHIMERX 2350-01 2350-02

Sac I continued

CONCENTRATION or VOLUME ACTIVITY	UNIT DEFINITION	REACTION BUFFER (As Provided)	ADDITIONAL ACTIVITIES and PURITY	SUPPLIER CATALOG No.
Streptomyces achromogenes continued				
8-12 U/μL; 40-60 U/μL	1 unit hydrolyzes 1 μg DNA/hr at 37°C in a 50 μL reaction volume using λ phage DNA as substrate	10 mM Bis Tris-Propane-HCl, 10 mM MgCl$_2$, 1 mM DTT, 0.1 mg/mL BSA, pH 6.5	>90% DNA fragments ligated and recut after 50-fold overdigestion	Fermentas ER1131 ER1132 ER1133
2000-20,000 U/mL		10 mM Tris-HCl, 0.1 mM EDTA, 1.0 mM DTT, 200 μg/mL BSA, 50% glycerol, pH 7.4		ICN 151450
	10 units does not produce unspecific cleavage products with 1 μg λ DNA/Hind III digest/16 hr at 37°C	10 mM Tris-HCl, 10 mM MgCl$_2$, 1 mM DTT, 100 μg/mL BSA, pH 7.9	>90% DNA fragments ligated and recut after 10-fold overdigestion	MINOTECH 122
8-12 U/μL >40 U/μL	1 unit digests completely 1 μg λ DNA substrate/hr at 37°C	33 mM Tris-acetate, 66 mM KOAc, 10 mM MgOAc, 0.5 mM DTT, pH 8.2		NBL Gene 011703 011706 011756
10 U/μL and 20-200 U/μL	1 unit hydrolyzes 1 μg λ DNA to completion/hr in a 50 μL total reaction volume	100 mM Tris-HCl, 100 mM NaCl, 100 mM MgCl$_2$, 100 mM β-MSH, 1.0 mg/mL BSA, pH 7.5		Oncor 110371 110372
8000-15,000 U/mL and 50,000-80,000 U/mL	1 unit completely digests 1 μg Hind III pre-digested λ DNA/hr at pH 7.5, 37°C in a 50 μL total assay mixture	100 mM Tris-acetate, 100 mM MgOAc, 500 mM KOAc, pH 7.5		Pharmacia 27-0970-01 27-0970-02 27-0970-18
10 U/μL and 40-80 U/μL	1 unit completely digests 1 μg EcoR I λ DNA fragments/hr at 37°C in a 50 mL reaction volume	Blue/White Cloning Qualified; 100 mM Tris-HCl, 70 mM MgCl$_2$, 500 mM KCl, 10 mM DTT, pH 7.5	90% ligation	Promega R6061 R6062 R6065 R4064
2000-20,000 U/mL	1 unit completely cleaves 1.0 μg λ EcoR I digest DNA/hr in a 50 μL reaction mixture	100 mM Bis Tris-Propane-HCl, 100 mM MgCl$_2$, 10 mM DTT, BSA, pH 7.0	100% cleavage, 100% ligation, 100% recleavage, 40% endonuclease (nickase) No detectable endonuclease (overdigestion)	Sigma R5268
4000-20,000 U/mL		50 mM KCl, 10 mM Tris-HCl, 0.1 mM EDTA, 1 mM DTT, 0.15% Triton X-100, 500 μg/mL BSA, 50% glycerol, pH 7.4		Stratagene 500910
2000 U and 10,000 U	1 unit completely digests 1 μg λ DNA/hr at 37°C in 50 μL of supplied buffer	100 mM Tris-HCl, 100 mM MgCl$_2$, 10 mM DTT, pH 7.5	Purity is confirmed by overdigestion, ligation-recutting and cloning test	TaKaRa 1078

Sac I continued

CONCENTRATION or VOLUME ACTIVITY	UNIT DEFINITION	REACTION BUFFER (As Provided)	ADDITIONAL ACTIVITIES and PURITY	SUPPLIER CATALOG NO.
Streptomyces achromogenes continued				
3-20 U/µL and 40-50 U/µL	1 unit completely digests 1 µg DNA/hr at 37°C	10 mM Tris-HCl, 10 mM MgCl$_2$, 1 mM DTT, pH 7.5	>90% ligation; >95% recut	Toyobo SAC-101 SAC-102 SAC-152
Streptomyces achromogenes ATCC 12767				
20,000 U/mL and 100,000 U/mL	1 unit completely digests 1 µg λ DNA (Hind III digest)/hr in a total reaction volume of 50 µL using the buffer provided	10 mM Bis Tris-Propane-HCl, 10 mM MgCl$_2$, 1 mM DTT, pH 7.0; add 100 µg/mL BSA	>95% DNA fragments ligated and recut after 20-fold overdigestion	NE Biolabs 156S 156L
5-20 U/µL and 30-100 U/µL	φ105 DNA substrate, 37°C	100 mM Tris-HCl, 100 mM MgCl$_2$, 10 mM DTT, pH 7.5	90% ligation and 100% recutting	Nippon Gene 319-00302 313-00305 311-00306- 311-02121

Sac II

RECOGNITION SEQUENCE
  5'...CCGC▼GG...3'
  3'...GG▲CGCC...5'

ISOSCHIZOMERS
  Sst II

OPTIMUM REACTION CONDITIONS
  pH 7.9

HEAT INACTIVATION
  65°C for 15-30 min.

METHYLATION EFFECTS
  Does not cleave m^{5}CCGCGG

NOTES
- EC 3.1.21.4
- A type II site specific deoxyribonuclease
- A large group of enzymes recognizing specific short DNA sequences. They cleave either within the recognition site, or at a short specific distance from it
- Sensitive to DNA contaminants and NaCl >50 mM
- Sites in l and φX174 DNA cleave at significantly lower rates than sites in other substrates
- Compatible cohesive ends: BsaO I, BsiE I, Ksp I, Mcr I, Sst I
- Requires Mg^{2+}

CONCENTRATION or VOLUME ACTIVITY	UNIT DEFINITION	REACTION BUFFER (As Provided)	ADDITIONAL ACTIVITIES and PURITY	SUPPLIER CATALOG NO.
Streptomyces achromogenes				
2500 U and 5000 U	1 unit completely digests 1 μg DNA substrate/hr at 37°C	10 mM Tris-HCl, 10 mM MgCl$_2$, 1 mM DTT, 0.1 mg/mL BSA, pH 7.5	>95% DNA fragments ligated and recut after 10-fold overdigestion	Adv Biotech AB-0398-a AB-0398-b
10 U/μL; 40 U/μL	1 unit completely cleaves 1 μg substrate DNA/hr	60 mM KCl, 20 mM Tris, 0.75 mM MgCl$_2$, pH 7.5	90-98% cleavage-ligation-recleavage	Am Allied L-26-02000 H-26-02000 L-26-10000 H-26-10000
3-20 U/μL	1 unit completely digests 1 μg Ad2 DNA/hr at 37°C in the buffer provided	100 mM Tris-HCl, 100 mM MgCl$_2$, 10 mM DTT, pH 7.5	>90% DNA fragments ligated and >95% recut	Amersham E 0221Y E 0221Z
	1 unit completely cleaves 1 μg Ad2 DNA/hr at pH 7.5, 37°C in a reaction volume of 50 μL	10 mM Tris-HCl, 10 mM MgCl$_2$, 1.0 mM DTT, 100 μg/mL BSA, pH 7.5		CHIMERX 2352-01 2352-02
	10 units does not produce unspecific cleavage products with 1 μg λ DNA/Hind III digest/16 hr at 37°C	10 mM Tris-HCl, 10 mM MgCl$_2$, 1 mM DTT, 100 μg/mL BSA, pH 7.9	>90% DNA fragments ligated and recut after 10-fold overdigestion	MINOTECH 123
6-16 U/μL	1 unit digests completely 1 μg λ DNA substrate/hr at 37°C	10 mM Tris-HCl, 10 mM MgCl$_2$, 1 mM DTT, pH 7.8		NBL Gene 012101 012103

Sac II continued

CONCENTRATION or VOLUME ACTIVITY	UNIT DEFINITION	REACTION BUFFER (As Provided)	ADDITIONAL ACTIVITIES and PURITY	SUPPLIER CATALOG No.
Streptomyces achromogenes continued				
10 U/µL and 20-40 U/µL	1 unit hydrolyzes 1 µg λ DNA to completion/hr in a 50 µL total reaction volume	330 mM Tris-acetate, 660 mM KOAc, 100 mM MgOAc, 5 mM DTT, 1 mg/mL BSA, pH 8		Oncor 110591 110592
5000-10,000 U/mL	1 unit completely digests 1 µg Ad2 DNA/hr at pH 7.5, 37°C in a 50 µL total assay mixture	100 mM Tris-acetate, 100 mM MgOAc, 500 mM KOAc, pH 7.5		Pharmacia 27-0910-01 27-0910-02
10 U/µL	1 unit completely digests 1 µg Ad2 DNA/hr at 37°C in a 50 mL reaction volume	Blue/White Cloning Qualified; 100 mM Tris-HCl, 100 mM MgCl$_2$, 500 mM NaCl, 10 mM DTT, pH 7.9	50% ligation	Promega R6221 R6222
4000-20,000 U/mL		50 mM KCl, 10 mM Tris-HCl, 0.1 mM EDTA, 1 mM DTT, 0.1% Triton X-100, 500 µg/mL BSA, 50% glycerol, pH 7.4		Stratagene 500920
1000 U and 5000 U	1 unit completely digests 1 µg λ DNA/hr at 37°C in 50 µL of supplied buffer and BSA	Genomic grade; 330 mM Tris-acetate, 100 mM MgOAc, 5 mM DTT, 660 mM KOAc, BSA, pH 7.9	Purity is confirmed by overdigestion, ligation-recutting, genome DNA digestion and cloning test	TaKaRa 1079
3-20 U/µL	1 unit completely digests 1 µg DNA/hr at 37°C	10 mM Tris-HCl, 10 mM MgCl$_2$, 1 mM DTT, pH 7.5	>90% ligation; >95% recut	Toyobo SAC-203 SAC-204
Streptomyces achromogenes ATCC 12767				
2000-30,000 U/mL		50 mM KCl, 10 mM Tris-HCl, 0.1 mM EDTA, 1 mM DTT, 200 µg/mL BSA, 50% glycerol, pH 7.4	No detectable Sac I, Sac III	ICN 153824
5-20 U/µL	λ DNA substrate, 37°C	100 mM Tris-HCl, 100 mM MgCl$_2$, 10 mM DTT, pH 7.5	95% ligation and 100% recutting	Nippon Gene 319-00922 313-00925 311-00926
Streptomyces lividens strain, carrying the cloned Sac II gene from Streptomyces achromogenes				
20,000 U/mL	1 unit completely digests 1 µg Ad2 DNA/hr in a total reaction volume of 50 µL using the buffer provided	20 mM Tris-acetate, 10 mM MgOAc, 50 mM KOAc, 1 mM DTT, pH 7.9	>95% DNA fragments ligated and recut after 10-fold overdigestion	NE Biolabs 157S 157L
Streptomyces lividens strain, carrying the cloned Sac II gene from Streptomyces achromogenes continued				
2000-20,000 U/mL	1 unit cleaves 95% λ DNA/hr in a 50 µL reaction mixture	200 mM Tris-acetate, 500 mM KOAc, 100 mM MgOAc, 10 mM DTT, pH 7.9	100% cleavage, 100% ligation, 100% recleavage, 10% endonuclease (nickase) No detectable endonuclease (overdigestion)	Sigma R4882

Sal I

RECOGNITION SEQUENCE
5'...G▼TCGAC...3'
3'...CAGCT▲G...5'

OPTIMUM REACTION CONDITIONS
pH 7.5-7.9

HEAT INACTIVATION
65°C for 15-30 min.

METHYLATION EFFECTS
Cleaves GTCGAm^{5}C
Does not cleave GTm^{5}CGAC, GTCGm^{6}AC or Ghm^{5}UCGAC

NOTES
- EC 3.1.21.4
- A type II site specific deoxyribonuclease
- A large group of enzymes recognizing specific short DNA sequences. They cleave either within the recognition site, or at a short specific distance from it
- Sensitive to DNA contaminants and NaCl <100 mM
- Supercoiled pBR322 and pUC require 10-fold overdigestion for complete digestion
- High glycerol or enzyme concentration, or ethylene glycol or DMSO solution results in star activity
- Compatible cohesive ends: *Ccr* I, *Pae*R7 I, *Xho* I
- Requires Mg^{2+}

CONCENTRATION or VOLUME ACTIVITY	UNIT DEFINITION	REACTION BUFFER (As Provided)	ADDITIONAL ACTIVITIES and PURITY	SUPPLIER CATALOG No.
Escherichia coli strain, carrying the cloned gene from Streptomyces albus G				
20,000 U/mL; 100,000 U/mL	1 unit completely digests 1 µg λ DNA (*Hind* III digest)/hr in a total reaction volume of 50 µL using the buffer provided	10 mM Tris-HCl, 10 mM MgCl$_2$, 150 mM NaCl, 1 mM DTT, pH 7.9; add 100 µg/mL BSA	>95% DNA fragments ligated and recut after 10-fold overdigestion	NE Biolabs 138S 138L
Streptomyces albus				
10 U/µL; 40 U/µL	1 unit completely cleaves 1 µg substrate DNA/hr	100 mM NaCl, 50 mM Tris, 5 mM MgCl$_2$, 2 mM β-MSH, pH 7.8	90-98% cleavage-ligation-recleavage	Am Allied L-05-01000 H-05-01000 L-05-02500 H-05-02500 L-05-10000 H-05-10000
10 U/µL and 40 U/µL	1 unit completely digests 1 µg substrate DNA/hr	50 mM Tris-HCl, 10 mM MgCl$_2$, 100 mM NaCl, 1 mM DTE, pH 7.5		Boehringer 348783 567663 1047612

Sal I continued

CONCENTRATION or VOLUME ACTIVITY	UNIT DEFINITION	REACTION BUFFER (As Provided)	ADDITIONAL ACTIVITIES and PURITY	SUPPLIER CATALOG NO.
Streptomyces albus continued				
8-12 U/μL; 40-60 U/μL	1 unit hydrolyzes 1 μg DNA/hr at 37°C in a 50 μL reaction volume using λ phage DNA as substrate	50 mM Tris-HCl, 10 mM $MgCl_2$, 100 mM NaCl, pH 7.5	>90% of the DNA fragments ligated and recut after 50-fold overdigestion	Fermentas ER0641 ER0642 ER0643
2000-20,000 U/mL		50 mM KCl, 5 mM KPO_4, 0.1 mM EDTA, 5 mM β-MSH, 500 μg/mL BSA, 50% glycerol, pH 7.4		ICN 152041
8000-15,000 U/mL and 50,000-80,000 U/mL	1 unit completely digests 1 μg Ad2 DNA/hr at pH 7.5, 37°C in a 50 μL total assay mixture	100 mM Tris-acetate, 100 mM MgOAc, 500 mM KOAc, pH 7.5		Pharmacia 27-0882-01 27-0882-01 27-0882-18
20,000 U/mL		Blue/white certified; 50 mM Tris-HCl, 10 mM $MgCl_2$, 100 mM NaCl, 1 mM DTT, pH 7.6	90% DNA fragments ligated and recut after 10-fold overdigestion	SibEnzyme E115 E116
Streptomyces albus G				
7500 U and 20,000 U	1 unit completely digests 1 μg DNA substrate/hr at 37°C	10 mM Tris-HCl, 10 mM $MgCl_2$, 100 mM NaCl, 1 mM DTT, 0.1 mg/mL BSA, pH 7.5	>90% DNA fragments ligated and recut after 10-fold overdigestion	Adv Biotech AB-0239-a AB-0239-b
8-12 U/μL	1 unit completely digests 1 μg λ DNA/hr at 37°C in the buffer provided	Genomic grade; 500 mM Tris-HCl, 100 mM $MgCl_2$, 10 mM DTT, 1000 mM NaCl, pH 7.5	>95% DNA fragments ligated and 100% recut	Amersham E 1080Y E 1080Z
>40 U/μL	1 unit completely digests 1 μg λ DNA/hr at 37°C in the buffer provided	Genomic grade; 500 mM Tris-HCl, 100 mM $MgCl_2$, 10 mM DTT, 1000 mM NaCl, pH 7.5	>95% DNA fragments ligated and 100% recut	Amersham E 1080YH E 1080ZH
10,000 U/mL	1 unit degrades 1 μg λ dam⁻ dcm⁻ DNA/hr at 37°C in a 0.05 mL mixture	50 mM Tris-HCl, 10 mM $MgCl_2$, 100 mM NaCl, 100 μg/mL BSA, pH 7.5	No detectable contaminating nuclease; 90% ligated and 95% recut after 10-fold overdigestion	AGS Heidelb A00560S A00560M
	1 unit completely cleaves 1 μg Ad2 DNA/hr at pH 7.5, 37°C in a reaction volume of 50 μL	50 mM Tris-HCl, 10 mM $MgCl_2$, 100 mM NaCl, 1.0 mM DTT, 100 μg/mL BSA, pH 7.5		CHIMERX 2370-01 2370-02
8-12 U/μL and 50 U/μL	1 unit completely digests 1 mg Ad2 DNA/hr at 37°C	1 M Tris-HCl, 100 mM $MgCl_2$, 1.5 M NaCl, pH 7.6	≥85% ligation; 100% cleavage	Life Technol 15217-011 15217-029 15217-037
	10 units does not produce unspecific cleavage products with 1 μg λ DNA/Hind III digest/16 hr at 37°C	10 mM Tris-HCl, 10 mM $MgCl_2$, 150 mM NaCl, 1 mM DTT, 100 μg/mL BSA, pH 7.9	>90% DNA fragments ligated and recut after 10-fold overdigestion	MINOTECH 124

Sal I continued

CONCENTRATION or VOLUME ACTIVITY	UNIT DEFINITION	REACTION BUFFER (As Provided)	ADDITIONAL ACTIVITIES and PURITY	SUPPLIER CATALOG No.
Streptomyces albus G continued				
6-12 U/µL	1 unit digests completely 1 µg λ DNA substrate/hr at 37°C	50 mM Tris-HCl, 100 mM NaCl, 10 mM MgCl$_2$, 1 mM DTT, pH 7.8		NBL Gene 010403 010406
5-20 U/µL 30-100 U/µL	λ DNA substrate, 37°C	1 M NaCl, 500 mM Tris-HCl, 100 mM MgCl$_2$, 10 mM DTT, pH 7.5	80% ligation and 100% recutting	Nippon Gene 318-00311 314-00313 312-00314 315-01781
10 U/µL and 15-80 U/µL	1 unit hydrolyzes 1 µg λ DNA to completion/hr in a 50 µL total reaction volume	500 mM Tris-HCl, 1 M NaCl, 100 mM MgCl$_2$, 10 mM DTE, 1.0 mg/mL BSA, pH 8		Oncor 110385 110386
10 U/µL and 40-80 U/µL	1 unit completely digests 1 µg Ad2 DNA/hr at 37°C in a 50 mL reaction volume	Blue/White Cloning Qualified; genome qualified; 60 mM Tris-HCl, 60 mM MgCl$_2$, 1.5 M NaCl, 10 mM DTT, pH 7.9	90% ligation	Promega R6051 R6052 R6055 R4054
2000-20,000 U/mL	1 unit cleaves 80% non-methylated λ DNA/hr in a 50 µL reaction mixture	Unique buffer + BSA	90% cleavage, 95% ligation, 80% recleavage, 20% endonuclease (nickase) No detectable endonuclease (overdigestion)	Sigma R0754
8000-15,000 U/mL		50 mM KCl, 5 mM KPO$_4$, 0.1 mM EDTA, 5 mM β-MSH, 200 µg/mL BSA, 50% glycerol, pH 7.4		Stratagene 500930 500931
3000 U and 15,000 U	1 unit completely digests 1 µg λ DNA/hr at 37°C in 50 µL of supplied buffer	Genomic grade; 500 mM Tris-HCl, 100 mM MgCl$_2$, 10 mM DTT, 1 M NaCl, pH 7.5	Purity is confirmed by overdigestion, ligation-recutting, genome DNA digestion and cloning test	TaKaRa 1080
3-20 U/µL and 40-50 U/µL	1 unit completely digests 1 µg DNA/hr at 37°C	50 mM Tris-HCl, 10 mM MgCl$_2$, 100 mM NaCl, 1 mM DTT, pH 7.5	>90% ligation; >95% recut	Toyobo SAL-101 SAL-102 SAL-152

SanD I

RECOGNITION SEQUENCE
5'...GG▼GWCCC...3'
3'...CCCWG▲GG...5'

NOTES
- EC 3.1.21.4
- A type II site specific deoxyribonuclease
- A large group of enzymes recognizing specific short DNA sequences. They cleave either within the recognition site, or at a short specific distance from it
- Requires Mg^{2+}

CONCENTRATION or VOLUME ACTIVITY	UNIT DEFINITION	REACTION BUFFER (As Provided)	ADDITIONAL ACTIVITIES and PURITY	SUPPLIER CATALOG No.
Streptomyces species				
				Stratagene

Sap I

RECOGNITION SEQUENCE
5'...GCTCTTC(N)$_1$▼...3'
3'...CGAGAAG(N)$_4$▲...5'

HEAT INACTIVATION
65°C for 20 min.

NOTES
- EC 3.1.21.4
- A type II site specific deoxyribonuclease
- A large group of enzymes recognizing specific short DNA sequences. They cleave either within the recognition site, or at a short specific distance from it
- Requires Mg^{2+}

CONCENTRATION or VOLUME ACTIVITY	UNIT DEFINITION	REACTION BUFFER (As Provided)	ADDITIONAL ACTIVITIES and PURITY	SUPPLIER CATALOG NO.
Saccharopolyspora species				
1000-5000 U/mL		150 mM NaCl, 10 mM Tris-HCl, 0.1 mM EDTA, 1 mM DTT, 200 µg/mL acetylated BSA, 50% glycerol, pH 7.5		ICN 159465
1000 U/mL	1 unit completely digests 1 µg λ DNA/hr in a total reaction volume of 50 µL using the buffer provided	20 mM Tris-acetate, 10 mM MgOAc, 50 mM KOAc, 1 mM DTT, pH 7.9	>95% DNA fragments ligated and recut after 2-fold overdigestion	NE Biolabs 569S 569L

Sau3A I

RECOGNITION SEQUENCE
5'...▼GATC...3'
3'...CTAG▲...5'

ISOSCHIZOMERS
Mbo I

OPTIMUM REACTION CONDITIONS
pH 7.5

HEAT INACTIVATION
65°C for 10-30 min.

METHYLATION EFFECTS
Cleaves Gm^6ATC and GAhm^5UC
Does not cleave GATm^5C, GATm^4C or GAThm^5C
Not blocked by *dam* methylation; *Dpn* II, *Nde* II and *Mbo* I are

NOTES
- EC 3.1.21.4
- A type II site specific deoxyribonuclease
- A large group of enzymes recognizing specific short DNA sequences. They cleave either within the recognition site, or at a short specific distance from it
- Activity is affected by adjacent DNA sequences
- Glycerol >10% or DMSO solution results in star activity
- 6 m*M* β-mercaptoethanol decreases activity 50%
- 30-fold enzyme excess may cause nicks in double-stranded DNA
- Compatible cohesive ends: *Bam*H I, *Bcl* I, *Bgl* II, *Bst*Y I, *Mbo* I, *Mfl* I, *Nde* II, *Xho* II
- Requires Mg^{2+}

CONCENTRATION or VOLUME ACTIVITY	UNIT DEFINITION	REACTION BUFFER (As Provided)	ADDITIONAL ACTIVITIES and PURITY	SUPPLIER CATALOG NO.
Staphylococcus aureus				
5000-15,000 U/mL	1 unit cleaves 98% λ DNA/hr in a 50 μL reaction mixture	Unique buffer + BSA	98% cleavage, 100% ligation, 30% recleavage, 60% endonuclease (nickase) No detectable endonuclease (overdigestion)	Sigma R0762
Staphylococcus aureus 3A				
10 U/μL; 40 U/μL	1 unit completely cleaves 1 μg substrate DNA/hr	60 m*M* KCl, 20 m*M* Tris, 0.75 m*M* MgCl$_2$, pH 7.5	90-98% cleavage-ligation-recleavage	Am Allied L-22-00200 L-22-01000 H-22-01000
8-12 U/μL	1 unit completely digests 1 μg λ DNA/hr at 37°C in the buffer provided	500 m*M* Tris-HCl, 100 m*M* MgCl$_2$, 10 m*M* DTT, 1000 m*M* NaCl, pH 7.5	>95% DNA fragments ligated and 100% recut	Amersham E 1082Y E 1082Z
>40 U/μL	1 unit completely digests 1 μg λ DNA/hr at 37°C in the buffer provided	500 m*M* Tris-HCl, 100 m*M* MgCl$_2$, 10 m*M* DTT, 1000 m*M* NaCl, pH 7.5	>95% DNA fragments ligated and 100% recut	Amersham E 1082ZH

Sau3A I continued

CONCENTRATION or VOLUME ACTIVITY	UNIT DEFINITION	REACTION BUFFER (As Provided)	ADDITIONAL ACTIVITIES and PURITY	SUPPLIER CATALOG No.
Staphylococcus aureus 3A continued				
2-10 U/μL	1 unit degrades 1 μg λ dam⁻ dcm⁻ DNA/hr at 37°C in a 0.05 mL mixture	50 mM Tris-HCl, 10 mM MgCl$_2$, 100 mM NaCl, 100 μg/mL BSA, pH 7.5	>90% ligated and recut after 25-fold overdigestion	AGS Heidelb A00570S A00570M
1-5 U/μL	1 unit completely digests 1 μg substrate DNA/hr	33 mM Tris-acetate, 10 mM MgOAc, 66 mM KOAc, 0.5 mM DTT, pH 7.9		Boehringer 709743 709751
	1 unit completely cleaves 1 μg unmethylated λ DNA/hr at pH 7.5, 37°C in a reaction volume of 50 μL	10 mM MES, 10 mM MgCl$_2$, 50 mM NaCl, 100 μg/mL BSA, pH 6.5		CHIMERx 2375-01 2375-02
		50 mM KCl, 10 mM Tris-HCl, 0.1 mM EDTA, 1.0 mM DTT, 250 μg/mL BSA, 50% glycerol, pH 7.5		ICN 153838
8-12 U/μL	1 unit completely digests 1 mg λ DNA/hr at 37°C	200 mM Tris-HCl, 50 mM MgCl$_2$, 500 M KCl, pH 7.4	≥95% ligation; 100% cleavage	Life Technol 15229-016 15229-024
2-7 U/μL 30-40 U/μL	1 unit digests completely 1 μg λ DNA substrate/hr at 37°C	33 mM Tris-acetate, 66 mM KOAc, 10 mM MgOAc, 0.5 mM DTT, pH 8.2		NBL Gene 010901 010903 010953
3-15 U/μL	λ DNA substrate, 37°C	500 mM NaCl, 100 mM Tris-HCl, 100 mM MgCl$_2$, 10 mM DTT, pH 7.5 (10X buffer)	90% ligation and 100% recutting	Nippon Gene 313-00322 311-00323 319-00324
3 U/μL	1 unit hydrolyzes 1 μg λ DNA to completion/hr in a 50 μL total reaction volume	100 mM Tris-acetate, 500 mM KOAc, 50 mM MgOAc, 1.0 mg/mL BSA, pH 7.5		Oncor 110391 110392
5000-15,000 U/mL	1 unit completely digests 1 μg λ DNA/hr at pH 7.5, 37°C in a 50 μL total assay mixture	100 mM Tris-acetate, 100 mM MgOAc, 500 mM KOAc, pH 7.5		Pharmacia 27-0913-01 27-0913-02
3-10 U/μL	1 unit completely digests 1 μg λ DNA/hr at 37°C in a 50 mL reaction volume	60 mM Tris-HCl, 60 mM MgCl$_2$, 500 mM NaCl, 10 mM DTT, pH 7.5	90% ligation	Promega R6191 R6192 R6195
5000-15,000 U/mL		50 mM KCl, 10 mM Tris-HCl, 0.1 mM EDTA, 1 mM DTT, 200 μg/mL BSA, 50% glycerol, pH 7.4		Stratagene 500940
200 U and 1000 U	1 unit completely digests 1 μg λ DNA/hr at 37°C in 50 μL of supplied buffer	500 mM Tris-HCl, 100 mM MgCl$_2$, 10 mM DTT, 1 M NaCl, pH 7.5	Purity is confirmed by overdigestion and ligation-recutting test	TaKaRa 1082

Sau3A I continued

CONCENTRATION or VOLUME ACTIVITY	UNIT DEFINITION	REACTION BUFFER (As Provided)	ADDITIONAL ACTIVITIES and PURITY	SUPPLIER CATALOG NO.
Staphylococcus aureus 3A continued				
3-20 U/µL	1 unit completely digests 1 µg DNA/hr at 37°C	10 mM Tris-HCl, 10 mM MgCl$_2$, 50 mM NaCl, 1 mM DTT, pH 7.5	>90% ligation; >95% recut	Toyobo S3A-101 S3A-102
4000 U/mL; 20,000 U/mL	1 unit completely digests 1 µg λ DNA/hr in a total reaction volume of 50 µL using the buffer provided	10 mM Bis Tris-Propane-HCl, 10 mM MgCl$_2$, 100 mM NaCl, 1 mM DTT, pH 7.0; add 100 µg/mL BSA	>95% DNA fragments ligated and recut after 20-fold overdigestion	NE Biolabs 169S 169L

Sau96 I

RECOGNITION SEQUENCE
5'...G▼GNCC...3'
3'...CCNG▲G...5'

ISOSCHIZOMERS
Asu I, Cfr13 I

OPTIMUM REACTION CONDITIONS
pH 7.5

HEAT INACTIVATION
Resistant; 65°C for 15 min.; 70°C for 30 min.

METHYLATION EFFECTS
Does not cleave GGNm^{5}CC, GGNCm^{5}C or GGNHm^{5}Chm^{5}C
Blocked by overlapping dcm methylation

NOTES
- EC 3.1.21.4
- A type II site specific deoxyribonuclease
- A large group of enzymes recognizing specific short DNA sequences. They cleave either within the recognition site, or at a short specific distance from it
- Requires Mg^{2+}

CONCENTRATION or VOLUME ACTIVITY	UNIT DEFINITION	REACTION BUFFER (As Provided)	ADDITIONAL ACTIVITIES and PURITY	SUPPLIER CATALOG NO.
Escherichia coli strain, carrying the Sau96 I gene from Staphylococcus aureus PS96				
10,000 U/mL	1 unit completely digests 1 µg λ DNA/hr in a total reaction volume of 50 µL using the buffer provided	20 mM Tris-acetate, 10 mM MgOAc, 50 mM KOAc, 1 mM DTT, pH 7.9	>95% DNA fragments ligated and recut after 5-fold overdigestion	NE Biolabs 165S 165L
5000-15,000 U/mL	1 unit cleaves 98% λ DNA/hr in a 50 µL reaction mixture	200 mM Tris-acetate, 500 mM KOAc, 100 mM MgOAc, 10 mM DTT, pH 7.9	100% cleavage, 100% ligation, 100% recleavage. No detectable endonuclease (overdigestion)	Sigma R6385

Sau96 I continued

CONCENTRATION or VOLUME ACTIVITY	UNIT DEFINITION	REACTION BUFFER (As Provided)	ADDITIONAL ACTIVITIES and PURITY	SUPPLIER CATALOG NO.
Staphylococcus aureus PS96				
3-20 U/μL	1 unit completely digests 1 μg λ DNA/hr at 37°C in the buffer provided	100 mM Tris-HCl, 100 mM MgCl$_2$, 10 mM DTT, pH 7.5	>90% DNA fragments ligated and >95% recut	Amersham E 0228Y E 0228Z
10 U/μL	1 unit completely digests 1 μg substrate DNA/hr	33 mM Tris-acetate, 10 mM MgOAc, 66 mM KOAc, 0.5 mM DTT, pH 7.9		Boehringer 651303 709735
5000-15,000 U/mL		50 mM KCl, 10 mM Tris-HCl, 0.1 mM EDTA, 1.0 mM DTT, 200 μg/mL BSA, 50% glycerol, pH 7.4		ICN 153825
3-10 U/μL	1 unit digests completely 1 μg λ DNA substrate/hr at 37°C	33 mM Tris-acetate, 66 mM KOAc, 10 mM MgOAc, 0.5 mM DTT, pH 8.2		NBL Gene 014601 014603
2-8 U/μL	λ (dcm⁻) DNA substrate, 37°C	500 mM NaCl, 100 mM Tris-HCl, 100 mM MgCl$_2$, 10 mM DTT, pH 7.5	90% ligation and 100% recutting	Nippon Gene 310-00332 314-00335 312-00336
10 U/μL	1 unit completely digests 1 μg λ DNA/hr at 37°C in a 50 mL reaction volume	100 mM Tris-HCl, 100 mM MgCl$_2$, 500 mM NaCl, 10 mM DTT, pH 7.9	90% ligation	Promega R6461 R6462
3-20 U/μL	1 unit completely digests 1 μg DNA/hr at 37°C	10 mM Tris-HCl, 10 mM MgCl$_2$, 1 mM DTT, pH 7.5	>90% ligation; >95% recut	Toyobo S96-101 S96-102

Sbf I

RECOGNITION SEQUENCE
5'...CCTGCA▼GG...3'
3'...GG▲ACGTCC...5'

HEAT INACTIVATION
65°C for 20 min.

NOTES
- EC 3.1.21.4
- A type II site specific deoxyribonuclease
- A large group of enzymes recognizing specific short DNA sequences. They cleave either within the recognition site, or at a short specific distance from it
- Requires Mg^{2+}

CONCENTRATION or VOLUME ACTIVITY	UNIT DEFINITION	REACTION BUFFER (As Provided)	ADDITIONAL ACTIVITIES and PURITY	SUPPLIER CATALOG No.
Streptomyces species Sb61				
5000 U/mL		Blue/white certified; 33 mM Tris-acetate, 10 mM MgOAc, 66 mM KOAc, 1 mM DTT, pH 7.9	90% DNA fragments ligated and recut after 10-fold overdigestion	SibEnzyme E101 E102

Sca I

RECOGNITION SEQUENCE
5'...AGT▼ACT...3'
3'...TCA▲TGA...5'

OPTIMUM REACTION CONDITIONS
pH 7.4-7.5

HEAT INACTIVATION
Resistant; 60°C for 15 min.; 65°C for 10-15 min.; 70°C for 30 min.

METHYLATION EFFECTS
Cleaves AGTAm^{5}CT
Does not cleave AGTm6ACT

NOTES
- EC 3.1.21.4
- A type II site specific deoxyribonuclease
- A large group of enzymes recognizing specific short DNA sequences. They cleave either within the recognition site, or at a short specific distance from it
- Low ionic strength, high pH or enzyme concentration, or Mn^{2+} solution may result in star activity
- Compatible cohesive ends: any blunt end
- Requires Mg^{2+}

Sca I continued

CONCENTRATION or VOLUME ACTIVITY	UNIT DEFINITION	REACTION BUFFER (As Provided)	ADDITIONAL ACTIVITIES and PURITY	SUPPLIER CATALOG No.
Streptomyces caespitosus				
8-12 U/µL	1 unit completely digests 1 µg λ DNA/hr at 37°C in the buffer provided	500 mM Tris-HCl, 100 mM MgCl$_2$, 10 mM DTT, 1000 mM NaCl, pH 7.5	>95% DNA fragments ligated and 100% recut	Amersham E 1084Y E 1084Z
8000 U/mL	1 unit degrades 1 µg λ dam$^-$ dcm$^-$ DNA/hr at 37°C in a 0.05 mL mixture	10 mM Tris-HCl, 5 mM MgCl$_2$, 100 mM NaCl, 100 µg/mL BSA, pH 7.2	No detectable contaminating nuclease; 90% ligated and 100% recut after 10-fold overdigestion	AGS Heidelb F00585S F00585M
10 U/µL and 40 U/µL	1 unit completely digests 1 µg substrate DNA/hr	50 mM Tris-HCl, 10 mM MgCl$_2$, 100 mM NaCl, 1 mM DTE, pH 7.5		Boehringer 775258 775266 1207636
	1 unit completely cleaves 1 µg λ DNA/hr at pH 7.5, 37°C in a reaction volume of 50 µL	50 mM Tris-HCl, 10 mM MgCl$_2$, 100 mM NaCl, 1.0 mM DTT, 100 µg/mL BSA, pH 7.5		CHIMERx 2378-01 2378-02
8-12 U/µL	1 unit hydrolyzes 1 µg DNA/hr at 37°C in a 50 µL reaction volume using λ phage DNA as substrate	10 mM Tris-HCl, 5 mM MgCl$_2$, 100 mM NaCl, 0.1 mg/mL BSA, pH 7.2, 37°C	>80% of the DNA fragments ligated and >90% recut after 50-fold overdigestion	Fermentas ER0431 ER0432
1000-8000 U/mL		50 mM KCl, 10 mM Tris-HCl, 0.1 mM EDTA, 1.0 mM DTT, 200 µg/mL BSA, 50% glycerol, pH 7.4		ICN 152048
8-12 U/µL	1 unit completely digests 1 mg λ DNA/hr at 37°C	500 mM Tris-HCl, 60 mM MgCl$_2$, 500 mM NaCl, 500 mM KCl, pH 7.4	≥65% ligation; 100% cleavage	Life Technol 15436-017
	10 units does not produce unspecific cleavage products with 1 µg λ DNA/16 hr at 37°C	10 mM Tris-HCl, 10 mM MgCl$_2$, 100 mM NaCl, 1 mM DTT, 100 µg/mL BSA, pH 7.9	>90% DNA fragments ligated and recut after 10-fold overdigestion	MINOTECH 125
8-12 U/µL	1 unit digests completely 1 µg λ DNA substrate/hr at 37°C	50 mM Tris-HCl, 100 mM NaCl, 10 mM MgCl$_2$, 1 mM DTT, pH 7.8		NBL Gene 012202 012205
8 U/µL	1 unit hydrolyzes 1 µg λ DNA to completion/hr in a 50 µL total reaction volume	500 mM Tris-HCl, 1 M NaCl, 100 mM MgCl$_2$, 10 mM DTE, 1.0 mg/mL BSA, pH 8		Oncor 110401 110402
5000-10,000 U/mL	1 unit completely digests 1 µg λ DNA/hr at pH 7.5, 37°C in a 50 µL total assay mixture	100 mM Tris-acetate, 100 mM MgOAc, 500 mM KOAc, pH 7.5		Pharmacia 27-0977-01 27-0977-02
8-12 U/µL and 40-80 U/µL	1 unit completely digests 1 µg λ DNA/hr at 37°C in a 50 mL reaction volume	100 mM Tris-HCl, 100 mM MgCl$_2$, 1.5 M KCl, pH 7.4	90% ligation	Promega R6211 R6212 R4214

Sca I continued

CONCENTRATION or VOLUME ACTIVITY	UNIT DEFINITION	REACTION BUFFER (As Provided)	ADDITIONAL ACTIVITIES and PURITY	SUPPLIER CATALOG No.
Streptomyces caespitosus continued				
5000-20,000 U/mL	1 unit completely cleaves 1.0 µg λ DNA/hr in a 50 µL reaction mixture	Unique buffer	100% cleavage, 90% ligation, 80% recleavage, 70% endonuclease (nickase) No detectable endonuclease (overdigestion)	Sigma R5007
4000-15,000 U/mL		50 mM KCl, 10 mM Tris-HCl, 0.1 mM EDTA, 1 mM DTT, 200 µg/mL BSA, 50% glycerol, pH 7.4		Stratagene 500960
1500 U and 7500 U	1 unit completely digests 1 µg λ DNA/hr at 37°C in 50 µL of supplied buffer	500 mM Tris-HCl, 100 mM MgCl$_2$, 10 mM DTT, 1 M NaCl, pH 7.5	Purity is confirmed by overdigestion and ligation-recutting test	TaKaRa 1084
3-20 U/µL	1 unit completely digests 1 µg DNA/hr at 37°C	50 mM Tris-HCl, 10 mM MgCl$_2$, 100 mM NaCl, 1 mM DTT, pH 7.5	>90% ligation; >95% recut	Toyobo SCA-103 SCA-104
Streptomyces caespitosus (H. Takahashi and R.J. Roberts)				
10,000 U/mL; 50,000 U/mL	1 unit completely digests 1 µg λ DNA/hr in a total reaction volume of 50 µL using the buffer provided	10 mM Tris-HCl, 10 mM MgCl$_2$, 100 mM NaCl, 1 mM DTT, pH 7.4	>90% DNA fragments ligated and recut after 5-fold overdigestion	NE Biolabs 122S 122L
Streptomyces caespitosus (H. Takahashi)				
3-15 U/µL	λ DNA substrate, 37°C	125 mM NaCl, 6 mM Tris-HCl, 6 mM MgCl$_2$, 6 mM β-MSH, pH 7.5	90% ligation and 100% recutting	Nippon Gene 313-01081 319-01083 317-01084

ScrF I

RECOGNITION SEQUENCE
  5'...CC▼NGG...3'
  3'...GGN▲CC...5'

ISOSCHIZOMERS
  Dsa V

OPTIMUM REACTION CONDITIONS
  pH 7.5

HEAT INACTIVATION
  65°C for 15-30 min.

METHYLATION EFFECTS
  Blocked by *dcm* methylation

NOTES
- EC 3.1.21.4
- A type II site specific deoxyribonuclease
- A large group of enzymes recognizing specific short DNA sequences. They cleave either within the recognition site, or at a short specific distance from it
- Partially cleaves DNA from dcm^+ *E. coli*
- ScrF I fragments are difficult to ligate with T4 DNA ligase, as are other single base 5'-extensions
- Requires Mg^{2+}

CONCENTRATION or VOLUME ACTIVITY	UNIT DEFINITION	REACTION BUFFER (As Provided)	ADDITIONAL ACTIVITIES and PURITY	SUPPLIER CATALOG No.
Streptococcus cremoris F				
10,000 U/mL	1 unit degrades 1 µg λ dam⁻ dcm⁻ DNA/hr at 37°C in a 0.05 mL mixture	10 m*M* Tris-HCl, 100 m*M* NaCl, 100 µg/mL BSA, pH 8.0	No detectable contaminating nuclease; 0% ligated and recut after 2-fold overdigestion	AGS Heidelb F00586S F00586M
10 U/µL	1 unit completely digests 1 µg substrate DNA/hr	10 m*M* Tris-HCl, 5 m*M* MgCl₂, 100 m*M* NaCl, 1 m*M* β-MSH, pH 8.0		Boehringer 1081292
1000-10,000 U/mL		50 m*M* KCl, 10 m*M* Tris-HCl, 0.1 m*M* EDTA, 1.0 m*M* DTT, 200 µg/mL BSA, 50% glycerol, pH 7.4		ICN 152051
5-20 U/µL	λ (dam⁻) DNA substrate, 37°C	1 *M* NaCl, 500 m*M* Tris-HCl, 100 m*M* MgCl₂, 10 m*M* DTT, pH 7.5	60% ligation and 95% recutting	Nippon Gene 318-00931 316-00932 310-00935
2000-10,000 U/mL	1 unit completely cleaves 1.0 µg λ DNA/hr in a 50 µL reaction mixture	200 m*M* Tris-acetate, 500 m*M* KOAc, 100 m*M* MgOAc, 10 m*M* DTT, pH 7.9	100% cleavage, 20% ligation, 80% recleavage No detectable endonuclease (overdigestion)	Sigma R7888
3-20 U/µL	1 unit completely digests 1 µg DNA/hr at 37°C	50 m*M* Tris-HCl, 10 m*M* MgCl₂, 100 m*M* NaCl, 1 m*M* DTT, pH 7.5		Toyobo SCR-101 SCR-102

ScrF I continued

CONCENTRATION or VOLUME ACTIVITY	UNIT DEFINITION	REACTION BUFFER (As Provided)	ADDITIONAL ACTIVITIES and PURITY	SUPPLIER CATALOG NO.
Streptococcus cremoris F (C. Daly)				
5000 U/mL	1 unit completely digests 1 µg λ DNA/hr in a total reaction volume of 50 µL using the buffer provided	20 mM Tris-acetate, 10 mM MgOAc, 50 mM KOAc, 1 mM DTT, pH 7.9	None of the DNA fragments ligated after 2-fold overdigestion	NE Biolabs 110S 110L

Sdu I

Recognition Sequence

$5'...G(^T_{AG})GC(^T_{AC})\blacktriangledown C...3'$
$3'...C\blacktriangle(^C_{TA})CG(^G_{TA})G...5'$

Notes
- EC 3.1.21.4
- A type II site specific deoxyribonuclease
- A large group of enzymes recognizing specific short DNA sequences. They cleave either within the recognition site, or at a short specific distance from it
- Requires Mg^{2+}

CONCENTRATION or VOLUME ACTIVITY	UNIT DEFINITION	REACTION BUFFER (As Provided)	ADDITIONAL ACTIVITIES and PURITY	SUPPLIER CATALOG NO.
Streptococcus durans RFL3				
8-12 U/µL	1 unit hydrolyzes 1 µg DNA/hr at 37°C in a 50 µL reaction volume using λ phage DNA as substrate	10 mM Tris-HCl, 3 mM MgCl₂, 150 mM NaCl, 0.1 mg/mL BSA, pH 7.2	>90% of the DNA fragments ligated and recut after 50-fold overdigestion	Fermentas ER0651 ER0652

SexA I

RECOGNITION SEQUENCE

5'...A▼CC(^A_T)GGT...3'
3'...TGG(^T_A)CC▲A...5'

NOTES
- EC 3.1.21.4
- A type II site specific deoxyribonuclease
- A large group of enzymes recognizing specific short DNA sequences. They cleave either within the recognition site, or at a short specific distance from it
- Requires Mg^{2+}

CONCENTRATION or VOLUME ACTIVITY	UNIT DEFINITION	REACTION BUFFER (As Provided)	ADDITIONAL ACTIVITIES and PURITY	SUPPLIER CATALOG NO.
Streptomyces exfoliatus				
10 U/μL	1 unit completely digests 1 μg substrate DNA/hr	10 mM Tris-HCl, 5 mM MgCl₂, 100 mM NaCl, 1 mM β-MSH, pH 8.0		Boehringer 1497995 1498002

SfaN I

RECOGNITION SEQUENCE

5'...GCATC(N)₅▼...3'
3'...CGTAG(N)₉▲...5'

HEAT INACTIVATION
65°C for 20 min.

NOTES
- EC 3.1.21.4
- A type II site specific deoxyribonuclease
- A large group of enzymes recognizing specific short DNA sequences. They cleave either within the recognition site, or at a short specific distance from it
- Requires Mg^{2+}

SfaN I continued

CONCENTRATION or VOLUME ACTIVITY	UNIT DEFINITION	REACTION BUFFER (As Provided)	ADDITIONAL ACTIVITIES and PURITY	SUPPLIER CATALOG No.
Streptococcus faecalis				
500-2000 U/mL		250 mM KCl, 10 mM Tris-HCl, 0.1 mM EDTA, 1.0 mM DTT, 200 μg/mL acetylated BSA, 50% glycerol, pH 7.4		ICN 159466
Streptococcus faecalis ATCC ND547				
500-2000 U/mL	1 unit completely digests 1 μg λ DNA/hr in a total reaction volume of 50 μL using the buffer provided	50 mM Tris-HCl, 10 mM MgCl$_2$, 100 mM NaCl, 1 mM DTT, pH 7.9	>95% DNA fragments ligated and recut after 2-fold overdigestion	NE Biolabs 172S 172L
Streptococcus faecalis N				
1000 U/mL		50 mM Tris-HCl, 10 mM MgCl$_2$, 100 mM NaCl, 1 mM DTT, pH 7.6	95% DNA fragments ligated and recut after 2-fold overdigestion	SibEnzyme E165 E166

Sfc I

RECOGNITION SEQUENCE
    5'...C▼TPuPyAG...3'
    3'...GAPyPuT▲C...5'

HEAT INACTIVATION
    65°C for 20 min.

NOTES
- EC 3.1.21.4
- A type II site specific deoxyribonuclease
- A large group of enzymes recognizing specific short DNA sequences. They cleave either within the recognition site, or at a short specific distance from it
- Enzyme concentrations ≥1 U/μg DNA are recommended
- Requires Mg^{2+}

CONCENTRATION or VOLUME ACTIVITY	UNIT DEFINITION	REACTION BUFFER (As Provided)	ADDITIONAL ACTIVITIES and PURITY	SUPPLIER CATALOG No.
Streptococcus faecalis				
1000-5000 U/mL		150 mM KCl, 10 mM Tris-HCl, 0.1 mM EDTA, 1.0 mM DTT, 400 μg/mL acetylated BSA, 50% glycerol, pH 7.4		ICN 159467

Sfc I continued

CONCENTRATION or VOLUME ACTIVITY	UNIT DEFINITION	REACTION BUFFER (As Provided)	ADDITIONAL ACTIVITIES and PURITY	SUPPLIER CATALOG NO.
Streptococcus faecium NEB 674				
1000 U/mL	1 unit completely digests 1 µg λ DNA/hr in a total reaction volume of 50 µL using the buffer provided	20 mM Tris-acetate, 10 mM MgOAc, 50 mM KOAc, 1 mM DTT, pH 7.9; add 100 µg/mL BSA	>95% DNA fragments ligated and recut after 2-fold overdigestion	NE Biolabs 561S 561L

Sfi I

RECOGNITION SEQUENCE
  5'...GGCC(N)$_4$▼NGGCC...3'
  3'...CCGGN▲(N)$_4$CCGG...5'
OPTIMUM REACTION CONDITIONS
  pH 7.9–8.0 and 50°C
HEAT INACTIVATION
  Resistant; 65°C for 2 min.; 85°C for 30 min.
METHYLATION EFFECTS
  Cleaves GGCC(N)$_5$GGCm^5 and GGm^{5}CC(N)$_5$GGm^{5}CC
  Does not cleave GGCm^{5}C(N)$_5$GGCC
  Blocked by overlapping *dcm* methylation

NOTES
- EC 3.1.21.4
- A type II site specific deoxyribonuclease
- A large group of enzymes recognizing specific short DNA sequences. They cleave either within the recognition site, or at a short specific distance from it
- Its 8-nucleotide recognition sequence makes *Sfi* I useful for producing large fragments from genomic DNA
- Requires Mg^{2+}

CONCENTRATION or VOLUME ACTIVITY	UNIT DEFINITION	REACTION BUFFER (As Provided)	ADDITIONAL ACTIVITIES and PURITY	SUPPLIER CATALOG NO.
Escherichia coli strain, carrying the cloned gene from Streptomyces fimbriatus				
10,000 U/mL	1 unit completely digests 1 µg Ad2 DNA/hr in a total reaction volume of 50 µL using the buffer provided	10 mM Tris-HCl, 10 mM MgCl$_2$, 50 mM NaCl, 1 mM DTT, pH 7.9; add 100 µg/mL BSA	>95% DNA fragments ligated and recut after 10-fold overdigestion	NE Biolabs 123S 123L
Streptomyces fimbriatus				
750 U and 2000 U	1 unit completely digests 1 µg DNA substrate/hr at 50°C	10 mM Tris-HCl, 10 mM MgCl$_2$, 50 mM NaCl, 1 mM DTT, 0.1 mg/mL BSA, pH 7.5	>95% DNA fragments ligated and recut after 10-fold overdigestion	Adv Biotech AB-0403-a AB-0403-b
8–12 U/µL	1 unit completely digests 1 µg Ad2 DNA/hr at 50°C in the buffer provided	100 mM Tris-HCl, 100 mM MgCl$_2$, 10 mM DTT, 500 mM NaCl, pH 7.5	>95% DNA fragments ligated and 100% recut	Amersham E 0102Y E 0102Z

Sfi I continued

CONCENTRATION or VOLUME ACTIVITY	UNIT DEFINITION	REACTION BUFFER (As Provided)	ADDITIONAL ACTIVITIES and PURITY	SUPPLIER CATALOG No.
Streptomyces fimbriatus continued				
10,000 U/mL	1 unit degrades 1 µg Ad2 DNA/hr at 50°C in a 0.05 mL mixture	10 mM Tris-HCl, 10 mM MgCl$_2$, 100 µg/mL BSA, pH 7.5	No detectable contaminating nuclease; 90% ligated and recut after 10-fold overdigestion	AGS Heidelb F00587S F00587M
10 U/µL and 40 U/µL	1 unit completely digests 1 µg substrate DNA/hr	10 mM Tris-HCl, 10 mM MgCl$_2$, 50 mM NaCl, 1 mM DTE, pH 7.5, 37 °C		Boehringer 1288016 1288024 1288032 1288059
	1 unit completely cleaves 1 µg Ad2 DNA/hr at pH 7.5, 37°C in a reaction volume of 50 µL	10 mM Tris-HCl, 10 mM MgCl$_2$, 50 mM NaCl, 1.0 mM DTT, 100 µg/mL BSA, pH 7.5		CHIMERx 2380-01 2380-02
2000-10,000 U/mL		50 mM KCl, 10 mM Tris-HCl, 0.1 mM EDTA, 1.0 mM DTT, 200 µg/mL BSA, 50% glycerol, pH 7.5		ICN 153839
8-12 U/µL	1 unit completely digests 1 mg Ad2 DNA/hr at 50°C	500 mM Tris-HCl, 100 mM MgCl$_2$, 500 mM NaCl, pH 8.0	≥65% ligation; 100% cleavage	Life Technol 15457-013
	10 units does not produce unspecific cleavage products with 1 µg Ad2 DNA/16 hr at 50°C	10 mM Tris-HCl, 10 mM MgCl$_2$, 50 mM NaCl, 1 mM DTT, 100 µg/mL BSA, pH 7.9	>90% DNA fragments ligated and recut after 10-fold overdigestion	MINOTECH 126
6-12 U/µL	1 unit digests completely 1 µg Ad2 DNA substrate/hr at 50°C	10 mM Tris-HCl, 50 mM NaCl, 10 mM MgCl$_2$, 1 mM β-MSH, 100 µg/mL BSA, pH 8.4		NBL Gene 015901 015903
4-20 U/µL	Ad2 DNA substrate, 50°C	500 mM NaCl, 100 mM Tris-HCl, 100 mM MgCl$_2$, 10 mM DTT, pH 7.5	80% ligation and 100% recutting	Nippon Gene 319-01441 315-01443 313-01444
2 U/µL and 5-15 U/µL	1 unit hydrolyzes 1 µg Ad2 DNA to completion/hr in a 50 µL total reaction volume	100 mM Tris-HCl, 500 mM NaCl, 100 mM MgCl$_2$, 100 mM β-MSH, 1.0 mg/mL BSA, pH 8		Oncor 110531 110532
5000-10,000 U/mL and 50,000-80,000 U/mL	1 unit completely digests 1 µg Ad2 DNA/hr at pH 7.5, 50°C in a 50 µL total assay mixture	100 mM Tris-acetate, 100 mM MgOAc, 500 mM KOAc, pH 7.5		Pharmacia 27-0975-01 27-0975-02 27-0975-18
10 U/µL and 40-80 U/µL	1 unit completely digests 1 µg Ad2 DNA/hr at 50°C in a 50 mL reaction volume	Blue/White Cloning Qualified; genome qualified; 60 mM Tris-HCl, 60 mM MgCl$_2$, 500 mM NaCl, 10 mM DTT, pH 7.2	90% ligation	Promega R6391 R6392 R4394

CONCENTRATION or VOLUME ACTIVITY	UNIT DEFINITION	REACTION BUFFER (As Provided)	ADDITIONAL ACTIVITIES and PURITY	SUPPLIER CATALOG No.
Streptomyces fimbriatus continued				
5000 U/mL		10 mM Tris-HCl, 10 mM MgCl$_2$, 50 mM NaCl, 1 mM DTT, pH 7.6	90% DNA fragments ligated and recut after 10-fold overdigestion	SibEnzyme E123 E124
2000-10,000 U/mL	1 unit completely cleaves 1.0 μg λ DNA/hr in a 50 μL reaction mixture	100 mM Tris-HCl, 500 mM NaCl, 100 mM MgCl$_2$, 10 mM DTT, BSA, pH 7.9	100% cleavage, 90% ligation, 100% recleavage, 0% endonuclease (nickase) No detectable endonuclease (overdigestion)	Sigma R8256
5000-25,000 U/mL		200 mM KCl, 10 mM Tris-HCl, 0.1 mM EDTA, 1 mM DTT, 200 μg/mL BSA, 50% glycerol, pH 7.5		Stratagene 501010 501011
3-20 U/μL	1 unit completely digests 1 μg DNA/hr at 50°C	10 mM Tris-HCl, 10 mM MgCl$_2$, 50 mM NaCl, 1 mM DTT, pH 7.5	>90% ligation; >95% recut	Toyobo SFI-101 SFI-102

Sfr274 I

RECOGNITION SEQUENCE
5'...C▼TCGAG...3'
3'...GAGCT▲C...5'

HEAT INACTIVATION
Resistant

NOTES
- EC 3.1.21.4
- A type II site specific deoxyribonuclease
- A large group of enzymes recognizing specific short DNA sequences. They cleave either within the recognition site, or at a short specific distance from it
- Requires Mg^{2+}

CONCENTRATION or VOLUME ACTIVITY	UNIT DEFINITION	REACTION BUFFER (As Provided)	ADDITIONAL ACTIVITIES and PURITY	SUPPLIER CATALOG NO.
Streptomyces fradiae 274				
10,000 U/mL		10 mM Tris-HCl, 10 mM $MgCl_2$, 1 mM DTT, pH 7.6	95% DNA fragments ligated and recut after 10-fold overdigestion	SibEnzyme E125 E126

Sfr303 I

RECOGNITION SEQUENCE
5'...CCGC▼GG...3'
3'...GG▲CGCC...5'

HEAT INACTIVATION
65°C for 20 min.

NOTES
- EC 3.1.21.4
- A type II site specific deoxyribonuclease
- A large group of enzymes recognizing specific short DNA sequences. They cleave either within the recognition site, or at a short specific distance from it
- Requires Mg^{2+}

CONCENTRATION or VOLUME ACTIVITY	UNIT DEFINITION	REACTION BUFFER (As Provided)	ADDITIONAL ACTIVITIES and PURITY	SUPPLIER CATALOG NO.
Streptomyces fradiae 303				
20,000 U/mL		10 mM Tris-HCl, 10 mM $MgCl_2$, 1 mM DTT, pH 7.6	95% DNA fragments ligated and recut after 10-fold overdigestion	SibEnzyme E127 E128

Sfu I

RECOGNITION SEQUENCE
  5'...TT▼CGAA...3'
  3'...AAGC▲TT...5'
ISOSCHIZOMERS
  Asu II

NOTES
- EC 3.1.21.4
- A type II site specific deoxyribonuclease
- A large group of enzymes recognizing specific short DNA sequences. They cleave either within the recognition site, or at a short specific distance from it
- Requires Mg^{2+}

CONCENTRATION or VOLUME ACTIVITY	UNIT DEFINITION	REACTION BUFFER (As Provided)	ADDITIONAL ACTIVITIES and PURITY	SUPPLIER CATALOG NO.
Streptomyces fulvissimus				
10 U/μL	1 unit completely digests 1 μg substrate DNA/hr	50 mM Tris-HCl, 10 mM $MgCl_2$, 100 mM NaCl, 1 mM DTE, pH 7.5		Boehringer 1243497

Sgf I

RECOGNITION SEQUENCE
  5'...GCGAT▼CGC...3'
  3'...CGC▲TAGCG...5'
OPTIMUM REACTION CONDITIONS
  pH 7.9
HEAT INACTIVATION
  65°C for 15 min.
METHYLATION EFFECTS
  Not blocked by dcm methylation

NOTES
- EC 3.1.21.4
- A type II site specific deoxyribonuclease
- A large group of enzymes recognizing specific short DNA sequences. They cleave either within the recognition site, or at a short specific distance from it
- High glycerol or enzyme concentration results in star activity
- Compatible cohesive ends: Pvu I
- Requires Mg^{2+}

Sgf I continued

CONCENTRATION or VOLUME ACTIVITY	UNIT DEFINITION	REACTION BUFFER (As Provided)	ADDITIONAL ACTIVITIES and PURITY	SUPPLIER CATALOG No.
Streptomyces griseoruber				
8-12 U/µL	1 unit completely digests 1 µg Ad2 DNA/hr at 37°C in a 50 mL reaction volume	Genome qualified; 100 mM Tris-HCl, 100 mM MgCl$_2$, 500 mM NaCl, 10 mM DTT, pH 7.9	90% ligation	Promega R7101 R7102
3-20 U/µL	1 unit completely digests 1 µg DNA/hr at 37°C		>50% ligation; >95% recut	Toyobo SGF-101 SGF-102

SgrA I

RECOGNITION SEQUENCE
$$5'...C(^A_G)\blacktriangledown CCGG(^T_C)G...3'$$
$$3'...G(^T_C)GGCC\blacktriangle(^A_G)C...5'$$

NOTES
- EC 3.1.21.4
- A type II site specific deoxyribonuclease
- A large group of enzymes recognizing specific short DNA sequences. They cleave either within the recognition site, or at a short specific distance from it
- Requires Mg^{2+}

CONCENTRATION or VOLUME ACTIVITY	UNIT DEFINITION	REACTION BUFFER (As Provided)	ADDITIONAL ACTIVITIES and PURITY	SUPPLIER CATALOG No.
Streptomyces griseus				
10 U/µL	1 unit completely digests 1 µg substrate DNA/hr	33 mM Tris-acetate, 10 mM MgOAc, 66 mM KOAc, 0.5 mM DTT, pH 7.9		Boehringer 1277014 1277022

Sin I

RECOGNITION SEQUENCE
5'...G▼G(^A_T)CC...3'
3'...CC(^T_A)G▲G...5'

ISOSCHIZOMERS
Ava II

METHYLATION EFFECTS
Does not cleave GG(^A_T)m^{5}CC

NOTES
- EC 3.1.21.4
- A type II site specific deoxyribonuclease
- A large group of enzymes recognizing specific short DNA sequences. They cleave either within the recognition site, or at a short specific distance from it
- Requires Mg^{2+}

CONCENTRATION or VOLUME ACTIVITY	UNIT DEFINITION	REACTION BUFFER (As Provided)	ADDITIONAL ACTIVITIES and PURITY	SUPPLIER CATALOG NO.
Salmonella infantis				
8000-12,000 U/mL		50 m*M* KCl, 10 m*M* Tris-HCl, 0.1 m*M* EDTA, 1.0 m*M* DTT, 500 µg/mL acetylated BSA, 50% glycerol, pH 7.4		ICN 159468
6-16 U/µL	1 unit digests completely 1 µg λ DNA substrate/hr at 37°C	10 m*M* Tris-HCl, 10 m*M* MgCl₂, 1 m*M* DTT, BSA, pH 7.8		NBL Gene 011201 011203
10 U/µL and 40-80 U/µL	1 unit completely digests 1 µg λ DNA/hr at 37°C in a 50 mL reaction volume	60 m*M* Tris-HCl, 60 m*M* MgCl₂, 60 m*M* NaCl, 10 m*M* DTT, pH 7.5	90% ligation	Promega R6141 R6142 R4144

Sma I

RECOGNITION SEQUENCE
5'...CCC▼GGG...3'
3'...GGG▲CCC...5'

ISOSCHIZOMERS
Xma I

OPTIMUM REACTION CONDITIONS
pH 7.5-7.9 and 30°C; activity is 2-fold greater at 30°C than 37°C ($T_{1/2}$ = 15 min)

HEAT INACTIVATION
65°C for 5-30 min.

METHYLATION EFFECTS
Cleaves Cm^5CCGGG
Does not cleave $m^5CCCGGG$, CCm^5CGGG or any m^4C-methylated sequence

NOTES
- EC 3.1.21.4
- A type II site specific deoxyribonuclease
- A large group of enzymes recognizing specific short DNA sequences. They cleave either within the recognition site, or at a short specific distance from it
- Produces blunt-end fragments; Xma I gives 5'-extensions
- Compatible cohesive ends: any blunt end
- Requires Mg^{2+} and KCl

CONCENTRATION or VOLUME ACTIVITY	UNIT DEFINITION	REACTION BUFFER (As Provided)	ADDITIONAL ACTIVITIES and PURITY	SUPPLIER CATALOG NO.
Escherichia coli strain, carrying the cloned gene from Serratia marcescens				
20,000 U/mL	1 unit completely digests 1 μg λ DNA (Hind III digest)/hr in a total reaction volume of 50 μL using the buffer provided	20 mM Tris-acetate, 10 mM MgOAc, 50 mM KOAc, 1 mM DTT, pH 7.9	>95% DNA fragments ligated and recut after 2-fold overdigestion	NE Biolabs 141S 141L
Serratia marcescens				
4500 U and 15,000 U	1 unit completely digests 1 μg DNA substrate/hr at 25°C	10 mM Tris-acetate, 10 mM MgOAc, 50 mM KOAc, 1 mM DTT, 0.1 mg/mL BSA, pH 7.9	>90% DNA fragments ligated and recut after 2-fold overdigestion	Adv Biotech AB-0314-a AB-0314-b
10 U/μL; 40 U/μL	1 unit completely cleaves 1 μg substrate DNA/hr	60 mM KCl, 20 mM Tris, 0.75 mM $MgCl_2$, pH 7.5	90-98% cleavage-ligation-recleavage	Am Allied L-23-01000 H-23-01000 L-23-05000 H-23-05000
10,000 U/mL	1 unit degrades 1 μg λ dam⁻ dcm⁻ DNA/hr at 25°C in a 0.05 mL mixture	33 mM Tris-acetate, 10 mM MgOAc, 66 mM KOAc, 100 μg/mL BSA, pH 7.9	No detectable contaminating nuclease; 90% ligated and 100% recut after 50-fold overdigestion	AGS Heidelb A00590S A00590M

Sma I continued

CONCENTRATION or VOLUME ACTIVITY	UNIT DEFINITION	REACTION BUFFER (As Provided)	ADDITIONAL ACTIVITIES and PURITY	SUPPLIER CATALOG NO.
Serratia marcescens continued				
10 U/μL and 40 U/μL	1 unit completely digests 1 μg substrate DNA/hr	33 mM Tris-acetate, 10 mM MgOAc, 66 mM KOAc, 0.5 mM DTT, pH 7.9		Boehringer 220566 656348 1047639
	1 unit completely cleaves 1 μg λ DNA/hr at pH 7.5, 25°C in a reaction volume of 50 μL	10 mM Tris-HCl, 10 mM MgCl$_2$, 15 mM KCl, 1.0 mM DTT, 100 μg/mL BSA, pH 8.0		CHIMERx 2390-01 2390-02
8-12 U/μL; 40-60 U/μL	1 unit hydrolyzes 1 μg DNA/hr at 37°C in a 50 μL reaction volume using λ phage DNA as substrate	33 mM Tris-acetate, 10 mM MgOAc, 66 mM KOAc, 0.1 mg/mL BSA, pH 7.9, 30°C	>90% of the DNA fragments ligated and recut after 50-fold overdigestion	Fermentas ER0661 ER0662 ER0663
5000-20,000 U/mL		50 mM KCl, 10 mM Tris-HCl, 0.1 mM EDTA, 1.0 mM DTT, 200 μg/mL BSA, 50% glycerol, pH 7.4		ICN 152059
8-12 U/μL and 50 U/μL	1 unit completely digests 1 mg Ad2 DNA/hr at 30°C	200 mM Tris-HCl, 50 mM MgCl$_2$, 500 M KCl, pH 7.4	≥65% ligation; 100% cleavage	Life Technol 15228-018 15228-034 15228-026
	10 units does not produce unspecific cleavage products with 1 μg λ DNA/16 hr at 25°C	20 mM Tris-acetate, 10 mM MgOAc, 50 mM KOAc, 1 mM DTT, 100 μg/mL BSA, pH 7.9	>90% DNA fragments ligated and recut after 10-fold overdigestion	MINOTECH 127
5-20 U/μL 30-100 U/μL	λ (dcm⁻) DNA substrate, 30°C	500 mM KOAc, 200 mM Tris-acetate, 100 mM MgOAc, 10 mM DTT, pH 7.9 (10X buffer)	90% ligation and 100% recutting	Nippon Gene 314-00352 318-00355 316-00356 318-02131
10 U/μL and 15-100 U/μL	1 unit hydrolyzes 1 μg λ DNA to completion/hr in a 50 μL total reaction volume	100 mM Tris-HCl, 200 mM KCl, 100 mM MgCl$_2$, 100 mM β-MSH, 1.0 mg/mL BSA, pH 8.0		Oncor 110411 110412
2000-10,000 U/mL and 50,000-80,000 U/mL	1 unit completely digests 1 μg λ DNA/hr at pH 7.5, 30°C in a 50 μL total assay mixture	100 mM Tris-acetate, 100 mM MgOAc, 500 mM KOAc, pH 7.5		Pharmacia 27-0942-01 27-0942-02 27-0942-18
10 U/μL and 40-80 U/μL	1 unit completely digests 1 μg λ DNA/hr at 25°C in a 50 mL reaction volume	Blue/White Cloning Qualified; genome qualified; 100 mM Tris-HCl, 70 mM MgCl$_2$, 500 mM KCl, 10 mM DTT, pH 7.8	90% ligation	Promega R6121 R6122 R6125 R4124

Sma I continued

CONCENTRATION or VOLUME ACTIVITY	UNIT DEFINITION	REACTION BUFFER (As Provided)	ADDITIONAL ACTIVITIES and PURITY	SUPPLIER CATALOG NO.
Serratia marcescens continued				
25,000 U/mL		Blue/white certified; 33 mM Tris-acetate, 10 mM MgOAc, 66 mM KOAc, 1 mM DTT, pH 7.9	50% DNA fragments ligated and recut after 2-fold overdigestion	SibEnzyme E177 E178
Serratia marcescens Sb				
8-12 U/μL	1 unit completely digests 1 μg λ DNA/hr at 30°C in the buffer provided	Genomic grade; 330 mM Tris-acetate, 100 mM MgOAc, 5 mM DTT, 660 mM KOAc, BSA, pH 7.9	>95% DNA fragments ligated and 100% recut	Amersham E 1085Y E 1085Z
	1 unit completely digests 1 μg λ DNA/hr at 30°C in the buffer provided	Genomic grade; 330 mM Tris-acetate, 100 mM MgOAc, 5 mM DTT, 660 mM KOAc, BSA, pH 7.9	>95% DNA fragments ligated and 100% recut	Amersham E 1085YH
8-12 U/μL	1 unit digests completely 1 μg λ DNA substrate/hr at 30°C	33 mM Tris-acetate, 66 mM KOAc, 10 mM MgOAc, 0.5 mM DTT, pH 8.2		NBL Gene 011003 011006
5000-20,000 U/mL	1 unit cleaves 98% λ DNA/hr in a 50 μL reaction mixture	200 mM Tris-acetate, 500 mM KOAc, 100 mM MgOAc, 10 mM DTT, pH 7.9	100% cleavage, 90% ligation, 100% recleavage, 50% endonuclease (nickase) No detectable endonuclease (overdigestion)	Sigma R4503
5000-25,000 U/mL		50 mM KCl, 10 mM Tris-HCl, 0.1 mM EDTA, 1 mM DTT, 200 μg/mL BSA, 50% glycerol, pH 7.4		Stratagene 501030
2000 U and 10,000 U	1 unit completely digests 1 μg λ DNA/hr at 30°C in 50 μL of supplied buffer and BSA	Genomic grade; 330 mM Tris-acetate, 100 mM MgOAc, 5 mM DTT, 660 mM KOAc, BSA, pH 7.9	Purity is confirmed by overdigestion, ligation-recutting, genome DNA digestion and cloning test	TaKaRa 1085
3-20 U/μL and 40-50 U/μL	1 unit completely digests 1 μg DNA/hr at 30°C		>90% ligation; >95% recut	Toyobo SMA-101 SMA-102 SMA-152

Smi I

RECOGNITION SEQUENCE
5'...ATTT▼AAAT...3'
3'...TAAA▲TTTA...5'
HEAT INACTIVATION
65°C for 20 min.

NOTES
- EC 3.1.21.4
- A type II site specific deoxyribonuclease
- A large group of enzymes recognizing specific short DNA sequences. They cleave either within the recognition site, or at a short specific distance from it
- Requires Mg^{2+}

CONCENTRATION or VOLUME ACTIVITY	UNIT DEFINITION	REACTION BUFFER (As Provided)	ADDITIONAL ACTIVITIES and PURITY	SUPPLIER CATALOG NO.
Streptococcus milleri S				
5000 U/mL		50 mM Tris-HCl, 10 mM MgCl$_2$, 100 mM NaCl, 1 mM DTT, pH 7.6	50% DNA fragments ligated and recut after 5-fold overdigestion	sibEnzyme E225 E226

SnaB I

RECOGNITION SEQUENCE
5'...TAC▼GTA...3'
3'...ATG▲CAT...5'
OPTIMUM REACTION CONDITIONS
pH 7.0
HEAT INACTIVATION
Resistant; 60°C for 15 min.
METHYLATION EFFECTS
Does not cleave TAm^{5}CGTA or Tm^{6}ACGTA

NOTES
- EC 3.1.21.4
- A type II site specific deoxyribonuclease
- A large group of enzymes recognizing specific short DNA sequences. They cleave either within the recognition site, or at a short specific distance from it
- Compatible cohesive ends: any blunt end
- Sensitive to NaCl >100 mM
- Requires Mg^{2+}

CONCENTRATION or VOLUME ACTIVITY	UNIT DEFINITION	REACTION BUFFER (As Provided)	ADDITIONAL ACTIVITIES and PURITY	SUPPLIER CATALOG No.
Sphaerotilus natans				
8-12 U/μL	1 unit completely digests 1 μg λ DNA/hr at 37°C in the buffer provided	Genomic grade; SnaB I basal buffer	90% DNA fragments ligated and >95% recut	Amersham E 1179Y
10 U/μL	1 unit completely digests 1 μg substrate DNA/hr	10 mM Tris-HCl, 10 mM MgCl$_2$, 50 mM NaCl, 1 mM DTE, pH 7.5		Boehringer 997480
	10 units does not produce unspecific cleavage products with 1 μg λ DNA/EcoR I digest/16 hr at 37°C	10 mM Bis Tris-propane-HCl, 10 mM MgCl$_2$, 1 mM DTT, 100 μg/mL BSA, pH 7.0	>90% DNA fragments ligated and recut after 10-fold overdigestion	MINOTECH 128
4-12 U/μL	1 unit digests completely 1 μg λ DNA substrate/hr at 37°C	10 mM Tris-HCl, 50 mM NaCl, 10 mM MgCl$_2$, 1 mM DTT, pH 7.8		NBL Gene 014301
2-10 U/μL	1 unit completely digests 1 μg EcoR I λ DNA fragments/hr at 37°C in a 50 mL reaction volume	60 mM Tris-HCl, 60 mM MgCl$_2$, 500 mM NaCl, 10 mM DTT, pH 7.5	90% ligation	Pharmacia Promega R6791 R6792 R6795
2000-10,000 U/mL	1 unit completely cleaves 1.0 μg λ DNA/hr in a 50 μL reaction mixture	200 mM Tris-acetate, 500 mM KOAc, 100 mM MgOAc, 10 mM DTT, BSA, pH 7.9	100% cleavage, 90% ligation, 95% recleavage, 50% endonuclease (nickase), 100% endonuclease (overdigestion)	Sigma R6509
2000-5000 U/mL		50 mM KCl, 10 mM Tris-HCl, 0.1 mM EDTA, 1 mM DTT, 200 μg/mL BSA, 50% glycerol, pH 7.4		Stratagene 501040
1000 U and 5000 U	1 unit completely digests 1 μg λ DNA/hr at 37°C in 50 μL of supplied buffer	Genomic grade; 10 mM Tris-HCl, 7 mM MgCl$_2$, 7 mM β-MSH, 50 mM NaCl, 0.01% BSA, pH 8.0	Purity is confirmed by overdigestion, ligation-recutting and genome DNA digestion test	TaKaRa 1179
Sphaerotilus natans ATCC 15291				
2000-10,000 U/mL		50 mM KCl, 10 mM Tris-HCl, 0.1 mM EDTA, 1 mM DTT, 200 μg/mL BSA, 50% glycerol, pH 7.4		ICN 153827
4000 U/mL	1 unit completely digests 1 μg T7 DNA/hr in a total reaction volume of 50 μL using the buffer provided	20 mM Tris-acetate, 10 mM MgOAc, 50 mM KOAc, 1 mM DTT, pH 7.9; add 100 μg/mL BSA	>95% DNA fragments ligated and recut after 20-fold overdigestion	NE Biolabs 130S 130L

Spe I

RECOGNITION SEQUENCE
5'...A▼CTAGT...3'
3'...TGATC▲A...5'

OPTIMUM REACTION CONDITIONS
pH 7.5; activity is enhanced 4-fold at 50°C

HEAT INACTIVATION
60°C for 15 min.; 65°C for 10-20 min.; 70°C for 30 min.

METHYLATION EFFECTS
Does not cleave m^{6}ACTAGT or Am^{5}CTAGT

NOTES
- EC 3.1.21.4
- A type II site specific deoxyribonuclease
- A large group of enzymes recognizing specific short DNA sequences. They cleave either within the recognition site, or at a short specific distance from it
- The 5'-CTAG extension produced is compatible with cohesive ends from: Avr II, Bln I, Nhe I, Xba I
- Low ionic strength or DMSO solution results in star activity
- Requires Mg^{2+}
- Useful for analysis of bacterial genomes since its recognition site has CTAG, the rarest tetranucleotide in most GC-rich genomes

CONCENTRATION or VOLUME ACTIVITY	UNIT DEFINITION	REACTION BUFFER (As Provided)	ADDITIONAL ACTIVITIES and PURITY	SUPPLIER CATALOG NO.
Sphaerotilus natans				
8-12 U/μL	1 unit completely digests 1 μg Ad2 DNA/hr at 37°C in the buffer provided	Genomic grade; 100 mM Tris-HCl, 100 mM MgCl$_2$, 10 mM DTT, 500 mM NaCl, pH 7.5	>95% DNA fragments ligated and 100% recut	Amersham E 1086Y E 1086Z
10,000 U/mL	1 unit degrades 1 μg Ad2 DNA/hr at 37°C in a 0.05 mL mixture	10 mM Tris-HCl, 100 mM NaCl, 100 μg/mL BSA, pH 8.0	No detectable contaminating nuclease; 90% ligated and 98% recut after 50-fold overdigestion	AGS Heidelb A00605S A00605M
10 U/μL and 40 U/μL	1 unit completely digests 1 μg substrate DNA/hr	50 mM Tris-HCl, 10 mM MgCl$_2$, 100 mM NaCl, 1 mM DTE, pH 7.5		Boehringer 1008943 1008951 1207644
3-15 U/μL	Ad2 DNA substrate, 37°C	500 mM NaCl, 100 mM Tris-HCl, 100 mM MgCl$_2$, 10 mM DTT, pH 7.5	80% ligation and 100% recutting	Nippon Gene 315-01541 311-01543 319-01544

Spe I continued

CONCENTRATION or VOLUME ACTIVITY	UNIT DEFINITION	REACTION BUFFER (As Provided)	ADDITIONAL ACTIVITIES and PURITY	SUPPLIER CATALOG No.
Sphaerotilus natans continued				
10 U/μL	1 unit completely digests 1 μg Ad2 DNA/hr at 37°C in a 50 mL reaction volume	Blue/White Cloning Qualified; genome qualified; 60 mM Tris-HCl, 60 mM MgCl$_2$, 500 mM NaCl, 10 mM DTT, pH 7.5	90% ligation	Promega R6591 R6592 R6595
300 U and 1500 U	1 unit completely digests 1 μg Ad2 DNA/hr at 37°C in 50 μL of supplied buffer	Genomic grade; 100 mM Tris-HCl, 100 mM MgCl$_2$, 10 mM DTT, 500 mM NaCl, pH 7.5	Purity is confirmed by overdigestion, ligation-recutting and genome DNA digestion test	TaKaRa 1086
3-20 U/μL	1 unit completely digests 1 μg DNA/hr at 37°C	10 mM Tris-HCl, 10 mM MgCl$_2$, 50 mM NaCl, 1 mM DTT, pH 7.5	>90% ligation; >95% recut	Toyobo SPE-101 SPE-102
Sphaerotilus species				
400 U and 1000 U	1 unit completely digests 1 μg DNA substrate/hr at 37°C	10 mM Tris-HCl, 10 mM MgCl$_2$, 50 mM NaCl, 1 mM DTT, 0.1 mg/mL BSA, pH 7.5	>90% DNA fragments ligated and recut after 10-fold overdigestion	Adv Biotech AB-0243-a AB-0243-b
10 U/μL; 40 U/μL	1 unit completely cleaves 1 μg substrate DNA/hr	60 mM KCl, 20 mM Tris, 0.75 mM MgCl$_2$, pH 7.5	90-98% cleavage-ligation-recleavage	Am Allied L-30-00200 L-30-01000 H-30-01000
	1 unit completely cleaves 1 μg Ad2 DNA/hr at pH 7.5, 37°C in a reaction volume of 50 μL	10 mM Tris-HCl, 10 mM MgCl$_2$, 50 mM NaCl, 1.0 mM DTT, 100 μg/mL BSA, pH 7.5		CHIMERx 2398-01 2398-02
8-12 U/μL	1 unit completely digests 1 mg Ad2 DNA/hr at 37°C	200 mM Tris-HCl, 50 mM MgCl$_2$, 500 M KCl, pH 7.4	≥95% ligation; 100% cleavage	Life Technol 15443-013 15443-047
3-10 U/μL	1 unit digests completely 1 μg λ DNA substrate/hr at 37°C	50 mM Tris-HCl, 100 mM NaCl, 10 mM MgCl$_2$, 1 mM DTT, pH 7.8		NBL Gene 016601 016603
8000-12,000 U/mL	1 unit completely digests 1 μg Ad2 DNA/hr at pH 7.5, 37°C in a 50 μL total assay mixture	100 mM Tris-acetate, 100 mM MgOAc, 500 mM KOAc, pH 7.5		Pharmacia 27-0875-01
1000-8000 U/mL	1 unit completely cleaves 1.0 μg λ DNA/hr in a 50 μL reaction mixture	100 mM Bis Tris-Propane-HCl, 100 mM MgCl$_2$, 10 mM DTT, pH 7.0	100% cleavage, 90% ligation, 90% recleavage, 10% endonuclease (nickase) No detectable endonuclease (overdigestion)	Sigma R5257
1000-8000 U/mL		50 mM KCl, 10 mM Tris-HCl, 0.1 mM EDTA, 1 mM DTT, 200 μg/mL BSA, 50% glycerol, pH 7.4		Stratagene 501050 501051

Spe I continued

CONCENTRATION or VOLUME ACTIVITY	UNIT DEFINITION	REACTION BUFFER (As Provided)	ADDITIONAL ACTIVITIES and PURITY	SUPPLIER CATALOG NO.
Sphaerotilus species ATCC 13923				
1000-10,000 U/mL		50 mM KCl, 10 mM Tris-HCl, 0.1 mM EDTA, 1 mM DTT, 200 µg/mL BSA, 50% glycerol, pH 7.4		ICN 153828
3000 U/mL; 15,000 U/mL	1 unit completely digests 1 µg Ad2 DNA/hr in a total reaction volume of 50 µL using the buffer provided	10 mM Tris-HCl, 10 mM MgCl₂, 50 mM NaCl, 1 mM DTT, pH 7.9; add 100 µg/mL BSA	>90% DNA fragments ligated and recut after 5-fold overdigestion	NE Biolabs 133S 133L

Sph I

RECOGNITION SEQUENCE
   5'...GCATG▼C...3'
   3'...C▲GTACG...5'
OPTIMUM REACTION CONDITIONS
   pH 7.5-7.9
HEAT INACTIVATION
   60°C for 15 min.; 65°C for 10-30 min.
METHYLATION EFFECTS
   Cleaves Ghm^{5}CATGhm^{5}C and GCATGm^{5}C
   Does not cleave GCm^{6}ATGC

NOTES
- EC 3.1.21.4
- A type II site specific deoxyribonuclease
- A large group of enzymes recognizing specific short DNA sequences. They cleave either within the recognition site, or at a short specific distance from it
- Cleavage leaves 3'-CATG extensions compatible with cohesive ends produced by Bbu I, Nla III, Nsp I, Nsp7524 I
- Requires Mg^{2+}

CONCENTRATION or VOLUME ACTIVITY	UNIT DEFINITION	REACTION BUFFER (As Provided)	ADDITIONAL ACTIVITIES and PURITY	SUPPLIER CATALOG NO.
Escherichia coli strain, carrying the cloned gene from Streptomyces phaeochromogenes				
5000 U/mL	1 unit completely digests 1 µg λ DNA/hr in a total reaction volume of 50 µL using the buffer provided	10 mM Tris-HCl, 10 mM MgCl₂, 50 mM NaCl, 1 mM DTT, pH 7.9	>95% DNA fragments ligated and recut after 10-fold overdigestion	NE Biolabs 182S 182L
Streptomyces phaeochromogenes				
200 U and 1000 U	1 unit completely digests 1 µg DNA substrate/hr at 37°C	10 mM Tris-HCl, 10 mM MgCl₂, 50 mM NaCl, 1 mM DTT, 0.1 mg/mL BSA, pH 7.5	>95% DNA fragments ligated and recut after 10-fold overdigestion	Adv Biotech AB-0526-a AB-0526-b

Sph I continued

CONCENTRATION or VOLUME ACTIVITY	UNIT DEFINITION	REACTION BUFFER (As Provided)	ADDITIONAL ACTIVITIES and PURITY	SUPPLIER CATALOG No.
Streptomyces phaeochromogenes continued				
10 U/μL; 40 U/μL	1 unit completely cleaves 1 μg substrate DNA/hr	100 mM NaCl, 50 mM Tris, 5 mM MgCl$_2$, 2 mM β-MSH, pH 7.8	90-98% cleavage-ligation-recleavage	Am Allied L-20-00200 L-20-00500 H-20-00500 L-20-01500 H-20-01500
8-12 U/μL	1 unit completely digests 1 μg λ DNA/hr at 37°C in the buffer provided	500 mM Tris-HCl, 100 mM MgCl$_2$, 10 mM DTT, 1000 mM NaCl, pH 7.5	>95% DNA fragments ligated and 100% recut	Amersham E 1180Y E 1180Z
2-10 U/μL	1 unit degrades 1 μg λ dam$^-$ dcm$^-$ DNA/hr at 37°C in a 0.05 mL mixture	50 mM Tris-HCl, 10 mM MgCl$_2$, 100 mM NaCl, 100 μg/mL BSA, pH 7.5	>90% ligated and recut after 10-fold overdigestion	AGS Heidelb A00610S A00610M
10 U/μL and 40 U/μL	1 unit completely digests 1 μg substrate DNA/hr	10 mM Tris-HCl, 10 mM MgCl$_2$, 50 mM NaCl, 1 mM DTE, pH 7.5		Boehringer 1026950 606120 1026534 1026542
	1 unit completely cleaves 1 μg λ DNA/hr at pH 7.5, 37°C in a reaction volume of 50 μL	10 mM Tris-HCl, 10 mM MgCl$_2$, 50 mM NaCl, 50 mM KCl, 1.0 mM DTT, 100 μg/mL BSA, pH 7.5		CHIMERx 2400-01 2400-02
1000-4000 U/mL		50 mM KCl, 10 mM Tris-HCl, 0.1 mM EDTA, 1.0 mM DTT, 200 μg/mL acetylated BSA, 50% glycerol, pH 7.4		ICN 152071
8-12 U/μL	1 unit completely digests 1 mg λ DNA/hr at 37°C	500 mM Tris-HCl, 60 mM MgCl$_2$, 500 mM NaCl, 50 mM KCl, pH 7.4	≥95% ligation; 100% cleavage	Life Technol 15413-016 15413-040
	10 units does not produce unspecific cleavage products with 1 μg λ DNA/16 hr at 37°C	10 mM Tris-HCl, 10 mM MgCl$_2$, 50 mM NaCl, 1 mM DTT, 100 μg/mL BSA, pH 7.9	>90% DNA fragments ligated and recut after 10-fold overdigestion	MINOTECH 129
5-12 U/μL	1 unit digests completely 1 μg λ DNA substrate/hr at 37°C	6 mM Tris-HCl, 150 mM NaCl, 6 mM MgCl$_2$, 10 mM β-MSH, 100 μg/mL BSA, pH 7.4		NBL Gene 013011 013002
2-8 U/μL	λ DNA substrate, 37°C	1 M NaCl, 500 mM Tris-HCl, 100 mM MgCl$_2$, 10 mM DTT, pH 7.5	95% ligation and 100% recutting	Nippon Gene 310-01091 316-01093 314-01094
5 U/μL	1 unit hydrolyzes 1 μg Ad2 DNA to completion/hr in a 50 μL total reaction volume	100 mM Tris-HCl, 500 mM NaCl, 100 mM MgCl$_2$, 100 mM β-MSH, 1.0 mg/mL BSA, pH 8		Oncor 110421 110422

Sph I continued

CONCENTRATION or VOLUME ACTIVITY	UNIT DEFINITION	REACTION BUFFER (As Provided)	ADDITIONAL ACTIVITIES and PURITY	SUPPLIER CATALOG NO.
Streptomyces phaeochromogenes continued				
1000-5000 U/mL	1 unit completely digests 1 µg λ DNA/hr at pH 7.5, 37°C in a 50 µL total assay mixture	100 mM Tris-acetate, 100 mM MgOAc, 500 mM KOAc, pH 7.5		Pharmacia 27-0951-01 27-0951-02
10 U/µL	1 unit completely digests 1 µg λ DNA/hr at 37°C in a 50 mL reaction volume	Blue/White Cloning Qualified; 100 mM Tris-HCl, 100 mM MgCl$_2$, 1.5 M KCl, pH 7.4	90% ligation	Promega R6261 R6262 R6265
5000 U/mL		Blue/white certified; 10 mM Tris-HCl, 10 mM MgCl$_2$, 50 mM NaCl, 1 mM DTT, pH 7.6	90% DNA fragments ligated and recut after 5-fold overdigestion	SibEnzyme E129 E130
2000-20,000 U/mL	1 unit completely cleaves 1.0 µg λ DNA/hr in a 50 µL reaction mixture	Unique buffer	100% cleavage, 90% ligation, 95% recleavage, 50% endonuclease (nickase) No detectable endonuclease (overdigestion)	Sigma R7135
1000-4000 U/mL		100 mM KCl, 10 mM Tris-HCl, 0.1 mM EDTA, 1 mM DTT, 400 µg/mL BSA, 50% glycerol, pH 7.4		Stratagene 501060 501061
400 U and 2000 U	1 unit completely digests 1 µg λ DNA/hr at 37°C in 50 µL of supplied buffer	500 mM Tris-HCl, 100 mM MgCl$_2$, 10 mM DTT, 1 M NaCl, pH 7.5	Purity is confirmed by overdigestion, ligation-recutting and cloning test	TaKaRa 1180
3-20 U/µL and 40-50 U/µL	1 unit completely digests 1 µg DNA/hr at 37°C	50 mM Tris-HCl, 10 mM MgCl$_2$, 100 mM NaCl, 1 mM DTT, pH 7.5	>90% ligation; >95% recut	Toyobo SPH-101 SPH-102 SPH-152

Spl I

RECOGNITION SEQUENCE
 5'...C▼GTACG...3'
 3'...GCATG▲C...5'

HEAT INACTIVATION
 Resistant; use phenol extraction

NOTES
- EC 3.1.21.4
- A type II site specific deoxyribonuclease
- A large group of enzymes recognizing specific short DNA sequences. They cleave either within the recognition site, or at a short specific distance from it
- Requires Mg^{2+}

CONCENTRATION or VOLUME ACTIVITY	UNIT DEFINITION	REACTION BUFFER (As Provided)	ADDITIONAL ACTIVITIES and PURITY	SUPPLIER CATALOG NO.
Spirulina platensis				
8-12 U/μL	1 unit completely digests 1 μg λ DNA/hr at 55°C in the buffer provided	Genomic grade; 500 mM Tris-HCl, 100 mM $MgCl_2$, 10 mM DTT, 1000 mM NaCl, BSA, pH 7.5	>95% DNA fragments ligated and 100% recut	Amersham E 1087Y E 1087Z
300 U and 1500 U	1 unit completely digests 1 μg λ DNA/hr at 55°C in 50 μL of supplied buffer and BSA	500 mM Tris-HCl, 100 mM $MgCl_2$, 10 mM DTT, 1 M NaCl, BSA, pH 7.5	Purity is confirmed by overdigestion, ligation-recutting, genome DNA digestion and cloning test	TaKaRa 1087

Spl II

RECOGNITION SEQUENCE
5'...C▼GTACG...3'
3'...GCATG▲C...5'

METHYLATION EFFECTS
Cleaves CGTm^{6}ACG

NOTES
- EC 3.1.21.4
- A type II site specific deoxyribonuclease
- A large group of enzymes recognizing specific short DNA sequences. They cleave either within the recognition site, or at a short specific distance from it
- Requires Mg^{2+} and BSA for full activity

CONCENTRATION or VOLUME ACTIVITY	UNIT DEFINITION	REACTION BUFFER (As Provided)	ADDITIONAL ACTIVITIES and PURITY	SUPPLIER CATALOG NO.
Spirilina platensis				
30 U and 1500 U	1 unit completely cleaves 1 μg DNA/hr in 50 μL reaction mixture	500 mM Tris-HCl, 100 mM MgCl$_2$, 10 mM DTT, 1 M NaCl, 0.1% BSA, pH 7.5		PanVera TAK 1087

Srf I

RECOGNITION SEQUENCE
5'...GCCC▼GGGC...3'
3'...CGGG▲CCCG...5'

NOTES
- EC 3.1.21.4
- A type II site specific deoxyribonuclease
- A large group of enzymes recognizing specific short DNA sequences. They cleave either within the recognition site, or at a short specific distance from it
- Blunt end ligations are enhanced over 16 hr by using decreased ATP and 15% PEG
- Requires Mg^{2+}

Srf I continued

CONCENTRATION or VOLUME ACTIVITY	UNIT DEFINITION	REACTION BUFFER (As Provided)	ADDITIONAL ACTIVITIES and PURITY	SUPPLIER CATALOG NO.
Streptomyces species				
5000–15,000 U/mL		200 mM KCl, 10 mM Tris-HCl, 1 mM EDTA, 10 mM β-MSH, 200 µg/mL BSA, 0.1% Triton X-100, 50% glycerol, pH 7.7		Stratagene 501064

Sse9 I

RECOGNITION SEQUENCE
5'...▼AATT...3'
3'...TTAA▲...5'

HEAT INACTIVATION
65°C for 20 min.

NOTES
- EC 3.1.21.4
- A type II site specific deoxyribonuclease
- A large group of enzymes recognizing specific short DNA sequences. They cleave either within the recognition site, or at a short specific distance from it
- Requires Mg^{2+}

CONCENTRATION or VOLUME ACTIVITY	UNIT DEFINITION	REACTION BUFFER (As Provided)	ADDITIONAL ACTIVITIES and PURITY	SUPPLIER CATALOG NO.
Sporosarcina species 9				
2000 U/mL	1 unit degrades 1 µg λ dam⁻ dcm⁻ DNA/hr at 37°C in a 0.05 mL mixture	10 mM Tris-HCl, 10 mM MgCl₂, 100 mM KCl, 100 µg/mL BSA, pH 8.5	No detectable contaminating nuclease; 95% ligated and recut after 5-fold overdigestion	ACS Heidelb F00613S F00613M
5000 U/mL		10 mM Tris-HCl, 10 mM MgCl₂, 1 mM DTT, pH 7.6	90% DNA fragments ligated and recut after 5-fold overdigestion	SibEnzyme E217 E218

Sse8387 I

RECOGNITION SEQUENCE
5'...CCTGCA▼GG...3'
3'...GG▲ACGTCC...5'

HEAT INACTIVATION
60°C for 15 min.

NOTES
- EC 3.1.21.4
- A type II site specific deoxyribonuclease
- A large group of enzymes recognizing specific short DNA sequences. They cleave either within the recognition site, or at a short specific distance from it
- Requires Mg^{2+} and BSA for full activity

CONCENTRATION or VOLUME ACTIVITY	UNIT DEFINITION	REACTION BUFFER (As Provided)	ADDITIONAL ACTIVITIES and PURITY	SUPPLIER CATALOG No.
Streptomyces species				
8-12 U/μL	1 unit completely digests 1 μg λ DNA/hr at 37°C in the buffer provided	Genomic grade; 100 mM Tris-HCl, 100 mM MgCl₂, 10 mM DTT, 500 mM NaCl, BSA, pH 7.5	>90% DNA fragments ligated and 100% recut	Amersham E 1183Y E 1183Z
Streptomyces species (TaKaRa strain 8387)				
400 U and 2000 U	1 unit completely cleaves 1 μg DNA/hr in 50 μL reaction mixture	100 mM Tris-HCl, 100 mM MgCl₂, 10 mM DTT, 500 mM NaCl, 0.01% BSA, pH 7.5		PanVera TAK 1183
400 U and 2000 U	1 unit completely digests 1 μg λ DNA/hr at 37°C in 50 μL of supplied buffer	Genomic grade; 100 mM Tris-HCl, 100 mM MgCl₂, 10 mM DTT, 500 mM NaCl, pH 7.5	Purity is confirmed by overdigestion, ligation-recutting, genome DNA digestion and cloning test	TaKaRa 1183

SseB I

RECOGNITION SEQUENCE
  5'...AGG▼CCT...3'
  3'...TCC▲GGA...5'

ISOSCHIZOMERS
  Stu I

OPTIMUM REACTION CONDITIONS
  pH 7.9

METHYLATION EFFECTS
  Cleaves m^{6}AATATT

NOTES
- EC 3.1.21.4
- A type II site specific deoxyribonuclease
- A large group of enzymes recognizing specific short DNA sequences. They cleave either within the recognition site, or at a short specific distance from it
- Requires Mg^{2+}

CONCENTRATION or VOLUME ACTIVITY	UNIT DEFINITION	REACTION BUFFER (As Provided)	ADDITIONAL ACTIVITIES and PURITY	SUPPLIER CATALOG NO.
Streptomyces species				
	50 units does not produce unspecific cleavage products with 1 μg λ DNA/*Hind* III digest/16 hr at 37°C	50 m*M* Tris-HCl, 10 m*M* MgCl$_2$, 100 m*M* NaCl, 1 m*M* DTT, 100 μg/mL BSA, pH 7.9	>90% DNA fragments ligated and recut after 10-fold overdigestion	MINOTECH 140

Ssp I

RECOGNITION SEQUENCE
  5'...AAT▼ATT...3'
  3'...TTA▲TAA...5'

OPTIMUM REACTION CONDITIONS
  pH 7.4-7.9

HEAT INACTIVATION
  60°C for 15 min.; 65°C for 10-30 min.; 80°C for 5 min.

NOTES
- EC 3.1.21.4
- A type II site specific deoxyribonuclease
- A large group of enzymes recognizing specific short DNA sequences. They cleave either within the recognition site, or at a short specific distance from it
- Low ionic strength, high glycerol or enzyme concentration, or pH >8.0 results in star activity
- Compatible cohesive ends: any blunt end
- Requires Mg^{2+}
- Inhibited by NaCl >50 mM
- AT-rich recognition sequence makes it useful for analysis of GC-rich bacterial genomes

CONCENTRATION or VOLUME ACTIVITY	UNIT DEFINITION	REACTION BUFFER (As Provided)	ADDITIONAL ACTIVITIES and PURITY	SUPPLIER CATALOG NO.
Escherichia coli strain, carrying the cloned gene from Sphaerotilus species ATCC 13925				
5000 U/mL	1 unit completely digests 1 μg λ DNA/hr in a total reaction volume of 50 μL using the buffer provided	100 mM Tris-HCl, 10 mM MgCl₂, 50 mM NaCl, 0.025% Triton X-100, pH 7.5	>95% DNA fragments ligated and recut after 10-fold overdigestion	NE Biolabs 132S 132L
Sphaerotilus natans				
8-12 U/μL	1 unit completely digests 1 μg λ DNA/hr at 37°C in the buffer provided	Genomic grade; 200 mM Tris-HCl, 100 mM MgCl₂, 10 mM DTT, 1000 mM KCl, pH 8.5	90% DNA fragments ligated and 100% recut	Amersham E 1185Y E 1185Z
10 U/μL and 40-80 U/μL	1 unit completely digests 1 μg λ DNA/hr at 37°C in a 50 mL reaction volume	Genome qualified; 60 mM Tris-HCl, 60 mM MgCl₂, 1 M NaCl, 10 mM DTT, pH 7.5	90% ligation	Promega R6601 R6602 R4604
500 U and 2500 U	1 unit completely digests 1 μg λ DNA/hr at 37°C in 50 μL of supplied buffer	200 mM Tris-HCl, 100 mM MgCl₂, 10 mM DTT, 1 M KCl, pH 8.5	Purity is confirmed by overdigestion, ligation-recutting and genome DNA digestion test	TaKaRa 1185
Sphaerotilus natans ATCC 13925				
2-8 U/μL	λ DNA substrate, 37°C	100 mM NaCl, 100 mM Tris-HCl, 50 mM MgCl₂, 10 mM β-MSH, pH 8.5	90% ligation and 95% recutting	Nippon Gene 313-01101 311-01102 315-01105

Ssp I continued

CONCENTRATION or VOLUME ACTIVITY	UNIT DEFINITION	REACTION BUFFER (As Provided)	ADDITIONAL ACTIVITIES and PURITY	SUPPLIER CATALOG NO.
Sphaerotilus species				
10,000 U/mL	1 unit degrades 1 μg λ dam⁻ dcm⁻ DNA/hr at 37°C in a 0.05 mL mixture	10 mM Tris-HCl, 100 mM NaCl, 100 μg/mL BSA, 1 mM DTT, pH 8.0	No detectable contaminating nuclease; 95% ligated and recut after 10-fold overdigestion	AGS Heidelb F00615S F00615M
10 U/μL and 40 U/μL	1 unit completely digests 1 μg substrate DNA/hr	50 mM Tris-HCl, 10 mM MgCl₂, 100 mM NaCl, 1 mM DTE, pH 7.5		Boehringer 972967 972975 1207652
8-12 U/μL	1 unit hydrolyzes 1 μg DNA/hr at 37°C in a 50 μL reaction volume using λ phage DNA as substrate	10 mM Tris-HCl, 10 mM MgCl₂, 50 mM NaCl, 0.1 mg/mL BSA, pH 7.5	>90% of the DNA fragments ligated and recut after 10-fold overdigestion	CHIMERx Fermentas ER0771 ER0772
	10 units does not produce unspecific cleavage products with 1 μg λ DNA/16 hr at 37°C	50 mM Tris-HCl, 10 mM MgCl₂, 100 mM NaCl, 1 mM DTT, 100 μg/mL BSA, pH 7.9	>90% DNA fragments ligated and recut after 10-fold overdigestion	MINOTECH 130
8-12 U/μL	1 unit digests completely 1 μg λ DNA substrate/hr at 37°C	50 mM Tris-HCl, 100 mM NaCl, 10 mM MgCl₂, 1 mM DTT, pH 7.8		NBL Gene 016701 016703
8000-12,000 U/mL	1 unit completely digests 1 μg λ DNA/hr at pH 7.5, 37°C in a 50 μL total assay mixture	100 mM Tris-acetate, 100 mM MgOAc, 500 mM KOAc, pH 7.5		Pharmacia 27-0873-01
25,000 U/mL		100 mM Tris-HCl, 10 mM MgCl₂, 50 mM NaCl, 0.025% Triton X-100, pH 7.6	90% DNA fragments ligated and recut after 10-fold overdigestion	SibEnzyme E041 E042
5000-12,000 U/mL		50 mM KCl, 10 mM Tris-HCl, 0.1 mM EDTA, 1 mM DTT, 200 μg/mL BSA, 50% glycerol, pH 7.4		Stratagene 501070
3-20 U/μL	1 unit completely digests 1 μg DNA/hr at 37°C	Ssp I	>90% ligation; >95% recut	Toyobo SSP-101 SSP-102
Sphaerotilus species ATCC 13925				
2000-10,000 U/mL		50 mM KCl, 10 mM Tris-HCl, 0.1 mM EDTA, 1 mM DTT, 200 μg/mL BSA, 50% glycerol, pH 7.4		ICN 153828
8-12 U/μL	1 unit completely digests 1 mg λ DNA/hr at 37°C	500 mM Tris-HCl, 60 mM MgCl₂, 500 mM NaCl, 500 mM KCl, pH 7.4	≥85% ligation; 100% cleavage	Life Technol 15458-011 15458-037

SspB I

RECOGNITION SEQUENCE
5'...T▼GTACA...3'
3'...ACATG▲T...5'

ISOSCHIZOMERS
Bsp1407 I, BsrG I

NOTES
- EC 3.1.21.4
- A type II site specific deoxyribonuclease
- A large group of enzymes recognizing specific short DNA sequences. They cleave either within the recognition site, or at a short specific distance from it
- Requires Mg^{2+}

CONCENTRATION or VOLUME ACTIVITY	UNIT DEFINITION	REACTION BUFFER (As Provided)	ADDITIONAL ACTIVITIES and PURITY	SUPPLIER CATALOG No.
Streptomyces species				
				Boehringer

Sst I

RECOGNITION SEQUENCE
  5'...GAGCT▼C...3'
  3'...C▲TCGAG...5'

ISOSCHIZOMERS
  Sac I

OPTIMUM REACTION CONDITIONS
  pH 7.9

HEAT INACTIVATION
  65°C for 10 min.

METHYLATION EFFECTS
  Does not cleave DNA when 5'-residue is 5-methylcytosine

NOTES
- EC 3.1.21.4
- A type II site specific deoxyribonuclease
- A large group of enzymes recognizing specific short DNA sequences. They cleave either within the recognition site, or at a short specific distance from it
- Recognition sequence specificity is altered by glycerol >5%
- Requires Mg^{2+}

CONCENTRATION or VOLUME ACTIVITY	UNIT DEFINITION	REACTION BUFFER (As Provided)	ADDITIONAL ACTIVITIES and PURITY	SUPPLIER CATALOG No.
Streptomyces stanford				
8-12 U/µL	1 unit completely digests 1 mg Ad2 DNA/hr at 37°C	500 mM Tris-HCl, 100 mM MgCl$_2$, 500 mM NaCl, pH 8.0	≥95% ligation; 100% cleavage	Life Technol 15222-011 15222-037 15222-029
	10 units does not produce unspecific cleavage products with 1 µg λ DNA/Hind III digest/16 hr at 37°C	10 mM Tris-HCl, 10 mM MgCl$_2$, 1 mM DTT, 100 µg/mL BSA, pH 7.9	>90% DNA fragments ligated and recut after 10-fold overdigestion	MINOTECH 131
5000-25,000 U/mL	1 unit completely cleaves 1.0 µg λ DNA/hr in a 50 µL reaction mixture	100 mM Tris-HCl, 500 mM NaCl, 100 mM MgCl$_2$, 10 mM DTT, pH 7.9	100% cleavage, 98% ligation, 100% recleavage, <1% endonuclease (nickase) No detectable endonuclease (overdigestion)	Sigma R4881

Sst II

RECOGNITION SEQUENCE
  5'...CCGC▼GG...3'
  3'...GG▲CGCC...5'

ISOSCHIZOMERS
  Sac II

NOTES
- EC 3.1.21.4
- A type II site specific deoxyribonuclease
- A large group of enzymes recognizing specific short DNA sequences. They cleave either within the recognition site, or at a short specific distance from it
- Recognition sequence specificity is altered by glycerol >5%
- Requires Mg^{2+}

CONCENTRATION or VOLUME ACTIVITY	UNIT DEFINITION	REACTION BUFFER (As Provided)	ADDITIONAL ACTIVITIES and PURITY	SUPPLIER CATALOG NO.
Streptomyces stanford				
8-12 U/µL	1 unit completely digests 1 mg Ad2 DNA/hr at 37°C	500 mM Tris-HCl, 100 mM $MgCl_2$, 500 mM NaCl, pH 8.0	≥95% ligation; 100% cleavage	Life Technol 45230-018
5000-25,000 U/mL	1 unit completely cleaves 1.0 µg λ DNA/hr in a 50 µL reaction mixture	100 mM Tris-HCl, 500 mM NaCl, 100 mM $MgCl_2$, 10 mM DTT, pH 7.9	100% cleavage, 95% ligation, 99% recleavage, 10% endonuclease (nickase) No detectable endonuclease (overdigestion)	Sigma R1760

Stu I

RECOGNITION SEQUENCE
   5′...AGG▼CCT...3′
   3′...TCC▲GGA...5′
ISOSCHIZOMERS
   Aat I
OPTIMUM REACTION CONDITIONS
HEAT INACTIVATION
   60°C for 15 min.; 65°C for 10-20 min.
METHYLATION EFFECTS
   Does not cleave AGGm^{5}CCT, AGGCm^{4}CT or AGGCm^{5}CT
   Blocked by overlapping *dcm* methylation

NOTES
- EC 3.1.21.4
- A type II site specific deoxyribonuclease
- A large group of enzymes recognizing specific short DNA sequences. They cleave either within the recognition site, or at a short specific distance from it
- Activity is 50% inhibited by 0.1 M NaCl
- Compatible cohesive ends: any blunt end
- Requires Mg^{2+}

CONCENTRATION or VOLUME ACTIVITY	UNIT DEFINITION	REACTION BUFFER (As Provided)	ADDITIONAL ACTIVITIES and PURITY	SUPPLIER CATALOG NO.
Streptomyces tubercidicus				
1750 U and 5000 U	1 unit completely digests 1 µg DNA substrate/hr at 37°C	10 mM Tris-HCl, 10 mM MgCl$_2$, 50 mM NaCl, 1 mM DTT, 0.1 mg/mL BSA, pH 7.5	>90% DNA fragments ligated and recut after 10-fold overdigestion	Adv Biotech AB-0249-a AB-0249-b
8-12 U/µL	1 unit completely digests 1 µg λ DNA/hr at 37°C in the buffer provided	100 mM Tris-HCl, 100 mM MgCl$_2$, 10 mM DTT, 500 mM NaCl, pH 7.5	>95% DNA fragments ligated and 100% recut	Amersham E 1088Y
10 U/µL and 40 U/µL	1 unit completely digests 1 µg substrate DNA/hr	10 mM Tris-HCl, 5 mM MgCl$_2$, 100 mM NaCl, 1 mM β-MSH, pH 8.0		Boehringer 753351 753360 1047680
	1 unit completely cleaves 1 µg unmethylated λ DNA/hr at pH 7.5, 37°C in a reaction volume of 50 µL	10 mM Tris-HCl, 10 mM MgCl$_2$, 50 mM NaCl, 1.0 mM DTT, 100 µg/mL BSA, pH 7.5		CHIMERX 2405-01 2405-02
8-12 U/µL	1 unit completely digests 1 mg λ DNA/hr at 37°C	500 mM Tris-HCl, 100 mM MgCl$_2$, 500 mM NaCl, pH 8.0	≥55% ligation; 90% cleavage	Life Technol 15438-013 15438-047
8-12 U/µL	1 unit digests completely 1 µg λ DNA substrate/hr at 37°C	10 mM Tris-HCl, 100 mM NaCl, 5 mM MgCl$_2$, 1 mM β-MSH, pH 8.3		NBL Gene 014702 014705

Stu I continued

CONCENTRATION or VOLUME ACTIVITY	UNIT DEFINITION	REACTION BUFFER (As Provided)	ADDITIONAL ACTIVITIES and PURITY	SUPPLIER CATALOG NO.
Streptomyces tubercidicus continued				
5-20 U/μL	λ (dcm⁻) DNA substrate, 37°C	500 mM NaCl, 100 mM Tris-HCl, 100 mM MgCl$_2$, 10 mM DTT, pH 7.5 (10X buffer)	80% ligation and 100% recutting	Nippon Gene 315-00365 311-00367 319-00368
5000-10,000 U/mL	1 unit completely digests 1 μg λ DNA/hr at pH 7.5, 37°C in a 50 μL total assay mixture	100 mM Tris-acetate, 100 mM MgOAc, 500 mM KOAc, pH 7.5		Pharmacia 27-0973-01 27-0973-02
10 U/μL	1 unit completely digests 1 μg λ DNA/hr at 37°C in a 50 mL reaction volume	60 mM Tris-HCl, 60 mM MgCl$_2$, 500 mM NaCl, 10 mM DTT, pH 7.5	90% ligation	Promega R6421 R6422
2000-25,000 U/mL	1 unit completely cleaves 1.0 μg λ DNA/hr in a 50 μL reaction mixture	100 mM Tris-HCl, 500 mM NaCl, 100 mM MgCl$_2$, 10 mM DTT, BSA, pH 7.9	100% cleavage, 90% ligation, 90% recleavage, 10% endonuclease (nickase) No detectable endonuclease (overdigestion)	Sigma R8013
4000-12,000 U/mL		50 mM KCl, 10 mM Tris-HCl, 0.1 mM EDTA, 1 mM DTT, 200 μg/mL BSA, 50% glycerol, pH 7.4		Stratagene 501080
500 U and 2500 U	1 unit completely digests 1 μg λ DNA/hr at 37°C in 50 μL of supplied buffer	100 mM Tris-HCl, 100 mM MgCl$_2$, 10 mM DTT, 500 mM NaCl, pH 7.5	Purity is confirmed by overdigestion, ligation-recutting and cloning test	TaKaRa 1088
Streptomyces tubercidicus (H. Takahashi)				
4000-25,000 U/mL		50 mM KCl, 10 mM Tris-HCl, 0.1 mM EDTA, 1 mM DTT, 200 μg/mL BSA, 50% glycerol, pH 7.4		ICN 153830
10,000 U/mL	1 unit completely digests 1 μg λ DNA/hr in a total reaction volume of 50 μL using the buffer provided	10 mM Tris-HCl, 10 mM MgCl$_2$, 50 mM NaCl, 1 mM DTT, pH 7.9	>95% DNA fragments ligated and recut after 50-fold overdigestion	NE Biolabs 187S 187L

Sty I

RECOGNITION SEQUENCE
$$5'...C\blacktriangledown C(^A_T)(^A_T)GG...3'$$
$$3'...GG(^T_A)(^T_A)C\blacktriangle C...5'$$

ISOSCHIZOMERS
*Eco*T14 I

OPTIMUM REACTION CONDITIONS
pH 7.9

HEAT INACTIVATION
65°C for 10-30 min.

NOTES
- EC 3.1.21.4
- A type II site specific deoxyribonuclease
- A large group of enzymes recognizing specific short DNA sequences. They cleave either within the recognition site, or at a short specific distance from it
- Inhibited by glycerol concentration >5%
- Requires Mg^{2+}

CONCENTRATION or VOLUME ACTIVITY	UNIT DEFINITION	REACTION BUFFER (As Provided)	ADDITIONAL ACTIVITIES and PURITY	SUPPLIER CATALOG NO.
Escherichia coli KM201 (pST27hsd⁺S-a)				
10 U/μL	1 unit completely digests 1 μg substrate DNA/hr	50 mM Tris-HCl, 10 mM $MgCl_2$, 100 mM NaCl, 1 mM DTE, pH 7.5		Boehringer 1047744 1047752
Escherichia coli strain, carrying pST27				
5000-20,000 U/mL	1 unit completely cleaves 1.0 μg λ DNA/hr in a 50 μL reaction mixture	500 mM Tris-HCl, 1 M NaCl, 100 mM $MgCl_2$, 10 mM DTT, BSA, pH 7.9	100% cleavage, 100% ligation, 98% recleavage, 40% endonuclease (nickase) No detectable endonuclease (overdigestion)	Sigma R7009
Escherichia coli WA921/pST27 hsd⁺				
	10 units does not produce unspecific cleavage products with 1 μg λ DNA/16 hr at 37°C	50 mM Tris-HCl, 10 mM $MgCl_2$, 100 mM NaCl, 1 mM DTT, 100 μg/mL BSA, pH 7.9	>90% DNA fragments ligated and recut after 10-fold overdigestion	MINOTECH 132
10,000 U/mL	1 unit completely digests 1 μg λ DNA/hr in a total reaction volume of 50 μL using the buffer provided	50 mM Tris-HCl, 10 mM $MgCl_2$, 100 mM NaCl, 1 mM DTT, pH 7.9; add 100 μg/mL BSA	>95% DNA fragments ligated and recut after 50-fold overdigestion	NE Biolabs 500S 500L
5-20 U/μL	λ DNA substrate, 37°C	1 M NaCl, 500 mM Tris-HCl, 100 mM $MgCl_2$, 10 mM DTT, pH 7.5	95% ligation and 100% recutting	Nippon Gene 315-00941 311-00943 319-00944
10 U/μL	1 unit completely digests 1 μg λ DNA/hr at 37°C in a 50 mL reaction volume	Blue/White Cloning Qualified; 100 mM Tris-HCl, 100 mM $MgCl_2$, 1 M NaCl, 10 mM DTT, pH 8.5	90% ligation	Promega R6481 R6482

Sty I continued

CONCENTRATION or VOLUME ACTIVITY	UNIT DEFINITION	REACTION BUFFER (As Provided)	ADDITIONAL ACTIVITIES and PURITY	SUPPLIER CATALOG NO.
Salmonella typhi				
5000-20,000 U/mL		50 mM KCl, 10 mM Tris-HCl, 0.1 mM EDTA, 1 mM DTT, 200 µg/mL BSA, 50% glycerol, pH 7.4		Stratagene 501090
Salmonella typhi (K. Mise)				
8-12 U/µL	1 unit completely digests 1 mg λ DNA/hr at 37°C	500 mM Tris-HCl, 100 mM MgCl$_2$, 1 M NaCl, pH 8.0	≥65% ligation; 100% cleavage	Life Technol 15442-015
Salmonella typhi 27 (E.S. Anderson)				
5000-30,000 U/mL		50 mM KCl, 10 mM Tris-HCl, 0.1 mM EDTA, 1 mM DTT, 200 µg/mL BSA, 50% glycerol, pH 7.4		ICN 153831

Sun I

RECOGNITION SEQUENCE

5'...C▼GTACG...3'
3'...GCATG▲C...5'

NOTES
- EC 3.1.21.4
- A type II site specific deoxyribonuclease
- A large group of enzymes recognizing specific short DNA sequences. They cleave either within the recognition site, or at a short specific distance from it
- Requires Mg^{2+}

CONCENTRATION or VOLUME ACTIVITY	UNIT DEFINITION	REACTION BUFFER (As Provided)	ADDITIONAL ACTIVITIES and PURITY	SUPPLIER CATALOG NO.
Synechococcus uniformis				
8-12 U/µL	1 unit completely digests 1 mg ΦX174 DNA/hr at 55°C	200 mM Tris-HCl, 50 mM MgCl$_2$, 500 M KCl, pH 7.4	≥85% ligation; 100% cleavage	Life Technol 15480-015

Swa I

RECOGNITION SEQUENCE
  5'...ATTT▼AAAT...3'
  3'...TAAA▲TTTA...5'

NOTES
- EC 3.1.21.4
- A type II site specific deoxyribonuclease
- A large group of enzymes recognizing specific short DNA sequences. They cleave either within the recognition site, or at a short specific distance from it
- Requires Mg^{2+}

CONCENTRATION or VOLUME ACTIVITY	UNIT DEFINITION	REACTION BUFFER (As Provided)	ADDITIONAL ACTIVITIES and PURITY	SUPPLIER CATALOG NO.
Staphylococcus warneri				
10 U/µL and 40 U/µL	1 unit completely digests 1 µg substrate DNA/hr	50 mM Tris-HCl, 10 mM MgCl₂, 100 mM NaCl, 1 mM DTE, pH 7.5		Boehringer 1371517 1371525 1371533

Tai I

RECOGNITION SEQUENCE
  5'...ACGT▼...3'
  3'...▲TGCA...5'
OPTIMUM REACTION CONDITIONS
  <10% activity at 37°C
METHYLATION EFFECTS
  Blocked by CG methylation

NOTES
- EC 3.1.21.4
- A type II site specific deoxyribonuclease
- A large group of enzymes recognizing specific short DNA sequences. They cleave either within the recognition site, or at a short specific distance from it
- Requires Mg^{2+}

Tai I continued

CONCENTRATION or VOLUME ACTIVITY	UNIT DEFINITION	REACTION BUFFER (As Provided)	ADDITIONAL ACTIVITIES and PURITY	SUPPLIER CATALOG NO.
Thermus aquaticus Cc1-331				
4-8 U/μL	1 unit hydrolyzes 1 μg DNA/hr at 37°C in a 50 μL reaction volume using λ phage DNA as substrate	10 mM Tris-HCl, 100 mM KCl, 10 mM $MgCl_2$, pH 8.5	>90% DNA fragments ligated and recut after 50-fold overdigestion	Fermentas ER1141 ER1142
				NE Biolabs

Taq I

RECOGNITION SEQUENCE
  5'...T▼CGA...3'
  3'...AGC▲T...5'
ISOSCHIZOMERS
  *Tth*HB8 I
OPTIMUM REACTION CONDITIONS
  65°C; <10% activity at 37°C
HEAT INACTIVATION
  Resistant; 85°C for 30 min.; 90°C for 5 min.
METHYLATION EFFECTS
  Cleaves Tm^5CGA and Thm^5CGA
  Does not cleave $TCGm^6A$
  Blocked by overlapping *dam* methylation

NOTES
- EC 3.1.21.4
- A type II site specific deoxyribonuclease
- A large group of enzymes recognizing specific short DNA sequences. They cleave either within the recognition site, or at a short specific distance from it
- Cuts single-stranded DNA 100-fold slower than double-stranded
- Partially cleaves DNA from *dam*+ *E. coli*
- Compatible cohesive ends: *Acc* I, *Acy* I, *Aha* II, *Asu* II, *Ban* III, *Bbi* II, *Bst*B I, *Cla* I, *Csp*45 I, *Fnu*4H I, *Hap* II, *Hin*1 I, *Hin*P I, *Hpa* II, *Mae* II, *Msp* I, *Nar* I, *Nsp* V, *Sfu* I, *Tth*HB8 I
- Requires Mg^{2+} and BSA for full activity

CONCENTRATION or VOLUME ACTIVITY	UNIT DEFINITION	REACTION BUFFER (As Provided)	ADDITIONAL ACTIVITIES and PURITY	SUPPLIER CATALOG NO.
Escherichia coli strain, carrying a *Taq* I overproducing plasmid				
5000-50,000 U/mL	1 unit completely cleaves 1.0 μg λ DNA/hr in a 50 μL reaction mixture	Unique buffer + BSA	≥95% cleavage, 100% ligation, 100% recleavage No detectable endonuclease (overdigestion)	Sigma R9507

Taq I continued

CONCENTRATION or VOLUME ACTIVITY	UNIT DEFINITION	REACTION BUFFER (As Provided)	ADDITIONAL ACTIVITIES and PURITY	SUPPLIER CATALOG NO.
Escherichia coli strain, carrying a *Taq*[a] I overproducing plasmid (F. Barany using an NEB clone)				
20,000 U/mL; 100,000 U/mL	1 unit completely digests 1 µg λ DNA/hr in a total reaction volume of 50 µL using the buffer provided	10 mM Tris-HCl, 10 mM MgCl$_2$, 100 mM NaCl, pH 8.4; add 100 µg/mL BSA	>95% DNA fragments ligated and recut after 30-fold overdigestion	NE Biolabs 149S 149L
Obtained from a cloned source				
10,000–40,000 U/mL		300 mM KCl, 10 mM Tris-HCl, 0.1 mM EDTA, 1 mM DTT, 200 µg/mL BSA, 50% glycerol, pH 7.4		Stratagene 501100 501101
Thermus aquaticus				
>40 U/µL	1 unit completely digests 1 µg dam⁻ λ DNA/hr at 65°C in the buffer provided	330 mM Tris-acetate, 100 mM MgOAc, 5 mM DTT, 660 mM KOAc, pH 7.9	>99% DNA fragments ligated and recut	Amersham E 0103XH
	1 unit completely cleaves 1 µg unmethylated λ DNA/hr at pH 7.5, 70°C in a reaction volume of 50 µL	10 mM Tris-HCl, 10 mM MgCl$_2$, 100 mM NaCl, 0.02% Triton X-100, 1.0 mM DTT, 100 µg/mL BSA, pH 8.5, 25°C		CHIMERX 2410-01 2410-02
5000–50,000 U/mL		300 mM KCl, 10 mM Tris-HCl, 0.1 mM EDTA, 1.0 mM DTT, 200 µg/mL BSA, 50% glycerol, pH 7.4		ICN 152095
10 U/µL and 15-50 U/µL	1 unit hydrolyzes 1 µg λ DNA to completion/hr in a 50 µL total reaction volume	330 mM Tris-acetate, 660 mM KOAc, 100 mM MgOAc, 5 mM DTT, 1.0 mg/mL BSA, pH 8		Oncor 110445 110441 110442
5000–10,000 U/mL and 50,000–80,000 U/mL	1 unit completely digests 1 µg λ DNA/hr at pH 7.5, 65°C in a 50 µL total assay mixture	100 mM Tris-acetate, 100 mM MgOAc, 500 mM KOAc, pH 7.5		Pharmacia 27-0956-01 27-0956-02 27-0956-18
Thermus aquaticus YT-1				
7500 U and 20,000 U	1 unit completely digests 1 µg DNA substrate/hr at 65°C	10 mM Tris-HCl, 10 mM MgCl$_2$, 50 mM NaCl, 1 mM DTT, 0.1 mg/mL BSA, pH 7.5	>90% DNA fragments ligated and recut after 10-fold overdigestion	Adv Biotech AB-0244-a AB-0244-b
10,000 U/mL	1 unit degrades 1 µg λ dam⁻ dcm⁻ DNA/hr at 70°C in a 0.05 mL mixture	33 mM Tris-acetate, 10 mM MgOAc, 66 mM KOAc, 100 µg/mL BSA, pH 7.9	No detectable contaminating nuclease; 95% ligated and 100% recut after 10-fold overdigestion	AGS Heidelb B00640S B00640M
10 U/µL and 40 U/µL	1 unit completely digests 1 µg substrate DNA/hr	10 mM Tris-HCl, 5 mM MgCl$_2$, 100 mM NaCl, 1 mM β-MSH, pH 8.0, 37°C		Boehringer 404128 567671 1175114 997471

Taq I continued

CONCENTRATION or VOLUME ACTIVITY	UNIT DEFINITION	REACTION BUFFER (As Provided)	ADDITIONAL ACTIVITIES and PURITY	SUPPLIER CATALOG No.
Thermus aquaticus YT-1 continued				
8-12 U/µL; 40-60 U/µL	1 unit hydrolyzes 1 µg DNA/hr at 37°C in a 50 µL reaction volume using λ phage DNA as substrate	10 mM Tris-HCl, 10 mM MgCl$_2$, 100 mM KCl, 0.1 mg/mL BSA, pH 8.5, 65°C	>90% of the DNA fragments ligated and recut after 50-fold overdigestion	Fermentas ER0671 ER0672 ER0673
8-12 U/µL and 50 U/µL	1 unit completely digests 1 mg ΦX174 DNA/hr at 65°C	500 mM Tris-HCl, 100 mM MgCl$_2$, 500 mM NaCl, pH 8.0	≥95% ligation; 100% cleavage	Life Technol 15218-019 15218-027 15218-035
	10 units does not produce unspecific cleavage products with 1 µg λ DNA/16 hr at 65°C	10 mM Tris-HCl, 10 mM MgCl$_2$, 100 mM NaCl, 10 mM β-MSH, 100 µg/mL BSA, pH 8.4	>90% DNA fragments ligated and recut after 10-fold overdigestion	MINOTECH 133
5-12 U/µL >40 U/µL	1 unit digests completely 1 µg λ DNA substrate/hr at 65°C	10 mM Tris-HCl, 100 mM NaCl, 5 mM MgCl$_2$, 1 mM β-MSH, pH 8.3		NBL Gene 011103 011106 011156
5-20 U/µL 30-100 U/µL	λ (dcm⁻) DNA substrate, 65°C	500 mM KOAc, 200 mM Tris-acetate, 100 mM MgOAc, 10 mM DTT, pH 7.9	90% ligation and 100% recutting	Nippon Gene 318-00372 312-00375 310-00376 316-03031
10 U/µL and 40-80 U/µL	1 unit completely digests 1 µg λ DNA/hr at 65°C in a 50 mL reaction volume	60 mM Tris-HCl, 60 mM MgCl$_2$, 1 M NaCl, 10 mM DTT, pH 7.5	90% ligation	Promega R6151 R6152 R6155 R4154
3-20 U/µL and 40-50 U/µL	1 unit completely digests 1 µg DNA/hr at 65°C	50 mM Tris-HCl, 10 mM MgCl$_2$, 100 mM NaCl, 1 mM DTT, pH 7.5	>90% ligation; >95% recut	Toyobo TAQ-101 TAQ-102 TAQ-152

Tfi I

RECOGNITION SEQUENCE
5'...G▼A(^A_T)TC...3'
3'...CT(^T_A)A▲G...5'

OPTIMUM REACTION CONDITIONS
10% activity at 37°C

HEAT INACTIVATION
Resistant

NOTES
- EC 3.1.21.4
- A type II site specific deoxyribonuclease
- A large group of enzymes recognizing specific short DNA sequences. They cleave either within the recognition site, or at a short specific distance from it
- Low ionic strength, high glycerol or enzyme concentration, or pH >8.0 results in star activity
- Requires Mg^{2+}

CONCENTRATION or VOLUME ACTIVITY	UNIT DEFINITION	REACTION BUFFER (As Provided)	ADDITIONAL ACTIVITIES and PURITY	SUPPLIER CATALOG NO.
Thermus filiformis				
1000-5000 U/mL		50 mM KCl, 10 mM Tris-HCl, 0.1 mM EDTA, 1 mM DTT, 200 µg/mL acetylated BSA, 50% glycerol, pH 7.4		ICN 159470
Thermus filiformis (D. Cowan)				
2000 U/mL	1 unit completely digests 1 µg λ DNA/hr in a total reaction volume of 50 µL using the buffer provided	50 mM Tris-HCl, 10 mM MgCl₂, 100 mM NaCl, 1 mM DTT, pH 7.9	>95% DNA fragments ligated and recut after 5-fold overdigestion	NE Biolabs 546S 546L

Tha I

RECOGNITION SEQUENCE
5'...CG▼CG...3'
3'...GC▲GC...5'

METHYLATION EFFECTS
Does not cleave DNA when either C residue is 5-methylcytosine

NOTES
- EC 3.1.21.4
- A type II site specific deoxyribonuclease
- A large group of enzymes recognizing specific short DNA sequences. They cleave either within the recognition site, or at a short specific distance from it
- Requires Mg^{2+}

CONCENTRATION or VOLUME ACTIVITY	UNIT DEFINITION	REACTION BUFFER (As Provided)	ADDITIONAL ACTIVITIES and PURITY	SUPPLIER CATALOG NO.
Thermoplasma acidophilum				
4-6 U/μL	1 unit completely digests 1 mg λ DNA/hr at 60°C	500 m*M* Tris-HCl, 100 m*M* MgCl$_2$, pH 8.0	≥65% ligation; 100% cleavage	Life Technol 15227-010

Tru1 I

RECOGNITION SEQUENCE
5'...T▼TAA...3'
3'...AAT▲T...5'

NOTES
- EC 3.1.21.4
- A type II site specific deoxyribonuclease
- A large group of enzymes recognizing specific short DNA sequences. They cleave either within the recognition site, or at a short specific distance from it
- Requires Mg^{2+}

CONCENTRATION or VOLUME ACTIVITY	UNIT DEFINITION	REACTION BUFFER (As Provided)	ADDITIONAL ACTIVITIES and PURITY	SUPPLIER CATALOG NO.
Thermus ruber RFL1				
8-12 U/μL	1 unit hydrolyzes 1 μg DNA/hr at 37°C in a 50 μL reaction volume using λ phage DNA as substrate	10 mM Tris-HCl, 10 mM MgCl₂, 100 mM KCl, 0.1 mg/mL BSA, pH 8.5, 65°C	>90% of the DNA fragments ligated and recut after 50-fold overdigestion	Fermentas ER0981 ER0982

Tru9 I

RECOGNITION SEQUENCE
5'...T▼TAA...3'
3'...AAT▲T...5'
ISOSCHIZOMERS
  Mse I
HEAT INACTIVATION
  Resistant

NOTES
- EC 3.1.21.4
- A type II site specific deoxyribonuclease
- A large group of enzymes recognizing specific short DNA sequences. They cleave either within the recognition site, or at a short specific distance from it
- Compatible cohesive ends: *Ase* I, *Asn* I, *Mae* I, *Mse* I, *Nde* I, *Vsp* I
- Requires Mg^{2+}

Tru9 I continued

CONCENTRATION or VOLUME ACTIVITY	UNIT DEFINITION	REACTION BUFFER (As Provided)	ADDITIONAL ACTIVITIES and PURITY	SUPPLIER CATALOG NO.
Thermus ruber 9				
10,000 U/mL	1 unit degrades 1 μg λ dam⁻ dcm⁻ DNA/hr at 65°C in a 0.05 mL mixture	10 mM Tris-HCl, 10 mM MgCl$_2$, 100 mM KCl, 100 μg/mL BSA, pH 8.5	No detectable contaminating nuclease; 95% ligated and recut after 5-fold overdigestion	AGS Heidelb F00642S F00642M
8-12 U/μL and 40 U/μL	1 unit completely digests 1 μg substrate DNA/hr	10 mM Tris-HCl, 10 mM MgCl$_2$, 50 mM NaCl, 1 mM DTE, pH 7.5, 37°C		Boehringer 1464817 1464825 1464833
8000-12,000 U/mL		50 mM NaCl, 10 mM Tris-HCl, 0.1 mM EDTA, 1 mM DTT, 500 μg/mL acetylated BSA, 50% glycerol, pH 7.4		ICN 159471
3-15 U/μL	1 unit digests completely 1 μg λ DNA substrate/hr at 65°C	10 mM Tris-HCl, 50 mM NaCl, 10 mM MgCl$_2$, 1 mM DTT, pH 7.8		NBL Gene 018901 018903
8-12 U/μL	1 unit completely digests 1 μg λ DNA/hr at 65°C in a 50 mL reaction volume	100 mM Tris-HCl, 100 mM MgCl$_2$, 1 M NaCl, 10 mM DTT, pH 7.8	90% ligation	Promega R7011 R7012
20,000 U/mL		10 mM Tris-HCl, 10 mM MgCl$_2$, 100 mM NaCl, 1 mM DTT, pH 8.5	95% DNA fragments ligated and recut after 20-fold overdigestion	SibEnzyme E199 E200
10,000-20,000 U/mL	1 unit completely cleaves 1.0 μg λ DNA/hr in a 50 μL reaction mixture	100 mM Tris-HCl, 500 mM NaCl, 100 mM MgCl$_2$, 10 mM DTT, BSA, pH 7.9, 25°C	100% cleavage, 95% ligation, 95% recleavage No detectable endonuclease (overdigestion)	Sigma R6638

Tsc I

RECOGNITION SEQUENCE
5'...ACGT▼...3'
3'...▲TGCA...5'

NOTES
- EC 3.1.21.4
- A type II site specific deoxyribonuclease
- A large group of enzymes recognizing specific short DNA sequences. They cleave either within the recognition site, or at a short specific distance from it
- Compatible cohesive ends: Aat II
- Requires Mg^{2+}

CONCENTRATION or VOLUME ACTIVITY	UNIT DEFINITION	REACTION BUFFER (As Provided)	ADDITIONAL ACTIVITIES and PURITY	SUPPLIER CATALOG No.
Thermus species 49				
4-10 U/μL	1 unit digests completely 1 μg λ DNA substrate/hr at 65°C	50 mM Tris-HCl, 50 mM NaCl, 10 mM MgCl₂, 1 mM DTT, pH 8.3		NBL Gene 012811 012802

Tse I

RECOGNITION SEQUENCE
5'...G▼CWGC...3'
3'...CGWC▲G...5'

NOTES
- EC 3.1.21.4
- A type II site specific deoxyribonuclease
- A large group of enzymes recognizing specific short DNA sequences. They cleave either within the recognition site, or at a short specific distance from it
- Requires Mg^{2+}

CONCENTRATION or VOLUME ACTIVITY	UNIT DEFINITION	REACTION BUFFER (As Provided)	ADDITIONAL ACTIVITIES and PURITY	SUPPLIER CATALOG No.
Thermus species 93170				
				NE Biolabs

Tsp45 I

RECOGNITION SEQUENCE
5'...▼GT(^C_G)AC...3'
3'...CA(^G_C)TG▲...5'

OPTIMUM REACTION CONDITIONS
10% activity at 37°C

HEAT INACTIVATION
Resistant

NOTES
- EC 3.1.21.4
- A type II site specific deoxyribonuclease
- A large group of enzymes recognizing specific short DNA sequences. They cleave either within the recognition site, or at a short specific distance from it
- Requires Mg^{2+}

CONCENTRATION or VOLUME ACTIVITY	UNIT DEFINITION	REACTION BUFFER (As Provided)	ADDITIONAL ACTIVITIES and PURITY	SUPPLIER CATALOG NO.
Thermus species (R.A.D Williams)				
4000 U/mL	1 unit completely digests 1 µg λ DNA/hr in a total reaction volume of 50 µL using the buffer provided	10 mM Bis Tris-Propane-HCl, 10 mM $MgCl_2$, 1 mM DTT, pH 7.0; add 100 µg/mL BSA	>95% DNA fragments ligated and recut after 4-fold overdigestion	NE Biolabs 583S 583L

Tsp509 I

RECOGNITION SEQUENCE
5'...▼AATT...3'
3'...TTAA▲...5'

OPTIMUM REACTION CONDITIONS
<10% activity at 37°C

HEAT INACTIVATION
Resistant

METHYLATION EFFECTS
Sensitive to methylation by *Eco*R I methylase

NOTES
- EC 3.1.21.4
- A type II site specific deoxyribonuclease
- A large group of enzymes recognizing specific short DNA sequences. They cleave either within the recognition site, or at a short specific distance from it
- Requires Mg^{2+}

CONCENTRATION or VOLUME ACTIVITY	UNIT DEFINITION	REACTION BUFFER (As Provided)	ADDITIONAL ACTIVITIES and PURITY	SUPPLIER CATALOG No.
***Thermus* species**				
4000-20,000 U/mL		50 m*M* KCl, 10 m*M* Tris-HCl, 0.1 m*M* EDTA, 1 m*M* DTT, 200 µg/mL acetylated BSA, 50% glycerol, pH 7.4		ICN 159472
5000 U/mL	1 unit completely digests 1 µg λ DNA/hr in a total reaction volume of 50 µL using the buffer provided	10 m*M* Bis Tris-Propane-HCl, 10 m*M* MgCl$_2$, 1 m*M* DTT, pH 7.0	>95% DNA fragments ligated and recut after 10-fold overdigestion	NE Biolabs 576S 576L
***Thermus* species E**				
4-10 U/µL	1 unit digests completely 1 µg λ DNA substrate/hr at 65°C			NBL Gene 018811 018802
				Toyobo

TspR I

RECOGNITION SEQUENCE
5'...NNCAGTGNN▼...3'
3'...▲NNGTCACNN...5'

OPTIMUM REACTION CONDITIONS
10% activity at 37°C

HEAT INACTIVATION
Resistant

NOTES
- EC 3.1.21.4
- A type II site specific deoxyribonuclease
- A large group of enzymes recognizing specific short DNA sequences. They cleave either within the recognition site, or at a short specific distance from it
- Produces DNA fragments with a 9 base 3'-extension
- Requires Mg^{2+}

CONCENTRATION or VOLUME ACTIVITY	UNIT DEFINITION	REACTION BUFFER (As Provided)	ADDITIONAL ACTIVITIES and PURITY	SUPPLIER CATALOG NO.
Thermus species (R.A.D Williams)				
4000 U/mL	1 unit completely digests 1 μg λ DNA/hr in a total reaction volume of 50 μL using the buffer provided	20 mM Tris-acetate, 10 mM MgOAc, 50 mM KOAc, 1 mM DTT, pH 7.9; add 100 μg/mL BSA	>95% DNA fragments ligated and recut after 10-fold overdigestion	NE Biolabs 582S 582L

Tth111 I

RECOGNITION SEQUENCE
 5'...GACN▼NNGTC...3'
 3'...CTGNN▲NCAG...5'

OPTIMUM REACTION CONDITIONS
 <10% activity at 37°C

HEAT INACTIVATION
 Resistant; use ethanol precipitation

NOTES
- EC 3.1.21.4
- A type II site specific deoxyribonuclease
- A large group of enzymes recognizing specific short DNA sequences. They cleave either within the recognition site, or at a short specific distance from it
- Low ionic strength, high glycerol or enzyme concentration, or pH >8.0 results in star activity
- Produces an ambiguous single base extension which is difficult to ligate
- Requires Mg^{2+}

CONCENTRATION or VOLUME ACTIVITY	UNIT DEFINITION	REACTION BUFFER (As Provided)	ADDITIONAL ACTIVITIES and PURITY	SUPPLIER CATALOG NO.
Thermus thermophilus 111				
8-12 U/μL	1 unit completely digests 1 μg λ DNA/hr at 65°C in the buffer provided	200 m*M* Tris-HCl, 100 m*M* MgCl$_2$, 10 m*M* DTT, 1000 m*M* KCl, pH 8.5	>90% DNA fragments ligated and 100% recut	Amersham E 1090Y
10,000 U/mL	1 unit degrades 1 μg λ dam⁻ dcm⁻ DNA/hr at 65°C in a 0.05 mL mixture	33 m*M* Tris-acetate, 10 m*M* MgOAc, 66 m*M* KOAc, 100 μg/mL BSA, pH 7.9	No detectable contaminating nuclease; 30% ligated and 90% recut after 2-fold overdigestion	AGS Heidelb F00645S F00645M
	1 unit completely cleaves 1 μg *Hind* III fragments of λ DNA/hr at pH 7.5, 65°C in a reaction volume of 50 μL	50 m*M* Tris-HCl, 10 m*M* MgCl$_2$, 100 m*M* NaCl, 1.0 m*M* DTT, 100 μg/mL BSA, pH 7.5		CHIMERx 2420-01 2420-02
8-12 U/μL	1 unit completely digests 1 μg *Eco*R I λ DNA fragments/hr at 65°C in a 50 mL reaction volume	60 m*M* Tris-HCl, 60 m*M* MgCl$_2$, 500 m*M* NaCl, 10 m*M* DTT, pH 7.0	0% ligation	Pharmacia Promega R6841 R6842
5000 U/mL		33 m*M* Tris-acetate, 10 m*M* MgOAc, 66 m*M* KOAc, 1 m*M* DTT, pH 7.9	10% DNA fragments ligated and recut after 2-fold overdigestion	SibEnzyme E097 E098
500 U and 2500 U	1 unit completely digests 1 μg λ DNA/hr at 65°C in 50 μL of supplied buffer	200 m*M* Tris-HCl, 100 m*M* MgCl$_2$, 10 m*M* DTT, 1 *M* KCl, pH 8.5	Purity is confirmed by overdigestion and ligation-recutting test	TaKaRa 1090

Tth111 I continued

CONCENTRATION or VOLUME ACTIVITY	UNIT DEFINITION	REACTION BUFFER (As Provided)	ADDITIONAL ACTIVITIES and PURITY	SUPPLIER CATALOG No.
Thermus thermophilus 111 (T. Oshima)				
2000-20,000 U/mL		200 mM KCl, 10 mM Tris-HCl, 0.1 mM EDTA, 1 mM DTT, 200 µg/mL BSA, 50% glycerol, pH 7.8	No detectable Tth111 II	ICN 153832
4000 U/mL	1 unit completely digests 1 µg λ DNA (Hind III digest)/hr in a total reaction volume of 50 µL using the buffer provided	20 mM Tris-acetate, 10 mM MgOAc, 50 mM KOAc, 1 mM DTT, pH 7.9	10% DNA fragments ligated after 2-fold overdigestion	NE Biolabs 185S 185L

TthHB8 I

RECOGNITION SEQUENCE
    5'...T▼CGA...3'
    3'...AGC▲T...5'
ISOSCHIZOMERS
    Taq I
HEAT INACTIVATION
    Resistant; use phenol extraction

NOTES
- EC 3.1.21.4
- A type II site specific deoxyribonuclease
- A large group of enzymes recognizing specific short DNA sequences. They cleave either within the recognition site, or at a short specific distance from it
- Requires Mg^{2+}

CONCENTRATION or VOLUME ACTIVITY	UNIT DEFINITION	REACTION BUFFER (As Provided)	ADDITIONAL ACTIVITIES and PURITY	SUPPLIER CATALOG No.
Thermus thermophilus HB8				
8-12 U/µL	1 unit completely digests 1 µg dam⁻ λ DNA/hr at 65°C in the buffer provided	500 mM Tris-HCl, 100 mM MgCl₂, 10 mM DTT, 1000 mM NaCl, pH 7.5	>95% DNA fragments ligated and 100% recut	Amersham E 1092Y
500 U and 2500 U	1 unit completely digests 1 µg λ DNA/hr at 65°C in 50 µL of supplied buffer	500 mM Tris-HCl, 100 mM MgCl₂, 10 mM DTT, 1 M NaCl, pH 7.5	Purity is confirmed by overdigestion and ligation-recutting test	TaKaRa 1092

Van91 I

RECOGNITION SEQUENCE
   5'...CCA(N)$_4$▼NTGG...3'
   3'...GGTN▲(N)$_4$ACC...5'

ISOSCHIZOMERS
   PflM I

METHYLATION EFFECTS
   Does not cleave CCANNNCm^{5}CTGG
   Blocked by overlapping dcm methylation

NOTES
- EC 3.1.21.4
- A type II site specific deoxyribonuclease
- A large group of enzymes recognizing specific short DNA sequences. They cleave either within the recognition site, or at a short specific distance from it
- Requires Mg^{2+}

CONCENTRATION or VOLUME ACTIVITY	UNIT DEFINITION	REACTION BUFFER (As Provided)	ADDITIONAL ACTIVITIES and PURITY	SUPPLIER CATALOG No.
Vibrio anguillarum RFL91				
12,000 U/mL	1 unit degrades 1 μg λ dam$^-$ dcm$^-$ DNA/hr at 37°C in a 0.05 mL mixture	10 mM Tris-HCl, 100 mM NaCl, 100 μg/mL BSA, pH 8.0	No detectable contaminating nuclease; 90% ligated and 100% recut after 50-fold overdigestion	AGS Heidelb F00485S F00485M
5 U/μL	1 unit completely digests 1 μg substrate DNA/hr	10 mM Tris-HCl, 5 mM MgCl$_2$, 100 mM NaCl, 1 mM β-MSH, pH 8.0, 37°C		Boehringer 1379275
8-12 U/μL	1 unit hydrolyzes 1 μg DNA/hr at 37°C in a 50 μL reaction volume using λ phage DNA as substrate	10 mM Tris-HCl, 10 mM MgCl$_2$, 100 mM KCl, 0.1 mg/mL BSA, pH 8.5, 37°C	>90% of the DNA fragments ligated and recut after 50-fold overdigestion	Fermentas ER0711 ER0712
200 U and 1000 U	1 unit completely digests 1 μg λ DNA/hr at 37°C in 50 μL of supplied buffer	200 mM Tris-HCl, 100 mM MgCl$_2$, 10 mM DTT, 1 M KCl, pH 8.5	Purity is confirmed by overdigestion and ligation-recutting test	TaKaRa 1193

Vha464 I

RECOGNITION SEQUENCE
5'...C▼TTAAG...3'
3'...GAATT▲C...5'

HEAT INACTIVATION
65°C for 20 min.

NOTES
- EC 3.1.21.4
- A type II site specific deoxyribonuclease
- A large group of enzymes recognizing specific short DNA sequences. They cleave either within the recognition site, or at a short specific distance from it
- High concentration T4 DNA ligase gives more efficient ligation
- Requires Mg^{2+}

CONCENTRATION or VOLUME ACTIVITY	UNIT DEFINITION	REACTION BUFFER (As Provided)	ADDITIONAL ACTIVITIES and PURITY	SUPPLIER CATALOG NO.
Vibrio harveyi 464				
15,000 U/mL		10 mM Tris-HCl, 10 mM MgCl₂, 50 mM NaCl, 1 mM DTT, pH 7.6	40% DNA fragments ligated and recut after 5-fold overdigestion	SibEnzyme E135 E136

Vne I

RECOGNITION SEQUENCE
5'...G▼TGCAC...3'
3'...CACGT▲G...5'

HEAT INACTIVATION
65°C for 20 min.

NOTES
- EC 3.1.21.4
- A type II site specific deoxyribonuclease
- A large group of enzymes recognizing specific short DNA sequences. They cleave either within the recognition site, or at a short specific distance from it
- Requires Mg^{2+}

Vne I continued

CONCENTRATION or VOLUME ACTIVITY	UNIT DEFINITION	REACTION BUFFER (As Provided)	ADDITIONAL ACTIVITIES and PURITY	SUPPLIER CATALOG NO.
Vibrio nereis 18				
20,000 U/mL		50 mM Tris-HCl, 10 mM MgCl$_2$, 100 mM NaCl, 1 mM DTT, pH 7.6	90% DNA fragments ligated and recut after 10-fold overdigestion	SibEnzyme E137 E138

Vsp I

RECOGNITION SEQUENCE
5'...AT▼TAAT...3'
3'...TAAT▲TA...5'

HEAT INACTIVATION
65°C for 20 min.

NOTES
- EC 3.1.21.4
- A type II site specific deoxyribonuclease
- A large group of enzymes recognizing specific short DNA sequences. They cleave either within the recognition site, or at a short specific distance from it
- Compatible cohesive ends: *Ase* I, *Asn* I, *Mae* I, *Mse* I, *Nde* I, *Tru*9 I
- Requires Mg^{2+}

CONCENTRATION or VOLUME ACTIVITY	UNIT DEFINITION	REACTION BUFFER (As Provided)	ADDITIONAL ACTIVITIES and PURITY	SUPPLIER CATALOG NO.
Vibrio species				
10,000 U/mL	1 unit degrades 1 μg λ dam⁻ dcm⁻ DNA/hr at 37°C in a 0.05 mL mixture	10 mM Tris-HCl, 100 mM NaCl, 100 μg/mL BSA, pH 8.0	No detectable contaminating nuclease; 80% ligated and 100% recut after 50-fold overdigestion	AGS Heidelb F00073S F00073M
8-12 U/μL	1 unit hydrolyzes 1 μg DNA/hr at 37°C in a 50 μL reaction volume using λ phage DNA as substrate	50 mM Tris-HCl, 10 mM MgCl$_2$, 100 mM NaCl, 0.1 mg/mL BSA, pH 7.5, 37°C	>70% of the DNA fragments ligated and >90% recut after 50-fold overdigestion	Fermentas ER0911 ER0912
8-12 U/μL	1 unit completely digests 1 mg λ DNA/hr at 37°C	500 mM Tris-HCl, 100 mM MgCl$_2$, 500 mM NaCl, pH 8.0	≥95% ligation; 100% cleavage	Life Technol 15485-014
8-12 U/μL	1 unit completely digests 1 μg λ DNA/hr at 37°C in a 50 mL reaction volume	60 mM Tris-HCl, 60 mM MgCl$_2$, 1.5 M NaCl, 10 mM DTT, pH 7.9	70% ligation	Promega R6851 R6852

Vsp I continued

CONCENTRATION or VOLUME ACTIVITY	UNIT DEFINITION	REACTION BUFFER (As Provided)	ADDITIONAL ACTIVITIES and PURITY	SUPPLIER CATALOG NO.
Vibrio species continued				
10,000-15,000 U/mL		100 mM KCl, 10 mM Tris-HCl, 1 mM EDTA, 1 mM DTT, 200 µg/mL BSA, 50% glycerol, pH 7.4		Stratagene 501120
Vibrio species 343				
40,000 U/mL		10 mM Tris-HCl, 10 mM MgCl$_2$, 100 mM NaCl, 1 mM DTT, pH 8.5	95% DNA fragments ligated and recut after 10-fold overdigestion	SibEnzyme E139 E140

Xba I

RECOGNITION SEQUENCE
  5'...T▼CTAGA...3'
  3'...AGATC▲T...5'
OPTIMUM REACTION CONDITIONS
  pH 7.9-8.0
HEAT INACTIVATION
  60°C for 15 min.; 65°C for 5-30 min.
METHYLATION EFFECTS
  Does not cleave TCTAGm^{6}A, Tm^{5}CTAGA or Thm^{5}CTAGA
  Blocked by overlapping dam methylation

NOTES
- EC 3.1.21.4
- A type II site specific deoxyribonuclease
- A large group of enzymes recognizing specific short DNA sequences. They cleave either within the recognition site, or at a short specific distance from it
- Partially cleaves DNA from dam$^+$ E. coli
- Low ionic strength, high glycerol or enzyme concentration, or DMSO solution results in star activity
- Compatible cohesive ends: Avr II, Bln I, Nhe I, Spe I
- Requires Mg^{2+}

CONCENTRATION or VOLUME ACTIVITY	UNIT DEFINITION	REACTION BUFFER (As Provided)	ADDITIONAL ACTIVITIES and PURITY	SUPPLIER CATALOG NO.
Escherichia coli strain, carrying the cloned gene from Xanthomonas badrii				
20,000 U/mL; 100,000 U/mL	1 unit completely digests 1 µg λ DNA (dam–/Hind III digest)/hr in a total reaction volume of 50 µL using the buffer provided	10 mM Tris-HCl, 10 mM MgCl$_2$, 50 mM NaCl, 1 mM DTT, pH 7.9; add 100 µg/mL BSA	>95% DNA fragments ligated and recut after 20-fold overdigestion	NE Biolabs 145S 145L

Xba I continued

CONCENTRATION or VOLUME ACTIVITY	UNIT DEFINITION	REACTION BUFFER (As Provided)	ADDITIONAL ACTIVITIES and PURITY	SUPPLIER CATALOG NO.
Xanthomonas badrii				
7500 U and 20,000 U	1 unit completely digests 1 μg DNA substrate/hr at 37°C	10 mM Tris-HCl, 10 mM MgCl$_2$, 50 mM NaCl, 1 mM DTT, 0.1 mg/mL BSA, pH 7.5	>90% DNA fragments ligated and recut after 2-fold overdigestion	Adv Biotech AB-0245-a AB-0245-b
10 U/μL; 40 U/μL	1 unit completely cleaves 1 μg substrate DNA/hr	100 mM NaCl, 50 mM Tris, 5 mM MgCl$_2$, 2 mM β-MSH, pH 7.8	90-98% cleavage-ligation-recleavage	Am Allied L-09-02000 H-09-02000 L-09-10000 H-09-10000
8-12 U/μL	1 unit completely digests 1 μg dam$^-$ λ DNA/hr at 37°C in the buffer provided	Genomic grade; 100 mM Tris-HCl, 100 mM MgCl$_2$, 10 mM DTT, 500 mM NaCl, BSA, pH 7.5	100% DNA fragments ligated and recut	Amersham E 1093Y E 1093Z
	1 unit completely digests 1 μg dam$^-$ λ DNA/hr at 37°C in the buffer provided	Genomic grade; 100 mM Tris-HCl, 100 mM MgCl$_2$, 10 mM DTT, 500 mM NaCl, BSA, pH 7.5	100% DNA fragments ligated and recut	Amersham E 1093ZH
10,000 U/mL	1 unit degrades 1 μg λ dam$^-$ dcm$^-$ DNA BssH II-fragments/hr at 37°C in a 0.05 mL mixture	50 mM Tris-HCl, 10 mM MgCl$_2$, 100 mM NaCl, 100 μg/mL BSA, pH 7.5	No detectable contaminating nuclease; 90% ligated and 80% recut after 50-fold overdigestion	AGS Heidelb A00650S A00650M
10 U/μL and 40 U/μL	1 unit completely digests 1 μg substrate DNA/hr	50 mM Tris-HCl, 10 mM MgCl$_2$, 100 mM NaCl, 1 mM DTE, pH 7.5, 37°C		Boehringer 674257 674265 674273 1047663
	1 unit completely cleaves 1 μg EcoR I fragments of unmethylated λ DNA/hr at pH 7.5, 65°C in a reaction volume of 50 μL	50 mM Tris-HCl, 10 mM MgCl$_2$, 100 mM NaCl, 1.0 mM DTT, 100 μg/mL BSA, pH 7.5		CHIMERx 2430-01 2430-02
8-12 U/μL; 40-60 U/μL	1 unit hydrolyzes 1 μg DNA/hr at 37°C in a 50 μL reaction volume using λ phage DNA as substrate	33 mM Tris-acetate, 10 mM MgOAc, 66 mM KOAc, 0.1 mg/mL BSA, pH 7.9, 37°C	>90% of the DNA fragments ligated and recut after 50-fold overdigestion	Fermentas ER0681 ER0682 ER0683
5000-20,000 U/mL		50 mM KCl, 10 mM Tris-HCl, 0.1 mM EDTA, 1 mM DTT, 200 μg/mL acetylated BSA, 50% glycerol, pH 7.4		ICN 197010
8-12 U/μL and 50 U/μL	1 unit completely digests 1 mg Ad2 DNA/hr at 37°C	500 mM Tris-HCl, 100 mM MgCl$_2$, 500 mM NaCl, pH 8.0	≥65% ligation; 100% cleavage	Life Technol 15226-012 15226-038 15226-020

Xba I continued

CONCENTRATION or VOLUME ACTIVITY	UNIT DEFINITION	REACTION BUFFER (As Provided)	ADDITIONAL ACTIVITIES and PURITY	SUPPLIER CATALOG No.
Xanthomonas badrii continued				
	10 units does not produce unspecific cleavage products with 1 μg λ DNA/16 hr at 37°C	10 mM Tris-HCl, 10 mM MgCl$_2$, 50 mM NaCl, 1 mM DTT, 100 μg/mL BSA, pH 7.9	>90% DNA fragments ligated and recut after 10-fold overdigestion	MINOTECH 134
6-16 U/μL	1 unit digests completely 1 μg λ DNA substrate/hr at 37°C	50 mM Tris-HCl, 100 mM NaCl, 10 mM MgCl$_2$, 1 mM DTT, pH 7.8		NBL Gene 012004 012007
10 U/μL and 20-300 U/μL	1 unit hydrolyzes 1 μg λ DNA to completion/hr in a 50 μL total reaction volume	100 mM Tris-HCl, 500 mM NaCl, 100 mM MgCl$_2$, 100 mM β-MSH, 1.0 mg/mL BSA, pH 8		Oncor 110461 110462 110465
8000-15,000 U/mL and 50,000-80,000 U/mL	1 unit completely digests 1 μg EcoR I pre-digested Ad2 DNA/hr at pH 7.5, 37°C in a 50 μL total assay mixture	100 mM Tris-acetate, 100 mM MgOAc, 500 mM KOAc, pH 7.5		Pharmacia 27-0948-01 27-0948-02 27-0948-18
10 U/μL and 40-80 U/μL	1 unit completely digests 1 μg Bal II unmethylated λ DNA fragments/hr at 37°C in a 50 mL reaction volume	Blue/White Cloning Qualified; genome qualified; 60 mM Tris-HCl, 60 mM MgCl$_2$, 1.5 M NaCl, 10 mM DTT, pH 7.9	90% ligation	Promega R6181 R6182 R6185 R4184
25,000 U/mL		Blue/white certified; 50 mM Tris-HCl, 10 mM MgCl$_2$, 100 mM NaCl, 1 mM DTT, pH 7.6	90% DNA fragments ligated and recut after 10-fold overdigestion	SibEnzyme E141 E142
2000-20,000 U/mL	1 unit completely cleaves 1.0 μg λ DNA/hr in a 50 μL reaction mixture	100 mM Tris-HCl, 500 mM NaCl, 100 mM MgCl$_2$, 10 mM DTT, BSA, pH 7.9, 25°C	100% cleavage, 90% ligation, 100% recleavage, 10% endonuclease (nickase) No detectable endonuclease (overdigestion)	Sigma R7260
8000-30,000 U/mL		50 mM KCl, 10 mM Tris-HCl, 0.1 mM EDTA, 1 mM DTT, 200 μg/mL BSA, 50% glycerol, pH 7.4		Stratagene 501130 501131
3000 U and 15,000 U	1 unit completely digests 1 μg N^6-methyladenine free λ DNA/hr at 37°C in 50 μL of supplied buffer and BSA	Genomic grade; 100 mM Tris-HCl, 100 mM MgCl$_2$, 10 mM DTT, 500 mM NaCl, BSA, pH 7.5	Purity is confirmed by overdigestion, ligation-recutting, genome DNA digestion and cloning test	TaKaRa 1093
3-20 U/μL and 40-50 U/μL	1 unit completely digests 1 μg DNA/hr at 37°C	10 mM Tris-HCl, 10 mM MgCl$_2$, 50 mM NaCl, 1 mM DTT, pH 7.5	>90% ligation; >95% recut	Toyobo XBA-101 XBA-102 XBA-152

Xba I continued

CONCENTRATION or VOLUME ACTIVITY	UNIT DEFINITION	REACTION BUFFER (As Provided)	ADDITIONAL ACTIVITIES and PURITY	SUPPLIER CATALOG NO.
Xanthomonas badrii ATCC 1672				
5-20 U/μL 30-100 U/μL	λ (dcm⁻) DNA substrate, 37°C	500 mM NaCl, 100 mM Tris-HCl, 100 mM MgCl$_2$, 10 mM DTT, pH 7.5	90% ligation and 100% recutting	Nippon Gene 317-00381 313-00383 317-00386 312-02151

Xcm I

RECOGNITION SEQUENCE
    5'...CCA(N)$_5$▼(N)$_4$TGG...3'
    3'...GGT(N)$_4$▲(N)$_5$ACC...5'
HEAT INACTIVATION
    65°C for 20 min.

NOTES
- EC 3.1.21.4
- A type II site specific deoxyribonuclease
- A large group of enzymes recognizing specific short DNA sequences. They cleave either within the recognition site, or at a short specific distance from it
- Requires Mg^{2+}

CONCENTRATION or VOLUME ACTIVITY	UNIT DEFINITION	REACTION BUFFER (As Provided)	ADDITIONAL ACTIVITIES and PURITY	SUPPLIER CATALOG NO.
Xanthomonas campestris				
1-5 U/μL	1 unit completely digests 1 μg λ DNA/hr at 37°C in the buffer provided	100 mM Tris-HCl, 100 mM MgCl$_2$, 10 mM DTT, 500 mM NaCl, pH 7.5	>20% DNA fragments ligated and 95% recut	Amersham E 0222Y E 0222Z
1000-5000 U/mL		50 mM KCl, 10 mM Tris-HCl, 0.1 mM EDTA, 1 mM DTT, 200 μg/mL acetylated BSA, 50% glycerol, pH 7.4		ICN 159473
Xanthomonas campestris (NEB 497)				
1000 U/mL	1 unit completely digests 1 μg λ DNA/hr in a total reaction volume of 50 μL using the buffer provided	10 mM Tris-HCl, 10 mM MgCl$_2$, 50 mM NaCl, 1 mM DTT, pH 7.9	>20% DNA fragments ligated and >95% recut after 5-fold overdigestion	NE Biolabs 533S 533L

Xho I

RECOGNITION SEQUENCE
5'...C▼TCGAG...3'
3'...GAGCT▲C...5'

ISOSCHIZOMERS
PaeR7 I

OPTIMUM REACTION CONDITIONS
pH 7.9-8.0

HEAT INACTIVATION
Resistant, use ethanol precipitation; 65°C for 15-20 min.

METHYLATION EFFECTS
Does not cleave CTm^{5}CGAG, CTCGm^{6}AG or m^{5}CTCGAG

NOTES
- EC 3.1.21.4
- A type II site specific deoxyribonuclease
- A large group of enzymes recognizing specific short DNA sequences. They cleave either within the recognition site, or at a short specific distance from it
- Very sensitive to DNA contaminants
- Compatible cohesive ends: Ava I, PaeR7 I, Sal I
- Requires Mg^{2+}

CONCENTRATION or VOLUME ACTIVITY	UNIT DEFINITION	REACTION BUFFER (As Provided)	ADDITIONAL ACTIVITIES and PURITY	SUPPLIER CATALOG NO.
Escherichia coli strain, carrying the cloned gene from *Xanthomonas holcicola*				
20,000 U/mL	1 unit completely digests 1 µg λ DNA/hr in a total reaction volume of 50 µL using the buffer provided	10 m*M* Tris-HCl, 10 m*M* MgCl₂, 50 m*M* NaCl, 1 m*M* DTT, pH 7.9; add 100 µg/mL BSA	>95% DNA fragments ligated and recut after 50-fold overdigestion	NE Biolabs 146S 146L
Xanthomonas holcicola (Xanthomonas campestris)				
10,000 U and 20,000 U	1 unit completely digests 1 µg DNA substrate/hr at 37°C	10 m*M* Tris-HCl, 10 m*M* MgCl₂, 100 m*M* NaCl, 1 m*M* DTT, 0.1 mg/mL BSA, pH 7.5	>90% DNA fragments ligated and recut after 10-fold overdigestion	Adv Biotech AB-0246-a AB-0246-b
10 U/µL; 40 U/µL	1 unit completely cleaves 1 µg substrate DNA/hr	100 m*M* NaCl, 50 m*M* Tris, 5 m*M* MgCl₂, 2 m*M* β-MSH, pH 7.8	90-98% cleavage-ligation-recleavage	Am Allied L-08-04000 H-08-04000 L-08-20000 H-08-20000
8-12 U/µL	1 unit completely digests 1 µg λ DNA/hr at 37°C in the buffer provided	Genomic grade; 500 m*M* Tris-HCl, 100 m*M* MgCl₂, 10 m*M* DTT, 1000 m*M* NaCl, pH 7.5	>95% DNA fragments ligated and 100% recut	Amersham E 1094Y E 1094Z
	1 unit completely digests 1 µg λ DNA/hr at 37°C in the buffer provided	Genomic grade; 500 m*M* Tris-HCl, 100 m*M* MgCl₂, 10 m*M* DTT, 1000 m*M* NaCl, pH 7.5	>95% DNA fragments ligated and 100% recut	Amersham E 1094ZH

Xho I continued

CONCENTRATION or VOLUME ACTIVITY	UNIT DEFINITION	REACTION BUFFER (As Provided)	ADDITIONAL ACTIVITIES and PURITY	SUPPLIER CATALOG No.
Xanthomonas holcicola (Xanthomonas campestris) continued			
10,000 U/mL	1 unit degrades 1 μg λ dam⁻ dcm⁻ DNA/hr at 37°C in a 0.05 mL mixture	10 mM Tris-HCl, 100 mM NaCl, 100 μg/mL BSA, 1 mM DTT, pH 8.0	No detectable contaminating nuclease; 80% ligated and recut after 50-fold overdigestion	AGS Heidelb F00660S F00660M
10 U/μL and 40 U/μL	1 unit completely digests 1 μg substrate DNA/hr	50 mM Tris-HCl, 10 mM MgCl₂, 100 mM NaCl, 1 mM DTE, pH 7.5, 37°C		Boehringer 899194 703770 703788
	1 unit completely cleaves 1 μg λ DNA/hr at pH 7.5, 37°C in a reaction volume of 50 μL	50 mM Tris-HCl, 10 mM MgCl₂, 100 mM NaCl, 1.0 mM DTT, 100 μg/mL BSA, pH 7.5		CHIMERx 2440-01 2440-02
8-12 U/μL; 40-60 U/μL	1 unit hydrolyzes 1 μg DNA/hr at 37°C in a 50 μL reaction volume using λ phage DNA as substrate	10 mM Tris-HCl, 10 mM MgCl₂, 100 mM KCl, 0.1 mg/mL BSA, pH 8.5, 37°C	>90% of the DNA fragments ligated and recut after 50-fold overdigestion	Fermentas ER0691 ER0692 ER0693
		50 mM KCl, 20 mM Tris-HCl, 0.1 mM EDTA, 1.0 mM DTT, 100 μg/mL BSA, 50% glycerol, pH 7.5		ICN 153840
8-12 U/μL	1 unit completely digests 1 mg Ad2 DNA/hr at 37°C	500 mM Tris-HCl, 100 mM MgCl₂, 500 mM NaCl, pH 8.0	≥95% ligation; 100%-cleavage	Life Technol 15231-012 15231-020
	10 units does not produce unspecific cleavage products with 1 μg λ DNA/16 hr at 37°C	10 mM Tris-HCl, 10 mM MgCl₂, 150 mM NaCl, 1 mM DTT, 100 μg/mL BSA, pH 7.9	>90% DNA fragments ligated and recut after 10-fold overdigestion	MINOTECH 135
6-14 U/μL	1 unit digests completely 1 μg λ DNA substrate/hr at 37°C	50 mM Tris-HCl, 100 mM NaCl, 10 mM MgCl₂, 1 mM DTT, pH 7.8		NBL Gene 011304 011307
10 U/μL	1 unit hydrolyzes 1 μg λ DNA to completion/hr in a 50 μL total reaction volume	100 mM Tris-HCl, 1 M NaCl, 100 mM MgCl₂, 100 mM β-MSH, 1.0 mg/mL BSA, pH 8		Oncor 110471 110472
8000-15,000 U/mL and 50,000-80,000 U/mL	1 unit completely digests 1 μg EcoR I pre-digested λ DNA/hr at pH 7.5, 37°C in a 50 μL total assay mixture	100 mM Tris-acetate, 100 mM MgOAc, 500 mM KOAc, pH 7.5		Pharmacia 27-0950-01 27-0950-02 27-0950-18
8-12 U/μL and 40-80 U/μL	1 unit completely digests 1 μg EcoR I λ DNA fragments/hr at 37°C in a 50 mL reaction volume	Blue/White Cloning Qualified; genome qualified; 60 mM Tris-HCl, 60 mM MgCl₂, 1.5 M NaCl, 10 mM DTT, pH 7.9	90% ligation	Promega R6161 R6162 R6165 R4164

Xho I continued

CONCENTRATION or VOLUME ACTIVITY	UNIT DEFINITION	REACTION BUFFER (As Provided)	ADDITIONAL ACTIVITIES and PURITY	SUPPLIER CATALOG No.
Xanthomonas holcicola (Xanthomonas campestris) continued				
5000-20,000 U/mL	1 unit completely cleaves 1.0 μg λ DNA/hr in a 50 μL reaction mixture	100 mM Tris-HCl, 500 mM NaCl, 100 mM MgCl$_2$, 10 mM DTT, pH 7.9, 25°C	100% cleavage, 95% ligation, 100% recleavage, 10% endonuclease (nickase) No detectable endonuclease (overdigestion)	Sigma R6379
10,000-60,000 U/mL		50 mM KCl, 10 mM Tris-HCl, 0.1 mM EDTA, 1 mM DTT, 200 μg/mL BSA, 50% glycerol, pH 7.4		Stratagene 501140 501141
5000 U and 25,000 U	1 unit completely digests 1 μg λ DNA/hr at 37°C in 50 μL of supplied buffer	Genomic grade; 500 mM Tris-HCl, 100 mM MgCl$_2$, 10 mM DTT, 1 M NaCl, pH 7.5	Purity is confirmed by overdigestion, ligation-recutting, genome DNA digestion and cloning test	TaKaRa 1094
3-20 U/μL and 40-50 U/μL	1 unit completely digests 1 μg DNA/hr at 37°C	50 mM Tris-HCl, 10 mM MgCl$_2$, 100 mM NaCl, 1 mM DTT, pH 7.5	>90% ligation; >95% recut	Toyobo XHO-101 XHO-102 XHO-152
Xanthomonas holcicola ATCC 13461				
5-20 U/μL 30-100 U/μL	λ DNA substrate, 37°C	1 M NaCl, 500 mM Tris-HCl, 100 mM MgCl$_2$, 10 mM DTT, pH 7.5	80% ligation and 100% recutting	Nippon Gene 312-00392 316-00395 314-00396 319-02161

Xho II

RECOGNITION SEQUENCE
  5'...Pu▼GATCPy...3'
  3'...PyCTAG▲Pu...5'

METHYLATION EFFECTS
  Not blocked by overlapping *dam* methylation

NOTES
- EC 3.1.21.4
- A type II site specific deoxyribonuclease
- A large group of enzymes recognizing specific short DNA sequences. They cleave either within the recognition site, or at a short specific distance from it
- Compatible cohesive ends: *Bam*H I, *Bcl* I, *Bgl* II, *Bst*Y I, *Mbo* I, *Mfl* I, *Nde* II, *Sau*3A I
- Requires Mg^{2+}

CONCENTRATION or VOLUME ACTIVITY	UNIT DEFINITION	REACTION BUFFER (As Provided)	ADDITIONAL ACTIVITIES and PURITY	SUPPLIER CATALOG NO.
Xanthomonas holcicola (Xanthomonas campestris)				
1-5 U/μL	1 unit completely digests 1 μg substrate DNA/hr	10 mM Tris-HCl, 10 mM $MgCl_2$, 1 mM DTE, pH 7.5, 37°C		Boehringer 742929 742937
10 U/μL	1 unit completely digests 1 μg pBR322 DNA/hr at 37°C in a 50 mL reaction volume	100 mM Tris-HCl, 100 mM $MgCl_2$, 500 mM NaCl, 10 mM DTT, pH 7.9	90% ligation	Promega R6811 R6812 R6815
500-3000 U/mL		50 mM KCl, 10 mM Tris-HCl, 0.1 mM EDTA, 1 mM β-MSH, 200 μg/mL BSA, 0.01% Triton X-100, 50% glycerol, pH 7.4		Stratagene 501150
Xanthomonas holcicola ATCC 13461				
500-10,000 U/mL		50 mM KCl, 10 mM Tris-HCl, 0.1 mM EDTA, 10 mM β-MSH, 0.01% Triton X-100, 100 μg/mL BSA, 50% glycerol, pH 7.5		ICN 153833

Xma I

RECOGNITION SEQUENCE
   5'...C▼CCGGG...3'
   3'...GGGCC▲C...5'

ISOSCHIZOMERS
   Sma I

HEAT INACTIVATION
   65°C for 20 min.

METHYLATION EFFECTS
   Cleaves CCm^{5}CGGG
   Does not cleave m^{4}CCCGGG, m^{5}CCCGGG, Cm^{4}CCGGG or CCm^{4}CGGG

NOTES
- EC 3.1.21.4
- A type II site specific deoxyribonuclease
- A large group of enzymes recognizing specific short DNA sequences. They cleave either within the recognition site, or at a short specific distance from it
- Xma I produces fragments with 5'-extensions; Sma I makes blunt ended fragments
- Compatible cohesive ends: Acc III, Age I, Bca77 I, BsaW I, BseA I, BsiM I, BspE I, BsrF I, Bsu23 I, Cfr9 I, Cfr10 I, Kpn2 I, Mro I, NgoA IV, NgoM I, PinA I, SgrA I
- GC-rich recognition sequence makes this enzyme useful for genomic DNA analysis
- Requires Mg^{2+}

CONCENTRATION or VOLUME ACTIVITY	UNIT DEFINITION	REACTION BUFFER (As Provided)	ADDITIONAL ACTIVITIES and PURITY	SUPPLIER CATALOG No.
Escherichia coli, carrying the Xma I gene from Xanthomonas malvacearum				
1 U/μL	1 unit completely digests 1 μg λ DNA/hr at 37°C in the buffer provided	Genomic grade; 100 mM Tris-HCl, 100 mM MgCl$_2$, 10 mM DTT, pH 7.5	>95% DNA fragments ligated and 95% recut	Amersham E 0223Y E 0223Z
1000 U/mL; 5000 U/mL	1 unit completely digests 1 μg λ DNA/hr in a total reaction volume of 50 μL using the buffer provided	10 mM Bis Tris-Propane-HCl, 10 mM MgCl$_2$, 1 mM DTT, pH 7.0	>95% DNA fragments ligated and recut after 10-fold overdigestion	NE Biolabs 180S 180L
500-5000 U/mL	1 unit completely cleaves 1.0 μg λ DNA/hr in a 50 μL reaction mixture	100 mM Bis Tris-Propane-HCl, 100 mM MgCl$_2$, 10 mM DTT, pH 7.0, 25°C	100% cleavage, 100% ligation, 95% recleavage No detectable endonuclease (overdigestion)	Sigma R5006
Xanthomonas malvacearum				
3000 U/mL	1 unit degrades 1 μg λ dam⁻ dcm⁻ DNA/hr at 37°C in a 0.05 mL mixture	10 mM Tris-HCl, 10 mM MgCl$_2$, 100 μg/mL BSA, pH 7.5	No detectable contaminating nuclease; 100% ligated and 95% recut after 10-fold overdigestion	AGS Heidelb F00675S F00675M

Xma I continued

CONCENTRATION or VOLUME ACTIVITY	UNIT DEFINITION	REACTION BUFFER (As Provided)	ADDITIONAL ACTIVITIES and PURITY	SUPPLIER CATALOG No.
Xanthomonas malvacearum continued				
2000-6000 U/mL	1 unit completely digests 1 μg λ DNA/hr at pH 7.5, 37°C in a 50 μL total assay mixture	100 mM Tris-acetate, 100 mM MgOAc, 500 mM KOAc, pH 7.5		Pharmacia 27-0877-01
1-5 U/μL	1 unit completely digests 1 μg Ad2 DNA/hr at 37°C in a 50 mL reaction volume	Blue/White Cloning Qualified; 60 mM Tris-HCl, 60 mM MgCl$_2$, 500 mM NaCl, 10 mM DTT, pH 7.5		Promega R6491 R6492 R6495
2000 U/mL		33 mM Tris-acetate, 10 mM MgOAc, 66 mM KOAc, 1 mM DTT, pH 7.9	90% DNA fragments ligated and recut after 10-fold overdigestion	SibEnzyme E233 E234
Xanthomonas malvacearum ATCC 9924				
500-50,000 U/mL		50 mM KCl, 10 mM Tris-HCl, 0.1 mM EDTA, 1 mM DTT, 200 μg/mL BSA, 50% glycerol, pH 7.4		ICN 153834

Xma III

RECOGNITION SEQUENCE
  5'...C▼GGCCG...3'
  3'...GCCGG▲C...5'

HEAT INACTIVATION
  Partially resistant at 65°C for 10 min.

METHYLATION EFFECTS
  Does not cleave DNA when the middle residue is 5-methyl cytosine

NOTES
- EC 3.1.21.4
- A type II site specific deoxyribonuclease
- A large group of enzymes recognizing specific short DNA sequences. They cleave either within the recognition site, or at a short specific distance from it
- Cleavage rates vary greatly among different Xma III sites
- Requires Mg^{2+}

CONCENTRATION or VOLUME ACTIVITY	UNIT DEFINITION	REACTION BUFFER (As Provided)	ADDITIONAL ACTIVITIES and PURITY	SUPPLIER CATALOG No.
Obtained from a cloned source				
5000-20,000 U/mL		100 mM KCl, 20 mM Tris-HCl, 0.1 mM EDTA, 10 mM β-MSH, 500 µg/mL BSA, 50% glycerol, pH 7.6		stratagene 501172
Xanthomonas malvacearum				
5000-20,000 U/mL		100 mM KCl, 20 mM Tris-HCl, 0.1 mM EDTA, 10 mM β-MSH, 500 µg/mL acetylated BSA, 50% glycerol, pH 7.6		ICN 159474
Xanthomonas malvacearum M (Xanthomonas campestris)				
2-8 U/µL	1 unit completely digests 1 mg Ad2 DNA/hr at 25°C	100 mM Tris-HCl, 80 mM $MgCl_2$, pH 8.2	≥95% ligation; 90% cleavage	Life Technol 15414-014

XmaC I

RECOGNITION SEQUENCE
5'...C▼CCGGG...3'
3'...GGGCC▲C...5'

ISOSCHIZOMERS
*Cfr*9 I, *Psp*A I, *Sma* I, *Xma* I

NOTES
- EC 3.1.21.4
- A type II site specific deoxyribonuclease
- A large group of enzymes recognizing specific short DNA sequences. They cleave either within the recognition site, or at a short specific distance from it
- Requires Mg^{2+}

CONCENTRATION or VOLUME ACTIVITY	UNIT DEFINITION	REACTION BUFFER (As Provided)	ADDITIONAL ACTIVITIES and PURITY	SUPPLIER CATALOG NO.
Xanthomonas malvacearum strain C				
				Boehringer

Xmn I

RECOGNITION SEQUENCE
5'...GAANN▼NNTTC...3'
3'...CTTNN▲NNAAG...5'

HEAT INACTIVATION
65°C for 20 min.

NOTES
- EC 3.1.21.4
- A type II site specific deoxyribonuclease
- A large group of enzymes recognizing specific short DNA sequences. They cleave either within the recognition site, or at a short specific distance from it
- Low ionic strength, high glycerol or enzyme concentration, or pH > 8.0 results in star activity
- Sensitive to NaCl >80 m*M*
- Requires Mg^{2+}
- *Eco*R I sites that have been cleaved and filled in with Klenow generate *Xmn* I sites after ligation

Xmn I continued

CONCENTRATION or VOLUME ACTIVITY	UNIT DEFINITION	REACTION BUFFER (As Provided)	ADDITIONAL ACTIVITIES and PURITY	SUPPLIER CATALOG NO.
Xanthomonas manihotis				
10,000 U/mL	1 unit degrades 1 μg λ dam⁻ dcm⁻ DNA/hr at 37°C in a 0.05 mL mixture	10 mM Tris-HCl, 100 mM NaCl, 100 μg/mL BSA, pH 8.0	No detectable contaminating nuclease; 65% ligated and 90% recut after 10-fold overdigestion	AGS Heidelb A00685S A00685M
10 U/μL	1 unit completely digests 1 μg λ DNA/hr at 37°C in a 50 mL reaction volume	10 mM Tris-HCl, 100 mM glutamate, 1 mM DTT, 0.1 mM EDTA, pH 7.5	80% ligation	Promega R7271 R7272 R7273
Xanthomonas manihotis 7AS1				
5000-10,000 U/mL		50 mM KCl, 10 mM Tris-HCl, 0.1 mM EDTA, 1 mM DTT, 200 μg/mL BSA, 50% glycerol, pH 7.5		Stratagene 501170
Xanthomonas manihotis 7AS1 (B.-C. Lin)				
2000-20,000 U/mL		50 mM KCl, 10 mM Tris-HCl, 0.1 mM EDTA, 1 mM DTT, 200 μg/mL BSA, 50% glycerol, pH 7.4		ICN 153835
6000 U/mL	1 unit completely digests 1 μg λ DNA/hr in a total reaction volume of 50 μL using the buffer provided	10 mM Tris-HCl, 10 mM MgCl₂, 50 mM NaCl, 1 mM DTT, pH 7.9; add 100 μg/mL BSA	>95% DNA fragments ligated and recut after 25-fold overdigestion	NE Biolabs 194S 194L

Zsp2 I

RECOGNITION SEQUENCE
5'...ATGCA▼T...3'
3'...T▲ACGTA...5'

HEAT INACTIVATION
65°C for 20 min.

NOTES
- EC 3.1.21.4
- A type II site specific deoxyribonuclease
- A large group of enzymes recognizing specific short DNA sequences. They cleave either within the recognition site, or at a short specific distance from it
- Requires Mg^{2+}

CONCENTRATION or VOLUME ACTIVITY	UNIT DEFINITION	REACTION BUFFER (As Provided)	ADDITIONAL ACTIVITIES and PURITY	SUPPLIER CATALOG NO.
Zoogloea species 2				
5000 U/mL		10 mM Tris-HCl, 10 mM $MgCl_2$, 1 mM DTT, pH 7.6	90% DNA fragments ligated and recut after 10-fold overdigestion	SibEnzyme E145 E146

Chapter 7. DNA Methyltransferases

Alu I Methylase

METHYLATION SEQUENCE
5'...AGCm^{5}T...3'
3'...TCm^{5}GA...5'

SYNONYMS
Modification methylase, restriction-modification system

OPTIMUM REACTION CONDITIONS
pH 7.5 and 37°C

NOTES
- EC 2.1.1.73
- A site-specific DNA-methyltransferase (cytosine-specific)
- Forms "restriction-modification systems" in conjunction with restriction enzymes of similar site specificity listed under EC 3.1.21.3, 3.1.21.4 or 3.1.21.5
- Modifies the cytosine residue (C^5) in the methylation sequence
- Alu I methylated DNA is *mcr*B sensitive
- EDTA or high salt concentrations inhibit contaminating endo- and exonucleases

CONCENTRATION or VOLUME ACTIVITY	UNIT DEFINITION	REACTION BUFFER (As Provided)	ADDITIONAL ACTIVITIES and PURITY	SUPPLIER CATALOG NO.
Arthrobacter luteus				
3000-10,000 U/mL		50 m*M* KCl, 10 m*M* Tris-HCl, 0.1 m*M* EDTA, 1 m*M* DTT, 200 µg/mL acetylated BSA, 50% glycerol, pH 7.5		ICN 159389
5000 U/mL	1 unit protects 1 µg λ DNA/hr at 37°C in a reaction mixture of 10 µL against cleavage by Alu I restriction endonuclease	50 m*M* KCl, 10 m*M* Tris-HCl, 0.1 m*M* EDTA, 1 m*M* DTT, 200 µg/mL BSA, 50% glycerol, pH 7.5	No detectable contaminating endo- and exonuclease	NE Biolabs 220S 220L
				TaKaRa

*Bam*H I Methylase

METHYLATION SEQUENCE
  5'...GGATCm^{5}C...3'
  3'...CCm^{5}TAGG...5'

SYNONYMS
  Modification methylase, restriction-modification system

OPTIMUM REACTION CONDITIONS
  pH 7.5 and 37°C

NOTES
- EC 2.1.1.73
- A site-specific DNA-methyltransferase (cytosine-specific)
- Forms "restriction-modification systems" in conjunction with restriction enzymes of similar site specificity listed under EC 3.1.21.3, 3.1.21.4 or 3.1.21.5
- Modifies the internal cytosine residue (C^5) in the methylation sequence
- EDTA or high salt concentrations inhibit contaminating endo- and exonucleases

CONCENTRATION or VOLUME ACTIVITY	UNIT DEFINITION	REACTION BUFFER (As Provided)	ADDITIONAL ACTIVITIES and PURITY	SUPPLIER CATALOG No.
Bacillus amyloliquefaciens H				
3000–10,000 U/mL		50 mM Tris-HCl, 10 mM EDTA, 1 mM DTT, 200 µg/mL acetylated BSA, 50% glycerol, pH 7.5		ICN 159397
Escherichia coli strain, carrying the cloned modification gene from *Bacillus amyloliquefaciens* H				
4000 U/mL	1 unit protects 1 µg λ DNA/hr at 37°C in a reaction mixture of 10 µL against cleavage by *Bam*H I restriction endonuclease	50 mM Tris-HCl, 10 mM EDTA, 1 mM DTT, 200 µg/mL BSA, 50% glycerol, pH 7.5	No detectable contaminating endo- and exonuclease	NE Biolabs 223S 223L
Unspecified				
				TaKaRa

Bsu15 I Methylase

METHYLATION SEQUENCE
5'...ATCGAm^{6}T...3'
3'...TAm^{6}GCTA...5'

SYNONYMS
Modification methylase, restriction-modification system

OPTIMUM REACTION CONDITIONS
pH 8.0 and 37°C

NOTES
- EC 2.1.1.72
- A site-specific DNA-methyltransferase (adenine-specific)
- Forms "restriction-modification systems" in conjunction with restriction enzymes of similar site specificity listed under EC 3.1.21.3, 3.1.21.4 or 3.1.21.5
- Modifies the internal adenine residue in the recognition sequence, yielding N^6-methyladenine
- EDTA or high salt concentrations inhibit contaminating endo- and exonucleases

CONCENTRATION or VOLUME ACTIVITY	UNIT DEFINITION	REACTION BUFFER (As Provided)	ADDITIONAL ACTIVITIES and PURITY	SUPPLIER CATALOG No.
Escherichia coli, carrying the cloned *Bsu*15 IM gene from *Bacillus subtilis* 15				
8–12 U/μL	1 unit protects 1 μg λ DNA/hr at 37°C in a 50 μL reaction buffer against cleavage by complementary restriction endonuclease	10 m*M* KPO$_4$, 100 m*M* KCl, 1 m*M* EDTA, 7 m*M* β-MSH, 50% glycerol, 0.2 mg/mL BSA, pH 7.4	No detectable other methylases <0.02% 5'-, 3'-exonuclease, nonspecific endonuclease, phosphatase contamination	Fermentas EM0271 EM0272

Cla I Methylase

METHYLATION SEQUENCE
$5'...ATCGAm^6T...3'$
$3'...TAm^6GCTA...5'$

SYNONYMS
Modification methylase, restriction-modification system

OPTIMUM REACTION CONDITIONS
pH 7.5 and 37°C

NOTES
- EC 2.1.1.72
- A site-specific DNA-methyltransferase (adenine-specific)
- Forms "restriction-modification systems" in conjunction with restriction enzymes of similar site specificity listed under EC 3.1.21.3, 3.1.21.4 or 3.1.21.5
- Modifies the internal adenine residue (N^6) in the methylation sequence
- EDTA or high salt concentrations inhibit contaminating endo- and exonucleases

CONCENTRATION or VOLUME ACTIVITY	UNIT DEFINITION	REACTION BUFFER (As Provided)	ADDITIONAL ACTIVITIES and PURITY	SUPPLIER CATALOG No.
Caryophanon latum				
3000-10,000 U/mL		5 mM β-MSH, 50 mM Tris-HCl, 10 mM EDTA, 200 μg/mL acetylated BSA, 50% glycerol, pH 7.5		ICN 159427
10,000 U/mL	1 unit protects 1 μg λ DNA/hr at 37°C in a reaction mixture of 10 μL against cleavage by Cla I restriction endonuclease	50 mM Tris-HCl, 10 mM EDTA, 5 mM β-MSH, 200 μg/mL BSA, 50% glycerol, pH 7.5	No detectable contaminating endo- and exonuclease	NE Biolabs 218S 218L
				TaKaRa

dam Methylase

METHYLATION SEQUENCE
5'...GAm^{6}TC...3'
3'...CTAm^{6}G...5'

SYNONYMS
Modification methylase, restriction-modification system

OPTIMUM REACTION CONDITIONS
pH 7.5 and 37°C

NOTES
- EC 2.1.1.72
- A site-specific DNA-methyltransferase (adenine-specific)
- Forms "restriction-modification systems" in conjunction with restriction enzymes of similar site specificity listed under EC 3.1.21.3, 3.1.21.4 or 3.1.21.5
- Modifies the adenine residue (N^6) in the methylation sequence
- EDTA or high salt concentrations inhibit contaminating endo- and exonucleases

CONCENTRATION or VOLUME ACTIVITY	UNIT DEFINITION	REACTION BUFFER (As Provided)	ADDITIONAL ACTIVITIES and PURITY	SUPPLIER CATALOG No.
Escherichia coli strain, with the plasmid pTP166 carrying the *dam* modification gene of *E. coli* (M. Marinus)				
5000–20,000 U/mL	1 unit protects 1 µg λ DNA/hr at 37°C in a reaction mixture of 10 µL against cleavage by *Mbo* I restriction endonuclease	50 mM KCl, 50 mM Tris-HCl, 10 mM EDTA, 1 mM DTT, 200 µg/mL BSA, 50% glycerol, pH 7.5	No detectable contaminating endo- and exonuclease	NE Biolabs 222S 222L
Escherichia coli, with *dam* modification				
5000–20,000 U/mL		50 mM KCl, 50 mM Tris-HCl, 10 mM EDTA, 1 mM DTT, 200 µg/mL acetylated BSA, 50% glycerol, pH 7.5		ICN 159428

EcoR I Methylase

METHYLATION SEQUENCE
5'...GAAm^{6}TTC...3'
3'...CTTAm^{6}AG...5'

SYNONYMS
Modification methylase, restriction-modification system

OPTIMUM REACTION CONDITIONS
pH 8.0 and 37°C

NOTES
- EC 2.1.1.72
- A site-specific DNA-methyltransferase (adenine-specific)
- Forms "restriction-modification systems" in conjunction with restriction enzymes of similar site specificity listed under EC 3.1.21.3, 3.1.21.4 or 3.1.21.5
- Modifies the adenine residue (N^6) in the methylation sequence
- Does not restrict GAm^{6}ATTC
- 50% activity in the presence of 4 mM MgCl$_2$
- EDTA or high salt concentrations inhibit contaminating endo- and exonucleases
- Useful in protecting internal EcoR I restriction sites of DNA fragments when constructing cDNA or genomic libraries using EcoR I linkers

CONCENTRATION or VOLUME ACTIVITY	UNIT DEFINITION	REACTION BUFFER (As Provided)	ADDITIONAL ACTIVITIES and PURITY	SUPPLIER CATALOG No.
Escherichia coli				
5000-40,000 U/mL		200 mM NaCl, 100 mM KPO$_4$, 0.1 mM EDTA, 10 mM β-MSH, 200 μg/mL acetylated BSA, 50% glycerol, pH 7.4		ICN 159441
				TaKaRa
Escherichia coli RY13				
20-40 U/μL	1 unit gives >90% protection to 1 μg λ DNA against digestion with EcoR I	100 mM KPO$_4$, 200 mM NaCl, 1 mM EDTA, 2 mM DTT, 50% glycerol, 100 μg/mL BSA, pH 7.4	No detectable endonuclease, exonuclease	NBL Gene 021208 021210
40-100 U/mL	37°C	100 mM, 10 mM β-MSH, 200 M NaCl, 0.1 mM EDTA, 0.2 mg/mL BSA, 50% glycerol, pH 7.4		Nippon Gene 316-01211 312-01213
Escherichia coli strain, carrying the cloned modification gene from *Escherichia coli* RY13				
40,000 U/mL	1 unit protects 1 μg λ DNA/hr at 37°C in a reaction mixture of 10 μL against cleavage by EcoR I restriction endonuclease	200 mM NaCl, 100 mM KPO$_4$, 0.1 mM EDTA, 10 mM β-MSH, 200 μg/mL BSA, 50% glycerol, pH 7.4	No detectable contaminating endo- and exonuclease	NE Biolabs 211S 211L
Escherichia coli, recombinant				
20-40 U/μL	1 unit protects, by greater than 90%, 1 μg λ DNA against cleavage by the restriction endonuclease EcoR I/hr at pH 8.0, 37°C	1 M Tris-HCl, 100 mM EDTA, pH 8.0 and BSA and S-adenosylmethionine		Promega M4511 M4512

FnuD II Methylase

METHYLATION SEQUENCE
$$5'...Cm^5GCG...3'$$
$$3'...GCGCm^5...5'$$

SYNONYMS
Modification methylase, restriction-modification system

NOTES
- EC 2.1.1.73
- A site-specific DNA-methyltransferase (cytosine-specific)
- Forms "restriction-modification systems" in conjunction with restriction enzymes of similar site specificity listed under EC 3.1.21.3, 3.1.21.4 or 3.1.21.5
- Modifies the external cytosine residue (C^5) in the methylation sequence
- EDTA or high salt concentrations inhibit contaminating endo- and exonucleases

CONCENTRATION or VOLUME ACTIVITY	UNIT DEFINITION	REACTION BUFFER (As Provided)	ADDITIONAL ACTIVITIES and PURITY	SUPPLIER CATALOG NO.
Escherichia coli strain, carrying the cloned modification gene from *Fusobacterium nucleatum* D				
10,000 U/mL	1 unit protects 1 µg λ DNA/hr at 37°C in a reaction mixture of 10 µL against cleavage by *Fnu*D II or *Bst*U I restriction endonuclease	50 m*M* KCl, 10 m*M* Tris-HCl, 0.1 m*M* EDTA, 1 m*M* DTT, 200 µg/mL BSA, 50% glycerol, pH 7.5	No detectable contaminating endo- and exonuclease	NE Biolabs 227S 227L
Fusobacterium nucleatum D				
5000-15,000 U/mL		50 m*M* KCl, 10 m*M* Tris-HCl, 0.1 m*M* EDTA, 1 m*M* DTT, 200 µg/mL acetylated BSA, 50% glycerol, pH 7.5		ICN 159444

Hae III Methylase

METHYLATION SEQUENCE
   5'...GGCm^{5}C...3'
   3'...CCm^{5}GG...5'

SYNONYMS
   Modification methylase, restriction-modification system

NOTES
- EC 2.1.1.73
- A site-specific DNA-methyltransferase (cytosine-specific)
- Forms "restriction-modification systems" in conjunction with restriction enzymes of similar site specificity listed under EC 3.1.21.3, 3.1.21.4 or 3.1.21.5
- Modifies the internal cytosine residue (C^5) in the methylation sequence
- EDTA or high salt concentrations inhibit contaminating endo- and exonucleases
- Protects DNA against cleavage by *Not* I and *Sfi* I

CONCENTRATION or VOLUME ACTIVITY	UNIT DEFINITION	REACTION BUFFER (As Provided)	ADDITIONAL ACTIVITIES and PURITY	SUPPLIER CATALOG NO.
Escherichia coli strain, carrying the cloned modification gene from *Haemophilus aegyptius*				
5000 U/mL	1 unit protects 1 µg λ DNA/hr at 37°C in a reaction mixture of 10 µL against cleavage by *Hae* III restriction endonuclease	50 mM KCl, 50 mM Tris-HCl, 10 mM EDTA, 1 mM DTT, 200 µg/mL BSA, 50% glycerol, pH 7.5	No detectable contaminating endo- and exonuclease	NE Biolabs 224S 224L
Haemophilus aegyptius				
1000-5000 U/mL		50 mM KCl, 10 mM Tris-HCl, 10 mM EDTA, 1 mM DTT, 200 µg/mL acetylated BSA, 50% glycerol, pH 7.5		ICN 159446
Unspecified				
				TaKaRa

Hap II Methylase

METHYLATION SEQUENCE
$5'...CCm^5GG...3'$
$3'...GGCm^5C...5'$

SYNONYMS
Modification methylase, restriction-modification system

NOTES
- EC 2.1.1.73
- A site-specific DNA-methyltransferase (cytosine-specific)
- Forms "restriction-modification systems" in conjunction with restriction enzymes of similar site specificity listed under EC 3.1.21.3, 3.1.21.4 or 3.1.21.5
- Modifies the internal cytosine residue (C^5) in the methylation sequence, yielding C^5-methyl cytosine
- EDTA or high salt concentrations inhibit contaminating endo- and exonucleases

CONCENTRATION or VOLUME ACTIVITY	UNIT DEFINITION	REACTION BUFFER (As Provided)	ADDITIONAL ACTIVITIES and PURITY	SUPPLIER CATALOG NO.
Unspecified				
				TaKaRa

Hha I Methylase

METHYLATION SEQUENCE
$5'...GCm^5GC...3'$
$3'...CGCm^5G...5'$

SYNONYMS
Modification methylase, restriction-modification system

NOTES
- EC 2.1.1.73
- A site-specific DNA-methyltransferase (cytosine-specific)
- Forms "restriction-modification systems" in conjunction with restriction enzymes of similar site specificity listed under EC 3.1.21.3, 3.1.21.4 or 3.1.21.5
- Modifies the internal cytosine residue (C^5) in the methylation sequence
- EDTA or high salt concentrations inhibit contaminating endo- and exonucleases

Hha I Methylase continued

CONCENTRATION or VOLUME ACTIVITY	UNIT DEFINITION	REACTION BUFFER (As Provided)	ADDITIONAL ACTIVITIES and PURITY	SUPPLIER CATALOG NO.
Escherichia coli strain, carrying the cloned modification gene from Haemophilus haemolyticus				
25,000 U/mL	1 unit protects 1 μg λ DNA/hr at 37°C in a reaction mixture of 10 μL against cleavage by Hha I restriction endonuclease	150 mM NaCl, 50 mM Tris-HCl, 10 mM EDTA, 5 mM β-MSH, 200 μg/mL BSA, 50% glycerol, pH 7.5	No detectable contaminating endo- and exonuclease	NE Biolabs 217S 217L
Haemophilus haemolyticus				
5000-25,000 U/mL		50 mM NaCl, 50 mM Tris-HCl, 10 mM EDTA, 5 mM β-MSH, 200 μg/mL acetylated BSA, 50% glycerol, pH 7.5		ICN 159407

Hind III Methylase

SYNONYMS
  Modification methylase, restriction-modification system

NOTES
- EC 2.1.1.73
- A site-specific DNA-methyltransferase (cytosine-specific)
- Forms "restriction-modification systems" in conjunction with restriction enzymes of similar site specificity listed under EC 3.1.21.3, 3.1.21.4 or 3.1.21.5
- EDTA or high salt concentrations inhibit contaminating endo- and exonucleases

CONCENTRATION or VOLUME ACTIVITY	UNIT DEFINITION	REACTION BUFFER (As Provided)	ADDITIONAL ACTIVITIES and PURITY	SUPPLIER CATALOG NO.
Unspecified				TaKaRa

Hpa II Methylase

METHYLATION SEQUENCE
5'...CCm^{5}GG...3'
3'...GGCm^{5}C...5'

SYNONYMS
Modification methylase, restriction-modification system

NOTES
- EC 2.1.1.73
- A site-specific DNA-methyltransferase (cytosine-specific)
- Forms "restriction-modification systems" in conjunction with restriction enzymes of similar site specificity listed under EC 3.1.21.3, 3.1.21.4 or 3.1.21.5
- Recognizes the same sequence as Msp I methylase, but modifies the internal cytosine residue (C^5) in the methylation sequence
- EDTA or high salt concentrations inhibit contaminating endo- and exonucleases

CONCENTRATION or VOLUME ACTIVITY	UNIT DEFINITION	REACTION BUFFER (As Provided)	ADDITIONAL ACTIVITIES and PURITY	SUPPLIER CATALOG NO.
Escherichia coli strain, carrying the cloned *Hpa* II M gene from *Haemophilus influenzae*				
8-12 U/μL	1 unit protects 1 μg λ DNA/hr at 37°C in a 50 μL reaction buffer against cleavage by complementary restriction endonuclease	10 mM KPO$_4$, 1 mM EDTA, 7 mM β-MSH, 50% glycerol, 0.2 mg/mL BSA, pH 7.5	No detectable other methylases <0.02% 5'-, 3'-exonuclease, nonspecific endonuclease, phosphatase contamination	Fermentas EM0241 EM0242
Escherichia coli strain, carrying the cloned modification gene from *Haemophilus parainfluenzae*				
1000-5000 U/mL	1 unit protects 1 μg λ DNA/hr at 37°C in a reaction mixture of 10 μL against cleavage by *Hpa* II restriction endonuclease	50 mM Tris-HCl, 10 mM EDTA, 5 mM β-MSH, 200 μg/mL BSA, 50% glycerol, pH 7.5	No detectable contaminating endo- and exonuclease	NE Biolabs 214S 214L
Haemophilus parainfluenzae				
1000-5000 U/mL		50 mM Tris-HCl, 10 mM EDTA, 5 mM β-MSH, 200 μg/mL acetylated BSA, 50% glycerol, pH 7.5		ICN 159447

Msp I Methylase

METHYLATION SEQUENCE
5′...Cm^{5}CGG...3′
3′...GGCCm^5...5′

SYNONYMS
Modification methylase, restriction-modification system

NOTES
- EC 2.1.1.73
- A site-specific DNA-methyltransferase (cytosine-specific)
- Forms "restriction-modification systems" in conjunction with restriction enzymes of similar site specificity listed under EC 3.1.21.3, 3.1.21.4 or 3.1.21.5
- Recognizes the same sequence as *Hpa* II methylase, but modifies the external cytosine residue (C^5) in the methylation sequence
- EDTA or high salt concentrations inhibit contaminating endo- and exonucleases

CONCENTRATION or VOLUME ACTIVITY	UNIT DEFINITION	REACTION BUFFER (As Provided)	ADDITIONAL ACTIVITIES and PURITY	SUPPLIER CATALOG NO.
Escherichia coli strain, carrying the cloned modification gene from *Moraxella* species				
5000 U/mL	1 unit protects 1 μg λ DNA/hr at 37°C in a reaction mixture of 10 μL against cleavage by *Msp* I restriction endonuclease	50 m*M* NaCl, 50 m*M* Tris-HCl, 10 m*M* EDTA, 5 m*M* β-MSH, 200 μg/mL BSA, 50% glycerol, pH 7.5	No detectable contaminating endo- and exonuclease	NE Biolabs 215S 215L
Moraxella species				
4000-20,000 U/mL		50 m*M* KCl, 10 m*M* Tris-HCl, 0.1 m*M* EDTA, 1 m*M* DTT, 200 μg/mL acetylated BSA, 50% glycerol, pH 8.0		ICN 159453

Mva I Methylase

METHYLATION SEQUENCE

$5'...CCm^4(^A_T)GG...3'$
$3'...GG(^T_A)Cm^4C...5'$

SYNONYMS

Modification methylase, restriction-modification system

NOTES

- EC 2.1.1.73
- A site-specific DNA-methyltransferase (cytosine-specific)
- Forms "restriction-modification systems" in conjunction with restriction enzymes of similar site specificity listed under EC 3.1.21.3, 3.1.21.4 or 3.1.21.5
- Modifies the internal cytosine residue in the methylation sequence to yield (N^4)-methyl cytosine
- EDTA or high salt concentrations inhibit contaminating endo- and exonucleases

CONCENTRATION or VOLUME ACTIVITY	UNIT DEFINITION	REACTION BUFFER (As Provided)	ADDITIONAL ACTIVITIES and PURITY	SUPPLIER CATALOG No.
Escherichia coli strain, carrying the cloned *Mva* IM gene from *Micrococcus varians* RFL19				
4-8 U/μL	1 unit protects 1 μg λ DNA/hr at 37°C in a 50 μL reaction buffer against cleavage by complementary restriction endonuclease	10 mM Tris-HCl, 50 mM NaCl, 0.1 mM EDTA, 1 mM DTT, 50% glycerol, 0.2 mg/mL BSA, pH 7.4	No detectable other methylases <0.02% 5'-, 3'-exonuclease, nonspecific endonuclease, phosphatase contamination	Fermentas EM0251 EM0252

Pst I Methylase

METHYLATION SEQUENCE
5'...CTGCAm^{6}G...3'
3'...GAm^{6}CGTC...5'

SYNONYMS
Modification methylase, restriction-modification system

OPTIMUM REACTION CONDITIONS
pH 7.5 and 37°C

NOTES
- EC 2.1.1.72
- A site-specific DNA-methyltransferase (adenine-specific)
- Forms "restriction-modification systems" in conjunction with restriction enzymes of similar site specificity listed under EC 3.1.21.3, 3.1.21.4 or 3.1.21.5
- Modifies the adenine residue (N^6) in the methylation sequence
- EDTA or high salt concentrations inhibit contaminating endo- and exonucleases

CONCENTRATION or VOLUME ACTIVITY	UNIT DEFINITION	REACTION BUFFER (As Provided)	ADDITIONAL ACTIVITIES and PURITY	SUPPLIER CATALOG NO.
Escherichia coli, carrying the cloned modification gene from *Providencia stuartii*				
1000-5000 U/mL	1 unit protects 1 μg λ DNA/hr at 37°C in a reaction mixture of 10 μL against cleavage by *Pst* I restriction endonuclease	50 mM NaCl, 50 mM Tris-HCl, 10 mM EDTA, 10 mM β-MSH, 200 μg/mL BSA, 50% glycerol, pH 7.5	No detectable contaminating endo- and exonuclease	NE Biolabs 216S
Providencia stuartii				
1000-5000 U/mL		50 mM Tris-HCl, 50 mM KCl, 10 mM EDTA, 10 mM β-MSH, 200 μg/mL acetylated BSA, 50% glycerol, pH 7.5		ICN 159464

Sss I (CpG) Methylase

METHYLATION SEQUENCE
5'...Cm^{5}G...3'
3'...GCm^5...5'

SYNONYMS
Modification methylase, restriction-modification system

Applications
- blocking restriction endonuclease cleavage
- studying CpG methylation-dependent gene expression
- probing sequence-specific contacts within the major groove of DNA
- altering the physical properties of DNA
- uniform [^{3}H]-labeling DNA
- decreasing the number of sites cut by class II enzymes, yielding an apparent increase in specificity

NOTES
- EC 2.1.1.73
- A site-specific DNA-methyltransferase (cytosine-specific)
- Forms "restriction-modification systems" in conjunction with restriction enzymes of similar site specificity listed under EC 3.1.21.3, 3.1.21.4 or 3.1.21.5
- Methylates all cytosine residues (C^5) within the dinucleotide methylation sequence
- EDTA or high salt concentrations inhibit contaminating endo- and exonucleases

CONCENTRATION or VOLUME ACTIVITY	UNIT DEFINITION	REACTION BUFFER (As Provided)	ADDITIONAL ACTIVITIES and PURITY	SUPPLIER CATALOG NO.
Escherichia coli strain, carrying the methylase gene from *Spiroplasma* sp. strain MQ1				
2000 U/mL	1 unit protects 1 µg λ DNA/hr at 37°C in a reaction mixture of 20 µL against cleavage by *Bst*U I restriction endonuclease	10 m*M* Tris-HCl, 0.1 m*M* EDTA, 1 m*M* DTT, 200 µg/mL BSA, 50% glycerol, pH 7.4		NE Biolabs 226S 226L

Taq I Methylase

METHYLATION SEQUENCE
5'...TCGAm^6...3'
3'...Am^{6}GCT▲...5'

SYNONYMS
Modification methylase, restriction-modification system

NOTES
- EC 2.1.1.72
- A site-specific DNA-methyltransferase (adenine-specific)
- Forms "restriction-modification systems" in conjunction with restriction enzymes of similar site specificity listed under EC 3.1.21.3, 3.1.21.4 or 3.1.21.5
- Modifies the adenine residue (N^6) in the methylation sequence
- EDTA or high salt concentrations inhibit contaminating endo- and exonucleases

CONCENTRATION or VOLUME ACTIVITY	UNIT DEFINITION	REACTION BUFFER (As Provided)	ADDITIONAL ACTIVITIES and PURITY	SUPPLIER CATALOG No.
Escherichia coli strain, carrying the cloned modification gene from Thermus aquaticus				
10,000 U/mL	1 unit protects 1 μg λ DNA/hr at 65°C in a reaction mixture of 10 μL against cleavage by *Taq* I restriction endonuclease	100 m*M* NaCl, 10 m*M* Tris-HCl, 0.1 m*M* EDTA, 1 m*M* DTT, 200 μg/mL BSA, 50% glycerol, pH 7.4	No detectable contaminating endo- and exonuclease	NE Biolabs 219S 219L
Thermus aquaticus				
4000-20,000 U/mL		100 m*M* NaCl, 10 m*M* Tris-HCl, 0.1 m*M* EDTA, 1 m*M* DTT, 200 μg/mL acetylated BSA, 50% glycerol, pH 7.4		ICN 159469

TthHB8 I Methylase

SYNONYMS
Modification methylase, restriction-modification system

NOTES
- EC 2.1.1.73
- A site-specific DNA-methyltransferase (cytosine-specific)
- Forms "restriction-modification systems" in conjunction with restriction enzymes of similar site specificity listed under EC 3.1.21.3, 3.1.21.4 or 3.1.21.5
- EDTA or high salt concentrations inhibit contaminating endo- and exonucleases

CONCENTRATION or VOLUME ACTIVITY	UNIT DEFINITION	REACTION BUFFER (As Provided)	ADDITIONAL ACTIVITIES and PURITY	SUPPLIER CATALOG NO.
Thermus thermophilus HB8				
				TaKaRa

Part 4. Unclassified Enzymes

Chapter 8. Enzymes Not Assigned an EC Number

Unclassified Carbohydrases

BIOLOGICAL SOURCE	SPECIFIC ACTIVITY	UNIT DEFINITION	PREPARATION FORM	SUPPLIER CATALOG No.
Galactomannase				
Effective in breaking down cellulosic and hemicellulosic polymers associated with plant cell walls like galactomannans, β-glucans, xylans and other cell wall polymers. Used in coffee processing				
Aspergillus niger	0.2 U/mg	1 unit releases 1 μmol reducing sugar equivalents (as maltose)/min at pH 4.0, 37°C using galactomannan as substrate	Lyophilized	Fluka 48235
Aspergillus niger	10,800 LBGV U/g	Locust bean gum viscosity is a relative activity unit reflecting the reduction in viscosity of a standard locust bean gum solution at pH 4.8, 22°C	Powder; also contains cellulase, hemicellulase, β-glucanase, xylanase optimum pH = 2.5-4.5, T = 65-70°C	Karlan 4000
Lacto-*N*-biosidase				
Specifically hydrolyzes type 1 chain oligosaccharides, producing lacto-N-biose (Galb-3GlcNAc). Does not hydrolyze type 2 chain oligosaccharides, α-2,3-sialyllacto-N-tetraose, or lact-N-fucopentaose I and II				
Streptomyces species 142		1 unit hydrolyzes 1 μmol PA-lacto-*N*-tetraose/min at pH 5.5, 37°C	50 mM NaOAC buffer solution containing 0.05% Brij 58 optimum pH = pH 5.5	PanVera TAK 4456
Streptomyces species 142	100 μU/vial	1 unit hydrolyzes 1 μmol PA-lacto-*N*-tetraose/min at pH 5.5, 37°C	50 mM NaOAc buffer solution containing 0.05% Brij 58 optimum pH = pH 5.5	TaKaRa 4456
Meicelase				
Very active against β-1,4-glucan, β-1,3-glucan and β-1,4-xylan				
Trichoderma viride	8,000,000 U/g	Unit is based on filter paper decomposing activity (modified Sotoyama method)	Powder optimum pH = 3.0-7.0, T = 20-60°C	Karlan 6250
Naringinase				
Penicillium decumbens	300-600 U/g solid	1 unit liberates 1.0 μmol reducing sugar (as glucose) from naringin/min at pH 4.0, 40°C	≤150 U/g β-glucosidase (1 unit liberates 1 μmol glucose from salicin/min at pH 5.25, 37°C)	Sigma N1385
Textile and laundry				
Bacillus stearothermophilus				EnzymeDev Enzeco[R] Exsizease 2.5X
Bacillus subtilis				EnzymeDev Enzeco[R] LTA 5000

BIOLOGICAL SOURCE	SPECIFIC ACTIVITY	UNIT DEFINITION	PREPARATION FORM	SUPPLIER CATALOG No.
Textile and laundry *continued*				
Unspecified				SpecialtyEnz Denifade
Unspecified				SpecialtyEnz Desize 3X
Unspecified				SpecialtyEnz Softzyme
Unspecified				SpecialtyEnz Stonezyme
α-Xylosidase				
Bacillus species 693-1	0.2 U/mg	1 unit releases 1 μmol *p*-nitrophenol from *p*-nitrophenyl-α-D-xyloside/min at pH 7.5, 45°C	Lyophilized optimum pH = 7.5	Seikagaku 120460-1

BIOLOGICAL SOURCE	SPECIFIC ACTIVITY	UNIT DEFINITION	PREPARATION FORM	SUPPLIER CATALOG NO.
rec A protein				
Escherichia coli strain KM 1842	≥0.25 U/mg protein; 200 µg/vial	1 unit hydrolyzes 1.0 µmol UTP/min at pH 6.2, 37°C in the presence of single stranded calf thymus DNA	Solution containing 10% glycerol, 20 mM Tris-HCl, 0.1 mM DTT, 0.1 mM EDTA, pH 7.5	Sigma R4005
Ribonuclease C				
Human plasma	40,000–100,000 U/mg protein	1 unit produces acid soluble oligonucleotides equivalent to a ΔA_{260} of 2.0/15 min at pH 7.5, 25°C in a 1.0 mL reaction volume with polycytidylic acid (5′) as substrate	Solution containing 20% glycerol, 0.05 M KPO$_4$, 2 M KCl, pH 8	Sigma R7007
λ-Terminase Recognizes a sequence of 100 Bp at the cos region of λ DNA, cleaving the cos site to generate a 5′ protruding end with 12 bases or 4-5 bases. Linearizes the cosmid or λ vector so that clones can be mapped				
Escherichia coli JM109, carrying a plasmid enclosing the A gene and Nu gene of λ phage	8 U/µL; 200 U	1 unit digests 50% of 0.5 µg *Pst* I-digested pHC79 DNA/2 hr at 30°C in 25 µL of the basal buffer	Solution containing 10 mM Tris-HCl, 1 mM EDTA, 10 mM β-MSH, 50% glycerol, pH 8.0; no detectable non-specific nuclease	TaKaRa 2810
Escherichia coli JM109, carrying a plasmid enclosing the A gene and Nu gene of λ phage		1 unit digests 50% of 0.5 µg *Pst* I-digested pHC79 DNA/2 hr at 30°C in 25 µL basal buffer	Solution containing 10 mM Tris-HCl, 1 mM EDTA, 10 mM β-MSH, 50% glycerol, pH 8.0; no detectable non-specific nuclease	PanVera TAK 2810
Escherichia coli, recombinant			10X Reaction buffer: 130 mM Tris-HCl, 5 mM EDTA, 30 mM MgCl$_2$, 50 mM DTT, 500 mM KOAc, 20 mM spermidine, 40 mM putrescine, 10 mM ATP, 0.1% Triton X-100, pH 8.0 <1% DNase, <3% RNase, ≥90% supercoiled plasmid	Promega M1671 M1672
A and Nu1 genes of bacteriophage λ		1 unit linearizes a minimum of 70% of 1 µg of cosmid DNA vector pHC79/30 min at pH 7.8, RT	Solution containing 50% glycerol, 50 mM Tris-HCl, 0.1 M NaCl, 0.1 mM EDTA, 1 mM DTT, 0.1% Triton X-100, pH 7.5; no detectable exogenous RNase, endonuclease or ss-exonuclease	Epicentre LT4450 LT44200
Escherichia coli JM109, carrying a plasmid enclosing the A gene and Nu gene of λ phage	10 U/µL	1 unit digests 50% 0.5 µg *Pst* I-digested pHC79 DNA/2 hr at 30°C in 25 µL of basal buffer	10 mM Tris-HCl, 1 mM EDTA, 10 mM β-MSH, 50% glycerol, pH 8.0; no detectable non-specific nuclease	Amersham E 2810Y

Unclassified Proteases

BIOLOGICAL SOURCE	SPECIFIC ACTIVITY	UNIT DEFINITION	PREPARATION FORM	SUPPLIER CATALOG No.
Alkaline protease				
Unspecified				SpecialtyEnz
Ecarin				
Prothrombin activator				
Echis carinatus venom	50 U/vial	1 unit activates prothrombin to produce 1 unit of amidolytic activity at pH 8.4, 37°C; 1 amidolytic unit hydrolyzes 1.0 μmol N-p-tosyl-Gly-Pro-Arg-p-nitroanilide/min at pH 8.4, 37°C	Contains thimerosal and lactose	Sigma E0504
Gelsolin				
An actin severing protein found in mammalian cells and blood plasma. Recently reported to decrease the viscosity of cystic fibrosis sputum samples in vitro				
Bovine plasma	20-100 U/mg protein	1 unit reduces the viscosity difference between actin solution and buffer by 50% at 28°C in a 1 mL mixture	Lyophilized containing 5% protein; balance NaCl, Tris buffer salt and EGTA	Sigma G8032
Metalloendopeptidase				
Specifically cleaves on the amino-terminal side of lysine residues. Purified from the fruiting bodies of a higher Basidiomycete				
Grifola frondosa	10 U/vial	1 unit changes A_{366} by 1.0/30 min at pH 10.0, 37°C using azocasein as substrate	Lyophilized	ICN 151616
Grifola frondosa	2270 U/mg protein	1 unit changes A_{366} by 1.0/30 min at pH 10.0, 37°C using azocasein as substrate	Homogeneous by SDS-PAGE; lyophilized; no detectable protease	ICN 320841
Grifola frondosa	2000 U/mg protein	1 unit increases A_{366} by 1.0 O.D. unit/30 min at pH 10.0, 37°C	Lyophilized optimum pH = 10-10.5	Seikagaku 100965-1
Grifola frondosa	4 U/vial	1 unit hydrolyzes casein to produce peptide equivalent to 1.0 μmol tyrosine/min at pH 11.0, 30°C		Sigma M4394
Mutanolysin				
Proteolytic enzyme used for cell lysis				
Streptomyces globisporus	50000 U/mg	1 unit decreases A_{600} by 0.001/min at pH 6.8, 37°C, Streptococcus faecalis cell walls as substrate	Lyophilized <0.00001% protease	Fluka 70017
Streptomyces globisporus ATCC 21553	4000-8000 U/mg protein	1 unit/mL produces a ΔA_{600} of 0.01/min at pH 6.0, 37°C using a suspension of S. faecalis cell walls as substrate	Chromatographically purified; lyophilized containing Ficoll and Na succinate buffer salts	Sigma M9901
Neutral protease				
Unspecified				SpecialtyEnz

Unclassified Proteases

BIOLOGICAL SOURCE	SPECIFIC ACTIVITY	UNIT DEFINITION	PREPARATION FORM	SUPPLIER CATALOG NO.
Prolase **Used in food processing**				
Papaya	0.3-0.5 U/mg solid	1 unit hydrolyzes casein producing the color equivalent of 1.0 µmol tyrosine/min at pH 7.5, 37°C	Crude No detectable DNase, RNase	ICN 101024
Unspecified				SpecialtyEnz Prolase
TEV protease **Has a 7 AA recognition site making it a highly site-specific protease. Used in the removal of affinity tags from fusion proteins. Purified as fusion protein s/rTEV**				
Escherichia coli, expressing the TEV protease gene	10 U/µL	1 unit cleaves ≥95% of 3 µg control substrate/hr at pH 8.0, 30°C	>95% single-band purity; 1 *M* Tris-HCl, 10 m*M* EDTA, pH 8.0 and 100 m*M* DTT; no detectable non-specific protease optimum pH = 6.0-8.5	Life Technol 10127-017

Unclassified Macerating and Lytic Enzymes

BIOLOGICAL SOURCE	SPECIFIC ACTIVITY	UNIT DEFINITION	PREPARATION FORM	SUPPLIER CATALOG NO.
Acutase				
Agkistrodon acutus (Formosan Hundred-pace snake) venom	60 U/mg	NIH units obtained by direct comparison to an NIH thrombin reference standard	Lyophilized containing 10% protein; balance NaOAc, pH 7.0	Sigma A2669
Agkistrodon acutus (Formosan Hundred-pace snake) venom	100-400 NIH U/mg protein	NIH units obtained by direct comparison to an NIH thrombin reference standard	Lyophilized containing 10% protein; balance NaOAc; unaffected by heparin concentration up to 50 USP U/mL	Sigma A4818
Driselase[R] **Used in preparation of protoplasts**				
Basidiomycetes species	1 U/mg	1 unit hydrolyzes 1 µmol OBR-hydroxyethylcellulose/min at pH 4.8, 30°C	Powder; also contains cellulase, pectinase, xylanase, dextranase, laminarinase, amylase, protease, macerating and lytic activity	Fluka 44585
Lysopeptase **Best suited for hydrolysis of *Staphylococci* cell walls. Hydrolyzes glycyl-glycine or alanyl-alanine linkages in the bridge peptide of cell walls. Also hydrolyzes *N*-acetyl-muramyl-*L*-alanine bonds at a slow rate**				
Cytophaga B-30	>2000 U/mg protein	1 unit decreases turbidity by 50% in *S. epidermidis* cells/hr at 37°C	Lyophilized	ICN 320831
Macerozyme **Useful for separating cells**				
Rhizopus species	3000 maceration U/g	1 unit measures macerating activity on potato tuber disks as determined by weight loss at pH 5.0, 40°C	Powder	ICN 152340
Rhizopus species	3000 maceration U/g	Unit is based on the volume of single cells released from potato tuber disks	Powder optimum pH = 5.0-6.0, T = 40-50°C	Karlan Macerozyme R10 6152
Streptolysin S				
Streptococcus pyogenes	2000-10,000 U/mg protein	1 unit causes 50% lysis of a 2% red blood cell suspension in PO_4 buffered saline/45 min at pH 7.4, 37°C	Lyophilized containing 3% protein; balance is core RNA, PO_4 buffer salts and NaCl	Sigma S2888
Yeast Lytic Enzyme **Composed of β-1,3-glucanase, protease, hemicellulase, amylase and pectinase. Lyses cell walls and membranes. Used for isolation of yeast spores and spheroplasts, spheroplast transformation and generation of DNA in agarose plugs**				
Arthrobacter luteus	1800 U/mg	1 unit decreases A_{800} by 0.001/min at pH 7.5, 25°C with brewers yeast as substrate	Powder	Fluka Lyticase 62979

Unclassified Macerating and Lytic Enzymes

BIOLOGICAL SOURCE	SPECIFIC ACTIVITY	UNIT DEFINITION	PREPARATION FORM	SUPPLIER CATALOG No.
Yeast lytic enzyme *continued*				
Arthrobacter luteus	1500 U/mg	1 unit decreases A_{800} by 0.001/min at pH 7.5, 25°C with brewers yeast as substrate	Lyophilized	Fluka Lyticase 62980
Arthrobacter luteus	500 U/mg	1 unit decreases A_{800} by 0.001/min at pH 7.5, 25°C with brewers yeast as substrate	Powder	Fluka Lyticase 62982
Arthrobacter luteus	20,000 U/g	1 unit produces a ΔA_{800} of 0.001/min at pH 7.5, 25°C using a brewers yeast suspension as substrate in a 3 mL reaction mixture	Lyophilized	ICN 153526
Arthrobacter luteus	100,000 U/g	1 unit produces a ΔA_{800} of 0.001/min at pH 7.5, 25°C using a brewers yeast suspension as substrate in a 3 mL reaction mixture	Lyophilized	ICN 190123
Arthrobacter luteus	> 70,000 lytic U/g	1 unit causes a ΔA_{800} of 0.001/min at pH 7.5, 25°C using a suspension of brewers yeast as substrate in a 3 mL reaction mixture		ICN 360941 360942 360943 360944
Arthrobacter luteus	5000 lytic U/g	1 unit causes a ΔA_{800} of 0.001/min at pH 7.5, 25°C using a suspension of brewers yeast as substrate in a 3 mL reaction mixture		ICN 360951 360952 360953 360954
Arthrobacter luteus	4000-10,000 U/mg protein	1 unit produces a ΔA_{800} of 0.001/min at pH 7.5, 25°C using brewers yeast as substrate in a 3 mL reaction mixture	Partially purified; powder containing 20% protein; balance $(NH_4)_2(SO_4)$ and stabilizer	Sigma Lyticase L5263
Arthrobacter luteus	5000-20,000 U/mg protein	1 unit produces a ΔA_{800} of 0.001/min at pH 7.5, 25°C using brewers yeast as substrate in a 3 mL reaction mixture	Partially purified; lyophilized containing 20% protein; balance primarily KPO_4 buffer salts	Sigma Lyticase L5763
Arthrobacter luteus	200-1000 U/mg solid	1 unit produces a ΔA_{800} of 0.001/min at pH 7.5, 25°C using brewers yeast as substrate in a 3 mL reaction mixture	Lyophilized	Sigma Lyticase L8012
Arthrobacter luteus	100,000 U/g	1 unit decreases A_{800} of brewers yeast suspension by 30%/2 hr at 25°C	Lyophilized optimum pH = 7.5, T = 35°C for lysis of viable yeast; pH = 6.5, T = 45°C for hydrolysis of yeast glucan	Seikagaku Zymolase 100T 120493-1

Unclassified Macerating and Lytic Enzymes

BIOLOGICAL SOURCE	SPECIFIC ACTIVITY	UNIT DEFINITION	PREPARATION FORM	SUPPLIER CATALOG NO.
Yeast lytic enzyme *continued*				
Arthrobacter luteus	20,000 U/g	1 unit decreases A_{800} of brewers yeast suspension by 30%/2 hr at 25°C	Lyophilized optimum pH = 7.5, T = 35°C for lysis of viable yeast; pH = 6.5, T = 45°C for hydrolysis of yeast glucan	Seikagaku Zymolase 20T 120491-1
Achromobacter species	15,000 U/g	1 unit decreases A_{660} by 0.01/min at pH 7.0, 40°C	Powder optimum pH = 7, T = 55°C	Amano YL-15
Achromobacter species	5000 U/g lytic	1 unit produces a ΔA_{800} of 0.001/min at pH 7.5, 25°C using a brewers yeast suspension as substrate in a 3 mL reaction mixture	Lyophilized	ICN 150214
Arthrobacter species	>70,000 U/g lytic	1 unit produces a ΔA_{800} of 0.001/min at pH 7.5, 25°C using a brewers yeast suspension as substrate in a 3 mL reaction mixture	Lyophilized	ICN 152270
Oerskovia xanthineolytica	7000 U/mg protein	1 unit decreases A_{600} of yeast suspension by 10%/30 min at 30°C	Solution	Boehringer Lyticase 1372467
Oerskovia xanthineolytica	>10,000 U/mg protein	1 unit decreases A_{600} by 10% of a yeast suspension/30 min at 30°C	Solution containing 50 mM KPO$_4$ and 50% glycerol, pH 7.5	Calbiochem Lyticase 440485
Rhizoctonia solani	150,000 U/g	1 unit decreases turbidity 15%/min in a 1 mL reaction mixture		ICN 153527
Rhizoctonia solani	≥150,000 U/g lytic activity; ≥2000 U/g endo-β-1,3-glucanase activity		Powder optimum pH = 7.0, T = 35°C for lysis; pH = 5.0, T = 45°C for endo-glucanase	Wako Kitalase 118-00371 114-00373
Unspecified	20,000 lytic U/g		Purified	ICN Zymolase 20T 360921
Unspecified	100,000 lytic U/g		Purified	ICN Zymolase 100T 360931 360932

BIOLOGICAL SOURCE	SPECIFIC ACTIVITY	UNIT DEFINITION	PREPARATION FORM	SUPPLIER CATALOG No.
Bakery protease				
Sulfite replacement in crackers				
Aspergillus and plant				Danisco GRINDAMYL™ PR 34
Unidentified				
Unspecified				SpecialtyEnz Cracker Special
Unspecified				SpecialtyEnz Starzyme LHS
Unspecified				SpecialtyEnz Starzyme XC
Unspecified				SpecialtyEnz Enpezyme

Unclassified Miscellaneous Enzymes

BIOLOGICAL SOURCE	SPECIFIC ACTIVITY	UNIT DEFINITION	PREPARATION FORM	SUPPLIER CATALOG No.
Calculase **Softens hard dental plaque**				
Unspecified				ICN 100386
Cholesterol dehydrogenase				
Nocardia species	≥5 U/mg	1 unit produces 1 μmol NADH/min at pH 8.5, 25°C	Powder optimum pH = 10.0, pI = 4.2/5.3/6.8, T = 30°C, MW = 53,000 Da; stable pH 6.5-7.5 [37°C, 15 min], T < 35°C [pH 5.0]	Amano Cholesterol dehydrogenase
Diacetinase **Diacetyl glycerol ester + 2 water ↔ Glycerol + 2 fatty acid** **Used in determination of lipase activity in biological fluids**				
Bacillus subtilis	>20 U/mg protein; >100 U/mL	1 unit releases 1.0 μmol fatty acid from tributyrin/min at 37°C	Solution	Genzyme 6304
Glutaminic-γ-amino butyric acid-ase (GABase)				
Pseudomonas fluorescens	0.4 U/mg protein	1 unit converts 1.0 μmol GABA to succinic semialdehyde and then to succinate/min at pH 8.6, 20°C with a stoichiometric reduction of 1.0 μmol NADP	Partially purified; lyophilized	ICN 100657
α-Hemolysin **A channel forming protein similar to complement and perforin, penetrating the cell membrane and creating a defined size pore.** **Shown to stimulate cellular phospholipase activity**				
Staphylococcus aureus	10,000-80,000 U/mg protein	1 unit lyses 50% of a 1% suspension of rabbit red blood cells in PO_4 buffered saline containing 1% BSA/30 min at pH 7.0, 37°C	60% protein; balance primarily Na citrate buffer	Sigma H9395
Vibrio parahaemolyticus	400+ U/mg protein	1 unit lyses 50% of a 1% suspension of human red blood cells in PO_4 buffered saline/2 hrs at pH 7.0, 37°C	Kanagawa α-hemolysin	Sigma H3142
Luciferase				
Recombinant			Lyophilized, containing luciferin, EDTA, MgOAc, tricine buffer and BSA	Wako 123-03921

Unclassified Miscellaneous Enzymes

BIOLOGICAL SOURCE	SPECIFIC ACTIVITY	UNIT DEFINITION	PREPARATION FORM	SUPPLIER CATALOG NO.
Mashazyme				
Used for food processing				
				Specialty Enzymes
N-Acetyltransferase				
Found in human liver, acetylating a large number of hydrazine and arylamine derivatives. Mutations in the NAT-2 gene are associated with slow acetylator phenotype, resulting in drug toxicity & increased cancer risk				
Human, recombinant			Purified	PanVera NAT-1 NAT-2
Oligosaccharide dehydrogenase				
Staphylococcus species	100-300 U/mg solid	1 unit oxidizes 1 μmol maltose/min at pH 7.5, 37°C	Lyophilized. Activated by Ca^{2+}; inhibited by Zn^{2+}, EDTA optimum pH = 8.0, pI = 9.54, MW = 80,000 Da (gel filtration); K_M = 0.54 mM [β-D-glucose], 1.36 mM [maltose], 17 μM [PMS]; stable pH 6.5-8.5, T < 45°C [pH 7, 10 min]	Asahi ODH T-36
Peptide N-fatty Acylase				
Deacylates acyl α-amino acids, acyl ω-amino acids and acyl peptides. Removes the acyl group from acyl peptides without affecting the peptide moiety. Synonymous with the polymyxin acylase discovered by Kimura *et al.*				
Pseudomonas species	≥0.1 U/mg protein		Crude MW = 80,000 Da [gel filtration], 20,000 and 60,000 Da [SDS-PAGE]	Wako 164-16081
Pseudomonas species	≥10 U/mg protein		Purified	Wako 161-16091
Theophylline oxidase				
Microbial	>2 U/mg protein	1 unit reduces 3 μmol ferricyanide/min at pH 7.5, 30°C in the presence of theophylline	Solution optimum pH = 7.5, T = 30°C	GDS Tech TR-100

Part 5. Indexes

Index A. Enzyme Supplier Index

Enzyme Supplier Index

Aalto
USA: Aalto Scientific, Ltd.
Carlsbad, CA 92008
Phone: 619-431-7922
Fax: 619-431-6942

ABM-RP
United Kingdom: Rhone-Poulenc ABM Brewing & Enzymes Group
Stockport, Cheshire SK6 1PQ
Phone: 061-910-1500
Fax: 061-910-1529

Adv Biotech
United Kingdom: Advanced Biotechnologies Ltd.
Leatherhead, Surrey KT22 7BA
Phone: 01372-360-123
Fax: 01372-363-263
E-mail: adbio@advbtech.demon.co.uk

Australia: Integrated Sciences
Willoughby, New South Wales 2068
Phone: 00-612-417-7866
Fax: 00-612-417-5066

Austria: Handesgesell Schaft mbH
A-1210 Wien
Phone: 00-43-1292-3527
Fax: 00-43-1292-1361

Canada: Diamed Lab Supplies Inc.
Mississauga
Phone: 001-905-625-6021
Fax: 001-905-625-6280

Denmark: Kebo Lab A/S
DK-2620 Albertslund
Phone: 00-45-4386-8788
Fax: 00-45-4386-8790

Finland: Kebo Lab Oy
02270 Espoo
Phone: 00-358-0804-551
Fax: 00-358-0804-552

Germany: Dianova GmbH
20354 Hamburg
Phone: 00-40-40-323-074
Fax: 00-49-40-322-190

Israel: Getter (2,000) Ltd.
Ramat Gan 52511
Phone: 00-972-3576-1555
Fax: 00-972-3752-3620

Italy: Societa Italiana Chimici
00162 Rome
Phone: 00-39-64-424-4078
Fax: 00-39-64-429-0775

Japan: Cosmo Bio Co. Ltd.
Tokyo 135
Phone: 00-81-356-32-9605
Fax: 00-81-356-32-9618

Norway: Kebo Lab
N-0901 Oslo
Phone: 00-47-22-32-90-00-00
Fax: 00-47-22 32-90-00-40

South Africa: Southern Cross Biotechnology (Pty.) Ltd.
Cape Town
Phone: 00-27-21-615-166
Fax: 00-27-21-617-734

Spain: Comercial Rafer
50008 Zaragoza
Phone: 00-34-76-2374-00
Fax: 00-34-76-2171-52

Sweden: Kebo Lab. AB.
S-16394 Spanga
Phone: 00-46-8-521 3400
Fax: 00-46-8-621-3470

Adv Biofact
USA: Advance Biofactures Corp.
Lynbrook, NY 11563
Phone: 516-593-7000
Fax: 516-593-7039

Adv Immuno
USA: Advanced ImmunoChemical Inc.
Long Beach, CA 90803
Phone: 310-434-4676
Gratis phone: 800-788-0034
Fax: 310-494-3776

Amano
Japan: Amano Pharmaceutical Co., Ltd.
Nagoya
Phone: 052-211-3032
Fax: 052-211-3054

United Kingdom: Amano Enzyme Europe Ltd.
Phone: 0908-374117
Fax: 0908-378886

USA: Amano Enzyme USA Co., Ltd.
Troy, VA 22974
Phone: 804-589-8278
Gratis phone: 800-446-7652
Fax: 804-589-8270

USA: Amano Enzyme USA Co., Ltd.
Hinsdale, IL 60521
Phone: 708-850-9255
Fax: 708-850-9261

USA: Amano Enzyme USA Co., Ltd.
Lombard, IL
Phone: 708-953-1891
Gratis phone: 800-446-7652
Fax: 708-953-1895

Ambion
USA: Ambion, Inc.
Austin, TX 78744-1832
Phone: 512-445-6979
Gratis phone: 800-888-8804
Fax: 512-445-7139

Australia: Bresatec Ltd.
Adelaide, S.A. 5000
Phone: 61-8-2342644
Fax: 61-8-2342699

Australia: Medos Co. Pty. Ltd.
Burwood, Victoria 3125
Phone: 61-3-8089077
Fax: 61-3-8080926

Canada: PDI Bioscience, Inc.
Aurora
Gratis phone: 800-661-4556
Fax: 800-661-4557

Czechoslovakia: BioVendor
61600 Brno
Phone: 0049-69-6772127
Fax: 0049-69-6301-5937

France: CliniSciences
92120 Montrouge
Phone: 33-1-42-53-14-53
Fax: 33-1-46-56-97-33

Germany: ITC Biotechnology GmbH
D-69020 Heidelberg
Phone: 49-6221-303907
Fax: 49-6221-303511

Hong Kong: TWC
Shatin
Phone: 85-2-649-9988
Fax: 85-2-635-0379

Italy: Celbio Srl
20016 Pero MI
Phone: 39-2-38103171
Fax: 39-2-38101465

Japan: Funakoshi Co. Ltd.
Tokyo
Phone: 81-3-5684-1620
Fax: 81-3-5684-1775

Korea: BMS Bio-Medical & Science Co., Ltd.
Seoul 135-080
Phone: 82-02-569-6902
Fax: 82-02-553-9670

Enzyme Supplier Index

Singapore: Omega Medical & Scientific Pte. Ltd.
Singapore 0315
Phone: 65-2789688
Fax: 65-2782852

Spain: AMS Biotechnology (Espana) SL
28007 Madrid
Phone: 34-91-5515403
Fax: 34-91-4334545

Sweden: Intermedica
104 25 Stockholm
Phone: 46-8-7495940
Fax: 46-8-7495105

Switzerland: AMS Biotechnology (Europe Ltd.)
6934 Bioggio Lugano
Phone: 49-69-779099
Fax: 41-91-591785

Taiwan 106 R.O.C: Taiwan Ivy Corp.
Taipei
Phone: 886-02-700-2286
Fax: 886-02-700-3625

The Netherlands: Sanbio BV
5400 AM Uden
Phone: 31-4132-51115
Fax: 31-4132-66605

United Kingdom: AMS Biotechnology (UK) Ltd.
Witney Oxon OX8 7GE
Phone: 44-0993-706500
Fax: 44-0993-706006

Am Allied

USA: American Allied Biochemical, Inc.
Aurora, CO 80014
Phone: 303-755-7137
Gratis phone: 800-873-3018
Fax: 303-755-7135

Am Labs

USA: American Laboratories, Inc.
Omaha, NE 68127
Phone: 402-339-2494
Gratis phone: 800-445-5989
Fax: 402-339-0801

Amersham

USA: Amersham Life Science, Inc.
Arlington Heights, IL 60005
Phone: 847-593-6300
Gratis phone: 800-341-7543
Fax: 847-437-1640

AMRESCO

USA: AMRESCO Inc.
Solon, OH 44139-4300
Phone: 216-349-1199
Gratis phone: 800-448-4442
Fax: 216-349-1182

Australia: Astral Scientific
Gymea
Phone: +61-2-5402055
Fax: +61-2-5402051

Austria: Bio Trade
A 1230 Vienna
Phone: +43-1-889-18-19
Fax: +43-1-889-18-19-20

Brazil: Genoma Biotechnologia Brasil
Belo Horizonte/MG
Phone: +55-31-275-1240
Fax: +55-31-337-9324

Brazil: Sellex, Inc.
05013-001- Sao Paulo - SP
Phone: +55-11-872-2015
Fax: +55-11-872-1024

Canada: Inter Medico
Toronto
Phone: 905-470-2520
Gratis phone: 800-387-9643
Fax: 905-470-2381

Canada: Inter Medico
Markham
Gratis phone: 800-263-1626

Canada: Inter Medico
Gratis phone: 800-268-1150

Chile: Biocronogen Ltda.
Correo 47, LaFlorida
Phone: +56-2-287-3802
Fax: +56-2-287-3802

Czech Republic
(Czech Republic, Hungary & Poland):
Bio Vendor s.r.o.
602 00 Brno
Phone: +42-5-4221-6367-71
Fax: +42-5-4221-6364

Denmark: Medinova Scientific ApS
DK-2900 Hellerup
Phone: +45-31-56-2000
Fax: +45-31-56-1942

Egypt: Medicopharmatrade
Agouza
Phone: +20-2-3445774
Fax: +20-2-3465623

Finland: Finnzymes Oy
SF 02201 Espoo
Phone: 358-0-584-121
Fax: 358-0-584-12200

France: Interchim
75013 Paris
Phone: +33-47-07-99-72
Fax: +33-45-35-63-28

France: Interchim
03103 Montlucon
Phone: +33-70-03-88-55
Fax: +33-70-03-82-60

Germany: Biometra
37079 Gottingen
Phone: 05-51-50-686-0
Fax: 05-51-50-686-66

Greece: Diachel Diagnostic

Chemical Instrumentation Ltd.
116 34 Athens
Phone: +30-1-7235523/7243911
Fax: +30-1-7219874

Hong Kong (China & Hong Kong): BCH Medical Supplies Co.
Chai Wan
Phone: 852-8983008
Fax: 852-8983041

India: Biogene India
New Delhi 110015
Phone: +91-11-5451426
Fax: +91-11-5451426

Israel: Tamar
Jerusalem, 91350
Phone: 972-2-6520279
Fax: 972-2-6527318

Italy: Prodotti Gianni
20138 Milano
Phone: +39-2-5097-2
Fax: +39-2-5097-358

Japan: Cosmo Bio Co., Ltd.
Tokyo 135
Phone: +81-3-5632-9630
Fax: +81-3-5632-9624

Korea: KDR Biotech Co. Ltd.
Seoul 138-240
Phone: +82-2-203-0707
Fax: +82-2-424-7259

Mexico: Biomedix
Mexico, D.F. CP 08400
Phone: +52-5-654-4245
Fax: +52-5-654-4245

Republic of Ireland: Labkem Ltd.
Baldoyle, Dublin 13
Phone: 353-1-8391212
Fax: 353-1-8391521

Enzyme Supplier Index

Singapore & Malaysia: All Eight Marketing Service
Singapore 1336
Phone: +65-288-6388
Fax: +65-284-9805

Slovakia: Bio Vendor Ltd.
949 01 Nitra
Phone: +42-87-31538
Fax: +42-87-31538

South Africa: Anatech Instruments
Sloane Park 2152
Phone: +27-11-792-3300
Fax: +27-11-792-3363

Spain: Interlabo S. A.
28007 Madrid
Phone: +34-1-551-14-91
Fax: +34-1-552-94-03

Sweden: A. B. Labassco
S-431 53 Moindal
Phone: +46-31-706-30-00
Fax: +46-31-706-30-30

Switzerland: P.H. Stehelin & Cie AG
CH-4003 Basel
Phone: +41-61-272-3924
Fax: +41-61-271-3907

Switzerland (Yugoslav Distributor): Lightning Instrumentation
Modern City B, CH-1012
Phone: +41-21-728-30-66
Fax: +41-21-728-30-67

Taiwan, R.O.C.: Protech Technology Enterprise Co., Ltd.
Taipei
Phone: 886-2-381-0844
Fax: 886-2-311-8524

Thailand: Gibthai Co., Ltd.
Bangkok 10310
Phone: +66-2-2748331-3
Fax: +66-2-2748336

The Netherlands: Integro B.V.
1506 VH Zaandam
Phone: +31-75-631-05-53
Fax: +31-75-6170564

United Kingdom: Biometra Ltd.
Kent ME160LS
Phone: 01622 678872
Fax: 01622 752774

USA: Midwest Scientific
Valley Park, MO 63088
Phone: 314-225-9997
Fax: 314-225-9998

USA: Continental Lab Products
San Diego, CA 92126
Phone: 619-549-7800
Gratis phone: 800-456-7741
Fax: 619-549-7865

Venezuela: Proserv Bioelectronica C.A.
Caracas
Phone: +58-2-267-6635
Fax: +58-2-267-6635

AGS Heidelb

Germany: Angewandte Gentechnologie Systeme GmbH
W-69123 Heidelberg
Phone: 0049-6221-831023
Fax: 0049-6221-840610
E-mail: AGS_Heidelberg@T-Online.de

Asahi

Japan: Asahi Chemical Industry Co., Ltd.
(Asahi Kasei Kogyo Kabushiki Kaisha)
Tokyo 108
Phone: +81-3-5476-8219
Fax: +81-3-5476-8277

ART

USA: Athens Research and Technology, Inc.
Athens, GA 30604
Phone: 706-546-0207
Fax: 706-546-7395

BASF

Germany: BASF Aktiengesellschaft Marketing Feed-Enzymes
D-67056 Ludwigshafen/Rhein
Phone: +49-621-60-99551
Fax: +49-621-60-40109

Beckman

USA: Beckman
OEM Diagnostic Products
Carlsbad, CA 92008
Phone: 619-438-9151
Fax: 619-438-6389

Biocatalysts

United Kingdom: Biocatalysts Limited
Pontypridd
Phone: +44-1443-843712
Fax: +44-1443-841214

BIOMOL

USA: BIOMOL Research Laboratories, Inc.
Plymouth Meeting, PA 19462
Phone: 610-941-0430
Fax: 610-941-9252

Australia: Sapphire Bioscience Pty. Ltd.
Alexandria
Phone: +61-2-313-4139
Fax: +61-2-669-2562

Belgium: Sanvertech N.V.
B-2530 Boechout
Phone: +32-3-454-00-66
Fax: +32-3-454-18-88

Denmark: SMS Gruppen
DK-2970 Hoersholm
Phone: +45-4-286-4400
Fax: +45-4-286-4881

France: Tebu
F-78610 Le Perray en Yvelines
Phone: +33-1-3-484-6252
Fax: +33-1-3-484-9357

Germany: BIOMOL Feinchemikalien GmbH
D-22769 Hamburg 50
Phone: 49-40-8532-600
Fax: 49-40-8511-929

Israel: Ornat Biochemicals & Lab Equipment
Ness-Ziona 74031
Phone: +972-8-940-6530
Fax: +972-8-940-6498
E-mail: ornatbio@netvision.net.il

Italy: D.B.A. Italia SRL
20090 Segrate
Phone: +39-2-2-692-2300
Fax: +39-2-2-692-6058

Japan: Funakoshi Co., Ltd.
Tokyo
Phone: +81-3-5684-1622
Fax: +81-3-5684-1633
E-mail: funa@gol.com

Korea: Jaesae Yang Heng Corp.
Seoul
Phone: +82-2-945-2431
Fax: +82-2-945-2434

Norway: Ing F. Heidenreich A/S
N-0401 Oslo
Phone: +47-22-220411
Fax: +47-22-221150
E-mail: ifh@powertech.no

Spain: Quimigranel S.A.
28020 Madrid
Phone: +34-1-556-1614
Fax: +34-1-555-0374

Enzyme Supplier Index

Switzerland: Anawa Trading SA
CH-8602 Wangen/Zurich
Phone: +41-1-833-0555
Fax: +41-1-833-0575
E-mail: hassler@anawa.ch

Taiwan: Hong Jing Co., Ltd.
Taipei
Phone: +886-2-393-0185
Fax: +866-2-356-0943

The Netherlands: Sanvertech B.V.
Breda
Phone: +31-76-522-62-57
Fax: +31-76-521-71-66

United Kingdom: Affiniti Research Products Ltd.
Mamhead
Phone: +44-1-62-689-1010
Fax: +44-1-62-689-1090
E-mail: 100337.1606@compuserve.com

Biozyme

USA: Biozyme Laboratories International Limited
San Diego, CA 92131-1029
Phone: 619-549-4484
Gratis phone: 800-423-8199
Fax: 619-549-0138

India: J. Mitra & Bros. Pvt. Limited
New Delhi-110020
Phone: 11-6818971
Fax: 11-6810203

Japan: Funakoshi Company Limited
Tokyo 113
Phone: 03-5684-1622
Fax: 03-5684-1633

Korea: Bision Business Company
Seoul
Phone: 02-586-9150
Fax: 02-586-9151

Switzerland: Dr. Rudolf Streuli AG
Zurich
CH-8032 Zurich
Phone: 01-26-12345
Fax: 01-26-12344

Taiwan 103, R.O.C.: Flow Science Limited
Taipei
Phone: 02-555-9700
Fax: 02-555-9695

Thailand: Biotechnical Company, Ltd.
Bangkok 10900
Phone: 2-5795490
Fax: 662-5611761

United Kingdom: Biozyme Laboratories Limited
Gwent NP4 9RL
Phone: 01495-790678
Fax: 01495-791780

Boehringer

Germany: Boehringer-Mannheim GmbH
D-68305 Mannheim
Phone: +49-621-759-85-68
Fax: +49-621-759-40-83

Australia: Boehringer Mannheim Australia Pty. Ltd.
Castle Hill
Phone: 02-899-7999
Fax: 02-899-7893

Austria: Boehringer Mannheim GmbH,
Wien
A-1210 Wien
Phone: 0222-277-87
Fax: 0222-277-87-17

Belgium: Boehringer Mannheim Belgium
1120 Bruxelles/Brussels
Phone: 02-247-49-30
Fax: 02-247-46-80

Brazil: BioAgency Ltda.
01227-000 Sao Paulo-SP
Phone: 55-11-66-3565
Fax: 55-11-825-2225

Canada: Boehringer Mannheim Canada
Laval
Phone: 514-686-7050
Fax: 514-686-7009

Chile: Boehringer Mannheim de Chile Ltda.
Santiago
Phone: 00-56-2-22-33-737
Fax: 00-56-2-22-32-049

China: Boehringer Mannheim China Ltd.
Kwai Chung, N.T.
Phone: 86-21-4164320 (China)
Gratis phone: 852-24857596 (Hong Kong)
Fax: 852-24180728

Czech Republic: B.M. - COMP, spolecnost s.r.o.
290 01 Podebrady
Phone: 0324-45-54, 58-71-2
Fax: 0324-45-53

Denmark: Boehringer Mannheim Ercopharm a-s
3490 Kvistgaard
Phone: 49-13-82-32
Fax: 49-13-80-62

Finland: Oriola Oy Prolab
FIN-02101 Espoo
Phone: 90-429-2342
Fax: 90-429-3117

France: Boehringer Mannheim France S.A.
F-38242 Meylan Cedex
Phone: 76-76-30-86
Fax: 76-76-46-90

Greece: Kekis S.A.
115 26 Athens
Phone: 01-64-96-683-6
Fax: 01-69-17-479

India: Boehringer Mannheim India Ltd.
Bombay 400093
Phone: 22-837-0794
Fax: 22-837-9906

Indonesia: Boehringer Mannheim Indonesia
Division of PT Rajawall Nusindo
Jakarta 12950
Phone: 62-21-520-2820
Fax: 62-21-520-2844 or -2829

Israel: Agentek (1987) Ltd.
Tel Aviv 61580
Phone: 972-3-6-49-31-11
Fax: 972-3-6-48-12-57

Italy: Boehringer Mannheim Italia SpA Biochimici
20162 Milano
Phone: 02-270-96209
Fax: 02-270-96250

Japan: Boehringer Mannheim K.K.
Tokyo 105
Phone: 03-3432-3155
Fax: 03-3434-4917

Malaysia: Boehringer Mannheim Malaysia
Sdn Bhd
Selangor Darul Ehsan
Phone: 60-03-755-5039
Fax: 60-03-755-5418

Mexico: Farmaceuticos Lakeside SA de CV
14386 Mexico, D.F.
Phone: 5-227-8967, -61
Fax: 5-227-8950

Enzyme Supplier Index

New Zealand: Boehringer Mannheim N.Z. Ltd.
Mr. Wellington
Phone: 09-276-4157
Fax: 09-276-8917

Norway: Medinor Produkter A/S
0611 Oslo
Phone: 22-07-65-00
Fax: 22-07-65-05

Poland: Hand-Prod Sp. z o.o.
01-113 Warszawa
Phone: +48-22-36-06-77-87
Fax: +48-22-37-42-35

Portugal: Boehringer Mannheim de Portugal, Lda.
2796 Linda-a-Velha
Phone: 01-417-1717
Fax: 01-417-1313

Singapore 168730: Boehringer Mannheim Singapore Pte. Ltd.
Singapore
Phone: 65-272-9200
Fax: 65-371-6500

South Africa: Boehringer Mannheim (South Africa) Pty. Ltd.
Randburg 2125
Phone: 011-886-2400
Fax: 011-886-2962

South Korea: Bio-Medical & Science Co. Ltd.
Kangnam-ku, Seoul #135-080
Phone: 02-569-6902
Fax: 02-553-9670

Spain: Boehringer Mannheim S.A.
E-08006 Barcelona
Phone: 93-201-44-11
Fax: 93-201-30-04

Sweden: Boehringer Mannheim Scandinavia AB
S-161 26 Bromma
Phone: 08-98-81-50
Fax: 08-98-44-42

Switzerland: Boehringer Mannheim (Schweiz) AG
CH-6343 Rotkreuz
Phone: 0-41/7-99-61-61
Fax: 0-41/7-99-65-45

Taiwan, R.O.C.: Formo Industrial Co., Ltd.
Taipei
Phone: 02-736-7125
Fax: 02-736-2647

Thailand: Boehringer Mannheim Group Thailand
Bankok 10310
Phone: 66-2-274-0708-13
Fax: 66-2-274-0736

The Netherlands: Boehringer Mannheim BV
NL-1300 BA Almere
Phone: 036-53-94-911
Fax: 036-53-94-231

Turkey: Dr. Sevgen Laboratuar Teknolojisi
Feneryolu/Istanbul
Phone: 1-349-81-76-79
Fax: 1-349-81-80

United Kingdom: Boehringer Mannheim UK
(Diagnostics & Biochemicals) Limited
East Sussex BN7 1LG
Phone: 01-273-480-444
Fax: 01-273-480-266

USA: Boehringer Mannheim Corporation
Biochemical Products
Indianapolis, IN 46209-1000
Gratis phone: 800-428-5433
Fax: 800-428-2883

Calbiochem
USA: Calbiochem-Novabiochem International
La Jolla, CA 92039-2087
Phone: 619-450-9600
Gratis phone: 800-854-3417
Fax: 800-776-0999 or 619-453-3552

Argentina: Biodynamics SRL
Buenos Aires
Phone: 541-383-3000
Fax: 541-384-7316

Australia
(Australia & New Zealand): Calbiochem-Novabiochem Pty.
Alexandria
Phone: +61-2-318-0322
Fax: +61-2-319-2440

Austria: R.u.P. Margaritella Ges.m.b.H. Bio-Trade
1230 Wien-Rodaun
Phone: 431-889-1819
Fax: 431-889-181920

Belgium: Euro Biochem SCRL
B-1301 Bierges
Phone: +32-10-41-24-55
Fax: +32-10-41-26-13

Brazil: Genoma Biotecnologia
Belo Horizonte MG 30140-062
Phone: +55-31-275-1240
Fax: +55-31-291-0168

Brazil: CB-Tech Do Brasil
Campinas SP 13044-370
Phone: +55-19-230-5961
Fax: +55-19-230-7465

Canada: Rose Scientific Ltd.
Edmonton
Phone: 403-438-5110
Gratis phone: 800-661-9289
Fax: 403-462-5776

Canada: InterSciences Inc.
Markham
Phone: 905-940-1831
Gratis phone: 800-661-6431
Fax: 905-940-1832

China: Fudan Biotechnology
Shanghai 200438
Phone: +86-21-65242386
Fax: +86-21-65242384

Denmark: Bie & Berntsen Ltd.
Roedovre DK-2610
Phone: 4544-94-88-22
Fax: 4544-94-27-09

Finland: Ya-Kemia OY
FIN-00700 Helsinki
Phone: 358-0350-9250
Fax: 358-0350-92555

France: France Biochem
92190 Meudon
Phone: 331-4626-7870
Fax: 331-4534-2520

Germany: Calbiochem-Novabiochem GmbH
D-65796 Bad Soden
Phone: +49-6196-63955 or 56
Fax: +49-6196-62361

Greece: Biodata Hellas Ltd.
11362 Athens
Phone: 1-884-06-13
Fax: 1-884-06-14

India: Biobusiness Development Agency
110018 New Delhi
Phone: 11-559-68-20
Fax: 11-559-68-20

Israel: Megapharm Ltd.
Hod Ha'Sharon 45105
Phone: 972-9-904514 or -904596
Fax: 972-9-904514

Enzyme Supplier Index

Italy: Inalco S.P.A.
20139 Milano
Phone: 02-55213005
Fax: 02-5694518

Japan: Iwai Chemicals Company Ltd.
Phone: 03-3241-0376
Fax: 03-3270-2425

Japan: Nacalai Tesque, Inc.
Phone: 06-381-8121
Fax: 06-381-8137

Japan: Wako Pure Chemical Industries, Ltd.
Tokyo
Phone: 03-3270-8571
Fax: 03-3242-6501

Japan: Shigematsu & Co., Ltd.
Phone: 06-231-6146
Fax: 06-231-6149

Japan: Calbiochem-Novabiochem Japan Ltd.
Tokyo 108
Phone: 03-5443-0281
Fax: 03-5443-0271

Japan: Kokusan Chemical Works Ltd.
Phone: 03-3241-0271
Fax: 03-3241-6263

Japan: Wako Pure Chemical Industries, Ltd.
Osaka
Phone: 06-203-3741
Fax: 06-229-1298

Korea: Koram Biotech Corp.
Seoul 135-080
Phone: 822-556-0311
Fax: 822-556-0828

Mexico: Control Tecnico y Representaciones
Mexico City
Phone: 525-399-2840
Fax: 525-399-2870

Mexico: Control Tecnico y Representaciones
Monterrey
Phone: 528-371-6050 or -370-1571
Fax: 528-373-2891

Norway: Bio Test A/S
Rygge N-1580
Phone: 4769-26-1777
Fax: 4769-26-1760

Portugal: Biocontec
2775 Carcavelos
Phone: 351-1458-1641
Fax: 351-1458-1636

South Africa: Noristan Ltd.
0127 Silverton
Phone: 12-8421000
Fax: 12-8038396

Spain: AMS Biotecnologia (Espana) S.L.
28007 Madrid
Phone: +34-1-551-54-03
Fax: +34-1-433-45-45

Sweden: Labkemi AB
135 70 Stockholm
Phone: +46-8-742-02-50
Fax: +46-8-742-72-99

Switzerland: Juro Supply AG
CH-6000 Lucerne 5
Phone: 041-410-16-51
Fax: 041-410-45-64

Taiwan 112: Cashmere Scientific Co.
Taipei
Phone: 886-2-821-3004 or -541-6188
Gratis phone: 080-291137
Fax: 886-2-821-7749

Taiwan 625: Cashmere Scientific Co.
Chai-Yi
Phone: 886-5-345-1214
Fax: 886-5-345-3548

The Netherlands: Omnilabo International B.V.
4800 DX Breda
Phone: 076-5795-795
Gratis phone: 06-099-7775
Fax: 076-5876-236

The Netherlands: Omnilabo International B.V.
4800 DX Breda
Phone: 076-5795-795
Gratis phone: 06-099-7775
Fax: 076-5876-236

Turkey: Biomar
41410 Gebze/Kocaeli
Phone: 262-646-4570
Fax: 262-646-4572

United Kingdom: Calbiochem-Novabiochem (UK) Ltd.
Beeston
Phone: 44-1159-430-840
Fax: 44-1159-430-951

Calzyme

USA: Calzyme Laboratories, Inc.
San Luis Obispo, CA 93401
Phone: 805-541-5754
Gratis phone: 800-523-9127
Fax: 805-541-8301

United Kingdom: Calzyme Laboratories, Inc.
Birmingham, B13 8LS
Phone: 021-449-7058
Fax: 021-442-4347

Can Inova

Canada: Canadian Inovatech Inc.
Abbotsford
Phone: 604-857-9080
Gratis phone: 800-665-3447
Fax: 604-857-2679

Cayman

USA: Cayman Chemical
Ann Arbor, MI 48108
Phone: 313-662-6756
Gratis phone: 800-364-9897
Fax: 313-662-6896

Cedar Lane

Canada: Cedar Lane Laboratories Ltd.
Hornby
Phone: 905-878-8891
Gratis phone: 800-268-5058
Fax: 905-878-7800

Argentina: Biocientifica S.A.
1232 Buenos Aires
Phone: 01-942-3654
Fax: 01-942-0988

Australia: Bioscientific Pty. Ltd.
Gymea, N.S.W.
Phone: 02-521-2177
Gratis phone: 008-251-437
Fax: 02-542-3037

Austria: Szabo
A-1172 Vienna
Phone: 0222-409-3961-0
Fax: 0222-409-3961-7

Bahrain: Gulf Pharmacy & General Store
Manama
Phone: 973-254431
Fax: 973-276129

Denmark: Trichem APS
DK-2830 Virum
Phone: +45-45-85-82-83
Fax: +45-45-85-95-93

Finland: Immuno Diagnostic OY
Fin-13131 Hameenlinna
Phone: 17-22758
Fax: 17-22039

Enzyme Supplier Index

France: Tebu
F-78610 Le Perray en Yvelines
Phone: 1-34-84-62-52
Fax: 1-34-84-93-57

Germany: Camon Labor Service GmbH
D-65205 Wiesbaden
Phone: 49-611-702846
Fax: 49-611-713782

Greece: Biodynamics S.A.
115 21 Athens
Phone: 0030-1-64-49-421
Fax: 0030-1-64-42-266

India: Hysel India
New Delhi-110 025
Phone: 91-11-6846565
Fax: 91-11-6846565

Israel: Tarom Applied Technologies Ltd.
Tel Aviv, 61292
Phone: 972-3-5377871
Fax: 972-3-5377868

Italy: Celbio S.r.l.
20016-Pero, MI
Phone: 02-38103171
Fax: 02-38101465

Japan: Wako Pure Chemical
Osaka 541
Phone: 06-203-3740
Fax: 06-229-1298

Japan: Dainippon Pharmaceutical Co., Ltd.
Osaka 564
Phone: 06-386-2164
Fax: 06-337-1606

Japan: Cosmo Bio
Tokyo 103
Phone: 81-3-3663-0723
Fax: 81-3-3663-0725

Korea: Medilab Korea Co.
Seoul, 138-50
Phone: 02-424-6367-9
Fax: 02-412-6535

Norway: Ingenior F. Heidenreich
N-0401 Oslo 4
Phone: 22-22-0411
Fax: 22-22-1150

Saudi Arabia: Baaboud Medical Supplies
Jeddah 21411
Phone: 6421437
Fax: 6422877

Spain: Atom S.A.
08024 Barcelona
Phone: 34-3-284-79-04
Fax: 34-3-210-82-55

Sweden: Dakopatts AB
S-125 21 Alvsjo
Phone: 46-8-99-60-00
Fax: 46-8-99-60-65

Switzerland: Bioreba AG
4153 Reinach BL 1
Phone: 061-712-11-15
Fax: 061-712-11-17

Taiwan: Cold Spring Biotechnology Co., Ltd.
Taipei Hsien
Phone: 886-2-694-0066
Fax: 886-2-694-3204

Thailand: Rapport Co. Ltd.
Bangkok 10240
Phone: 3796889-90
Fax: 662-379-7072

The Netherlands: Sanbio BV
5400 AM Uden
Phone: 31-4132-51115
Fax: 31-4132-66605

United Kingdom: VH Bio Ltd
Gosforth
Phone: +44-91-492-0022
Fax: +44-91-410-0916

USA: Accurate Chemical & Scientific Corporation
Westbury, NY 11590
Phone: 516-333-2221
Gratis phone: 800-645-6264
Fax: 516-997-4948

USA: Accurate Chemical & Scientific Corporation
San Diego, CA
Phone: 619-296-9945
Gratis phone: 800-255-9378

Chemicon

USA: Chemicon International Inc.
Temecula, CA 92590
Phone: 909-676-8080
Gratis phone: 800-437-7500
Fax: 909-676-9209

United Kingdom: Chemicon International, Ltd.
Harrow HA1 2JR
Phone: 0181-863-0415
Fax: 0181-863-0416

CHIMERx

USA: CHIMERx
Madison, WI 53704
Phone: 608-244-6319
Gratis phone: 800-626-7833
Fax: 608-244-6318

ChrHansens

Denmark: Chr. Hansen's Laboratory Danmark A/S
DK-2970 Horsholm
Phone: +45-45-76-76-76
Fax: +45-45-76-56-33

Australia: Chr. Hansen's Laboratory
Greenacres
Phone: 8-2618122
Fax: 8-2617254

Argentina: Laboratorio Chr. Hansen S.A.I.C.
1878 Quilmes
Phone: 1-2531746
Fax: 1-2571514

Brazil: Tres Coroas
Industria e Comercio Ltda.
CEP 06300-Carapicuiba-SP
Phone: 11-4296944

Brazil: "HA-LA" do Brasil Chr. Hansen Industria e Comercio Ltda.
CEP 13.270-Valinhos-SP
Phone: 192-713655
Fax: 192-713376

Canada: Chr. Hansen's Laboratory Ltd.
Mississauga
Phone: 416-625-2560
Fax: 416-625-8157

France: Societe Boll
F-91292 Arpajon Cedex
Phone: 1-60840727
Fax: 1-60841594

Germany: Chr. Hansen's Laboratorium GmbH
D-2400 Lubeck 1
Phone: 451-480070
Fax: 451-4800777

Ireland: Chr. Hansen's Laboratory Ireland Limited
Little Island
Phone: 21-353500
Fax: 21-353912

Italy: Caglio Italiano "Chr. Hansen" SpA
1-20094 Corsico (Milano)
Phone: 2-4478341
Fax: 2-4472497

Enzyme Supplier Index

New Zealand: Chr. Hansen's Laboratory P/L
Hamilton
Phone: 71-80795
Fax: 71-393798

Spain: CHL-LACTA S.L.
E-28028 Madrid
Phone: 1-255-6607
Gratis phone: 1-255-1019

Sweden: Axel Bergmark & Co. AB
S-40242 Goteborg 12
Phone: 31-922030
Fax: 31-920923

United Kingdom: Chr. Hansen's Laboratory Ltd.
Reading RG2 OQL
Phone: 734-861056
Fax: 734-312937

USA: Chr. Hansen's Laboratory, Inc.
Milwaukee, WI 53214-4298
Phone: 414-476-3630
Gratis phone: 800-558-0802
Fax: 414-476-2313

Yugoslavia: Chr. Hansen's Laboratory Doo
Dj. Djakorica bb
Phone: 022-54-855
Fax: 022-53-355

CLONTECH

USA: CLONTECH Laboratories, Inc.
Palo Alto, CA 94303-4230
Phone: 415-424-8222
Gratis phone: 800-662-2566
Fax: 415-424-1064
E-mail: orders@clontech.com

Austria: BIO-TRADE
A-1230 Wien-Rodaun
Phone: 43-1-889-1819
Fax: 43-1-889-181920

Brazil: BIO AGENCY
01227-000 Sao Paulo, SP
Phone: 55-11-663-565
Fax: 55-11-825-2225

Canada: Bio/Can Scientific, Inc.
Mississauga
Phone: 905-828-2455
Fax: 905-828-9422

China: Clontech China
Shanghai 200032
Phone: 86-21-549-2222
Fax: 86-21-534-1650

Germany: ITC Biotechnology GmbH
69020 Heidelberg
Phone: 49-6221-303907
Fax: 49-6221-303511

Greece: Bio+Analytica
114, 71 Athens
Phone: 30-1-643-6138
Fax: 30-01-646-2748

Hong Kong: Letrich Industries Co., Ltd.
Tsuen Wan, N.T.
Phone: 852-416-8132
Fax: 852-412-1863

India: KPS Scientific Products Co.
New Delhi-110 024
Phone: 91-11-6425156
Fax: 91-11-6425156

Italy: Technogenetics S.P.A.
Milan
Phone: 39-2-95258282
Fax: 39-2-9521361

Mexico: APCO, Inc.
03020 Mexico, D.F.
Phone: 525-519-3463
Fax: 525-538-1884

Spain: ITISA
28034 Madrid (Fuencarral)
Phone: 34-1-358-92-08
Fax: 34-1-358-97-54

Sweden: Intermedica
104 25 Stockholm
Phone: 46-8-749-5940
Fax: 46-8-749-5105

Switzerland: P.H. Stehelin & Cie AG
4003 Basel
Phone: 41-61-272-39-24
Fax: 41-61-271-39-07

Turkey: Ermanak Miskciyan
80000 Karakoy-Istanbul
Phone: 90-216-385-8321
Fax: 90-216-385-4649

United Kingdom: Cambridge Bioscience
Cambridge
Phone: 44-223-31-6855
Fax: 44-223-60-732

Collaborative

USA: BD Labware / Collaborative Biomedical
Bedford, MA 01730
Phone: 617-275-0004
Gratis phone: 800-343-2035
Fax: 617-275-0043

Cortex

USA: Cortex Biochem, Inc.
San Leandro, CA 94577
Phone: 510-568-2228
Gratis phone: 800-888-7713
Fax: 510-568-2467

Crystal

USA: Crystal Chem, Inc.
Chicago, IL 60660
Phone: 312-764-3549
Fax: 312-764-8666

Danisco

DENMARK: Danisco Ingredients
DK-8220 Brabrand
Phone: +45-86-43-50-00
Fax: +45-86-25-10-77

Diagnostic

Canada: Diagnostic Chemicals Limited
Charlottetown
Phone: 902-566-1396
Gratis phone: 800-565-0265
Fax: 902-566-2498
E-mail: orders@dclchem.com

Australia: ScimaR
Templestowe
Phone: 842-3386
Fax: 842-3407

Canada: Xymotech Biosystems
Mt. Royal
Phone: 514-738-1386
Gratis phone: 800-663-1392
Fax: 514-738-8499

USA: Diagnostic Chemicals Limited (USA)
La Jolla, CA 92037
Phone: 619-546-2926
Gratis phone: 800-794-0899
Fax: 619-452-8541

USA: Diagnostic Chemicals Limited (USA)
Oxford, CT 06478
Phone: 203-881-2020
Gratis phone: 800-325-2436
Fax: 203-888-1143

Difco

USA: Difco Laboratories
Detroit, MI 48232-7-58
Phone: 313-462-8500
Gratis phone: 800-521-0851
Fax: 313-462-8517

Elastin

USA: Elastin Products Company, Inc.
Owensville, MO 65066
Phone: 314-437-2193
Fax: 314-437-4632

Enzyme Supplier Index

EnzymeDev
USA: Enzyme Development Corporation
New York, NY 10121-0034
Phone: 212-736-1580
Fax: 212-279-0056

EnzymeRes
USA: Enzyme Research Laboratories, Inc.
South Bend, IN 46601
Phone: 219-288-2268
Gratis phone: 800-729-5270
Fax: 219-288-2272

Epicentre
USA: Epicentre Technologies
Madison, WI 53713
Phone: 608-258-3080
Gratis phone: 800-284-8474
Fax: 608-258-3088

Fermentas
Lithuania: Fermentas AB
2028 Vilnius
Phone: 370-2-641279
Fax: 370-2-643436

Austria: GATT
A-6010 Innsbruck
Phone: 0512-58-30-17
Fax: 0512-58-91-96

Canada: MBI Fermentas Inc.
Flamborough
Gratis phone: 800-959-5673
Fax: 800-472-8322
E-mail: info@fermentas.com

Czech Republic: Biogen sro
CZ-12820 Praha 2
Phone: 2-29-5265
Fax: 2-29-3228

Denmark: Life Science Denmark ApS
DK-3480 Fredensborg
Phone: 48-47-50-51
Fax: 48-47-50-61

Finland: L-Team International Oy
FIN-02630 Espoo
Phone: 0-502-3600
Fax: 0-502-3636

France: Euromedex
Etablissement Pharmaceutique-
Autorisation Ministerielle
F-67460 Souffelweyersheim
Phone: +88-18-07-22
Fax: +88-18-07-25

Germany: MBI Fermentas GmbH
D-68789 St. Leon-Rot
Phone: 06227-55853
Fax: 06227-53694

Hungary: Biological Research Centre
Hungarian Academy of Science
H-6701 Szeged
Phone: 62-432-232
Fax: 62-433-484

Israel: Eisenberg Bros. Lt.
ISR-54030 Givat Shmuel
Phone: 03-532-1715
Fax: 03-532-5696

Italy: Dasit s.p.a.
20010 Cornaredo, Milan
Phone: 02-93991-1
Fax: 02-93991390

Japan: Cosmo Bio Co., Ltd.
Tokyo 135
Phone: 3-5632-9630
Fax: 3-5632-9624

Poland: PPH ABO
Tarnowska-Boreysza
80-255 Gdansk
Phone: +058-41-21-43
Fax: +058-31-57-25

Russia: Gnii Genetika
113545 Moscow
Phone: +095-315-0501
Fax: +095-315-0065

South Korea: Dae Myung Medical Co.
Seoul
Phone: 82-2-458-5835
Fax: 82-2-452-1221

Spain: Quimigranel S.A.
28020 Madrid
Phone: 9-1-556-16-14
Fax: 9-1-555-03-74

Sweden: AB Labassco
S-433 85 Partille
Phone: 031-36-30-00
Fax: 031-44-60-64

Switzerland: Gebrueder Maechler AG
CH-4003 Basel
Phone: 061-272-3065
Fax: 061-271-3907

The Netherlands: ITK Diagnostics BV
NL-1420 AB Uithoorn
Phone: 0-2975-68893
Fax: 0-2975-63485

Turkey: Prizma Laboratuar Urunleri
Sanayi Ve Tic Ltd. Sti.
81010 Kadikoy Istanbul
Phone: 216-418-24-14
Fax: 216-418-24-13

United Kingdom: Immunogen
International Limited
Sunderland
Phone: 0191-516-0554
Fax: 0191-516-8415

USA: MBI Fermentas Inc.
Buffalo, NY 14202
Gratis phone: 800-340-9026
Fax: 800-472-8322
E-mail: info@fermentas.com

5′→3′
USA: Five Prime Three Prime Inc.
Boulder, CO 80303
Phone: 303-440-3705
Gratis phone: 800-533-5703
Fax: 303-440-0835

Fluka
Switzerland: Fluka Chemie AG
CH-9470 Buchs
Phone: 081-755-25-11
Gratis phone: 155-66-33
Fax: 081-756-54-49

Australia: Fluka Chemicals
Castle Hill NSW 2154
Phone: 008-800-097
Fax: 008-800-096

Austria: Sigma-Aldrich Handels GmbH
A1110 Wien
Phone: 43-222-74040-644
Fax: 43-222-74040-643

Belgium: Fluka Chemie
B-2880 Bornem
Phone: 03-899-13-01
Fax: 03-899-13-11

Brazil: Sigma-Aldrich
01239-010 Sao Paulo, SP
Phone: 55-11-231-1866
Fax: 55-11-257-9079

Czech Republic: Sigma-Aldrich Ltd.
186 00 Praha 8
Phone: 42-2-242-25-285
Fax: 42-2-242-24-031

France: Fluka Chimie
F-38297 St. Quentin Fallavier Cedex
Phone: 74-82-28-00
Gratis phone: 05-21-14-08
Fax: 05-03-10-52

Enzyme Supplier Index

Germany: Fluka Chemie
D-89231 Neu-Ulm
Phone: 0731-973-03
Gratis phone: 0131-2341
Fax: 0131-84-77-66

Hungary: Sigma-Aldrich Kft
1067 Budapest
Phone: 36-1-269-1288
Fax: 36-1-153-3391

India: Sigma-Aldrich Corporation
Hyderabad 500 033
Phone: 091-40-244-739
Fax: 091-40-244-794

Israel: Sigma Israel Chemical Co. Ltd.
Holon 58100
Phone: 972-3-559-6610
Fax: 972-3-559-6596

Italy: Fluka Chimica
I-20151 Milano
Phone: 02-33417-310
Gratis phone: 02-38010737

Japan: Fluka Fine Chemicals
Tokyo
Phone: 81-03-3255-4787
Fax: 81-03-3258-0157

Mexico: Sigma-Aldrich Quimica
S.A. de C.V.
14210 Mexico, D.F.
Phone: 52-5-631-3671
Fax: 52-5-631-3780

South Korea: Sigma-Aldrich
Seoul
Phone: 82-2-783-5211
Fax: 82-2-783-5011

Spain: Fluka Quimica
E-28100 Alcobendas (Madrid)
Phone: 91-661-99-77
Gratis phone: 900-101-376
Fax: 91-661-96-42

United Kingdom: Fluka Chemicals
Gillingham
Phone: 0747-82-30-97
Gratis phone: 0800-26-23-00
Fax: 0800-56-57-50

USA: Fluka Chemical Corp.
Milwaukee, WI 53233
Phone: 414-273-3850
Gratis phone: 800-358-5287
Fax: 414-273-4979

GDS Tech

USA: GDS Technology, Inc.
Elkhart, IN 46515
Phone: 219-264-7384
Fax: 219-262-0109

Genencor

USA: Genencor International, Inc.
Rochester, NY 14618
Phone: 716-256-5200
Gratis phone: 800-847-5311
Fax: 716-244-9988

Finland: Genencor International, Oy
SF-00241 Helsinki
Phone: +358-0-134-411
Fax: +358-0-1344-1319

Japan: Genencor International Japan Ltd.
Saitama 336
Phone: +81-48-861-8330
Fax: +81-48-861-8346

USA: Genencor International, Inc.
Cedar Rapids, IA 52404
Phone: 319-363-9601
Fax: 319-363-1840

USA: Genencor International, Inc.
South San Francisco, CA 94080
Phone: 415-742-7500
Fax: 415-742-7220

Genzyme

United Kingdom: Genzyme Diagnostics
West Malling
Phone: +44-1732-220022
Fax: +44-1732-220024/5

USA: Genzyme Corporation
Cambridge, MA 02139-1562
Phone: 617-252-7500
Gratis phone: 800-332-1042
Fax: 617-252-7759

Hayashibara

Japan: Hayashibara Biochemical Laboratories, Inc.
Okayama 700
Phone: 086-224-4311
Fax: 086-225-5630

Hindustan

India: Hindustan Antibiotics Limited
Pune 411 018
Phone: 091-212-776511-12-13
Fax: 091-212-772327

ICN

USA: ICN Pharmaceuticals Inc.
Costa Mesa, CA 92626
Gratis phone: 800-854-0530
Fax: 800-334-6999

Argentina: ICN Argentina
Provincia Buenos Aires
Phone: 541-203-42-40
Fax: 541-203-56-81

Australia: ICN Pharmaceuticals Australasia Pty. Ltd.
Seven Hills NSW 2147
Phone: 61-02-838-7422
Fax: 61-02-838-7390

Belgium: ICN Pharmaceuticals, NV/SA
B-1733 Asse-Relegem
Phone: 32-02-4660000
Fax: 32-02-4662642

Canada: ICN Pharmaceuticals Canada Inc.
Montreal H4M 1V1
Phone: 514-744-6792
Fax: 514-744-6272

France: ICN Pharmaceuticals France
91893 Orsay Cedex
Phone: 1-60-19-37-37
Fax: 1-60-19-34-60

Germany: ICN Pharmaceuticals GmbH
D-53340 Meckenheim
Phone: 49-22-25/88-05-0
Fax: 49-22-25/88-05-81

Italy: ICN Pharmaceuticals, S.r.l.
20090 Opera (MI)
Phone: 02-57601041
Fax: 02-57601610

Mexico: ICN Pharmaceuticals Mexico
09080 Mexico, D.F.
Phone: 525-670-0739
Fax: 525-581-4938

Russia: ICN Galenika
121151 Moscow
Phone: 010-7095-243-5022
Fax: 010-7095-241-7820

Singapore: ICN Far East
Singapore 9152
Phone: 65-782-9534
Fax: 65-787-2205

Spain: ICN Hubber, S.A.
Barcelona
Phone: 93-688-05-44
Fax: 343-688-04-01

The Netherlands: ICN Pharmaceuticals BV
2700 AJ Zoetermeer
Phone: 06-022-74-16
Fax: 06-022-74-89

Enzyme Supplier Index

U.A.E: ICN Middle East
Deira-Dubai
Phone: 971-4-622603 (620-570)
Fax: 61-2-838-7390

United Kingdom: ICN Pharmaceuticals Ltd.
Thame
Phone: +44-844-213366
Fax: +44-844-214455

USA: ICN Pharmaceuticals Inc.
Aurora, OH 44202
Phone: 216-562-1500
Gratis phone: 800-854-0530
Fax: 216-562-1987

Yugoslavia: ICN Galenika
Belgrade, Zemun
Phone: 381-11-199-642
Fax: 381-11-108-354

ImmunoVis

USA: ImmunoVision Inc.
Springdale, AR 72764
Phone: 501-751-7005
Gratis phone: 800-541-0960
Fax: 501-751-7002

Canada: Farmington Biotechnologies
Brampton
Phone: 416-452-0206
Fax: 416-458-8292

Japan: Veritas Corporation
Tokyo
Phone: 03-435-1558
Fax: 03-435-1526

Luxembourg L-2543: Cellon Sarl
Phone: 352-49-59-75
Fax: 40-35-07

Imperial

United Kingdom: Imperial Biotechnology Limited
London SE1 6LN
Phone: +44-0-171-620-0800
Fax: +44-0-171-620-1010

USA: Imperial Biotechnology U.S., Inc.
St. Louis, MO 63102
Gratis phone: 800-746-4428
Fax: 314-982-3625

Intergen

USA: Intergen Company
Purchase, NY 10577
Phone: 914-694-1700
Gratis phone: 800-431-4505
Fax: 914-694-1429

Japan: International Reagents Corporation
Kobe
Phone: 81-78-231-4053
Fax: 81-78-232-0557

United Kingdom: Intergen/Europe
Phone: 44-1-491-613792
Fax: 44-1-491-613594

Iogen

Canada: Iogen Corporation
Ottawa
Phone: 613-733-9830
Fax: 613-733-5127

Karlan

USA: Karlan Research Products Corporation
Santa Rosa, CA 95403
Phone: 707-576-1225
Gratis phone: 800-231-9186
Fax: 707-576-1349

Latoxan

France: Latoxan
05150 Rosans
Phone: 33-92-66-60-64
Fax: 33-92-66-63-40

Japan: Funakoshi Company Ltd.
Tokyo 113
Phone: 03-5684-1622
Fax: 03-5684-1633

USA: Accurate Chemical and Scientific Corporation
Westbury, NY 11590
Phone: 516-333-2221
Gratis phone: 800-645-6264
Fax: 516-997-4948

USA: Molecular Probes Inc.
Eugene, OR 97402
Phone: 503-344-3007
Fax: 503-344-6504

Life Technol

USA: Life Technologies, Inc., GibcoBRL
Gaithersburg, MD 20884-9980
Phone: 301-840-8000
Gratis phone: 800-828-6686
Fax: 301-670-1394

Argentina: Neutro Quimica, S.A.
Buenos Aires
Phone: 54-1-822-61-64
Fax: 54-1-823-04-06

Australia: Life Technologies-GibcoBRL
Melbourne
Phone: 03-562-8245
Gratis phone: 1-800-331-627
Fax: 03-562-7773

Bangladesh: Tradeworths Ltd.
Dakka
Phone: 2-235646
Fax: 2-833645

Brazil: INCIBRAS Biotecnologia
Sao Paulo
Phone: 55-11-536-05-75
Fax: 55-11-531-32-10

Canada: Life Technologies-GibcoBRL
Burlington
Phone: 905-335-2255
Fax: 905-335-5455

Chile: Bios Chile I.G.S.A
Santiago
Phone: 56-2-238-18-78
Fax: 56-2-239-42-50

Colombia: ARC Quimicos, Ltda.
Santa Fe de Bogota
Phone: 57-1-244-37-89
Fax: 57-1-244-37-89

Hong Kong: Life Technologies-GibcoBRL
Twuen Wan
Phone: 2407-8450
Fax: 2408-2280

India: Life Technologies-GibcoBRL
New Delhi
Phone: 91-11-647-4701
Fax: 91-11-647-4718

Indonesia: PT Elo Karsa Utama
Jakarta
Phone: 21-739-2565
Fax: 21-739-2856

Japan: Life Technologies-GibcoBRL
Tokyo
Phone: 03-3663-8240
Fax: 03-3663-8242

Latin America: Life Technologies-GibcoBRL
Phone: 301-840-4027
Fax: 301-258-8238

Enzyme Supplier Index

Malaysia: Bio-Diagnostics Sdn Bhd
Selangor
Phone: 7754063
Fax: 60-3-7772610

Mexico: ACCESOLAB
Accesorios Para Laboratorios, S.A.
Mexico City
Phone: 52-5-250-08-05
Fax: 52-5-255-55-20

New Zealand: Life Technologies-GibcoBRL
Auckland
Phone: 09-579-3024
Gratis phone: 0-800-657-565
Fax: 09-579-3119

People's Republic of China: Life Technologies-GibcoBRL
Phone: 22-231-0770
Fax: 22-335-7985

Philippines: Medical Test Systems, Inc.
Makati, M.M.
Phone: 8166821
Fax: 63-2-8166350

Scotland: Life Technologies-GibcoBRL
Paisley
Gratis phone: 0-800-269210
Fax: 0141-814-6317

Singapore: SPD Scientific Pte., Ltd.
Singapore
Phone: 473-3720
Fax: 473-2503

South Korea: KDR Biotech Co., Ltd.
Phone: 2-203-0707
Fax: 2-203-0606

Switzerland
Gratis phone: 0800-269-210
Fax: 041-887-1167

Taiwan R.O.C.: Life Technologies-GibcoBRL
Phone: 2-652-2380
Fax: 2-652-2381

Thailand: Gibthai Company Ltd.
Bangkok
Phone: 2-2748331
Fax: 2-274-8336

Uruguay: METEC, S.A
Montevideo
Phone: 59-8-2-63-66-88
Fax: 59-8-2-63-41-45

USA: Life Technologies-GibcoBRL
Grand Island, NY
Gratis phone: 800-828-6686
Fax: 800-331-2286

Venezuela: BioDan, C.A.
Caracas
Phone: 58-2-752-11-33
Fax: 58-2-752-41-13

Midland

USA: Midland Certified Reagent Company
Midland, TX 79701
Phone: 915-694-7950
Gratis phone: 1-800-247-8766
Fax: 915-694-2387
E-mail: orders@oligos.com

MINOTECH

Greece: MINOTECH Molecular Biology Products
IMBB - Foundation for Research and Technology - Hellas
Heraklio 71110
Phone: 081-210235
Fax: 081-210234
E-mail: saitakis@nefeli.imbb.forth.gr

Hungary: Agricultural Biotechnology Center
HO-2101 Godollo
Phone: +36-28-320521
Fax: +36-28-330602

Switzerland
CII-1724 Praroman
Phone: +41-37-331850
Fax: +41-37-333588

USA: Biotechnology Trading Co. Inc.
Hannover, NH 03755
Phone: 603-643-8334
Fax: 603-643-4371
E-mail: john.n.vournakis@dartmouth.edu

Nacalai

Japan: Nacalai Tesque, Inc.
Kyoto 600
Phone: 81-0-75-371-5631
Fax: 81-0-75-371-5543

Nagase

Japan: Nagase Biochemicals Limited
Kyoto 620
Phone: 81-773-27-5801
Fax: 81-773-27-2040

Japan: Nagase & Company Limited
Tokyo 103
Phone: +81-3-3665-3364
Fax: +81-3-3665-3940

USA: Nagase America Corporation
New York, NY 10110
Phone: 212-354-3140
Fax: 212-398-0687

NBL Gene

United Kingdom: NBL Gene Sciences Limited
Cramlington
Phone: +44-01670-73-29-92
Fax: +44-01670-73-04-54
E-mail: 100660.53@compuserve.com

NE Biolabs

USA: New England BioLabs
Beverly, MA 01915-5599
Phone: 508 927-5054
Gratis phone: 800-632-5227
Fax: 508-921-1350
E-mail: info@neb.com

Argentina: Migliore Laclaustra S.R.L.
1055 Buenos Aires
Phone: 1-3729045
Fax: 1-3729045

Australia: Genesearch Pty. Ltd.
Arundel
Phone: 075 94-0299
Gratis phone: 1-800-074-278
Fax: 075-94-0562
E-mail: Genesearch@eworld.com

Brazil: BioAgency Ltda.
Sao Paulo-SP-CEP: 01227-000
Phone: 011-66-3565
Fax: 011-825-2225

Canada: New England Biolabs, Ltd.
Mississauga
Phone: 905-672-3370
Gratis phone: 800-387-1095
Fax: 905-672-3414
E-mail: info@ca.neb.com

Chile: Genetica y Tecnologia Ltda.
Santiago
Phone: 02-633-5269
Fax: 02-639-3124
E-mail: leaton@abello.dic.uchile.cl

China: Union Friendship Medical Technology LTD
Huaying Li
Phone: 86-010-2020011-3117
Fax: 86-010-7787601

Denmark: Medinova Scientific ApS
DK 2900 Hellerup
Phone: 31-56-20-00
Fax: 31-56-19-42
E-mail: medinova@inet.uni-c.dk

Enzyme Supplier Index

Finland: Finnzymes OY
FIN-02201 Espoo
Phone: 0-420-8077
Fax: 0-420-8653

France: Ozyme
78180 Montigny-le-Bretonneux
Phone: 1-34-60-24-24
Fax: 1-30-45-50-35

Germany: New England Biolabs GmbH
65820 Schwalbach/Taunus
Phone: 06196-3031
Gratis phone: 0130-83-30-31
Fax: 06196-83639
E-mail: info@de.neb.com

Greece: BioLine Scientific
104 37 Athens
Phone: 01-5226547
Fax: 01-5244744

Hong Kong: TWC Biosearch International
Hong Kong
Phone: 649-9988
Fax: 635-0379

India: Biotech India
Varanasi-221005
Phone: 542-311473
Fax: 542-313474

Israel: Gamidor Ltd.
Savyon 56915
Phone: 972-3-5351205
Fax: 972-3-5346573
E-mail: gamidor@gold.goldnet.co.il

Italy: CelBio S.r.l.
20016 Pero (MI)
Phone: 02-38103171
Fax: 02-38101465

Japan: Daiichi Pure Chemicals Co. Ltd.
Tokyo 103
Phone: 03-3552-1531
Fax: 03-5820-9409
E-mail: 100202.3653@compuserve.com

Malaysia: Research Instruments Sdn. Bhd.
47400 Petaling Jaya, Selangor
Phone: 03-718-3600
Fax: 03-718-3618

Mexico: Uniparts, S.A.
03020 Mexico, D.F.
Phone: 5-519-3463
Fax: 5-538-1884

New Zealand: Biolab Scientific Ltd.
Auckland 9
Phone: 09-418-3039
Fax: 09-480-3430

Norway: Ing. F. Heidenreich A.S.
N-0401 Oslo 4
Phone: 47-22-22-04 11
Fax: 47-22-22-11-50
E-mail: ifh@powertech.no

Portugal: Quimigranel Ltda.
1801 Lisbon CODEX
Phone: 01-868-15 64
Fax: 01-868-37-10

Singapore: Research Biolabs Pte. Ltd.
Singapore 1646
Phone: 4457927
Fax: 4483966
E-mail: yapsm@singnet.com.sg

South Korea: Koram Biotech Corp.
Seoul
Phone: 02-556-0311
Fax: 02-556-0828

Spain: Landerdiagnostico S.A.
28015 Madrid
Phone: 01-594-08-06
Fax: 01-448-71-92

Sweden: In Vitro AB
S-17136 Solna
Phone: 08-734-8300
Fax: 08-730-4640

Switzerland: BioConcept
4123 Allschwil 3
Phone: 061-481-47-13
Fax: 061-481-71-12

Taiwan R.O.C.: Taigen Bioscience Corp.
Taipei
Phone: 02-8802913
Fax: 02-8802916

Thailand: Theera Trading Co., Ltd.
Bangkokyai Bangkok 10600
Phone: 02-4125672
Fax: 02-4123244

The Netherlands: Westburg B.V.
NL-3830 AE Leusden
Phone: 033-4950094
Gratis phone: 0800-1-9815
Fax: 033-951222
E-mail: info_westburg@e-mail.com

Turkey: Rafigen Tic.
81030 Kadaky Istanbul
Phone: 0216-4113780
Fax: 0216-4113780

United Kingdom: New England Biolabs (U.K.), Ltd.
Hitchin
Phone: 01462-420616
Gratis phone: 0800-31-84 86
Fax: 01462-421057
E-mail: info@uk.neb.com

Nippon Gene

Japan: Nippon Gene Co., Ltd.
Toyama 930
Phone: +81-764-51-6548
Fax: +81-764-51-6547

Novo Nordisk

Denmark: Novo Nordisk A/S
2880 Bagsvaerd
Phone: +45-4444 8888
Fax: +45-4444-1021

Switzerland: Novo Nordisk Ferment Ltd.
4243 Dittingen
Phone: +41-61-7656111
Fax: +41-61-7656333

USA: Novo Nordisk Biochem of North America
Franklinton, NC 27525
Gratis phone: 800-879-6686
Fax: 919-494-3401

Oncogene

France: Oncogene Science SA
75019 Paris
Phone: 33-1-44-52-73-73
Fax: 33-1-44-52-91-53

Australia: Amrad Pharmacia Biotech
Boronia Vic 3155
Phone: 613-887-3909
Fax: 613-887-3912

Austria: Bio-Trade
A1230- Wien-Rodaun
Phone: 43-1-889-1819
Fax: 43-1-889-1819-20

Canada: Cedar Lane Laboratories Ltd.
Hornby
Phone: 905-878-8891
Fax: 905-878-7800

Croatia: HEBE spo
58000 Split
Phone: 385-58-56-37-76
Fax: 385-58-56-37-76

Denmark: AH Diagnostics
8200 Aarhus N
Phone: 45-86-10-10-55
Fax: 45-86-16-15-33

Enzyme Supplier Index

Egypt: Medico Pharma Trade
Agouza-Giza
Phone: 20-2-344-5774
Fax: 20-2-346-5623

Germany: dianova
20354 Hamburg
Phone: 49-40-450670
Fax: 49-40-322190

Greece: Biodynamics, SA
Athens, 11471
Phone: 30-1-644-8632
Fax: 30-1-644-2266

India: OSB Agencies
Delhi, 110031
Phone: 91-11-224-9973
Fax: 91-11-221-6736

Israel: Tarom Applied Technologies
Tel-Aviv 67778
Phone: 972-3-537-7871
Fax: 972-3-537-7868

Italy: Genzyme Srl
20092 Cinisello B-MI
Phone: 39-2-612-76-21
Fax: 39-2-66-01-19-23

Japan: CosmoBio Co. Ltd.
Tokyo 135
Phone: 81-3-5632-9620
Fax: 81-3-5632-9619

Korea: Kormed Corp.
135-010 Seoul
Phone: 82-5-540-4663
Fax: 82-5-544-4539

Singapore: Scimed (Asia) Pte. Ltd.
Singapore 2260
Phone: 65-266-1884
Fax: 65-266-3086

South Africa: Weil Organization Pty. Ltd.
Kelvin 2054
Phone: 27-011-444-4330
Fax: 27-011-444-5457

Spain: Itisa
28049-Madrid
Phone: 34-1-358-9208
Fax: 34-1-358-9754

Sweden: Novakemi ab
Stockholm
Phone: 46-8-390-490
Fax: 46-8-659-1559

Switzerland: PH Stehelin & CIE AG
CH-4003 Basel
Phone: 41-61-272-3924
Fax: 41-61-271-3907

Taiwan: Crystal Ray Ltd.
Panchiao City, Taipei
Phone: 886-2-968-5666
Fax: 886-2-966-7337

Taiwan 112: Cashmere Scientific Co.
Taipei
Phone: 886-2-821-3004
Fax: 886-2-821-7749

Taiwan, R.O.C.: Unimed Healthcare Inc.
Taipei
Phone: 886-2-720-2215
Fax: 886-2-723-3666

The Netherlands: Sanbio BV
5405 PB Uden
Phone: 31-4-1325-1115
Fax: 31-4-1326-6605

Turkey: Bio-Kem
Findikzade, Istanbul
Phone: 90-1-534-0103
Fax: 90-1-525-9832

United Kingdom: Cambridge BioScience
CB58LA Cambridge
Phone: 44-223-316-855
Fax: 44-223-607-32

USA: ONCOGENE Science, Inc.
San Diego, CA 92122
Phone: 619-546-4850
Gratis phone: 800-662-2616
Fax: 617-658-0692

USA: Oncogene Research Products
Cambridge, MA 02142
Phone: 617-557-9333
Gratis phone: 800-662-2616
Fax: 617-577-8015
E-mail: resprod@oncogene.com

USA: ONCOGENE Science, Inc.
Uniondale, NY 11553-3649
Phone: 516-222-0023
Fax: 516-222-0114

Venezuela: Vargas Scientific Supplies
Chacao, Caracas
Phone: 58-2-261-6256
Fax: 58-2-261-7280

Oncor

France: Appligene/Oncor
F-67402 Illkirch
Phone: +33-88-67-22-67
Fax: +33-88-67-19-45

Argentina: Tecnolab S.A.
Buenos Aires
Phone: 011-54-1553-4727
Fax: 011-54-1553-3331

Australia: CSL Biosciences
Victoria 3052
Phone: 011-61 3-9389-1389
Fax: 011-61 3-9389-1646

Brazil: Fairuz
Sao Paulo
Phone: 011-0192-571-461
Fax: 011-0192-390-461

Canada: P.D.I. Bioscience
Aurora
Phone: 905-713-1201
Fax: 905-713-1205

Canada: Vector Biosystems
Toronto
Phone: 416-920-8446
Fax: 416-920-1685

Hong Kong: Gene Company Ltd.
Chai Wan
Phone: 011-852-896-6283
Fax: 011-852-515-9371

Israel: Tal-Ron
Rehovat 76349
Phone: 011-972-8947-2563
Fax: 011-972-8947-1156

Japan: Cosmo Bio Co., Ltd.
Tokyo 135
Phone: 011-81-356-329-630
Fax: 011-81-356-329-624

Korea: Koaman/Seoul Hyo Sung
Seoul
Phone: 011-82-2-556-3863
Fax: 011-82-2-557-4096

Kuwait: Tareq Co.
13066 Safat
Phone: 011-965-243-6045
Fax: 011-965-243-7700

Mexico: Apco, Inc.
03020 Mexico, D.F.
Phone: 011-525-519-3463
Fax: 011-525-538-1884

Enzyme Supplier Index

Taiwan R.O.C.: Unimed Healthcare, Inc.
Taipei
Phone: 011-886-272-02215
Fax: 011-886-272-33666

USA: Oncor
Gaithersburg, MD 20877
Phone: 301-963-3500
Gratis phone: 800-776-6267
Fax: 301-926-6129

USA: Labs U.S.A.
San Diego, CA 92130-2054
Phone: 619-259-2626
Fax: 619-259-1342

OYC

JAPAN: Oriental Yeast Co. Ltd.
Tokyo 103
Phone: +81-3-3663-8218
Fax: +81-3-3663-8238

PanVera

USA: PanVera Corporation
Madison, WI 53711
Phone: 608-233-5050
Gratis phone: 800-791-1400
Fax: 608-233-3007

Pharmacia

Sweden: Pharmacia Biotech Inc.
Uppsala
Phone: 46-(0)18-16-50-00
Fax: 46-(0)18-14-38-20

Australia: Pharmacia Biotech Inc.
Phone: 008-25-22-65
Fax: 3-887-39-12

Australia: AMRAD Pharmacia Biotech
North Ryde, NSW 2113
Fax: 02-367-4250

Austria: Pharmacia Biotech Inc.
Phone: 01-68-66-250
Fax: 01-68-66-25-211

Belgium: Pharmacia Biotech
1130 Brusselles
Phone: 02-727-42-51
Fax: 02-727-42-99

Brazil: Pharmacia Biotech do Brasil Ltda.
05011-001 Sao Paulo
Phone: 55-11-872-68-33
Fax: 55-11-873-04-17

Canada: Pharmacia Biotech Inc.
Quebec H9X 3V1
Gratis phone: 800-463-5800
Fax: 800-567-1008

China: Pharmacia Ltd.
Beijing
Phone: 01-256-56-03
Fax: 01-256-56-03

Denmark: Pharmacia Biotech AS
DK-3450 Allerod
Phone: 48-14-10-00
Fax: 48-14-10-06

Eastern Europe: Pharmacia Biotech Ges.m.b.H.
A-1152 Wien
Phone: 43-0-1982-38-26
Fax: 43-0-1985-38-27

Finland: Pharmacia Biotech OY
02630 ESPOO
Phone: 90-5021-077
Fax: 80-52-01-56

France: Pharmacia Biotech S.A.
78051 St. Quentin, Yvelines, Cedex
Phone: 1-30-64-34-00
Fax: 1-30-43-44-45

Germany: Pharmacia Biotech Europe GmbH
79111 Freiburg
Phone: 07-61-49-030
Fax: 07-61-49-03-246

Holland: Pharmacia Biotech Benelux
NL-4700 BJ Roosendaal
Phone: 31-1-650-80-400
Fax: 31-1-650-80-401

Hong Kong: Pharmacia Biotech Far East Ltd.
Hong Kong
Phone: 852-811-86-93
Fax: 852-811-52-70

India: Pharmacia Biotech Far East Ltd.
Madras 600 035
Phone: 44-453-622
Fax: 44-476-722

Italy: Pharmacia Biotech S.p.A.
20093 Cologne Monzese, Milano
Phone: 02-273-221
Fax: 02-273-02-212

Japan: Pharmacia Biotech K.K.
Tokyo 103
Phone: 03-34-92-94-97
Fax: 03-34-92-94-82

Korea: Pharmacia Dong-il Co. Ltd.
Seoul
Fax: 02-541-2163

Malaysia: Pharmacia Biotech Asean
47500 Petaling
Phone: 603-735-39-72
Fax: 603-735-46-72

Middle East: Pharmacia Export S.A.
GF-116610 Glyfada
Phone: 30-1-96-27-396
Fax: 30-1-96-29-963

New Zealand
Phone: 800-73-38-93
Fax: 9-815-06-23

Norway: Pharmacia Biotech AS
2001 Lillestrom
Phone: 63-89-23-10
Fax: 63-89-23-10

Portugal: Kabi Pharmacia Laboratorios SA
2795 Linda-a-Velha
Phone: 01-417-24-72
Fax: 01-417-01-45

Russia & CIS
Phone: 095-95-61-137
Fax: 095-95-61-137

Spain: Pharmacia Biotech S.A.
Barcelona
Phone: 93-589-07-01
Fax: 93-589-34-73

Sweden: Pharmacia Biotech Norden AB
191 27 Sollentuna
Phone: 08-623-85-00
Fax: 08-623-00-69

Switzerland: Pharmacia Biotech AG
8600 Dubendorf
Phone: 01-802-81-50
Fax: 01-802-81-51

Taiwan: Pharmacia Biotech Far East Ltd.
Taipei 11125
Phone: 02-831-5310
Fax: 02-831-5311

United Kingdom: Pharmacia Biotech Ltd.
St. Albans
Phone: 0727-814000
Fax: 0727-814001

USA: Pharmacia Biotech Inc.
Piscataway, NJ 08855-1327
Gratis phone: 800-526-3593
Fax: 800-329-3593

Primalco

Finland: Primalco Ltd. Biotec
FIN-05200 Rajamaki
Phone: +358-0-1331-1
Fax: +358-0-1331-236

Enzyme Supplier Index

Promega

USA: Promega Corporation
Madison, WI 53711-5399
Phone: 608-274-4330
Gratis phone: 800-356-9526
Fax: 608-273-6967

Argentina: Biodynamics, Srl.
Buenos Aires
Phone: 011-54-1-381-7962
Fax: 011-54-1-382-1986

Australia: Promega Corporation
Annandale NSW 2038
Phone: 02-565-1100
Gratis phone: 800-22-5123
Fax: 800-626-017

Austria: Bender MedSystems
Wien 1121
Phone: 1-801-05-615
Fax: 111-43-1-80105-488

Brazil: Genoma Biotecnologia do Brasil Ltda.
Belo Horizonte MG 30140
Phone: 011-0055-31-275-1240
Fax: 011-0055-31-337-9324

Canada: Fisher Scientific, Ltd.
Nepean
Gratis phone: 800-267-7424
Fax: 1-613-226-8639

Chile: FERMELO, Ltda.
Santiago
Phone: 011-56-2-220-1801
Fax: 011-56-2-220-4804

China: Sino-American Biotechnology Co. (SABC)
LuoYang
Phone: 011-86-379-3935-607
Fax: 011-86-379-3938-113

China: Promega Beijing
Beijing 100080
Phone: 011-86-1-256-3159
Fax: 011-86-1-255-9444

China: Sino-American Biotechnology Co. (SABC)
Beijing 100088
Phone: 011-86-1-202-0359
Fax: 011-86-1-202-0354

China: Sino-American Biotechnology Co. (SABC)
Shanghai 200032
Phone: 011-86-21-404-2371
Fax: 011-86-21-404-3824

China: Shanghai Promega Biological Products, Ltd.
Shanghai 200233
Phone: 011-86-21-470-0892
Fax: 011-86-21-470-0176

Denmark: Bie & Berntsen A-S
Roedovre DK-2610
Phone: 011-45-44-948-822
Fax: 011-45-44-942-709

Egypt: Lab Technology
Cairo 11351
Phone: 011-20-2-245-1785
Fax: 011-20-2-242-8366

Finland: YA-Kemia Oy
Helsinki 00510
Phone: 011-358-0-708-0900
Fax: 011-358-0-708-0437

France: Promega France
Charbonnieres les Bains 69260
Phone: 011-33-78-87-81-74
Gratis phone: 05-48-79-99
Fax: 011-33-78-87-81-10

France
Paris 75015
Phone: 011-33-14-533-6717
Fax: 05-38-39-83

Germany: SERVA Feinbiochemica GmbH & Co.
Heidelberg D-69115
Phone: 011-49-6221-502122
Fax: 011-49-6221-502113

Greece: BioAnalytica, S.A.
Athens
Phone: 011-30-1-64-36-138
Fax: 011-30-1-64-62-748

Hong Kong
Kowloon
Phone: 011-8522-751-9488
Fax: 011-8522-751-9217
E-mail: techcomp@HongKong.Super.NET

Hungary: East Port Scientific
Budapest
Phone: 011-36-1-251-0344
Fax: 011-36-1-183-3787

India: HYSEL INDIA
New Delhi 1100
Phone: 011-91-11-6846565
Fax: 011-91-11-6846565

Indonesia: PT Diastika Biotekindo
Jakarta
Phone: 011-62-21-723-1018
Fax: 011-62-21-726-0320

Ireland: Medical Supply Co. Ltd.
Dublin 9
Phone: 011-353-1-8426644
Fax: 011-353-1-8426747

Israel: ORNAT Biochemicals & Laboratory Equipment
Ness-Ziona 70400
Phone: 011-972-8-406930
Fax: 011-972-8-406498

Italy: GENENCO (Life Science) M-Medical, Srl.
Florence 50132
Phone: 011-39-55-5001871
Fax: 011-39-55-5002255

Japan: Seikagaku Corporation
Tokyo 103
Phone: 011-81-3-3270-0536
Fax: 011-81-3-3242-5335

Japan: Promega KK
Tokyo 103
Phone: 011-81-3-3669-7981
Fax: 011-81-3-3669-7982

Korea: Seoulin Scientific Co., Ltd.
Seoul 135-270
Phone: 011-82-2-3461-5911
Fax: 011-82-2-575-5572

Malaysia: Filtarite Company
Petaling Jaya, Selangor 47500
Phone: 011-60-3-735-1559
Fax: 011-60-3-735-1306

Mexico: UNIPARTS, S.A.
Mexico, D.F.
Phone: 011-52-5-519-3463
Fax: 011-52-5-538-1884

New Zealand: DADE Diagnostics Pty. Ltd.
Auckland
Phone: 011-64-9-570-3200
Fax: 011-64-9-570-3201

Norway: Bionor Research Products
Skien 3701
Phone: 011-47-35-54-1999
Fax: 011-47-35-53-7130

Poland: University of Gdansk
Kladki 24
Phone: 011-48-58-31-28-07
Fax: 011-48-58-31-28-07

Portugal: Biocontec Biotechnologia Ambiente, Lda.
Carcavelos
Phone: 011-351-1-458-1641
Fax: 011-351-1-458-1636
E-mail: biocontec@individual.puug.pt

Russia: Institute of General Genetics
Moscow 117809
Phone: 011-7095-135-42-06
Fax: 011-7095-135-42-06
E-mail: Bion@glas.apc.org

Enzyme Supplier Index

Singapore: Research Instruments
Phone: 011-65-775-7284
Fax: 011-65-775-9228

South Africa: Zotos House
Johannesburg 2195
Phone: 011-27-11-476-9457
Fax: 011-27-11-678-4826

Spain: Ingelheim Diagnostica y
Tecnologia, S.A.
Barcelona 08017
Phone: 011-34-3-404-52-14
Fax: 011-34-3-404-54-85

Sweden/Iceland: SDS Scandinavian
Diagnostic Services
Falkenberg S-311 35
Phone: 011-46-346-83050
Fax: 011-46-346-84840

Switzerland: Catalys AG
Wallisellen CH-8304
Phone: 011-41-1-830-70-37
Fax: 011-41-1-830-55-78

Taiwan: Pan Asia Hospital Supply Co., Ltd.
Taipei
Phone: 011-886-2-733-6877
Fax: 011-886-2-735-0741

Thailand: Diagnostic Biotechnology Co., Ltd.
Bangkok
Phone: 011-66-2-284-0950
Fax: 011-66-2-295-3632

The Netherlands: Promega BNL
Leiden 2331 BB
Phone: 06-0221910
Fax: 06-0226545

Turkey: ERMANAK MISKCIYAN
80000 Karakoy, Istanbul
Phone: 011-90-2-216-385-8321
Fax: 011-90-2-216-385-4649

United Kingdom: Promega UK
Southampton SO1
Phone: 011-44-1-703-760225
Gratis phone: 0800-378994
Fax: 011-44-1-703-767014

USA: Fisher Scientific
Pittsburgh, PA 15219
Gratis phone: 1-800-866-7000
Fax: 1-800-926-1166

Venezuela: Vargas Scientific Supplies, C.A.
Caracas
Phone: 011-58-2-265-0891
Fax: 011-58-2-263-0924

ProZyme

USA: ProZyme, Inc.
San Leandro, CA 94577-1258
Phone: 510-638-6900
Gratis phone: 800-457-9444
Fax: 510-638-6919
E-mail: info@prozyme.com

Randox

United Kingdom: Randox Laboratories Ltd.
Crumlin
Phone: +44-1849-422-413
Fax: +44-1849-422-413

Canada: Randox Laboratories
Mississauga
Phone: 905-564-9372
Fax: 905-564-8979

France: Randox Laboratories
Paris Nord II, BP 50037
Phone: +33-1-48657617
Fax: +33-1-48659821

Germany: Randox Laboratories
47807 Krefeld
Phone: 02151-310444
Fax: 02151-310444

Portugal: Irlandox Laboratories Quimica Analitica Ltda
4000 Porto
Phone: +351-25104278
Fax: +351-2580200

Spain
Gratis phone: 900-98-4494

USA: Randox Laboratories (California)
California
Phone: 310-531-3593
Fax: 310-531-0302

Rohm

Germany: Rohm GmbH
D-64275 Darmstadt
Phone: +49-6151-184774
Fax: +49-6151-184120

Singapore: ROHM Enzymes
Singapore 609916
Phone: +65-562-8387
Fax: +65-562-8397

USA: ROHM Enzymes
Somerset, NJ 08873-6821
Phone: 908-560-6322
Fax: 908-560-6356

SBI

USA: Systems Bio-Industries, Inc.
Waukesha, WI 53187-1609
Phone: 414-547-5531
Fax: 414-547-0587

Scripps

USA: Scripps Laboratories
San Diego, CA 92121-2904
Phone: 619-546-5800
Fax: 619-546-5812

Belgium: C. Canion
B-1301 Bierges
Phone: 010-41-35-56
Fax: 010-41-26-13

Italy: Inalco SpA
I-20139 Milano
Phone: 02-55213005
Fax: 02-5694518

Japan: Scripps Laboratories Japan, Inc.
Tokyo, 112
Phone: 03-3818-7631
Fax: 03-3818-4453

Switzerland: MILAN ANALYTICA AG
1634 La Roche
Phone: 037-332751
Fax: 037-332703

Thailand: Clinical Diagnostics Ltd.
Bangkok 10300
Phone: 02-281-2820
Fax: 02-280-3623

USA: Latin America Business & Strategies, Inc. (LABS)
San Diego, CA 92130-2054
Phone: 619-259-2626
Fax: 619-259-1342

Seikagaku

Japan: Seikagaku Corporation
Chuo-ku, Tokyo 103
Phone: 81-3-3245-1951
Fax: 81-3-3242-5335

Australia: Sapphire Bioscience Pty. Ltd.
Alexandria, NSW 2015
Phone: 02-313-4139
Fax: 02-669-2562

Austria: R.u.P. Margaritella Ges.m.b.H.
A-1230 Vienna
Phone: 1-889-18-19
Fax: 1-889-18-19-20

Canada: PDI Joldon
Aurora, Ontario L4G 3W2
Phone: 1-800-661-4556

Enzyme Supplier Index

Denmark: Bie & Berntsen Ltd.
DK-2610 Roedovre
Phone: 044-948822
Fax: 044-942709

Finland: YA-Kemia Oy
SF-00510 Helsinki
Phone: 0-708081
Fax: 0-753-8642

France: OSI
78312 Maurepas Cedex
Phone: 1-30-13-24-00
Fax: 1-30-13-24-24

France: Coger
75015 Paris
Phone: 1-45-33-67-17
Gratis phone: 05-38-39-83
Fax: 1-45-32-71-04

Germany: Medac GmbH
D-22083 Hamburg
Phone: 040-22-65-5-0
Fax: 040-22-65-5-123

Holland: Brunschwig Chemie B.V.
NL-1070 BE Amsterdam
Phone: 020-611-31-33
Fax: 020-613-75-96

Holland: Sanbio B.V.
5400 AM Uden
Phone: 04132-51115
Fax: 04132-66605

Italy: DBA Italia s.r.l.
20090 Segrate
Phone: 02-26411973
Fax: 02-2640540

Italy: Chebios s.r.l.
00162 Roma
Phone: 06-420506
Fax: 06-44290724

Korea: Chem-Bio Service Co., Ltd
Seoul 137-071
Phone: 02-588-5961
Fax: 02-588-5965

Spain: Quimica Farmaceutica Bayer, S.A.
28008 Madrid
Phone: 91-542-51-00
Fax: 91-542-05-22

Sweden: LabKemi Skandinavien AB
S-13570 Tyreso
Phone: 08-742-0250
Fax: 08-742-7299

Switzerland: Bio-Science Products AG
CH-6020 Emmenbrucke
Phone: 041-555875
Fax: 041-558322

Taiwan R.O.C.: Cheng Chin Trading Co., Ltd.
Taipei 10008
Phone: 02-3313607
Fax: 02-3822197

Taiwan R.O.C.: Cold Spring Biotechnology Co., Ltd.
221 Taipei Hsien
Phone: 02-694-0066
Fax: 02-694-3204

United Kingdom: AMS Biotechnology (UK) Ltd.
Witney
Phone: 0993-706500
Fax: 0993-706006

USA: Seikagaku America, Inc.
Ijamsville, MD 21754
Gratis phone: 800-237-4512
Fax: 301-424-6961

SibEnzyme

Russia: SibEnzyme Ltd.
Novosibirsk-90 630090
Phone: 383-235-3350
Fax: 383-232-8831
E-mail: info@siben.nsk.su

Sigma

USA: Sigma Chemical Company
St. Louis, MO 63178-9916
Phone: 314-771-5750
Gratis phone: 800-325-8070
Fax: 800-325-5052

Australia: Sigma-Aldrich Pty., Limited
Castle Hill NSW 2154
Phone: 899-9977
Gratis phone: 008-800-097
Fax: 008-800-096

Austria: Sigma-Aldrich Handels GmbH
A-1110 Wien
Phone: 0222/740-40-644
Fax: 0222/740-40-643

Belgium: Sigma-Aldrich N.V./S.A.
B-2880 Bornem
Phone: 03-8991301
Gratis phone: 0800-14747
Fax: 0800-14745

Brazil: Sigma-Aldrich Chemical Representacoes Ltda.
01239-010 Sao Paulo, SP
Phone: 011-231-1866
Fax: 011-257-9079

Czech Republic: Sigma-Aldrich s.r.o.
186 00 Praha 8
Phone: 02/2422-5285
Fax: 02/2422-4031

France: Sigma-Aldrich Chimie S.a.r.l.
38297 St. Quentin Fallavier Cedex
Phone: 74-82-28-00
Gratis phone: 05-21-14 08
Fax: 05-03-10-52

Germany: Sigma-Aldrich Chemie GmbH
D-82041 Deisenhofen
Phone: 089/6513-0
Gratis phone: 0130-5155
Fax: 0130-6490

Hungary: Sigma-Aldrich Kft
1067 Budapest
Phone: 06-1-269-1288
Fax: 06-1-153-3391

India: Sigma-Aldrich Corporation
Hyderabad 500 033
Phone: 040-244-739
Fax: 040-244-794

India: Sigma-Aldrich Corporation
New Delhi-110 029
Phone: 11-688-6872
Fax: 11-688-6873

Italy: Sigma-Aldrich S.r.l.
20151 Milano
Phone: 02-33417-310
Gratis phone: 1678-27018
Fax: 02-38010737

Japan: Sigma-Aldrich Japan K.K.
Tokyo 103
Phone: 03-3258-0155
Gratis phone: 0120-070406
Fax: 0120-676788

Korea: Sigma-Aldrich Korea
Seoul
Gratis phone: 080-023-7111
Fax: 080-023-8111

Mexico: Sigma-Aldrich Quimica, S.A. de C.V.
14210 Mexico, D.F.
Phone: 5-631-3671
Gratis phone: 91-800-00753
Fax: 5-631-3780

Enzyme Supplier Index

Poland: Sigma-Aldrich Sp. z.o.o
61-663 Poznan
Phone: 48-61-232-481
Fax: 48-61-232-781

Spain: Sigma-Aldrich Quimica S.A.
Madrid
Phone: 34-1-6619977
Gratis phone: 900-101376
Fax: 34-1-6619642

Sweden: LabKemi
Stockholm
Phone: 08-742-02-50
Fax: 08-742-72-99

Switzerland: Sigma Chemie
CH-9470 Buchs
Phone: 081-755-2721
Gratis phone: 155-00-20
Fax: 081-756-7420

The Netherlands: Sigma-Aldrich
N.V./S.A.
Phone: 00-32-3-8991301
Gratis phone: 06-0224748
Fax: 06-0224745

United Kingdom: Sigma-Aldrich
Company Ltd.
Poole
Phone: 01202-733114
Gratis phone: 0800-373731
Fax: 0800-378785

SpecialtyEnz

USA: Specialty Enzymes &
Biochemicals
Yorba Linda, CA 92687
Phone: 909-613-1660
Fax: 909-613-1663

India: Specialty Enzymes and
Biochemicals Co.
Thane (West) 400 601
Phone: 91-22-534-3119
Fax: 91-22-534-3445

SPL

USA: Scientific Protein Labs (Viobin
Corporation)
Waunakee, WI 53597-0158
Phone: 608-849-5944
Fax: 608-849-4053

USA: Viobin Corporation
Monticello, IL 61856
Phone: 217-762-2561
Fax: 217-762-2489

STC Labs

Canada: STC Laboratories Inc.
Div. of Export Packers Co. Ltd.
Brampton
Phone: 905-792-9700
Fax: 905-792-7421

Stratagene

USA: Stratagene Cloning Systems
La Jolla, CA 92037
Phone: 619-535-5400
Gratis phone: 800-424-5444
Fax: 619-535-0045

Argentina: Tecnolab F.A.
1427 Buenos Aires
Phone: 0115415550010
Fax: 0115415533331

Australia: Integrated Sciences
Willoughby NSW 2068
Phone: 011-61-2-417-7866
Gratis phone: 1-800-252-204
Fax: 011-61-2-417-5066

Austria: Chemomedica, Creutzberg &
Co.
1013 Wien
Phone: 011-43-1-533-2666
Fax: 011-43-1-535-330658

Brazil: Instrucom
04616-Sao Paulo
Phone: 011-55-11-530-7833
Fax: 011-55-11-530-0895

Canada: PDI Bioscience
Aurora, Ontario L4G 3W2
Phone: 905-713-1201
Fax: 905-713-1205

Canada: PDI Bioscience
Richmond, BC V6X 1X5
Phone: 604-270-2107
Fax: 604-270-2184

China: Cold Springs Biotech. (Beijing)
Co. Ltd. 1001012
Beijing
Phone: 86-10-4992996, 4928555
Fax: 011-86-1-492-8555

Czechoslovakia: Chemos
Czechoslovakia
CS 102 00 Praha 10-Hostivar
Phone: 011-42-2-7862198
Fax: 011-42-2-7862198

Denmark: AH Diagnostics
8200 Aarhus N
Phone: 011-45-8-6101055
Fax: 011-45-8-6161533

Finland: Kebo Finland
02150 Espoo
Phone: 011-358-0-43542003
Fax: 011-358-0-80455200

France: Ozyme
78180 Montigny-Le-Bretonneux
Phone: 011-33-1-34-60-2424
Fax: 011-33-1-34-60-9212
E-mail:
100726,2670@compuserve.com.fr

Germany: Stratagene GmbH
D-6900 Heidelberg
Phone: 011-49-6221-400634
Gratis phone: 0130-840911
Fax: 011-49-6221-400639

Hong Kong: Line Analytics Limited
North Point
Phone: 011-852-5785839
Fax: 011-852-2807-2674

Hungary: Kvalilex Scientific
Technological Trading Ltd.
H-1136 Budapest
Phone: 011-3611111700
Fax: 011-3611111700

India: Genetix
New Delhi-110 015
Phone: 011-91-11-5421714
Fax: 011-91-11-5467637

Iran: Farda Beh Co.
Tehran 14
Phone: 011-98-21-893292
Fax: 011-98-21-8801251

Ireland: Brownes Ltd.
Dublin 18
Phone: 01135312953401
Fax: 01135312953818

Israel: Getter (2000) Ltd.
Tel Aviv 61162
Phone: 011-972-3-7521687
Fax: 011-972-3-7523620

Italy: Eppendorf s.r.l.
20138 Milan
Phone: 011-39-2-5801-3409
Fax: 011-39-2-5801-3438

Japan: Funakoshi Co. Ltd.
Tokyo 113
Phone: 011-81-3-5684-1622
Fax: 011-81-3-5684-1633

Japan: Toyobo Co. Ltd.
Tokyo 103
Phone: 011-81-3-3660-4819
Fax: 011-81-3-3660-4887

Japan: Toyobo Co. Ltd.
Osaka 530
Phone: 011-81-6-348-3785
Fax: 011-81-6-348-3322

Enzyme Supplier Index

Korea: Koram Biotech. Corp.
Seoul 135-080
Phone: 011-82-2 5560311
Fax: 011-82-2-5560828

Malaysia: Interscience Sdn Bhd
47301 Petaling Jaya
Phone: 0116037031888
Fax: 0116037038047

Mexico: Bioselect
Fabricantes Y Distribuidores de Material Y Equipo de Laboratorio
02060 Mexico, D.F.
Phone: 011-525-341-7764
Fax: 011-525-341-7403
E-mail: bioselec@mail.internet.com.mx

New Zealand: Lab Supply Pierce (NZ) Ltd.
Auckland 10
Phone: 011-64-9 443-5867
Fax: 011-64-9-444-7314

Norway: A H Diagnostics AS
0566 Oslo
Phone: 011-47-22-71-5090
Fax: 011-47-22-71-9290

Portugal: Biocontec
2775 Carcavelos
Phone: 011-351-1-458-1641
Fax: 011-351-1-458-1636

Russia: DRG International, Inc.
Moscow 12 36 10
Phone: 011-7-095-253-1904
Fax: 011-7-095-253-1082

Singapore: ITS Science & Medical Pte. Ltd.
Singapore
Phone: 011652730898
Fax: 011652730810

South Africa (RSA): Whitehead Scientific Supplies
Capetown
Phone: 011-27-21-9811560
Fax: 011-27-21-9815789

South America: LABS
San Diego, CA 92130-2054
Phone: 619-259-2626
Fax: 619-259-1342

Spain: Cultek, S.L.
28034 Madrid
Phone: 011-34-1-729-03-33
Fax: 011-34-1-358-17-61

Sweden: AH Diagnostics AB
127 40 Skarholmen
Phone: 011-46-8-680-0845
Fax: 011-46-8-680-0435

Switzerland: Stratagene GmbH
CH-8035 Zurich
Phone: 011-41-1-3641106
Fax: 011-41-1-3657707

Taiwan, R.O.C.: Cold Springs Biotechnology Co.
Taipei Hsien
Phone: 011-886-2-695-9990
Fax: 011-886-2-695-9963

Taiwan, R.O.C.: Unimed Health Care Inc.
Taipei
Phone: 011-886-2-720-2215
Fax: 011-886-2-723-3666

Thailand: ITS (Thailand) Co., Ltd.
Bangkok 10320
Phone: 011-6623080611
Fax: 011-6623080612

The Netherlands: Westburg B.V.
NL-3830 AE LEUSDEN
Phone: 011-31-33-4950094
Gratis phone: 011-31-78-119815
Fax: 011-31-33-4951222

Turkey: Genomed Ltd.
80200 Istanbul
Phone: 011-90-212-248-2000
Fax: 011-90-212-231-4767

United Kingdom: Stratagene LTD.
Cambridge, CB4 4GF
Phone: 011-44-223-420955
Gratis phone: 011-44-800-585370
Fax: 011-44-223-420234

TaKaRa

Japan: Takara Shuzo Co., Ltd.
Biomedical Group
Shiga, 520-21
Phone: +81 775-43-7247
Fax: +81-775-43-9254
E-mail: trkbiosm@yo.rim.or.jp

France: Takara Biomedical Europe S.A.
92230 Gennevilliers
Phone: +33-1-41-47-01-14
Fax: +33-1-47-92-18-80

Korea: BOHAN Biomedical Inc.
Seoul 135-270
Phone: +82-2-577-2002
Fax: +82-2-577-3691

Taiwan: Cheng Chin Trading Co., Ltd.
Taipei
Phone: +886-2-331-3111
Fax: +886-2-382-2197

Taiwan: Protech Technology Enterprise Co., Ltd.
Taipei
Phone: +886-2-381-0844
Fax: +886-2-311-8524

USA: PanVera Corporation
Madison, WI 53711
Phone: 608-233-5050
Gratis phone: 800-791-1400
Fax: 608-233-3007
E-mail: info@panvera.com

Toyobo

Japan: Toyobo
Osaka 530
Phone: 06-348-3786
Fax: 06-348-3833

Unitika

Japan: Unitika Ltd.
Osaka 541
Phone: 06-281-5021
Fax: 06-281-5256

USA: SchweizerHall
Piscataway, NJ 08854
Phone: 908-981-8200
Fax: 908-981-8282

UOP

USA: UOP
Des Plaines, IL 60017-5017
Phone: 708-391-2127
Gratis phone: 1-800-348-0832
Fax: 708-391-3804

Vital

USA: Vital Products, Inc.
St. Louis, MO 63131-3927
Phone: 314-966-6996
Fax: 314-966-5695

Wako

Japan: Wako Pure Chemical Industries, Ltd.
Osaka 541
Phone: +81-6-203-3741
Fax: +81-6-222-1203

Germany: WAKO Chemicals GmbH
D-41468 Neuss
Phone: 49-2131-3-11-0
Fax: 49-2131-31-11-00

Enzyme Supplier Index

USA: Wako Chemicals USA, Inc.
Richmond, VA 23237
Phone: 804-271-7677
Gratis phone: 800-992-WAKO
Fax: 804-271-7791

Worthington
USA: Worthington Biochemical Corporation
Freehold, NJ 07728
Phone: 908-462-3838
Gratis phone: 800-445-9603
Fax: 800-368-3108 or 908-308-4453

Australia: ScimaR
Templestowe
Phone: 03-842-3386
Gratis phone: 1-800-639-364
Fax: 03-842-3407

Belgium: Bayer Belgium S.A.-N.V.
Diagnostic Division
1050 Brussels
Phone: 0297-280-666
Fax: 0297-284-165

Belgium: GESTIMED s.p.r.l./b.v.b.a.
1160 Brussels
Phone: 02-672-26-02
Fax: 02-672-26-02

Denmark: MEDINOVA Acientific ApS
DK-2900 Hellerup
Phone: 31-56-20-00
Fax: 31-56-19-42

Finland: YA-KEMIA OY
FIN-00510 Helsinki
Phone: 358-0-708-081
Fax: 358-0-708-0437

France: SERLABO
94866 Bonneuil-Sur-Marne
Phone: 1-49-80-50-00
Fax: 1-49-80-15-00

Germany: CellSystems
53424 Remagen 2
Phone: 02228-8057
Fax: 02228-8555

Hong Kong: Line Analytics, Ltd.
North Point
Phone: 578-5839
Fax: 807-2674

Israel: Enco Scientific Services, Ltd.
Petah Tikva
Phone: 3-9349922
Fax: 3-9349876

Italy: D.B.A. Italia, s.r.l.
Milano 20090 Segrate
Phone: 39-2-26922300
Fax: 39-2-26926058

Japan: Funakoshi Company, Ltd.
Tokyo 113
Phone: 3-5684-1622
Fax: 3-5684-1633

Korea: LRS Laboratories
Seoul 136-062
Phone: 02-924-8697
Fax: 02-924-8696

Korea: BMS Bio-Medical & Science Co., Ltd.
Seoul 135-080
Phone: 589-6902-4
Fax: 553-9670

Norway: INSTRUNOR A/S
N-3705 Skien
Phone: 3-501800
Fax: 3-502320

South Africa: STERILAB SERVICES cc
Kempton Park 1620
Phone: 27-11-974-5641
Fax: 27-11-395-1083

Spain: Eurofinsa, S.A.
28046 Madrid
Phone: 1-415-9312
Fax: 1-314-6829

Spain: LabClinics
E-08026 Barcelona
Phone: 3-446-4713
Fax: 3-348-1039

Sweden: AB Kemila-Preparat
S-19127 Sollentuna
Phone: 468-35-90-40
Fax: 468-35-05-45

Switzerland: Bioconcept
CH-4123 Allschwil
Phone: 61-481-47-13
Fax: 61-481-71-12

Taiwan R.O.C.: Protech Technology
Taipei
Phone: 2-381-0844
Fax: 2-311-8524

Taiwan R.O.C.: Johnson & Annie Company, Ltd.
Taipei 10099
Phone: 2-221-4949
Fax: 2-701-3970

The Netherlands: BAYER B.V.
3641 RT Mijdrecht
Phone: 02979-80692
Fax: 02979-80672

The Netherlands: ARISTOFORMA-INTERCHEMA
1437 CJ Rozenburg/Aalsmeer
Phone: 020-653-4567
Fax: 020-653-4433

United Kingdom: Lorne Laboratories, Ltd.
Twyford, Reading
Phone: 0734-342400
Fax: 0734-342788

Index B. General Index of EC Numbers and Enzyme Names, Synonyms and Abbreviations

1

1.1.1.1, *14*
1.1.1.2, *17*
1.1.1.5, *18*
1.1.1.6, *19*
1.1.1.8, *21*, *756*
1.1.1.10, *23*
1.1.1.14, *23*
1.1.1.18, *24*
1.1.1.19, *17*
1.1.1.21, *25*
1.1.1.22, *25*
1.1.1.26, *26*
1.1.1.27, *27*, *759*
1.1.1.28, *38*
1.1.1.30, *39*
1.1.1.33, *17*
1.1.1.35, *41*, *758*
1.1.1.37, *42*
1.1.1.40, *47*
1.1.1.41, *47*
1.1.1.42, *48*
1.1.1.44, *49*
1.1.1.47, *50*
1.1.1.48, *51*
1.1.1.49, *52*, *758*
1.1.1.50, *58*, *759*
1.1.1.51, *59*, *759*
1.1.1.53, *61*
1.1.1.55, *17*
1.1.1.67, *62*
1.1.1.72, *63*
1.1.1.83, *63*
1.1.1.95, *64*
1.1.1.119, *65*
1.1.1.122, *66*
1.1.1.159, *67*
1.1.1.176, *68*
1.1.1.204, *86*
1.1.2.3, *68*
1.1.2.4, *69*
1.1.3.4, *70*, *757*
1.1.3.6, *75*
1.1.3.9, *78*
1.1.3.13, *79*
1.1.3.15, 81
1.1.3.17, *82*
1.1.3.21, *83*
1.1.3.22, *85*
1.1.99.3, *88*
1.1.99.10, *89*
1.1.99.11, *89*
1.2.1.1, *90*, *96*
1.2.1.2, *91*
1.2.1.4, *92*
1.2.1.5, *92*
1.2.1.12, *93*, *758*
1.2.1.16, *96*
1.2.1.46, *96*
1.2.1.51, *97*
1.2.3.3, *98*
1.2.3.4, *100*
1.2.4.1, *100*
1.2.4.2, *101*
1.3.1.14, *101*
1.3.1.15, *114*
1.3.3.5, *102*
1.3.3.6, *103*
1.4.1.1, *104*
1.4.1.3, *104*
1.4.1.4, *107*
1.4.1.9, *108*
1.4.1.20, *109*
1.4.3.2, *81*, *110*
1.4.3.3, *111*
1.4.3.4, *113*
1.4.3.6, *114*
1.4.3.10, *115*
1.4.3.11, *115*
1.4.3.14, *116*
1.5.1.3, *117*, *129*
1.5.1.5, *118*
1.5.1.7, *118*
1.5.1.11, *119*
1.5.3.1, *120*
1.5.99.1, *121*
1.6.2.4, *122*
1.6.4.2, *122*
1.6.5.3, *128*
1.6.6.1, *124*
1.6.6.2, *124*
1.6.8.1, *125*
1.6.99.1, *125*
1.6.99.2, *127*
1.6.99.3, *128*
1.6.99.7, *129*
1.7.3.3, *129*
1.8.1.4, *133*
1.8.3.1, *135*
1.8.3.3, *136*
1.8.4.7, *86*
1.9.3.1, *137*
1.9.6.1, *137*
1.10.3.1, *138*, *174*
1.10.3.2, *139*
1.10.3.3, *139*
1.11.1.1, *141*
1.11.1.6, *142*, *757*
1.11.1.7, *147*, *759*
1.11.1.8, *155*
1.11.1.9, *155*
1.11.1.10, *157*
1.12.1.2, *91*
1.13.11.3, *158*
1.13.11.12, *159*
1.13.11.31, *160*
1.13.11.33, *161*
1.13.11.34, *162*

General Index

1.13.12.4, *163*
1.13.12.7, *164*
1.14.13.1, *166*
1.14.13.2, *167*
1.14.13.16, *168*
1.14.13.22, *169*
1.14.13.39, *169*
1.14.13.41, *170*
1.14.14.1, *122*
1.14.14.3, *171*
1.14.17.1, *173*
1.14.18.1, *138*, *174*
1.14.99.1, *175*, *759*
1.15.1.1, *176*
1.16.3.1, *180*
1.18.1.2, *181*

2

2.1.1.6, *181*
2.1.1.8, *182*
2.1.1.28, *183*
2.1.1.72, See Chapter 7, *1222-1238*
2.1.1.73, See Chapter 7, *1222-1238*
2.1.1.77, *184*
2.1.1.80, *184*
2.1.2.5, *185*
2.1.3.2, *185*
2.1.3.3, *186*
2.1.3.6, *186*
2.2.1.1, *187*
2.2.1.2, *188*
2.3.1.6, *188*
2.3.1.7, *189*
2.3.1.8, *190*
2.3.1.16, *191*, *758*
2.3.1.28, *192*
2.3.1.43, *193*
2.3.1.55, *193*

2.3.2.2, *194*
2.3.2.13, *196*
2.4.1.1, *196*
2.4.1.5, *197*
2.4.1.7, *198*
2.4.1.8, *199*
2.4.1.11, *199*
2.4.1.13, *200*
2.4.1.14, *200*
2.4.1.17, *201*
2.4.1.19, *202*
2.4.1.22, *203*, *745*
2.4.1.24, *204*
2.4.1.25, *204*
2.4.1.38, *205*, *745*
2.4.1.90, *203*
2.4.1.186, *199*
2.4.2.1, *206*
2.4.2.4, *207*
2.4.2.5, *206*
2.4.2.6, *207*
2.4.2.8, *208*
2.4.2.10, *209*, *670*, *759*
2.4.2.30, *209*
2.4.2.31, 701*699*
2.4.99.1, *210*
2.4.99.2, *211*
2.4.99.4, *211*
2.5.1.18, *212*
2.6.1.1, *42*, *214*
2.6.1.2, *216*
2.6.1.57, *214*
2.7.1.1, *218*, *758*
2.7.1.2, *222*
2.7.1.6, *223*
2.7.1.11, *223*
2.7.1.12, *225*
2.7.1.15, *225*
2.7.1.19, *226*
2.7.1.23, *226*

2.7.1.30, *227*
2.7.1.32, *230*
2.7.1.37, *231*
2.7.1.38, *236*
2.7.1.40, *759*
2.7.1.78, *241*
2.7.1.90, *244*
2.7.1.95, *245*
2.7.1.107, *246*
2.7.1.112, *247*, *352*
2.7.1.117, *248*
2.7.1.120, *248*
2.7.1.123, *248*
2.7.1.135, *248*
2.7.2.1, *249*
2.7.2.2, *186*, *250*
2.7.2.3, *250*, *758*
2.7.3.2, *252*
2.7.3.3, *257*
2.7.4.3, *257*
2.7.4.4, *259*
2.7.4.6, *260*
2.7.4.8, *261*
2.7.6.1, *262*
2.7.7.1, *262*
2.7.7.6, *263*
2.7.7.7, *270*, *756*
2.7.7.8, *286*
2.7.7.9, *287*
2.7.7.10, *288*
2.7.7.12, *289*
2.7.7.31, *289*
2.7.7.49, *292*, *756*
2.7.7.50, *296*
2.8.1.1, *297*
2.8.3.5, *298*
2.8.3.10, *680*

3

3.1.1.1, *299*
3.1.1.2, *299*
3.1.1.3, *301*, *753*
3.1.1.4, *193*, *312*
3.1.1.5, *193*, *299*, *316*
3.1.1.6, *299*, *317*
3.1.1.7, *318*
3.1.1.8, *320*
3.1.1.11, *317*, *323*, *746*, *747*
3.1.1.13, *323*
3.1.1.20, *327*
3.1.1.23, *299*
3.1.1.28, *299*
3.1.1.34, *328*
3.1.2.2, *299*
3.1.2.6, *329*
3.1.2.16, *680*
3.1.3.1, *330*, *657*
3.1.3.2, *341*
3.1.3.5, *344*
3.1.3.6, *345*
3.1.3.8, *346*
3.1.3.9, *347*, *657*
3.1.3.11, *347*
3.1.3.16, *348*
3.1.3.26, *351*
3.1.3.48, *349*, *352*
3.1.4.1, *353*
3.1.4.10, *361*
3.1.4.12, *362*
3.1.4.17, *363*
3.1.4.2, *355*
3.1.4.3, *356*
3.1.4.4, *359*
3.1.4.37, *365*
3.1.6.1, *366*, *739*
3.1.6.9, *367*
3.1.6.10, *368*
3.1.11.1, *369*
3.1.11.2, *370*
3.1.11.3, *372*
3.1.11.5, *373*
3.1.11.6, *374*
3.1.15.1, *375*
3.1.16.1, *376*
3.1.21.1, *378*, *383*, *756*
3.1.21.3, See Chapter 7, *1222-1238*
3.1.21.4, See Chapter 6, *762-1220*, Chapter 7, *1222-1238*
3.1.21.5, See Chapter 7, *1222-1238*
3.1.22.1, *383*
3.1.25.1, *385*
3.1.26.4, *386*
3.1.27.1, *388*
3.1.27.3, *389*
3.1.27.4, *392*
3.1.27.5, *393*, *756*
3.1.27.8, *398*
3.1.30.1, *399*
3.1.30.2, *404*
3.1.31.1, *405*
3.2.1.1, *407*, *738*, *750*, *751*, *753*
3.2.1.2, *417*, *738*, *739*, *751*
3.2.1.3, *419*, *452*, *738*, *745*, *750*
3.2.1.4, *422*, *741*, *742*, *743*, *744*, *752*, *753*
3.2.1.6, *429*, *738*, *740*, *741*, *742*, *750*, *751*, *752*
3.2.1.6$_{39}$, *752*
3.2.1.8, *432*, *738*, *739*, *740*, *741*, *742*, *744*, *747*
3.2.1.10, *434*
3.2.1.11, *435*
3.2.1.14, *436*, *752*
3.2.1.15, *438*, *742*, *743*, *744*, *746*, *747*, *751*, *752*
3.2.1.17, *436*, *441*
3.2.1.18, *445*
3.2.1.20, *452*
3.2.1.21, *455*, *739*, *740*, *741*, *742*
3.2.1.22, *457*
3.2.1.23, *458*
3.2.1.24, *463*
3.2.1.25, *464*, *747*
3.2.1.26, *464*
3.2.1.28, *466*
3.2.1.31, *467*, *739*
3.2.1.32, *471*
3.2.1.35, *472*
3.2.1.36, *476*
3.2.1.37, *477*
3.2.1.39, *477*
3.2.1.40, *478*, *741*
3.2.1.41, *479*, *486*, *738*, *739*, *745*
3.2.1.45, *480*
3.2.1.48, *434*
3.2.1.49, *481*
3.2.1.51, *482*
3.2.1.52, *419*, *483*
3.2.1.61, *485*
3.2.1.62, *499*
3.2.1.63, *485*, *499*
3.2.1.68, *486*
3.2.1.70, *487*
3.2.1.74, *488*, *742*, *743*
3.2.1.78, *488*, *742*, *743*, *744*, *746*, *747*, *752*
3.2.1.81, *490*
3.2.1.87, *752*
3.2.1.91, *492*, *741*, *742*, *743*
3.2.1.96, *744*, *745*
3.2.1.97, *495*
3.2.1.99, *496*, *742*, *746*, *752*
3.2.1.103, *497*
3.2.1.108, *499*
3.2.1.111, *485*, *499*
3.2.1.113, *500*
3.2.1.114, *501*
3.2.1.115, *746*, *752*
3.2.1.123, *502*

3.2.1.132, *503*
3.2.2.5, *503*
3.2.2.17, *504*
3.2.2.20, *506*
3.2.2.21, *506*
3.2.3.1, *507*
3.3.1.1, *507*
3.4.11.1, *508*, *511*, *749*
3.4.11.2, *510*
3.4.11.5, *511*
3.4.11.10, *512*
3.4.13.9, *514*
3.4.14.1, *514*, *524*
3.4.14.2, *515*, *524*
3.4.14.3, *524*
3.4.14.4, *524*
3.4.14.5, *515*, *524*
3.4.15.1, *516*
3.4.16.1, *517*
3.4.17.1, *519*
3.4.17.2, *521*
3.4.17.16, *523*
3.4.19.1, *524*
3.4.19.3, *525*
3.4.19.9, *526*
3.4.21.1, *527*, *748*
3.4.21.2, *554*
3.4.21.4, *531*, *748*
3.4.21.5, *541*
3.4.21.6, *547*
3.4.21.7, *549*
3.4.21.9, *550*
3.4.21.12, *552*
3.4.21.19, *552*, *624*
3.4.21.20, *554*
3.4.21.21, *555*, *748*
3.4.21.22, *556*, *748*
3.4.21.25, *556*
3.4.21.26, *557*
3.4.21.27, *558*

3.4.21.34, *558*
3.4.21.35, *559*
3.4.21.36, *561*, *565*
3.4.21.37, *565*
3.4.21.38, *566*
3.4.21.40, *560*
3.4.21.50, *567*
3.4.21.59, *568*
3.4.21.60, *547*
3.4.21.62, *556*, *569*, *748*
3.4.21.64, *573*
3.4.21.66, *576*
3.4.21.68, *579*
3.4.21.69, *576*
3.4.21.70, *577*
3.4.21.71, *578*
3.4.21.72, *578*
3.4.21.73, *579*
3.4.21.74, *580*
3.4.22.1, *581*
3.4.22.2, *583*, *749*
3.4.22.3, *587*
3.4.22.6, *588*
3.4.22.8, *589*, *607*
3.4.22.14, *590*
3.4.22.15, *581*, *591*
3.4.22.16, *591*
3.4.22.17, *592*
3.4.22.32, *593*, *749*
3.4.22.33, *594*
3.4.23.1, *595*, *749*
3.4.23.4, *598*, *749*
3.4.23.5, *599*
3.4.23.15, *598*, *601*
3.4.23.18, *602*, *603*, *750*, *752*
3.4.23.19, *603*
3.4.23.20, *604*
3.4.23.21, *604*, *751*
3.4.23.23, *605*
3.4.23.24, *606*

3.4.23.27, *606*
3.4.24.3, *607*, *617*, *748*, *750*
3.4.24.11, *617*
3.4.24.13, *578*
3.4.24.20, *618*
3.4.24.25, *618*
3.4.24.26, *618*
3.4.24.27, *618*, *751*
3.4.24.28, *620*, *618*, *748*, *750*, *751*, *752*, *753*
3.4.24.29, *618*, *624*
3.4.24.31, *618*, *625*
3.4.24.32, *627*
3.4.24.33, *628*
3.4.24.40, *629*
3.5.1.1, *629*
3.5.1.2, *631*
3.5.1.4, *299*, *632*
3.5.1.5, *632*
3.5.1.11, *636*
3.5.1.13, *299*, *637*
3.5.1.14, *638*
3.5.1.15, *639*
3.5.1.24, *640*
3.5.1.52, *640*, *744*, *745*
3.5.1.54, *725*
3.5.2.5, *642*
3.5.2.6, *643*
3.5.2.10, *646*
3.5.2.14, *648*
3.5.3.1, *649*
3.5.3.3, *650*
3.5.3.12, *186*
3.5.3.15, *651*
3.5.4.1, *652*
3.5.4.3, *652*
3.5.4.4, *653*
3.5.4.6, *655*
3.5.4.9, *118*
3.5.4.21, *655*

3.6.1.1, *657*
3.6.1.3, *659*
3.6.1.5, *660*
3.6.1.9, *661*
3.6.1.32, *662*
3.8.1.2, *662*

4

4.1.1.1, *663*
4.1.1.2, *664*
4.1.1.3, *664*
4.1.1.4, *665*
4.1.1.15, *666*
4.1.1.17, *667*
4.1.1.18, *668*
4.1.1.19, *669*
4.1.1.22, *669*
4.1.1.23, *209, 670, 759*
4.1.1.25, *671*
4.1.1.31, *672*
4.1.1.39, *674*
4.1.1.53, *674*
4.1.2.10, *675*
4.1.2.11, *676*
4.1.2.13, *676*
4.1.3.3, *679*
4.1.3.6, *680*
4.1.3.7, *42, 681*
4.1.3.34, *680*
4.1.99.1, *682*
4.2.1.1, *682*
4.2.1.2, *685*
4.2.1.11, *685*
4.2.1.17, *687, 758*
4.2.1.24, *687*
4.2.1.74, *687*
4.2.2.1, *688*
4.2.2.2, *689, 746, 747*

4.2.2.3, *690, 746*
4.2.2.4, *690*
4.2.2.5, *692*
4.2.2.7, *693*
4.2.2.8, *694*
4.2.2.10, *695*
4.3.1.1, *697*
4.3.1.3, *697*
4.3.1.4, *185*
4.3.1.5, *698*
4.3.2.1, *699*
4.3.2.2, *699*
4.4.1.5, *700*
4.4.1.11, *700*
4.6.1.1, *701*

5

5.1.3.1, *702*
5.1.3.2, *702*
5.1.3.3, *703*
5.2.1.8, *705*
5.3.1.1, *705, 758*
5.3.1.5, *707*
5.3.1.6, *708*
5.3.1.8, *708*
5.3.1.9, *709*
5.3.2.1, *711*
5.3.4.1, *711*
5.3.99.3, *713, 759*
5.3.99.4, *713, 759*
5.3.99.5, *714*
5.4.2.1, *715*
5.4.2.2, *716*
5.4.2.4, *715*
5.4.99.2, *717*
5.99.1.2, *717*
5.99.1.3, *719*

6

6.1.1.21, *720*
6.2.1.1, *720*
6.2.1.3, *721, 723*
6.2.1.4, *722*
6.2.1.20, *723*
6.3.1.2, *724*
6.3.4.3, *118*
6.3.4.6, *725*
6.3.5.1, *726*
6.4.1.1, *726*
6.5.1.1, *727*
6.5.1.2, *731*
6.5.1.3, *733*

A

AAO, *139*
Aat I, *1186*
Aat II, *762, 1198*
Acc I, *763, 938, 946, 1027, 1067, 1079, 1191*
Acc II, *765, 862, 920, 998*
Acc III, *766, 804, 880, 1062, 1215*
*Acc*16 I, *767*
*Acc*65 I, *768, 1042*
*Acc*113 I, *769*
*Acc*B1 I, *770*
*Acc*B2 I, *770*
*Acc*B7 I, *771*
*Acc*BS I, *771, 910*
Accelase, *508*
Acetate
   -CoA ligase, *720*
   kinase, *249*
   thiokinase, *720*
Acetic-ester acetylhydrolase, *317*
Acetinase, *317*

Acetoacetate
  carboxy-lyase, *665*
  decarboxylase, *665*
Acetoin dehydrogenase, *18*
Acetoin:NAD⁺ oxidoreductase, *18*
Acetokinase, *249*
Acetyl activating enzyme, *720*
Acetylcholine acetylhydrolase, *318*
Acetylcholinesterase, *318*
Acetyl-CoA
  c-acyltransferase, *191*
  synthetase, *720*
Acetyl-CoA:
  acetyl-CoA c-acyltransferase, *191*
  carnitine O-acetyltransferase, *189*
  chloramphenicol 3-O-acetyltransferase, *192*
  choline O-acetyltransferase, *188*
  kanamycin $N^{6'}$-acetyltransferase, *193*
  orthophosphate acetyltransferase, *188*
β-N-Acetyl-D-hexosaminide N-acetylhexosaminohydrolase, *483*
Acetylesterase, *299*, *317*
α-N-Acetylgalactosaminidase, *481*
Acetylglucosaminidase
  α-N-, *481*
  β-D-, *419*
  endo-β-N-, *492*
β-N-Acetyl
  glucosaminylglycopeptide β-1,4-galactosyltransferase, *205*, *745*
  hexosaminidase, *483*
N-Acetyl
  lactosamine synthase, *203*
  muramidase, *441*
  neuraminate lyase, *679*
  neuraminate pyruvate-lyase, *679*
  neuraminic acid aldolase, *679*
  transferase, *1250*
AChE, *318*
Achromopeptidase, *627*
Aci I, *772*, *938*, *1067*, *1079*
Acid

phosphatase, *341*
phosphomonoesterase, *341*
protease, *749*
Acid:CoA ligase (AMP-forming), *721*
Acl I, *773*
*Acl*N I, *773*
AcN I, *774*
ACOD, *103*
Acs I, *774*
Actinidain, *590*
Actinidin, *590*
Activated Christmas Factor, *556*
Actomyosin, *662*
Acutase, *1245*
Acy I, *775*, *818*, *938*, *943*, *1015*, *1030*, *1036*, *1067*, *1070*, *1079*, *1191*
Acylamidase, *632*
Acylamide amidohydrolase, *632*
Acylase, *632*, *638*, *639*
Acyl
  coenzymeA oxidase, *103*
  [acyl-carrier-protein] synthetase, *723*
  activating enzyme, *720*, *721*
  amino-acid-releasing enzyme, *524*
  aminoacyl-Peptidase, *524*
  carnitine hydrolase, *299*
  choline acylhydrolase, *320*
Acyl-CoA
  oxidase, *103*, *721*
  synthetase, *103*, *721*
Acyl-CoA:
  acetyl-CoA C-acyltransferase, *191*
  oxygen 2-oxidoreductase, *103*
Acylglycerol lipase, *299*
N-Acyl-L-
  amino acid amidohydrolase, *638*
  aspartate amidohydrolase, *639*
Acylneuraminyl hydrolase, *445*
N-Acylpeptide hydrolase, *524*
ADA, *653*
Adenosine
  aminohydrolase, *653*

deaminase, *206*, *653*
diphosphatase, *660*
triphosphatase, *164*, *252*, *662*, *659*
5'-triphosphatase, *659*
Adenosylhomocysteinase, *507*
S-Adenosyl-L-
  homocysteine hydrolase, *507*
  methionine:catechol O-methyltransferase, *181*
  methionine:histamine N-*tele*-methyltransferase, *182*
  methionine:phenylethanolamine N-methyltransferase, *183*
  methionine:protein L-glutamate O-methyltransferase, *184*
  methionine:protein-L-isoaspartate O-methyltransferase, *184*
Adenyl cyclase, *701*
Adenylate
  cyclase, *701*
  kinase, *257*, *721*
5'-Adenylic acid deaminase, *655*
Adenylosuccinate lyase, *699*
Adenylpyrophosphatase, *659*
Adenylylcyclase, *701*
Adjuzyme^R, *420*
ADH, *14*, *17*
AdK, *257*
ADPase, *See* Adenosine diphosphatase
ADP-ribosyl cyclase, *503*
ADP-ribosyltransferase, *209*
Adrenalin oxidase, *113*
Adrenodoxin reductase, *181*
Aeromondysin, *618*
Afa I, *776*, *1134*
Afe I, *776*
Afl II, *777*, *830*, *894*, *1071*
Afl III, *778*, *1083*
AgarACE™, *490*
1, β-Agarase, *490*
Agarose 3-glycanohydrolase, *490*
Age I, *779*, *796*, *804*, *944*, *1114*, *1215*

AGH, *452*
Aha II, *775, 847, 938, 943, 1015, 1030, 1070, 1079, 1191*
Aha III, *950*
Ahd I, *780*
AK, *249*
AlaDH, *104*
Alanine
   aminopeptidase, *510*
   aminotransferase, *216*
   dehydrogenase, *104*
   transaminase, *216*
L-Alanine:
   2-oxoglutarate aminotransferase, *216*
   NAD^+ oxidoreductase (deaminating), *104*
Alcalase^R, *571*
   Novo, *569*
AlcDH, *14*
Alcohol
   dehydrogenase, *14*, ($NADP^+$), *17*
   oxidase, *79*
Alcohol:
   NAD^+ oxidoreductase, *14*
   $NADP^+$ oxidoreductase, *17*
   oxygen oxidoreductase, *79*
Aldehyde
   dehydrogenase, ($NADP^+$) and [$NAD(P)^+$], *92*
   monooxygenase, *171*
   reductase, (NADPH), *14, 17, 25*
Aldehyde:
   $NAD(P)^+$ oxidoreductase, *92*
   $NADP^+$ oxidoreductase, *92*
Alditol:$NAD(P)^+$ 1-oxidoreductase, *25*
AlDH, *92, 129*
Aldoketomutase, *700*
Aldolase, *676*
Aldomax, *421*
Aldose 1-
   dehydrogenase, D-threo-, *66*
   aldose dehydrogenase, (2S,3R)-, *66, 703*
Aldose
   mutarotase, *703*
   reductase, *25*
Aldose:NAD^+ 1-oxidoreductase
   D-threo-, *66*
Alginate lyase I, *690*
Ali-esterase, *299*
Alkaline
   lipase, *301*
   phosphatase, *330, 657*
   phosphomonoesterase, *330*
   protease, *569, 620, 1243*
   proteinase, *572*
Alkanal monooxygenase (FMN-linked), *171*
Alkanal,reduced-FMN:oxygen oxidoreductase (1-hydroxylating, luminescing), *171*
Alkylated-DNA glycohydrolase (releasing methyladenine and methylguanine), *506*
Allantoin amidohydrolase, *642*
Allantoinase, *642*
Allophanate hydrolase, *725*
ALOD, *80*
ALP, *330*
Alphalase, *738, 750*
ALT, *98, 216*
Alu I, *780, 1222*
   methylase, *1222*
Alw I, *774, 783*
*Alw*21 I, *784, 823, 1128*
*Alw*26 I, *785, 876*
*Alw*44 I, *786, 792*
*Alw*N I, *787*
*Ama*87 I, *788, 878*
Ambazyme, *419, 420, 745*
AMG, *419*
Amidase, *299, 561, 632*
Amine oxidase, *113, 114*
Amine:oxygen oxidoreductase (deaminating), *113, 114*
Amino acid
   arylamidase, *510*
   oxidase, D-, *111*, L-, *81, 110*
Amino-acid:oxygen oxidoreductase
   (deaminating), D-, *111*
   (deaminating), L-, *110*
Aminoacylase, *638*
   II, *639*
Aminoacyltransferase, *194, 196*
Aminolevulinate
   dehydratase, *687*
   hydro-lyase (adding 5-aminolevulinate and cyclizing), 5-, *687*
Amino-oligopeptidase, *510*
Aminopeptidase, *512, 524, 591, 607, 749*
   cytosol, *508, 511*
   N-formylmethionine (fMet), *524*
   leucine, *508*
   membrane/microsomal/bound, *510*
   proline, *511, 515*
   pyroglutamyl, *525*
Ammodytoxin, *312*
AMP
   aminase, *655*
   aminohydrolase, *655*
   deaminase, *655*
cAMP-dependent protein kinase, *231*
Ampligase^R, *732*
AMP-lyase
   N^6-(1,2-dicarboxyethyl)AMP, *699*
Amylase, *70, 753, 1245*
   α-, *407, 419, 452, 455, 738, 750, 751, 754, 755*
   bacterial α-, *407*
   fungal α-, *407*
   β-, *417, 738, 739, 751*
   γ-, *419*
   Bacillus macerans, *202*
   fungal, *417*
   maltogenic, *204*
   saccharogenic, *417*
Amyloglucosidase, *See* Glucoamylase
Amylopectin 6-glucanohydrolase, *479*
Amylophosphorylase, *196*
Amylozyme, *409, 432, 738, 739, 750*
Ancrod, *580*

General Index

Angiotensin
   converting enzyme, *516*
   forming enzyme, *601*
Angiotensinogenase, *601*
AO, *139*
Aoc I, *788, 809, 855*
Aoc II, *886*
AOD, *110*
AOO, *80*
*Aor*51H I, *776, 789, 968*
Aos I, *808*
AP, *330*
Apa I, *790, 883*
*Apa*L I, *786, 792*
Apo I, *793*
Apy I, *915, 986*
Apyrase, *660*
Aquazym^R, *414*
Arabinan
   1,5-α-L-arabinohydrolase, 1,5-α-L-, *496*
   endo-1,5-α-L-arabinosidase, *496*
Arabinanase, *742, 746, 752*
   endo-1,5-α-L-, *496*
Arabinase, *496, 740, 744*
Arachidonate
   5-lipoxygenase, *162*
   12-lipoxygenase, *160*
   15-lipoxygenase, *161*
Arachidonate:oxygen
   5-oxidoreductase, *162*
   12-oxidoreductase, *160*
   15-oxidoreductase, *161*
Arachidonyl-CoA synthetase, *721*
Arg-C, *559*
Arginase, *649*
Arginine
   amidinohydrolase, L-, *642*
   amidinase, *649*
   carboxy-lyase, L-, *669*
   decarboxylase, *669*
   kinase, *257*
Arginine,NADPH:oxygen oxidoreductase
   L-, (nitric-oxide-forming), *169*
Arginine:NAD$^+$ oxidoreductase (L-arginine forming)
   N^2-(D-1-carboxyethyl)-L-, *119*
Argininosuccinate lyase, *699*
Arginosuccinase, *699*
Arginylendopeptidase, *559*
Aromatic alcohol dehydrogenase
   NADP$^+$-dependent, (R)- and (S)-, *17*
Aromatic-amino-acid transaminase, *214*
ARS, *366*
Aryl-acylamidase, *299, 637*
Aryl-acylamide amidohydrolase, *637*
Arylamidase, *510*
Arylesterase, *299*
Arylsulfatase, *366, 737, 739*
Aryl-sulfate sulfohydrolase, *366*
Asc I, *794, 900, 1058*
L-Ascorbate oxidase, *139*
L-Ascorbate:oxygen oxidoreductase, *139*
Ase I, *795, 1086, 1196, 1206*
*Asi*A I, *796*
Asn I, *796, 1086, 1196, 1206*
ASOD, *139*
ASOM, *139*
Asp I, *797, 803*
*Asp*700 I, *798*
*Asp*718 I, *768, 798, 814, 1042*
Asparaginase, *629*
Aspartase, *697*
Aspartate
   aminotransferase, *214*
   ammonia-lyase, *697*
   carbamoyltransferase, *185*
   transaminase, *42, 214*
   transcarbamylase, *185*
L-Aspartate:2-oxoglutarate aminotransferase, *214*
Aspartic proteinase, *See* Proteinase
Aspartoacylase, *639*
Aspartyl protease, *See* Protease

*Asp*E I, *799*
*Asp*H I, *800, 823, 1128*
*Asp*LE I, *800*
*Asp*S9 I, *801, 802, 1127*
Aspergillopepsin, *602, 603, 750*
Aspergillopeptidase A, *602*
Asperzyme^R, *416*
Assimilatory nitrate reductase, *124*
AST, *98, 214*
Asu I, *802, 936, 1150*
Asu II, *775, 866, 882, 938, 943, 1030, 1036, 1048, 1070, 1079, 1105, 1163, 1191*
*Asu*HP I, *802*
*Asu*NH I, *803*
ATPase. *See* Adenosine triphosphatase
ATP
   diphosphatase, *660*
   diphosphohydrolase, *660*
   kinase, *252*
   monophosphatase, *659*
   phosphohydrolase, *659*
   pyrophosphate-lyase (cyclizing), *701*
ATP:(d)GMP phosphotransferase, *261*
ATP:1,2-diacylglycerol 3-phosphotransferase, *246*
ATP:3-phospho-D-glycerate 1-phosphotransferase, *250*
ATP:5′-dephosphopolynucleotide 5′-phosphotransferase, *241*
ATP:acetate phosphotransferase, *249*
ATP:AMP phosphotransferase, *257*
ATP:carbamate phosphotransferase, *250*
ATP:choline phosphotransferase, *230*
ATP:creatine N-phosphotransferase, *252*
ATP:D-fructose 6-phosphate 1-phosphotransferase, *223*
ATP:D-galactose 1-phosphotransferase, *223*

ATP:D-gluconate 6-phosphotransferase, *225*
ATP:D-glucose 6-phosphotransferase, *222*
ATP:D-hexose 6-phosphotransferase, *218*
ATP:D-ribose 5-phosphotransferase, *225*
ATP:D-ribose-5-phosphate
　pyrophosphotransferase, *262*
　1-phosphotransferase, *226*
ATP:glycerol 3-phosphotransferase, *227*
ATP:kanamycin 3′-O-phosphotransferase, *245*
ATP:L-arginine N-phosphotransferase, *257*
ATP:NAD$^+$ 2′-phosphotransferase, *226*
ATP:nicotinamide-nucleotide
　adenylyltransferase, *262*
ATP:nucleoside-
　diphosphate phosphotransferase, *260*
　phosphate phosphotransferase, *259*
ATP:phosphorylase-*b* phosphotransferase, *236*
ATP:protein
　O-phosphotransferase (calmodulin-dependent), *248*
　phosphotransferase, *231*
　tyrosine O-phosphotransferase, *247*
ATP:pyruvate 2-O-phosphotransferase, *237*
ATP-
　citrate (*pro-S*)-lyase, *42*
　dependent deoxyribonuclease, *371*
　urea amidolyase, *725*
Ats I, *803*
Aureolysin, *618*, *624*
Ava I, *788*, *804*, *828*, *878*, *975*, *1211*
Ava II, *806*, *838*, *967*, *1012*, *1165*
Ava III, *990*, *1101*, *1128*
Avi II, *767*, *809*
Avr II, *808*, *837*, *888*, *1091*, *1171*, *1207*
Awamorin, *602*
Axy I, *788*, *809*, *855*, *946*

B

*Bac*77 I, *766*
Bacillolysin, *618*
Bacterase, *411*
Bacterial leucyl aminopeptidase, *510*
Bacterial luciferase. See Luciferase
Bake P, *425*
Bakery enzymes. See Chapter 8, *1248*
Bal I, *810*, *1060*, *1064*
*Bam*H I, *811*, *825*, *834*, *904*, *949*, *1052*, *1088*, *1148*, *1214*
　methylase, *1223*
BAN, *411*
Ban I, *770*, *814*
Ban II, *815*, *991*, *1000*
Ban III, *775*, *817*, *845*, *938*, *943*, *1030*, *1067*, *1070*, *1079*, *1191*
Batroxobin, *580*
Bbe I, *818*, *992*, *1005*
Bbi I, *1070*, *1079*
Bbi II, *775*, *818*, *938*, *943*, *1015*, *1030*, *1067*, *1191*
*Bbr*P I, *819*, *973*
Bbs I, *820*, *823*, *852*
Bbu I, *821*, *1173*
Bbv I, *822*
Bbv II, *820*, *840*
*Bbv*12 I, *823*
*Bbv*16 II, *823*
BC, *320*
*Bca*77 I, *1215*
Bcg I, *824*
Bcl I, *811*, *825*, *834*, *870*, *949*, *997*, *1052*, *1088*, *1148*, *1214*
Bcn I, *827*
Bco I, *788*, *804*, *828*, *878*
Benzamidase, *638*
Benzenediol:oxygen oxidoreductase, *139*
　1,2-, *138*
Benzoylcholinesterase, *320*
Bet I, *850*
Bfa I, *829*, *1049*, *1086*
Bfm I, *830*, *919*
Bfr I, *777*, *830*
Bgl I, *831*, *949*
Bgl II, *811*, *825*, *834*, *1052*, *1088*, *1148*, *1214*
Bile salt hydrolase, *640*
Bilirubin oxidase, *102*
Bilirubin:oxygen oxidoreductase, *102*
Bioprase, *572*
BioTouch^R, *741*
Biozyme, *415*, *417*
Bisphosphoglycerate mutase, *715*
Bln I, *837*, *888*, *1091*, *1171*, *1207*
Blp I, *838*, *888*
*Bme*18 I, *838*, *1012*
Bmy I, *839*, *1128*
Bpi I, *823*, *840*
Bpm I, *841*, *1004*
*Bpu*14 I, *842*, *930*
*Bpu*1102 I, *838*, *842*, *888*, *931*, *994*
*Bpu*A I, *843*
Brewers
　Fermex^R, *738*
　Mylase^R, *738*
Bromelain, *593*, *594*, *749*
Bromoperoxidase, *147*
Bsa I, *844*
*Bsa*29 I, *845*
*Bsa*A I, *845*, *909*
*Bsa*B I, *846*, *855*, *864*, *865*
*Bsa*H I, *775*, *847*, *938*, *1015*, *1067*, *1079*
*Bsa*J I, *848*
*Bsa*M I, *848*
*Bsa*O I, *849*, *915*, *1141*
*Bsa*W I, *804*, *850*, *1215*
Bsc I, *845*, *850*, *938*

*Bsc*4 I, *851*
*Bsc*91 I, *852*
*Bsc*B I, *852*
*Bsc*C I, *853*
*Bsc*F I, *853*
*Bse*1 I, *854, 899*
*Bse*3D I, *854*
*Bse*8 I, *855*
*Bse*21 I, *809, 855*
*Bse*118 I, *856, 900*
*Bse*A I, *804, 856, 880, 1215*
*Bse*C I, *845, 857*
*Bse*D I, *858*
*Bse*N I, *854, 859, 899*
*Bse*P I, *859, 900*
*Bse*P1 I, *1109*
*Bse*R I, *860*
Bsg I, *861*
Bsh I, *861*
*Bsh*1236 I, *765, 862, 998*
*Bsh*1285 I, *863, 915*
*Bsh*1365 I, *855, 864*
*Bsh*N I, *770, 865, 972*
*Bsi*B I, *846, 855, 865*
*Bsi*C I, *842, 866, 930, 938, 1067, 1079*
*Bsi*E I, *849, 867, 915, 1141*
*Bsi*HKA I, *823, 868, 1128*
*Bsi*L I, *868, 905, 986*
*Bsi*M I, *804, 869, 880, 1215*
*Bsi*Q I, *825, 870, 811, 834*
*Bsi*S I, *870*
*Bsi*W I, *871*
*Bsi*X I, *845, 872, 938, 1067, 1079*
*Bsi*Y I, *851, 872*
*Bsi*Z I, *801, 802, 873, 1127*
Bsl I, *851, 872, 874*
Bsm I, *848, 875*
*Bsm*A I, *785, 876*
*Bsm*B I, *877*

*Bsm*F I, *878*
*Bso*B I, *788, 878*
*Bso*F I, *879, 997, 1003*
*Bsp*13 I, *880*
*Bsp*19 I, *880*
*Bsp*50 I, *765*
*Bsp*68 I, *881*
*Bsp*106 I, *845, 882, 938, 1067, 1079*
*Bsp*119 I, *842, 882, 930*
*Bsp*120 I, *790, 883, 958, 1095*
*Bsp*143 I, *884*
*Bsp*143 II, *770, 885, 914, 1005*
*Bsp*1286 I, *839, 886, 1128*
*Bsp*1407 I, *887, 898, 1183*
*Bsp*1720 I, *888*
*Bsp*A2 I, *888*
*Bsp*C I, *889*
*Bsp*D I, *817, 845, 882, 889, 938, 1067, 1079*
*Bsp*E I, *766, 804, 880, 890, 1215*
*Bsp*H I, *891, 1083, 1133*
*Bsp*L I, *892*
*Bsp*LU11 I, *893*
*Bsp*M I, *869, 893*
*Bsp*M II, *766, 856, 890, 1045, 1062*
*Bsp*T I, *894*
*Bsp*X I, *845, 938*
Bsr I, *854, 859, 895, 899*
*Bsr*B I, *771, 896, 910*
*Bsr*BR I, *855, 896*
*Bsr*D I, *854, 897*
*Bsr*F I, *804, 856, 898, 900, 935, 1215*
*Bsr*G I, *898, 1183*
*Bsr*S I, *854, 899*
*Bss*A I, *856, 900*
*Bss*C I, *1007*
*Bss*H II, *900, 1058*
*Bss*HI I, *859*
*Bss*K I, *902*

*Bss*S I, *903, 905*
*Bss*T1 I, *903, 994*
Bst I, *811, 825, 834, 904, 1052*
*Bst*2B I, *905*
*Bst*2U I, *905*
*Bst*71 I, *822, 906*
*Bst*98 I, *906*
*Bst*1107 I, *907*
*Bst*B I, *775, 842, 908, 930, 938, 943, 1030, 1048, 1067, 1070, 1079, 1103, 1191*
*Bst*BA I, *909*
*Bst*D102 I, *910*
*Bst*DE I, *910*
*Bst*DS I, *911*
*Bst*E II, *911, 918, 975, 980*
*Bst*F5 I, *913*
*Bst*H2 I, *770, 914*
*Bst*HP I, *914*
*Bst*MC I, *915*
*Bst*N I, *868, 905, 915, 986*
*Bst*NS I, *916*
*Bst*O I, *905, 917*
*Bst*P I, *911, 918*
*Bst*SF I, *919*
*Bst*SN I, *920*
*Bst*U I, *765, 862, 920, 998*
*Bst*X I, *921*
*Bst*X2 I, *924*
*Bst*Y I, *811, 825, 834, 924, 1052, 1148, 1214*
*Bst*Z I, *925, 969, 1095*
*Bsu*6 I, *926*
*Bsu*15 I, *845, 882, 926, 938*
    methylase, *1224*
*Bsu*23 I, *804, 1215*
*Bsu*36 I, *788, 809, 855, 927, 974*
*Bsu*R I, *928, 1007*
β-Bungarotoxin, *312*
Butyrylcholine esterase, *320*

C

C4S, *367*
C6S, *368*
CA, *682*
Ca^{2+}/calmodulin
  kinase II, *248*
  -dependent protein kinase, *248*
*Cac*8 I, *929*
Calcineurin, *348, 349, 705*
Calculase, *1249*
Caldesmon kinase, *248*
Calpain, *592*
CaM kinase II, *248*
Canavanase, *649*
Candidapepsin, *606*
CANP, *592*
CAPALASE^{TM}, *302*
Carbamate kinase, *250*
Carbamoyl phosphate:
  L-ornithine carbamoyltransferase, *186*
  L-aspartate carbamoyltransferase, *185*
Carbamylaspartotranskinase, *185*
N-Carbamylsarcosine amidohydrolase, *648, 655*
Carbohydrases
  mixtures, See Chapter 5, *738-747, 750-755*
  unclassified, See Chapter 8, *1240-1241*
Carbonate
  dehydratase, *682*
  hydro-lyase, *682*
Carbonic anhydrase, *682*
Carboxycathepsin, *516*
α-Carboxylase, *663*
Carboxylesterase, *299*
Carboxypeptidase, *517, 519, 521, 523, 526*
  carboxylic-ester hydrolase, *299*
Carboxypolypeptidase, *519*
Carboxytripeptidase, *515*
Carnitine O-acetyltransferase, *189*

Carotene oxidase, *159*
Casein kinase, *231*
CAT, *192*
Catalase, *129, 141, 142, 143, 757*
Catechol
  O-methyltransferase, *181*
  oxidase, *138, 174*
Catecholase, *174*
Cathepsin
  A, *517, 599*
  B, *581, 591*
  C, *514*
  D, *599*
  G, *554*
  H, *591*
  L, *581, 591*
Cau II, *827*
Cbi I, *842, 930*
*Cci*N I, *930*
Ccr I, *1143*
CE, *323*
Cel II, *842, 888, 931, 946, 994*
Cellobiase, *453, 740*
Cellobiohydrolase, *741, 742, 743*
Celluclast^R, *426*
Cellulase, *422, 423, 696, 741, 742, 743, 744, 752, 753*
Cellulizer, *428*
Cellulose 1,4-β-cellobiosidase, *492*
Celluzyme^R, *424*
Cephalosporinase, *643*
Ceramide glycanase, *480*
Cereflo^R, *429*
Ceremix^R, *750*
Ceruloplasmin, *180*
Cfo I, *800, 932, 1013*
Cfr I, *993, 957, 969*
*Cfr*9 I, *766, 934, 1215, 1218*
*Cfr*10 I, *804, 856, 898, 900, 935, 1215*
*Cfr*13 I, *801, 802, 936, 1127, 1150*
*Cfr*19 I, *766*

*Cfr*42 I, *937*
Chillproof, *584*
Chitinase, *436, 752*
Chitodextrinase, *436*
Chitosan N-acetylglucosaminohydrolase, *503*
Chitosanase, *503*
Chloramphenicol
  acetyltransferase, *192*
  O-acetyltransferase, *192*
Chloride peroxidase, *157*
Chloride:hydrogen-peroxide oxidoreductase, *157*
Chloroperoxidase, *157*
ChO, *75*
Cholesterol
  dehydrogenase, *1249*
  esterase, *323*
  oxidase, *75, 323*
Cholesterol:oxygen oxidoreductase, *75*
Cholesterol-binding proteinase, *577*
Cholesteryl ester synthase, *323*
Choline
  acetylase, *188*
  acetyltransferase, *188*
  O-acetyltransferase, *188*
  esterase, *158, 316, 318*
  kinase, *230*
  oxidase, *82, 359*
  phosphatase, *359*
Choline:oxygen 1-oxidoreductase, *82*
Cholinesterase, *318, 320*
Choloyl
  glycine hydrolase, *640*
  taurine hydrolase, *640*
Chondro
  -4-sulfatase, *367*
  -6-sulfatase, *368*
Chondroitin
  ABC eliminase, *690*
  ABC lyase, *690*
  AC eliminase, *692*

AC lyase, *692*
sulfate lyase, *692*
Chondroitinase, *690, 692*
Christmas Factor, activated, *556*
Chymogen™, *598*
Chymopapain, *588*
Chymosin, *598, 749*
Chymotrypsin, *527, 547, 549, 550, 552, 557, 558, 560, 561, 565, 566, 568-569, 577, 578, 580*
Citrase, *680*
Citratase, *680*
Citrate
  aldolase, *680*
  CoA-transferase, *680*
  condensing enzyme, *681*
  lyase, *680*
  (*si*)-synthase, *42, 681*
[Citrate oxaloacetate-lyase ((*pro*-3S)-CH$_2$COO$^-$→acetate)], *680*
[Citrate oxaloacetate-lyase ((*pro*-3S)-CH$_2$COO$^-$→ acetyl-CoA), *681*
[Citrate-(*pro*-3S)-lyase], *680*
Citridesmolase, *680*
Citritase, *680*
Citrogenase, *681*
Citrulline phosphorylase, *186*
Citrus acetylesterase, *317*
Citryl-CoA lyase, *680*
CK, *252*
CL, *680*
Cla I, *775, 817, 845, 857, 872, 882, 889, 926, 938, 943, 1030, 1067, 1070, 1079, 1191*
  methylase, *1225*
Clostridopeptidase
  A, *607*
  B, *589*
Clostripain, *589, 607*
CMP-*N*-acetylneuraminate:

β-*D*-galactoside α-2,3-*N*-acetylneuraminyltransferase, *211*
β-*D*-galactosyl-1,4-*N*-acetyl-β-*D*-glucosamine α-2,6-*N*-acetylneuraminyltransferase, *210*
CNPase, *365*
CO, 77, 137
Co-Adjuzyme^R, *410*
Coagulation Factor
  Va, *541, 576*
  VIIa, *555, 566, 576, 748*
  IXa, *556, 748*
  X activating enzyme, *745*
  Xa, *541, 547, 748*
  XIa, *556, 558, 566*
  XIIa, *558, 566*
  XIIIa, *196*
Cocaine esterase, *299*
Cocoonase, *531*
COD, *82*
Colicin, *378*
Colipase, *301*
Collagenase, *531, 607, 748, 752*
Combizyme, *750, 751*
COMT, *181*
CON, *75*
Condensing enzyme, *681*
Conjugase, *526*
COO, *77*
Cookerzyme^R, *413*
Copper oxidase, *174*
Corolase^R, *572, 577, 603*
Correndonuclease II, *385*
Cortisone reductase, *61*
Cottage cheese coagulator, *598*
Cpase Y, *517*
CPK, *252*
Cpo I, *940, 941*
Cracker Special, *1248*
Creatinase, *650*
Creatine
  amidinohydrolase, *121, 650*
  kinase, *164, 227, 252*

phosphokinase, *252*
Creatininase, *646*
Creatinine
  amidohydrolase, *121, 646*
  deaminase, *655*
  deiminase, *648, 655*
  iminohydrolase, *655*
Cresolase, *174*
Crotalase, *580*
Crotonase, *687*
CS, *681*
Csp I, *940, 941*
*Csp*6 I, *776, 942, 1086*
*Csp*45 I, *775, 842, 930, 938, 943, 1030, 1067, 1070, 1079, 1103, 1191*
*Csp*A I, *944*
Cucumisin, *556*
*Cvi*B III methylase, *1282*
*Cvi*J I, *944*
Cvn I, *788, 809, 855, 945, 946, 974*
3′,5′-Cyclic AMP synthetase, *701*
Cyclic-CMP phosphodiesterase, *365*
2′,3′-Cyclic-nucleotide 3′-phosphodiesterase, *365*
3′,5′-Cyclic-nucleotide 5′-nucleotidohydrolase, *363*
3′,5′-Cyclic-nucleotide phosphodiesterase, *363*
Cyclodextrin glucanotransferase, *202*
Cyclohexanone monooxygenase, *169*
Cyclohexanone,NADPH:oxygen oxidoreductase (6-hydroxylating, 1,2-lactonizing), *169*
Cyclomaltodextrin glucanotransferase, *202*
Cyclooxygenase, *175, 759*
Cyclopentanone monooxygenase, *168*
Cyclopentanone,NADPH:oxygen oxidoreductase (5-hydroxylating, lactonizing), *168*
Cyclophilin, *705*

Cysteine protease. *See* Protease
Cytochrome
    a_3, *137*
    c reductase, *128*
    oxidase, *137*
Cytochrome-c oxidase, *137*
Cytocuprein, *176*
Cytosine
    aminohydrolase, *652*
    deaminase, *652*
Cytosol aminopeptidase, *508*, *511*

D

dam Methylase, *1226*
Dde I, *910*, *946*
Deamido-NAD$^+$:L-glutamine amido-ligase (AMP-forming), *726*
Deaminase, *655*
Deamizyme, *655*
Debitrase, *508*, *516*
Debranching enzyme, *479*, *486*
Deep Vent$_R^R$, *277*
Dehydropeptidase II, *638*
Denazyme, *603*
Denifade, *1241*
Deoxin, *757*
4-Deoxy-β-D-gluc-4-enuronosyl-(1,3)-*N*-acetyl-D-galactosamine-
    4-sulfate 4-sulfohydrolase, *367*
    6-sulfate 6-sulfohydrolase, *368*
Deoxy-D-cyclobutadipyrimidine polynucleotidodeoxyribohydrolase, *504*
Deoxyguanylate kinase, *261*
Deoxynucleoside-triphosphate:DNA deoxynucleotidyltransferase
    (DNA-directed), *270*
    (RNA-directed), *292*
Deoxyribonuclease, *369*, *372*, *378*, *383*, *385*, *399*, *573*, *756*

Deoxyribopyrimidine endonucleosidase, *504*
Dephosphophosphorylase kinase, *236*
Depol, *418*, *424*, *434*, *478*, *738*, *739*, *740*, *743*, *751*
Desize, *412*, *413*, *1241*
Dextranase, *435*
Dextransucrase, *197*
Dextrin
    6-glucanohydrolase, α-, *479*
    6-α-D-glucanohydrolase, *434*
    endo-1,6- α-glucosidase, α-, *486*
    glucanohydrolase, α-, *486*
    glycosyltransferase, *204*
Dextrinase, *432*
Dextrozyme™, *745*
DI, *125*
Diacetinase, *1249*
Diacetyl reductase, *18*
N,N'-Diacetylchitobiosyl β-*N*-acetylglucosaminidase, *492*
Diacylglycerol
    kinase, *sn*-1,2-, *244*
    lipase, *328*
Diamine oxidase, *114*
Diamino oxhydrase, *114*
Diaphorase, *125*, *133*
Diastase, *407*
Diglyceride
    kinase, *246*
    lipase, *328*
Dihydrofolate reductase, *117*, *129*
Dihydrolipoamide dehydrogenase, *133*
Dihydrolipoamide:NAD$^+$ oxidoreductase, *133*
Dihydroorotate dehydrogenase, *101*
(*S*)-Dihydroorotate:NAD$^+$ oxidoreductase, *101*
Dihydropteridine reductase, *129*
Dihydroxyacetonetransferase, *188*

3,4-Dihydroxyphenethylamine, Ascorbate:oxygen oxidoreductase (β-hydroxylating), *173*
Dinucleotide nucleotidohydrolase, *661*
Dioxygenase, *175*
Dipeptide hydrolase, *516*
Dipeptidyl
    aminopeptidase, *512*, *513*
    arylamidase II, *515*
    carboxypeptidase I, *516*
    -peptidase, *512*, *514*, *515*, *524*
    transferase, *514*
Diphenol oxidase, *138*
O-Diphenolase, *138*
Dispase, *617*, *620*, *748*, *752*
Disproportionating enzyme, *204*
DNA
    gyrase, *719*
    helicase, *369*
    joinase, *727*, *731*
    ligase, *727*, *731*, *915*, *979*, 1*155*, 1*205*
    nucleotidylexotransferase, *289*
    nucleotidyltransferase, *270*, *292*
    polymerase, *270*, *271*, *370*, *756*, *915*, *1218*
    repair enzyme, *727*, *731*
    topoisomerase, *717*, *719*
DNA methyltransferases. *See* Chapter 7
    Alu I, *1222*
    *Bam*H I, *1223*
    *Bsu*15 I, *1224*
    Cla I, *1225*
    dam, *1226*
    *Eco*R I, *1227*
    *Fnu*D II, *1228*
    Hae III, *1229*
    Hap II, *1230*
    Hha I, *1230*
    *Hin*d III, *1231*
    Hpa II, *1232*
    Msp I, *1233*
    Mva I, *1234*
    Pst I, *1235*
    Sss I (CpG), *1236*

Taq I, *1237*
*Tth*HB8 I, *1238*
DNA-3-methyladenine glycosidase, *506*
DNA-directed
    DNA polymerase, *270*
    RNA polymerase, *263*
DNase. See Deoxyribonuclease
Donor:hydrogen-peroxide oxidoreductase, *147*
Dopamine
    β-hydroxylase, *173*
    β-monooxygenase, *173*
Dox, *750*
DPN
    hydrolase, *503*
    kinase, *226*
DPNase, *503*
Dpn I, *948*, *1052*
Dpn II, *811*, *825*, *834*, *949*, *1052*, *1148*
Dra I, *950*
Dra II, *852*, *981*
Dra III, *953*
Drd I, *955*, *956*
Driselase[R], *1245*
Dsa I, *911*, *955*, *1058*
Dsa V, *1155*
DsbA, *711*
*Dse*D I, *956*

E

Eae I, *925*, *933*, *957*, *958*, *969*, *1095*
Eag I, *925*, *958*, *969*, *1095*
*Eam*1104 I, *926*, *959*, *1047*
*Eam*1105 I, *960*
Ear I, *926*, *959*, *961*, *1047*
Ecarin, *1243*
*Ecl*136 II, *962*
*Ecl*HK I, *963*
*Ecl*X I, *963*, *969*, *1095*
*Eco*24 I, *964*, *991*, *1000*
*Eco*31 I, *965*
*Eco*32 I, *966*, *987*
*Eco*47 I, *806*, *838*, *967*, *1012*
*Eco*47 III, *776*, *789*, *968*
*Eco*52 I, *925*, *958*, *969*, *1095*
*Eco*57 I, *971*
*Eco*64 I, *770*, *972*
*Eco*72 I, *973*
*Eco*81 I, *788*, *809*, *855*, *927*, *974*
*Eco*88 I, *788*, *804*, *878*, *975*
*Eco*91 I, *911*, *975*
*Eco*105 I, *920*, *976*
*Eco*130 I, *903*, *977*, *989*, *994*
*Eco*147 I, *977*
*Eco*255 I, *769*, *978*
*Eco*ICR I, *979*
*Eco*N I, *979*
Econase[R], *425*, *426*, *430*, *431*, *432*, *433*, *742*, *743*
*Eco*O65 I, *911*, *980*
*Eco*O109 I, *952*, *981*
Ecopen, *426*
Ecopulp[R], *433*, *747*
*Eco*R I, *983*, *1072*, *1218*, *1227*
    methylase, *1200*, *1227*
*Eco*R II, *905*, *915*, *986*, *1074*
*Eco*R V, *966*, *987*
Ecostone[R], *425*, *426*, *427*, *741*, *742*
*Eco*T14 I, *903*, *989*, *994*, *1188*
*Eco*T22 I, *990*, *1101*, *1128*
*Eco*T38 I, *991*, *1000*
Ege I, *992*
Ehe I, *992*, *993*
Elastase, *561*, *565*, *577*, *578*, *754*, *755*
Emulsin, *455*
Endo
    -1,3(4)-β-glucanase, *429*
    -1,3-β-glucanase, *429*
    -1,4-β-glucanase, *422*
    -1,4-β-xylanase, *432*
Endodeoxyribonuclease (pyrimidine dimer), *385*
Endoglucanase, *429*
Endoglycoceramidase, *502*
Endoglycosidase
    D, *492*, *641*
    F, *492*, *641*, *744*, *745*
    F1, *492*
    F2, *492*
    H, *492*, *641*
Endoglycosylceramidase, *502*
Endonuclease, *369*
    I, *E. coli*, *378*
    II, T4, *378*
    II, T7, *378*
    III, *E. coli*, *385*
    V, *E. coli*, *385*
    V, T4, *385*
    S_1, *Aspergillus*, *399*
    micrococcal, *405*
    Serratia marcescens, *404*
    single-stranded-nucleate, *399*
    spleen, *406*
Endopectin lyase, *746*
Endopeptidase, *591*
    24.11, *617*
    E, pancreatic, *577*
    K, *573*
    glutamyl, *552*, *624*
    neutral, *510*
    proline-specific, *557*
    Rhizopus niveus, *604*
    serine, *517*, *576*, *578*
Endoperoxide isomerase, *713*
Endoproteinase
    Arg-C, *559*, *589*
    Asp-*N*-, *628*
    Glu-C, *552*
    Lys-C, *567*
Endoribonuclease
    I, *393*
    V, *398*

H (calf thymus), *386*
Endoxylanase, *432*
Enolase, *685*
Enoyl hydrase, *687*
Enoyl-CoA hydratase, *687*
  2-, *758*
Enpezyme, *1248*
Enterokinase, *550*
Enteropeptidase, *550*
Envo, *426*
Enzeco[R], *146, 300, 303, 304, 327, 411, 412, 412, 413, 420, 422, 423, 430, 433, 439, 442, 459, 471, 585, 602, 603, 621, 623, 742, 749, 750, 753, 1240*
Enzoguard[R], *309, 570, 743*
Enzopharm[R], *750*
D-Enzyme, *204*
Enzyme-thiol transhydrogenase (oxidized-glutathione), *86*
Epidermal growth-factor-binding protein, *560*
Epidioxy-15-hydroxyprosta-5,13-dienoate
  6-isomerase, (5Z,13E)-(15S)-9α,11α-, *713*
  E-isomerase, (5Z,13E)-(15S)-9α,11α-, *713*
  thromboxane-A$_2$-isomerase, (5Z,13E)-(15S)-9α,11α-, *714*
Erh I, *903, 994*
Erythrocuprein, *176*
Erythrose-4-phosphate isomerase, *702*
Esp I, *842, 931, 946*
Esp3 I, *877, 994*
Esp4 I, *777*
Esp1396 I, *995*
Esperase[R], *571*
Esterase, *299, 561, 583*
  B-, *299*
  C-, *317*
Euphorbain, *556*
Euro[R] G, *743*
Exo-cellobiohydrolase, *492*
Exodeoxyribonuclease
  I, *369*
  III, *370*
  V, *373*
  V, deoxyribonuclease (ATP-dependent), *373*
  VII, *374*
  λ-induced, *372*
  T2- and T4-induced, *369*
Exoenzyme C3, *209*
Exoglycosidase, *477, 492*
Exonuclease
  I, *369*
  III, *370*
  IV, *369*
  V, *373*
  3'-, *376*
  5'-, *353*
  3'-to-5'-,double-strand specific, *369*
  Haemophilus influenzae, *370*
  λ, *372*
  Lactobacillus, *375*
  Micrococcus luteus, *375*
  spleen, *376*
  T$_4$, *372*
  T$_5$, *372*
  T$_7$, *372*
  venom, *375*
Exopeptidase, *508, 523*
Extracellular proteinase, *629*

F

Factor. See Coagulation Factor
Fatty acid thiokinase (long chain), *721*
Fau I, *996*
FauND I, *996*
Fba I, *825, 997*
Fe(II):oxygen oxidoreductase, *180*
Fermcolase[R], *144*
Ferredoxin:NADP$^+$
  oxidoreductase, *181*
  reductase, *181*
Ferrihemoprotein P-450 reductase, *122*
Ferrocytochrome:nitrate oxidoreductase, *137*
Ferrocytochrome-c:oxygen oxidoreductase, *137*
Ferroxidase, *108*
Fibrinase, *549*
Fibrinogenase, *541*
Fibrinoligase, *196*
Fibrinolysin, *549, 579*
Ficain, *587*
Ficin, *587*
Fin I, *878*
Finase[R], *346*
Finizym[R], *429*
Firefly luciferase. See Luciferase
Flavorpro, *619, 622*
Flavourzyme™, *749*
FMN reductase, *125*
Fnu4H I, *879, 997, 1003, 1191*
FnuD II, *765, 862, 920, 998, 1076*
  methylase, *1228*
Fok I, *913, 999*
Formaldehyde dehydrogenase, *96, 646, 650*
  glutathione, *90*
Formaldehyde:NAD$^+$ oxidoreductase, *96*
  glutathione-formylating, *90*
Formate
  dehydrogenase, *91*
  hydrogenlyase, *91*
Formate:NAD$^+$ oxidoreductase, *91*
Formate-tetrahydrofolate ligase, *118*
Formic dehydrogenase, *90*
Formimino-L-glutamic acid transferase, *185*
Formiminotetrahydrofolate
  cyclodeaminase, *185*
5-Formiminotetrahydrofolate:L-glutamate N-formiminotransferase, *185*
N-Formylmethionine (fMet)
  aminopeptidase, *524*

Fresh-N^R, *413*
FriO I, **991**, **1000**
β-Fructofuranosidase, **464**
β-D-Fructofuranoside fructohydrolase, **464**
Fructotransferase, **465**
D-Fructose 1,6-bisphosphate
    1-phosphohydrolase, **347**
    D-glyceraldehyde-3-phosphate-lyase, **676**
Fructose 5-dehydrogenase, **89**
D-Fructose dehydrogenase, **89**
D-Fructose:(acceptor) 5-oxidoreductase, **89**
Fructose-1,6-bisphosphate
    triosephosphate-lyase, **676**
D-Fructose-1,6-diphosphatase, **347**
Fructose-6-phosphate kinase, **223**, **244**
Fructose-bisphosphatase, **347**
Fructose-bisphosphate aldolase, **676**
Fruit bromelain, **594**
Fse I, **1001**
Fsp I, **767**, **1002**
Fsp II, **908**
*Fsp*4H I, **997**, **1003**
2-α-L-Fucopyranosyl-β-D-galactoside
    fucohydrolase, **485**
L-Fucose dehydrogenase, **66**
Fucosidase
    II, almond emulsin, **485**, **499**
    1,2-α-L-, **485**, **499**
    1,3-α-L-, **485**, **499**
    α-L-, **480**
α-L-Fucoside fucohydrolase, **482**
3-α-L-Fucosyl-N-acetylglucosaminyl-
    glycoprotein fucohydrolase, **499**
Fumarase, **685**
Fumarate hydratase, **685**
Fumaric aminase, **697**
Funcelase, **429**, **431**
Fungamyl^R, **410**

G

GABase, **1249**
GAD, **666**
GAH, **458**
αGal, **457**
βGal, **458**, **497**
Galacto
    kinase, **223**
    mannase, **1240**
Galactosaminide N-
    acetylgalactosaminohydrolase
    α-N-acetyl-D-, **481**
Galactose
    dehydrogenase, **51**
    oxidase, **78**
D-Galactose:
    NAD$^+$ 1-oxidoreductase, **51**
    oxygen 6-oxidoreductase, **78**
Galactose-1-phosphate uridylyltransferase, **288**
Galactosidase
    α-, **457**
    β-, **458**
    β-1,3-, **458**
    endo-β-, **497**
Galactoside galactohydrolase
    α-D-, **457**
    β-D-, **458**
β-Galactoside
    α-2,3-sialyltransferase, **211**
    α-2,6-sialyltransferase, **210**
Galactosyl transferase, **203**
D-Galactosyl-N-acetyl-α-D-galactosamine D-
    galactosyl-N-acetyl-
    galactosaminohydrolase, **495**
GAM, **419**
Gamanase^{TM}, **489**
GalO, **78**
GAO, **79**

GAPDH, **93**
GDH, **19**
Gelase^{TM}, **490**, **491**
Gelsolin, **1243**
Gentiobiase, **455**
GK, **227**, **328**
Glc6PDH, **52**
GlcDH, **65**
GlcK, **222**
GlDH, **104**
Glu-C, **552**
Glucan
Glucan
    1,4-α-Glucosidase, **419**, **452**
    1,4-β-Glucosidase, **488**, **742**, **743**
    1,6-α-Glucosidase, **487**
    3(4)-glucanohydrolase, 1,3-(1,3;1,4)-β-D-, **429**
    4-α-D-(1,4-α-D-glucano)-transferase (cyclizing), 1,4-α-D-, **202**
    4-glucanohydrolase, 1,3-1,4-α-D-, **485**, 1,4-(1,3;1,4)-β-D-, **422**
    6-α-glucosyltransferase, 1,4-α-, **204**
    6-glucanohydrolase, 1,6-α-D-, **435**
    cellobiohydrolase, 1,4-β-D-, **492**
    endo-1,3,β-D-glucosidase **477**
    glucanohydrolase, 1,3-β-D-, **477**, 1,4-α-D-, **407**, 1,4-β-D-, **488**, 1,6-α-D-, **487**
    glucohydrolase, 1,4-α-D-, **419**
    glucosidase, **741**
    maltohydrolase, 1,4-α-D-, **417**
Glucan, 1,4-α-D-:1,4-α-D-glucan
    (D-glucose) 6-α-D-glucosyltransferase, **204**
    4-α-D-glycosyltransferase, **204**
Glucanase, **477**
    β-, **422**, **429**, **738**, **740**, **741**, **742**, **750**, **751**
    β-1,3-, **429**, **752**, **1245**
    endo-1,3-β-, **477**
Glucanotransferase
    4-α-, **204**
Glucoamylase, **419**, **738**, **745**, **750**
    See also Amyloglucosidase
Glucodextranase, **487**

General Index

Glucoinvertase, *452*
Glucokinase, *218, 222*
Gluconate
  2-dehydrogenase, *88*
  dehydrogenase (NAD(P) independent), *88*
  kinase, *225*
D-Gluconate:(acceptor) 2-oxidoreductase, *88*
Gluconokinase, *225*
Glucosaminide β-1,4-galactosyltransferase
  β-N-acetyl-D-, *205*
α-D-Glucose 1,6-phosphomutase, *716*
Glucose 1-dehydrogenase, *50*
  NADP$^+$, *65*
D-Glucose 6-phosphate ketol-isomerase, *709*
D-Glucose 6-phosphate:NADP$^+$ 1-oxidoreductase, *52*
Glucose
  dehydrogenase, *50, 65, 89*
  isomerase, *707*
  oxidase, *70, 142, 703, 757*
  oxyhydrase, *70*
  phosphomutase, *716*
D-Glucose:
  (acceptor) 1-oxidoreductase, *89*
  NADP$^+$ 1-oxidoreductase, *65*
β-D-Glucose:
  NAD(P)$^+$ 1-oxidoreductase, *50*
  oxygen 1-oxidoreductase, *70*
Glucose-1-phosphate uridylyltransferase, *287*
Glucose-6-phosphatase, *347, 657*
Glucose-6-phosphate
  1-dehydrogenase, *52*
  dehydrogenase, *52, 218, 758*
  isomerase, *709*
  phosphohydrolase, *347*
Glucosidase
  β-, *455, 741, 742*
  endo-1,3-β-D-, *475*
  endo-1,6-α-, *479*
  exo-1,4-α-, *419*
  exo-1,4-β-, *488*
  exo-1,6-β-, *487*
  lysosomal, α-, *419*
  oligo-1,6-, *434*
Glucoside glucohydrolase
  α-D-, *452*
  β-D-, *455*
Glucosidosucrase, *452*
Glucosphingosine glucosylhydrolase, *480*
Glucosylceramidase, *480*
(Glucosyl, 1,4-α-D-)$_n$:orthophosphate α-D-glucosyltransferase, *196*
D-Glucosyl-N-acylsphingosine glucohydrolase, *480*
Glucox, *70, 757*
Glucozyme, *419, 421*
Glucuronate reductase, *17*
β-Glucuronidase, *467, 739*
β-D-Glucuronoside glucuronosohydrolase, *467*
Glucuronosyltransferase, *201*
Gluczyme, *421*
αGluD, *452*
Glutamate
  1-carboxy-lyase, *666*
  decarboxylase, *666*
  dehydrogenase, (NADP$^+$), *107*, [NAD(P)$^+$], *104*, L-, *104*
  formiminotransferase, *185*
  formyltransferase, *185*
  oxidase, *115*
L-Glutamate:
  ammonia ligase (ADP-forming), *724*
  NAD(P)$^+$ oxidoreductase (deaminating), *104*
  NADP$^+$ oxidoreductase (deaminating), *107*
  oxygen oxidoreductase (deaminating), *115*
Glutamate-ammonia ligase, *724*
Glutamic dehydrogenase, *104, 107*
  L-, *104, 107*
Glutamic
  -alanine transaminase, *216*
  -aspartic transaminase, *214*
  -oxaloacetic transaminase, *42, 214*
  -pyruvic transaminase, *216*
Glutaminase, *631*
Glutamine
  amidohydrolase, *631*
  synthetase, *724*
Glutaminic-γ-amino butyric acid-ase, *1249*
Glutamyl
  carboxypeptidase, lysosomal γ-, *526*
  endopeptidase, *552*
  hydrolase, γ-, *526*
  transpeptidase, γ-, *194*
(Glutamyl, 5-L-)-peptide:amino-acid 5-glutamyltransferase, *194*
γ-Glutamyltransferase, *194*
Glutathione
  hydrolase, S-(2-hydroxyacyl), *329*
  oxidase, *136*
  peroxidase, *143, 155*
  reductase (NADPH), *122*
  S-alkyltransferase, *212*
  S-aralkyltransferase, *212*
  S-aryltransferase, *212*
  S-transferase, *212*
  sulfhydryl oxidase, *136*
  transferase, *212*
Glutathione:
  hydrogen-peroxide oxidoreductase, *155*
  oxygen oxidoreductase, *136*
Glu-X carboxypeptidase, *524*
N-Glycanase, *640*
D-Glyceraldehyde 3-phosphate:NAD$^+$ oxidoreductase (phosphorylating), *93*
Glyceraldehyde-3-phosphate
  dehydrogenase, *758*, phosphorylating, *93*
  ketol-isomerase, D-, *705*
sn-Glycero-3-phosphocholine glycerophosphohydrolase, *355*
Glycerokinase, *227*
sn-Glycerol 3-phosphate:
  NAD$^+$ 2-oxidoreductase, *21*

oxygen 2-oxidoreductase, *83*
Glycerol
   dehydrogenase, *19, 21*, NADP$^+$, *63*
   kinase, *83, 227*
Glycerol:
   NAD$^+$ 2-oxidoreductase, *19*
   NADP$^+$ oxidoreductase, *63*
Glycerol-3-phosphate
   dehydrogenase, *758*, NAD$^+$, *21*
   oxidase, *83*
Glycerophosphatase, *330, 341*
α-Glycerophosphate dehydrogenase, *756*
Glycerophosphocholine phosphodiesterase, *355*
Glycerophosphorylcholine phosphodiesterase, *355*
Glycocholase, *640*
Glycogen
   6-glucanohydrolase, *486*
   (starch) synthase, *199*
   synthase α kinase, *231*
Glycogenase, *407, 417*
Glycogenin glucosyltransferase, *199*
Glycolaldehydetransferase, *187*
Glycolate oxidase, *81*
Glycolate:NAD$^+$ oxidoreductase, *26*
Glycopeptidase, *640*
Glycopeptide
   α-N-acetylgalactosaminidase, *495*
   -(N-acetyl-β-D-glucosaminyl)-L-asparagine amidohydrolase N-linked, *640*
   β-1,4-galactosyl-transferase β-N-acetylglucosaminyl, *743*
   N-glycosidase, *640*
   -D-mannosyl-N^4-(N-acetyl-D-glucosaminyl)$_2$-asparagine 1,4-N-acetyl-β-glucosaminohydrolase, *492*
Glycoprotein 4-β-galactosyl-transferase, *205*
N-Glycosidase, *640, 744, 745*
Glycosylceramidase, *499*
Glyoxalase

I, *700*
II, *329*
Glyoxylate reductase, *26*
Gly-Pro naphthylamidase, *515*
cGMP-dependent protein kinase, *231*
GO, *70*
GOD, *70*
Good Earth, *427*
GOT, *42, 98, 214*
GPCP, *355*
G3PDH, *21*
GPDH, *21*
GPO, *83, 328*
GPOS, *85*
GPT, *98, 216*
GR, *122*
GRINDAMYL™, *74, 408, 411, 489, 759, 1248*
GRINDAZYM™, *744*
GRS, *467*
GS, *724*
Gsu I, *841, 1004*
γGT, *194*
cGTase, *202*
GTP:mRNA guanylyltransferase, *296*
Guanase, *652*
Guanine
   aminase, *652*
   aminohydrolase, *652*
   deaminase, *652*
   phosphoribosyltransferase, *208*
Guanylate kinase, *296*
Guanyloribonuclease, *389*
Guanylyltransferase, *294*
G-Zyme^R, *316, 413, 420, 707*

H

Hae II, *770, 885, 914, 1005*
Hae III, *861, 1007*

   methylase, *1229*
Hageman Factor (activated), *566*
2-Haloacid
   dehalogenase, *662*
   halihydrolase, *662*
Hannilase^R, 599
Hap II, *775, 938, 943, 1010, 1030, 1070, 1079, 1191*
   methylase, *1230*
Hemicellulase, *488, 696, 742, 743, 744, 746, 747, 749, 752, 1245*
   See also 3.2.1.78
Hemocuprein, *176*
α-Hemolysin, *1249*
Heparin
   eliminase, *693*
   lyase, *693*
Heparinase, *693*
Heparin-sulfate
   eliminase, *694*
   lyase, *694*
Heparitinase, *694*
Heparitin-sulfate lyase, *694*
Hesperidinase, *741*
Hexokinase, *218, 758*
Hexomax™, *465*
Hexosaminidase, *483*
Hexose-1-phosphate uridylyltransferase, *289*
Hexosediphosphatase, *347*
Hexosephosphate isomerase, *709*
Hga I, *1011*
*Hgi*A I, *784, 800, 868, 1128*
*Hgi*C I, *814, 865, 972*
*Hgi*D I, *938, 1067, 1079*
*Hgi*E I, *838, 1012*
*Hgi*H I, *814*
*Hgi*J II, *964*
Hha I, *800, 932, 1013, 1017, 1027*
   methylase, *1230*
Hidelase, *757*

Hin1 I, *775, 938, 943, 1015, 1030, 1067, 1070, 1079, 1191*
Hin2 I, *1016*
Hin6 I, *800, 1013, 1017*
Hinc II, *1018, 1020*
Hind II, *1018, 1020*
Hind III, *1021*
   methylase, *1231*
Hinf I, *1024*
Hinf II, *1026*
HinP I, *943, 1036, 1067, 1070, 1079, 1191*
HinP1 I, *775, 800, 938, 1013, 1027, 1030*
Hippuricase, *638*
Histaminase, *114*
Histamine *N*-methyltransferase, *182*
Histidase, *697*
Histidinase, *697*
Histidine
   α-deaminase, *697*
   ammonia-lyase, L-, *697*
   carboxy-lyase, L-, *669*
   decarboxylase, *669*
L-Histidine:tRNA^{His} ligase (AMP-forming), *720*
Histidine-tRNA ligase, *720*
Histidyl-tRNA synthetase, *720*
Histozyme, *638*
HK$_t$, *218*
HK^{TM}, *331*
HNMT, *182*
Hot Tub^{TM}, *286*
Hpa I, *914, 1028*
Hpa II, *775, 870, 938, 943, 1010, 1016, 1030, 1067, 1070, 1079, 1191*
   methylase, *1232, 1233*
Hph I, *802, 1032*
HRP, *129*
HSD
   3α, *58*
   7α, *67*
   12α, *68*

HSDH, *58, 59, 61, 67, 68*
Hsp92 I, *1033*
Hsp92 II, *1034*
HspA I, *800, 1035*
Human prostate-specific antigen, *560*
Hurain, *556*
Hyaluroglucosidase, *688*
Hyaluronate
   3-glycanohydrolase, *476*
   4-glycanohydrolase, *472*
   lyase, *688*
Hyaluronidase, *472, 476, 531, 688*
Hyaluronoglucosaminidase, *472*
Hyaluronoglucuronidase, *476*
Hydridase^{TM}, *388*
Hydrogen dehydrogenase, *91*
Hydrogen-peroxide:hydrogen-peroxide oxidoreductase, *142*
Hydrolases. See Chapter 4, *299-663*
Hydroxamate transferase, *632*
(S)-2-Hydroxy acid:oxygen 2-oxidoreductase, *81*
Hydroxy-acid oxidase
   A, *81*
   B, *81*
   (S)-2-, *81*
β-Hydroxyacyl dehydrogenase, *41*
Hydroxyacyl-CoA dehydrogenase
   3-, *41, 758*
   β-, *41*
Hydroxyacyl-CoA hydro-lyase
   (3S)-3-, *687*
Hydroxyacyl-CoA: NAD$^+$ oxidoreductase
   (S)-3, *41*
Hydroxyacylglutathione hydrolase, *329*
S-(Hydroxyalkyl)glutathione lyase, *212*
Hydroxyalkyl-protein kinase, *231*
Hydroxyaryl-protein kinase, *245*
Hydroxybenzoate
   3-monooxygenase, 4-, *167*
   hydrolyase, *158*, p-, *167*

4-Hydroxybenzoate,NADPH:oxygen oxidoreductase (3-hydroxylating), *167*
(R)-3-Hydroxybutanoate:NAD$^+$ oxidoreductase, *39*
Hydroxybutyrate dehydrogenase, *39*
4-Hydroxymandelonitrile hydroxybenzaldehyde-lyase, *676*
Hydroxymandelonitrile lyase, *676*
Hydroxynitrile lyase, *675, 676*
Hydroxyprostaglandin dehydrogenase, *58*
Hydroxysteroid dehydrogenase, *59*
   3α-, *58, 759*
   3α-, B-specific, *58*
   3α-20β-, *61*
   3α (or 20β)-, *61*
   3 (or 17)β-, *59, 759*
   7α-, *67*
   12α-, *68*
   β-, *59*
   (R)-20-, *61*
Hydroxysteroid:
   NAD(P)$^+$ oxidoreductase, 3α-(B-specific), *60*, 3 (or 17) β-, *59*
   NAD$^+$ 7-oxidoreductase, 7α-, *67*
   NAD$^+$ oxidoreductase, 3α (or 20β)-, *61*
   NADP$^+$ 12-oxidoreductase, 12α-, *68*
Hypoxanthine
   oxidase, *85*
   phosphoribosyltransferase, *208*
   -guanine phosphoribosyltransferase, *208*

I

ICDH, *47, 48*
I-*Ceu* I, *1035*
I-*Cla* I, *1036*
Icosa-5,8,11,14-tetraenoate,hydrogen-donor:oxygen oxidoreductase (5Z,8Z,11Z,14Z)-, *175*
L-Iditol 2-dehydrogenase, *23*
L-Iditol: NAD$^+$ 2-oxidoreductase, *23*

I-*Dmo* I, **1037**
IgA protease, **578**
IgA-specific metalloendopeptidase
  bacterial, **578**
Imidodipeptidase, **513**
IMP pyrophosphorylase, **208**
IMP:pyrophosphate phospho-D-ribosyltransferase, **208**
IndiAge, **742**, **743**
Inorganic pyrophosphatase, **657**
Inosine phosphorylase, **206**
Inositol 2-dehydrogenase, **24**
  myo-, **24**
myo-Inositol hexakisphosphate 3-phosphohydrolase, **346**
Inositol:NAD$^+$ 2-oxidoreductase, **24**
  myo-, **24**
myo-Inositol-hexakisphosphate 6-phosphohydrolase, **351**
Invertase, **464**
Iodide peroxidase, **155**
Iodide:hydrogen-peroxide oxidoreductase, **155**
Iodinase, **155**
Iodotyrosine deiodase, **155**
I-*Ppo* I, **1038**
I-*Sce* I, **1039**
Isoamylase, **486**
Isocitrate dehydrogenase
  NAD$^+$, **47**
  NADP$^+$, **48**
Isocitrate:
  NAD$^+$ oxidoreductase (decarboxylating), **47**
  NADP$^+$ oxidoreductase (decarboxylating), **48**
Isocitric dehydrogenase, **47**
  NADP, **48**
Isomaltase, **434**
Isomerases. See Chapter 4, **702-719**
Iso-*Nae* I, **1040**
IsoTherm™, **272**

Ita I, **997**, **1003**, **1040**
ITALASE™, **302**

J

Jo-1 antigen, **720**

K

Kallikrein, **558**, **559**, **560**, **566**, **579**
Kanamycin
  6'-N-acetyltransferase, **193**
  kinase, **245**
KaryMAX, **755**
Kas I, **992**, **1041**
Keratanase, **497**
Keratan-sulfate
  1,4-β-D-galactanohydrolase, **497**
  endo-1,4-β-galactosidase, **497**
α-Ketoacid carboxylase, **663**
3-Ketoacyl-CoA thiolase, **191**
α-Ketoglutarate dehydrogenase, **101**
α-Ketoglutaric dehydrogenase, **101**
β-Ketoglutaric-isocitric carboxylase, **47**
Ketomax^R, **707**
Ketone-aldehyde mutase, **700**
β-Keto-reductase, **41**
β-Ketothiolase, **191**
Kininase II, **516**
T-Kininogenase, **560**
Kininogenin, **558**
Kitalase, **1247**
Klenow fragment. See DNA polymerase
Kpn I, **768**, **1042**
*Kpn*2 I, **804**, **880**, **1045**, **1215**
KSHRASE™, **302**
Ksp I, **1046**, **1141**
*Ksp*22 I, **1046**

*Ksp*632 I, **926**, **959**, **1047**
*Kzo*9 I, **1048**

L

Laccase, **139**
Lactaldehyde reductase (NADPH), **17**
β-Lactam hydrolase, **643**
β-Lactamase, **643**
Lactase, **458**, **499**
Lactate
  2-monooxygenase, **163**
  oxidase, **163**
  oxidative decarboxylase, **163**
Lactate dehydrogenase, **27**, **237**, **721**, **759**
  cytochrome, D-, **69**, L-, **68**
  cytochrome b_2, L-, **68**
  D-, **38**
  L-, **27**
  NAD-, **29**
  NAD-dependent, **27**
(R)-Lactate:
  NAD$^+$ oxidoreductase, **38**
  ferricytochrome c 2-oxidoreductase, (S)-, **68**, (R)-, **69**
  oxygen 2-oxidoreductase (decarboxylating), (S)-, **163**
Lactic acid dehydrogenase, **27**, **38**, **68**, **69**
  L-, **29**
Lactic dehydrogenase, **27**
  L-, **27**
Lacto-N-biosidase, **1240**
Lactoperoxidase, **147**, **157**
Lactose
  galactohydrolase, **499**
  synthase, **203**, **745**
Lactoylglutathione
  lyase, **700**
  methylglyoxal-lyase (isomerizing), (R)-S-, **700**
Lactozym^R, **462**
Laminarinase, **429**, **477**, **752**

Laminex^R, *431*
LAP, *147*, *508*
Laundry enzymes, *1240*, *1241*
Lecitase™, *315*
Lecithinase
  A, *312*
  B, *316*
  C, *356*
  D, *359*
Lecithin-cholesterol acyltransferase, *193*
Leucine
  aminopeptidase, *509*
  dehydrogenase, *108*
L-Leucine:NAD$^+$ oxidoreductase (deaminating), *108*
Leucolysin, *617*
Leucyl
  aminopeptidase, *508*, *511*
  peptidase, *506*
  t-RNA synthetase, *257*
LeuDH, *108*
Leukotriene-A$_4$ synthase, *162*
Liberase™, *755*
Ligases. See Chapter 4, *720-735*
Lignoceroyl-CoA synthase, *721*
Lilipase, *310*
Limit dextrinase, *479*
Linoleate:oxygen 13-oxidoreductase, *159*
Lipase, *299*, *301*, *754*, *755*
  A, *301*
  B, *301*
  clearing factor, *328*
Lipid hydroperoxidase, *155*
Lipoamide
  dehydrogenase, *133*
  reductase (NADH), *133*
Lipoic dehydrogenase, *133*
Lipolase™, *303*
Lipomod, *303*, *305*, *306*, *307*, *315*
LipoPDE II, *359*
Lipophosphodiesterase
  I, *356*
  II, *359*
Lipoprotein lipase, *83*, *328*
Lipoxidase, *159*
Lipoxygenase, *159*
  5-, *162*
  12-, *160*
  15-, *161*
  P4, *161*
Lipoyl dehydrogenase, *133*
Lipozyme^R, *306*
Liquipanol^R, *585*
Long-chain-enoyl-CoA hydratase, *687*
Long-chain-fatty-acid:[acyl-carrier-protein] ligase (AMP-forming), *723*
Long-chain-fatty-acid-CoA ligase, *721*, *723*
LO, *163*
LOX, *163*
LPO, *147*
Lsp I, *842*, *930*, *1048*
LU, *171*
Luciferase, *164*, *171*, *1249*, *1310*
  bacterial, *171*
  firefly, *164*
Lumafast™, *309*
Lyases. See Chapter 4, *663-701*
LYPL, *316*
Lys-C, *567*
Lysine
  carboxy-lyase, L-, *668*
  decarboxylase, *668*
  oxidase, L-, *116*
  α-oxidase, L-, *116*
  -2-oxoglutarate reductase, *118*
Lysine:
  NAD$^+$ oxidoreductase (L-lysine-forming) N^6-(L-1,3-dicarboxypropyl)-L-, *118*
  oxygen 2-oxidoreductase (deaminating), *116*
Lysolecithinase, *316*
Lysopeptase, *1245*
2-Lysophosphatidylcholine acylhydrolase, *316*
Lysophospholipase, *193*, *299*, *316*

Lysosomal
  carboxypeptidase A, *517*
  tyrosine carboxypeptidase, *517*
Lysozyme, *441*, *536*
Lysyl endopeptidase, *567*
  α-, *552*
Lytic
  enzymes, *1245*
  metalloendopeptidase, β-, *627*
  proteinase, *Myxobacter* α-, *551*
Lyticase, *1245-1247*

M

Macerating and lytic enzymes. See Chapter 8, *1245-1247*
Macerozyme, *1245*
Mae I, *829*, *1049*, *1086*, *1196*, *1206*
Mae II, *775*, *938*, *943*, *1030*, *1036*, *1050*, *1067*, *1070*, *1079*, *1191*
Mae III, *911*, *1050*
Malate dehydrogenase, *42*
  decarboxylating, D-, *63*
  oxaloacetate-decarboxylating (NADP$^+$), *47*
(S)-Malate hydro-lyase, *685*
(S)-Malate:NAD$^+$ oxidoreductase, *42*
  decarboxylating, *63*
(S)-Malate:NADP$^+$ oxidoreductase (oxaloacetate-decarboxylating), *47*
Malic
  dehydrogenase, *42*
  enzyme, *47*
Maltase, *452*
  acid, *419*
Maltase-glucoamylase, *452*
Maltogenic amylase, *204*
Maltose phosphorylase, *199*
Maltose:orthophosphate 1-β-D-glucosyltransferase, *199*
Mam I, *846*, *855*, *1051*

Mandelonitrile
  benzaldehyde-lyase, *675*
  lyase, *675*
Mannan
  endo-1,4-β-mannosidase, *488*
  mannanohydrolase, 1,4-β-D-, *488*
Mannanase, *464, 747*
  endo-1,4-β-, *488*
Mannase, *464*
Mannitol
  2-dehydrogenase, *62*
  dehydrogenase, *62*
D-Mannitol:NAD$^+$ 2-oxidoreductase, *6623*
Mannose-6-phosphate isomerase, *708*
Mannosidase
  1A, *500*
  1B, *500*
  II, *501*
  α-, *463*
  α-1,6-, *501*
  β-, *464*
  endo-1,4-β-, *464*
Mannoside mannohydrolase
  α-D-, *463*
  β-D-, *464*
Mannosyl-glycoprotein endo-β-N-acetylglucosamidase, *492*
Mannosyl-oligosaccharide
  1,2-α-mannosidase, *500*
  1,3-1,6-α-mannosidase, *501*
  α-D-mannohydrolase, 1,2-α-, *500*, 1,3-(1,6-)-, *501*
MAO, *113*
Mashazyme, *1250*
Mash-zyme, *750*
MAX-LIFE, *411*
Mbo I, *811, 825, 834, 853, 884, 948, 949, 1052, 1088, 1148, 1214*
Mbo II, *1054*
Mcr I, *849, 863, 1141*
McrBC I, *1055*
MDH, *42*

Megadex™, *413*
Meganuclease, *1039*
Megateriopeptidase, *620*
Meicelase, *1240*
Melibiase, *457*
Membrane
  alanyl aminopeptidase, *510*
  aminopeptidase I, *510*
  Pro-X carboxypeptidase, *523*
Menadione reductase, *127*
Metalloendopeptidase, *1243*
  bacterial, *620, 625*
  IgA-specific, bacterial, *578*
  lytic, β-, *627*
  peptidyl-Asp, *628*
  peptidyl-Lys, *618*
Methanol oxidase, *79*
Methenyltetrahydrofolate cyclohydrolase, *118*
L-Methionase, *700*
Methionine
  γ-lyase, *700*
  methanethiol-lyase (deaminating), L-, *700*
(R)-2-Methyl-3-oxopropanoyl-CoA CoA-carbonylmutase, *717*
Methyl-accepting chemotaxis protein O-methyltransferase, *184*
Methylbutyrase, *299*
Methylenetetrahydrofolate dehydrogenase
  5,10-, *118*
  NADP$^+$, *118*
5,10-Methylenetetrahydrofolate:NADP$^+$ oxidoreductase, *118*
Methylglyoxalase, *700*
N-Methylhydantoin
  amidohydrolase, *648*
  hydrolase, *655*
N-Methylhydantoinase (ATP-hydrolyzing), *648*
N-Methylimidazolidine-2,4-dione amidohydrolase (ATP-hydrolyzing), *648*

Methylmalonyl-CoA mutase, *717*
Methyltransferases. *See* DNA methyltransferases
Mevaldate reductase (NADPH), *17*
Mfe I, *1056*
Mfl I, *811, 825, 834, 924, 1052, 1057, 1148, 1214*
Microbial collagenase, *607*
Microperoxidase, *147*
Microsomal aminopeptidase, *510*
Microtubule-associated protein 2 kinase, *248*
Miscellaneous enzymes
  mixtures, *See* Chapter 5, *757-759*
  unclassified, *See* Chapter 8, *1249-1250*
Mixtures
  Carbohydrases. *See* Chapter 5, *738-747*
  Carbohydrases and proteases. *See* Chapter 5, *750-752*
  Carbohydrases, lipases and proteases. *See* Chapter 5, *753*
  Carbohydrases, lipases, nucleic acid-active enzymes and proteases. *See* Chapter 5, *754-755*
  Nucleic acid-active enzymes. *See* Chapter 5, *756*
  Proteases. *See* Chapter 5, *748-749*
  Miscellaneous enzymes. *See* Chapter 5, *757-759*
MK, *257*
Mlu I, *900, 1058*
MluN I, *1060*
Mly113 I, *992*
Mnl I, *1061*
Modification methylases. *See* Chapter 7, *1222-1238*
Monoamine oxidase, *113*
Monobutyrase, *299*
Monophenol
  monooxygenase, *138, 174*
  oxidase, *174*

Monophenol, L-dopa:oxygen oxidoreductase, *174*
Monophosphatidylinositol phosphodiesterase, *361*
Monosialoganglioside sialyltransferase, *211*
Mph1103 I, *1062*
mRNA
   capping enzyme, *296*
   guanylyltransferase, *296*
Mro I, *766, 804, 869, 880, 1062, 1215*
MroN I, *1063*
MroX I, *1064*
Msc I, *810, 1064*
Mse I, *1065, 1086, 1196, 1206*
Msl I, *1066*
Msp I, *775, 938, 943, 1030, 1036, 1067, 1079, 1191*
   methylase, *1232, 1233*
MspA1 I, *1070*
MspC I, *1071*
MspR9 I, *902, 1072*
Mst II, *788, 946*
Mucorpepsin, *605*
Multifect^R, *431, 433, 439, 442, 569, 570, 571, 743*
Multifresh^R-*408*
Multizyme^R, *751*
Mun I, *1072*
Muramidase, *441*
Muscle phosphorylase *a* and *b*, *196*
Mutanolysin, *1243*
Mutarotase, *703*
Mva I, *905, 915, 986, 1074*
   methylase, *1234*
Mva1269 I, *1075*
Mvn I, *765, 862, 998, 1076*
Mwo I, *1076*
Mycodextranase, *485*
Mycolysin, *618, 625*
Myeloperoxidase, *147*

myo-Inositol 2-dehydrogenase, *24*
myo-Inositol:NAD$^+$ 2-oxidoreductase, *24*
Myokinase, *257*
Myosin
   ATP phosphohydrolase (actin-translocating), *662*
   ATPase, *662*
   -light-chain kinase, *248*
Myrosinase, *507*

N

N. *Bst*SE Nickase, *918*
N-Acetyllactosamine synthase, *203*
NAD(P)$^+$, 5,6,7,8-tetrahydropteridine, *129*
NAD(P)$^+$-arginine ADP-ribosyltransferase, *701*
NAD(P)H dehydrogenase
   FMN, *125*
   quinone, *127*
NAD(P)H:
   6,7-dihydropteridine oxidoreductase, *129*
   FMN oxidoreductase, *125*
   nitrate oxidoreductase, *124*
   (quinone-acceptor) oxidoreductase, *127*
NAD$^+$
   ADP-ribosyltransferase, *209*
   glycohydrolase, *503*
   kinase, *226*
   nucleosidase, *503*
   pyrophosphorylase, *262*
   synthase (glutamine-hydrolyzing), *726*
   synthetase (glutamine-hydrolyzing), *726*
NAD$^+$:poly(adenine-diphosphate-D-ribosyl)-acceptor ADP-D-ribosyl-transferase, *209*
NADase, *503*
NADH
   dehydrogenase, *128*
   oxidase, *141*
   peroxidase, *141*
NADH:
   (acceptor) oxidoreductase, *128*
   hydrogen-peroxide oxidoreductase, *141*
   nitrate oxidoreductase, *124*

NAD-lactate dehydrogenase, *27*
NADH-
   dependent dehydrogenase, *133*
   FMN oxidoreductase, *128*
NADP-cytochrome reductase, *122*
NADPH
   dehydrogenase, *125*
   -dependent dehydrogenase, *133*
   diaphorase, *125*
NADPH:
   (acceptor) oxidoreductase, *125*
   ferricytochrome oxidoreductase, *122*
   oxidized-glutathione oxidoreductase, *122*
NADPH-
   ferrihemoprotein reductase, *122*
   FMN oxidoreductase, *125*
   P450 Oxidoreductase, *122*
Nae I, *1077, 1089*
NAG, *481*
Nagapsin^R, *636*
Nagarse, *569*
   proteinase, *569*
NAGlc, *483*
NAH, *483*
NANA, *679*
Nar I, *775, 818, 938, 943, 992, 993, 1030, 1036, 1041, 1067, 1070, 1079, 1191*
Naringinase, *1240*
Natugrain, *740*
Natuphos^R, *346*
Nci I, *827, 1081*
Nco I, *880, 1083*
Nde I, *996, 1086, 1196, 1206*
Nde II, *811, 825, 834, 1052, 1088, 1148, 1214*
Neomycin
   phosphotransferase II, *245*
   -kanamycin phosphotransferase, *245*
Neprilysin, *617*
Nerve growth factor
   γ-subunit, *560*
Neuraminidase, *445*

Neuron specific enolase, **685**
Neurotoxin C3
    botulinum, **209**
Neutral
    endopeptidase, **510**, **617**
    protease, **602**, **620**
    proteinase, **747**
Neutralase™, **626**
Neutrase^R, 626
Newlase, **605**
N-Glycosidase F, **742**
*Ngo*A IV, **804**, **1089**, **1215**
*Ngo*M I, **804**, **1077**, **1089**, **1090**, **1215**
Nhe I, **803**, **808**, **1091**, **1171**, **1207**
Nicking enzyme, **378**
Nicking-closing enzyme, **717**
Nicotinamide-nucleotide
    adenylyltransferase, **262**
Nitrate reductase
    cytochrome, **137**
    assimilatory, NADH, NAD(P)H, **124**
Nitric-oxide synthase, **169**
Nla II, **825**
Nla III, **821**, **1093**, **1173**
Nla IV, **892**, **1094**
9°N$_m$™, **274**
Noradrenaline *N*-methyltransferase, **183**
NOS, **169**
Not I, **925**, **930**, **958**, **969**, **1095**, **1229**
Notexin Np, **312**
Novamyl™, **205**
Novozym^R, **144**
Nru I, **881**, **1098**
*Nru*G I, **1100**
Nsi I, **990**, **1101**, **1128**
Nsp I, **821**, **916**, **1102**, **1173**
Nsp II, **886**
Nsp III, **804**
Nsp V, **775**, **842**, **930**, **938**, **943**, **1030**, **1067**, **1070**, **1079**, **1103**, **1191**
*Nsp*7524 I, **938**, **1104**, **1173**
*Nsp*7524 V, **1067**, **1079**, **1105**
*Nsp*7542 I, **1103**
*Nsp*B II, **1105**
*Nsp*H II, **806**
NT, **124**
Nuclease
    5'-, **105**
    P1, **399**
    S1, **399**, **404**
    S7, **405**
    Azotobacter, **405**
    Bacillus subtilis, **376**
    *Bal*31, **399**
    Chlamydomonas, **406**
    Lactobacillus acidophilus, **376**
    micrococcal, **405**
    mung bean, **399**
    N. crassa, **399**
    potato, **405**
    salmon testis, **376**
    Serratia marcescens, **404**
    silkworm, **405**
Nucleic acid-active enzymes
    mixtures, See Chapter 5, **754-756**
    unclassified. See Chapter 8, **1242**
Nucleoside
    2',3'-cyclic phosphate 2'-nucleotidohydrolase, **365**
    5'-diphosphate kinase, **260**
    deoxyribosyltransferase, **207**
    diphosphokinase, **164**
    monophosphate kinase, **259**
    phosphorylase, **206**
    ribosyltransferase, **206**
Nucleoside-
    diphosphate kinase, **260**
    phosphate kinase, **259**
Nucleoside-triphosphate:
    DNA deoxynucleotidylexotransferase, **289**
    RNA nucleotidyltransferase (DNA-directed), **263**
Nucleotidase
    3'-, **345**
    5'-, **206**, **344**
Nucleotide pyrophosphatase, **661**
Nun II, **938**

O

D-Octopine
    dehydrogenase, **119**
    synthase, **119**
Oligoglucan-branching glycosyltransferase, **204**
Oligoglycosylglucosylceramide
    glycohydrolase, **502**
Oligonucleate 5'-nucleotidohydrolase, **353**
Oligosaccharide
    dehydrogenase, **1250**
    glycopeptidase, *N*-, **640**
Ophio-amino-acid oxidase, **110**
Orgotein, **176**
Ornithine
    carbamoyltransferase, **186**
    carboxy-lyase, *L*-, **667**
    decarboxylase, **667**
    transcarbamylase, **186**
Orotate phosphoribosyltransferase, **209**, **670**
Orotate reductase
    NADH, **101**
    NADPH, **114**
Orotidine-5'-phosphate:pyrophosphate phospho-α-D-ribosyltransferase, **209**
Orotidine-5'-monophosphate
    decarboxylase, **670**
Orotidine-5'-phosphate
    carboxy-lyase, **670**
    decarboxylase, **209**, **670**
    pyrophosphorylase, **209**
    decarboxylase, **759**
    pyrophosphorylase, **759**
Orotidylic acid phosphorylase, **209**

Orthophosphate:oxaloacetate carboxy-lyase
(phosphorylating), **672**
Orthophosphoric-monoester
phosphohydrolase
  acid optimum, **341**
  alkaline optimum, **330**
Oxalate
  carboxy-lyase, **664**
  decarboxylase, **664**
  oxidase, **100**
Oxalate:oxygen oxidoreductase, **100**
Oxaloacetate
  carboxy-lyase, **664**
  decarboxylase, **664**
  β-decarboxylase, **664**
  transacetase, **681**
Oxalosuccinate decarboxylase, **48**
OxaN I, **946**
Oxidoreductases. See Chapter 4, **14-181**
2-Oxo-acid carboxy-lyase, **663**
3-Oxoacid CoA-transferase, **298**
3-Oxoacyl-CoA thiolase, **758**
Oxoglutarate decarboxylase, **101**
Oxoglutarate dehydrogenase
  lipoamide, **101**
  2-, **101, 133**
2-Oxoglutarate:lipoamide 2-oxidoreductase
  (decarboxylating and acceptor-
  succinylating), **101**
Oxoisomerase, **709**
5-Oxoprolyl-peptidase, **525**
OxyGO^R, **72**

P

p34^{cdc2}/cyclinB, **231**
p60^{c-src} enzyme, **247**
Pac I, **1106**
Pae I, **11611073**
PaeR7 I, **1107, 1143, 1211**

PaeRT I, **1211**
Pal I, **1007, 1108**
Palatase^R, **310**
Palmitoyl-CoA
  hydrolase, **299**
  synthase, **721**
Panazyme, **602, 748**
Pancreatic ribonuclease, **393**
Pancreatin, **754, 755**
Pancreatopeptidase E, **561**
Pancrelipase, **301**
Panol^R, **585**
PAO, **114**
PAP, **341**
Papain, **583, 587, 588, 589, 590, 591, 749**
Particle-bound aminopeptidase, **510**
Pau I, **859, 1109**
PDE, **353, 363, 375**
PDH, **97**
Pectate
  lyase, **689**
  transeliminase, **689**
Pectin
  demethoxylase, **323**
  depolymerase, **438**
  lyase, **695, 696**
  methoxylase, **323**
  methylesterase, **323**
  pectylhydrolase, **323**
Pectinase, **438, 477, 695, 742, 743, 744, 746, 751, 752, 1245**
Pectinesterase, **317, 323, 696, 746, 747**
Pectinex™, **439, 746, 747**
Pectin-transeliminase, **746, 747**
Pectolyase, **695, 746**
PEKTOLASE™, **440**
Penicillinase, **643**
Penicillin
  acylase, **636**
  amidase, **636**
  amidohydrolase, **636**

Penicillopepsin, **604**
Pentosanase, **432, 738, 739, 751**
PEPC, **672**
PEP-carboxylase, **42**
Pepsin, **561, 595, 604, 749**
Pepsinogen, **595**
Peptidase, **749**
  A, **604, 617**
  D, **513**
  E, **510**
  P, **516**
  S, **508**
  γ-, **513**
  pyroglutamyl, **525**
  pyrrolidone-carboxylate, **525**
  serine-type, **515**
Peptide
  -N^4-(N-acetyl-β-glucosaminyl)asparagine
    amidase, **640**
  N-fatty acylase, **1250**
Peptidoglycan N-acetylmuramoylhydrolase, **441**
Peptidyl
  arginine deiminase, **651**
  -Asp metalloendopeptidase, **628**
  -dipeptidase A, **516**
  -Lys metalloendopeptidase, **618**
  proline cis-trans-isomerase, **705**
  prolyl isomerase, **705**
  -tripeptidase, **607**
Peroxidase, **78, 102, 142, 143, 147, 148, 175, 227, 323, 759**
Pfl23 II, **1109**
PflF I, **803**
PflM I, **771, 995, 1110, 1204**
PFK, **223**
PGK, **250**
Pfu DNA ligase, **727**
3PGDH, **64**
PGlc I, **709**
PGlcM, **716**
6PGluDH, **49**

PGlyM, 715
PGM, *714, 716*
3PGPK, *250*
Phasaropepsin, *606*
Phaseolin, *517*
Phenolase, *138, 174*
Phenylalanine
　ammonia-lyase, *698*
　carboxy-lyase, L-, *674*
　deaminase, *698*
　decarboxylase, *674*
　dehydrogenase, *109*, L-, *109*
L-Phenylalanine:NAD$^+$ oxidoreductase
　(deaminating), *109*
Phenylethanolamine N-methyltransferase, *183*
Phenylpyruvate
　keto-enol-isomerase, *711*
　tautomerase, *711*
PHI, *709*
pHlozyme™, *408*
Phosphatase, prostatic acid, *341*
Phosphate acetyltransferase, *190*
Phosphatidase, *312*
Phosphatidolipase, *312*
Phosphatidylcholine
　2-acylhydrolase, *312*
　cholinephosphohydrolase, *356*
　phosphatidohydrolase, *359*
Phosphatidylcholine:sterol O-
　acyltransferase, *193*
1-Phosphatidyl-D-myo-inositol
　inositolphosphohydrolase (cyclic-
　phosphate-forming), *361*
Phosphatidylinositol
　phosphodiesterase, 1-, *361*
　phospholipase C, *361*
Phosphoacylase, *190*
6-Phospho-D-gluconate: NADP$^+$ 2-
　oxidoreductase (decarboxylating), *49*
Phospho-D-glycerate
　carboxy-lyase (dimerizing), 3-, *674*
　hydro-lyase, 2-, *685*
Phosphodiesterase, *363, 375*
　I, *353*
　5'-, *353*
　hog kidney, *375*
　spleen, *376, 404406*
　venom, *375*
Phosphoenol transphosphorylase, *237*
Phosphoenolpyruvate
　carboxylase, *42, 672*
　kinase, *237*
6-Phosphofructokinase
　(pyrophosphate), *244*
Phosphoglucomutase, *716*
Phosphogluconate dehydrogenase
　(decarboxylating), *49*
Phosphogluconic acid dehydrogenase, *49*
6-Phosphogluconic
　carboxylase, *49*
　dehydrogenase, *49*
Phosphoglucose isomerase, *709*
Phosphoglycerate
　2,3-phosphomutase, D-, *715*
　dehydratase, 2-, *685*
　dehydrogenase, *64*, 3-, *64*
　kinase, *93, 250*, 3-, *250*
　mutase, *715*
　phosphomutase, *715*
3-Phosphoglycerate:NAD$^+$ 2-
　oxidoreductase, *64, 65*
3-Phosphoglyceric phosphokinase, *250, 758*
Phosphoglyceromutase, *715*
Phosphohexoisomerase, *708*
Phosphohexokinase, *223*
Phosphohexomutase, *708, 709*
Phosphohexose isomerase, *709*
Phospholipase
　A$_2$, *193, 312*
　B, *316*
　C, *301, 356*
　D, *82, 357*
Phospholipid-cholesterol acyltransferase, *193*
Phosphomannose isomerase, *708*
Phosphomonoesterase, *330, 331, 341*
Phosphopentoisomerase, *708*
Phosphopentokinase, *226*
Phosphoprotein
　phosphatase, *348*
　phosphohydrolase, *348*
Phosphopyruvate hydratase, *685*
Phosphoriboisomerase, *708*
Phosphoribosylpyrophosphate synthetase, *262*
Phosphoribulokinase, *226*
Phosphoribulose epimerase, *702*
Phosphorylase, *196, 236*
　b kinase kinase, *231*
　kinase, *236*
Phosphosaccharomutase, *709*
Phosphotransacetylase, *190*
Phosphotriose isomerase, *705*
Phosphotyrosine phosphatase, *352*
Photinus luciferin:oxygen 4-oxidoreductase
　(decarboxylating, ATP-hydrolyzing), *164*
Photinus-luciferin 4-monooxygenase (ATP-
　hydrolyzing), *164*
Phylloquinone reductase, *127*
Physaropepsin, *605*
Phytase, *346, 351*
Phytate 6-phosphatase, *351*
*Pin*A I, *796, 804, 1114, 1215*
Pinguinain, *594*
PK, *237*
PKA, *231*
PKC, *231*
PKG, *231*
PL, *301*
PLA$_2$, *312*
Plasma
　amine oxidase, *114*
　kallikrein, *558, 566, 579*

Plasmid-Safe™, *374*
Plasmin, *541, 549, 579*
Plasminogen activator, *579*
PLC, *356*
PLD, *359*
Ple I, *1114*
*Ple*19 I, *1115*
PI-*Psp* I, *1111*
PI-*Sce* I, *1112*
PI-*Tli* I, *1113*
*Pma*C I, *819, 973, 1116*
Pme I, *1116*
*Pme*55 I, *1117*
PMI, *708*
Pml I, *973, 1118*
PNGase F, *492, 640*
PNMT, *183*
PNPase, *286*
PO, *98*
POD, *147, 328*
Poly(1,4-(*N*-acetyl-β-*D*-glucosaminide))
   glycanohydrolase, *436*
Poly(1,4-α-*D*-galacturonide)
   glycanohydrolase, *438*
   lyase, *689*
Poly(A) polymerase, *263*
Poly(ADP-ribose) synthase, *209*
Poly(β-*D*-1,4-mannuronide) lyase, *690*
Poly(β-*D*-mannuronate) lyase, *690*
Poly(deoxyribonucleotide):poly(deoxyribonucleotide) ligase
   AMP-forming, *727*
   AMP-forming, NMN-forming, *731*
Poly(methoxy-*L*-galacturonide) lyase, *695*
Poly(ribonucleotide):poly(ribonucleotide)
   ligase (AMP-forming), *733*
Poly-
   β-glucosaminidase, *436*
   N-acetylglucosaminidase, 1,4-β-, *436*
Polydeoxyribonucleotide synthase
   ATP, *727*
   NAD$^+$, *731*
Polygalacturonase, *438, 696, 746, 747*
Polynucleotide
   5'-hydroxyl-kinase, *241*
   kinase, T4, *241*
   ligase, *727*, NAD$^+$, *731*
   phosphorylase, *286*
Polyol dehydrogenase, *23*
   NADP$^+$, *25*
Polyphenol oxidase, *138, 174*
Polyphosphorylase, *196*
Polyribonucleotide
   nucleotidyltransferase, *286*
   synthase (ATP), *733*
Polyribonucleotide:orthophosphate
   nucleotidyltransferase, *286*
Polysaccharidase, *752*
Porphobilinogen synthase, *687*
Post-proline cleaving enzyme, *557*
Postproline dipeptidyl aminopeptidase IV, *515*
PP-*348*
Ppase, *657*
*Ppu*10 I, *1119*
*Ppu*M I, *806, 1120*
Primafast^R, *425, 426*
Proctase, *602, 603*
Procaine esterase, *299*
PROK, *573*
Prolase, *1244*
Proleather, *623*
Prolidase, *513*
Proline
   aminopeptidase, *511*
   dipeptidase, *513*
   iminopeptidase, *511*
   -specific endopeptidase, *557*
Prolyl
   aminopeptidase, *511*
   oligopeptidase, *557*
Promod, *571, 584, 619, 622*

Promozyme^R, *479*
Pronase^R, *625*
Pronase component, *625*
Prostacyclin synthase, *713, 759*
Prostaglandin disulfide-isomerase, *711*
Prostaglandin synthase, *175*
   I, *713*
   E, *713*
   G/H, *175*
   H$_2$ E, *713, 759*
   endoperoxide, *175*
Prostate-specific antigen
   human, *560*
Protamex™, *749*
Protaminase, *521*
Protease, *749, 750, 751, 753, 602*
   C, *515, 517*
   D, submaxillary, *559*
   II, mast cell, *568*
   acid, *751*
   alkaline, *531, 569, 620, 1243*
   alkaline, serine, *748*
   alkalophilic, *531*
   anionic, *590*
   aspartyl, *595, 598, 599, 601, 602, 603, 604, 605, 606*
   collagenolytic, *607*
   cysteine, *525, 581, 583, 587, 588, 589, 590, 591, 592, 593*
   fungal, *604*
   IgA, *578*
   mixtures, See Chapter 5, *748-755*
   neutral, *592, 750, 751, 1243*
   neutral, Ca^{2+}-activated, *592*
   pancreatic, *583*
   sulfhydryl, *607*
   unclassified. See Chapter 8, *1243-1244*
Protein C (activated), *576*
Protein disulfide-isomerase, *711*
Protein kinase, *231*
   II (Ca^{2+}/CaM dependent), *231, 248*
   A, C, G, *231*
   cAMP, cGMP-dependent, *231*
   hydroxyalkyl, *231*

hydroxyaryl, *247*
serine (threonine), *231*
tau, *248*
tyrosine, *247*
Protein phosphatase, *346, 347*
ω-Protein, *717*
Proteinase, *751, 752*
   I, *567, 627*
   II, *618*
   A, B, *560, 602, 603, 606*
   E, *629*
   K, *331, 573*
   alkaline, *573, 629*
   aspartic, *603, 604, 606*
   cholesterol-binding, *577*
   collagenolytic, *607*
   fungal, *602*
   lytic, *552, 627*
   neutral, *592, 617, 618, 620, 624, 625, 750*
   Novo, bacterial, *569*
   serine, *552, 576*
   V8, *552*
Protein-arginine
   deiminase, *651*
   iminohydrolase, L-, *651*
Protein-β-aspartate O-methyltransferase, *184*
Protein-glutamate O-methyltransferase, *184*
Protein-glutamine γ-glutamyltransferase, *196*
Protein-glutamine:amine γ-glutamyltransferase, *196*
Protein-L-isoaspartate
   (D-Aspartate) O-methyltransferase, *184*
   O-methyltransferase, *184*
Protein-tyrosine
   kinase, *247, 352*
   -phosphatase, *349, 352*
   -phosphate phosphohydrolase, *352*
Prothrombase, *547*
Prothrombin activator, *1243*
Prothrombinase, *547*

Protocatechuate
   3,4-dioxygenase, *158*
   oxygenase, *158*
Protocatechuate:oxygen 3,4-oxidoreductase (decyclizing), *158*
Pro-X aminopeptidase, *511*
Ps-1, *567*
Pseudo
   catalase, *143*
   cholinesterase, *320*
   leucine aminopeptidase, *510*
   lysin, *618*
*Psh*A I, *1121*
*Psh*B I, *1121*
*Psp*5 II, *1122*
*Psp*124B I, *1123*
*Psp*1406 I, *773, 938, 1067, 1079, 1123*
*Psp*A I, *1124, 1218*
*Psp*E I, *1125*
*Psp*L I, *1125*
*Psp*N4 I, *1126*
*Psp*OM I, *1126*
*Psp*P I, *1127*
*Psp*PP I, *1127*
Pss I, *952*
Pst I, *1101, 1128*
   methylase, *1235*
Psychosine hydrolase, *480*
PTA, *190*
Pteroyl-poly-γ-glutamate hydrolase, *526*
PTK, *247*
Pullulanase, *479, 738, 739, 745*
Pulluzyme, *480*
Purafect^R, *570*
Purine nucleoside:orthophosphate ribosyltransferase, *206*
Purine-nucleoside phosphorylase, *206*
Putrescine
   oxidase, *115, 186*
   synthase, *187*

Putrescine:oxygen oxidoreductase (deaminating), *115*
Pvu I, *889, 1131, 1163*
Pvu II, *889, 1132*
Pyrimidine dimer DNA-glycosylase, *504*
Pyrimidine phosphorylase, *207*
Pyroglutamyl
   aminopeptidase, *525*
   -peptidase, *523*, I, *523*
Pyrophosphatase, *657*
Pyrophosphate:D-fructose 6-phosphate 1-phosphotransferase, *244*
Pyrophosphate
   -fructose-6-phosphate 1-phosphotransferase, *244*
   phosphohydrolase, *657*
Pyrrolidone-carboxylate peptidase, *523*
Pyruvate
   carboxylase, *726*
   decarboxylase, *100, 663*
   dehydrogenase, *100, 133*
   lipoamide, *100*, NADP$^+$, *97*
   kinase, *237, 721, 759*
   oxidase, *98*
Pyruvate:
   carbon-dioxide ligase (ADP-forming), *726*
   lipoamide 2-oxidoreductase (decarboxylating and acceptor-acetylating), *100*
   NADP$^+$ 2-oxidoreductase (CoA-acetylating), *97*
   oxygen 2-oxidoreductase (phosphorylating), *98*
Pyruvic
   carboxylase, *726*
   decarboxylase, *663*
   dehydrogenase, *100*
   -malic carboxylase, *47*
   oxidase G, *98*

Q

Quinone reductase, *127*

R

Rca I, *1083, 1133*
Ready-Lyse™, *445*
Rearrangase, S-S, *707119*
rec A protein, *1242*
Relaxing enzyme, *717*
Renin, *560, 598, 601*
Rennet, *598, 749*
Rennin, *598, 605*
Replitherm™, *282*
Restriction endonucleases. See Chapter 6, *761-1220*
Restriction-modification system. See Chapter 7
Retrotherm™, *756*
Reverse transcriptase, *292, 756*
Revertase, *292*
Reyonet, *146*
α-L-Rhamnosidase, *478, 741*
α-L-Rhamnoside rhamnohydrolase, *478*
Rhizopuspepsin, *604*
Rhodanese, *297*
Ribokinase, *225*
Ribonuclease, *389, 392, 393, 573, 754, 755*
   I, *393, 388, 390*
   IA, *393*
   II, *388*
   III, *392*
   V, *398*
   V1, *398*
   A, *393, 756*
   B, *393*
   C, *1242*
   H, *292, 369, 386*
   M, *393*
   N_1, *390*
   N_2, *388, 390*
   T1, *389*
   T2, *388*
   U2, *392*
   U3, *392*
   microbial, *389, 390, 392, 393*
   pancreatic, *393*
   *Phy*M, *392*
   plant, *388*
   venom, *393*
Ribonucleotide phosphohydrolase
   3'-, *345*
   5'-, *344*
Ribose 5-phosphate
   ketol-isomerase, D-, *708*
   isomerase, *708*
Ribose-phosphate pyrophosphokinase, *262*
D-Ribulose 1,5-diphosphate carboxylase, *674*
D-Ribulose 5-phosphate
   ketol-isomerase, *708*
   3-epimerase, *702*
   kinase, *226*
Ribulose-bisphosphate carboxylase, *674*
Ribulosebisphosphate carboxylase/oxygenase, *674*
Ribulose-phosphate 3-epimerase, *702*
Rma I, *1049*
RNA
   ligase (ATP), *733*
   nucleotidyltransferase (DNA-directed), *263*
   polymerase, *263, 383*
RNA-directed DNA polymerase, *292*
t-RNA synthetase
   leucyl, *257*
RNase. See Ribonuclease
Rohalase[R], *427, 471*
Rohament[R], *427, 439*
Rohapect[R], *439, 752*
Romase, *299*
Rsa I, *776, 942, 1134*
Rsr II, *940, 941, 1136*
*Rsp*X I, *891*
Rst II, *806*
RT, *292*
Rubisco, *674*
Rusticyanin, *180*
RX:glutathione *R*-transferase, *212*

S

Sac I, *962, 1138, 1184*
Sac II, *937, 1046, 1141, 1142, 1185*
Saccharase, *464*
Saccharopine dehydrogenase (NAD^+, L-lysine-forming), *118*
Sal I, *804, 1143, 1211*
Salicylate
   1-monooxygenase, *166*
   hydroxylase, *166*
Salicylate,NADH:oxygen oxidoreductase (1-hydroxylating, decarboxylating), *166*
*San*D I, *1146*
Sap I, *1147*
Sarcosine
   dehydrogenase, *121, 646, 650*
   Sarcosine oxidase, *120, 646, 650*, G, *120*
Sarcosine:
   (acceptor) oxidoreductase (demethylating), *121*
   oxygen oxidoreductase (demethylating), *120*
Sau I, *788, 946, 974*
*Sau*3A I, *811, 825, 834, 948, 949, 1052, 1088, 1148, 1214*
*Sau*96 I, *801, 802, 873, 936, 952, 1127, 1150*
Savinase[R], *571*
Sbf I, *1152*
Sca I, *769, 1152*
Scintillase, *584*
*Scr*F I, *902, 1155*
SCS, *722*
Scutelarin, *547*
SDH, *23*
Sdu I, *886, 1156*
Sealase, *727*

Sec I, **848**, **858**
Sedoheptulose 7-phosphate:D-glyceraldehyde 3-phosphate
  glyceronetransferase, **188**
  glycolaldehydetransferase, **187**
γ-Seminoprotein, **559**, **560**
Septizyme, **753**
Sequenase™, **276**
SequiTherm™, **286**
Serotonin, **580**
Serralysin, **629**
Serum
  kallikrein, **558**
  transaminase, **105**
SexA I, **1157**
SfaN I, **1157**
Sfc I, **919**, **1158**
Sfe I, **830**
Sfi I, **1095**, **1159**, **1229**
Sfr274 I, **1162**
Sfr303 I, **1162**
Sfu I, **775**, **842**, **930**, **938**, **943**, **1030**, **1036**, **1067**, **1070**, **1079**, **1191**
Sgf I, **1163**
SgrA I, **804**, **1164**, **1215**
Sialidase, **445**
  exo-α-, **445**
Sialyltransferase
  2,6-, **210**
  α-2,3-, **211**
Sin I, **806**, **838**, **1012**, **1165**
Sinigrinase, **507**
Site-specific DNA-methyltransferases. See Chapter 7
Sma I, **934**, **1166**, **1215**, **1218**
Smi I, **1169**
Sna I, **907**
SnaB I, **920**, **976**, **1169**
Sno I, **786**, **792**
SOD, **176**

Softline, **427**
Softzyme, **1241**
Solanain, **556**
Sorbitol dehydrogenase, **23**
SOX, **120**
SP, **198**
Spe I, **773**, **808**, **1091**, **1171**, **1207**
Speedase, **414**
Spezyme^R, **412**, **418**, **420**, **426**, **707**, **738**
Sph I, **822**, **1149**, **1229**
Sphingomyelin
  cholinephosphohydrolase, **362**
  phosphodiesterase, **362**
Sphingomyelinase, **362**
Spl I, **768**, **871**, **1176**
Spl II, **1177**
Srf I, **1177**
Sse9 I, **1178**
Sse8387 I, **1128**, **1179**
SseB I, **1180**
Ssp I, **1181**
SspB I, **1183**
Sss I (CpG) methylase, **1236**
Sst I, **1138**, **1141**, **1184**
Sst II, **1046**, **1141**, **1185**
Staphylokinase, **624**
Starzyme, **1248**
Stem bromelain, **593**, **594**
Steroid hydroperoxide, **155**
Sterol esterase, **323**
Steryl-ester acylhydrolase, **323**
Stonezyme, **1241**
Straub diaphorase, **133**
Streptodornase, **378**
Streptolysin S, **1245**
Stu I, **1127**, **1180**, **1186**
Sty I, **903**, **989**, **994**, **1188**
Subtilisin, **556**, **557**, **569**, **573**, **576**
Subtilopeptidase, **569**
Succinate

  arginine-lyase, N-(L-arginino), **699**
  thiokinase, **722**
Succinate:CoA ligase (GDP-forming), **722**
Succinate-CoA ligase (GDP-forming), **722**
Succinate-semialdehyde dehydrogenase [NAD(P)$^+$], **96**
Succinate-semialdehyde:NAD(P)$^+$ oxidoreductase, **96**
Succinyl-CoA
  synthetase (GDP-forming), **722**
  transferase, **298**
Succinyl-CoA:3-oxo-acid CoA-transferase, **298**
Sucrase-isomaltase, **534**
Sucrose
  6-glucosyltransferase, **197**
  α-glucosidase, **434**
  glucosyltransferase, **198**
  phosphate synthetase, **200**
  phosphate-UDP glucosyltransferase, **200**
  phosphorylase, **198**
  synthase, **200**
  synthetase, **200**
Sucrose:
  1,6-α-D-glucan 6-α-D-glucosyltransferase, **197**
  orthophosphate α-D-glucosyltransferase, **198**
Sucrose-phosphate synthase, **200**
Sucrose-UDP glucosyltransferase, **200**
Sulfatase, **366**
Sulfite oxidase, **135**
Sulfite:oxygen oxidoreductase, **135**
Sun I, **1189**
Superoxide dismutase, **85**, **143**, **176**
Superoxide:superoxide oxidoreductase, **176**
Swa I, **1190**
Sweetzyme^R, **707**
Swivelase, **717**

T

T4 DNA ligase, *727, 879, 915, 979, 1155, 1205*
T4 RNA ligase, *733*
Tabernamontanain, *556*
Tai I, *1050, 1190*
Tannase, *327*
Tannin acylhydrolase, *327*
Taq DNA ligase, *731*
Taq I, *775, 938, 943, 1030, 1036, 1126, 1067, 1070, 1079, 1191, 1203*
 methylase, *1237*
Tau-protein kinase, *246*
Tautomerase, *711*
TD, *671*
Termamyl^R, *412*
Terminal
 addition enzyme, *289*
 deoxyribonucleotidyltransferase, *289*
 transferase, *289*
λ-Terminase, *1242*
Tetrahydrofolate dehydrogenase, *117*
Tetrahydrofolate:NADP$^+$ oxidoreductase, 5,6,7,8-, *117*
TEV protease, *1244*
Textile enzymes
 unclassified, *1240-1241*
Tfi I, *1194*
TH, *170*
Tha I, *765, 862, 998, 1195*
Theophylline oxidase, *1250*
Thermitase, *576*
Thermolysin, *510, 625, 627*
Thioglucosidase, *507*
Thioglucoside glucohydrolase, *507*
Thiol-dependent carboxypeptidase, *526*
Thioredoxin, *711*
Thiosulfate
 cyanide transsulfurase, *297*
 sulfurtransferase, *297*
 thiotransferase, *297*
Thiosulfate:cyanide sulfurtransferase, *297*
Thrombin, *541, 576, 580, 592*
Thrombocytin, *580*
Thrombokinase, *547*
Thromboxane
 synthase, *714*
 -A synthase, *714*
Thymidine phosphorylase, *207*
Thymidine:orthophosphate deoxy-D-ribosyltransferase, *207*
Thymonuclease, *378*
TIM, *705*
TK, *247*
TOD, *113*
Tonin, *560*
Toxin
 Clostridium oedematiens β- and γ-, *356*
 Clostridium welchii α-, *356*
TPI, *705*
TPNH$_2$-Cytochrome c reductase, *122*
Transaldolase, *188*
Transaminase
 A, *214*
 serum, *105*
Transferases. See Chapter 4, *181-298*
Transglutaminase, *196*
Transketolase, *187*
Transphosphoribosidase, *208*
α,α-Trehalase, *466*
α,α-Trehalose glucohydrolase, *466*
Triacylglycerol
 acylhydrolase, *301*
 lipase, *301*
Triacylglycero-protein acylhydrolase, *328*
Tributyrase, *301*
Triglyceride lipase, *301*
Trihydroxy-5β-cholan-24-oylglycine amidohydrolase
 3α, 7α, 12α, *640*
Triosephosphate
 dehydrogenase, *93*
 isomerase, *705, 758*
 mutase, *705*
Triphosphatase, *659*
Triterpenol esterase, *323*
*Tru*1 I, *1196*
*Tru*9 I, *1086, 1196, 1206*
Trypsin, *530, 531, 541, 549, 550, 560, 561, 595, 607, 748, 754, 755*
Trypsinogen kinase, *602*
Tryptase, *568*
L-Tryptophan indole-lyase (deaminating), *682*
Tryptophanase, *682*
Tsc I, *1198*
Tse I, *1198*
*Tsp*45 I, *1199*
*Tsp*509 I, *1200*
*Tsp*A I, *986*
*Tsp*R I, *1201*
Tth DNA ligase, *727*
*Tth*111 I, *797, 803, 1202*
*Tth*HB8 I, *775, 938, 943, 1030, 1070, 1079, 1191*
 methylase, *1238*
Topoisomerase, DNA, *715, 717*
Tyraminase, *113*
Tyramine oxidase, *113*
Tyrosinase, *138, 174*
Tyrosine
 carboxy-lyase, L-, *671*
 decarboxylase, *671*
 hydroxylase, *170*
 N-hydroxylase, *170*
 N-monooxygenase, *170*
L-Tyrosine,NADPH:oxygen oxidoreductase (N-hydroxylating), *170*
Tyrosylprotein kinase, *247*

U

UDPGlcDH, *25*
UDPgalactose 4-epimerase, *702*
UDPgalactose:
   D-glucose 4-β-D-galactotransferase, *203*
   N-acetyl-β-D-glucosaminyl-glycopeptide β-1,4-galactosyltransferase, *205*
UDPgalactose-glucose galactosyltransferase, *203*
UDPgalactose-glycoprotein galactosyltransferase, *205*
UDPglucose
   4-epimerase, *702*
   6-dehydrogenase, *25*
   dehydrogenase, *25*
   pyrophosphorylase, *287*
UDPglucose:
   α-D-galactose-1-phosphate uridylyltransferase, *287*
   D-fructose 2-α-D-glucosyltransferase, *200*
   D-fructose-6-phosphate 2-α-D-glucosyltransferase, *200*
   glycogen 4-α-D-glucosyltransferase, *199*
   NAD^+ 6 oxidoreductase, *25*
UDPglucose-
   fructose glucosyltransferase, *200*
   fructose-phosphate glucosyltransferase, *200*
   glycogen glucosyltransferase, *199*
   hexose-1-phosphate uridylyltransferase, *289*
UDPglucuronate:(acceptor) β-D-glucuronosyltransferase, *201*
UDPglucuronosyltransferase, *201*
Ultrazym[R], *439*, *746*
Unclassified enzymes. See Chapter 8
UNG, *504*
Unsaturated acyl-CoA hydratase, *687*
Unspecific monooxygenases, *122*
Untwisting enzyme, *717*
UOD, *129*
Uracil-DNA glycosylase, *504*

Urate oxidase, *129*
Urate:oxygen oxidoreductase, *129*
Urea carboxylase, *725*
   hydrolyzing, *724*
Urea:carbon-dioxide ligase (ADP-forming), *725*
Urease, *632*
   ATP-hydrolyzing, *725*
Uricase, *129*
Uridine-5′-diphosphoglucose
   dehydrogenase, *25*
   pyrophosphorylase, *287*
Uridyl transferase, *289*
Urishiol oxidase, *139*
Urokinase, *579*
UTP:
   α-D-glucose-1-phosphate uridylyltransferase, *287*
   α-D-hexose-1-phosphate uridylyltransferase, *288*
UTPglucose-1-phosphate uridylyltransferase, *287*
UTPhexose-1-phosphate uridylyltransferase, *288*

V

Van91 I, *955*, *1204*
Venombin A, *580*
Vent$_R$[R], *278*
Veron[R], *410*, *572*, *585*, *603*
Vevozyme, *413*
Vha464 I, *1205*
Viscozyme[R], *742*
Vne I, *1205*
Vsp I, *1086*, *1196*, *1206*

W

Wheatzyme, *740*

X

Xaa-Pro-dipeptidyl-aminopeptidase, *515*
Xanthine
   dehydrogenase, *86*
   oxidase, *85*
Xanthine:oxygen oxidoreductase, *85*
Xba I, *808*, *1091*, *1171*, *1207*
Xcm I, *1210*
Xho I, *1107*, *1143*, *1211*
Xho II, *811*, *825*, *834*, *924*, *1052*, *1057*, *1088*, *1148*, *1214*
Xma I, *766*, *934*, *1166*, *1215*, *1218*
Xma II, *1095*
Xma III, *925*, *963*, *969*, *971*, *1217*
   methylase, *1095*
*Xma*C I, *1218*
Xml I, *889*
Xmn I, *798*, *1218*
Xor II, *889*
X-Pro dipeptidase, *513*
Xylan
   1,4-β-xylosidase, *477*
   endo-1,3-β-xylosidase, *471*
   xylanohydrolase, 1,3-β-D-, *471*, 1,4-β-D-, *432*
   xylohydrolase, 1,4-β-D-, *477*
Xylanase, *432*, *471*, *696*, *740*, *741*, *742*, *742*, *744*, *747*, *752*
   endo-1,3-β-, *471*
Xylitol:$NADP^+$ 4-oxidoreductase (L-xylulose-forming), *23*
Xylobiase, *477*
Xylose
   isomerase, *707*
   ketol-isomerase, D-, *707*
Xylosidase
   α-, *1241*
   β-, *477*
   exo-1,4-β-, *477*
L-Xylulose reductase, *23*

Y

Yeast lytic enzyme, **1245**, **1246**, **1247**

Z

Zsp2 I, **1220**
Zymolase, **1246**, **1247**
Zymolean^R, **748**